continued on back

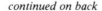

STATISTICAL
INFERENCE

STATISTICAL INFERENCE

VIJAY K. ROHATGI
Department of Mathematics and Statistics
Bowling Green State University
Bowling Green, Ohio

JOHN WILEY & SONS
New York · Chichester · Brisbane · Toronto · Singapore

Library of Congress Cataloging in Publication Data:

Rohatgi, V. K., 1939–
  Statistical inference.

  Includes index.
  1. Mathematical statistics. I. Title.

QA276.R624 1984 519.5 83-21848
ISBN 0-471-87126-5

To Bina and Sameer

PREFACE

This course in statistical inference is designed for juniors and seniors in most disciplines (including mathematics). No prior knowledge of probability or statistics is assumed or required. For a class meeting four hours a week this is a two-semester or three-quarter course. The mathematical prerequisite is modest. The prospective student is expected to have had a three-semester or four-quarter course in calculus. Whenever it is felt that a certain topic is not covered in a calculus course at the stated prerequisite level, enough supplementary details are provided. For example, a section on gamma and beta functions is included, as are supplementary details on generating functions.

There are many fine books available at this level. Then why another? Their titles notwithstanding, almost all of these books are in probability and statistics. Roughly the first half of these texts is usually devoted to probability and the second half to statistics. Statistics almost always means parametric statistics, with a discussion of nonparametric statistics usually relegated to the last chapter.

This text is on statistical inference. My approach to the subject separates this text from the rest in several respects. First, probability is treated here from a modeling viewpoint with strong emphasis on applications. Second, statistics is not relegated to the second half of the course—indeed, statistical thinking is encouraged and emphasized from the beginning. Formal language of statistical inference is introduced as early as Chapter 4 (essentially the third chapter, since Chapter 1 is introductory), immediately after probability distributions have been introduced. Inferential questions are considered along with probabilistic models, in Chapters 6 and 7. This approach allows and facilitates an early introduction to parametric as well as nonparametric techniques. Indeed, every attempt has been made to integrate the two: Empirical distribution function is introduced in Chapter 4, in Chapter 5 we show that it is unbiased and consistent, and in Chapter 10 we show that it is the maximum likelihood estimate of the population distribution function. Sign test and Fisher–Irwin test are introduced in Chapter 6, and inference concerning quantiles is covered in Section 8.5. There is not even a separate chapter entitled nonparametric statistical inference.

Apart from the growing importance of and interest in statistics, there are several reasons for introducing statistics early in the text. The traditional approach in which statistics follows probability leaves the reader with the false notion that probability and statistics are the same and that statistics is the mathematics of computing certain probabilities. I do not believe that an appreciation for the utility of statistics should be withheld until a large dose of probability is digested. In a traditional course, students who leave the course after one semester or one quarter learn little or no statistics. They are left with little understanding of the important role that statistics plays in scientific research. A short course in probability becomes just another hurdle to pass before graduation, and the students are deprived of the chance to use statistics in their disciplines. I believe that the design of this text alleviates these problems and enables the student to acquire an outlook approaching that of a modern mathematical statistician and an ability to apply statistical methods in a variety of situations.

There appears to be a reasonable agreement on the topics to be included in a course at this level. I depart a little from the traditional coverage, choosing to exclude regression since it did not quite fit into the scheme of "one sample, two sample, many sample" problems. On the other hand, I include the Friedman test, Kendall's coefficient of concordance, and multiple comparison procedures, which are usually not done at this level.

The guiding principles in my selection have been usefulness, interrelationship, and continuity. The ordering of the selections is dictated by need rather than relative importance.

While the topics covered here are traditional, their order, coverage, and discussion are not. Many other features of this text separate it from previous texts. I mention a few here:

(i) An unusually large number of problems (about 1450) and examples (about 400) are included. Problems are included at the end of each section and are graded according to their degree of difficulty; more advanced (and usually more mathematical) problems are identified by an asterisk. A set of review problems is also provided at the end of each chapter to test the student's ability to choose relevant techniques. Every attempt has been made to avoid the annoying and time-consuming practice of creating new problems by referring to earlier problems (often scores of pages earlier). Either completely independent problems are given in each section or relevant details (with cross references) are restated whenever a problem is important enough to be continued in a later section. The amount of duplication, however, is minimal and improves readability.

(ii) Sections with a significant amount of mathematical content are also identified by an asterisk. These sections are aimed at the more mathematically inclined students and may be omitted at first reading. This procedure allows us to encompass a much wider audience without sacrificing mathematical rigor. Needless to say, this is not a recipe book. The emphasis is on the how and why of all the techniques introduced here, in the hope that the student is challenged to think like a statistician.

(iii) Applications are included from diverse disciplines. Most examples and problems are application oriented. It is true that no attempt has been made to include "real life data" in these problems and examples but I hope that the student will be motivated enough to follow up this course with an exploratory data analysis course.

(iv) A large number of figures (about 150) and remarks supplement the text. Summaries of main results are highlighted in boxed or tabular form.

In a two-semester course, meeting four times a week, my students have been able to cover the first twelve chapters of the text without much haste. In the first semester we cover the first five chapters with a great deal of emphasis on Chapter 4, the introduction to statistical inference. In the second semester we cover all of Chapters 6 and 7 on models (but at increased pace and with emphasis on inferential techniques), most of Chapter 8 on random variables and random vectors (usually excluding Section 6, depending on the class composition), and all of Chapter 9 on large-sample theory.

In Chapter 10 on point and interval estimation, more time is spent on sections on sufficiency, method of moments, maximum likelihood estimation, and confidence intervals than on other sections. In Chapter 11 on testing hypotheses, we emphasize sections on Wilcoxon signed rank test, two-sample tests, chi-square test of goodness of fit, and measures of association. The point is that if the introductory chapter on statistical inference (Chapter 4) is covered carefully, then one need not spend much time on unbiased estimation (Section 10.3), Neyman–Pearson Lemma (Section 11.2), composite hypotheses (Section 11.3), or likelihood ratio tests (Section 11.4). Chapter 12 on categorical data is covered completely.

In a three-quarters course, the pace should be such that the first four chapters are covered in the first quarter, Chapters 5 to 9 in the second quarter, and the remaining chapters in the third quarter. If it is found necessary to cover Chapter 13 on k-sample problems in detail, we exclude the technical sections on transformations (Section 8.3) and generating functions (Section 8.6), and also sections on inference concerning quantiles (Section 9.8), Bayesian estimation (Section 10.6), and composite hypotheses (Section 11.3).

I take this opportunity to thank many colleagues, friends, and students who made suggestions for improvement. In particular, I am indebted to Dr. Humphrey Fong for drawing many diagrams, to Dr. Victor Norton for some numerical computations used in Chapter 4 and to my students, especially Lisa Killel and Barbara Christman, for checking many solutions to problems. I am grateful to the Literary Executor of the late Sir Ronald A. Fisher, F. R. S., to Dr. Frank Yates, F. R. S., and to Longman Group Ltd., London, for permission to reprint Tables 3 and 4 from their book *Statistical Tables for Biological, Agricultural and Medical Research* (6th edition, 1974). Thanks are also due to Macmillan Publishing Company, Harvard University Press, the Rand Corporation, Bell Laboratories, Iowa University Press, John Wiley & Sons, the Institute of Mathematical Statistics, Stanford University Press, Wadsworth Publishing Company, Biometrika Trustees, Statistica Neerlandica, Addison–Wesley Publishing Company, and the American Statistical Association for permission to use tables and to John Wiley & Sons for permission to use some diagrams.

I also thank the several anonymous reviewers whose constructive comments greatly improved presentation. Finally, I thank Mary Chambers for her excellent typing and Beatrice Shube, my editor at John Wiley & Sons, for her cooperation and support in this venture.

VIJAY K. ROHATGI

Bowling Green, Ohio
February 1984

CONTENTS

STATISTICAL INFERENCE

CHAPTER 1

Introduction

1.1 INTRODUCTION

Probabilistic statements are an integral part of our language. We use the expressions random, odds, chance, risk, likelihood, likely, plausible, credible, as likely as not, more often than not, almost certain, possible but not probable, and so on. All these words and phrases are used to convey a certain degree of uncertainty although their nontechnical usage does not permit sharp distinctions, say, between probable and likely or between improbable and impossible. One of our objectives in this course is to introduce (in Chapter 2) a numerical measure of uncertainty. Once this is done we can use the apparatus of mathematics to describe many physical or artificial phenomena involving uncertainty. In the process, we shall learn to use some of these words and phrases as technical terms.

The basic objective of this course, however, is to introduce techniques of statistical inference. It is hardly necessary to emphasize here the importance of statistics in today's world. Statistics is used in almost every field of activity. News media carry statistics on unemployment rate, inflation rate, batting averages, average rainfall, the money supply, crime rates. They carry the results of Gallop and Harris polls and many other polls. Most people associate statistics with a mass of numerical facts, or *data*. To be sure statistics does deal with the collection and description of data. But a statistician does much more. He or she is—or should be—involved in the planning and design of experiments, in collecting information, and in deciding how best to use the collected information to provide a basis for decision making. This text deals mostly with this latter aspect: the art of evaluating information to draw reliable inferences about the true nature of the phenomenon under study. This is called *statistical inference*.

1.2 STOCHASTIC MODELS

Frequently the objective of scientific research is to give an adequate mathematical description of some natural or artificial phenomenon. A *model* may be defined as a mathematical idealization used to approximate an observable phenomenon. In any such idealization, certain assumptions are made (and hence certain details are

1

ignored as unimportant). The success of the model depends on whether or not these assumptions are valid (and on whether the details ignored actually are unimportant).

In order to check the validity of a model, that is, whether or not a model adequately describes the phenomenon being studied, we take observations. The process of taking observations (to discover something that is new or to demonstrate something that is already known) is called an *experiment*.

A *deterministic model* is one which stipulates that the conditions under which an experiment is performed determine the outcome of the experiment. Thus the distance d traveled by an automobile in time t hours at a constant speed s kilometers per hour is governed by the relation $d = st$. Knowledge of s and t precisely determines d. Similarly, gravitational laws describe precisely what happens to a falling object, and Kepler's laws describe the behavior of planets.

A *nondeterministic* (or *stochastic*) model, on the other hand, is one in which past information, no matter how voluminous, does not permit the formulation of a rule to determine the precise outcome of an experiment. Many natural or artificial phenomena are *random* in the sense that the exact outcome cannot be predicted, and yet there is a predictable long-term pattern. Stochastic models may be used to describe such phenomena. Consider, for example, the sexes of newborns in a certain County Hospital. Let B denote a boy and G a girl. Suppose sexes are recorded in order of birth. Then we observe a sequence of letters B and G, such as

$$G\,B\,G\,G\,B\,G\,B\,B\,B\,G\,G\ldots$$

This sequence exhibits no apparent regularity. Moreover, one cannot predict the sex of the next newborn, and yet one can predict that in the long run the proportion of girls (or boys) in this sequence will settle down near $1/2$. This long-run behavior is called *statistical regularity* and is noticeable, for example, in all games of chance.

In this text we are interested only in experiments that exhibit the phenomena of randomness and statistical regularity. Probability models are used to describe such phenomena.

We consider some examples.

    Example 1. Measuring Gravity. Consider a simple pendulum with unit mass suspended from a fixed point O which swings only under the effect of gravity. Assume that the string is of unit length and is weightless. Let $t = t(\theta)$ be the period of oscillation when θ is the angle between the pendulum and the vertical (see Figure 1). It is shown in calculus[†] that when the pendulum goes from $\theta = \theta_0$ to $\theta = 0$ (corresponding to one fourth of a period)

$$\frac{t}{4} = \sqrt{\frac{1}{2g}} \int_0^{\theta_0} \frac{d\theta}{\sqrt{\cos\theta - \cos\theta_0}}$$

[†]Al Shenk, *Calculus and Analytic Geometry*, Scott-Foresman, Glenview Illinois, 1979, p. 544.

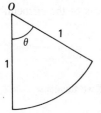

Figure 1

so that

$$t = 4\sqrt{\frac{1}{g}} \int_0^{\pi/2} \frac{dx}{\sqrt{1 - k^2 \sin x}}, \qquad k = \sin\frac{\theta_0}{2}.$$

Hence for small oscillations $k = 0$ and $t = 2\pi/\sqrt{g}$ gives approximately the period of oscillation. This gives a deterministic model giving g as a function of t, namely,

$$g = \frac{4\pi^2}{t^2}.$$

If t can be measured accurately, this formula gives the value of g. If repeated readings on t are taken, they will all be found different (randomness) and yet there will be a long-run pattern in that these readings will all concentrate near the true value of t (statistical regularity). The randomness may be due to limitations of our measuring device and the ability of the person taking the readings. To take into account these nondeterministic factors we may assume that t varies in some random manner such that

$$t = \frac{2\pi}{\sqrt{g}} + \varepsilon$$

where $2\pi/\sqrt{g}$ is the true value of t and ε is a random error that varies from reading to reading. A stochastic model will postulate that $t = 2\pi/\sqrt{g} + \varepsilon$ along with some assumptions about the random error ε. A statistician's job, then, is to estimate g based on, say, n readings $t_1, t_2, \ldots, t_n$ on t.   □

Example 2.   Ohm's Law.   According to Ohm's law for a simple circuit, the voltage V is related to the current I and the resistance R according to the formula $V = IR$. If the conditions underlying this deterministic relationship are met, this model predicts precisely the value of V given those of I and R. Such a description may be adequate for most practical purposes.

If, on the other hand, repeated readings of either I or R or both are found to vary, then V will also vary in a random manner. Given a pair of readings on I and R, V is still determined by $V = IR$. Since all pairs of readings on I and R are different, so will be the values of V. In a stochastic model the assumptions we make concerning the randomness in I and/or R determine the random behavior of V through the relation $V = IR$.   □

Example 3.   Radioactive Disintegration.   The deterministic model for radioactive disintegration postulates that the rate of decay of a quantity of radioactive element is proportional to the mass of the element. That is,

(1)
$$\frac{dm}{dt} = -\lambda m,$$

where $\lambda > 0$ is a constant giving the rate of decay and m is the mass of the element. Integrating (1) with respect to t we get:

$$\ln m = -\lambda t + c$$

where c is a constant. If $m = m_0$ at time $t = 0$, then $c = \ln m_0$ and we have:

(2) $$m = m_0 \exp(-\lambda t)$$

for $t \geq 0$. Given m_0 and λ, we know m as a function of t.

The exact number of decays in a given time interval, however, cannot be predicted with certainty because of the random nature of the time at which an element disintegrates. This forces us to consider a stochastic model. In a stochastic model one makes certain assumptions concerning the probability that a given element will decay in time interval $[0, t]$. These lead to an adequate description of the probability of exactly k decays in $[0, t]$.   □

Example 4. Rolling a Die. A die is rolled. Let X be the number of points (face value) on the upper face. No deterministic model can predict which of the six face values 1, 2, 3, 4, 5, or 6 will show up on any particular roll of the die (randomness) and yet there is a predictable long-run pattern. The proportion of any particular face value in a sequence of rolls will be about $\frac{1}{6}$ provided the die is not loaded.   □

It should be clear by now that in some experiments a deterministic model is adequate. In other experiments (Examples 3 and 4) we must use a stochastic model. In still other experiments, a stochastic model may be more appropriate than a deterministic model. Experiments for which a stochastic model is more appropriate are called *statistical* or *random* experiments.

DEFINITION 1. (RANDOM OR STATISTICAL EXPERIMENT). An experiment that has the following features is called a random or statistical experiment.

(i)   All possible outcomes of the experiment are known in advance.
(ii)   The exact outcome of any specific performance of the experiment is unpredictable (randomness).
(iii) The experiment can be repeated under (more or less) identical conditions.
(iv)   There is a predictable long-run pattern (statistical regularity).

Example 5. Some Typical Random Experiments. We list here some typical examples of random experiments.

(i)   Toss a coin and observe the up face.
(ii)   A light bulb manufactured at a certain plant is put to a lifetime test and the time at which it fails is recorded.
(iii) A pair of dice is rolled and the face values that show up are recorded.
(iv)   A lot consisting of N items containing D defectives ($D \leq N$) is sampled. An item sampled is not replaced, and we record whether the item selected is defective or nondefective. The process continues until all defective items are found.
(v)   The three components of velocity of an orbital satellite are recorded continuously for a 24-hour period.

(vi)   A manufacturer of refrigerators inspects its refrigerators for 10 types of defects. The number of defects found in each refrigerator inspected is recorded.

(vii) The number of girls in every family with five children is recorded for a certain town.

Problems for Section 1.2

In the following problems, state whether a deterministic or a nondeterministic model is more appropriate. Identify the sources of randomness. (In each case there is no clearly right or wrong answer; the decision you make is subjective.)

1. The time (in seconds) elapsed is measured between the end of a question asked of a person and the start of her or his response.

2. In order to estimate the average size of a car pool, a selection of cars is stopped on a suburban highway and the number of riders recorded.

3. On a graph paper, a line with equation $y = 3x + 5$ is drawn and the values of y for $x = 1, 3, 5, 7, 9$ are recorded.

4. Consider a binary communication channel that transmits coded messages consisting of a sequence of 0's and 1's. Due to noise, a transmitted 0 might be received as a 1. The experiment consists of recording the transmitted symbol (0 or 1) and the corresponding received symbol (0 or 1).

5. In order to estimate the average time patients spent at the emergency room of a county hospital from arrival to departure after service, the service times of patients are recorded.

6. A coin is dropped from a fixed height, and the time it takes to reach the ground is measured.

7. A roulette wheel is spun and a ball is rolled on its edge. The color (black or red) of the sector in which the ball comes to rest is recorded. (A roulette wheel consists of 38 equal sectors, marked 0, 00, 1, 2, ..., 36. The sectors 0 and 00 are green. Half of the remaining 36 sectors are red, the other half black.)

1.3   PROBABILITY, STATISTICS, AND INFERENCE[†]

There are three essential components of a stochastic model:

(i)   Indentification of all possible outcomes of the experiment.
(ii)   Identification of all events of interest.
(iii) Assignment of probabilities to these events of interest.

The most important as well as most interesting and difficult part of model building is the assignment of probabilities. Consequently a lot of attention will be devoted to it.

[†]This section uses some technical terms that are defined in Chapter 2. It may be read in conjunction with or after Chapter 2.

Consider a random experiment and suppose we have agreed on a stochastic model for it. This means that we have identified all the outcomes and relevant events and made an assignment of probabilities to these events. The word *population*, in statistics, refers to the collection of all outcomes along with the assignment of probabilities to events. The object in statistics is to say something about this population. This is done on the basis of a *sample*, which is simply a part of the population. It is clear that a sample is not just any part of the population. In order for our inferences to be meaningful, randomness should somehow be incorporated in the process of sampling. More will be said about this in later chapters.

At this stage let us distinguish between probability and statistics. In probability, we make certain assumptions about the population and then say something about the sample. That is, the problem in probability is: *Given a stochastic model, what can we say about the outcomes?*

In statistics, the process is reversed. The problem in statistics is: *Given a sample (set of outcomes), what can we say about the population (or the model)?*

> **Example 1. Coin Tossing.** Suppose the random experiment consists of tossing a coin and observing the outcome. There are two possible outcomes, namely, heads or tails. The stochastic model may be that the coin is fair, that is, not fraudulently weighted. We will see that this completely specifies the probability of a head ($= 1/2$) and hence also of tails ($= 1/2$). In probability we ask questions such as: Given that the coin is fair, what is the chance of observing 10 heads in 25 tosses of the coin? In statistics, on the other hand we ask: Given that 25 tosses of a coin resulted in 10 heads, can we assert that the coin is fair?   □

> **Example 2. Gasoline Mileage.** When a new car model is introduced, the automobile company advertises (an estimated) Environmental Protection Agency rating of fuel consumption (miles per gallon) for comparison purposes. The initial problem of determining the probability distribution of fuel consumption for this model is a statistical problem. Once this has been solved, the computation of the probability that a particular car will give, say, at least 38 miles per gallon is a problem in probability. Similarly, estimating the average gas mileage for the model is a statistical problem.   □

> **Example 3. Number of Telephone Calls.** The number of telephone calls initiated in a time interval of length t hours is recorded at a certain exchange. The initial problem of estimating the probability that k calls are initiated in an interval of length t hours is a problem in statistics. Once these probabilities have been well established for each $k = 0, 1, 2, \ldots$, the computation of the probability that more than j calls are initiated in a one-hour period is a probability problem.   □

Probability is basic to the study of statistics, and we devote the next two chapters to the fundamental ideas of probability theory. Some basic notions of statistics are introduced in Chapter 4. Beginning with Chapter 5, the two topics are integrated.

Statistical inference depends on the laws of probability. In order to ensure that these laws apply to the problem at hand, we insist that the sample be random in a certain sense (to be specified later). Our conclusions, which are based on the sample outcomes, are therefore as good as the stochastic model we use to represent the experiment. If we observe an event that has a small probability of occurring under the model, there are two possibilities. Either an event of such a small probability has actually occurred or the postulated model is not valid. Rather than accept the explanation that a rare event has happened, statisticians look for alternative explanations. They argue that events with low probabilities cast doubt on the validity of the postulated model. If, therefore, such an event is observed in spite of its low probability, then it provides evidence against the model. Suppose, for example, we assume that a coin is fair and that heads and tails are equally likely on any toss. If the coin is then tossed and we observe five heads in a row, we begin to wonder about our assumption. If the tossing continues and we observe 10 heads in a row, hardly anyone would argue against our conclusion that the coin is loaded. Probability provides the basis for this conclusion. The chance of observing 10 heads in a row in 10 tosses of a fair coin, as we shall see, is 1 in $2^{10} = 1024$, or less than .001. This is evidence against the model assumption that the coin is fair. We may be wrong in this conclusion, but the chance of being wrong is 1 in 1024. And that is a chance worth taking for most practical purposes.

CHAPTER 2

Probability Model

2.1 INTRODUCTION

We recall that a stochastic model is an idealized mathematical description of a random phenomenon. Such a model has three essential components: the set of all possible outcomes, events of interest, and a measure of uncertainty. In this chapter we begin a systematic study of each of these components. Section 2.2 deals with the first two components. We assume that the reader is familiar with the language and algebra of set theory.

The most interesting part of stochastic modeling, and the most difficult, is the assignment of probabilities to events. The rest of the chapter is devoted to this particular aspect of stochastic modeling. It is traditional to follow the axiomatic approach to probability due to Kolmogorov. In Section 2.3 we introduce the axioms of probability, and in Section 2.4 we study their elementary consequences. Section 2.5 is devoted to the uniform assignment of probability. This assignment of probability applies, in particular, to all games of chance. In Section 2.6 we introduce conditional probability: Given that the outcome is a point in A, what is the probability that it is also in B? The important special case when the occurrence of an event A does not affect the assignment of probability to event B requires an understanding of the concept of independence, introduced in Section 2.7. This concept is basic to statistics and probability.

Assuming a certain model (Ω, ζ, P) for the random experiment, Sections 2.2 through 2.7 develop computational techniques to determine the chance of a given event in any performance of the experiment. In statistics, on the other hand, the problem is to say something about the model given a set of observations usually referred to as a sample. In order to make probability statements, we have to be careful how we sample. Section 2.8 is devoted to sampling from a finite population. We consider only a special type of probability sample called a simple random sample.

2.2 SAMPLE SPACE AND EVENTS

Consider a random experiment. We recall that such an experiment has the element of uncertainty in its outcomes on any particular performance, and that it exhibits statistical regularity in its replications. Our objective is to build a mathematical model that will take into account randomness and statistical regularity. In Section 1.3 we noted that such a model has three components. In this section we identify all the outcomes of the experiment and the so-called *events* of interest. In mathematical terms this means that we associate with each experiment a set of outcomes that becomes a point of reference for further discussion.

DEFINITION 1. (SAMPLE SPACE). A sample space, denoted by Ω, associated with a random experiment is a set of points such that:

(i)   each element of Ω denotes an outcome of the experiment, and

(ii) any performance of the experiment results in an outcome that corresponds to exactly one element of Ω.

In general, many sets meet these requirements and may serve as sample spaces for the same experiment, although one may be more suitable than the rest. The important point to remember is that all possible outcomes must be included in Ω. It is desirable to include as much detail as possible in describing these outcomes.

*Example 1. **Drawing Cards from a Bridge Deck.*** A card is drawn from a well-shuffled bridge deck. One can always choose Ω to be the set of 52 points $\{AC, 2C, \ldots, QC, KC, AD, 2D, \ldots, KD, AH, 2H, \ldots, KH, AS, 2S, \ldots, KS\}$ where AC represents the ace of clubs, 2H represents the two of hearts, and so on. If the outcome of interest, however, is the suit of the card drawn, then one can choose the set $\Omega_1 = \{C, D, H, S\}$ for Ω. On the other hand, if the outcome of interest is the denomination of the card drawn, one can choose the set $\Omega_2 = \{A, 2, 3, \ldots, J, Q, K\}$ for Ω. If the outcome of interest is that a specified card is drawn, then we have to choose the larger set:

$$\Omega = \{AC, 2C, \ldots, KC, \ldots, AS, 2S, \ldots, KS\}.$$

If we choose Ω_1 as the sample space and then are asked the question, "Is the card drawn an ace of a black suit?", we cannot answer it since our method of classifying outcomes was too coarse. In this case the sample space Ω offers a finer classification of outcomes and is to be preferred over Ω_1 and Ω_2. □

*Example 2. **Coin Tossing and Die Rolling.*** A coin is tossed once. Then $\Omega = \{H, T\}$ where H stands for heads and T for tails. If the same coin is tossed twice, then $\Omega = \{HH, HT, TH, TT\}$. If instead we count the number of heads in our outcome, then we can choose $\{0, 1\}$ and $\{0, 1, 2\}$ as the respective sample spaces. It would, however, be preferable to choose $\Omega = \{HH, HT, TH, TT\}$ in the case of tossing a coin twice, since this sample space describes the outcomes in complete detail.

If a die is rolled once, then $\Omega = \{1, 2, 3, 4, 5, 6\}$. If the die is rolled twice (or two dice are rolled once) then

$$\Omega = \{(i, j): i = 1, 2, 3, 4, 5, 6, \quad j = 1, 2, 3, 4, 5, 6\}.$$

$$= \{(1, 1), (1, 2), \ldots, (1, 6), \ldots, (6, 1), (6, 2), \ldots, (6, 6)\}.$$ □

Example 3. Tossing a Coin Until a Head Appears. The experiment consists of tossing a coin until a head shows up. In this case

$$\Omega = \{\, H, TH, TTH, \ldots \,\}$$

is the countably infinite set of outcomes. □

Example 4. Choosing a Point. A point is chosen from the interval [0, 1]. Here $\Omega = \{x\colon 0 \leq x \leq 1\}$ and contains an uncountable number of points. If a point is chosen from the square bounded by the points $(0, 0)$, $(1, 0)$, $(1, 1)$, and $(0, 1)$, then $\Omega = \{(x, y)\colon 0 \leq x \leq 1,\ 0 \leq y \leq 1\}$. If the experiment consists of shooting at a circular target of radius one meter and center $(0, 0)$, then Ω consists of the interior of a circle of radius one meter and center $(0, 0)$, that is,

$$\Omega = \{(x, y)\colon x^2 + y^2 \leq 1\}.$$ □

Example 5. Life Length of a Light Bulb. The time until failure of a light bulb manufactured at a certain plant is recorded. If the time is recorded to the nearest hour, then $\Omega = \{0, 1, 2, \ldots\}$. If the time is recorded to the nearest minute, we may choose $\Omega = \{0, 1, 2, \ldots\}$ or may simply take $\Omega = \{x\colon x \geq 0\}$ for convenience. □

Example 6. Urn Model. Many real or conceptual experiments can be modeled after the so-called urn model. Consider an urn that contains marbles of various colors. We draw a sample of marbles from the urn and examine the color of the marbles drawn. By a proper interpretation of the terms "marble," "color," and "urn," we see that the following examples, and many more, are special cases.

(i) *Coin tossing.* We may identify the coin as a marble, and the two outcomes H and T as two colors. If the coin is tossed twice we may identify each toss as a marble and outcomes HH, HT, TH, TT as different colors. Alternatively, we could consider an urn with two marbles of colors H and T and consider tossing the coin twice as a two-stage experiment consisting of two draws with replacement.

(ii) *Opinion polls.* A group of voters (the sample) is selected from all the voters (urn) in a congressional district and asked their opinion on some issue. We may think of the voters as marbles and different opinions as different colors.

(iii) *Draft lottery.* Days of the year are identified as marbles and put in an urn. Marbles are drawn one after another without replacement to determine the sequence in which the draftees will be called up for service. □

We note the following features of a sample space.

(i) A sample space Ω corresponds to the universal set. Once selected, it remains fixed and all discussion corresponds to this sample space.

(ii) The outcomes or points of Ω may be numerical or categorical.

(iii) A sample space may contain a countable (finite or infinite) or uncountable number of elements.

DEFINITION 2.   (DISCRETE SAMPLE SPACE).   A sample space Ω is said to be discrete if it contains at most a countable number of elements.

DEFINITION 3.   (CONTINUOUS SAMPLE SPACE).   A sample space is said to be continuous if its elements constitute a continuum: all points in an interval, or all points in the plane, or all points in the k-dimensional Euclidean space, and so on.

The sample spaces considered in Examples 1, 2, 3, and 6 are all discrete; that considered in Example 4 is continuous. In Example 5 the sample space $\Omega = \{0, 1, 2, \ldots\}$ is discrete whereas $\Omega = [0, \infty)$ is continuous. Continuous sample spaces arise whenever the outcomes are measured on a continuous scale of measurement. This is the case, for example, when the outcome is temperature, time, speed, pressure, height, or weight.

In practice, owing to limitations of our measuring device, all sample spaces are discrete. Consider, for example, the experiment consisting of measuring the length of life of a light bulb (Example 5). Suppose the instrument (watch or clock) we use is capable of recording time only to one decimal place. Then our sample space becomes $\{0.0, 0.1, 0.2, \ldots\}$. Moreover, it is realistic to assume that no light bulb will last forever so that there is a maximum possible number of hours, say T, where T may be very large. With this assumption we are dealing with a finite sample space $\Omega = \{0.0, 0.1, 0.2, \ldots, T\}$. Mathematically, however, it is much more convenient to select the idealized version $\Omega = [0, \infty)$ as the sample space.

The next step in stochastic modeling is to identify the subsets of Ω that are of interest. For this purpose we need the concept of an event.

DEFINITION 4.   (EVENT).   Let Ω be a sample space. An event (with respect to Ω) is a set of outcomes. That is, an event is a subset of Ω.

It is convenient to include the empty set, denoted by $\varnothing$, and Ω in our collection of events. When Ω is discrete, every subset of Ω may be taken as an event. If, on the other hand, Ω is continuous, then some technical problems arise. Not every subset of Ω may be considered as an event. For example, if $\Omega = \mathbb{R} = (-\infty, \infty)$ then the fundamental objects of interest are intervals (open, closed, semiclosed). Thus, all intervals should be events, as also should all (countable) unions and intersections of intervals. In this book we assume that associated with every sample space Ω, there is a class of subsets of Ω, denoted by ζ[†]. Elements of ζ are referred to as events. As pointed out earlier, if Ω is discrete then ζ may be taken to be the class of all subsets of Ω. If Ω is an interval on the line or a set in the plane, or a subset of the k-dimensional Euclidean space $\mathbb{R}_k$, then ζ will be a class of sets having a well-defined length or area or volume, as the case may be.

Events are denoted by the capital letters A, B, C, D, and so on. Let $A \in \zeta$. We say that an event has happened if the outcome is an element of A. We use the methods of set theory to combine events. In the following, $A, B, C \in \zeta$ and we describe some typical events in terms of set operations. We assume that the reader

---

[†]See Section 2.3 for a formal definition and more details.

is familiar with the operations of union, intersection, and complementation denoted respectively by $\cup$, $\cap$, and $^-$.

Verbal Description of Event	Set
A or B (at least one of A or B)	$A \cup B$
A and B	$A \cap B$
Not A	$\bar{A}$
A but not B	$A \cap \bar{B} (= A - B)$
A but not B or C	$A \cap (\overline{B \cup C}) (= A \cap \bar{B} \cap \bar{C})$
A and B but not C	$A \cap B \cap \bar{C}$
A or B but not C	$(A \cup B) \cap \bar{C}$
At least one of A, B, or C	$A \cup B \cup C$
None of A, B, or C	$\overline{A \cup B \cup C} (= \bar{A} \cap \bar{B} \cap \bar{C})$
Exactly one of A, B, or C	$(A \cap \bar{B} \cap \bar{C}) \cup (\bar{A} \cap B \cap \bar{C}) \cup (\bar{A} \cap \bar{B} \cap C)$
At most one of A, B, or C	$(\overline{A \cup B \cup C}) \cup (A \cap \bar{B} \cap \bar{C}) \cup (\bar{A} \cap B \cap \bar{C})$ $\cup (\bar{A} \cap \bar{B} \cap C)$
Exactly two of A, B, or C	$(A \cap B \cap \bar{C}) \cup (B \cap C \cap \bar{A}) \cup (A \cap C \cap \bar{B})$
At most two of A, B, or C	$\overline{A \cap B \cap C}$
At least two of A, B, or C	$(A \cap B \cap \bar{C}) \cup (B \cap C \cap \bar{A}) \cup (A \cap C \cap \bar{B})$ $\cup (A \cap B \cap C)$
All three of A, B, and C	$A \cap B \cap C$

DEFINITION 5. (MUTUALLY EXCLUSIVE EVENTS). We say that two events A and B are mutually exclusive if $A \cap B = \varnothing$.

In the language of set theory, mutually exclusive events are disjoint sets. Mutual exclusiveness is therefore a set theoretic property.

   Example 7. Families with Four Children. Consider families with four children. A family is selected and the sexes of the four children are recorded. Writing b for boy and g for girl, we can choose

$$\Omega = \{\, bbbb, bbbg, bbgb, bgbb, gbbb, bbgg, bggb, ggbb, bgbg, gbbg, gbgb, bggg, gbgg,$$

$$ggbg, gggb, gggg \,\},$$

where the sex is recorded in order of the age of the child. The event A that the family has two girls is given by

$$A = \{\, bbgg, bggb, ggbb, bgbg, gbbg, gbgb \,\}.$$

Some other events are listed below.

$$B = \{\text{at most one boy}\} = \{ gggg, bggg, gggb, gbgg, ggbg \},$$

$$C = \{\text{at least three girls}\} = \{ gggg, gggb, ggbg, gbgg, bggg \}.$$

Note that

$$B = \{\text{no boys}\} \cup \{\text{exactly one boy}\}.$$

$$C = \{\text{three girls}\} \cup \{\text{four girls}\}.$$

Similarly,

$$D = \{\text{at least one girl}\} = \{\overline{\text{no girl}}\} = \Omega - \{ bbbb \}. \qquad \square$$

Example 8.   Length of Life of a Television Tube.   A television tube manufactured by a certain company is tested and its total time of service (life length or time to failure) is recorded. We take the idealized sample space $\Omega = \{ \omega : \omega \geq 0 \}$. The event A that the tube outlasts 50 hours is $\{ \omega : \omega > 50 \}$, the event B that the tube fails on or before 150 hours is $\{ \omega : 0 \leq \omega \leq 150 \}$ and the event C that the service time of the tube is at least 25 hours but no more than 200 hours is $\{ \omega : 25 \leq \omega \leq 200 \}$. Then

$$A \cap C = (50, \infty) \cap [25, 200] = (50, 200]$$

$$= \{\text{tube survives 50 hours but does not last more than 200 hours}\}$$

$$B \cup C = [0, 150) \cup [25, 200] = [0, 200]$$

$$= \{\text{tube does not last more than 200 hours}\}$$

$$= \overline{(200, \infty)},$$

and so on. $\qquad \square$

Example 9.   Picking a Point from Unit Square in the Plane.   The experiment consists of picking a point from the square bounded by the points $(0, 0)$, $(1, 0)$, $(1, 1)$, and $(0, 1)$ and recording its coordinates. We take

$$\Omega = \{ (x, y) : 0 \leq x \leq 1, 0 \leq y \leq 1 \}.$$

Clearly all subrectangles of Ω, their unions, and their intersections should be events, as should circles. Indeed, all subsets of Ω that have well-defined area should belong to ζ. The event A that the point selected lies within a unit distance from the origin is given by (see Figure 1):

$$A = \{ (x, y) : x^2 + y^2 < 1, 0 \leq x \leq 1, 0 \leq y \leq 1 \}.$$

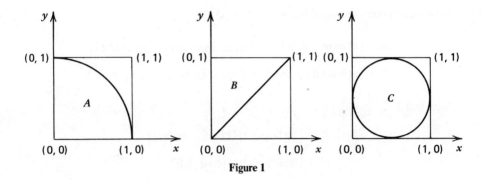

Figure 1

Similarly,

$B = \{$ Point selected lies on or above the diagonal joining $(0,0)$ and $(1,1)\}$

$= \{(x, y): 0 \le x \le y \le 1\},$

$C = \{$ Point selected lies in the inscribed circle with center $(\frac{1}{2}, \frac{1}{2})$ and radius $\frac{1}{2}\}$

$= \{(x, y): (x - \frac{1}{2})^2 + (y - \frac{1}{2})^2 < \frac{1}{4}, 0 \le x \le 1, 0 \le y \le 1\},$

and so on.   □

Problems for Section 2.2

1. For each of the following experiments, choose a sample space.
    (i)   A coin is tossed three times. Four times.
    (ii)   A die is rolled three times. Four times.
    (iii)   A bridge hand is drawn from a well-shuffled bridge deck.
    (iv)   A coin is tossed n times and the number of heads observed is recorded.
    (v)   A car door is assembled with a number of spot welds, say n welds to a door. Each door is checked and the number of defects recorded.
    (vi)   The three components of velocity of a missile are recorded, t minutes after the launch.
    (vii)   A lot of six items contains two defective items. Items are drawn one after another without replacement until both the defectives have been found. The number of draws required is recorded.
    (viii) A pointer is spun and the angle θ at which the point comes to rest is measured.
    (ix)   The age (in years), height, and weight of an individual are measured.
    (x)   Diastolic and systolic blood pressures are measured.
    (xi)   A point is selected from the triangle bounded by $(0,0)$, $(1,0)$, and $(1,1)$.
    (xii)   A point x is selected from $(0,1)$ and then a point y is selected from $(0, x)$.
    (xiii) Consider a rat in a T-maze. The rat is placed in the maze five times and each time it reaches the T bar it can turn to its right or left. The experiment consists of recording the direction of each turn.

2. Consider a committee of three persons who vote on an issue for or against with no abstentions allowed. Describe a suitable sample space and write as sets the following events.
    (i) A tie vote.   (ii) Two "for" votes.   (iii) More votes against than for.

3. Consider the population of all students at a state university. A student is selected. Let A be the event that the student is a male, B the event that the student is a psychology major, C the event that the student is under 21 years of age.
    (i) Describe in symbols the following events:
        (a) The student is a female, aged 21 years or over.
        (b) The student is a male under 21 years of age and is not a psychology major.
        (c) The student is either under 21 years of age or a male but not both.
        (d) The student is either under 21 or a psychology major and is a female.
    (ii) Describe in words the following events:
        (a) $A \cap (B - C)$.   (b) $(A \cup B \cup C) - (A \cap B) \cup (B \cap C) \cup (A \cap C)$.
        (c) $A \cup B \cup C - A \cap B$.   (d) $B - (A \cup C)$.
        (e) $(A \cup B \cup C) - A \cap B \cap C$.   (f) $A \cap B - C$.

4. Which of the following pairs of events are mutually exclusive?
    (i) Being a male or a smoker.
    (ii) Drawing a king or a black card from a bridge deck.
    (iii) Rolling an even-faced or odd-faced value with a die.
    (iv) Being a college student or being married.

5. Three brands of coffee, X, Y, and Z, are to be ranked according to taste by a judge. (No ties are allowed.) Define an appropriate sample space and describe the events "X preferred over Y," "X ranked best," "X ranked second best," in terms of sets.

6. Three transistor batteries are put on test simultaneously. At the end of 50 hours each battery is examined to see if it is operational or has failed. Write an appropriate sample space and identify the following events:
    (i) Battery 2 has failed.
    (ii) Batteries 2 and 3 have either both failed or are both operational.
    (iii) At least one battery has failed.

7. A point is selected from the unit square bounded by $(0,0)$, $(1,0)$, $(1,1)$, and $(0,1)$. Identify the following events:
    (i) The point lies on or above the diagonal joining $(1,0)$ and $(0,1)$.
    (ii) The point lies in the square bounded by $(\frac{1}{2},0)$, $(1,0)$, $(1,\frac{1}{2})$, and $(\frac{1}{2},\frac{1}{2})$.
    (iii) The point lies in the area bounded by $y = 3x$, $y = 0$, and $x = 1$.

8. Consider a 24-hour period. At some time x a switch is put into the ON position. Subsequently, at time y during the same 24 hour period, the switch is put into the OFF position. The times x and y are recorded in hours, minutes, and seconds on the time axis with the beginning of the time period taken as the origin. Let $\Omega = \{(x, y): 0 \leq x \leq y \leq 24\}$. Identify the following events:
    (i) The switch is on for no more than two hours.
    (ii) The switch is on at time t where t is some instant during the 24-hour period.
    (iii) The switch is on prior to time t_1 and off after time t_2 ($t_1 < t_2$ are two instants during the same 24-hour period).

9. A sociologist is interested in a comparative study of crime in five major metropolitan areas. She selected Chicago, Detroit, Philadelphia, Los Angeles, and Houston for her study but decides to use only three of the cities. Describe the following events as subsets of an appropriate sample space:
   (i) Chicago is not selected.
   (ii) Chicago is selected but not Detroit.
   (iii) At least two of Chicago, Detroit, and Los Angeles are selected.
   (iv) Philadelphia, Detroit, and Los Angeles are selected.

10. There are five applicants for a job of which two are to be chosen. Suppose the applicants vary in competence, 1 being the best, 2 the second best, and so on. The employer does not know these ranks. Write down an appropriate sample space for the selection of two applicants and identify the following events:
   (i) The best and one of the two poorest applicants are selected.
   (ii) The poorest and one of the two best are selected.
   (iii) At least one of the two poorest applicants is selected.

2.3 PROBABILITY AXIOMS

Consider a sample space Ω of a statistical experiment and suppose ζ is the class of events. We now consider what is usually the most difficult part of stochastic modeling, assigning probabilities to events in Ω. For completeness we need to be a little more specific about ζ.

DEFINITION 1*. A class ζ of subsets of Ω is said to be a sigma field (written "σ-field") if it satisfies the following properties:

   (i) $\Omega \in \zeta$,
   (ii) if $A \in \zeta$, then $\bar{A} \in \zeta$,
   (iii) if $A_1, A_2, \ldots \in \zeta$, then $\bigcup_{n=1}^{\infty} A_n \in \zeta$,
   (iv) if $A_1, A_2, \ldots \in \zeta$, then $\bigcap_{n=1}^{\infty} A_n \in \zeta$.

We note that (iv) follows easily from (ii) and (iii). We are not greatly concerned in this course with membership in class ζ. The reader needs to remember only that (a) with every Ω we associate a class of events ζ, (b) not every subset of Ω is an event, and (c) probability is assigned only to events (sets of class ζ).

Consider a random experiment with sample space Ω and let ζ be the class of events. Let $A \in \zeta$ be an event. How do we quantify the degree of uncertainty in A? Suppose the experiment is repeated[†] n times and let $f(A)$ be the *frequency* of A, that is, the number of times A happens. Due to statistical regularity, we feel that the *relative frequency* of A, namely, $f(A)/n$, will stabilize near some number p_A as $n \to \infty$. That is, we expect that

(1)
$$\lim_{n \to \infty} \frac{f(A)}{n} = p_A.$$

---

*Parts (sections, theorems, proofs, definitions) marked * are included for completeness; they are intended for mathematically inclined students.

[†]Here we need the condition that these repetitions are independent. For a technical definition of independence, see Section 2.7. Roughly speaking, it means that the relative frequency of what happens on any repetition does not affect the relative frequency of what happens on any other repetition.

It is therefore tempting to take (1) as a definition of the probability of event A. There are both technical and operational problems in doing so. How do we check in any physical situation that the limit in (1) exists? What should be the value of n before we assign to event A probability $f(A)/n$? The probability of an event A should not depend on the experimenter or on a particular frequency $f(A)$ that one observes in n repetitions. Moreover, in many problems it may be impractical to repeat the experiment.

Let us write for convenience

$$r(A) = \frac{f(A)}{n},$$

and examine properties of $r(A)$. We note that

  (i) $0 \le r(A) \le 1$,

 (ii) $r(A) = 0$ if and only if $f(A) = 0$, that is, if and only if A never occurs in n repetitions,

(iii) $r(A) = 1$ if and only if $f(A) = n$, that is, if and only if A occurs in each of the n repetitions,

(iv) if A and B are mutually exclusive, $r(A \cup B) = r(A) + r(B)$.

All these properties appear desirable in any measure of uncertainty. It is intuitively attractive also to desire that $r(A)$ converge to some p_A as $n \to \infty$.

Based on these considerations, it is reasonable to postulate that probability of an event satisfies the following axioms.

DEFINITION 2. Let Ω be a set of outcomes and $\mathfrak{F}$ be the associated collection of events. A set function P defined on $\mathfrak{F}$ is called a probability if the following axioms are satisfied:

Axiom I. $0 \le P(A) \le 1$ for all $A \in \mathfrak{F}$.

Axiom II. $P(\Omega) = 1$.

Axiom III. If $\{A_n\}$ is a sequence of mutually exclusive events ($A_i \cap A_j = \varnothing$ for $i \ne j$), then $P(\bigcup_{i=1}^{\infty} A_i) = \sum_{i=1}^{\infty} P(A_i)$.

Axiom III is much stronger than property (iv) of relative frequency and is the most useful from a computational point of view. In an experiment that terminates in a finite number of outcomes, it is hard to see why Axiom III is required. If the experiment is repeated indefinitely and we consider the combined experiment, there are events that can only be described by a countable number of sets. Consider, for example, repeated tossings of a fair coin until a head shows up. Let us write A_i for the event that a head shows up for the first time on the ith trial, $i = 1, 2, \ldots$, and let us write A for the event that a head will be observed eventually. Then

$$A = \bigcup_{i=1}^{\infty} A_i, \qquad A_i \cap A_j = \varnothing, \quad i \ne j,$$

and one needs Axiom III to compute $P(A)$ from $P(A_i)$, $i = 1, 2, \ldots$.

In Definition 2 we do not require $P(\varnothing) = 0$ since it follows as a consequence of Axioms I and III.

We first consider some examples.

Example 1.   Uniform Assignment.   We consider a sample space Ω that has a finite number, say n, of elements. Let $\Omega = \{\omega_1, \ldots, \omega_n\}$ and let ζ be the set of all subsets of Ω. Each one point set $\{\omega_j\}$ is called an *elementary event*. Suppose we define P on ζ as follows: For each j,

$$P\{\omega_j\} = p_j, \quad j = 1, 2, \ldots, n, \quad p_j \geq 0, \quad \sum_{j=1}^{n} p_j = 1.$$

Then for any $A \in \zeta$, $A = \bigcup_j \{\omega_{i_j}\}$ is a finite union of elementary events. Accordingly, we define

$$P(A) = \sum_j P\{\omega_{i_j}\}.$$

Then P on ζ satisfies Axioms I to III and hence is a probability. In particular, take

$$p_1 = p_2 = \cdots = p_n = \frac{1}{n}$$

so that

$$P(A) = \frac{n(A)}{n}$$

where $n(A) = $ number of elements in A. This assignment of probability is called the *uniform* assignment and is basic to the study of all games of chance such as coin tossing, die rolling, bridge and poker games, roulette and lotteries.   □

According to the uniform assignment, a problem of computing the probability of an event A reduces to that of counting the number of elements in A. It is for this reason that we devote Section 2.5 to the study of some simple counting methods.

We now take some specific examples of the uniform model.

Example 2.   Coin Tossing.   A coin is tossed and the up face is noted. Then $\Omega = \{H, T\}$. Let us define P by setting

$$P\{H\} = p = 1 - P\{T\}, \quad 0 \leq p \leq 1.$$

Then P defines a probability on (Ω, ζ), where $\zeta = \{\varnothing, \{H\}, \{T\}, \{H, T\}\}$. If the coin is fair, we take $p = \frac{1}{2}$ since $n(\Omega) = 2$.

If a fair coin is tossed three times, then Ω has 8 sample outcomes

$$\{HHH, HHT, HTH, THH, HTT, THT, TTH, TTT\}$$

and we take $p_1 = p_2 = \cdots = p_8 = \frac{1}{8}$. Thus

$$P(\text{exactly two heads}) = P\{ HHT, HTH, THH \} = \tfrac{3}{8},$$

$$P(\text{no heads}) = P\{ TTT \} = \tfrac{1}{8},$$

and so on.   □

Example 3. Round-off Error. Suppose the balances of checking accounts at a certain branch of a bank are rounded off to the nearest dollar. Then the actual round-off error equals the true value minus the recorded (rounded off) value. If an outcome ω is the actual round-off error in an account, then

$$\omega \in \Omega = \{ -.50, -.49, \ldots, -.01, 0, .01, \ldots, .50 \}.$$

Clearly, $n(\Omega) = 101$. Let us consider the assignment:

$$p_j = P\left\{ \frac{j}{100} \right\} = \frac{1}{101} \qquad \text{for } j = -50, -49, \ldots, 0, 1, \ldots, 49, 50.$$

Then P defines a probability. The probability that the round-off error in an account is at least 2 cents is given by

$$P\{ -50, \ldots, -2, 2, \ldots, 50 \} = 98/101.$$ □

We now consider some examples of experiments where Ω is countably infinite. This is typically the case when we do not know in advance how many repetitions are required before the experiment terminates. Let $\Omega = \{ \omega_1, \omega_2, \ldots \}$ be a countable set of points and let $\mathfrak{S}$ be the class of all subsets of Ω. Consider the assignment:

$$P\{ \omega_j \} = p_j, \qquad \text{where } p_j \geq 0 \quad \text{and} \quad \sum_{j=1}^{\infty} p_j = 1.$$

Set

$$P(A) = \sum_{\{ j:\, \omega_j \in A \}} p_j.$$

Then

$$0 \leq P(A) \leq 1, \qquad P(\Omega) = \sum_{j=1}^{\infty} p_j = 1,$$

and for any collection $\{ A_n \}$ of disjoint events,

$$P\left(\bigcup_{j=1}^{\infty} A_j \right) = P\left[\bigcup_{j=1}^{\infty} \left(\bigcup_i \{ \omega_{j_i} \} \right) \right] = \sum_{j=1}^{\infty} \sum_i P\{ \omega_{j_i} \}$$

$$= \sum_{j=1}^{\infty} P(A_j).$$

It follows that P defines a probability on $(\Omega, \mathfrak{S})$.

It should be clear that there is no analog of the uniform assignment of probability when Ω is countably infinite. Indeed, we cannot assign equal probability to a countably infinite number of points without violating Axiom III.

Example 4. **Tossing A Coin Until First Head.** Consider repeated tossings with a possibly loaded coin with $P\{H\} = p$, $0 \leq p \leq 1$. The experiment terminates the first time a head shows up. Here $\Omega = \{H, TH, TTH, TTTH, \ldots\}$ is countably infinite. For reasons that will become clear subsequently, consider the assignment

$$p_k = P\left\{\underbrace{TTT\ldots TH}_{k-1}\right\} = (1-p)^{k-1}p, \quad k = 1, 2, \ldots$$

Then

$$0 \leq p_k \leq 1, \quad k = 1, 2, \ldots.$$

Moreover,

$$P(\Omega) = \sum_{k=1}^{\infty} p_k = p \sum_{k=1}^{\infty} (1-p)^{k-1} = p \cdot \frac{1}{(1-(1-p))} = 1.$$

($\sum_{k=0}^{\infty} r^k$ is a geometric series with sum $1/r$ for $|r| < 1$. See Section 6.5 for some properties of this series.) It follows that P defines a probability on $(\Omega, \mathfrak{z})$. Let $A = \{$at least n throws are required$\}$. Then:

$$P(A) = \sum_{k=n}^{\infty} p_k = p \sum_{k=n}^{\infty} (1-p)^{k-1} = p(1-p)^{n-1} \sum_{k=0}^{\infty} (1-p)^k$$

$$= \frac{p(1-p)^{n-1}}{1-(1-p)} = (1-p)^{n-1}, n = 1, 2, \ldots.$$

In particular, if $p = 1/n$, then[†] $P(A) \doteq \lim_{n \to \infty} (1 - 1/n)^{n-1} = 1/e$.   □

Example 5. **Number of Arrivals at a Service Counter.** The number of arrivals at a service counter in a time interval of length t hours is recorded. Then $\Omega = \{0, 1, 2, \ldots\}$ and $\mathfrak{z}$ is the class of all subsets of Ω. A reasonable assignment of probability, as we shall see, is the following:

$$p_k = P\{k\} = \exp(-\lambda t)\frac{(\lambda t)^k}{k!}, \quad k = 0, 1, 2, \ldots$$

where $\lambda > 0$ is a constant. As usual, we postulate that $P(A) = \sum_{k \in A} p_k$. Then $0 \leq P(A) \leq 1$, and

$$P(\Omega) = \sum_{k=0}^{\infty} p_k = e^{-\lambda t} \sum_{k=0}^{\infty} \frac{(\lambda t)^k}{k!} = e^{-\lambda t}e^{\lambda t} = 1.$$

It follows that P defines a probability on $\mathfrak{z}$. If $A = \{$no more than 1 arrival in 15

[†]The notation $\doteq$ means "approximately equal to."

minutes}, then $t = \frac{1}{4}$ and

$$P(A) = P\{0,1\} = e^{-\lambda/4} + e^{-\lambda/4}\frac{(\lambda/4)}{1!}$$

$$= e^{-\lambda/4}\left(1 + \frac{\lambda}{4}\right). \qquad \square$$

Finally, we consider some examples where Ω is a continuous sample space.

Example 6. *Selecting a Point from [a, b].* Suppose a point is selected from the interval $[a, b]$. Then $\Omega = [a, b]$. The outcome $\omega \in [a, b]$. Since Ω contains uncountably many points, we cannot assign positive probability to every one point set $\{\omega\}$, $\omega \in \Omega$ without violating Axiom III. The events of interest here are subintervals of Ω since we are interested in the probability that the chosen point belongs to an interval $[c, d]$, say, where $[c, d] \subseteq [a, b]$. As mentioned in Section 2.2, subintervals of $[a, b]$, as also their unions and intersections, are events. (Indeed $\mathcal{F}$ is taken to be the smallest σ-field containing subintervals of Ω. This is called the Borel σ-field.)

How do we assign probability to events in $\mathcal{F}$? In this case probability is assigned by specifying a method of computation. For example, let f be a nonnegative integrable function defined on Ω such that $\int_a^b f(x)\, dx = 1$. For any subinterval I of Ω, define

$$P(I) = \int_I f(x)\, dx$$

and if $A \in \mathcal{F}$ is a disjoint union of subintervals I_k of Ω, define

$$P(A) = \int_A f(x)\, dx = \sum_k \int_{I_k} f(x)\, dx.$$

Then P defines a probability on Ω.

In particular, what is the analog of the uniform assignment of probability? For convenience, take $a = 0$, $b = 1$. If the points $\omega \in \Omega$ are to be equally likely (as in the case when Ω is finite), each one point set must be assigned the probability 0. Intuitively, the probability that the number selected will be in the interval $[0, \frac{1}{4}]$ should be $\frac{1}{4}$ and so also should be the probability that the number selected is in the interval $[\frac{1}{2}, \frac{3}{4}]$. Therefore a reasonable translation of the uniform assignment in the continuous case is to assign to every interval a probability proportional to its length. Thus we assign to $[c, d] \in \mathcal{F}$

$$P[c, d] = \frac{d - c}{1 - 0} = d - c, \qquad 0 \le c \le d \le 1.$$

This assignment is considered in detail in Section 7.2. $\qquad \square$

Example 7. *Length of Life of a Transistor.* The life length of a transistor (in hours) is measured. Then $\Omega = [0, \infty)$, and we take $\mathcal{F}$ to be the class of events containing all subintervals of Ω. Let us define

$$P(A) = \int_A \lambda^{-1} e^{-x/\lambda}\, dx$$

where $\lambda > 0$ is a fixed constant. If $A = (a, b) \subseteq \Omega$, then

$$P(a, b) = \int_a^b \lambda^{-1} e^{-x/\lambda} \, dx = e^{-a/\lambda} - e^{-b/\lambda}.$$

We leave the reader to show that P is a probability on (Ω, ζ).   □

Example 8. Two-Dimensional Uniform Model. Let us now construct a two-dimensional analog of the uniform model. Suppose $\Omega = \{(x, y): a \le x \le b, c \le y \le d\}$. Clearly all subrectangles of Ω and their unions have to be taken as events. Indeed ζ is taken to contain all subsets of Ω that have a well-defined area. These are sets over which one can integrate. To each $A \in \zeta$ we assign a probability proportional to its area. Thus

$$P(A) = \frac{\text{area of } A}{(b - a)(d - c)}.$$

This is a reasonable analog of the uniform assignment of probability considered in Example 6. According to this assignment, all sets of equal area irrespective of their shape or relative position are assigned the same probability, and in this sense all outcomes are equally likely   □

To summarize, a probability model consists of a triple (Ω, ζ, P) where Ω is some set of all outcomes, ζ is the set of events, and P is a probability defined on events. We call (Ω, ζ, P) a *probability model* or a *probability space* or a *probability system*. To specify this model, one lists all possible outcomes and then all events of interest. If Ω is discrete, it suffices to assign probability to each elementary event, that is, each one-point set. If Ω is an interval or some subset of $\mathbb{R}_k$, we specify a method of computing probabilities. This usually involves integration over intervals (in one dimension) or rectangles (in two or more dimensions).

Remark 1. Definition 2 says nothing about the method of assigning probabilities to events. Since probability of an event is not observable, it cannot be measured exactly. Thus any assignment we make will be an idealization, at best an approximation. Examples 1, 6, and 8 give some specific methods of assigning probabilities to events. (See also Remark 2.) In the final analysis the problem of assigning a specific probability in a practical situation is a statistical problem. If the assignment is poor, it is clear that our model will give a poor fit to the observations, and consequently our inferences are not going to be reliable.

Remark 2. In statistics the word "random" connotes the assignment of probability in a specific way. The word "random" corresponds to our desire to treat all outcomes as equally likely. In the case when Ω is finite this means (Example 1) that the assignment of probability is uniform in the sense that each elementary outcome is assigned probability $1/n(\Omega)$. Thus "a card is randomly selected from a bridge deck" simply means that each card has probability $\frac{1}{52}$ of being selected. We elaborate on this idea still further in Section 2.8.

If, on the other hand, $\Omega \subseteq \mathbb{R}_k$ then the phrase "a point is randomly selected" in Ω simply means that the assignment of probability to events in Ω is uniform. That is, the point selected may be any point of Ω and the probability that the selected point falls in some subset (event) A of Ω is proportional to the measure (length, area, or volume) of A as in Examples 6 and 8.

Remark 3. (Interpretation). The probability of an event is a measure of how likely the event is to occur. How, then, do we interpret a statement such as $P(A) = \frac{1}{2}$? The *frequentist* or objective view, that we will adhere to, is that even though $P(A)$ is not observable it is empirically verifiable. According to this view, $P(A) = \frac{1}{2}$ means that in a long series of (independent) repetitions of the experiment the proportion of times that A happens is about one-half. We thus apply to a single performance of the experiment a measure of the chance of A based on what would happen in a long series of repetitions. It should be understood that $P(A) = \frac{1}{2}$ does not mean that the number of times A happens will be one-half the number of repetitions. (See Sections 5.4 and 6.3.1.)

There is yet another interpretation of the probability of an event. According to this so-called *subjective* interpretation, $P(A)$ is a number between 0 and 1 that represents a person's assessment of the chance that A will happen. Since different people confronted with the same information may have different assessments of the chance of A, their (subjective) estimates of the probability of A will be different. For example, in horse racing, the odds against various horses participating in a certain race as reported by different odds makers are different, even though all the odds makers have essentially the same information about the horses. An important difference between the frequentist and subjectivist approach is that the latter can be applied to a wider class of problems. We emphasize that the laws of probability we develop here are valid no matter what interpretation one gives to the probability of an event. In a wide variety of problems, subjective probability has to be reassessed in the light of new evidence to which a frequency interpretation can be given. These problems are discussed in Section 2.6.

Remark 4. (Odds). In games of chance, probability is stated in terms of *odds*. For example, in horse racing a \$2 bet on a horse to win with odds of 2 to 1 gives a total return of (approximately) \$6 if the horse wins the race. We say that *odds against an event A are a to b if the probability of A is $b/(a + b)$*. Thus odds of 2 to 1 (against a horse) means that the horse has a chance of 1 in 3 to win. This means, in particular, that for every dollar we bet on the horse to win, we collect \$3 if the horse comes in first in the race (see Section 3.7).

Remark 5. For any event A we know that $0 \le P(A) \le 1$. Suppose $A \ne \varnothing$ or Ω. If $P(A)$ is near (or equal to) zero we say that A is possible but improbable. For example, if a point is selected at random from the interval $[0, 1]$, then the probability of every one-point set is zero. It does not mean, for example, that the event $\{.93\}$ is impossible but only that it is possible (since .93 is a possible value) but highly improbable (or has probability zero). In the relative frequency sense, it simply means that in a long series of independent selections of points from $[0, 1]$, the relative frequency of the event $\{.93\}$ will be near zero.

If $P(A) > \frac{1}{2}$ we say that A is more likely than not, and if $P(A) = \frac{1}{2}$ we say that A is as likely or probable as not (or that it has 50–50 chance). If $P(A)$ is close to 1 we say A is highly probable (or very likely), and so on.

Problems for Section 2.3

1. Which of the following set functions define probabilities? State the reasons for your answer.
   - (i) $\Omega = (0, \infty)$, $P(A) = 0$ if A is finite, and $P(A) = 1$ if A is an infinite set.
   - (ii) $\Omega = [0, 1]$, $P(A) = \int_A f(x) \, dx$ where $f(x) = 2x, 0 < x < 1$ and zero elsewhere.
   - (iii) $\Omega = \{1, 2, 3, \dots\}$, $P(A) = \sum_{x \in A} p(x)$ where $p(x) = (\frac{1}{3})^x$, $x = 1, 2, \dots$.
   - (iv) Let $\Omega = \{1, 2, 3, \dots, 2n, 2n + 1\}$ where n is a positive integer. For each $A \subseteq \Omega$ define P as follows:

$$P(A) = 0 \qquad \text{if } A \text{ has an even number of elements,}$$

   and

$$P(A) = 1 \qquad \text{if } A \text{ has an odd number of elements.}$$

   - (v)   Let $\Omega = \{2, 3, 4, 5, 6\}$. Define $p(y) = y/20$, $y \in \Omega$ and for $A \subseteq \Omega$, $P(A) = \sum_{y \in A} p(y)$.
   - (vi) Let $\Omega = \{2, 3, 4, 5, 6\}$. Define $P(A) = \sum_{y \in A} p(y)$, $A \subseteq \Omega$, where $p(y) = (5 - y)/10$, $y \in \Omega$.
   - (vii) Let $\Omega = (0, 1)$. For an event $A \subseteq \Omega$ define $P(A) = \int_A 6x(1 - x) \, dx$.

2. A pocket contains 4 pennies, 3 nickels, 5 dimes, and 6 quarters. A coin is drawn at random. Find the probability that the value of the coin (in cents) drawn is:
   - (i) One cent.   (ii) 25 cents.   (iii) Not more than 5 cents.
   - (iv) At least 5 cents.   (v) At most 10 cents.

3. A lot consists of 24 good articles, four with minor defects, and two with major defects. An article is chosen at random. Find the probability that
   - (i)   It has no defects.
   - (ii) It has no major defects.
   - (iii) It is either good or has minor defects.

4. An integer is chosen at random from integers 1 through 50. Find the probability that:
   - (i)   It is divisible by 3.   (ii) It is divisible by 4 and 6.
   - (iii) It is not larger than 10 but larger than 2.   (iv) It is larger than 6.

5. There are five applicants for a job. Unknown to the employer, the applicants are ranked from 1 (best) through 5 (worst). A random sample of two is selected by the employer. What is the probability that the employer selects:
   - (i)   The worst and one of the two best applicants.
   - (ii) At least one of the two best applicants.
   - (iii) The best applicant.

6. Assuming that in a four-child family, the 16 possible sex distributions of children *bbbb, bbbg, bbgb, bgbb, gbbb, ..., gggg* are equally likely, find the probability that
   - (i)   Exactly one child is a girl.
   - (ii) At least one child is a girl.
   - (iii) At most one child is a girl.
   - (iv) The family has more girls than boys.

7. In order to determine which of three persons will pay for coffee, each tosses a fair coin. The person with a different up face than the other two pays. That is, the "odd

man" who tosses a head when the other two toss tails, or vice versa, pays. Assuming that all the eight outcomes $HHH, HHT, HTH, THH, TTH, HTT, THT, TTT$ are equally likely, what is the probability that the result of one toss will differ from the other two?

8. A person has six keys in her key chain of which only one fits the front door to her house. She tries the keys one after another until the door opens. Let ω_j be the outcome that the door opens on the jth try, $j = 1, 2, \ldots, 6$. Suppose all the outcomes are equally likely. Find the probability that the door opens
    (i) On the sixth try.
    (ii) On either the first or the fourth or the sixth try.
    (iii) On none of the first four tries.

9. Two contestants play a game as follows. Each is asked to select a digit from 1 to 9. If the two digits match they both win a prize. (The two contestants are physically separated so that one does not hear what the other has picked.) What is the probability that the two contestants win a prize?

10. Three people get into an elevator on the first floor. Each can get off at any floor from 2 to 5. Find the probability that:
    (i) They all get off on the same floor.
    (ii) Exactly two of them get off on the same floor.
    (iii) They all get off on different floors.

11. Five candidates have declared for the position of mayor. Assume that the probability that a candidate will be elected is proportional to the length of his or her service on the city council. Candidate A has six years of service on the council; Candidate B has three years of service; C has four years; D, three years; and E, two years. What is the probability that a candidate with less than 4 years of service will be elected? More than 3 years?

12. Roulette is played with a wheel with 38 equally spaced slots, numbered $00, 0, 1, \ldots, 36$. In addition to the number, each slot is colored as follows: 0 and 00 slots are green; 1, 3, 5, 7, 9, 12, 14, 16, 18, 19, 21, 23, 25, 27, 30, 32, 34, 36 are red, and the remaining slots are black. The wheel is spun and a ball is rolled around its edge. When the wheel and the ball slow down, the ball falls into a slot giving the winning number and winning color. (Bets are accepted only on black or red colors and/or on any number or on certain combinations of numbers.) Find the probability of winning on:
    (i) Red.
    (ii) An even number (excluding zero).
    (iii) 1, 4, 7, 10, 13, 16, 19, 22, 25, 28, 31, 34.
    (iv) 1 or red or 1, 2, 3, 4, 5, 6.   (v) 1 and red and 1, 2, 3, 4, 5, 6.

13*. Consider a line segment ACB of length $a + b$ where $AC = a$ and $CB = b, a < b$. A point X is chosen at random from line segment AC and a point Y from line segment BC. Show that the probability that the segments AX, XY, and YB form a triangle is $a/(2b)$. (Hint: A necessary and sufficient condition for the three segments to form a triangle is that the length of each segment must be less than the sum of lengths of the other two.)

14*. The interval $(0, 1)$ is divided into three intervals by choosing two points at random. Show that the probability that the three-line segments form a triangle is $\frac{1}{4}$.

15*. The base and the altitude of a right triangle are obtained by picking points randomly from $[0, a]$ and $[0, b]$ respectively. Show that the probability that the area of the triangle so formed will be less than $ab/4$ is $(\frac{1}{2})(1 + \ln 2)$.

16*. A point X is chosen at random on a line segment AB.
   (i) Show that the probability that the ratio AX/BX is smaller than $a (a > 0)$ is $a/(1 + a)$.
   (ii) Show that the probability that the length of the shorter segment to that of the longer segment is $< \frac{1}{3}$ is $\frac{1}{2}$.

2.4 ELEMENTARY CONSEQUENCES OF AXIOMS

Consider a random experiment with set of outcomes Ω. In Section 2.3 we saw that it is not necessary to assign probability P to every event in Ω. If suffices to give a recipe or a method according to which the probability of any event can be computed. In many problems we either have insufficient information or have little interest in a complete specification of P. We may know the probability of some events of interest and may seek the probability of some related event or events. For example, if we know $P(A)$, what can we say about $P(\overline{A})$? If we know $P(A)$ and $P(B)$, what can we say about $P(A \cup B)$ or $P(A \cap B)$? In this section we derive some properties of a probability set function P to answer some of these questions. These results follow as elementary consequences of the three axioms and do not depend on the particular assignment P on ζ.

Suppose the proportion of female students at a certain college is .60. Then the proportion of male students at this college is .40. A similar result holds for the probability of an event and is proved in Proposition 2 below. Similarly if the proportion of students taking Statistics 100 is .05, that of students taking Mathematics 100 is .18, and that of students taking both Mathematics 100 and Statistics 100 is .03, then the proportion of students taking Mathematics 100 or Statistics 100 is $(.05 + .18 - .03) = .20$. A similar result holds for the set function P and is proved in Proposition 3.

We begin with a somewhat obvious result. Since the event $\varnothing$ has no elements it should have zero probability.

PROPOSITION 1. *For the empty set $\varnothing$, $P(\varnothing) = 0$.*

Proof. Since $\varnothing \cap \varnothing = \varnothing$ and $\varnothing \cup \varnothing = \varnothing$, it follows from Axiom III that

$$P(\varnothing) = P(\varnothing) + P(\varnothing) = 2P(\varnothing).$$

Hence $P(\varnothing) = 0$.   □

In view of Proposition 1 we call $\varnothing$ the *null* or *impossible event*. It *does not* follow that if $P(A) = 0$ for some $A \in \zeta$, then $A = \varnothing$. Indeed, if $\Omega = [0, 1]$, for example, then we are forced to assign zero probability to each elementary event $\{\omega\}$, $\omega \in \Omega$. Note also that if $P(A) = 1$, then it *does not* follow that $A = \Omega$.

PROPOSITION 2. *Let $A \in \mathfrak{F}$. Then*:

(1)
$$P(\overline{A}) = 1 - P(A).$$

Proof. Indeed, $A \cap \overline{A} = \varnothing$ and $A \cup \overline{A} = \Omega$, so that in view of Axioms II and III,

$$P(\Omega) = 1 = P(A) + P(\overline{A}).$$ □

Proposition 2 is very useful in some problems where it may be easier to compute $P(\overline{A})$.

Example 1. Rolling a Pair of Dice. Suppose a pair of fair dice is rolled. What is the probability that the sum of face values is at least 5? Here

$$\Omega = \{(i, j), i = 1, 2, \ldots, 6, j = 1, 2, \ldots, 6\}$$

has 36 outcomes. Since the dice are fair, all pairs are equally likely and

$$P\{(i, j)\} = \tfrac{1}{36} \qquad \text{for all } i, j.$$

By Proposition 2:

$$P\{(i, j): i + j \geq 5, i, j \in [1, 2, 3, 4, 5, 6]\}$$

$$= 1 - P\{(i, j): i + j \leq 4, i, j \in [1, 2, \ldots, 6]\}$$

$$= 1 - P\{(1, 1), (1, 2), (1, 3), (2, 1), (2, 2), (3, 1)\}$$

$$= 1 - \tfrac{6}{36} = \tfrac{5}{6}.$$

A direct computation involves counting the number of elements in the event

$$\{(i, j): i + j = 5, 6, 7, \ldots, 12, i, j \in [1, 2, \ldots, 6]\}$$

which, though not difficult, is time-consuming. □

PROPOSITION 3. *Let $A, B \in \mathfrak{F}$, A and B being not necessarily mutually exclusive. Then*

(2)
$$P(A \cup B) = P(A) + P(B) - P(A \cap B).$$

Proof. A Venn diagram (Figure 1) often helps in following the argument here. It suggests that the events $A \cap \overline{B}$, $A \cap B$, and $B \cap \overline{A}$ are disjoint, so that by Axiom III,

(3)
$$P(A \cup B) = P(A \cap \overline{B}) + P(B \cap \overline{A}) + P(A \cap B).$$

Since, however,

$$A = (A \cap B) \cup (A \cap \overline{B}), \quad B = (A \cap B) \cup (B \cap \overline{A})$$

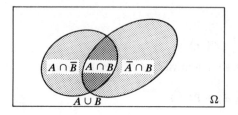

Figure 1. Venn diagram for $A \cup B$.

we have (again Axiom III):

(4) $$P(A \cap \bar{B}) = P(A - B) = P(A) - P(A \cap B),$$

(5) $$P(B \cap \bar{A}) = P(B - A) = P(B) - P(A \cap B).$$

Substituting for $P(A \cap \bar{B})$ and $P(B \cap \bar{A})$ from (4) and (5) into (3), we get (2). ☐

COROLLARY 1. *If $A, B, C \in \zeta$, then:*

(6) $$P(A \cup B \cup C) = P(A) + P(B) + P(C) - P(A \cap B) - P(B \cap C)$$
$$-P(A \cap C) + P(A \cap B \cap C).$$

Proof. Writing $A \cup B \cup C = (A \cup B) \cup C$ and applying Proposition 3 twice, we get (6). ☐

COROLLARY 2. *For $A_1, A_2, \ldots, A_n \in \zeta$*

(7) $$P\left(\bigcup_{i=1}^{n} A_i \right) = \sum_{i=1}^{n} P(A_i) - \sum_{i<j=2}^{n} P(A_i \cap A_j)$$
$$+ \sum_{i<j<k=3}^{n} P(A_i \cap A_j \cap A_k) + \cdots$$
$$+ (-1)^{n-1} P(A_1 \cap A_2 \cap \cdots \cap A_n).$$

Proof. The proof follows by mathematical induction. ☐

COROLLARY 3. *If A and $B \in \zeta$, then*

(8) $$P(A \cup B) \leq P(A) + P(B).$$

Proof. Since $P(A \cap B) \geq 0$, the result (8) follows from Proposition 3. ☐

COROLLARY 4. *If $A_1, A_2, \ldots, A_n \in \zeta$, then*

(9) $$P\left(\bigcup_{k=1}^{n} A_k \right) \leq \sum_{k=1}^{n} P(A_k).$$

Proof. Follows from (8) by induction on n. ☐

It should be clear by now that Axiom III is the only way to compute probabilities of compound events. The property of Axiom III is called *additivity* and that of Corollaries 3 and 4 is called *subadditivity*. The extension of (9) to countable number of events is given below.

In statistics, probability bounds such as (9) are often useful. For example, let $P(A) = .4$ and $P(B) = .2$ and suppose we want $P(A \cup B)$. If A and B are not mutually exclusive, we cannot compute $P(A \cup B)$ exactly unless we know $P(A \cap B)$. See relation (2). Corollary 3, however, tells us that $P(A \cup B) \leq .6$. In fact, even though we do not know $P(A \cup B)$ exactly, we can conclude that $.4 \leq P(A \cup B) \leq .6$. The inequality on the left follows from Proposition 4 (Corollary 2) below.

Example 2. Lottery. Consider a promotional lottery run by a mail-order magazine clearing house. Suppose 1 million entries are received. There are 10 prizes, and prizes are awarded by a random drawing. This means uniform assignment of probability so that each ticket has $10/10^6$ chance of winning a prize. Suppose someone sends in three entries, what is the probability of this person winning at least one prize? Let A_k be the event that the kth ticket wins a prize. Then the event that the person wins at least one prize is

$$A = A_1 \cup A_2 \cup A_3,$$

where A_1, A_2, A_3 are not mutually exclusive. We cannot apply Axiom III, but (6) is applicable and certainly Corollary 4 is applicable. In view of Corollary 4,

$$P(A) \leq P(A_1) + P(A_2) + P(A_3) = 3(10)^{-5} = .00003.$$

Thus even though we have not evaluated the exact probability, we are able to say that it is highly improbable that he or she will win a prize. □

Example 3. Random Selection from Freshman Class. The freshman class at a college has 353 students of which 201 are women, 57 are majoring in mathematics, and 37 mathematics majors are women. If a student is selected at random from the freshman class, what is the probability that the student will be either a mathematics major or a woman?

Each of the 353 students has probability $1/353$ of being selected. Let M be the event that the student is a mathematics major and W be the event that the student is a woman. Then

$$P(M) = 57/353, P(W) = 201/353 \quad \text{and} \quad P(M \cap W) = 37/353,$$

and from (2),

$$P(M \cup W) = P(M) + P(W) - P(M \cap W) = 57/353 + 201/353 - 37/353$$

$$= 221/353.$$ □

Let $A, B \in \zeta$ such that $A \subseteq B$. Is A more probable than B? The answer is no.

PROPOSITION 4. *Let $A, B \in \zeta$ such that $A \subseteq B$. Then*

(10) $$P(A) \leq P(B).$$

Proof. Indeed, since $A \subseteq B$, $A \cap B = A$ so that from (5)

$$P(B - A) = P(B) - P(A \cap B) = P(B) - P(A) \geq 0.$$ □

COROLLARY 1.   *For $A_n \in \zeta$, $n = 1, 2, \ldots$*

(11)
$$P\left(\bigcup_{n=1}^{\infty} A_n \right) \leq \sum_{n=1}^{\infty} P(A_n).$$

Proof.   We have

$$P\left(\bigcup_{n=1}^{\infty} A_n \right) = P\left(A_1 \cup \left(A_2 \cap \bar{A}_1 \right) \cup \left(A_3 \cap \bar{A}_1 \cap \bar{A}_2 \right) \cup \cdots \right)$$

$$\leq P(A_1) + P(A_2) + P(A_3) + \cdots \qquad \text{(Axiom III and Proposition 4)}$$

$$= \sum_{n=1}^{\infty} P(A_n). \qquad\qquad\qquad\qquad \square$$

COROLLARY 2. $\max\{ P(A), P(B) \} \leq P(A \cup B) \leq \min\{ P(A) + P(B), 1 \}$.

PROPOSITION 5.   *Let $A, B \in \zeta$, then*

(12)
$$P(A \cap B) \geq 1 - P(\bar{A}) - P(\bar{B})$$

and, more generally,

(13)
$$P\left(\bigcap_{n=1}^{\infty} A_n \right) \geq 1 - \sum_{n=1}^{\infty} P(\bar{A}_n).$$

Proof.   We have

$$\overline{A \cap B} = \bar{A} \cup \bar{B}$$

so that

$$P(A \cap B) = 1 - P(\overline{A \cap B}) \qquad \text{(Proposition 2)}$$

$$= 1 - P(\bar{A} \cup \bar{B})$$

$$\geq 1 - P(\bar{A}) - P(\bar{B}). \qquad \text{(Inequality (8))}$$

In the general case:

$$\overline{\bigcap_{n=1}^{\infty} A_n} = \bigcup_{n=1}^{\infty} \bar{A}_n$$

so that

$$P\left(\overline{\bigcap_{n=1}^{\infty} A_n} \right) = 1 - P\left(\bigcup_{n=1}^{\infty} \bar{A}_n \right) \qquad \text{(Proposition 2)}$$

$$\geq 1 - \sum_{n=1}^{\infty} P(\bar{A}_n). \qquad \text{(Inequality (11))} \qquad \square$$

COROLLARY 1. $\min\{P(A), P(B)\} \geq P(A \cap B) \geq \max\{0, 1 - P(\overline{A}) - P(\overline{B})\}$.

Example 4. Highly Probable Events. Suppose that in a metropolitan area 80 percent of the crimes occur at night and 90 percent of the crimes occur within the city limits. What can we say about the percentage of crimes that occurs within the city limits at night?

Let A be the event "crime at night" and B be the event "crime within city limits." Then $P(A) = .80$ and $P(B) = .90$, so that from Proposition 2, $P(\overline{A}) = .20$ and $P(\overline{B}) = .10$. In view of (12),

$$P(A \cap B) \geq 1 - .2 - .1 = .70$$

so that the probability of both a crime occurring at night and a crime occurring within the city limits is at least .70. Note that $P(A \cap B) \leq .8$.

It should be noted that the bounds (8) and (12) may yield trivial results. In Example 4, if we want $P(A \cup B)$, then the bound (8) yields $P(A \cup B) \leq .8 + .9 = 1.7$ which, though correct, is useless. Similarly, if $P(A) = .3$ and $P(B) = .1$, then (12) yields

$$P(A \cap B) \geq 1 - .7 - .9 = -.6$$

which, though true, is of no use.   □

Problems for Section 2.4

1. A house is to be given away at a lottery. Tickets numbered 001 to 500 are sold for \$200 each. A number is then randomly selected from 001 to 500 and the person with the corresponding ticket wins the house. Find the probability that the number drawn will be:
   (i)   Divisible by 5.
   (ii) Not divisible by 5.
   (iii) Larger than 200 and not divisible by 5.
   (iv) Divisible by 5 but not 10.
   (v)   Divisible by 5, 6, and 10.

2. The probability that a customer at a bank will cash a check is .72, the probability that he will ask to have access to his safety deposit box is .09, and the probability that he will do both is .003. What is the probability that a customer at this bank will either cash a check or ask to have access to his safety deposit box? What is the probability that he will do neither?

3. Let $P(A) = .12$, $P(B) = 0.89$, $P(A \cap B) = .07$. Find:
   (i) $P(A \cup B)$.   (ii) $P(A \cup \overline{B})$.   (iii) $P(A \cap \overline{B})$.   (iv) $P(\overline{A} \cup B)$.
   (v) $P(\overline{A} \cup \overline{B})$.   (vi) $P(\overline{A} \cap B)$.

4. Suppose A, B, and C are events in some sample space Ω and the following assignment is made:

   $$P(A) = .6, \qquad P(B) = .4, \qquad P(C) = .7.$$

   Find the following probabilities whenever possible. If not, find lower and upper bounds for the probability in question.
   (i) $P(\overline{B})$.   (ii) $P(A \cup C)$.   (iii) $P(B \cap C)$.   (iv) $P(\overline{A} \cup \overline{B})$.
   (v) $P(\overline{A} \cap \overline{B})$.   (vi) $P(A \cup B \cup \overline{C})$.   (vii) $P(A \cap C) + P(\overline{A} \cap C)$.

5. Consider a large lot of spark plugs. The company looks for three kinds of defects, say A, B, and C, in the plugs it manufactures. It is known that .1 percent of plugs have defects of type A, 1 percent of plugs have defects of type B, and .05 percent of plugs have defects of type C. Moreover, .05 percent of plugs have defects of type A and B, .02 percent have defects of type B and C, .03 percent have defects of type A and C, and .01 percent have all three defects. What is the probability that a sampled plug from the lot is defective, that is, has at least one of the three defects?

6. Let A and B be two events. Then $A \Delta B$ is the (disjoint) union of events $A - B$ and $B - A$. That is, $A \Delta B$ is the event that either A happens or B but not both. Show that:

$$P(A \Delta B) = P(A) + P(B) - 2P(A \cap B).$$

7. A number is selected at random from the unit interval $[0, 1]$. What is the probability that the number picked will be a rational number? (Hint: The set of rational numbers is countable.)

8. Let $\Omega = (0, \infty)$ and for each event A define $P(A) = \int_A e^{-\omega} \, d\omega$. What is the probability of the event:

$$E = \{ \omega \in (0, \infty) : |\omega - n| \leq .2 \quad \text{for some nonnegative integer } n \}?$$

9. Tom was interviewed by three prospective employers. On the basis of his interviews he estimates that his chances of being offered a job by the three employers are, respectively, .8, .9, .75. What can you say about his chances of being offered all three jobs?

10. A manufacturer of magnetic tapes for tape decks checks the tapes for two types of defects: improperly applied magnetic coating, and tear or wrinkles in the tape. It is known that about .06 percent of the tapes have coating defects, about .02 percent have wrinkles or tears, and about .015 percent have both defects. Find the probability that a customer will receive:
   (i) a defective tape, (ii) a nondefective tape.

11. It is estimated that a certain bill has an 80 percent chance of being passed by the House of Representatives, a 90 percent chance of being passed by the Senate and a 95 percent chance of being passed by at least one chamber.
   (i) What is the probability that the bill will come before the President for signatures?
   (ii) What is the probability that it will pass one of the chambers but not the other?
   (iii) What is the probability that it will be passed by neither chamber?

12. The probability that Trent will buy a car this year is .90, that he will buy a video cassette recorder is .80, that he will buy a home computer is .95, and that he will buy a car as well as a video cassette recorder is .75. What is your best estimate of the lower bound of the probability that he will buy all three?

13. In a sample survey of 100 students at a university, the students were asked the following question: Which of the three magazines A, B, or C do you read? The

following results were obtained:

55 read A   28 read A and B   22 read all three
42 read B   32 read B and C
69 read C   44 read A and C

If a student is selected at random from these 100 students, find the probability that she reads:

(i) Only A.   (ii) Only B.   (iii) Only C.   (iv) Only A or B.
(v) None of the three.   (vi) At least one.   (vii) At most one.
(viii) Exactly two.   (ix) At least two.

14. A study of a busy metropolitan airport yielded the following estimates of probabilities for the number of airplanes waiting to land at a peak hour:

Number of Airplanes Waiting to Land	1	2	3	4	5	6	7 or more
Probability	.05	.08	.12	.15	.20	.30	.10

Find the probability that
   (i) At least two planes are waiting to land.
   (ii) No more than three planes are waiting to land.
   (iii) At most five planes are waiting to land.
   (iv) More than two planes are waiting to land.

15. In a study of herpes victims it is found that 50 percent are women, 95 percent are white, 80 percent are 20–40 years of age, 55 percent have had at least four years of college, and 55 percent earn over $20,000 a year. Find bounds on the probability that a randomly chosen victim
   (i) Is white and aged 20–40 years.
   (ii) Has had four years of college and earns more than $20,000 a year.
   (iii) Is nonwhite or not aged 20–40 years.

2.5 COUNTING METHODS

Let Ω be a finite sample space with n elements. According to the uniform assignment of probability, if $A \subseteq \Omega$, then

$$P(A) = \frac{n(A)}{n(\Omega)} = \frac{n(A)}{n}$$

where $n(A) =$ the number of elements in A. This assignment reduces the problem of probability computation to that of counting. In this section we consider some rules that facilitate counting. Our subsequent development is not crucially dependent on the knowledge of these rules. Their knowledge does, however, help simplify computations.

We begin with the simplest and yet the most important rule, from which all the remaining rules will be derived.

Rule 1. Let $A = \{a_1, a_2, \ldots, a_m\}$ and $B = \{b_1, b_2, \ldots, b_n\}$ be two sets. Then the number of pairs (a_j, b_k) where $a_j \in A$ and $b_k \in B$ that can be formed is mn. That is, the product set $A \times B$ has mn elements.

Proof. Let us arrange the pairs in a rectangular array (or an $m \times n$ table), as follows.

$\backslash B$ $A\backslash$	b_1	b_2	$\cdots$	b_k	$\cdots$	b_n
a_1	(a_1, b_1)	(a_1, b_2)	$\cdots$	(a_1, b_k)	$\cdots$	(a_1, b_n)
a_2	(a_2, b_1)	(a_2, b_2)	$\cdots$	(a_2, b_k)	$\cdots$	(a_2, b_n)
$\vdots$	$\vdots$	$\vdots$		$\vdots$		$\vdots$
a_j	(a_j, b_1)	(a_j, b_2)	$\cdots$	(a_j, b_k)	$\cdots$	(a_j, b_n)
$\vdots$	$\vdots$	$\vdots$		$\vdots$		$\vdots$
a_m	(a_m, b_1)	(a_m, b_2)	$\cdots$	(a_m, b_k)	$\cdots$	(a_m, b_n)

The assertion of Rule 1 is now obvious. An alternative but frequently used method is to draw a tree diagram, such as in Figure 1. We first draw an element from A and then an element from B. The result of the first draw is exhibited by m branches emanating from a point. Each branch corresponds to one and only one element of A. The result of the second draw is represented by n branches emanating from $a_1, a_2, \ldots, a_m$. The total number of pairs is simply the total number of branches after the second draw, namely, mn. $\square$

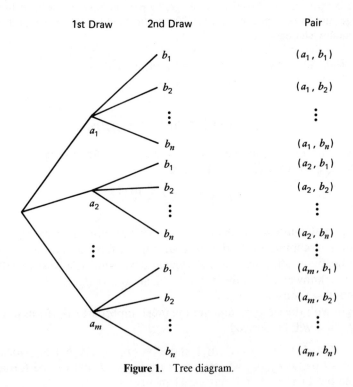

Figure 1. Tree diagram.

COROLLARY 1.   *Let $A_j = \{a_{j,1}, a_{j,2}, \ldots, a_{j,n_j}\}, j = 1,2,\ldots,k$. Then the total number of ordered k-tuples $(a_{1,i_1}, a_{2,i_2}, \ldots, a_{k,i_k})$ that can be formed by taking one element from each set (in the order $A_1, A_2, \ldots, A_k$) is given by $n_1 n_2 \cdots n_k$. That is, the number of elements in the product set $A_1 \times A_2 \times \cdots \times A_k$ is $n_1 n_2 \cdots n_k$.*

Proof.   For $k = 2$, this is simply Rule 1. Now proceed by induction. A tree diagram makes the result transparent.   □

COROLLARY 2.   (*Sampling with Replacement-Ordered Samples*).   *The number of ordered k-tuples $(a_{i_1}, \ldots, a_{i_k})$ where $a_{i_j} \in A = \{a_1, a_2, \ldots, a_n\}$ is n^k. (The term with replacement refers to the fact that the symbol drawn on any draw is replaced before the next draw is made.)*

Proof.   In Corollary 1, take $A_1 = A_2 = \cdots = A_k = A$.   □

Example 1.   Number of License Plates.   Suppose a license plate in a state consists of three letters of alphabet followed by two integers from 0 to 9. What is the total number of license plates possible?

We take $A_1 = A_2 = A_3 = \{A, B, \ldots, Z\}, A_4 = A_5 = \{0, 1, 2, \ldots, 9\}$ in Corollary 1 to Rule 1. Then $n_1 = n_2 = n_3 = 26, n_4 = n_5 = 10$ and the number of possible license plates is $(26)^3 (10)^2 = 1,757,600$.   □

Example 2.   Sex Distribution of Children.   Consider families with five children. Let us write b for a boy and g for a girl and write the sexes of children in order of their ages. How many different distributions are possible? Let us take $A = \{b, g\}, n = 2$ and $k = 5$ in Corollary 2 to Rule 1. The number of possible sex distributions is $2 \cdot 2 \cdot 2 \cdot 2 \cdot 2 = 2^5 = 32$.

Similarly, if we consider families with n children, then there are 2^n different sex distributions.   □

Often sampling is done *without replacement*, that is, an item drawn is not replaced before the next draw. There are two cases to consider: (i) when order matters, (ii) when order does not matter.

Rule 2.   (Ordered Samples Without Replacement).   The number of ordered k-tuples $(a_{i_1}, a_{i_2}, \ldots, a_{i_k})$, where $a_{i_j} \in A = \{a_1, a_2, \ldots, a_n\}$ and *repetition is not allowed* (that is, no two a_{i_j}'s are the same), is given by

$$_nP_k = n(n - 1) \cdots (n - k + 1), \qquad k = 1, 2, \ldots, n.$$

Proof.   Since repetition is not permitted at each draw, one loses exactly one element. Applying Corollary 1 to Rule 1 with $n_1 = n, n_2 = n - 1, \ldots, n_k = n - k + 1$, the assertion follows.   □

COROLLARY 1.   *The number of ordered n-tuples $(a_{i_1}, a_{i_2}, \ldots, a_{i_n})$ where $a_{i_j} \in A = \{a_1, a_2, \ldots, a_n\}$ and repetition is not allowed is $n(n - 1) \cdots 2 \cdot 1$.*

Proof.   Take $k = n$ in Rule 2.   □

The number $_nP_k$ is sometimes called the *number of permutations of k objects out of n* or the *number of permutations of n objects taken k at a time*. In the special case when $k = n$ (Corollary 1) $_nP_n$ represents a reordering or *permutation* of n objects. For convenience we write

$$_nP_n = n! = n(n-1)\cdots 2\cdot 1.$$

It follows that

$$_nP_k = n(n-1)\cdots(n-k+1) = \frac{n!}{(n-k)!}$$

for integers $0 \le k \le n$, and we set $_nP_k = 0$ for integers $k > n$. We adopt the convention that $0! = 1$.

> **Example 3. Management Committee.** From the five-member board of directors of a company, a three-member management committee is to be selected, consisting of a chairman, a treasurer, and a secretary. Since order matters, and sampling is without replacement (it matters whether a member is secretary or treasurer and a member cannot be both), we use Rule 2 with $n = 5$ and $k = 3$. The number of possible committees is $_5P_3 = 5\cdot 4\cdot 3 = 60$. □

> **Example 4. Photographer's Dilemma.** The photographer of a small-town newspaper is asked to take pictures at a swim meet. Some of the pictures are to be included along with the results of the meet in a forthcoming edition of the paper. The photographer takes 20 pictures, all of which turn out well. The editor tells the photographer to select two and to decide on the order in which they will appear in the paper. How many possible choices are there? Using Rule 2 with $n = 20$ and $k = 2$, we see that the photographer has $20(19) = 380$ possible choices. □

Let us now consider a case when the order in which the objects are drawn does not matter. In this context we speak of population and subpopulation. A population (subpopulation) of size n is a collection of n objects without regard to order. We say that two populations (or subpopulations) are different if and only if they have at least one element that is not common to both.

Rule 3. (Unordered Samples Without Replacement). The number of subpopulations of size k or the number of unordered subsets $\{a_{i_1},\ldots,a_{i_k}\}$ or the number of ways of selecting k distinct objects out of n objects in $A = \{a_1, a_2,\ldots,a_n\}$ is

$$\binom{n}{k} = \frac{n!}{k!(n-k)!}.$$

Proof. Let x be the number of subpopulations of size k drawn from n objects $\{a_1, a_2,\ldots,a_n\}$. We note that there are $_nP_k$ ordered k-tuples $(a_{i_1}, a_{i_2},\ldots,a_{i_k})$ where $a_{i_j} \in A$. Each of the $k!$ possible permutations of $(a_{i_1}, a_{i_2},\ldots,a_{i_k})$ leads to the same subpopulation $\{a_{i_1}, a_{i_2},\ldots,a_{i_k}\}$. Hence

$$x(k!) = _nP_k$$

so that

$$x = \frac{_nP_k}{k!} = \frac{n!}{(n-k)!k!} = \binom{n}{k}. \qquad \square$$

The number $\binom{n}{k}$ is called a *binomial coefficient* since it is the coefficient of x^k in the binomial expansion

$$(1 + x)^n = \sum_{k=0}^{n} \binom{n}{k} x^k.$$

The number $\binom{n}{k}$ is sometimes called the number of *combinations of k objects out of n*. We define

$$\binom{n}{0} = 1, \quad \text{and} \quad \binom{n}{k} = 0 \quad \text{for } k > n.$$

Note that

$$\binom{n}{k} = \binom{n}{n-k}, \quad 0 \le k \le n.$$

Example 5. Number of Bridge and Poker Hands. Consider a bridge deck in a game of cards. In a bridge hand, the order of cards in a hand is disregarded. It follows (with $n = 52$ and $k = 13$) that there are $\binom{52}{13} = 635{,}013{,}559{,}600$ different bridge hands. In a five-card poker game, if we disregard the order in which cards are dealt, the number of different hands is $\binom{52}{5} = 2{,}598{,}960$. □

Example 6. Number of Committees. Suppose a committee of five senators is to be selected from the 100 senators in U.S. Senate. The number of such committees is given by $\binom{100}{5}$. □

We note that taking a subpopulation of size k from a population of n objects is equivalent to partitioning the population into two parts of which the first contains k objects and the second $n - k$ objects. Suppose we wish to partition a population of n objects into r parts with $k_1, k_2, \ldots, k_r$ elements such that

$$k_1 + k_2 + \cdots + k_r = n, \quad k_j \ge 0.$$

Rule 4. Let $k_1, k_2, \ldots, k_r$ be nonnegative integers such that $\sum_{j=1}^{r} k_j = n$. The number of ways in which a population of n objects can be partitioned into r subpopulations such that the first contains k_1 elements, the second k_2 elements, and so on, is

$$\binom{n}{k_1, k_2, \ldots, k_r} = \frac{n!}{k_1! k_2! \cdots k_r!}.$$

Proof. We first select k_1 elements from n in $\binom{n}{k_1}$ ways; of the remaining $n - k_1$ elements we select k_2 elements in $\binom{n-k_1}{k_2}$ ways, and so on. By Corollary 1 to Rule 1, the number of such partitions is

$$\binom{n}{k_1}\binom{n-k_1}{k_2}\binom{n-k_1-k_2}{k_3}\cdots\binom{n-k_1-\cdots-k_{r-2}}{k_{r-1}}$$

$$= \frac{n!}{k_1!(n-k_1)!}\frac{(n-k_1)!}{k_2!(n-k_1-k_2)!}\cdots\frac{(n-k_1-\cdots-k_{r-2})!}{k_{r-1}!(n-k_1-\cdots-k_{r-2}-k_{r-1})!}$$

$$= \frac{n!}{k_1!k_2!\cdots k_r!} = \binom{n}{k_1,k_2,\ldots,k_r},$$

since $n - k_1 - \cdots - k_{r-1} = k_r$. (We note that we can make only $r - 1$ choices, the last one being determined automatically.) $\quad\square$

The numbers $\binom{n}{k_1,\ldots,k_r}$ are called *multinomial* coefficients since they appear as coefficients in the multinomial expansion

$$(a_1 + a_2 + \cdots + a_r)^n = \sum_{\substack{k_1,\ldots,k_r \\ k_1+\cdots+k_r=n}} \binom{n}{k_1,\ldots,k_r} a_1^{k_1} a_2^{k_2} \cdots a_r^{k_r}.$$

Example 7. Grade Distribution. A class of 12 students is to be divided in such a way that two students get A's, 2 get B's, 5 get C's and 3 get D's. Then the number of possible grade distributions is given by

$$\binom{12}{2,2,5,3} = \frac{12!}{2!2!5!3!} = 166,320. \qquad\square$$

The following diagram summarizes the results of this section.

Type of Sampling		Use	Typical Examples
With replacement (ordered)—infinite samples possible.		Rule 1	Repeated tossings of coin, die, etc.
Without replacement— finite samples possible.	Ordered	Rule 2	Selecting committees where members have designated jobs.
	Unordered	Rule 3	Selecting members of a committee.

Applications to Probability

We recall that when Ω is finite and the probability assignment is uniform

$$P(A) = \text{number of elements in } A / \text{number of elements in } \Omega.$$

We now consider some applications.

We can use the metaphor of balls and cells to describe the sampling process in a wide variety of applications. Thus k balls can be randomly distributed in n cells in n^k different ways. This follows from Rule 1, since we need to choose one cell for each of the k balls. Thus the experiment of throwing a die k times has 6^k possible outcomes. If we require that no cell has more than one ball, that is, there is no repetition, then Rule 2 gives the number of distributions to be $_nP_k$, $k \leq n$. If all possible distributions have the same probability n^{-k}, then the probability that no cell has more than one ball is given by $_nP_k/n^k$.

Example 8. Elevators, Subway Trains, and Birthdays.

(i) An elevator starts with $k = 5$ passengers and stops at $n = 6$ floors. The probability that no two passengers alight at the same floor is $_6P_5/6^5 = 5/54$.

(ii) A subway train has n cars and is boarded by k passengers ($k \leq n$). Suppose the passengers choose cars randomly. Then the probability that all passengers will board different cars is $_nP_k/n^k$. In particular, if $k = n = 6$, then this probability is $5/324 = .015432$, a rather improbable event.

(iii) Consider a class of k students. The birthdays of these students form a sample of size k from the population of $n = 365$ days in the year (assuming that all years are of equal length). Then the probability that all k birthdays will be different is $_{365}P_k/(365)^k$, and the probability that at least two students will have the same birthday is $\{1 - _{365}P_k/(365)^k\}$. The following table gives $p_k = _{365}P_k/(365)^k$ for some selected values of k.

k	20	23	25	30	35	60
p_k	.589	.493	.431	.294	.186	.006

We note that in a class of 23 students the probability is $1 - .493 = .507$ that at least two students will have a common birthday, whereas in a class of 35 students this probability is .814. □

Example 9. Four Bridge Players Holding an Ace Each.

A bridge deck of 52 cards can be partitioned into four hands of 13 cards each in $52!/(13!)^4$ ways. What is the probability that every player has an ace? We note that the four aces can be permuted (Rule 2) in 4! ways. Each permutation represents exactly one way of giving one ace to each player. The remaining 48 cards can be partitioned (by Rule 4) into four groups of 12 cards each in $48!/(12!)^4$ ways. By Rule 1, the number of distributions where each player holds an ace is $(4!)[48!/(12!)^4]$, so that the required probability is

$$\frac{4!48!}{(12!)^4} \cdot \frac{(13!)^4}{52!} = \frac{2197}{20825} = 0.1055.$$ □

Example 10. Selecting Line Segments Randomly to Form a Triangle.

Given five line segments of lengths 1, 3, 5, 7, 9 centimeters, What is the probability that three randomly selected line segments will form a triangle? Clearly the order of selection is not important. It follows from Rule 3 that there are $\binom{5}{3} = 10$ possible ways of selecting three line segments out of five, each of which has the same probability of being selected, namely $1/10$. Next we note that in order for any three line segments to form a triangle, the sum of any two line segments must be greater than the third. Since only three selections $\{3, 5, 7; \ 5, 7, 9; \ 3, 7, 9\}$ meet this requirement, the probability is $3/10$. □

Example 11. Absent-minded Professor. Consider a professor who signs five letters of references for one of his students and then puts them into five preaddressed envelopes at random. What is the probability that none of the letters is put in its correct envelope? By Rule 2 the number of elements in Ω is $5! = 120$ so that each outcome is assigned the probability of $1/120$. Let A_i be the event that the ith letter is placed in the correct envelope. Then the probability we seek is $1 - P(\bigcup_{i=1}^{5} A_i)$. From Corollary 2 to Proposition 2.4.3, we have:

$$P\left(\bigcup_{i=1}^{5} A_i\right) = \sum_{i=1}^{5} P(A_i) - \sum_{i<j=1}^{5} P(A_i \cap A_j) + \cdots + P(A_1 \cap \cdots \cap A_5).$$

To compute $P(A_i)$, we note that since the ith letter goes to its correct envelope, the number of random distributions of the remaining four letters is $4!$. Hence $P(A_i) = 4!/5! = 1/5 = .20$.

Exactly the same argument applies to $P(A_i \cap A_j)$, $P(A_i \cap A_j \cap A_k)$, and so on. We have:

$$P(A_i \cap A_j) = 3!/5!, P(A_i \cap A_j \cap A_k) = 2!/5!,$$

$$P(A_i \cap A_j \cap A_k \cap A_l) = 1/5! = P(A_1 \cap A_2 \cap A_3 \cap A_4 \cap A_5).$$

It follows that

$$P\left(\bigcap_{i=1}^{5} A_i\right) = 5(1/5) - \binom{5}{2}(1/20) + \binom{5}{3}(1/60) - \binom{5}{4}(1/120) + (1/120) = 19/30$$

so that the required probability is $11/30$.

If there are n envelopes and n letters, then the probability that at least one letter is put into its correct envelope is given by

$$\binom{n}{1}\frac{(n-1)!}{n!} - \binom{n}{2}\frac{(n-2)!}{n!} + \binom{n}{3}\frac{(n-3)!}{n!} + \cdots \mp \frac{1}{n!}$$

$$= 1 - \frac{1}{2!} + \frac{1}{3!} + \cdots \mp \frac{1}{n!} \rightarrow 1 - e^{-1} \quad \text{as } n \rightarrow \infty.$$

It follows that the probability that none of the letters is put into its correct envelope $\rightarrow$ $e^{-1} = .3679$ as $n \rightarrow \infty$. □

Example 12. Parking Tickets. In some cities, overnight street parking is not allowed. A person who got 10 tickets for overnight street parking noticed that all the tickets were issued on Mondays, Tuesdays, and Fridays. Would she be justified in renting a garage only for those nights? (That is, do the police have a system?) Let us assume randomness, that is, that the police have no system and randomly patrol the city. Then the total number of distributions of 7 days ($n = 7$) to 10 tickets ($k = 10$) is 7^{10} (each ticket is assigned one of the seven days). Of these, there are 3^{10} ways of assigning the three days to each ticket. Hence, under the assumption that the police do not have a system, the probability of being ticketed 10 times on Mondays, Tuesdays, and Fridays is $(3/7)^{10} = .0002$. In other words, if the streets were patrolled randomly then it is highly improbable that in a selection of 10 tickets all will be issued on Mondays, Tuesdays, and Fridays. This small probability casts doubt on our assumption that the streets are patrolled randomly.

Suppose that in 10 tickets none is given on Saturdays. Is there sufficient evidence to conclude that no tickets are issued on Saturdays? Proceeding exactly as above, the probability that all 10 tickets are issued on Sundays through Fridays, assuming randomness, is $(6/7)^{10} = .2141$. Since this probability is not small, the evidence does not justify the conclusion that no tickets are issued on Saturdays. □

Problems for Section 2.5

1. A restaurant menu lists three soups, eight meat dishes for the main course, ten desserts, and seven beverages. In how many ways can a full meal be ordered consisting of one soup, one meat dish, one dessert, and one beverage?

2. A nationwide fast-food chain used to advertise that it could serve hamburgers in 256 ways. What is the number of choices of "fixings" (lettuce, ketchup, onion, pickles, etc.) that the chain offers?

3. A Social Security number consists of nine digits. How many different Social Security numbers can there be?

4. Find the number of different four-digit numbers larger than 2000 that can be formed with digits 1, 2, 3, 4 if
    (i) No repetitions are allowed.
    (ii) Repetitions are allowed.

5. Find the number of different four-letter words that can be formed from letters in the word LOVE when
    (i) No repetitions are allowed.
    (ii) Repetitions are allowed.

6. In baseball's World Series, the first team that wins four games out of seven possible games wins the Series. Find the number of ways in which the Series can end.

7. In how many different ways can you permute letters of the word INSTRUMENTS?

8. Find the number of different seven-digit telephone numbers that can be formed if
    (i) The first digit must be 3 and second 2.
    (ii) The first digit must be 3 and no repetitions are allowed.

9. Each permutation of the digits 1, 2, 3, 4, and 5 determines a five-digit number. If numbers corresponding to all possible permutations are listed in increasing order of magnitude beginning with 12345, what is the 96th number in this list?

10. In how many different ways can seven people be seated around a circular table?

11. A professor tells exactly three jokes a semester in his course. If it is his policy never to repeat the same combination of three jokes,
     (i) How many semesters will eight jokes last him?
     (ii) What is the minimum number of different jokes he will tell in 35 semesters? 20?

12. A young woman has four suitors. A gypsy tells the woman that she will be married twice and that both her husbands will come from this group of four men. How many marital histories are possible for this woman?

13. Ten friends meet at a party and shake hands. How many handshakes are exchanged if each person shakes hands with every other person once and only once?

14. Find out how many ways a student can answer a 20-problem true/false examination if
     (i) She marks half the questions true and the other half false.
     (ii) She marks no two consecutive answers the same.

15. What is the possible number of bridge hands that can consist of seven spades, three hearts, two diamonds, and one club?

16. A die is tossed ten times. Find the number of ways in which 1 appears two times, 2 appears three times, 3 appears one time, 4 appears two times, and 5 appears two times.

17. In a survey of students at a state university, 20 students were asked to respond to the following question: "Are you in favor of holding classes on Saturdays? Answer yes, no, no opinion." What is the number of ways in which 5 of the respondents can answer yes, 12 answer no, and 3 answer no opinion?

18*. Find the number of unordered sets (subpopulations) $\{a_{i_1}, a_{i_2}, \ldots, a_{i_k}\}$ where $a_{i_j} \in A = \{a_1, a_2, \ldots, a_n\}$ when repetition is allowed. (The answer gives the rule for unordered sampling with replacement and completes the diagram on page 38.)

19. The automobile license plates issued in a certain state have three digits followed by three letters. A license plate number is selected at random, and the person whose license plate matches the number wins a new car. If you have three cars registered in this state, what is your chance of winning the new car?

20. From six positive and 10 negative numbers, four are picked at random (without replacement) and multiplied. What is the probability that the product is negative?

21. A five-person Credentials Committee is to be chosen from a group of nine men and six women delegates to a national convention. If all the members of the committee selected are men, do women have reason to complain?

22. A lot of 100 fuses is known to contain 20 defective fuses. If a sample of 15 fuses contains only one defective fuse, is there reason to doubt the randomness of the selection?

23. In a game recently run by a well-known fast-food restaurant chain, each customer is given a card with 12 squares. The customer is allowed to scratch exactly three squares to reveal whether a square is marked yes or sorry. Three yeses win a prize. Suppose five squares are marked yes and seven marked sorry. What is the probability that a customer will win a prize?

24. A closet contains six pairs of shoes. Four shoes are selected at random. What is the probability that among the four shoes there will be:
    (i) Exactly one pair, (ii) At least one pair.

25. (i) A married couple and four of their friends enter a row of seats in a concert hall. What is the probability that the wife will sit next to her husband if all possible seating arrangements are equally likely?
    (ii) In part (i), suppose the six people go to a restaurant after the concert and sit at a round table. What is the probability that the wife will sit next to her husband?

26. In a five-card poker game, find the probability that a hand will have:
    (i) A royal flush (ace, king, queen, jack, and ten of the same suit).
    (ii) A straight flush (five cards in a sequence, all of the same suit; ace is high but A, 2, 3, 4, 5 is also a sequence) excluding a royal flush.
    (iii) Four of a kind (four cards of the same face value).
    (iv) A full house (three cards of the same face value x and two cards of the same face value y).
    (v) A flush (five cards of the same suit excluding cards in a sequence).
    (vi) A straight (five cards in a sequence).
    (vii) Three of a kind (three cards of the same face value and two cards of different face values).
    (viii) Two pairs.
    (ix) A single pair.

27*. Consider a town with N people. A person sends two letters to two separate people, each of whom is asked to repeat the procedure. Thus for each letter received, two letters are sent out to separate persons chosen at random (irrespective of what happened in the past). What is the probability that in the first n stages the person who started the chain letter game will not receive a letter?

28. Consider a town with N people. A person tells a rumor to a second person, who in turn repeats it to a third person, and so on. Suppose that at each stage the recipient of the rumor is chosen at random from the remaining $N - 1$ people. What is the probability that the rumor will be repeated n times
    (i) Without being repeated to any person.
    (ii) Without being repeated to the originator.

29. A lot of 15 items contains two defective items. Two items are drawn at random (without replacement). Find the probability that
    (i) Neither is defective. (ii) At least one is not defective.

30. In order to raise money for its Pee Wee League baseball team, a club runs a lottery at its annual picnic. There is exactly one grand prize. Each participant draws a ticket from a box; the winning number is to be announced after all participants have drawn their tickets. Two customers, Jones and Paul, arrive at the same time and Jones is allowed to draw a ticket first. Paul complains that this is unfair since if Jones has drawn the grand prize, Paul has no chance. Is Paul justified in complaining?

31. A local bank has a time-and-temperature service that is repeated over the telephone every 15 seconds. The members of a local swimming team are advised to call the service before coming for morning practice. If the temperature is below 60°F there is no practice. On a certain day 30 people showed up for practice. All claimed that

they called the time and temperature number during the 30-minute period 6:00 a.m. to 6:30 a.m. and none of the callers received a busy signal. Do you have any reason to doubt their claim? (Assume that if two or more calls are received in any 15-second period during which the report is being given, the callers receive a busy signal.)

32. There were four accidents in a town during a seven-day period. Would you be surprised if all four occurred on the same day? Each of the four occurred on a different day?

33. Consider the lottery of Example 2.4.2, where 1 million entries are received and 10 prizes are to be awarded by a random drawing. What is the probability that a person who sends in three entries will win at least one prize?

2.6 CONDITIONAL PROBABILITY

Let (Ω, ζ, P) be a probability model for some random experiment. We recall that P is defined on events, that is, on elements of ζ, and for $A \in \zeta$, $P(A)$ may be thought of as a measure of our belief in the uncertainty of A prior to a performance of the experiment. Often, however, some information becomes available while the experiment is in progress. For example, suppose we know that about 50 percent of people over age 40 are overweight and suppose that the percentage of people who are overweight and have heart disease is 25. A person is randomly chosen from an over-40 age group and is found to be overweight. How do we use this information to compute the probability that the person has heart disease? How do we modify our model to take this information into account?

Formally, suppose $A \in \zeta$ and we are told that A has happened. What is the probability that B has also happened? That is, suppose the outcome is a point in A. What is the probability that it is also in B or, to put it alternatively, what is the probability that event A was "caused" by B? Since the outcome $\omega \in A$, we know that $\omega \notin \overline{A}$, so that all subsets of $\overline{A}$ should get zero probability. Moreover, $\omega \in A$ and $\omega \in B$ if and only if $\omega \in A \cap B$, so that we should assign to B the proportional mass $P(A \cap B)/P(A)$. This is the proportion of event A "occupied" by B (see Figure 1). We are led to the following definition.

DEFINITION 1.   (CONDITIONAL PROBABILITY).   Let $A \in \zeta$ be such that $P(A) > 0$. Then the conditional probability of $B \in \zeta$ given A, written $P\{B|A\}$, is defined by

$$P\{B|A\} = \frac{P(A \cap B)}{P(A)}.$$

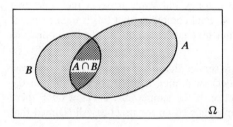

Figure 1.   Part of A occupied by B, $A \cap B$.

In the example above, the (conditional) probability that a randomly selected person over 40 years of age will have heart disease, given that he or she is overweight, is $(\frac{1}{4})/(\frac{1}{2}) = (\frac{1}{2})$.

Are we justified in calling $P\{B|A\}$ a probability? Does $P\{\cdot|A\}$ satisfy the three probability axioms? Clearly $0 \le P\{B|A\} \le 1$ for all $B \in \mathcal{S}$ since $0 \le P(A \cap B) \le P(A)$. Moreover, $P\{A|A\} = 1$. Finally, if $\{A_n\}$ is any sequence of mutually exclusive events, then

$$P\left\{\bigcup_{n=1}^{\infty} A_n | A\right\} = \frac{P\left(A \cap \bigcup_{n=1}^{\infty} A_n\right)}{P(A)} = \frac{P\left(\bigcup_{n=1}^{\infty} (A \cap A_n)\right)}{P(A)}$$

$$= \frac{\sum_{n=1}^{\infty} P(A \cap A_n)}{P(A)} = \sum_{n=1}^{\infty} P\{A_n | A\}$$

since $\{A_n \cap A\}$ are also mutually exclusive. We have thus shown that for each fixed $A \in \mathcal{S}$ with $P(A) > 0$, $P\{\cdot|A\}$ defines a probability on $(A, \mathcal{S})$. It follows that it has all the properties of a probability described in Section 2.3.

Remark 1. One can also motivate the definition of conditional probability in terms of relative frequency. If the experiment is repeated n (independent) times, we look at all those cases where the outcome is in A. If the outcome is also in B for a proportion $p_{B|A}$ times, then we should assign to B the probability $p_{B|A}$. The proportion of times A occurs is approximately $P(A)$ while the proportion of these when B also occurs is about $P(A \cap B)$, so that

$$p_{B|A} = \frac{P(A \cap B)}{P(A)}.$$

Example 1. Distribution of Students by Sex and Declared Major. The distribution of 1237 students at a community college by sex and declared major is given below

Sex	Arts and Sciences (A)	Business (B)	Music (C)	Total
Male (M)	127	383	40	550
Female (F)	380	242	65	687
Total	507	625	105	1237

If a business major is randomly selected from this college, what is the probability that the student is a female? Using uniform assignment of probability, each student has the same chance, $1/1237$, of being selected. Hence

$$P\{F|B\} = \frac{P(B \cap F)}{P(B)} = \frac{242}{625}.$$

Similarly,
$$P\{C|F\} = 65/687, \qquad P\{F|C\} = 65/105. \qquad \square$$

Example 2. Lifetime of a Light Bulb. The lifetime of a light bulb is recorded in hours of use. Take $\Omega = [0, \infty)$. Let P be defined on the events in Ω by

$$P(A) = \int_A \frac{1}{1200} e^{-(1/1200)x} \, dx.$$

The probability of the event A that a bulb will last at least 1500 hours is then given by

$$P(A) = P[1500, \infty) = \int_{1500}^{\infty} (1/1200) e^{-(1/1200)x} \, dx = e^{-1.25},$$

whereas the probability of the event B that a bulb will last 3000 hours given that it has lasted 1500 hours is given by

$$P\{B|A\} = \frac{P(A \cap B)}{P(A)} = \frac{P([1500, \infty) \cap [3000, \infty))}{P[1500, \infty)}$$

$$= \frac{\int_{3000}^{\infty} (1/1200) e^{-(1/1200)x} \, dx}{\int_{1500}^{\infty} (1/1200) e^{-(1/1200)x} \, dx} = \frac{e^{-2.5}}{e^{-1.25}} = e^{-1.25}. \qquad \square$$

Example 3. Meeting for Lunch. Suppose two friends a and b decide to meet for lunch at a preassigned place. Due to traffic conditions, the two can only be sure of arriving between 12:00 noon and 1:00 p.m. Let $\Omega = \{(x, y): 0 \le x \le 1, 0 \le y \le 1\}$ where x and y (in hours) are the respective times measured from 12 noon at which a and b arrive. Then for any event A

$$P(A) = \frac{\text{area of } A}{1} = \text{area of } A.$$

Suppose the two friends further agree to wait no more than 15 minutes from the time of arrival or until the end of the hour. Let A be the event that they meet, and let B be the event that b arrives later than a. Then

$$A = \{(x, y): |x - y| \le 1/4, \qquad 0 \le x \le 1, 0 \le y \le 1\}.$$

Therefore (see Figure 2)

$$P(A) = \text{area of } A = 1 - 2\{(3/4)(3/4)(1/2)\} = 7/16.$$

Now

$$P\{B|A\} = \frac{P(A \cap B)}{P(A)} = \frac{1}{2}$$

since B is the area above the main diagonal. $\qquad \square$

Remark 2. Given the information that A has happened does not necessarily translate into a larger probability for another event B. Indeed, if $A \cap B = \emptyset$,

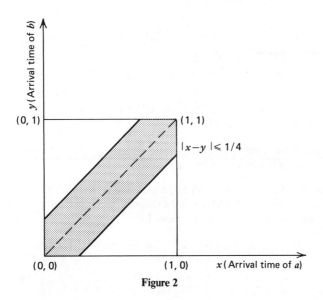

Figure 2

then $P\{B|A\} = 0 \le P(B)$. Here the knowledge of A tells us that B cannot happen. On the other hand, if $B \subseteq A$, then the fact that A has happened cannot decrease the chances of B. Indeed, since $A \cap B = B$,

$$P\{B|A\} = \frac{P(B)}{P(A)} \ge P(B).$$

If $B \supseteq A$, then trivially $P\{B|A\} = 1$.

Remark 3. The conditional probability $P\{B|A\}$ remains undefined for $P(A) = 0$.

As a simple consequence of Definition 1, we note that

(1) $\qquad\qquad P(A \cap B) = P(A)P\{B|A\}, \qquad \text{if } P(A) > 0,$

and,

(2) $\qquad\qquad P(A \cap B) = P(B)P\{A|B\}, \qquad \text{if } P(B) > 0.$

Relations (1) and (2) are special cases of the so-called *multiplication rule* of probability.

PROPOSITION 1. (MULTIPLICATION RULE). *Let $A_1, A_2, \ldots, A_n$ be events such that $P(A_1 \cap A_2 \cap \cdots \cap A_{n-1}) > 0$. Then*

(3) $\quad P\left(\bigcap_{j=1}^{n} A_j\right) = P(A_1)P\{A_2|A_1\}P\{A_3|A_1 \cap A_2\} \cdots P\{A_n|A_1 \cap \cdots \cap A_{n-1}\}.$

Proof. We note that

$$A_1 \supseteq (A_1 \cap A_2) \supseteq \cdots \supseteq (A_1 \cap A_2 \cap \cdots \cap A_{n-1})$$

so that

$$P(A_1) \geq P(A_1 \cap A_2) \geq \cdots \geq P\left(\bigcap_{j=1}^{n-1} A_j\right) > 0$$

and all the conditional probabilities in (3) are well defined. The proof is now completed by mathematical induction using (1) or (2).   □

Proposition 1 is often useful in computing $P(\cap_{j=1}^{n} A_j)$ since evaluation of conditional probability is easier in many models.

Example 4. *Lot Inspection.*   A lot contains 100 items of which 10 are defective. Items are chosen one after another and examined to see if they are defective. Suppose two items are chosen without replacement. Let

$$A_1 = \{\text{first item defective}\}, \quad A_2 = \{\text{second item defective}\}.$$

Then

$$P(A_1) = 10/100 \quad \text{and} \quad P\{A_2 | A_1\} = 9/99$$

so that

$$P(A_1 \cap A_2) = (10/100)(9/99) = 1/110.$$

Here a direct evaluation of $P(A_1 \cap A_2)$, though possible, is cumbersome but $P\{A_2 | A_1\}$ is easily obtained. A tree diagram often helps. In Figure 3 we write (x, y) at each stage

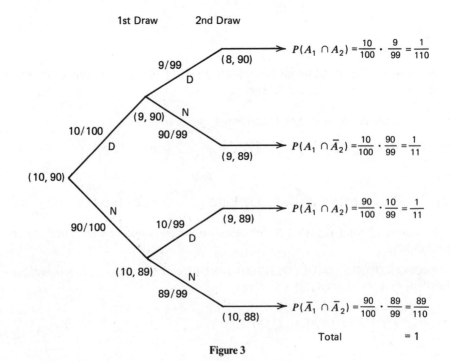

Figure 3

to indicate the contents of the lot: x defectives (D), y nondefectives (N), and along the branches we write conditional probabilities. Thus, if

$$A_3 = \{\text{third item defective}\},$$

then

$$P\{A_3|A_1 \cap A_2\} = 8/98$$

$$P\{A_3|A_1 \cap \bar{A}_2\} = 9/98,$$

and so on. In particular

$$P(A_1 \cap A_2 \cap A_3) = (10/100)(9/99)(8/98)$$

which can be read off from Figure 3 by following the branches labeled D, D, D.   □

We now apply the multiplication rule to compute the probability of an event A.

PROPOSITION 2.   (TOTAL PROBABILITY RULE).   *Let* $\{H_n\}$, $n = 1, 2, \ldots$ *be a sequence of events such that*

$$H_i \cap H_j = \varnothing, \quad i \neq j, \quad \text{and} \quad \bigcup_{n=1}^{\infty} H_n = \Omega.$$

Suppose $P(H_n) > 0$ *for all* $n \geq 1$. *Then for any event* A

(4)
$$P(A) = \sum_{j=1}^{\infty} P(H_j) P\{A|H_j\}.$$

Proof.   Clearly

$$A = A \cap \Omega = A \cap \left(\bigcup_{n=1}^{\infty} H_n \right) = \bigcup_{n=1}^{\infty} (A \cap H_n)$$

and

$$(A \cap H_i) \cap (A \cap H_j) = \varnothing \quad \text{for } i \neq j.$$

By axiom of countable additivity of P we have

$$P(A) = \sum_{n=1}^{\infty} P(A \cap H_n)$$

$$= \sum_{n=1}^{\infty} P(H_n) P\{A|H_n\} \quad [\text{by (1), since } P(H_n) > 0].$$ □

Proposition 2 is very useful. Often $P(A)$ is difficult to compute directly and it is easier to evaluate $P\{A|H_j\}$.

Example 5. Making a Sale.   Consider a local manufacturer of mattresses who consistently runs a sales campaign in the local media. In order to evaluate the effectiveness of the campaign, the manufacturer keeps a record of whether or not a customer has seen the advertisement and whether or not the customer makes a purchase. On the basis of this record, it is estimated that the probability that a customer has seen their advertisement is .6, the conditional probability that a sale is completed given that the customer has seen their advertisement is .7, and the conditional probability that a sale is completed given that the customer has not seen their advertisement is .2. What is the probability that a randomly selected customer will make a purchase?

Let A be the event that the customer has seen their advertisement and S be the event that a customer makes a purchase. Then

$$P(S) = P(A \cap S) + P(\bar{A} \cap S)$$

$$= P(A)P\{S|A\} + P(\bar{A})P\{S|\bar{A}\}$$

$$= (.6)(.7) + (.4)(.2) = .50.$$

A tree diagram often helps. As before, we write conditional probabilities along the branches in Figure 4. We note that the probabilities along the two branches emanating from any node or vertex (such as A or $\bar{A}$) must add to 1. ☐

Example 6. Polya's Urn Model. Consider an urn that contains r red and g green marbles. A marble is drawn at random and its color noted. Then the marble, together with $c > 0$ marbles of the same color, are returned to the urn. Suppose n such draws are made from the urn. What is the probability of selecting a red marble at any draw?

This so-called Pólya's urn model can be used to study the spread of epidemics. The red marbles may correspond to people in a population not having some contagious disease, and the green marbles to people having a certain disease.

Let R_j be the event that the jth marble drawn is red, and G_j the event that the jth marble drawn is green, $j = 1, 2, \ldots, n$. It is clear that R_j and G_j are mutually exclusive. At the kth draw there are $r + g + (k-1)c$ marbles in the urn, and we assume that each is equally likely. We note that for $j \geq 2$

$$P(R_j) = P(R_{j-1} \cap R_j) + P(G_{j-1} \cap R_j)$$

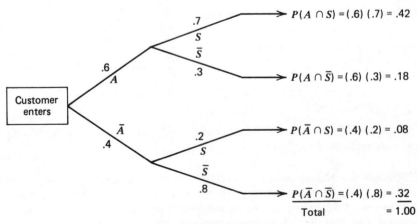

Figure 4

and

$$P(R_1) = \frac{r}{g+r}, \qquad P(G_1) = \frac{g}{g+r}.$$

In particular,

$$P(R_1 \cap R_2) = P(R_1)P\{R_2|R_1\}$$

$$= \frac{r}{r+g}\frac{r+c}{r+g+c}$$

and

$$P(G_1 \cap R_2) = P(G_1)P\{R_2|G_1\}$$

$$= \frac{g}{g+r}\frac{r}{r+g+c}$$

so that

$$P(R_2) = \frac{r(r+c) + gr}{(r+g)(r+g+c)} = \frac{r}{r+g}.$$

Consequently

$$P(G_2) = \frac{g}{r+g}.$$

By induction, it can be shown that

$$P(R_j) = \frac{r}{r+g} \qquad \text{for } j = 1, 2, \ldots, n.$$

It is possible to modify the drawing scheme to make it a more realistic epidemic model. For example we may postulate that c marbles of the color drawn and d marbles of the opposite color are added to the urn (along with the marble drawn) at each draw. The integers c and d may be chosen any way we please. They may even be negative (in which case the procedure terminates after a finite number of draws). The particular case $d = 0$ and $c = -1$ corresponds to drawing without replacement when the procedure terminates after $g + r$ draws. The important point to remember is that the drawing procedure is a convenient way to specify conditional probabilities, which are used to compute some basic probabilities of interest. □

Bayes's Rule

Let Ω be a sample space and let H_j be mutually exclusive events such that

$$\bigcup_{j=1}^{\infty} H_j = \Omega \quad \text{and} \quad P(H_j) > 0, \qquad j = 1, 2, \ldots.$$

We can think of H_j's as the "causes" (or the hypotheses) which lead to the outcome of an experiment. The probabilities $P(H_j), j = 1, 2, \ldots$ are called *prior probabilities*. Suppose the experiment results in an outcome from event A where $P(A) > 0$. What is the probability that the observed event was due to cause H_j?

In other words, we seek the conditional probability $P\{H_j|A\}$. These probabilities are frequently known as *posterior* probabilities. The information that A has occurred leads us to reassess the probability $P(H_j)$ assigned to H_j.

PROPOSITION 3. (BAYES'S RULE). *Let* $\{H_j\}$ *be mutually exclusive events such that* $P(H_j) > 0$ *for* $j = 1, 2, \ldots,$ *and* $\Omega = \bigcup_{j=1}^{\infty} H_j$. *Let* A *be an event with* $P(A) > 0$. *Then for* $j = 1, 2, \ldots$

$$(5) \qquad P\{H_j|A\} = \frac{P(H_j)P\{A|H_j\}}{\sum\limits_{k=1}^{\infty} P(H_k)P\{A|H_k\}}.$$

Proof. By definition

$$(6) \qquad P\{H_j|A\} = \frac{P(A \cap H_j)}{P(A)}$$

$$= \frac{P(H_j)P\{A|H_j\}}{\sum\limits_{k=1}^{\infty} P(H_k)P\{A|H_k\}},$$

where we have used (4).   □

We need not memorize formula (5). It gives nothing new, being simply an alternative way of writing the definition of conditional probability. In applications, we simply recreate formula (5) from the definition.

Example 7. Medical Diagnosis. In medical diagnosis a doctor observes that a patient has one or more of certain specified symptoms $A = \{S_1, S_2, \ldots, S_l\}$ and is faced with the problem of deciding which of the several possible diseases $\{D_1, D_2, \ldots, D_k\}$ is the most probable cause of the observed symptoms. Suppose it has been estimated from medical records that $P(D_j) = p_j$, $p_j > 0$, $j = 1, 2, \ldots, k$. We assume that the diseases never occur together in the same person. We assume further that the conditional probability $P\{A|D_j\}$ can also be estimated from medical records. Then for $j = 1, 2, \ldots, k$ Bayes's rule gives

$$P\{D_j|A\} = \frac{P(D_j)P\{A|D_j\}}{\sum\limits_{j=1}^{k} P(D_j)P\{A|D_j\}}$$

which is the probability that a person with one or more symptoms $A = \{S_1, \ldots, S_l\}$ has disease D_j. The most probable cause of illness then is the disease D_j, for which

$$P\{D_j|A\} > P\{D_i|A\} \qquad \text{for } i = 1, 2, \ldots, k, \quad i \neq j.$$

In particular, let $P(D_1) = .40$, $P(D_2) = .25$, $P(D_3) = .35$, and $P\{A|D_1\} = .8$, $P\{A|D_2\} = .6$, $P\{A|D_3\} = .9$. Then

$$P(A) = .4(.8) + .25(.6) + .35(.9) = .785$$

and

$$P\{D_1|A\} = \frac{.4(.8)}{.785} = .4076, \quad P\{D_2|A\} = \frac{.25(.6)}{.785} = .1911,$$

$$P\{D_3|A\} = \frac{.35(.9)}{.785} = .2473.$$

It follows that a patient who exhibits one or more of the symptoms $\{S_1, S_2, \ldots, S_l\}$ is most likely to have disease D_1. If no further information is available, the patient should be treated for disease D_1.   □

Example 8. Drilling for Oil.   Suppose the probability of a successful strike when drilling for oil is .05. (This can be estimated from past records.) Let A be a set of conditions of the rock. Assume further that the conditional probability of rock condition A when there is oil below is estimated to be .8 and that the conditional probability of rock condition A in the absence of oil is .5. If condition of the rock at a site is observed to be A, what is the chance of striking oil?

Let E the event that there is oil under the ground. Then

$$P(E) = .05 \quad \text{and} \quad P(\bar{E}) = .95.$$

Also

$$P\{A|E\} = .80 \quad \text{and} \quad P\{A|\bar{E}\} = .5.$$

It follows from Bayes's Rule that

$$P\{E|A\} = \frac{P(E)P\{A|E\}}{P(E)P\{A|E\} + P(\bar{E})P\{A|\bar{E}\}}$$

$$= \frac{.05(.80)}{.05(.80) + .95(.5)} = \frac{.04}{.515} = .0777.$$

Thus, the presence of rock condition A at a site does increase the chances of striking oil at the site. It pays to do a geological survey of the site before exploring for oil.   □

Problems for Section 2.6

1. The following table gives the percentage distribution of employees at an automobile plant by sex and marital status.

Marital Status	Female F	Male M	Total
Married, W	26	32	58
Unmarried, $\bar{W}$	22	20	42
Total	48	52	100

An employee is randomly selected from this plant. Find:
(i) $P\{W|F\}$.   (ii) $P\{W|\bar{F}\}$.   (iii) $P\{\bar{W}|F\}$.   (iv) $P\{\bar{W}|F\}$.

2. In a certain town, the probability that it will rain on June 1 is .50, and the probability that it will rain on June 1 and June 2 is .30. What is the probability that it will rain on June 2 given that June 1 was a rainy day?

3. From a lot containing five defective and five nondefective items, a sample of five is drawn (without replacement). Given that the sample contains at least three defectives, what is the probability that it contains exactly two nondefectives?

4. A player is dealt a five-card hand from a 52-card bridge deck. A flush consists of five cards of the same suit.
   (i) Given that all the cards in the hand are black, what is the probability that the hand has a flush?
   (ii) Given that the hand dealt has four clubs and one card of some other suit, and the player is allowed to discard one card in order to draw another, what is the probability that she will be successful in completing a flush? (Assume that she discards the card that is not a club.)
   (iii) If the hand dealt has two cards that are not clubs and the player discards them to draw two new cards, what is the probability that she will complete a club flush?

5. Consider an electronic assembly with two subsystems A and B. Suppose:

$$P(A \text{ fails}) = .1, \quad P(B \text{ fails alone}) = .05, \quad P(A \text{ and } B \text{ fail}) = .08.$$

Find:
   (i) $P\{A \text{ fails}|B \text{ has failed}\}$.   (ii) $P\{A \text{ fails alone}\}$.

6. The diameter of an electrical cable is recorded. Let $\Omega = [0,1]$, and for any event A, $P(A) = \int_A 6x(1 - x)\, dx$. Let A be the event that the diameter is $\leq 1/2$, B the event that diameter is not less than $1/3$ but less than $2/3$. What is $P\{A|B\}$?

7. If 80 percent of all graduate students at a university receive some kind of financial support, and if 10 percent of graduate students who are on support also work part-time, what is the probability that a randomly selected graduate student will be on support as well as working part-time?

8. Let A and B be two events such that $P(A) > 0$, and $P(A) + P(B) > 1$. Show that $P\{B|A\} \geq 1 - \{P(\bar{B})/P(A)\}$.

9. Suppose 45 percent of the students at a university are male and 55 percent are female. If 50 percent of the females and 40 percent of the males smoke, what is the probability that a smoker is male?

10. Consider a lot of light bulbs that contains 30 percent of Brand A bulbs, 45 percent of Brand B bulbs, and 25 percent of Brand C bulbs. It is known that 2 percent of all Brand A bulbs, 3 percent of all Brand B bulbs, and 1 percent of all Brand C bulbs are defective. What is the probability that a randomly selected bulb from the lot will be defective?

11. Consider a bicyclist who leaves a point P (see map), choosing one of the roads PR_1, PR_2, PR_3 at random. At each subsequent crossroad he again chooses a road at random.
    (i) What is the probability that he will arrive at point A?
    (ii) What is the conditional probability that he will arrive at A via road PR_3?

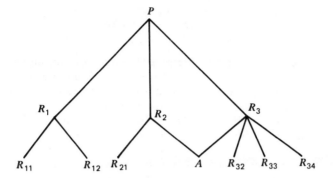

12. It is known from insurance records that a driver who has had a driver education program at high school or through the American Automobile Association has an 85 percent chance of not having an accident during the first year of driving, whereas a new driver who has not had such a program has a 60 percent chance of not having an accident during the first year of driving. If 80 percent of all new drivers undergo a driver education program,
    (i) What is the probability that a new driver will finish the first year of driving without an accident?
    (ii) What is the probability that a new driver who is involved in an accident during the first year of driving has had a driver education?

13. Five percent of patients suffering from a certain disease are selected to undergo a new treatment that is believed to increase the recovery rate from 30 percent to 50 percent. A person is randomly selected from these patients after the completion of the treatment and is found to have recovered. What is the probability that the patient received the new treatment?

14. Three identical cards, one red on both sides, one black on both sides, and the third red on one side and black on the flip side, are placed in a box. A card is then randomly picked from the box and tossed on the table. (Assume that the color on the underside of the card is not observed.)
    (i) What is the probability that the up face is red?
    (ii) Given that the up face is red, what is the probability that the down face is also red?

15. Four roads lead away from the county jail. A prisoner has escaped from the jail and selects a road at random. If road I is selected, the probability of escaping is 1/8; if road II is selected, the probability of success is 1/6; if road III is selected, the probability of escaping is 1/4; and if road IV is selected, the probability of success is 9/10.
    (i) What is the probability that the prisoner will succeed in escaping?
    (ii) If the prisoner succeeds, what is the probability that the prisoner escaped by using road IV? Road I?

16. The geographical distribution of students at a private college is given as: 50 percent from the South, 10 percent from the East, 20 percent from the Midwest and 20 percent from the West. If 60 percent of the Easterners, 30 percent of the southerners,

10 percent of the Midwesterners, and 5 percent of the Westerners wear ties, find the probability that a student wearing a tie comes from the
(i) East.   (ii) Midwest.   (iii) South.   (iv) West.

17. A diagnostic test for a certain disease is 95 percent accurate in that if a person has the disease, it will detect it with a probability of .95, and if a person does not have the disease, it will give a negative result with a probability of .95. Suppose only 0.5 percent of the population has the disease in question. A person is chosen at random from this population. The test indicates that this person has the disease. What is the (conditional) probability that he or she does have the disease?

2.7 INDEPENDENCE

We now consider an important special case of the multiplication rule (2.6.1) (or (2.6.2) or Proposition 2.6.1) when the occurrence of an event A does not affect the assignment of probability to B. This means that $P\{B|A\} = P(B)$, provided $P\{B|A\}$ is well defined, that is, $P(A) > 0$. By multiplication rule, we see that

(1) $$P(A \cap B) = P(A)P\{B|A\} = P(A)P(B).$$

We note that this relation is symmetric in A and B. We are therefore led to the following definition.

DEFINITION 1. (PAIRWISE INDEPENDENCE). Let A and B be two events. We say that A and B are (pairwise) independent if

$$P(A \cap B) = P(A)P(B).$$

If A and B are not independent, we say that they are *dependent*. If A and B are independent then trivially

$$P\{A|B\} = P(A) \quad \text{and} \quad P\{B|A\} = P(B)$$

provided the conditional probabilities are well defined. We note that if $P(A) = 0$, then for any event B

$$0 \leq P(A \cap B) \leq P(A) = 0$$

so that

$$P(A \cap B) = P(A)P(B)$$

and A is independent of any other event.

The reader needs to distinguish between independence and mutually exclusive events. Independence is a property of probability P whereas mutual exclusion is a set theoretic property. If A and B are mutually exclusive events with $P(A) > 0$ and $P(B) > 0$, then $P(A \cap B) = 0 \neq P(A)P(B)$ so that A and B cannot be independent. In fact, $P\{A|B\} = 0 = P\{B|A\}$ so that if A occurs, B cannot and if B occurs, A cannot. Thus A and B are strongly dependent when $A \cap B = \varnothing$.

If A and B are independent, it means that the occurrence of A does not influence the probability of B, that is, one cannot say anything about the occurrence of B from the knowledge that A has occurred. Therefore, if A and B are independent, so should be A and $\overline{B}$ and, by symmetry, $\overline{A}$ and B, and $\overline{A}$ and $\overline{B}$. We leave the reader to prove the following result.

PROPOSITION 1. *Two events A and B are independent if and only if*

(i) *A and $\overline{B}$ are independent, or*

(ii) *$\overline{A}$ and B are independent, or*

(iii) *$\overline{A}$ and $\overline{B}$ are independent.*

In practice, it is often easy to conclude from physical consideration of the experiment whether or not two events are independent, but the only way to be sure is to check that (1) holds.

Example 1. **Sampling With Replacement.** A lot contains 1000 items of which five are defective. Two items are randomly drawn as follows. The first item drawn is tested and replaced in the lot before the second item is selected. Let A_i be the event that ith item drawn is defective, $i = 1, 2$. Then A_1 and A_2 are independent events and

$$P(A_1 \cap A_2) = P(A_1)P(A_2) = (5/1000)(5/1000) = 1/40{,}000.$$

On the other hand, if the first item is not replaced in the lot than A_1 and A_2 are dependent events. Moreover,

$$P(A_1 \cap A_2) = P(A_1)P\{A_2|A_1\} = (5/1000)(4/999)$$

$$= 1/49{,}950. \qquad \square$$

Example 2. **Families with Two Children.** Consider families with two children and assume that the four possible sex distributions bb, bg, gb, gg are equally likely. Let A be the event that a randomly selected family has children of both sexes and B the event that the family has at most one girl. Then

$$P(A) = P\{bg, gb\} = 2/4, P(B) = P\{bb, bg, gb\} = 3/4, \quad P(A \cap B) = 2/4.$$

It follows that A and B are dependent. This is by no means obvious from the experiment.

Indeed, if we consider families with three children, then

$$P(A) = P\{bbg, bgb, gbb, bgg, gbg, ggb\} = 6/8,$$

$$P(B) = P\{bbb, bbg, bgb, gbb\} = 4/8,$$

$$P(A \cap B) = P\{bbg, bgb, gbb\} = 3/8,$$

so that

$$P(A \cap B) = P(A)P(B)$$

and A and B are independent. Thus it is not always obvious whether or not two events A and B are independent. $\square$

Example 3. System Reliability. Consider a system that has two components or subsystems connected in series, as shown in Figure 1. The system functions if and only if both the components C_1 and C_2 are operational. If the two components operate independently then the reliability of the system as measured by the probability that it will survive to time t is given by

$$P(\text{system survives to time } t) = P(C_1 \text{ and } C_2 \text{ survive to time } t)$$

$$= P(C_1 \text{ survives to time } t) P(C_2 \text{ survives to time } t).$$

$\square$

If in Example 3 we have more than two components in the system, how do we go about computing system reliability? Clearly, we need to extend the concept of independence to more than two events. Suppose we have three events A, B, and C. If they are pairwise independent, then

$$(2) \qquad P(A \cap B) = P(A)P(B), \quad P(B \cap C) = P(B)P(C),$$

$$P(A \cap C) = P(A)P(C).$$

Is the requirement (2) enough to say A, B, and C are independent? If not, it does not help us solve the problem of Example 3 with three components. For that purpose we need to say something about $P(A \cap B \cap C)$. Suppose we require that

$$(3) \qquad P(A \cap B \cap C) = P(A)P(B)P(C).$$

It turns out that (3) does not imply (2) and (2) does not imply (3), so we must require both. We are led to the following definition.

DEFINITION 2. (MUTUAL INDEPENDENCE). The n events $A_1, A_2, \ldots, A_n$ are mutually independent if the following $2^n - n - 1$ relations hold:

$$P(A_i \cap A_j) = P(A_i)P(A_j), \qquad 1 \le i < j \le n,$$

$$P(A_i \cap A_j \cap A_k) = P(A_i)P(A_j)P(A_k), \qquad 1 \le i < j < k \le n,$$

$$\vdots$$

$$P(A_1 \cap A_2 \cap \cdots \cap A_n) = P(A_1)P(A_2) \cdots P(A_n).$$

Figure 1. System in series.

Example 4. ***Examples where (2) ⇏ (3) and (3) ⇏ (2).*** Suppose four identical slips of cards marked *aaa, abb, bab, bba* respectively are placed in a box. The contents of the box are well shuffled and a card is drawn at random. Define the events

$$A = \{\text{First letter on card is } a\}$$

$$B = \{\text{Second letter is } a\}$$

$$C = \{\text{Third letter is } a\}.$$

Then $P(A) = P(B) = P(C) = 1/2$,

$$P(A \cap B) = P(B \cap C) = P(A \cap C) = 1/4$$

so that A, B, C are pairwise independent. Since, however,

$$P(A \cap B \cap C) = \tfrac{1}{4} \neq P(A)P(B)P(C),$$

A, B, C are not mutually independent. Thus (2) ⇏ (3).

We next consider an example where (3) ⇏ (2). Suppose a pair of fair dice is rolled so that

$$P\{(i, j)\} = 1/36 \qquad \text{for all } i, j = 1, 2, \ldots, 6.$$

Let

$$A = \{\text{First die shows 3, 4, or 5}\}$$

$$B = \{\text{First die shows 1, 2, or 3}\}$$

$$C = \{\text{Sum of face values is 9}\}.$$

It is clear that $A, B,$ and C are not independent. In fact A, B and B, C and C, A are pairwise dependent. However,

$$P(A \cap B \cap C) = P\{(3,6)\} = \tfrac{1}{36} = P(A)P(B)P(C). \qquad \square$$

Fortunately, it is not very often that we need to check that given events $A_1, A_2, \ldots, A_n$ are independent. Often it will be a part of our hypothesis or will follow from the manner in which the experiment is set up (see page 61). In that case we can use the multiplication rule for independent events

$$P\left(\bigcap_{j=1}^{k} A_{i_j}\right) = \prod_{j=1}^{k} P\left(A_{i_j}\right)$$

for any subset $A_{i_1}, A_{i_2}, \ldots, A_{i_k}$ of $A_1, A_2, \ldots, A_n$.

Example 5.   System Reliability.   Consider a system with k subsystems connected in series (Example 3). The system operates successfully if and only if all the subsystems operate successfully. We assume that the subsystems operate independently.

Let A be the event that the system operates successfully and A_i, the event that the ith subsystem operates successfully. Then the reliability of A is given by

$$P(A) = P\left(\bigcap_{i=1}^{k} A_i\right) = \prod_{i=1}^{k} P(A_i)$$

since A_i's are independent. In particular, if $k = 5$ and $P(A_i) = .95$ for each i, then

$$P(A) = (.95)^5 = .7738.$$

Next we consider a system connected in *parallel* (as shown in Figure 2). The system operates successfully if and only if at least one of the subsystems successfully operates. It follows that $A = \bigcup_{i=1}^{k} A_i$ so that the reliability of A is given by

$$P(A) = P\left(\bigcup_{i=1}^{k} A_i\right).$$

In order to use the independence of A_i we write:

$$P(A) = 1 - P\left(\bigcap_{i=1}^{k} \overline{A}_i\right) = 1 - \prod_{i=1}^{k} P(\overline{A}_i)$$

$$= 1 - \prod_{i=1}^{k} [1 - P(A_i)].$$

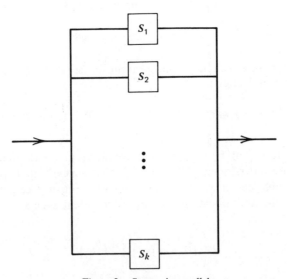

Figure 2.   System in parallel.

In particular, if $k = 5$, and $P(A_i) = .90$ for each i, then
$$P(A) = 1 - (.1)^5 = .99999.$$ □

Independent Trials

Recall that events A and B are independent if and only if $P(A \cap B) = P(A)P(B)$. We emphasize that for more than two events to be independent, it is not enough to check pairwise independence. We have to show that the multiplication rule holds for every pair, every triple, and so on.

We now wish to make precise what we mean by "independent repetitions of an experiment" that was used in Section 2.3 [see (2.3.1) and the following footnote].

For convenience, any performance of an experiment is referred to as a *trial*. If the same experiment is repeated n times we speak of n trials of the experiment.

Suppose we perform n experiments with respective sample spaces $\Omega_1, \Omega_2, \ldots, \Omega_n$. We can think of the n experiments as a single combined experiment with sample space Ω where[†] Ω consists of outcomes $(\omega_1, \omega_2, \ldots, \omega_n)$ with $\omega_i \in \Omega_i, i = 1, 2, \ldots, n$. This means that the first experiment results in outcome ω_1, the second in outcome ω_2, and so on. We require further that if A_i is an event in Ω_i, then $\{(\omega_1, \omega_2, \ldots, \omega_n)\} \in \Omega, \omega_i \in A_i\}$ is an event in Ω.

DEFINITION 3. The n experiments with sample spaces $\Omega_1, \Omega_2, \ldots, \Omega_n$ are said to be independent if any events $A_1, A_2, \ldots, A_n$ whatever with $A_i \subseteq \Omega_i, i = 1, 2, \ldots, n$ are (mutually) independent (in the sense of Definition 2).

According to Definition 3, the probability of any event in any sample space Ω_i is unaffected by what happens in any other sample space $\Omega_j (i \neq j)$. We emphasize that we say nothing in Definition 3 about events in the same sample space. Thus if A_i, B_i are events in Ω_i they may or may not be independent. But any event B_i in Ω_i is independent of A_j or B_j in Ω_j, and so on.

Example 6. *Tossing a Coin and Rolling a Die.* A fair die is rolled and a fair coin is tossed independently. Then we have two experiments with

$$\Omega_1 = \{1, 2, 3, 4, 5, 6\} \quad \text{and} \quad \Omega_2 = \{H, T\}.$$

On Ω_1 we assign uniform probability $P\{i\} = 1/6$ for $i = 1, 2, \ldots, 6$, and on Ω_2 also we assign uniform probability $P\{H\} = P\{T\} = 1/2$. The combined experiment has sample space

$$\Omega = \{(i, H), (i, T): i = 1, 2, \ldots, 6\},$$

where we assign the equal probabilities

$$P\{(i, H)\} = P\{(i, T)\} = 1/12, \quad i = 1, 2, \ldots, 6.$$

Let $A = \{$even face value on die$\} \subseteq \Omega_1$, and $B = \{$heads on the coin$\} \subseteq \Omega_2$. Then A and B are independent. However, if $C = \{$odd face value on die$\} \subseteq \Omega_1$, then A and C are dependent but C and B are independent. □

[†]In the language of set theory, it means that Ω is the product space $\Omega = \Omega_1 \times \Omega_2 \times \cdots \times \Omega_n = \{(\omega_1, \omega_2, \ldots, \omega_n): \omega_i \in \Omega_i, i = 1, 2, \ldots, n\}$.

The case of most interest in statistics is the case when $\Omega_1 = \Omega_2 = \cdots = \Omega_n$. That is, when the same experiment is repeated n times. If the repetitions are independent in the sense of Definition 3, we speak of n independent trials of the experiment (with sample space Ω_1). In practice it may almost be impossible to check that a model with independent trials is appropriate. Fortunately, we do not have to do so if we are careful enough to set up the experiment in such a way that there is no interaction between the trials. Given, however, that the trials are independent, we can always use the multiplication rule on the combined sample space Ω.

Example 7. Coin Tossing. A coin is tossed 20 (independent) times. Each toss is considered a trial. Since the trials are independent, the probability that 20 trials produce 20 heads is p^{20}, where $p = P$ (heads on any trial), $0 < p < 1$.

Suppose a coin is tossed 10 times and 9 heads are observed. Is there evidence to suggest that the coin is biased? If the coin were fair, the probability of observing 9 heads in 10 tosses is $\binom{10}{9}(\frac{1}{2})^{10}$. (We choose 9 trials for 9 heads out of 10 trials in $\binom{10}{9}$ ways and each of these sequences of 9 heads and one tail, by independence, has the same chance $(\frac{1}{2})^{10}$.) This probability is small ($= .0098$). Note that if we were to reject the claim that the coin is fair when we observe 9 heads in 10 trials, we will certainly do so when we observe all 10 heads. Hence the event of interest is not "exactly 9 heads" but "at least nine heads." The probability of this event is

$$\left\{\binom{10}{9} + 1\right\}\left(\frac{1}{2}\right)^{10} = .0107.$$

Thus in a large number of sequences of 10 trials with a fair coin, about 1 percent of the trials will produce at least 9 heads. We can hardly dismiss 9 heads on 10 trials as a mere coincidence, and we have to conclude that the coin is not fair.   □

Example 8. Tossing a Pair of Fair Coins. A pair of fair coins is tossed n (independent) times. Each trial has sample space $\Omega_1 = \{HH, HT, TH, TT\}$ on which we assign uniform probability $P\{\omega\} = 1/4$, $\omega \in \Omega_1$. Let $A = \{$exactly one head$\}$, $B = \{$at least one head$\}$. On any given trial, say trial 1, A and B are dependent events since $P(A) = 1/2$, $P(B) = 3/4$ and $P(A \cap B) = 2/4 \neq P(A)P(B)$. But event A on trial 1 is independent of event B on trial 2 or of event A on trial 2, and so on. Thus

$$P(A \text{ on trial 1 and } B \text{ on trial 2}) = P(A)P(B) = 3/8.$$ □

Example 9. Drug Testing. A doctor studying cures for a certain disease tries a new drug on n randomly selected patients suffering from the disease. In this case the administration of the drug to each patient may be considered a trial. It is reasonable to assume that the trials are independent since the probability that the first patient is cured does not affect (and is not affected by) the probability that the second patient is cured, and so on.

Suppose that the doctor tries the new drug on eight patients and finds that seven patients are cured. It is known that the traditional treatment is only 40 percent effective. Is there evidence to suggest that the new drug is more effective? If the new drug is not effective, than the probability of success is .4. Under this assumption, the probability that 7 or more patients will be cured by the new drug is

$$\binom{8}{7}(.4)^7(.6)^1 + \binom{8}{8}(.4)^8 = .0085.$$

(The argument used here is the same as in Example 7.) Since it is highly improbable to get seven or more cured out of eight patients with the traditional treatment, we have to conclude that our assumption that the probability of success with new drug is .4 is not reasonable. It does not mean that our assumption is necessarily false but only that it is more reasonable to conclude that the new drug is more effective than the one presently in use.   □

Example 10. Quality Control. To keep control on the quality of flash bulbs that a company produces, n randomly selected bulbs are inspected from a day's production. Each inspection is considered a trial. Since the probability that the first bulb, say, is defective does not affect the probability that the ith bulb ($1 \le i \le n$) is defective, and so on, it is reasonable to postulate that the trials are independent.   □

Problems for Section 2.7

1. Which of the following pairs of events are independent?
    (i)   Smoking and lung cancer.
    (ii)   Rain on a particular day and rain on the day following the first.
    (iii) Drinking alcohol and having a car accident.
    (iv)   High-school grades and college performance.
    (v)   Winning a state lottery and buying a car.
    (vi)   Being a male and having blue eyes.
    (vii) Being a girl and playing baseball.
    (viii) Being six feet tall and weighing over 160 pounds.

2. Which of the following pairs of events are mutually exclusive?
    (i)   Being under 20 and earning $20,000 year.
    (ii)   Being a college student and owning a car.
    (iii) Rolling ones on two successive rolls of a die.
    (iv)   Rolling a sum of 6 or a sum of 12 with a pair of dice.
    (v)   Rolling an odd face value and rolling an even face value on a single roll of a die.

3. A fair die is rolled twice. Let A_1, A_2, A_3 be the events

$$A_1 = \{5 \text{ or } 6 \text{ on the first roll}\},$$

$$A_2 = \{5 \text{ on the second roll}\},$$

$$A_3 = \{\text{both rolls produce the same face value}\}.$$

Are A_1, A_2, A_3 pairwise independent?

4. A pair of fair dice is rolled. Let A be the event that the first die shows 4, 5, or 6, let B be the event that the second die shows 1, 2, or 3, and let C be the event that the sum of face values is 7. Are A, B, and C independent events? Are they pairwise independent?

5. Let A and B be events such that $P(A) = .3$, $P(A \cup B) = .8$, and $P(B) = p$.
   (i) For what value of p are A and B independent?
   (ii) What is a necessary condition for A and B to be mutually exclusive?

6. An executive has three telephones on her desk, each with a different number, and none is an extension. During a 15-second interval the probability that phone 1 rings is .35, that phone 2 rings is .60, and that phone 3 rings is .10. Find the probability that during this interval of time
   (i) All three phones ring.   (ii) At least one phone rings.
   (iii) At most two phones ring.   (iv) None of the phones ring.

7. Show that an event A is independent of itself if and only if $P(A) = 0$ or 1.

8. Let A, B, and C be independent events. Show that A and $B \cup C$ are independent and so also are A and $B-C$.

9. Let A and B be independent events. Suppose A and B occur with probability .1 and neither A nor B occur with probability .4. Find $P(A)$ and $P(B)$.

10. A possibly biased coin with $P(\text{heads}) = p$, $0 < p < 1$, is tossed repeatedly until for the first time one observes a run of three heads or three tails. What is the probability that the game will end at the seventh toss?

11. A lot of five identical batteries is life tested. The probability assignment is assumed to be

$$P(A) = \int_A (1/\lambda) e^{-x/\lambda} \, dx$$

for any event $A \subseteq [0, \infty)$ where $\lambda > 0$ is a known constant. Thus the probability that a battery fails after time t is given by

$$P(t, \infty) = \int_t^\infty (1/\lambda) e^{-x/\lambda} \, dx, t \geq 0.$$

If the times to failure of the batteries are independent, what is the probability that at least one battery will be operating after t_0 hours?

12. On $\Omega = (a, b)$, $-\infty < a < b < \infty$, each subinterval is assigned a probability proportional to the length of the interval. Find a necessary and sufficient condition for two events to be independent.

13. It is estimated that the percentages of A, B, O, and AB blood types in a certain population are, respectively, 40, 20, 30, and 10. Three individuals are picked (independently) from this population. (Assume that the population is very large.) Find the probability that:
   (i) They are all type AB.
   (ii) They are all of the same blood type.

(iii) They are all different blood types.

(iv) At least two are blood type A.

(v) None is blood type A.

14. In the World Series of baseball, two teams play a sequence of 4 to 7 games. The first team to win 4 games wins the Series, and no further games are played. If the two teams are evenly matched and the outcomes of different games are independent, find the probability that the series will terminate at the end of:

(i) The fourth game. (ii) The fifth game. (iii) The sixth game.

(iv) The seventh game.

15. A process is in control if it produces at most 5 percent defectives. Random samples of 20 items are taken periodically, and if the sample contains at least two defectives, the process is stopped. What is the probability that the process will be stopped when in fact it is producing only 5 percent defectives?

16. In People vs. Collins, an elderly woman was assauited from behind and robbed on June 18, 1964 in Los Angeles. The victim claimed that she saw a young blond woman run from the scene. Another witness said that he saw a Caucasian woman with dark blond hair and ponytail run out of the alley and enter a yellow automobile driven by a black male with mustache and beard. Within a few days of the crime an interracial couple, Janet and Malcolm Collins, was arrested on the strength of these descriptions and charged with the crime. At the trial a mathematician using the product rule for independent events with probabilities

Event	Probability
A. Yellow car	.10
B. Man with mustache	.25
C. Woman with ponytail	.10
D. Woman with blond hair	.33
E. Black man with beard	.10
F. Interracial couple in a car	.001

testified that the chance was 1 in 12 million that a couple selected at random would possess these characteristics. The probability was overwhelming, he argued, that the accused were guilty because they matched the unusual description. The jury convicted. Do you agree with the mathematician's calculations and conclusion? If not, why not? (On appeal the conviction was reversed. The Supreme Court of California ruled that the trial court should not have admitted the probabilistic evidence.)

17*. In People vs. Collins of Problem 16, the court computed the conditional probability that at least one other couple exists matching these characteristics, given that at least one such couple exists, as

$$\frac{1 - (1 - p)^N - Np(1 - p)^{N-1}}{1 - (1 - p)^N}$$

where p is the probability of selecting at random a couple with the characteristics of the accused and N is the total number of couples in the area. Derive this result.

Letting $p = 1/N$ and $N \to \infty$ show that this conditional probability approaches $(e - 2)/(e - 1) \doteq .418$. (The court argued that even if one accepts the prosecution's figures that $p = 1/12,000,000$, the probability is .42 that the couple observed by the witnesses could be duplicated by at least one other couple with the same characteristics.)

18*. A game of craps is played with a pair of fair dice as follows. A player rolls the dice. If a sum of 7 or 11 shows up, the player wins; if a sum of 2, 3, or 12 shows up, the player loses. Otherwise the player continues to roll the pair of dice until the sum is either 7 or the first number rolled. In the former case the player loses and in the latter the player wins.
  (i)   Find the probability that the player wins on the nth roll.
  (ii) Find the probability that the player wins the game.
  (iii) What is the probability that the game ends on:
       (a) The first roll?   (b) Second roll?   (c) Third roll?

19. Consider a pair of fair dice, A and B. Die A is tossed once. If the face value is i, $i \in \{1, 2, 3, 4, 5, 6\}$, then die B is tossed i times and the sum of the face values on the i tosses of die B is taken.
  (i)   Find the probability that the sum is 6.
  (ii) Find the probability that die B was rolled j times ($j = 1, 2, 3, 4, 5, 6$) given that the sum of face values is 6.

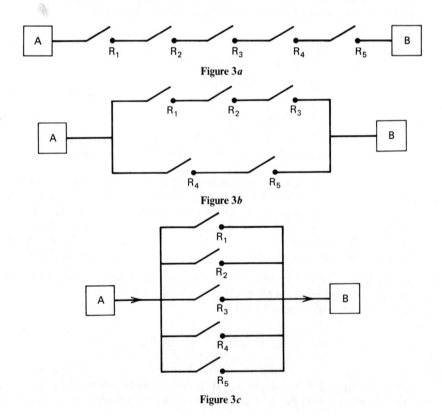

Figure 3a

Figure 3b

Figure 3c

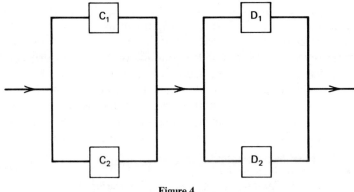

Figure 4

(iii) Suppose die A is loaded in such a way that $P\{i\} = p_i > 0$, $i = 1, 2, \ldots, 6$, $\sum_{i=1}^{6} p_i = 1$ and die B is fair. Find the probabilities in parts (i) and (ii) above.

(iv) Suppose die A is fair but die B is loaded with $P\{i\} = q_i > 0$, $\sum_{i=1}^{6} q_i = 1$. Repeat parts (i) and (ii).

20. Consider five relays connected as shown in Figures $3a$, $3b$ and $3c$. A message will be transmitted from Terminal A to Terminal B: if the relays R_1 to R_5 in Figure $3a$ are all closed; if the relays R_1 to R_3 or R_4 and R_5 in Figure $3b$ are closed; if at least one of the relays $R_1, R_2, \ldots, R_5$ in Figure $3c$ is closed. Suppose all the relays function independently and have the same probability .8 of being closed. Find the probability (in each case) that the message will be transmitted.

21. Consider two parallel systems connected in a series as in Figure 4. Assuming that all the four components operate independently and that the probability that a C-type component functions is p_1 while that for a D-type component is p_2, what is the reliability of the system?

2.8 SIMPLE RANDOM SAMPLING FROM A FINITE POPULATION

The reader has no doubt heard the phrase "random sample" as applied to samples taken by the Gallup and Harris organizations. For the Gallup poll, for example, a sample of size of about 1500 is taken from the entire voting age population of citizens of the United States. How can one make inferences about the election results, for example, on the basis of such a small sample?

We recall that a population is a group of objects about which some information is desired. Thus we speak of the population of voting age citizens of a country, the population of heights of entering freshman, the population of family incomes in a city, and so on. A sample is a part of the population under study, and the objective in statistics is to make inferences about the population from the sample. For example, one may be interested in the proportion of eligible voters who are registered to vote or in the average age of all eligible voters. Clearly no amount of sampling is going to lead to conclusions that are 100 percent accurate unless the sample consists of the entire population, that is, unless we take a *census*. A census, however, is frightfully expensive and time-consuming. Moreover, in many instances (for example, in life testing of batteries) the items in the sample are destroyed. If an automobile manufacturer has to test-drive every car it produces in order to report the average miles per gallon of gasoline that its cars give, then there will be nothing to sell. A carefully chosen sample, on the other hand, can give fairly reliable results. It is clear that not just any sample will do if we have to make probability statements about our inferences. For example, if you were the buyer for a large department store, you would not just look at a couple of shirts from the top of a box. You would do well to sample each box in order to evaluate the quality of the shipment. The two shirts on top may not be representative of the entire box. The ones at the bottom, for example, may be irregular.

In statistics we take *probability samples*, that is, some random or probability device is used in the selection process. This allows us to evaluate the performance of our procedure.

The simplest form of a probability sample is called a *simple random sample*.

DEFINITION 1. (SIMPLE RANDOM SAMPLE). A simple random sample of size n from a population of N objects is a sample chosen in such a way that every collection of n objects in the population has the same chance of being selected.

We note that there are N^n different ordered samples of size n from a population of N objects if samples are taken with replacement and $N^P n$ ordered samples if sampling is done without replacement. In sampling without replacement and without regard to order, there are $\binom{N}{n}$ samples of size n. It follows that in simple random sampling all possible samples have the same chance, namely, $1/N^n$ in ordered sampling with replacement and $1/N^P n$ in ordered sampling without replacement. Similarly, in unordered, without-replacement sampling, each sample has probability $1/\binom{N}{n}$ of being selected. For large N and small n/N, the ratio $N^P n/N^n$ is close to one, so that for large populations (and relatively small-sized samples) we expect the two methods of sampling to be equivalent for all practical purposes.

Typical examples of random sampling with replacement are coin tossing and die rolling. Random sampling without replacement occurs typically in lotteries, successive drawing of cards from a deck, and opinion surveys.

In this section we consider only the case when the population is finite and sampling is done without replacement (and without regard to order). The case when the population is infinite and also its special case when the population is finite and sampling is done with replacement will be considered in Section 3.8.

How do we go about taking a simple random sample? (Unless otherwise stated, we will assume that all samples are simple random samples and we will drop the adjective "simple" from here on.) In theory taking, a random sample is equivalent to drawing n marbles from an urn that contains N marbles, each corresponding to one and only one element of the population. In practice, this is hardly, if ever, done. One uses, instead, a table of random numbers (such as Table A1 in the

Appendix). A table of random numbers is a list of ten digits $0, 1, 2, \ldots, 9$ with the following properties.

(i) The digit in any position in the list has the same chance, namely $1/10$ of being 0, or 1, or 2,..., or 9.

(ii) The digits in different positions are independent.

The random digits are arranged in groups of five into numbered rows in order to facilitate the use of the table. In view of (i) and (ii), we note that any pair of digits in the table has the same chance, namely $1/100$, of being 00, or 01, or 02,..., or 99; in general, any k-tuple of digits in the table has the same chance, namely, $1/10^k$ of being $00\ldots0$, or $00\ldots01$, or $00\ldots02,\ldots$, or $99\ldots9$.

In order to use Table A1, we take the following steps.

(i) Find the number of digits in $N - 1$ where N is the population size.

(ii) Label (in any manner whatever) the N elements in the population from $00\ldots0$ to $N - 1$ where each label has the same number of digits as $N - 1$.

(iii) Enter Table A1 in any place whatever, and systematically read through the table either across or down in groups of k where k is the number of digits in $N - 1$.

(iv) Choose as many k-digit numbers that are less than or equal to $N - 1$ as desired. (We drop duplicates of random numbers already selected.)

Example 1. Random Sample of Voters. A random sample of 10 voters is to be taken from a voters list consisting of 4832 names. Since $N - 1 = 4831$ has four digits we look for four-digit random numbers from 0000 to 4831. For example, entering Table A1 on line 21 and reading across, we get the random numbers

41849	84547	46850	52326	34677
58300	74910	64345	19325	81549
46352	33049	69248	93460	45305
07521	61318	31855	14413	70951

If we agree to take only the first four digits of each five-digit group, and we take only the numbers less than or equal to 4831, then our sample consists of voters labeled

4184	4685	3467	1932	4635
3304	4530	0752	3185	1441

We leave the reader to show that each of 4832 voters has the same probability $1/4832$ of being included in the sample. □

Example 2. Choosing Five of the Fifty States. Suppose a random sample of five is to be taken from 50 states of the United States. Let us list the states alphabetically and number them from 00 for Alabama to 49 for Wyoming. Suppose we begin with the

00	Alabama	25	Montana
01	Alaska	26	Nebraska
02	Arizona	27	Nevada
03	Arkansas	28	New Hampshire
04	California	29	New Jersey
05	Colorado	30	New Mexico
06	Connecticut	31	New York
07	Delaware	32	North Carolina
08	Florida	33	North Dakota
09	Georgia	34	Ohio
10	Hawaii	35	Oklahoma
11	Idaho	36	Oregon
12	Illinois	37	Pennsylvania
13	Indiana	38	Rhode Island
14	Iowa	39	South Carolina
15	Kansas	40	South Dakota
16	Kentucky	41	Tennessee
17	Louisiana	42	Texas
18	Maine	43	Utah
19	Maryland	44	Vermont
20	Massachusetts	45	Virginia
21	Michigan	46	Washington
22	Minnesota	47	West Virginia
23	Mississippi	48	Wisconsin
24	Missouri	49	Wyoming

42nd row of Table A1. We get the following two-digit numbers:

$$81 \quad 61 \quad 61 \quad 87 \quad 11 \quad 53 \quad 34 \quad 24 \quad 42 \quad 76 \quad 75 \quad 12.$$

Ignoring numbers above 49 and duplicates, we obtain the following five distinct numbers,

$$11 \quad 34 \quad 24 \quad 42 \quad 12,$$

which correspond to Idaho, Ohio, Missouri, Texas, and Illinois. This is our random sample of five states. □

Remark 1. It is customary to use a different page and a different line of Table A1 for each application. This can be done by some random procedure.

Remark 2. A random sample is used to represent the population. One cannot simply look at a sample and determine whether the sample resembles the population or not. All one can do is to examine the method used in selecting the sample. If, for example, the method used systematically excludes some elements of the population, the selection procedure is biased. It is therefore important to ascertain how the sample was chosen.

Remark 3.   Simple random sampling is not the only method of obtaining probability samples. Other probability samples such as *stratified random samples* or *cluster samples* are often used and may be more efficient. (A stratified random sample divides the population into several strata. Simple random samples are then taken from each strata. The combined sample is called a stratified random sample. In cluster sampling, groups of elements are chosen randomly.) The Gallup election year poll, for example, uses a multistage cluster sample rather

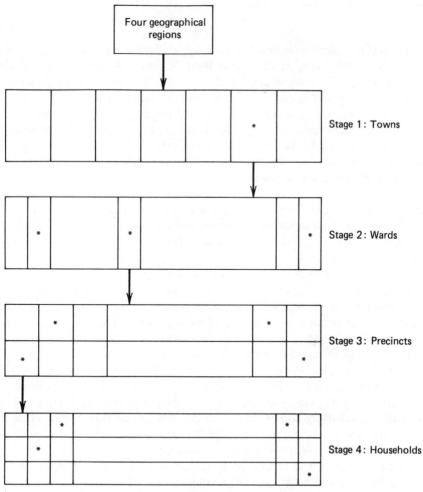

Figure 1

**TABLE 1.   Sample Sizes for Prescribed Levels of Confidence and Margins of Error
(in Simple Random Sampling)**

True Proportion in Population, p	Margin of Error ($\pm$)	Sample Size at Confidence Level		
		90%	95%	99%
.1 (or .9)	3%	271	385	664
	4%	153	217	373
	5%	98	139	239
.3 (or .7)	3%	632	897	1548
	4%	356	505	871
	5%	228	323	557
.5	3%	758	1068	1842
	4%	423	601	1037
	5%	271	385	664

than a simple random sample. The country is divided into four geographical regions: Northeast, South, Midwest, and West. Within each region, all population centers of similar size are grouped together and in the first stage, simple random sample of these towns is selected. Each town is divided into wards, which are further subdivided into precincts. At the second stage, a random sample of wards is selected. A simple random sample of precincts is selected at the third stage, from the selected wards. The fourth stage consists of drawing a simple random sample of households from the selected precincts. Finally, some members of the households selected are interviewed.

Remark 4.   (Sample Size).   We have not yet answered the question that is most often asked about sampling, or more specifically, about public opinion polls. How does a sample of 1500 to 1800 voters from a population of 135 million eligible voters allow us to make such accurate predictions? The first point to remember is that no sample short of a census will produce results that are 100 percent accurate. The object, therefore, is to come as close as possible to 100 percent accuracy. We now need to make two choices. First, how accurate do we want to be in our predictions? The level of accuracy desired (known as confidence level) is usually taken to be large (90, 95 or 99 percent). The second choice, given the first, is that of the margin of error. Within how many percentage points of the true figure do we want to be? In sample surveys the margin of error is usually taken to be between 3 to 5 percent. *The size of the sample depends on these two choices and not on the population size* (provided the population size is large and the sample size is relatively small). In particular, this means that pretty much the same sample size will be required whether the population consists of 150 million people or 500,000 people, provided the confidence level and the margin of error desired are the same. The second point to remember is that in simple random sampling the margin of error for a given confidence level is proportional to $1/\sqrt{n}$ and not $1/n$ where n is the sample size. This means that a doubling of the sample size does not halve the margin of error. Since even sample sizes of 1 or 2 million will produce a certain (however small) margin of error, the time and cost considerations demand that we choose as small a sample as is consistent with our requirements. Most surveys determine the sample size that gives a margin of error of 3 to 5 percent with 95 percent accuracy. If p is the true proportion of voters in a population favoring candidate X (p will not be known and we need to make a guess at it), then the following table gives the sample size required in simple random sampling for several selected levels of confidence and margins of errors associated with the sample proportion favoring X as an estimate or guess of p.

The various concepts such as confidence level, margin of error, and the procedure to determine sample size will all be discussed in detail in later chapters. In particular, see Chapter 4 and Example 9.6.3.

Remark 5. Having settled the question of sample size, how can a sample survey be so accurate? Probability samples are representative of the population being sampled. The situation is similar to blood testing where we take a few drops of blood as representative of the condition of the blood of a person. People show much more variation; nevertheless, as a totality they do conform to certain patterns of behavior. If, therefore, a sample is broadly representative of the population being sampled, we should get fairly reliable results. This is accomplished by probability samples.

Problems for Section 2.8

1. Which of the indicated samples in the following cases are simple random samples?
    (i) In order to estimate the proportion of publishing statisticians with Ph.D. degrees, a statistical society sends a questionnaire to all its members and bases its estimate on the replies received.
    (ii) A radio station asks its listeners to call in after a radio debate between two candidates for a mayoral election. It predicts the winner on the basis of the calls received.
    (iii) In order to evaluate the efficacy of a new drug, a doctor decides to prescribe it to every fourth of his willing patients suffering from the disease. The remaining patients serve as controls for comparison purposes.
    (iv) A mathematics professor chooses a sample of 25 students from her large lecture class consisting of 285 students. She uses a table of random numbers as follows: The last three digits of each group of five numbers is used, beginning with the second line on a randomly selected page of the table and skipping every other line.
    (v) A telephone survey is taken in a certain town to ascertain the sentiment of voters on the Equal Rights Amendment. A random sample of telephone numbers is selected from the telephone directory, and these numbers are called until there are 150 respondents.
    (vi) Every tenth name in a telephone book is picked.

2. In each case in Problem 1, identify the population.

3. The grade distribution of 83 students enrolled in Elementary Statistics is given below.

Student	Grade	Student	Grade	Student	Grade
1	73	29	98	57	64
2	84	30	96	58	62
3	13	31	82	59	73
4	29	32	84	60	75
5	99	33	72	61	84
6	62	34	55	62	92
7	54	35	53	63	89
8	46	36	12	64	76
9	68	37	82	65	75
10	72	38	72	66	62
11	73	39	70	67	61
12	61	40	63	68	58
13	48	41	34	69	59
14	96	42	94	70	67
15	84	43	82	71	92
16	75	44	83	72	94
17	89	45	75	73	83
18	37	46	65	74	72
19	15	47	62	75	76
20	26	48	63	76	73
21	38	49	56	77	61
22	95	50	59	78	54
23	98	51	69	79	67
24	100	52	72	80	70
25	75	53	71	81	72
26	64	54	75	82	71
27	83	55	75	83	35
28	68	56	36		

Use the table of random numbers (Table A1) to select 10 random samples of size 10 each.

4. In Problem 3, use Table A1 to select a random sample of size 20. Begin with line 3, read across, and choose the first two digits of each group of 5 digits.

5. What is the probability that five consecutive digits in a random sample of size 5 from a random digit table will be all different?

6. A lot of 20 light bulbs contains three defective bulbs. Find the (smallest) sample size of the random sample from the lot such that the probability that the sample contains at least one defective light bulb will exceed:
   (i) 1/4.   (ii) .4.   (iii) .95.

7. A class has five sophomores, three juniors, and two seniors.
   (i) What is the smallest sample size required so that with a probability of at least .6 the sample will contain all three juniors?
   (ii) What is the smallest sample size needed so that with probability at least .45 the sample will contain all three juniors and both seniors?

8. From a population of N elements a random sample of size n is taken. Show that
   (i) in without-replacement sample, the probability that *any fixed* element of the population will be included in the random sample is n/N, $(n < N)$.
   (ii) In with-replacement sampling the corresponding probability is $1 - \{(N - 1)/N\}^n$.

9. In 1936, Literary Digest magazine predicted an overwhelming victory (by a margin of 57 percent to 43 percent) for Landon over Roosevelt. This prediction was based on a mail-in poll conducted by the magazine. It mailed out 10 million questionnaires of which 2.4 million were returned. These names came from the club's membership roster, from telephone books, and from other lists. However, Roosevelt won the election by 62 percent to 38 percent. Where did the magazine go wrong in its prediction? (The same year Gallup predicted with a sample of only 3000 what the Literary Digest's predictions were going to be. Moreover, with another sample of 50,000 people Gallup was able to predict correctly that Roosevelt would win, although he forecast that the margin would be 56 percent for Roosevelt to 44 percent for Landon.)

10. In the 1948 election three independent polls (Gallup, Crossley, and Roper) predicted well ahead of the election that Dewey would win over Truman by a margin of about 5 percent. In the election, however, Truman won over Dewey by a margin of about 5 percent. The method used by three pollsters is called *quota sampling*. In quota sampling, each interviewer has a fixed quota of subjects to interview. Moreover, the number of subjects falling into certain categories such as sex, age, race, and economic status are also fixed. The interviewer was free to select anyone he or she wished to. What went wrong?

11. Before the 1980 presidential election, Jimmy Carter and Ronald Reagan had a televised debate, on October 28. The ABC Television network announced that it would conduct a poll of viewers immediately following the debate. Viewers were asked to call in one of two special telephone numbers, one for those favoring Carter and the other for those favoring Reagan. The viewers were told that their local telephone company would bill them 50 cents for the call. The results were reported by ABC News later that night. Callers favored Reagan over Carter by a 2–1 margin. The poll was roundly criticized by the media and experts. Why?
   (About 736,000 calls were received by 12:53 a.m. of October 29, when the ABC poll closed. At any time until then, about 5,000 calls could be received by either number. In the 1980 election Reagan won over Carter by a huge margin.)

2.9 REVIEW PROBLEMS

1. Let f be as defined below, and define P by $P(A) = \int_A f(x)\, dx$ where A is an event. Does P define a probability?
   (i) $f(x) = (x + \theta/2)/\theta^2$, $0 \le x \le \theta$, and zero elsewhere.

(ii) $f(x) = 3x^2/2\theta^3$ if $0 \le x \le \theta$, $= 3(x - 2\theta)^2/2\theta^3$ if $\theta \le x \le 2\theta$, and zero elsewhere.

(iii) $f(x) = 3(\theta + x)^2/2\theta^3$ for $-\theta \le x \le 0$, $= 3(\theta - x)^2/2\theta^3$ for $0 \le x \le \theta$, and zero elsewhere.

(iv) $f(x) = 1/x^2$, $x \ge 1$, and zero elsewhere.

(v) $f(x) = \theta e^{\theta x}$, $x < 0$, and zero elsewhere.

In each case where the answer is yes, find $P(A)$ where $A = (-1, 1)$.

2. Let $\Omega = \{0, 1, 2, \dots\}$. For $A \subseteq \Omega$, define $P(A)$ as follows:

   (i) $P(A) = 1$ if A contains the element 0 and $P(A) = 0$ otherwise.

   (ii) $P(A) = \dfrac{1}{2} \displaystyle\sum_{x \in A} \left\{ \dfrac{e^{-\lambda}\lambda^x}{x!} + p(1 - p)^x \right\}$; $\lambda > 0$, $0 < p < 1$ are constants.

   Does P define a probability?

3. Let A, B, and C be three events, and suppose that $A \cap B \subseteq C$. Show that $P(\bar{C}) \le P(\bar{A}) + P(\bar{B})$.

4. The Supreme Court of the United States has nine Justices (a Chief Justice and eight Associate Justices). Suppose all Justices vote randomly (for or against) on a particular issue before the Court.

   (i) Find the probability that a majority votes in favor of the plaintiff.

   (ii) If a specified judge abstains, find the probability of a tied vote.

   (iii) If there are four women and five men Justices, find the probability that they vote along sex lines. (A unanimous vote on the issue may be excluded.)

5. Suppose n books numbered 1 through n are arranged in a random order on a shelf. What is the probability that books numbered $1, 2, \dots, r$ $(r < n)$ appear as neighbors in that order?

6. A movie theatre can accommodate $n + k$ people. Suppose n seats are occupied. Find the probability that $r(\le n)$ given seats are occupied.

7. Let A and B be two independent events such that $P(A) > 0$, $P(B) > 0$. For any event C show that $P\{C|A\} = P(B)P\{C|A \cap B\} + P\{\bar{B}\}P\{C|A \cap \bar{B}\}$.

8. In a multiple choice test with k choices to each question, a candidate either knows the answer with probability p, $0 \le p \le 1$, or does not know the answer with probability $1 - p$. If he knows the answer, he puts down the correct answer with probability .9, whereas if he guesses, the probability of his guessing correctly is $1/k$. Find the conditional probability that the candidate knew the answer to a question, given that he has answered the question correctly. Show that this probability tends to 1 as $k \to \infty$.

9. A biased coin with probability p, $0 < p < 1$, of heads is tossed until a head appears for the first time. What is the probability that the number of tosses required is even? Odd?

10. A pond contains red and golden fish. There are 3000 red and 7000 golden fish, of which respectively 200 and 500 are tagged.

    (i) Find the probability that a random sample of 100 red and 200 golden fish will show 15 and 20 tagged fish respectively.

    (ii) Find the probability that a random sample of 300 fish will show 15 red tagged and 20 green tagged fish.

11. In Problem 2.3.8 we considered a person who has six keys in her key chain, only one of which fits the door to her house. We assumed that the outcomes ω_j, $j = 1, 2, \ldots, 6$, that the door opens on the jth try, are equally likely. Suppose instead that she tries the keys one at a time, choosing at random from those keys that have not yet been tried. What is the probability that she opens the door
    (i)   On the third try?
    (ii) On either the first or the third try?
    (iii) On none of the first four tries?

12. A man named in a paternity suit is required to take a blood test. Suppose that the child has a blood type that could only have come from his or her father. If the man does not have the type in question, he is not the father; if he does have the type in question, the chance of his being the father is increased. Let q be the proportion of people in the population with the blood type in question. Let p be the probability that the mother's assertion of the man's paternity is true.
    (i) Find the probability that the alleged father has the blood type in question.
    (ii) Find the posterior probability of paternity given that the alleged father has the blood type in question in terms of p and q. In particular, if $q = .4$, and $p = .6$, find the posterior probability of paternity.

13. Suppose the proportion of people in a population with a certain characteristic A (such as a certain blood type or a certain latent palm print) is p, $0 < p < 1$. From physical evidence at the scene of a crime it is known that the criminal has characteristic A. A person with characteristic A is caught and accused of the crime. Let π be the prior probability that the accused is guilty. Find the posterior probability that the accused is guilty for values of $p = .01, .1, .2, .3, .5, .9$, and $\pi = .01, .1, .2, .3, .5, .9$.

14. Blackjack, or Twenty-one, is a game requiring two players and a bridge deck. Cards with a face value of 2 to 10 have a value equal to the face value on the card, while face cards (kings, queens, and jacks) are all valued 10. An ace is valued either a 1 or an 11. The dealer deals two cards to each player and to herself or himself. The players can draw more cards. The players bet against the dealer. A player wins if he or she has a hand with value that exceeds the value of the dealer's hand without going over 21. If the face value of the two initial cards dealt is 21 it is called blackjack. If the cards are dealt from a well-shuffled deck of 52 cards, what is the chance of drawing a blackjack? (There are several variations of the game.)

15. In an office pool, 17 players contribute a dollar each. Seventeen slips of the same size are prepared, each with a name corresponding to one of the participants. An eighteenth blank slip is added. All the slips are put in a hat and shuffled thoroughly. Slips are drawn one after another until the blank slip shows up. The person whose name appears on the *next* slip wins.
    (i) What is the chance of a particular participant winning the pool if nobody wins when the last slip drawn is blank?
    (ii) What is the chance of a participant winning the pool if the procedure is repeated every time the last slip drawn is blank?

16. Suppose a polygraph test is 95 percent reliable in the sense that the conditional probability is .95 that the test says yes when the person is telling the truth and the conditional probability is also .95 that the test says no when the person is lying.

Suppose that the prior probability that a person is lying is .005. What is the posterior probability that the person is lying given that the test states that he or she is lying?

17. Let $A_1, A_2, \ldots, A_n$ be n events. We have seen in Corollary 2 to Proposition 2.4.3 how to compute $P(\bigcup_{i=1}^n A_i)$, given $P(A_i)$, $P(A_i \cap A_j)$, $P(A_i \cap A_j \cap A_k), \ldots$, and so on. Show that the following bounds hold:

$$P\left(\bigcup_{i=1}^n A_i\right) \le \sum_{i=1}^n P(A_i)$$

$$P\left(\bigcup_{i=1}^n A_i\right) \ge \sum_{i=1}^n P(A_i) - \sum_{i<j} P(A_i \cap A_j)$$

$$P\left(\bigcup_{i=1}^n A_i\right) \le \sum_{i=1}^n P(A_i) - \sum_{i<j} P(A_i \cap A_j)$$

$$+ \sum_{i<j<k} P(A_i \cap A_j \cap A_k)$$

$$\ge \cdots$$

$$\le \cdots$$

where the inequality changes direction each time we add an additional term in the expansion of $P(\bigcup_{i=1}^n A_i)$.

In particular, let:

$$P(A_1) = .5, \quad P(A_2) = .7, \quad P(A_3) = .4, \quad P(A_4) = .6,$$

$$P(A_1 \cap A_2) = .3, \quad P(A_1 \cap A_3) = .35, \quad P(A_2 \cap A_3) = .25,$$

$$P(A_1 \cap A_4) = .4, \quad P(A_2 \cap A_4) = .45, \quad P(A_3 \cap A_4) = .3,$$

$$P(A_1 \cap A_2 \cap A_3) = .2, \quad P(A_1 \cap A_2 \cap A_4) = .25, \quad P(A_1 \cap A_3 \cap A_4) = .20,$$

$$P(A_2 \cap A_3 \cap A_4) = .15.$$

Find the best possible bounds for $P(A_1 \cup A_2 \cup A_3 \cup A_4)$.

18. An Elementary Statistics course has 123 students, numbered 1 to 123. Select a random sample of size 23 as follows. Begin with line 24 of Table A1, selecting the first three digits of each group of five-digit numbers.

19. Suppose A and B are independent events with $P(A) = 1/4$ and $P(B) = 3/4$. Find:
   (i) $P(A \cup B)$,   (ii) $P\{A|A \cup B\}$.   (iii) $P\{B|A \cup B\}$.

20. Suppose current flows through three switches A_1, A_2, and A_3 to a heater and back to the battery as shown in the diagram. Suppose further that the switches operate

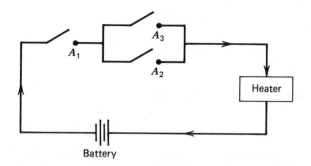

independently and the probability that switch A_1 is on (closed) is .9, that switch A_2 is on is .8, and that switch A_3 is on is .85. Find the probability that the heater is on.

21. A nuclear power plant has a fail-safe mechanism that consists of eight protective devices $A_1, A_2, \ldots, A_8$ which operate independently. An accident will take place if at least seven devices fail. Suppose the probability that any single device fails is .1. Find the probability of an accident. Find the probability that at least five out of the eight devices fail.

22. Two mechanics working for the same company have the same type of citizen band equipment in their vans. The range of their system is 30 miles. Mechanic A traveling towards base from the West is within 35 miles of the base at 5 p.m., and Mechanic B traveling towards base from the South is within 40 miles of the base at 5 p.m. Find the probability that they can communicate with each other at 5 p.m.

23. A bomber is to drop a bomb inside a half-mile square field in order to destroy a munitions area in the field. At each of the four corners of the field is a building that will be destroyed if the bomb falls within 1/4 mile of it. Find the probability that:
    (i)   None of the buildings is destroyed.
    (ii) At least one building is destroyed.
    (iii) Exactly one building is destroyed.
    (iv) At least two buildings are destroyed.

CHAPTER 3

Probability Distributions

3.1 INTRODUCTION

Suppose (Ω, ζ, P) is the probability model of a random experiment. This means that Ω is a set of all possible outcomes of the experiment, ζ is a collection of events of interest and P is an assignment of probabilities to events in ζ. Since it is inconvenient to work with sets (for example, one cannot perform arithmetical operations on elements of ζ) and set functions we consider an alternative model for the random experiment expressed in terms of point functions. In this new model the outcomes of Ω are converted into numerical outcomes by associating with each outcome $\omega \in \Omega$ a real number $X(\omega)$. Thus X defined on Ω is a real valued function and is known as a random variable. Suppose that instead of observing ω we observe $X(\omega)$. Then we have a new set of outcomes $\{X(\omega), \omega \in \Omega\}$ which may be taken to be $\mathbb{R}$, the real line. Subsets of $\mathbb{R}$ are events, although not every set of real numbers can be regarded as an event. Only those subsets E of $\mathbb{R}$ are taken to be events for which we can compute $P(\omega: X(\omega) \in E)$. In particular, all intervals, which are the fundamental objects of interest in this new sample space, are events.

How do we compute probabilities for events on $\mathbb{R}$? This is done by associating with each event $E \subseteq \mathbb{R}$ the probability $P_X(E) = P(\omega: X(\omega) \in E)$. This new model $(\mathbb{R}, \mathcal{B}, P_X)$ where $\mathcal{B}$ is the collection of events on $\mathbb{R}$ is based entirely on the model (Ω, ζ, P). This description, however, is not satisfactory. We need still to simplify computation of probabilities. This is done by replacing P_X by a point function on $\mathbb{R}$ which allows us to use our knowledge of calculus.

Formal definition of a random variable is given in Section 3.2. We also consider the case when each observation is a k-tuple of real numbers. In Section 3.3 we consider some (equivalent) methods of describing a random variable. The case of several random variables is considered in Section 3.4. In Section 3.5 we consider conditional models. In order to specify the joint distribution of two random variables X and Y it is sufficient to specify the conditional distribution of $X(Y)$ given $Y(X)$ and the marginal distribution of $Y(X)$. Both these concepts are introduced in Section 3.5. The important special case of independence of X and Y is considered in Section 3.6. In this case it suffices to specify the marginal distributions of X and Y in order to specify the joint distribution of (X, Y) Section 3.7 deals with some numerical characteristics of the distribution of a random variable. In Section 3.8 we define what we mean by a random sample in sampling from an infinite population.

3.2 RANDOM VARIABLES AND RANDOM VECTORS

Suppose we have selected a probabilistic model for a random phenomenon. This means that we have chosen a sample space Ω, specified events in this sample space, and have made an assignment of probability to these events. Usually we are not interested in all of Ω but only some event or events in Ω. The last chapter was devoted to the computation of probabilities of such events. The reader may have noticed that in many instances we are interested in a much smaller set of outcomes whose members are real numbers. For example, when we toss a coin n times we are mostly interested in the number of heads. We hardly, if ever, keep a record of what happens on any particular trial. Hence what we observe is a number between 0 and n (both inclusive) rather than a particular n length sequence of heads and tails.

Similarly, if two persons start walking from the same point at the same time along perpendicular lines, then the sample space Ω consists of all ordered pairs of positive real numbers (x, y) corresponding to their distances from the starting point, say O. If X and Y represent their speeds in miles per hour and if we are interested in the distance d_t between the two after t hours, then the random variable of interest is

$$d_t = t\sqrt{X^2 + Y^2}.$$

To every point (outcome) (x, y) in Ω we assign a positive real number $t\sqrt{x^2 + y^2}$ that represents the distance between the two walkers after t hours. (See Figure 1.)

In both these examples we are interested *not* in an outcome ω in the original sample space, but some real number assigned to ω. Often the outcome ω itself is a real number. We formalize this assignment in the following definition.

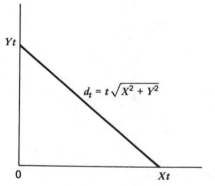

Figure 1

DEFINITION 1.   (RANDOM VARIABLE).   Let Ω be a sample space with a class of events $\mathfrak{F}$. Any rule (assignment or recipe) X that assigns to each $\omega \in \Omega$ a real number $X(\omega)$ is called a random variable.

Remark 1.   Suppose X is a random variable on Ω. Then $X(\omega)$ takes values on the real line $\mathbb{R}$, and in effect we have a new space $\mathcal{X} = \{ X(\omega): \omega \in \Omega \} \subseteq \mathbb{R}$ which may be taken to be $\mathbb{R}$. What are the events in $\mathbb{R}$? Since the fundamental objects of interest on $\mathbb{R}$ are intervals, every interval is taken to be an event. So also are unions, differences and intersections of intervals. It is therefore reasonable to insist that each interval on $\mathbb{R}$ be assigned a probability consistent with the assignment P to events on Ω. This is done by assigning to an interval $I \subseteq \mathbb{R}$ the probability $P\{\omega: X(\omega) \in I\}$. In order for this assignment to be meaningful, we must insist that the sets $\{\omega: X(\omega) \in I\}$ for each interval on $\mathbb{R}$ are themselves events in $\mathfrak{F}$. Definition 1 therefore should have stated that a real valued function X defined on Ω is a random variable if for each interval I,

$$\{\omega: X(\omega) \in I\} \text{ is in } \mathfrak{F}.$$

However, in all cases to be considered in this text, this condition will be satisfied and we will make no attempt to check that this in fact is the case in any specific situation. Thus for all practical purposes Definition 1 suffices. In terms of calculus, a random variable is simply a real valued function on Ω.

Remark 2.   Random variables will be denoted by capital letters X, Y, Z, U, V W, etc (with or without subscripts). The value $X(\omega)$ assumed by X is denoted by x. That is, we write x for $X(\omega)$.

Remark 3.   Recall (Remark 1) that we assign $P\{X(\omega) \in I\}$ to each interval I on $\mathbb{R}$. This induces a new probability P_X on the intervals (or events) on $\mathbb{R}$ which is called the (*induced*) *distribution* of X. We will suppress the subscript X and write $P(I)$ for $P_X(I) = P\{\omega: X(\omega) \in I\}$. We will not show here that P_X does indeed define a probability on the set of events in $\mathbb{R}$.

Remark 4.   Every random variable X partitions Ω into disjoint sets $\{X = x_1\}, \{X = x_2\}, \ldots$ (see Figure 2).

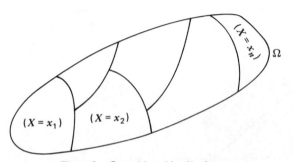

Figure 2. Ω partitioned by X values.

Some typical examples of random variables are: number of children in a family, number of rooms in a house, family income, the length of life of a piece of equipment, the error in rounding off a number to the nearest integer, the number of errors in a typed page, the electrical current passing a given point, and so on.

Example 1. Tossing a Fair Coin. A fair coin is tossed 5 times. Then Ω consists of all 32 sequences of length 5 of heads and tails. Let X be the number of heads in 5 tosses. Then X takes values $0, 1, 2, 3, 4, 5$. The event $\{X = 1\}$ is the event that 1 head (and hence 4 tails) shows up. This happens if and only if ω is any one of the 5 outcomes

$$HTTTT, THTTT, TTHTT, TTTHT, TTTTH.$$

Thus

$$P(X = 1) = \sum_{\omega} P\{\omega : X(\omega) = 1\} = \frac{5}{2^5} = \frac{5}{32}.$$

The new outcome space is $\mathscr{X} = \{0, 1, 2, 3, 4, 5\} \subset \mathbb{R}$ and all subsets of $\mathscr{X}$ are events of interest.

Yet another random variable of interest is Y, the numerical value of the difference between number of heads and number of tails. The random variable Y takes values in $\mathscr{Y} = \{1, 3, 5\}$. Thus

$$\{Y = 3\} = \{4 \text{ heads and 1 tail or 4 tails and 1 head}\}$$

$$= \left\{ \begin{array}{l} HHHHT, HHHTH, HHTHH, HTHHH, THHHH, \\ TTTTH, TTTHT, TTHTT, THTTT, HTTTT \end{array} \right\}.$$

Clearly

$$P(Y = 3) = \sum_{\omega} P\{\omega : Y(\omega) = 3\} = \frac{10}{32} \qquad \square$$

Example 2. Rolling a Fair Die. A fair die is rolled once. In this case $\Omega = \{1, 2, 3, 4, 5, 6\}$. We note that the outcome is a real number. If we let X denote the face value, then $X(\omega) = \omega$ is the identity function. On the other hand we may simply be interested in whether the outcome is an odd or an even number. In that case we can define

$$Y(\omega) = 1 \qquad \text{if } \omega \text{ is even, and zero otherwise}$$

Then Y takes values 0 (when $\omega = 1$, 3, or 5) and 1 (when $\omega = 2$, 4, or 6).

If the die is rolled twice, then $\Omega = \{(i, j): i, j = 1, 2, 3, 4, 5, 6\}$. Several random variables of interest are:

$$X(\omega) = \text{Sum of face values in } \omega.$$

$$Y(\omega) = \text{Difference in face values in } \omega.$$

$$Z(\omega) = \text{Absolute value of the difference in face values in } \omega.$$

$$U(\omega) = \text{Larger face value in } \omega.$$

$$V(\omega) = \text{Smaller face value in } \omega.$$

In each case the new sample space may be taken to be $\mathbb{R}$. In effect, though, a subset of $\mathbb{R}$ suffices. For example, X takes values in $\mathscr{X} = \{2, 3, \ldots, 12\}$, Y takes values in $\mathscr{Y} = \{-5, -4, \ldots, -1, 0, 1, \ldots, 5\}$, Z takes values in $\mathscr{Z} = \{0, 1, \ldots, 5\}$, U takes values in $\mathscr{U} = \{1, 2, \ldots, 6\}$ and V takes values in $\mathscr{V} = \{1, 2, \ldots, 6\}$.

The event $\{X = 5)$ happens if and only if $\omega \in \{(1, 4), (4, 1), (2, 3), (3, 2)\}$. The event $\{U = 3\}$ occurs if and only if $\omega \in \{(1, 3), (2, 3), (3, 1), (3, 2), (3, 3)\}$, and so on. Thus

$$P(X = 5) = \frac{4}{6^2}, \qquad P(U = 3) = \frac{5}{6^2}.$$

Suppose we roll the die until a face value of 6 is observed. In this case, since we are only interested in the number W of trials necessary to observe a 6 and since W takes values in $\mathscr{W} = \{1, 2, 3, \ldots\}$, we may take $\mathscr{W}$ to be the sample space. The events of interest are all subsets of $\mathscr{W}$. Thus

$$\{W \geq 5\} = \{(\omega_1, \omega_2, \omega_3, \omega_4, 6), (\omega_1, \ldots, \omega_5, 6), \ldots :$$

$$\omega_i \in \{1, 2, 3, 4, 5\}, i = 1, 2, \ldots\}$$

so that

$$P(W \geq 5) = \sum_{k=4}^{\infty} \left(\frac{5}{6}\right)^k \left(\frac{1}{6}\right) = \left(\frac{5}{6}\right)^4$$

$$= P(\text{first four rolls do not produce a } 6). \qquad \square$$

Example 3.   Life Length of a Battery.   Consider a battery which is put into use at time $t = 0$. The time at which the battery fails is recorded. Here $\Omega = [0, \infty)$ and the events of interest are subintervals of Ω. A random variable of interest is X, the time to failure of the battery. In that case $X(\omega) = \omega$ so that X is the identity function. How do we assign probabilities to events in this case? We need to compute probabilities such as $P(t_1 \leq X \leq t_2)$ where $t_1, t_2 \in [0, \infty)$, $t_1 < t_2$. This will be done in Section 3.3.

Another random variable of interest may be Y defined by

$$Y(\omega) = \begin{cases} 1 & \text{if } X(\omega) < t \\ 0 & \text{if } X(\omega) \geq t. \end{cases}$$

The random variable Y tells us whether or not the battery has failed prior to time t where $t > 0$ is known. Yet another random variable of interest is

$$Z(\omega) = \begin{cases} X(\omega) & \text{if } X(\omega) < t \\ t & \text{if } X(\omega) \geq t. \end{cases}$$

If the battery fails prior to time t, Z is the time to failure, otherwise Z takes the value t. Then Z takes values in the new sample space $[0, t]$.

We need to compute $P(Z = t)$ and probabilities of the form $P(t_1 \leq Z \leq t_2)$, $0 \leq t_1 \leq t_2 < t$. But this can be done by going back to Ω provided an assignment P has somehow been made there. Indeed,

$$P(t_1 \leq Z \leq t_2) = P(t_1 \leq X \leq t_2) \qquad \text{for } 0 \leq t_1 \leq t_2 < t$$

and

$$P(Z = t) = P(X \geq t). \qquad \qquad \square$$

Example 4.   Distance Traveled by an Automobile.   When we say that an automobile travels at a constant speed of 55 miles per hour, it means that in every time interval of length t hours it travels $55t$ miles. If we let d_t denote the distance traveled in $[0, t]$, then $d_t = 55t$. In this deterministic model, given t we know precisely what d_t is. In a stochastic model we assume that the speed V is not constant but random (due to road conditions, weather conditions, and degree of traffic congestion). If we let X_t denote the distance traveled in $[0, t]$, then X_t takes values in $[0, \infty)$ and X_t is a random variable. Given t, we no longer know precisely what X_t is. In fact we are no longer interested in the event $\{ X_t = x \}$, $x > 0$ but in events such as $\{ X_t > x \}$ or $\{x_1 < X_t < x_2\}$. The probabilities of such events will depend on the randomness assumption on V. In the case of a deterministic model we were interested in the recipe or formula to compute d_t given t. In the stochastic case our interest shifts from a formula for X_t to the distribution of X_t. $\square$

Remark 5.   In Examples 1 and 2 we note that Ω is at most countable so that the random variables of interest take values in a finite or countably infinite set. In these cases it is easier to take $\mathscr{X} = \{ x \in \mathbb{R}: X(\omega) = x, \omega \in \Omega \}$ as the new sample space, and it is sufficient to assign probabilities to events such as $(X = x)$ for $x \in \mathscr{X}$. We note that

$$P(X = x) = P\{ \omega: X(\omega) = x \} = \sum_{\{ \omega: X(\omega) = x \}} P\{ \omega \}.$$

In Example 3, $\Omega = [0, \infty)$ is uncountably infinite. The random variables defined on Ω may take either at most a countable number of values (such as Y) or an uncountable number of values (such as X or Z). In the latter case we need to assign probability to each interval on $\mathbb{R}$ according to

$$P(t_1 \leq X \leq t_2) = P\{ \omega: t_1 \leq X(\omega) \leq t_2 \}.$$

We note that if $\Omega \subseteq \mathbb{R}$, $\{ \omega: t_1 \leq X(\omega) \leq t_2 \}$ may not, in general, be an interval on $\mathbb{R}$. Nevertheless, it is an event with a well defined probability.

To summarize, a random variable is a real valued function defined on the sample space Ω. It is a device to transform original outcomes $\omega \in \Omega$, whenever necessary, into real numbers, which enables us to use familiar properties of the real-number system. In the process, it also filters out needless information and often leads us to consider a much smaller subspace $\mathscr{X} \subseteq \mathbb{R}$.

So far we have considered only one random variable defined on Ω. Quite frequently, however, it is necessary to consider several random variables on Ω. For example, for a collection of individuals we measure X as the height in inches, Y the weight in pounds, and Z the body temperature. We call (X, Y, Z) a *three-dimensional random variable* or simply a *random vector* defined on Ω. In this case our outcome is a point (x, y, z) in the three-dimensional space. In general, let $X_1, X_2, \ldots, X_k$ be k random variables, $k \geq 2$, defined on Ω. We say that $(X_1, X_2, \ldots, X_k)$ is a *random vector* on Ω or that $(X_1, X_2, \ldots, X_k)$ is a *k-dimensional random variable*. Clearly each ω is assigned a k-tuple $(X_1(\omega), X_2(\omega), \ldots, X_k(\omega))$ in $\mathbb{R}_k$. We shall write $(x_1, x_2, \ldots, x_k)$ for a value assumed by $(X_1, X_2, \ldots, X_k)$.

Remarks similar to 1, 3, 4, and 5 apply also to random vectors. We content ourselves with some examples.

Example 5. Rolling a Fair Die. Let us consider again Example 2. Suppose a fair die is rolled twice. Then $\Omega = \{(i, j): i, j = 1, 2, 3, 4, 5, 6\}$. If we let $X(\omega) = X(i, j) = i$, and $Y(\omega) = Y(i, j) = j$, then X and Y represent the face values on the two rolls. Thus (X, Y) is a somewhat trivial example of a two-dimensional random variable. On the other hand, if $X(\omega) = i + j$, $Y(\omega) = |i - j|$, $Z(\omega) = \min\{i, j\}$, $U(\omega) = \max\{i, j\}$ for $\omega = (i, j)$, then (X, Y, Z, U) is a random vector. We note that (X, Y, Z, U) takes values in the discrete set $\{(x, y, z, u): x \in \{2, 3, \ldots, 12\}, y \in \{0, 1, \ldots, 5\}, z$ and $u \in \{1, 2, \ldots, 6\}\}$. We need only assign probabilities to such points and this is easily done just as in the one-dimensional case. For example:

$$P\{X = 3, Y = 1, Z = 1, U = 2\} = P\{(1, 2), (2, 1)\} = 2/6^2. \qquad \square$$

Example 6. Length and Width of a Board. Suppose the length and the width of a wooden board obtained from a cutting process are both random. Let X denote the length and Y the width of a randomly selected board. Then (X, Y) is a random vector. In this case the fundamental events of interest to which we need to assign probabilities are rectangles of the form $\{x_1 < X \leq x_2, y_1 < Y \leq y_2\}$ in the plane. Once this is done, it will be possible to compute probabilities of events such as: the area of a wood board from this process is at most $x_0 (= (XY \leq x_0))$ or of $(XY \geq x_1)$, and so on. Consider, for example, the assignment

$$P(A) = \frac{\text{Area of } (A \cap S)}{\text{Area of } S} = \frac{\text{Area of } (A \cap S)}{2}$$

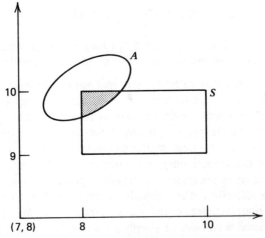

Figure 3

where $S = \{8 \leq x \leq 10, 9 \leq y \leq 10\}$, and A is any set in plane with a well-defined area (see Figure 3). In particular:

$$P(8 \leq X \leq 9, 9 \leq Y \leq 9.5) = 1(0.5)/2 = 1/4.$$

We will see in later sections how to compute probabilities such as $P(XY \leq x_0)$. ☐

Problems for Section 3.2

1. Consider the classification of three-children families according to the sex of the first, second, and the third child. Let X be the number of girls. List possible values of X. Describe the following events in terms of X:
   (i) At most two girls.   (ii) Exactly two girls.   (iii) Fewer than two girls.
   (iv) At least one girl.
   In each case list the simple events that make up the event.

2. A lot contains 10 items of which 4 are defective. A random sample of 5 is taken from the lot. Let X be the number of defectives in the lot. List possible values of X. Describe the following events:
   (i) $(X \leq 3)$.   (ii) $(X \geq 4)$.   (iii) $(X = 0)$.

3. (i) A fair coin is tossed until a head appears for the first time. Let X be the number of trials needed. What are the possible values of X? Write the events $(X = 1)$, $(X \geq 2)$ as subsets of Ω. What are their probabilities?
   (ii) If the coin is tossed until a head appears for the first time, or until 10 consecutive tails appear, let Y be the number of trials required.
       (a) What are the possible values of Y?
       (b) Write the event $(Y = 10)$ as a subset of Ω.
       (c) What probability will you assign to this event?

4. Let X be the height of a randomly selected freshman from a certain university.
   (i) What are the possible values of X?
   (ii) If Y is the weight of the freshman selected, what are the possible values of Y?
   (iii) What are the possible values of (X, Y)?

5. In each of the following examples define some random variables of interest.
   (i) The experiment consists of drawing a random sample of 20 students from a large Calculus class.
   (ii) The experiment consists of recording the opening stock prices of 50 selected stocks.
   (iii) The experiment consists of recording the height of mercury in a barometer at every hour on the hour.
   (iv) The experiment consists of recording the number of customers who enter a store during each 15-minute interval starting at 10 a.m. (opening time).

6. Consider families with four children. Define X by $X = 1$ if the first born was a girl and zero otherwise, and Y by $Y = 1$ if the fourth child was a girl and zero otherwise. Write down the events $(X = 0, Y = 0)$, $(X = 1, Y = 0)$, $(X = 0, Y = 1)$, $(X = 1, Y = 1)$ as events in the sample space $(\Omega, \mathfrak{I})$. What probabilities will you assign to these events if boys and girls are equally likely?

7. A committee of four is selected from a group of 10 students consisting of 4 seniors, 2 juniors, 2 sophomores, and 2 freshman. Let X be the number of seniors and Y the

number of juniors in the committee. Describe the events $(X = 1, Y = 0)$, $(X = 3, Y = 1)$, $(X = 2, Y = 1)$, $(X = 1, Y = 1)$. What probabilities will you assign to these events?

8. Suppose $\Omega = [0, 1]$. Let $\mathcal{F}$ the set of events in Ω and suppose P on $\mathcal{F}$ is defined as follows. To each subinterval $(a, b]$ of Ω, P assigns probability

$$P(a, b] = \text{Area of the rectangle of width } (a, b] \text{ and height 1.}$$

Define X on Ω be setting $X(\omega) = 2\omega + 1$.
   (i) What are the possible values of X?
   (ii) Find $P(X \le 2)$, $P(X > 1)$, $P(a < X \le b)$.

9. Let $\Omega = [-2, 1]$ and $\mathcal{F}$ the set of events on Ω. On $\mathcal{F}$ let P be defined as follows: To each subinterval of $(a, b]$ of Ω, P assigns probability

$$P(a, b] = \text{Area of the region bounded by } (a, b] \text{ and the line } y = \frac{2x + 4}{9}.$$

Let Y be defined as follows:

$$Y(\omega) = 3\omega - 1.$$

   (i) What are the possible values of Y?
   (ii) Find $P(-3 < Y \le 1)$, $P(Y \le 0)$, $P(Y > -1)$.

10. Let $\Omega = \{(\omega_1, \omega_2): 0 < \omega_1 < \omega_2 < 1\}$. Suppose P on $\mathcal{F}$ is defined as follows: For $A \subseteq \Omega, A \in \mathcal{F}, P(A) = \iint\limits_{A} 2 \, d\omega_1 \, d\omega_2$. Let

$$X(\omega_1, \omega_2) = \omega_1 \quad \text{and} \quad Y(\omega_1, \omega_2) = \omega_2, \quad 0 < \omega_1 < \omega_2 < 1.$$

Find $P(X < 1/2)$, $P(Y > 1/2)$, $P(1/3 < X < Y < 5/8)$.

3.3 DESCRIBING A RANDOM VARIABLE

We began with the following description of a random experiment. Identify Ω, the set of outcomes of the experiment, $\mathcal{F}$, the events of interest, and define probability P on $\mathcal{F}$. Then the triple $(\Omega, \mathcal{F}, P)$ gives a stochastic description of the experiment. In Section 3.2 we saw how we transform Ω into a more convenient set of outcomes whose elements are real numbers. We indicated how to identify events on this sample space and how to compute probabilities of these events. The mathematical description of the model is complete once we know X, the events of interest, and the induced distribution P_X of X. This situation is different from that in calculus. There we identify the domain set D and the recipe f which associates with each $x \in D$ some real number $f(x)$. Knowledge of D and f gives a complete description of the model. In statistics the description is not complete until one specifies in addition the distribution of X.

The specification of P_X through P on $(\Omega, \mathfrak{z})$ is not satisfactory since one frequently has to compute P_X for an incredibly large set of events according to the formula

$$P_X(A) = P\{\omega: X(\omega) \in A\},$$

where A is an event on the new sample space, $A \subseteq \mathbb{R}$. Fortunately it is not necessary to do so. In this section we study two equivalent ways to specify P_X. For this purpose it is convenient to consider two separate cases according to whether X takes values in a countable subset $\mathfrak{X}$ of $\mathbb{R}$ or values in an interval on $\mathbb{R}$. In the former case if $\mathfrak{X} = \{x_1, x_2, \ldots\}$, it is sufficient to specify $P(X = x_j) = P\{\omega: X(\omega) = x_j\}$ for each j. In the latter case it suffices to specify some nonnegative function f on $\mathbb{R}$ such that

$$P_X(A) = \int_A f(x)\, dx$$

for each event $A \subseteq \mathbb{R}$.

DEFINITION 1.   A random variable X on $(\Omega, \mathfrak{z})$ is said to be of the discrete type if it takes values in a countable set $\mathfrak{X} = \{x_1, x_2, \ldots\}$, that is, if the set of all distinct $X(\omega)$, as ω varies over Ω, is at most countable. The distribution P_X of X is specified by the numbers

$$P_X\{x_j\} = P\{\omega: X(\omega) = x_j\}$$

and $\{P_X\{x_j\}\}$ is known as the probability function of X. For any event A in $\mathfrak{X}$, that is, any subset A of $\mathfrak{X}$, we have

$$P_X(A) = \sum_{x_j \in A} P_X\{x_j\} = \sum_{x_j \in A} P\{\omega: X(\omega) = x_j\}.$$

DEFINITION 2.   A random variable X is said to be of the continuous type if it takes all values over some interval, that is, the set $\{X(\omega), \omega \in \Omega\}$ is an interval on $\mathbb{R}$. The distribution of X is then specified by specifying some nonnegative integrable function f on $\mathbb{R}$ such that

$$P_X(A) = \int_A f(x)\, dx,$$

for every event $A \subseteq \mathbb{R}$. The function f is known as the density function of X.

In most cases X is either of the discrete type or the continuous type. But mixed types can and do occur in practice. In this case there is a countable set of points $x_1, x_2, \ldots$ and a nonnegative function f such that

$$(1) \qquad\qquad P_X(A) = \sum_{x_j \in A} P_X\{x_j\} + \int_A f(x)\, dx,$$

for every event A. We will restrict our discussion almost exclusively to discrete and continuous cases and content ourselves with a few examples of the mixed type.

In the discrete case $1 \geq P_X\{x_j\} \geq 0$ for all j and

$$(2) \qquad \sum_{j=1}^{\infty} P_X\{x_j\} = P_X(\mathcal{X}) = P(\Omega) = 1.$$

In the continuous case $f(x) \geq 0$ for all $x \in \mathbb{R}$ and

$$(3) \qquad \int_{-\infty}^{\infty} f(x)\, dx = P_X(\mathbb{R}) = P(\Omega) = 1.$$

Often we need the probability $P(X \leq x)$ that is cumulated up to and including a real number x. This is just a function (of x) on $\mathbb{R}$ and is given a special name.

DEFINITION 3. (DISTRIBUTION FUNCTION). The distribution function of a random variable X on Ω, denoted by F, is defined by the relation

$$F(x) = P\{\omega: -\infty < X(\omega) \leq x\} = P(X \leq x), \qquad x \in \mathbb{R}.$$

The basic reason why the distribution function is important is that it specifies and is completely specified by the distribution of X. Indeed, given P_X for all intervals I, we get F by choosing $I = (-\infty, x]$. Conversely, suppose F is given. Then for any $x_1, x_2 \in \mathbb{R}$, $x_1 < x_2$, we have

$$P\{\omega: X(\omega) \leq x_2\} = P\{\omega: X(\omega) \leq x_1\} + P\{\omega: x_1 < X(\omega) \leq x_2\}$$

from additivity of P on $(\Omega, \mathcal{F})$. Hence

$$P\{\omega: X(\omega) \in (x_1, x_2]\} = P(X \leq x_2) - P(X \leq x_1)$$

so that

$$(4) \qquad P_X((x_1, x_2]) = F(x_2) - F(x_1).$$

In the special case when $x_2 = \infty$, we get

$$(5) \qquad P(X > x_1) = 1 - F(x_1).$$

The probability $P(X > x_1)$ is called a *tail probability* and is often used to compare distribution functions. (See also Remark 8, Section 3.7.)

Thus a description of the model is complete the moment we specify F. Later we will show that this is equivalent to specifying the probability or density function of X.

What are some properties of F? What does it look like? Here are some simple properties.

PROPOSITION 1.   *Let F be a distribution function. Then*

(i) $0 \le F(x) \le 1$ *for all* $x \in \mathbf{R}$,

(ii) $x_1 < x_2 \Rightarrow F(x_1) \le F(x_2)$ *so that F is nondecreasing,*

(iii)   *F is right continuous, that is,* $\lim_{x \to x_0^+} F(x) = F(x_0)$ *for all* $x_0 \in \mathbf{R}$,

(iv) $\lim_{x \to \infty} F(x) = 1$ *and* $\lim_{x \to -\infty} F(x) = 0$.

Proof.   Part (i) follows from definition of F and part (ii) from (4). Part (iii) is proved in Lemma 1 below, and part (iv) follows from (iii) by letting $x \to \pm \infty$.

LEMMA 1.   *For any distribution function f and any real number* x_0

$$\lim_{x \to x_0^+} F(x) = F(x_0) \tag{6}$$

and

$$\lim_{x \to x_0^-} F(x) = F(x_0) - P(X = x_0) = P(X < x_0) \tag{7}$$

Proof*.   Since F is nondecreasing and $0 \le F \le 1$, the existence of the limits follows at once from calculus. (Any bounded monotone sequence has a limit.) We need only show that the limits have values indicated in (6) and (7). We prove only (6). The proof of (7) is similar. The idea of the proof is that as $x \to x_0^+$, the interval $(x_0, x]$ shrinks. If P is "continuous" then we should have $P(x_0, x] \to 0$ as $x \to x_0^+$.

Let $x > x_0$. Then

$$F(x) = P(X \le x_0) + P(x_0 < X \le x)$$

so that

$$\lim_{x \to x_0^+} F(x) = F(x_0) + \lim_{x \to x_0^+} P(x_0 < X \le x).$$

Let $\cdots > x_n > x_{n-1} > \cdots > x_0$ such that $x_n \to x_0$. We need only show that $P(x_0 < X \le x_n) \to 0$. Choose $x_n = x_0 + 1/n$ and set $p_n = P(x_0 < X \le x_0 + 1/n)$. Let $A_n = (x_0 + 1/(n+1) < X \le x_0 + 1/n)$. Then A_n's are disjoint and

$$\bigcup_{n=1}^{\infty} A_n = (x_0 < X \le x_0 + 1).$$

Moreover,

$$P\left(\bigcup_{n=1}^{\infty} A_n \right) = p_1 = \sum_{n=1}^{\infty} P(A_n) = \sum_{n=1}^{\infty} (p_n - p_{n+1}).$$

The series converges, since $\sum_{n=1}^{\infty} P(A_n) = p_1$. Hence

$$p_1 = \sum_{n=1}^{\infty} (p_n - p_{n+1}) = \lim_{N \to \infty} \sum_{n=1}^{N} (p_n - p_{n+1}) = \lim_{N \to \infty} (p_1 - p_{N+1})$$

$$= p_1 - \lim_{N \to \infty} p_{N+1}$$

so that $\lim_{N \to \infty} p_N = 0$ as asserted. $\square$

Next we show that there is a 1–1 correspondence between the distribution function of a random variable X and its probability (or density) function. This is simple. In the discrete case

$$(8) \qquad F(x) = \sum_{x_j \le x} P_X\{x_j\} = \sum_{x_j \le x} P(X = x_j),$$

and in the continuous case

$$(9) \qquad F(x) = \int_{-\infty}^{x} f(t)\, dt$$

gives the distribution function for X. Conversely, suppose F is given. Then from Lemma 1

$$(10) \qquad P(X = x) = F(x) - F(x^-).$$

If X is of discrete type taking values $\mathscr{X} = \{x_1, x_2, \ldots\}$ then

$$(11) \qquad P(X = x_j) = F(x_j) - F(x_j^-)$$

gives the probability function of X. In the continuous case we use the fundamental theorem of calculus to conclude that

$$(12) \qquad f(x) = \frac{dF(x)}{dx}$$

for all x where F is differentiable. (See also Remark 3.)

In a more advanced course it is shown that every function F defined on $\mathbb{R}$ and satisfying (i) to (iv) of Proposition 1 is a distribution function of *some* random variable. Moreover, from remarks following Definition 3 we note that F is determined by, and determines uniquely, a distribution P_X on the set of events in $\mathbb{R}$ through the correspondence (4) or, equivalently, $P_X((-\infty, x]) = F(x)$.

The following result is now somewhat obvious.

PROPOSITION 2.

(i) *Given a set of nonnegative real numbers p_k such that $\sum_{k=1}^{\infty} p_k = 1$, then $\{p_k\}$ is the probability function of some discrete type random variable X.*

(ii) *Given a nonnegative real valued function f that is integrable on $\mathbb{R}$ and satisfies $\int_{-\infty}^{\infty} f(x)\, dx = 1$, then f is the density function of some random variable X.*

Proof. All we need to do is to construct a distribution function F. This is easy. In the continuous case for example set

$$F(x) = \int_{-\infty}^{x} f(y)\, dy,$$

then F satisfies all the four properties of Proposition 1. $\qquad\qquad\square$

Remark 1.   We emphasize that the distribution of X is uniquely determined by specifying either the distribution function or the density (or probability) function. In the following the phrase "X has distribution given by" will be followed by either the distribution function or the density (or probability) function. When we ask the reader to find the distribution of X, it is sufficient to find either the density (or probability) function or the distribution function of X.

Remark 2.   We note that if Ω is a discrete set, that is, a countable set of points, then any random variable X defined on Ω is necessarily discrete. However, if Ω contains an interval on $\mathbb{R}$, a random variable on Ω may still be discrete. In the discrete case, $P(X = x_j)$ represents the probability that X takes the value x_j. If $P(X = x_j) > 0$ for some x_j we say that X (or the corresponding distribution function) has a *jump of size* $P(X = x_j)$ at x_j and call x_j a *jump point*. If X takes values in $\{x_1, x_2, \ldots\}$ then its distribution function F given by (8), namely,

$$F(x) = \sum_{x_j \leq x} P_X\{x_j\} = \sum_{x_j \leq x} P\{X = x_j\},$$

has jumps of size

$$P(X = x_j) = P(X \leq x_j) - P(X < x_j) = F(x_j) - F(x_j^-)$$

at each $x_j, j = 1, 2, \ldots$. Figure 1 represents the probability function and distribution function of a typical discrete type random variable.

In analogy to mechanics, we can think of a total mass of one unit distributed over the real line such that the mass distribution consists of point masses at $x_1, x_2, \ldots$.

Remark 3.   In the continuous case we can get the distribution function from the density function by the relation (9) and, conversely, the density function from F by the relation (12). We note that f is not unique and may be defined arbitrarily at values of x where $F'(x)$ is not defined. Also, F is everywhere continuous (in

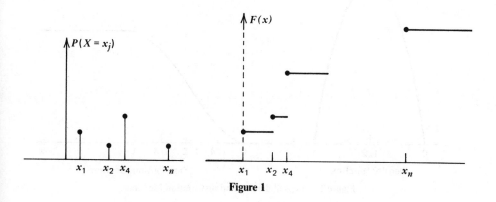

Figure 1

fact, absolutely continuous) and from (10) $P(X = x) = 0$ for all $x \in \mathbb{R}$. (From (10) F is continuous at x if and only if $P(X = x) = 0$.) Thus for any $a, b \in \mathbb{R}$, $a < b$

$$P(a < X < b) = P(a \leq X < b) = P(a < X \leq b)$$

$$= P(a \leq X \leq b)$$

whenever X is of the continuous type. The fact that $P(X = x) = 0$ for each $x \in \mathbb{R}$ does not mean that X cannot take the value x. From a relative frequency viewpoint it simply means that if the associated experiment is repeated n times, the relative frequency of the event $(X = x)$ will tend to zero as $n \to \infty$. In fact $f(x)$ does not represent the probability of any event. From the mean value theorem if f is continuous at x then we can interpret $f(x) \Delta x \doteq P(X \in [x, x + \Delta x])$ for small Δx and $f(x)$ as the probability per unit of length at x. Figure 2 illustrates a typical density and associated distribution function.

In analogy to mechanics we can think of a total mass of one unit continuously spread over the real line in such a way that the probability associated with any set of points is the mass of the set. Then $f(x)$ represents the mass density at x. Geometrically we can therefore interpret $P(A) = \int_A f(x)\, dx$ as the area bounded by the set A and the curve f. This interpretation frequently simplifies computations of probabilities.

Remark 4.   Remark 3 exhibits the fact that $P(A) = 0 \nRightarrow A = \varnothing$ and $P(B) = 1 \nRightarrow B = \Omega$. Thus $P(A) = 0$ does not mean that A cannot happen, only that A has probability zero. A relative frequency interpretation of this statement may be given as in Remark 3. If $P(A) = 0$ we say that A is possible but highly improbable.

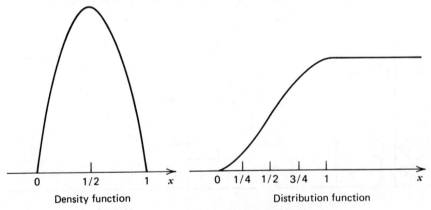

Density function   Distribution function

Figure 2.   Typical density and distribution functions.

Remark 5. We note that we can define, equivalently, a continuous type random variable to be one whose distribution function F has the representation (9) for some nonnegative function f on $\mathbb{R}$. This means that F is absolutely continuous. We emphasize that F may be continuous without being absolutely continuous so that continuity of F is not sufficient for a random variable X to be of the continuous type.

Remark 6. When X is a discrete type random variable taking values on equally spaced points, say integers, we will prefer to draw a *probability histogram* of the probability function of X. This is done by drawing rectangles centered at each x value with heights proportional to the probability at x so that the area of a rectangle with center x and width 1 represents $P(X = x)$. Then the total area of the rectangles—that is, the area under the histogram—is one, and in analogy with the continuous case (Remark 3), we can use areas to represent probabilities of intervals (or events).

Remark 7. In practice no real-world experiment or measurement can possibly yield a continuous type random variable. The assumption that X is of continuous type is a simplifying assumption leading to an idealized description of the model. If X measures height of a tree in feet, X takes discrete integer values. If X is measured to the nearest inch, it still takes discrete integer values but the corresponding rectangles in the histogram get narrower and the histogram has a much smoother appearance than in the previous case (see Figure 3). If we shrink the length of the base by making units of measurement smaller and smaller, the histogram will tend towards a smooth curve representing the (idealized) density of X.

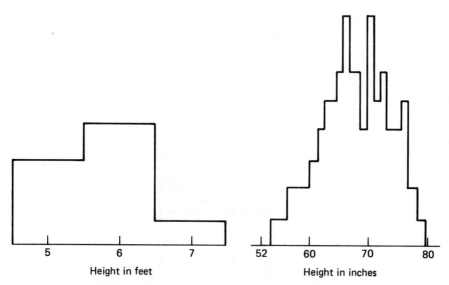

Figure 3. Histogram of the same data in feet and in inches.

Example 1. Tossing a Fair Coin. Let X be the number of heads in one toss of a fair coin. Then the probability function of X is given by

$$P(X = 0) = P\{T\} = 1/2$$

$$P(X = 1) = P\{H\} = 1/2,$$

and the distribution function by

$$F(x) = \begin{cases} 0, & x < 0 \\ 1/2, & 0 \le x < 1 \\ 1, & x \ge 1. \end{cases}$$

We note that F has jumps at $x = 0$ and at $x = 1$. Figure 4a gives the graph of F, and Figure 4b gives the probability histogram of X.   □

Example 2. Life-Testing. Consider an experiment that consists of life-testing a piece of equipment such as a battery or a light bulb or a transistor. Let us take the starting time of the test to be time zero and measure the length of life of the piece of equipment until it fails. Then we can take $\Omega = [0, \infty)$. Many random variables can be defined on Ω. Suppose X denotes the life length (failure time) of the piece. Then X maps Ω on to itself so that $\mathscr{X} = \Omega = [0, \infty) \subseteq \mathbb{R}$. Some events of interest in Ω are

$\{X = t\}$:   failure at time t

$\{X \le t\}$:   failure at or prior to time t

$\{X > t\}$:   failure after time t *or* survival to time t

$\{t_1 < X \le t_2\}$:   survival to time t_1 but failure at or prior to time t_2.

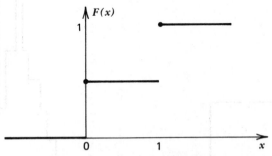

Figure 4a.   Graph of F, Example 1.

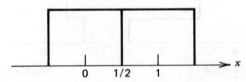

Figure 4b.   Probability histogram of X, Example 1.

Suppose X has density function

$$f(x) = e^{-x}, \quad x > 0, \quad \text{and zero elsewhere.}$$

Then the distribution function of X is given by

$$F(x) = \begin{cases} 0, & \text{if } x \le 0 \\ \int_0^x e^{-y} \, dy = 1 - e^{-x}, & \text{if } x > 0. \end{cases}$$

Suppose we terminate testing at time t_0. Let Y be the time to failure if the equipment piece fails prior to time t_0, and let $Y = t_0$ if it survives time t_0. Then Y is a mixed type random variable with a jump at t_0. The distribution function of Y is given by

$$F_Y(y) = \begin{cases} 0, & y \le 0 \\ 1 - e^{-y}, & 0 < y \le t_0 \\ 1, & y \ge t_0. \end{cases}$$

Figures 5a and 5b give the distribution function and density function of X and Figure 6 shows the distribution function of Y.

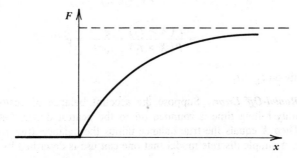

Figure 5a. $F(x) = 1 - e^{-x}, x > 0$ and zero elsewhere.

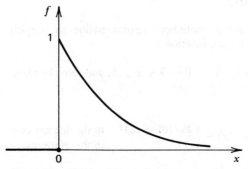

Figure 5b. $f(x) = e^{-x}, x > 0$ and zero elsewhere.

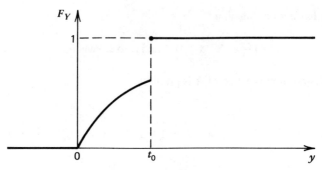

Figure 6. Graph of $F_Y(y)$.

Probability computations are somewhat easier if F is known. Thus

$$P(X > t) = 1 - P(X \le t) = 1 - F(t) = 1 - (1 - e^{-t}) = e^{-t},$$

$$P(t_1 < X < t_2) = F(t_2) - F(t_1) = e^{-t_1} - e^{-t_2},$$

and

$$P\{X > t_2 | X > t_1\} = \frac{P(X > t_2 \text{ and } X > t_1)}{P(X > t_1)}$$

$$= \frac{P(X > t_2)}{P(X > t_1)} = \frac{e^{-t_2}}{e^{-t_1}} = e^{-(t_2 - t_1)}$$

which depends only on $t_2 - t_1$.   □

Example 3. Round-Off Error. Suppose the account balance of customers of a bank at the monthly billing time is rounded off to the nearest dollar. Let X be the round-off error. Then X equals the true balance minus the balance after round-off to the nearest dollar. A simple discrete model that one can use is described by

$$P\left(X = \frac{j}{100}\right) = 1/101, \quad j = -50, -49, \ldots, -1, 0, 1, \ldots, 50.$$

One can, however, use a continuous approximation to simplify computation by assuming that X has density function

$$f(x) = 1 \quad \text{if } -.5 \le x \le .5, \text{ and zero elsewhere.}$$

Thus:

$$P(X \ge .25) = \begin{cases} 26/101 = .257 & \text{in the discrete case} \\ .25 & \text{in the continuous case} \end{cases}$$

and for most practical purposes we get nearly the same result.   □

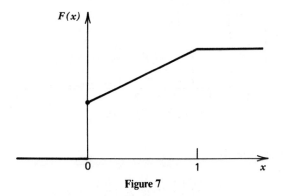

Figure 7

Example 4.   A Mixed Distribution.   Consider the function (see Figure 7)

$$F(x) = \begin{cases} 0, & x < 0 \\ (x+1)/2, & 0 \le x < 1 \\ 1, & 1 \le x. \end{cases}$$

We note that $0 \le F \le 1$, F is right continuous, nondecreasing, and satisfies $F(-\infty) = 0$, $F(\infty) = 1$. Hence F is a distribution function. Note that $F(0^-) = 0$ and $F(0^+) = 1/2$ so that F is not continuous at 0 and hence it is not the distribution function of a continuous type random variable. There is a jump of size

$$F(0^+) - F(0^-) = 1/2$$

at $x = 0$, and F is continuous for $x > 0$. Thus F is the distribution function of a mixed type random variable X described by

$$P(X = x) = 1/2, \quad \text{for } x = 0,$$

and for $0 < x \le 1$, X has density f given by $f(x) = (d/dx)F(x) = 1/2$, and zero elsewhere. We note that necessarily

$$P(X = 0) + \int_0^1 f(x)\, dx = 1. \qquad \Box$$

Example 5.   Waiting for the First Boy.   Consider a couple who would like a boy. Suppose they have children until they have their first boy. Let X be the number of children they have before the birth of the boy. Then X takes values in $\mathscr{X} = \{0, 1, 2, \dots\}$. Assume that boys and girls are equally likely and that the sex at each birth is independent of that on any other birth. Then for $x \ge 0$, x integral:

$$P(X = x) = P(x \text{ girls followed by one boy})$$

$$= \left(\frac{1}{2}\right)^x \frac{1}{2},$$

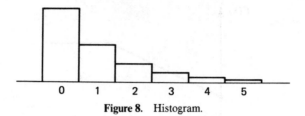

Figure 8. Histogram.

due to independence. Clearly

$$\sum_{x=0}^{\infty} \frac{1}{2}\left(\frac{1}{2}\right)^{x} = \frac{1}{2}\sum_{x=0}^{\infty}\left(\frac{1}{2}\right)^{x} = 1$$

so that $P(X = x) = (1/2)^{x+1}$, $x = 0,1,\ldots$ defines the probability function of X and the distribution function is given by $F(x) = 0$ for $x < 0$ and for $x \geq 0$

$$F(x) = \sum_{0 \leq j \leq [x]} \left(\frac{1}{2}\right)^{j+1} = 1 - \left(\frac{1}{2}\right)^{[x]+1}$$

where $[x]$ is the largest integer $\leq x$. The graph of probability histogram and distribution function are given in Figures 8 and 9. Thus

$$P(X > x_1) = (1/2)^{[x_1]+1}$$

and

$$P(x_1 < X \leq x_2) = F(x_2) - F(x_1) = (1/2)^{[x_1]+1} - (1/2)^{[x_2]+1}.$$

In particular,

$$P(X > 5) = 1/64 \quad \text{and} \quad P(3 < X \leq 9) = 63/1024. \qquad \square$$

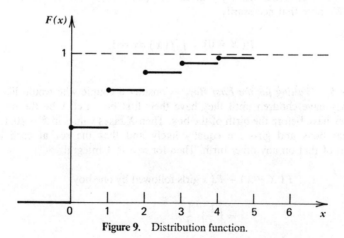

Figure 9.   Distribution function.

Example 6. *Volume of a Sphere.* Consider a sphere of radius X, where X has distribution function

$$F(x) = \begin{cases} 0, & x < 0 \\ 3x^2 - 2x^3, & 0 \le x < 1 \\ 1, & 1 \le x. \end{cases}$$

Let us find the probability that the volume of the sphere is larger than v_0, say. We first note that F is continuous everywhere and differentiable except at $x = 0$ and $x = 1$. Hence for $x \ne 0$ or 1 we get the density of X on differentiating F to be

$$f(x) = 6x(1 - x), \quad 0 < x < 1, \quad \text{and zero for } x < 0 \text{ or } x > 1.$$

At $x = 0$ and $x = 1$ we can define f arbitrarily. We choose $f(0) = 0$, $f(1) = 0$ so that

$$f(x) = 6x(1 - x), \quad 0 \le x \le 1, \quad \text{and zero elsewhere.}$$

Now the volume V of a sphere of radius X is given by $V = (4/3)\pi X^3$. Clearly $V \le (4/3)\pi$ and for $0 \le v_0 < (4/3)\pi$

$$P(V > v_0) = P\left(\frac{4}{3}\pi X^3 > v_0\right) = P\left(X^3 > \frac{3}{4}\frac{v_0}{\pi}\right) = 1 - F\left(\left(\frac{3v_0}{4\pi}\right)^{1/3}\right)$$

$$= 1 - 3\left(\frac{3v_0}{4\pi}\right)^{2/3} + \frac{3v_0}{2\pi}. \qquad \square$$

Remark 8. We emphasize that the distribution function F and the density function f are defined for all real numbers x. The probability function of a discrete random variable taking only values $x_1, x_2, \ldots$ with positive probability may also be defined for all $x \in \mathbb{R}$ by writing $P(X = x) = 0$ if $x \notin \{x_1, x_2, \ldots\}$. This is what we will do at least in the beginning. Sometimes however, we will specify $P(X = x)$ for only those x for which $P(X = x) > 0$ to conserve space. In the continuous case, we will specify f for all $x \in \mathbb{R}$. This is done to avoid mistakes, since in the continuous case f is usually given by a formula and there is a tendency to substitute the expression for f in the integral without regard to where it is positive and where it is zero.

We conclude this section with the following definition.

DEFINITION 4. A random variable X is said to have a symmetric distribution about a point α if

$$P(X \ge \alpha + x) = P(X \le \alpha - x) \qquad \text{for all } x.$$

If F is the distribution function of X, then F is symmetric about α if

$$F(\alpha - x) = 1 - F(\alpha + x) + P(X = \alpha + x).$$

In particular, if X is of the continuous type with density function f, then X is symmetric about α if and only if

$$f(\alpha - x) = f(\alpha + x) \qquad \text{for all } x.$$

If $\alpha = 0$, we say that f (or F or X) is symmetric. Examples of symmetric distributions are easy to come by. The density functions

$$f(x) = \frac{1}{\sigma\sqrt{2\pi}} \exp\left\{ -\frac{(x-\theta)^2}{2\sigma^2} \right\}, \qquad x \in \mathbb{R}$$

and

$$f(x) = \frac{1}{\pi} \frac{1}{1 + (x-\theta)^2}, \qquad x \in \mathbb{R}$$

are both symmetric about θ. The probability function

$$P(X = 1) = 1/2 = 1 - P(X = 0)$$

is symmetric about $1/2$.

Problems for Section 3.3

1. Are the following functions probability functions?
   (i) $p(x) = \dfrac{x-6}{5}$, $x = 7, 8, 9$, and zero elsewhere.
   (ii) $p(x) = x/21$, $x = 1, 2, 3, 4, 5, 6$, and zero elsewhere.
   (iii) $p(x) = x^2/55$, $x = 1, 2, 3, 4, 5$, and zero elsewhere.
   (iv) $p(x) = (x^2 - 2)/50$, $x = 1, 2, 3, 4, 5$, and zero elsewhere.

2. For each of the following functions find the constant c so that p is a probability function.
   (i) $p(x) = c(1/4)^x$, $x = 1, 2, \ldots$, and zero elsewhere.
   (ii) $p(x) = c(1/3)^x(2/3)^{15-x}$, $x = 0, 2, \ldots, 15$, and zero otherwise.
   (iii) $p(x) = c2^x$, $x = 1, 2, \ldots, 11$, and zero elsewhere.
   In each case write down the distribution function.

3. Consider an urn containing 5 red and 3 black marbles. A random sample of size 3 is selected (without replacement). Let X denote the number of black marbles drawn. Find the probability function of X. Find $P(1 \le X \le 3)$.

4. Suppose X has probability function given by:

x	-2	-1	0	1	2	4
$P(X = x)$	.3	.1	.15	.2	.1	.15

Find $P(X < 0)$, $P(X \text{ even})$, $P\{X = 0 | X \le 0\}$, $P\{X \ge 2 | X > 0\}$.

5. For what values of c are the following probability functions?

(i)
$$f(x) = \begin{cases} .2 + \dfrac{x}{20}, & x = 0, \pm 2 \\ c, & x = \pm 1, 3 \\ 0 & \text{otherwise?} \end{cases}$$

(ii) $p(x) = (2x + 1)c$, $x = 0, 1, \ldots, 6$, and zero otherwise?

(iii) $p(x) = xc$, $x = 1, 2, \ldots, N$, and zero elsewhere?

In each case find the distribution function and compute $P(X < 3)$, $P(2 < X < 5)$. Find the smallest x for which $P(X \le x) > 0.5$. Find the largest x for which $P(X \le x) < 0.5$.

6. A lot of 10 items contains 2 defectives. Let X denote the number of defectives in a random sample of size 3 (without replacement). Find the probability function of X, its distribution function, and $P(0 < X < 2)$, $P(X > 1)$ and $P(X < 2)$.

7. The number of arrivals at a supermarket in any period of length t hours has probability function $P(X(t) = k) = (8t)^k e^{-8t}/k!$, $k = 0, 1, \ldots$, and zero otherwise. Find $P(X(.25) \ge 1)$, $P(X(.25) < 2)$, $P(X(.25) = 1)$.

8. Which of the following functions are density functions:
   (i) $f(x) = x(2 - x)$, $0 < x < 2$, and zero elsewhere.
   (ii) $f(x) = x(2x - 1)$, $0 < x < 2$, and zero elsewhere.
   (iii) $f(x) = \dfrac{1}{\lambda} \exp\{-(x - \theta)/\lambda\}$, $x > \theta$, and zero elsewhere.
   (iv) $f(x) = (1/2)\exp\{-|x|\}$, $x \in \mathbb{R}$.
   (v) $f(x) = \sin x$, $0 < x < \pi/2$, and 0 elsewhere.
   (vi) $f(x) = 1 - |1 - x|$, $0 \le x \le 2$, and zero elsewhere.
   (vii) $f(x) = 0$ for $x < 0$, $= (x + 1)/9$ for $0 \le x < 1$, $= 2(2x - 1)/9$ for $1 \le x < 3/2$, $= 2(5 - 2x)/9$ for $3/2 \le x < 2$, $= 4/27$ for $2 \le x < 5$, and zero elsewhere.
   (viii) $f(x) = 1/[\pi(1 + x^2)]$, $x \in \mathbb{R}$.

9. In Problem 8, compute the distribution functions for the densities in parts (iii) to (viii).

10. For what values of c are the following functions density functions?
    (i) $f(x) = cx(2 - x)$, $0 < x < 2$, and zero elsewhere.
    (ii) $f(x) = 0$ if $x \le -1$, $= c(x + 1)$ if $-1 < x \le 2$, $= 3c$ if $2 < x \le 5$, and zero elsewhere.
    (iii) $f(x) = cx^4(1 - x)$, $0 \le x \le 1$, and zero elsewhere.
    (iv) $f(x) = c/(a^2 + x^2)$, $|x| \le a$, and zero elsewhere.
    (v) $f(x) = c/x^5$, $x \ge 1000$, and zero elsewhere.

11. Are the following functions distribution functions? If so, find the corresponding density or probability functions.
    (i) $F(x) = 0$ for $x \le 0$, $= x/2$ for $0 \le x < 1$, $= 1/2$ for $1 \le x < 2$, $= x/4$ for $2 \le x < 4$ and $= 1$ for $x \ge 4$.
    (ii) $F(x) = 0$ if $x < -\theta$, $= \dfrac{1}{2}\left(\dfrac{x}{\theta} + 1\right)$ if $|x| \le \theta$, and 1 for $x > \theta$ where $\theta > 0$.
    (iii) $F(x) = 0$ if $x < 0$, and $= 1 - (1 + x)\exp(-x)$ if $x \ge 0$.

(iv) $F(x) = 0$ if $x < 1$, $= (x - 1)^2/8$ if $1 \le x < 3$, and 1 for $x \ge 3$.

(v) $F(x) = 0$ if $x < 0$, and $= 1 - e^{-x^2}$ if $x \ge 0$.

12. For what values of c are the following functions distribution functions of continuous type random variables?

(i) $F(x) = c[\arctan x + \pi/2\}$, $x \in \mathbb{R}$

(ii) $F(x) = 0$, if $x < 1$, $= c(x - 1)^3$ if $1 < x \le 3$, $= 1$ if $x \ge 3$.

(iii) $F(x) = 0$, if $x < 0$, $= cx^2$ if $0 \le x < r$, $= 1$ if $x \ge r$.

(iv) $F(x) = c \arccos [(r - x)/r]$ if $0 \le x \le 2r$, $= 0$ if $x < 0$ and $= 1$ if $x \ge 2r$.

13. The length X of a randomly selected call (in minutes) at an exchange is a random variable whose density can be approximated by

$$f(x) = (1/10)e^{-x/10}, \qquad x > 0, \text{ and zero elsewhere.}$$

Find the probability that a call will last:

(i) More than 10 minutes. (ii) Between 15 and 25 minutes. (iii) Less than 12 minutes.

14. A leading manufacturer of color television sets offers a four-year warranty of free replacement if the picture tube fails during the warranty period. It is estimated that the length of life of the picture tubes used by this manufacturer has density function $f(t) = (1/10)e^{-t/10}$, $t > 0$, and zero elsewhere. What percentage of sets will have picture tube failures during warranty period? During the first year?

15. Let X be the radius of the opening of a finely calibrated tube. Suppose X has density function

$$f(x) = \begin{cases} 10, & 2.05 \le x \le 2.15 \\ 0 & \text{otherwise.} \end{cases}$$

Find the probability that the cross-sectional area πX^2 of the opening is greater than 14.25.

16. Consider the random variable X of Example 6, which has density function

$$f(x) = 6x(1 - x), \qquad 0 \le x \le 1, \text{ and zero elsewhere.}$$

Let $S = 4\pi X^2$ be the surface area of the sphere (of radius X). Find $P(S > s_0)$.

17. Consider a street car line of length l. Passengers board the street car at a random point X in the interval $[0, l]$ and it is assumed that the probability that a passenger will board the street car in a small neighborhood of x is proportional to $x(l - x)^3$. That is, suppose X has density $f(x) = cx(l - x)^3$, $0 \le x \le l$, and zero elsewhere. Find c and $P(X > l/2)$.

18. The yearly incomes are usually modeled by the distribution function

$$F(x) = \begin{cases} 1 - (x_0/x)^\alpha, & x \ge x_0 \\ 0, & x < x_0 \end{cases}$$

where $\alpha > 0$ and $x_0 > 0$. What proportion of incomes are in excess of x_1? What proportion of incomes are between x_2 and x_3 $(x_2 < x_3)$?

19. Let X_1, X_2, X_3 be random variables with distributions given by

$$f_1(x) = 1, \quad 0 < x < 1, \quad \text{and zero otherwise,}$$

$$f_2(x) = \frac{1}{\lambda}\exp\left(-\frac{x}{\lambda}\right), \quad x > 0, \quad \text{and zero otherwise,}$$

$$f_3(x) = p(1-p)^{x-1}, \quad 0 < p < 1, \quad x = 1, 2, \ldots, \quad \text{and zero otherwise,}$$

respectively. An auxiliary experiment is performed that assigns equal probability to X_1, X_2, and X_3, and an observation is taken (on X_1, X_2, or X_3) depending on the result of the auxiliary experiment.
   (i) Find the probability that the observation is greater than 2.
   (ii) If a value greater than 2 is observed, what is the probability that is it was on X_1? On X_2? On X_3?

20. Suppose $P(X \geq x)$ is given for a random variable X (of the continuous type) for all x. How will you find the corresponding density function? In particular find the density function in each of the following cases:
   (i) $P(X \geq x) = 1$ if $x \leq 0$, and $P(X \geq x) = 1 - e^{-\lambda x}$ for $x > 0$; $\lambda > 0$ is a constant.
   (ii) $P(X \geq x) = 1$ if $x < 0$, and $= (1 + x/\lambda)^{-\lambda}$, for $x \geq 0$, $\lambda > 0$ is a constant.
   (iii) $P(X \geq x) = 1$ if $x \leq 0$, and $= 3/(1 + x)^2 - 2/(1 + x)^3$ if $x > 0$.
   (iv) $P(X > x) = 1$ if $x \leq x_0$, and $= (x_0/x)^\alpha$ if $x > x_0$; $x_0 > 0$ and $\alpha > 0$ are constants.

21. Let f be a symmetric density. Let $a > 0$. Show that:
   (i) $F(0) = 1/2$, where F is the distribution function.
   (ii) $P(X \geq a) = 1/2 - \int_0^a f(x)\, dx$.
   (iii) $F(-a) + F(a) = 1$.
   (iv) $P(|X| > a) = 2F(-a) = 2[1 - F(a)]$.
   (v) $P(|X| \leq a) = 2F(a) - 1$.

3.4 MULTIVARIATE DISTRIBUTIONS

Consider a random vector $(X_1, X_2, \ldots, X_k)$, $k \geq 2$, defined on (Ω, ζ, P). How do we describe the model in terms of the new sample space, namely, $\mathbb{R}_k$ (or a subset of $\mathbb{R}_k$) and the set of events $\mathscr{E}$ on this new outcome space? Our object now is to study some equivalent ways of describing the induced distribution of $(X_1, X_2, \ldots, X_k)$ on $\mathscr{E}$. The development follows closely the univariate case (Section 3.3). If Ω is a discrete set, then the new sample space is also discrete, and $\mathscr{E}$ is the class of all sets. In this case it suffices to specify the so-called *joint probability function*, as we shall see. In the case when $\mathbb{R}_k$ or a rectangle in $\mathbb{R}_k$ is the new sample space, the fundamental objects of interest are rectangles. Let us take $k = 2$ for convenience. Let (X, Y) be a random vector defined on (Ω, ζ), and let $\mathbb{R}_2$ be the new space of outcomes. Then the events of interest are of the type $I_1 \times I_2$ where I_1, I_2 are intervals on $\mathbb{R}$. Actually the set of events of interest in $\mathbb{R}_2$ is much larger; it includes all sets in the plane that have a well defined area such as circles, ellipses, triangles, parallelograms, and so on. The induced distribution $P_{X,Y}$ of (X, Y) is defined just as in the one-dimensional case by the relation

$$(1) \qquad P_{X,Y}(A) = P\{\omega : (X(\omega), Y(\omega)) \in A\}$$

for every event A. In particular if $A = I_1 \times I_2$ where I_1, I_2 are intervals on $\mathbb{R}$, then

$$(2) \qquad P_{X,Y}(I_1 \times I_2) = P(X \in I_1, Y \in I_2)$$

is the probability that the outcome (X, Y) falls in $I_1 \times I_2$. To give a simpler description of $P_{X,Y}$ we proceed as in Section 3.3.

DEFINITION 1. *(JOINT PROBABILITY FUNCTION).* A random vector (X, Y) is said to be of the discrete type if there exists a countable set $\mathcal{X} \subseteq \mathbb{R}_2$ such that $P((X, Y) \in \mathcal{X}) = 1$. The distribution $P_{X,Y}$ is specified by the numbers

$$P_{X,Y}\{(x_i, y_j)\} = P\{\omega : X(\omega) = x_i, Y(\omega) = y_j\}$$

where $(x_i, y_j) \in \mathcal{X}$, $i, j = 1, 2, \ldots$ and $\{P_{X,Y}\{(x_i, y_j)\}\}$ is known as the joint probability function of (X, Y). For any subset A of $\mathcal{X}$ we have

$$P_{X,Y}(A) = \sum_{(x_i, y_j) \in A} P_{X,Y}\{(x_i, y_j)\} = \sum_{(x_i, y_j) \in A} P(X = x_i, Y = y_j).$$

DEFINITION 2. *(JOINT DENSITY FUNCTION).* A random vector (X, Y) is said to be of the continuous type with joint density function $f(x, y)$ if for every event $A \subseteq \mathbb{R}_2$ the joint distribution of (X, Y) is given by

$$P_{X,Y}(A) = P((X, Y) \in A) = \iint_A f(x, y) \, dx \, dy.$$

In the discrete case, $1 \geq P(X = x_i, Y = y_j) \geq 0$ for all (x_i, y_j) and

$$(3) \qquad \sum_i \sum_j P(X = x_i, Y = y_j) = 1.$$

In the continuous case, $f(x, y) \geq 0$ for all $(x, y) \in \mathbb{R}_2$ and

$$(4) \qquad \int_{-\infty}^{\infty} \int_{-\infty}^{\infty} f(x, y) \, dx \, dy = 1.$$

We will mostly deal with either the discrete case or the continuous case, although mixed cases do arise in practice where one variable is a discrete type and the other a continuous type. Often we need the probability $P(X \leq x, Y \leq y)$, which is a function of $(x, y) \in \mathbb{R}_2$.

DEFINITION 3. (JOINT DISTRIBUTION FUNCTION). The joint distribution function of (X, Y) is defined by the relation

$$F(x, y) = P(X \leq x, Y \leq y) = P\{\omega : X(\omega) \leq x, Y(\omega) \leq y\}$$

for all $(x, y) \in \mathbb{R}_2$.

Just as in the univariate case F determines and is completely determined uniquely by $P_{X,Y}$, similarly F determines and is completely determined by the joint density or joint probability function. It is therefore sufficient to specify F or the joint density (or probability) function of (X, Y) to complete the description of the model.

The joint distribution function has properties similar to a univariate distribution function. These are stated below. A result similar to Proposition 3.3.2 holds, but we will skip the details.

PROPOSITION 1. *Let F be the joint distribution function of (X, Y). Then*

(i) $0 \leq F(x, y) \leq 1$ *for all (x, y),*
(ii) *F is nondecreasing in each argument,*
(iii) *F is right continuous in each argument,*
(iv) $F(-\infty, y) = 0 = F(x, -\infty)$ *and* $F(\infty, \infty) = 1$, *and*
(v)   *for every* $x_1 < x_2, y_1 < y_2$ *the inequality*

(5) $$F(x_2, y_2) - F(x_2, y_1) + F(x_1, y_1) - F(x_1, y_2) \geq 0$$

   holds.

Proof. We need only prove part (v), since proofs of parts (i) to (iv) are similar to the proof of Proposition 3.3.1. The inequality (5) simply says (see Figure 1) that $P((X, Y) \in (x_1, x_2] \times (y_1, y_2]) \geq 0$. Indeed,

$$F(x_2, y_2) - F(x_2, y_1) + F(x_1, y_1) - F(x_1, y_2)$$

$$= P(X \leq x_2, Y \leq y_2) - P(X \leq x_2, Y \leq y_1)$$

$$+ P(X \leq x_1, Y \leq y_1) - P(X \leq x_1, Y \leq y_2)$$

$$= P(X \leq x_2, y_1 < Y \leq y_2) - P(X \leq x_1, y_1 < Y \leq y_2)$$

$$= P(x_1 < X \leq x_2, y_1 < Y \leq y_2) \geq 0$$

as asserted.   □

The reader will recognize that in the multivariate case joint distribution functions have an additional property, namely, (5). It turns out that the conditions of Proposition 1 are also sufficient for a function F to be a joint distribution

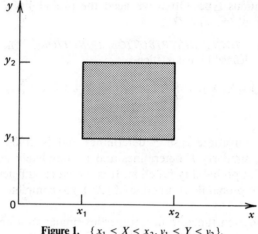

Figure 1. $\{x_1 < X \leq x_2, y_1 < Y \leq y_2\}$.

function. Often a function of two variables satisfies (i)-(iv) of Proposition 1 but not (v). In that case F cannot be a joint distribution function (Problem 7; see also Example 6). Proposition 1 can be extended to more than two variable case.

In the discrete case

$$F(x, y) = \sum_{x_i \leq x} \sum_{y_j \leq y} P_{X,Y}\{(x_i, y_j)\} = \sum_{x_i \leq x} \sum_{y_j \leq y} P(X = x_i, Y = y_j)$$

and in the continuous case

(6) $$F(x, y) = \int_{-\infty}^{x} \int_{-\infty}^{y} f(u, v) \, dv \, du.$$

Similarly, if F is known we can obtain $P_{X,Y}\{(x_i, y_j)\}$ or $f(x, y)$ from it. For example, if the joint density exists then it can be obtained from the joint distribution function according to the formula

(7) $$f(x, y) = \frac{\partial^2 F(x, y)}{\partial x \partial y},$$

for all (x, y) where F has a mixed partial derivative.

One can show (as noted above) that any function $p(x, y)$ satisfying

(i) $$p(x, y) \geq 0, \quad \text{(ii)} \quad \sum_x \sum_y p(x, y) = 1$$

is the joint probability function of some random vector (X, Y). Similarly, any function f satisfying

(i) $\quad\quad\quad f(x, y) \geq 0,$ (ii) $\displaystyle\int_{-\infty}^{\infty}\int_{-\infty}^{\infty} f(x, y)\, dy\, dx = 1$

is the joint density function of some random vector (X, Y).

For the bivariate case

$$P(A) = \iint_A f(x, y)\, dx\, dy$$

represents the volume of the solid formed by the surface f and the set A in the (x, y) − plane (see Figure 2).

Example 1. Lot Inspection. Consider a large lot that has two types of defective items. Let us call these defects of type A and type B. Suppose an item is drawn at random. Define (X, Y) as follows. Let

$$X = \begin{cases} 1 & \text{if the item has type } A \text{ defect,} \\ 0 & \text{otherwise,} \end{cases}$$

and

$$Y = \begin{cases} 1 & \text{if the item has type } B \text{ defect,} \\ 0 & \text{otherwise.} \end{cases}$$

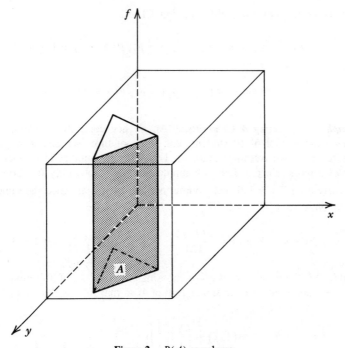

Figure 2. $P(A)$ as volume.

There are four possibilities: the item drawn is not defective, or has both type A and type B defects, or only defects of type A, or only defects of type B. Suppose the probability function of (X, Y) is as summarized in the bivariate table:

Y \ X	0	1
0	p_1	$p_3 - p_1$
1	$p_2 - p_1$	$1 + p_1 - p_2 - p_3$

Thus

$$P(X = 0, Y = 0) = p_1 = \text{proportion of items without any defect}$$

$$P(X = 1, Y = 0) = p_3 - p_1 = \text{proportion of items with type } A \text{ defect only}$$

and so on. Consequently, the distribution function of (X, Y) is given by

$$F(x, y) = \begin{cases} 0, & x < 0 \text{ or } y < 0 \\ p_1, & 0 \le x < 1, \text{ and } 0 \le y < 1 \\ p_2, & 0 \le x < 1, \text{ and } y \ge 1 \\ p_3, & x \ge 1, \text{ and } 0 \le y < 1 \\ 1, & x \ge 1, \text{ and } y \ge 1. \end{cases}$$

Figure 3 shows the joint probability function. Clearly

$$P(X = 0) = p_1 + (p_2 - p_1) = p_2, \, P(X = 1) = 1 - p_2$$

and

$$P(Y = 0) = p_1 + (p_3 - p_1) = p_3, \, P(Y = 1) = 1 - p_3. \qquad \square$$

Example 2. Selecting a Committee. An ad hoc committee of three is selected randomly from a pool of 10 students consisting of three seniors, three juniors, two sophomores, and two freshmen students. Let X be the number of seniors and Y the number of juniors selected. Let us compute the joint probability function of (X, Y). Clearly there are $\binom{10}{3} = 120$ such committees and each is assigned the same probability $1/120$. Now

$$P(X = i, Y = j) = \frac{n(i, j)}{120}, \qquad i, j = 0, 1, 2, 3, \quad i + j \le 3$$

where $n(i, j)$ is the number of ways of choosing i seniors (out of 3), j juniors (out of 3) and $3 - i - j$ sophomores or freshmen (out of 4) so that from Rules 1 and 3 of Section 2.5

$$n(i, j) = \binom{3}{i}\binom{3}{j}\binom{4}{3 - i - j}.$$

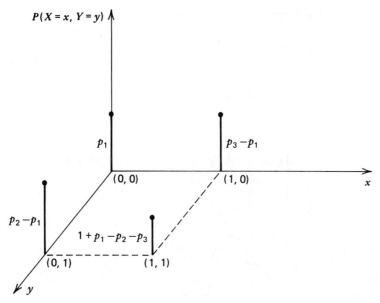

Figure 3

It follows that the joint probability function of (X, Y) is given by

X Y	0	1	2	3
0	4/120	18/120	12/120	1/120
1	18/120	36/120	9/120	0
2	12/120	9/120	0	0
3	1/120	0	0	0

It is easy to write the joint distribution function of (X, Y). Clearly,

$$P(0 < X \le 2, Y = 3) = \sum_{x=1}^{2} P(X = x, Y = 3) = 0,$$

$$P(0 < X \le 2, Y = 1) = 45/120,$$

$$P(X \ge 1) = \sum_{x=1}^{3} \sum_{y=0}^{3} P(X = x, Y = y) = \frac{85}{120},$$

and so on. Figure 4 shows the joint probability function of (X, Y). □

Example 3. Tensile Strength. The tensile strengths X and Y of two kinds of nylon fibers have a joint density function proportional to $xy \exp\{-(x + y)/\lambda\}$, $x, y > 0$.

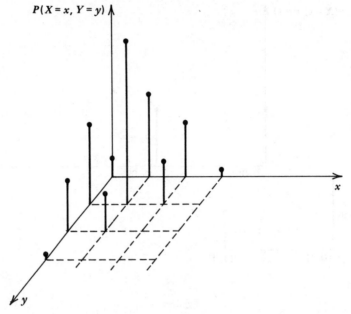

Figure 4. Joint probability function of (X, Y).

Then the joint density function is given by:

$$f(x, y) = \begin{cases} cxy \exp\left\{-\dfrac{x+y}{\lambda}\right\}, & x > 0, y > 0 \\ 0 & \text{otherwise.} \end{cases}$$

Since f is a density function,

$$\int_{-\infty}^{\infty}\int_{-\infty}^{\infty} f(x, y)\, dx\, dy = 1$$

and it follows, on integration by parts, that $c = 1/\lambda^4$. The joint distribution function of (X, Y) is given by

$$F(x, y) = 0 \quad \text{if} \quad x \le 0 \quad \text{or} \quad y \le 0$$

and

$$F(x, y) = \int_0^x \int_0^y \frac{1}{\lambda^4} uv \exp\left\{-\frac{u+v}{\lambda}\right\} du\, dv$$

$$= \int_0^x \frac{1}{\lambda^2} u \exp\left(-\frac{u}{\lambda}\right) du \int_0^y \frac{1}{\lambda^2} v \exp\left(-\frac{v}{\lambda}\right) dv$$

$$= \left\{1 - e^{-x/\lambda} - \left(\frac{x}{\lambda}\right) e^{-x/\lambda}\right\}\left\{1 - e^{-y/\lambda} - \left(\frac{y}{\lambda}\right) e^{-y/\lambda}\right\},$$

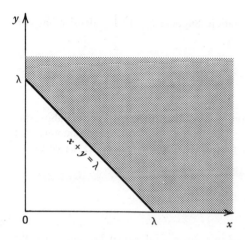

Figure 5. $\{(x, y): x + y > \lambda\}$.

for $x > 0, y > 0$. Suppose we want to find $P(X + Y > \lambda)$. By definition:

$$P(X + Y > \lambda) = \iint\limits_{x+y>\lambda} f(x, y) \, dx \, dy.$$

This means that we need to integrate $f(x, y)$ over the shaded region in Figure 5. Note, however, that

$$P(X + Y > \lambda) = 1 - P(X + Y \leq \lambda)$$

so that

$$P(X + Y > \lambda) = 1 - \int_0^\lambda \left(\int_0^{\lambda-y} \frac{1}{\lambda^4} xy \exp\left\{ -\frac{x+y}{\lambda} \right\} dx \right) dy$$

$$= 1 - \int_0^\lambda \frac{1}{\lambda^2} y e^{-y/\lambda} \left(\int_0^{\lambda-y} \frac{1}{\lambda^2} x e^{-x/\lambda} \, dx \right) dy$$

$$= 1 - \int_0^\lambda \frac{1}{\lambda^2} y e^{-y/\lambda} \left[1 - e^{-(\lambda-y)/\lambda} - \frac{1}{\lambda}(\lambda - y) e^{-(\lambda-y)/\lambda} \right] dy$$

$$= 1 - \frac{1}{\lambda^2} \int_0^\lambda \left[y e^{-y/\lambda} - y e^{-1} - \left(y \frac{(\lambda - y)}{\lambda} \right) e^{-1} \right] dy$$

$$= 1 - \frac{1}{\lambda^2} \left[-\lambda^2 e^{-1} + \lambda^2 (1 - e^{-1}) - \lambda^2 e^{-1} + \frac{\lambda^2}{3} e^{-1} \right] = \frac{8}{3e}$$

$$\doteq .981. \qquad \qquad \square$$

Example 4. Radar Screen. Let (X, Y) be the rectangular coordinates of a point on a circle of radius r and center $(0, 0)$. For example, (X, Y) may be the coordinates of a spot on the circular screen of a radar station. Suppose (X, Y) has density function

$$
f(x, y) = \begin{cases} \dfrac{1}{\pi r^2}, & x^2 + y^2 \le r^2 \\ 0 & \text{elsewhere.} \end{cases}
$$

Geometrically, f is just the circular section of a cylinder of height $1/(\pi r^2)$ above the circle with center at the origin and radius r in the (x, y)-plane. The probability that $(X, Y) \in A$ for any subset of the circle is proportional to the area of set A (see Figure 6). That is:

$$
P(A) = \frac{\text{area of } A}{\pi r^2}.
$$

The probability that a spot will be observed in a square of diagonal $2r$, centered at the origin, is $2r^2/\pi r^2 = 2/\pi$ (Figure 7).

We have from Figure 8 $P(X > Y) = 1/2$. Directly,

$$
P(X > Y) = \iint_{x>y} \frac{1}{\pi r^2} \, dx \, dy = \frac{1}{\pi r^2} \int_{-r}^{r} \left(\int_{y}^{\sqrt{r^2-y^2}} dx \right) dy
$$

$$
= \frac{1}{\pi r^2} \int_{-r}^{r} \left(\sqrt{r^2 - y^2} - y \right) dy = \frac{2}{\pi r^2} \int_{0}^{r} \sqrt{r^2 - y^2} \, dy = \frac{2}{\pi r^2} \left(\frac{r^2 \pi}{4} \right)
$$

$$
= 1/2. \qquad\qquad \square
$$

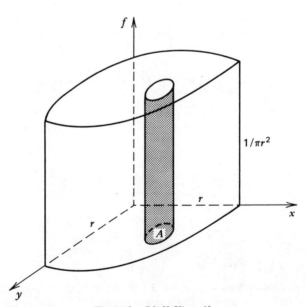

Figure 6. $P\{(X, Y) \in A\}$.

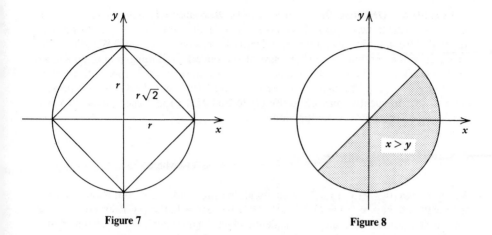

Figure 7   **Figure 8**

Example 5.   A Trivariate Density.   Let (X, Y, Z) have joint density function given by

$$f(x, y, z) = \begin{cases} \dfrac{1}{243} xy \exp\left\{-\dfrac{x + y + z}{3}\right\}, & x > 0, y > 0, z > 0 \\ 0 & \text{elsewhere.} \end{cases}$$

Let us find $P(X < Y < Z)$. We have

$$P(X < Y < Z) = \iiint\limits_{x<y<z} f(x, y, z)\, dz\, dy\, dx$$

$$= \int_{x=0}^{\infty}\int_{y=x}^{\infty}\int_{z=y}^{\infty} \frac{1}{243} xy \exp\left\{-\frac{x + y + z}{3}\right\} dz\, dy\, dx$$

$$= \frac{1}{243}\int_0^{\infty} xe^{-x/3}\left(\int_x^{\infty} ye^{-y/3}\left(\int_y^{\infty} e^{-z/3}\, dz\right) dy\right) dx$$

$$= \frac{1}{243}\int_0^{\infty} xe^{-x/3}\left(\int_x^{\infty} ye^{-y/3}\cdot 3e^{-y/3}\, dy\right) dx$$

$$= \frac{1}{81}\int_0^{\infty} xe^{-x/3}\left[\frac{ye^{-2y/3}}{-2/3}\bigg|_x^{\infty} + \frac{3}{2}\int_x^{\infty} e^{-2y/3}\, dy\right] dx$$

$$= \frac{1}{81}\int_0^{\infty} xe^{-x/3}\left[\frac{3}{2} xe^{-2x/3} + \frac{9}{4} e^{-2x/3}\right] dx$$

$$= \frac{1}{81}\int_0^{\infty}\left[\frac{3}{2} x^2 e^{-x} + \frac{9}{4} xe^{-x}\right] dx$$

$$= \frac{1}{81}\left[\frac{3}{2}(2) + \frac{9}{4}(1)\right] \quad \text{(integration by parts)}$$

$$= (1/81)(21/4) = 7/108. \qquad \square$$

Example 6. Obtaining Density from a Joint Distribution Function.   This example serves to illustrate two points: How do we show that a function F is a distribution function? If so, how do we find the joint density in the continuous case? We will restrict ourselves to the continuous case. In view of the remarks following Proposition 1, we need to check conditions (i) through (v) in order to show that F is a joint distribution function. Although (i) through (iv) are easily checked, (v) is not always easy to check. A quicker way, frequently, may be to use (7) to find $f(x, y)$ and check to see if f is a density function. Consider the function

$$F(x, y) = \begin{cases} 1 - \exp\{-x - y\}, & x > 0, y > 0 \\ 0 & \text{elsewhere.} \end{cases}$$

We note that $0 \le F(x, y) \le 1$, F is nondecreasing and right continuous in each argument, and $F(-\infty, y) = F(x, -\infty) = 0$, $F(\infty, \infty) = 1$. But (v) of Proposition 1 is not satisfied. Rather than find a rectangle for which (5) does not hold, we assume that F is a distribution function. In that case (7) gives the density function as

$$f(x, y) = \frac{\partial^2 F(x, y)}{\partial x \partial y} = \frac{\partial}{\partial x}(e^{-x}e^{-y}) = -e^{-x-y}, \qquad x, y > 0$$

and since $f < 0$ for $x, y > 0$ we conclude that f cannot be a density function. Hence F is not a joint distribution function.

Consider the function

$$F(x, y) = \begin{cases} 0, & x < 0 \;\text{ or }\; y < 0 \\ xy, & 0 \le x < 1, \;\; 0 \le y < 1 \\ y, & 1 \le x, \;\; 0 \le y < 1 \\ x, & 1 \le y, \;\; 0 \le x < 1 \\ 1, & x \ge 1 \;\text{ and }\; y \ge 1. \end{cases}$$

Differentiating partially with respect to y and then with respect to x, we get

$$f(x, y) = \begin{cases} 1, & 0 < x < 1, 0 < y < 1 \\ 0 & \text{elsewhere.} \end{cases}$$

Since $f \ge 0$ and $\int_0^1\int_0^1 f(x, y)\, dy\, dx = 1$, we see that f is a density function so that F must be a distribution function.   □

Problems for Section 3.4

1. A fair die is rolled twice. Let

$$X = \text{sum of face values}$$

$$Y = \text{absolute value of the difference in face values.}$$

(i) Find the joint probability function of (X, Y).
(ii) Find $P(X \ge 8, Y \le 2)$, $P(X \le 6, Y \le 3)$.

2. An urn contains two black, one white, and three red marbles. A random sample of two marbles is drawn (a) with replacement, and (b) without replacement. Let

$$X_k = \begin{cases} 1 & \text{if the marble drawn on } k\text{th draw is white} \\ 0 & \text{otherwise,} \end{cases}$$

for $k = 1, 2$. Find the joint probability function of (X_1, X_2) in each case, and also find the joint distribution functions.

3. Four fair coins are rolled once. Let

$$X = \text{number of heads}$$

$$Y = \text{number of heads less the number of tails.}$$

(i) Find the joint distribution of (X, Y).
(ii) Find $P(X \geq 3, Y \geq 0)$, $P(X \geq 2, Y \leq 1)$.

4. Consider the following joint probability function of (X, Y)

X Y	5	6
1	.15	.10
2	.20	.20
3	.30	.05

(i) Find $P(X + Y \geq 8)$, $P(X + Y \leq 7)$
(ii) Find $P(6 \leq XY \leq 13)$, $P(XY < 8)$.

5. An urn contains five identical slips numbered 1 through 5. One slip is drawn and let X be the number on the slip. Then all the slips numbered *less* than X are also removed from the urn and another slip is drawn from the remaining slips. Let Y be the number on the second slip drawn.
(i) Find the joint probability function of (X, Y).
(ii) Find $P(X \leq 3, Y \leq 4)$, $P(X + Y \leq 7)$, $P(Y - X \geq 2)$. (If $X = 5$, take $Y = 0$.)

6. Which of the following functions are joint density functions?
(i) $f(x, y) = e^{-(x+y)}$, $x > 0$, $y > 0$, and zero elsewhere.
(ii) $f(x, y) = xy$, $0 \leq x \leq 2$, $0 \leq y \leq 1$, and zero elsewhere.
(iii) $f(x, y) = (1 + xy)/4$, $|x| \leq 1$, $|y| \leq 1$, and zero elsewhere.
(iv) $f(x, y) = 8x^2y$, $0 \leq y \leq x \leq 1$, and zero elsewhere.
(v) $f(x, y) = (1 - e^{-x/\lambda_1})(1 - e^{-y/\lambda_2})$, $x > 0$, $y > 0$, and zero elsewhere.
(vi) $f(x, y) = x\exp\{-x(1 + y)\}$, $0 < x < \infty$, $0 < y < \infty$, and zero otherwise.
(vii) $f(x, y) = e^{-y}$, $0 < x < y < \infty$, and zero elsewhere.

7. Show that the following functions are *not* joint distribution functions.
(i) $F(x, y) = 0$, $x + 2y < 1$, and 1 elsewhere.
(ii) $F(x, y) = 1 - x\exp(-x - y)$, $x > 0$, $y > 0$, and zero elsewhere.
(iii) $F(x, y) = 0$, $x + y \leq 0$, and 1 elsewhere.
(iv) $F(x, y) = 1 - (1 - e^{-x})e^{-y}$, $x > 0$, $y > 0$, and zero elsewhere.
(v) $F(x, y) = 1 - \exp\{-x^2 - y^2\}$, $x > 0$, $y > 0$, and zero elsewhere.

8. Suppose (X, Y) is of the continuous type with joint density function f given below. Find c.
   (i) $f(x, y) = c(1 - x^2 - y^2)$, $x^2 + y^2 \leq 1$, and zero elsewhere.
   (ii) $f(x, y) = cxe^{-y}$, $0 < x < y < \infty$, and zero elsewhere.
   (iii) $f(x, y) = c(x + y)$, $0 < x < 1, 0 < y < 1, 0 < x + y < 1$, and zero elsewhere.
   (iv) $f(x, y) = cxy$, $0 < y < x < 1$, and zero otherwise.
   (v) $f(x, y) = c \exp\{-(x + y)\}$, $0 \leq x \leq y < \infty$, and zero otherwise.
   (vi) $f(x, y) = cg(x)g(y)$, $-\infty < x \leq y < \infty$, where g is a density function on $(-\infty, \infty)$.
   In each case find the distribution function.

9. A point (X, Y) is selected at random from the unit circle according to joint density function f where f is proportional to the inverse of the distance of (X, Y) from the center of the circle. What is the joint density function of (X, Y)? Find $P(1/2 < X < 1, 0 < Y < 3/4)$.

10. Let X be the lifetime of an electronic tube and Y that of the replacement installed after the failure of the first tube. Suppose (X, Y) have joint density $f(x, y) = \exp(-x - y)$, $x, y > 0$ and zero elsewhere. Find $P(X + Y \geq 2)$, $P(X \leq 5, Y \leq 7)$, $P(X/Y \geq 1)$.

11. Is the function

$$f(x, y, z, u) = \begin{cases} \exp(-u), & 0 < x < y < z < u \\ 0 & \text{elsewhere} \end{cases}$$

a joint density function? If so, find $P(X \leq 7)$ where (X, Y, Z, U) is a random variable with density f.

12. Is the function

$$f(x_1, x_2, \ldots, x_n) = \begin{cases} c \exp\left\{-\sum_{j=1}^{n} x_j\right\}, & x_1, x_2, \ldots, x_n > 0 \\ 0 & \text{elsewhere} \end{cases}$$

a joint density function? If so, find c. Also find

$$P(0 \leq X_1 \leq 1, \ldots, 0 \leq X_n \leq 1) \quad \text{and} \quad P(1 \leq X_1 < X_2 < \cdots < X_n < \infty).$$

13. Consider an L.P. gas tank that has a random amount Y in supply at the beginning of the month. Suppose a random amount X is used up during the month without being resupplied. Suppose X and Y have joint density function

$$f(x, y) = \begin{cases} 2/k^2, & 0 \leq x \leq y \leq k \\ 0 & \text{elsewhere} \end{cases}$$

where k is a known positive constant.
   (i) Show that f is a density function.
   (ii) Find the distribution function of (X, Y).
   (iii) Find $P(X \leq k/2)$, $P(X \leq k/2, Y \geq 3k/4)$.

14. Consider an electronic system that has three types of components operating simultaneously. The lifetimes X, Y, Z of the three components have joint distribution function given by

$$F(x, y, z) = \begin{cases} (1 - e^{-x^2})(1 - e^{-y^2})(1 - e^{-z^2}), & x > 0, y > 0, z > 0, \\ 0 & \text{elsewhere} \end{cases}$$

Find $P(X > Y > Z)$. (See Problem 19 for a general result.)

15. Show that the function defined by

$$f(x, y, z, u) = \frac{24}{(1 + x + y + z + u)^5}, \qquad x > 0, y > 0, z > 0, u > 0$$

and zero elsewhere is a joint density function.
(i) Find $P(X > Y > Z > U)$.
(ii) Find $P(X + Y + Z + U \geq 1)$.

16. Let $f_1, f_2, \ldots, f_k$ be probability density functions on (a, b) where a may be $-\infty$ and b may equal ∞. Is

$$f(x_1, \ldots, x_k) = \prod_{j=1}^{k} f_j(x_j)$$

a density function of some random vector $(X_1, X_2, \ldots, X_k)$? If so, find the distribution function of $(X_1, X_2, \ldots, X_k)$. Find $P(X_1 \leq x, X_2 \leq x, \ldots, X_k \leq x)$ when $f_1 = f_2 = \cdots = f_k$.

17. Let A be a closed triangle in the plane with vertices $(0, 0)$, $(0, \sqrt{2})$ and $(\sqrt{2}, \sqrt{2})$. Let $F(x, y)$ denote the area of the intersection of A with set $\{(x_1, y_1): x_1 \leq x, y_1 \leq y\}$. Show that F defines a distribution function. Find the corresponding joint density function. Find $P(X + Y < 1)$, $P(Y - X > 1)$.

18. Suppose (X, Y) are jointly distributed with density

$$f(x, y) = g(x)g(y), x \in \mathbf{R}, y \in \mathbf{R}$$

where g is a density function on $\mathbf{R}$. Show that $P(X \geq Y) = P(X \leq Y) = 1/2$.

19. Suppose (X, Y, Z) are jointly distributed with density

$$f(x, y, z) = \begin{cases} g(x)g(y)g(z), & x > 0, y > 0, z > 0 \\ 0 & \text{elsewhere.} \end{cases}$$

Find $P(X > Y > Z)$. Hence find the probability that $(x, y, z) \notin \{X > Y > Z\}$ or $\{X < Y < Z\}$. (Here g is density function on $\mathbf{R}$.)

20. Let f and g be (univariate) density functions with respective distribution functions F and G. Let $0 \leq \alpha \leq 1$ and define f_α on $\mathbf{R}_2$ by:

$$f_\alpha(x, y) = f(x)g(x)[1 + \alpha(2F(x) - 1)(2G(x) - 1)].$$

Show that f_α is a joint density function for each α.

21*. Let (X, Y) have joint density function f and joint distribution function F. Suppose that:

$$f(x_1, y_1)f(x_2, y_2) \le f(x_1, y_2)f(x_2, y_1)$$

holds for $x_1 \le a \le x_2$ and $y_1 \le b \le y_2$. Show that:

$$F(a, b) \le F_1(a)F_2(b).$$

3.5 MARGINAL AND CONDITIONAL DISTRIBUTIONS

Let $(X_1, X_2, \ldots, X_k)$ be a random vector. Does the knowledge of the joint distribution of $(X_1, X_2, \ldots, X_k)$ lead to the joint distribution of any subset of $\{X_1, X_2, \ldots, X_k\}$? Conversely, does the knowledge of the marginal distributions of $X_1, X_2, \ldots, X_k$ lead to the joint distribution of $(X_1, X_2, \ldots, X_k)$? The answer to the first question is yes. Indeed, writing $F_{X_1, X_2, \ldots, X_k}$ for the joint distribution function of $(X_1, X_2, \ldots, X_k)$, we have:

$$(1) \quad F_{X_{i_1}, X_{i_2}, \ldots, X_{i_j}}\left(x_{i_1}, x_{i_2}, \ldots, x_{i_j}\right) = \lim_{\substack{x_i \to \infty \\ i \ne i_1, \ldots, i_j}} F_{X_1, X_2, \ldots, X_k}(x_1, x_2, \ldots, x_k)$$

for any subset $\{i_1, i_2, \ldots, i_j\} \subseteq \{1, 2, \ldots, k\}$. Moreover, if $(X_1, X_2, \ldots, X_k)$ is of the continuous type, then

$$(2) \quad f_{X_{i_1}, \ldots, X_{i_j}}\left(x_{i_1}, \ldots, x_{i_j}\right) = \int_{-\infty}^{\infty} \cdots \int_{-\infty}^{\infty} f_{X_1, \ldots, X_k}(x_1, \ldots, x_k) \prod_{\substack{i=1 \\ i \ne i_1, \ldots, i_j}}^{k} dx_i$$

and in the discrete case

$$(3) \quad P\left(X_{i_1} = x_{i_1}, \ldots, X_{i_j} = x_{i_j}\right) = \sum_{x_1} \cdots \sum_{\substack{x_k \\ x_i \ne x_{i_1}, \ldots, x_{i_j}}} P(X_1 = x_1, \ldots, X_k = x_k).$$

The distribution function in (1) is called the *joint (marginal) distribution function* of $(X_{i_1}, \ldots, X_{i_j})$. The density function on the left side of (2) is called the *joint (marginal) density function* of $(X_{i_1}, \ldots, X_{i_j})$ and the function on the left side of (3) is known as the *joint (marginal) probability function* of $(X_{i_1}, \ldots, X_{i_j})$.

Suppose $k = 2$. For convenience, let us write (X, Y) for (X_1, X_2). In the discrete case, the marginal distributions of (X, Y) are easy to comprehend. In terms of a bivariate table, these are column and row totals that translate to the

X Y	x_1	x_2	$\cdots$	$P(Y = y_j)$
y_1 y_2 $\vdots$				
$P(X = x_i)$				1

graph in Figure 1. Basically it means that if we move a plane from left to right and another plane from right to left, both parallel to the $(x, P(X = x_i, Y = y_j))$-plane until they meet the $(x, P(X = x_i, Y = y_j))$-plane, the marginal distribution of X is the total mass accumulated at various values of X in the $(x, P(X = x_i, Y = y_j))$-plane. A similar geometrical interpretation may be given to the marginal probability function of Y.

In the continuous case, the marginals of X and Y are the integrals of $f(x, y)$ over all values of y and x respectively. In geometric terms, the marginal density of X is the impression left on $(x, f(x, y))$-plane as we condense the solid enclosed by $f(x, y)$ by moving planes parallel to the $(x, f(x, y))$-plane from left to right and right to left until they meet the $(x, f(x, y))$-plane.

In what follows we restrict ourselves to the $k = 2$ case. For convenience we use the notation (X, Y) instead of (X_1, X_2). We now consider the second question raised in the first paragraph of this section. Does the knowledge of the marginal distributions of X and Y yield their joint distribution? The answer, in general, is no. A special case where the answer is yes will be considered in Section 3.6. The

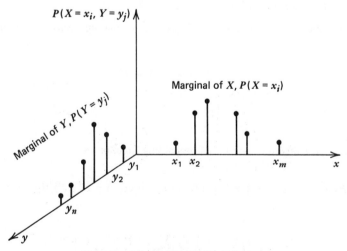

Figure 1.   Marginal probability functions.

fact that the answer in general is no should not be surprising; the marginal distributions of X and Y provide no information on the dependence structure between X and Y. This information is provided by the conditional distribution defined below.

Suppose a value of X is observed. Given $X = x$, what is the distribution of a future value of Y? This looks like a problem in conditional probability. In the discrete case, given $X = x_i$, the probability that $Y = y_j$ is given by

$$P\{Y = y_j | X = x_i\} = \frac{P\{(X = x_i) \cap (Y = y_j)\}}{P(X = x_i)}$$

(4)
$$= \frac{P(X = x_i, Y = y_j)}{P(X = x_i)}$$

by the definition of conditional probability, provided that $P(X = x_i) > 0$. Here $X = x_i$ is fixed and y_j varies. We note that

(5)
$$P\{Y = y_j | X = x_i\} \geq 0$$

and

(6)
$$\sum_j P\{Y = y_j | X = x_i\} = \sum_j \frac{P(X = x_i, Y = y_j)}{P(X = x_i)} = 1$$

by definition of the marginal probability function of X. Hence $P\{Y = y_j | X = x_i\}$, $j = 1, 2, \ldots$ defines a probability function.

DEFINITION 1. Let (X, Y) be a discrete type random vector. Then the function defined by

$$P\{Y = y_j | X = x_i\} = \begin{cases} \dfrac{P(X = x_i, Y = y_j)}{P(X = x_i)} & \text{if } P(X = x_i) > 0 \\ 0 & \text{otherwise,} \end{cases}$$

for $j = 1, 2, \ldots$ is called the conditional probability function of Y given $X = x_i$. We define $P\{Y = y_j | X = x_i\} = 0$ if $P(X = x_i) = 0$.

Since $P\{Y = y_j | X = x_i\}$ is a probability function it has all the properties of a probability function. Thus

(7)
$$P\{Y \in A | X = x_i\} = \sum_{y_j \in A} P\{Y = y_j | X = x_i\}.$$

In particular, if $A = (-\infty, y]$, then

(8)
$$P\{Y \le y | X = x_i\} = \sum_{y_j \le y} P\{Y = y_j | X = x_i\}$$

is called the *conditional distribution function of Y given X = x_i.*

In geometric terms, $P\{Y = y_j | X = x_i\}$ is the cross section of $P(X = x_i, Y = y_j)$ in the $X = x_i$ plane adjusted to give a total mass of one. The adjustment factor is $P(X = x_i)$.

Example 1. Coin Tossing. Consider three independent tossings of a fair coin. Let

$$X = \text{number of heads,}$$

and

$$Y = \text{number of tails before the first head.}$$

Then X takes values 0, 1, 2, 3, and Y takes values 0, 1, 2, 3. The joint distribution of (X, Y) is given in the following table.

Y \ X	0	1	2	3	$P(Y = y)$
0	0	1/8	2/8	1/8	4/8
1	0	1/8	1/8	0	2/8
2	0	1/8	0	0	1/8
3	1/8	0	0	0	1/8
$P(X = x)$	1/8	3/8	3/8	1/8	1

The row totals give the marginal distribution of Y and the column totals give the marginal distribution of X. Let us compute the conditional distribution of X given $Y = y$ for $y = 0, 1, 2, 3$ respectively. We have

$$P\{X = x | Y = 0\} = \begin{cases} 1/4, & \text{if } x = 1 \text{ or } 3 \\ 1/2, & \text{if } x = 2 \\ 0 & \text{otherwise,} \end{cases}$$

$$P\{X = x | Y = 1\} = \begin{cases} 1/2, & \text{if } x = 1 \text{ or } 2 \\ 0 & \text{otherwise,} \end{cases}$$

$$P\{X = x | Y = 2\} = \begin{cases} 1, & \text{if } x = 1 \\ 0 & \text{elsewhere,} \end{cases}$$

and

$$P\{X = x | Y = 3\} = \begin{cases} 1, & \text{if } x = 0 \\ 0 & \text{elsewhere.} \end{cases}$$

In particular,

$$P\{ X \geq 2 | Y = 0\} = 3/4,$$

$$P\{ X + Y \geq 2 | Y = 1\} = P\{ X \geq 1 | Y = 1\} = 1,$$

and so on. The conditional distribution function of X given $Y = 0$, for example, is given by

$$F(x|0) = \begin{cases} 0, & \text{if } x < 1 \\ 1/4, & \text{if } 1 \leq x < 2 \\ 3/4, & \text{if } 2 \leq x < 3 \\ 1, & \text{if } x \geq 3. \end{cases}$$

In Figure 2, we plot the marginal distribution of Y and the conditional distribution of X, given $Y = 0$.   □

Example 2.   *Quality Control.*   Every manufacturer of major appliances such as washers, dryers, and refrigerators is concerned about the quality of the merchandise before shipping. Suppose a manufacturing firm subjects its finished product to a final inspection that consists of looking for two kinds of defects, those in the finish (scratches, dents, and flaws in porcelain finish) and those that can be classified as mechanical. A random sample of 100 pieces of the appliance is examined. Let X be the number of defects of the first kind, and Y the number of defects of the second kind.

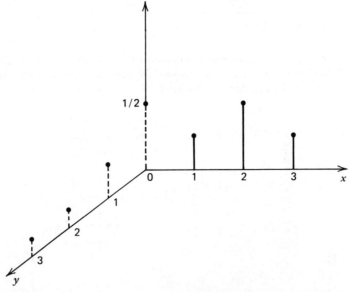

Figure 2.   Marginal distribution of Y (dotted line); conditional distribution of X given $Y = 0$ (solid line).

Suppose the following results are obtained.

	Number of Defects X				
Y	0	1	2	3	
0	50	8	2	1	61
1	11	8	2	1	22
2	5	3	1	0	9
3	6	1	0	0	7
4	1	0	0	0	1
	73	20	5	2	100

In a situation such as this example, the joint distribution of (X, Y) will be estimated by the manufacturer after repeated sampling over a period of time. Suppose this table reflects the long-run joint distribution of defects. Then the probability function is obtained by dividing through by 100. The marginal distributions are then given by

x	0	1	2	3	4
$P(X = x)$	.73	.20	.05	.02	–
$P(Y = x)$	.61	.22	.09	.07	.01

and, in particular,

$$P\{Y = y | X = 1\} = \begin{cases} 8/20, & y = 0,1 \\ 3/20, & y = 2 \\ 1/20, & y = 3 \\ 0 & \text{elsewhere.} \end{cases}$$

Thus

$$P\{1 \le Y \le 3 | X = 1\} = 8/20 + 3/20 + 1/20 = .6.$$

Similarly,

$$P\{X + Y \le 3 | X = 2\} = P\{Y \le 1 | X = 2\} = (2/5) + (2/5) = .8. \qquad \square$$

Let us now consider the case when X and Y have a joint density f. Let f_1, f_2 be the marginal densities of X and Y respectively. Since $P(X = x) = 0$ for every x we can no longer follow the above development in order to define the conditional distribution of Y given $X = x$. However, we can imitate (4) and set for $f_1(x) > 0$

$$(9) \qquad g(y|x) = \frac{f(x, y)}{f_1(x)}, \qquad \text{for } y \in \mathbb{R}.$$

Then

$$(10) \qquad g(y|x) \geq 0 \quad \text{and} \quad \int_{-\infty}^{\infty} g(y|x)\, dy = \int_{-\infty}^{\infty} \frac{f(x, y)\, dy}{f_1(x)} = 1,$$

and $g(y|x)$ is a density function.

DEFINITION 2. Let (X, Y) be a continuous type random vector with joint density function f and let f_1 be the marginal density of X. Then the conditional density of Y given $X = x$ is defined by

$$g(y|x) = \frac{f(x, y)}{f_1(x)}, \qquad y \in \mathbb{R}$$

provided $f_1(x) > 0$.

We recall that a density may be changed on a finite set without affecting its integral, so that $g(\cdot|x)$ is not strictly unique. When we speak of the conditional density, we simply mean that a version is chosen and held fixed. It is clear that the roles of X and Y may be reversed to define the conditional density function of X given $Y = y$.

We now provide a justification for Definition 2. The reader may skip the following discussion. It is not needed in our subsequent investigation.

Let $h > 0$ and suppose $P(x - h < X \leq x + h) > 0$. By the definition of conditional probability, we have

$$(11) \quad P\{Y \leq y | x - h < X \leq x + h\} = \frac{P(x - h < X \leq x + h, Y \leq y)}{P(x - h < X \leq x + h)}.$$

Let F be the joint distribution function of (X, Y), f the joint density, and f_1 the marginal density of X. Note that since F_1 (the distribution function of X) is continuous,

$$P(x - h < X \leq x + h) = \{F_1(x + h) - F_1(x - h)\} \downarrow 0 \text{ as } h \downarrow 0.$$

Moreover, $(x - h < X \leq x + h) \downarrow (X = x)$ as $h \downarrow 0$. Let us write

$$\lim_{h \downarrow 0} P\{Y \leq y | x - h < X \leq x + h\} = P\{Y \leq y | X = x\},$$

provided that the limit exists. Then

$$(12) \qquad P\{Y \leq y | X = x\} = \lim_{h \downarrow 0} \frac{F(x + h, y) - F(x - h, y)}{F_1(x + h) - F_1(x - h)}.$$

Both numerator and denominator tend to 0 as $h \downarrow 0$ and we have a typical case of L'Hôpital's rule. Dividing numerator and denominator on the right side of (12) by $2h$ and noting that if f_1 is continuous at x, then

$$\lim_{h \to 0} \frac{1}{2h} \{ F_1(x + h) - F_1(x - h) \} = f_1(x)$$

and

$$\lim_{h \to 0} \frac{1}{2h} \{ F(x + h, y) - F(x - h, y) \} = \int_{-\infty}^{y} f(x, v) \, dv,$$

we have

(13) $$P\{ Y \le y | X = x \} = \int_{-\infty}^{y} \frac{f(x, v)}{f_1(x)} \, dv$$

provided $f_1(x) > 0$.

We note that $P\{ Y \le y | X = x \}$ defines a distribution function as a function of y for fixed x. This is called the *conditional distribution function of Y given X = x*. Differentiating with respect to y we have

(14) $$\frac{d}{dy} P\{ Y \le y | X = x \} = \frac{f(x, y)}{f_1(x)}$$

provided f is continuous at (x, y), f_1 at x, and $f_1(x) > 0$. This is precisely (9).

Sometimes the notation $f_{Y|X}(y|x)$ is used to denote the conditional density function and $F_{Y|X}(y|x)$ to denote the conditional distribution function of Y given $X = x$. For convenience we will not use this subscript notation unless necessary. Whenever we write $g(y|x)$ [or $h(x|y)$] it will mean the conditional density of Y given $X = x$ (or of X given $Y = y$): the first letter within the parentheses is the variable, the letter following the vertical bar is fixed (and represents the given value of the random variable).

Geometrically, $g(y|x)$ represents the cross section of $f(x, y)$ in the $X = x$ plane, adjusted so as to give a total area of 1 under the curve.

Since $g(y|x)$ is a density function, we can compute $P\{ Y \in A | X = x \}$ from the formula

(15) $$P\{ Y \in A | X = x \} = \int_A g(y|x) \, dy.$$

It follows from the definition of conditional probability or density function that

(16) $$P(X = x_i, Y = y_j) = P(X = x_i) P\{ Y = y_j | X = x_i \}$$

$$= P(Y = y_j) P\{ X = x_i | Y = y_j \},$$

and

$$(17) \qquad f(x, y) = f_1(x)g(y|x) = f_2(y)h(x|y)$$

where f_2 is the marginal density of Y and $h(x|y)$ the conditional density of X given $Y = y$.

Relations (16) and (17) answer the second question we raised. In order to obtain the joint distribution of (X, Y) we need one marginal and one conditional distribution (given the marginal). We note that for $f_2(y) > 0$

$$(18) \qquad h(x|y) = \frac{f_1(x)g(y|x)}{f_2(y)} = \frac{f_1(x)g(y|x)}{\displaystyle\int_{-\infty}^{\infty} f_1(x)g(y|x)\, dx}$$

which gives the continuous analog of the Bayes rule (Proposition 2.6.3).

Often, one of X and Y is continuous type and the other discrete type. That is, (X, Y) is the mixed type. It should be easy for the reader to modify the above discussion to include such a possibility. We will consider some examples of this type here and later in Sections 10.6 and 10.7.

Remark 1. The reader should carefully distinguish between two types of problems involving conditional probability. The problem of finding, say, $P\{X \in B | Y \in A\}$ where $P\{Y \in A\} > 0$ is a simple problem of conditional probability to be treated by methods of Section 2.6. The problem of finding $P\{X \in B | Y = y\}$, however, is a problem of conditional probability distribution to be treated by methods of this section. In the case when Y is discrete and $P(Y = y) > 0$, however, this distinction is not necessary.

    Example 3. Target Practice. In target practice, a marksman aims a gun at a certain target, say, the origin of a rectangular coordinate system, but due to random factors the hit-point can be any point (X, Y) in a circle of radius r with center at the origin. Suppose (X, Y) have joint density

$$f(x, y) = \frac{1}{\pi r^2}, \qquad \text{if } x^2 + y^2 \le r^2, \text{ and zero elsewhere.}$$

(See also Example 3.4.4.) Let us compute f_1, the marginal density of X. We have:

$$f_1(x) = \int_{-\infty}^{\infty} f(x, y)\, dy = \int_{x^2+y^2 \le r^2} (1/\pi r^2)\, dy$$

$$= (1/\pi r^2) \int_{-\sqrt{r^2-x^2}}^{\sqrt{r^2-x^2}} dy = \frac{2\sqrt{r^2 - x^2}}{\pi r^2}$$

for $|x| \le r$, and zero elsewhere. By symmetry:

$$f_2(y) = \frac{2\sqrt{r - y^2}}{\pi r^2}, \qquad |y| \le r, \text{ and zero elsewhere.}$$

Consequently, for fixed $x \in (-r, r)$,

$$g(y|x) = \frac{f(x, y)}{f_1(x)} = \left(2\sqrt{r^2 - x^2}\right)^{-1}, \qquad |y| \leq \sqrt{r^2 - x^2}, \text{ and zero elsewhere.}$$

Thus

$$P\{Y \geq r/2 | X = r/2\} = \int_{r/2}^{r} g(y|r/2)\, dy$$

$$= \int_{r/2}^{r} \left(2\sqrt{3r^2/4}\right)^{-1} dy = \frac{r}{2} \cdot \frac{1}{(\sqrt{3})r} = \frac{1}{2\sqrt{3}}.$$

Figure 3 shows the marginal distribution of X and the conditional distribution of Y given $X = x$.

In the same target model, suppose the marksman has more skill so that the joint density function is more concentrated towards the target, that is, the center of the circle. The joint density function

$$f(x, y) = \frac{2}{\pi r^4}\left\{r^2 - (x^2 + y^2)\right\}, \qquad 0 \leq x^2 + y^2 \leq r^2, \text{ and zero otherwise}$$

describes such a density function.

The marginal densities are easily computed. Indeed,

$$f_1(x) = f_2(x) = \frac{2}{\pi r^4} \int_{-\sqrt{r^2-x^2}}^{\sqrt{r^2-x^2}} (r^2 - x^2 - y^2)\, dy$$

$$= \frac{8}{3\pi r^4}(r^2 - x^2)^{3/2},$$

for $|r| \leq x$ and zero elsewhere. The conditional density of Y given $X = x$ for $x \in (-r, r)$ is given by

$$g(y|x) = \frac{3}{4}\frac{r^2 - x^2 - y^2}{(r^2 - x^2)^{3/2}}, \qquad |y| \leq \sqrt{r^2 - x^2}, \text{ and zero elsewhere.} \qquad \square$$

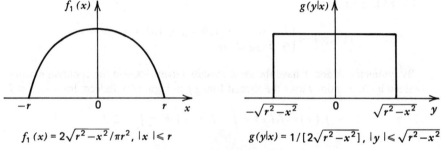

$$f_1(x) = 2\sqrt{r^2-x^2}/\pi r^2, \ |x| \leq r \qquad\qquad g(y|x) = 1/[2\sqrt{r^2-x^2}], \ |y| \leq \sqrt{r^2-x^2}$$

Figure 3. Graphs of f_1 and g.

*Example 4. **Sale Price and Volume of Sales.*** Suppose the selling price p of an article varies over the interval $(0, 1)$ depending on the final production costs (determined from an initial run). The sales volume V is also a random variable depending on the selling price. Let the joint density be given by:

$$f(p, v) = \begin{cases} 12p^4(1 - p)v \exp(-vp), & 0 < p < 1, v > 0 \\ 0 & \text{elsewhere.} \end{cases}$$

Then

$$f_1(p) = 12p^4(1 - p)\int_0^\infty ve^{-vp}\, dv = 12p^4(1 - p)(1/p^2)$$

$$= 12p^2(1 - p), \text{ for } 0 < p < 1,$$

and zero elsewhere, and

$$f_2(v) = 12v\int_0^1 (p^4 - p^5)e^{-vp}\, dp \qquad \text{for } v > 0$$

and zero elsewhere. It is easy to compute f_2 explicitly on integration by parts. We leave the details to the reader. It follows that for $p \in (0, 1)$,

$$g(v|p) = \begin{cases} vp^2e^{-vp}, & v > 0 \\ 0 & \text{elsewhere.} \end{cases}$$

Hence

$$P\{V > 10|p\} = p^2\int_{10}^\infty ve^{-vp}\, dv = p^2\left[-\left.\frac{ve^{-vp}}{p}\right|_{10}^\infty + \frac{1}{p}\int_{10}^\infty e^{-vp}\, dv \right]$$

$$= \left[p(10e^{-10p}) + e^{-10p} \right] = (1 + 10p)e^{-10p}.$$

In particular, if $p = 1/2$, then

$$P\{V > 10|p = 1/2\} = 6e^{-5} = .0404. \qquad \square$$

*Example 5. **Additives in Gasoline.*** Let X and Y be the proportions of two different additives in a sample taken from a certain brand of gasoline. Suppose the joint density of (X, Y) is given by:

$$f(x, y) = \begin{cases} 2, & 0 \le x \le 1, 0 \le y \le 1, 0 \le x + y \le 1, \\ 0 & \text{elsewhere.} \end{cases}$$

By symmetry, X and Y have the same marginal densities, and the common density is obtained by integrating f over the shaded triangle in Figure 4b. Indeed, for $0 \le x \le 1$

$$f_1(x) = f_2(x) = \int_{-\infty}^\infty f(x, y)\, dy = \int_0^{1-x} 2\, dy$$

$$= 2(1 - x),$$

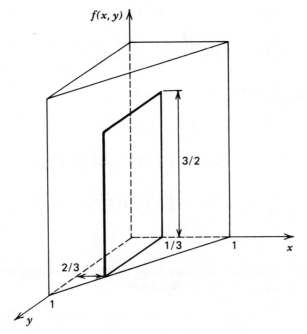

Figure 4a. Graph of $g(y|\frac{1}{3})$.

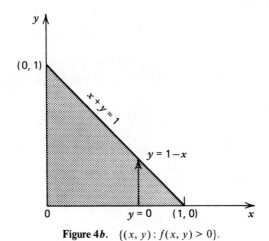

Figure 4b. $\{(x, y) : f(x, y) > 0\}$.

131

and $f_1(x) = 0$ elsewhere. The conditional density of Y given $X = x$ is given by

$$g(y|x) = \begin{cases} \dfrac{2}{2(1-x)} = \dfrac{1}{1-x}, & 0 \le y \le 1-x \\ 0 & \text{elsewhere}, \end{cases}$$

for $0 \le x \le 1$. In particular, if $x = 1/3$, then

$$g\left(y\left|\dfrac{1}{3}\right.\right) = \begin{cases} \dfrac{1}{1-1/3} = \dfrac{3}{2}, & 0 \le y \le 2/3 \\ 0 & \text{elsewhere}, \end{cases}$$

and

$$P\left\{\dfrac{1}{2} \le Y \le \dfrac{7}{8}\left|X = \dfrac{1}{3}\right.\right\} = \int_{1/2}^{7/8} g\left(y\left|\dfrac{1}{3}\right.\right) dy = \dfrac{3}{2}\int_{1/2}^{2/3} dy$$

$$= (3/2)(1/6) = 1/4. \qquad \square$$

Example 6. A Mixed Distribution: Store Holdups. The weekly number of store holdups in a certain city has probability function

$$P_\lambda(X = x) = e^{-\lambda}\lambda^x/x!, \qquad x = 0, 1, 2, \ldots, \text{ and zero elsewhere.}$$

We write P_λ instead of P to indicate that the distribution depends on λ. Once λ is specified we can compute probability of events defined in terms of X. Here $\lambda > 0$ measures the intensity of the incidence of holdups. Suppose the intensity varies from week to week according to the density function

$$f_2(\lambda) = 10e^{-10\lambda}, \qquad \lambda > 0, \text{ and zero elsewhere.}$$

Let us call this random variable Λ. Then $P_\lambda(X = x)$ gives the conditional distribution of X given $\Lambda = \lambda$, and the joint distribution of X and Λ is given by

$$P_\lambda(X = x)f_2(\lambda) = \left(\dfrac{e^{-\lambda}\lambda^x}{x!}\right)(10e^{-10\lambda}), \qquad x = 0, 1, 2, \ldots; \quad \lambda > 0$$

and zero elsewhere. The unconditional distribution of X is given by

$$P(X = x) = \int_{-\infty}^{\infty} P_\lambda(X = x)f_2(\lambda)\, d\lambda = \int_0^\infty \dfrac{e^{-\lambda}\lambda^x}{x!}10e^{-10\lambda}\, d\lambda$$

$$= \dfrac{10}{x!}\int_0^\infty \lambda^x e^{-11\lambda}\, d\lambda$$

for $x = 0, 1, 2, \ldots$. Integrating by parts repeatedly, we get[†]

$$P(X = x) = \dfrac{10}{x!}\dfrac{1}{(11)^{x+1}}\cdot(x!) = \dfrac{10}{(11)^{x+1}}, \qquad x = 0, 1, \ldots.$$

[†] This is a gamma integral, to be studied in Section 7.3.

Thus

$$P(X \geq 5) = 10 \sum_{x=5}^{\infty} \frac{1}{(11)^{x+1}} = 10 \cdot \frac{1}{11^6} \sum_{x=0}^{\infty} \frac{1}{11^x}$$

$$= (10/11^6)(11/10) = 1/11^5 = 6.21 \times 10^{-6}$$

where we have used the fact that $\sum_{0}^{\infty} r^x = 1/(1 - r)$, $0 < |r| < 1$ (see Section 6.5), so that five or more holdups in a week are highly improbable. If $\lambda = .1$ for a week, for example, then

$$P_\lambda(X \geq 5) = \sum_{x=5}^{\infty} e^{-.1} \frac{(.1)^x}{x!}$$

$$= 1 - e^{-.1}\left\{1 + .1 + \frac{(.1)^2}{2} + \frac{(.1)^3}{6} + \frac{(.1)^4}{24}\right\}$$

$$= 1 - .9999998 = .0000002 = 2 \times 10^{-7},$$

whereas if $\lambda = 10$, then

$$P_\lambda(X \geq 5) = \sum_{x=5}^{\infty} e^{-10} \frac{10^x}{x!} = 1 - .02925 = .97075.$$

The conditional probability $P_\lambda(X \geq 5)$ fluctuates between 0 and 1 depending on the value of λ. The unconditional probability $P(X \geq 5)$ is in some sense an average probability that takes into account the distribution of Λ. Since large values of Λ are improbable it is not surprising that $P(X \geq 5)$ is small.

Suppose we know that in some week $X = 2$. What can we say about Λ? The information about Λ, given $X = x$ is provided by the conditional density $g(\lambda|x)$. For $x \geq 0$ (integral) we have

$$g(\lambda|x) = \frac{P_\lambda(X = x)f_2(\lambda)}{P(X = x)} = \frac{(e^{-\lambda}\lambda^x/x!)10e^{-10\lambda}}{10/(11)^{x+1}}$$

$$= \frac{11}{x!}(11\lambda)^x e^{-11\lambda}, \qquad \lambda > 0$$

and

$$g(\lambda|x) = 0 \qquad \text{elsewhere.}$$

Thus

$$P\{.1 \leq \Lambda \leq .3|X = 2\} = \frac{11}{2!} \int_{.1}^{.3} (11\lambda)^2 e^{-11\lambda} \, d\lambda$$

$$= \frac{1}{2} \int_{1.1}^{3.3} u^2 e^{-u} \, du \qquad (u = 11\lambda)$$

$$= \frac{1}{2}[-u^2 e^{-u} - 2ue^{-u} - 2e^{-u}]_{1.1}^{3.3} = 0.541$$

and

$$P\{.5 \leq \Lambda | X = 2\} = \frac{11}{2} \int_{.5}^{\infty} (11\lambda)^2 e^{-11\lambda} \, d\lambda$$

$$= \frac{1}{2} \left[-e^{-u}(u^2 + 2u + 2) \right]_{5.5}^{\infty} = .0066.$$

If $X = 2$ is observed then it is improbable that Λ is large, that is, $\Lambda \geq .5$, and much more probable that Λ is small, $.1 \leq \Lambda \leq .3$. $\quad\square$

We conclude this section with some remarks concerning the more than two variate case. Let $(X_1, X_2, \ldots, X_k)$, $k \geq 2$, be a random vector defined on $(\Omega, \mathcal{S}, P)$. We have seen in (1) through (3) how to obtain the joint marginal distribution of a subset of $X_1, X_2, \ldots, X_k$ from the joint distribution of $(X_1, X_2, \ldots, X_k)$. We can also define the conditional distribution in a similar manner. Indeed, suppose $(X_1, \ldots, X_k)$ is of discrete type, and we want to find the (joint) conditional distribution of $(X_1, X_2, \ldots, X_j)$ given $X_{j+1}, \ldots, X_k$, for $j < k$. Then:

$$P\{ X_1 = x_1, \ldots, X_j = x_j | X_{j+1} = x_{j+1}, \ldots, X_k = x_k \}$$

$$= \frac{P\{ X_1 = x_1, \ldots, X_j = x_j, X_{j+1} = x_{j+1}, \ldots, X_k = x_k \}}{P(X_{j+1} = x_{j+1}, \ldots, X_k = x_k)}$$

provided $P(X_{j+1} = x_{j+1}, \ldots, X_k = x_k)$, the marginal probability function of $(X_{j+1}, \ldots, X_k)$, is positive at $(x_{j+1}, \ldots, x_k)$. In the continuous case the conditional density of $(X_1, X_2, \ldots, X_j)$ given $X_{j+1} = x_{j+1}, \ldots, X_k = x_k$ is defined by

$$g(x_1, \ldots, x_j | x_{j+1}, \ldots, x_k) = \frac{f(x_1, \ldots, x_j, x_{j+1}, \ldots, x_k)}{f_{X_{j+1}, \ldots, X_k}(x_{j+1}, \ldots, x_k)}$$

for all $(x_1, \ldots, x_j)$ provided $f_{X_{j+1}, \ldots, X_k}(x_{j+1}, \ldots, x_k) > 0$. Here f is the joint density of $(X_1, \ldots, X_k)$ and $f_{X_{j+1}, \ldots, X_k}$ that of $(X_{j+1}, \ldots, X_k)$. It suffices to consider an example.

Example 7. *Choosing Points from a Line Segment.* A point X_1 is chosen from $[0, 1]$ such that the probability that X_1 belongs to a small neighborhood of x is proportional to the length of the neighborhood. That is, X_1 has density

$$f_1(x) \alpha c \, \Delta x$$

for x in a neighborhood of length Δx. Since f_1 has to be a density on $[0, 1]$, $c = 1$ and

$$f_1(x) = 1, \qquad 0 \leq x \leq 1, \text{ and zero otherwise.}$$

A point X_2 is now chosen from $[0, X_1]$ according to the density

$$f(x_2 | x_1) = \frac{1}{x_1}, \qquad 0 \leq x_2 \leq x_1, \text{ and zero elsewhere.}$$

Another point X_3 is chosen from the interval $[0, x_2]$ according to the density

$$f(x_3|x_1, x_2) = \frac{1}{x_2}, \qquad 0 \le x_3 \le x_2, \text{ and zero elsewhere,}$$

and so on. Suppose $X_1, X_2, \ldots, X_n$ are n such points. Then the joint density of $(X_1, X_2, \ldots, X_n)$ is given by:

$$f(x_1, x_2, \ldots, x_n) = f_1(x_1)f(x_2|x_1)f(x_3|x_1, x_2) \cdots f(x_n|x_1, \ldots, x_{n-1}).$$

This follows from an obvious extension of the product rule (17). Hence

$$f(x_1, x_2, \ldots, x_n) = \begin{cases} \dfrac{1}{x_1} \dfrac{1}{x_2} \cdots \cdot \dfrac{1}{x_{n-1}}, & 0 \le x_n \le x_{n-1} \le \cdots \le x_1 \le 1 \\ 0 & \text{elsewhere.} \end{cases}$$

Let $k < n$. Then the joint marginal density of $(X_1, \ldots, X_k)$ is given by

$$f_{X_1, \ldots, X_k}(x_1, \ldots, x_k) = \begin{cases} \dfrac{1}{x_1 x_2 \cdots x_{k-1}}, & 0 \le x_k \le \cdots \le x_1 \le 1, \\ 0 & \text{elsewhere} \end{cases}$$

and the conditional density of $(X_{k+1}, \ldots, X_n)$ given $X_1 = x_1, \ldots, X_k = x_k$ is given by:

$$f_{X_{k+1}, \ldots, X_n | X_1, \ldots, X_k}(x_{k+1}, \ldots, x_n | x_1, \ldots, x_k)$$

$$= \frac{1}{x_k x_{k+1} \cdots x_n}, 0 \le x_n \le \cdots \le x_{k+1} \le x_k,$$

and zero elsewhere. $\qquad\qquad\qquad\qquad\qquad\qquad\qquad\qquad\qquad\qquad$ □

Problems for Section 3.5

1. Consider a fair die that is rolled three times. Let X be the number of times the face value is 1 or 2, and Y the number of times the face value is 4, 5, or 6.
   (i) Find the joint distribution of (X, Y).
   (ii) Find the two marginal distributions.
   (iii) Find the conditional distribution of X given $Y = y$ and that of Y given $X = x$.
   (iv) Find $P(1 \le X \le 3 | Y = 1)$.

2. A fruit basket contains five oranges, two apples, and four bananas. The fruit are dropped in a sack and a random sample of 4 is taken (without replacement). Let X be the number of apples and Y be the number of bananas in the sample.
   (i) Find the joint probability function of (X, Y).
   (ii) Find the two marginal distributions.
   (iii) Find the conditional distribution of Y given $X = x$ (where $x = 0$ or 1 or 2).
   (iv) Find $P\{Y \le y | X = 1\}$ for $y = 0, 1, 2, 3, 4$.
   (v) Find $P\{1 < Y < 3 | X = 0\}$.

3. From a table of random numbers, a number X is drawn from the integers 1, 2, 3, 4, 5, 6. Suppose $X = x$. Then a second number Y is drawn from the integers x, $x + 1, \ldots, 9$.
   (i) Find the joint probability function of (X, Y).
   (ii) Find the two marginal distributions.
   (iii) Find the conditional distribution of X given $Y = y$.
   (iv) Find $P\{3 < X \le 9 | Y = 8\}$

4. Let (X, Y) have joint probability function given by

$$P(X = x, Y = y) = \begin{cases} \dfrac{3x + 2y}{72}, & x = 1, 2, 3, y = 0, 1, 2 \\ 0 & \text{elsewhere.} \end{cases}$$

   (i) Find the marginal distributions of X and Y.
   (ii) Find the conditional distribution of X given $Y = y$.
   (iii) Find $P\{X \le 2 | Y = 1\}$.

5. From a group of three Republicans, four Democrats and two Independents, a committee of three is selected. Let X be the number of Republicans selected and Y, the number of Democrats.
   (i) Find the joint distribution of (X, Y).
   (ii) Find the two marginal distributions.
   (iii) Find the two conditional distributions.
   (iv) Find $P(X = 3)$, $P(Y = 3)$. If the committee consists of either three Republicans or three Democrats, do the Independents have reason to complain? If the committee consists only of Democrats and Republicans, do the Independents have reason to complain?
   (v) Find $P\{Y = 3 | X = 0\}$, $P\{X = 3 | Y = 0\}$.

6. For the following joint density functions, compute (a) the marginal distributions of X and Y, (b) the two conditional distributions.
   (i) $f(x, y) = 2(2x + 3y)/5, 0 < x < 1, 0 < y < 1$, and zero elsewhere.
   (ii) $f(x, y) = (x + y)/3, 0 < x < 2, 0 < y < 1$, and zero elsewhere.
   (iii) $f(x, y) = 1/2, 0 < x < 2, 0 < y < x$, and zero elsewhere.
   (iv) $f(x, y) = 8xy, 0 < x < y < 1$, and zero otherwise.
   (v) $f(x, y) = x \exp\{-x(1 + y)\}, x > 0, y > 0$, and zero otherwise.
   (vi) $f(x, y) = (\pi^2/8)\sin(\pi(x + y)/2), 0 < x < 1, 0 < y < 1$, and zero otherwise.
   (vii) $f(x, y) = 1/y, 0 < x < y < 1$, and zero elsewhere.

7. Let X be the proportion of households that will respond to a mail-order solicitation and let Y be the proportion of households that will respond to a solicitation on the phone. Suppose (X, Y) have joint probability density function given by:

$$f(x, y) = \frac{(3x + 5y)}{4}, \qquad 0 \le x \le 1, 0 < y < 1, \text{ and zero elsewhere.}$$

   (i) Find the two marginal densities.
   (ii) Find $P(X > .25)$, $P(Y \le .75)$.
   (iii) Find the two conditional densities.
   (iv) Find $P\{X \ge .75 | Y \le .25\}$, $P\{X + Y \le .50 | X \ge .25\}$

8. The joint distribution of (X, Y), where

$X = $ total time spent by a customer in a bank,

$Y = $ time spent by the customer in the bank waiting to be served,

is modeled by joint density function

$$f(x, y) = \exp(-x), 0 < y < x < \infty, \text{ and zero elsewhere.}$$

(i)   Find the two marginal distributions.
(ii) Find the two conditional distributions.
(iii) Find $P\{Y > 5 | X < 10\}$, $P\{X > 10 | Y \le 2\}$.
(iv) Note that $X - Y$ is the service time of the customer. Find

$$P\{X - Y \ge 2 | X = 10\}, P\{X - Y \ge 2 | 1 \le X \le 10\}.$$

9. A point X is chosen from the interval $[0, 1]$ according to the density function f_1 (given below). Another point Y is chosen from the interval $[X, 1]$ (so that $X \le Y \le 1$) according to the density function $g(y|x)$. Let

$$f_1(x) = 1, \qquad 0 \le x \le 1, \text{ and zero elsewhere,}$$

and for $0 \le x < 1$

$$g(y|x) = \frac{1}{1 - x} \qquad \text{if } x \le y \le 1, \text{ and zero elsewhere.}$$

(i)   Find the joint density of X and Y and the marginal density of Y.
(ii) Find the conditional density of X given $Y = y$.
(iii) Find $P\{1/4 \le X \le 1/2 | Y = 3/8\}$.

10. The surface tension X and the acidity Y of a certain chemical compound are jointly distributed with density function

$$f(x, y) = \frac{3 - x - y}{3}, \qquad 0 \le x \le 1, 0 \le y \le 2, \text{ and zero otherwise.}$$

(i)   Find the two marginal densities, and the two conditional densities.
(ii) Find $P\{X \le .92 | Y = 1\}$, $P\{1 \le X + Y \le 2 | X = .1\}$.

11*. Customers arrive at a service counter (a theater ticket booth or a bank teller's or a retail counter), stand on line for service, and depart after service. Suppose $X(t)$, the number of arrivals in a time interval of length t, has probability function

$$P(X(t) = k) = \frac{e^{-\lambda t}(\lambda t)^k}{k!}, \qquad k = 0, 1, \dots, \quad \text{and zero elsewhere}$$

where $\lambda > 0$ is a fixed constant. Let T be the time it takes to service a single customer and let N be the number of additional customers who arrive during the service time T of a customer. Suppose T has density function

$$f_1(t) = \lambda e^{-t\lambda}, \qquad t > 0, \text{ and zero elsewhere.}$$

(i)   Find the joint distribution of T and N and the marginal distribution of N.
(ii) Find the conditional distribution of T given $N = n$.
(iii) Find $P(N \geq 2)$, $P\{T \leq 2 | N = 2\}$. [Hint: $P\{N = n | T = t\} = P\{X(t) = n\}$
      $= e^{-\lambda t}(\lambda t)^n / n!.$]

12*. A new drug is designed to reduce the effects of a certain disease. Suppose the
      proportion X of patients who respond favorably to the new drug is a random
      variable that has density function

$$f_1(x) = 6x(1 - x), \qquad 0 < x < 1, \text{ and zero elsewhere.}$$

The number of patients Y who show improvement from a total of n patients
administered the drug has probability function

$$P\{Y = y | X = x\} = \binom{n}{y} x^y (1 - x)^{n-y}, y = 0, 1, \ldots, n.$$

(i)   Find the marginal probability function of Y.
(ii) Find the conditional density of X given $Y = y$.
(iii) Find $P\{0 < X < .1 | Y = n\}$.
      [Hint: $\int_0^1 x^{\alpha-1}(1 - x)^{\beta-1} dx = B(\alpha, \beta) = (\alpha - 1)!(\beta - 1)!/(\alpha + \beta - 1)!$
      for $\alpha, \beta > 0$ integral. See Section 7.3.]

13.   Let f given below be the joint density of $(X_1, X_2, \ldots, X_n)$.
      (i) $f(x_1, x_2, \ldots, x_n) = e^{-x_n}, 0 < x_1 < x_2 < \cdots < x_n < \infty$, and zero elsewhere.
      (ii) $f(x_1, x_2, \ldots, x_n) = (1/n!)\exp(-\max x_i), x_1 > 0, \ldots, x_n > 0$, and zero other-
            wise.
      In each case find
      (iii) the joint marginal density of $(X_3, X_4, \ldots, X_n)$, and
      (iv) the joint conditional density of (X_1, X_2) given $X_3 = x_3, \ldots, X_n = x_n$.
      In (i) also find
      (v) $P\{X_2 - X_1 > a | X_3 = x_3, \ldots, X_n = x_n\}$, for $a > 0$ fixed.

14.   For distribution functions F, F_1, F_2 show that

$$\max(F_1(x) + F_2(y) - 1, 0) \leq F(x, y) \leq \min(F_1(x), F_2(y)),$$

for all $x, y \in \mathbf{R}$ if and only if F_1, F_2 are the marginal distribution functions of F.

15.   Let f_1, f_2 be density functions with corresponding distribution functions F_1 and F_2.
      Let α be a constant, $|\alpha| \leq 1$. Define

$$f_\alpha(x_1, x_2) = f_1(x_1)f_2(x_2)\{1 + \alpha[2F_1(x_1) - 1][2F_2(x_2) - 1]\},$$

for all x_1, x_2. Then f_α is a joint density function (Problem 3.4.20). Find
(i)   The two marginal densities.
(ii) The two conditional densities.

3.6   INDEPENDENT RANDOM VARIABLES

A basic problem in probability is to determine the joint distribution of the random vector $(X_1, X_2, \ldots, X_k)$ given the marginal distributions of $X_1, X_2, \ldots, X_k$. In Section 3.5 we showed that in general we cannot get the joint distribution from the marginal distributions alone. We need additional information on the dependence structure of $X_1, X_2, \ldots, X_k$. If this information is available in the form of conditional distributions, then we can get the joint distribution. Thus, in the discrete case

$$(1) \quad P(X_1 = x_1, X_2 = x_2, \ldots, X_k = x_k)$$

$$= P(X_1 = x_1)P\{X_2 = x_2, \ldots, X_k = x_k | X_1 = x_1\}$$

$$= P(X_1 = x_1)P\{X_2 = x_2 | X_1 = x_1\}$$

$$\times P\{X_3 = x_3, \ldots, X_k = x_k | X_1 = x_1, X_2 = x_2\}$$

$$= P(X_1 = x_1)P\{X_2 = x_2 | X_1 = x_1\} \cdots$$

$$\times P\{X_k = x_k | X_1 = x_1, \ldots, X_{k-1} = x_{k-1}\}.$$

This follows by a simple extension of the product rule (3.5.16). An analog of (1) holds also for the continuous case. An important special case is the case of *independence* which we now study. In this case the probability that $X_j \in A_j$ is unaffected by the values of the remaining random variables.

DEFINITION 1.   Let $X_1, X_2, \ldots, X_k$ be random variables defined on (Ω, ζ, P). Then $X_1, X_2, \ldots, X_k$ are said to be independent if for all events $A_1, A_2, \ldots, A_k$ (in $\mathbb{R}$)

$$P(X_1 \in A_1, X_2 \in A_2, \ldots, X_k \in A_k) = \prod_{j=1}^{k} P(X_j \in A_j).$$

Definition 1 says that all the events $\{\omega: X_j(\omega) \in A_j\}$ are independent on ζ and is a natural extension of the definition of independence of events. Definition 1 can be simplified and written equivalently in terms of distribution or density (probability) functions.

DEFINITION 2.   Let F be the joint distribution function of $(X_1, \ldots, X_k)$ and let F_j be the marginal distribution function of $X_j, j = 1, 2, \ldots, k$. Then $X_1, X_2, \ldots, X_k$ are independent if

$$F(x_1, x_2, \ldots, x_k) = \prod_{j=1}^{k} F_j(x_j), \quad x_1, \ldots, x_k \in \mathbb{R}.$$

DEFINITION 3. If $X_1, \ldots, X_k$ are discrete type random variables then $X_1, \ldots, X_k$ are independent if

$$P(X_1 = x_1, \ldots, X_k = x_k) = \prod_{j=1}^{k} P(X_j = x_j)$$

for all $x_1, x_2, \ldots, x_k \in \mathbb{R}$.

If $(X_1, \ldots, X_k)$ have a joint density function f with marginal densities $f_j, j = 1, \ldots, k$ then $X_1, \ldots, X_k$ are independent if

$$f(x_1, \ldots, x_k) = \prod_{j=1}^{k} f_j(x_j)$$

for all $x_1, x_2, \ldots, x_k \in \mathbb{R}$.

Definition 3 is the easiest to use in practice. We leave the reader to show that Definitions 2 and 3 are equivalent. Definition 1 is also equivalent to Definition 2. Indeed, according to Definition 1

$$P(X_1 \in (-\infty, x_1], \ldots, X_k \in (-\infty, x_k]) = \prod_{j=1}^{k} P(X_j \in (-\infty, x_j])$$

which is the statement of Definition 2. Conversely, according to Definition 2:

$$(2) \qquad F(x_1, \ldots, x_k) = \prod_{j=1}^{k} F_j(x_j)$$

so that either

$$(3) \qquad P(X_1 = x_1, \ldots, X_k = x_k) = \prod_{j=1}^{k} P(X_j = x_j)$$

(in the discrete case) or

$$(4) \qquad f(x_1, \ldots, x_k) = \prod_{j=1}^{k} f_j(x_j)$$

(in the continuous case) for all $x_1, \ldots, x_k$. Now we sum both sides of (3) or integrate both sides of (4), as the case may be, over all $x_1, \ldots, x_k$ such that $x_j \in A_j, j = 1, \ldots, k$. This leads to Definition 1.

Example 1. Family Incomes. Let X and Y represent the family incomes above $\$I$ of workers in two different trades. Suppose (X, Y) have joint density

$$f(x, y) = \frac{I^2}{x^2 y^2}, \quad x \geq I, y \geq I, \quad \text{and zero elsewhere.}$$

Then X and Y are independent. Ineed, by symmetry

$$f_1(x) = f_2(x) = \frac{I^2}{x^2} \int_I^\infty \frac{1}{y^2} dy = \frac{I}{x^2}, \ x \geq I, \text{ and zero elsewhere}$$

so that

$$f(x, y) = f_1(x) \cdot f_2(y).$$

We note that $f(x, y)$ can be written as a product of a function of x alone and a function of y alone and the domains of X and Y separate. ☐

Example 2. Distribution of Children by Sex. Consider families with three children. Let X be the number of boys in a randomly selected family and Y the number of girls. Since $X + Y = 3$, X and Y cannot be independent. Indeed, the joint distribution of (X, Y) is given (assuming boys and girls are equally likely) by

Y＼X	0	1	2	3	$P(Y = y)$
0	0	0	0	1/8	1/8
1	0	0	3/8	0	3/8
2	0	3/8	0	0	3/8
3	1/8	0	0	0	1/8
$P(X = x)$	1/8	3/8	3/8	1/8	1

Since

$$P(X = 0, Y = 0) = 0 \neq 1/8 \cdot 1/8,$$

X and Y are not independent. We note that to prove independence of X, Y we need to check (3) [or (4)] for all pairs, but to disprove independence we need only find one pair (x, y) for which (3) or (4) does not hold.

On the other hand, if we define U, V, Z as follows:

$$U = \begin{cases} 1 & \text{eldest child boy} \\ 0 & \text{otherwise} \end{cases}$$

$$V = \begin{cases} 1 & \text{middle child boy} \\ 0 & \text{otherwise} \end{cases}$$

and

$$Z = \begin{cases} 1 & \text{youngest child boy} \\ 0 & \text{otherwise} \end{cases}$$

then U, V, Z are independent. This follows since

$$P(U = u, V = v, Z = z) = \left(\frac{1}{2}\right)^3 = P(U = u)P(V = v)P(Z = z)$$

for $u, v, z \in \{0, 1\}$ and $P(U = u, V = v, Z = z) = 0$ otherwise. ☐

Example 3. Checking Independence. Suppose X and Y have joint density given by

$$f(x) = \begin{cases} \dfrac{1 + xy}{4}, & |x| < 1, |y| < 1, \\ 0 & \text{elsewhere.} \end{cases}$$

The form of f does not suggest that we can write it as a product of two functions f_1 and f_2 where f_1 is a function of x alone and f_2 that of y alone. Formally, for $|x| < 1$

$$f_1(x) = \frac{1}{4} \int_{-1}^{1} (1 + xy) \, dy = \frac{1}{2},$$

and $f_1(x) = 0$ elsewhere. By symmetry:

$$f_2(y) = 1/2, \ |y| < 1, \quad \text{and zero elsewhere.}$$

Thus $f(x, y) \neq f_1(x)f_2(y)$ and X and Y are not independent. Note, however, that for $0 < x < 1, 0 < y < 1,$

$$P(X^2 \leq x, Y^2 \leq y) = P(|X| \leq \sqrt{x}, |Y| \leq \sqrt{y})$$

$$= \frac{1}{4} \int_{-\sqrt{x}}^{\sqrt{x}} \int_{-\sqrt{y}}^{\sqrt{y}} (1 + uv) \, dv \, du$$

$$= (\sqrt{x})(\sqrt{y})$$

$$= P(X^2 \leq x).P(Y^2 \leq y)$$

so that X^2 and Y^2 are independent. □

Independence is a very strong condition. If $X_1, X_2, \ldots, X_k$ are independent, that is, if (2) holds, then the random variables in any subset of $\{X_1, X_2, \ldots, X_k\}$ are also independent. (Contrast relation (2) with the independence of events $A_1, A_2, \ldots, A_k$ where we had to insist that the product relation holds for all subcollections of $A_1, A_2, \ldots, A_k$.) But we cannot go backwards. Independence of X_1 and X_2, X_2 and X_3, and X_1 and X_3 does not imply that of X_1, X_2, X_3.

PROPOSITION 1. *Let $X_1, X_2, \ldots, X_k$ be independent. Then*

(i) $\{X_{i_1}, \ldots, X_{i_j}\} \subset \{X_1, \ldots, X_k\}$ *are also independent,*

(ii) *any functions $f_1(X_1), \ldots, f_k(X_k)$ are also independent, and*

(iii) *any functions $f(X_1, \ldots, X_j)$, $g(X_{j+1}, \ldots, X_k)$ of disjoint subsets are also independent.*

Proof. (i) By independence of $(X_1, \ldots, X_k)$

$$F(x_1, \ldots, x_k) = \prod_{j=1}^{k} F_j(x_j)$$

in an obvious notation. Now let $x_i \to \infty$ for all $i = 1, \ldots, k$, $i \neq i_1$ or $i_2, \ldots, i_j$. Then by definition of a joint marginal distribution function and properties of distribution functions in one dimension, we have

$$F_{X_{i_1},\ldots,X_{i_j}}(x_{i_1},\ldots,x_{i_j}) = \prod_{l=1}^{j} F_{i_l}(x_{i_l}).$$

(ii) We take $k = 2$ for convenience. Now the event $\{f_1(X_1) \le x_1, f_2(X_2) \le x_2\}$ can be rewritten as $\{X_1 \in A_1, X_2 \in A_2\}$ where A_1 depends only on f_1 and A_2 only on f_2. Since X_1, X_2 are independent,

$$P(f_1(X_1) \le x_1, f_2(X_2) \le x_2) = P(X_1 \in A_1, X_2 \in A_2)$$

$$= P(X_1 \in A_1)P(X_2 \in A_2)$$

$$= P(f_1(X_1) \le x_1)P(f_2(X_2) \le x_2)$$

since $X_j \in A_j$ if and only if $f_j(X_j) \le x_j, j = 1, 2$.

(iii) By (i), $(X_1,\ldots,X_j)$ and $(X_{j+1},\ldots,X_k)$ are independent. Hence the result follows from (ii). $\qquad\square$

What does independence say about conditional distributions? If (X_1, X_2) are independent, the probability that $X_1 \in A_1$ is the same regardless of what happens to X_2 so that the conditional distribution of X_1 given $X_2 = x_2$ should not depend on x_2. This is true. Indeed, in the continuous case

$$g(x_2|x_1) = \frac{f(x_1, x_2)}{f_1(x_1)} = \frac{f_1(x_1)f_2(x_2)}{f_1(x_1)} = f_2(x_2).$$

The same argument holds in the discrete case as also in the case when $k > 2$.

Example 4. Pairwise Independence vs. Independence. Suppose (X, Y, Z) have joint probability function

$$P(X = x, Y = y, Z = z) = \begin{cases} 1/4, & \text{if } (x, y, z) \in A \\ 0 & \text{elsewhere,} \end{cases}$$

where $A = \{(1,0,0),(0,1,0),(0,0,1),(1,1,1)\}$. Then

$$P(X = 0) = P(X = 1) = 1/2,$$

$$P(Y = 0) = P(Y = 1) = 1/2,$$

$$P(Z = 0) = P(Z = 1) = 1/2,$$

$$P(X = x, Y = y) = P(X = x, Z = y) = P(Y = x, Z = y) = 1/4,$$

for $x, y \in \{0,1\}$ and zero elsewhere. It follows that X, Y, Z are *pairwise* independent. That is, X and Y are independent, Y and Z are independent, and X and Z are independent. But (X, Y, Z) are not independent since, for example,

$$P(X = 1, Y = 0, Z = 0) = 1/4 \neq P(X = 1)P(Y = 0)P(Z = 0) = 1/8. \quad \square$$

Example 5. Ohm's Law. Consider an electrical circuit for which the Ohm's law $V = IR$ holds where V is the voltage, I the current and R the resistance. Suppose I and R are independent random variables with respective density functions

$$h_I(i) = h(i) = 2i, \quad 0 \leq i \leq 1, \text{ and zero elsewhere,}$$

and

$$g_R(r) = g(r) = r^2/9, \quad 0 \leq r \leq 3, \text{ and zero elsewhere.}$$

Then the joint density of I and R is given by

$$f(i, r) = h(i)g(r) = (2i)\frac{r^2}{9}, \quad 0 \leq i \leq 1, \quad 0 \leq r \leq 3, \text{ and zero elsewhere.}$$

Thus (see Figure 1):

$$P(V \leq 2) = \iint\limits_{ir \leq 2} f(i, r) \, dr \, di$$

$$= \int_{r=0}^{2}\int_{i=0}^{1}\frac{2}{9}ir^2 \, di \, dr + \int_{r=2}^{3}\int_{i=0}^{2/r}\frac{2}{9}ir^2 \, di \, dr$$

$$= (2/9)[(1/2)(8/3) + 2(3 - 2)]$$

$$= (20/27). \quad \square$$

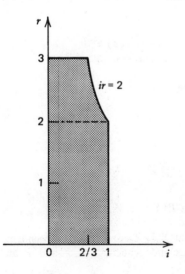

Figure 1. $\{(i, r) : ir \leq 2\}$.

Example 6. *Lifetimes of Television Picture Tubes.* Let X be the lifetime of a television tube manufactured by a certain company and let Y and Z be the lifetimes of successive replacement tubes installed after the failure of the first tube. Suppose X, Y, Z are independent with common density function

$$f_1(x) = (1/\lambda)e^{-x/\lambda}, \qquad x > 0, \text{ and zero elsewhere.}$$

What is $P(X + Y + Z \geq 3\lambda)$?

By independence of X, Y, and Z, their joint density is given by

$$f(x, y, z) = \begin{cases} (1/\lambda^3)\exp\left\{-\dfrac{x+y+z}{\lambda}\right\}, & x > 0, \quad y > 0, \quad z > 0, \\ 0 & \text{elsewhere.} \end{cases}$$

Hence:

$$P(X + Y + Z \geq 3\lambda) = 1 - P(X + Y + Z < 3\lambda)$$

where

$$P(X + Y + Z < 3\lambda) = \iiint\limits_{x+y+z<3\lambda} \frac{1}{\lambda^3}\exp\left\{-\frac{x+y+z}{\lambda}\right\} dz\, dy\, dx$$

$$= \frac{1}{\lambda^3} \iint\limits_{x+y<3\lambda} \left(\int_0^{3\lambda-x-y} e^{-z/\lambda}\, dz\right) e^{-(x+y)/\lambda}\, dy\, dx$$

$$= \frac{1}{\lambda^2} \int_0^{3\lambda} \left(\int_0^{3\lambda-x} \{e^{-(x+y)/\lambda} - e^{-3}\}\, dy\right) dx$$

$$= \frac{1}{\lambda^2} \int_0^{3\lambda} \left[e^{-x/\lambda}\lambda\{1 - e^{-3+x/\lambda}\} - (3\lambda - x)e^{-3}\right] dx$$

$$= \frac{1}{\lambda}\left[\lambda(1 - e^{-3}) - 3\lambda e^{-3}\right] - \frac{1}{\lambda^2}e^{-3}\frac{9\lambda^2}{2}$$

$$= 1 - 4e^{-3} - (9/2)e^{-3}.$$

It follows that

$$P(X + Y + Z \geq 3\lambda) = 1 - \left(1 - (17/2)e^{-3}\right) = (17/2)e^{-3} = 0.4232. \qquad \square$$

Example 7. *Jury Discrimination: Cassell vs. Texas.* In Cassell vs. Texas [399 U.S. 282 (1950)], black plaintiff Cassell was convicted of murder and appealed the decision on the grounds of jury discrimination. His argument was based on the fact that blacks were underrepresented on the juries in Dallas County. Although blacks formed 15.8 percent of the population in Dallas County, in a sample of 21 juries examined only 17 of the 252 jurors (that is, 6.7 percent) were black. Moreover, no jury ever had more than one black on it.

The State of Texas argued that blacks formed only 6.5 percent of poll-tax payers. The Supreme Court accepted this explanation of the difference between the 15.8 percent and the 6.7 percent figures. It reversed the conviction, however, on the basis of nonstatistical aspects of the case. Three judges noted that it was quite improbable that no more than one black would every appear on a jury. Any such limit, it observed, would be unconstitutional.

There are at least two kinds of jury discrimination problems here. Suppose we accept the figure of 6.5 percent advanced by the State of Texas and accepted by the Supreme Court. Then p = proportion of blacks = .065. Let us first consider the problem of underrepresentation. Assuming that p = .065, how likely is it to observe 17 or fewer blacks in a random sample of size 252? If X denotes the number of blacks in a sample of size 252, then this probability is given by

$$(5) \qquad P(X \le 17) = \sum_{k=0}^{17} \binom{252}{k}(.065)^k(.935)^{252-k}$$

The argument used here is as follows. The probability of exactly k blacks and hence $(252 - k)$ nonblacks by independence of selections is $(.065)^k(.935)^{252-k}$. Since there are $\binom{252}{k}$ ways of choosing k blacks and $(252 - k)$ non-blacks, and since each selection has the same chance, it follows that (5) holds. (See Section 6.3 for more details.) A probability such as the one in (5) is not easy to evaluate exactly. In Chapter 9 we consider an approximation for the sum on the right side of (5), which yields

$$P(X \le 17) \doteq .3859.$$

It is hardly reasonable to claim underrepresentation.

The second problem here is that of limitation. That is, discrimination by arranging never to have more than one black to a jury. Let $X_1, X_2, \ldots, X_{21}$ be the number of blacks on the 21 juries. Then

$$P(X_i \le 1) = P(X_i = 0) + P(X_i = 1) = (1 - .065)^{12} + \binom{12}{1}(.065)(.935)^{11}$$

$$= .818.$$

Assuming that X_i's are independent, we have

P(Each of 21 juries has at most one black)

$$= P(X_1 \le 1, X_2 \le 1, \ldots, X_{21} \le 1) = [P(X_i \le 1)]^{21} = (.818)^{21} = .015$$

which is considerably smaller than $P(X \le 17)$. Thus, under the assumption that blacks selected for each jury formed a random sample of size 12 from a population that is 6.5 precent black, there is only 1.5 percent chance that of the 21 juries examined each will have at most one black on it. Whether this is sufficient evidence to conclude that there was discrimination (due to limitation of blacks) in jury selection is a subjective judgment depending on the margin of error that one is willing to allow.

Although the Supreme Court accepted the 6.5 percent figure, it is interesting to compare the two figures: 6.7 percent blacks on juries and 15.8 percent blacks in the population. How likely is it to obtain a random sample of size 252 that has 6.7 percent blacks from a population that has 15.8 percent blacks? Using the approximation mentioned above, one can show that this probability is approximately .0001. In technical language to be explained in Section 4.4, this means that the two proportions (sample and population) are significantly different at level .0001. The general methods of comparing sample and population proportions are discussed in Sections 4.4, 6.3, and 9.7.   □

Problems for Section 3.6

1. Given the joint distribution

	X		
Y	-1	0	1
0	$1/6$	$1/3$	$1/6$
1	$1/6$	$-$	$1/6$

Are X and Y independent?

2. If (X, Y) have joint probability function given by

	X			
Y	0	1	2	3
1	0	$3/8$	$3/8$	0
3	$1/8$	0	0	$1/8$

are they independent?

3. Let (X, Y, Z) have probability function defined by

$$P(X = x, Y = y, Z = z) = \begin{cases} 3/16, & (x, y, z) = (0,0,0) \text{ or } (0,1,1) \\ & \text{or } (1,0,1) \text{ or } (1,1,0) \\ 1/16, & (x, y, z) = (0,0,1) \text{ or } (0,1,0) \\ & \text{or } (1,0,0) \text{ or } (1,1,1), \\ 0, & \text{elsewhere.} \end{cases}$$

(i) Are X, Y, Z independent?
(ii) Are X, Y, Z pairwise independent?

4. Suppose X_1, X_2 are independent having the same probability function given by

$$P(X_i = \pm 1) = 1/2.$$

Let $X_3 = X_1 X_2$. Show that X_1, X_2, X_3 are not independent but are pairwise independent.

5. Let A and B be events of positive probability and let X and Y denote the indicator functions of A and B respectively. That is, $X(\omega) = I_A(\omega) = 1$ if $\omega \in A$ and zero otherwise. Show that X and Y are independent if and only if A and B are independent events.

6. Let X and Y be independent random variables with probability functions given by

$$P(X = k) = P(Y = k) = 1/N, \quad k = 1, 2, \ldots, N, \text{ and zero elsewhere.}$$

Find $P(X + Y \le N/2)$ and $P(X + Y \le N)$.

7. A retail outlet has two stores. In each store the daily demand for a particular item has probability function

$$P(X = k) = \frac{6k^2}{N(N + 1)(2N + 1)}, \quad k = 1, 2, \ldots, N, \text{ and zero otherwise}$$

where N is known. If the demands X_1, X_2 are independent, find $P(X_1 \ge X_2)$, $P(X_1 - X_2 \ge 5)$, $P(X_1 + X_2 \ge N)$. [Hint: $\sum_{k=1}^{N} k^2 = N(N + 1)(2N + 1)/6$.]

8. A fair die is rolled twice. Let X be the sum of face values and Y the absolute value of the difference in face values on the two rolls. Are X and Y independent?

9. The random vector (X, Y) has joint density function f given below. Are X and Y independent?

   (i) $f(x, y) = \left(\dfrac{1}{\alpha + 1} + \dfrac{1}{\beta + 1} \right)(x^\alpha + y^\beta)$, $0 < x < 1$, $0 < y < 1$, and zero otherwise.

   (ii) $f(x, y) = xy$, $0 \le x \le 2$, $0 \le y \le 1$, and zero elsewhere.

   (iii) $f(x, y) = 4y(x - y)e^{-(x+y)}$, $0 \le y \le x \le \infty$, and zero elsewhere.

   (iv) $f(x, y) = 24y(1 - x)$, $0 \le y \le x \le 1$, and zero otherwise.

   (v) $f(x, y) = (1/2u^2v)$, $1 \le u \le \infty$, $1/u \le v \le u$, and zero elsewhere.

   (vi) $f(x, y) = 6x^2y$, $0 \le x \le 1$, $0 \le y \le 1$, and zero otherwise.

   (vii) $f(x, y) = 2 \exp\{-(x + y)\}$, $0 \le y \le x \le \infty$, and zero elsewhere.

   (viii) $f(x, y) = 8xy$, $0 \le y \le x \le 1$, and zero otherwise.

10. Let X, Y, and Z be the three perpendicular components of the velocity of a molecule of gas in a container. Suppose X, Y, and Z are independent with common density function given by

$$f(x) = 1/2a, \quad x \in (-a, a), \text{ and zero elsewhere.}$$

Let E be the kinetic energy of a molecule of mass m. Then $E = m(X^2 + Y^2 + Z^2)/2$. Find $P(E \le t)$ for $t > 0$ and, in particular, $P(E \le a/2)$.

11. Suppose two buses, A and B operate on a route. A person arrives at a certain bus stop on this route at time 0. Let X and Y be the arrival times of buses A and B respectively at this bus stop. Suppose X and Y are independent and have density functions given respectively by

$$f_1(x) = \frac{1}{a}, \quad 0 \le x \le a, \text{ and zero elsewhere,}$$

$$f_2(y) = \frac{1}{b}, \quad 0 \le y \le b, \text{ and zero otherwise.}$$

What is the probability that bus A will arrive before bus B?

12. A leading manufacturer of TV guarantees its picture tubes for four years. Suppose the time to failure of its picture tubes has density function

$$f(t) = \frac{1}{5} \exp\left(-\frac{t}{5}\right), \qquad 0 < t < \infty, \text{ and zero otherwise.}$$

If the time to failure of the tubes are independent, what is the probability that of the four television sets that are sold by a retailer:
(i) All the picture tubes will fail during the guarantee period.
(ii) Only two will fail during the guarantee period.

13. Consider a simple electrical circuit for which Ohm's law $V = IR$ holds where V is the voltage, I the current, and R the resistance. Suppose I and V are independent random variables with density functions, respectively,

$$f_1(i) = 2\exp(-2i), \qquad 0 < i < \infty, \text{ and zero elsewhere}$$

and

$$f_2(v) = e^{-v}, \qquad 0 < v < \infty, \text{ and zero elsewhere.}$$

Find $P(R > r_0)$ where r_0 is a fixed positive number.

14. The length X of time (in days) after the end of a certain advertising campaign that an individual is able to remember the name brand of the product being advertised has density function

$$f(x) = (1/4)x^{-1/2}\exp\{-(1/2)\sqrt{x}\}, \qquad x > 0, \text{ and zero elsewhere.}$$

Suppose five individuals who viewed the campaign are independently selected 15 days after the campaign ended.
(i) Find the probability that none will remember the name brand.
(ii) Find the probability that only one will remember the name brand.
What can you say about the success of the advertising campaign?

15. Consider two batteries, one of Brand A and the other of Brand B. Brand A batteries have a length of life with density function

$$f(x) = 3\lambda x^2\exp(-\lambda x^3), \qquad x > 0, \text{ and zero elsewhere}$$

whereas Brand B batteries have a length of life with density function given by

$$g(y) = 3\mu y^2\exp(-\mu y^3), \qquad y > 0, \text{ and zero elsewhere.}$$

Brand A and Brand B batteries are put to a test. What is the probability that Brand B battery will outlast Brand A? In particular, what is the probability if $\lambda = \mu$?

16. Let X and Y be independent random variables of the continuous type. Find $P(X > Y)$ in terms of the densities f and g of X and Y respectively. In particular, if $f = g$ find $P(X > Y)$. (Problem 15 is a special case.)

17. Let X be the length (inches) and Y the width (inches) of a rectangle where X and Y are independent random variables with respective density functions

$$f(x) = 1/2, \qquad 8 \le x \le 10, \text{ and zero otherwise}$$

and

$$g(y) = 1/2, \qquad 7 \le y \le 9, \text{ and zero elsewhere.}$$

What is the probability that
   (i) The perimeter of the rectangle will be at least 36 inches?
   (ii) The area of the rectangle will be between 60 and 80 square inches?

18. (i) Let (X, Y) have joint density f. Show that X and Y are independent if and only if for some constant $k > 0$ and nonnegative functions f_1 and f_2

$$f(x, y) = kf_1(x)f_2(y)$$

for all $x, y \in \mathbb{R}$.
   (ii) Let $A = \{f_X(x) > 0\}$, $B = \{f_Y(y) > 0\}$, and f_X, f_Y are marginal densities of X and Y respectively. Show that if X and Y are independent then $\{f > 0\} = A \times B$.

3.7 NUMERICAL CHARACTERISTICS OF A DISTRIBUTION—EXPECTED VALUE

Averages are part of our day-to-day life. We hear of the Dow Jones industrial average, the average rainfall in a city in a year, the average temperature in July, the average age of workers in a plant, the average faculty compensation, and so on. We think of height relative to an average. A person is tall for his or her age if he or she is taller than the average height for the group and short if he or she is shorter than the average for that group. Most people are comfortable with the Dow Jones industrial average as a measure of general movement in stock prices. The average gasoline mileages announced by the Environmental Protection Agency have become a standard basis of comparison in automobile advertising.

Our aim in this section is to consider some numerical characteristics of a population distribution. The most common average used in statistics is the *mean* or *expected value* or *mathematical expectation*.

Let X be a random variable defined on (Ω, ζ, P) and let g be a real valued function defined on $\mathbb{R}$. Define $g(X)$ by

(1) $$g(X)(\omega) = g(X(\omega)), \qquad \omega \in \Omega.$$

In all cases considered in this text $g(X)$ is a random variable.

DEFINITION 1. Suppose $g(X)$ is a discrete type random variable. If $\sum_{j=1}^{\infty}|g(x_j)|P(X = x_j) < \infty$, then we define the mean of $g(X)$ or the expected value of

$g(X)$ by

$$\mathscr{E}g(X) = \sum_{j=1}^{\infty} g(x_j) P(X = x_j).$$

If $g(X)$ is of the continuous type and f is the density function of X, then we define

$$\mathscr{E}g(X) = \int_{-\infty}^{\infty} g(x)f(x)\, dx$$

provided that $\int_{-\infty}^{\infty} |g(x)| f(x)\, dx < \infty$.

According to Definition 1, $\mathscr{E}g(X)$ exists provided $\mathscr{E}|g(X)| < \infty$. Some special cases of interest arise when we take $g(x) = x$, $g(x) = x - c$, $g(x) = |x|^{\alpha}$, and we give these special names.

$g(x)$	$\mathscr{E}g(X)$	Notation	Name Given
x	$\mathscr{E}X$	μ	First moment or mean of X.
$x - c$	$\mathscr{E}(X - c)$		First moment of X about c.
$|x|^{\alpha}$	$\mathscr{E}|X|^{\alpha}$		Absolute moment of X of order α, $\alpha > 0$.
x^k	$\mathscr{E}X^k$	m_k	Moment of order k, $k \geq 0$ integral.
$(x - c)^k$	$\mathscr{E}(X - c)^k$		Moment of order k about c, $k \geq 0$ integral.
$(x - \mu)^2$	$\mathscr{E}(X - \mu)^2$	σ^2	Variance
$(x - \mu)^k$	$\mathscr{E}(X - \mu)^k$	μ_k	Central moment of order k, $k \geq 0$ integral.

Example 1.   *Average Face Value in Rolling a Die.* Suppose a fair die is rolled once. What is the average face value observed? Clearly X has probability function

$$P(X = k) = 1/6, \qquad k = 1, 2, 3, 4, 5, 6, \text{ and zero elsewhere.}$$

Since X takes only a finite number of values, $\mathscr{E}X$ exists and equals

$$\sum_{k=1}^{6} kP(X = k) = \sum_{1}^{6} \frac{k}{6} = 3.5. \qquad \square$$

Example 2.   *Average Number of Telephone Calls.* Let X_t be the number of telephone calls initiated in a time interval of length t and suppose X_t has probability function

$$P(X_t = k) = e^{-\lambda t} \frac{(\lambda t)^k}{k!}, \qquad k = 0, 1, \ldots, \text{ and zero elsewhere.}$$

Then

$$\mathscr{E} X_t = \sum_{k=0}^{\infty} k P(X_t = k) = \sum_{k=0}^{\infty} k e^{-\lambda t} \frac{(\lambda t)^k}{k!}$$

$$= e^{-\lambda t} \sum_{k=1}^{\infty} \frac{(\lambda t)^k}{(k-1)!} = e^{-\lambda t} \sum_{l=0}^{\infty} \frac{(\lambda t)^{l+1}}{l!}$$

$$= e^{-\lambda t}(\lambda t e^{\lambda t}) = \lambda t.$$

Note that X_t takes only nonnegative integer values with positive probabilities so that $\sum_{k=0}^{\infty} |k| P(X_t = k) = \sum_{k=0}^{\infty} k P(X_t = k) = \lambda t < \infty$. □

Example 3. *Average Length of Life of a Light Bulb.* A manufacturer of light bulbs of a certain wattage approximates that the time to failure X of light bulbs has density function

$$f(x) = \frac{1}{\lambda} e^{-x/\lambda}, \qquad x > 0, \text{ and zero elsewhere.}$$

Then

$$\mathscr{E} X = \int_{-\infty}^{\infty} x f(x)\, dx = \int_0^{\infty} \frac{1}{\lambda} x e^{-x/\lambda}\, dx$$

$$= \frac{1}{\lambda} \left[\frac{x e^{-x/\lambda}}{-1/\lambda} \Big|_0^{\infty} + \lambda \int_0^{\infty} e^{-x/\lambda}\, dx \right] = \int_0^{\infty} e^{-x/\lambda}\, dx = \lambda.$$ □

Example 4. *Expected Value Need Not Exist.* Consider the density function

$$f(x) = \frac{1}{2x^2}, \qquad |x| > 1, \text{ and zero elsewhere.}$$

Then $\mathscr{E} X$ does not exist. In fact

$$\int_{-\infty}^{\infty} |x| f(x)\, dx = \int_{|x|>1} \frac{|x|}{2x^2}\, dx = \int_{|x|>1} \frac{1}{2|x|}\, dx = \infty.$$

According to Definition 1, $\mathscr{E} X$ does not exist, that is, it is undefined. □

The following results and its corollaries are useful.

PROPOSITION 1. *Suppose $\mathscr{E} X$ exists, that is, $\mathscr{E}|X| < \infty$. Let a, b be real numbers. Then $\mathscr{E}|aX + b| < \infty$ and*

(2) $$\mathscr{E}(aX + b) = a\mathscr{E} X + b.$$

Proof. The proof for both the cases being similar, we consider only the discrete case. By the triangular inequality

$$\mathscr{E}|aX + b| \le \mathscr{E}|aX| + \mathscr{E}|b|.$$

Now

$$\mathscr{E}|b| = \sum_{j=1}^{\infty} |b| P(X = x_j) = |b| \sum_{j=1}^{\infty} P(X = x_j) = |b|$$

and

$$\mathscr{E}|aX| = \sum_{j=1}^{\infty} |ax_j| P(X = x_j) = |a| \sum_{j=1}^{\infty} |x_j| P(X = x_j) = |a|\mathscr{E}|X|$$

Hence:

$$\mathscr{E}|aX + b| \le |a|\mathscr{E}|X| + |b| < \infty.$$

Moreover,

$$\mathscr{E}(aX + b) = \sum_{j=1}^{\infty} (ax_j + b) P(X = x_j) = a \sum_{j=1}^{\infty} x_j P(X = x_j) + b \sum_{j=1}^{\infty} P(X = x_j)$$

$$= a\mathscr{E}X + b. \qquad \square$$

COROLLARY 1. *If $\mathscr{E}X$ exists then $\mathscr{E}(X - \mathscr{E}X) = 0$.*

Proof. Take $a = 1$ and $b = -\mathscr{E}X$ in (2). $\qquad \square$

COROLLARY 2. *Let $g_1, g_2, \ldots, g_k$ be real valued functions such that $g_1(X), \ldots, g_k(X)$ are random variables. If $\mathscr{E}|g_j(X)| < \infty$ for all j, then $\mathscr{E}\left| \sum_{j=1}^{k} g_j(X) \right| < \infty$ and*

(3)
$$\mathscr{E}\left(\sum_{j=1}^{k} g_j(X) \right) = \sum_{j=1}^{k} \mathscr{E}g_j(X).$$

Proof. Use the triangular inequality to show the existence of $\mathscr{E}|\sum_{j=1}^{k} g_j(X)|$ and then split the left side of (3) into k separate sums (or integrals) to get the right side of (3). $\qquad \square$

Remark 1. The term "expected value" is somewhat misleading in that $\mathscr{E}X$ is not necessarily a value of X to be expected when the experiment is performed. Thus in Example 1, the expected face value when a fair die is rolled once is 3.5, which is not a possible value of X. From a relative frequency point of view, where we think of probabilities as limits of observable relative frequencies in repeated trials, the expected value can be thought of as the limit of the averages $(X_1 + X_2 + \cdots + X_n)/n$ (as $n \to \infty$) where $X_1, X_2, \ldots, X_n$ are the n observations on X when the experiment is repeated n times (see Section 3.8). We show in Section 5.4

that for large n the average $(X_1 + \cdots + X_n)/n$ is likely to be near $\mathscr{E}X$. In Example 1 this means that if the die is rolled n times then the average of the observed face values, for large n, is likely to be near 3.5.

Remark 2. The concept of expected value is analogous to the concept of center of gravity in mechanics. If we distribute a unit mass along the real line at points $x_1, x_2, \ldots$, and if $P(X = x_i)$ is the mass at x_i, then $\mathscr{E}X$ represents the center of gravity. The same analogy holds in the continuous case. In geometric terms, this means that (in the continuous case) if we were to draw the density function f of X on a cardboard of uniform density, cut it along the curve, and balance it on an edge perpendicular to the x-axis at the point $x = \mathscr{E}X$, it will be in an equilibrium state (see Figure 1).

Many applications of probability involve minimization of expected loss or maximization of expected profit. We consider an example.

Example 5. An Inventory Problem. Stores order seasonal merchandise well in advance. Any article not sold is heavily discounted or remaindered at the end of the season. Suppose a store orders summer suits at a cost of $50 a suit which it sells at $150 per suit. At the end of the season, suppose the unsold suits are sold to a discount outlet at $35 per suit. From past experience it is known that the probability function of the number of suits the store can sell to its customers is given by

$$P(X = x) = e^{-\lambda}\frac{\lambda^x}{x!}, \qquad x = 0, 1, 2, \ldots, \text{ and zero elsewhere.}$$

How many suits must be ordered to maximize the expected profit?

Let z be the number of suits ordered and stocked. Let $\varphi(z)$ be the profit when z suits are stocked. Then

$$\varphi(z) = \begin{cases} (150 - 50)X + (35 - 50)(z - X) & \text{if } 0 \le X \le z, \\ (150 - 50)z & \text{if } z < X. \end{cases}$$

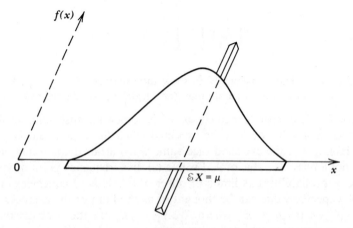

Figure 1

We note that $\varphi(z)$ is a random variable which is a function of X. In view of the definition of $\mathscr{E}g(X)$ we have

$$\mathscr{E}\varphi(z) = \sum_{x=0}^{z} \{100x - 15(z - x)\} P(X = x) + 100zP(X > z)$$

$$= \sum_{x=0}^{z} \{115x - 15z\} \frac{e^{-\lambda}\lambda^x}{x!} + 100z\left\{1 - \sum_{x=0}^{z} e^{-\lambda}\frac{\lambda^x}{x!}\right\}$$

$$= 100z + \sum_{x=0}^{z} 115(x - z)e^{-\lambda}\frac{\lambda^x}{x!}.$$

To find the value of z for which $\mathscr{E}\varphi(z)$ is maximum we compute $\mathscr{E}\varphi(z + 1) - \mathscr{E}\varphi(z)$. As long as $\mathscr{E}\varphi(z + 1) - \mathscr{E}\varphi(z) > 0$, $\mathscr{E}\varphi(z)$ is increasing as a function of z and the value of z at which $\mathscr{E}\varphi(z + 1) - \mathscr{E}\varphi(z)$ changes sign gives us the maximum. We have

$$\mathscr{E}\varphi(z + 1) = 100(z + 1) + 115 \sum_{x=0}^{z+1} (x - z - 1)e^{-\lambda}\frac{\lambda^x}{x!}$$

$$= 100z + 100 + 115 \sum_{x=0}^{z} (x - z - 1)e^{-\lambda}\frac{\lambda^x}{x!}$$

$$= 100 + \mathscr{E}\varphi(z) - 115 \sum_{x=0}^{z} e^{-\lambda}\frac{\lambda^x}{x!}$$

so that

$$\mathscr{E}\varphi(z + 1) - \mathscr{E}\varphi(z) = 100 - 115 \sum_{x=0}^{z} e^{-\lambda}\frac{\lambda^x}{x!}$$

It follows that

$$\mathscr{E}\varphi(z + 1) - \mathscr{E}\varphi(z) \begin{cases} > 0 & \text{if } \sum_{x=0}^{z} e^{-\lambda}\frac{\lambda^x}{x!} < \frac{100}{115} \\ < 0 & \text{if } \sum_{x=0}^{z} e^{-\lambda}\frac{\lambda^x}{x!} > \frac{100}{115}. \end{cases}$$

Thus the expected profit is maximum for the smallest value of z for which $\mathscr{E}\varphi(z + 1) - \mathscr{E}\varphi(z) = 0$ or negative. This value clearly depends on λ and can be computed once λ is specified.

Suppose $\lambda = 5$. In this case we want to find the smallest z for which either

$$\sum_{x=0}^{z} e^{-5}\frac{5^x}{x!} = \frac{100}{115} = .8696 \quad \text{or} \quad > .8696$$

Table A4 in the Appendix gives $\sum_{x=0}^{z} e^{-\lambda} \lambda^x / x!$ for various values of λ and z. For $\lambda = 5$ we see that

$$\sum_{x=0}^{7} e^{-5} \frac{5^x}{x!} = .867 \quad \text{and} \quad \sum_{x=0}^{8} e^{-5} \frac{5^x}{x!} = .932$$

so that we must choose $z = 8$. The following table gives the values of z for some selected values of λ.

λ	5	6	7	8	9	10	11
z	8	9	10	11	12	14	15

For large λ one can use normal approximation (Section 9.4). The probability function of X used here is the well known Poisson distribution, studied in detail in Section 6.5. The argument used to find the maximum of $\mathscr{E}\varphi(z)$ is typical of discrete random variables and will be used often. ☐

Remark 3. In this book we will use the terms "the mean of X" and "the mean of F" interchangeably where F is the distribution of X.

3.7.1 Variance of a Random Variable

The center of gravity of a mass distribution provides no information about how the mass is spread about this center. In mechanics, a measure of this spread (dispersion) is provided by the moment of inertia. In probability, $\mathscr{E}X$ provides information only about the center or location of the distribution but says nothing about how the distribution of X is spread about $\mathscr{E}X$. Indeed, it is possible to have many widely different distributions with the same mean or expected value. We now study a measure of variation, or dispersion, or spread that is analogous to the moment of inertia.

What we want is a measure that should take into account the deviations $X - \mu$ where $\mu = \mathscr{E}X$. Since $\mathscr{E}(X - \mu) = 0$ it is clear that we cannot use $\mathscr{E}(X - \mu)$ as a measure of spread. This suggests using $\mathscr{E}|X - \mu|$ as a measure of average dispersion. Unfortunately, the *mean absolute deviation* $\mathscr{E}|X - \mu|$ is not mathematically tractable; manipulation with absolute values is not always easy. Yet another alternative is to consider $\mathscr{E}|X - \mu|^2$.

DEFINITION 2. *(VARIANCE).* The variance of X (or of the distribution of X) is defined by

$$\mathscr{E}(X - \mu)^2 = \begin{cases} \sum_{j=1}^{\infty} (x_j - \mu)^2 P(X = x_j), & \text{if } X \text{ is discrete type} \\ \int_{-\infty}^{\infty} (x - \mu)^2 f(x) \, dx, & \text{if } X \text{ is continuous type with density } f \end{cases}$$

provided $\mathscr{E}|X|^2 < \infty$, and is denoted by σ^2. The (positive) square root σ is called the standard deviation of X.

It is usually convenient to compute the variance from the relation

$$(4) \qquad\qquad \sigma^2 = \mathscr{E}X^2 - \mu^2.$$

Indeed, in view of Corollary 2 to Proposition 1

$$\sigma^2 = \mathscr{E}(X - \mu)^2 = \mathscr{E}(X^2 - 2\mu X + \mu^2) = \mathscr{E}X^2 - 2\mu\mathscr{E}X + \mu^2$$

$$= \mathscr{E}X^2 - \mu^2.$$

Remark 4.   The following interpretation maybe given to the variance. Suppose you are asked to predict a value of X. If you predict the value to be c and X is observed, suppose you lose $(X - c)^2$. Then your best predictor is $c = \mathscr{E}X$ in the sense that it minimizes your average loss. Indeed, the average loss given by

$$\mathscr{E}(X - c)^2 = \mathscr{E}(X - \mu - c + \mu)^2$$

$$= \mathscr{E}(X - \mu)^2 - 2(c - \mu)\mathscr{E}(X - \mu) + (c - \mu)^2$$

$$= \sigma^2 + (c - \mu)^2 \geq 0$$

is minimized for $(c - \mu)^2 = 0$, that is, for $c = \mu$. Thus, if a fair die is tossed once, your best guess of the face value that will turn up is 3.5 even though you know in advance that a face value of 3.5 is impossible. (See Remark 1.)

The following proposition is very useful.

PROPOSITION 2.   *Suppose $\mathscr{E}|X|^2 < \infty$.*

(i)   *We have*

$$(5) \qquad\qquad \mathrm{var}(X) = 0 \quad \text{if and only if} \quad P(X = c) = 1$$

for some constant c.

(ii)   *For any real numbers a and b, we have*

$$(6) \qquad\qquad \mathrm{var}(aX + b) = a^2 \mathrm{var}(X)$$

Proof.   (i)   Suppose there is a c for which $P(X = c) = 1$. Then $\mathscr{E}X = c$ and

$$\mathscr{E}(X - \mu)^2 = (c - c)^2 . 1 = 0.$$

Conversely, suppose var$(X) = 0$. Then $P(X = \mu) = 1$. Indeed, in the discrete case suppose $P(X = x) > 0$ for some $x \neq \mu$. Then

$$\text{var}(X) = \sum_{x_j} (x_j - \mu)^2 P(X = x_j) = (x - \mu)^2 P(X = x)$$

$$+ \sum_{x_j \neq x} (x_j - \mu)^2 P(X = x_j)$$

$$> 0.$$

This contradiction shows that there cannot be an $x \neq \mu$ for which $P(X = x) > 0$. Hence $P(X = \mu) = 1$. A similar proof may be given in the continuous case.

(ii) First we note that $\mathscr{E}|X|^2 < \infty$ implies, by triangular inequality, that $\mathscr{E}|aX + b|^2 < \infty$ so that var$(aX + b)$ exists. Now

$$\text{var}(aX + b) = \mathscr{E}\{aX + b - \mathscr{E}(aX + b)\}^2$$

$$= \mathscr{E}\{aX + b - a\mathscr{E}X - b\}^2 \qquad \text{(Proposition 1)}$$

$$= a^2 \mathscr{E}(X - \mathscr{E}X)^2 = a^2 \text{var}(X).$$

$\square$

COROLLARY 1. *If $\mathscr{E}X^2 < \infty$, and $Z = (X - \mu)/\sigma$, then $\mathscr{E}Z = 0$ and var$(Z) = 1$.*

Proof. We have

$$\mathscr{E}Z = \mathscr{E}\left\{\frac{X - \mu}{\sigma}\right\} = \frac{1}{\sigma}\mathscr{E}(X - \mu) = 0$$

and from Proposition 2(b)

$$\text{var}(Z) = \text{var}\left[\frac{X - \mu}{\sigma}\right] = \frac{1}{\sigma^2}\text{var}(X - \mu) = \frac{\sigma^2}{\sigma^2} = 1. \qquad \square$$

Remark 5. The random variable Z defined in Corollary 1 is called the *standardized X*. Given a random variable X we standardize it by subtracting from it the mean and dividing this deviation by the standard deviation. This changes the origin to μ and the scale to standard deviation units. Standardization is useful in comparing two or more distributions. Suppose for example that a student scores 80 on a calculus test where the mean was 70 and the standard deviation 4, and she scores 75 on a statistics test where the mean was 60 and the standard deviation 5. At first glance it appears that the student did better on her calculus test since her score was 80. Converting to standard units we note that her standard score on the calculus test was $(80-70)/4 = 2.5$ whereas here standard score on the statistics test was $(75-60)/5 = 3$, so that her relative position in statistics class is better.

Remark 6. We note that $\text{var}(X) > 0$ for all nontrivial random variables. Moreover, if the distribution of X is concentrated near $\mathscr{E}X$ then σ^2 will be small. On the other hand, if values of X are widely dispersed than σ^2 will be large. We will see (Remark 9) that a small value of σ^2 means the probability is small that X will deviate much from its mean. But a large value of σ^2 does not necessarily mean the probability is large that X will be far from $\mathscr{E}X$ (Problem 21).

Remark 7. We defined $\mathscr{E}X^k$ as the moment of order k and $\mathscr{E}(X - \mu)^k$ as the moment of order k about the mean (also called the kth central moment of X). We note that if $\mathscr{E}|X|^n < \infty$ for some n, then $\mathscr{E}|X|^k < \infty$ for $k < n$. Indeed, in the discrete case for $k < n$

$$\mathscr{E}|X|^k = \sum_j |x_j|^k P(X = x_j) = \left(\sum_{|x_j|^k < 1} + \sum_{|x_j|^k \geq 1} \right) |x_j|^k P(X = x_j)$$

$$\leq \sum_{|x_j|^k < 1} P(X = x_j) + \sum_{|x_j|^k \geq 1} |x_j|^k P(X = x_j)$$

$$\leq P(|X|^k < 1) + \mathscr{E}|X|^n < \infty.$$

Remark 8. Probabilities of the type $P(|X| > n)$ or $P(X > n)$ or $P(X < -n)$ are called tail probabilities. The behavior of the tail probability determines whether $\mathscr{E}|X|^k < \infty$ or not. In fact, if $\mathscr{E}|X|^k < \infty$ then

$$n^k P(|X| > n) \to 0 \qquad \text{as } n \to \infty,$$

and a partial converse also holds. We have:

$$\infty > \int_0^\infty |x|^k f(x)\, dx = \lim_{n \to \infty} \int_{|x| \leq n} |x|^k f(x)\, dx$$

so that $\int_{|x| > n} |x|^k f(x)\, dx \to 0$ as $n \to \infty$. Since

$$\int_{|x| > n} |x|^k f(x)\, dx) \geq n^k (P|X| > n)$$

we see that $n^k P(|X| > n) \to 0$ as $n \to \infty$. In the converse direction, $n^k P(|X| > n) \to 0$ as $n \to \infty$ implies that $\mathscr{E}|X|^{k-\delta} < \infty$ for all $0 < \delta < k$, but we will skip the proof.

Example 6. *Variance Computation.* Let X have density function

$$f(x) = \frac{1}{\pi}, \qquad 0 \leq x \leq \pi, \text{ and zero elsewhere.}$$

Then

$$\mathscr{E}X = \frac{1}{\pi}\int_0^\pi x\, dx = \frac{\pi}{2}, \mathscr{E}X^2 = \frac{1}{\pi}\int_0^\pi x^2\, dx = \frac{\pi^2}{3}$$

and

$$\sigma^2 = \text{var}(X) = \mathscr{E}X^2 - (\mathscr{E}X)^2 = \frac{\pi^2}{12}, \quad \text{and } \sigma = \frac{\pi}{2\sqrt{3}}.$$

Suppose Y has density

$$g(x) = \frac{1}{2}\sin x, \qquad 0 \le x \le \pi, \text{ and zero elsewhere.}$$

Then

$$\mathscr{E}Y = \frac{1}{2}\int_{-\infty}^\infty yg(y)\, dy = \frac{1}{2}\int_0^\pi y\sin y\, dy$$

$$= \frac{1}{2}\left[-y\cos y\Big|_0^\pi + \int_0^\pi \cos y\, dy\right] = \frac{\pi}{2}$$

and

$$\mathscr{E}Y^2 = \frac{1}{2}\int_0^\pi y^2\sin y\, dy = \frac{1}{2}\left[-y^2\cos y\Big|_0^\pi + 2\int_0^\pi y\cos y\, dy\right]$$

$$= \frac{1}{2}\left[\pi^2 + 2y\sin y\Big|_0^\pi - 2\int_0^\pi \sin y\, dy\right]$$

$$= \frac{1}{2}[\pi^2 - 4].$$

It follows that

$$\text{var}(Y) = \frac{\pi^2}{2} - 2 - \left(\frac{\pi}{2}\right)^2 = \frac{\pi^2}{4} - 2.$$

Figure 2 shows the graphs of f and g. The density g assigns more probability near the mean $\pi/2$ than the density f. It is not surprising then that $\text{var}(X) > \text{var}(Y)$.   □

Example 7. Computing Variance and Higher Order Moments. Suppose X has density function

$$f(x) = 3x^2, \qquad 0 \le x \le 1, \text{ and zero elsewhere.}$$

Clearly, $\mathscr{E}X^k$ exists for any k and

$$\mathscr{E}X^k = \int_0^1 x^k(3x^2)\, dx = \frac{3}{k+3}.$$

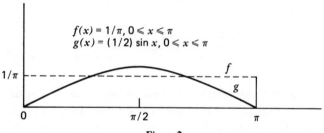

$f(x) = 1/\pi, 0 \leqslant x \leqslant \pi$
$g(x) = (1/2) \sin x, 0 \leqslant x \leqslant \pi$

Figure 2

Hence:

$$\mathscr{E}X = 3/4, \mathscr{E}X^2 = 3/5 \quad \text{and} \quad \text{var}(X) = (3/5) - (3/4)^2 = 3/80. \qquad \square$$

3.7.2 Chebyshev's Inequality[†]

We now obtain a very useful inequality that gives an operational meaning to the definition of variance. Let us introduce the distance $|X - \mathscr{E}X|$. Let $\varepsilon > 0$. What is the chance that $|X - \mathscr{E}X| \geq \varepsilon$?

THEOREM 1. *Suppose $\mathscr{E}X^2 < \infty$ and $\mathscr{E}X = \mu$, var$(X) = \sigma^2$, $0 < \sigma^2 < \infty$. Then for every $\varepsilon > 0$:*

(7)
$$P(|X - \mathscr{E}X| \geq \varepsilon) \leq \frac{\sigma^2}{\varepsilon^2}.$$

Proof. We give the proof only in the continuous case. We have

$$\text{var}(X) = \sigma^2 = \mathscr{E}(X - \mu)^2 = \int_{-\infty}^{\infty} (x - \mu)^2 f(x)\, dx$$

$$= \int_{|x-\mu| < \varepsilon} (x - \mu)^2 f(x)\, dx + \int_{|x-\mu| \geq \varepsilon} (x - \mu)^2 f(x)\, dx$$

$$\geq \int_{|x-\mu| \geq \varepsilon} (x - \mu)^2 f(x)\, dx$$

$$\geq \int_{|x-\mu| \geq \varepsilon} \varepsilon^2 f(x)\, dx = \varepsilon^2 P(|X - \mu| \geq \varepsilon).$$

Since $\varepsilon > 0$, it follows that (7) holds. $\qquad \square$

COROLLARY 1. (CHEBYSHEV'S INEQUALITY). *For any $k \geq 1$,*

(8)
$$P(|X - \mu| \geq k\sigma) \leq 1/k^2,$$

[†]Recent work has shown that although this inequality is well known in the literature as Chebyshev's inequality it is also due to I. J. Bienayme.

or, equivalently,

(9) $$P(|X - \mu| < k\sigma) \geq 1 - 1/k^2.$$

Proof. Take $\varepsilon = k\sigma$ in (7). $\hspace{4cm}$ □

Remark 9. From (7) we see that the smaller the σ, the smaller the probability $P(|X - \mathscr{E}X| \geq \varepsilon)$. That is, if σ is small then the probability that X will deviate from μ by more than ε is small. From (8), the larger the k, the smaller the $P(|X - \mu| \geq k\sigma)$. Thus

$$P(|X - \mu| \geq 5\sigma) \leq 1/25 = .04$$

and

$$P(|X - \mu| \geq 10\sigma) \leq 1/100 = .01.$$

If σ is large however, it does not follow that $P(|X - \mathscr{E}X| \geq \varepsilon)$ will be large. (See Problem 21.)

Remark 10. The result (7) or (8) applies to all distributions for which $\mathscr{E}|X|^2 < \infty$. It is not surprising, therefore, that it provides a fairly crude bound for tail probabilities. We hasten to add, though, that the inequality is sharp. That is, it cannot be improved unless X is further restricted. Nevertheless, we will find the inequality to be very useful and use it often.

Remark 11. Chebyshev's inequality, in fact, is a property of any finite set of real numbers. In Problem 28 we ask the reader to imitate the proof of Theorem 1 to show that if $x_1, x_2, \ldots, x_n$ is any finite set of real numbers, and

$$\bar{x} = \sum_{i=1}^{n} \frac{x_i}{n}, s^2 = \sum_{i=1}^{n} \frac{(x_i - \bar{x})^2}{n - 1}$$

then at least a fraction $(1 - 1/k^2)$ of the n numbers are included in the interval $(\bar{x} - ks, \bar{x} + ks)$.

Example 8. **Choosing a Point on (0, 1).** A point is chosen from the interval $(0, 1)$ according to the density

$$f(x) = 1, \quad 0 < x < 1, \text{ and zero elsewhere.}$$

Then $\mathscr{E}X = 1/2$, var$(X) = 1/12$, and according to Chebyshev's inequality

$$P(|X - 1/2| \geq k/\sqrt{12}) \leq 1/k^2.$$

For $k = 2$, $P(X \geq 1/2 + 2/\sqrt{12}$ or $X \leq 1/2 - 2/\sqrt{12}) \leq .25$, and for $k = 3$, $P(X \geq 1/2 + 3/\sqrt{12}$ or $X \leq 1/2 - 3/\sqrt{12}) \leq .11$. However, by direct integration

$$P(X \geq 1/2 + 2/\sqrt{12} \quad \text{or} \quad X \leq 1/2 - 2/\sqrt{12}) = 0$$

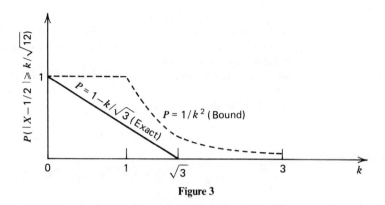

Figure 3

since $2/\sqrt{12} \doteq .58$ so that the Chebyshev's inequality considerably overestimates the probability in the tail. In Figure 3 we compare the exact probability $P(|X - 1/2| \geq k\sigma)$ with Chebyshev's bound. Note that

$$P(|X - 1/2| \geq k/\sqrt{12}) = 1 - 2k/\sqrt{12} \qquad \text{for } k \leq \sqrt{12}/2 = \sqrt{3}. \qquad \square$$

Example 9.   Chebyshev's Inequality Is Sharp.   Let X have probability function

$$P(X = \pm 1) = \frac{1}{2k^2}, \qquad P(X = 0) = 1 - \frac{1}{k^2}, \qquad k > 1$$

and

$$P(X = x) = 0, \qquad \text{for } x \neq 0, \pm 1.$$

Then

$$\mathscr{E}X = 0, \qquad \text{var}(X) = \mathscr{E}X^2 = \frac{1}{k^2}$$

and

$$P(|X - \mu| \geq k\sigma) = P\left(|X| \geq k\frac{1}{k}\right) = P(X = 1, \text{ or } X = -1) = \frac{1}{k^2}$$

$\square$

so that equality holds in (8).

Example 10.   Some Typical Applications of Chebyshev's Inequality.   According to (9),

$$P(\mu - k\sigma < X < \mu + k\sigma) \geq 1 - \frac{1}{k^2} \qquad \text{for } k > 1.$$

This inequality has three parameters μ, σ, and k. Given μ, σ, and k, it says that $X \in (\mu - k\sigma, \mu + k\sigma)$ with probability at least $1 - 1/k^2$. In practice μ or σ are hardly if ever known. The object, then, is to find bounds for μ or σ that will cover the unknown μ or σ with a given probability, say $1 - \alpha$, $0 < \alpha < 1$.

Suppose μ is unknown but σ is known and we want to find $a(X)$ and $b(X)$ so that $P(a(X) < \mu < b(X)) \geq 1 - \alpha$.

We choose $1 - \alpha = 1 - 1/k^2$ so that $k = 1/\sqrt{\alpha} > 1$. Then

$$P\left(\mu - \alpha^{-1/2}\sigma < X < \mu + \alpha^{-1/2}\sigma\right) \geq 1 - \alpha$$

or, equivalently,

$$P\left(X - \alpha^{-1/2}\sigma < \mu < X + \alpha^{-1/2}\sigma\right) \geq 1 - \alpha.$$

Thus $a(X) = X - \sigma\alpha^{-1/2}$ and $b(X) = X + \sigma\alpha^{-1/2}$. Similarly, if μ is known, one can show that

$$P\left(0 < \sigma \leq |X - \mu|\sqrt{\alpha}\right) \leq \alpha. \qquad \square$$

Remark 12. Sometimes we will use (8) with strict inequality and (9) with $<$ replaced by $\leq$. It is clear that (8) holds with "$\geq k\sigma$" replaced by "$> k\sigma$" and (9) holds with "$< k\sigma$" replaced by "$\leq k\sigma$."

3.7.3 Some Additional Measures of Location and Dispersion

We have seen that moments such as $\mathscr{E}X$ and $\text{var}(X)$ do not always exist. In such cases the theory developed in the earlier sections is inapplicable. We now define some measures of location and dispersion which always exist.

DEFINITION 3. (QUANTILE OF ORDER p).   Let $0 < p < 1$. A quantile of order p of a random variable X (or the distribution P_X) is any number ζ_p such that

$$P\left(X \leq \zeta_p\right) \geq p \quad \text{and} \quad P\left(X \geq \zeta_p\right) \geq 1 - p,$$

or, equivalently,

$$P\left(X < \zeta_p\right) \leq p \leq P\left(X \leq \zeta_p\right).$$

For $p = 1/2$ we call $\zeta_{.5}$ a median of X.

If X has distribution function F, then ζ_p is a quantile of order p if it satisfies

$$F\left(\zeta_p\right) - P\left(X = \zeta_p\right) \leq p \leq F\left(\zeta_p\right).$$

In particular, if F is continuous at ζ_p then ζ_p is a solution of $F(\zeta_p) = p$. However, it may still not be unique unless F is monotone. Graphically, Figure 4 illustrates computation of ζ_p. If the graph of F is flat at p then ζ_p is not unique. A median is frequently used as a center of the distribution and the difference $\zeta_{.75} - \zeta_{.25}$, called the *interquartile range*, is used as a measure of spread or dispersion. Here $\zeta_{.25}$ is called the *first quartile* and $\zeta_{.75}$ the *third quartile* of X. We note that $\mathscr{E}X$ (if it exists) is influenced by extremely high or low outcomes. In such cases a median

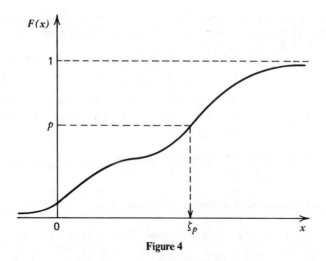

Figure 4

may be preferable as a measure of location. Such is the case, for example, when X is family income or annual wages.

Yet another measure of location is *mode*. A *mode* of a random variable (or a distribution) is its *most probable value*. In the case when X is of the discrete type, a mode of X is that value of X at which the probability function $P(X = x)$ has its largest value. If, on the other hand, X is of the continuous type, then $P(X = x) = 0$ for all x. In this case a mode is that value of X for which $f(x)$ is maximum. A mode may or may not exist.

A distribution is said to be *unimodal* if the probability or density function has a unique maxima (Figure 5*a*) and bimodal if the probability or density function has two modes (Figure 5*b*). Most of the distributions studied in this text are unimodal.

Example 11. Number of Arrivals at a Bank.   Suppose the number of arrivals X at a bank in any 15-minute interval has probability function

$$P(X = x) = \frac{5^x e^{-5}}{x!}, \qquad x = 0, 1, 2, \ldots, \text{ and zero elsewhere.}$$

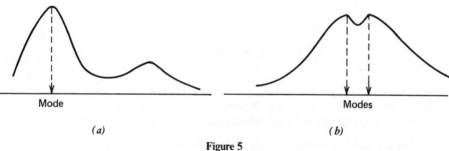

Mode   Modes

(*a*)   (*b*)

Figure 5

This is the so-called Poisson probability function encountered in Example 5, for which the distribution function is given in Table A4 in the Appendix. The table below abridged from Table A4 gives $P(X \leq x)$.

x	0	1	2	3	4	5	6	7	8	9
$P(X \leq x)$	.007	.040	.125	.265	.440	.616	.762	.867	.932	.968
$P(X = x)$	.007	.033	.085	.140	.175	.176	.146	.105	.065	.036
x	10	11	12	13	14					

$P(X \leq x)$	.986	.995	.998	.999	1.000
$P(X = x)$	.018	.009	.003	.001	.001

From the definition of median, we see that since

$$P(X \leq 5) = .616 \geq .5 \quad \text{and} \quad P(X \geq 5) \geq .56 \geq .5,$$

$\zeta_{.5} = 5$ is the unique median of X. Similarly,

$$\zeta_{.25} = 3 \quad \text{and} \quad \zeta_{.75} = 6$$

so the interquartile range is $6 - 3 = 3$. The modal value is 5 and the distribution is unimodal. We note that $\mathscr{E}X = 5$ so that $\mathscr{E}X = \zeta_{.5} = \text{Mode}(X)$. ☐

Example 12. Survival Times of Patients with Advanced Cancer of the Bladder. Suppose the survival times of patients who have advanced cancer of the bladder can be modeled by the density function

$$f(x) = \frac{1}{\lambda} \exp\left(-\frac{x}{\lambda}\right), \quad x \geq 0, \quad \text{and zero otherwise.}$$

Then $\mathscr{E}X = \lambda$ so that the average time a patient survives is λ. We note that

$$F(x) = \int_{-\infty}^{x} f(t)\, dt = \int_{0}^{x} \frac{1}{\lambda} e^{-t/\lambda}\, dt = 1 - e^{-x/\lambda},$$

for $x \geq 0$. Since F is continuous, given $0 < p < 1$, ζ_p is given by

$$F(\zeta_p) = 1 - e^{-\zeta_p/\lambda} = p.$$

It follows that

$$\zeta_p = \lambda \ln\left(\frac{1}{1-p}\right),$$

so that $\zeta_{.5} = \lambda \ln 2$, $\zeta_{.25} = \lambda \ln(4/3)$, $\zeta_{.75} = \lambda \ln 4$. Thus about 50 percent of the patients will die by time $\lambda \ln 2$, and about 25 percent will die by the time $\lambda \ln(4/3)$. Since f is a decreasing function of x it follows that the mode is at $x = 0$. ☐

Applications of median will appear quite often in this text.

Problems for Section 3.7

1. A used-car salesman sells 0, 1, 2, 3, 4, 5, or 6 cars each week with equal probability. Find the average number of cars sold by the salesman each week. If the commission on each car sold is $150, find the average weekly commission.

2. Find the expected value for the following probability functions.
    (i) $p(x) = (x + 1)/21$, $x = 0, 1, 2, 3, 4, 5$, and zero elsewhere.
    (ii) $p(x) = \binom{4}{x} \left(\frac{1}{2} \right)^4$, $x = 0, 1, 2, 3, 4$, and zero elsewhere.
    (iii) $p(x) = (x^2 + 1)/60$, $x = 1, 2, 3, 4, 5$, and zero elsewhere.
    In each case find the variance.

3. A lot of six items contains three defectives. Find the average number of defectives in samples of size 3 from the lot. Find the probability that the number of defectives in a sample of size 3 will be larger than the average number of defectives.

4. A manufacturer of a small appliance examines samples of his product for quality control. He counts the number of mechanical defects in each appliance. The distribution of the number of defects X is given by

x	0	1	2	3
$p(x)$	9/10	1/20	1/40	1/40

    (i) Find the average number of defects per appliance.
    (ii) If it costs $2 to correct each defect and 50¢ to examine each appliance for defects, what is the average quality control cost per appliance?

5. An insurance policy pays $a if an event A occurs. Suppose $P(A) = p$, $0 \le p \le 1$. What should the company charge as premium in order to ensure an expected profit of 5 percent of the amount insured?

6. A recent game run by a major soft drink bottling company advertised the following odds:

Prize	25¢	50¢	$1	$25	$100	$500
Odds	1–25	1–500	1–1,000	1–25,000	1–100,000	1–500,000

    Suppose that in order to enter you buy a two-liter bottle costing $1. What is your expected loss assuming that the soft drink you bought is of no value to you?

7. The percentage of additive in a brand of gasoline has density function

    $$f(x) = 20 x^3 (1 - x), \qquad 0 < x < 1, \text{ and zero elsewhere.}$$

    (i) Find the mean percentage of the additive.
    (ii) If the profit is given by $P = 15 + 5X$ where X is the percentage of additive find $\mathcal{E}P$.

8. The percentage of a certain additive in gasoline determines its specific gravity, which in turn determines its price. Suppose X, the percentage of the additive, has density

function

$$f(x) = 6x(1 - x), \qquad 0 < x < 1, \text{ and zero elsewhere.}$$

If $X < .75$, the gasoline sells as low test for \$1.50 per gallon, and if $X > .80$, the gasoline sells as high test for \$1.90 per gallon. Otherwise, it sells as regular for \$1.75 per gallon. Find the expected revenue per gallon of gasoline sold.

9*. The newspaper boys or girls in a large city pay 18¢ for a newspaper that they sell for 25¢. Unsold newspapers, however, can only be returned for 9¢ each. A newspaper girl has observed that her daily demand has approximately the probability function given by

Number of customers, X	35	36	37	38	39	40	41	42	43
Probability $P(X = x)$	.05	.10	.10	.10	.15	.15	.15	.15	.05

How many papers must she stock to maximize her expected profit?

10. A manufacturer of heat pumps offers a five-year warranty of free repair or replacement if the condenser fails. The time to failure of the condenser it uses has density function

$$f(t) = (1/8)e^{-t/8}, \qquad t > 0, \text{ and zero elsewhere.}$$

(i) Find the average time to failure of a condenser.
(ii) Find the probability that a condenser will fail before the mean time to failure.
(iii) If the profit per sale is \$550 and the replacement or repair cost of a condenser is \$200, find the expected profit.

11. The acidity of a compound X depends on the proportion Y of a certain chemical present in the compound. Suppose

$$X = (\alpha + \beta Y)^2$$

where α, β are constants. If the density of Y can be assumed to be

$$g(y) = 2y, \qquad 0 \le y \le 1, \text{ and zero elsewhere}$$

find the average acidity of the compound. Also find var(X).

12. It costs a company $\$C_1$ in fixed costs (irrespective of the volume of its production) and a variable manufacturing cost of $\$C_2$ per piece of equipment it produces. All the pieces it produces are sold at $\$C_3$ each. The density function of the number of pieces it produces is assumed to be

$$f(x) = \frac{1}{\alpha}e^{-x/\alpha}, \qquad x > 0, \text{ and zero elsewhere.}$$

(For convenience it is assumed that the number of pieces produced has a continuous distribution.)
(i) Find the expected profit.
(ii) If the company has $\$C_4$ in cash reserves, find the probability that it will go bankrupt (assuming that it cannot borrow).

13. A manufacturer produces a perishable item. He has found that the number of units X of the item ordered each week is a continuous random variable with density function

$$f(x) = 1/3, \quad 3 \le x \le 6, \text{ and zero elsewhere.}$$

Each unit of item sold brings a profit of $800 while each unit that is not sold during the week must be destroyed at a loss of $200. Items once sold by the manufacturer are not returned for credit. Suppose p is the number of units manufactured each week. What is the expected weekly profit? What should be the optimum value of p in order to maximize the expected weekly profit?

14. In each of the following cases, compute $\mathscr{E}X$, var(X), and $\mathscr{E}X^n$ (for $n \ge 0$, an integer) whenever they exist.
    (i) $f(x) = 1, -1/2 \le x \le 1/2$, and zero elsewhere.
    (ii) $f(x) = e^{-x}, x \ge 0$, and zero elsewhere.
    (iii) $f(x) = (k - 1)/x^k, x \ge 1$, and zero elsewhere; $k > 1$ is a constant.
    (iv) $f(x) = 1/[\pi(1 + x^2)], -\infty < x < \infty$.
    (v) $f(x) = 6x(1 - x), 0 < x < 1$, and zero elsewhere.
    (vi) $f(x) = xe^{-x}, x \ge 0$, and zero elsewhere.
    (vii) $P(X = x) = p(1 - p)^{x-1}, x = 1, 2, \ldots$, and zero elsewhere; $0 < p < 1$.
    (viii) $P(X = x) = 1/N, x = 1, 2, \ldots, N$, and zero elsewhere; $N \ge 1$ is a constant.

15. Suppose X has density function

$$f(x) = 6x(1 - x), \quad 0 < x < 1, \text{ and zero elsewhere.}$$

Find
    (i) $\mathscr{E}(1/X)$. (ii) $\mathscr{E}(1 - X)$. (iii) $\mathscr{E}X(1 - X)$.

16. The length of life X of a light bulb has density function

$$f(x) = (1/\beta)\exp(-x/\beta), \quad x > 0, \text{ and zero elsewhere.}$$

    (i) Find $\mathscr{E}X$, and var(X).
    (ii) Find $P(X > \mathscr{E}X)$.
    (iii) If it is known that one-half of the bulbs fail in 1500 hours, find the probability that a randomly chosen bulb will survive for 2500 hours.

17. Suppose it takes on the average 1.5 days for a letter to reach its destination with a standard deviation of .2 days. If you want to be 95 percent sure that your letter will reach its destination on time, how early should you mail it?

18. For the density

$$f(x) = (1/\beta)\exp(-x/\beta), \quad x > 0, \text{ and zero elsewhere}$$

of Problem 16, find exactly $P(|X - \mu| \le k\sigma)$ and compare your result with the bound for this probability as given by Chebyshev's inequality.

19. The number of cars crossing a bridge during a certain 15-minute period of time each day has mean 16 and variance 16. What is the maximum probability that more than

25 cars will cross the bridge in this 15-minute period on a randomly selected day? What is the probability that between 10 and 22 (both inclusive) cars will cross the bridge in any such 15-minute time interval?

20. The annual rainfall in a certain locality has a mean of 50 inches with a standard deviation of 10 inches. What is the probability that in a particular year the total rainfall will be within 1.5 standard deviation of the mean? What is the probability that the rainfall will be no more than 12 inches from the mean?

21. Construct an example to show that if σ^2 is large it does not necessarily follow that $P(|X - \mu| > \varepsilon)$ is large.

    [Hint: Take $P(X = c) = 1 - 1/n$ and $P(X = c \pm n^k) = 1/(2n)$ where c is a constant and $k \geq 1$. Then $\mathscr{E}X = c$ and $\text{var}(X) = n^{2k-1}$. Find $P(|X - c| \leq \varepsilon)$ for $\varepsilon > 0$.]

22. Imitate the proof of Chebyshev's inequality to show that for a nonnegative random variable X for which $0 < \mathscr{E}X < \infty$, the inequality

$$P(X > \varepsilon) \leq \frac{\mathscr{E}X}{\varepsilon}$$

holds for all $\varepsilon > 0$. In particular if $\mathscr{E}|X|^k < \infty$ for some k, then

$$P\left(|X| > \varepsilon\left(\mathscr{E}|X|^k\right)^{1/k}\right) \leq 1/\varepsilon^k$$

For $k = 2$ replacing X by $X - \mathscr{E}X$ we get the Chebyshev's inequality.

23*. Let X be a random variable such that $P(a \leq X \leq b) = 1$ where $-\infty < a < b < \infty$. Show that $\text{var}(X) \leq (b - a)^2/4$.

    [Hint: $1 = P(a \leq X \leq b) = P\left(|X - (a + b)/2| \leq \dfrac{b - a}{2}\right)$. Use the result of Problem 22.]

24. For each of the following distributions, find a (the) mode if it exists.
    (i) $P(X = x) = p(1 - p)^{x-1}$, $x = 1, 2, \ldots$, and zero elsewhere; $0 < p < 1$.
    (ii) $f(x) = 1/(\sigma\sqrt{2\pi})\exp\{-(x - \mu)^2/2\sigma^2\}$, $-\infty < x < \infty$; $\sigma > 0$, $\mu \in \mathbb{R}$.
    (iii) $f(x) = (1/\beta^2)xe^{-x/\beta}$, $x > 0$, and zero elsewhere; $\beta > 0$.
    (iv) $f(x) = \dfrac{(\alpha + \beta - 1)!}{(\alpha - 1)!(\beta - 1)!}x^{\alpha-1}(1 - x)^{\beta-1}$, $0 < x < 1$, and zero otherwise; $\alpha, \beta \geq 1$, $\alpha + \beta > 2$, α, β integral.
    (v) $f(x) = (\theta/\pi)\{\theta^2 + (x - \alpha)^2\}^{-1}$, $-\infty < x < \infty$; $\theta > 0$, $\alpha \in \mathbb{R}$.
    (vi) $f(x) = (12/\alpha^4)(\alpha - x)^2$, $0 < \alpha < x$, and zero elsewhere.
    (vii) $f(x) = 2xe^{-x^2}$, $x > 0$, and zero elsewhere.

25*. Suppose $\mathscr{E}|X - c| < \infty$ for some constant c. Let m be a median of X.
    (i) If X has density function f, show that
    (a) $\mathscr{E}|X - c| = \mathscr{E}|X - m| + 2\displaystyle\int_m^c (c - x)f(x)\,dx$.
    (b) $\mathscr{E}|X - c|$ is minimized by choosing $c = m$.

(ii) If X is of the discrete type, then show that

$$\mathscr{E}|X - c| = \begin{cases} \mathscr{E}|X - m| + 2 \sum_{m \le x_j \le c} (c - x_j)P(X = x_j) \\ \qquad + \dfrac{c - m}{2}\left\{ P(X \le m) - \dfrac{1}{2}\right\}, \\ \qquad\qquad\qquad\qquad \text{if } c > m, \\ \mathscr{E}|X - m| + 2 \sum_{c \le x_j \le m} (x_j - c)P(X = x_j) \\ \qquad + \dfrac{m - c}{2}\left\{ P(X > m) - \dfrac{1}{2}\right\}, \\ \qquad\qquad\qquad\qquad \text{if } c \le m, \end{cases}$$

and $\mathscr{E}|X - c|$ is minimized by choosing $c = m$.

$$[\text{Hint: } |X - c| - |X - m| = \begin{cases} c - m & \text{if } X < c \\ 2(X - c) + (c - m) & \text{if } c \le X \le m \\ m - c & \text{if } X > m \end{cases}$$

for $c \le m$. Similar result for $c > m$. Compute $\mathscr{E}|X - c|$.]

26. For each of the following distributions, find a (the) median.
   (i) $f(x) = 1/(\sigma\sqrt{2\pi})\exp\{-(x - \mu)^2/2\sigma^2\}$; $-\infty < x < \infty$, $\mu \in \mathbb{R}$.
   (ii) $P(X = x) = p(1 - p)^{x-1}$, $x = 1, 2, \ldots$, and zero elsewhere; $0 < p < 1$.
   (iii) $f(x) = (\theta/\pi)\{\theta^2 + (x - \alpha)^2\}^{-1}$, $-\infty < x < \infty$; $\theta > 0$, $\alpha \in \mathbb{R}$.
   (iv) $f(x) = 1/x^2$, $x \ge 1$, and zero elsewhere.
   (v) $P(X = k) = 1/N$, $k = 1, 2, \ldots, N$, and zero elsewhere.
   (vi) $f(x) = 2x\exp(-x^2)$, $x > 0$, and zero elsewhere.

27. Find the quantile of order p $(0 < p < 1)$ for the following distributions.
   (i) $f(x) = 1/x^2$, $x \ge 1$, and zero elsewhere.
   (ii) $f(x) = 2x\exp(-x^2)$, $x \ge 0$, and zero otherwise.
   (iii) $f(x) = 1/\theta$, $0 \le x \le \theta$, and zero elsewhere.
   (iv) $P(X = x) = \theta(1 - \theta)^{x-1}$, $x = 1, 2, \ldots$, and zero otherwise; $0 < \theta < 1$.
   (v) $f(x) = (1/\beta^2)x\exp(-x/\beta)$, $x > 0$, and zero otherwise; $\beta > 0$.
   (vi) $f(x) = (3/b^3)(b - x)^2$, $0 < x < b$, and zero elsewhere.

28*. Let $x_1, x_2, \ldots, x_n$ be n real numbers. Set $\bar{x} = \sum_{i=1}^{n} x_i/n$ and $s^2 = \sum_{i=1}^{n} (x_i - \bar{x})^2/(n - 1)$. Let $k \ge 1$. Show that the fraction of $x_1, \ldots, x_n$ included in the interval $(\bar{x} - ks, \bar{x} + ks)$ is at least $1 - 1/k^2$.

29. If X is symmetric about α, show that α is the median of X.

3.8 RANDOM SAMPLING FROM A PROBABILITY DISTRIBUTION

Consider the disintegration of radioactive elements (Example 1.2.3). Recall that the rate of decay of a quantity of a radioactive element is proportional to the mass of the element. That is,

$$(1) \qquad\qquad \frac{dm}{dt} = -\lambda m,$$

where $\lambda > 0$ is a constant giving the rate of decay and m is the mass of the element. Integrating with respect to t we see that

$$\ln m = -\lambda t + c$$

where c is a constant. If $m = m_0$ at time $t = 0$ then $c = \ln m_0$ and

$$(2) \qquad\qquad m = m_0\exp(-\lambda t), \, t \geq 0.$$

Given m_0 and λ we know m as a function of t. The constant λ is the *parameter* of the relationship (2) in the sense that for any particular choice of $\lambda > 0$ we obtain a specific exponential relationship between m and t.

If, on the other hand, the time X at which the element disintegrates is considered to be random we would be interested in the probability that a given element will decay before time t. In order to compute this probability we need to have a suitable model for the density of X. Note that the fraction of the original mass m_0 that decays in $[0, t]$ is given by

$$\frac{m_0 - m_0 e^{-\lambda t}}{m_0} = 1 - e^{-\lambda t}$$

which may be taken as $P(X \leq t)$ for $t > 0$. That is, we take

$$(3) \qquad\qquad P(X \leq t) = 1 - e^{-\lambda t}, \, t > 0,$$

which gives the density function of X as

$$(4) \qquad\qquad f(t) = \lambda e^{-\lambda t}, \, t > 0.$$

Define $f(t) = 0$ for $t \geq 0$. Again, $\lambda > 0$ is a *parameter* of this stochastic or nondeterministic model of radioactive decay in the sense that a particular choice of $\lambda > 0$ gives a specific density of X and a specific probability $P(X \leq t)$. (See also Section 4.2.) For now, we think of parameters as numerical quantities associated with a probability distribution. Thus, the mean and the variance (if they exist) and also the quantiles of a distribution are all parameters.

We recall that the basic difference between probability theory and statistical inference is that in the former one specifies a model and studies its consequences (what is the probability that an element will decay by time t?) whereas in the latter the model is not completely known. The object, then, is to determine some properties of the model (underlying distribution of X) on the basis of observed data. In the example considered above this means, in particular, to make a guess at the unknown value of λ (estimation) or perhaps to settle the validity of a less precise statement about λ such as $\lambda \leq \lambda_0$ (testing of hypothesis). Often, however, even the functional form of the distribution of X is not known. (See Chapter 4 for more details.)

The basic statistical technique used in statistical inference is to replicate the experiment under (more or less) identical conditions, thus obtaining what is called a *random sample*. In the example under consideration this means that we take repeated and independent observations on X, say $x_1, x_2, \ldots, x_n$. These observations may be regarded as values of random variables $X_1, X_2, \ldots, X_n$ respectively where $X_1, X_2, \ldots, X_n$ have the same distribution and are independent.

DEFINITION 1.   Two random variables X and Y are said to be identically distributed if they have the same distribution function.

We emphasis that if X and Y are identically distributed it does not follow that $P(X = Y) = 1$. (See Problem 3.) It is clear that the X_i's need to have the same distribution if they are to represent replications of the same experiment. Independence is used to specify the joint distribution of the sample.

DEFINITION 2.   Let $X_1, X_2, \ldots, X_n$ be independent and identically distributed with common distribution function F. We say that $X_1, X_2, \ldots, X_n$ is a random sample from F.

If $X_1, X_2, \ldots, X_n$ is a random sample from a distribution function F, then the joint distribution of $X_1, \ldots, X_n$ is given by

$$g(x_1, \ldots, x_n) = \prod_{j=1}^{n} f(x_j),$$

if F has density function f and by

$$P(X_1 = x_1, \ldots, X_n = x_n) = \prod_{j=1}^{n} P(X_1 = x_j)$$

when $X_1, X_2, \ldots, X_n$ are of the discrete type with common probability function $P(X_1 = x_j)$. Sometimes the term *population* is used to describe the universe from which the sample is drawn; the population may be conceptual (see Remark 3 below). Often F is referred to as the population distribution.

Remark 1.   In sampling from a probability distribution, randomness is inherent in the phenomenon under study. The sample is obtained by independent replications. In sampling from a finite population, on the other hand, randomness is a consequence of the sampling design. We recall that in sampling from a finite population of N elements, a sample of size n is said to be a (simple) random sample if each of the $\binom{N}{n}$ possible samples has the same chance of being selected (in sampling without replacement and without regard to order).

Remark 2.   Sampling from a probability distribution is sometimes referred to as sampling from an infinite population, since one can obtain samples of any size one wants even if the population is finite (by sampling with replacement).

Remark 3. In sampling from a finite population, the term population is meaningful in that it refers to some measurable characteristic or characteristics of a group of individuals or items. When the population is not finite it is often difficult to identify an underlying population. This happens, for example, in coin tossing or rolling a die, or when measurements involve a manufacturing process (such as life length of a light bulb or net weight of coffee in 10 oz. coffee jars). In these cases the sample is generated by an ongoing random phenomenon.

Remark 4. We can also think of a random sample as an outcome of a (composite) random experiment $\mathscr{E}$ whose outcomes are n-tuples $(x_1, x_2, \ldots, x_n)$. Then x_i may be thought of as the outcome of a component random experiment $\mathscr{E}_i$ of $\mathscr{E}$. In general, given a composite random experiment $\mathscr{E}$ with components $\mathscr{E}_i$ and the joint probability distribution of $\mathscr{E}$, we can always compute the probability distribution of $\mathscr{E}_i$ (by taking the marginal distribution). In the case of random sampling from a probability distribution, however, the probability distribution of $\mathscr{E}$ determines and is completely determined by that of any $\mathscr{E}_i$. (See also Section 2.7)

Example 1. **Number of Girls in a Family of Five Children.** Assuming that both sexes are equally likely, the number of girls X in a randomly selected family of five children has probability function

$$P(X = x) = \binom{5}{x}\left(\frac{1}{2}\right)^5, \qquad x = 0, 1, 2, 3, 4, 5, \text{ and zero elsewhere}$$

(see Section 6.3). A random sample of n families (with five children) showed $x_1, x_2, \ldots, x_n$ girls. Then the probability of observing this sample is given by

$$P(X_1 = x_1, X_2 = x_2, \ldots, X_n = x_n) = \prod_{j=1}^{n} P(X = x_j)$$

$$= \prod_{j=1}^{n} \left\{ \binom{5}{x_j}\left(\frac{1}{2}\right)^5 \right\},$$

for $0 \le x_j \le 5, j = 1, 2, \ldots, n$. In a random sample of size $n = 3$, the probability that $x_1 + x_2 + x_3 \le 1$ is given by

$$P(X_1 + X_2 + X_3 \le 1) = P(X_1 = X_2 = X_3 = 0) + P(X_1 = 1, X_2 = X_3 = 0)$$

$$+ P(X_2 = 1, X_1 = X_3 = 0) + P(X_3 = 1, X_1 = X_2 = 0)$$

$$= \left(\frac{1}{2}\right)^{15} + 3\left[\binom{5}{1}\left(\frac{1}{2}\right)^{15}\right] = \frac{16}{2^{15}} = .0005.$$

That is, in a sample of three families the probability is .0005 that one will observe a total of at most one girl. The probability that each of the three sampled families will have at most one girl is given by

$$P(X_1 \le 1, X_2 \le 1, X_3 \le 1) = [P(X_1 \le 1)]^3 = \left\{ \left(\frac{1}{2}\right)^5 + \binom{5}{1}\left(\frac{1}{2}\right)^5 \right\}^3 = \frac{216}{2^{15}}$$

$$= .0066.$$ □

Example 2. Reliability of Electrical Parts. In studying the reliability of electrical parts, the lifetime X (in hours) of a given part can be assumed to be a random variable with density function

$$f(x) = \frac{\beta}{\theta}\left(\frac{x}{\theta}\right)^{\beta-1} \exp\left\{-\left(\frac{x}{\theta}\right)^{\beta}\right\}, \quad x \ge 0, \quad \text{and zero elsewhere.}$$

A random sample of n parts is put to test. Then the joint distribution of the sample $X_1, X_2, \ldots, X_n$ is given by:

$$g(x_1, x_2, \ldots, x_n) = \prod_{j=1}^{n} f(x_j) = \left(\frac{\beta}{\theta}\right)^n \prod_{j=1}^{n} \left[\left(\frac{x_j}{\theta}\right)^{\beta-1} \exp\left\{-\left(\frac{x_j}{\theta}\right)^{\beta}\right\}\right]$$

$$= \left(\frac{\beta}{\theta}\right)^n \exp\left\{-\sum_{j=1}^{n}\left(\frac{x_j}{\theta}\right)^{\beta}\right\} \prod_{j=1}^{n}\left(\frac{x_j}{\theta}\right)^{\beta-1}$$

for $x_1 \ge 0, x_2 \ge 0, \cdots, x_n \ge 0$, and zero elsewhere. Here $\theta > 0$, $\beta > 0$ are parameters either or both of which may be unknown. The probability that each part will survive t hours is given by:

$$P(X_1 > t, X_2 > t, \ldots, X_n > t) = \int_t^{\infty} \cdots \int_t^{\infty} g(x_1, \ldots, x_n) \, dx_n \ldots dx_1$$

$$= \prod_{j=1}^{n} P(X_1 > t) = [P(X_1 > t)]^n.$$

Now

$$P(X_1 > t) = \int_t^{\infty} \frac{\beta}{\theta}\left(\frac{x}{\theta}\right)^{\beta-1} \exp\left\{-\left(\frac{x}{\theta}\right)^{\beta}\right\} \, dx$$

$$= \int_{(t/\theta)^{\beta}}^{\infty} e^{-y} \, dy \quad \left(y = \left(\frac{x}{\theta}\right)^{\beta}\right)$$

$$= \exp\left\{-\left(\frac{t}{\theta}\right)^{\beta}\right\},$$

so that

$$P(X_1 > t, X_2 > t, \ldots, X_n > t) = \exp\left\{-n\left(\frac{t}{\theta}\right)^{\beta}\right\}.$$

If β, θ are known, this probability can be evaluated. If either θ or β or both are unknown, then the problem (of statistical inference) may be to estimate the unknown parameters from the data $x_1, x_2, \ldots, x_n$ or to estimate the probability $P(X_1 > t, \ldots, X_n > t)$. $\square$

Sample Statistics

In practice we often observe a function of the sample. When we toss a coin n times we hardly, if ever, keep track of the result of each trial. Rather, we count the number of heads (or equivalently the number of tails) in n trials. Let $X_1, X_2, \ldots, X_n$ be a random sample. Then any function $(X_1, X_2, \ldots, X_n)$ that can be computed from the sample without the knowledge of values of the unknown parameters is given a special name.

DEFINITION 3. Let $(X_1, X_2, \ldots, X_n)$ be a random vector. Then a statistic $\phi(X_1, X_2, \ldots, X_n)$ is an observable function of the observation vector $(X_1, X_2, \ldots, X_n)$. That is, the statistic ϕ is directly computable from the data vector $(X_1, X_2, \ldots, X_n)$. A statistic ϕ may be univariate or multivariate.

We note that a statistic is an alternative name given to a random variable (or random vector) when we have a sample of observations. In practice $X_1, X_2, \ldots, X_n$ will be a random sample, that is, $X_1, X_2, \ldots, X_n$ will be independent and identically distributed.

Example 3. Sample Mean and Sample Variance. Let $X_1, X_2, \ldots, X_n$ be a random sample from a distribution function F. Then two sample statistics of great importance in statistics are the *sample mean* defined by

$$(4) \qquad \overline{X} = \sum_{i=1}^{n} \frac{X_i}{n}$$

and the *sample variance* defined by

$$(5) \qquad S^2 = \sum_{i=1}^{n} \frac{(X_i - \overline{X})^2}{n - 1}.$$

The statistic $(\overline{X}, S^2)$ is usually referred to as a joint statistic. Note that we divide by $(n - 1)$ and not n in (5) for reasons that will become clear later. Moreover, both $\overline{X}$ and S^2 are always finite irrespective of whether the population mean and variance do or do not exist. Some other statistics of interest are $\max(X_1, X_2, \ldots, X_n)$, $\min(X_1, X_2, \ldots, X_n)$, median, and mode of the sample.

In particular, let X denote the number of children in a family. Suppose a random sample of 10 families shows the following values of X:

$$5 \; 0 \; 3 \; 2 \; 1 \; 4 \; 3 \; 2 \; 2 \; 3$$

Then

$$\overline{x} = \frac{5 + 0 + 3 + 2 + 1 + 4 + 3 + 2 + 2 + 3}{10} = \frac{25}{10} = 2.5$$

and

$$s^2 \doteq \frac{\sum\limits_{i=1}^{n}(x_i - \bar{x})^2}{n-1} = \frac{\begin{aligned}(2.5)^2 + (2.5)^2 + (.5)^2 + (.5)^2 + (1.5)^2 \\ + (1.5)^2 + (.5)^2 + (.5)^2 + (.5)^2 + (.5)^2\end{aligned}}{10-1}$$

$$= \frac{18.5}{9} = 2.06.$$

Sometimes the following computational formula for S^2 is useful

$$(6) \qquad S^2 = \sum_{i=1}^{n} \frac{(X_i - \bar{X})^2}{n-1} = \frac{\sum\limits_{i=1}^{n} X_i^2 - n\bar{X}^2}{n-1} = \frac{\sum\limits_{i=1}^{n} X_i^2 - \left(\sum\limits_{i=1}^{n} X_i\right)^2/n}{n-1}.$$

The positive square root S of S^2 is known as the *sample standard deviation*. We note that for a specific set of observed values $(x_1, x_2, \ldots, x_n)$ of $(X_1, X_2, \ldots, X_n)$ we use the notation $\bar{x}$ and s^2 for the corresponding values of $\bar{X}$ and S^2. This is consistent with our convention of using lower-case letters to denote values of a random variable.   □

We remark that sample statistics are simply the numerical characteristics of the sample just as parameters are numerical characteristics of the population. However, sample statistics are random variables and vary from sample to sample, whereas parameters are fixed constants. The following table summarizes some facts about parameters and statistics.

Numerical Characteristics	Population Parameters (fixed constants)	Sample Statistics (random variables)
Mean	$\mu = \mathscr{E}X$	$\bar{X} = \sum\limits_{i=1}^{n} X_i/n$
Variance	$\sigma^2 = \mathscr{E}(X - \mu)^2$	$S^2 = \sum\limits_{i=1}^{n}(X_i - \bar{X})^2/(n-1)$
Standard deviation	σ	S
kth Moment	$m_k = \mathscr{E}X^k$	$\sum\limits_{j=1}^{n} X_j^k/n$

According to Problem 3.7.28, an analog of the Chebyshev's inequality holds for each sample $x_1, x_2, \ldots, x_n$.

Example 4.   Proportion of Defective Parts In a Sample.   The length X (in centimeters) of a manufactured part may be considered to be a random variable. A random sample of 25 parts showed a mean of 1.01 cm with a standard deviation of .01 cm. If a manufactured part is considered defective when its length is $\leq .99$ cm or ≥ 1.03 cm, what proportion of the sample parts is defective?

In the absence of any information about the distribution of X we cannot compute the probability $P(.99 < \bar{X} \le 1.03)$ or $P(\bar{X} \le .99$ or $\bar{X} \ge 1.03)$. We can, however, use Chebyshev's inequality to get an upper bound on the proportion of defective parts in the sample. From Problem 3.7.28 we recall that the proportion of sample values included in the interval $(\bar{x} - ks, \bar{x} + ks)$ is at least $1 - 1/k^2$. Since $\bar{x} - ks = 1.01 - k(.01) = .99$ we see that $k = 2$. Hence at most a proportion $1/k^2 = .25$ of the 25 parts in the sample is defective. $\quad\square$

Example 5. Reliability of Electrical Parts (Example 2). In Example 2 the lifetime in hours of an electrical part was assumed to have density function

$$f(x) = \frac{\beta}{\theta}\left(\frac{x}{\theta}\right)^{\beta-1} \exp\left\{-\left(\frac{x}{\theta}\right)^{\beta}\right\}, \qquad x \ge 0, \text{ and zero elsewhere.}$$

What is the probability that a random sample of size $n = 2$ will produce an average life-length of t hours or more? Here we wish to evaluate $P(\bar{X} \ge t)$ where $\bar{X} = (X_1 + X_2)/2$. For convenience we take $\beta = 1$. We have

$$P(\bar{X} \ge t) = P(X_1 + X_2 \ge 2t) = \iint\limits_{x_1+x_2 \ge 2t} g(x_1, x_2)\, dx_2\, dx_1$$

$$= \iint\limits_{x_1+x_2 \ge 2t} \left(\frac{1}{\theta}\right)^2 \exp\left\{-\frac{x_1 + x_2}{\theta}\right\} dx_2\, dx_1$$

$$= 1 - (1/\theta)^2 \int_{x_1=0}^{2t} e^{-x_1/\theta}\left(\int_{x_2=0}^{2t-x_1} e^{-x_2/\theta}\, dx_2\right) dx_1 \quad \text{(See Figure 1)}$$

$$= 1 - (1/\theta)\int_0^{2t} e^{-x_1/\theta}\left[1 - e^{-(2t-x_1)/\theta}\right] dx_1$$

$$= 1 - (1/\theta)\int_0^{2t}\left[e^{-x_1/\theta} - e^{-2t/\theta}\right] dx_1 = 1 - (1 - e^{-2t/\theta}) + (2t/\theta)e^{-2t/\theta}$$

$$= (1 + 2t/\theta)e^{-2t/\theta}. \qquad\qquad \square$$

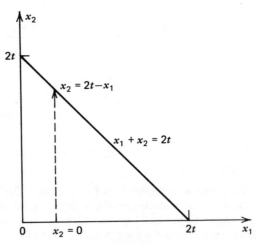

Figure 1

Remark 5. (Sample Median and Sample Mode). Let $x_1, x_2, \ldots, x_n$ be a random sample. Then a *sample* mode is a value that occurs most often. In Example 3, there are two sample modes. Family sizes 2 and 3 occur three times each and are modal family sizes. A *sample median* is a value that divides the sample into two equal parts, each part containing at least half the values. In order to compute a sample median we arrange $x_1, x_2, \ldots, x_n$ in increasing order, say, $x_{(1)} \le x_{(2)} \le \cdots \le x_{(n)}$. If n is an odd integer then there is exactly one middle value, namely $x_{((n+1)/2)}$, which then is a median. If n is even, then there are two middle values $x_{(n/2)}$ and $x_{((n/2)+1)}$. Then any x, $x_{(n/2)} < x < x_{((n/2)+1)}$ is a median. Usually one takes the average $(x_{(n/2)} + x_{(n/2+1)})/2$ as a median. In Example 3, the ordered arrangement is

$$0\ 1\ 2\ 2\ 2\ 3\ 3\ 3\ 4\ 5.$$

Since $n = 10$, the two middle values are 2 and 3 and we take $(2 + 3)/2 = 2.5$ as the sample median. A more formal discussion of sample quantiles appears in Section 8.5.

Remark 6. (Infinite Number of Random Variables). We conclude this section with some comments concerning an infinite number of random variables. In the later chapters we shall be often concerned with some properties of an infinite sequence of random variables. The question that we need to answer is the following: How do we specify a probability model for this sequence of random variables? That is done by specifying the joint distribution of $(X_1, X_2, \ldots, X_n)$ for every positive integer n. That is, in order to specify a probability model for an infinite sequence $\{X_n\}$ we need to specify the joint probabilities

$$P(X_1 \in A_1), \quad P((X_1, X_2) \in A_2), \quad P((X_1, X_2, X_3) \in A_3), \ldots$$

where $A_1, A_2, A_3, \ldots$ respectively are events in $\mathbb{R}_1, \mathbb{R}_2, \mathbb{R}_3, \ldots$. We emphasize that we cannot arbitrarily define the successive joint distributions. If we know the joint distributions of $(X_1, X_2, \ldots, X_{n+1})$, then the (marginal) joint distribution of $(X_1, \ldots, X_n)$ is obtained by letting X_{n+1} take on all possible values. Hence the successive joint distributions must satisfy this *consistency* property. In the special case when $X_1, X_2, \ldots$ are independent and identically distributed, the case of most interest to us, the specification of the marginal distribution F of X_1 suffices since in that case the joint distribution function of $(X_1, \ldots, X_n)$ for every n is the product of the marginal distribution functions of $X_1, \ldots, X_n$ each of which is F.

Problems for Section 3.8

1. Transistors have a life length that has the density function

$$f(x) = (1/\lambda)\exp\left(-\frac{x}{\lambda}\right), \quad x > 0, \text{ and zero elsewhere.}$$

A random sample of 5 transistors is taken and their time to failure is recorded. What is the joint density function of the sample?

2. A random sample of size n is taken from the probability function

$$P(X = x) = e^{-\lambda}\frac{\lambda^x}{x!}, \qquad x = 0, 1, \ldots, \text{ and zero elsewhere.}$$

What is the joint probability function of the sample?

3. A fair coin is tossed once. Let X be the number of heads and Y the number of tails. Are X and Y identically distributed? Is $P(X = Y) = 1$?

4. A fair coin is tossed eight times. Each time a head is observed we write 1, and 0 (zero) otherwise. Suppose the observations are 1, 1, 1, 0, 1, 1, 1, 1. What is the probability of observing the sequence $(1, 1, 1, 0, 1, 1, 1, 1)$? What is the probability of observing as many as seven heads? Based on this computation, do you have any reason to doubt the fairness of the coin?

5. A sample of size 2 is taken from the density function

$$f(x) = 1, \qquad 0 \le x \le 1, \text{ and zero elsewhere.}$$

What is the probability that $\overline{X}$ is at least .9?

6. A sample of size 2 is taken from probability function

$$P(X = 1) = p = 1 - P(X = 0), \qquad 0 < p < 1.$$

(i) Find $P(\overline{X} \le p)$. (ii) Find $P(S^2 \ge .5)$.

[Hint: For a sample of size $n = 2$, $S^2 = \sum_{i=1}^{2} (X_i - \overline{X})^2/(2 - 1) = (X_1 - X_2)^2/2$.]

7. The number of defective refrigerators per week produced at a plant is recorded for 26 weeks with the following results:

$$2, 4, 3, 9, 3, 1, 0, 7, 2, 4, 1, 1, 1, 0, 2, 3, 4, 1, 2, 0, 1, 1, 0, 0, 2, 4.$$

Find the average number of defective refrigerators in the sample. Find also the sample variance.

8. A simple random sample of 65 independent grocery stores in a large metropolitan area gave the following information on the number of employees per store.

Number of employees	1	2	3	4	5	6	8	9	10
Number of stores	6	10	10	7	4	5	7	8	8

Find the sample mean number of employees per store and the sample standard deviation.

9. In Problem 2 take $n = 2$. Find $P(\overline{X} \ge 1)$.

10. Let $x_1, x_2, \ldots, x_n$ be n real numbers. Show that:

$$\max|x_i - \overline{x}| \le \frac{(n-1)s}{\sqrt{n}}.$$

[Hint: $x_i - \overline{x} = -\sum_{j \ne i}^{n} (x_j - \overline{x})$.]

11. Let $x_1, x_2, \ldots, x_n$ be a sample of size n. Write $y_i = ax_i + b$, $i = 1, 2, \ldots, n$. Show that

$$\bar{y} = a\bar{x} + b \quad \text{and} \quad s_y^2 = a^2 s_x^2$$

where $\bar{y}$ is the mean of $y_1, \ldots, y_n$, s_y^2 is the sample variance of the y's and s_x^2 is the sample variance of the x's.

12. The weights (in kilograms) of five bags of navy beans were recorded to be: 1.1, 1.3, 0.9, 1.2, 1.0.
    (i)   What is the sample mean and the sample variance?
    (ii) What is the sample mean and the sample standard deviation if the weights were recorded in pounds?
    [Hint: One kilogram $\doteq$ 2.2 pounds. Use the result of Problem 11.]

13. A random sample of 25 I.Q. scores given to entering freshman at a state university showed a mean of 125 with a standard deviation of 13.2.
    (i)   Find an interval in which at least 75 percent of the I.Q. scores lie.
    (ii) Give an upper bound for the proportion of students in the sample whose I.Q. score on the test is at least 150.
    [Hint: Use Chebyshev's inequality of Problem 3.7.28.]

14. Let $x_1, x_2, \ldots, x_n$ be a set of n real numbers. Let $x_{(n)} = \max\{x_1, x_2, \ldots, x_n\}$ and $x_{(1)} = \min_n\{x_1, x_2, \ldots, x_n\}$. Show that for any set of real numbers $a_1, a_2, \ldots, a_n$ such that $\sum_{i=1}^{n} a_i = 0$ the following inequality holds.

$$\left| \sum_{i=1}^{n} a_i x_i \right| \le \frac{1}{2} (x_{(n)} - x_{(1)}) \sum_{i=1}^{n} |a_i|.$$

[Hint: $\sum_{i=1}^{n} a_i x_i = \sum_{i=1}^{n} a_i(x_i - c)$ for any constant c. Use triangular inequality and choose c appropriately.]

15. Show that X has a symmetric distribution if and only if X and $-X$ are identically distributed.

16*. Let X_1, X_2 be a random sample from a continuous distribution which is symmetric about θ. That is, if f is the common density of the X_i's, then

$$f(\theta + x) = f(\theta - x) \quad \text{for all } x.$$

(i)   Show that

$$P(X_1 + X_2 \le 2\theta + x) = P(X_1 + X_2 \ge 2\theta - x)$$

for all x, so that the distribution of $X_1 + X_2$ is symmetric about 2θ. (By induction one can show that the distribution of $X_1 + \cdots + X_n$ is symmetric about $n\theta$.)

(ii) Show that

$$P(X_1 - X_2 \le x) = P(X_1 - X_2 \ge -x)$$

for all $x \in \mathbb{R}$ so that the distribution of $X_1 - X_2$ is symmetric about 0.

3.9 REVIEW PROBLEMS

1. Which of the following functions are density functions? Find the corresponding distribution function whenever the function defined below is a density function.
   (i) $f(x) = \sin x$, $0 \le x \le \pi/2$, and zero elsewhere.
   (ii) $f(x) = (1/2)|\sin x|$, $|x| < \pi/2$, and zero elsewhere.
   (iii) $f(x) = x/4$ for $0 \le x \le 1$, $= 5/8$ for $1 \le x < 2$, $= (3 - x)/2$ for $2 \le x < 3$, and zero elsewhere.
   (iv) $f(x) = (1/24)x(1 - x/12)$, $0 \le x \le 12$, and zero otherwise.
   (v) $f(x) = (1/4\theta)$ if $0 < x < \theta/2$ or $3\theta/2 \le x \le 2\theta$, $= 3/(4\theta)$ if $\theta/2 < x < 3\theta/2$, and zero elsewhere.
   (vi) $f(x) = (1/2\theta)\cos(x/\theta)$, $-\pi\theta/2 \le x \le \pi\theta/2$ and zero otherwise.

2. Are the following functions distribution functions? If so, find the corresponding probability or density functions.
   (i) $F(x) = 0$, $x < 0$, and $= 1 - \exp(-x^2/2\sigma^2)$ if $x \ge 0$; $\sigma > 0$.
   (ii) $F(x) = 0$ if $x \le 0$, $= \ln x$ if $0 < x \le 2$, and $= 1$ if $x > 0$.
   (iii) $F(x) = 0$ if $x < 1$, $= \ln x/\ln 2$ if $1 \le x \le 2$, and $= 1$ if $x > 2$.
   (iv) $F(x) = 1 - \exp(-x^\beta)$ if $x \ge 0$, and zero elsewhere; $\beta > 0$.
   (v) $F(x) = 1 - (2x + 1)e^{-2x}$, for $x \ge 0$ and zero otherwise.
   (vi) $F(x) = 0$ if $x < 0$, $= 1/2$ if $0 \le x < 1/2$, $= x$ if $1/2 \le x < 1$, and $= 1$ if $x > 1$.

3. The number of arrivals X at a bank in any 15-minute period has probability function

$$P(X = k) = \frac{(2.5)^k e^{-2.5}}{k!}, \quad k = 0, 1, 2, \dots.$$

   Find:
   (i) $P(X \ge 1)$.   (ii) $P(X < 2)$.   (iii) $P\{X \ge 1 | X < 2\}$.

4. The storage capacity of jet fuel at an airport is 1.5 million gallons. Suppose the daily consumption (in million gallons) of jet fuel at this airport has density function given by

$$f(x) = 1.5e^{-1.5x}, \quad x \ge 0, \text{ and zero elsewhere.}$$

   How often will the fuel run out by the end of the day if the tank is filled at the beginning of each day?

5. The percentage X of antiknock additive in a particular gasoline has density function

$$f(x) = cx^2(1 - x), \quad 0 \le x \le 1, \text{ and zero elsewhere.}$$

   (i) Find c, $\mathscr{E}X$ and var(X).
   (ii) Find $P(X \ge 1/2)$.

6. The manufacturer of liquid fertilizer makes 89¢ profit on each gallon of fertilizer it sells in the summer of the year it is manufactured. Amount not sold by the end of the summer is stored at a cost of 25¢ per gallon. Suppose the demand X for his fertilizer

has density function

$$f(x) = \frac{2}{10^6}, \qquad 10^6 \le x \le (1.5)10^6, \text{ and zero otherwise.}$$

At what level should he stock in order to maximize his average profit? What is the maximum average profit?

7. Let $f_1, f_2, \ldots, f_k$ be probability density functions on the interval (a, b).

   (i)   Show that $\displaystyle\sum_{i=1}^{k} f_i$ cannot be a density function on the interval (a, b).

   (ii) Show that $\displaystyle\sum_{i=1}^{k} \alpha_i f_i,\, 0 \le \alpha_i \le 1,\, \sum_{i=1}^{k} \alpha_i = 1$, is a density function on (a, b).

8. Let X be a bounded random variable. That is, suppose there exists a constant $c > 0$ such that $P(|X| \le c) = 1$. Show that $\mathscr{E}|X| < \infty$ and $|\mathscr{E}X| \le c$.

9. Suppose X has symmetric distribution about a point c and $\mathscr{E}|X| < \infty$. Show $\mathscr{E}X = c$.

10. Suppose $P(X \ge Y) = 1$. Suppose $\mathscr{E}|X| < \infty$ and $\mathscr{E}|Y| < \infty$. Show that $\mathscr{E}X \ge \mathscr{E}Y$ and $\mathscr{E}X = \mathscr{E}Y$ if and only if $P(X = Y) = 1$.

11. Let X be the amount (in thousands of dollars) of itemized deductions on a randomly selected federal income tax return. Suppose X has density function

$$f(x) = \frac{(\alpha - 1)5^{\alpha - 1}}{x^\alpha}, \qquad \text{if } x \ge 5, \text{ and zero otherwise.}$$

   Here $\alpha > 1$ is a constant.
   (i)   Find mode if it exists.
   (ii) Find the quantile of order $p,\, 0 < p < 1$.
   (iii) Find the median if $\alpha = 2$.

12. Find the mean and the variance of the density functions in part (iv), (v) and (vi) of Problem 1.

13. Compare the probability bounds obtained from Chebyshev's inequality in Problem 12 with the exact results.

14. Suppose X has probability function given by

$$P(X = a + bk) = c, \qquad k = 1, 2, \ldots N,$$

   where $c > 0$ is a constant. Find the constant c and the mean and the variance of X.

15. A manufacturer counts major and minor defects in sump pumps it manufactures for quality control purposes. Suppose only four characteristics are inspected. Let X be

the number of minor defects and Y that of major defects in a sump pump. Suppose (X, Y) has joint probability function given by

Y \ X	0	1	2	3	4
0	5/36	1/36	2/36	2/36	3/36
1	1/36	2/36	2/36	3/36	0
2	2/36	2/36	3/36	0	0
3	2/36	3/36	0	0	0
4	3/36	0	0	0	0

   (i) Find the two marginal distributions and the conditional distributions.
   (ii) Find $P\{X \geq 1 | Y \leq 3\}$ and $P\{X \leq 2 | Y = 1\}$.
   (iii) Find $P(Y > X)$ and $P(Y \geq X)$.

16. Let (X, Y) have joint density function

$$f(x, y) = 8xy, \qquad 0 < y < x < 1, \text{ and zero elsewhere.}$$

   (i) Are X and Y independent?
   (ii) Find $P(|X - Y| > 1/2)$.
   (iii) Find the marginal density of Y and the conditional density of X given $Y = y$.
   (iv) Find $P\{1/4 < X < 1/2 | Y = 3/8\}$ and $P\{1/4 < X < 1/2 | 1/4 < Y < 3/8\}$.

17. Let X and Y be the times to failure of two batteries and suppose X and Y are independent with respective densities

$$f_1(x) = e^{-x}, \qquad x \geq 0, \text{ and zero elsewhere,}$$

$$f_2(y) = ye^{-y}, \qquad y \geq 0, \text{ and zero elsewhere.}$$

   Find $P(X - Y > 0)$ and $P(|X - Y| > 2)$

18. Three points X_1, X_2, and X_3 are randomly selected from the interval $(0, 1)$
   (i) Find the probability that exactly two of the points selected will be $\geq .60$.
   (ii) Find the probability that $X_1 + X_2 > X_3$.

19. A random sample of size 2 is taken from the density function

$$f(x) = \lambda e^{-\lambda x}, \qquad x > 0, \text{ and zero otherwise.}$$

   What is the probability that the sample variance will be greater than $\lambda^2/2$?

20. A random sample of size two is taken from the probability function

$$P(X = 1) = p = 1 - P(X = 0), \qquad 0 < p < 1.$$

   What is the probability that the sample variance will be at least as large as the sample mean?

CHAPTER 4

Introduction to
Statistical Inference

4.1 INTRODUCTION

Given a random experiment, we have constructed a stochastic model for the experiment by identifying a set of all outcomes Ω, the events of interest A, $A \in \zeta$, and by assigning to each event A a probability $P(A)$. We then translated the problem to the real line (or $\mathbb{R}_n$) by defining a random variable (or a random vector) X on Ω. We saw that X induces a probability distribution P_X on events E in this new sample space through the correspondence

$$P_X(E) = P(\omega: X(\omega) \in E).$$

We showed that P_X can be specified simply by specifying the distribution function, or the density or probability function, of X.

Suppose we have n observations $x_1, x_2, \ldots, x_n$ on some random variable X. What information does it contain about the distribution F of X? How does it help us to make a guess at F or some numerical function of F? The answer depends on the type of assumptions we are willing to make about F. If we make little or no assumptions about F, then the family of possible distributions of X is called nonparametric. On the other hand, if we assume that the form of F is known except for some numerical values of a finite number of constants (parameters) $\theta_1(F), \theta_2(F), \ldots, \theta_k(F)$, then the family of possible distributions of X is called parametric. Section 4.2 gives formal definitions of these two concepts and provides some typical examples of parametric and nonparametric inference problems.

Often the problem of inference is to make a guess at the numerical value of some constant(s) $\theta(F)$. In Section 4.3 we consider the problem of estimating $\theta(F)$ by a point estimate. We explore some desirable properties of a point estimate of $\theta(F)$. Sometimes it is preferable to estimate $\theta(F)$ by an interval (or a set) of values that includes the true $\theta(F)$ with a large prespecified probability. This so-called problem of confidence estimation is also considered in Section 4.3.

In earlier chapters we often computed $P(T(X_1, X_2, \ldots, X_n) \in A)$ where $T(X_1, X_2, \ldots, X_n)$ is a sample statistic. The numerical value of this probability was then used to assess the strength of argument that the sample data provides for or against our underlying model assumptions. This, however, is not enough. We also need to contend with the probability $P(T(X_1, X_2, \ldots, X_n) \in A)$ computed under the assumption that the model is false. Suppose, for example, that P_0 and P_1 are the two contending probability models for the underlying distribution of the X_i's. How do we choose between P_0 and P_1? This so-called problem of hypothesis testing is the subject of investigation in Section 4.4. The close relation between confidence estimation and hypothesis testing is also explored.

In Section 4.5 we consider some simple graphic methods of looking at the data. Methods of estimating the distribution function and probability or density function are also introduced. The shape of the graph of the estimate of probability or density function can often be used to see if the sample could have come from a known family of distributions.

4.2 PARAMETRIC AND NONPARAMETRIC FAMILIES

Let $(X_1, X_2, \ldots, X_n)$ be a random vector from a joint distribution function G. A specification of the stochastic model governing the observations requires a specification of the set $\mathfrak{G}$ of *all possible joint distribution functions G* for the X_i's. How is the family $\mathfrak{G}$ selected? As much as one would like $\mathfrak{G}$ to be large in order that no possible underlying distribution for the X_i's is excluded, it is not realistic to expect that one realization $(x_1, x_2, \ldots, x_n)$ will tell us anything about the specific $G \in \mathfrak{G}$. Since the object is usually to predict future performance when replications of the same experiment are made, it is reasonable to limit $\mathfrak{G}$ by assuming that $X_1, X_2, \ldots, X_n$ are independent, identically distributed, and have a common distribution function F. That is, $X_1, \ldots, X_n$ is a random sample from F so that $G(x_1, \ldots, x_n) = \prod_{j=1}^{n} F(x_j)$. It is then sufficient to consider univariate distribution functions $F \in \mathfrak{F}$ say. If we restrict $\mathfrak{F}$ still further by specifying the form of F, we should be able to get fairly detailed information on the specific $F \in \mathfrak{F}$ from the sample $(x_1, \ldots, x_n)$. However, taking $\mathfrak{F}$ too small may lead to exclusion of the true F and hence wrong conclusions.

It is convenient to distinguish between parametric and nonparametric families of distributions because the methods used in two cases differ greatly. In a *parametric model*, we specify the form of the distribution (through its density or probability function) with the exception of the *index* or *parameter* θ where θ belongs to a *parameter set* Θ that is a subset of a finite dimensional Euclidean space. Thus we get all distributions in $\mathfrak{F}$ by varying at most a finite number of scalar parameters whose values completely determine a unique member of $\mathfrak{F}$.

Example 1. Model for Coin Tossing. Suppose a coin is tossed n times. Let $X_i = 1$ if the ith toss results in a head and zero otherwise, and let $P(X_i = 1) = p = 1 - P(X_i = 0)$. If we know n, then a specification of p determines precisely one distribution for $Y = \sum_{i=1}^{n} X_i$, the number of heads in n trials. Here p is the parameter that is scalar, and

$\Theta = \{0 \leq p \leq 1\}$ is the parameter set. On the other hand, if we make m observations $Y_1, Y_2, \ldots, Y_m$ on the Y's without the knowledge of n and p, then the parameters are n and p so that $\theta = (n, p)$ and $\Theta = \{(n, p): n = 1, 2, \ldots, 0 \leq p \leq 1\}$. If we specify the numerical values of both n and p, we know precisely the distribution that governs the Y's. □

Example 2. Survival Times of Patients with Advanced Lung Cancer. Suppose 100 patients with advanced lung cancer are followed until their death. Let $X_1, X_2, \ldots, X_{100}$ be their survival times. If we assume that the survival times can be modeled according to the density function

$$f(x) = \theta^{-1} e^{-x/\theta}, \qquad x > 0, \text{ and zero elsewhere}; \qquad \theta > 0,$$

then the parameter θ is scalar with $\theta \in \Theta = (0, \infty)$. All possible distributions for the X_i's can be obtained by varying θ in Θ, and a specification of the numerical value of θ completely specifies the underlying density of the X_i's. □

It is clear, then, that we need a dictionary of some well-known and frequently used models as well as a listing of their properties, which will help facilitate our choice of an appropriate parametric model for the random phenomenon under study. This is done in Chapters 6 and 7.

Often, however, we do not have enough information to rely on a specific parametric model. In that case we may consider a much wider class of distributions. In a *nonparametric model*, the family of underlying distributions $\mathfrak{F}$ for X is so wide that $\mathfrak{F}$ cannot be indexed by a finite number of numerical parameters. Thus a specification of a finite number of numerical parameters does not uniquely determine a member of $\mathfrak{F}$.

The terms "nonparametric" and "distribution-free" are used interchangeably. It should be noted, however, that "nonparametric" does not mean that there is no parameter and distribution-free does not mean that no distributions are involved.

Example 3. Family of Distributions with Finite Mean. Let $\mathfrak{F}$ be the family of all distribution functions on the real line that have finite mean. This is a very large class of distribution functions, and no matter how many scalar parameters are specified we cannot in general specify the specific F in $\mathfrak{F}$ that governs the observations. Hence $\mathfrak{F}$ is classified as a nonparametric family. □

Example 4. Symmetric Distributions. Let $\mathfrak{F}$ be the set of all distributions on the real line that are symmetric about their median. Again $\mathfrak{F}$ is a nonparametric family since it cannot be characterized by a finite set of scalars. □

In nonparametric statistics, the most that we will assume (unless otherwise specified) is that the underlying distribution is of the continuous type. This is a mathematically convenient assumption that allows us to assume that the observations can be arranged in increasing (or decreasing) order with no ties [for continuous-type random variables $P(X = Y) = 0$]. This family is so wide that in general the procedures used will not be as sharp and powerful as those developed for a narrowly defined parametric family. At times, however, the nonparametric procedures are almost as good as the parametric ones. The point is that before we apply any parametric procedure we must make sure that the underlying assumptions are satisfied, or at least reasonably satisfied. If it is felt that there is not enough information available to choose a specific parametric model, it is advisable to use a nonparametric procedure.

Problems for Section 4.2

In each of the following cases, determine whether you will classify the family of distributions as parametric or nonparametric.

1. $\mathfrak{F} = \{f : f \text{ is a density satisfying } f(x) = f(-x)\}$.

2. $\mathfrak{F} = \{P_\lambda(x) = e^{-\lambda}\lambda^x/x!; \ x = 0, 1, \ldots; \lambda > 0\}$.

3. $\mathfrak{F} = \{F : F \text{ is a continuous distribution function or has only jump discontinuities}\}$.

4. $\mathfrak{F} = \{f : f \text{ is a density with unique median}\}$.

5. $\mathfrak{F} = \{f_\theta(x) = 1/\theta, \ -\theta/2 < x < \theta/2, \text{ zero otherwise}, \theta > 0\}$.

6. $\mathfrak{F} = \{f_{\mu, \sigma^2}(x) = (1/\sigma\sqrt{2\pi})\exp\{-(x-\mu)^2/2\sigma^2\}, x \in \mathbb{R}; \mu \in \mathbb{R}, \sigma > 0\}$.

7. $\mathfrak{F} = \{f_{\mu, \theta}(x) = (1/\theta)\exp\{-(x-\mu)/\theta\}, x > \mu, \text{ and zero elsewhere}, \mu \in \mathbb{R}, \theta > 0\}$.

8. $\mathfrak{F} = \{P_N(x) = 1/N, x = 1, 2, \ldots, N; N \geq 1\}$.

9. $\mathfrak{F} = \{f : f \text{ is unimodal density}\}$.

10. $\mathfrak{F} = \{f_{\alpha, \beta}(x) = x^{\alpha-1}e^{-x/\beta}/((\alpha-1)!\beta^\alpha), x > 0, \text{ and zero elsewhere}; \alpha \geq 1 \text{ is an integer}, \beta > 0\}$.

In the following problems construct an appropriate statistical model for the experiment. Identify $\mathfrak{F}$, the parameter, and the parameter set.

11. A die is tossed repeatedly. The object is to decide whether the die is fair or loaded. A record is kept of the number of 1's, 2's, 3's, 4's, 5's, and 6's that show up in n trials.

12. A manufacturer produces 1-inch wood screws, some of which go unslotted. A random sample of n screws is taken from a very large batch to see whether the overall proportion of unslotted screws is sufficiently small.

13. A manufacturer of 12-volt automobile batteries takes a sample of size n from a large lot and measures the length of life of each battery.
    (i) The object is to estimate the overall proportion of batteries that do not meet the given specification (length of life greater than a prespecified number of hours).
    (ii) The object is to estimate the average length of life of batteries it produces.

4.3 POINT AND INTERVAL ESTIMATION

Suppose a decision has been made concerning the model, whether nonparametric or parametric. How do we estimate a parameter or some parameters of the model? How accurate is this estimate? That is, how do we measure the precision of our estimate? In order to motivate discussion, suppose we wish to find the average survival times of patients with advanced lung cancer. Let X denote the survival time of a randomly selected patient, and suppose F is the distribution function of X. We take a sample of n patients and follow them until their death. Let $X_1, X_2, \ldots, X_n$ be the survival times of the n patients. How do we use the observations $X_1, X_2, \ldots, X_n$ to estimate $\mu_F = \mathscr{E}_F X$?[†] This very much depends on the assumptions we are willing to make about F. If we assume only that $F \in \mathfrak{F}$ where $\mathfrak{F}$ is the family of all continuous distribution functions on the real line with finite mean, then $\mathfrak{F}$ is a nonparametric family of distributions. In this case the form of F is unknown and hence cannot be used. If, on the other hand, it is felt that the density

$$f(x) = \frac{1}{\beta} \exp\left(-\frac{x}{\beta}\right), \qquad x > 0, \text{ and zero elsewhere}$$

adequately describes the distribution, then $\mathfrak{F}$ is a parametric family with parameter β belonging to the parameter set $\Theta = (0, \infty)$. In this case $\mu_F = \mathscr{E}_F X = \beta$ so that the problem reduces to estimating β. Once β is known, the distribution F is completely known. (In the nonparametric case, knowledge of μ_F tells us little about the distribution function F.)

We have so far been using the word "estimate" in its dictionary sense. Let us now make a formal definition.

DEFINITION 1. (POINT ESTIMATE). Let $(X_1, X_2, \ldots, X_n)$ be a random vector with joint distribution function $F \in \mathfrak{F}$. Let $\theta = \theta(F)$, $F \in \mathfrak{F}$ be the (scalar or vector) unknown parameter of interest which takes values in $\Theta = \{\theta(F): F \in \mathfrak{F}\}$. Any statistic $T(X_1, X_2, \ldots, X_n)$ is said to be a (point) estimate of θ if it can take values only in Θ.

Some remarks are in order. First, we note that the statistic T must be observable (that is, computable from the sample). Hence it cannot be a function of any unknown parameter. Next, we note that frequently a distinction is made between an *estimate* and an *estimator*. This distinction is the same as that between a random variable and a value assumed by it. An estimator is the statistic used to estimate the parameter, and a numerical value of the estimator is called an estimate. For notational convenience, we will not follow this distinction. Third, even though all statistics that take values in Θ are possible candidates for estimates of θ, it is clear that we have to develop criteria that indicate which estimates are "good" and which may be rejected out of hand. Certainly estimates that are constants (not based on the observations) should not be admissible under any reasonable criteria (except for pedagogical reasons).

Example 1. Point Estimate for Median of a Distribution. Let $X_1, X_2, \ldots, X_n$ be a random sample from a continuous distribution that is symmetric about its midpoint. This means that $X_1, X_2, \ldots, X_n$ are independent random variables, each having a common continuous distribution function $F(x) = P(X \le x)$, which is assumed to be symmetric about its midpoint or median m_F (assumed unique):

$$F(m_F) = 1/2 \quad \text{and} \quad F(m_F + x) = 1 - F(m_F - x) \qquad \text{for all } x \in \mathbb{R}.$$

[†]See Remark 2 below for an explanation of the notation $\mathscr{E}_F$.

This is clearly a nonparametric situation with $\theta = m_F \in \Theta = \mathbb{R}$. Some reasonable estimates of m_F are

   (i) median of $X_1, X_2, \ldots, X_n$,

   (ii) $(\min X_i + \max X_i)/2$.

Of course there are infinitely many estimates to choose from since every real valued function of the sample is an estimate. However, often the choice narrows down to a few or at least to a manageable subset of the set of all estimates.

If, on the other hand, it is determined that the model can be narrowed down to the family of parametric density:

$$f(x) = \frac{1}{\sqrt{2\pi}} \exp\left\{-\frac{(x-\theta)^2}{2}\right\}, x \in \mathbb{R}$$

where $\theta \in \mathbb{R}$ is unknown, then the choice of estimates of θ (the median) is focused owing to the knowledge of the form of F or its density. Some reasonable estimates of θ are: (i) $\bar{X} = \sum_{i=1}^{n} X_i/n$, (ii) median of $X_1, X_2, \ldots, X_n$. The reason for choosing $\bar{X}$ is that θ is also the mean of f. Moreover, as we shall see in Section 5.4 for large n, $\bar{X}$ is likely to be near θ. The estimate $\bar{X}$ has several other important properties. The choice of median of $X_1, \ldots, X_n$ as an estimate derives from the fact that θ is also the median of f. $\square$

Example 2. Estimating the Mean of a Distribution. Let $X_1, X_2, \ldots, X_n$ be a random sample from a distribution function $F \in \mathfrak{F}$ where $\mathfrak{F}$ is the class of all distributions on the real line with finite mean μ_F. Suppose nothing else is assumed about F. Then the family $\mathfrak{F}$ is nonparametric. A reasonable estimate of μ_F is the sample mean

$$\bar{X} = \sum_{i=1}^{n} \frac{X_i}{n}$$

We will show in Section 5.4 that for large n, $\bar{X}$ is likely to be near μ, moreover $\mathscr{E}\bar{X} = \mu_F$ (see Section 5.3). On the other hand, if it is felt, say, that the uniform distribution on $[0, \theta]$, $\theta > 0$, adequately represents F, then the knowledge of the form of F can be used to construct other reasonable estimates of the mean $(= \theta/2)$. For example, in addition to $\bar{X}$ two other reasonable estimates are the sample median and one half the largest observation. All these estimates tend to be near θ for large n. $\square$

Remark on Notation 1. Let $\theta \in \Theta$ be the unknown parameter to be estimated based on the sample $X_1, X_2, \ldots, X_n$ of size n. Estimates will usually be denoted by $\hat{\theta}(X_1, X_2, \ldots, X_n) = \hat{\theta}$ (with or without subscripts depending on the number of estimates under consideration). Whenever a Greek or other letter appears with a "hat" above it, such as $\hat{\alpha}$, $\hat{\beta}$, $\hat{\gamma}$, it is assumed to be an estimate of the corresponding parameter (without hat).

We now turn our attention to two related problems in point estimation. How do we judge the worth or precision of an estimate? What are some desirable properties that we may require our estimate to possess? Requiring an estimate to possess a certain desirable property restricts the search to a smaller set of estimates, whereas a measure of precision of an estimate helps us to choose between two (or more) estimates having the same property. Neither provides a constructive procedure for finding estimates.

Precision usually means some measure of "nearness." How close is the estimate to the parameter? If an estimate gives, *on the average*, the true parameter value, we call this property of the estimate unbiasedness.

DEFINITION 2.   An estimate $T(X_1, \ldots, X_n)$ for a parameter $\theta \in \Theta$ is said to be unbiased for θ if $\mathscr{E}_\theta |T| < \infty$ and

$$\mathscr{E}T(X_1, \ldots, X_n) = \theta \qquad \text{for all } \theta \in \Theta.$$

Remark On Notation 2.   When we subscript $\mathscr{E}$ by θ, it means that the expected value is to be computed under the density (or probability) function when θ is the true value of the parameter. The same remarks apply to the notation P_θ for probability computations. The notation $\mathscr{E}_F$ (or P_F) has a similar interpretation.

DEFINITION 3.   An estimate $T(X_1, \ldots, X_n)$ is said to be biased for θ if $\mathscr{E}_\theta T \neq \theta$ for some $\theta \in \Theta$. The quantity

$$b(T, \theta) = \mathscr{E}_\theta T(X_1, \ldots, X_n) - \theta$$

is called the bias of T.

Bias is a systematic error (in the same direction). Unbiasedness of T says that T is correct on the average, that is, the mean of distribution of T is θ. If we take a large number of random samples and for each sample construct the corresponding value of $T(x_1, \ldots, x_n)$, then this collection of T values will have a mean approximately equal to θ.

Remark 3.   Let $\hat{\theta}$ be unbiased for θ. Then $a\hat{\theta} + b$ is unbiased for $a\theta + b$. (See equation (3.7.2).) This simple observation frequently allows us to construct unbiased estimates from biased ones. Thus, if

$$\mathscr{E}_\theta \hat{\theta} = a\theta + b, \quad a \neq 0$$

then

$$\mathscr{E}_\theta \left(\frac{\hat{\theta} - b}{a} \right) = \theta.$$

In general, however, if $\hat{\theta}$ is unbiased for θ and φ is a real valued function on Θ, then $\varphi(\hat{\theta})$ is *not* unbiased for $\varphi(\theta)$.

Remark 4.   To find an unbiased estimate for a parameter, one begins with the computation of the first few moments to see if the parameter is linearly related to any moment (or moments). If so, then an unbiased estimate is easily obtained. (See Remark 3.)

Is unbiasedness necessarily a desirable characteristic of an estimate? Let T_1, T_2 be two estimates where T_1 is unbiased for θ but T_2 is not. Suppose the possible values of T_1 are widely scattered around θ whereas those of T_2 are concentrated near θ_1 (which is close to θ). Then T_2 may be preferable to T_1. (See Figure 1.) It is clear, then, that one needs also to look at the variability of the estimate around θ to make a choice.

The variance of an estimate T measures the dispersion of T around $\mathscr{E}_\theta T$. If T is unbiased, its variance gives a good measure of its precision. If T is biased, we should still be looking at the variability of T around θ as a measure of its precision.

DEFINITION 4.   The mean square error (MSE) of an estimate T for θ is defined to be $\mathrm{MSE}_\theta(T) = \mathscr{E}_\theta(T(X_1,\ldots,X_n) - \theta)^2$.

We note (see Remark 3.7.4) that:

(1) $$\mathrm{MSE}_\theta(T) = \mathscr{E}_\theta(T - \mathscr{E}_\theta T + \mathscr{E}_\theta T - \theta)^2 = \mathrm{var}_\theta(T) + [b(T,\theta)]^2$$

so that

(2) $\mathrm{MSE}_\theta(T) > \mathrm{var}_\theta(T)$ for some $\theta \in \Theta$ unless $b(T,\theta) = 0$, for all θ

that is, T is unbiased.

Smaller MSE means greater precision. In comparing two estimates T_1 and T_2 of θ we choose the one with smaller MSE. Thus, if

(3) $$\mathrm{MSE}_\theta(T_1) \le \mathrm{MSE}_\theta(T_2) \text{ for all } \theta \in \Theta$$

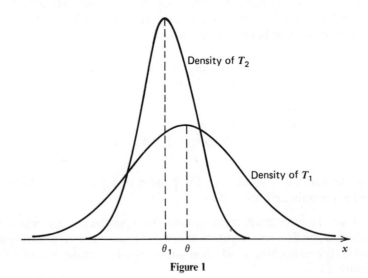

Figure 1

with strict inequality for at least one θ, then the estimate T_1 is preferred over T_2. Often this comparison is made by considering the ratio $\mathrm{MSE}_\theta(T_2)/\mathrm{MSE}_\theta(T_1)$.

DEFINITION 5. The relative efficiency (RE) of T_1 in relation to T_2 is defined by the ratio

$$\mathrm{RE}_\theta(T_1, T_2) = \frac{\mathrm{MSE}_\theta(T_2)}{\mathrm{MSE}_\theta(T_1)}.$$

Clearly, T_1 is preferred over T_2 if $\mathrm{RE}_\theta(T_1, T_2) \geq 1$ with strict inequality for at least one θ.

Example 3. Estimating the Size of a Population. A box contains N (unknown) items marked 1 through $N(\geq 2)$. We make one selection from the box. How do we estimate N? This is a problem one encounters, for example, in estimating the number of taxis in a town. In World War II this problem was encountered by the Allies who wanted to estimate enemy production of tanks.

Let X be the observed number. Then

$$P(X = x) = \frac{1}{N}, \quad x = 1, 2, \ldots, N; \qquad \Theta = \{N : N \geq 2, N \text{ integral}\}.$$

An obvious estimate of N is X, call it $\hat{N}_1$. Clearly $X \leq N$. We have:

$$\mathscr{E}_N \hat{N}_1 = \sum_{k=1}^{N} k \cdot \frac{1}{N} = \frac{N+1}{2}$$

so that $\hat{N}_1$ is biased. The bias is given by

$$b(\hat{N}_1, N) = \frac{N+1}{2} - N = \frac{1}{2} - \frac{N}{2} < 0.$$

Hence $\hat{N}_1$ underestimates N, which is to be expected since we know that $X \leq N$. The MSE of $\hat{N}_1$ is given by:

$$\mathrm{MSE}_N(\hat{N}_1) = \mathscr{E}(\hat{N}_1 - N)^2 = \mathscr{E}\hat{N}_1^2 - 2N\mathscr{E}\hat{N}_1 + N^2$$

$$= \sum_{k=1}^{N} k^2 \frac{1}{N} - 2N\left(\frac{N+1}{2}\right) + N^2$$

$$= \frac{(N+1)(2N+1)}{6} - N = \frac{(2N-1)(N-1)}{6}$$

where we have used the fact that

$$\sum_{k=1}^{N} k^2 = N\frac{(N+1)(2N+1)}{6}.$$

(A method of proving this last result for readers who are unfamiliar with it is suggested in Problem 6.2.1.)

We note that although $\hat{N}_1$ is not unbiased, $\hat{N}_2 = 2\hat{N}_1 - 1$ is. In fact, since

$$\mathscr{E}_N \hat{N}_1 = \frac{(N+1)}{2}$$

is a linear function of N, we have:

$$\mathscr{E}_N(\hat{N}_2) = \mathscr{E}_N(2\hat{N}_1 - 1) = 2\mathscr{E}_N\hat{N}_1 - 1 = N.$$

Moreover,

$$\text{MSE}_N(\hat{N}_2) = \text{var}_N(2\hat{N}_1 - 1) = 4\,\text{var}(\hat{N}_1) \qquad [(3.7.6)]$$

$$= 4\{\text{MSE}_N(\hat{N}_1) - [b(\hat{N}_1, N)]^2\} \qquad [\text{from (1)}]$$

$$= \frac{4(2N-1)(N-1)}{6} - 4 \cdot \frac{(N-1)^2}{4} = \frac{(N^2-1)}{3} > \text{MSE}_N(\hat{N}_1).$$

Suppose, now, that we have n observations $X_1, X_2, \ldots, X_n$ on X. How do we estimate N? Since $\mathscr{E}X = (N+1)/2$, and $\overline{X} = \sum_{i=1}^{n} X_i/n$, it is plausible that

$$\mathscr{E}_N \overline{X} = \mathscr{E}_N X = \frac{N+1}{2}$$

so that $\mathscr{E}_N(2\overline{X} - 1) = N$. Then $\hat{N}_3 = 2\overline{X} - 1$ will be an unbiased estimate of N. Another estimate of N that we can use is $\hat{N}_4 = \max(X_1, X_2, \ldots, X_n)$. Since $\hat{N}_4 \leq N$, we note that $\hat{N}_4$ underestimates N and is a biased estimate of N. We consider the performance of these and some other estimates of N in Examples 6.2.1 and 10.4.3. $\qquad \square$

Example 4. Estimating the Average Number of Colds in a Year. Let X denote the number of colds that a person has in a year. Suppose the distribution of X can be adequately represented by the probability function

$$P(X = x) = e^{-\lambda}\frac{\lambda^x}{x!}, \, x = 0, 1, 2, \ldots$$

where $\lambda > 0$. Then $\Theta = \{\lambda > 0\}$ and we wish to estimate λ. Let us compute $\mathscr{E}X$. We have from Example 3.7.2 with $t = 1$ that $\mathscr{E}X = \lambda$. It follows that X is an unbiased estimate of λ. We leave the reader to show that the precision of X as an estimate of λ as measured by its variance is given by

$$\text{var}(X) = \mathscr{E}X^2 - (\mathscr{E}X)^2 = \lambda.$$

We note that the estimate X takes the value 0 (which is *not* in Θ) with probability $e^{-\lambda} > 0$.

Suppose we take a random sample of size n. Let X_i be the number of colds that the ith person in the sample has in a year, $i = 1, 2, \ldots, n$. How do we estimate λ? Since

$\lambda = \mathscr{E} X_i$, we have said in Example 2 that $\bar{X}$ is a reasonable estimate of λ. Is $\bar{X}$ unbiased? That is, is $\mathscr{E}_\lambda \bar{X} = \lambda$? Since $\bar{X} = \sum_{i=1}^n X_i/n$ and each X_i has mean λ, it is plausible that

$$\mathscr{E}\bar{X} = \sum_{i=1}^n \mathscr{E}\frac{X_i}{n} = \lambda.$$

This, however, is not a proof. The result is true, as we show in Section 5.3.   □

Example 5.   Estimating θ^2.   Let Y be the tensile strength of a cable of diameter X mm. Suppose Y is proportional to X^2. That is, $Y = cX^2$ for some constant c. Let X have density function given by

$$f(x) = 1/\theta, \quad 5 < x < 5 + \theta, \text{ and zero elsewhere.}$$

What is an unbiased estimate of θ^2?

We note that if X is observed, then

$$\mathscr{E} X = \int_5^{5+\theta} x\frac{1}{\theta}\,dx = 5 + \frac{\theta}{2},$$

$$\mathscr{E} X^2 = \int_5^{5+\theta} x^2\frac{1}{\theta}\,dx = \frac{\theta^2}{3} + 5\theta + 25.$$

It follows that

$$\mathscr{E}[2(X - 5)] = \theta, \quad \text{and} \quad \mathscr{E}\{3(X^2 - 5\theta - 25)\} = \theta^2$$

so that

$$\mathscr{E}\{3(X^2 - 25 - 10(X - 5))\} = \theta^2.$$

Hence:

$$T(X) = 3(X^2 - 10X + 25) = 3(X - 5)^2$$

is an unbiased estimate of θ^2.

On the other hand, if Y is observed, then we can rewrite T in terms of Y as

$$T_1(Y) = 3\left(\frac{Y}{\sqrt{c}} - 5\right)^2$$

which is unbiased for θ^2.   □

Remark 5.   In Examples 3, 4, and 5 we considered only parametric families and usually samples of size 1. This is because we have not yet defined expectations in the case of random vectors. We will show later, for example, that $\bar{X}$ is always an unbiased estimate of μ (Example 2) and that its variance gets smaller as n gets larger (which is to be expected).

Remark 6. Sometimes Θ is an open interval but the estimate of θ takes values in the closure of Θ (see Example 4). This awkward situation can sometimes be avoided by taking the parameter set to be the closure of Θ.

Often it is desirable to estimate a parameter θ to within some prespecified deviation, ε. This cannot be done with certainty because $T(X_1, \ldots, X_n)$ is a random variable. However, one can compute

$$(4) \qquad\qquad P_\theta\{|T(X_1, \ldots, X_n) - \theta| \leq \varepsilon\}$$

as a function of θ, ε, and n, and use it as a measure of closeness of T to θ.

It is intuitively appealing to desire that as the sample size increases, the bias decreases. This need not, however, be the case unless we also demand that the distribution of T tends to concentrate near θ with increasing sample size. This property is called *consistency*: an estimate T is said to be *consistent for θ* if for every $\varepsilon > 0$

$$(5) \qquad\qquad P_\theta(|T(X_1, \ldots, X_n) - \theta| > \varepsilon) \to 0$$

as $n \to \infty$. We emphasize that consistency is a property of a sequence of estimates and is a large sample property. We note that in view of a simple extension of Chebyshev's inequality (Problem 3.7.22) as $n \to \infty$ we have

$$P_\theta(|T(X_1, \ldots, X_n) - \theta| > \varepsilon) \leq \frac{\mathrm{MSE}_\theta(T)}{\varepsilon^2} \to 0$$

provided $\mathrm{MSE}_\theta(T) \to 0$ as $n \to \infty$. Some further aspects of consistency are explored in Section 9.5. We content ourselves here with an example that demonstrates the amount of work we still need to do to use these concepts with ease.

> **Example 6. Estimating Average Length of Life of a Fuse.** Let X be the length of life (in hours) of a fuse manufactured by a certain process. Suppose the distribution of X is given by the density function
>
> $$f(x) = (1/\beta)\exp\left(-\frac{x}{\beta}\right), \qquad x > 0 \text{ and zero elsewhere.}$$
>
> Then $\mathscr{E}X = \beta$ (Example 3.7.5) and X is an unbiased estimate of β, the average life length. Suppose we take a random sample $X_1, X_2, \ldots, X_n$ of size n. Then the sample mean $\overline{X} = \sum_{i=1}^{n} X_i/n$ is a natural choice for an estimate of β. We will show that $\overline{X}$ is in fact unbiased for β (Section 5.3). Is $\overline{X}$ consistent? That is, does
>
> $$P(|\overline{X} - \beta| > \varepsilon) \to 0$$
>
> as $n \to \infty$ for every $\varepsilon > 0$? In order to check this, we need to compute $P(|\overline{X} - \beta| > \varepsilon)$. This means that we need to compute the distribution of $\overline{X}$. This is done in Chapter 8. Alternatively, we could use Chebyshev's inequality and have
>
> $$P(|\overline{X} - \beta| > \varepsilon) \leq \mathrm{var}(\overline{X})/\varepsilon^2.$$

If $\mathrm{var}(\overline{X}) \to 0$ (and it does) as $n \to \infty$ we are through. The computation of $\mathrm{var}(\overline{X})$ is done in Section 5.3, and a general result concerning the consistency of $\overline{X}$ for μ is proved in Section 5.4.

What about some other estimates of β? Let $X_{(1)} = \min(X_1, X_2, \ldots, X_n)$. That is, $X_{(1)}(\omega) = \min\{X_1(\omega), X_2(\omega), \ldots, X_n(\omega)\}$ for $\omega \in \Omega$. Then $X_{(1)}$ is a random variable (Section 8.5) and we will see that $\mathscr{E} X_{(1)} = \beta/n$ so that $nX_{(1)}$ is also an unbiased estimate of β. Is $nX_{(1)}$ consistent? We again need the distribution of $nX_{(1)}$. We will show (Section 8.5) that $nX_{(1)}$ has the same distribution as X, that is, $nX_{(1)}$ has density f given above, so that

$$P\big(|nX_{(1)} - \beta| > \varepsilon\big) = \int_0^{\beta-\varepsilon} \frac{1}{\beta} e^{-x/\beta}\, dx + \int_{\beta+\varepsilon}^{\infty} \frac{1}{\beta} e^{-x/\beta}\, dx$$

$$= 1 - \exp(-(\beta - \varepsilon)/\beta) + \exp(-(\beta + \varepsilon)/\beta).$$

Hence $nX_{(1)}$ is not consistent for β.   □

Example 6 demonstrates the types of problems a statistician encounters. These include judgmental as well as mathematical problems. For example, which of the two estimates $\overline{X}$ and $nX_{(1)}$ is preferable? Since both estimates are unbiased, we could use variance as a measure of their precision. We will see that $\mathrm{var}(\overline{X}) < \mathrm{var}(nX_{(1)})$ so that $\overline{X}$ is preferred if smaller variance is desirable. Both $\overline{X}$ and $nX_{(1)}$ are easily computed but in order to compute $\overline{X}$ we need the whole sample, whereas $nX_{(1)}$ can be computed the moment the first of the fuses under test fails. If time is of the essence, then $nX_{(1)}$ will have to merit serious consideration.

We have thus far given some basic definitions and some properties that we may want our estimate to possess. We have not considered any methods of estimation. In Chapter 10 we consider some methods of estimation and properties of estimates they produce. Until then, what we have learned so far will be sufficient.

We emphasize that we will concentrate mainly on some time-tested common methods of estimation. In any particular situation, the choice of an estimate in particular and of a statistical procedure in general is governed by both statistical factors (such as validity of the procedure, reliability of the sampling procedure used, and so on) and nonstatistical factors (such as simplicity, practical and economic implications, and so on).

The problems at the end of this section deal primarily with the case of one observation even though in practice one usually has a sample of size $n > 1$. We do this to avoid distributional problems of the type mentioned in Example 6. The case $n > 1$ is considered in the following chapters.

Confidence Interval Estimation

It should be clear by now that a point estimate is meaningless without some measure of its precision. One such measure we suggested above is to use the probability, $P(|T - \theta| \le \varepsilon)$, that T does not deviate from θ by more than a prespecified ε. Since $|T - \theta|$ is the *error* in estimation, ε is an upper bound for error. Rather than specify the error in advance, we may wish to specify a lower bound for the probability: What is the error ε such that

(6) $P_\theta(|T - \theta| \leq \varepsilon) \geq 1 - \alpha$ for all $\theta \in \Theta$?

Rewriting (6) as

(7) $P_\theta(T - \varepsilon \leq \theta \leq T + \varepsilon) \geq 1 - \alpha$ for $\theta \in \Theta$,

we see that the question may be posed equivalently as follows. What is the interval that contains the true θ with a probability of at least $1 - \alpha$? From (7) we see that $[T - \varepsilon, T + \varepsilon]$ is such an interval where ε is computed from (6). An interval such as $[T - \varepsilon, T + \varepsilon]$ is usually called a $1 - \alpha$ *level confidence interval* for θ. The assertion is that the range of values $[T - \varepsilon, T + \varepsilon]$ contains the true (but unknown) θ with probability at least $1 - \alpha$. The quantity $1 - \alpha$ is a measure of the degree of credibility of the interval.

Let $[T - \varepsilon, T + \varepsilon]$ be a $1 - \alpha$ level confidence interval for θ so that (7) holds. Then any interval $I(T)$ that contains $[T - \varepsilon, T + \varepsilon]$ is also a $1 - \alpha$ level confidence interval for θ. Or, equivalently, any $1 - \alpha$ level confidence interval is also a $1 - \alpha'$ level confidence interval for $\alpha' > \alpha$. The length of a confidence interval is used as a measure of its precision. In general, we want our confidence intervals to be as small as possible. Since the confidence interval is our estimate of the interval in which θ is located, the smaller the interval the more precise the information about θ. These considerations motivate the following definition.

DEFINITION 6. Let $X_1, X_2, \ldots, X_n$ be a random sample from a distribution function $F \in \mathfrak{F}$ and let $\theta = \theta(F)$ be a numerical parameter. Let $l(X_1, \ldots, X_n)$ and $u(X_1, \ldots, X_n)$ be two statistics such that for all $F \in F$

$$P_F(l(X_1, \ldots, X_n) \leq \theta \leq u(X_1, \ldots, X_n)) = 1 - \alpha, \quad 0 < \alpha < 1.$$

Then the random interval $[l(X_1, \ldots, X_n), u(X_1, \ldots, X_n)]$ is known as a minimum level $1 - \alpha$ confidence interval for θ. The quantity $1 - \alpha$ is called the confidence level. If $P_F(\theta \leq u(X_1, \ldots, X_n)) = 1 - \alpha$ we call u a minimum level $1 - \alpha$ upper confidence bound, and if $P_F(l(X_1, \ldots, X_n) \leq \theta) = 1 - \alpha$, we call l a minimum level $1 - \alpha$ lower confidence bound for θ.

When T is a discrete type random variable, then in Definition 6 not all levels $1 - \alpha$ are attainable (see Example 11 below). In such cases, we replace the equality in Definition 6 with the inequality " $\geq$ ".

In the following discussions we drop the adjective minimum from the phrase "minimum level $1 - \alpha$ confidence interval" for θ whenever the statistic T has a continuous type distribution.

In many problems an interval $[l, u]$ is preferable to a point estimate for θ. A point estimate gives a single numerical value for θ for each sample that is either right or wrong, but we will never know since θ is unknown. An interval estimate, on the other hand gives a range of possible values of θ (for each sample). The probability $1 - \alpha$ is not associated with any specific confidence interval.

It is asserted that about $100(1 - \alpha)$ percent of statements of the form "$[l(X_1, \ldots, X_n), u(X_1, \ldots, X_n)] \ni \theta$" should be correct. That is, a proportion $1 - \alpha$ of all such intervals will contain the true θ. It is in this sense that we say we are $100(1 - \alpha)$ percent confident that any specific interval will contain the true θ.

It is clear that in any given case there will be more than one (often infinitely many) confidence intervals at level $1 - \alpha$. Moreover, the confidence interval depends on the choice of the statistic T. As mentioned above, the length of a confidence interval is taken to be a measure of its precision.

We choose, if possible, a confidence interval that has the least length among all minimum level $1 - \alpha$ confidence intervals for θ. This is a very difficult problem, however. Instead, we concentrate on all confidence intervals based on some statistic T that have minimum level $1 - \alpha$ and choose one which has the least length. Such an interval, if it exists, is called *tight*. If the distribution of T is symmetric about θ, it is easy to see (Problem 12) that the length of the confidence interval based on T is minimized by choosing an equal tails confidence interval. That is, we choose ε in

$$P(|T - \theta| \le \varepsilon) = 1 - \alpha$$

by taking

$$P(T - \theta \ge \varepsilon) = \frac{\alpha}{2} \quad \text{and} \quad P(T - \theta \le -\varepsilon) = \frac{\alpha}{2}$$

(see Figure 2). We will often choose equal tails confidence intervals for convenience even though the distribution of T may not be symmetric.

Example 7. Upper Confidence Bound for Tensile Strength of Copper Wire. The manufacturer of a copper wire estimates that the tensile strength X, in kilograms, of copper wire he produces has density function

$$f(x) = (x/\theta^2)e^{-x/\theta}, \qquad x > 0 \text{ and zero otherwise}; \quad \theta > 0.$$

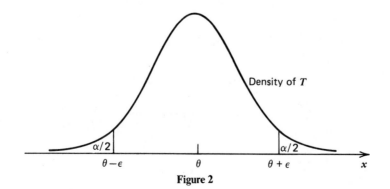

Figure 2

The average tensile strength is given by

$$\mathscr{E}X = \int_0^\infty \frac{x^2}{\theta^2} e^{-x/\theta}\, dx = 2\theta.$$

A potential customer can use the wire only if the average tensile strength is not less than θ_0. The customer decides to put a random sample of n pieces of copper wire produced by the manufacturer to a test. Let $x_1, x_2, \ldots, x_n$ be the tensile strengths of the n pieces in the sample. On the basis of this sample the customer would like to construct a lower confidence bound $l(x_1, \ldots, x_n)$ for the average tensile strength 2θ at level $1 - \alpha$. If $l \geq \theta_0$ then the customer should buy copper wire from this manufacturer. The problem, therefore, is to find a statistic $l(X_1, \ldots, X_n)$ such that

$$P_\theta\big(\mathscr{E}X \geq l(X_1, \ldots, X_n)\big) = 1 - \alpha \qquad \text{for all } \theta > 0.$$

Since $\overline{X}$ is a good estimate of $\mathscr{E}X$, we take $T(X_1, \ldots, X_n) = \overline{X}$ and find $c > 0$ such that

$$P_\theta(\overline{X} \leq c) = 1 - \alpha \qquad \text{for all } \theta.$$

We have

$$1 - \alpha = P_\theta(\overline{X} \leq c) = \int \cdots \int\limits_{x_1 + \cdots + x_n \leq nc} \left(\prod_{i=1}^n \frac{x_i}{\theta^2} \right) \exp\left(-\sum_{i=1}^n \frac{x_i}{\theta} \right) \prod_{i=1}^n dx_i.$$

Having found c as a function of θ we hope that we can invert the inequality $\overline{X} \leq c$ to yield a lower bound for 2θ.

Let us illustrate the computation for the $n = 2$ case. Anticipating a change of variable in the multiple integral, let us write $c = k\theta$ and find $k > 0$ from

$$1 - \alpha = P_\theta\left(\frac{\overline{X}}{\theta} \leq k \right) = \iint\limits_{x_1 + x_2 \leq 2k\theta} \frac{x_1 x_2}{\theta^4} e^{-(x_1 + x_2)/\theta}\, dx_1\, dx_2$$

$$= \int_{x_2=0}^{2k\theta} \left(\int_{x_1=0}^{2k\theta - x_2} \frac{x_1}{\theta^2} e^{-x_1/\theta}\, dx_1 \right) \frac{x_2}{\theta^2} e^{-x_2/\theta}\, dx_2$$

$$= \int_0^{2k\theta} \left(\int_0^{2k - x_2/\theta} y_1 e^{-y_1}\, dy_1 \right) \frac{x_2}{\theta^2} e^{-x_2/\theta}\, dx_2 \qquad \left(y_1 = \frac{x_1}{\theta} \right)$$

$$= \int_0^{2k\theta} \left[-\left(2k - \frac{x_2}{\theta} \right) e^{-2k + x_2/\theta} + 1 - e^{-2k + x_2/\theta} \right] \frac{x_2}{\theta^2} e^{-x_2/\theta}\, dx_2$$

$$= \int_0^{2k\theta} \frac{x_2}{\theta^2} \left[-\left(2k - \frac{x_2}{\theta} \right) e^{-2k} + e^{-x_2/\theta} - e^{-2k} \right] dx_2$$

$$= \int_0^{2k} y_2 \left[-(2k - y_2) e^{-2k} + e^{-y_2} - e^{-2k} \right] dy_2 \qquad \left(y_2 = \frac{x_2}{\theta} \right)$$

$$= \left\{ -2k \frac{(2k)^2}{2} + \frac{(2k)^3}{3} - \frac{(2k)^2}{2} \right\} e^{-2k} + \left\{ -2k e^{-2k} + 1 - e^{-2k} \right\}$$

$$= \left\{ -\frac{4k^3}{3} - 2k^2 - 2k - 1 \right\} e^{-2k} + 1.$$

Thus k is a solution of

$$\alpha = e^{-2k}(4k^3 + 6k^2 + 6k + 3)/3.$$

We note that $g(k) = e^{-2k}(4k^3 + 6k^2 + 6k + 3)/3$ is strictly decreasing on $(0, \infty)$ since $g'(k) < 0$ for all k. Hence $g(k) = \alpha$ has a unique positive solution. Given α, the equation can be solved numerically. In the following table we give the solutions for some selected values of α.

α	.01	.02	.05	.10
k	5.02	4.54	3.88	3.34

For a given α let $k_1 > 0$ be a solution of $g(k) = \alpha$. Then for all $\theta > 0$

$$1 - \alpha = P_\theta\left(\overline{X} \le k_1\theta\right) = P_\theta\left(2\theta \ge \frac{2\overline{X}}{k_1}\right).$$

It follows that the minimum level $1 - \alpha$ lower confidence bound for $\mathscr{E}X = 2\theta$ based on $\overline{X}$ is given by $l(X_1, X_2) = (X_1 + X_2)/k_1$. $\square$

Example 8.   Confidence Interval for θ.   Suppose X has density function

$$f(x) = \frac{1}{\theta}, \qquad 0 < x < \theta, \text{ and zero elsewhere.}$$

Let us find a $1 - \alpha$ level confidence interval for θ based on a single observation. Given α, $0 < \alpha < 1$, our object is to find two random variables $l(X)$ and $u(X)$ such that for all $\theta > 0$

$$P(l(X) \le \theta \le u(X)) = 1 - \alpha.$$

We note that f is symmetric about the mean $\theta/2$. Hence we can minimize the length of the confidence interval by choosing $c = c(\alpha)$ such that

(8) $$P\left(-c \le X - \frac{\theta}{2} \le c\right) = 1 - \alpha.$$

In fact, $c = \theta(1 - \alpha)/2$. (See Figure 3.) We can rewrite (8) as follows:

$$1 - \alpha = P\left(-c \le X - \frac{\theta}{2} \le c\right) = P\left(-\theta\frac{1-\alpha}{2} + \frac{\theta}{2} \le X \le \theta\frac{1-\alpha}{2} + \frac{\theta}{2}\right)$$

$$= P\left(\frac{X}{1 - \alpha/2} \le \theta \le \frac{X}{\alpha/2}\right)$$

so that $[X/(1 - \alpha/2), X/(\alpha/2)]$ is a $1 - \alpha$ level confidence interval for θ.

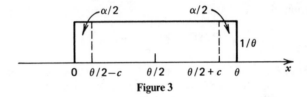

Figure 3

If, for example, the observation is 1.524 and $\alpha = .05$, then the confidence interval for θ is $[1.563, 60.96]$. The assertion is *not* that the interval $[1.563, 60.96]$ contains unknown θ with probability .95. Indeed this interval, being fixed, either contains θ or it does not. Since θ is unknown, we will never know which of these two statements is correct. The assertion is that in sampling from density f, about 95 percent intervals of this type will contain the true θ. In Table 1 we demonstrate the result of taking 50 samples of size 1 from density f where θ is known to be 1. Confidence intervals that do not contain $\theta = 1$ are identified with an asterisk.

We note that 48 out of 50 intervals, or 96 percent, do contain the true value of θ.

□

TABLE 1.   95 Percent Confidence Intervals for θ

Sample	Confidence Interval	Sample	Confidence Interval
.59	[.605, 23.6]	.09	[.092, 3.6]
.96	[.985, 38.4]	.46	[.472, 18.4]
.78	[.800, 31.2]	.56	[.574, 22.4]
.16	[.164, 6.4]	.25	[.256, 10.0]
.87	[.892, 34.8]	.91	[.933, 36.4]
.13	[.133, 5.2]	.26	[.267, 10.4]
.73	[.749, 29.2]	.77	[.790, 30.4]
.41	[.421, 16.4]	.55	[.564, 22.0]
.74	[.759, 29.6]	.72	[.738, 28.8]
.55	[.564, 22.0]	.50	[.513, 20.0]
.72	[.738, 28.8]	.96	[.985, 38.4]
.47	[.482, 18.8]	.47	[.482, 18.8]
.43	[.441, 17.2]	.07	[.072, 2.8]
.31	[.318, 12.4]	.75	[.769, 30.0]
.56	[.574, 22.4]	.02	[.021, 0.8]*
.22	[.226. 8.8]	.36	[.369, 14.4]
.90	[.923, 36.0]	.68	[.697, 27.2]
.78	[.800, 30.2]	.17	[.174, 6.8]
.66	[.677, 26.4]	.52	[.533, 20.8]
.18	[.185, 7.2]	.01	[.010, 0.4]*
.73	[.749, 29.2]	.70	[.718, 28.0]
.43	[.441, 17.2]	.64	[.656, 25.6]
.58	[.595, 23.2]	.61	[.626, 24.4]
.11	[.112, 4.4]	.18	[.185, 7.2]
.16	[.164, 6.4]	.64	[.656, 25.6]

The inequality (6) invites comparison with Chebyshev's inequality. Indeed if $\mathscr{E}_\theta T = \theta$ then

$$P_\theta(|T - \theta| \leq \varepsilon) \geq 1 - \frac{\text{var}_\theta(T)}{\varepsilon^2}.$$

Choosing $\varepsilon = k\sqrt{\text{var}_\theta(T)}$, we get

$$P_\theta\left(|T - \theta| \leq k\sqrt{\text{var}_\theta(T)}\right) \geq 1 - (1/k^2).$$

Letting $1 - \alpha = 1 - 1/k^2$ we get $k = 1/\sqrt{\alpha}$ so that for all $\theta \in \Theta$

$$(9) \qquad P_\theta\left(\theta - \sqrt{\frac{\text{var}_\theta(T)}{\alpha}} \leq T \leq \theta + \sqrt{\frac{\text{var}_\theta(T)}{\alpha}}\right) \geq 1 - \alpha.$$

If we are able to write the event

$$(10) \qquad \left(\theta - \sqrt{\frac{\text{var}_\theta(T)}{\alpha}} \leq T \leq \theta + \sqrt{\frac{\text{var}_\theta(T)}{\alpha}}\right)$$

as

$$(l(T) \leq \theta \leq u(T))$$

then $[l(T), u(T)]$ is a $1 - \alpha$ level confidence interval for θ.

We note, however, that $[l(T), u(T)]$ is not a minimum level $1 - \alpha$ confidence interval for θ. The level is *at least* $1 - \alpha$. Since the Chebyshev's inequality is crude, this method provides a rough and ready but fast confidence interval for the parameter. The length of such a confidence interval will usually be greater than the minimum level $1 - \alpha$ confidence interval. The advantage, however, is that we do not need the underlying distribution. All we need is $\text{var}_\theta(T)$.

Example 9. *Confidence Interval for θ of Example 8 Using Chebyshev's Inequality.* Let X have density

$$f(x) = 1/\theta, \qquad 0 < x < \theta, \text{ and zero elsewhere.}$$

Then $\mathscr{E}X = \theta/2$ and $\text{var}(X) = \theta^2/12$. Choosing $T(X) = 2X$ in (9) so that T is an unbiased estimate of θ we have

$$\text{var}_\theta(T(X)) = \text{var}_\theta(2X) = 4\,\text{var}_\theta(X) = \theta^2/3.$$

Hence (10) reduces to

$$\left(\theta - \theta(3\alpha)^{-1/2} \leq 2X \leq \theta + \theta(3\alpha)^{-1/2}\right).$$

Note, however, that for $0 < \alpha < 1/3$, $1 - (3\alpha)^{-1/2} < 0$ and $1 + (3\alpha)^{-1/2} > 2$. It follows that for $0 < \alpha < 1/3$ Chebyshev's inequality leads to the trivial confidence interval $[X, \infty]$. The length of this confidence interval is infinite, whereas the length of the confidence interval constructed in Example 8 was finite. If, on the other hand, $\alpha > 1/3$ then the confidence interval is given by $[2X/\{1 + (3\alpha)^{-1/2}\}, 2X/\{1 - (3\alpha)^{-1/2}\}]$. We leave the reader to show that the length of this confidence interval is greater than the length of the confidence interval constructed in Example 8.   □

Example 10.   Confidence Interval for Average Life Length.   Let X be the length of life (in hours) of a fuse manufactured by a certain process. Suppose the distribution of X is given by the density

$$f(x) = \frac{1}{\beta} \exp\left(-\frac{x}{\beta}\right), \qquad x > 0, \text{ and zero elsewhere.}$$

Then $\mathscr{E}X = \beta$ and we seek a $1 - \alpha$ level confidence interval for β.

Let $X_1, X_2, \ldots, X_n$ be a random sample of size n from f. We need to find $l(X_1, X_2, \ldots, X_n)$ and $u(X_1, X_2, \ldots, X_n)$ such that

$$P_\beta\big(l(X_1, X_2, \ldots, X_n) \le \beta \le u(X_1, X_2, \ldots, X_n)\big) = 1 - \alpha \qquad \text{for all } \beta > 0$$

where $0 < \alpha < 1$. We have said that $\overline{X}$ is a good estimate of $\mathscr{E}X_1 = \beta$. We therefore choose $T(X_1, X_2, \ldots, X_n) = \overline{X}$. Let us find c_1 and c_2, $0 < c_1 < c_2$, such that for all $\beta > 0$

$$1 - \alpha = P_\beta\big(c_1 \le \overline{X} \le c_2\big) = P_\beta\big(nc_1 \le X_1 + X_2 + \cdots + X_n \le nc_2\big).$$

$$= \int \cdots \int_{nc_1 \le x_1 + \cdots + x_n \le nc_2} \frac{1}{\beta^n} \exp\left\{-\sum_{i=1}^n \frac{x_i}{\beta}\right\} \prod_{i=1}^n dx_i.$$

We first note that this last equation has infinitely many solutions, each pair of solutions leading to a minimum level $1 - \alpha$ confidence interval. Next we note that the density f is not symmetric about β or any other point. We therefore assign equal probability to each tail and choose c_1 and c_2 from the equations

$$\alpha/2 = \int \cdots \int_{0 < x_1 + \cdots + x_n < nc_1} \frac{1}{\beta^n} \exp\left(-\sum_{i=1}^n \frac{x_i}{\beta}\right) \prod_{i=1}^n dx_i$$

and

$$\alpha/2 = \int \cdots \int_{x_1 + \cdots + x_n > nc_2} \frac{1}{\beta^n} \exp\left(-\sum_{i=1}^n \frac{x_i}{\beta}\right) \prod_{i=1}^n dx_i.$$

We now have a complicated problem of multiple integration. In Chapter 8 we show how to reduce these integrals to univariate integrals. We can then find c_1 and c_2 and invert the inequalities $c_1 \le \overline{X} \le c_2$ to yield a $1 - \alpha$ level confidence interval for β.

Let us show how this can be done in the case when we have two observations. We use a simple trick here that is equivalent to a change of variables in the multiple integral. Instead of finding c_1, c_2 such that

$$\frac{\alpha}{2} = P(\overline{X} < c_1) \quad \text{and} \quad \frac{\alpha}{2} = P(\overline{X} > c_2)$$

let us find k_1, k_2 such that

$$\frac{\alpha}{2} = P\left(\frac{\overline{X}}{\beta} < k_1\right) \quad \text{and} \quad \frac{\alpha}{2} = P\left(\frac{\overline{X}}{\beta} > k_2\right).$$

In view of Example 3.8.5.

$$P\left(\frac{\overline{X}}{\beta} < k_1\right) = P(X_1 + X_2 < 2k_1\beta)$$

$$= \int_{0 < x_1 + x_2 < 2k_1\beta} (1/\beta^2) \exp\left[-\frac{x_1 + x_2}{\beta}\right] dx_1 \, dx_2$$

$$= 1 - (1 + 2k_1)e^{-2k_1}$$

and

$$P\left(\frac{\overline{X}}{\beta} > k_2\right) = P(X_1 + X_2 > 2k_2\beta) = (1 + 2k_2)e^{-2k_2}.$$

It follows that k_1, k_2 are solutions of the following equations

$$1 - \frac{\alpha}{2} = (1 + 2k_1)e^{-2k_1} \quad \text{and} \quad \frac{\alpha}{2} = (1 + 2k_2)e^{-2k_2}.$$

Given α, these equations can be solved numerically giving values of k_1, k_2. Note that the function $g(x) = (1 + 2x)e^{-2x}$ for $x > 0$ is strictly decreasing since $g'(x) < 0$ for all $x > 0$. It follows that the equations

$$1 - \frac{\alpha}{2} = g(k_1) \quad \text{and} \quad \frac{\alpha}{2} = g(k_2)$$

have unique solutions in $(0, \infty)$. In the following table we have numerically computed k_1 and k_2 for some selected values of α.

α	.01	.02	.05	.10
k_1	0.0517	0.0743	0.1211	0.1777
k_2	3.7151	3.3192	2.7858	2.3719

Armed with the values of k_1 and k_2 we have for all $\beta > 0$

$$1 - \alpha = P_\beta(\beta k_1 \leq \overline{X} \leq \beta k_2)$$

$$= P_\beta\left(\frac{\overline{X}}{k_2} \leq \beta \leq \frac{\overline{X}}{k_1}\right)$$

so that $[\overline{X}/k_2, \overline{X}/k_1]$ is a $1 - \alpha$ level confidence interval for β.

For the case when we have exactly one observation, the problem is considerably simpler. In that case we find c_1 and c_2, $0 < c_1 < c_2$, such that

$$1 - \alpha = P_\beta(c_1 \le X \le c_2) = \int_{c_1}^{c_2} \frac{1}{\beta} e^{-x/\beta} \, dx = e^{-c_1/\beta} - e^{-c_2/\beta}.$$

Once again we choose c_1 and c_2 by assigning equal probability to each tail. Thus we choose c_1 and c_2 from

$$\frac{\alpha}{2} = \int_0^{c_1} \frac{1}{\beta} e^{-x/\beta} \, dx \quad \text{and} \quad \frac{\alpha}{2} = \int_{c_2}^{\infty} \frac{1}{\beta} e^{-x/\beta} \, dx$$

so that

$$c_1 = -\beta \ln\left(1 - \frac{\alpha}{2}\right) \quad \text{and} \quad c_2 = -\beta \ln \frac{\alpha}{2}.$$

It follows that $[-X/\ln(\alpha/2), -X/\ln(1 - \alpha/2)]$ is a $1 - \alpha$ level confidence interval for β.

In particular, if $\alpha = .10$, then $\ln(\alpha/2) = -2.996$, and $\ln(1 - \alpha/2) = -.051$ so that a 90 percent confidence interval is given by $[X/2.996, X/.051]$. Intervals of this form will include the true (unknown) value of β for 90 percent of the values of X that we would obtain in repeated random sampling from the density f. If, in particular, $X = 207.5$ is observed, then the confidence interval takes the value $[69.26, 4068.63]$. $\quad\square$

Example 11. Confidence Interval for Probability of Heads. Consider a possibly biased coin with $p = P(\text{heads})$, $0 < p < 1$. Rather than estimate p on the basis of n tosses of the coin, suppose we wish to estimate the interval in which p is located with probability $1 - \alpha$. For $i = 1, 2, \ldots, n$ let $X_i = 1$ if the ith trial results in a head and zero otherwise. Then $\sum_{i=1}^n X_i/n$ is the proportion of successes in n independent tosses of the coin. We have said that $\overline{X}$ is a good estimate of $\mathscr{E}X_i = p$. Therefore, we take $T(X_1, \ldots, X_n) = \overline{X}$. Our object is to find two statistics l and u such that

$$P_p(l(X_1, \ldots, X_n) \le p \le u(X_1, \ldots, X_n)) = 1 - \alpha, \text{ for all } p \in (0,1).$$

Since $(X_1, \ldots, X_n)$ assumes only 2^n values, both l and u assume at most 2^n values each. Clearly not all levels $1 - \alpha$ are attainable. That is, given α, $0 < \alpha < 1$, there may be no minimum level $1 - \alpha$ confidence interval for p. If n is large, it is possible to use the normal approximation to the distribution of $T(X_1, \ldots, X_n) = \overline{X}$ to obtain confidence interval with approximate level $1 - \alpha$. This is done in Section 9.6.

For small n, the procedure for finding l and u for an observed value of $\overline{X}$ is a fairly involved numerical problem. (As noted earlier, when T is a discrete type random variable we seek confidence intervals of *level at least* $1 - \alpha$.) There are two choices. One can use Chebyshev's inequality (Section 6.3) or use extensive tables and graphs that have been developed for this purpose. These tables are more accurate. For example, Table 2 below gives the equal tails confidence interval for p at level at least .95 for $n = 10$.

TABLE 2.   Equal Tails Confidence Interval for p at Level $\geq$ .95, $n = 10$

$\bar{x}$	0	.1	.2	.3	.4	.5	.6	.7	.8	.9	1.0
l	0	.002	.025	.067	.122	.187	.262	.348	.444	.555	.589
u	.311	.445	.556	.652	.738	.813	.878	.933	.975	.998	1.000

Table A-23 in *Experimental Statistics* by M. G. Natrella (National bureau of Standards Handbook No. 91, 1963) gives confidence intervals for p for $n \leq 30$.

Suppose, for example, that in 10 independent tosses of the coin in question, 8 heads are observed. Then $\bar{x} = 8/10 = .8$, and from Table 2 we see that a 95 percent confidence interval for p is given by $[.444, .975]$. $\square$

Problems for Section 4.3

1. In the following problems a random sample $X_1, X_2, \ldots, X_n$ of size n is taken from the given distribution. Give some reasonable estimates of the parameters specified in each case.

    (i) $f(x) = 1/\theta, 0 \leq x \leq \theta$, and zero elsewhere; $\theta \in (0, \infty)$.

    (ii) $P(X_i = 1) = \theta = 1 - P(X_i = 0)$, and zero elsewhere; $0 \leq \theta \leq 1$.

    (iii) $f(x) = 1/(2\theta)$, $-\theta \leq x \leq \theta$, and zero elsewhere; $\theta > 0$.

    (iv) F is a continuous distribution function. Estimate quantile of order p, $0 < p < 1$.

    (v) $P(X = x) = 1/N$, $x = 1, 2, \ldots, N$, and zero elsewhere; $N \geq 1$ is an integer. Estimate N.

    (vi) F is any distribution function on the real line with finite second moment. Estimate $m_2(F) = \mathscr{E}_F X^2$. Also, $\sigma_F^2 = \text{var}_F(X)$.

    (vii) $f(x) = (1/\beta)\exp(-x/\beta)$, $x > 0$, and zero elsewhere; $\theta = P_\beta(X > t) = \exp(-t/\beta)$. Estimate θ.

    (viii) $P(X_i = 1) = \theta = 1 - P(X_i = 0)$, and zero elsewhere; $0 \leq \theta \leq 1$. Estimate θ^2, $P(X = 1) - P(X = 0)$, $\theta - \theta^2$.

2. In the following cases find an unbiased estimate of θ based on one observation.

    (i) $f(x|\theta) = 2(\theta - x)/\theta^2$, $0 \leq x \leq \theta$, and zero elsewhere; $\theta > 0$.

    (ii) $f(x|\theta) = (1/2)\exp\{-|x - \theta|\}$, $x \in \mathbb{R}$; $\theta \in \mathbb{R}$.

    (iii) $f(x|\theta) = 1/(\alpha\theta + \beta)$, $0 < x < \alpha\theta + \beta$, and zero elsewhere; α, β known, $\alpha \neq 0$, $\alpha\theta + \beta > 0$.

    (iv) $P(X = 0) = 1 - P(X = 1) = 1 - \theta$, and zero elsewhere; $0 \leq \theta \leq 1$.

    (v) $f(x|\theta) = \exp\{-(x - \theta)\}$, $x > \theta$, and zero elsewhere.

3. For the density $f(x) = 2(\theta - x)/\theta^2$, $0 \leq x \leq \theta$, and zero elsewhere, consider the biased estimate X for θ. Find the relative efficiency of the estimate X of θ relative to the unbiased estimate found in Problem 2(i).

4. Let $k > 0$ be a known constant. Find an unbiased estimate of θ^k based on a sample of size one from the density $f(x) = 1/\theta$, $0 \leq x \leq \theta$, and zero elsewhere; $\theta > 0$.

5. Let X be the length of life of a certain brand of fuse manufactured by a certain company. Suppose X has density function $f(x) = (1/\beta)\exp(-x/\beta)$, $x > 0$, and zero elsewhere. Find an unbiased estimate of $3\beta + 4\beta^2$.

6. Let X have density function $f(x) = 1$, $\theta < x < \theta + 1$, and zero elsewhere.
   (i)   Show that X is a biased estimate of θ.
   (ii) Find the bias and the mean square error of the estimate X.
   (iii) Find an unbiased estimate of θ.
   (iv) Find the relative efficiency of the unbiased estimate computed in (iii) relative to the biased estimate X.

7. The content X of a metal in an alloy has density function given by $f(x) = 2x/\theta^2$, $0 \le x \le \theta$, and zero elsewhere. Find an unbiased estimate of the average profit $15 + 3\theta$.

8. Let X have density function

$$f(x) = \begin{cases} 1/2, & 0 < x < 1, \\ \dfrac{1}{2\beta}\exp\left[-\dfrac{x-1}{\beta}\right], & x > 1, \\ 0 & \text{elsewhere.} \end{cases}$$

   (i)   Is X an unbiased estimate for β? Find its bias and the mean square error.
   (ii) Find an unbiased estimate for β and its variance.
   (iii) Compute the relative efficiency of the estimate in (i) relative to that in (ii).

9. Find a $(1 - \alpha)$ level equal tails confidence interval for the parameter θ in each of the following cases. (Take a sample of one observation.)
   (i) $f(x) = \theta/x^2$, $0 < \theta \le x < \infty$, and zero elsewhere.
   (ii) $f(x) = \exp\{-(x - \theta)\}$, $x > \theta$, and zero elsewhere.
   (iii) $f(x) = \theta x^{\theta - 1}$, $0 < x < 1$, and zero elsewhere.
   (iv) $f(x) = (1/\pi)[1 + (x - \theta)^2]^{-1}$, $x \in \mathbb{R}$.
   (v) $f(x) = (1/2)\exp\{-|x - \theta|\}$, $x \in \mathbb{R}$.

10. In Examples 6 and 10 the length of life of a fuse has density function

$$f(x) = \frac{1}{\beta}\exp\left(-\frac{x}{\beta}\right), \qquad x > 0, \text{ and zero elsewhere.}$$

   Use Chebyshev's inequality to find a confidence interval for β at level at least $1 - \alpha$.

11. In Problem 9 (excluding part iii), apply Chebyshev's inequality (wherever applicable) to find confidence intervals for the parameter θ.

12*. Suppose the distribution of T is continuous and symmetric about θ and

$$P_\theta(T - a \le \theta \le T + b) = 1 - \alpha$$

   for all θ. Show that the length $b + a$ of the $1 - \alpha$ level confidence interval $[T - a, T + b]$ for θ is minimized for $a = b$.
   [Hint: Minimize $L(a, b) = a + b$ subject to $\int_{\theta - b}^{\theta + a} f(t)\, dt = 1 - \alpha$.]

13*. Two points X_1, X_2 are randomly and independently picked from the interval $[0, \theta]$ in order to estimate θ by an interval with confidence level $1 - \alpha$. If $\bar{X}$ is used as an estimate of $\mathscr{E}X = \theta/2$, find the confidence interval.

14. Let X be a single observation from the density

$$f(x) = \begin{cases} x/\theta^2, & 0 < x < \theta \\ (2\theta - x)/\theta^2, & \theta \le x < 2\theta, \theta > 0. \end{cases}$$

(i) Show that for $\alpha < 1/2$, $X/\sqrt{2\alpha}$ is a minimum level $1 - \alpha$ upper confidence bound for θ.

(ii) Show that a minimum level $1 - \alpha$ confidence interval for θ is given by $[X/(2 - \sqrt{\alpha}), X/\sqrt{\alpha}]$.

15. In Problem 3, show that a lower $1 - \alpha$ level confidence bound for θ is given by $X/(1 - \sqrt{\alpha})$.

16. Let X be an observation from density function

$$f(x) = \begin{cases} (\theta + x)/\theta^2, & -\theta \le x \le 0 \\ (\theta - x)/\theta^2, & 0 \le x \le \theta, \end{cases} \quad \theta > 0.$$

Show that $|X|/(1 - \sqrt{\alpha})$ is a minimum level $1 - \alpha$ lower confidence bound for θ.

4.4 TESTING HYPOTHESES

A stochastic model is based on certain assumptions. When we say that the observable quantity X associated with the random phenomenon under study can be modeled by a density function $f(x, \theta)$, this is a theoretical assumption about the phenomenon. Any such assumption is called a *statistical hypothesis*. The reader has already encountered the terms "random variables are independent," "random variables are identically distributed," "F is a symmetric distribution function," "the coin is fair," and so on. All these assumptions are examples of statistical hypotheses.

Let $(X_1, \ldots, X_n)$ be a random vector with joint distribution function $F_\theta, \theta \in \Theta$. The family $\mathfrak{F} = \{F_\theta: \theta \in \Theta\}$ may be parametric (in which case θ is a finite dimensional numerical vector) or nonparametric (in which case $\mathfrak{F}$ cannot be characterized by a finite dimensional vector of scalers). The problem of statistical inference that we have considered so far is that of point (or interval) estimation of some numerical parameter(s) of F_θ. We now consider the problem of hypothesis testing where we are not asked to make a guess at the numerical value of a parameter but rather to settle some statement about a parameter such as "the median is nonnegative" or "the variance is $\ge \sigma_0^2$."

Formally, let $\Theta_0 \subseteq \Theta$ and $\Theta_1 = \Theta - \Theta_0$ be specified and let $\mathfrak{F}_0 = \{F_\theta: \theta \in \Theta_0\} \subseteq \mathfrak{F}$ and $\mathfrak{F}_1 = \mathfrak{F} - \mathfrak{F}_0$. The problem we wish to consider now may be stated as follows: Given a data vector $(x_1, x_2, \ldots, x_n)$, is the true distribution of $(X_1, X_2, \ldots, X_n)$ from $\mathfrak{F}_0$ or from $\mathfrak{F}_1$? Before we give formal definitions of the concepts involved, we consider some examples.

Example 1. Is a Given Coin Fair? Before engaging in a coin tossing game against an adversary (who supplies the coin for the game) we should like to be reasonably sure that the coin is fair. Of course, the correct thing to do is to assume that the coin is fair until there is evidence to the contrary. This assumption is therefore the hypothesis to be tested and we call it the *null hypothesis*. Note that the specification of the null hypothesis forces us to think about the choice of a rival conjecture. What happens if we reject the null hypothesis? We may postulate as the rival conjecture that the coin is not fair, or that it is biased in favor of heads, or that it is biased in favor of tails. The rival conjecture is called the *alternative hypothesis* and its specification depends very much on the use to which we wish to put the coin.

How do we go about testing the null hypothesis that the coin is fair? We do what comes naturally, toss it, say, n times. Let

$$X_i = \begin{cases} 1 & \text{if the } i \text{th toss results in heads} \\ 0 & \text{otherwise.} \end{cases}$$

Then $X_1, X_2, \ldots, X_n$ are independent and identically distributed random variables. Let $P(\text{Heads}) = p$, $0 \le p \le 1$. Then

$$P(X_i = 1) = 1 - P(X_i = 0) = p, \qquad 0 \le p \le 1.$$

Here p is the parameter and $\Theta = [0, 1]$. The null hypothesis translates to the statement $H_0: p = 1/2$ so that $\Theta_0 = \{1/2\}$. The alternative hypothesis that the coin is not fair translates to the statement $H_1: p \ne 1/2$ so that $\Theta_1 = [0, 1/2) \cup (1/2, 1]$. We note that under H_0 the common distribution of the X's is completely specified, whereas under H_1 the common distribution is specified except for the value of p.

The outcome is a point in $\mathfrak{X} = \{(x_1, \ldots, x_n): x_i = 0 \text{ or } 1, 1 \le i \le n\}$. Under what circumstances are we willing to reject H_0? Probability theory enables us to compute the probability of observing the sample on the basis of the given null and alternative hypotheses. Thus the probability of observing the sample outcome $x_1, x_2, \ldots, x_n$ when $P(X_i = 1) = p$ is given by

$$P_p(X_1 = x_1, X_2 = x_1, \ldots, X_n = x_n) = p^{x_1}(1 - p)^{1 - x_1}$$

$$\cdot p^{x_2}(1 - p)^{1 - x_2} \cdots p^{x_n}(1 - p)^{1 - x_n}$$

$$= p^{\sum_{i=1}^{n} x_i}(1 - p)^{n - \sum_{i=1}^{n} x_i}$$

where we have written for convenience

$$P_p(X = x) = p^x(1 - p)^{1 - x}, \qquad x = 0 \text{ or } 1.$$

In particular,

$$P_{H_0}(X_1 = x_1, X_2 = x_2, \ldots, X_n = x_n) = \left(\frac{1}{2}\right)^n$$

and

$$P_{H_1}(X_1 = x_1, X_2 = x_2, \ldots, X_n = x_n) = p^{\sum_{i=1}^{n} x_i}(1 - p)^{n - \sum_{i=1}^{n} x_i}, \qquad p \ne 1/2.$$

We note that $P_p(X_1 = x_1, \ldots, X_n = x_n)$ depends only on $\sum_{i=1}^n x_i$, the total number of heads in n tosses. This means that if we are willing to reject H_0 if our sample is such that $\sum_{i=1}^n x_i = s$, then we should be willing to reject H_0 for every sample where the total number of successes is s no matter what the individual x_i's are. Suppose we observe a sample that is very improbable from the point of view of H_0. That is, suppose that $P_{H_0}(\sum_{i=1}^n X_i = s)$ is "small." If in spite of its small probability we actually observe the outcome $(x_1, x_2, \ldots, x_n)$ for which $\sum_{i=1}^n x_i = s$ then this is a strong argument against H_0. We find it hard to believe that something so improbable could happen. Realizing that the probability computation was made under the belief that H_0 is correct, we start doubting the basis of our computation.

To be more specific, let us take $\Theta_0 = \{1/2\}$, $\Theta_1 = \{3/4\}$, and $n = 6$. Then we need to choose between H_0: $p = 1/2$ and H_1: $p = 3/4$ on the basis of a random sample of size 6. We note that

$$P_p(X_1 = x_1, \ldots, X_6 = x_6) = p^{\sum_{i=1}^6 x_i}(1 - p)^{6 - \sum_{i=1}^6 x_i}$$

$$= \begin{cases} (1/2)^6, & \text{under } H_0: p = 1/2 \\ (3/4)^s(1/4)^{6-s}, & \text{under } H_1: p = 3/4 \end{cases}$$

where $s = \sum_{i=1}^6 x_i$ is the number of heads observed. As noted above, the probability of observing $(x_1, \ldots, x_6)$ depends only on the number of heads observed in six tosses and not on any particular of the 2^6 sequences of 0's and 1's observed. Thus if we decide to reject H_0 when the sequence $HTTTTT$(or 1, 0, 0, 0, 0, 0) is observed, we should also reject it when the sequence $THTTTT$ is observed, and so on. The following table gives the probabilities associated with the observations $s = 0, 1, \ldots, 6$ under both H_0 and H_1.

Note that in computing the probability of $S = 1$ we have to add the probability of each of the six possible sequences of outcomes that result in exactly one heads. Similarly for $S = 2$, there are $\binom{6}{2} = 15$ such sequences, and so on. (See Section 6.3 for more details.)

Now it is tempting to argue as follows. If we observe s successes in six trials, we check to see from the above table if such a value of s is more probable under H_0 or under H_1. We choose H_0 if the observed s is more probable under H_0, and H_1 if the

TABLE 1. Probability of the Outcome[a]

Outcomes	$p = 1/2$	$p = 3/4$
0	$1/64 = .0156$	$1/4096 = .0002$
1	$6/64 = .0938$	$18/4096 = .0044$
2	$15/64 = .2344$	$135/4096 = .0330$
3	$20/64 = .3125$	$540/4096 = .1318$
4	$15/64 = .2344$	$1215/4096 = .2966$
5	$6/64 = .0938$	$1458/4096 = .3560$
6	$1/64 = .0156$	$729/4096 = .1780$

[a]The total in second column does not add up to 1 due to round-off error.

observed s is more probable under H_1. This leads to the following rule:

$$\text{Accept } H_0 \quad \text{if } S = 0, 1, 2, 3$$

$$\text{Accept } H_1 \quad \text{if } S = 4, 5, 6.$$

What are the consequences of this rule? That is, what is the cost of being wrong? Suppose the observed s falls in $C = \{4, 5, 6\}$ (C is called the *critical* region of the rule) so that we reject H_0. In that case the probability of being wrong is given by

$$P\{S = 4, 5, \text{ or } 6 | H_0 \text{ true}\} = P\{S = 4, 5, \text{ or } 6 | p = 1/2\} = .348.$$

The probability $P\{\text{reject } H_0 | H_0 \text{ true}\}$ is called the *probability of type I error* (and the error in rejecting H_0 when it is true is called a type I error). There is yet another possible error, that of accepting H_0 when H_1 is true (H_0 is false). In our case

$$P\{S = 0, 1, 2, \text{ or } 3 | p = 3/4\} = .1694,$$

which is called the probability of type II error.

The reader can easily come up with several other reasonable rules for the problem under consideration. How do we choose amongst different rules for the same problem? Clearly we should like both probabilities of errors to be as close to zero as possible. Unfortunately, however, both these probabilities cannot be controlled simultaneously for a fixed sample size. The usual procedure is to control $P(\text{type I error})$ at a preassigned level and then find a rule that minimizes $P(\text{type II error})$.

We will see in Chapter 11 that the rule considered above turns out to be the best in a certain sense. See also Examples 4 and 5. □

Example 2. MPG Rating of a New Car Model. A new automobile model has just been introduced and the manufacturer claims that it gives an average of 48 miles per gallon (mpg) in highway driving. A consumer protection agency decides to test this claim. It takes a random sample of 15 cars. Each car is then run on a stretch of highway and it is found that the sample average mileage per gallon of gasoline is 42. This figure is less than that claimed by the manufacturer. Is it, however, low enough to doubt the manufacturer's claim? Could the difference between the claimed average and the observed sample average be due to chance?

Let μ be the true (unknown) average for the model. The correct thing to do is to assume that the manufacturer's claim is right so that the null hypothesis that we wish to put to test is H_0: $\mu = 48$ mpg. The alternative hypothesis is H_1: $\mu < 48$ mpg. Let X_i represent the mpg of the ith car $i = 1, 2,, \ldots, 15$, in the test run. We are given that for a sample of size $n = 15$, the sample average $\bar{x}$ is 42. If the observed value of $\bar{X}$ were at least 48 there would be no reason to question the validity of H_0. Since the observed value of $\bar{X}$ is 42, it is natural to ask the following question: How probable is it to observe a value of $\bar{X}$ as small as 42 in random sampling from a population with $\mu = 48$? (We have said that $\bar{X}$ is a good estimate of μ so we expect large values of $\bar{X}$ under H_0 rather than under H_1). If this probability is "small" the evidence against H_0 is strong, and if the probability is "large" the evidence against H_0 is weak.

The problem, therefore, is to compute $P_{H_0}(\bar{X} \leq 42)$. For an exact computation of this probability, we need to know the distribution of $\bar{X}$. This means that we have to make certain assumptions about the underlying distribution of the X_i's. If we believe that the family of possible densities of X_i is $\{f(x, \mu), \mu > 0\}$ where f is completely specified when μ is specified, then

$$P_{H_0}(\overline{X} \le 42) = P_{H_0}(X_1 + \cdots + X_{15} \le 15(42))$$

$$= \int \cdots \int_{x_1 + \cdots + x_{15} \le 630} f(x_1, \mu_0) \ldots f(x_{15}, \mu_0)\, dx_1 \ldots dx_{15}.$$

This last multiple integral can be evaluated to give $P_{H_0}(\overline{X} \le 42)$. On the other hand, if n is large it may be possible to approximate $P_{H_0}(\overline{X} \le 42)$.

If we are not willing to assume that the family of distributions of X_i is parametric but only that it is continuous and symmetric about μ, then we cannot compute $P_{H_0}(\overline{X} \le 42)$ since we do not know f under H_0. According to our assumption, X_i has a density that is symmetric about the median $\mu = 48$ under H_0. Thus, under H_0, $P(X_i \ge 48) = P(X_i \le 48) = 1/2$ so that we expect about half the sample X_i's to be smaller than 48 and about half larger than 48. If, however, we observe a relatively large number of X_i's smaller than 48, then we would suspect that the median is smaller than 48. If we know each of the 15 values of X_i we could simply count the number of X_i's that are smaller than 48 and base our decision whether to accept or reject H_0 on the magnitude of this number. This nonparametric approach leads to the well known sign test, which will be discussed in detail in Section 6.3. □

Example 3. Response Times for Two Stimuli. In order to compare the lengths of response time for two different stimuli in cats, both stimuli are applied to each of n cats and the response times are recorded in seconds. The object is to determine if there is any difference in mean response times for the two stimuli. Let $(X_1, Y_1), \ldots, (X_n, Y_n)$ be the response times for the n subjects and suppose $D_i = X_i - Y_i$. Let us now choose a probabilistic model for the problem. Suppose we are only willing to assume that $D_i = X_i - Y_i$, $i = 1, 2, \ldots, n$ is a random sample from a continuous distribution $F \in \mathfrak{F}$ where $\mathfrak{F}$ is the class of all continuous distribution functions on the real line. Then $\mathfrak{F}$ is a nonparametric family. Let $p = P(X_i > Y_i)$.

In view of Problem 3.6.16 the null hypothesis of no difference between the stimuli (that is, X_i and Y_i are identically distributed) translates to

$$H_0: P(X_i > Y_i) = P(X_i < Y_i) \qquad \text{or } p = 1/2$$

and the alternative hypothesis is

$$H_1: P(X_i > Y_i) \ne P(X_i < Y_i) \qquad \text{or } p \ne 1/2.$$

If, on the other hand, we are willing to narrow down $\mathfrak{F}$ further by assuming that the common distribution F of the D_i's is given by the density

$$f(d) = \frac{1}{\sigma\sqrt{2\pi}} \exp\left\{-\frac{(d - \mu)^2}{2\sigma^2}\right\}, \, d \in \mathbb{R},$$

Then $\mathfrak{F}$ is a parametric family and H_0 translates to

$$H_0: \mu = 0 \, (\sigma^2 \text{ unknown})$$

and H_1 to

$$H_1: \mu \neq 0 \ (\sigma^2 \text{ unknown}).$$

The point is that the procedure we use depends on the assumptions we make concerning $\mathfrak{F}$. The more we know about $\mathfrak{F}$ the better we should be able to do. The nonparametric case of this example is discussed in Section 6.3 and the parametric case in Section 8.7.   □

We are now ready to formalize the notions introduced in the above examples. Let $X_1, X_2, \ldots, X_n$ be the data vector (often $X_1, X_2, \ldots, X_n$ will be a random sample) with joint distribution function F_θ, $\theta \in \Theta$. Let $\mathfrak{F} = \{ F_\theta: \theta \in \Theta \}$.

DEFINITION 1.   A statistical hypothesis is an assertion about the joint distribution F_θ of $(X_1, \ldots, X_n)$. The statement being tested that $F_\theta \in \mathfrak{F}_0 = \{ F_\theta: \theta \in \Theta_0 \}$ is called the null hypothesis, and we abbreviate by writing it as $H_0: \theta \in \Theta_0$. The rival statement that we suspect is true instead of H_0 is called the alternative hypothesis and we abbreviate by writing $H_0: \theta \in \Theta_1 = \Theta - \Theta_0$. The hypothesis under test is called nonparametric if $\mathfrak{F}_0$ is a nonparametric family and parametric if $\mathfrak{F}_0$ is a parametric family.

Usually the null hypothesis is chosen to correspond to the smaller or simpler subset Θ_0 of Θ and is a statement of "no difference." In all cases we consider the null hypothesis will be of the form $\theta = \theta_0$, $\theta \geq \theta_0$, or $\theta \leq \theta_0$. Note that the equality sign always appears in the null hypothesis, whereas the strict inequality sign will always appear in H_1. If the distribution of $(X_1, \ldots, X_n)$ is completely specified by a hypothesis we call it a *simple hypothesis*, otherwise the hypothesis is called *composite*. Thus, whenever Θ_0 or Θ_1 consists of exactly one point the corresponding hypothesis is simple, otherwise it is composite.

Let $\mathfrak{X}$ be the set of all possible values of $(X_1, X_2, \ldots, X_n)$. Then $\mathfrak{X} \subseteq R_n$.

DEFINITION 2.   (TEST).   A rule (decision) that specifies a subset C of $\mathfrak{X}$ such that

$$\text{if } (x_1, x_2, \ldots, x_n) \in C, \text{ reject } H_0$$

$$\text{if } (x_1, x_2, \ldots, x_n) \notin C, \text{ accept } H_0$$

is called a test of H_0 against H_1 and C is called the critical region of the test. A test statistic is a statistic that is used in the specification of C.

One can make two types of errors.

	Reject	
True	H_0	H_1
H_0	Error (type I)	Correct action
H_1	Correct action	Error (type II)

DEFINITION 3.   (TWO TYPES OF ERRORS).

> Type I error: Reject H_0 when H_0 is true.
> Type II error: Accept H_0 when H_0 is false.

For a simple null hypothesis H_0: $\theta = \theta_0$ against a simple alternative H_1: $\theta = \theta_1(\theta_1 > \theta_0)$, the two types of error probabilities are exhibited in Figure 1.

Ideally, we should like both types of error to have probability 0, but this is impossible except in trivial cases. We therefore have to content ourselves by trying to keep these probabilities at an acceptably small level. It is customary, however, to fix the probability of type I error at a preassigned (small) level α, $0 < \alpha < 1$, and then try to minimize the probability of type II error.

DEFINITION 4.   (SIZE OF A TEST).   A test of the null hypothesis H_0: $\theta \in \Theta_0$ against H_1: $\theta \in \Theta_1$ is said to have size α, $0 \le \alpha \le 1$, if

$$\sup_{\theta \in \Theta_0} P_\theta \left(\text{Reject } H_0 \right) = \alpha.$$

The chosen size α is often unattainable. Indeed in many problems only a countable number of levels α in $[0, 1]$ are attainable. In that case we usually take the largest level less than α that is attainable. One also speaks of a critical region C of *significance level* α if

$$P_\theta(C) \le \alpha \qquad \text{for all } \theta \in \Theta_0.$$

If $\sup_{\theta \in \Theta_0} P_\theta(C) = \alpha$ then the level and the size of C both equal α. On the other hand, if $\sup_{\theta \in \Theta_0} P_\theta(C) < \alpha$, then the size of C is smaller than its significance level α. A chosen significance level α is usually attainable.

If H_0 is a simple hypothesis, then it is clear that $P_{H_0}(C)$ is the size of the critical region C, which may or may not equal a given significance level α.

The choice of a specific value for α, however, is completely arbitrary. Indeed it is not strictly a statistical problem. Considerations such as the possible conse-

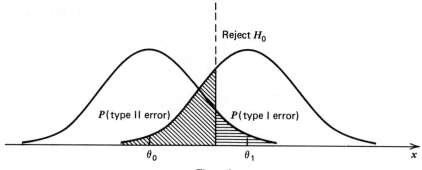

Figure 1

quences of rejecting H_0 falsely, the economic and practical implications of rejecting H_0, and so on, should influence the choice of α. We therefore follow an alternative approach, wherever possible, by reporting the so-called P-value of the observed test statistic. This is the smallest level α at which the observed sample statistic is significant.

Before we give a formal definition, let us return to Example 1. Suppose $S = 5$ is observed. Then under H_0, $P(S = 5) = .0938$. The probability of interest, however, is not $P(S = 5)$ but $P(S \geq 5)$ since the probability $P_{H_0}(S = 6)$ is even smaller, and if we reject H_0 when $S = 5$ is observed we should do so also when a more extreme value of S is observed. This motivates Definition 5 below.

Suppose the appropriate critical region for testing H_0 against H_1 is one-sided. That is, suppose that C is of the form $\{T \geq c_1\}$ or $\{T \leq c_2\}$ where T is the test statistic.

DEFINITION 5. (P-VALUE). The probability of observing under H_0 a sample outcome at least as extreme as the one observed is called the P-value. If t_0 is the observed value of the test statistic and the critical region is in the right tail then the P-value is $P_{H_0}(T \geq t_0)$. If the critical region is in the left tail then the P-value is $P_{H_0}(T \leq t_0)$. The smaller the P-value, the more extreme the outcome and the stronger the evidence against H_0.

In most problems considered in this text P-value is a well defined quantity. If the critical region C is two-sided, that is, if C is of the form $(T \geq t_1$ or $T \leq t_2)$, then we will double the one-tailed P-value and report it as the P-value even if the distribution of T is not symmetric. The P-value is variously referred to as the descriptive level or the effective level or the level attained.

Whenever the P-value is well defined, it can be interpreted as the smallest level at which the observed value of the test statistic is significant. Reporting a P-value conveys much more information than simply reporting whether or not an observed value of T is statistically significant at level α. Indeed, if the level α is preassigned and p_0 is the P-value associated with t_0, then t_0 is significant at level α if $p_0 \leq \alpha$. Often the statistician does not have much feeling about the economic consequences or the practical implications of the decision to be reached. Reporting the P-value permits each individual to choose his or her level of significance.

In many instances we do not have readily available tables of the null distribution (that is, distribution under H_0) of the test statistic. Even in such cases it is desirable to report the interval in which the P-value lies.

Let us now turn our attention to the probability of a type II error. In order to assess the performance of a test under the alternative hypothesis, we make the following definition.

DEFINITION 6. (POWER FUNCTION). Let C be the critical region of a test of hypothesis H_0 against H_1. Then the function

$$\beta(\theta) = P_\theta (\text{Reject } H_0) = P_\theta(C)$$

as a function of θ is called the power function of the test.

We note that if $\theta \in \Theta_0$ then $\beta(\theta)$ is the probability of a type I error, and if $\theta \in \Theta_1$ then $\beta(\theta)$ equals $1 - P$ (type II error). Thus minimization of a type II error probability is equivalent to the maximization of $\beta(\theta)$ for $\theta \in \Theta_1$ subject to the (significance) level requirements, namely, that

$$\beta(\theta) \leq \alpha \qquad \text{for } \theta \in \Theta_0.$$

In the case when $\Theta_0 = \{\theta_0\}$, H_0 is a simple null hypothesis and $\beta(\theta_0)$ is the size of the test. If $\Theta_1 = \{\theta_1\}$ then

$$\beta(\theta_1) = 1 - P \text{ (type II error)} = 1 - P_{\theta_1} \left(\text{Reject } H_1\right).$$

Figure 2 shows some typical power functions.

If H_1 is composite, the following definition provides an optimality criterion for selecting a test.

DEFINITION 7.   (UNIFORMLY MOST POWERFUL TEST).   A test with critical region C of the null hypothesis $H_0 \colon \theta \in \Theta_0$ is said to be uniformly most powerful of its size $\sup_{\theta \in \Theta_0} P_\theta(C)$ if it has the maximum power among all critical regions C' of significance level $\sup_{\theta \in \Theta_0} P(C)$. That is, C is best (uniformly most powerful) if for all tests C' with $P_\theta(C') \leq \sup_{\theta \in \Theta_0} P_\theta(C)$, $\theta \in \Theta_0$ the inequality

$$P_\theta(C) \geq P_\theta(C')$$

holds for each $\theta \in \Theta_1$.

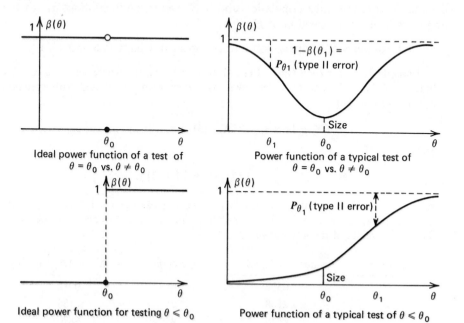

Figure 2.   Some typical power functions.

Uniformly most powerful tests do not exist for many hypothesis-testing problems. Even when they do exist, they are often not easy to find. In the rest of this section we consider some examples to illustrate the concepts that we have introduced. We will restrict ourselves mostly to simple cases when we have samples of size 1 or 2 in order to avoid distributional problems. In Chapter 11 we study some general methods of finding good tests. In this section we will appeal to intuition in finding reasonable test procedures.

Remark 1. It is important to remember that the P-value is computed under H_0 whereas the power is (essentially) computed under H_1. An event that has small probability under H_0 does not necessarily have large probability under H_1. That is, an event that is rare under H_0 does not have to be relatively frequent under H_1. Thus, in order to properly evaluate the strength of evidence against H_0 on the basis of the magnitude of the P-value, we also need to look at the power of the test.

Remark 2. Inferential statements are not probability statements in the sense that once a sample is taken and the decision is made on the basis of this sample then there is no probability attached to the inference. The statement is either true or false and we will never know that unless we know θ. As pointed out in the case of confidence intervals (see remarks following Definition 4.3.6), all these statements can be given a "long run" interpretation. Thus, "a test has significance level α" does not mean that when the test is applied to a specific sample and we reject H_0, the probability of rejecting H_0 falsely is at most α. In relative frequency terms it simply means that if the same test is performed on a large number, say N, of samples from the same population and if H_0 were true then about αN of the samples will result in rejection of H_0.

Several examples of nonparametric tests appear in Chapters 6 and 11.

Example 4. Is the Coin Fair? Let us return to Example 1 where we toss a coin six times. Let $X_1, X_2, \ldots, X_6$ be independent and identically distributed with common probability function

$$P(X_i = 1) = 1 - P(X_i = 0) = p, \quad 0 \le p \le 1.$$

Then

$$\Theta = [0,1], \Theta_0 = \{1/2\}, \Theta_1 = [0,1/2) \cup (1/2,1]$$

and $H_0: p = 1/2$ is a simple parametric hypothesis whereas $H_1: p \ne 1/2$ is a composite hypothesis. We can use Table 1 in Example 1 to compute the P-values associated with the various values of the test statistic S.

Outcome, s	P-value $= 2P_{H_0}(S \le s)$	Outcome	P-value $= 2P_{H_0}(S \ge s)$
0	.0312	6	.0312
1	.2188	5	.2188
2	.6876	4	.6876

TABLE 2. Power Function of $C = \{0, 6\}$

p	.9	.8	.7	.6	.5	.4	.3	.2	.1
$\beta(p)$	.5314	.2622	.1185	.0508	.0312	.0508	.1185	.2622	.5314

If $\alpha = .05$ is given, then the critical region is $C = \{S = 0 \text{ or } 6\}$ with actual level (.0312) considerably less than .05. The situation becomes much worse if $\alpha = .20$. In this case, again $C = \{S = 0 \text{ or } 6\}$ with P-value .0312.

For the power function we need to specify a test. Suppose we take the critical region $C = \{S = 0 \text{ or } 6\}$ with size .0312. Then

$$\beta(p) = P\{S = 0 \text{ or } S = 6|p\}$$

is a function of p and can be computed for various values of p. In Table 1 of Example 1, this was done for $p = 3/4$ and by symmetry the same result holds for $p = 1/4$. Thus

$$\beta(1/4) = \beta(3/4) = P_{p=3/4}(S = 0 \text{ or } S = 6) = .0002 + .1780$$

$$= .1782.$$

In Table 2 we give some other values of $\beta(p)$. As we might expect, the larger the deviation $|p - 1/2|$, the larger the power. (See Figure 3.)   □

Example 5. Testing Simple vs. Simple in Example 1.   In Example 1 let us look at the problem of testing $H_0: p = 1/2$ against $H_1: p = 3/4$ a little more carefully. Let S be the number of heads in six tosses. The following table gives values of $P(S = s)$ under H_0, H_1 and the ratio $P(S = s|H_1)/P(S = s|H_0)$ for $s = 0, 1, \ldots, 6$.

s	0	1	2	3	4	5	6
$P_{H_0}(S = s)$	1/64	6/64	15/64	20/64	15/64	6/64	1/64
$P_{H_1}(S = s)$	1/4096	18/4096	135/4096	540/4096	1215/4096	1458/4096	729/4096
$P_{H_1}(S = s)/$ $P_{H_0}(S = s)$	1/64	3/64	9/64	27/64	81/64	243/64	729/64

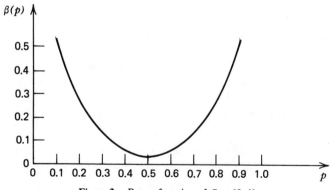

Figure 3.   Power function of $C = \{0, 6\}$.

We first note that if S is used as the test statistic then there are only a finite number of levels α attainable ranging from $\alpha = 0$ (we never reject H_0 no matter what S is observed) to $\alpha = 1$ (we always reject H_0 no matter what S is observed). Suppose we choose $\alpha = 1/64 = .0156$ to be the size of the test. Then there are only two critical regions $C_1 = \{S = 0\}$ and $C_2 = \{S = 6\}$ with size $\alpha = 1/64$. Which is better of the two for testing H_0 against H_1? We note that

$$P_{H_0}(C_1) = 1/64, \quad P_{H_1}(C_1) = 1/4096, \quad \beta_1 = 1 - P_{H_1}(C_1) = 4095/4096$$

and

$$P_{H_0}(C_2) = 1/64, \quad P_{H_1}(C_2) = 729/4096, \quad \beta_2 = 1 - P_{H_1}(C_2) = 3367/4096.$$

If we choose C_1 over C_2 then even though both have the same size we are faced with the prospect of rejecting H_0 when H_0 is true more often than rejecting H_0 when it is false (H_1 is true). On the other hand, if we choose C_2, then the probability of rejecting H_0 when it is true is much smaller than the probability of rejecting H_0 when it is false. This is clearly a desirable feature of a critical region. Indeed, C_2 is the best critical region of size $\alpha = 1/64$ since it has larger power. In fact

$$729/4096 = P_{H_1}(C_2) > P_{H_1}(C_1) = 1/4096.$$

This result is consistent with our intuition. We expect a larger number of successes under H_1 than under H_0.

The ratio $P_{H_1}(S = s)/P_{H_0}(S = s)$ also provides us a tool to determine the best critical region for a given value of size α. For example, we note that the ratio obtains its maximum value for $s = 6$ which is in the best critical region C_2. The ratio tells us that in relative terms, among all values of s the value $s = 6$ occurs most frequently under H_1 than under H_0.

Let us take another example. Suppose we choose $\alpha = 7/64$. Then the only possible critical regions of size $\alpha = 7/64$ are

$$C_3 = \{S = 0 \text{ or } 1\}, \quad C_4 = \{S = 0 \text{ or } 5\}, \quad C_5 = \{S = 1 \text{ or } 6\}, \quad C_6 = \{S = 5 \text{ or } 6\}.$$

We note that the ratio $P_{H_1}(S = s)/P_{H_0}(S = s)$ has two largest values for $s = 5$ and $s = 6$ so that C_6 should be the best critical region. The following table shows the power for each critical region.

Critical Region	C_3	C_4	C_5	C_6
Power	19/4096	1459/4096	747/4096	2187/4096

In terms of power, therefore, we see that C_6 is the best critical region and C_3 the worst. (There are several other critical regions of significance level $\alpha = 7/64$: $C_7 = \{S = 0 \text{ or } 6\}$, $C_8 = \{S = 0\}$, $C_9 = \{S = 6\}$, $C_{10} = \{S = 1\}$, $C_{11} = \{S = 5\}$, and $C_{12} = \{$always accept $H_0\}$. We leave the reader to check that C_7 to C_{12} also have smaller power than C_6.) $\qquad\square$

Example 6. Time to Failure of an Electric Tube. The time to failure, in thousands of hours, of an electric tube produced at a certain plant has density function given by

$$f(x) = \frac{1}{\theta} \exp\left(-\frac{x}{\theta}\right), \quad x > 0, \quad \text{and zero elsewhere.}$$

The mean life is θ thousand hours. In order to determine if the process is in control, a sample of size n is taken and the tubes are put to a test. Let $X_1, X_2, \ldots, X_n$ be the n failure times. The process is assumed to be in control if $\theta \geq 5$ so that the null hypothesis that the process is in control is $H_0: \theta \geq 5$ and the alternative is $H_1: \theta < 5$. Both H_0 and H_1 are composite hypotheses. Let $\overline{X} = \sum_{i=1}^n X_i/n$ be the sample mean. We have said that $\overline{X}$ is a reasonable estimate of θ and therefore a reasonable candidate for the test statistic. Now large values of $\overline{X}$ tend to support H_0 whereas small values of $\overline{X}$ tend to support H_1. It is therefore reasonable to reject H_0 if $\overline{X}$ is small, that is, if $\overline{X} \leq k$ for some constant k. The question is, how small should $\overline{X}$ be before we start doubting H_0? In other words, what should be the value of k? This depends on the amount of error that we are willing to allow in rejecting H_0 when in fact it is true.

Let us compute the power function of the critical region $C = \{\overline{X} \leq k\}$. We have for all $\theta > 0$,

$$\beta(\theta) = P_\theta(\overline{X} \leq k) = P_\theta(X_1 + X_2 + \cdots + X_n \leq nk)$$

$$= \int \cdots \int_{x_1 + \cdots + x_k \leq nk} \left(\frac{1}{\theta}\right)^n \exp\left\{-\sum_{i=1}^n \frac{x_i}{\theta}\right\} \prod_{i=1}^n dx_i.$$

Although it is not difficult to evaluate this multiple integral, we will not do so here. In Chapter 8 we show how to find the distribution of $\sum_{i=1}^n X_i$ when X_i are independent and identically distributed with density f. Once this is done, it is easy to compute $P_\theta(\overline{X} \leq k)$ with the help of a univariate integral. We content ourselves here with the case when $n = 2$. In that case:

$$\beta(\theta) = P_\theta(\overline{X} \leq k) = P_\theta(X_1 + X_2 \leq 2k)$$

$$= 1 - P_\theta(X_1 + X_2 \geq 2k)$$

$$= 1 - (1 + 2k/\theta)\exp(-2k/\theta) \qquad \text{(Example 3.8.5)}.$$

Suppose k is fixed. This fixes the critical region $\overline{X} \leq k$. We note that $\beta'(\theta) < 0$ so that $\beta(\theta)$ is a decreasing function of θ. Hence

$$\sup_{\theta \in \Theta_0} \beta(\theta) = \sup_{\theta \geq 5} \beta(\theta) = \beta(5) = 1 - \left(1 + \frac{2k}{5}\right)\exp\left(-\frac{2k}{5}\right)$$

is the size of the test. If the size is preassigned to be α, $0 < \alpha < 1$, we can choose k from

$$\alpha = 1 - \left(1 + \frac{2k}{5}\right)\exp\left(-\frac{2k}{5}\right).$$

On the other hand suppose we observe $X_1 = 3$, $X_2 = 4$. Then the observed value of the test statistic $\overline{X}$ is $\overline{x} = (3 + 4)/2 = 3.5$. Let us find the associated P-value. We note that the critical region is in the left tail so that the P-value is given by

$$P\{\overline{X} \leq 3.5 | H_0\} = \sup_{\theta \in \Theta_0} P_\theta\{\overline{X} \leq 3.5\},$$

$$= 1 - \left(1 + \frac{7}{5}\right)\exp\left(-\frac{7}{5}\right) = 1 - .5918 = .4082$$

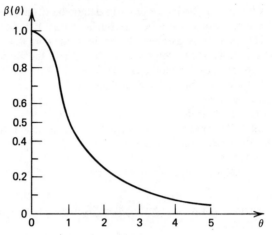

Figure 4. Power function of $\{X_1 + X_2 \le 2\}$.

and we can reject the null hypothesis only at levels greater than or equal to .4082. Clearly the evidence against the null hypothesis is not strong.

In particular for $k = 1$, $\alpha = .0616$ the power function of the size $\alpha = .0616$ test with critical region $C = \{\overline{X} \le 1\}$ is given below for some selected values of θ.

θ	4	3	2	1	0.5	0
$\beta(\theta)$	.0902	.1443	.2642	.5040	.9084	1

Figure 4 shows the graph of $\beta(\theta)$ for $0 \le \theta \le 5$.   □

Example 7. Impurities in a Chemical Product.   The percentage of impurities per batch in a certain chemical product is a random variable X with density function

$$f(x) = \theta x^{\theta - 1}, \quad 0 \le x \le 1, \quad \text{and zero elsewhere;} \quad \theta > 0.$$

Here $\Theta = (0, \infty)$. Let $\theta_0 \in (0, \infty)$ and suppose we wish to test H_0: $\theta \le \theta_0$ against H_1: $\theta > \theta_0$ on the basis of a sample of size $n = 2$. Consider the critical region $C = \{(x_1, x_2): x_1 x_2 \ge k, x_1, x_2 \ge 0\}$ where k (< 1) is a fixed positive constant.

This critical region, as we see in Chapter 11, is the best critical region for testing H_0 against H_1 and is arrived at by considering the ratio $f_{\theta_1}(x_1) f_{\theta_1}(x_2) / (f_{\theta_0}(x_1) f_{\theta_0}(x_2))$, for $\theta_1 > \theta_0$, of the joint density of (X_1, X_2) under H_1 and under H_0. The reason for considering this so-called *likelihood ratio* is similar to the one given for considering the ratio $P(S = s|H_1) / P(S = s|H_0)$ in Example 5, but we will not go into it here.

The test statistic here is $X_1 X_2$, and both H_0 and H_1 are composite hypotheses. The size of C is given by:

$$\sup_{\theta \in \Theta_0} P_\theta(C) = \sup_{\theta \le \theta_0} \iint_{x_1 x_2 \ge k} \theta^2 (x_1 x_2)^{\theta - 1} \, dx_2 \, dx_1.$$

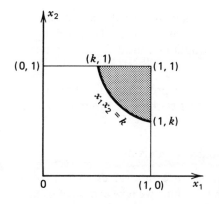

Figure 5. $\{(x_1, x_2): x_1 x_2 \geq k\}$.

In view of Figure 5 we have

$$\beta(\theta) = \theta^2 \iint_{x_1 x_2 \geq k} (x_1 x_2)^{\theta-1} \, dx_2 \, dx_1 = \theta^2 \int_k^1 x_1^{\theta-1} \left(\int_{k/x_1}^1 x_2^{\theta-1} \, dx_2 \right) dx_1$$

$$= \theta \int_k^1 x_1^{\theta-1} \left[1 - \left(\frac{k}{x_1} \right)^{\theta} \right] dx_1 = 1 - k^{\theta}(1 - \theta \ln k).$$

Since $k \leq 1$, so that $1 - \theta \ln k > 0$ and k^{θ} is a decreasing function of θ, it follows that $\beta(\theta) = P_{\theta}(C)$ is an increasing function of θ. Hence

$$\sup_{\theta \leq \theta_0} P_{\theta}(C) = 1 - k^{\theta_0}(1 - \theta_0 \ln k)$$

gives the size of C with power function

$$\beta(\theta) = P_{\theta}(C) = 1 - k^{\theta}(1 - \theta \ln k).$$

In particular, if $\theta_0 = 1$ and $k = .8$, then the size of C is given by

$$\beta(1) = 1 - .8(1 - \ln .8) = .2 + .8 \ln .8 = .0215$$

and the power function for some selected values of $\theta(> 1)$ is given below.

θ	2	3	4	10
$\beta(\theta)$	.0744	.1453	.2248	.653

$\square$

Remark 3.   The discerning reader will notice that we have used a dual approach in selecting critical regions: (i) in terms of likelihood, judging whether an observed event has "too small" a probability under a given hypothesis; (ii) in terms of discrepancy from expectations. In some examples (Examples 5 and 7 and also Problem 6) and in Remark 1 [under (i)] we have indicated that we need to look at the ratio of the likelihoods under the two hypotheses. This is done in detail in Sections 11.2 and 11.4. For most distributions considered in this book, the two approaches lead to the same conclusion.

Remark 4. It often helps to graph the density or probability function of the test statistic under both H_0 and H_1 (whenever possible) in order to find a reasonable critical region.

Relation Between Hypothesis Testing and Confidence Intervals

There is a close relationship between the confidence interval estimation of a parameter θ and a test of hypothesis concerning θ, which we now explore. Given a confidence interval at minimum level $1 - \alpha$ for a parameter θ, we can easily construct a size α test of hypothesis H_0: $\theta = \theta_0$ against one-sided or two-sided alternatives.

PROPOSITION 1. *Let $[l(\hat{\theta}), u(\hat{\theta})]$ be a minimum level $1 - \alpha$ confidence interval for θ. Then the test with critical region*

$$C = \{ \text{Reject } \theta = \theta_0 \text{ if } \theta_0 \notin [l(\hat{\theta}), u(\hat{\theta})] \}$$

is a size α test.

Proof. Indeed,

$$P_{\theta_0}(C) = 1 - P_{\theta_0}(l(\hat{\theta}) \le \theta_0 \le u(\hat{\theta}))$$

$$= 1 - (1 - \alpha) = \alpha,$$

as asserted. $\qquad\qquad\square$

According to Proposition 1, the $1 - \alpha$ level confidence interval $[l(\hat{\theta}), u(\hat{\theta})]$ is the acceptance region of a size α test of $\theta = \theta_0$.

If, however, $[l(\hat{\theta}), u(\hat{\theta})]$ has level at least $1 - \alpha$, that is, if

$$P_{\theta_0}(l(\hat{\theta}) \le \theta_0 \le u(\hat{\theta})) \ge 1 - \alpha$$

then

$$P_{\theta_0}(C) \le 1 - (1 - \alpha) = \alpha$$

and C is a rejection region of $\theta = \theta_0$ of significance level α.

In the converse direction, given the acceptance region of a test of $\theta = \theta_0$ of size α, we can construct a minimum level $1 - \alpha$ confidence interval for θ.

PROPOSITION 2. *Let $C(\theta_0)$ be the rejection region of a size α test of H_0: $\theta = \theta_0$ and let $S(\theta_0) = \overline{C}(\theta_0)$ be the corresponding acceptance region. Then the set of all values of $\theta \in \Theta$ whose acceptance regions include the observation $(x_1, \ldots, x_n)$ is a $(1 - \alpha)$ level confidence set for θ. That is, the set*

$$A(x_1, x_2, \ldots, x_n) = \{ \theta: (x_1, \ldots, x_n) \in S(\theta) \}$$

is a $(1 - \alpha)$ level confidence set for θ.

Proof.   We have:

$$\theta \in A(x_1,\ldots,x_n) \Leftrightarrow (x_1, x_2,\ldots,x_n) \in S(\theta)$$

so that

$$P_\theta(\theta \in A(X_1,\ldots,X_n)) = P_\theta((X_1,\ldots,X_n) \in S(\theta)) = 1 - \alpha,$$

for all $\theta \in \Theta$. It follows that A has confidence level $1 - \alpha$.   □

Example 8.   Example 4.3.8 Revisited.   In Example 4.3.8 we showed that a $(1 - \alpha)$ level confidence interval for θ based on a single observation X from the density function

$$f(x) = \frac{1}{\theta}, \qquad 0 < x < \theta, \qquad \text{and zero otherwise,}$$

is given by $[X/(1 - \alpha/2), X/(\alpha/2)]$. It follows from Proposition 1 that the test with critical region

$$C = \left\{ \text{Reject } \theta = \theta_0 \text{ if } \theta_0 \notin \left[\frac{2X}{2 - \alpha}, \frac{2X}{\alpha} \right] \right\}$$

is a size α test for testing $H_0: \theta = \theta_0$ against $H_1: \theta \neq \theta_0$. In particular, if $\alpha = .05$ and $X = 1.524$ is observed, then a 95 percent confidence interval for θ is given by $[1.563, 60.96]$. Since, for example, $\theta_0 = 1.5$ does not belong to this interval, we reject $H_0: \theta = 1.5$ at level .05.   □

Example 9.   Confidence Bound for Mean Time to Failure.   In Example 6, the time to failure of an electric tube was assumed to have density function

$$f(x) = \frac{1}{\theta} \exp\left(-\frac{x}{\theta} \right), \qquad x > 0, \qquad \text{and zero elsewhere.}$$

Suppose we wish to test $H_0: \theta \geq 5$ against $H_1: \theta < 5$ on the basis of one observation. Since a large observation tends to support H_0, a reasonable critical region is given by $C = \{x \leq a\}$ for some constant a. If the size of C is to be preassigned at α, $0 < \alpha < 1$, then we choose a to satisfy

$$\alpha = \sup_{\theta \geq 5} \beta(\theta) = \sup_{\theta \geq 5} P_\theta(C)$$

$$= \sup_{\theta \geq 5} \int_0^a (1/\theta) e^{-x/\theta}\, dx = \sup_{\theta \geq 5} [1 - e^{-a/\theta}]$$

$$= 1 - e^{-a/5}$$

so that $a = 5\ln(1 - \alpha)^{-1}$. Hence a size α critical region is $C = \{X \leq 5\ln(1 - \alpha)^{-1}\}$.

According to Proposition 2 the set

$$A(x) = \{\theta: x \in S(\theta)\}$$

where $S(\theta) = \overline{C}(\theta)$ gives a $(1 - \alpha)$ level confidence set for θ. In our case $C(\theta) = \{ X \leq a(\theta)\}$ where $a = a(\theta)$ is chosen from

$$\alpha = \sup_{\theta' \geq \theta} \beta(\theta') = 1 - e^{-a/\theta}.$$

Thus

$$a(\theta) = \theta \ln(1 - \alpha)^{-1} \quad \text{and} \quad S(\theta) = \left\{ X \geq \theta \ln(1 - \alpha)^{-1}\right\}.$$

Since

$$X \in S(\theta) \quad \text{if and only if} \quad \theta \in A(X),$$

we have:

$$X \in S(\theta) \Leftrightarrow \theta \leq \frac{X}{\ln(1 - \alpha)^{-1}}$$

and

$$A(X) = \left\{ \theta : \theta \leq \frac{X}{\ln(1 - \alpha)^{-1}}\right\}$$

is a $(1 - \alpha)$ level confidence interval for θ. The quantity $X/\ln(1 - \alpha)^{-1}$ is a $(1 - \alpha)$ level upper confidence bound for θ.

In particular, if $\alpha = .05$ and $X = 3$ is observed than we cannot reject H_0: $\theta \geq 5$ since the critical region of size .05 test is $C = \{ x \leq 5\ln(1/.95)\} = \{ x \leq .2565\}$. We note that the 95 percent level upper confidence bound for θ in this case is $3/\ln(1/.95) = 59.49$, which is greater than $\theta = 5$. $\qquad\square$

Problems for Section 4.4

1. In each case below identify the null and an appropriate alternative hypothesis.
    (i)   The mean age of entering freshman at a certain state university is 18.
    (ii)   The proportion of blue-collar workers in Detroit is .40.
    (iii) College graduates earn more money on the average than nongraduates.
    (iv)   American tourists spend more money in Switzerland than do other tourists.
    (v)   Females are more in favor of the Equal Rights Amendment than males.
    (vi)   The median income of college professors is not the same as that of dentists.
    (vii) Blood storage does not effect triglyceride concentration.
    (viii) A multiple-choice examination consists of 20 questions. Each question has five possible answers, only one of which is correct. A student gets 16 correct answers. Is he or she guessing the answers?
    (ix)   Two brands of denture cleansers are not equally effective in preventing accumulation of stain on dentures.
    (x)   Method A of training is superior to Method B.
    (xi)   A given die is loaded.

(xii)   There are sex differences in brand preference for a particular soft drink.
(xiii) The average birth weight of babies in a certain hospital is less than 7.5 pounds.
(xiv)   There is no sex difference in average birth weight of babies born in a certain hospital.

2. A lady claims that she is able to tell by taste alone whether a cup of tea has been made with tea bags or with loose tea leaves. In order to test her claim, she is asked to taste eight pairs of cups of tea. One cup of each pair is made with a tea bag and the other with loose tea leaves of the same type of tea. Suppose she is able to identify correctly in six cases.
   (i)   State the null and alternative hypotheses.
   (ii) What is the P-value associated with the observed number of correct identifications?
   (iii) Find the size of the following critical regions: $(X \geq 5)$, $(X \geq 6)$, $(X \geq 7)$, $(X \geq 8)$, where X is the number of correct identifications.

3. It is believed that a typist makes many more errors, on the average, during the afternoon than in the morning. The number of errors, X, per page has approximately the probability function

$$P(X = k) = e^{-\lambda}\frac{\lambda^k}{k!}, k = 0,1,2,\ldots$$

where $\mathscr{E}X = \lambda > 0$. Suppose further that during the morning session the average number of errors λ is known to be 1.25. In order to test this belief a random sample of eight pages (known to have been typed either in a morning or in an afternoon session) was checked and found to contain a total of 19 errors.
   (i)   State the null and alternative hypotheses.
   (ii) Find a test of H_0 against H_1 that has significance level $\alpha = .10$.
   (iii) Find the P-value associated with the observed number of errors in eight pages. Would you reject H_0 at level $\alpha = .01$?
         [Hint: Use the result that $S_8 = X_1 + X_2 + \cdots + X_8$ has the distribution $P(S_8 = s) = e^{-8\lambda}(8\lambda)^s/s!, s = 0,1,2,\ldots$. Table A4 in the Appendix gives the cumulative probabilities for S_8.]

4. According to a genetic theory, about 40 percent of a certain species have a particular characteristic. In order to test the theory a random sample of seven is taken and it is found that only one individual in the sample has the characteristic.
   (i)   What are the null and alternative hypotheses?
   (ii) What is a reasonable test statistic?
   (iii) What is the P-value associated with the outcome?

5. Let X be a single observation from probability function

$$P(X = x) = e^{-\lambda}\frac{\lambda^x}{x!}, \qquad x = 0,1,2,\ldots, \text{ and zero elsewhere.}$$

Consider the test with critical region $C = \{X \geq 4\}$ for testing $H_0: \lambda = 1$ against $H_1: \lambda = 2$. Find the size of the test, the probability of type II error, and the power of the test.

6. In Problem 5 suppose a sample of size 2 is taken to test H_0: $\lambda = 1$ against H_1: $\lambda = 2$. Consider the critical region

$$C = \left\{ \frac{P_{H_1}(X_1 = x_1) P_{H_1}(X_2 = x_2)}{P_{H_0}(X_1 = x_1) P_{H_0}(X_2 = x_2)} \geq 16e^{-2} \right\}.$$

Find the size and the power of this critical region.
[Hint: Simplify and show that the critical region C can be written as $\{(x_1, x_2): x_1 + x_2 \geq 4\}$.]

7. A random sample of two observations is taken from the distribution with density function

$$f(x) = \frac{1}{\theta} \exp\left(-\frac{x}{\theta}\right), \qquad x \geq 0, \text{ and zero elsewhere.}$$

The null hypothesis H_0: $\theta = 2$ is rejected against the alternative H_1: $\theta < 2$ if the observed values x_1, x_2 of X_1, X_2 fall in the set. $\{(x_1, x_2): x_1 + x_2 < 1.5, x_1 \geq 0, x_2 \geq 0\}$. Find the size of the test and its power function at $\theta = 1.8, 1.6, \dots, .2$.

8. In Problem 7 suppose we wish to test H_0: $\theta = 2$ against H_1: $\theta = 1$. Consider the critical region $C = \{(x_1, x_2): \frac{f(x_1, 1) f(x_2, 1)}{f(x_1, 2) f(x_1, 2)} > 2\}$ where $f(x, \theta) = (1/\theta) \exp(-x/\theta)$ for $x \geq 0$ and zero elsewhere. Find the size of this critical region and the power at $\theta = 1$.

9. A random variable X has probability function

x	1	2	3	4	5	6
Under H_0	1/6	1/6	1/6	1/6	1/6	1/6
Under H_1	2/15	1/6	1/5	1/5	1/6	2/15

A sample of size one is taken. If $x \in \{3, 4\}$, we reject H_0 and accept it otherwise.
(i)   Find the probability of type I error.
(ii) Find the power of the test and the probability of type II error.

10. In Problem 2 suppose the lady does have certain ability to distinguish between the two methods of making tea. In particular, suppose that the probability that she is able to correctly identify the method is $p = 2/3$.
(i)   Find the probability of type II error and the power of each of the four critical regions.
(ii) Given $\alpha = .0352$, find the best critical region of size α.

11. A gambler believes that a certain die is loaded in favor of the face value 6. In order to test his belief, he rolls the die five times. Let X be the number of times a six shows up in five rolls and consider the rejection regions $C_i = \{ X \geq i \}$, $i = 0, 1, 2, \dots, 5$.
(i)   State the null and alternative hypotheses.
(ii) Find the size of each of the six critical regions.
(iii) Find the power of each of the six critical regions at $p = P(6) = 1/5$.

12. In Problem 11 suppose Y is the number of rolls needed to get the first 6. Consider the critical regions $D_i = \{Y \le i\}$, $i = 1, 2, \ldots$ based on the test statistic Y.
    (i) Find the size of each D_i, $i \ge 1$.
    (ii) Find the power of each critical region when $p = P(6) = 1/5$.
    (iii) Suppose in two experiments the average number of rolls needed to get a 6 was 4.5. Find the associated P-value. Would you accept H_0 at 10 percent level of significance?

13. The life length of a fuse has density function

$$f(x) = \frac{1}{\theta} \exp\left(-\frac{x}{\theta}\right), \qquad x > 0, \text{ and zero elsewhere.}$$

    Find a size α test of $H_0: \theta = \theta_0$ against $H_1: \theta \ne \theta_0$ of the form $C = \{x: x < c_1 \text{ or } x > c_2\}$, where $c_1 < c_2$. (Choose c_1, c_2 such that each tail has probability $\alpha/2$.) What is the power function of your test? Find a $1 - \alpha$ level confidence interval for θ.

14. Let X have density function

$$f(x) = \exp\{-(x - \theta)\}, \qquad x \ge \theta, \text{ and zero elsewhere.}$$

    Find a size α test of $H_0: \theta \le \theta_0$ against $H_1: \theta > \theta_0$ based on one observation. Find a $1 - \alpha$ level lower confidence bound for θ. (Hint: Graph f under H_0 and H_1.)

15. Let X have density function

$$f(x) = \frac{1}{2} \exp\{-|x - \theta|\}, x \in \mathbb{R}.$$

    Find a size α test of $H_0: \theta = 1$ against $H_1: \theta \ne 1$ based on a single observation. Find the power function of your test. (Hint: Graph f under H_0 and H_1.)

16. Let X have density function

$$f(x) = \frac{\theta}{x^2}, \qquad 0 < \theta \le x < \infty, \text{ and zero elsewhere.}$$

    Find a size α test of $H_0: \theta \ge \theta_0$ against $H_1: \theta < \theta_0$ based on one observation. Find the power function of your test. Find a $1 - \alpha$ level upper confidence bound for θ. (Hint: Graph f under H_0 and H_1.)

17. The content X of a metal in any alloy has density function given by $f(x) = 2x/\theta^2$, $0 \le x \le \theta$, and zero elsewhere. It is required that $H_0: \theta = 4$ be tested against $H_1: \theta > 4$. A sample of size one produced the observation $x = 3$.
    (i) What is the P-value associated with the sample? What is the P-value if $x = 3.5$ is observed?
    (ii) Find a size α test of H_0 against H_1.
    (iii) Find the power function of your size α test.
    (iv) Find a minimum level $1 - \alpha$ lower confidence bound for θ.
    [Hint: For part (ii), graph f under H_0 and H_1.]

18. (i) For the following densities, find a size $\alpha(< 1/2)$ test of H_0: $\theta = \theta_0$ against H_1: $\theta > \theta_0$ based on a sample of size 1.

(a)

$$f(x) = \begin{cases} x/\theta^2, & 0 < x \le \theta \\ (2\theta - x)/\theta^2, & \theta \le x \le 2\theta \\ 0 & \text{otherwise.} \end{cases}$$

(b) $f(x) = 2(\theta - x)/\theta^2$, $0 < x < \theta$, and zero otherwise.

(c) $f(x) = |x|/\theta^2$, $0 \le |x| < \theta$, and zero elsewhere.

(d) $f(x) = (\theta - |x|)/\theta^2$, $0 \le |x| < \theta$, and zero elsewhere.

(ii) Find the power function of each test in part (i) for $\theta > \theta_0$.

(iii) Find the minimum level $1 - \alpha$ lower confidence bound for θ in each case of part (i) from your size α test.

4.5 FITTING THE UNDERLYING DISTRIBUTION

Most of the work we have done so far assumes that we know the probability distribution of the random variables of interest except perhaps for the numerical value of some parameter or parameters. This means that we specify a parametric model for the observable phenomenon. If the conditions describing the experiment are precise enough to suggest that the distribution can be closely approximated by a known distribution $F = F_\theta$ (except perhaps for θ), then the problem is simply to estimate θ from the data and then use a test to check if there is agreement between the mathematical model and the data. Often, however, one has little or no prior information on F. This may be because the conditions describing the physical experiment are too vague or incomplete to suggest any particular model. In that case we can try to estimate either the distribution function or the density function.

Before making a specification of F, it is a good practice to look at the data in a graphical from in order to see if they could have come from one of a known family of distributions. A list of some well known distributions and their properties is useful for this purpose. This is provided in Chapters 6 and 7.

One of the simplest ways of looking at the data is to make a *stem-and-leaf plot*. It is best introduced with the help of an example.

Example 1. Weights of Students in a Statistics Class. A sample of 39 students was taken from a large class in an elementary statistics course at a state university and their weights (in pounds) were recorded as follows:

182	134	118	96	92	192	147	168	154	183	89	125	159	116	112
125	157	174	165	179	114	116	125	174	152	147	173	165	143	118
124	137	121	115	118	126	134	132	160						

Each weight, regarded as a three-digit number, is arranged in a display by designating the first two digits (the hundredth-place and the tenth-place digits) as the *stem* and the remaining digit as the *leaf*. The distinct stem values are then listed vertically and

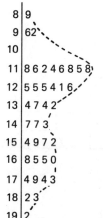

```
 8 | 9
 9 | 6 2
10 |
11 | 8 6 2 4 6 8 5 8
12 | 5 5 5 4 1 6
13 | 4 7 4 2
14 | 7 7 3
15 | 4 9 7 2
16 | 8 5 5 0
17 | 4 9 4 3
18 | 2 3
19 | 2
```

Figure 1. Stem-and-Leaf plot for weights of students (in pounds).

the leaves are recorded on the lines corresponding to their stem values. Thus, the weight 96 has a stem value of 9 and leaf value of 6, which is recorded on the horizontal line corresponding the stem value 9, and so on. Figure 1 gives a stem-and-leaf plot of the data on weights. □

A stem and leaf plot for the data on weight in Example 1 contains all the information on weights of students in the sample. It is quick, simple, easy to draw and quite effective. No hand drawing is required. The shape of the plot suggests the form of the underlying density function. (One can imagine or even draw a curve through the leaves appearing on the extreme right of each stem value.)

The classical method of presenting data is to construct a *histogram*. We first encountered a histogram in Section 3.3 where we drew histograms of probability functions of integer valued random variables by erecting bars or rectangles on intervals $(x - 1/2, x + 1/2)$ of unit length with midpoints x, where x is an integer. The height of the rectangle on base $(x - 1/2, x + 1/2)$ was taken to be proportional to $P(X = x)$.

In order to draw a histogram for a set of continuous measurements, the range (largest observation–smallest observation) is partitioned into intervals of prefer-ably equal length. When the observations are unbounded (the range is infinite) we take unbounded intervals on one or both extremes as required. The number of observations in each interval is called the *frequency* of the interval, or *class frequency*. Rectangles are then drawn with class intervals as base and height proportional to the class frequency so that the area of a rectangle on a class interval is proportional to the frequency of that class interval. A stem-and-leaf plot is in effect a histogram for class intervals of length equal to the difference between two consecutive stem values. The frequency of each class interval is simply the length of the row for the corresponding stem value. In Figure 1, intervals are of length 10 each, the frequency of interval $[80, 90)$ is 1, that of $[90, 100)$ is 2, and so on.

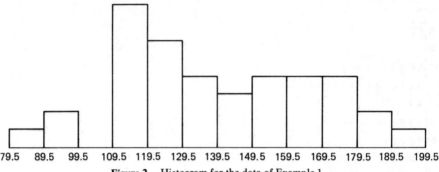

Figure 2.   Histogram for the data of Example 1.

Formally we select k, $5 \le k \le 15$, intervals, the actual number depending on the number of observations, the nature of the data, and convenience. Then the width of each interval is approximately equal to (range$/k$). The first interval begins somewhere below the smallest observation. In order to avoid ambiguities, each interval begins and ends half way between two successive possible values of the variable.

In Example 1, range $= 192 - 89 = 103$. If we choose $k = 15$, then width $103/15 \doteq 7$, and if we choose $k = 12$, then width $= 103/12 \doteq 9$, and so on. For the stem-and-leaf plot of Figure 1 we chose 12 intervals each of width 10. These were $(79.5, 89.5)$, $(89.5, 99.5), \ldots, (189.5, 199.5)$. The end points of a class interval are called *class boundaries*. The frequency of each interval is already known from Figure 1, and one can draw the histogram as shown in Figure 2.

In practice, however, it is preferable to draw a *relative frequency histogram* since the relative frequency of an interval estimates the probability in that interval.

Example 2.   Relative Frequency Histogram for Example 1.   Suppose we chose $k = 7$ intervals. Then class width is approximately 15. Let us choose intervals beginning with say 88. Since the measurements are approximated to nearest pound, we begin each interval half way between two possible values of the measurement. Thus the intervals

TABLE 1.   Frequency Distribution of Weights of Example 1

Class Interval	Frequency	Cumulative Frequency	Relative Frequency	Cumulative Relative Frequency
(87.5, 102.5)	3	3	.08	.08
(102.5, 117.5)	5	8	.13	.21
(117.5, 132.5)	10	18	.26	.47
(132.5, 147.5)	6	24	.15	.62
(147.5, 162.5)	5	29	.13	.75
(162.5, 177.5)	6	35	.15	.90
(177.5, 192.5)	4	39	.10	1.00

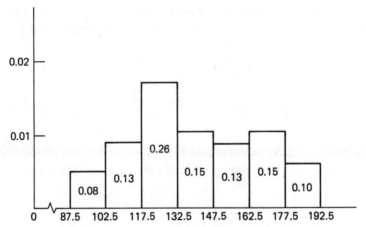

Figure 3. Relative frequency histogram for Table 1.

have boundaries

$$(87.5, 102.5) \qquad (102.5, 117.5) \qquad (117.5, 132.5) \qquad (132.5, 147.5)$$
$$(147.5, 162.5) \qquad (162.5, 177.5) \qquad (177.5, 192.5)$$

Table 1 gives the relevant information.

The relative frequency histogram is given in Figure 3.   □

Remark 1. It is clear that there is a certain amount of arbitrariness involved in the construction of a histogram. Nevertheless, unless one selects the number of intervals (and hence size of intervals) rather poorly, a histogram gives a rough idea of the shape of the underlying distribution.

Remark 2. It is not necessary, although it is often desirable, to choose equal length intervals. If one chooses intervals of different lengths, one can still draw a histogram for the data. The point to remember is that it is the area of the rectangle that represents (or is proportional to) the frequency or relative frequency of the interval and not the height of the rectangle.

Remark 3. When X is of the discrete type, it is easy to construct a frequency distribution for the data. All we need to do is to count the frequency of each of the possible values of X.

Let us now turn our attention to the theory underlying the histogram estimation of a density function. In the discrete case, the relative frequency of $\{ X = x_j \}$ estimates $P(X = x_j)$, $j = 1, 2, \ldots$. In the continuous case, let the range be partitioned into k class intervals $I_j, j = 1, 2, \ldots, k$ and suppose that $I_j = (a_{j-1}, a_j)$, $j = 1, 2, \ldots, k$ where $a_0 < a_1 < \cdots < a_k$. Define the function $\hat{f}$ by

$$\hat{f}(x) = \begin{cases} f_j / [n(a_j - a_{j-1})] & \text{for } a_{j-1} < x \le a_j, j = 1, 2, \ldots, k \\ 0 & \text{otherwise} \end{cases}$$

where f_j is the frequency of the class interval I_j and $n = \sum_{j=1}^{k} f_j$ is the total number of observations. The function $\hat{f}$ defined on $(-\infty, \infty)$ is called a *relative frequency histogram*. We note that

(1) $$\hat{f}(x) \geq 0 \text{ for all } x \in (-\infty, \infty),$$

(2) $$\int_{-\infty}^{\infty} \hat{f}(x)\, dx = \int_{a_0}^{a_k} \hat{f}(x)\, dx = \sum_{j=1}^{k} \int_{a_{j-1}}^{a_j} \frac{f_j}{n(a_j - a_{j-1})}\, dx = \sum_{j=1}^{n} \frac{f_j}{n} = 1.$$

Hence $\hat{f}$ defines a density function and is used as an estimate of the underlying density function f. In particular, an estimate of the probability of an event A is given by

$$\hat{P}(A) = \int_A \hat{f}(x)\, dx.$$

If, for example, $A = \bigcup_j I_j$ then

$$\hat{P}(A) = \sum_j \int_{I_j} \hat{f}(x)\, dx = \sum_j \frac{f_j}{n}.$$

We will see that $\hat{P}(A)$ is an unbiased and consistent estimate of $P(A)$ (see Problem 5.4.6 and also Problem 5.6.11).

Yet another way to look at the data is to draw the *sample or empirical distribution function*. Suppose we observe the sample $X_1 = x_1$, $X_2 = x_2, \ldots, X_n = x_n$. Then we can construct a distribution function from $x_1, x_2, \ldots, x_n$ by assigning to each x_i a probability mass of $1/n$. This leads to the following definition.

DEFINITION 1.   (*SAMPLE OR EMPIRICAL DISTRIBUTION FUNCTION*).
Let $x_1, x_2, \ldots, x_n$ be the observed values of the random sample $X_1, X_2, \ldots, X_n$ from some distribution function F. Then the function

$$\hat{F}_n(x) = \frac{\text{no. of } x_j\text{'s in the sample} \leq x}{n}$$

defined for each $x \in \mathbb{R}$ is called the empirical distribution function.

We first note that $0 \leq \hat{F}_n(x) \leq 1$ for all x, and, moreover, that $\hat{F}_n$ is right continuous and nondecreasing. Clearly, $\hat{F}_n(-\infty) = 0$ and $\hat{F}_n(\infty) = 1$. It follows that $\hat{F}_n$ is a distribution function. Moreover, $\hat{F}_n$ is a discrete type distribution function even though F may be continuous. We will see (Examples 5.4.3 and 10.3.8) that $\hat{F}_n$ is a consistent and unbiased estimate of F. Note that $\hat{F}_n(x)$ is the relative frequency of the event $\{X \leq x\}$.

Example 3. **Empirical Distribution Function for Weights of Students (Example 1).**
Let X_i be the weight of the ith student. Then the 39 weights of Example 1 are the 39 observations $x_1, x_2, \ldots, x_{39}$. Clearly, $\hat{F}_{39}(x) = 0$ for $x < 89$, $\hat{F}_{39}(89) = 1/39$, $\hat{F}_{39}(92) = 2/39$, $\hat{F}_{39}(96) = 3/39, \ldots \hat{F}_{39}(192) = 1$, and $\hat{F}_{39}(x) = 1$ for $x > 192$. Figure 4 shows the graph of $\hat{F}_{39}(x)$.   □

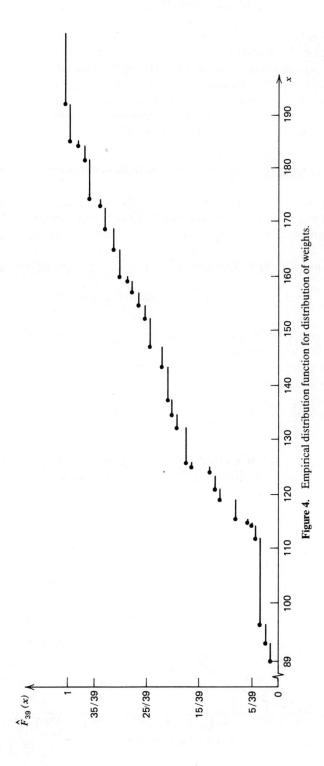

Figure 4. Empirical distribution function for distribution of weights.

235

Unless n is large, $\hat{F}_n$ looks a bit silly as an estimate of F when F is continuous since $\hat{F}_n$ goes up by jumps of size $1/n$ whereas F increases continuously. But when n is large, $\hat{F}_n$ is tedious to draw since it amounts to listing all the n observations in an increasing order of magnitude. Moreover, the step size $1/n$ is small when n is large. For large n it is usual to group the observations in a frequency distribution. How, then, do we use $\hat{F}_n$ to compute an estimate of F from a grouped frequency distribution? Suppose the intervals are $I_j = (a_{j-1}, a_j), j = 1, 2, \ldots, k$. Then

$$\hat{F}_n(a_j) = \frac{\text{no. of observations} \leq a_j}{n} = \text{relative frequency of } (-\infty, a_j].$$

Let us graph the points $(a_0, \hat{F}_n(a_0) = 0)$, $(a_1, \hat{F}_n(a_1))$, $(a_2, \hat{F}_n(a_2)), \ldots,$ $(a_k, \hat{F}_n(a_k) = 1)$ and join the successive points by line segments. The resulting plot is known as an *ogive* and provides an estimate of F from a grouped frequency distribution.

Let $\hat{F}$ denote the ogive function. Then $0 \leq \hat{F}(x) \leq 1$ for all $x \in \mathbb{R}$,

$$\hat{F}(x) = \int_{-\infty}^{x} \hat{f}(u) \, du = \int_{a_0}^{x} \hat{f}(u) \, du,$$

and $\hat{F}$ is right continuous and nondecreasing on $\mathbb{R}$. Since $\hat{F}(-\infty) = 0$ and $\hat{F}(\infty) = 1$, $\hat{F}$ is a distribution function.

We note that the slope of the line joining the points $(a_{j-1}, \hat{F}_n(a_{j-1}))$ and $(a_j, \hat{F}_n(a_j))$ is given by

$$\frac{\hat{F}_n(a_j) - \hat{F}_n(a_{j-1})}{a_j - a_{j-1}} = \frac{\hat{F}(a_j) - \hat{F}(a_{j-1})}{a_j - a_{j-1}} = \frac{f_j/n}{a_j - a_{j-1}} = \hat{f}(x)$$

for $x \in (a_{j-1}, a_j]$. It follows that $\hat{F}'(x) = \hat{f}(x)$ for $x \in (a_{j-1}, a_j)$.

For the frequency distribution of Table 1, Figure 5 shows the plot of the ogive.

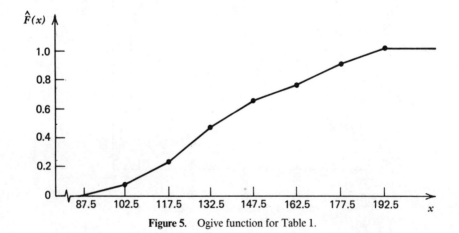

Figure 5. Ogive function for Table 1.

So far we have answered the following questions: What does the underlying density function f or the distribution function F look like? How do we estimate f or F? The next question that one needs to answer is: Does f or F look like a known or specified distribution? One obvious way to proceed is to plot the specified f (or F) and its estimate $\hat{f}$ (or $\hat{F}_n$ or $\hat{F}$) on the same paper and see how good the agreement is. If the agreement between $\hat{f}$ and f is good, it is reasonable to conclude that f is the underlying density. Since the eye can be deceptive, we will have to devise a more scientific procedure to check the fit. This is done in Section 9.9 and Chapter 11. Nevertheless, this is a good first step to follow.

Example 4. **Random Sampling from the Interval (0, 1).** A random sample of 80 points is picked from the interval $(0, 1)$ so that the underlying density is $f(x) = 1$, $0 < x < 1$, and zero elsewhere. The data are as follows.

.59	.72	.47	.43	.31	.56	.22	.90
.96	.78	.66	.18	.73	.43	.58	.11
.78	.16	.09	.46	.56	.25	.91	.26
.16	.77	.55	.72	.50	.96	.47	.07
.87	.75	.02	.36	.68	.17	.52	.01
.13	.70	.64	.61	.18	.64	.44	.59
.73	.44	.64	.50	.67	.70	.29	.88
.41	.18	.81	.33	.90	.59	.13	.43
.74	.53	.84	.06	.73	.30	.22	.15
.55	.59	.96	.74	.97	.16	.98	.97

Let us construct a relative frequency histogram using 10 equal classes with $a_0 = .005$ and $a_{10} = 1.005$. We expect each interval to contain eight observations. Table 2 gives the relative frequency histogram.

In Figure 6 we plot the histogram estimate $\hat{f}$ of f and f, and in Figure 7 we plot the ogive function estimate $\hat{F}$ of F (the distribution function corresponding to f) and F. It

TABLE 2. **Relative Frequency Histogram and Ogive**

Class Interval	Frequency	Relative Frequency	$\hat{f}$	$\hat{F}^a$
$(0.005, 0.105)$	5	.063	0.625	.063
$(0.105, 0.205)$	11	.138	1.375	.200
$(0.205, 0.305)$	6	.075	0.750	.275
$(0.305, 0.405)$	3	.038	0.375	.313
$(0.405, 0.505)$	11	.138	1.375	.450
$(0.505, 0.605)$	11	.138	1.375	.588
$(0.605, 0.705)$	9	.113	1.125	.700
$(0.705, 0.805)$	11	.138	1.375	.838
$(0.805, 0.905)$	6	.075	0.750	.913
$(0.905, 1.005)$	7	.088	0.875	1.000
Total	80	1.004^b		

^{a} This column only gives the values of $\hat{F}$ at upper class boundaries.
^{b} Total does not add to 1 owing to round-off error.

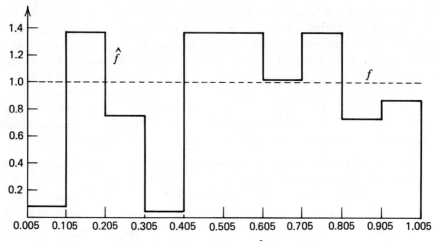

Figure 6.   Histogram estimate $\hat{f}$ of f, and f.

is not at all clear from these figures whether the agreement between $\hat{f}$ and f or $\hat{F}$ and F should be considered good or not. We will have to apply some test of goodness of fit. This is done in Section 9.9. $\square$

Finally, we note that the relative frequency histogram can be used to estimate moments of the underlying distribution and the ogive function can be properly used to estimate the quantiles of F. For example, the mean of $\hat{f}(x)$ by definition is given by

$$\int_{a_0}^{a_k} x\hat{f}(x)\,dx = \sum_{j=1}^{k} \int_{a_{j-1}}^{a_j} \frac{xf_j}{n(a_j - a_{j-1})}\,dx = \sum_{j=1}^{k} \frac{f_j}{n}\left(\frac{a_j + a_{j-1}}{2}\right) = \sum_{j=1}^{k} \frac{f_j x_j}{n}$$

where $x_j = (a_{j-1} + a_j)/2$ is called the class mark or the midpoint of class interval I_j. Similarly, the variance of $\hat{f}$ can be used to estimate var(X). However,

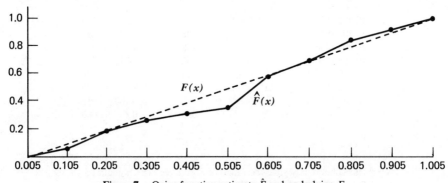

Figure 7.   Ogive function estimate $\hat{F}$ and underlying F.

the estimate for variance that is frequently used is given by

$$(n-1)^{-1}\left\{\sum_{j=1}^{k} f_j x_j^2 - \left(\sum_{j=1}^{k} f_j x_j\right)^2 / n\right\}.$$

From the ogive function one can estimate the quantile of order p, $0 < p < 1$ by simply reading off from the graph of $\hat{F}$ the abscissa of the point of intersection of $y = p$ and $y = \hat{F}(x)$.

Problems for Section 4.5

1. The failure times of 50 transistor batteries of a certain kind in hours are as follows:

    13.8 41.2 39.7 42.6 15.9 12.8 23.7 12.6 13.8 16.7 26.8 24.7
    21.4 32.6 16.8 39.3 19.6 18.7 16.8 15.9 20.4 41.2 33.7 34.8
    39.6 43.4 28.6 18.8 21.4 16.3 30.6 19.3 18.7 31.4 38.7 34.2
    29.6 24.6 20.2 21.3 28.7 18.7 14.6 16.3 29.7 17.6 33.7 31.9
    29.1 28.2

    (i) Construct a frequency distribution using 11 intervals of equal length beginning with 12.55.
    (ii) Draw the relative frequency histogram and the ogive for the data.
    (iii) Find an estimate for the mean and the median of the underlying distribution of failure times.

2. The high-school grade point averages of a random sample of 60 entering freshmen are given below.

2.87	2.76	2.63	3.67	4.00	2.93
2.42	3.25	4.00	3.80	3.30	3.00
3.89	3.56	3.68	2.32	2.50	2.68
3.57	3.17	3.90	3.13	2.60	2.58
2.90	3.69	3.18	3.20	2.85	2.87
2.25	2.17	2.84	3.00	4.00	3.83
2.82	2.69	2.77	2.92	3.17	3.54
3.82	3.17	3.77	2.26	2.69	2.87
3.42	3.30	3.25	3.50	2.87	2.79
2.83	2.28	2.37	2.42	3.18	3.81

    Construct a relative frequency histogram and an ogive for the data.

3. A die was tossed 120 times with the following results.

Face value	1	2	3	4	5	6
Frequency	17	18	24	18	21	22

    How will you estimate the probability function $P(X = x)$, $x = 1, 2, \ldots, 6$ of X, the face value that shows up on a toss? Do the data indicate that the die is fair? [Compare your estimate of $P(X = x)$ and $P(X = x) = 1/6$, $x = 1, 2, \ldots, 6$ by means of a graph.]

4. For the following data, compute the empirical distribution function and plot it.

$$4 \quad 2 \quad 8 \quad 2 \quad 7 \quad 4 \quad 6 \quad 3 \quad 7 \quad 5$$

What is your estimate of $P(X \leq 3)$? Estimate $P(X > 4)$.

5. An urn contains five slips numbered 1 through 5. A random sample of size 18 is taken with replacement with the following results:

$$1 \quad 5 \quad 2 \quad 3 \quad 4 \quad 3 \quad 5 \quad 4 \quad 2 \quad 2 \quad 1 \quad 1 \quad 3 \quad 1 \quad 2 \quad 5 \quad 4 \quad 4$$

Compute and plot the empirical distribution function. Plot also the underlying distribution function on the same paper. Estimate

$$P(1 < X < 5), P(X > 3), P(X < 3).$$

6. A random sample of 60 infants yielded the following heights (in inches):

18.2	21.4	22.6	17.4	17.6	16.7	17.1	21.4	20.1	17.9	16.8	23.1
22.3	21.7	19.6	18.4	17.7	19.3	18.4	18.6	17.8	16.9	21.4	20.6
19.8	18.7	17.5	17.8	18.3	18.9	19.6	20.6	18.7	18.3	18.8	21.4
20.9	21.8	22.6	22.1	21.4	22.3	21.4	23.2	21.6	22.4	19.6	18.6
19.9	20.7	21.8	22.2	21.5	21.1	19.6	18.9	20.8	19.6	20.4	23.0

(i) Construct an appropriate frequency distribution with eight classes of equal length.
(ii) Construct the relative frequency histogram and the ogive.
(iii) Find estimates of the mean and the median heights of infants.

7. The following random sample of 100 observations is taken from the interval $(0, 1)$:

.50	.24	.89	.54	.34	.89	.92	.17	.32	.80
.06	.21	.58	.07	.56	.20	.31	.17	.41	.38
.88	.61	.35	.06	.90	.13	.23	.60	.93	.90
.83	.24	.16	.43	.60	.59	.82	.27	.13	.22
.27	.65	.49	.37	.32	.19	.89	.86	.93	.51
.46	.34	.75	.66	.54	.38	.36	.75	.86	.05
.21	.41	.67	.13	.05	.93	.05	.50	.70	.44
.89	.73	.57	.38	.50	.47	.22	.32	.81	.62
.04	.83	.26	.57	.63	.03	.76	.34	.84	.30
.61	.78	.36	.68	.58	.62	.62	.27	.41	.55

(i)   Construct a frequency distribution with 10 class intervals of equal length.
(ii) Construct the relative frequency histogram and the ogive.
(iii) Estimate the mean and the quantiles of order $p = .25, .5$, and $.75$.
(iv) What is your guess of f and F?

4.6 REVIEW PROBLEMS

1. The breaking strength (in ounces) of a random sample of 50 threads is given below:

33.2	22.6	15.8	18.9	19.7	38.5	14.6	29.8	38.2	27.6
28.4	25.6	24.7	29.2	32.6	29.0	17.8	16.9	18.4	31.6
24.8	29.3	28.7	25.4	31.3	34.8	18.7	19.4	28.3	31.1
20.9	32.5	37.1	22.1	26.8	22.7	21.3	33.4	29.8	23.6
25.3	27.4	21.3	19.1	29.3	30.1	23.1	34.2	36.1	18.9

(i)   Construct a frequency distribution with 5 intervals beginning with 13.95 ounces.
(ii) Obtain the histogram estimate of the density function of breaking strength and also the ogive estimate of the distribution function.
(iii) Estimate the mean breaking strength of the population of threads from which the sample was taken. (There are two ways to do so, either by using the sample observations or by using the frequency distribution.)

2. Suppose that a student's score X on a certain examination has density function

$$f(x) = \frac{(x + \theta/2)}{\theta^2}, \qquad 0 \le x \le \theta, \text{ and zero elsewhere.}$$

(i)   Find an unbiased estimate of θ based on a single observation. Find the variance of your estimate.
(ii) Consider also the estimate $\hat{\theta} = X$. Find the bias and mean square error of $\hat{\theta}$.
(iii) Find the efficiency of your unbiased estimate relative to $\hat{\theta}$.

3. The time X (in hours) between arrivals of customers in a bank is a random variable with density function

$$f(x) = \exp\{-(x - \theta)\}, \qquad x > \theta, \text{ and zero elsewhere.}$$

(i)   Show that the critical region $\{x: f_{H_0}(x) = 0\}$ is the best size 0 critical region for testing $H_0: \theta = \theta_0$ against $H_1: \theta = \theta_1$ ($\theta_0 > \theta_1$) based on a sample of size 1.
(ii) Find the best critical region of size zero for testing H_0 against H_1 [of part (i)] based on a sample of size 2.

4. Let $X_1, X_2, \ldots, X_n$ be a random sample from a distribution function F. Let c be a given constant. How will you estimate $P_F(X_1 > c)$ without the knowledge of F?

In particular, in Problem 1, what is your estimate of the proportion of thread that has breaking strength greater than 28 ounces?

5. An urn contains N identical slips numbered 1 through N. A random sample of two slips is taken with replacement in order to estimate N by a confidence interval of

level at least $1 - \alpha$. Find an equal tails confidence interval based on $\max(X_1, X_2)$ where X_1 and X_2 are the numbers on the first and the second slip drawn. Hence find a test of $H_0: N = N_0$ against $H_1: N \neq N_0$ of significance level α. [Hint: $\max(X_1, X_2) \leq k$ if and only if $X_1 \leq k$ and $X_2 \leq k$.]

In particular, if $x_1 = 8$, $x_2 = 27$, and $\alpha = .10$, what is your interval estimate of N?

6. Let X have density

$$f(x) = \begin{cases} 3x^2/(2\theta^3), & 0 \leq x \leq \theta, \\ 3(x - 2\theta)^2/(2\theta^3), & \theta \leq x \leq 2\theta, \\ 0 & \text{elsewhere.} \end{cases}$$

(i) Find a size α ($< 1/2$) test of $H_0: \theta = \theta_0$ against $H_1: \theta < \theta_0$ based on a sample of size 1.

(ii) Find the power function of your test.

(iii) Find an upper confidence bound at minimum level $1 - \alpha$ for θ.

7. Let X have density

$$f(x) = \begin{cases} 3(\theta + x)^2/(2\theta^3), & -\theta \leq x \leq 0 \\ 3(\theta - x)^2/(2\theta^2), & 0 \leq x \leq \theta \\ 0 & \text{otherwise.} \end{cases}$$

(i) Find a size α test of $H_0: \theta = \theta_0$ against $H_1: \theta < \theta_0$ based on a single observation.

(ii) Find the power function of your test.

(iii) Find an upper minimum level $1 - \alpha$ confidence bound for θ.

8. A chemist has taken a sample of size 2 from a chemical compound and tests each sample for proportion of impurity. On the basis of this analysis she would like to determine if the compound came from Supplier A or Supplier B. It is known that the proportion of impurity is a random variable having density $f_A(x) = 2x$, $0 \leq x \leq 1$ and zero otherwise if the compound comes from Supplier A and density $f_B(x) = 3x^2$, $0 < x < 1$ and zero otherwise if it comes from Supplier B. Let x_1, x_2 be the sample proportions of impurities. The chemist chooses to reject H_0 that X has density f_A if

$$\frac{f_B(x_1)f_B(x_2)}{f_A(x_1)f_A(x_2)} \geq .95.$$

Find the size of her test and its power.

9. Two observations were taken from the uniform density

$$f(x) = \frac{1}{\theta}, \qquad 0 \leq x \leq \theta, \text{ and zero elsewhere}$$

in order to test the null hypothesis $\theta = \theta_0$ against $H_1: \theta \neq \theta_0$.

(i) Let $\bar{X}$ be the test statistic.

(a) Find a size α test of H_0.

(b) Find the power of your test.

(c) Find the minimum level $1 - \alpha$ equal tails confidence interval for θ.

   (ii) Take $T(X_1, X_2) = \max(X_1, X_2)$ as the test statistic.
   - (a) Find a size α test of H_0.
   - (b) Find the power of your test.
   - (c) Find the minimum level $1 - \alpha$ equal tails confidence interval for θ.
   (iii) Compare the powers of your tests in (i) and (ii). (You may assume that $\alpha < 1/2$.)

10. After the April 20, 1980, helicopter attempt to rescue American hostages held in Iran, it was claimed that the eight helicopters used in the rescue mission had each flown 20 missions without a malfunction. On the day of the rescue mission, however, three of the eight helicopters involved in the rescue failed. Let $p = P$ (failure on a single flight). The assertion that no failures occurred in previous missions should not be taken to mean that $p = 0$ but only that p is very small, say $p = 1/25$. Test H_0: $p = 1/25$ against H_1: $p > 1/25$ to see if there is any inconsistency between the observed failure rate and the asserted failure rate.

   Do the same problem with H_0: $p = .1$ and with H_0: $p = .01$. (You may assume that the failure of any helicopter on a mission is independent of the failure of any other helicopter on the mission.)

11. From Table A1 in the Appendix, the following sample of 105 random digits is taken:

24122	66591	27699	06494	14845	46672	61958
77100	90899	75754	61196	30231	92962	61773
41839	55382	17267	70943	78038	70267	30532

   (i) Find the histogram estimate of the distribution of digits $0, 1, \ldots, 9$.
   (ii) Compute the empirical distribution function. Graph the empirical distribution function along with the theoretical distribution function.

CHAPTER 5

More on Mathematical Expectation

5.1 INTRODUCTION

In Chapter 4 we saw that in order to compute the numerical value of a confidence interval or to compute the critical region of a level α test we often need to evaluate a multiple integral. This integration can be reduced to a univariate integral provided we know the distribution of the statistic $T(X_1, X_2, \ldots, X_n)$. This will be done in Chapter 8. When the parameter of interest is μ we often took $T(X_1, X_2, \ldots, X_n) = \overline{X}$ as an estimate or a test statistic. How do we evaluate the performance of $\overline{X}$ as an estimate for μ without the knowledge of the distribution of $\overline{X}$? How do we construct a confidence interval for μ based on $\overline{X}$ without the knowledge of the distribution of $\overline{X}$? For this purpose we first need to define the expected value of $T(X_1, X_2, \ldots, X_n)$ with respect to the joint distribution of $(X_1, X_2, \ldots, X_n)$. This is done in Section 5.2. In Section 5.3 we show how to compute $\mathscr{E}T$ and $\mathrm{var}(T)$ when T is a linear function of $X_1, X_2, \ldots, X_n$. This allows us to compute $\mathscr{E}T$ and $\mathrm{var}(T)$ from the knowledge of $\mathscr{E}X_j$, $\mathrm{var}(X_j)$, and $\mathrm{cov}(X_i, X_j)$. We do not need to know the joint distribution of $X_1, X_2, \ldots, X_n$. In Section 5.4 we give a precise formulation of the well known law of averages and prove a weak version of this result. This so-called weak law of large numbers allows us to conclude, for example, that $\overline{X}$ is a consistent estimate for μ so that for large n, $\overline{X}$ is likely to be near μ. In Section 5.5 we define expected values with respect to a conditional distribution. Suppose X, Y are jointly distributed dependent random variables. How do we predict Y given $X = x$? More precisely, for what function ϕ does the estimate $\phi(x)$ of Y have minimum mean square error $\mathscr{E}(Y - \phi(x))^2$? The answer, as we show in Section 5.5, is that $\phi(x)$ is the mean of the condition distribution of Y given $X = x$.

5.2 MOMENTS IN THE MULTIVARIATE CASE

Let X be the length and Y the width of a rectangle and suppose that X and Y are random variables. What is the average area of the rectangle so formed? If X is the total time that a customer spends in a bank and Y the time spent waiting in line, what is the average service time of the customer at this bank? To answer these questions we need to define expected value of a real valued function of a random vector.

Let $(X_1, X_2, \ldots, X_k)$ be a random vector with known joint probability (or density) function. Let $g(x_1, \ldots, x_k)$ be a real valued function defined on $\mathbb{R}_k$ and assume that g itself is a random variable.

DEFINITION 1. If $(X_1, X_2, \ldots, X_k)$ is of the discrete type and $\sum_{x_1} \cdots \sum_{x_k} |g(x_1, \ldots, x_k)| P(X_1 = x_1, \ldots, X_k = x_k) < \infty$ then we define the expected value (mathematical expectation) of g by:

$$\mathscr{E}g(X_1, \ldots, X_k) = \sum_{x_1} \cdots \sum_{x_k} g(x_1, \ldots, x_k) P(X_1 = x_1, \ldots, X_k = x_k).$$

If, on the other hand, $(X_1, \ldots, X_k)$ has joint density function f and

$$\int_{-\infty}^{\infty} \cdots \int_{-\infty}^{\infty} |g(x_1, \ldots, x_k)| f(x_1, \ldots, x_k) \prod_{j=1}^{k} dx_j < \infty$$

then we define the expected value of g by:

$$\mathscr{E}g(X_1, \ldots, X_k) = \int_{-\infty}^{\infty} \cdots \int_{-\infty}^{\infty} g(x_1, \ldots, x_k) f(x_1, \ldots, x_k) \prod_{j=1}^{k} dx_j.$$

Some immediate consequences of Definition 1 are listed below. The reader is asked to supply the proofs. (*In the following, whenever we write $\mathscr{E}g(X_1, \ldots, X_k)$ for any function g we assume that g is a random variable and the expected value exists in the sense of Definition 1.*)

I. Let a and b be constants. Then

$$\mathscr{E}\{ag(X_1, \ldots, X_k) + b\} = a\mathscr{E}g(X_1, \ldots, X_k) + b.$$

II. Let $g_1, g_2, \ldots, g_n$ be real valued functions of $X_1, \ldots, X_k$. Then

$$\mathscr{E}\{g_1(X_1, \ldots, X_k) + \cdots + g_n(X_1, \ldots, X_k)\}$$
$$= \mathscr{E}g_1(X_1, \ldots, X_k) + \cdots + \mathscr{E}g_n(X_1, \ldots, X_k).$$

III. Let $X_1, X_2, \ldots, X_k$ be independent random variables and $g_1, g_2, \ldots, g_k$ be real valued functions of $X_1, X_2, \ldots, X_k$ respectively. Then

$$\mathscr{E}\prod_{j=1}^{k} g_j(X_j) = \prod_{j=1}^{k} \mathscr{E}g_j(X_j).$$

Since III is an important result we give a proof. We restrict attention to the continuous case. The proof for the discrete case (or mixed case) is similar. We have

$$\mathscr{E}\prod_{j=1}^{k} g_j(X_j) = \int_{-\infty}^{\infty} \cdots \int_{-\infty}^{\infty} \prod_{j=1}^{k} g_j(x_j) f(x_1,\ldots,x_k) \prod_{j=k}^{1} dx_j$$

$$= \int_{-\infty}^{\infty} \cdots \int_{-\infty}^{\infty} \prod_{j=1}^{k} \left[g_j(x_j) f_j(x_j) \right] \prod_{j=k}^{1} dx_j$$

where f_j is the marginal density of X_j and we have used the independence of $X_1,\ldots,X_k$. It follows that

$$\mathscr{E}\prod_{j=1}^{k} g_j(X_j) = \int_{-\infty}^{\infty} g_1(x_1) f_1(x_1) \, dx_1 \int_{-\infty}^{\infty} g_2(x_2) f_2(x_2) \, dx_2 \cdots$$

$$\times \int_{-\infty}^{\infty} g_k(x_k) f_k(x_k) \, dx_k$$

$$= \prod_{j=1}^{k} \mathscr{E} g_j(X_j)$$

as asserted. We emphasize that the *converse of property III is not true.*
    Choosing

$$g(x_1,\ldots,x_k) = \left(x_j - \mathscr{E}X_j \right)^n, \qquad n \geq 1, \text{ integral}$$

we see that (in the continuous case):

$$\mathscr{E}\left(X_j - \mathscr{E}X_j \right)^n = \int_{-\infty}^{\infty} \cdots \int_{-\infty}^{\infty} \left(x_j - \mathscr{E}X_j \right)^n \prod_{i=1}^{k} f_i(x_i) \prod_{i=k}^{1} dx_i$$

$$= \int_{-\infty}^{\infty} \left(x_j - \mathscr{E}X_j \right)^n f_j(x_j) \, dx_j \int_{-\infty}^{\infty} \cdots \int_{-\infty}^{\infty} \prod_{\substack{i=k \\ i \neq j}}^{k} f_i(x_i) \prod_{\substack{i=k \\ i \neq j}}^{1} dx_i$$

$$= \int_{-\infty}^{\infty} \left(x_j - \mathscr{E}X_j \right)^n f_j(x_j) \, dx_j$$

which is the nth central moment of the marginal distribution of X_j. Thus $\mathscr{E}X_j$ is the mean and $\mathscr{E}(X_j - \mathscr{E}X_j)^2$ is the variance of the marginal distribution of X_j. In other words, we can compute moments of each X_j either from the marginal distribution of X_j or from the joint distribution of $X_1,\ldots,X_k$ whichever happens to be convenient.

Remark 1.   Since $Y = g(X_1, X_2,\ldots,X_k)$ is a (one-dimensional) random variable, it has a distribution that can be computed from the joint distribution of

$X_1, X_2, \ldots, X_k$. This is done in Chapter 8. Suppose f_Y is the density function of Y if Y is of the continuous type [or $P(Y = y_j)$ is the probability function of Y if it is of the discrete type]. If $\mathscr{E}|Y| < \infty$ then

$$\mathscr{E}Y = \begin{cases} \displaystyle\int_{-\infty}^{\infty} y f_Y(y)\, dy, & \text{continuous case,} \\[2mm] \displaystyle\sum_{j=1}^{\infty} y_j P(Y = y_j), & \text{discrete case.} \end{cases}$$

Does $\mathscr{E}Y$ coincide with $\mathscr{E}g(X_1, X_2, \ldots, X_k)$ as computed in Definition 1? This is true but we will not prove it here. The point is that if the joint distribution of $X_1, \ldots, X_k$ is known then we can compute $\mathscr{E}g(X_1, X_2, \ldots, X_k)$ directly from Definition 1 without computing the distribution of $Y = g(X_1, X_2, \ldots, X_k)$.

Example 1.   Mean Service Time at a Bank.   Let X be the total time (minutes) that a customer spends at a bank, and Y the time she spends waiting in line before reaching the teller. Assuming that the customer gets into a line immediately after entering the bank, the service time is $X - Y$ and the mean service time is given by $\mathscr{E}(X - Y)$. We can use either Definition 1 or property II with $g_1(X, Y) = X$ and $g_2(X, Y) = Y$ to compute $\mathscr{E}(X - Y)$.

Suppose X, Y have joint density

$$f(x, y) = \lambda^2 e^{-\lambda x}, \qquad 0 \le y \le x < \infty, \text{ and zero elsewhere.}$$

Then, from Definition 1:

$$\mathscr{E}(X - Y) = \int_0^{\infty}\left[\int_0^x (x - y)\lambda^2 e^{-\lambda x}\, dy\right] dx$$

$$= \lambda^2 \int_0^{\infty} e^{-\lambda x}\left[x^2 - \frac{x^2}{2}\right] dx = \frac{\lambda^2}{2}\int_0^{\infty} x^2 e^{-\lambda x}\, dx = \frac{1}{\lambda}$$

on integration by parts. Alternatively, the marginal of X is given by

$$f_1(x) = \lambda^2 e^{-\lambda x}\int_0^x dy = \begin{cases} \lambda^2 x e^{-\lambda x}, & x \ge 0 \\ 0, & x < 0, \end{cases}$$

and that of Y by

$$f_2(y) = \lambda^2 \int_y^{\infty} e^{-\lambda x}\, dx = \begin{cases} \lambda e^{-\lambda y}, & y \ge 0 \\ 0, & y < 0. \end{cases}$$

Hence

$$\mathscr{E}(X - Y) = \mathscr{E}X - \mathscr{E}Y \qquad \text{(property II)}$$

$$= \frac{2}{\lambda} - \frac{1}{\lambda} = \frac{1}{\lambda} \qquad \text{(integration by parts).} \qquad \square$$

Example 2. Average Amount of Money Spent on a Commodity. The wholesale price X of a certain commodity and the total sales Y have a joint distribution given (approximately) by

$$f(x, y) = \begin{cases} xe^{-xy}, & \theta < x < \theta + 1, y > 0 \\ \\ 0 & \text{otherwise.} \end{cases} \quad (\theta > 0)$$

What is the average amount of money spent on this commodity?

Since the total amount of money spent is XY, the average amount of money spent is

$$\mathscr{E}(XY) = \int_\theta^{\theta+1} \int_0^\infty xy(xe^{-xy}) \, dy \, dx$$

$$= \int_\theta^{\theta+1} x^2 \left[\frac{-ye^{-xy}}{x} \Big|_0^\infty + \int_0^\infty \frac{e^{-xy}}{x} \, dy \right] dx$$

$$= \int_\theta^{\theta+1} x \left[\frac{1}{x} \right] dx = 1. \qquad \square$$

Example 3. Families with Three Children. Consider families with three children. Let X denote the number of boys and Y the excess of boys over girls. Then X and Y have joint probability function given by

	X	Number of Boys			
Y	0	1	2	3	
-3	1/8	0	0	0	1/8
Excess $\quad -1$	0	3/8	0	0	3/8
1	0	0	3/8	0	3/8
3	0	0	0	1/8	1/8
	1/8	3/8	3/8	1/8	1

We note that

$$\mathscr{E}XY = \Sigma_i \Sigma_j x_i y_j P(X = x_i, Y = y_j)$$

$$= (1)(-1)(3/8) + (2)(1)(3/8) + (3)(3)(1/8) = 1.5.$$

A look at the table shows that X and Y increase or decrease together and, moreover, that all nonzero values are along the diagonal. This suggests that X and Y have a linear relationship. Indeed, $Y =$ (number of boys less number of girls) $= X - (3 - X) = 2X - 3$. The measurement of the extent to which X and Y vary together is the subject of discussion below. $\qquad \square$

Covariance and Correlation

Let X and Y be two random variables with joint distribution function F and marginal distribution functions F_1 and F_2 respectively. We recall that X and Y are independent if and only if

$$F(x, y) = F_1(x)F_2(y)$$

for all $x, y \in \mathbb{R}$. We saw that (stochastic) independence of X and Y simply means that the knowledge of one of X or Y provides no new probabilistic information about the other variable. Therefore, if X and Y are dependent we feel that X and Y somehow vary together and the knowledge of one variable may allow us to say something about the other. If X is the shoe size and Y the height of an individual and if we know that X is large, we feel pretty confident that Y is also large. Similarly, when the price of a commodity goes up we expect that the demand will go down. If a person has a few drinks we do not hesitate to recommend that she or he drive carefully or not drive at all, because we feel that the greater the amount of alcohol in the blood the poorer the motor coordination.

How do we measure this connection or association between X and Y? How do X and Y vary jointly? Any such measure should clearly depend on the joint distribution of X and Y. Moreover, if X and Y increase together then $(X - \mathscr{E}X)(Y - \mathscr{E}Y)$ should be positive whereas if X decreases while Y increases (and conversely) then the product should be negative. Hence the average value of $(X - \mathscr{E}X)(Y - \mathscr{E}Y)$ provides a measure of association or joint variation of X and Y.

DEFINITION 2. (COVARIANCE). If $\mathscr{E}\{(X - \mathscr{E}X)(Y - \mathscr{E}Y)\}$ exists, we call it the covariance between X and Y.

If covariance exists, we write:

(1) $$\sigma_{XY} = \text{cov}(X, Y) = \mathscr{E}\left[(X - \mu_1)(Y - \mu_2)\right]$$

where $\mu_1 = \mathscr{E}X$ and $\mu_2 = \mathscr{E}Y$. Note that

$$\sigma_{XX} = \text{cov}(X, X) = \mathscr{E}(X - \mu_1)^2 = \sigma_X^2$$

and

$$\sigma_{XY} = \sigma_{YX}.$$

For computational purposes, the following formula is sometimes useful:

(2) $$\sigma_{XY} = \mathscr{E}XY - \mathscr{E}X\mathscr{E}Y.$$

For the proof of (2), we note that

$$\mathscr{E}(X - \mu_1)(Y - \mu_2) = \mathscr{E}\{XY - \mu_1 Y - \mu_2 X + \mu_1\mu_2\}$$
$$= \mathscr{E}XY - \mu_1\mu_2$$

in view of property II.

PROPOSITION 1.   *If X and Y are independent,*

(3) $\text{cov}(X, Y) = 0.$

Proof.   If X and Y are independent, it follows from property III that

$$\mathscr{E}XY = \mathscr{E}X\mathscr{E}Y = \mu_1\mu_2$$

and the result follows from (2).   □

Remark 2.   The converse of Proposition 1 is false. If $\text{cov}(X, Y) = 0$, it does not follow that X and Y are independent. For, let X be symmetric with $\mathscr{E}X = 0$ and $\mathscr{E}|X|^3 < \infty$. Let $Y = X^2$. Then

$$\mathscr{E}\{(X - \mu_1)(Y - \mu_2)\} = \mathscr{E}X^3 = 0$$

but X and Y are strongly dependent. For some exceptions, see Problem 12 and Section 7.7.

Example 4.   Covariance Between Service Time and Total Time Spent.   For the random variables X and Y of Example 1, we have

$$\mathscr{E}XY = \lambda^2 \int_0^\infty \left(\int_0^x xye^{-\lambda x}\, dy \right) dx$$

$$= \lambda^2 \int_0^\infty xe^{-\lambda x} \cdot \frac{x^2}{2}\, dx$$

$$= \frac{\lambda^2}{2} \cdot \frac{6}{\lambda^4} \qquad \text{(integration by parts)}$$

$$= 3/\lambda^2.$$

Since $\mathscr{E}X = 2/\lambda$ and $\mathscr{E}Y = 1/\lambda$ we have from (2)

$$\text{cov}(X, Y) = \frac{3}{\lambda^2} - \left(\frac{2}{\lambda}\right)\left(\frac{1}{\lambda}\right) = \frac{1}{\lambda^2} > 0$$

which is not surprising, since the greater the service time Y the greater the total time spent X.   □

Example 5.   Covariance Between Stock and Sales.   The weight X in tons of a bulk item stocked by a supplier at the beginning of every month and the weight Y of this item sold by the supplier during the month have a joint probability density function given by

$$f(x, y) = \frac{1}{10x}, \qquad 0 \le y \le x \le 10, \text{ and zero elsewhere.}$$

In order to compute $\text{cov}(X, Y)$, we compute:

$$\mathscr{E}XY = \iint xyf(x, y)\, dx\, dy = \int_0^{10}\left(\int_y^{10} xy\frac{1}{10x}\, dx\right) dy = \frac{1}{10}\int_0^{10} y(10 - y)\, dy$$

$$= \frac{1}{10}\left(\frac{10^3}{2} - \frac{10^3}{3}\right) = \frac{100}{6},$$

$$\mathscr{E}X = \int_0^{10}\left(\int_y^{10} x\frac{1}{10x}\, dx\right) dy = \frac{1}{10}\int_0^{10}(10 - y)\, dy = 10 - \frac{10}{2} = 5,$$

$$\mathscr{E}Y = \int_0^{10}\left(\int_0^x y\frac{1}{10x}\, dy\right) dx = \int_0^{10}\frac{1}{10x}\cdot\frac{x^2}{2}\, dx = \frac{1}{20}\cdot\frac{10^2}{2} = \frac{5}{2}.$$

Hence:

$$\text{cov}(X, Y) = \mathscr{E}XY - \mathscr{E}X\mathscr{E}Y = \frac{100}{6} - 5\left(\frac{5}{2}\right) = \frac{25}{6}. \qquad \square$$

The following result will be quite useful.

PROPOSITION 2. (CAUCHY–SCHWARZ INEQUALITY). *If $\mathscr{E}X^2 < \infty$, $\mathscr{E}Y^2 < \infty$ then σ_{XY} exists and*

(4) $$\sigma_{XY}^2 \leq \text{var}(X)\text{var}(Y).$$

Equality in (4) holds if and only if there exist $\alpha, \beta, c \in \mathbb{R}$, α, β not both zero such that $P(\alpha X + \beta Y = c) = 1$.

Proof. Since $(a - b)^2 \geq 0$, we have

$$(x - \mu_1)(y - \mu_2) \leq \frac{(x - \mu_1)^2 + (y - \mu_2)^2}{2}$$

and it follows that σ_{XY} exists whenever σ_X^2 and σ_Y^2 do.

Let $\alpha, \beta \in \mathbb{R}$. Then for all $\alpha, \beta \in \mathbb{R}$,

(5) $$\mathscr{E}\left(\alpha(X - \mu_1) + \beta(Y - \mu_2)\right)^2 = \alpha^2\sigma_X^2 + \beta^2\sigma_Y^2 + 2\alpha\beta\sigma_{XY} \geq 0.$$

If X is degenerate $\sigma_X = 0$ and (4) holds trivially. Hence we may assume that $\sigma_X^2 > 0$. Choose $\alpha = -\sigma_{XY}/\sigma_X^2$ in (5). Then

$$\frac{\sigma_{XY}^2}{\sigma_X^2} + \beta^2\sigma_Y^2 - 2\beta\frac{\sigma_{XY}^2}{\sigma_X^2} \geq 0 \qquad \text{for all } \beta \in \mathbb{R}$$

and in particular with $\beta = 1$,

$$\sigma_Y^2 \geq \frac{\sigma_{XY}^2}{\sigma_X^2}$$

which is (4). Equality holds in (4) if and only if there exist α and $\beta \in \mathbb{R}$ (not both zero) such that

$$P\{\alpha(X - \mu_1) + \beta(Y - \mu_2) = 0\} = 1.$$

That is, if and only if

$$P(\alpha X + \beta Y = c) = 1$$

for some α, β, not both zero, and $c \in \mathbb{R}$. ☐

Remark 3. In Proposition 2, $\mathscr{E}X^2 < \infty$, $\mathscr{E}Y^2 < \infty$ is a sufficient but not a necessary condition for the existence of cov(X, Y). For example, if X and Y are independent then

$$\text{cov}(X, Y) = \mathscr{E}(X - \mu_1)\mathscr{E}(Y - \mu_2)$$

exists (and equals 0) provided $\mathscr{E}|X| < \infty$ and $\mathscr{E}|Y| < \infty$. In fact cov(X, Y) may exist but $\mathscr{E}X^2$ or $\mathscr{E}Y^2$ may be infinite even when X and Y are dependent (Problems 14 and 15).

The following result points out an unsatisfactory feature of covariance as a measure of the joint variation of X and Y.

PROPOSITION 3. *Let a, b, c, and d be real numbers. Then*

(6) $$\text{cov}(aX + b, cY + d) = ac\,\text{cov}(X, Y).$$

Proof. By definition

$$\text{cov}(aX + b, cY + d) = \mathscr{E}\{[aX + b - (a\mathscr{E}X + b)][cY + d - (c\mathscr{E}Y + d)]\}$$

$$= \mathscr{E}\{[a(X - \mathscr{E}X)][c(Y - \mathscr{E}Y)]\} = ac\,\text{cov}(X, Y). \qquad ☐$$

COROLLARY 1. Var($aX + b$) = a^2var(X).

COROLLARY 2. *Let $\mathscr{E}X = \mu_1$, var(X) = σ_1^2, $\mathscr{E}Y = \mu_2$, var(Y) = σ_2^2. Then*

(7) $$\text{cov}\left(\frac{X - \mu_1}{\sigma_1}, \frac{Y - \mu_2}{\sigma_2}\right) = \frac{\text{cov}(X, Y)}{\sigma_1 \sigma_2}.$$

Proof. Choose $a = \dfrac{1}{\sigma_1}, b = -\dfrac{\mu_1}{\sigma_1}, c = \dfrac{1}{\sigma_2}, d = -\dfrac{\mu_2}{\sigma_2}$ in (6). ☐

The covariance between X and Y is positive if X and Y tend to increase or decrease together and negative if X and Y tend to move in opposite directions. Unfortunately, the covariance is sensitive to the scales of measurement used (Proposition 3). Therefore, it is not possible to interpret what a numerical value of cov(X, Y) means. For example, suppose X and Y are measured in meters and cov(X, Y) = 0.001. If we change the scale to centimeters by writing $U = 100X$ and $V = 100Y$, then from (6):

$$\text{cov}(U, V) = (100)^2(.001) = 10.$$

A simple change in units of measurement leads to a completely different impression about the strength of relationship between X and Y. It is clear, therefore, that we need to devise a measure that does not depend on the scale. This is accomplished by measuring both X and Y in standard units, as in Corollary 2.

DEFINITION 3.   (CORRELATION COEFFICIENT).   The correlation coefficient between X and Y, denoted by $\rho = \rho(X, Y)$ is the covariance between standardized X and standardized Y. That is,

$$\rho = \text{cov}\left(\frac{X - \mu_1}{\sigma_1}, \frac{Y - \mu_2}{\sigma_2}\right) = \frac{\text{cov}(X, Y)}{\sigma_1\sigma_2}.$$

We note that cov(X_1, X_2) = $\rho\sigma_1\sigma_2$ and, moreover, in view of Proposition 2:

(8) $|\rho| \leq 1$

with equality if, and only if with probability one, X and Y lie on the same line. In fact, $\rho = 1$ if and only if $Y = (\sigma_2/\sigma_1)X$ + constant (the line has positive slope) with probability one and $\rho = -1$ if and only if $Y = -(\sigma_2/\sigma_1)X$ + constant (the line has negative slope) with probability one. Hence ρ may properly be called the *coefficient of linear correlation* between X and Y. A value of $|\rho|$ close to 1 indicates a linear relationship between X and Y, whereas in-between values of ρ indicate a departure from a linear relationship. If $\rho < 0$, the trend of the (X, Y) values is downward to the right and if $\rho > 0$, it is upward to the right. If $\rho = 0$ ($\rho = 0$ if and only if cov(X, Y) = 0), we say that X and Y are *uncorrelated*. It is clear that ρ and cov(X, Y) have the same sign.

The specific way in which ρ measures the straight line dependence is considered in Example 5.3.2. Care is needed in interpretation of the numerical value of ρ, as the following example illustrates.

Example 6.   Uniform Distribution.   Let X have density

$$f(x) = 1, \qquad 0 < x < 1, \text{ and zero elsewhere.}$$

This is the uniform density on $(0,1)$. Let $Y = X^k$ for $k > 0$. Then X and Y are dependent. We compute ρ. Clearly, for $n > -1$

$$\mathscr{E}X^n = \int_0^1 x^n(1)\, dx = \frac{1}{n + 1}.$$

In particular, $\mathscr{E}X = \frac{1}{2}$, $\mathscr{E}Y = 1/(k+1)$,

$$\sigma_1^2 = \mathrm{var}(X) = \mathscr{E}X^2 - (\mathscr{E}X)^2 = \frac{1}{12},$$

$$\sigma_2^2 = \mathrm{var}(Y) = \mathscr{E}Y^2 - (\mathscr{E}Y)^2 = \frac{1}{2k+1} - \frac{1}{(k+1)^2} = \left(\frac{k}{k+1}\right)^2 \frac{1}{(2k+1)},$$

and

$$\mathrm{cov}(X,Y) = \mathscr{E}(XY) - \mathscr{E}X\mathscr{E}Y = \frac{1}{k+2} - \frac{1}{2(k+1)} = \frac{k}{2(k+1)(k+2)}.$$

It follows that

$$\rho = \frac{\mathrm{cov}(X,Y)}{\sigma_1\sigma_2} = \frac{\sqrt{6k+3}}{k+2}.$$

If $k = 2$, then $\rho = .968$ and if k is large, ρ is small. In fact $\rho \to 0$ as $k \to \infty$. Thus, ρ may be small even though X and Y are strongly dependent and ρ may be large (close to 1) even though the relationship between X and Y is not linear.   □

Example 7. *Correlation Between Service Time and Total Time.* In Example 4 we saw that $\mathrm{cov}(X,Y) = 1/\lambda^2$.

Using the marginal densities

$$f_1(x) = \lambda^2 x e^{-\lambda x}, \qquad x \geq 0, \text{ and zero elsewhere,}$$

and

$$f_2(y) = e^{-\lambda y}, \qquad y \geq 0, \text{ and zero elsewhere,}$$

as computed in Example 1 we have

$$\mathrm{var}(X) = \frac{2}{\lambda^2} \quad \text{and} \quad \mathrm{var}(Y) = \frac{1}{\lambda^2}.$$

It follows that

$$\rho = \frac{\mathrm{cov}(X,Y)}{\sigma_1\sigma_2} = \frac{1/\lambda^2}{\sqrt{2/\lambda^4}} = \frac{1}{\sqrt{2}} = .71.   \qquad □$$

Remark 4. We emphasize that ρ is not a measure of how strongly two random variables X and Y depend on each other. It only measures the extent to which they are associated or vary together. It should not be used to exhibit a causal (cause and effect) relationship between X and Y. If X and Y are, for instance, positively correlated it does not follow that a change in $X(Y)$ causes a change in

$Y(X)$. Frequently X and Y may appear to be highly correlated but they may not be directly associated. The high correlation may be due to a third variable correlated with both X and Y.

Problems for Section 5.2

1. Let X and Y be numbers of customers at two branches of a chain store. Assuming X and Y are independent and identically distributed, find the average number of customers in the two stores. Find $\mathscr{E}(XY)$.

2. Assume that the round-off error in rounding off a number of the form $N.NN$ to the nearest integer has density function

$$f(x) = 1, \qquad -.5 \le x \le .5, \text{ and zero elsewhere.}$$

    If five numbers are independently rounded off to their nearest integers, what is the average total round-off error?

3. The six words in the sentence I DO NOT LIKE THIS GLASS are written on six identical slips and put in a box. A slip is then selected at random. Let X be length of the word selected and Y the number of I's in the selected word. What is the average number of letters other than I in the word selected?

4. Let X be the proportion of carbon monoxide emission per minute by a car without a PCV (positive crankcase ventilation) valve, and let Y be the proportion of carbon monoxide emission per minute by the same car under the same conditions after a PCV valve is installed. Suppose X, Y have joint density function given by

$$f(x, y) = 2 \qquad 0 \le y \le x \le 1,$$

    and

$$f(x, y) = 0 \qquad \text{elsewhere.}$$

    Find the average reduction in the proportion of emission per minute due to the PCV valve.

5. From a group of five Republicans, three Democrats, and two Independents, a committee of three is selected at random. Let X be the number of Republicans and Y the number of Democrats in the committee. Find $\mathscr{E}(X + Y), \mathscr{E}(X - Y), \mathscr{E}(XY)$.

6. Suppose the joint distribution of the fractional part Y of a number $X (\ge 1)$ resulting from a computer operation involving a random number generator is given by

$$f(x, y) = \begin{cases} \dfrac{2(k-1)(k-2)}{3k-4}\left(\dfrac{x+y}{x^k}\right), & 1 \le x < \infty, 0 \le y \le 1 \\ 0 & \text{elsewhere} \end{cases}$$

    where $k > 2$ is a fixed known number. Find
    (i) $\mathscr{E}(X + Y)$,   (ii) $\mathscr{E}(X - Y)$,   (iii) $\mathscr{E}(X^\alpha Y^\beta)$ whenever it exists.

7. In the following cases, state whether you would expect a positive, negative, or no correlation.
   (i)   Height and weight of students in a class.
   (ii)   SAT scores and grade point average in freshman year.
   (iii) Ages of husband and wife.
   (iv)   Education level and income.
   (v)   Grades in psychology and statistics.
   (vi)   Diastolic and systolic blood pressures.
   (vii) Nicotine content and tar in a cigarette.
   (viii) Length of skirt and sense of humor.
   (ix)   Hat size and batting average.
   (x)   Age of premature infant and weight.
   (xi)   Age of an appliance and its trade-in value.

8. Let X have density given below and let $Y = X^\alpha$ where $\alpha \geq 2$ is an integer. Find the covariance between X and Y and the correlation coefficient.

$$f(x) = e^{-x}, \qquad x > 0, \text{ and zero elsewhere.}$$

   Show that $\rho \to 0$ as $\alpha \to \infty$.

9. Let X be the number of heads and Y the difference in absolute value between number of heads and number of tails in three independent tosses of a fair coin. Find the correlation coefficient between X and Y.

10. Let X and Y have the joint density given below. Find the correlation coefficient between X and Y.
    (i) $f(x, y) = \dfrac{1 + xy}{4}$, $|x| < 1$, $|y| < 1$, and zero elsewhere.
    (ii) $f(x, y) = x + y$, $0 < x < 1$, $0 < y < 1$, and zero elsewhere.
    (iii) $f(x, y) = x^2 + (xy/3)$, $0 < x < 1$, $0 < y < 2$, and zero elsewhere.
    (iv) $f(x, y) = 3y(1 + x)^{-6}\exp\{-y/(1 + x)\}$, $x, y \geq 0$, and zero elsewhere.
    (v) $f(x, y) = (4/5)(x + 3y)\exp(-x - 2y)$, $x, y \geq 0$, and zero elsewhere.
    (vi) $f(x, y) = 1$, $0 < |y| < x < 1$ and zero elsewhere.
    (vii) $f(x, y) = e^{-y}$, $0 < x < y < \infty$, and zero elsewhere.

11*. Let X and Y be identically distributed random variables with probability function

$$P(X = k) = P(Y = k) = \frac{1}{N}, \qquad k = 1, 2, \ldots, N; \; N > 1.$$

    (i) Show that $\rho = 1 - \dfrac{6\mathscr{E}(X - Y)^2}{N^2 - 1}$.
    (ii) Show that $\rho = 1$ if and only if $P(X = Y) = 1$ and $\rho = -1$ if and only if $P(Y = N + 1 - X) = 1$.
    [Hint: Use $\Sigma_1^N k = N(N + 1)/2$, $\Sigma_1^N k^2 = N(N + 1)(2N + 1)/6$. See Problem 6.2.1 for proof. Note that $\mathscr{E}XY = (\mathscr{E}X^2 + \mathscr{E}Y^2 - \mathscr{E}(X - Y)^2)/2$.]

12*. Let X and Y have probability functions

$$P(X = x_1) = p_1 = 1 - P(X = x_2)$$

and

$$P(Y = y_1) = p_2 = 1 - P(Y = y_2).$$

Show that X and Y are independent if and only if $\rho(X, Y) = 0$.
[Hint: Let $U = aX + b$, $V = cY + d$. From Proposition 3, $\rho = 0$ implies $\rho(U, V) = 0$. Use (2) to show that U and V are independent by choosing $a = 1$, $b = -x_1$ and $c = 0$, $d = -y_1$.]

13. Let X and Y have common mean 0, common variance 1, and correlation coefficient ρ. Show that

$$\mathscr{E}\{\max(X^2, Y^2)\} \le 1 + \sqrt{1 - \rho^2}\,.$$

[Hint: $\max(x, y) = (1/2)\{|x + y| + |x - y|\}$. Use Cauchy–Schwarz inequality.]

14*. In this problem, we ask the reader to show that $\mathrm{cov}(X, Y)$ exists but $\mathscr{E}X^2$ is not finite (see Remark 2). Suppose X is the length of time between successive demands for auto parts at a store with density

$$h(x|y) = ye^{-yx}, \qquad x \ge 0, \text{ and zero elsewhere,}$$

where $1/y$ is the mean demand. Suppose Y itself is a random variable with density

$$f_2(y) = ye^{-y}, \qquad y \ge 0, \text{ and zero elsewhere.}$$

(i) Show that $\mathrm{cov}(X, Y)$ exists and find $\mathrm{cov}(X, Y)$.
(ii) Show that the unconditional density of X is

$$f_1(x) = \frac{2}{(1 + x)^3}, \qquad x \ge 0, \text{ and zero elsewhere}$$

so that $\mathrm{var}(X)$ does not exist.

15*. Show with the help of the following joint density that $\mathscr{E}(XY)$ may exist but even $\mathscr{E}Y$ may not:

$$f(x, y) = x\exp\{-x(1 + y)\}, \qquad x \ge 0, y \ge 0, \text{ and zero elsewhere.}$$

Show that
(i) $\mathscr{E}XY = 1$.   (ii) $\mathscr{E}Y$ does not exist.

16. Let X be any random variable. Show that X is independent of itself if and only if $P(X = c) = 1$ for some constant c.

5.3   LINEAR COMBINATIONS OF RANDOM VARIABLES

Let $(X_1, X_2, \dots, X_m)$ be a random vector and let $a_1, a_2, \dots, a_m$ be real numbers (not all zero). In statistics, linear combinations of the form $S = \sum_{j=1}^{m} a_j X_j$ occur quite often and we will find it convenient to compute the mean and the variance of S. For example, if $X_1, \ldots, X_m$ is a random sample from a distribution function F, we have said that the statistic $\overline{X} = m^{-1}\sum_{j=1}^{m} X_j$, which is the sample mean, is an intuitively appealing estimate of $\mu = \mu_F$. What are some properties of this estimate? What is its precision as measured by its variance? How do we use Chebyshev's inequality to construct a confidence interval estimate for μ based on $\overline{X}$?

THEOREM 1. *Let* $X_1, X_2, \ldots, X_m$ *and* $Y_1, Y_2, \ldots, Y_n$ *be random variables with* $\mathscr{E} X_i = \mu_i$ *and* $\mathscr{E} Y_i = \xi_i$. *Set*

$$U_m = \sum_{j=1}^{m} a_j X_j, \qquad V_n = \sum_{j=1}^{n} b_j Y_j,$$

for real numbers $a_1, \ldots, a_m$ *and* $b_1, \ldots, b_n$. *Then*

(i) $$\mathscr{E} U_m = \sum_{j=1}^{m} a_j \mu_j, \qquad \mathscr{E} V_n = \sum_{j=1}^{n} b_j \xi_j,$$

(ii) $$\mathrm{var}(U_m) = \sum_{j=1}^{m} a_j^2 \mathrm{var}(X_j) + 2\sum\sum_{i<j} a_i a_j \mathrm{cov}(X_i, X_j),$$

$$\mathrm{var}(V_m) = \sum_{j=1}^{m} b_j^2 \mathrm{var}(Y_j) + 2\sum\sum_{i<j} b_i b_j \mathrm{cov}(Y_i, Y_j),$$

where $\sum\sum_{i<j}$ *denotes the double sum over all* i *and* j *with* $i < j$, *and*

(iii) $$\mathrm{cov}(U_m, V_n) = \sum_{i=1}^{m} \sum_{j=1}^{n} a_i b_j \mathrm{cov}(X_i, Y_j).$$

[*In parts* (*ii*) *and* (*iii*) *we have assumed that* $\mathrm{var}(X_i)$ *and* $\mathrm{var}(Y_j)$ *are finite for all* i *and* j.]

Proof. Part (i) follows trivially from properties I and II of Section 5.2. It suffices to prove part (iii) since part (ii) is a special case of (iii).

By definition of covariance, using part (i), we have:

$$\mathrm{cov}(U_m, V_m) = \mathscr{E}\{(U_m - \mathscr{E} U_m)(V_n - \mathscr{E} V_n)\}$$

$$= \mathscr{E}\left\{\left[\sum_{i=1}^{m} a_i(X_i - \mu_i)\right]\left[\sum_{j=1}^{n} b_j(Y_j - \xi_j)\right]\right\}$$

$$= \mathscr{E}\left\{\sum_{i=1}^{m} \sum_{j=1}^{n} a_i b_j(X_i - \mu_i)(Y_j - \xi_j)\right\}$$

$$= \sum_{i=1}^{m} \sum_{j=1}^{n} a_i b_j \mathrm{cov}(X_i, Y_j)$$

in view of properties I and II of Section 5.2 and the definition of covariance.

To obtain part (ii) we take $m = n$, $a_i = b_i$, $X_i = Y_i$ and we have

$$\text{var}(U_m) = \text{cov}(U_m, U_m)$$

$$= \sum_{i=1}^{m} \sum_{j=1}^{m} a_i a_j \text{cov}(X_i, X_j)$$

$$= \sum_{i=1}^{m} a_i^2 \text{cov}(X_i, X_i) + \sum\sum_{i \neq j} a_i a_j \text{cov}(X_i, X_j)$$

$$= \sum_{i=1}^{m} a_i^2 \text{var}(X_i) + 2\sum\sum_{i<j} a_i a_j \text{cov}(X_i, X_j). \qquad \square$$

The following corollary is very useful.

COROLLARY 1. *Let* $X_1, \ldots, X_m$ *be uncorrelated random variables. If* $\text{var}(X_i)$ *is finite for each* i, *then*

$$\text{var}\left(\sum_{j=1}^{m} a_j X_j\right) = \sum_{j=1}^{m} a_j^2 \text{var}(X_j).$$

This corollary follows from part (ii) of Theorem 1 by noting that $\text{cov}(X_i, X_j) = 0$ for all $i \neq j$, $i, j = 1, 2, \ldots, m$. An important special case in random sampling from a distribution with finite mean μ and finite variance σ^2 is especially useful and will be used repeatedly. This is the case when $a_i = 1/n$. In that case $U_n = \sum_{i=1}^{n} (1/n) X_i = \overline{X}$ and we have the following important result.

Let $X_1, X_2, \ldots, X_n$ be a random sample from a distribution with finite mean μ. Then
(i) $\mathscr{E} \overline{X} = \mu$.
If $\text{var}(X_i) = \sigma^2$ is finite, then
(ii) $\text{var}(\overline{X}) = \sigma^2/n$.

We note that independence is not required for part (i) and only " uncorrelated" is required instead of independence for part (ii). The result says that the *sample mean is an unbiased estimate of the population mean and the precision of the sample mean as measured by its variance is given by* $\text{var}(\overline{X}) = \sigma^2/n$. Precision increases (that is, variance decreases) as n increases, and $\to 0$ as $n \to \infty$.

We next show that S^2 is unbiased for σ^2. That is,

$$\boxed{\mathscr{E} S^2 = \sigma^2}$$

In fact,

$$\mathscr{E}S^2 = \frac{1}{n-1}\mathscr{E}\left(\sum_{i=1}^{n} X_i^2 - n\bar{X}^2\right) = \frac{1}{n-1}\left(n\mathscr{E}X_1^2 - n\mathscr{E}\bar{X}^2\right)$$

$$= \frac{1}{n-1}\left[n\mathscr{E}X_1^2 - n\left(\operatorname{var}\bar{X} + (\mathscr{E}\bar{X})^2\right)\right]$$

$$= \frac{1}{n-1}\left[n\left(\mathscr{E}X_1^2 - (\mathscr{E}\bar{X})^2\right) - n \cdot \frac{\sigma^2}{n}\right] = \sigma^2.$$

Example 1. Sampling from a Finite Population. Consider a finite population Π of N distinct elements $\omega_1, \omega_2, \ldots, \omega_N$. Frequently our object is to estimate the population mean

$$(1) \qquad\qquad \mu = \sum_{i=1}^{N} \frac{\omega_i}{N}$$

or the population variance

$$(2) \qquad\qquad \sigma^2 = \sum_{i=1}^{N} \frac{(\omega_i - \mu)^2}{N}.$$

We take a simple random sample $X_1, X_2, \ldots, X_n$ of size $n \ (\leq N)$ without replacement. Then $X_1, X_2, \ldots, X_n$ has joint probability function

$$(3) \qquad P(X_1 = x_1, \ldots, X_n = x_n) = \frac{1}{N(N-1)\ldots(N-n+1)}$$

where $x_1, \ldots, x_n \in \Pi$. It is then natural to consider $\bar{X}$ as an estimate of μ. How good is $\bar{X}$ as an estimate of μ? We note that the marginal distribution of each X_j is given by

$$(4) \qquad\qquad P(X_j = \omega_i) = \frac{1}{N}, i = 1, 2, \ldots, N$$

so that $\mathscr{E}X_j = \mu$ for each j. It follows from Theorem 1 (i) that $\bar{X}$ is unbiased for μ, that is,

$$(5) \qquad\qquad \mathscr{E}\bar{X} = \frac{1}{n}\sum_{i=1}^{n}\mathscr{E}X_i = \mu.$$

What about the precision of $\bar{X}$? We have from Theorem 1 (ii)

$$\operatorname{var}(\bar{X}) = \frac{1}{n^2}\left\{\sum_{j=1}^{n}\operatorname{var}(X_j) + 2\sum_{i<j}\operatorname{cov}(X_i, X_j)\right\}$$

so that

(6)
$$\text{var}(\overline{X}) = \frac{1}{n^2}\{n\,\text{var}(X_j) + n(n-1)\text{cov}(X_i, X_j)\}.$$

Now

(7)
$$\text{var}(X_j) = \mathscr{E}X_j^2 - (\mathscr{E}X_j)^2 = \sum_{i=1}^{N} \frac{\omega_i^2}{N} - \mu^2 = \sum_{i=1}^{N} \frac{(\omega_i - \mu)^2}{N} = \sigma^2.$$

Moreover, since (6) holds for all n, it holds in particular for $n = N$ but in that case $\overline{X} \equiv \mu$ and $\text{var}(\overline{X}) = 0$. Substituting $n = N$ in (6), using (7) we get[†]

$$0 = \frac{1}{N}\{\sigma^2 + (N-1)\text{cov}(X_i, X_j)\}$$

so that

(8)
$$\text{cov}(X_i, X_j) = -\frac{\sigma^2}{N-1}.$$

Consequently, for $1 < n \le N$

(9)
$$\text{var}(\overline{X}) = \frac{1}{n}\left\{\sigma^2 - \frac{n-1}{N-1}\sigma^2\right\} = \frac{N-n}{N-1}\cdot\frac{\sigma^2}{n} < \frac{\sigma^2}{n}.$$

On the other hand, if we sample with replacement then $X_1,\ldots,X_n$ are independent and identically distributed with common probability function given by (4), and we have seen that $\text{var}(\overline{X}) = \sigma^2/n$. Since in with replacement sampling we can take samples of any size whatever it is frequently referred to as sampling from an infinite population (or random sampling from a probability distribution). We note that the variance of $\overline{X}$ in sampling from a finite population equals $(N-n)/(N-1)$ times the variance of $\overline{X}$ in sampling from an infinite population. The fraction $(N-n)/(N-1)$ is therefore called *finite population correction*. If $n = N$, this factor vanishes and $\text{var}(\overline{X}) = 0$.

When n/N is small and N large, the correction factor may be dropped, since in that case:

$$\frac{N-n}{N-1} = \frac{1-n/N}{1-1/N} \doteq 1.$$

How about the sample variance S^2 as an estimate of σ^2? Since

$$S^2 = \frac{\displaystyle\sum_{i=1}^{n} X_i^2 - n\overline{X}^2}{n-1}$$

[†] The distribution of (X_i, X_j) is independent of n. This follows from (3). See also Problem 6.2.6.

it follows from Theorem 1 that

$$\mathscr{E}S^2 = \frac{\sum_{i=1}^{n} \mathscr{E}X_i^2 - n\mathscr{E}\overline{X}^2}{n-1}$$

$$= \frac{n\mathscr{E}X_1^2 - n\left[\mathrm{var}(\overline{X}) + \mu^2\right]}{n-1}$$

$$= \frac{n\left(\mathscr{E}X_1^2 - \mu^2\right) - \sigma^2(N-n)/(N-1)}{(n-1)} = \frac{N\sigma^2}{(N-1)},$$

so that S^2 is nearly unbiased for σ^2. In fact, $(N-1)S^2/N$ is unbiased for σ^2.   □

Example 2. Variance Reduction: ρ^2 As a Measure of Straight-Line Association.
Suppose $\mathrm{cov}(X, Y) = \rho\sigma_1\sigma_2$. Then the least value of $\mathrm{var}(Y + aX)$ is given by $(1 - \rho^2)\sigma_2^2$. Indeed, from Theorem 1:

$$\phi(a) = \mathrm{var}(Y + aX) = \sigma_2^2 + 2a\rho\sigma_1\sigma_2 + a^2\sigma_1^2.$$

It follows that

$$\phi'(a) = 2\rho\sigma_1\sigma_2 + 2a\sigma_1^2,$$

$$\phi''(a) = 2\sigma_1^2 > 0,$$

so that $\phi'(a) = 0$ yields a minimum, namely, $a = -\rho\sigma_2/\sigma_1$. The minimum value of $\phi(a)$ is given by

$$\min_a \phi(a) = \sigma_2^2 - 2\rho^2\sigma_2^2 + \rho^2\sigma_2^2 = \sigma_2^2(1 - \rho^2).$$

Of the total variation σ_2^2 in Y, a proportion ρ^2 is accounted for by the straight-line dependence of Y on X. By symmetry, the same result holds for the variance of X.   □

Example 3*. Waiting Times in Sampling. A fair die is rolled repeatedly. How many rolls will be needed on the average before we observe all six face values? In the birthday problem [Example 2.5.8(iii)], how many students need to be sampled on the average up to the time when our sample contains n different birthdays? These are all special cases of the following problem: A population contains N distinct elements which are sampled *with replacement*. Let S_n be the sample size needed to sample n distinct objects. What is $\mathscr{E}S_n$? To use Theorem 1 we need to write S_n as a sum of random variables. We note that the first draw produces the first element of the sample. Let X_1 be the waiting time for the second new element: this is the number of draws from the second draw up to and including the draw at which the second new element appears. Let X_j be the waiting time for the $(j + 1)$st new element. Then for $n \geq 2$ (see Figure 1)

$$S_n = 1 + X_1 + \cdots + X_{n-1}$$

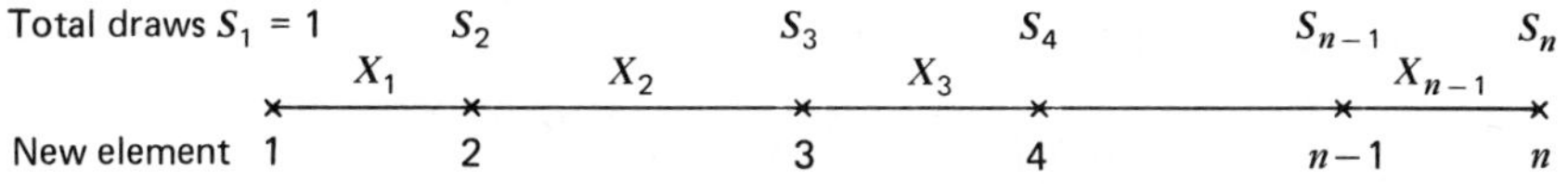

Total draws $S_1 = 1$ S_2 S_3 S_4 S_{n-1} S_n

X_1 X_2 X_3 X_{n-1}

New element 1 2 3 4 $n-1$ n

Figure 1

is the (minimum) sample size required for n distinct elements. It follows that

$$\mathscr{E}S_n = 1 + \mathscr{E}X_1 + \cdots + \mathscr{E}X_{n-1}.$$

To compute $\mathscr{E}X_k$, we note that the sample contains k different elements so that the probability of drawing a new one at each draw is $p_k = (N - k)/N$. Since X_k equals one (last success) plus the number of failures preceding that success, it has a (geometric) probability function (see also Section 6.5) given by

$$P(X_k = x + 1) = p_k \cdot (1 - p_k)^x, \qquad x = 0, 1, \ldots.$$

Then

$$\mathscr{E}X_k = \sum_{x=0}^{\infty} (x + 1) p_k (1 - p_k)^x = p_k \sum_{x=1}^{\infty} x(1 - p_k)^{x-1}$$

$$= \frac{1}{p_k} = \frac{N}{N - k}. \qquad \left(\text{Differentiate } \sum_{x=1}^{\infty} (1 - p_k)^x = \frac{1 - p_k}{p_k}. \right)$$

It follows that

$$\mathscr{E}S_n = 1 + \frac{N}{N-1} + \cdots + \frac{N}{N-(n-1)} = N \sum_{k=1}^{n} \frac{1}{N+1-k}.$$

In particular, if $n = N$, then

$$\mathscr{E}S_n = N \sum_{k=1}^{N} \frac{1}{N+1-k}$$

is the expected number of draws necessary to exhaust the entire population. In the case of tossing a fair die $N = 6$, we need, on the average

$$\mathscr{E}S_6 = 6\left\{ \tfrac{1}{6} + \tfrac{1}{5} + \cdots + 1 \right\} = 14.7$$

tosses to observe all six face values. For $N = 6$ and $n = 4$ we have

$$\mathscr{E}S_4 = 6\left\{ \tfrac{1}{6} + \tfrac{1}{5} + \tfrac{1}{4} + \tfrac{1}{3} \right\} = 5.7$$

so we need about five to six tosses to observe four distinct face values and it takes about nine additional tosses to observe the remaining two face values.

We can approximate $\mathscr{E}S_n$ by

$$\mathscr{E}S_n \doteq N \int_{N-n+1/2}^{N+1/2} \frac{1}{x} \, dx = N \ln \frac{N+1/2}{N-n+1/2}.$$

If $N = 365$ and $n = 25$, then

$$\mathscr{E}S_n \doteq 365 \ln \frac{365.5}{340.5} = 25.86$$

so that we need to sample only 25 to 26 students on the average to observe 25 different birthdays.   □

Example 4.   Confidence Interval for Population Mean μ.   In Section 4.3 we saw that in order to compute the numerical value of the minimum level $1 - \alpha$ confidence interval for the population mean μ for a given sample, we either need to know the distribution of $\overline{X}$ or we have to evaluate a multiple integral to compute c_1 and c_2 such that

$$P(\overline{X} < c_2) = \frac{\alpha}{2} = P(\overline{X} < c_1).$$

On the other hand, if we know σ then we can compute a rough confidence interval estimate for μ at level at least $1 - \alpha$ by using Chebyshev's inequality. Indeed, since $\mathscr{E}\overline{X} = \mu$ and $\text{var}(\overline{X}) = \sigma^2/n$ we have from Chebyshev's inequality

$$P(|\overline{X} - \mu| \leq k\sigma/\sqrt{n}) \geq 1 - \frac{1}{k^2}.$$

Choosing $k = 1/\sqrt{\alpha}$ we get

$$1 - \alpha \leq P\left(|\overline{X} - \mu| \leq \frac{1}{\sqrt{\alpha n}} \sigma\right) = P\left(\overline{X} - \frac{\sigma}{\sqrt{\alpha n}} \leq \mu \leq X + \frac{\sigma}{\sqrt{\alpha n}}\right),$$

which holds for all μ. Hence the confidence interval of level at least $1 - \alpha$ is given by $[\overline{X} - \sigma/\sqrt{\alpha n}, \overline{X} + \sigma/\sqrt{\alpha n}]$. If σ is unknown and n is large we can use S as an estimate of σ to get an approximate confidence interval $[\overline{X} - S/\sqrt{\alpha n}, \overline{X} + S/\sqrt{\alpha n}]$. For large n, however, we will see in Chapter 9 that it is possible to do much better by using an approximation for the distribution of $\overline{X}$.   □

Remark 1.   It is important to remember that we do not need to know the joint distribution of $X_1, X_2, \ldots, X_m$ to apply Theorem 1. We need only know the means of X_i, $i = 1, 2, \ldots, m$ in order to compute the mean of any linear function $\sum_{j=1}^{m} a_j X_j$ of $(X_1, \ldots, X_m)$. Similarly, if we know the variances $\text{var}(X_j)$ and covariances $\text{cov}(X_i, X_j)$, $i \neq j$, $i, j = 1, 2, \ldots, m$ we can compute $\text{var}(\sum_{j=1}^{m} a_j X_j)$.

Example 5.   Empirical Distribution Function As an Unbiased Estimate of the Population Distribution Function.   Let $X_1, X_2, \ldots, X_n$ be a random sample from distribution function F. Let $\hat{F}_n(x)$ be the empirical distribution function of the sample. Define $Y_i = 1$ if $X_i \leq x$ and $Y_i = 0$ if $X_i > x$. Then $Y_1, Y_2, \ldots, Y_n$ are independent and identically distributed with common distribution

$$P(Y_i = 1) = P(X_i \leq x) = F(x) = 1 - P(Y_i = 0).$$

Moreover, $\mathscr{E}Y_i = F(x)$ and $\operatorname{var}(Y_i) = F(x) - [F(x)]^2$. We note that

$$\overline{Y} = \sum_{i=1}^{n} \frac{Y_i}{n} = \frac{\text{number of } X_i\text{'s} \le x}{n} = \hat{F}_n(x).$$

It follows that

$$\mathscr{E}\overline{Y} = \mathscr{E}\hat{F}_n(x) = F(x)$$

for all $x \in \mathbb{R}$ so that $\hat{F}_n(x)$ is an unbiased estimate of $F(x)$. $\square$

Example 6.   Estimating Covariance and Correlation Coefficient.   Given a random sample $(x_1, y_1), (x_2, y_2), \ldots, (x_n, y_n)$ of n observations on (X, Y), how do we estimate $\operatorname{cov}(X, Y)$ and $\rho(X, Y)$? We have shown that $S_1^2 = \sum_{i=1}^{n} (X_i - \overline{X})^2/(n-1)$ and $S_2^2 = \sum_{i=1}^{n} (Y_i - \overline{Y})^2/(n-1)$ are unbiased estimates of $\sigma_1^2 = \operatorname{var}(X)$ and $\sigma_2^2 = \operatorname{var}(Y)$ respectively. This suggests that we consider

$$S_{11} = \frac{\sum_{i=1}^{n} (X_i - \overline{X})(Y_i - \overline{Y})}{n-1}$$

as an estimate of $\operatorname{cov}(X, Y)$. We show that S_{11} is an unbiased estimate. In fact

$$(n-1)\mathscr{E}S_{11} = \mathscr{E}\sum_{i=1}^{n} (X_i - \overline{X})(Y_i - \overline{Y}) = \mathscr{E}\left\{ \sum_{i=1}^{n} X_iY_i - n\overline{X}\,\overline{Y} \right\}.$$

$$= \sum_{i=1}^{n} \mathscr{E}(X_iY_i) - n\mathscr{E}(\overline{X}\,\overline{Y}).$$

$$= n[\operatorname{cov}(X, Y) + \mathscr{E}X\mathscr{E}Y] - \frac{1}{n}\mathscr{E}\sum_{i=1}^{n}\sum_{j=1}^{n} X_iY_j$$

$$= n[\operatorname{cov}(X, Y) + \mathscr{E}X\mathscr{E}Y] - \frac{1}{n}\left[\sum_{i=1}^{n} \mathscr{E}X_iY_i + n(n-1)\mathscr{E}X\mathscr{E}Y \right]$$

$$= (n-1)[\operatorname{cov}(X, Y) + \mathscr{E}X\mathscr{E}Y] - (n-1)\mathscr{E}X\mathscr{E}Y$$

$$= (n-1)\operatorname{cov}(X, Y).$$

It follows that $\mathscr{E}S_{11} = \operatorname{cov}(X, Y)$ and S_{11} is unbiased.

In order to estimate $\rho(X, Y)$ we recall that

$$\rho(X, Y) = \operatorname{cov}(X, Y)/\sigma_1\sigma_2.$$

It is therefore natural to consider

$$R = \frac{S_{11}}{S_1S_2} = \frac{\sum_{i=1}^{n} (X_i - \overline{X})(Y_i - \overline{Y})}{\sqrt{\sum_{i=1}^{n} (X_i - \overline{X})^2 \sum_{i=1}^{n} (Y_i - \overline{Y})^2}}$$

as an estimate of ρ. The estimate R, however, is not unbiased for ρ. We note that S_{11} and R are consistent estimates of $\text{cov}(X, Y)$ and $\rho(X, Y)$ respectively (Problem 9.10.24). In Example 11.4.3 and Section 11.9 we show how the statistic R is used in testing $H_0: \rho = 0$. The statistic S_{11} is called the *sample covariance* and R is called the *sample correlation coefficient*.   □

Example 7.   Proportion of Faculty in Favor of Semester System.   Suppose that at a certain state university a random sample of 65 faculty members is taken out of a total of 565 faculty members. Each member in the sample is asked the question, Are you in favor of or against the semester system? The object is to estimate the true proportion of faculty members who are in favor of the semester system. Suppose 35 of the respondents are found to be in favor of the proposition.

Let p be the true proportion of faculty members who are in favor of the semester system. Let $X_i = 1$ if the ith member of the sample responds with a yes and zero otherwise. Then

$$\sum_{i=1}^{65} X_i = \text{number in the sample who responded with a yes}$$

and according to Example 1, $\hat{p} = \overline{X}$ is an unbiased estimate of p with variance

$$\sigma_{\hat{p}}^2 = \frac{N - n}{N - 1} \frac{\sigma^2}{n}$$

where $N = 565$, $n = 65$, $\sigma^2 = \text{var}(X_i) = p - p^2$. Hence our point estimate of p is $\hat{p} = 35/65 = .5385$ and the variance of the estimate $\hat{p}$ is given by

$$\sigma_{\hat{p}}^2 = \frac{500}{564} \cdot \frac{p(1 - p)}{65}.$$

Since p is unknown, we have two choices to estimate $\sigma_{\hat{p}}^2$: we can either replace p by $\hat{p}$ in the expression for $\sigma_{\hat{p}}^2$ or use the unbiased estimate $(N - 1)S^2/N$ for σ^2 (see Example 1). Writing $\hat{\sigma}_{\hat{p}}^2$ for the estimate of $\sigma_{\hat{p}}^2$ when we replace p by $\hat{p}$ we get

$$\hat{\sigma}_{\hat{p}}^2 = \frac{500}{564} \frac{(35/65)(30/65)}{65} = .00339$$

and

$$\hat{\sigma}_{\hat{p}} = .0582.$$

If we want a confidence interval estimate for p at level at least .95 say, we use Chebyshev's inequality (as in Example 4). We have for all $0 \le p \le 1$ with $k = 1/\sqrt{.05}$

$$.95 \le P\left(|\hat{p} - p| \le \frac{1}{\sqrt{.05}} \sigma_{\hat{p}}\right) = P\left(\hat{p} - \frac{\hat{\sigma}_{\hat{p}}}{\sqrt{.05}} \le p \le \hat{p} + \frac{\sigma_{\hat{p}}}{\sqrt{.05}}\right).$$

Since $\sigma_{\hat{p}}$ is unknown we use $\hat{\sigma}_{\hat{p}}$ instead to get an approximate confidence interval for p as

$$\left[.5385 - \frac{.0582}{\sqrt{.05}}, \quad .5385 + \frac{.0582}{\sqrt{.05}}\right] = [.2782, .7988].$$

On the other hand, if we use $(N-1)S^2/N$ to estimate σ^2, then

$$(n-1)S^2 = \sum_{i=1}^{65} X_i^2 - 65\overline{X}^2 = \sum_{i=1}^{65} X_i - 65\left(\sum_{i=1}^{65} \frac{X_i}{65}\right)^2$$

so that

$$\hat{\sigma}_{\hat{p}}^2 = \frac{N-n}{N-1} \cdot \frac{(N-1)s^2}{N} \cdot \frac{1}{n}$$

$$= \frac{500}{565}\left(35 - \frac{35^2}{65}\right)\left(\frac{564}{565}\right)\frac{1}{(64)(65)} = .00343,$$

and $\hat{\sigma}_{\hat{p}} = .0586$. There is hardly any difference between the two estimates for $\sigma_{\hat{p}}$.

The important point to remember here is that the sampling is without replacement from a finite population so that we need to use finite population correction in the computation of $\sigma_{\hat{p}}^2$. If we do not do so, we will get a less precise confidence interval. Note also that in sampling from a finite population we do have a natural lower bound for p. Indeed, $p \geq \sum_{i=1}^{n} X_i/N$. In sampling from an infinite population we do not have any such lower bound for p. □

Approximation of Mean and Variance*

Let X, Y be jointly distributed random variables. If g is a linear function of X and Y, Theorem 1 shows us how to compute the mean and the variance of g. For most practical purposes this result suffices. In some problems, however, g is not linear. Suppose we wish to estimate the coefficient of variation σ/μ by, say, $S/\overline{X}$. Is $S/\overline{X}$ unbiased? What is the variance of $S/\overline{X}$? Or, if the mean length of life of a piece of equipment is $1/\mu$, we know that $\overline{X}$ is unbiased for $1/\mu$. It is then natural to consider $1/\overline{X}$ as an estimate of μ. How do we find the variance of $1/\overline{X}$ as an estimate of μ? Theorem 1 clearly does not apply here.

One possibility is to compute the distribution of $g(X, Y)$. This is done in Chapter 8, where we see that the computation of distribution of g is often not simple. Even if we are able to compute the distribution of g, the computation of variance or mean square error of g may not be easy. In any case, the procedure is long and complicated.

A second possibility, which is at least less time-consuming, is to compute $\mathscr{E}g(X, Y)$ or $\text{var}(g(X, Y))$ directly from the distribution of (X, Y).

In the case of most interest, however, X and Y will be some statistics based on one or more samples. Neither of the above two procedures, in such cases, guarantees any success. We explore the third possibility, that of finding approximations for $\mathscr{E}g(X, Y)$ and $\mathrm{var}(g(X, Y))$. We consider mainly the univariate case.

THEOREM 2. *Let g be a real valued function that is twice differentiable at $x = \theta$. Then*

$$(10) \qquad \mathscr{E}_\theta g(X) \doteq g(\theta) + \mathscr{E}_\theta (X - \theta) g'(\theta) + \tfrac{1}{2} g''(\theta) \mathscr{E}_\theta (X - \theta)^2$$

and

$$(11) \qquad \mathscr{E}_\theta (g(X) - g(\theta))^2 \doteq \left[g'(\theta) \right]^2 \mathscr{E}_\theta (X - \theta)^2.$$

Outline of Proof. Expanding g in a Taylor series about $x = \theta$ in two terms we have

$$(12) \quad g(X) = g(X - \theta + \theta) = g(\theta) + (X - \theta) g'(\theta) + \frac{(X - \theta)^2}{2!} g''(\theta) + R(X, \theta)$$

where R is the remainder. If $\mathscr{E}R(X, \theta)$ is discarded, then

$$\mathscr{E}_\theta g(X) \doteq g(\theta) + (\mathscr{E}_\theta X - \theta) g'(\theta) + \tfrac{1}{2} \mathscr{E}(X - \theta)^2 g''(\theta)$$

which is (10).

If we expand g in a Taylor series only to one term, then

$$g(X) = g(\theta) + (X - \theta) g'(\theta) + R_1(X, \theta)$$

and discarding R_1 we get

$$\mathscr{E}_\theta (g(X) - g(\theta))^2 = \mathrm{MSE}_\theta (g(X)) \doteq \left[g'(\theta) \right]^2 \mathscr{E}_\theta (X - \theta)^2. \qquad \square$$

COROLLARY 1. *If $\mathscr{E}_\theta X = \theta$ and $\mathscr{E}_\theta (X - \theta)^2 = \sigma^2$ then*

$$(13) \qquad \mathscr{E}_\theta g(X) \doteq g(\theta) + \tfrac{1}{2} \sigma^2 g''(\theta)$$

and

$$(14) \qquad \mathrm{var}_\theta (g(X)) \doteq \left[g'(\theta) \right]^2 \sigma^2.$$

It is possible to improve the approximation for $\mathscr{E}_\theta (g(X) - g(\theta))^2$ by using (12) and higher-order moments of X. Similarly, an extension to two (or more) variables can be obtained by using Taylor expansion for two (or more) variables. For example, if $\mathscr{E}X = \mu_1$, $\mathscr{E}Y = \mu_2$, $\mathrm{var}(X) = \sigma_1^2$, $\mathrm{var}(Y) = \sigma_2^2$, then under appropriate conditions on g,

$$(15) \quad \mathscr{E}g(X, Y) \doteq g(\mu_1, \mu_2)$$

$$+ \frac{1}{2} \left\{ \left. \frac{\partial^2 g(\mu_1, \mu_2)}{\partial x^2} \right|_{x = \mu_1} \sigma_1^2 + \left. \frac{\partial^2 g(\mu_1, \mu_2)}{\partial y^2} \right|_{y = \mu_2} \sigma_2^2 \right\}$$

and

(16)
$$\text{var}(g(X, Y)) \doteq \left[\frac{\partial g}{\partial x}\right]^2_{x=\mu_1} \sigma_1^2 + \left[\frac{\partial g}{\partial y}\right]^2_{y=\mu_2} \sigma_2^2.$$

We omit the details.

Remark 2. We emphasize that $\mathscr{E}_\theta g(X)$ or $\text{var}_\theta(g(X))$ may not exist. These approximations do not assert the existence of mean and/or variance of $g(X)$. For example, $\mathscr{E}(S/\overline{X})$ does not exist for some common distributions.

Example 8.   Estimating p^2 where p is Probability of a Success.   Let $X_1, X_2, \dots, X_n$ be a random sample from probability function $P(X_i = 1) = p = P(X_i = 0)$. This is the coin tossing experiment with a head (or success) identified as 1 and a tail as 0. Then $\overline{X}$ is the proportion of heads (successes) in n trials and $\mathscr{E}\overline{X} = p$, $\text{var}(\overline{X}) = p(1 - p)/n$. (Here $\mu = \mathscr{E}X_i = p$, $\sigma^2 = \mathscr{E}X_i^2 - (\mathscr{E}X_i)^2 = p(1 - p)$.) Suppose we wish to estimate $g(p) = p^2$. Consider $g(\overline{X}) = \overline{X}^2$ as an estimate of p^2. We note that

$$g'(p) = 2p \quad \text{and} \quad g''(p) = 2$$

so that from (13)

$$\mathscr{E}g(\overline{X}) \doteq p^2 + \frac{1}{2}\text{var}(\overline{X})(2) = p^2 + \frac{p(1 - p)}{n}$$

and

$$\text{var}(g(\overline{X})) \doteq [2p]^2\text{var}(\overline{X}) = 4p^2p\frac{1 - p}{n} = 4p^3\frac{1 - p}{n}.$$

For large n, $p(1 - p)/n$ is nearly zero and we see that $g(\overline{X}) = \overline{X}^2$ is nearly unbiased for p^2. We note that $\mathscr{E}\overline{X}^2 = \text{var}(\overline{X}) + (\mathscr{E}\overline{X})^2 = p(1 - p)/n + p^2$, so that the Taylor approximation gives an exact result for $\mathscr{E}g(\overline{X})$ in this case.   □

Example 9.   Estimating the Gravitational Constant.   In Example 1.2.1 we considered a simple pendulum with unit mass suspended from a fixed point O (see Figure 2). Under the assumption that the pendulum swings only under the effect of gravity we showed that the period of oscillation $T = T(\theta)$ as the pendulum goes from $\theta = \theta_0$ to $\theta = 0$ is given by

$$T \doteq \frac{4}{\sqrt{2g}} \int_0^{\theta_0} \frac{d\theta}{\cos\theta - \cos\theta_0}$$

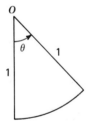

Figure 2

and for a small oscillation

$$T \doteq 2\pi/\sqrt{g}$$

approximately. Suppose $T = 2\pi/\sqrt{g} + \varepsilon$ where ε is the random error with $\mathscr{E}(\varepsilon) = 0$. Then $\mathscr{E}T = 2\pi/\sqrt{g}$. How do we estimate g based on observable values $T_1, T_2, \ldots, T_n$ of T? We note that $g = c/T^2$ where $c = 4\pi^2$. There are two obvious possibilities. For each observed T_i, we get an observation g_i on g as c/T_i^2 and the average $\hat{g}_1 = \bar{g} = \sum_{i=1}^n g_i/n$ is then an estimate of g. (Here we are using the same notation for the random variable c/T^2 and the gravitational constant g.)

The second estimate can be obtained by first estimating $\mathscr{E}T$ by $\bar{T}$ so that

$$\hat{g}_2 = \frac{c}{\bar{T}^2}$$

also estimates g.

We have

$$\mathscr{E}\hat{g}_1 = \mathscr{E}g_i = c\mathscr{E}\frac{1}{T_i^2}$$

and

$$\mathscr{E}\hat{g}_2 = c\mathscr{E}\frac{1}{\bar{T}^2}.$$

Let $\mathscr{E}T_i = \mu$ and $\mathrm{var}(T_i) = \sigma^2$. In view of (13),

$$\mathscr{E}\hat{g}_1 \doteq c\left(\frac{1}{\mu^2} + \frac{3}{\mu^4}\sigma^2\right)$$

and

$$\mathscr{E}\hat{g}_2 \doteq c\left(\frac{1}{\mu^2} + \frac{3}{\mu^4}\frac{\sigma^2}{n}\right).$$

Since $\mu = \mathscr{E}T_i = 2\pi\sqrt{1/g}$ we have

$$\mathscr{E}\hat{g}_1 \doteq g + \frac{3g^2\sigma^2}{4\pi^2},$$

$$\mathscr{E}\hat{g}_2 \doteq g + \frac{3g^2\sigma^2}{4\pi^2 n}.$$

An application of (14) yields

$$\mathrm{var}(\hat{g}_1) = \mathrm{var}(\hat{g}_2) = \frac{g^3\sigma^2}{\pi^2 n}.$$

As $n \to \infty$ we note that $\hat{g}_2$ is nearly unbiased for g but $\hat{g}_1$ is positively biased. $\quad\square$

Problems for Section 5.3

Problems 22–25 Refer to the Subsection.

1. A student takes a five-answer multiple choice test consisting of 10 questions. Suppose he or she guesses at every question. What is the average number of his or her correct answers?

2. Let $X_1, X_2, \ldots, X_n$ be random variables with common mean μ and common variance σ^2. Find:
   (i) $\mathscr{E}(X_1 + X_2)$.   (ii) $\mathscr{E}(X_1 - X_2)$.   (iii) $\mathscr{E}\{(X_1 + X_2)/2 - \mu\}$.
   (iv) $\mathscr{E}\{((X_1 + X_2)/2 - \mu)/\sigma\}$.   (v) $\mathscr{E}\{\bar{X} - \mu\}$.   (vi) $\mathscr{E}\{(\bar{X} - \mu)/\sigma\}$.

3. In Problem 2, assume that the random variables, in addition, are uncorrelated. Find the variance in each case.

4. Let $\text{var}(X) = 2$, $\text{var}(Y) = 9$, and $\text{cov}(X, Y) = 2$. Find:
   (i) $\text{var}(X - Y)$.   (ii) $\text{var}(3X + 9Y)$.   (iii) $\text{var}(X + 4Y)$.
   (iv) $\text{var}(2X - 5Y)$.

5. A folding 4-foot ruler is constructed by joining four 12-inch pieces together. Suppose the pieces are chosen independently from a population with a mean length of 12 inches and standard deviation σ. What is the mean length of the ruler and what is the variance of the length of a ruler constructed in this fashion? If it is desired that the standard deviation of the length be .05 inches, what should be the value of σ?

6. The profit that a company makes each quarter is a random variable with mean \$100,000 and standard deviation \$20,000. If the profits each quarter are assumed independent and identically distributed, what is the average total profit for the year and the variance of total yearly profit?

7. Suppose X, Y have joint density given by:

   $$f(x, y) = 6y^2, \qquad 0 < |y| < x, 0 < x < 1, \text{ and zero elsewhere.}$$

   Find the variance of $Z = 3X + 2Y - 5$.

8. Suppose X, Y, Z are random variables such that $\text{cov}(X, Y) = 2$, $\text{cov}(X, Z) = -2$, $\text{var}(X) = 6$, $\text{var}(Y) = 4$, $\text{var}(Z) = 9$ whereas Y and Z are uncorrelated. Find $\text{cov}(U, V)$ where $U = X - 3Y + 2Z$ and $V = -3X + 4Y + Z$.

9. Suppose X and Y are two independent estimates of the true weight μ of an object. Assume X and Y are unbiased with finite variances σ_1^2 and σ_2^2 respectively.
   (i) Show that for each α, $\alpha X + (1 - \alpha)Y$ is also unbiased for μ.
   (ii) How will you choose α to minimize the variance of $\alpha X + (1 - \alpha)Y$?

10. Let X, Y have joint density given by

   $$f(x, y) = 24xy, \qquad 0 \le x + y \le 1, x \ge 0, y \ge 0, \text{ and zero elsewhere.}$$

   Find:
   (i) $\mathscr{E}(X + Y)$.   (ii) $\mathscr{E}(X^2 + 3Y)$.   (iii) $\mathscr{E}(X + 4Y)^2$.
   (iv) $\text{var}(2X + 3Y)$.   (v) $\text{var}(X^2 + 3Y)$.

11. Let X, Y and Z be jointly distributed random variables.
    (i)   Show that $\text{cov}(X + Z, Y) = \text{cov}(X, Y) + \text{cov}(Z, Y)$.
    (ii) Find the correlation coefficient between $X + Y$ and $X + Z$ when X, Y, Z are uncorrelated and have the same variance.
    (iii) Find $\rho(X + Y, X - Y)$ when X and Y are uncorrelated.
    (iv) Suppose $\mathscr{E}X = \mathscr{E}Y$ and $\mathscr{E}X^2 = \mathscr{E}Y^2$. Show that $\text{cov}(X - Y, X + Y) = 0$.

12. Let $X_1, X_2, \ldots, X_n$ be a random sample from a distribution with finite variance. Show that:
    (i) $\text{cov}(X_i - \bar{X}, \bar{X}) = 0$ for each i, and
    (ii) $\rho(X_i - \bar{X}, X_j - \bar{X}) = -1/(n - 1)$, $i \neq j$, $i, j = 1, \ldots, n$.

13. Let $X_1, X_2, \ldots, X_n$ be a random sample from a distribution with $\mathscr{E}X_i = 0 = \mathscr{E}X_i^3$. Show that $\text{cov}(\bar{X}, S^2) = 0$.
    [Hint: Show that $\text{cov}(\bar{X}, (X_i - \bar{X})^2) = \text{cov}(\bar{X}, X_i^2 - 2X_i\bar{X} + \bar{X}^2) = 0$ for each i.]

14. Let $X_1, X_2, \ldots, X_n$ be dependent random variables with $\rho(X_i, X_j) = \rho$ for $i \neq j$, $i, j = 1, 2, \ldots, n$. Show that $-(n - 1)^{-1} \leq \rho \leq 1$, when $\text{var}(X_i) = \text{var}(X_j)$, for all i, j.

15. Let $X_1 X_2, \ldots, X_{m+n}$ be independent and identically distributed random variables with common finite variance σ^2. Let $S_k = \Sigma_{j=1}^k X_j$, $k = 1, 2, \ldots, m + n$. Find $\rho(S_n, S_{m+n} - S_m)$ for $n > m$.

16*. A box contains D defective and $N - D$ nondefective items. A sample of size n is taken without replacement. Let X be the number of defective items in the sample. Show that

$$\mathscr{E}X = \frac{nD}{N} \quad \text{and} \quad \text{var}(X) = nD\frac{(N - D)(N - n)}{N^2(N - 1)}.$$

[Hint: Let $Y_i = 1$ if ith item drawn is defective and $Y_i = 0$ otherwise. Then $X = \Sigma_{i=1}^n Y_i$. Show first $P(Y_i = 1) = D/N$ for $i = 1, 2, \ldots, n$ and $P(Y_i = 1, Y_j = 1) = D(D - 1)/N(N - 1)$ for $i \neq j$.]

17*. Let $A_1, \ldots, A_k$ be disjoint events such that $\bigcup_{j=1}^k A_j = \Omega$ and $P(A_i) = p_i$, $0 < p_i < 1$, $\Sigma_{i=1}^k p_i = 1$. In n independent trials of the experiment, let $X_1, X_2, \ldots, X_k$ denote the number of times that $A_1, A_2, \ldots, A_k$ occur, respectively. Then $\Sigma_{j=1}^k X_j = n$. Show that $\text{cov}(X_i, X_j) = -np_i p_j$ for $i \neq j$. [Hint: Let $U_l = 1$ if lth trial results in A_i and zero otherwise. Let $V_m = 1$ if mth trial results in A_j and zero otherwise. Then $X_i = \Sigma_{l=1}^n U_l$ and $X_j = \Sigma_{m=1}^n V_m$. Use Theorem 1 (iii).]

18*. Consider a dichotomous experiment with a probability of success p. The experiment is repeated independently until r successes are observed. Let Y be the number of trials required to observe r successes. Show that $\mathscr{E}Y = r/p$ and $\text{var}(Y) = r(1 - p)/p^2$. [Hint: Define $X_i = $ trial number on which ith success is observed, $i = 1, 2, \ldots, r$ and $W_i = X_i - X_{i-1}$, $i = 1, 2, \ldots, r$, with $X_0 = 0$. Then W_i are independent identically distributed random variables and $Y = X_r = \Sigma_{i=1}^r (X_i - X_{i-1}) = \Sigma_{i=1}^r W_i$. The probability function of W_i is given by $P(W_i = w) = p(1 - p)^{w-1}$, $w = 1, 2, \ldots$.]

19*. A box contains N items of which D_j are of jth type, $j = 1, 2, \ldots, k$, $\sum_{j=1}^{k} D_j = N$. A random sample of size n is taken without replacement. Let X_i be the number of items of type i, $i = 1, \ldots, k$. Then $\sum_{j=1}^{k} X_j = n$. Show that for $i \neq j$, $\mathrm{cov}(X_i, X_j) = -nD_i D_j (N - n)/[N^2(N - 1)]$. [Hint: Let $U_l = 1$ if lth draw is of type i and zero elsewhere, and let $V_m = 1$ if mth draw is an item of type j and zero elsewhere. Then $\mathscr{E}U_l = D_i/N$, $\mathscr{E}V_m = D_j/N$ and $\mathscr{E}U_l V_m = D_i D_j/N(N - 1)$, for $l \neq m$, $\mathscr{E}U_l V_l = 0$.]

20. Let $X_1, X_2, \ldots, X_m$ and $Y_1, \ldots, Y_n$ be independent random samples from distributions with means μ_1 and μ_2 and variances σ_1^2 and σ_2^2 respectively. Show that:
(i) $\mathscr{E}(\overline{X} - \overline{Y}) = \mu_1 - \mu_2$.
(ii) $\mathrm{var}(\overline{X} - \overline{Y}) = \dfrac{\sigma_1^2}{m} + \dfrac{\sigma_2^2}{n}$.

21. In Problem 20, let S_1^2, S_2^2 be the two sample variances. Show that

$$\mathscr{E}\left(\frac{S_1^2}{m} + \frac{S_2^2}{n} \right) = \frac{\sigma_1^2}{m} + \frac{\sigma_2^2}{n}.$$

In particular, if $\sigma_1 = \sigma_2 = \sigma$ show that

$$S_p^2 = \frac{(m - 1)S_1^2 + (n - 1)S_2^2}{m + n - 2}$$

is unbiased for the common variance σ^2. In this case, two other unbiased estimates are S_1^2 and S_2^2. Which of the three do you think should be preferred and why?

22*. Let $X_1, X_2, \ldots, X_n$ be a random sample from a distribution with finite mean μ and finite variance σ^2. It is required that $g(\mu)$ be estimated by $g(\overline{X})$. Find approximate expressions for $\mathscr{E}g(\overline{X})$ and $\mathrm{var}(g(\overline{X}))$. In particular, consider the function $g(x) = x^2$. Obtain expressions for $\mathscr{E}\overline{X}^2$ and $\mathrm{var}(\overline{X}^2)$.

23*. Let $X_1, X_2, \ldots, X_n$ be a random sample from probability function

$$P(X_i = 1) = p = 1 - P(X_i = 0).$$

An estimate of $\mathrm{var}(X_i) = p(1 - p)$ is required. Consider the estimates (i) $\hat{\sigma}_1^2 = S^2$, and (ii) $\hat{\sigma}_2^2 = \overline{X}(1 - \overline{X})$. Compute approximate expressions for the means and the variances.
[Hint: Note that S^2 is unbiased for σ^2. Write $(n - 1)S^2 = \sum_{i=1}^{n} X_i^2 - n(\sum_{i=1}^{n} X_i/n)^2 = S_n - S_n^2/n$ where $S_n = \sum_{i=1}^{n} X_i$. Then $S^2 = 1/[n(n - 1)] - S_n(n - S_n) = [n/(n - 1)]\overline{X}(1 - \overline{X})$.]

24*. According to Ohm's law, the voltage V in a simple circuit can be expressed as $V = IR$, where I is the current and R the resistance of the circuit. Suppose I and R are independent random variables. Use (15) and (16) to show that

$$\mathscr{E}V = \mathscr{E}I\mathscr{E}R \text{ and } \mathrm{var}(V) \doteq (\mathscr{E}R)^2 \mathrm{var}(I) + (\mathscr{E}I)^2 \mathrm{var}(R).$$

(Note that $\mathscr{E}V = \mathscr{E}I\mathscr{E}R$ by independence of I and R.)

25*. Frequently it is of interest to estimate μ_1/μ_2 where μ_1, μ_2 are means of two random variables X and Y with variances σ_1^2 and σ_2^2, respectively. Let $X_1, \ldots, X_m$ and $Y_1, \ldots, Y_n$ be independent random samples. Consider the estimate $\overline{X}/\overline{Y}$. Show, using (15) and (16), that

$$\mathcal{E}\frac{\overline{X}}{\overline{Y}} \doteq \frac{\mu_1}{\mu_2}\left(1 + \frac{\sigma_2^2}{n\mu_2^2}\right)$$

and

$$\operatorname{var}\frac{\overline{X}}{\overline{Y}} \doteq \left(\frac{\mu_1}{\mu_2}\right)^2\left\{\frac{\sigma_1^2}{m\mu_1^2} + \frac{\sigma_2^2}{n\mu_2^2}\right\}.$$

In the special case when $m = n$, consider the estimate $n^{-1}\sum_{i=1}^{n}(X_i/Y_i)$. Find approximations for its mean and the variance.

5.4 THE LAW OF LARGE NUMBERS

We defined probability of an event A as a measure of its uncertainty, that is, as a measure of how likely A is to occur. Our intuitive notion of probability is based on the assumption that it is a long-term relative frequency of occurrence of A. That is, if the experiment is repeated n times under identical conditions and A occurs $N(A)$ times, then $N(A)/n$ should be close to $P(A)$, the probability of A. When we say that $P(A)$ is the probability of A on a particular trial we are applying to this particular trial a measure of its chance based on what would happen in a long series of trials. Thus when a fair coin is tossed five times and shows five heads, the probability of a tail on the next toss is still $1/2$. The so-called (and generally misinterpreted) *law of averages* does not demand that the sixth, seventh, or eightth trials produce tails to balance the five straight heads. If a surgeon performs a delicate surgery where the chances of success are only 1 in 10, it should be no comfort to the tenth patient that the previous nine operations were failures. If a couple has had six girls in a row, the probability that the seventh child will be a girl is still $1/2$. It is true that the probability of seven straight girls $(= 1/2^7 = .008)$ is small but given that the first six children born to them were girls, the probability of the seventh being a girl is $1/2$.

We now give a precise formulation of the law of averages (it is called *the law of large numbers*). We consider only a weak version of the law of large numbers.

Law of Large Numbers

Let $\{X_n\}$ be a sequence of independent random variables with a common distribution and let $\mathcal{E}X_n = \mu$ exist. Then for every $\varepsilon > 0$ as $n \to \infty$

$$P\{|\overline{X} - \mu| > \varepsilon\} \to 0,$$

that is, the probability that the sample average will differ from the population expectation by less than ε (arbitrary) approaches one as the sample size $n \to \infty$.

Unfortunately, even this version is beyond our scope. We prove the following result instead.

THEOREM 1.   *Let* $\{X_n\}$ *be a sequence of random variables. Suppose* $\operatorname{var}(X_j) = \sigma_j^2$ *is finite for each* $j = 1, 2, \ldots$ *. If* $n^{-2}\operatorname{var}(\Sigma_{j=1}^n X_j) \to 0$ *as* $n \to \infty$ *then for every* $\varepsilon > 0$ *as* $n \to \infty$

$$(1) \qquad P\left(\left| \sum_{j=1}^n \frac{X_j - \mathscr{E}X_j}{n} \right| > \varepsilon \right) \to 0.$$

Proof.   From Chebyshev's inequality

$$P\left(\left| \sum_{j=1}^n \frac{X_j - \mathscr{E}X_j}{n} \right| > \varepsilon \right) \le \frac{\operatorname{var}\left(\sum_{j=1}^n X_j \right)}{n^2 \varepsilon^2} \to 0$$

as $n \to \infty$.   □

COROLLARY 1.   *If* X_n's *are uncorrelated and identically distributed with common finite variance* σ^2, *and* $\mathscr{E}X_j = \mu$, *then for every* $\varepsilon > 0$ *as* $n \to \infty$ $P(|\overline{X} - \mu| > \varepsilon) \to 0$.

Proof.   Since X_n's are uncorrelated, it follows from Theorem 5.3.1 that

$$\operatorname{var}\left(\sum_{j=1}^n X_j \right) = \sum_{j=1}^n \operatorname{var}(X_j) = \sigma^2 \sum_{j=1}^n 1 = n\sigma^2$$

and hence

$$n^{-2}\operatorname{var}\left(\sum_{j=1}^n X_j \right) = n^{-2}(n\sigma^2) = \frac{\sigma^2}{n} \to 0$$

as $n \to \infty$.   □

In particular, if the X_i's are independent and identically distributed (as in the case of random sampling with finite variance, then the sample mean $\overline{X}$ is likely to be close to μ for large n. That is, $\overline{X}$ is a consistent estimate of μ. This provides yet another justification for using sample mean as an estimate of μ. (We do not need the existence of variance in this special case but the proof is beyond our scope.) In the rest of this text, when we speak of the law of large numbers we refer to Corollary 1 above.

Remark 1.   The law of large numbers asserts only that for every particular sufficiently large n the deviation $|\overline{X} - \mu|$ is likely to be small. It does not imply that $|\overline{X} - \mu|$ remains small for all large n.

Remark 2.   The law of large numbers as proved above makes no assertion that $P(\overline{X} \to \mu) = 1$. It asserts simply that $\overline{X}$ is likely to be near μ for every fixed sufficiently large n and not that $\overline{X}$ is bound to stay near μ if n is increased.

Actually, it is true that with probability one $|\overline{X} - \mu|$ becomes small and remains small in random sampling from a distribution with finite mean, but we will not prove it here or use it subsequently. This is the so-called strong law of large numbers.

Example 1. *Relative Frequency of an Event A is likely to be Near P(A).* Consider an event with probability $P(A)$. Suppose in n independent trials of the experiment, A occurs $N(A)$ times. Then the relative frequency of A is $R(A) = N(A)/n$. We show that $N(A)/n$ is likely to be near $P(A)$. That is, $R(A)$ is a consistent estimate of $P(A)$.

Define $X_i = 1$ if A occurs on the ith trial and $X_i = 0$ otherwise. Since the trials are independent, X_i's are independent and, moreover, $\mathscr{E}X_i = P(A)$, $\text{var}(X_i) = P(A)[1 - P(A)]$. Also

$$N(A) = \sum_{i=1}^{n} X_i$$

so that $R(A) = N(A)/n = \sum_{i=1}^{n} X_i/n$. By the law of large numbers, for every $\varepsilon > 0$ as $n \to \infty$

$$P\big(|R(A) - P(A)| > \varepsilon\big) = P\big(|\overline{X} - P(A)| > \varepsilon\big) \to 0$$

as asserted.   □

Example 2. *Error in Estimation and Sample Size Determination.* Let $X_1, X_2, \ldots, X_n$ be a random sample from a distribution with mean μ and variance σ^2. According to the law of large numbers

$$P\big(|\overline{X} - \mu| > \varepsilon\big) = P\big(|S_n - n\mu| > n\varepsilon\big)$$

$$= P\big(S_n > n(\mu + \varepsilon) \quad \text{or} \quad S_n < n(\mu - \varepsilon)\big)$$

$$\to 0$$

as $n \to \infty$ for every $\varepsilon > 0$ where $S_n = \sum_{i=1}^{n} X_i$. In other words,

$$P\big(n(\mu - \varepsilon) \le S_n \le n(\mu + \varepsilon)\big) \to 1$$

so that the law of large numbers tells us that S_n grows at rate n. We can use the law of large numbers either to find an error bound in estimating μ by $\overline{X}$ or to find the sample size n required given an error bound. Thus the error in estimation is at most ε with probability at least $1 - (\sigma^2/n\varepsilon^2)$. Indeed,

$$P\big(|\overline{X} - \mu| \le \varepsilon\big) \ge 1 - \frac{\sigma^2}{n\varepsilon^2} \qquad \text{(Chebyshev's inequality)}.$$

On the other hand, if we want the error to be at most ε with a probability of at least $1 - \alpha$, then we choose n from

$$1 - \alpha = 1 - \frac{\sigma^2}{n\varepsilon^2},$$

giving $n = \sigma^2/(\alpha\varepsilon^2)$. We will see that we can do much better with the help of the central limit theorem (see Section 9.3). $\quad\square$

Example 3. The Empirical Distribution Function. Let $X_1,\ldots,X_n$ be a random sample from a distribution function F. We recall that the empirical distribution function of the sample is defined by

$$\hat{F}_n(x) = \frac{\text{number of } X_i\text{'s} \le x}{n}.$$

Let

$$Y_i = \begin{cases} 1 & \text{if } X_i \le x \\ 0 & \text{otherwise.} \end{cases}$$

In Example 5.3.6 we saw that $Y_1, Y_2,\ldots, Y_n$ are independent and identically distributed with

$$\mathscr{E}Y_i = P(X_i \le x) = F(x)$$

and

$$\text{var}(Y_i) = \mathscr{E}Y_i^2 - (\mathscr{E}Y_i)^2 = F(x) - (F(x))^2.$$

Since $\bar{Y} = \hat{F}_n(x)$ it follows by the law of large numbers that for every $\varepsilon > 0$ and each $x \in \mathbb{R}$

$$P(|\bar{Y} - F(x)| > \varepsilon) \to 0$$

as $n \to \infty$. Thus $\hat{F}_n(x)$ is likely to be near $F(x)$ for large n, that is, $\hat{F}_n(x)$ is a consistent estimate of $F(x)$.

A similar argument can be used to show that the relative frequency histogram estimate of an event A is a consistent estimate of the probability $P(A)$. (See Problems 6 and 5.6.11.) $\quad\square$

Example 4. Evaluating an Integral. Let f be a continuous function from $[0,1]$ to $[0,1]$, and suppose that we wish to evaluate $\int_0^1 f(x)\,dx$ numerically. Let $X_1, Y_1, X_2, Y_2,\ldots$ be independent random variables with density function

$$g(x) = 1, \quad 0 \le x \le 1 \text{ and zero elsewhere.}$$

Define

$$Z_i = \begin{cases} 1 & \text{if } f(X_i) > Y_i \\ 0 & \text{otherwise,} \end{cases}$$

for $i = 1, 2,\ldots$. Clearly Z_i are also independent and

$$\mathscr{E}Z_i = P(f(X_i) > Y_i) = \iint\limits_{f(x)>y} (1)(1)\,dx\,dy$$

$$= \int_0^1 \left(\int_0^{f(x)} dy \right) dx = \int_0^1 f(x)\,dx.$$

By the law of large numbers, since $\mathscr{E}Z_i^2 < \infty$ $(\mathscr{E}Z_i^2 = \mathscr{E}Z_i)$ it follows that

$$P\left(\left|\bar{Z} - \int_0^1 f(x)\, dx\right| > \varepsilon\right) \to 0$$

as $n \to \infty$ so that $\bar{Z}$ is likely to be near $\int_0^1 f(x)\, dx$ for large n. Thus, using a random number generator to generate (X_i, Y_i) and some computing device to compute $\bar{Z}$, we can use $\bar{Z}$ to approximate the integral $\int_0^1 f(x)\, dx$. $\qquad\square$

A special case of the law of large numbers will be encountered in Section 6.3. In statistical applications, the law of large numbers is frequently used to prove consistency of a sequence of estimates for a parameter.

Problems for Section 5.4

1. Let $X_1, X_2, \ldots$ be a sequence of independent random variables with distribution given below. Does the sufficient condition of Theorem 1 hold in each case? (If $\mathrm{var}(S_n)/n^2 \nrightarrow 0$, it does not follow that the law of large numbers does not hold.)
   (i) $P(X_k = \pm 2^k) = 1/2, k = 1, 2, \ldots$
   (ii) $P(X_k = \pm 2^k) = 1/2^{2k+1}, P(X_k = 0) = 1 - 2^{-2k}, k = 1, 2, \ldots$
   (iii) $P(X_k = \pm \sqrt{k}) = 1/2, k = 1, 2, \ldots$
   (iv) $P(X_k = \pm k) = 1/2\sqrt{k}, P(X_k = 0) = 1 - 1/\sqrt{k}, k = 1, 2, \ldots$.

2. Suppose $X_1, X_2, \ldots, X_n$ is a random sample with finite mean μ and finite variance σ^2. Show that

$$P\left(\left|\frac{2\sum\limits_{j=1}^{n} jX_j}{n(n+1)} - \mu\right| > \varepsilon\right) \to 0$$

as $n \to \infty$. That is, $2\sum_{j=1}^{n} X_j/n(n+1)$ is a consistent estimate for μ. (Hint: Use Theorem 5.3.1 and Theorem 1 of this chapter.)

3. Suppose $\{X_n\}$ is a sequence of random variables with common finite variance σ^2. Suppose also that for all $i \neq j$, $\mathrm{cov}(X_i, X_j) < 0$. Show that the law of large numbers holds for $\{X_n\}$.

4. Suppose $X_1, X_2, \ldots$ are independent, identically distributed with $\mathscr{E}|X_i|^{2k} < \infty$ for some positive integer k. Show that

$$\overline{X^j} = \sum_{i=1}^{n} \frac{X_i^j}{n}$$

the unbiased estimate of $\mathscr{E}X_1^j$ (for $1 \leq j \leq k$) is likely to be near $\mathscr{E}X_1^j$ for large n. That is, show that

$$P\left(\left|\overline{X^j} - \mathscr{E}X_1^j\right| > \varepsilon\right) \to 0$$

as $n \to \infty$ for every $\varepsilon > 0$.

5*.   Let $\{X_n\}$ be such that X_j is independent of X_k for $k \neq j + 1$ or $k \neq j - 1$. Suppose $0 < \text{var}(X_n) < c$ for all n and some constant c. Show that $\{X_n\}$ obeys the law of large numbers.

6.   Let $x_1, x_2, \ldots, x_n$ be n observations from a known density function f. Let us construct a histogram of the data by subdividing the real line in disjoint intervals (a_j, a_{j+1}), $j = 0, 1, \ldots$. Let f_j be the class frequency of (a_j, a_{j+1}) and $\hat{f}(x) = f_j/[n(a_{j+1} - a_j)]$, $x \in (a_j, a_{j+1})$ be the histogram estimate of f. Let $\hat{P}(A) = \int_A \hat{f}(x)\, dx$ be the histogram estimate of $P(A)$. Show that $P(|\hat{P}(A) - \int_A f(x)\, dx| > \varepsilon) \to 0$ as $n \to \infty$ for every $\varepsilon > 0$, whenever A is a union of intervals (a_j, a_{j+1}). (Hint: Use Example 1.)

7*.   Let $X_1, X_2, \ldots$ be independent identically distributed with density function

$$f(x) = 1, \qquad 0 \le x \le 1, \text{ and zero elsewhere.}$$

Show that for every $\varepsilon > 0$ as $n \to \infty$

$$P\left(\left|e\left(\prod_{i=1}^{n} X_i\right)^{1/n} - 1\right| > \varepsilon\right) \to 0.$$

$[(\prod_{i=1}^{n} X_i)^{1/n}$ is called the geometric mean of $X_1, \ldots, X_n.]$

8.   In random sampling from the following distributions, check to see if the sufficient condition for the law of large numbers is satisfied. If so, state what the law says.
   (i) $f(x) = 1, 0 \le x \le 1$, and zero elsewhere.
   (ii) $f(x) = (1/\pi)[1/(1 + x^2)], x \in \mathbb{R}$.
   (iii) $f(x) = 1/x^2, x \ge 1$, and zero elsewhere.
   (iv) $f(x) = (1/\lambda)e^{-x/\lambda}, x > 0$, and zero elsewhere.
   (v) $f(x) = 6x(1 - x), 0 \le x \le 1$, and zero elsewhere.
   (vi) $f(x) = \frac{1}{2}e^{-|x|}, x \in \mathbb{R}$.
   (vii) $f(x) = 2(\theta - x)/\theta^2, 0 < x < \theta$, zero elsewhere.

9.   In estimating μ by $\bar{X}$ suppose we wish to be 95 percent sure that $\bar{X}$ will not deviate from μ by more than:
   (i) $\sigma/5$.   (ii) $\sigma/10$.   (iii) $\sigma/20$.
   Find the minimum sample size required in each case.

10.   Let $X_1, X_2, \ldots$ be a sequence of independent, identically distributed random variables and suppose that $\mathscr{E}X_i = \mu > 0$ and $\text{var}(X_i) = \sigma^2$ is finite. Show that $P(S_n \ge a) \to 1$ as $n \to \infty$ for every finite constant a where $S_n = \sum_{k=1}^{n} X_k, n = 1, 2, \ldots$.

5.5* CONDITIONAL EXPECTATION

Let (X, Y) be jointly distributed. In many practical problems it is desirable to predict Y based on X and the knowledge of the relationship between X and Y as described by the joint distribution of X and Y. For example, it may be desirable (or even critical) to control a certain parameter Y in a chemical process, but Y may not be directly observable. If an observation on X is available instead, how do we predict Y? Suppose the object is to minimize the mean square error. Recall that $\mathscr{E}(Y - c)^2$ is minimized if and only if we choose $c = \mathscr{E}Y$. If nothing is known about X (or if X and Y are independent), our best estimate (predictor) of Y to minimize MSE is simply $\mathscr{E}Y$. If, on the other hand, information on X becomes available during the experiment, the distribution of Y changes to a conditional distribution that should also affect our choice of an estimate for Y. Heuristically, if $X = x$ is observed, Y is now being sampled from a subpopulation where the distribution is the conditional distribution of Y given $X = x$. If we therefore wish to estimate Y to minimize MSE, this estimate should be the expected value of the conditional distribution.

Let X, Y be jointly distributed. We recall that

$$P\{Y = y \mid X = x\} = \frac{P(X = x, Y = y)}{P(X = x)}$$

whenever $P(X = x) > 0$, defines a probability mass function for each fixed x (as a function of y) provided (X, Y) is of the discrete type. In the continuous case

$$g(y \mid x) = \frac{f(x, y)}{f_1(x)}, \qquad \text{for } f_1(x) > 0, y \in \mathbb{R},$$

defines a probability density function for each fixed x (as a function of y) where f is the joint and f_1 the marginal density of (X, Y) and X respectively.

Clearly, then, for fixed x we can define the expected value of Y in this subpopulation just as we defined $\mathscr{E}g(X)$.

DEFINITION 1.   Suppose X and Y are discrete type random variables. We define the conditional expectation of Y given $X = x$ by:

$$\mathscr{E}\{Y \mid X = x\} = \sum_y yP\{Y = y \mid X = x\}$$

and that of X given $Y = y$ by:

$$\mathscr{E}\{X \mid Y = y\} = \sum_x xP\{X = x \mid Y = y\}$$

provided the sums converge absolutely.

If X and Y have a joint density function f, then we define

$$\mathscr{E}\{Y \mid X = x\} = \int_{-\infty}^{\infty} yg(y \mid x)\, dy$$

and

$$\mathscr{E}\{X \mid Y = y\} = \int_{-\infty}^{\infty} xh(x \mid y)\, dx$$

provided the integrals converge absolutely. Here g and h are respectively the conditional densities of Y given $X = x$ and X given $Y = y$.

If one of X and Y is a discrete type while the other is a continuous type, we define $\mathscr{E}\{X|Y = y\}$ and $\mathscr{E}\{Y|X = x\}$ in an analogous manner. Similarly we can define $\mathscr{E}\{\phi(X)|Y = y\}$ or $\mathscr{E}\{\phi(Y)|X = x\}$ for any function ϕ.

Example 1. Sampling Inspection. A lot contains 10 items, two of type A, four of type B, and four of type C. A sample of size two is drawn without replacement. Let

$$X = \text{number of type } A \text{ items in the sample.}$$

$$Y = \text{number of type } B \text{ items in the sample.}$$

Then the joint distribution of (X, Y) is given by the probability function:

Y＼X	0	1	2	$P(Y = y)$
0	6/45	8/45	1/45	15/45
1	16/45	8/45	0	24/45
2	6/45	0	0	6/45
$P(X = x)$	28/45	16/45	1/45	1

Clearly

$$P\{Y|X = 0\} = \begin{cases} 6/28, & y = 0 \\ 16/28, & y = 1 \\ 6/28, & y = 2 \end{cases}$$

so that

$$\mathscr{E}\{Y|X = 0\} = 0(6/28) + 1(16/28) + 2(6/28) = 1.$$

Similarly,

$$P\{X|Y = 0\} = \begin{cases} 6/15, & x = 0 \\ 8/15, & x = 1 \\ 1/15, & x = 2 \end{cases}$$

and

$$\mathscr{E}\{X|Y = 0\} = 0(6/15) + 1(8/15) + 2(1/15) = 10/15.$$

Also:

$$\mathscr{E}\{Y|X = 1\} = 1/2, \qquad \mathscr{E}\{Y|X = 2\} = 0$$

$$\mathscr{E}\{X|Y = 1\} = 1/3, \qquad \mathscr{E}\{X|Y = 2\} = 0$$

so that

$$\mathscr{E}\{Y|X = x\} = \begin{cases} 1, & x = 0 \\ 1/2, & x = 1 \\ 0, & x = 2 \end{cases}$$

and

$$\mathscr{E}\{X|Y=y\} = \begin{cases} 10/15, & y = 0 \\ 1/3, & y = 1 \\ 0, & y = 2. \end{cases}$$ □

Example 2. Standby-Redundant System. Consider a standby-redundant system. Each component in such a system operates continuously until it fails, and then it is immediately replaced by a standby component which operates until it fails. Components can fail while in use.

Let X, Y respectively denote the failure times of the original and the standby component as measured from time 0 when the system is switched on. Then $Y \geq X$. Suppose (X, Y) have joint density

$$f(x, y) = \begin{cases} (1/\alpha\beta)\exp\left\{-\left(\frac{x}{\alpha}\right) - \frac{(y-x)}{\beta}\right\}, & y \geq x, x > 0 \\ 0 & \text{elsewhere} \end{cases}$$

where $\alpha \neq \beta$ and $\alpha > 0$, $\beta > 0$. It is easy to check that the marginal densities of X and Y are given, respectively, by

$$f_1(x) = \frac{1}{\alpha}\exp\left(-\frac{x}{\alpha}\right), \qquad x > 0, \text{ and zero elsewhere}$$

and

$$f_2(y) = \frac{1}{\beta - \alpha}\left\{\exp\left(-\frac{y}{\beta}\right) - \exp\left(-\frac{y}{\alpha}\right)\right\}, \qquad y > 0, \text{ and zero elsewhere.}$$

Then

$$g(y|x) = \frac{f(x, y)}{f_1(x)} = \frac{1}{\beta}\exp\{-(y-x)\beta\}, \qquad y \geq x > 0, \text{ and zero elsewhere.}$$

It follows that

$$\mathscr{E}\{Y|X=x\} = \frac{1}{\beta}\int_{-\infty}^{\infty} yg(y|x)\, dy = (1/\beta)\int_x^{\infty} y\exp\left\{-\frac{y-x}{\beta}\right\} dy$$

$$= \beta + x \qquad \text{(on integration by parts).}$$

Similarly,

$$h(x|y) = \frac{f(x, y)}{f_2(y)} = \left(\frac{\beta - \alpha}{\alpha\beta}\right)\frac{\exp\{-(x/\alpha) - (y-x)/\beta\}}{\{\exp(-y/\beta) - \exp(-y/\alpha)\}},$$

for $0 < x \leq y$, and zero for $x > y$. Hence

$$\mathscr{E}\{X|Y=y\} = \int_0^y x\left(\frac{1}{\alpha} - \frac{1}{\beta}\right) \frac{\exp\{-x/\alpha - (y-x)/\beta\}}{\{\exp(-y/\beta) - \exp(-y/\alpha)\}} dx$$

$$= \left(\frac{1}{\alpha} - \frac{1}{\beta}\right)(e^{-y/\beta} - e^{-y/\alpha})^{-1}e^{-y/\beta}\int_0^y xe^{-x(1/\alpha-1/\beta)} dx$$

$$= \frac{\beta - \alpha}{\alpha\beta}[1 - e^{-y(1/\alpha-1/\beta)}]^{-1}\left\{\frac{-ye^{-y(1/\alpha-1\beta)}}{(1/\alpha - 1/\beta)} + \frac{1 - e^{-y(1/\alpha-1/\beta)}}{(1/\alpha - 1/\beta)^2}\right\},$$

on integration by parts. It follows that

$$\mathscr{E}\{X|Y=y\} = \left(\frac{1}{\alpha} - \frac{1}{\beta}\right)^{-1} - y(e^{y(1/\alpha-1/\beta)} - 1)^{-1}. \qquad \square$$

We note that as x varies, $\mathscr{E}\{Y|X=x\}$ varies, and we can think of $\mathscr{E}\{Y|X\}$ as a random variable that assumes the value $\mathscr{E}\{Y|X=x\}$ when $X=x$. Similarly, $\mathscr{E}\{X|Y\}$ is a random variable with distribution $P(Y=y)$ or $f_2(y)$ depending on whether Y is a discrete or continuous type. We can therefore speak of the expected value of $\mathscr{E}\{X|Y\}$, and so on.

THEOREM 1. *If $\mathscr{E}|X| < \infty$, then $\mathscr{E}\{X|Y=y\}$ is finite for every y for which the conditional distribution of X given $Y=y$ is defined. Moreover,*

(1) $$\mathscr{E}X = \mathscr{E}\{\mathscr{E}\{X|Y\}\}.$$

Proof. We shall prove the result in the discrete case only. A similar proof holds for the continuous case. Suppose $P(Y=y) > 0$. Then $P\{X=x|Y=y\}$ is defined, and moreover:

$$P(X=x|Y=y) = \frac{P(X=x, Y=y)}{P(Y=y)} \leq \frac{P(X=x)}{P(Y=y)}.$$

It follows that

$$\sum_x |x|P\{X=x|Y=y\} \leq \sum_x |x|\frac{P(X=x)}{P(Y=y)} = [P(Y=y)]^{-1}\mathscr{E}|X| < \infty.$$

In that case,

$$\mathscr{E}X = \sum_x xP(X=x) = \sum_x \sum_y xP(X=x, Y=y)$$

$$= \sum_x x\sum_y P(Y=y)P\{X=x|Y=y\}$$

$$= \sum_y P(Y=y)\sum_x xP\{X=x|Y=y\}$$

$$= \sum_y P(Y=y)\mathscr{E}\{X|Y=y\}$$

$$= \mathscr{E}\{\mathscr{E}\{X|Y\}\}$$

as asserted. The interchange of summation signs is justified because of absolute convergence. $\qquad\square$

Similarly, if $\mathscr{E}|Y| < \infty$ then

$$(2) \qquad\qquad \mathscr{E}Y = \mathscr{E}\{\mathscr{E}\{Y|X\}\}.$$

Both (1) and (2) are very useful, as we shall see. We emphasize that $\mathscr{E}\{X|Y\}$ may be finite without $\mathscr{E}X$ being finite.

We first state some properties of conditional expectations. For convenience we shall write $\mathscr{E}\{X|y\}$ for $\mathscr{E}\{X|Y = y\}$. In the following we assume that the indicated expectations exist.

PROPOSITION 1.

(i) *If X and Y are independent and $\mathscr{E}|X| < \infty$ then*

$$(3) \qquad\qquad \mathscr{E}\{X|y\} = \mathscr{E}X,$$

and if $\mathscr{E}|Y| < \infty$, then

$$(4) \qquad\qquad \mathscr{E}\{Y|x\} = \mathscr{E}Y.$$

(ii) *Let $\phi(X, Y)$ be a function of X and Y. Then*

$$(5) \qquad\qquad \mathscr{E}\{\phi(X,Y)|y\} = \mathscr{E}\{\phi(X,y)|y\}$$

and

$$(6) \qquad\qquad \mathscr{E}\{\phi(X,Y)|x\} = \mathscr{E}\{\phi(x,Y)|x\}.$$

Moreover,

$$(7) \quad \mathscr{E}\{\alpha_1\phi_1(X,Y) + \alpha_2\phi_2(X,Y)|x\} = \alpha_1\mathscr{E}\{\phi_1(x,Y)|x\} + \alpha_2\mathscr{E}\{\phi_2(x,Y)|x\}$$

for any functions ϕ_1 and ϕ_2 and real numbers α_1, α_2.

(iii) *If ψ is a function of X and ϕ a function of X and Y, then*

$$(8) \qquad\qquad \mathscr{E}\{\psi(X)\phi(X,Y)|x\} = \psi(x)\mathscr{E}\{\phi(x,Y)|x\}.$$

We leave it to the reader to supply a proof of Proposition 1.

*Example 3. **Sum of Random Number of Independent Random Variables.*** Let $X_1, X_2, \ldots$ be independent and identically distributed random variables and let N be a positive integer valued random variable that is independent of the X's. Set $S_N = \sum_{j=1}^N X_j$. What is $\mathscr{E}S_N$? We note that S_N has N terms where N itself is a random variable. We cannot apply Theorem 5.3.1 to compute $\mathscr{E}S_N$.

Random variables such as S_N arise often in practice. For example, suppose the number of customers per day who shop at a store is a random variable N. Let X be a consumer's demand for a certain item. Let $X_1, \ldots, X_n, \ldots$ be independent and identically distributed and suppose N is independent of X. Then the expected consumer demand for this item at this store per day is $\mathscr{E}(X_1 + X_2 + \cdots + X_N)$.

Suppose $\mathscr{E}|X_i| < \infty$ and $\mathscr{E}N < \infty$. Now $S_N = S_n$ when $N = n$ is observed. We use (1) and have

$$\mathscr{E}S_N = \mathscr{E}\{\mathscr{E}\{S_N|N\}\} = \sum_{n=1}^{\infty} \mathscr{E}\{S_n|N = n\}P(N = n)$$

$$= \sum_{n=1}^{\infty} \mathscr{E}S_n P(N = n) \qquad (N \text{ and } S_n \text{ are independent—Proposition 1})$$

$$= \sum_{n=1}^{\infty} (n\mathscr{E}X_i)P(N = n) \qquad (\text{Theorem 5.3.1})$$

$$= \mathscr{E}X_i \mathscr{E}N.$$

Actually, we first need to check that $\mathscr{E}|S_N| < \infty$ before we can apply (1). But this is easily done by replacing S_N by $|S_N|$ using triangular inequality and the fact that $\mathscr{E}|X_i| < \infty$ and $\mathscr{E}N < \infty$.

Note how Theorem 5.3.1 is modified when N is a random variable. We simply replace n by $\mathscr{E}N$ in $\mathscr{E}S_n = n\mathscr{E}X_i$. This result is called *Wald's equation* and has many applications. □

Example 4. $\mathscr{E}\{Y|X\}$ May Exist but $\mathscr{E}Y$ May Not. Consider the joint density function

$$f(x, y) = x \exp\{-x(1 + y)\}, \qquad x \geq 0, y \geq 0, \text{ and zero elsewhere,}$$

of Problem 5.2.15. We note that

$$f_1(x) = e^{-x}, \qquad x \geq 0, \text{ and zero elsewhere,}$$

and

$$f_2(y) = \frac{1}{(1 + y)^2}, \qquad y \geq 0, \text{ and zero elsewhere.}$$

Clearly, $\mathscr{E}Y = \int_0^{\infty} y/(1 + y)^2 \, dy$ does not exist (that is, is not finite), but

$$\mathscr{E}\{Y|x\} = \int_{-\infty}^{\infty} yg(y|x) \, dy = \int_0^{\infty} yx \exp(-xy) \, dy$$

$$= x(1/x^2) = 1/x \qquad (\text{integration by parts}). \qquad □$$

Example 5. Standby-Redundant System. In Example 2 we showed that

$$\mathscr{E}\{Y|x\} = \beta + x$$

and X has marginal density

$$f_1(x) = \frac{1}{\alpha} \exp\left(-\frac{x}{\alpha}\right), \qquad x > 0, \text{ and zero elsewhere.}$$

In view of Theorem 1

$$\mathscr{E}Y = \mathscr{E}\{\mathscr{E}\{Y|X\}\} = \mathscr{E}(\beta + X) = \beta + \mathscr{E}X = \beta + \alpha. \qquad \square$$

Let us now prove the result that we alluded to in the beginning of this section.

PROPOSITION 2.   *The estimate of Y based on X, say $\phi(X)$, which minimizes the MSE $\mathscr{E}(Y - \phi(X))^2$, is given by:*

$$\phi(X) = \mathscr{E}\{Y|X\}.$$

(The function $\mathscr{E}\{Y|X\}$ is also called the regression function of Y on X.)

Proof.   We have

$$\mathscr{E}(Y - \phi(X))^2 = \mathscr{E}\{\mathscr{E}\{(Y - \phi(X))^2|X\}\}$$

$$= \mathscr{E}\{\mathscr{E}\{[Y - \mathscr{E}\{Y|X\} + \mathscr{E}\{Y|X\} - \phi(X)]^2|X\}\}$$

$$= \mathscr{E}\{\mathscr{E}\{[Y - \mathscr{E}\{Y|X\}]^2|X\} + \mathscr{E}[\mathscr{E}\{Y|X\} - \phi(X)]^2$$

which is minimized by choosing $\phi(X) = \mathscr{E}\{Y|X\}$. $\qquad \square$

By taking $\phi(X) = \mathscr{E}Y$ and remembering that $\mathscr{E}Y = \mathscr{E}\{\mathscr{E}\{Y|X\}\}$, we get the following useful result:

$$(9) \qquad\qquad \operatorname{var}(Y) = \mathscr{E}(\operatorname{var}\{Y|X\}) + \operatorname{var}(\mathscr{E}\{Y|X\}).$$

Thus the unconditional variance of Y is the sum of the expectation of its conditional variance (that is, variance of the conditional distribution) and the variance of its conditional expectation.

Example 6. $P(X < Y)$ for Independent X and Y.   Let X and Y be independent random variables of the continuous type. Then $P(X < Y)$ is of great interest in statistics. For example, let X and Y be lifetimes of two components operating independently. Then $P(X < Y)$ is the probability that the second component (with lifetime Y) survives the first component (with lifetime X). Now, if I_A is the indicator function of set A, then

$$P(X < Y) = \mathscr{E}I_{[X<Y]} = \mathscr{E}\{\mathscr{E}\{I_{[X<Y]}|Y\}\}$$

by Theorem 1. We note that

$$\mathscr{E}\{I_{[X<Y]}|y\} = \mathscr{E}\{I_{[X<y]}|y\} \qquad \text{[Proposition 1 (ii)]}$$

$$= \mathscr{E}I_{[X<y]} \qquad \text{[Proposition 1 (i)]}$$

$$= F_X(y)$$

where F_X is the distribution function of X. Hence

$$P(X < Y) = \mathscr{E}\{F_X(Y)\} = \int_{-\infty}^{\infty} F_X(y)f_Y(y)\,dy$$

where f_Y is the density function of Y. In particular, if $f_X = f_Y$ then

$$P(X < Y) = \int_{-\infty}^{\infty} F_X(y)f_X(y)\,dy = \frac{1}{2}$$

on integration by parts.

More generally,

$$P(X - Y \le z) = \mathscr{E}\{\mathscr{E}\{I_{[X-Y\le z]}|Y\}\}$$

$$= \mathscr{E}\{F_X(Y + z)\} = \int_{-\infty}^{\infty} F_X(y + z)f_Y(y)\,dy,$$

which gives the distribution function of $X - Y$. □

Example 7. Trapped Miner. Consider a miner trapped in a mine with three doors. The first door leads to a tunnel with no escape and returns him to his cell in the mine in one hour. The second door also leads to a tunnel with no escape and returns him to his cell in the mine in three hours. The third door takes him to safety in one hour. If the miner at all times chooses randomly from the untried doors, what is the expected length of time until he reaches safety?

If we let X denote the time he takes to make good his escape and let Y_1 denote the door he chooses initially, then

$$\mathscr{E}X = \mathscr{E}\{\mathscr{E}\{X|Y_1\}\}$$

$$= \mathscr{E}\{X|Y_1 = 1\}P(Y_1 = 1) + \mathscr{E}\{X|Y_1 = 2\}P(Y_1 = 2)$$

$$+ \mathscr{E}\{X|Y_1 = 3\}P(Y_1 = 3)$$

$$= \frac{1}{3}\sum_{j=1}^{3}\mathscr{E}\{X|Y_1 = j\}.$$

Clearly, $\mathscr{E}\{X|Y_1 = 3\} = 1$. Now

$$\mathscr{E}\{X|Y_1 = 1\} = 1 + P\{Y_2 = 2|Y_1 = 1\}\mathscr{E}\{X|Y_1 = 1, Y_2 = 2\}$$

$$+ P\{Y_2 = 3|Y_1 = 1\}\mathscr{E}\{X|Y_1 = 1, Y_2 = 3\}$$

where Y_2 denotes the door he chooses on the second try. Then

$$\mathscr{E}\{X|Y_1 = 1\} = 1 + (1/2)(3 + 1) + (1/2)(1) = 7/2.$$

This follows since the miner having chosen door 1 on the first try returns to the mine after one hour and then chooses door 2 or door 3. If he chooses door 2, then he spends three hours in the tunnel, returns, and tries door 3, which leads him to safety in one additional hour. On the other hand, if he chooses door 3 on the second try, he reaches safety in an hour. Similarly,

$$\mathscr{E}\{X|Y_1 = 2\} = 3 + P\{Y_2 = 1|Y_1 = 2\}\mathscr{E}\{X|Y_1 = 2, Y_2 = 1\} + P\{Y_2 = 3|Y_1 = 2\}$$
$$\cdot\mathscr{E}\{X|Y_1 = 2, Y_2 = 3\} = 3 + (1/2)(1+1) + (1/2)(1) = 9/2.$$

It follows that

$$\mathscr{E}X = (1/3)\{(7/2) + (9/2) + 1\} = 3.$$

An alternative approach would be to evaluate $\mathscr{E}X$ directly, bypassing the conditional expectation approach. We note that the only possible door sequences are 123, 213, 23, 13, 3 with respective total times required to reach safety (in hours) 5, 5, 4, 2 and 1. Hence

$$\mathscr{E}X = (1/6)5 + (1/6)5 + (1/6)4 + (1/6)2 + (1/3)1 = 3.$$

For a somewhat simpler (but unrealistic) model, see Problem 3.   □

Example 8. **Estimating Proportions of Two Chemicals in a Compound.** Let X and Y denote the proportions of two different chemicals in a sample from a chemical mixture with joint density function

$$f(x, y) = \begin{cases} 2 & 0 \le x + y \le 1, 0 \le x \le 1, 0 \le y \le 1, \\ 0 & \text{elsewhere.} \end{cases}$$

Then (see Figure 1)

$$f_1(x) = \int_0^{1-x} 2\,dy = \begin{cases} 2(1-x), & 0 \le x \le 1 \\ 0 & \text{elsewhere} \end{cases}$$

and

$$f_2(y) = \int_0^{1-y} 2\,dx = 2(1-y), \qquad 0 \le y \le 1, \text{ and zero elsewhere.}$$

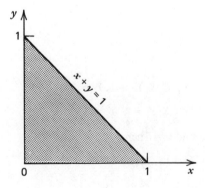

Figure 1. $\{(x, y): f(x, y) > 0\}$.

If X is not observed then our best estimate of Y is $\mathscr{E}Y = 2\int_0^1 y(1 - y)\,dy = 1/3$. If it is known that $X = 1/6$, say, then our best estimate of Y from Proposition 2 is

$$\mathscr{E}\left\{Y|x = \frac{1}{6}\right\} = \int_{-\infty}^{\infty} yg\left(y|\frac{1}{6}\right) dy = \int_0^{5/6} y\frac{1}{(1 - 1/6)}\,dy = \frac{6}{5}\cdot\frac{1}{2}\left(\frac{5}{6}\right)^2$$

$$= 5/12.$$

Let us find var(Y) using (9). First we note that

$$g(y|x) = \frac{1}{1 - x}, \qquad 0 \le y \le 1 - x, x \ge 0, \text{ and zero elsewhere.}$$

Hence:

$$\mathscr{E}\{Y|x\} = \int_0^{1-x} y\frac{1}{1 - x}\,dy = \frac{1 - x}{2}$$

and

$$\mathscr{E}\{Y^2|x\} = \frac{(1 - x)^2}{3}$$

so that

$$\text{var}\{Y|x\} = \frac{(1 - x)^2}{12}.$$

It follows that

$$\mathscr{E}(\text{var}\{Y|X\}) = \mathscr{E}\left\{\frac{(1 - X)^2}{12}\right\} = \left(\frac{1}{12}\right)\left(\frac{1}{2}\right) = \frac{1}{24}$$

and

$$\text{var}(\mathscr{E}\{Y|X\}) = \text{var}\left(\frac{1 - X}{2}\right) = \frac{\text{var}(X)}{4} = \frac{1}{72}.$$

Hence from (9),

$$\text{var}(Y) = (1/24) + (1/72) = 1/18.$$

In this case it is simpler to compute var(Y) directly from $f_2(y)$.

To summarize what we have done in this example, the best predictor of Y when no information on X is available is $\mathscr{E}Y = 1/3$. This estimate minimizes $\mathscr{E}(Y - c)^2$. If, on the other hand, it is known that $X = 1/6$, then our best predictor of Y is $\mathscr{E}\{Y|x = 1/6\} = 5/12$ in the sense that it minimizes $\mathscr{E}\{Y - \phi(X)\}^2$. In each case we minimize the average of the squared deviation between Y and its estimate. $\qquad\square$

Problems for Section 5.5

1. Each word in the sentence I DO NOT LIKE THIS GLASS is written on a slip and the six slips are returned to a box. A slip is then randomly selected. Let X be the length of the word on the slip drawn and Y the number of I's in the selected word. Find $\mathscr{E}\{X|Y\}$, $\mathscr{E}\{Y|X\}$, var$\{X|Y\}$, var$\{Y|X\}$.

2. Let X, Y be jointly distributed with probability function

Y \ X	0	1	2	$P(Y = y)$
0	1/60	4/60	9/60	14/60
1	2/60	6/60	12/60	20/60
2	3/60	8/60	3/60	14/60
3	4/60	2/60	6/60	12/60
$P(X = x)$	10/60	20/60	30/60	1

Find $\mathscr{E}\{X|y\}$, $\text{var}\{X|y\}$, $\mathscr{E}\{Y|x\}$, $\text{var}\{Y|x\}$.

3. A prisoner has just escaped from jail. There are three roads that lead away from the jail, A, B, and C. Of the three, C leads to freedom immediately and the other two lead back to the prison gate in 15 minutes. If at all times the prisoner selects roads A, B, and C with probability $1/2$, $1/3$, and $1/6$, what is the expected time that it will take to make good the escape?

4. A fair coin is tossed three independent times. Let X be the number of heads and Y the number of head runs, where a head run is defined to be a consecutive occurrence of at least two heads. Check that the joint probability function X and Y is given by:

Y \ X	0	1	2	3	$P(Y = y)$
0	1/8	3/8	1/8	0	5/8
1	0	0	2/8	1/8	3/8
$P(X = x)$	1/8	3/8	3/8	1/8	1

Find $\mathscr{E}\{Y|x\}$ and $\mathscr{E}\{X|y\}$.

5. Let X be the sum of face values and Y the numerical difference between face values when a pair of dice is tossed once. The joint probability function of (X, Y) is given below.

Y \ X	2	3	4	5	6	7
0	1/36	0	1/36	0	1/36	0
1	0	2/36	0	2/36	0	2/36
2	0	0	2/36	0	2/36	0
3	0	0	0	2/36	0	2/36
4	0	0	0	0	2/36	0
5	0	0	0	0	0	2/36
	1/36	2/36	3/36	4/36	5/36	6/36

Y \ X	8	9	10	11	12	
0	1/36	0	1/36	0	1/36	6/36
1	0	2/36	0	2/36	0	10/36
2	2/36	0	2/36	0	0	8/36
3	0	2/36	0	0	0	6/36
4	2/36	0	0	0	0	4/36
5	0	0	0	0	0	2/36
	5/36	4/36	3/36	2/36	1/36	1

Find $\mathscr{E}\{X|Y\}$, $\mathscr{E}\{Y|X\}$, var$\{X|Y\}$ and var$\{Y|X\}$.

6. Compute $\mathscr{E}\{X|Y\}$ and $\mathscr{E}\{Y|X\}$ for each of the following bivariate densities:
   (i)* $f(x, y) = 4y(x - y)\exp\{-(x + y)\}$, $x \geq y \geq 0$, and zero elsewhere.
   (ii) $f(x, y) = 2$, $0 \leq y \leq x \leq 1$, and zero elsewhere.
   (iii) $f(x, y) = (x + y)/3$, $0 < x < 1$, $0 < y < 2$, and zero elsewhere.
   (iv) $f(x, y) = (1/8)(x^2 - y^2)\exp(-x)$, $0 \leq x < \infty$, $|y| < x$, and zero elsewhere.

7. For the joint density

$$f(x, y) = (1/2)\, ye^{-xy}, \qquad 0 < y < 2, 0 < x < \infty, \text{ and zero elsewhere}$$

   of X and Y, show that $\mathscr{E}\{\exp(X/2)|y\} = 2y/(2y - 1)$, for $y > 1/2$.

8. Let (X, Y) have joint density given by

$$f(x, y) = \begin{cases} 6xy(2 - x - y), & 0 < x < 1, 0 < y < 1 \\ 0 & \text{elsewhere.} \end{cases}$$

   Show that the best estimate of an X value given $Y = y$ is given by $(5 - 4y)/(8 - 6y)$.

9. Let (X, Y) have joint density function given below.
   (i) $f(x, y) = y^{-1}e^{-y}$, $0 < x < y$, $0 < y < \infty$, and zero otherwise.
   (ii) $f(x, y) = y^{-1}e^{-x/y}e^{-y}$, $0 < x < \infty$, $0 < y < \infty$, and zero elsewhere.
   (iii) $f(x, y) = x/36$, $x \geq 0$, $y \geq 0$, $0 \leq x + y \leq 6$, and zero elsewhere.
   (iv) $f(x, y) = x\exp\{-x(1 + y)\}$, $x > 0$, $y > 0$, and zero elsewhere.
   Find, in each case, $\mathscr{E}\{X|Y\}$ and $\mathscr{E}\{X^2|Y\}$.

10. Find $\mathscr{E}XY$ by conditioning on X or Y in each of the following cases:
    (i) $f(x, y) = x\exp\{-x(1 + y)\}$, $x > 0$, $y > 0$, and zero elsewhere.
    (ii) $f(x, y) = 2$, $0 \leq y \leq x \leq 1$, and zero elsewhere.
    (iii) $f(x, y) = (x + y)/3$, $0 < x < 1$, $0 < y < 2$, and zero elsewhere.

11. Let T be the failure time of a component that is tested until failure at various voltage levels. The failure time is found to have a mean directly proportional to the square of the voltage. If the voltage has density

$$f(v) = 1/\theta, \qquad 0 \leq v \leq \theta, \text{ and zero elsewhere,}$$

    find the average time to failure of the component.

12. A point X is chosen from the interval $[0, 1]$ according to density

$$f(x) = 1, \qquad 0 \leq x \leq 1, \text{ and zero elsewhere.}$$

Another point Y is chosen from the interval $[0, X]$ according to the density

$$g(y|x) = \frac{1}{x}, \qquad 0 \leq y \leq x, \text{ and zero elsewhere.}$$

Find $\mathscr{E}\{Y^k|x\}$ and $\mathscr{E}\{Y^k\}$, where $k > 0$ is a fixed constant.

13. Let X and Y be independent continuous random variables.
   (i) Show that $P(X + Y \leq z) = \int_{-\infty}^{\infty} F_1(z - y)f_2(y)\,dy$ where F_1 is the distribution function of X and f_2 the density of Y.
   (ii) Show that $P(XY \leq z) = \int_{-\infty}^{0}[1 + F_1(z/y)]f_2(y)\,dy + \int_{0}^{\infty}F_1(z/y)f_2(y)\,dy.$
   (iii) Show that $P(X/Y \leq z) = \int_{-\infty}^{0}[1 - F_1(zy)]f_2(y)\,dy + \int_{0}^{\infty}F_1(zy)f_2(y)\,dy.$

14. Let $X_1, X_2, \ldots$ be independent identically distributed and let N be a nonnegative integer valued random variable which is independent of the X's. Show that

$$\text{var}\left(\sum_{i=1}^{N} X_i\right) = \mathscr{E}N\,\text{var}(X_i) + (\mathscr{E}X)^2\text{var}(N)$$

provided $\mathscr{E}X^2 < \infty$ and $\mathscr{E}N^2 < \infty$. [Hint: Use (9). Alternatively, proceed directly as in Example 3.]

15. Show that the MSE $\mathscr{E}\{Y - \mathscr{E}\{Y|X\}\}^2$ of the predictor $\mathscr{E}\{Y|X\}$ of Y is given by $\mathscr{E}(\text{var}\{Y|X\})$.

16*. Consider a battery that is installed in a process and operates until it fails. In that case it is immediately replaced by another battery, and so on. The process continues without interruption. Let F be the lifetime distribution of the battery. Let $N(t)$ be the number of batteries that fail by time t. Show that the battery in use at any time t tends to have a longer life length than an ordinary battery. That is, show that

$$P(X_{N(t)+1} > x) \geq 1 - F(x) \text{ for all } x.$$

[Hint: Let $S_{N(t)} = \sum_{j=1}^{N(t)}X_j$. Then $X_{N(t)+1} = S_{N(t)+1} - S_{N(t)}$ is the length of the battery in use. $S_{N(t)}$ is the time at which the last battery was replaced prior to t.

Condition on $S_{N(t)}$. Then $P(X_{N(t)+1} > x) = \mathscr{E}\{P\{X_{N(t)+1} > x|S_{N(t)}\}\}$. Show that $P\{X_{N(t)+1} > x|S_{N(t)} = t - s\} = 1$ if $s > x$ and $= (1 - F(x))/(1 - F(s))$ if $s \leq x$.]

17*. Let $X_1, X_2, \ldots, X_n$ be exchangeable random variables. Show that

$$\mathscr{E}\{X_j|X_1 + \cdots + X_n = z\} = \mathscr{E}\{X_k|X_1 + \cdots + X_n = z\}$$

for $j \neq k$, and hence conclude that

$$\mathscr{E}\left\{\frac{X_1 + \cdots + X_k}{X_1 + \cdots + X_n}\right\} = \frac{k}{n}, \qquad 1 \leq k \leq n.$$

Given $X_1 + \cdots + X_n = z$, what is the best predictor of X_1? ($X_1, X_2, \ldots, X_n$ are exchangeable if their joint distribution $F(x_1, x_2, \ldots, x_n)$ remains unchanged under any permutation of the arguments $x_1, x_2, \ldots, x_n$.)

18. Let X, given $\Lambda = \lambda$, have probability function

$$P(X = x|\lambda) = e^{-\lambda}\lambda^x/x!, \qquad x = 0, 1, 2, \ldots$$

and suppose Λ has density function

$$f(x) = e^{-\lambda}, \qquad \lambda > 0, \text{ and zero elsewhere.}$$

Find $\mathscr{E}\{\exp(-\Lambda)|X = 1\}$.

19*. Let (X, Y) be jointly distributed. Suppose X is observable and we wish to predict Y by $\phi(X)$ in order to minimize $\mathscr{E}|Y - \phi(X)|$. What is the best predictor $\phi(X)$? (Hint: Use Problem 3.7.25.)

5.6   REVIEW PROBLEMS

1. Two particles have lifetimes X and Y with joint density function given by

$$f(x, y) = 2e^{-(x+y)}, \qquad 0 \le x \le y, \text{ and zero elsewhere.}$$

   (i)   Find $\mathscr{E}(Y - X)$ and $\text{var}(Y - X)$.
   (ii) Find $\mathscr{E}\{Y|x\}$ and $\mathscr{E}\{X|y\}$.
   (iii) Verify that $\mathscr{E}\{\mathscr{E}\{Y|X\}\} = \mathscr{E}Y$ by using part (ii).

2. Let X and Y have joint density given by

$$f(x, y) = (6/5)(x + y^2), \qquad 0 < x < 1, 0 < y < 1, \text{ and zero elsewhere.}$$

   (i)   Find $\text{cov}(X, Y)$ and $\rho(X, Y)$
   (ii) Find $\text{var}(2X + 7Y)$.
   (iii) Find $\mathscr{E}\{X|y = 1/2\}$.

3. A random sample of 40 students is taken from a small college student body that has 450 students. Each student in the sample is asked if he or she is for or against Saturday morning classes. Suppose eight students are in favor of the proposal. Find an estimate of the true proportion of students at this college favoring the proposal. Find an estimate for the variance of your estimate and a 95 percent confidence interval for the true proportion using Chebyshev's inequality.

4. In Examples 5.2.1, 5.2.4, and 5.2.7, the joint distribution of the total time X that a customer spends in a bank and the time Y spent standing in line was assumed to have density function

$$f(x, y) = \lambda^2 e^{-\lambda x}, \qquad 0 \le y \le x < \infty, \text{ and zero elsewhere.}$$

What is the variance of the service time $X - Y$?

5. Let $X \geq 1$ be a real number and let Y be its fractional part. Suppose X and Y are jointly distributed with density function

$$f(x, y) = 4\frac{x + y}{5x^3}, \qquad 0 \leq y \leq 1, 1 \leq x < \infty, \text{ and zero otherwise.}$$

   (i) Find $\mathscr{E}\{Y|X\}$ and $\mathscr{E}Y$.
   (ii) Find $\text{var}\{Y|X\}$.

6. Consider a randomly selected family with six children. Let $X_i = 1$ if the ith oldest child is a girl and zero otherwise. If it is known that the family has five girls, what is $\mathscr{E}\{X_1|Y = 5\}$ where Y is the number of girls in a family with six children?

7. Let (X, Y) be the coordinates of a spot on the circular section of a radar station and suppose (X, Y) has joint density function

$$f(x, y) = \frac{1}{\pi r^2}, \qquad x^2 + y^2 \leq r^2, \text{ and zero elsewhere.}$$

   (i)   Find $\mathscr{E}\{X|Y\}$, $\mathscr{E}X$, and $\text{var}\{X|Y\}$.
   (ii) Find $\mathscr{E}\{X + Y\}$, $\mathscr{E}(XY)$, and $\rho(X, Y)$.
   (iii) Find $\text{var}(X + Y)$.
   (See also Examples 3.4.4 and 3.5.3.)

8. The surface tension X and the acidity Y of a certain chemical compound are jointly distributed with density function

$$f(x, y) = \frac{3 - x - y}{3}, \qquad 0 \leq x \leq 1, 0 \leq y \leq 2, \text{ and zero elsewhere.}$$

   (i)   Find $\mathscr{E}\{X|Y\}$ and $\text{var}\{X|Y\}$.
   (ii) Find $\text{cov}(X, Y)$ and $\rho(X, Y)$.
   (iii) Find $\text{var}(X + Y)$.
   (See also Problem 3.5.10.)

9. The best mean square error predictor of Y is $\mathscr{E}Y$ when nothing is known about X. The best mean square predictor of Y given $X = x$ is given by the conditional mean $\mathscr{E}\{Y|X = x\}$ (Proposition 5.5.2). Often one does not know how to find $\mathscr{E}\{Y|x\}$. In that case, one may be content to use the best linear predictor $a + bX$ where a and b are chosen to minimize the mean square error $\mathscr{E}(Y - a - bX)^2$.
   (i) Show that

$$b = \frac{\text{cov}(X, Y)}{\text{var}(X)} \quad \text{and} \quad a = \mathscr{E}Y - b\mathscr{E}X$$

   so that the best linear predictor of Y is given by

$$\mathscr{E}Y + \frac{\rho\sigma_2}{\sigma_1}(X - \mathscr{E}X)$$

   where ρ is the correlation coefficient and σ_1, σ_2 are the standard deviations of X and Y respectively.

(ii)   Show that the minimum mean square error is $\sigma_2^2(1 - \rho^2)$. (See also Example 5.3.2.) [The predictor in (i) is called the linear regression of Y on X.]

10.   Suppose one word is randomly selected from the sentence IT IS NOT YET TIME TO GO HOME. Let Y be the number of letters in the word selected.
   (i)   What is your best estimate of the length of the word selected in the sense of minimum variance (or mean square error)?
   (ii)   Let X be the number of letters E in the word selected. Suppose you know that $X = 1$. What is your best estimate of Y in the sense of minimum mean square error?
   (iii)   Suppose you wish to use a linear predictor $a + bX$ for Y. What is your best predictor?
   (iv)   Find the mean square error in each case.

11.   Let $\hat{f}(x)$ be the histogram estimate of $f(x)$ for $x \in (a_{j-1}, a_j), j = 1, 2, \ldots, .$
   (i)   Show that for $x \in (a_{j-1}, a_j)$:

$$\mathscr{E}\hat{f}(x) = \int_{a_{j-1}}^{a_j} \frac{f(u)\, du}{(a_j - a_{j-1})} = \frac{F(a_j) - F(a_{j-1})}{a_j - a_{j-1}}$$

   where F is the distribution function corresponding to f, and
   (ii)   as $n \to \infty$

$$\hat{f}(x) \to \mathscr{E}\hat{f}(x).$$

   By the mean value theorem $\int_{a_{j-1}}^{a_j} f(x)\, dx = (a_j - a_{j-1})f(x_j)$ for some $x_j \in (a_{j-1}, a_j)$ so that $\hat{f}(x)$ is an unbiased and consistent estimate of $f(x)$ for some x_j in (a_{j-1}, a_j).

12.   A process produces a proportion p of defective items where p is unknown. A random sample of size n is to be taken in order to estimate p by the sample proportion of defective items. If the error in estimation is not to exceed .01 with probability at least .90, how large should the sample size be?

13*.   The area X of triangle ABC is given by the relation $X = (ab/2)\sin \gamma$ where γ is a random variable having uniform distribution on $(0, \theta)$. Find approximate expressions for the mean and variance of X.

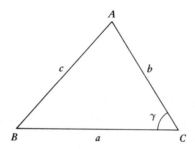

14*.   The altitude of the peak of a mountain in terms of the angle of inclination α and the distance X on the slope is given by $Y = X \sin \alpha$. Suppose X and α are independent

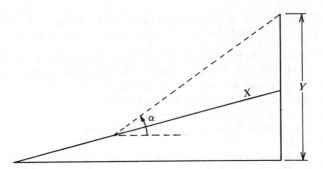

random variables. Find approximate expressions for the mean and variance of Y in terms of those of X and α.

15. An urn contains 10 identical slips numbered 0 to 9. Choosing a slip at random is equivalent to choosing a random digit from 0 to 9. Suppose n independent drawings are made with replacement.
    (i) What does the law of large numbers say about the occurrence of 2's in n drawings?
    (ii) How large should n be so that with a probability of at least .94, the relative frequency of 2's will be $\geq .09$ and $\leq .11$?

CHAPTER 6

Some Discrete Models

6.1 INTRODUCTION

A number of random experiments can often be modeled by the same stochastic model. The conditions underlying the experiment often fit a pattern for which a specific model may be applicable. In Chapters 6 and 7 we consider some examples of probability models that are very useful in practical applications of probability and statistics.

In this chapter we consider some discrete integer-valued stochastic models. Section 6.2 deals with the discrete uniform model, which has many applications, such as simple random sampling (Section 2.8).

Several models result when we seek answers to certain questions concerning a sequence of independent repetitions of a random experiment. Suppose for each repetition we are interested only in whether an event A does or does not occur. Each occurrence of A is usually called a success. Let $p = P(A)$ and set $X_i = 1$ if the ith trial is a success and $X_i = 0$ otherwise. Typical examples are coin tossing games, observing the sex of newborns in a hospital, observing whether or not a new drug is effective, and so on. Suppose there are n repetitions (trials). What is the probability of observing k successes, $k = 0, 1, 2, \ldots, n$? The random variable of interest here is $S_n = \sum_{k=1}^{n} X_k$ and we seek $P(S_n = k)$ for $k = 0, 1, \ldots, n$. The model $\{ P(S_n = k), k = 0, 1, \ldots n \}$ is called a binomial model and is the subject of discussion in Section 6.3. We also consider inferential questions concerning p for the case when n is small.

If, on the other hand, we seek the distribution of the number of trials necessary to achieve the first (rth) success, then the resulting model is known as a geometric (negative binomial) model. This is studied in Section 6.5.

In the binomial case, if n is large and p is small then one can use an approximating model called Poisson model. Poisson models are used to characterize rare events. Typical examples are the number of accidents at an intersection during a certain period, and the number of telephone calls received during a 15-minute interval. Section 6.6 deals with Poisson models.

Section 6.4 deals with the hypergeometric model that arises in practice in sampling (without replacement) from a finite population. This is the case, for example, in lot inspection. A lot contains N items of which D are defective. A sample of size n is taken and inspected. What is the probability that the sample contains X defectives? A typical application of the hypergeometric model is in estimating the size of an animal population by the following capture-recapture procedure. A sample of size D is captured, tagged, and then released into the population. After some time a new sample of size n is taken, and the number of tagged animals is counted.

In Section 6.7 we consider two multivariate models that are generalizations of binomial and hypergeometric models.

In spite of the length of this chapter, it can and should be covered rather quickly. The reader should concentrate mostly on the model assumptions and applications and on the inferential techniques developed in each section. The reader should already be familiar with routine summations to show whether a function is a probability function or with the computation of moments of a probability function. These are included here for completeness and easy accessibility.

6.2   DISCRETE UNIFORM DISTRIBUTION

Consider an experiment that can terminate in N mutually exclusive events $A_1, A_2, \ldots, A_N$, all equally likely. Let $\Omega = \bigcup_{k=1}^{N} A_k$ be the sample space of the experiment. Clearly $P(A_k) = 1/N$, $k = 1, 2, \ldots, N$. Let us define a random variable X on Ω as follows:

$$X(\omega) = x_k \text{ if and only if } \omega \in A_k.$$

Then the probability function of X is given by

$$(1) \qquad P(X = x_k) = P(A_k) = \frac{1}{N}, k = 1, 2, \ldots, N.$$

A random variable X with probability function (1) is said to have a *uniform distribution* on $\{x_1, x_2, \ldots, x_N\}$. For example, in a single throw with a fair die, if we let X be the face value on the side that shows up then X has uniform distribution on $\{1, 2, 3, 4, 5, 6\}$ with probability function

$$P(X = k) = 1/6, \qquad k = 1, 2, \ldots, 6.$$

For a random variable with probability function (1) we have

$$\mathscr{E}X^k = \sum_{k=1}^{N} x_i^k \frac{1}{N} = \frac{1}{N} \sum_{i=1}^{k} x_i^k, \qquad (k \geq 0 \text{ integral}).$$

In particular,

$$\mathscr{E}X = \sum_{i=1}^{N} \frac{x_i}{N}, \ \text{var}(X) = \frac{\sum_{i=1}^{N} x_i^2}{N} - \left(\frac{\sum_{i=1}^{N} x_i}{N}\right)^2,$$

and so on. In the special case when $x_i = i$, $i = 1, 2, \ldots, N$

$$\mathscr{E}X = \frac{N+1}{2}, \ \text{var}(X) = \frac{N^2 - 1}{12}.$$

The generation of samples from a uniform distribution is facilitated by the use of a table of random numbers (Table A1 in the Appendix), as we saw in Chapter 2. The most important application of the uniform distribution, as we have seen, arises in sampling from a finite population. Indeed (in unordered sampling without replacement), a sample is a simple random sample of size n from a population of N objects if each of the $\binom{N}{n}$ possible samples has the same probability of being drawn, namely, $1/\binom{N}{n}$. That is, the distribution of samples is uniform with probability function

$$P(X = k) = \frac{1}{\binom{N}{n}}, \quad k = 1, 2, \ldots, \binom{N}{n}$$

where $X = k$ if and only if the kth of the $\binom{N}{n}$ possible samples $S_1, S_2, \ldots, S\binom{N}{n}$ is chosen.

Another application of uniform distribution arises in sampling from a probability distribution. Let $X_1, X_2, \ldots, X_n$ be a random sample from a probability distribution function F. Then the empirical distribution function $\hat{F}_n(x)$ as defined in Section 4.5 is simply the distribution function of the uniform distribution on $X_1, X_2, \ldots, X_n$. The sample mean $\overline{X}$ is the mean and the sample variance S^2 is simply $(n/(n-1))$ times the variance of this distribution.

Some special cases of uniform distribution are important. The uniform distribution on one point, say c, has probability function

$$P(X = c) = 1.$$

A random variable X with this property will be referred to as a *degenerate* random variable. We will say that X is *degenerate at c* to mean that $P(X = c) = 1$. Writing

(2)
$$\varepsilon(x) = \begin{cases} 0 & x < 0 \\ 1 & x \geq 0, \end{cases}$$

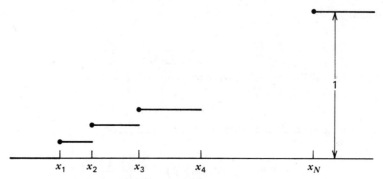

Figure 1.   Distribution function corresponding to (1) where $x_1 \leq x_2 \leq \cdots \leq x_N$.

we note that if X is degenerate at c then its distribution function is given by

$$F(x) = \varepsilon(x - c), \qquad x \in \mathbb{R}.$$

It is clear that the distribution function of X with probability function (1) is given by

$$F(x) = \frac{1}{N} \sum_{k=1}^{N} \varepsilon(x - x_k), \qquad x \in \mathbb{R}$$

which is a step function with graph as shown in Figure 1.

Another special case of great interest is the $N = 2$ case with

$$P(X = 0) = P(X = 1) = 1/2$$

which is discussed in greater generality and more detail in the next section.

    Example 1.   The Taxi Problem: Estimating the Number of Taxis in a Town. In Example 4.3.3 we considered the problem of estimating N the number of taxis in a town (or the size of a finite population) on the basis of a single observation. We now consider the problem of estimating N on the basis of a random sample of size n.

    Consider an urn that contains N marbles numbered 1 through N. Suppose n marbles are drawn with replacement. Let M_n be the largest number drawn. Then M_n is a random variable and can be used to estimate N. How good is this estimate? Clearly $M_n \leq N$. Moreover,

$$P(M_n = k) = P(M_n \leq k) - P(M_n \leq k - 1)$$

$$= \left(\frac{k}{N}\right)^n - \left(\frac{k-1}{N}\right)^n$$

since

$$P(M_n \leq k) = P(\text{all the } n \text{ numbers drawn are} \leq k) = (k/N)^n$$

by independence of the n draws. The mean of M_n is given by

$$\mathscr{E}M_n = N^{-n} \sum_{k=1}^{N} \left\{ k\left[k^n - (k-1)^n \right] \right\}$$

$$= N^{-n} \sum_{k=1}^{N} \left\{ k^{n+1} - (k-1+1)(k-1)^n \right\}$$

$$= N^{-n} \sum_{k=1}^{N} \left\{ \left[k^{n+1} - (k-1)^{n+1} \right] - (k-1)^n \right\}$$

$$= N^{-n} \left\{ N^{n+1} - \sum_{k=1}^{N} (k-1)^n \right\}.$$

Now

$$\sum_{k=1}^{N} (k-1)^n = 1^n + 2^n + \cdots + (N-1)^n$$

and for large N we can approximate it by the integral

$$\int_0^N y^n \, dy = \frac{N^{n+1}}{n+1}.$$

Hence:

$$\mathscr{E}M_n \doteq N^{-n} \left\{ N^{n+1} - \frac{N^{n+1}}{n+1} \right\} = \frac{nN}{n+1}.$$

It follows that

$$\mathscr{E}\frac{(n+1)M_n}{n} \doteq N$$

so that $\hat{N}_1 = (n+1)M_n/n$ is a nearly unbiased estimate of N for large N.

Suppose that a town has N taxis numbered 1 through N. A person standing at a street corner notices the taxi numbers on $n = 8$ taxis that pass by. Suppose these numbers (arranged in increasing order) are

$$124, 212, 315, 628, 684, 712, 782, 926.$$

An obvious estimate for N is $\hat{N} = M_8 = 926$. Since N is large we can use the estimate

$$\hat{N}_1 = \frac{n+1}{n} M_n = \frac{9}{8}(926) \doteq 1042.$$

There are many other simple estimates of N that can be easily constructed.

For example, the sample median is $(628 + 684)/2 = 656$. It follows that if we use the sample median to estimate the population median we can use $\hat{N}_2 = 656 + 655 = 1311$ as an estimate of N. Yet another estimate can be obtained by noting that there are 123 taxis with numbers smaller than the smallest observed number ($= 124$). Assuming that there are as many taxis with numbers larger than the largest observed number we arrive at $\hat{N}_3 = 926 + 123 = 1059$ as an estimate of N. Which of these estimates do we choose? That depends on what we desire of our estimate. Do we want our estimate to be unbiased and have the least variance? Do we want our estimate to be within, say, 50 of the true N with 90 percent confidence? Some of these questions will be answered in general in Chapter 10 and in particular in Example 10.3.4. See also Problems 3, 4, 7, 8, and 9 in the following problems section. $\square$

Problems for Section 6.2

1. Let X have uniform distribution on $\{1, 2, \ldots, N\}$. Show that

    $$\mathscr{E}X = \frac{N+1}{2}, \quad \mathrm{var}(X) = \frac{N^2-1}{12}, \quad \mathscr{E}X^3 = N\left(\frac{N+1}{2}\right)^2,$$

    $$\mathscr{E}X^4 = \frac{1}{30}(N+1)(2N+1)(3N^2+3N-1).$$

    (If you do not already know the sums of powers of the first N integers, use the following procedure to find them. Let $f_s(n) = \sum_{k=1}^n k^s = a_{s+1}\binom{n}{s+1} + a_s\binom{n}{s}$ $+ \cdots + a_1\binom{n}{1}$. Then $f_s(1) = 1 = a_1$. Solve for $a_2, \cdots, a_{s+1}$ by substituting $n = 2, 3, \ldots$ successively. Show that: $f_1(n) = n(n+1)/2$, $f_2(n) = n(n+1)(2n+1)/6$, $f_3(n) = [n(n+1)/2]^2$, $f_4(n) = n(n+1)(2n+1)(3n^2+3n-1)/30$.)

2. An urn contains N marbles numbered 1 through N. Two marbles are drawn one after the other as follows. If the first marble is numbered k (k fixed, $1 \le k \le N$), it is kept. Otherwise it is returned to the urn. A second marble is then drawn. Find the probability that the second marble drawn is numbered $j (\ne k), 1 \le j \le N$.

3. Consider the following procedure to estimate N in sampling from the uniform distribution on $\{1, 2, \ldots, N\}$. Let $x_1, x_2, \ldots, x_n$ be the n observed numbers in the sample of size n and assume, without loss of generality, that $x_1 < x_2 < \cdots < x_n$.

    Estimate the gaps between 1 to x_1, x_1 to $x_2, \ldots, x_{n-1}$ to x_n. Then estimate N by

    $$\hat{N}_4 = x_n + \text{average gap}.$$

    Show that $\hat{N}_4 = \frac{n+1}{n}M_n - 1$, where M_n is the largest observed number ($= x_n$ here). ($\hat{N}_4$ is essentially N_1 in Example 1.) The estimate $\hat{N}_4$ is called the *gap estimate* of N.

4*. In Example 1 we assumed sampling with replacement in computing the distribution of M_n. (If N is large then the probability that an observation appears twice is $1/N^2$ and is small. For large N, sampling with and without replacement are approximately the same.) Suppose N is not large and sampling is without replacement.
   (i) Show that:

$$P(M_n \le k) = \frac{k(k-1)\ldots(k-n+1)}{N(N-1)\ldots(N-n+1)} = \frac{\binom{k}{n}}{\binom{N}{n}}, k = n, n+1,\ldots,N$$

and

$$P(M_n = k) = \frac{\binom{k-1}{n-1}}{\binom{N}{n}}, k = n, n+1,\ldots,N.$$

(ii) Show that:

$$\mathscr{E}M_n = n\frac{N+1}{n+1},$$

$$\mathrm{var}(M_n) = n\frac{(N-n)(N+1)}{(n+2)(n+1)^2}.$$

Part (ii) shows that the gap estimate of Problem 3 is unbiased. [Hint: Use $\binom{n}{r}$ $+\binom{n}{r-1} = \binom{n+1}{r}$ to show that $\sum_{\nu=0}^{r}\binom{\nu+k-1}{\nu} = \binom{r+k}{k}$, which may then be used to prove (ii).]

5. Consider a roulette wheel whose circumference is divided into 38 arcs, or sectors, of equal length. The sectors are numbered $00, 0, 1, \ldots, 36$. The wheel is spun and a ball is rolled on the circumference. The ball comes to rest in one of the 38 sectors. Assume that the sector number so chosen has uniform distribution. Players betting \$1 on any of the 38 numbers receive \$36 if their number shows up, and they lose their \$1 if it does not. What is a player's expected gain?

6*. A sample of size n is chosen without replacement from the finite population of N objects $\{x_1, x_2, \ldots, x_N\}$. Show that the probability distribution of X_i, the object selected on the ith draw, is uniform:

$$P(X_i = x_j) = \frac{1}{N}, \qquad j = 1, 2, \ldots, N; i = 1, 2, \ldots, n.$$

Also show that

$$P(X_i = y_i, X_j = y_j) = \frac{1}{N(N-1)}$$

for $y_i, y_j = x_1, \ldots, x_N, y_i \ne y_j$ for $i \ne j, i, j = 1, 2, \ldots, n$.

(Thus, although X_i, X_j are not independent, their marginal distribution is uniform on $x_1, \ldots, x_N$ exactly as in the case of sampling with replacement.)

7*. Let $X_1, X_2, \ldots, X_n$ be a sample from the uniform distribution on $\{1, 2, \ldots, N\}$. Let $N_n = \min\{X_1, X_2, \ldots, X_n\}$.
   (i) If $X_1, X_2, \ldots, X_n$ are independent identically distributed (that is, if the sampling is with replacement), find the distribution of N_n and find $\mathscr{E}N_n$. Find $\mathscr{E}(N_n + M_n)$ and obtain an unbiased estimate of N.
   (ii) If the sampling is without replacement and X_j denotes the outcome of the jth draw, find the distribution of N_n and $\mathscr{E}N_n$.

8. A random sample of five is taken from the uniform distribution on $\{1, 2, \ldots, N\}$. The maximum observation in the sample was 500. Is there evidence to doubt that $N = 1500$?

6.3 BERNOULLI AND BINOMIAL DISTRIBUTIONS

In a wide variety of random phenomena the outcome can be categorized into one of two mutually exclusive categories, usually but not necessarily referred to as success and failure. Such is the case, for example, when a coin is tossed (heads success, tails failure), or when a dairy farmer observes the sex of a newly born calf (male, female), or when we observe the outcome of a delicate operation by a surgeon (successful or unsuccessful), and so on. Let Ω be a sample space and $A \subseteq \Omega$ be an event with $P(A) = p$, $0 < p < 1$. On Ω let us define a random variable X which is the indicator function of A, that is,

$$X(\omega) = \begin{cases} 1 & \omega \in A \\ 0 & \omega \in \overline{A}. \end{cases}$$

Then the probability function of X is given by

(1) $$P(X = 1) = p, \; P(X = 0) = 1 - p$$

and we say that X is a *Bernoulli random variable* or that (1) is a *Bernoulli probability function*. Clearly the distribution function of X is given by

(2) $$F(x) = \begin{cases} 0, & x < 0 \\ p, & 0 \le x < 1 \\ 1, & 1 \le x \end{cases}$$

and we see that

$$F(x) = p\varepsilon(x) + (1 - p)\varepsilon(x - 1)$$

where $\varepsilon(x) = 1$ if $x \ge 0$ and $\varepsilon(x) = 0$ if $x < 0$, as defined in equation (2) of Section 6.2. Moreover,

(3) $$\mathscr{E}X = p \quad \text{and} \quad \text{var}(X) = p(1 - p).$$

Any performance of an experiment that results in either a success or a failure is called a *Bernoulli trial*. Let us now consider the combined experiment consisting of n independent Bernoulli trials with a constant probability p of success.

DEFINITION 1. An experiment with the following properties is called a binomial experiment:
1. It consists of n trials.
2. Each trial is dichotomous, that is, the outcome falls into one of two mutually exclusive categories: a success (S) or a failure (F).
3. The probability of success $P(S) = p$ remains constant from trial to trial.
4. The trials are independent.

We shall be interested in the random variable

$$X = \text{the number of successes in } n \text{ trials.}$$

Example 1. Effectiveness of a Drug. A new drug that is effective in 30 percent of its applications is administered to 10 randomly chosen patients. What is the probability that it will be effective in five or more of these patients?

Let us check to see if this experiment meets the requirements of Definition 1.

(i)   The experiment consists of $n = 10$ trials.
(ii)   The trials are dichotomous. Either the drug is effective (S) or ineffective (F) on a particular patient.
(iii) The probability of success on each trial is .3.
(iv)   The trials are independent since the patients are randomly chosen.

It is therefore reasonable to conclude that this is a binomial experiment. Let X be the number of successes. Then we wish to find $P(X \geq 5)$. This computation will be made in Example 3.   □

Example 2. Probability of Successful Flight on a Jetliner. A jumbo jet has four engines that operate independently. Suppose that each engine has a probability 0.002 of failure. If at least two operating engines are needed for a successful flight, what is the probability of completing a flight?

Here the experiment consists of four trials that are dichotomous: either an engine is operating (S) or it has failed (F). Trials are independent since the engines operate independently. Finally, $p = P(S) = 1 - .002 = .998$ is the same for each engine. If we let $X =$ number of engines operating during the flight then

$$P(\text{successful flight}) = P(X \geq 2).$$

This computation is made in Example 4.   □

Let us now compute the probability function of X. Clearly $P(0 \leq X \leq n) = 1$. The sample space consists of all n-length sequences of the letters S and F. Thus

$$\Omega = \{(x_1 x_2 \ldots x_n): x_i \in \{S, F\}, \quad i = 1, \ldots, n\}$$

and Ω has 2^n outcomes. However, each sequence does not have the same probability (unless $p = 1/2$). Let $0 \le k \le n$ be an integer. Then

$$P(X = k) = P\{(x_1 x_2 \ldots x_n): x_i \in \{S, F\}, 1 \le i \le n,$$

exactly k of the x_i's are S and $(n - k)$ are $F\}$.

Consider a typical sample point from the event $(X = k)$. One such sample point is

$$\underbrace{SSS\ldots SS}_{k} \ \underbrace{FFF\ldots F}_{n-k}.$$

Another is

$$\underbrace{FFF\ldots FF}_{n-k} \ \underbrace{SSS\ldots SS}_{k},$$

and yet another is

$$\underbrace{F}_{1} \ \underbrace{SS\ldots SS}_{k} \ \underbrace{FF\ldots FF}_{n-k-1}.$$

In view of independence,

$$P\left(\underbrace{SS\ldots S}_{k} \ \underbrace{FF\ldots F}_{n-k} \right) = \underbrace{pp\ldots p}_{k} \ \underbrace{(1-p)(1-p)\ldots(1-p)}_{n-k} = p^k (1-p)^{n-k}.$$

Similarly,

$$P\left(\underbrace{FF\ldots F}_{n-k} \ \underbrace{SS\ldots S}_{k} \right) = (1-p)^{n-k} p^k = p^k (1-p)^{n-k},$$

and

$$P\left(\underbrace{FSS\ldots S}_{k} \ \underbrace{FF\ldots F}_{n-k-1} \right) = (1-p) p^k (1-p)^{n-k-1} = p^k (1-p)^{n-k}.$$

It is clear, therefore, that every sample point from $(X = k)$ has the same probability, namely $p^k (1-p)^{n-k}$, and we have

$$P(X = k) = p^k (1-p)^{n-k} \sum 1$$

where summation is over those sample points having exactly k S's and $(n - k)$ F's. It follows that

$$P(X = k) = (\text{number of sample points with } k \ S\text{'s}, (n - k) \ F\text{'s}) p^k (1-p)^{n-k}.$$

Since the number of sample points with k S's and $(n - k)$ F's is simply the number of ways of choosing k places for the S's (and hence $(n - k)$ places for the F's), we have

(4) $$P(X = k) = \binom{n}{k} p^k (1 - p)^{n-k}, k = 0, 1, 2, \ldots, n.$$

Binomial Probability Function

$$P(X = k) = \binom{n}{k} p^k (1 - p)^{n-k}, k = 0, 1, \ldots, n.$$

Let us show that (4) does actually define a probability function. Indeed, $P(X = k) \geq 0$ for every k and, moreover, by binomial theorem:

$$1 = [p + (1 - p)]^n = \sum_{k=0}^{n} \binom{n}{k} p^k (1 - p)^{n-k}.$$

Since the kth term in the binomial expansion $[p + (1 - p)]^n$ gives the probability of k successes, the distribution defined by (4) is called a *binomial distribution* and X is called a *binomial random variable*. Given n and p we can compute any probability concerning X. Thus:

$$P(X = 0) = \binom{n}{0} p^0 (1 - p)^{n-0} = (1 - p)^n,$$

$$P(X = 1) = np(1 - p)^{n-1},$$

$$P(X = n) = p^n,$$

$$P(X \geq l) = \sum_{k=l}^{n} P(X = k) = \sum_{k=l}^{n} \binom{n}{k} p^k (1 - p)^{n-k},$$

$$P(X \leq l) = \sum_{k=0}^{l} \binom{n}{k} p^k (1 - p)^{n-k},$$

$$P(k \leq X \leq l) = P(X \leq l) - P(X \leq k - 1)$$

$$= \sum_{j=k}^{l} \binom{n}{j} p^j (1 - p)^{n-j},$$

$$P(k < X < l) = P(k + 1 \leq X \leq l - 1)$$

$$= \sum_{j=k+1}^{l-1} \binom{n}{j} p^j (1 - p)^{n-j}.$$

It is also useful to remember some frequently used terms:

Term	Event
at least k successes no less than k successes }	$X \geq k$
at most k successes no more than k successes }	$X \leq k$
fewer than k successes	$X \leq k - 1$
more than k successes	$X \geq k + 1$

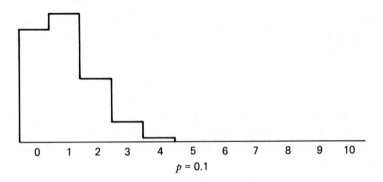

$p = 0.1$

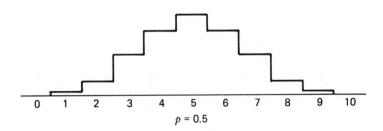

$p = 0.5$

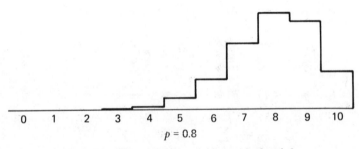

$p = 0.8$

Figure 1. Histogram for $n = 10$, $p = .1$, .5 and .8.

The distribution function of X is easy to compute. Indeed, for $x \in \mathbb{R}$

$$(5) \qquad F(x) = \sum_{k=0}^{n} \varepsilon(x - k)\binom{n}{k} p^k (1 - p)^{n-k}$$

$$= \begin{cases} 0, & x < 0 \\ (1 - p)^n, & 0 \le x < 1 \\ (1 - p)^n + np(1 - p)^{n-1}, & 1 \le x < 2 \\ \vdots \\ \sum_{k=0}^{n-1} \binom{n}{k} p^k (1 - p)^{n-k}, & n - 1 \le x < n \\ 1, & x \ge n. \end{cases}$$

Figure 1 shows the probability histograms of binomial probability functions for $n = 10, p = .1, .5, .8$.

Example 3. *Effectiveness of a Drug (Continuation of Example 1).* In Example 1 we had $n = 10$, $p = .3$ and we had to find $P(X \ge 5)$. From (4) we have

$$P(X \ge 5) = \sum_{k=5}^{10} \binom{10}{k}(.3)^k(.7)^{10-k} = .15. \qquad \square$$

Example 4. *Successful Flight on a Jetliner (Continuation of Example 2).* In Example 2 we had $n = 4$, $p = .998$ and

$$P(\text{successful flight}) = P(X \ge 2) = \sum_{k=2}^{4} \binom{4}{k}(.998)^k(.002)^{4-k}$$

$$= 1 - \sum_{k=0}^{1} P(X \le k) = 1 - (.002)^4 - 4(.998)(.002)^3$$

$$= 1 - .00000000001 - .000000031936$$

$$\doteq .99999997. \qquad \square$$

A table of cumulative probability under binomial distribution frequently helps in computations. Table A3 in the Appendix gives the cumulative probabilities $\sum_{j=0}^{k}\binom{n}{j}p^j(1 - p)^{n-j}$ for some selected values of n and p, and $k = 0, 1, 2, \ldots, n$. For larger values of n one uses the normal approximation discussed in Chapter 9 (or the Poisson approximation, discussed in Section 6.6).

Cumulative probabilities for $p > .5$ can be extracted from Table A3 by the simple observation that

$$P(X \le k) = P(n - X \ge n - k)$$

$$= P(\text{number of failures} \ge n - k)$$

$$= \sum_{j=n-k}^{n} \binom{n}{j}(1 - p)^j p^{n-j}$$

$$= 1 - \sum_{j=0}^{n-k-1} \binom{n}{j}(1 - p)^j p^{n-j}.$$

Thus if we want the probability of *at most* k successes in n trials with $p > .5$, then we look for the probability of *at least* $n - k$ failures each with probability $(1 - p)(< .5)$. Similarly,

$$P(X \ge k) = P(n - X \le n - k) = \sum_{j=0}^{n-k} \binom{n}{j}(1 - p)^j p^{n-j}.$$

The following relations are frequently useful:

$$P(X = k) = P(X \le k) - P(X \le k - 1), \quad k \ge 1.$$

$$P(X \ge k) = 1 - P(X \le k - 1), \quad k \ge 1.$$

$$P(X > k) = P(X \ge k + 1), \quad k \ge 0.$$

$$P(k \le X \le l) = P(X \le l) - P(X \le k - 1), \quad k \ge 1, l \ge k.$$

$$P(X < k) = P(X \le k - 1), \quad k \ge 1.$$

Table A3 can also be used for finding values of k corresponding to a given probability in the tail of the distribution. Thus, given n and p we can find the *smallest* k such that

$$P(X \ge k) \le \alpha$$

for a given α, $0 < \alpha < 1$, or the *largest* k such that

$$P(X \le k) \le \alpha, \quad 0 < \alpha < 1.$$

The following examples illustrate the use of Table A3.

Example 5. Effectiveness of a Drug. Experience has shown that with the present treatment, only 25 percent of all patients afflicted with leukemia will be alive 10 years after the onset of the disease. A researcher has developed a new drug which is tried on 15 randomly selected patients. It is found that 10 of these patients are still alive 10 years after the onset. Does the evidence indicate that the new drug is preferable to the present treatment?

The null hypothesis is H_0: $p = .25$ to be tested against H_1: $p > .25$. Let X be the number of patients in the sample of 25 who are still alive 10 years after the onset. Then the P-value is

$$P_{H_0}(X \geq 10) = 1 - P_{H_0}(X \leq 9) = 1 - .9992 = .0008 \qquad \text{(Table A3)}.$$

We conclude that the null hypothesis is not supported by the evidence. □

Example 6. An Application to Quality Control. A large lot of light bulbs is supposed to contain only 10 percent defectives. A sample of size 25 is randomly selected from the lot and tested. How many defective bulbs should we observe in the sample before we conclude that the lot does not meet the specifications? Suppose the level of significance is $\alpha = .10$.

Here $n = 25$ and H_0: $p = .10$. Let X be the number of defectives in the sample. We wish to find the smallest k such that

$$P_{H_0}(X \geq k) \leq .10.$$

Since Table A3 gives only $P(X \leq k)$, we rewrite this inequality as

$$P_{H_0}(X \leq k - 1) = 1 - P_{H_0}(X \geq k) \geq 1 - .10 = .90.$$

From Table A3, the smallest k satisfying this inequality is given by $k - 1 = 4$, so that $k = 5$. If we observe five or more defective bulbs in a sample of size 25, we should reject the lot and conclude that $p > .1$. □

In the beginning of this section we assumed $0 < p < 1$. Frequently we will include the (trivial) extreme cases $p = 0$ and $p = 1$ and assume that $0 \leq p \leq 1$.

Example 7. Estimating p. In practice p is unknown. To estimate p we note that the *most probable* value of k corresponds to the value of p given by

$$\frac{d}{dp}P(X = k) = \binom{n}{k}\left\{ kp^{k-1}(1 - p)^{n-k} - (n - k)p^k(1 - p)^{n-k-1} \right\}$$

$$= \binom{n}{k}p^{k-1}(1 - p)^{n-k-1}\{ k(1 - p) - (n - k)p \} = 0$$

so that $p = k/n$. Since $(d^2/dp^2)P(X = k) < 0$, $p = k/n$ is a maximum. Thus $\hat{p} = k/n$ is that value of p which makes the outcome $X = k$ most probable. Therefore, $\hat{p} = X/n$ is a reasonable estimate of p. Note that X/n is the proportion of successes in n trials and would have been the choice for estimating p on intuitive grounds. Further properties of this estimate are considered later.

In particular, if 10 successes are observed in 30 trials, then our estimate of p would be $\hat{p} = 10/30 = 1/3$. □

Example 8. *Reading Table A3 when p > .5.* A random sample of 15 observations is taken from the density function

$$f(x) = \begin{cases} 1, & 0 < x < 1 \\ 0 & \text{otherwise.} \end{cases}$$

What is the probability that five of these observations will be smaller than .6? Note that $n = 15$ and $p = P(Y < .6)$ where Y is a random variable with density f. Then $p = .6$. If X is the number of observations that are smaller than .6, then X has a binomial distribution with $n = 15$ and $p = .6$. Since $p > .5$, we cannot directly read $P(X = 5)$ from Table A3. But

$$P(X = 5) = P(15 - X = 10) = P(\text{ten failures in 15 trials with } 1 - p = .4)$$

$$= .9907 - .9662 = .0245.$$

Again,

$$P(6 \le X \le 9) = P(6 \le n - X \le 9)$$

$$= .9662 - .4032 = .5630$$

and

$$P(X \le 9) = P(n - X \ge 6) = 1 - P(n - X \le 5)$$

$$= 1 - .4032 = .5368. \qquad \square$$

6.3.1 Some Properties of Binomial Distribution

For later use, we will need some moments of the binomial distribution. We have

$$(6) \quad \mathscr{E}X = \sum_{k=0}^{n} k \binom{n}{k} p^k (1 - p)^{n-k} = \sum_{k=1}^{n} \frac{n!}{(k-1)!(n-k)!} p^k (1 - p)^{n-k}$$

$$= \sum_{k=0}^{n-1} np \binom{n-1}{k} p^{k-1} (1 - p)^{n-1-k} = np(p + (1 - p))^{n-1} = np.$$

A simpler method to compute moments of X uses the binomial expansion:

$$(7) \qquad (x + y)^n = \sum_{k=0}^{n} \binom{n}{k} x^k y^{n-k}.$$

Differentiating both sides of (7) with respect to x we get

$$(8) \qquad n(x + y)^{n-1} = \sum_{k=1}^{n} k \binom{n}{k} x^{k-1} y^{n-k},$$

(9)
$$n(n-1)(x+y)^{n-2} = \sum_{k=2}^{n} k(k-1)\binom{n}{k} x^{k-2} y^{n-k},$$

(10) $\quad n(n-1)(n-2)(x+y)^{n-3} = \sum_{k=3}^{n} k(k-1)(k-2)\binom{n}{k} x^{k-3} y^{n-k},$

and so on. Let us substitute $x = p$ and $y = 1 - p$. We have

$$\sum_{k=1}^{n} k\binom{n}{k} p^{k-1}(1-p)^{n-k} = n,$$

$$\sum_{k=2}^{n} k(k-1)\binom{n}{k} p^{k-2}(1-p)^{n-k} = n(n-1),$$

$$\sum_{k=3}^{n} k(k-1)(k-2)\binom{n}{k} p^{k-3}(1-p)^{n-k} = n(n-1)(n-2).$$

It follows that

$$\mathscr{E}X = np, \mathscr{E}X(X-1) = n(n-1)p^2$$

and

$$\mathscr{E}X(X-1)(X-2) = n(n-1)(n-2)p^3.$$

Hence:

(11)
$$\operatorname{var}(X) = \mathscr{E}X^2 - (\mathscr{E}X)^2 = \mathscr{E}X(X-1) + \mathscr{E}X - (\mathscr{E}X)^2$$

$$= n(n-1)p^2 + np - (np)^2 = np(1-p)$$

and

$$\mathscr{E}X^3 = \mathscr{E}X(X-1)(X-2) + 3\mathscr{E}X(X-1) + \mathscr{E}X$$

$$= n(n-1)(n-2)p^3 + 3n(n-1)p^2 + np.$$

Mean and Variance of Binomial

$$\mu = \mathscr{E}X = np, \quad \sigma^2 = \operatorname{var}(X) = np(1-p)$$

It also follows that

(12)
$$\mathscr{E}\frac{X}{n} = p \quad \text{and} \quad \operatorname{var}\left(\frac{X}{n}\right) = \frac{p(1-p)}{n}$$

where X/n is the proportion of successes in n trials.

Next we note that the number of successes in n independent trials can be looked upon as the sum

$$X = \sum_{k=1}^{n} X_k$$

where $X_k = 1$ if the kth trial results in a success, and $X_k = 0$ otherwise. Thus X is the sum of n independent and identically distributed Bernoulli random variables $X_1, \ldots, X_n$, with $\mathscr{E}X_k = p$ and $\mathrm{var}(X_k) = p(1 - p)$.

Similarly, if Y_1, Y_2 are two independent binomial random variables with parameters (n_1, p) and (n_2, p) the distribution of $Y_1 + Y_2$ is also binomial $(n_1 + n_2, p)$ since we can look upon $Y_1 + Y_2$ as the number of successes in $n_1 + n_2$ independent Bernoulli trials with probability of success p.

Returning to X/n, the proportion of successes in n trials, we note that it can be looked upon as a sample average $\bar{X} = n^{-1}\sum_{k=1}^{n} X_k$. Hence it must obey the weak law of large numbers studied in Section 5.4.

In fact, according to Chebyshev's inequality, for every $\varepsilon > 0$

$$P\left(\left|\frac{X}{n} - p\right| \geq \varepsilon\right) \leq \frac{p(1 - p)}{n\varepsilon^2}.$$

We note, however, that $4p(1 - p) < 1$ unless $p = 1/2$, in which case $p(1 - p) = 1/4$. [In fact, $4p(1 - p) - 1 = -(1 - 2p)^2 < 0$ for all $p \neq 1/2$.] It follows that

$$(13) \qquad P\left(\left|\frac{X}{n} - p\right| \geq \varepsilon\right) \leq \frac{1}{4n\varepsilon^2}.$$

Letting $n \to \infty$ we see that

$$P\left(\left|\frac{X}{n} - p\right| \geq \varepsilon\right) \to 0.$$

This special case of the weak law of large numbers (Section 5.4) is usually called the *Borel law of large numbers*.

The Borel Law of Large Numbers

Let A be any event. Then $R(A)$, the relative frequency of A in n independent and identical trials of the experiment satisfies

$$P\{|R(A) - P(A)| > \varepsilon\} \to 0$$

as $n \to \infty$. That is, as n increases, the probability that the relative frequency of A deviates from $P(A)$ by more than any given $\varepsilon > 0$ tends toward zero.

Remark 1. We emphasize once again that the law of large numbers as formulated here simply asserts that for large n the relative frequency of A in n

trials is likely to be close to $P(A)$. It says that for every $\varepsilon > 0$, there is a sufficiently large n such that the probability that $R(A)$ lies in the interval $(p - \varepsilon, p + \varepsilon)$ can be made as close to one as we wish. It does not assert the existence or the value of the limit $\lim_{n \to \infty} R(A)$ for any particular sequence of trials in the ordinary sense of convergence of a sequence of real numbers. It is true, however, that

$$P\left(\lim_{n \to \infty} R(A) = P(A) \right) = 1$$

but we will not attempt to prove this stronger result here.

Remark 2. Consider the special case when $p = 1/2$. Since $\mathscr{E}X = n/2$ and $\text{var}(X) = n/4$ we expect, on the average, $n/2$ successes, and according to Chebyshev's inequality, the number of successes will be in the interval $[(n - 3\sqrt{n})/2, (n + 3\sqrt{n})/2]$ with a probability of at least $8/9$. For $n = 100$ this means that $P(35 \le X \le 65) \ge 8/9$. (This is a rather crude bound since the actual probability is .9991.) The law of large numbers does not say that in a large number of trials the number of successes and the number of failures will be approximately the same in the sense that their difference is near zero. The assertion concerns only the proportion of successes (or failures). Thus in 100,000 tosses of a fair coin we may observe 51,000 heads and 49,000 tails. The proportion of heads and proportion of tails are nearly $1/2$ and yet the difference between the number of heads and the number of tails is 2000. Similarly, it is customary to expect that in a coin-tossing game with a fair coin one player will lead the other roughly half the time. This is not true either.

Remark 3. The law of large numbers gives an estimate of the rate of convergence. It gives an upper bound for probability $P(|X/n - p| \ge \varepsilon)$ as $1/(4n\varepsilon^2)$, which is quite crude. A much more precise bound for this probability is obtained by using the normal approximation (Section 9.4). If we take $\varepsilon = .02$, then

$$P\left(\left| \frac{X}{n} - p \right| \ge .02 \right) \le \frac{1}{4n(.0004)} = \frac{625}{n}.$$

For $n = 10,000$, the upper bound is .06; for $n = 100,000$, the upper bound is .006.

Example 9. *An Application to Opinion Polls.* A pollster wishes to estimate the proportion of voters who favor the Equal Rights Amendment based on a random sample of size n. She wants to use the sample proportion of those who favor ERA as an estimate of the true proportion p. If the pollster wants the estimate to be within 3 percent of the true p, with probability $\ge .95$, how large a sample should she take?

The pollster wants to use X/n to estimate the unknown p such that

$$P\left(\left| \frac{X}{n} - p \right| < .03 \right) \ge .95,$$

or,

$$P(|X/n - p| \geq .03) \leq .05.$$

Taking $\varepsilon = .03$ and $1/(4n\varepsilon^2) = .05$ in (13) we get

$$n = \frac{1}{4(0.5)(.03)^2} \doteq 5556.$$

The pollster should therefore take a sample of size at least 5556. At the expense of belaboring the point we emphasize that if the pollster does take a sample of size 5556 and computes the relative frequency of respondents who favor ERA, then this number will either be within 3% of the true p or not. Since p is unknown we will never know this for sure. The point is that if, say, 1000 samples were taken, each of size 5556, then for about 95 percent of these samples the relative frequency would be within 3 percent of the true p.   □

Example 10. **Chebyshev's Bounds for Proportion of Successes.**   Consider n tossings with a fair coin. Let X/n be the proportion of successes (heads). Clearly, $\mathscr{E}(X/n) = 1/2$ and $\text{var}(X/n) = 1/(4n)$. By Chebyshev's inequality

$$P\left(\left|\frac{X}{n} - \frac{1}{2}\right| \leq k\frac{1}{\sqrt{4n}}\right) \geq 1 - \frac{1}{k^2}.$$

Let us choose $k = 5$ so that $1 - 1/k^2 = 24/25 = .96$. Hence with a probability of at least .96, the proportion of successes will be between $(\frac{1}{2} - 5/2\sqrt{n})$ and $(\frac{1}{2} + 5/2\sqrt{n})$.

n	Bounds for Proportion of Successes	Maximum Difference Between No. of Successes and No. of Failures
100	[.25, .75]	50
625	[.4, .6]	125
10,000	[.475, .525]	500
1,000,000	[.4975, .5025]	5000 □

6.3.2   Statistical Inference for the Binomial Distribution

In almost all our earlier discussions of the binomial distribution it was assumed that both p, the probability of success, and n, the number of trials, are known. We have mostly concentrated on the computation of probabilities such as

$$P(X \geq k) = \sum_{r=k}^{n} \binom{n}{r} p^r (1-p)^{n-r}$$

either by direct computation or by the use of Table A3. That is, given n and p, the question we have so far answered is this: Assuming that the binomial model conditions hold, what is the probability of observing a certain number of successes in n trials? We now reverse this process and ask questions concerning p based on an observed number of successes. For example, if we observe five heads in 100 tosses of a coin, we ask what is a reasonable estimate of p? Can conclude that the coin is fair?

We first consider the point estimation of p. Suppose in n independent and identical Bernoulli trials with $P(\text{success}) = p$ we observe X successes. Suppose further that p is unknown. Consider the statistic

$$\hat{p} = \frac{X}{n}.$$

We note that $0 \le \hat{p} \le 1$ so that $\hat{p}$ is an estimate of p; it maps the sample space into the parameter set $\Theta = [0, 1]$. Moreover:

(i) $\mathscr{E}\hat{p} = \mathscr{E}X/n = p$; $\hat{p}$ is *unbiased*.
(ii) $P(|\hat{p} - p| \ge \varepsilon) \to 0$ as $n \to \infty$ so that $\hat{p}$ *is consistent* for p. The probability that $\hat{p}$ will deviate from p by more than any fixed amount $\varepsilon > 0$ can be made as small as we please by taking n sufficiently large.
(iii) $\hat{p} = k/n$ maximizes $P(X = k)$.
(iv) $\text{var}(\hat{p}) = p(1 - p)/n$; the variance of $\hat{p}$ can be made as close to 0 as we please by taking n sufficiently large.

Thus $\hat{p}$ is a reasonable point estimate of p. Some other properties of $\hat{p}$ will be considered later. Note, however, that $\hat{p}$ is a statistic (random variable); its value varies from sample to sample. The probability that $\hat{p}$ equals p is usually small. Indeed,

$$P(\hat{p} = p) = P\left(\frac{X}{n} = p\right) = P(X = np)$$

is zero when np is not an integer. Even when np is an integer this probability is small. For $n = 20$, $p = .4$,

$$P(\hat{p} = p) = P(X = 20(.4)) = P(X = 8) = .18,$$

and for $n = 100$, $p = .5$

$$P(\hat{p} = p) = P(X = 50) = .0796.$$

The estimate $\hat{p}$, therefore, and any other point estimate of p for that matter, is of limited value. In many applications what we want are bounds that will include the unknown value of p with a large probability. In other words, we want a confidence interval for p.

Let $0 < \alpha < 1$. Then we wish to find random variables $l(X)$ and $u(X)$ such that

(14) $$P(l(X) \le p \le u(X)) = 1 - \alpha.$$

Since X is a discrete type random variable, minimum level $1 - \alpha$ confidence intervals $[l(X), u(X)]$ do not exist for all α. We therefore replace the equality sign in (14) by " $\geq$ " and seek confidence intervals for p at level at least $1 - \alpha$. (See Section 4.3.) By Chebyshev's inequality we have for $\varepsilon > 0$

$$(15) \qquad P\left(|\hat{p} - p| \leq \varepsilon\sqrt{\frac{p(1-p)}{n}}\right) \geq 1 - \frac{1}{\varepsilon^2}.$$

Since $p(1 - p) \leq \frac{1}{4}$ we have

$$(16) \qquad P\left(|\hat{p} - p| \leq \varepsilon\frac{1}{2\sqrt{n}}\right) \geq P\left(|\hat{p} - p| \leq \varepsilon\sqrt{\frac{p(1-p)}{n}}\right) \geq 1 - \frac{1}{\varepsilon^2}.$$

Rewriting (16), we have

$$(17) \qquad P\left(\hat{p} - \frac{\varepsilon}{2\sqrt{n}} \leq p \leq \hat{p} + \frac{\varepsilon}{2\sqrt{n}}\right) \geq 1 - \frac{1}{\varepsilon^2}.$$

Let us now compare (14) and (17). We have

$$1 - \alpha = 1 - 1/\varepsilon^2$$

so that $\varepsilon = 1/\sqrt{\alpha}$ and

$$(18) \qquad l(X) = \hat{p} - \frac{1}{2\sqrt{n\alpha}}, \qquad u(X) = \hat{p} + \frac{1}{2\sqrt{n\alpha}}.$$

Point and Interval Estimates for p

The proportion of successes $\hat{p} = X/n$ is a good point estimate for p.

The interval $\left[\hat{p} - \frac{1}{2\sqrt{n\alpha}}, \hat{p} + \frac{1}{2\sqrt{n\alpha}}\right]$ is a $100(1 - \alpha)$ percent confidence interval for p.

Example 11. Busing to Achieve Racial Balance. Of 100 people interviewed, 79 registered opposition to busing students to achieve racial balance. Let p be the true proportion in the population of those who oppose busing. Then a point estimate of p based on this sample of size $n = 100$ is given by

$$\hat{p} = X/n = 79/100 = 0.79,$$

so (we estimate) that more than half the population is not in favor of busing. Suppose that the true p is .5. What is the probability of observing $\hat{p} \geq .79$ in a sample of size 100 in random sampling from a population where in fact $p = .5$? We have

$$P(\hat{p} \geq .79) = P(X \geq 79) = 0^+$$

from Table A3 with $n = 100$ and $p = .5$. It is therefore reasonable to conclude the true proportion of those opposed to busing in this population exceeds $.5$.

Suppose, instead, that we want a 95 percent confidence interval for p. Then $1 - \alpha = .95$, $\alpha = .05$, and a 95 percent confidence interval is given by

$$\left[\hat{p} - \frac{1}{2\sqrt{n\alpha}} , \hat{p} + \frac{1}{2\sqrt{n\alpha}} \right] = [.79 - .22, .79 + .22] = [.57, 1.0]$$

(the largest value of p is 1). This confidence interval has width $.43$. It is clear that the sample size is too small (and the interval is too crudely constructed) to give a narrow width. We note, however, that $p = .5$ is *not* included in this interval. $\square$

Example 12. **An Improved Confidence Interval for** p. The interval $[\hat{p} - (1/2\sqrt{n\alpha}), \hat{p} + (1/2\sqrt{n\alpha})]$ is a somewhat crude confidence interval for p. The length of a confidence interval is taken as a measure of its accuracy. The smaller the length the more informative is the confidence interval. The length of the above confidence interval is given by

(19)
$$L_1(X) = u(X) - l(X) = \frac{1}{\sqrt{n\alpha}},$$

which does not even depend on the observations. This feature, however, allows us to choose the sample size to guarantee a fixed length of d (> 0), say. In fact, by choosing

$$n \geq \frac{1}{d^2\alpha}$$

the confidence interval $[\hat{p} - (1/2\sqrt{n\alpha}), \hat{p} + (1/2\sqrt{n\alpha})]$ will be a $100(1 - \alpha)\%$ confidence interval with width at most d.

It is possible to obtain an improved confidence interval by manipulating the inequality

(20)
$$|\hat{p} - p| \leq \varepsilon\sqrt{\frac{p(1 - p)}{n}}.$$

Clearly, (20) holds if and only if

$$n(\hat{p} - p)^2 \leq \varepsilon^2 p(1 - p),$$

or, if and only if

(21)
$$(n + \varepsilon^2)p^2 - (2n\hat{p} + \varepsilon^2)p + n\hat{p}^2 \leq 0.$$

The left side of (21) is a quadratic equation in p so that the inequality holds (see Figure 2) if and only if p lies between the two roots of the equation

$$f(p) = (n + \varepsilon^2)p^2 - (2n\hat{p} + \varepsilon^2)p + n\hat{p}^2 = 0.$$

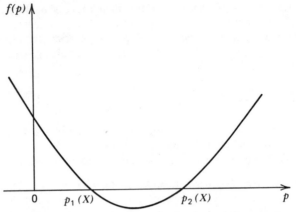

Figure 2. Graph of $f(p) = (n + \varepsilon^2)p^2 - (2n\hat{p} + \varepsilon^2)p + n\hat{p}^2$.

The two roots are given by

$$p_1(X) = \frac{2n\hat{p} + \varepsilon^2 - \sqrt{[2n\hat{p} + \varepsilon^2]^2 - 4n(n + \varepsilon^2)\hat{p}^2}}{2(n + \varepsilon^2)}$$

and

$$p_2(X) = \frac{2n\hat{p} + \varepsilon^2 + \sqrt{[2n\hat{p} + \varepsilon^2]^2 - 4n(n + \varepsilon^2)\hat{p}^2}}{2(n + \varepsilon^2)}.$$

Choosing $\varepsilon = 1/\sqrt{\alpha}$, we see that $[p_1(X), p_2(X)]$ is a $100(1 - \alpha)$ percent confidence interval for p. The length of this interval is given by

$$L_2(X) = p_2(X) - p_1(X) = \frac{\sqrt{[2n\hat{p} + \varepsilon^2]^2 - 4n(n + \varepsilon^2)\hat{p}^2}}{(n + \varepsilon^2)}$$

which is a random variable. Therefore, we cannot compare L_1 and L_2 directly. If n is large, however, then we can replace $\hat{p}$ by p in the expression for L_2. In that case:

$$L_2(X) \doteq \frac{\sqrt{\varepsilon^4 + 4np\varepsilon^2 - 4np^2\varepsilon^2}}{n + \varepsilon^2}$$

$$= \frac{\varepsilon}{n + \varepsilon^2}\sqrt{\varepsilon^2 + 4np(1 - p)} \le \frac{\varepsilon}{n + \varepsilon^2}\sqrt{n + \varepsilon^2} = \frac{\varepsilon}{\sqrt{n + \varepsilon^2}}$$

$$< \frac{\varepsilon}{\sqrt{n}} = L_1(X).$$

Thus, at least for large n, the $100(1 - \alpha)$ percent confidence interval $[p_1(X), p_2(X)]$ should be narrower than the $100(1 - \alpha)$ percent confidence interval given in (18).

Returning to the data of Example 11 for comparison purposes, we see that with $n = 100$, $\hat{p} = .79$, $\alpha = .05$,

$$\frac{\sqrt{(2n\hat{p} + 1/\alpha)^2 - 4n(n + 1/\alpha)\hat{p}^2}}{2(n + 1/\alpha)} = \frac{\sqrt{(158 + 20)^2 - 400(100 + 20)(.79)^2}}{2(100 + 20)}$$

$$= .1732$$

so that

$$p_1(X) = .7417 - .1732 = .5685 \quad \text{and} \quad p_2(X) = .7417 + .1732 = .9149.$$

Thus a 95 percent confidence interval is [.5685, .9149]. This confidence interval has width $2(.1732) = .3464$, which is much smaller than the width of the confidence interval considered in Example 11. In this sense, it is an improved confidence interval.   □

We note that the confidence intervals considered here are generally applicable and are used for *small or moderate sample sizes*. As we shall see, for large samples we can do much better.

Finally, we consider the problem of testing hypotheses concerning p. *The methods of this section are exact and apply to the case when n is small or moderate.*

Example 13.   Change of Voting Preference. Congressional District 5 in Wood County has consistently voted 60 per cent or more Republican in the past 20 years in Congressional elections. It is suspected that the depressed economic conditions during 1981–82 may have changed the percentage of voters who will vote for the Republican candidate for Congress in 1982. In a random sample of 25 voters from this district 12 voters indicated that they plan to vote Republican. Have the depressed economic conditions changed voting preference?

Here the null hypothesis to be tested is H_0: $p \geq .60$ where p = true proportion of voters in the district who plan to vote Republican. This is a one-sided hypothesis to be tested against the alternative that H_1: $p < .60$. Let X be the number of voters in the sample who favor Republicans. Since $X = 12$ is observed, our estimate of p is $\hat{p} = 12/25 = .48$, which is less than .60. The question is, How probable is such a small value of $\hat{p}$ in random sampling from a population where $p \geq .60$? In other words, we seek the probability

$$P(X \leq 12 \text{ when } p \geq .60).$$

However, this cannot be done unless p is completely specified. We note that

$$P(X \leq 12 \text{ when } p = .60) \geq P(X \leq 12 \text{ when } p \geq .60)$$

since if $p > .60$ we would expect more successes (that is more voters in favor of the Republican candidate). Now

$$P(X \leq 12 \text{ when } p = .60) = .154.$$

Thus, unless the level of significance α is at least .154 we have to conclude that there is not enough evidence to support the rejection of H_0. If we reject H_0 when X is at most 12, what is the probability of a type II error? This depends on the true p. Suppose

$p = .5$. Then the probability of a Type II error is given by

$$P(\text{accept } H_0 \text{ when } p = .5) = P(X \geq 13 \text{ when } p = .5) = 1 - .5 = .5.$$

The smaller the p the smaller this error. The power at $p = .5$ is $1 - P(\text{type II error}) = .5$.

Given α, $0 < \alpha < 1$, the level of significance, how do we find the critical region of a level α test of H_0: $p \geq .60$ against H_1: $p < .60$? Since larger values of X support H_0, a reasonable critical region is given by $\{X \leq c\}$ where c is the largest integer satisfying

$$P_{H_0}(X \leq c) \leq \alpha.$$

Since, as pointed out above,

$$P_{H_0}(X \leq c) \leq P\{X \leq c | p = .60\}$$

it suffices to choose the largest c satisfying

$$P\{X \leq c | p = .60\} \leq \alpha.$$

The values of c for some selected values of α are given below.

α	.01	.05	.15	.20
c	8	10	11	12

The probability of a type II error if we choose $\alpha = .05$ say, so that $c = 10$, is given by $P_{H_1}(X \geq 11)$. For some selected values of $p(< .60)$, the type II error probabilities are given below.

p	.5	.4	.3	.2	.1
$P(\text{type II error})$	.788	.414	.008	.006	0^+
Power	.212	.586	.902	.994	1^-

The procedure for testing H_0: $p \leq p_0$ against H_1: $p > p_0$ or for testing $p = p_0$ against $p \neq p_0$ is similar. We summarize the results below.

Testing Binomial p

Let X have a binomial distribution with parameters n and p and suppose that $X = x_0$ is the observed number of successes.

H_0	H_1	Reject H_0 at Level α if	P-Value
$p \geq p_0$	$p < p_0$	$x_0 \leq c$ where c is the largest integer such that $P\{X \leq c | p_0\} \leq \alpha$	$P\{X \leq x_0 | p_0\}$
$p \leq p_0$	$p > p_0$	$x_0 \geq c$ where c is the smallest integer such that $P\{X \geq c | p_0\} \leq \alpha$	$P\{X \geq x_0 | p_0\}$
$p = p_0$	$p \neq p_0$	$x_0 \leq c_1$ or $x_0 \geq c_2$ where c_1 is the largest integer such that $P\{X \leq c_1 | p_0\} \leq \alpha/2$ and c_2 the smallest integer such that $P\{X \geq c_2 | p_0\} \leq \alpha/2$	2(smaller of the two tail probabilities)

In the case when H_0: $p = p_0$ is to be tested against H_1: $p \neq p_0$ we have used the conventional equal-tails approach. Thus, if the level of significance α is given, we distribute it equally in the two tails to compute the critical values c_1 and c_2. If α is not given and we decide to reject with probability $P\{ X \leq x_0 | p = p_0 \}$ in the left tail, presumably we will allow the same probability of error in the right tail. The critical value in the right tail is computed by taking the smallest c such that

$$P\{ X \geq c | p = p_0 \} \leq P\{ X \leq x_0 | p = p_0 \}.$$

In the special case when $p_0 = \frac{1}{2}$, $c = n - x_0$.

Example 14. Sex Ratio of Births at a Hospital. It is desired to test that the sex ratio of births at a hospital is .5. A random sample of 20 births showed 13 girls. Let X be the number of female births in the sample and let p be the probability that the sex of a newly born child is female. Then we wish to test H_0: $p = p_0 = .5$ against H_1: $p \neq .5$. Here $x_0 = 13$ so the smaller tail probability is given by:

$$P\{ X \geq 13 | p = .5 \} = 1 - P\{ X \leq 12 | p = .5 \} = 1 - .868 = .132.$$

The P-value is $2(.132) = .264$, and there is not enough evidence to reject H_0. To find the corresponding rejection region in the left tail we choose the largest c such that:

$$P\{ X \leq c | p = .5 \} \leq .132.$$

From Table A3, $c = 7$. (Since $p_0 = .5$, $c = n - x_0 = 20 - 13 = 7$.) The size .264 rejection region then is $\{0, 1, \ldots, 7, 13, 14, \ldots, 20\}$.

On the other hand, if we are given the level of significance $\alpha = .10$, say, we reject H_0 if the observed $x_0 \in \{0, 1, \ldots, c, n - c, n - c + 1, \ldots, n\}$ where c is chosen from (the largest value)

$$2P\{ X \leq c | p = 1/2 \} \leq .10.$$

From Table A3, $c = 5$ so that the rejection region is $\{0, 1, 2, 3, 4, 5, 15, 16, 17, 18, 19, 20\}$. The size of the corresponding test is $2(.021) = .042$. Since $x_0 = 13$ does not fall into this region, we cannot reject H_0 at level .10. $\square$

6.3.3 An Application of Binomial Distribution: Sign Test

Suppose a new diet is designed to lower cholesterol levels. How do we go about testing whether the diet is effective or not? One way to do so would be to take a random sample of n subjects and put them on this diet for, say, six months. Let X_i be the cholesterol level of the ith subject before he or she goes on the diet, and let Y_i be his or her cholesterol level after six months. If the diet is effective on the ith subject, $X_i - Y_i$ will be positive. The number of pairs (X_i, Y_i) for which $X_i - Y_i > 0$ is then a measure of the effectiveness of the diet. The larger this number, the more plausible it is that the diet is effective. The null hypothesis that we wish to test is

$$H_0: P(X > Y) \leq P(X < Y) \qquad (X \text{ usually smaller than } Y, \text{ diet is ineffective})$$

against the alternative

$$H_1: P(X > Y) > P(X < Y) \qquad (X \text{ usually larger than } Y, \text{ diet is effective}).$$

We will assume in the following that (X, Y) is jointly of the continuous type so that $P(X = Y) = 0$, that is, ties occur with zero probability. In that case, under H_0

$$1 = P(X > Y) + P(X < Y) \leq 2P(X < Y)$$

or

$$P(X - Y < 0) \geq \tfrac{1}{2},$$

and under H_1

$$P(X - Y < 0) < \tfrac{1}{2}.$$

Let S be the number of pairs for which $X_i < Y_i$. Then S has a binomial distribution with parameters n and $p = P(X < Y)$ where $p \geq \tfrac{1}{2}$ under H_0. The test (for small samples) that we use is the one-sided test described in Section 6.3.2. We reject H_0 if S is small (or equivalently, the number of pairs with $X_i - Y_i > 0$ large), that is,

Reject H_0 if $S \leq c$ with level of significance given by

$$P\left(S \leq c \text{ when } p = \tfrac{1}{2}\right) = \frac{1}{2^n} \sum_{k=0}^{c} \binom{n}{k}.$$

If α is known we choose the largest c such that

$$\frac{1}{2^n} \sum_{k=0}^{c} \binom{n}{k} \leq \alpha.$$

A similar procedure applies for testing $p \leq \tfrac{1}{2}$ or $p = \tfrac{1}{2}$. Since the test statistics counts the number of negative (or positive) signs in $X_1 - Y_1, X_2 - Y_2, \ldots, X_n - Y_n$, the resulting test is called a *sign test*.

The Sign Test

To test for jointly continuous type random variables X and Y, the null hypothesis

$$H_0: P(X - Y < 0) = P(X - Y > 0) = \tfrac{1}{2}$$

against one-sided or two-sided alternatives, we reject H_0 if:

$$0 \leq S \leq c \qquad \text{if the alternative is } H_1: P(X - Y < 0) < \tfrac{1}{2};$$
$$n - c \leq S \leq n \qquad \text{if the alternative is } H_1: P(X - Y < 0) > \tfrac{1}{2};$$

$$\left.\begin{array}{l} 0 \leq S \leq c \\ \text{or} \\ n - c \leq S \leq n \end{array}\right\} \quad \text{if the alternative is } H_1: P(X - Y < 0) \neq \tfrac{1}{2}.$$

The levels of significance are given, respectively, by

$$\frac{1}{2^n} \sum_{k=0}^{c} \binom{n}{k}, \quad \frac{1}{2^n} \sum_{k=n-c}^{n} \binom{n}{k} \quad \text{and} \quad \frac{1}{2^{n-1}} \sum_{k=0}^{c} \binom{n}{k}.$$

We note that

$$\frac{1}{2^n} \sum_{k=n-c}^{n} \binom{n}{k} = \frac{1}{2^n} \sum_{k=0}^{c} \binom{n}{k}.$$

The same tests can be used in some other situations:

(i) Suppose we sample from a continuous symmetric random variable X with mean $\mu(=\text{median})$. Then $P(X - \mu > 0) = P(X - \mu < 0) = \tfrac{1}{2}$ and we can test $\mu = \mu_0$ against one-sided or two-sided alternatives. In this case we only need to count S, the number of negative signs in $X_1 - \mu_0, X_2 - \mu_0, \ldots, X_n - \mu_0$. Thus, $\mu = \mu_0$ is rejected against $\mu < \mu_0$ if $S \geq c$ at size $\alpha = (1/2^n)\sum_{k=c}^{n}\binom{n}{k}$. If the random variable X does not have a symmetric distribution, we can still use the same procedure to test $\tilde{\mu} = \text{med}(X) = \tilde{\mu}_0$.

(ii) In many applications X's and Y's are independent. This happens, for example, when we need to compare two treatments or drugs to ascertain which is better. In this case the two drugs cannot, in general, be administered to the same individual. Suppose we choose n patients who are administered the first drug and choose n other patients who receive the second drug. We assume that the patients are selected in n pairs in such a way that patients in each pair are matched as closely as possible for similar characteristics. In each pair we randomly select one to receive the first treatment and the other to receive the second treatment. Let $X_1, X_2, \ldots, X_n$ be the responses of patients who receive the first drug and $Y_1, Y_2, \ldots, Y_n$ be the responses of those who receive the second drug. Here the X's and the Y's are independent. Under the null hypothesis that both treatments are equally effective, we note that $P(X > Y) = P(X < Y) = \frac{1}{2}$ and we can use the sign test as above. All we need to do is to count S, the numbers of pairs (X_i, Y_i) where $X_i < Y_i$. The test is performed exactly as before.

It should be noted that to perform the sign test we do not need the numerical values of $X_i - Y_i$, $i = 1, 2, \ldots, n$. We only need to know the sign of $X_i - Y_i$ for each i. Also, even though we have assumed that $P(X_i - Y_i) = 0$, ties do occur in practice. *In case of ties we remove the tied pairs from consideration and perform the test with $n*$ observations, where*

$$n* = n \text{ minus the number of tied pairs.}$$

We remark that sign test loses information in not using the magnitudes of $X_i - Y_i$, $i = 1, 2, \ldots, n$. A test that also takes account of these magnitudes is discussed in Section 11.5.

Example 15. Effectiveness of a New Diet Pill. To test the effectiveness of a new diet pill, nine randomly selected subjects were weighed before they went on the pill for six months. Their weights before and after the program in kilograms were recorded as follows:

Subject	1	2	3	4	5	6	7	8	9
Initial weight, X	80	82	75	90	98	87	100	107	103
Final weight, Y	79	83	75	81	95	86	101	105	100
Sign of $X - Y$	+	−	tie	+	+	+	−	+	+

The null hypotheses that we wish to test is that the diet pill in ineffective, that is,

$$H_0 : p = P(X > Y) = P(X < Y) = \tfrac{1}{2}$$

against

$$H_1 : P(X > Y) > \tfrac{1}{2} \quad \text{(diet pill is effective)}$$

$$\left(\text{or equivalently: } p = P(X < Y) < \tfrac{1}{2} \right).$$

Let S = number of negative signs in $X_i - Y_i$, $i = 1, 2, \ldots, 9$. Since there is one tie (Subject 3), we drop that observation and use the remaining $n* = 8$ pairs. We note that $S = 2$ and the P-value is given by

$$P_{H_0}(S \le 2) = \frac{1}{2^8} \sum_{k=0}^{2} \binom{8}{k} = \frac{(1 + 8 + 28)}{2^8} = \frac{37}{256} = .1446.$$

If the level of significance is .10, we will have to conclude that the evidence does not support rejection of H_0.

If $\alpha = .10$ is given, then the critical region is $\{0, 1, 2, \ldots, c\}$, where c is the largest integer such that:

$$P\{ S \le c | p = \tfrac{1}{2} \} \le .10,$$

that is,

$$\sum_{k=0}^{c} \binom{8}{k}\left(\frac{1}{2}\right)^{8} \leq .10.$$

From Table A3, $c = 1$ and the critical region $\{0, 1\}$ has size .0352. □

Example 16. Waiting Time for a Bus. A Metropolitan Area Road Transport Authority claims that the average waiting period on a well traveled route is 20 minutes. A random sample of 12 passengers showed waiting times of 25, 15, 19, 16, 21, 24, 18, 18, 24, 28, 15, 11 minutes. Does evidence support the claim?

Here the null hypothesis that we wish to test is H_0: $\mu = 20$ against the alternative H_1: $\mu \neq 20$ where μ is the average waiting time. We assume that the waiting time X has a symmetric distribution about μ. There are seven negative signs in $X_i - 20$, $i = 1, 2, \ldots, 12$. Under H_0,

$$P(S \geq 7) = \sum_{k=7}^{12} \binom{12}{k}\left(\frac{1}{2}\right)^{12} = .3871.$$

The P-value, therefore, is $2(.3871) = .7742$ and there is hardly any evidence to reject the claim.

If, on the other hand, α is given to be, say, .10, then the rejection region is $\{0, 1, \ldots, c, 12 - c, \ldots, 12\}$, where c is the largest integer such that

$$2\sum_{k=0}^{c} \binom{12}{k}\left(\frac{1}{2}\right)^{12} \leq .10.$$

It follows that $c = 2$. The rejection region $\{0, 1, 2, 10, 11, 12\}$ has size $2(.0193) = .0386$. Since $S = 7$ is not in the critical region, we cannot reject H_0 at a 10 percent level of significance. □

Problems for Section 6.3

1. In each of the following cases, state whether the random variable X can or cannot be assumed to have a binomial distribution. Give reasons for your answer.
   (i) In a statistics class of 25 students, four are selected at random. Let X be the number of seniors in the sample.
   (ii) About 5 percent of students registered for a statistics course are absent from the first class in Winter semester (which starts in mid-January) because they have colds. Let X be the number of students with colds who will be absent from the first class in this statistics course in the coming Winter semester out of a total of 35 in the course.
   (iii) Let X be the number of babies having blonde hair out of 20 (independent) births at the county hospital.
   (iv) Let X be the number of patients cured of a certain infection out of 15 who were treated with penicillin.
   (v) Twenty-five people were asked to try two different brands of detergent. Let X be the number of those who prefer Brand A over Brand B.

(vi) A random sample of 50 persons from a large population is asked the question, Do you smoke? Let X be the number of those who smoke.

2. A well-known baseball player has a lifetime batting average of .300. If he comes to bat five times in a game, find the probability that:
   (i) He will get at least one hit.
   (ii) He will get more than three hits.
   (iii) He will get no hits.

3. About 10 percent of mail solicitations result in a donation. Find the probability that of a randomly selected 20 people who receive a mail solicitation five or more will donate some money.

4. From past experience it is known that about 90 percent of all stolen cars are recovered. Find the probability that at least six of 10 stolen cars will be recovered.

5. A binomial random variable X has mean μ and variance σ^2.
   (i) Find the distribution of X. In particular, let $\mu = 10$ and $\sigma^2 = 6$. Find $P(X \geq 10)$.
   (ii) Do $\mu = 5$, $\sigma^2 = 4$ and $\mu = 5$, $\sigma^2 = 6$ define binomial random variables?

6. For $n = 25$ and $p = .3$ and $p = .8$, use Table A3 to find:
   (i) $P(X > 4)$. (ii) $P(5 < X < 9)$. (iii) $P(X \leq 6)$.
   (iv) $P(4 < X \leq 8)$.

7. For $p = .2$ and $n = 25$, use Table A3 to find:
   (i) The smallest k such that $P(X \geq k) \leq .20$.
   (ii) The largest k such that $P(X \leq k) \leq .35$.
   (iii) The largest k such that $P(X \geq k) \geq .30$.
   (iv) The smallest k such that $P(X \leq k) \geq .40$.

8. It is known that in thoroughbred horse racing the favorites win about one-third of the time. In a nine-game card, a person bets on the favorite in every race. Find the probability that he or she will collect:
   (i) At least once. (ii) Never. (iii) At most 5 times.
   (iv) At least 8 times.

9. It is known that a standard cure for leukemia is successful 30 percent of the time. A new cure is tried on 20 patients and found to be successful in nine cases. Is there any evidence to suggest that the new cure is better than the standard cure?

10. Two fair dice are thrown n times. Let X be the number of throws in which the number on the first die is smaller than that on the second. What is the distribution of X? How large should n be in order that the probability that X is at least 1 is $\geq \frac{1}{2}$? Repeat the problem with $X =$ number of throws in which both face values are the same.

11. What is the probability of observing a face value of 1 or 2 at least seven times in one roll of 10 fair dice? In four rolls of 10 fair dice?

12. In repeated rolling of a fair die, find the minimum number of trials necessary in order for the probability of at least one ace (face value 1) to be:
    (i) $\geq \frac{1}{2}$. (ii) $\geq 3/4$. (iii) $\geq 9/10$. (iv) $\geq .95$.

13. You are asked to toss a fair coin a certain number of times independently. A prize is promised if you throw exactly seven heads. The catch is that at the outset you have to choose the number of tosses you will make. How many tosses should you choose to make to maximize your probability of winning the prize? What is you chance of winning the prize? (See also Problem 20.)

14. About 95 percent of a potato crop is good. The remaining 5 percent have rotten centers which cannot be detected without cutting open the potato. A sample of 25 potatoes is taken. What is the probability that at most 20 of these are good? What is the probability that more than half are rotten?

15. In a five-answer multiple choice test that has 25 questions, what is the probability of getting 8 or more correct answers by guessing? If the instructor decides that no grade will be passing unless the probability of getting or exceeding that grade by guessing is $< .05$, what is the minimum passing grade (that is, the number of correct answers)?

16. In n tossings with a biased coin with $P(\text{head}) = p$, $0 < p < 1$, find the probability of obtaining:
    (i) An even number of heads.   (ii) An odd number of heads.
    For what value of p is the probability of an even number of heads equal to that of an odd number of heads?

17. In n independent Bernoulli trials with constant probability p of success, show that

$$P\left\{X_j = 1 \,\middle|\, \sum_{j=1}^{n} X_j = k\right\} = k/n \qquad (0 \le k \le n, \text{ fixed})$$

where $X_j = 1$ if jth trial is a success and $X_j = 0$ otherwise. (This result provides yet another justification for using $\hat{p} = k/n$ as an estimate for p.)

18*. An urn contains N marbles numbered 1 through N. A marble is drawn from the urn. If x is the number drawn, then a coin with probability $p(x) = x/N$ of heads is tossed n independent times. Let Y be the number of heads observed. Find:
    (i) $P\{Y = k\}$, for $k = 0, 1, \ldots, n$.   (ii) $\mathscr{E}Y$.
    (iii) $P\{X = x \mid Y = k\}$ where X denotes the number on the marble drawn.

19*. Let X be the number of successes and Y the number of failures in n independent Bernoulli trials with $P(\text{success}) = p$. Let $U = X - Y$. Show that the distribution of U is given by:

$$P(U = k) = \binom{n}{\dfrac{n+k}{2}} p^{(n+k)/2}(1 - p)^{(n-k)/2}$$

where $k = -n, 2 - n, 4 - n, \ldots, n - 4, n - 2, n$. Moreover,

$$\mathscr{E}U = n(2p - 1), \quad \text{var}(U) = 4np(1 - p)$$

and for $\varepsilon > 0$

$$P\left(\left|\frac{U}{n} - (2p - 1)\right| > \varepsilon\right) \to 0 \text{ as } n \to \infty.$$

20*. For the binomial distribution, show that the most probable number of successes (for given p and n) is k where

$$(n + 1)p - 1 < k \leq (n + 1)p$$

unless $P(X = k - 1) = P(X = k)$, in which case $k = (n + 1)p$. [Hint: Show that

$$\frac{P(X = k)}{P(X = k - 1)} = 1 + \frac{(n + 1)p - k}{k(1 - p)}$$

so that $P(X = k) > P(X = k - 1)$ if and only if $k < (n + 1)p$. This result can be used to estimate n for given p and is not inconsistent with Example 7.]

21*. Let X be the number of successes in n independent Bernoulli trials, each with a constant probability of success equal to p, $0 < p < 1$. Then X has a binomial distribution. Derive the probability function of X as follows. Let $f(n, k) = P(X = k)$.

(i) Show that

$$f(n, k) = pf(n - 1, k - 1) + (1 - p)f(n - 1, k), \qquad k = 1, 2, \ldots, n.$$

(Hint: The last trial is either a success or a failure.)

(ii) Show by induction that

$$f(n, k) = \binom{n}{k} p^k (1 - p)^{n - k}, \qquad k = 1, 2, \ldots, n - 1.$$

For $k = 0$, $f(n, 0) = (1 - p)^n$ by direct computation and for $k = n$, $f(n, k) = p^n$ by direct computation. Clearly $f(n, k) = 0$ for $k > n$. Take $f(0, 0) = 1$. [Hint: $\binom{n}{r} = \binom{n - 1}{r} + \binom{n - 1}{r - 1}$ for $1 \leq r \leq n - 1$.]

(iii) Let $\mathscr{E}X = h(n)$. Use the result in part (i) to deduce that

$$h(n) - h(n - 1) = p, \qquad \text{for } n = 1, 2, \ldots$$

with $h(0) = 0$. Conclude, by summing, that

$$h(n) = \mathscr{E}X = np.$$

[Hint: $h(n) = \sum_{k=0}^{n} kf(n, k)$.]

(iv) Let $\mathscr{E}X(X - 1) = g(n)$. Use the result in part (i) to show that

$$g(n) - g(n - 1) = 2(n - 1)p^2, \qquad n = 1, 2, \ldots$$

with $g(0) = 0$, $g(1) = 0$. Hence show that $g(n) = n(n - 1)p^2$, $n = 2, 3, \ldots$ and deduce that $\text{var}(X) = np(1 - p)$. [Hint: $g(n) = \sum_{k=2}^{n} k(k - 1)f(n, k)$.]

22*. Show that for a binomial random variable X

$$P(X \geq k) = n\binom{n - 1}{k - 1} \int_0^p x^k (1 - x)^{n - k} \, dx$$

(Hint: Integrate by parts. Alternatively, differentiate both sides with respect to p, simplify the right side, and then integrate. This formula gives a method of computing the distribution function of X in terms of an incomplete beta function. See Chapter 7, Sections 7.3 and 7.5.)

23*. If $k \geq np$, show that the probability function of a binomial random variable satisfies

$$P(X \geq k) \leq P(X = k)\frac{(k+1)(1-p)}{k+1-(n+1)p}$$

and if $k \leq np$, then

$$P(X \leq k) \leq P(X = k)\frac{(n-k+1)p}{(n+1)p-k}.$$

[Hint: $P(X = k)/P(X = k - 1) = p(n + 1 - k)/k(1 - p) < 1$ if and only if $k > (n + 1)p$. Hence the ratio decreases monotonically as k increases. Conclude that

$$\frac{P(X = k)}{P(X = k - 1)} \leq \frac{(n-r)p}{(r+1)(1-p)} \text{ for } k \geq r + 1$$

and hence that

$$\frac{P(X = r + v)}{P(X = r)} \leq \left\{\frac{(n-r)p}{(r+1)(1-p)}\right\}^{v}, \qquad v = 1, 2, \dots .$$

Sum over $v = 0$ to $n - r$, noting that this sum is bounded by the sum of an infinite geometric series.]

24*. Let N be a binomial random variable with parameters $q, 0 < q < 1$, and M. Let the conditional distribution of X, given $N = n$, be binomial with parameters p and n. Find the unconditional distribution of X and its mean and variance.

25*. As an application of Problem 24, let N_1 be the (random) number of contacts that an insurance salesman makes. Suppose $N = N_1 - 1$ is binomial with parameters $M - 1$ and $q, 0 < q < 1$. Let X be the number of sales that result. Suppose the conditional distribution of X given $N_1 = n$ is binomial with parameters n and p. Find the unconditional distribution of X, its mean and its variance. If each sale results in a commission of $\$C$, find the expected commission and its variance.

26. A *venire* is an order by a judge to a sheriff instructing him to summon persons to serve as jurors; a person called to jury duty on such an order is called a venireman. Find the probability of at most five nonwhites in a venire of 25 when the probability of selecting a single nonwhite is p. Compute this probability for $p = .20, .15, .10,$ and $.05$. In each case find the probability of selecting 20 such venires independently. Comment on your results.

27. A random sample of 20 students at a large state university was asked to respond to the question "Do you favor conversion to a semester system?" There were 13 Yeses and 7 Noes.
    (i) Estimate p, the true proportion of students who favor conversion.
    (ii) If the students were evenly divided on this question, what is the probability of observing as many as 13 Yeses in the sample?

- (iii) What is the mean and variance of your estimator $\hat{p}$ of p in part (i) if the true proportion of Yeses is .6?
- (iv) Find $P\{|\hat{p} - p| \le .1|p = .6\}$, the probability that your estimate $\hat{p}$ will be within .1 of the true value $p = .6$.
- (v) Find Chebyshev's inequality upper bound for $P(|\hat{p} - p| \ge .1)$.
- (vi) Find a 90 percent confidence interval for p based on Chebyshev's inequality. Also compute the improved 90 percent confidence interval for p. Compare the lengths of the two intervals.

28. A manufacturer of missiles wishes to estimate the success rate of its missiles. A random sample of 18 missiles was fired, and 16 firings were classified as successful. If the manufacturer wishes to estimate the true success rate p of its missiles by $\hat{p}$, the observed proportion of successes, what is his estimate?

    Find a 95 percent confidence interval for p based on Chebyshev's inequality. Also find the improved 95 percent confidence interval for p.

    Suppose he wishes to estimate p by $\hat{p}$ to within .02 with 95 percent confidence, how many missiles need he test-fire?

29. A major soft drink company wishes to estimate the proportion of shoppers who can identify a new soft drink introduced by the company in order to assess the success of its advertising campaign. In a random sample of 25 shoppers, only 10 were able to identify the soft drink.
    - (i) What estimate of p, the true proportion of shoppers who can identify the drink, should it use?
    - (ii) What is the probability of observing as small a proportion as the one observed if the true p is .50?
    - (ii) Find a 95 percent confidence interval for p.
    - (iv) If the company wishes to estimate p by $\hat{p}$ to within .05 with 90 percent confidence, how large a sample should it take?
    - (v) Find $P\{|\hat{p} - p| \le .2|p = .5\}$.

30*. A survey of students at a junior high school is conducted to determine the percentage of those who smoke regularly. A random sample of 25 students showed that 10 smoked regularly.
    - (i) If you estimate the true proportion of smokers by the sample proportion $\hat{p}$, find $P\{\hat{p} \le .4|p = .5\}$.
    - (ii) Find $P\{|\hat{p} - p| \le .1|p = .5\}$.
    - (iii) Find a 90 percent confidence interval for the true p.
    - (iv) If you wish to estimate p to within .02 with 95 percent confidence, how large a sample size must be taken?

31. A random sample of 100 people showed that 40 believe in UFO's. Find a 95 percent confidence interval for the true proportion of the population which believes in UFO's.

32. A coin is tossed 15 times and four heads are observed.
    - (i) Is there evidence to suggest that the coin is biased in favor of tails? What is the P-value of your test if you were to reject H_0: $p = 1/2$? (Here p = probability of a head.)
    - (ii) Take $\alpha = .05$. Find the critical region for testing H_0 at level $\alpha = .05$. What is the size of this critical region?
    - (iii) Find the power of the test in (ii) at $p = .4, p = .3, p = .2, p = .1$.

33. A random sample of 20 faculty members from the faculty of a large state university showed that eight had published during the last five years. Is it reasonable to assert that about half the faculty at this university has published during the last five years?

   What will be your conclusion if the level of significance $\alpha = .10$? Find the power of your level .10 critical region at $p = .3, .4$ and $p = .6, .7$.

34. In a random sample of 100 people, 25 had blue eyes. Does the evidence support the claim that the proportion of blue-eyed people is $> .20$? Find the critical region of a level .05 test.

   Find the power of this critical region at $p = .3, .4$, and .5.

35. It is estimated that 25 percent of all automobiles in a certain state do not meet minimum safety standards. The State Auto Safety Council launched a six-month advertising campaign to encourage motorists to have their autos checked. In order to check the effectiveness of its campaign, the council examined a random sample of 100 cars and found that 18 did not meet the minimum safety standards. Is there evidence to suggest that the campaign has had some effect?

   What will your conclusion be if $\alpha = .1$?

36. A well-known seed company claims that 80 percent of its coriander seeds germinate. A random sample of 100 seeds is planted and 25 fail to germinate. Is there evidence that the company's claim is false? Find the critical region of your test if α is given to be .08. Find the power of your test at $p = .7$, $p = .6$, and $p = .5$.

37. Find the critical region of a level α test in each of the following cases:
   (i) $H_0: p = .10$ against $H_1: p > .10$ with $n = 25$, $\alpha = .05$.
   (ii) $H_0: p = .30$ against $H_1: p \neq .30$ with $n = 20$, $\alpha = .10$.
   (iii) $H_0: p = .30$ against $H_1: p < .30$ with $n = 20$, $\alpha = .10$.
   (iv) $H_0: p = .25$ against $H_1: p \neq .25$ with $n = 100$, $\alpha = .14$.

38. A manufacturer of light bulbs claims that 90 percent of its bulbs have a life length of 2000 hours or more. A consumer takes a sample of size 15 and decides to reject the manufacturer's claim if he finds two or more bulbs in the sample that fail before 2000 hours.
   (i) What is the size of the test?
   (ii) Find the probability of a type II error if the true proportion of defectives is .20. Repeat for $p = .30$, $p = .40$, $p = .50$.

39. It is claimed that 25 percent of the marbles in an urn are numbered 50 or less. A random sample of 100 marbles (drawn with replacement) from the urn produced only 10 marbles numbered 50 or less. Do you have any reason to doubt the claim? At what level of significance would you reject the claim?

40. A panel of 12 wine tasters $A, B, \ldots, L$ was asked to test two varieties of wine: Cabernet Sauvignon and Pinot Noir. Each member of the panel ranks the two wines according to his or her preference. Rank 2 is more preferred than rank 1. The following rankings were awarded.

Panel Member	A	B	C	D	E	F	G	H	I	J	K	L
Cabernet Sauvignon	2	2	1	2	2	2	1	2	2	1	1	2
Pinot Noir	1	1	2	1	1	1	2	1	1	2	2	1

(i)   Is there evidence to suggest that the two varieties are equally preferred?

(ii) If a significance level of $\alpha = .10$ is given, what is the critical region of your test for testing $H_0: p = 1/2$ against $H_1: p \neq 1/2$? What is the size of your critical region? (Here $p = P$ (Cabernet Sauvignon preferred over Pinot Noir).)

41. Ten chickens are fed an experimental diet for four weeks and their weight gains in ounces are recorded as follows:

$$80 \quad 75 \quad 65 \quad 84 \quad 40 \quad 60 \quad 49 \quad 50 \quad 38 \quad 39$$

It is claimed that the median weight gained is 48 ounces. Test the claim at $\alpha = .10$ level of significance.

42. Thirteen randomly selected volunteers A, B, ..., M are administered a pill designed to lower systolic blood pressure with the following results.

Volunteer	A	B	C	D	E	F	G
Before	115	135	140	130	135	150	122
After	120	128	142	112	111	150	110

Volunteer	H	I	J	K	L	M
Before	135	138	190	180	99	110
After	135	126	180	160	103	108

(i)   Is there evidence to suggest that the pill is not effective?

(ii) If a significance level of $\alpha = .15$ is given, find the critical region of your test.

43. A course is taught by two methods: the classroom instruction method A and the self-study method B. Ten students each are selected randomly and carefully matched according to their background and ability prior to instruction. At the end of the course the 10 pairs of scores are recorded as follows:

$$(96, 99) \quad (112, 110) \quad (115, 111) \quad (98, 103) \quad (95, 90)$$
$$(110, 95) \quad (98, 100) \quad (111, 92) \quad (97, 99) \quad (98, 93)$$

Do the data indicate any significant difference between the two methods at 10 percent level?

44. Eleven students are given a course designed to improve their I.Q. on a standard test. Their scores before and after the course are recorded as follows:

I.Q. before 95 110 120 98 99 95 89 93 95 116 95
I.Q. after   98 114 124 93 99 106 106 95 103 124 111

What can you conclude about the effectiveness of the course?

45. Twenty-five drivers are randomly selected and tested for reaction times. After three alcoholic drinks, the same test is administered. It is found that 18 had slower

reaction times, 5 drivers registered no change, and 2 had faster reaction times. Is there evidence to suggest that the reaction times remain unaffected due to drinking?

46. Given p, how will you estimate n based on the number of successes X?

47. Two proofreaders independently read a manuscript and find X_1, X_2 typographical errors. Let X_{12} be the number of errors found by both proofreaders. How will you estimate the total number of errors in this manuscript? (Hint: Use the law of large numbers.)

6.4 HYPERGEOMETRIC DISTRIBUTION

In estimating the size of an animal population, the following capture-recapture procedure is frequently used. A sample of size D is captured, tagged, and then released into the population. After awhile a new catch of n animals is made, and the number of tagged animals in the sample is noted. Let X be this number. Assuming that the two catches can be considered as random samples from the population of all animals, X is a random variable and provides an estimate of the total population of animals. Indeed, if we assume that there were no births or deaths, then we expect the proportion of tagged animals in the sample to be approximately the same as that in the population. If N is the true size of the animal population then

$$\frac{X}{n} = \frac{D}{N}$$

so that an estimate of N is $\hat{N} = nD/X$. We will see later that this estimate maximizes the probability of the observed value of X.

The distribution of X is called a *hypergeometric distribution*. Let us compute the probability function of X. Figure 1 shows the sampling scheme. First note that there are $\binom{N}{n}$ samples of size n each of which is assigned the same probability, $1/\binom{N}{n}$. To find $P(X = k)$ we need only count the number of samples containing exactly k tagged animals [and hence $(n - k)$ untagged animals]. Since the k tagged animals must be chosen from D and $(n - k)$ untagged animals from

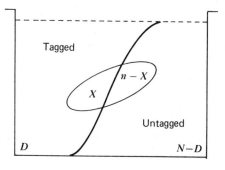

Figure 1. Sampling scheme.

$(N - D)$, the number of such samples is $\binom{D}{k}\binom{N-D}{n-k}$ (by Rules 1 and 3 of Section 2.5). It follows that

(1)
$$P(X = k) = \frac{\binom{D}{k}\binom{N-D}{n-k}}{\binom{N}{n}}.$$

Note that

$$0 \leq k \leq D \quad \text{and} \quad 0 \leq n - k \leq N - D,$$

so that

$$\max(0, n + D - N) \leq k \leq \min(D, n).$$

Hypergeometric Probability Function

$$P(X = k) = \frac{\binom{D}{k}\binom{N-D}{n-k}}{\binom{N}{n}},$$

where $k = \max(0, n + D - N), \ldots, \min(D, n)$.

Let us verify that (1) is indeed a probability function. Clearly, $P(X = k) \geq 0$ since the binomial coefficients are all ≥ 0. We need the following identity:

(2)
$$\sum_{k=0}^{m} \binom{a}{k}\binom{b}{m-k} = \binom{a+b}{m}$$

where we have used the convention that $\binom{a}{k} = 0$ whenever $k > a$; a, b and m are positive integers. Consider the identity

(3)
$$(1 + x)^a (1 + x)^b = (1 + x)^{a+b}, \quad x \in \mathbb{R}.$$

The coefficients of x on both sides must be the same (why?). Equating the coefficients of x^m on both sides of (3), using binomial expansions, we get (2). We take $a = D$, $b = N - D$, and $m = n$ in (2) to get

$$\sum_{k=0}^{n} \binom{D}{k}\binom{N-D}{n-k} = \binom{N}{n}$$

so that $\sum_{k=0}^{n} P(X = k) = 1$. [Again we have used the convention that $\binom{a}{k} = 0$ whenever $a < k$.] Hence (1) defines a probability function.

Hypergeometric distribution arises in many practical problems where the population being sampled has elements of two types: D "defectives" (tagged or successes) and $N - D$ "nondefectives" (untagged or failures). We take a random sample of n and seek the probability that the sample contains exactly k defectives. The experiment looks much like binomial except that we are sampling from a finite population *without replacement* so that the trials are not independent.

Example 1. **Binomial or Hypergeometric?** To highlight the differences between the two distributions we consider a box containing D defectives and $N - D$ nondefectives. We take a sample of size n from the box. What is the probability that the sample contains k defectives? That depends on whether we sample *with replacement* or *without replacement* (unordered sample).

If we sample *with replacement* (so that we can take samples of any size we desire and in this sense we are essentially sampling from an infinite population) then, considering each draw a trial and identifying success as a defective item drawn, the trials are independent, dichotomous, and the probability of success on each trial is $p = P(D) = D/N$. Hence

$$P(X = k) = \binom{n}{k}\left(\frac{D}{N}\right)^k \left(1 - \frac{D}{N}\right)^{n-k}, \qquad k = 0, 1, \ldots, n$$

where X is the number of successes (defectives) in n trials (draws). This is binomial distribution.

If we sample *without replacement* (so that $1 \le n \le N$) then we can either sample one after another or make a single draw of size n. Since the two procedures are identical (see Section 2.8) we assume that the draws are made one after another so that we have n trials or draws. *The trials are dependent so we can no longer use the binomial distribution.* The distribution of X in this case is given by (1).

We show that the probability of success on any draw is still D/N. Define events A_k, $k = 1, 2, \ldots, n$ where

$$A_k = \{\text{Success or defective on the } k\text{th draw}\}.$$

Then A_k happens if and only if the kth draw is a success and in the first $(k - 1)$ draws there are j defectives and $k - 1 - j$ nondefectives, $j = 0, 1, \ldots, \min(k - 1, D - 1)$. Hence

$$P(A_k) = \sum_{j=0}^{\min(k-1, D-1)} \frac{\binom{D}{j}\binom{N-D}{k-1-j}}{\binom{N}{k-1}} \frac{D-j}{N-k+1}$$

$$= \frac{D}{N} \sum_{j=0}^{k-1} \frac{\binom{D-1}{j}\binom{N-D}{k-1-j}}{\binom{N-1}{k-1}} = \frac{D}{N}$$

in view of (2). Since k is arbitrary, the result follows.

We emphasize, however, that the conditional probability of a defective on kth draw given information about the $(k-1)$th draw is no longer D/N since the trials are dependent. Thus in the binomial case (with replacement sampling),

$$P(\text{defective on any draw}) = P\{\text{defective on } k\text{th}|\text{defective on } (k-1)\text{th}\}$$

$$= \frac{D}{N}$$

whereas in sampling without replacement

$$P(\text{defective on any draw}) = \frac{D}{N},$$

but

$$P\{\text{defective on } k\text{th}|\text{defective on } (k\text{-1})\text{th}\} = \frac{D(D-1)}{N(N-1)} \neq \frac{D}{N}. \qquad \square$$

Binomial	Hypergeometric
1. Dichotomous trials	1. Dichotomous trials
2. Trials independent (sampling with replacement)	2. Trials dependent (sampling without replacement)
3. Probability of success same on any trial	3. Probability of success same on any trial
4. Number of trials is n.	4. Number of draws is n.

Example 2.   Racial Bias in Jury Selection.   A jury of 12 persons was randomly selected from a group of 30 potential jurors, of whom 20 were white and the remaining nonwhite (black, Hispanic, and other minorities). Suppose the jury contained only three nonwhites. Is there any reason to believe that the selection was biased?

Let us compute the probability of selecting as few as three nonwhites in a random sample of size 12 from a population of size 30. (If the selection is deemed biased when three nonwhites were selected, it will certainly be biased if less than three nonwhites were selected.) Here $N = 30$, $D = 10$, $n = 12$ so that if X is the number of nonwhites selected then

$$P(X \leq 3) = \sum_{k=0}^{3} \frac{\binom{10}{k}\binom{20}{12-k}}{\binom{30}{12}}$$

$$= \frac{\binom{20}{12} + \binom{10}{1}\binom{20}{11} + \binom{10}{2}\binom{20}{10} + \binom{10}{3}\binom{20}{9}}{\binom{30}{12}}$$

$$= \frac{3502}{10005} = .35.$$

Since the probability of selecting three or fewer nonwhites in a random sample of 12 from a population of 30 that has 10 nonwhites is .35, the evidence is not strong to suggest any bias in jury selection. $\qquad \square$

Example 3.   Estimating N.   Let us now justify that $\hat{N} = nD/X$ is a reasonable estimate of N. The idea here is that if $X = k$ is observed we estimate N by that number which maximizes $P(X = k)$. That is, we seek the value of N for which $X = k$ is most probable. Consider $P(X = k)$ as a function of N. Write $p_k(N) = \binom{D}{k}\binom{N-D}{n-k}/\binom{N}{n}$ and consider the ratio

$$R_k(N) = \frac{p_k(N)}{p_k(N-1)} = \frac{(N-D)(N-n)}{N(N-D-n+k)}.$$

We note that

$$R_k(N) > 1 \qquad \text{if and only if} \quad Nk < Dn$$

so that $p_k(N)$ first increases and then decreases, reaching a maximum when N is the largest integer $< nD/k$. Therefore, a value of N near nD/k makes $X = k$ the most probable value. The estimate $\hat{N} = nD/X$ is therefore a reasonable estimate of N (it is called the *maximum likelihood estimate* of N). If nD/X is integral then both $nD/X - 1$ and nD/X maximize $P(X = k)$. Note, however, that

$$\max(0, n + D - N) \le X \le \min(D, n)$$

so that unless $n + D > N$, $P(X = 0) > 0$ and moments of $\hat{N}$ are infinite. In particular, $\hat{N}$ is not unbiased for N. The estimates $[(n + 1)(D + 1)/(X + 1)] - 1$ or $(n + 2)(D + 2)/(X + 2)$ have been suggested instead. Neither has infinite moments.

In particular, let $k = 2$, $n = 3$, and $D = 3$. Then $\hat{N} = 3(3)/2 = 4.5$ so that we choose $N = 4$ as our estimate of N. The probability function of X gives

k	$P(X = k)$
2	3/4
3	1/4

so that $k = 2$ does have the larger probability.

In many problems N is known and the object is to estimate D. For example, the hypergeometric distribution has been used as a model for the number of children attacked by an infectious disease when a fixed number N are exposed to the disease. In this case the interest is in estimating D, given N, n and $X = k$. We apply the same argument as above, except that we consider $P(X = k)$ as a function of D. Consider the ratio

$$P_k(D) = \frac{\binom{D}{k}\binom{N-D}{n-k}}{\binom{D-1}{k}\binom{N-D+1}{n-k}} = \frac{D(N-D+1-n+k)}{(D-k)(N-D+1)}.$$

Note that

$$P_k(D) > 1 \quad \text{if and only if} \quad D < k\frac{N+1}{n}.$$

Hence $X = k$ is most probable for $D = \hat{D}$ where $\hat{D}$ is the largest integer $< k(N + 1)/n$. If $k(N + 1)/n$ is integral then either $(k(N + 1)/n) - 1$ or $k(N + 1)/n$ is our estimate of D.   □

Example 4. Sampling Inspection.   A manufacturer buys silicon chips in lots of 25. In order to check the quality of each lot he put a randomly chosen sample of size 4 from each lot to test. Let X be the number of defective chips in the sample. The following criterion is used: If $X \leq 1$, accept the lot; reject otherwise. Find the probability that a lot containing five defective chips will be accepted.

Here $N = 25$, $n = 4$. If D is the actual number of defectives in a lot, then X has hypergeometric distribution:

$$P(X = k) = \frac{\binom{D}{k}\binom{25 - D}{4 - k}}{\binom{25}{4}}, \qquad k = 0, 1, 2, \ldots, 4.$$

Hence the probability of accepting a lot that has D defectives is given by

$$P\{\text{accept the lot}\,|\,D \text{ defectives}\} = \sum_{k=0}^{1} \frac{\binom{D}{k}\binom{25 - D}{4 - k}}{\binom{25}{4}} = \frac{\binom{25 - D}{4}}{\binom{25}{4}} + \frac{\binom{D}{1}\binom{25 - D}{3}}{\binom{25}{4}}.$$

In particular, when $D = 5$,

$$P\{\text{accepting the lot}\,|\,D = 5\} = \frac{\binom{20}{4} + \binom{5}{1}\binom{20}{3}}{\binom{25}{4}} = \frac{4845 + 5700}{12650} = \frac{10545}{12650} = .8336.$$

Hence the manufacturer has an 83 percent chance of accepting a lot containing five defective chips. In fact, even if he adopts the rule:

Accept the lot if and only if $X = 0$,

he has about a 38.3 percent chance of accepting a lot with five defective chips.   □

Yet another question of interest is the following. Suppose a lot containing D defectives is acceptable. The manufacturer takes a random sample of size n and puts the items to test. When should a lot be rejected if the manufacturer is willing to allow an error of at most 100α percent of rejecting an acceptable (good) lot? In this case, we need to find the smallest c such that

$$P\{\,X \geq c\,|\,D \text{ defectives in the lot}\} \leq \alpha.$$

The lot is rejected if we observe $c, c + 1, \ldots, \min(n, D)$ defectives in the sample and accepted otherwise. For example, let us take $N = 20$, $D = 2$, $n = 3$, and $\alpha = 0.5$. Then we need to choose the smallest c such that

$$\sum_{k=c}^{2} \frac{\binom{2}{k}\binom{18}{3 - k}}{\binom{20}{3}} \leq .05.$$

These probabilities are given below:

c	$P(X \geq c)$
2	.0158
1	.2842
0	1

It follows that $c = 2$.

Finally, we show that we can approximate hypergeometric probabilities by binomial probabilities.

Example 5. *Approximating Hypergeometric by Binomial.* We note that the computation of $P(X = k)$ from (1) becomes cumbersome when N is large. If N is very large (infinite) it does not matter whether we sample with or without replacement. The trials will then be almost independent and we should expect to approximate $P(X = k)$ by a binomial probability. Indeed, if N is large then $p \doteq D/N$ and we should be able to approximate (1) by

$$\binom{n}{k}\left(\frac{D}{N}\right)^{k}\left(1 - \frac{D}{N}\right)^{n-k}.$$

Clearly, we cannot let $N \to \infty$ and D fixed since then $D/N \to 0$. So we need $N \to \infty$, $D \to \infty$ such that $D/N \to p, 0 < p < 1$. To see what other conditions we need in order that

$$\frac{\binom{D}{k}\binom{N-D}{n-k}}{\binom{N}{n}} \to \binom{n}{k}p^{k}(1-p)^{n-k}$$

let us rewrite (1) as

$$P(X = k) = \frac{D!}{k!(D-k)!}\frac{(N-D)!}{(n-k)!(N-D-n+k)!}\cdot\frac{n!(N-n)!}{N!}$$

$$= \frac{n!}{k!(n-k)!}\cdot\frac{D!}{(D-k)!}\cdot\frac{(N-D)!}{(N-D-n+k)!}\cdot\frac{(N-n)!}{N!}$$

$$= \binom{n}{k}\frac{D(D-1)\ldots(D-k+1)}{N(N-1)\ldots(N-k+1)}$$

$$\times\frac{(N-D)(N-D-1)\ldots(N-D-n+k+1)}{(N-k)(N-k-1)\ldots(N-n+1)}$$

$$= \binom{n}{k}\left\{\left(\frac{D}{N}\right)\frac{(D-1)/N}{(N-1)/N}\ldots\frac{(D-k+1)/N}{(N-k+1)/N}\right\}$$

$$\times\left\{\frac{(N-D)/N}{(N-k)/N}\cdot\frac{(N-D-1)/N}{(N-k-1)/N}\ldots\frac{(N-D-n+k+1)/N}{(N-n+1)/N}\right\}$$

Let $N \to \infty$, $D \to \infty$ *such that* $D/N \to p$, $0 < p < 1$. The middle term then $\to p^k$. The third term will converge to $(1 - p)^{n-k}$ provided n is *small compared with* N (note that $k \le n$). Then

$$P(X = k) \doteq \binom{n}{k}\left(\frac{D}{N}\right)^k \left(1 - \frac{D}{N}\right)^{n-k}.$$

The approximation is considered adequate when $N > 50$ and $n/N \le .10$. Since binomial probability can be approximated by Poisson or normal distribution, as we shall show, hypergeometric probabilities can also be approximated by Poisson and normal probabilities.

Let us show how good the binomial approximation is. We first apply binomial approximation to the data of Examples 2 and 4 even though the conditions for validity of the approximation are not satisfied. In Example 2, $p \doteq D/N = 10/30$ and $n = 12$ so that

$$P(X \le 3) \doteq \sum_{k=0}^{3} \binom{12}{k}\left(\frac{1}{3}\right)^k \left(\frac{2}{3}\right)^{12-k} = .3931.$$

In Example 4, $p \doteq D/N = 5/25 = .20$ and $n = 4$, so that

$$P(X \le 1) \doteq \sum_{k=0}^{1} \binom{4}{k}(.2)^k (.8)^{4-k} = .8192.$$

Neither approximation is that bad even though $N < 50$ and $n/N > .1$ in each case.

It should be noted that in (1) D and n can be interchanged without changing the distribution of X. This gives us another binomial approximation for $P(X = k)$, but with $p = n/N$.

Approximating
Hypergeometric Probability, $P(X = k)$

Use	When
I. $\binom{n}{k}\left(\dfrac{D}{N}\right)^k \left(1 - \dfrac{D}{N}\right)^{n-k}$	$D \ge n, n/N \le .1, N > 50$
II. $\binom{D}{k}\left(\dfrac{n}{N}\right)^k \left(1 - \dfrac{n}{N}\right)^{D-k}$	$D < n, D/N \le .1, N > 50$

It turns out that there are two more approximations possible due to another symmetry (see Problem 9). But we are not going to pursue the subject any further.

In Table 1 we give the exact probability under hypergeometric and the corresponding binomial approximation. □

TABLE 1. Binomial Approximation to Hypergeometric[a]

				$P(X = k)$	
N	n	D	k	Exact	Binomial Approximation
50	5	15	3	.127775	.13230
50	10	5	2	.209840	.20480
80	6	10	1	.402762	.38468
80	5	4	2	.017546	.02060
100	8	10	3	.028341	.03307
100	12	8	4	.006203	.00870

[a]Hypergeometric probabilities are taken from G. J. Lieberman and D. B. Owen, *Tables of Hypergeometric Probability Distribution*, Stanford University Press, 1961, with permission.

We now compute the mean and the variance of a hypergeometric distribution. We will need the following identities.

(4)
$$\sum_{k=1}^{n} k \binom{a}{k}\binom{b}{n-k} = a\binom{a+b-1}{n-1},$$

(5)
$$\sum_{k=2}^{n} k(k-1)\binom{a}{k}\binom{b}{n-k} = a(a-1)\binom{a+b-2}{n-2}.$$

To prove (4) we note that

$$\sum_{k=1}^{n} k\binom{a}{k}\binom{b}{n-k} = \sum_{k=1}^{n} a\binom{a-1}{k-1}\binom{b}{n-k}$$

$$= a\sum_{l=0}^{n-1}\binom{a-1}{l}\binom{b}{n-1-l} = a\binom{a+b-1}{n-1}$$

in view of (2). To prove (5) we apply the same procedure. Indeed,

$$\sum_{k=2}^{n} k(k-1)\binom{a}{k}\binom{b}{n-k} = \sum_{k=2}^{n} a(a-1)\binom{a-2}{k-2}\binom{b}{n-k}$$

$$= a(a-1)\sum_{l=0}^{n-2}\binom{a-2}{l}\binom{b}{n-2-l}$$

$$= a(a-1)\binom{a+b-2}{n-2} \qquad [\text{using (2)}].$$

Thus

$$(6) \quad \mathscr{E} X = \sum_{k=0}^{n} kP(X=k) = \sum_{k=1}^{n} k \frac{\binom{D}{k}\binom{N-D}{n-k}}{\binom{N}{n}} = \frac{D\binom{N-1}{n-1}}{\binom{N}{n}} = \frac{Dn}{N}$$

in view of (4), and

$$\mathscr{E} X(X-1) = \sum_{k=2}^{n} \frac{k(k-1)\binom{D}{k}\binom{N-D}{n-k}}{\binom{N}{n}}$$

$$= \frac{D(D-1)\binom{N-2}{n-2}}{\binom{N}{n}} = D(D-1) \cdot \frac{n(n-1)}{N(N-1)},$$

in view of (5). Hence

$$(7) \qquad \operatorname{var}(X) = \mathscr{E} X(X-1) + \mathscr{E} X - (\mathscr{E} X)^2$$

$$= \frac{D(D-1)n(n-1)}{N(N-1)} + \frac{Dn}{N} - \frac{D^2 n^2}{N^2}$$

$$= n \frac{D}{N}\left[\frac{(D-1)(n-1)}{N-1} + \frac{N-Dn}{N}\right]$$

$$= n\left(\frac{D}{N}\right)\left(\frac{N-D}{N}\right)\left(\frac{N-n}{N-1}\right).$$

Mean and Variance of Hypergeometric Distribution

$$\mathscr{E} X = n\left(\frac{D}{N}\right)$$

$$\operatorname{var}(X) = n\left(\frac{D}{N}\right)\left(\frac{N-D}{N}\right)\left(\frac{N-n}{N-1}\right)$$

If we let $D \to \infty$, $N \to \infty$ such that $D/N \to p$, $0 < p < 1$, then

$$\mathscr{E} X \to np \quad \text{and} \quad \operatorname{var}(X) \to np(1-p)$$

provided $n/N \to 0$. These are the mean and variance of the limiting binomial distribution. Whenever n/N is not small,

$$\operatorname{var}(X) \doteq \frac{N - n}{N - 1} np(1 - p).$$

The factor $(N - n)/(N - 1)$ is called (see Example 5.3.1) the *finite population correction factor* for the variance in sampling without replacement. If N is large compared with n, this factor is approximately 1 since, in that case, sampling with or without replacement are about the same.

Example 6. *Election Reversal.* This novel application of hypergeometric distribution is due to Finkelstein and Robbins.[†] Consider an election with two candidates, A and B, where A receives a votes and B receives b votes, $a > b$. Suppose n of the total $N = a + b$ votes cast were irregular (illegal). Candidate B challenges the result of the election, claiming that if these n votes were to be removed at random the outcome may result in a reversal of the election result. Let us compute the probability of a reversal.

Suppose these n votes are removed at random from the N votes cast. Let X be the number of these votes for Candidate A and hence $n - X$ for Candidate B. Then a reversal occurs if and only if

$$a - X \le b - (n - X),$$

that is, if and only if

$$X \ge \frac{(n + a - b)}{2}.$$

Since X is a random variable that is not observed [all we know is the triple (a, b, n)] there is no way to check if

$$X \ge \frac{(n + a - b)}{2}.$$

We first note that X has a hypergeometric distribution with $N = a + b$, $D = a$, and $n = n$, so that:

$$P(X = k) = \frac{\binom{a}{k}\binom{b}{n-k}}{\binom{a+b}{n}}, \qquad k = 0, 1, 2, \ldots, a.$$

Hence

$$P(\text{reversal}) = P\left(X \ge \frac{n + a - b}{2}\right)$$

$$= \sum_{k = \left[\frac{n+a-b}{2}\right]+1}^{a} \frac{\binom{a}{k}\binom{b}{n-k}}{\binom{a+b}{n}}.$$

[†]M. O. Finkelstein and H. E. Robbins, Mathematical Probability in Election Challenges, *Columbia Law Review*, 1973, 241–248.

If for given a, b, and n this probability is "small," we have to conclude that reversal is improbable (and no new election is called for).

Suppose $a = 510$, $b = 490$, and $n = 75$. Then $n/N = 75/1000 < .1$ and $p \doteq a/N = .51$. Since N is large we use the binomial approximation. We have

$$P(\text{reversal}) \doteq \sum_{k=48}^{75} \binom{75}{k} (.51)^k (.49)^{75-k}$$

$$= P(X_1 \geq 48)$$

where X_1 is a binomial random variable with $p = .51$ and $n = 75$. From binomial tables of Romig[†] we have

$$P(\text{reversal}) \doteq .0158.$$

Thus a margin of victory of as small as 20 votes out of 1000 votes cast has about 1.6 percent chance of reversal if 75 votes are randomly voided. Whether a probability of .0158 is small or not is a subjective decision to be made by the courts.

We emphasize that the crucial assumption here is that of randomness. This means that each voter has the same chance of casting an irregular vote. If there is any evidence of fraud or election irregularities, or if there is reason to believe that other factors made it more likely that the illegal votes were cast for one of the two candidates, then this assumption may not be valid.

The analysis can be extended to more than two candidates. Another (normal) approximation to the probability of reversal is considered later.   □

Applications of Hypergeometric Distribution to Statistical Inference

The hypergeometric distribution arises in many hypothesis-testing problems. We consider two applications.

Let X and Y be independent random variables of the continuous type with distribution functions F and G, respectively. Suppose we wish to test the null hypothesis

$$H_0 : F(x) = G(x) \qquad \text{for all } x \in \mathbb{R}$$

against one of the two one-sided alternatives $H_1 : F(x) \geq G(x)$ or $H_1 : F(x) \leq G(x)$, or against the two-sided alternative $H_1 : F(x) \neq G(x)$. (Here $F(x) \geq G(x)$ or $F(x) \leq G(x)$ or $F(x) \neq G(x)$ simply means that strict inequality holds on a set of positive probability.)

Let $X_1, X_2, \ldots, X_m$ and $Y_1, Y_2, \ldots, Y_n$ be the two random samples. Let us combine the two samples into one sample of size $m + n$ and order the observations in increasing order of magnitude. For convenience, assume that $m + n = 2l$, $l \geq 1$. Then the combined sample can be divided into two equal parts. Exactly l of the observations lie in the lower half and the remaining l in the upper half. Let U

[†] H. G. Romig, *50–100 Binomial Tables*, Wiley, New York, 1952.

be the number of observations on X (that is, the number of X_i's) that are in the lower half. Then U is a random variable with hypergeometric distribution (under H_0), as we shall show.

Under H_0, $F = G$ so that we expect the X's and Y's to be evenly spread out in the combined sample. If we observe a large value of U, it would be reasonable to suspect that X's tend to be smaller than the Y's so that $P(X > x) \le P(Y > x)$ or $F(x) \ge G(x)$. Thus a large value of U supports the alternative $F(x) \ge G(x)$. Similarly, a small value of U supports the alternative $F(x) \le G(x)$. A test based on U is called a *median test* since U is the number of X's that are smaller than the sample median (which may be taken to be the average of the two middle values).

Let us now compute the distribution of U under $H_0 : F(x) = G(x)$ for all $x \in \mathbb{R}$. Since X and Y are of the continuous type, $P(X_i = Y_j) = 0$ for all i, j and we can assume that equality does not hold for any of the mn pairs (X_i, Y_j). The total number of samples of size l from a population of size $2l$ is $\binom{2l}{l}$. Let us now count the number of those samples (of size l) in which exactly u values of X (and hence $l - u$ values of Y) appear below the sample median. This number is $\binom{m}{u}\binom{n}{l-u}$. Hence

(8)
$$P\{U = u | F = G\} = \frac{\binom{m}{u}\binom{n}{l-u}}{\binom{m+n}{l}}, \qquad u = 0, 1, \ldots, m$$

where $2l = m + n$. The same argument applies in the case when $m + n = 2l + 1$ is an odd integer. (In this case there is a unique middle value, the $(l + 1)$st value in the ordered arrangement, which is the sample median.)

Median Test for Testing $H_0 : F = G$

Order the combined sample of $m + n$ observations in increasing order of magnitude. Let u_0 be the observed number of X's that are smaller than the sample median.

H_1	Reject H_0 at Level α if	P-value
$F(x) \ge G(x)$	$u_0 \ge c$ where c is the smallest integer such that $P_{H_0}(U \ge c) \le \alpha$	$P_{H_0}(U \ge u_0)$
$F(x) \le G(x)$	$u_0 \le c$ where c is the largest integer such that $P_{H_0}(U \le c) \le \alpha$	$P_{H_0}(U \le u_0)$
$F(x) \ne G(x)$	$u_0 \ge c_1$ or $u_0 \le c_2$ where c_1, c_2 are determined as above with $\alpha/2$ probability in each tail	2(smaller tail probability)

Example 7. Comparing Two Sections of a Statistics Course. Two large sections of an elementary statistics course are taught by two instructors. At the end of the quarter a standard test is given to the two sections. It is desired to compare the performance of the students in the two sections. Since the classes were very large, a random sample of 10 students each was chosen and their scores were recorded as follows:

Section A:	95	85	98	102	74	65	92	72	78	61
Section B:	73	84	95	107	106	64	73	75	89	94

Let us analyze the data to see if there is any difference in student performance. Arranging the grades in increasing order of magnitude, we have:

Grade	Section		Grade	Section
61	A		85	A
64	B		89	B
65	A		92	A
72	A		94	B
73	B		95	A
73	B		95	B
74	A		98	A
75	B		102	A
78	A		106	B
84	B		107	B

The sample median here is $(84 + 85)/2 = 84.5$, and there are five students from Section A who have a score less than 84.5. Hence $U = 5$. Under H_0 that the distribution of scores in two sections is the same:

$$P\{U \leq 5\} = \sum_{k=0}^{5} \frac{\binom{10}{k}\binom{10}{10-k}}{\binom{20}{10}} = .6719$$

and

$$P(U \geq 5) = .6719$$

so that in random sampling there is a 67 percent chance of observing a value of U as small as (or as large as) 5. There is no evidence therefore to reject H_0: $F = G$. $\square$

Another application of hypergeometric distribution is in testing the equality of two proportions. Let X be binomial with parameters n and p_1, and let Y be binomial with parameters n and p_2. Suppose X and Y are independent and we wish to test H_0: $p_1 = p_2$ against, say, H_1: $p_1 < p_2$. Let $k = k_1 + k_2$ be the number of successes in $m + n$ trials. Under H_0, $p_1 = p_2 = p$ so that:

$$P(X + Y = k) = \binom{m + n}{k} p^k (1 - p)^{m+n-k}.$$

Moreover,

$$P(X = k_1) = \binom{m}{k_1} p^{k_1} (1 - p)^{m - k_1}$$

and

$$P(Y = k_2) = \binom{n}{k_2} p^{k_2} (1 - p)^{n - k_2}.$$

Hence

(9) $$P\{ X = k_1 | X + Y = k \} = \frac{P(X = k_1, Y = k_2)}{P(X + Y = k)}$$

$$= \frac{\binom{m}{k_1}\binom{n}{k_2}}{\binom{m + n}{k}}, \qquad k_1 = 0, 1, \ldots, m,$$

which is hypergeometric. We note that under H_0 the conditional distribution of X given $X + Y = k$ is independent of the common value $p_1 = p_2$ (which is unknown).

It is instructive to derive (9) directly as follows. The k successes in $m + n$ trials can be divided into two mutually exclusive categories, those happening in the first m trials and those happening in the last n trials. Hence the conditional distribution of the number of successes in the first m trials (the last n trials) given $X + Y = k$ has a hypergeometric distribution given by (9).

Returning now to the problem of testing $H_0 : p_1 = p_2$ against $H_1 : p_1 < p_2$, we note that given k, a large value of X tends to support H_0 and a small value of X tends to support H_1. It is therefore reasonable to reject H_1 when k_1 is large (that is, the observed value of X is large). This is the *Fisher–Irwin test* for comparing two unknown probabilities.

Fisher–Irwin Test of $H_0 : p_1 = p_2$

H_1	Reject H_0 at Level α if	P-value
$p_1 < p_2$	$k_1 \leq c$ where c is the largest integer such that $P_{H_0}\{ X \leq c | X + Y = k \} \leq \alpha$	$P_{H_0}\{ X \leq k_1 | X + Y = k \}$
$p_1 > p_2$	$k_1 \geq c$ where c is the smallest integer such that $P_{H_0}\{ X \geq c | X + Y = k \} \leq \alpha$	$P_{H_0}\{ X \geq k_1 | X + Y = k \}$
$p_1 \neq p_2$	$k_1 \leq c_1$ or $k_1 \geq c_2$ where c_1 and c_2 are determined as above with $\alpha/2$ probability in each tail	2(smaller tail probability)

Example 8.   Comparing Attitudes Towards ERA.   Independent random samples of 10 male and 10 female students at a junior college were asked whether or not they are in favor of the Equal Rights Amendment to the State Constitution. If six men and eight women favor the amendment, is there evidence to support the claim that the proportion of female students favoring ERA is more than the proportion of male students favoring ERA?

If p_1, p_2 respectively are the proportions of males and females favoring ERA at the college, then we wish to test $H_0: p_1 = p_2$ against $H_1: p_2 > p_1$. We have $k = 6 + 8 = 14$, $k_1 = 6$, $m = 10 = n$. We reject H_0 if X is too small. Under H_0,

$$P\{X \le 6 | X + Y = 14\} = \sum_{j=4}^{6} \frac{\dbinom{10}{j}\dbinom{10}{14-j}}{\dbinom{20}{14}} = .3142.$$

Since the observed event is quite probable under H_0, we cannot reject H_0. That is, the evidence does not support the claim that $p_2 > p_1$.   □

Problems for Section 6.4

1. Ten vegetable cans, all of the same size, have lost their labels. It is known that four contain tomatoes and six contain peas.
   (i) What is the probability that in a random sample of four cans all contain tomatoes?
   (ii) What is the probability that two or more contain peas?

2. An urn contains nine marbles of which three are red, three green, and three yellow. A sample of size 4 is taken (without replacement).
   (i) Find the probability that the sample will contain marbles of:
       (a) Only two colors.   (b) All three colors.
   (ii) What is the expected number of colors in the sample?
   (iii) What is the probability distribution of red marbles in the sample?
   (iv) How large a sample must you take so that the probability is at least .90 that it will contain all three red marbles?

3. A lot of 10 light bulbs contains two defective bulbs. A random sample of five bulbs is inspected from the lot.
   (i) What is the probability that the sample contains at least one defective?
   (ii) What is the minimum number of bulbs that should be examined so that the probability of finding at least one defective is at least 1/2? At least .80?

4. A shipment of 50 items is received. The shipment is accepted if a random sample of five items inspected shows fewer than two items defective.
   (i) If the shipment contains five defectives, what is the probability that it will be accepted?
   (ii) What is the probability of accepting a shipment containing 15 defectives? 30 defectives?

5. From a group of 10 (five men and five women) a committee of four was selected randomly. All committee members selected were men. Is there evidence to question the randomness of the selection?

6. Suppose 40 cards marked $1, 2, \ldots, 40$ are randomly arranged in a row. Find the distribution of the number of even integers in the first 25 positions.

7. There are 58 Republicans and 42 Democrats in the U.S. Senate. A committee of three Senators is to be chosen at random. Find the probability that:
   (i) All members of the committee are from the same party.
   (ii) The committee consists of at least one democrat.
   (iii) The committee consists of at least one member of each party.

8. From a deck of bridge cards, five cards are drawn at random. Find the probability that:
   (i) All five cards are of red suits.   (ii) All five cards are face cards.
   (iii) All four aces are drawn.

9. Let us write $p(k, n; N, D)$ for

$$P(X = k) = \frac{\binom{D}{k}\binom{N - D}{n - k}}{\binom{N}{n}}, \qquad k = 0, 1, \ldots, D.$$

Show that D and n can be interchanged without changing $P(X = k)$. That is, show that
   (i) $p(k, n; N, D) = p(k, D; N, n)$,
   (ii) $p(n - k, N - D; N, n) = p(D - k, N - n; N, D)$.

10. Consider the estimate $\hat{D} = X(N + 1)/n$ of D. Is $\hat{D}$ unbiased for D? (That is, check if $\mathscr{E}\hat{D} = D$.) Find $\text{var}(\hat{D})$. (See also Problem 13.)

11. Is it possible for a hypergeometric random variable X that $\mathscr{E}X = \text{var}(X)$?

12. Let X and Y be random variables such that X is binomial with parameters m and p and

$$P\{Y = y | X = x\} = \binom{n}{y - x} p^{y - x}(1 - p)^{n - y + x}, \qquad y = x, x + 1, \ldots, x + n.$$

   (i) Find the unconditional distribution of Y.
   (ii) Find the conditional distribution of X given $X = y$.

13. Show that:
   (i) $\hat{D} = NX/n$ is an unbiased estimate of D.
   (ii) X/n is an unbiased estimate of D/N so that estimation of D/N, the fraction of defectives, does not require knowledge of N.
   In particular, suppose that a random sample of 100 voters in a county showed 29 Republicans. What is an unbiased estimate of the proportion of Republican voters in the county? If it is known that the county has 6000 voters, what is an unbiased estimate of the number of Republican voters?

14. Find an upper bound for $P(|X/n - D/N| \le \varepsilon)$, where X is hypergeometric and $\varepsilon > 0$.

15*. Let $p(k, N; n, D)$ be the probability of drawing k defectives in a random sample of size n from a population of N elements containing D defectives. Since sampling without replacement is equivalent to sampling one after another, consider the last draw. It produces either a defective or a nondefective item. Show that

$$p(k, N; n, D) = \frac{D - k + 1}{N - n + 1} p(k - 1, N; n - 1, D)$$

$$+ \frac{N - D - (n - 1 - k)}{N - n + 1} p(k, N; n - 1, D).$$

Use this relation to show that

$$p(k, N; n, D) = \frac{\binom{D}{k}\binom{N - D}{n - k}}{\binom{N}{n}},$$

and also to derive the mean and variance of $p(k, N; n, D)$.

16. In order to study the efficacy of a drug to control hyperactivity in children under age 8, ten children chosen randomly were given the drug, and ten different children, also chosen randomly (and independently of the first group), were given a placebo. All 20 children were then given a test which was scored on an ordinal scale with the following results (higher score indicates higher level of hyperactivity).

Drug group	10	15	8	9	7	11	6	14	11	4
Placebo group	7	18	16	9	11	6	14	12	10	18

Is there evidence to indicate that the drug is effective in controlling hyperactivity? (The scores being ordinal here, we can only compare the median scores of each group by using the median test.)

17. Two different brands of automobile batteries are to be compared to determine if Brand X has a longer life length than Brand Y. Independent random samples of size 8 from Brand X batteries and 7 from Brand Y batteries were put to a test under identical conditions and yielded the following life lengths (in 1000 hours).

Brand X	11.1	15.8	19.3	18.8	17.3	16.5	12.7	20.9
Brand Y	12.1	13.2	16.5	8.9	10.3	11.2	9.8	

Use the median test to see if the data support the claim that Brand X batteries last longer.

18. To compare the average weight gains of pigs fed two different rations, nine pigs were fed on ration A for 30 days and six pigs were fed on ration B for 30 days. The gains in weight (in pounds) were recorded as follows.

Ration A	55	38	60	46	39	40	52	49	32
Ration B	52	48	25	19	45	36			

Is there evidence to suggest that the average gain due to ration A is larger?

19. In order to compare pollution levels in two rivers, independent samples were taken from each river at several different locations to determine the quantity of oxygen dissolved in the water. The following results were recorded.

River 1:	3.5	4.6	9.4	7.3	6.6	8.4	7.6
River 2:	4.2	5.8	7.4	6.2	8.7		

    How will you test the null hypothesis that the two rivers are equally polluted? Find the *P*-value of your test.

20. A random sample of five boys and six girls (ages 15–18) are asked for their opinion on legalized abortion on demand. The sample was taken from a large group of boys and girls who had assembled for a rock concert. It is believed that girls are more favorably disposed to legalized abortion than boys. Two boys and five girls were found to be in favor of legalized abortion. Is there evidence to support the belief? If the level of significance α is given to satisfy $.04 \leq \alpha \leq .10$, what is the critical region of your test?

21. A new treatment is believed to prolong the life of persons who have suffered a coronary attack. To test the efficacy of the treatment 20 patients are selected so that they are similar in age, general health, severity of the coronary attack, and so on. A random sample of 10 of these 20 patients is given the new treatment while the remaining 10 serve as controls. After two years, six of the patients who received the treatment were still alive whereas only three of the control-group patients survived. Is there evidence to suggest that the treatment is effective?

22. In Problem 21 suppose that the actual survival times of the 20 patients were recorded, in years, as follows:

Treatment group	1.9	2.7	5.5	1.7	.5	1.8	6.9	11.8	18.0	7.3
Control group	1.7	2.9	16.3	14.7	1.4	.5	1.9	1.1	.6	1.8

    Suppose the only information that is available for each patient is whether or not he or she is still alive after five years. Is there any evidence to suggest that the treatment prolongs the life of patients?

23. Of 25 income tax returns audited in a small town, 10 were from low and middle income families and 15 from high income families. Two of the low income families and four of the high income families were found to have underpaid their taxes. Are the two proportions of families who underpaid taxes the same?

24. A candidate for a congressional seat checks her progress by taking a random sample of 20 voters each week. Last week, six reported to be in her favor. This week nine reported to be in her favor. Is there evidence to suggest that her campaign is working?

25. Two large sections of a statistics course are taught during the spring semester, Section 1 meeting in the morning and Section 2 in the late afternoon. Students were randomly assigned to these sections and the same professor taught both courses using the same text. To compare the performance of the students, in order to

determine if there were any differences due to the time of the day the classes met, random samples of nine students each were selected. Section 1 had four A's and Section 2 had two A's. Is the proportion of A students in Section 1 significantly larger than that in Section 2?

6.5 GEOMETRIC AND NEGATIVE BINOMIAL DISTRIBUTIONS: DISCRETE WAITING-TIME DISTRIBUTION

Suppose that a couple decides to have children until they have a girl. What is the average number of issues needed before they have the first girl? What is the probability that they have three boys before they have a girl? Assuming that boys and girls are equally likely, that is, the probability that the sex of a newborn is male is $1/2$, it is intuitively clear that the average number of issues needed for the first girl is $1/(1/2) = 2$. How do we justify this? More important, how do we compute the required probability, namely, that the fourth child is the first girl? This probability is not $1/2$. It does not equal the conditional probability that the fourth is a girl, given that the first three children were boys. By independence, that probability is always $1/2$ no matter what the sex distribution of the first three of their children.

We note that the experiment looks much like a binomial:

(i)   There is dichotomy, either girl (success) or boy (failure).

(ii)   Trials are independent.

(iii)   The probability of success on each trial is $1/2$.

The number of trials, however, is not fixed in advance. Indeed the number of trials is a random variable which equals the number of times the experiment is repeated until a success occurs. We recall that in a binomial experiment, the number of trials is fixed in advance and the random variable of interest is the number of successes in n trials.

Let X be the number of trials needed for the first success. Then X is said to have a *geometric* distribution.

Binomial or Geometric

Assumptions: (i)   Dichotomous trials.
            (ii)   Probability of success p is same on each trial.
            (iii)   Trials are independent.

Binomial	Geometric
(iv) Number of trials is fixed in advance. Random variable of interest is the number of successes in n trials.	(iv) Number of trials is a random variable. Random variable of interest is the number of trials needed for the first success.

Let us compute the distribution of X. We note that $X \geq 1$ and for any integer $k \geq 1$,

(1) $P(X = k) = P(\underbrace{FF...F}_{k-1}S) = \underbrace{(1-p)(1-p)...(1-p)}_{k-1}p$ (independence)

$\qquad\qquad = p(1-p)^{k-1}, \qquad k = 1, 2, \dots .$

Geometric Probability Function

$$P(X = k) = p(1-p)^{k-1}, \qquad k = 1, 2, \dots; \quad 0 < p < 1.$$

First we show that (1) defines a probability function. Indeed $P(X = k) \geq 0$ and, moreover,

$$\sum_{k=1}^{\infty} P(X = k) = \sum_{k=1}^{\infty} p(1-p)^{k-1} = p \sum_{k=0}^{\infty} (1-p)^k = \frac{p}{1 - (1-p)} = 1.$$

Here we have used the following result. For $0 < |r| < 1$

(2) $$\sum_{k=0}^{\infty} r^k = \frac{1}{1-r}.$$

To prove (2), let

(3) $$S_n = \sum_{k=0}^{n} r^k = 1 + r + \cdots + r^n.$$

Multiplying both sides of (3) by r and subtracting from (3), we get

$$S_n - rS_n = \left(1 + r + \cdots + r^n\right) - \left(r + r^2 + \cdots + r^{n+1}\right) = 1 - r^{n+1}$$

so that

(4) $$S_n = \frac{1 - r^{n+1}}{1 - r}.$$

The result (4) holds for any $r \in \mathbb{R}, r \neq 1$. If, however, $0 < |r| < 1$, then $r^{n+1} \to 0$ as $n \to \infty$ and letting $n \to \infty$ in (4) we have

$$\sum_{k=0}^{\infty} r^k = \lim_{n \to \infty} S_n = \lim_{n \to \infty} \frac{1 - r^{n+1}}{1 - r} = \frac{1}{1 - r}.$$

We also need the following identities: For $0 < r < 1$

(5)
$$\sum_{k=1}^{\infty} kr^{k-1} = \frac{1}{(1-r)^2},$$

(6)
$$\sum_{k=2}^{\infty} k(k-1)r^{k-2} = \frac{2}{(1-r)^3}.$$

To prove (5) we differentiate both sides of (2) with respect to r. We have

$$\sum_{k=1}^{\infty} kr^{k-1} = \left(\frac{1}{1-r}\right)^2.$$

To prove (6), we differentiate (5) with respect to r to obtain

$$\sum_{k=2}^{\infty} k(k-1)r^{k-2} = \frac{2}{(1-r)^3}.$$

Using (5) and (6) we have

$$\mathscr{E}X = \sum_{k=1}^{\infty} kp(1-p)^{k-1} = \frac{p}{[1-(1-p)]^2} = \frac{1}{p},$$

and

$$\mathscr{E}X(X-1) = \sum_{k=2}^{\infty} k(k-1)p(1-p)^{k-1}$$

$$= p(1-p)\frac{2}{[1-(1-p)]^3} = \frac{2(1-p)}{p^2}.$$

Hence

$$\operatorname{var}(X) = \mathscr{E}X(X-1) + \mathscr{E}X - (\mathscr{E}X)^2 = \frac{2(1-p)}{p^2} + \frac{1}{p} - \frac{1}{p^2}$$

$$= \frac{1}{p^2}\{2 - 2p + p - 1\} = \frac{1-p}{p^2}.$$

Mean and Variance of Geometric Distribution

$$\mathscr{E}X = 1/p, \qquad \text{var}(X) = (1-p)/p^2, \qquad 0 < p < 1.$$

We are now ready to answer both the questions posed in the beginning of the section. Assuming $p = P(\text{girl}) = \frac{1}{2}$,

$$\mathscr{E}X = \frac{1}{p} = \frac{1}{\frac{1}{2}} = 2$$

so that the average number of births needed for the first girl is 2. Moreover,

$$P(\text{first girl born is the fourth child}) = P(X = 4) = \frac{1}{2}\left(1 - \frac{1}{2}\right)^3 = \frac{1}{16}.$$

Example 1. Multiple Choice Examinations. A student takes a multiple-choice oral examination. The grade is based on the number of questions asked until he gets one correct answer. Suppose the student guesses at each answer and there are five choices for each answer. In this case the probability of a correct guess (success) is $p = \frac{1}{5}$. The trials may be assumed to be independent. If X is the number of questions (trials) required for the first correct answer, then

$$P(X = k) = \frac{1}{5}\left(1 - \frac{1}{5}\right)^{k-1}, \qquad k = 1, 2, \ldots.$$

The average number of questions required for the first correct answer is

$$\mathscr{E}X = \frac{1}{p} = \frac{1}{\frac{1}{5}} = 5.$$

The probability that he gets a correct answer on or before the 10th trial is

$$P(X \le 10) = \sum_{k=1}^{10} \frac{1}{5}\left(\frac{4}{5}\right)^{k-1} = \frac{1}{5} \cdot \sum_{k=0}^{9} \left(\frac{4}{5}\right)^k = \frac{1}{5} \cdot \frac{1 - (4/5)^{10}}{1 - (4/5)} = 1 - \left(\frac{4}{5}\right)^{10}$$

$$= .8926. \qquad \qquad \square$$

In general, for $k \ge 1$

$$P(X \ge k) = \sum_{j=k}^{\infty} p(1-p)^{j-1} = p\left\{(1-p)^{k-1}\left[1 + (1-p) + \cdots\right]\right\}$$

$$= p(1-p)^{k-1} \cdot \frac{1}{1 - (1-p)} = (1-p)^{k-1}$$

so that[†]

(7) $$P(X > k) = P(X \geq k + 1) = (1 - p)^k$$

and

(8) $$P(X \leq k) = 1 - (1 - p)^k.$$

We note that for integers $k, l \geq 1$, we have from (7)

(9) $$P\{X > k + l | X > l\} = \frac{P(X > k + l, X > l)}{P(X > l)}$$

$$= \frac{P(X > k + l)}{P(X > l)} = \frac{(1 - p)^{k+l}}{(1 - p)^l} = (1 - p)^k$$

$$= P\{X > k\}$$

where we have used the fact that

$$X > k + l \quad \text{implies} \quad X > l$$

so that $\{X > k + l\} \subseteq \{X > l\}$. Hence $\{X > k + l, X > l\} = \{X > k + l\} \cap \{X > l\} = \{X > k + l\}$. Equation (9) has the following interpretation. Given that the first success happens after the lth trial (that is, no success was observed in the first l trials), the conditional probability that the first success will be observed after an additional k trials depends only on k. This is the so-called *lack of memory property* of a geometric law. The process forgets the initial string of l failures. This is a characterizing property of a geometric probability function. That is, any positive integer-valued random variable satisfying (9) must have a geometric distribution (Problem 28).

> **Example 2. Number of Trials Required to Achieve a Success.** In many applications, one is interested not in the average number of trials but the number of trials required to observe the first success with a given probability. That is, for given α, $0 < \alpha < 1$, usually small, we wish to find n such that

$$P(X \leq n) \geq 1 - \alpha.$$

In view of (8) we have

$$1 - (1 - p)^n \geq 1 - \alpha$$

so that

$$(1 - p)^n \leq \alpha.$$

[†]Alternatively, if more than k trials are needed for one success, there must be no successes in k trials. Hence $P(X > k) = P(Y = 0)$, where Y has binomial (k, p) distribution. See discussion following relation (12) below.

Taking logarithms and solving for n we see that

$$n \geq \frac{\ln \alpha}{\ln(1 - p)} \qquad [\ln \alpha \text{ as well as } \ln(1 - p) < 0].$$

Thus, if $p = \frac{1}{2}$ and $\alpha = .05$ so that we want to be 95 percent sure of achieving a success in n trials with a fair coin, then we must choose

$$n \geq \frac{\ln \frac{1}{20}}{\ln \frac{1}{2}} = \frac{\ln 2 + \ln 10}{\ln 2} = 1 + \frac{\ln 10}{\ln 2} = 4.32.$$

Hence with a fair coin we are 95 percent certain to achieve a success in five (or more) trials. $\qquad \square$

Example 3. Estimating p. Yet another question of interest is to estimate p. Suppose we need n trials to achieve a success. Then

$$P(X = n) = p(1 - p)^{n-1}.$$

The most probable value of n is obtained by maximizing $P(X = n)$ with respect to p. Thus

$$\frac{d}{dp} P(X = n) = -p(n - 1)(1 - p)^{n-2} + (1 - p)^{n-1}$$

$$= (1 - p)^{n-2}[-p(n - 1) + 1 - p] = 0$$

gives $\hat{p} = 1/n$ as our estimate of p. Note that

$$\frac{d^2}{dp^2} P(X = n) = -(n - 2)(1 - p)^{n-3}(1 - np) - n(1 - p)^{n-2} < 0$$

so that $p = 1/n$ is a maximum. In particular, if we need 90 tosses to achieve a success, then our estimate of p is $\frac{1}{90}$.

Note, however, that the estimate $\hat{p} = 1/X$ is not unbiased. In fact,

$$\mathscr{E}\hat{p} = \mathscr{E}\frac{1}{X} = \sum_{k=1}^{\infty} \frac{1}{k} p(1 - p)^{k-1} = p(1 - p)^{-1} \sum_{k=1}^{\infty} \frac{(1 - p)^k}{k}.$$

It is well known that the Taylor series expansion of $\ln(1 + t)$ for $|t| < 1$ is given by

$$-\ln(1 + t) = \sum_{k=1}^{\infty} (-1)^k \frac{t^k}{k}$$

so that on replacing t by $-t$ we have

$$\ln\frac{1}{1 - t} = \sum_{k=1}^{\infty} \frac{t^k}{k}, \qquad |t| < 1.$$

Hence with $t = 1 - p, 0 < p < 1$, we have

$$\mathscr{E}\hat{p} = p(1 - p)^{-1} \sum_{k=1}^{\infty} \frac{(1 - p)^k}{k} = p(1 - p)^{-1} \ln\frac{1}{p}$$

and it follows that $\hat{p} = 1/X$ is not an unbiased estimate of p. It can be shown (Problem 29) that there does not exist an unbiased estimate of p.   □

Negative Binomial Distribution*

An obvious generalization of the geometric distribution is obtained if we ask the following question. Let $r \geq 1$ be an integer. How many Bernoulli trials do we need to achieve r successes? Let Y be the number of trials required. Then Y is a random variable. Clearly $Y \geq r$. What is the probability function of Y?

To find $P(Y = k)$ for $k \geq r$ note that $Y = k$ if and only if kth trial results in the rth success so that:

(10)

$P(Y = k) = P((r - 1)$ successes in the first $(k - 1)$ trials

and the kth trial a success)

$= p \cdot P((r - 1)$ successes in $(k - 1)$ trials)   (independence)

$= p\binom{k - 1}{r - 1}p^{r-1}(1 - p)^{k-r}$ (binomial distribution)

$= \binom{k - 1}{r - 1}p^r(1 - p)^{k-r}, \quad k = r, r + 1, \ldots.$

This is the *negative binomial* (or Pascal) *distribution*.

Negative Binomial Probability Function

$$P(Y = k) = \binom{k - 1}{r - 1}p^r(1 - p)^{k-r}, \quad k = r, r + 1, \ldots.$$

For $r = 1$ we get the geometric distribution. The name "negative binomial" derives from the fact that the terms in the binomial series expansion with negative exponent, namely:

(11) $1 = p^r[1 - (1 - p)]^{-r} = p^r \sum_{k=0}^{\infty} \binom{-r}{k}(-(1 - p))^k$

give the probabilities $P(Y = k + r)$, $k = 0, 1, 2, \ldots$. Here

$$(12) \quad \binom{-r}{k} = \frac{(-r)(-r-1)\ldots(-r-k+1)}{k!} = (-1)^k \binom{r+k-1}{k}$$

for any $r > 0$ and $k \geq 0$ integral. (In our case $r > 0$ is an integer.)

The computation of probabilities for a negative binomial distribution is facilitated by the use of the following observation. If n or less trials are needed for the rth success, then the number of successes in n trials must be at least r. That is, if X is negative binomial with parameters r and p, and Y is binomial with parameters n and p, then we have the following relation.

Relation Between Negative Binomial and Binomial

$$P(X \leq n) = P(Y \geq r)$$

$$P(X > n) = P(Y < r)$$

Consequently, we can use Table A3 for binomial probabilities to compute probabilities under negative binomial (or geometric) distribution. Figure 1 shows some histograms for negative binomial distribution.

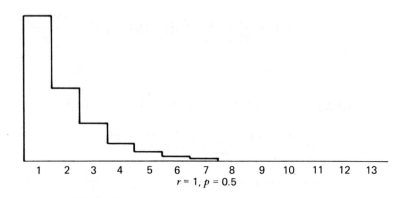

$$r = 1, p = 0.5$$

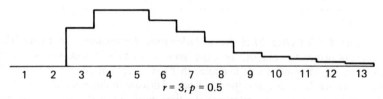

$$r = 3, p = 0.5$$

Figure 1. Histogram for negative binomial.

The mean and the variance are easily computed. We have

$$\mathscr{E}X = \sum_{k=r}^{\infty} k \binom{k-1}{r-1} p^r (1-p)^{k-r} = rp^r \sum_{k=r}^{\infty} \binom{k}{r} (1-p)^{k-r}$$

$$= rp^r \sum_{l=0}^{\infty} \binom{l+r}{r} (1-p)^l = rp^r \sum_{l=0}^{\infty} (-1)^l \binom{-r-1}{l} (1-p)^l \quad (l = k - r)$$

$$= rp^r \sum_{l=0}^{\infty} \binom{-r-1}{l} [(-1)(1-p)]^l = rp^r [1 - (1-p)]^{-r-1} = \frac{r}{p}$$

in view of (12) and (11). Similarly,

$$\mathscr{E}X(X+1) = p^r \sum_{k=r}^{\infty} k(k+1) \binom{k-1}{r-1} (1-p)^{k-r}$$

$$= r(r+1) p^r \sum_{k=r}^{\infty} \binom{k+1}{r+1} (1-p)^{k-r}$$

$$= r(r+1) p^r \sum_{l=0}^{\infty} \binom{l+r+1}{l} (1-p)^l$$

$$= r(r+1) p^r \sum_{l=0}^{\infty} \binom{-r-2}{l} [(-1)(1-p)]^l = \frac{r(r+1)}{p^2}.$$

Hence

$$\mathrm{var}(X) = \mathscr{E}X^2 - (\mathscr{E}X)^2 = \mathscr{E}X(X+1) - \mathscr{E}X - (\mathscr{E}X)^2$$

$$= \frac{r(r+1)}{p^2} - \frac{r}{p} - \left(\frac{r}{p}\right)^2 = \frac{r(1-p)}{p^2}.$$

Mean and Variance of Negative Binomial Distribution

$$\mathscr{E}X = r/p, \qquad \mathrm{var}(X) = r(1-p)/p^2, \qquad 0 < p < 1.$$

Example 4. Selecting Mice for a Laboratory Experiment. In many laboratory experiments involving animals, the experimenter needs a certain number of animals. Suppose an experimenter needs five mice that have a certain disease. A large collection of mice is available to him, and the incidence of disease in this population is 30 percent. If the experimenter examines mice one at a time until he gets the five mice needed, then the number of mice X that must be examined has a negative binomial distribution with

$p = .3$ and $r = 5$. The expected number of examinations required is $r/p = 5/.3 = 16.67$. The probability that the number of examinations required is more than 25 is given by

$$P(X > 25) = P(Y < 5)$$

where Y is binomial with parameters $n = 25$ and $p = .3$. From Table A3

$$P(X > 25) = P(Y \le 4) = .0905$$

so that there is only about a 9 percent chance that more than 25 examinations will be required to get five mice that have the disease. $\qquad\square$

Example 5. The World Series of Baseball. In the World Series, two baseball teams A and B play a series of at most seven games. The first team to win four games wins the series. Assume that the outcomes of the games are independent. Let p be the (constant) probability for team A to win each game, $0 < p < 1$. Let X be the number of games needed for A to win. Then $4 \le X \le 7$. Writing

$$A_k = \{\, A \text{ wins on the } k \text{th trial}\,\}, \qquad k = 4, 5, 6, 7$$

we note that $A_k \cap A_l = \varnothing$, $k \ne l$, so that

$$P(A \text{ wins}) = P\left(\bigcup_{k=4}^{7} A_k \right) = \sum_{k=4}^{7} P(A_k)$$

where

$$P(A_k) = P(\text{4th success on } k \text{th trial})$$

$$= \binom{k-1}{3} p^3 (1 - p)^{k-4} (p).$$

Hence,

$$P(A \text{ wins}) = p^4 \sum_{k=4}^{7} \binom{k-1}{3} (1 - p)^{k-4}.$$

In particular, let $p = 1/2$. Then $P(A \text{ wins}) = 1/2$. The probability that A will win in exactly six games is $\binom{5}{3}(\tfrac{1}{2})^6 = \tfrac{5}{32}$. The probability that A will win in six games or less is $11/32$.

Given that A has won the first two games, the conditional probability of A winning the series equals

$$\sum_{k=2}^{5} \binom{k-1}{1} \left(\frac{1}{2}\right)^2 \left(\frac{1}{2}\right)^{k-2} = \left(\frac{1}{4} + \frac{2}{8} + \frac{3}{16} + \frac{4}{32} \right) = \frac{13}{16}.$$

Given that A has won two and lost three of the first five games, the probability that A will win is $\tfrac{1}{4}$. $\qquad\square$

The negative binomial distribution is frequently referred to as a *waiting time distribution* since it may be interpreted as the *probability distribution of the waiting time for the rth success*. Let $S_k = X_1 + \cdots + X_k$ be the number of trials necessary for the kth success, $k = 1, 2, \ldots, r$. Then we can think of S_r as the sum of r independent and identically distributed random variables $X_1, \ldots, X_r$ each having a geometric distribution with parameter p (see Figure 2). Each X_i is a waiting time for one success. Thus X_1 is the waiting time for first success, X_2 is the waiting time for an additional success, and so on. Conversely, the sum of r independent identically distributed geometric random variables can be imagined as the number of trials necessary for the rth success and, hence, must have a negative binomial distribution. A formal proof is given in Chapter 8. A continuous analog of negative binomial (and geometric) distribution is considered in Section 7.4.

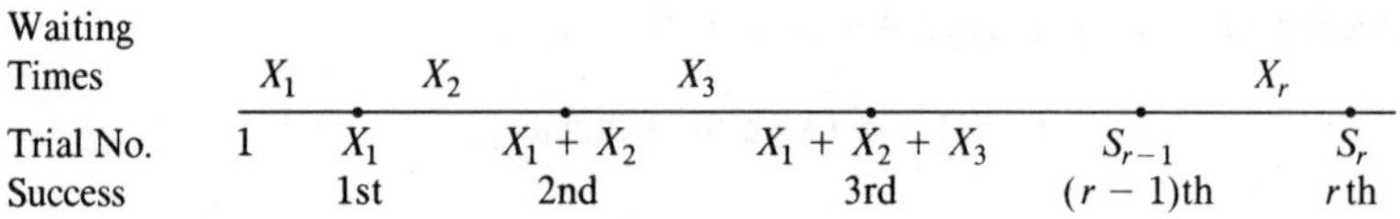

Waiting Times	X_1	X_2	X_3		X_r
Trial No.	1 $\quad X_1$	$X_1 + X_2$	$X_1 + X_2 + X_3$	S_{r-1}	S_r
Success	1st	2nd	3rd	$(r-1)$th	rth

Figure 2

Finally, we consider briefly an analogue of the negative binomial distribution in sampling without replacement. Considering the urn model with w white and b black marbles, if we sample *with replacement* and ask for the distribution of the number of draws X needed for the rth black marble, then X has the negative binomial distribution

$$(13) \quad P(X = k) = \binom{k-1}{r-1}\left(\frac{b}{b+w}\right)^r\left(\frac{w}{b+w}\right)^{k-r}, \quad k = r, r+1, \ldots .$$

If we sample *without replacement* one after the other, then in order that the kth draw $(k \geq r)$ be the rth black marble drawn, the kth draw must produce a black marble whereas in the previous $k - 1$ draws there must have been $(r - 1)$ black marbles. Hence:

$$(14) \qquad P(X = k) = \frac{\binom{b}{r-1}\binom{w}{k-r}}{\binom{b+w}{k-1}} \cdot \frac{b-r+1}{b+w-k+1},$$

$k = r, r+1, \ldots, b + w$.

Rewriting (14) we see that

$$(15) \qquad P(X = k) = \binom{k-1}{r-1}\frac{\binom{b+w-k}{b-r}}{\binom{b+w}{b}}$$

A simple argument now shows that $P(X = k)$ as given by (15) or (14) approaches

$$\binom{k-1}{r-1}\left(\frac{b}{b+w}\right)^r\left(\frac{w}{b+w}\right)^{k-r}$$

as $b + w \to \infty$ provided that $r/(b + w) \to 0$ and $k/(b + w) \to 0$ (Problem 24). This is precisely the probability given in (13) as we expect, since when $b + w \to \infty$ sampling with or without replacement is the same. [Although (15) looks much like the hypergeometric probability (6.4.1), it is not the same.]

In the case of dichotomous trials we have considered four probability distributions. Let us summarize the results.

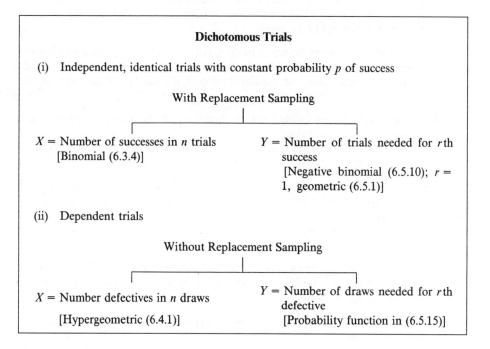

Problems for Section 6.5

Problems 16–27 refer to the Subsection on negative binomial distribution.

1. A fair die is rolled until an ace shows up.
   (i) Find the probability of the first ace showing up on the sixth trial.
   (ii) Find the probability that an ace shows up in at most six trials.
   (iii) Find the number of trials required so that the probability of getting an ace is at least $1/2$.

2. About 5 percent of potatoes grown by a farmer have rotten centers. In order to test whether a potato is rotten, one has to cut it open. Find the probability that:
   (i) The ninth potato cut is the first rotten potato.
   (ii) Twenty-five or more potatoes have to be destroyed to find the first rotten potato.

3. A test yields a positive reaction with probability .7. What is the probability that the first positive reaction occurs after nine negative reactions?

4. About 90 percent of apples grown by a farmer are good. It is not possible to detect a rotten apple without cutting the apple. If 15 good apples are needed to bake a pie, what is the probability that in a random selection 20 or more apples must be cut open?

5. Three people are having coffee in a diner. They each toss a coin to determine the odd person, who is then asked to pay. Tosses are continued until a decision is reached. Let p be the probability of success on each coin. Find the probability that fewer than n tossings are necessary to reach a decision.

   In particular, if $p = 1/2$ find the minimum number of tossings required to reach a decision in 90 percent of the cases.

6. In a series of independent Bernoulli trials, it is known that the average number of trials needed for a success is 10. Find the distribution of the number of trials needed to achieve a success. Also find the distribution of the number of successes in 10 trials.

7. (i) In successive tossings with a fair die, find the probability of an ace before an even-numbered face value.
   (ii) Answer part (i) if the die is loaded such that $P(1) = p$, $P(2) = P(4) = P(6) = p_1$, with $0 < p + 3p_1 < 1, p, p_1 > 0$.

8. Consider a series of independent Bernoulli trials. If the mean number of successes is five and the mean number of trials needed to achieve a success is also five, find the distribution of:
   (i) The number of successes.
   (ii) The number of trials needed to achieve a success.

9. A motorist encounters n consecutive lights. Each of the lights is equally likely to be green or red. Find the probability distribution of the number of green lights passed by the motorist before being stopped by a red light.

10. The number of times X that a light switch operates before being discarded has probability function

$$P(X = k) = p(1 - p)^{k-1}, \qquad k = 1, 2, \ldots; \quad 0 < p < 1.$$

   (i) Find the probability that the number of times the switch operates is an even number.
   (ii) Find the probability that the number of times the switch operates is an integral multiple of 3.

11. A lot of television tubes is tested by randomly testing tubes one after another until a defective tube is found. The number of tubes tested determines whether or not the lot is accepted. Let k be the critical number, that is, the lot is accepted if k or more tubes are tested for the first defective.
   (i) Find the probability that a lot containing a proportion p of defectives will be accepted.

(ii) If $k = 5$ and $p = .10$ find the probability of accepting the lot.

(iii) What should be the value of k if we want to be 90 percent confident of rejecting a lot that has 25 percent defectives?

12. Families with two children are required for a psychological study. The following sampling scheme is used. Randomly selected families (with two children) are interviewed sequentially until the first family with two boys appears for an interview. No more families are interviewed after that.

(i) What is the average number of families selected using this procedure?

(ii) What is the probability that more than 10 families will be selected for this study?

(iii) What is the probability that four families will be selected?

13. A pair of identically loaded coins are thrown together independently. Let p be the probability of a head on each coin. What is the distribution of number of throws required for both coins to show head simultaneously? What is the distribution of number of throws required for at least one of the coins to show a head?

14. A fair die is rolled repeatedly. What is the probability that a face value of 2 will show up before a face value of 5?

15. A pair of fair dice is rolled repeatedly. Find the probability that a sum of face values of 7 appears before a sum of face values of 4.

16. A baseball player has a season batting average of .300 going into the World Series. He needs 10 more hits to reach a season total of 150 hits. What is the probability that 12 or less times at bat are required to reach his goal? Does he have a reasonable chance of reaching his goal this season?

17. A student takes a five-answer multiple-choice oral test. His grade is determined by the number of questions required in order for him to get five correct answers. A grade of A is given if he requires only five questions; a grade of B is given if he requires 6 or 7 questions; a grade of C is given if he requires 8 or 9 questions; he fails otherwise. Suppose the student guesses at each answer. What is the probability of his getting an A? B? C?

18. In repeated tossings of a pair of fair dice, find the probability that the sum of face values is 7 *twice* before a sum of 4. What is the probability of two successive sums of 7 before a sum of 4? [Hint: P(sum of 7 before a sum of 4) $= \frac{2}{3}$ from Problem 15.]

19*. The game of craps is played with two fair dice as follows. A person throws the pair of dice. He wins on the first throw if he gets a sum of 7 or 11, loses if he throws 2, 3, or 12, and continues to throw otherwise until he throws a sum of 7 or the first number thrown. He wins if he throws the first number thrown and loses if 7 appears. (See also Problem 2.7.18.)

(i) Find the probability of winning for the player.

(ii) Find the expected number of throws required before the game ends.

20. In a sequence of independent Bernoulli trials with probability p of success, find the probability that n_1 successes will occur before n_2 failures.

21*. Let $q(n, r; p)$ be the probability of observing the rth success on the nth trial with constant probability p of success. By considering the result of the nth trial, show that

$$q(n, r; p) = pq(n - 1, r - 1; p) + pq(n - 1, n - r; 1 - p).$$

Hence show (by induction) that

$$q(n, r; p) = \binom{n-1}{r-1} p^r (1 - p)^{n-r}, \qquad n = r, r + 1, \dots .$$

Find $\mathscr{E}X$ and var(X) for the distribution $q(n, r; p)$ by using the recursive relation $[q(n, r; p) = 0$ for $n < r]$.

22. There are N marbles numbered 1 through N in an urn. Marbles are drawn one after another *with replacement*. Let X be the number of drawings until the first marble is drawn again. Find the distribution and expected value of X.

23. Consider a biased coin with probability p of heads, $0 < p < 1$. The coin is tossed once. If a head shows up it is tossed until another head shows up. Let X be the number of trials needed to stop. Find the distribution of X, its mean and its variance.

24. Show that the probability function (14) or (15) $\to$ negative binomial probability function as $b + w \to \infty$. That is, show that

$$\binom{k-1}{r-1} \frac{\binom{b+w-k}{b-r}}{\binom{b+w}{b}} \to \binom{k-1}{r-1} \left(\frac{b}{b+w}\right)^r \left(\frac{w}{b+w}\right)^{k-r}$$

where k and r are fixed constants.

25. (i) Show that for $n \geq r$

$$P\{ X_1 = k \mid X_1 + \cdots + X_r = n \} = \frac{\binom{n-k-1}{r-2}}{\binom{n-1}{r-1}}, \qquad k = 1, 2, \dots, n + 1 - r,$$

where $X_1, X_2, \dots, X_r$ are independent identically distributed geometric random variables with common parameter p. In particular, $P\{ X_1 = k \mid X_1 + X_2 = n \}$ is uniform on $\{1, 2, \dots, n - 1\}$.

(ii) Show that for any $l \geq 2$, the conditional distribution of $X_1 + \cdots + X_{l-1}$ given $X_1 + \cdots + X_l = n$ is given by

$$P\{ X_1 + \cdots + X_{l-1} = k \mid X_1 + \cdots + X_l = n \} = \frac{\binom{k-1}{l-2}}{\binom{n-1}{l-1}},$$

$k = l - 1, \dots, n.$

26*. Let X have a negative binomial distribution with parameters r and p.
   (i) Given r and p, for what value of k is $P(X = k)$ maximum? That is, what is the most probable value of X?
   (ii) Show that $\hat{p} = r/X$ is the estimate of p that maximizes $P(X = k)$.
   (iii) Show that $\hat{p} = r/X$ is not unbiased for p but that $\hat{p}_1 = (r - 1)/(X - 1)$ is unbiased for p (provided $r \geq 2$). (See Example 3.)
   (iv) Given p, find the estimate of r that maximizes $P(X = k)$.
   (v) Given p, find an unbiased estimate of r.
   [Hint: For (i) consider $P(X = k)/P(X = k - 1)$. For (ii) consider $P(X = k)$ as a function of p. For (iv) consider $P(X = k)$ as a function of r and use an argument similar to that of (i).]

27*. Since (15) defines a probability distribution, it follows that

$$\sum_{k=r}^{b+w} \binom{k-1}{r-1}\binom{b+w-k}{b-r} = \binom{b+w}{b}$$

and on rewriting with $l = k - r$,

$$\sum_{l=0}^{w} \binom{r+l-1}{l}\binom{b-r+w-l}{w-l} = \binom{b+w}{b}.$$

Use this identity to show that

$$\mathscr{E}X = r\frac{b+w+1}{b+1}, \quad \mathscr{E}X(X+1) = \frac{r(r+1)(b+w+1)(b+w+2)}{(b+1)(b+2)},$$

and

$$\mathrm{var}(X) = \frac{rw(b+w+1)(b+1-r)}{(b+1)^2(b+2)}.$$

28*. Show that the lack of memory property characterizes a geometric distribution. That is, let X be a (positive) integer valued random variable with $0 < P(X = 1) < 1$. If for all integers $k, l \geq 1$

$$P\{X > k + l \mid k > l\} = P\{X > l\}$$

then X must have a geometric distribution. (Hint: Iterate $P\{X > k + l\} = P\{X > l\}P\{X > k\}$.)

29*. Show that if X has a geometric distribution with parameter p, then there does not exist an unbiased estimate of $\mathscr{E}X = 1/p$. [Hint: If $h(X)$ is unbiased for $1/p$, then $\mathscr{E}h(X) = 1/p$ for all p, $0 < p < 1$. Compare coefficients of various powers of p on both sides.]

6.6 POISSON DISTRIBUTION

Another probability distribution which is closely connected to the binomial distribution often arises in practice. This is the distribution of the number of occurrences of a rare event, that is an event with small probability p, in n

independent trials. It is called the *Poisson distribution*. Typical examples are the number of customers that enter a store during a 15-minute interval, the number of machine failures per day in a plant, the number of printing errors per page of a book, the number of winning tickets among those purchased in a state lottery, the number of motorists arriving at a gas station per hour, the number of telephone calls at an exchange per hour, the number of errors that a typist makes per page, the number of radioactive particles that decay per unit time, and so on. In all these examples, the event of interest is rare (per unit of time, space, or volume).

To find the probability function of the number of rare events we need to make certain assumptions. Let $X(\Delta)$ be the number of events that occur during an interval Δ. We assume that

(i) The events are independent; this means that if $\Delta_1, \Delta_2, \ldots$ are disjoint intervals then the random variables $X(\Delta_1), X(\Delta_2), \ldots$ are independent.

(ii) The distribution of $X(\Delta)$ depends only on the length of Δ and not on the time of occurrence.

(iii) The probability that exactly one event occurs in a small interval of length Δt equals $\lambda \Delta t + o(\Delta t)$ where[†] $o(\Delta t) \to 0$ as length $\Delta t \to 0$, and $\lambda > 0$.

(iv) The probability that two or more events occur in a small interval of length Δt is $o(\Delta t)$.

In view of (iii) and (iv), the probability of zero events in a small interval of length Δt is $1 - \lambda \Delta t + o(\Delta t)$.

Consider the interval $[0, t]$ and let $X(t)$ be the number of events in $[0, t]$. Let us subdivide $[0, t]$ in n disjoint intervals each of length t/n. Let $\Delta_1, \Delta_2, \ldots, \Delta_n$ be these intervals. Then:

$$X(t) = \sum_{k=1}^{n} X(\Delta_k)$$

where $X(\Delta_k)$, $k = 1, 2, \ldots, n$ are independent and $X(t)$ has a binomial distribution with probability function

$$P(X(t) = k) = \binom{n}{k}\left[\frac{\lambda t}{n} + o\left(\frac{t}{n}\right)\right]^k \left[1 - \frac{\lambda t}{n} + o\left(\frac{t}{n}\right)\right]^{n-k}.$$

We need the following lemma, which describes a property of exponential functions.

LEMMA 1. *Let us write $f(x) = o(x)$ if $f(x)/x \to 0$ as $x \to 0$. Then*

(1)
$$\lim_{n \to \infty} \left\{1 + \frac{a}{n} + o\left(\frac{1}{n}\right)\right\}^n = e^a, \qquad a \in \mathbb{R}.$$

[†]We write $f(x) = o(x)$ if $f(x)/x \to 0$ as $x \to 0$. In particular, $f(x) = o(1)$ as $x \to 0$ if $f(x) \to 0$ as $x \to 0$.

Proof. Expanding f into a Taylor series about $x = 0$, we have:

$$f(x) = f(0) + f'(\theta x) \quad (0 < \theta < 1)$$

$$= f(0) + xf'(0) + x\{f'(\theta x) - f'(0)\}.$$

If f' is continuous at 0, then as $x \to 0$

$$f(x) = f(0) + xf'(0) + o(x).$$

In particular, let $f(x) = \ln(1 + x)$ so that $f'(x) = 1/(1 + x)$, which is continuous at 0. Then, as $x \to 0$

$$\ln(1 + x) = x + o(x)$$

Hence for large n,

$$n\ln\left\{1 + \frac{a}{n} + o\left(\frac{1}{n}\right)\right\} = n\left\{\frac{a}{n} + o\left(\frac{1}{n}\right) + o\left[\frac{a}{n} + o\left(\frac{1}{n}\right)\right]\right\}$$

$$= a + o(1)$$

so that

$$\lim_{n \to \infty} \left\{1 + \frac{a}{n} + o\left(\frac{1}{n}\right)\right\}^n = e^a$$

as asserted. $\qquad\square$

For fixed k we now let $n \to \infty$ in the expression for $P(X(t) = k)$ above. We have:

$$P(X(t) = k) = \lim_{n \to \infty} \binom{n}{k}\left[\frac{\lambda t}{n} + o\left(\frac{t}{n}\right)\right]^k\left[1 - \frac{\lambda t}{n} + o\left(\frac{t}{n}\right)\right]^{n-k}$$

$$= \frac{(\lambda t)^k}{k!} \lim_{n \to \infty} \frac{n(n-1) \cdots (n-k+1)}{n^k}$$

$$\times \left[1 + o\left(\frac{n}{\lambda t} \cdot \frac{t}{n}\right)\right]^k\left[1 - \frac{\lambda t}{n} + o\left(\frac{t}{n}\right)\right]^n\left[1 - \frac{\lambda t}{n} + o\left(\frac{t}{n}\right)\right]^{-k}$$

$$= \frac{(\lambda t)^k}{k!}e^{-\lambda t}$$

since as $n \to \infty$

$$\frac{n(n-1)\cdots(n-k+1)}{n^k} = (1)\left(1-\frac{1}{n}\right)\cdots\left(1-\frac{k-1}{n}\right) \to 1,$$

$$\left[1 + o\left(\frac{n}{\lambda t}\frac{t}{n}\right)\right]^k = \left[1 + o\left(\frac{1}{\lambda}\right)\right]^k \to 1,$$

$$\left[1 - \frac{\lambda t}{n} + o\left(\frac{t}{n}\right)\right]^{-k} \to 1,$$

and from Lemma 1,

$$\left[1 - \frac{\lambda t}{n} + o\left(\frac{t}{n}\right)\right]^n \to e^{-\lambda t}.$$

Hence:

$$(2) \qquad P(X(t) = k) = e^{-\lambda t}\frac{(\lambda t)^k}{k!}, \qquad k = 0,1,2,\ldots$$

where λ is called the *intensity* or the *rate of occurrence*.

A similar argument shows that if X is a binomial random variable with parameters p and n then

$$(3) \qquad P(X = k) = \binom{n}{k}p^k(1-p)^{n-k} \to e^{-\lambda}\frac{\lambda^k}{k!}$$

provided $n \to \infty$, $p \to 0$ in such a way that $np \to \lambda$. It is clear that (3) is a special case of (2) with $t = 1$.

We note that in view of (3), binomial probabilities can be approximated by Poisson probabilities. *The approximation is good provided p is small ($p < .1$) and $np = \lambda$ is moderate*[†] (≤ 10). In Table 1 we compare the binomial distribution with parameters n, p and the Poisson distribution with parameter $\lambda = np$ for $n = 50$, $p = .1$ and for $n = 100$, $p = .01$. Table A4 in the Appendix gives cumulative Poisson probabilities for some selected values of λ. Figure 1 shows histograms of (3) for some selected values of λ.

Clearly, $P(X = k) > 0$ and, moreover,

$$\sum_{k=0}^{\infty} P(X = k) = e^{-\lambda}\sum_{k=0}^{\infty}\frac{\lambda^k}{k!} = e^{-\lambda} \cdot e^{\lambda} = 1$$

so that $P(X = k)$ does define a probability function. Let us compute the mean

[†] For $np > 10$ it is more convenient to use the normal approximation given in Chapter 9.

TABLE 1. Poisson Approximation

k	$n = 50, p = .1$ $P(X = k)$ Binomial	Poisson	$n = 100, p = .01$ $P(X = k)$ Binomial	Poisson
0	0.0052	0.0067	0.3660	0.3679
1	0.0286	0.0337	0.3697	0.3679
2	0.0779	0.0843	0.1849	0.1839
3	0.1386	0.1403	0.0610	0.0613
4	0.1809	0.1755	0.0149	0.0153
5	0.1849	0.1755	0.0029	0.0031
6	0.1541	0.1462	0.0005	0.0005
7	0.1076	0.1044	0.0001^-	0.0001^-
8	0.0643	0.0653	0.0000^+	0.0000^+
9	0.0333	0.0363		
10	0.0152	0.0181		
11	0.0061	0.0082		
12	0.0022	0.0035		
13	0.0007	0.0013		
14	0.0002	0.0005		
15	0.0001^-	0.0001^-		
16	0.0000^+	0.0000^+		

and the variance of X. Since

$$e^\lambda = \sum_{k=0}^{\infty} \frac{\lambda^k}{k!}$$

we have on successive differentiation with respect to λ

$$e^\lambda = \sum_{k=1}^{\infty} k\frac{\lambda^{k-1}}{k!} \quad \text{and} \quad e^\lambda = \sum_{k=2}^{\infty} k(k-1)\frac{\lambda^{k-2}}{k!}.$$

Consequently,

$$(4) \qquad \mathscr{E}X = \sum_{k=0}^{\infty} kP(X = k) = e^{-\lambda} \sum_{k=1}^{\infty} \frac{k\lambda^k}{k!} = \lambda e^{-\lambda} \sum_{k=1}^{\infty} \frac{k\lambda^{k-1}}{k!} = \lambda$$

and

$$\mathscr{E}X(X-1) = \sum_{k=1}^{\infty} k(k-1)e^{-\lambda}\frac{\lambda^k}{k!} = e^{-\lambda}\lambda^2 \sum_{k=2}^{\infty} k(k-1)\frac{\lambda^{k-2}}{k!} = \lambda^2.$$

It follows that

$$(5) \qquad \text{var}(X) = \mathscr{E}X(X-1) + \mathscr{E}X - (\mathscr{E}X)^2 = \lambda^2 + \lambda - (\lambda)^2 = \lambda.$$

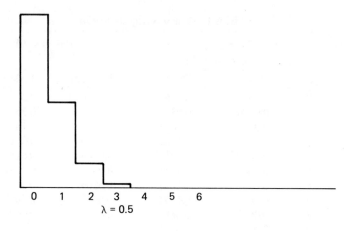

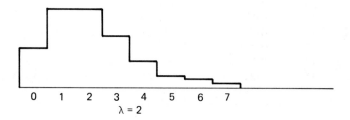

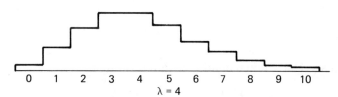

Figure 1. Histogram for Poisson, $\lambda = .5, 2, 4$.

Mean and Variance of Poisson Distribution

$$\mathscr{E}X = \lambda = \mathrm{var}(X)$$

Example 1. Probability of Hitting a Target. The probability of hitting a target with a single shot is .002. In 4000 shots, find the probability of hitting the target more than two times.

If we identify success by a hit, then $p = P(\mathrm{success}) = .002$ is small and the number of trials $n = 4000$ is large. Moreover, $\lambda = np = 8 \leq 10$. We can therefore apply the

Poisson approximation. If X denotes the number of hits, then

$$P(X > 2) = 1 - P(X \le 2)$$

$$\doteq 1 - \left\{ e^{-8} + 8e^{-8} + \frac{8^2}{2!} e^{-8} \right\} = 1 - 41e^{-8}$$

$$= .9862. \qquad \square$$

Example 2. Probability of Winning in a State Lottery. Suppose M of the N lottery tickets issued by a state lottery are winning tickets. Then the probability of buying a winning ticket is $p = M/N$. Usually N is very large and p is small. If a person buys n tickets then the number of winning tickets X in the n purchased has, approximately, a Poisson distribution with parameter $\lambda = np = nM/N$ and

$$P(X = k) \doteq e^{-\lambda} \frac{\lambda^k}{k!}, \qquad k = 0, 1, 2, \ldots .$$

Moreover,

$$\text{probability of winning} = P(X \ge 1) = 1 - P(X = 0) \doteq 1 - e^{-\lambda}.$$

If we wish to guarantee that

$$P(X \ge 1) \ge 1 - \alpha, \qquad 0 < \alpha < 1,$$

then

$$1 - e^{-\lambda} \ge 1 - \alpha$$

so that

$$\lambda \ge \ln \frac{1}{\alpha} .$$

That is,

$$\frac{nM}{N} \ge \ln \frac{1}{\alpha}$$

so that the purchaser needs to buy at least $(N/M)\ln(1/\alpha)$ tickets to be $100(1 - \alpha)$ percent confident of having at least one winning ticket. If $M/N = 1/10,000$ and $\alpha = .05$, then $n \ge 29957.3$, so that in a lottery where 1 out of 10,000 tickets is a winning ticket a player needs to buy 29,958 tickets or more in order to be 95 percent confident of owning a winning ticket. $\qquad \square$

Example 3. Automobile Accidents at an Intersection. The number of automobile accidents at a certain intersection in a month has approximately a Poisson distribution with a mean of 4. What is the probability that no accidents occur in a week?

Assuming 4 weeks to a month, we note that the weekly rate of accidents is $\lambda = 1$. If X denotes the number of accidents in a week at that intersection, then:

$$P(X = 0) = e^{-\lambda} = e^{-1} = .3679.$$

Also,

$$P(X > 5) = 1 - e^{-\lambda}\left(1 + \lambda + \frac{\lambda^2}{2} + \frac{\lambda^3}{3!} + \frac{\lambda^4}{4!} + \frac{\lambda^5}{5!}\right)$$

$$= 1 - 2.7167e^{-1} = .0006.$$

Thus the event that more than five accidents occur at this intersection during a week is a rare event. If more than five accidents are actually observed, we will have to suspect that the weekly rate $\lambda \neq 1$ (that is, it is larger).   □

Estimation of λ and tests of hypotheses concerning λ are considered in Chapters 10 and 11.

Problems for Section 6.6

1. The number of cars crossing a bridge during a certain interval has approximately a Poisson distribution with $\lambda = 4$. Find the probability that during a randomly chosen (similar) time interval:
   (i) No cars will cross the bridge.
   (ii) Exactly three cars will cross the bridge.
   (iii) No more than five cars will cross the bridge.

2. The number of common colds that a randomly chosen student at a Midwestern University has in a given academic year is a Poisson random variable with a mean of 2.6. Find the probability that a randomly chosen student has had:
   (i) More than 5 colds during the school year.
   (ii) At most 5 colds during the school year.
   (iii) At least one cold.

3. The number of customers X that enter a store during a one-hour interval has a Poisson distribution. Experience has shown that $P(X = 0) = .111$. Find the probability that during a randomly chosen one-hour interval
   (i) Five or more customers enter the store.
   (ii) Exactly one customer enters the store.
   (iii) No more than five customers enter the store.

4. Suppose a large number N of raisin buns of equal size are baked from a batch of dough that contains n well-mixed raisins. Let X be the number of raisins in a randomly selected bun. Find the probability that a randomly selected bun will contain:
   (i) No raisins.   (ii) At least one raisin.
   (iii) At least five raisins.   (iv) An even number of raisins (that is, $0, 2, 4, \ldots$).

5. A 500-page book has 1000 misprints. Find the probability of
   (i) More than two misprints on a page.   (ii) No misprint on a page.
   (iii) Less than two misprints on a page.

6. A well known television personality has committed suicide in New York. During the week following the suicide, a psychologist observes that there were 15 suicides in New York compared with an average of 10 per week. The psychologist attributes the extra suicides to the suggestive effect of the television personality's suicide. Do the data support the psychologist's explanation?

7. Timothy receives, on the average, five telephone calls a day. Find the probability that he will receive:
   (i) Exactly five calls on a randomly selected day.
   (ii) Fewer than five calls.
   (iii) More than five calls.

8. A life insurance company estimates the probability is .00001 that a person in the 50-to-60 age group dies during a year from a certain rare disease. If the company has 200,000 policyholders in this age group, find the probability that, due to death from this disease, the company must pay off during a year:
   (i) More than two claims.   (ii) More than five.
   (iii) Less than four.

9. A car salesman sells on the average λ cars per day.
   (i) Use Chebyshev's inequality to show that the number of cars X sold by him on a randomly chosen day satisfies

$$P(0 < X < 2\lambda) \geq 1 - \frac{1}{\lambda}.$$

   (ii) Suppose $\lambda = 3$. Compare the bound in (i) with the exact probability from Table A4.
   (iii) If $\lambda = 3$, find the probability that on a given day he will sell:
   (a) At most five cars.   (b) Exactly five cars.

10. In a lengthy manuscript, it is found that only one-third of the pages contain no typing errors. Assuming that the number of errors per page has a Poisson distribution, what percentage of pages have:
   (i) Exactly one error?   (ii) More than one error?

11. An intersection has averaged two accidents per week. Recognizing that the frequency of accidents at this intersection is intolerable, the city administration decided to install a traffic light at this intersection at considerable expense to the taxpayers. To check if the traffic light reduced the number of accidents, a week was randomly selected and showed one accident.
   (i) Would you say that the new lights are effective?
   (ii) What would be your conclusion if there was one accident over a period of two weeks?
   (iii) One accident over a period of four weeks?
   (Assume that the number of accidents at this intersection has, approximately, a Poisson distribution.)

12. Let Y have Poisson distribution with mean $\lambda = 5$. Find a simple upper bound for $P(Y > 8)$. Compare your result with the exact probability from Table A4.

13. Suppose customers arrive at a store at the rate of λ per hour. Then the distribution of $X(t)$, the number of customers arriving in $[0, t]$, has a Poisson distribution with

mean λt. Find the probability that less than r customers will arrive in time interval $[0, x]$.

14. Let X have a Poisson distribution with parameter λ. Show that for $n \geq 1$:
    (i) $P(X = n) = (\lambda/n)P(X = n - 1)$. (ii)* $P(X \geq n) < \lambda^n/n!$.
    In particular, use (ii) to show that $\mathscr{E}X < e^{\lambda}$.
    [Hint: For a positive integer valued random variable X, $\mathscr{E}X = \sum_{n=1}^{\infty} P(X \geq n)$.]

15. Find the most probable value of k for a Poisson random variable. That is, find k for which $P(X = k)$ is maximum for a given λ. [Hint: Consider the ratio $P(X = k)/P(X = k - 1)$.]

16. Find λ to maximize $P(X = k)$ for a Poisson random variable X.

17*. By Chebyshev's inequality,

$$P\left(|X - \lambda| \leq \sqrt{\lambda/\alpha}\right) \geq 1 - \alpha, \qquad 0 < \alpha < 1.$$

(i) Show that

$$P\left(X + \frac{1}{2\alpha} - \sqrt{\frac{X}{\alpha} + \frac{1}{4\alpha^2}} \leq \lambda \leq X + \frac{1}{2\alpha} + \sqrt{\frac{X}{\alpha} + \frac{1}{4\alpha^2}}\right) \geq 1 - \alpha.$$

[Hint: Consider $f(\lambda) = (X - \lambda)^2 - (\lambda/\alpha) \leq 0$. Find the points at which the parabola $f(\lambda)$ intersects the λ-axis (see Figure 2).]

(ii) In particular, let $\alpha = .05$, and suppose $X = 5$ is the observation. Find a 95 percent confidence interval for λ. Also find a 75 percent confidence interval for λ.

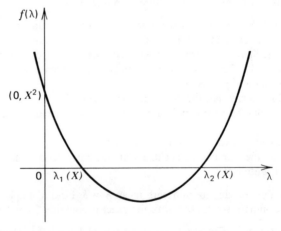

Figure 2. $f(\lambda) = (x - \lambda)^2 - \dfrac{\lambda}{\alpha} \leq 0$ for $\lambda_1(X) \leq \lambda \leq \lambda_2(X)$.

18*.   (Poisson as a limit of hypergeometric.) Show that

$$\frac{\dbinom{D}{k}\dbinom{N-D}{n-k}}{\dbinom{N}{n}} \to e^{-\lambda}\frac{\lambda^k}{k!}$$

as $n, N, D \to \infty$ such that $n/N \to 0$ and $nD/N \to \lambda$.

19*.   Player A observes a Poisson random variable X with parameter λ. If the outcome is x, he asks Player B to toss a coin x times. Let p be the probability of a head (success) on any trial. Let Y be the number of heads that B observes.
   (i)   Find the probability distribution of Y, its mean and variance.
   (ii) Suppose Player B tells you that he has observed k successes. What is the probability that he tossed the coin x times?

20*.   (Poisson as a limit of negative binomial.) Let X have a negative binomial distribution with probability function

$$P(X = n + r) = \binom{r+n-1}{n}p^r(1-p)^n, \qquad n = 0, 1, 2, \ldots .$$

[Substitute $k = n + r$ in (6.5.10).] Show that as $p \to 1$ and $r \to \infty$ in such a way that $r(1-p) \to \lambda > 0$ fixed,

$$P(X = n + r) \to e^{-\lambda}\frac{\lambda^n}{n!}.$$

21*.   By considering the ratio $r_k = \binom{n}{k}p^k(1-p)^{n-k}/[e^{-\lambda}\lambda^k/k!]$ as a function of k, show that r_k reaches its maximum when k is the largest integer not exceeding $n + 1 - \lambda(1-p)/p$.

6.7   MULTIVARIATE HYPERGEOMETRIC AND MULTINOMIAL DISTRIBUTIONS

We have considered some models that are appropriate for experiments where the outcomes can be put into two mutually exclusive categories. We recall that binomial distribution arises in sampling with replacement and that hypergeometric arises in sampling without replacement. In many applications, each trial results in an outcome that can be put into more than two categories of interest. This happens, for instance, when a pollster asks the respondents if they "strongly agree," "agree," "disagree," "strongly disagree," or are "neutral" on a certain issue.

To build a stochastic model for this problem, we consider an urn that contains k types of marbles. Let $n_j \, (> 0)$ be the number of marbles of type $j, j = 1, 2, \ldots, k$, and let $\sum_{j=1}^{k} n_j = N$.

First consider the case when we sample *without replacement* and draw a random sample of size n. Let X_j be the number of type j marbles in the sample.

Then $0 \le X_j \le n$ for $j = 1, 2, \ldots, k$ and $\sum_{j=1}^{k} X_j = n$ so that $X_1, X_2, \ldots, X_k$ are dependent random variables. To find the joint probability function of $X_1, X_2, \ldots, X_k$ we first note that there are $\binom{N}{n}$ samples of size n which are all equally likely. The event $\{ X_1 = x_1, X_2 = x_2, \ldots, X_k = x_k \}$ happens if and only if the sample contains x_1 type 1 marbles, x_2 type 2 marbles, $\ldots$, x_k type k marbles such that $x_1 + x_2 + \cdots + x_k = n$ (see Figure 1). Since there are $\binom{n_j}{x_j}$ ways of choosing the type j marbles, $j = 1, 2, \ldots, k$, it follows (Rule 1, Section 2.5) that

$$(1) \qquad P(X_1 = x_1, X_2 = x_2, \ldots, X_k = x_k) = \frac{\binom{n_1}{x_1}\binom{n_2}{x_2} \cdots \binom{n_k}{x_k}}{\binom{N}{n}},$$

where

$$x_j = 0, 1, \ldots, n_j, \quad \sum_{j=1}^{k} x_j = n, \quad \text{and} \quad \sum_{j=1}^{k} n_j = N.$$

Since $X_k = n - X_1 - \cdots - X_{k-1}$ we can rewrite (1) as

$$(2) \quad P(X_1 = x_1, X_2 = x_2, \ldots, X_{k-1} = x_{k-1}) = \frac{\left[\prod_{j=1}^{k-1} \binom{n_j}{x_j} \right] \binom{n_k}{n - \sum_{j=1}^{k-1} x_j}}{\binom{N}{n}}$$

where $0 \le x_j \le n_j$, $j = 1, 2, \ldots, k - 1$, such that $\sum_{j=1}^{k-1} x_j \le n$. For $k = 2$ we get the hypergeometric distribution (6.4.1) with $D = n_1$. The probability function (1) [or (2)] is known as a *multivariate hypergeometric probability function*. To show that (1) [or (2)] is a probability function we need only equate the coefficients of x^n

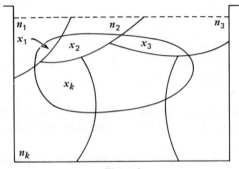

Figure 1

on both sides of the identity:

(3) $$(1 + x)^N = (1 + x)^{n_1}(1 + x)^{n_2} \cdots (1 + x)^{n_k}$$

to get

(4) $$\binom{N}{n} = \sum_{\substack{x_1 \cdots x_k \\ x_1 + \cdots + x_k = n}} \prod_{j=1}^{k} \binom{n_j}{x_j}.$$

Next we consider the case when we sample *with replacement*. Set $p_j = n_j/N$, $j = 1, 2, \ldots, k$ so that $\sum_{j=1}^{k} p_j = 1$. In this case, the trials (draws) are independent each with probability p_j of drawing a type j marble. By independence, the probability of drawing x_1 marbles of type 1, x_2 of type 2,..., x_k of type k, in a *specific order* (say first x_1 of type 1, then x_2 of type 2,..., finally x_k of type k) is $p_1^{x_1} p_2^{x_2} \ldots p_k^{x_k}$. Since there are $(n!/x_1! x_2! \ldots x_k!)$ ways of choosing x_1 marbles of type 1,..., x_k of type k (Rule 4; Section 2.5) it follows that

(5) $$P(X_1 = x_1, X_2 = x_2, \ldots, X_k = x_k) = \frac{n!}{x_1! x_2! \ldots x_k!} \prod_{j=1}^{k} p_j^{x_j},$$

where $0 \le x_j \le n$, $\sum_{j=1}^{k} x_j = n$, and $\sum_{j=1}^{k} p_j = 1$. We may rewrite (5) as

(6) $P(X_1 = x_1, \ldots, X_{k-1} = x_{k-1})$

$$= \frac{n!}{x_1! \ldots x_{k-1}! (n - x_1 - \cdots - x_{k-1})!} \prod_{j=1}^{k-1} p_j^{x_j} \left(1 - \sum_{j=1}^{k-1} p_j\right)^{n - \sum_{j=1}^{k-1} x_j}$$

where $0 \le x_j \le n$, $j = 1, 2, \ldots, k - 1$, and $\sum_{j=1}^{k-1} x_j \le n$. In view of the multinomial expansion

$$1 = (p_1 + p_2 + \cdots + p_k)^n = \sum_{\substack{x_1 \cdots x_k \\ x_1 + \cdots + x_k = n}} \frac{n!}{x_1! \ldots x_k!} \prod_{j=1}^{k} p_j^{x_j},$$

(5) [or (6)] is called a *multinomial probability function*. For $k = 2$ we get the binomial probability function (6.3.4) with $p = p_1$ and $1 - p = p_2$.

Multivariate Hypergeometric Distribution

$$P(X_1 = x_1, X_2 = x_2, \ldots, X_k = x_k) = \frac{\prod_{j=1}^{k} \binom{n_j}{x_j}}{\binom{N}{n}},$$

where $0 \le x_j \le n_j$, $\sum_{j=1}^{k} x_j = n$, and $\sum_{j=1}^{k} n_j = N$.

Multinomial Distribution

$$P(X_1 = x_1, X_2 = x_2, \ldots, X_k = x_k) = \frac{n!}{\displaystyle\prod_{j=1}^{k} x_j!} \prod_{j=1}^{k} p_j^{x_j},$$

where $0 \le x_j \le n$, $\sum_{j=1}^{k} x_j = n$, and $\sum_{j=1}^{k} p_j = 1$.

If N is large, samplings with or without replacement are essentially equivalent and we would expect (5) to approximate (1). Indeed, let $N \to \infty$, $n_1, n_2, \ldots, n_k \to \infty$ such that $n_j/N \to p_j$, $j = 1, 2, \ldots, k$. Then, after a little simplification:

$$\binom{N}{n}^{-1} \prod_{j=1}^{k} \binom{n_j}{x_j}$$

$$= \frac{n!}{x_1! \ldots x_k!} \cdot \frac{n_1(n_1 - 1) \cdots (n_1 - x_1 + 1) \cdots n_k(n_k - 1) \cdots (n_k - x_k + 1)}{N(N - 1) \cdots (N - x_1 + 1) \cdots (N - n + x_k) \cdots (N - n + 1)}$$

$$\to \frac{n!}{x_1! \ldots x_k!} p_1^{x_1} \cdots p_k^{x_k}$$

provided $n/N \to 0$.

Example 1. Randomness in Jury Selection. A panel of 20 prospective jurors contains eight white men, four white women, five nonwhite men, and three nonwhite women. A jury of six is selected at random and is found to contain three white men, one white woman, one nonwhite man and one nonwhite woman. Is there any reason to doubt the randomness of the selection?

Here $N = 20$, $n = 6$, $n_1 = 8$, $n_2 = 4$, $n_3 = 5$, $n_4 = 3$. In an obvious notation:

$$P(X_1 = 3, X_2 = 1, X_3 = 1, X_4 = 1) = \frac{\binom{8}{3}\binom{4}{1}\binom{5}{1}\binom{3}{1}}{\binom{20}{6}}.$$

Thus in random selection, 8.67 percent of samples of size 6 out of 20 jurors with eight white men, four white women, five nonwhite men, and three nonwhite women will contain three white men, one white woman, one nonwhite man, and one nonwhite woman. What we need is the probability $P\{X_1 \ge 3, X_2 \le 1, X_3 \le 1, X_4 \le 1 \,|\, X_1 + X_2 + X_3 + X_4 = 6\}$. We leave the reader to show that this probability is .1803. Thus, there is not enough evidence to doubt the randomness of the selection. □

Example 2. Four Brands of Toothpaste. It is known that the current market shares of toothpaste of four brands, A_1, A_2, A_3, A_4, are respectively 40, 15, 20, and 25 percent. What is the probability that in a random sample of 20 customers of toothpaste six will buy brand A_1, four will buy brand A_2, and five each will buy brands A_3 and A_4?

Assuming that the customers buy independently, the required probability is given by (5) with $n = 20$, $p_1 = .4$, $p_2 = .15$, $p_3 = .20$, $p_4 = .25$, $x_1 = 6$, $x_2 = 4$, $x_3 = x_4 = 5$. The required probability is:

$$\frac{20!}{6!\,4!\,5!\,5!}(.4)^6(.15)^4(.2)^5(.25)^5 = 0.007.$$ □

Suppose that $(X_1, X_2,\ldots,X_k)$ has a multivariate hypergeometric distribution. What is the marginal distribution of X_j for each j? What is the joint distribution of (X_i, X_j) for $i \neq j$? To find the distribution of X_j we need only concentrate on the count of those marbles that are type j (and hence also those that are *not* type j). But this is precisely the hypergeometric model situation, and we have

$$P(X_j = x_j) = \frac{\binom{n_j}{x_j}\binom{N - n_j}{n - x_j}}{\binom{N}{n}}, \qquad 0 \leq x_j \leq n_j$$

for $j = 1, 2,\ldots,k$. Hence each X_j has a hypergeometric distribution with

$$\mathscr{E}X_j = n\frac{n_j}{N} \quad \text{and} \quad \text{var}(X_j) = n\frac{n_j}{N}\left(\frac{N - n_j}{N}\right)\left(\frac{N - n}{N - 1}\right).$$

To find the distribution of (X_i, X_j) we have three classes of marbles, type i, type j, and remaining types. It follows that

$$P(X_i = x_i, X_j = x_j) = \frac{\binom{n_i}{x_i}\binom{n_j}{x_j}\binom{N - n_i - n_j}{n - x_i - x_j}}{\binom{N}{n}}$$

where $0 \leq x_i \leq n_i$, $0 \leq x_j \leq n_j$, $x_i + x_j \leq n$ so that the joint distribution of (X_i, X_j) is multivariate hypergeometric. The same result holds for the joint distribution of any subset of $\{X_1,\ldots,X_k\}$.

We now compute $\text{cov}(X_i, X_j)$, $i \neq j$ for a multivariate hypergeometric distribution.[†] We have:

$$\mathscr{E}(X_i X_j) = \sum_{\substack{x_i=0 \\ x_i+x_j \leq n}}^{n_i} \sum_{x_j=0}^{n_j} x_i x_j \binom{n_i}{x_i}\binom{n_j}{x_j}\binom{N - n_i - n_j}{n - x_i - x_j} \Big/ \binom{N}{n}$$

$$= \sum_{x_i=0}^{n_i} x_i \binom{n_i}{x_i}\left\{ \sum_{x_j=0}^{\min(n_j,\,n-x_i)} x_j \binom{n_j}{x_j}\binom{N - n_i - n_j}{n - x_i - x_j} \Big/ \binom{N}{n}\right\}.$$

[†]See Problem 5.3.19 for an alternative derivation.

Now the quantity in the braces can be rewritten as

$$\sum_{x_j=0}^{\min(n_j,\,n-x_i)}\left\{x_j\binom{n_j}{x_j}\binom{N-n_i-n_j}{n-x_i-x_j}\Bigg/\binom{N-n_i}{n-x_i}\right\}\left\{\binom{N-n_i}{n-x_i}\Bigg/\binom{N}{n}\right\}$$

$$=\left\{n_j\frac{n-x_i}{N-n_i}\right\}\left\{\binom{N-n_i}{n-x_i}\Bigg/\binom{N}{n}\right\}$$

in view of (6.4.6). Hence:

$$\mathscr{E}(X_iX_j)=\left(\frac{n_j}{N-n_i}\right)\sum_{x_i=0}^{n_i}x_i(n-x_i)\binom{n_i}{x_i}\binom{N-n_i}{n-x_i}\Bigg/\binom{N}{n}$$

$$=n_in_j\sum_{x_i=1}^{n_i}\binom{n_i-1}{x_i-1}\binom{N-n_i-1}{n-x_i-1}\Bigg/\binom{N}{n}$$

$$=n_in_j\binom{N-2}{n-2}\Bigg/\binom{N}{n}=\frac{n(n-1)}{N(N-1)}n_in_j$$

in view of (6.4.2). It follows that

$$\mathrm{cov}(X_i,X_j)=\frac{n(n-1)}{N(N-1)}n_in_j-n^2\left(\frac{n_i}{N}\right)\left(\frac{n_j}{N}\right)=-\frac{N-n}{N-1}n\frac{n_i}{N}\frac{n_j}{N}.$$

A similar argument applies when $(X_1,X_2,\ldots,X_k)$ has a multinomial distribution. We leave the reader to show that the marginal distribution of each X_j is binomial with parameters n_j, p_j so that

$$\mathscr{E}X_j=n_jp_j\quad\text{and}\quad\mathrm{var}(X_j)=n_jp_j(1-p_j).$$

Moreover, (X_i,X_j) for $i\neq j$ has a joint multinomial distribution with parameters n, p_i, p_j, $(1-p_i-p_j)$ so that

$$P(X_i=x_i,X_j=x_j)=\frac{n!}{x_i!x_j!(n-x_i-x_j)!}p_i^{x_i}p_j^{x_j}(1-p_i-p_j)^{n-x_i-x_j},$$

where $0\leq x_i\leq n$, $0\leq x_j\leq n$, $x_i+x_j\leq n$, and p_i, $p_j>0$, $0<p_i+p_j<1$. Indeed, every subset of $\{X_1,X_2,\ldots,X_k\}$ has a joint marginal multinomial distribution (Problem 9). Also, for $i\neq j$,

$$\mathscr{E}(X_iX_j)=\sum_{\substack{x_i=0\\x_i+x_j\leq n}}^{n}\sum_{x_j=0}^{n}x_ix_j\frac{n!}{x_i!x_j!(n-x_i-x_j)!}p_i^{x_i}p_j^{x_j}(1-p_i-p_j)^{n-x_i-x_j}$$

$$=\sum_{x_i=0}^{n}x_i\binom{n}{x_i}p_i^{x_i}\sum_{x_j=0}^{n-x_i}x_j\binom{n-x_i}{x_j}p_j^{x_j}(1-p_i-p_j)^{n-x_i-x_j}$$

$$=\sum_{x_i=0}^{n}x_i\binom{n}{x_i}p_i^{x_i}\left\{(n-x_i)p_j(1-p_i)^{n-x_i-1}\right\}$$

in view of (6.3.8). Hence

$$\mathscr{E}(X_i X_j) = n(n-1)p_i p_j \sum_{x_i=1}^{n-1} \binom{n-2}{x_i-1} p_i^{x_i-1}(1-p_i)^{n-x_i-1}$$

$$= n(n-1)p_i p_j \quad \text{(binomial expansion)}$$

and it follows[†] that

$$\text{cov}(X_i, X_j) = n(n-1)p_i p_j - (np_i)(np_j) = -np_i p_j.$$

Moments of Multivariate Hypergeometric and Multinomial Distributions

	Multivariate Hypergeometric	Multinomial
$\mathscr{E}X_j$	nn_j/N	np_j
$\text{var}(X_j)$	$nn_j(N-n_j)(N-n)/[N^2(N-1)]$	$np_j(1-p_j)$
$\text{cov}(X_i, X_j),\ i \neq j$	$-nn_i n_j(N-n)/[N^2(N-1)]$	$-np_i p_j$

Estimation of $(p_1, p_2, \ldots, p_{k-1})$ for a multinomial distribution is easy and is considered in Problem 11. Multinomial distribution has applications whenever the data is categorical. In particular, this is the case whenever the data are grouped in a (finite) number of class intervals. A most common application of multinomial distribution is in the analysis of contingency tables. Problems of this type are considered in Section 9.9 and Chapter 12.

Problems for Section 6.7

1. In a game of bridge, what is the probability that each of the four bridge players holds an ace?

2. A committee of college faculty members consists of three full, five associate, and two assistant professors. A subcommittee of six is randomly selected. What is the probability that the subcommittee is composed of three full, two associate, and one assistant professor?

3. Consider a class consisting of 16 students of which six are seniors, seven juniors, and three sophomores. A random selection of four was made. Find the probability that
   (i) All four are seniors.   (ii) None is a senior.
   (iii) The sample contains at least one each of seniors, juniors, and sophomores.

4. A psychiatric test is given to n children. The test consists of asking the children to pick a number from 1, 2, 3, 4, 5, 6. Assuming that the choices were random, find the

---

[†]See Problem 5.3.17 for an alternative derivation.

joint distribution of (X_1, X_2) where X_1 is the number of children who picked 4 and X_2 the number of children who picked 6. In particular, what is the probability that in a class of 25 children who were given this test, ten picked 4 and nine picked 6?

5. A target or dartboard is divided into three concentric rings A_1, A_2, and A_3. If the probabilities of a single shot hitting these rings are respectively $1/8$, $4/8$, and $3/8$, and five shots are fired in all, what is the probability that three hit A_1 and one hits A_2?

6. Eight identical cans containing fruit have lost their labels. If three contain peaches, four contain apricots, and one contains pineapple, what is the probability that a random sample of 4 will contain:
   (i) The pineapple can.   (ii) All three peach cans.
   (iii) Two peach cans and one apricot can.

7. About 50 percent of the students at a large state university live in dormitories, 20 percent live in fraternities and sororities, 20 percent live off campus (in town), and 10 percent are commuters. A committee of eight is randomly selected to represent the students on the Board of Trustees. If the selections are independent, what is the probability that all four groups are represented? (Do not evaluate.) What is the probability that the committee consists of four dormitory residents, two fraternity and sorority students, and one each of commuter and off-campus students?

8. Grass seed bags sold in lawn-and-garden stores are mixtures of several types of grass. A particular mixture contains 60 percent Kentucky blue grass, 25 percent red fescue, 10 percent annual rye grass, and 5 percent inert seeds. Let X_1, X_2, X_3, and X_4 denote the number of seeds of each type, respectively, in a random sample of nine seeds drawn from a 10-pound bag.
   (i) Find $P(X_1 = 4, X_2 = 2, X_3 = 2)$.   (ii) Find $P(X_4 = 1)$.

9. Suppose $(X_1, X_2, \ldots, X_k)$ has a multinomial distribution. Show that every subset of $\{X_1, X_2, \ldots, X_k\}$ has a joint marginal multinomial distribution.

10. Suppose $n \to \infty$ and $p_j \to 0$ such that $np_j \to \lambda_j$. Show that

$$\frac{n!}{\prod\limits_{j=1}^{k} x_j!} \prod_{j=1}^{k} p_j^{x_j} \to \exp\left\{-\sum_{j=1}^{k} \lambda_j\right\} \prod_{j=1}^{k} \left(\frac{\lambda_j^{x_j}}{x_j!}\right).$$

11. Estimate $p_1, p_2, \ldots, p_{k-1}$ in a multinomial distribution such that $P(X_1 = x_1, \ldots, X_{k-1} = x_{k-1})$ is most probable.

12. A loaded die is rolled n times. Suppose $P(\text{ace}) = p_1$, $P(\text{two}) = p_2$, $0 < p_1 < 1$, $0 < p_2 < 1$, and $p_1 + p_2 < 1$. Find the probability distribution of the number of times a face value of 2 shows up given that k of the trials produced aces. Find the mean of this distribution.

13. Let (X_1, X_2) have a multivariate hypergeometric distribution with parameters n, N, n_1, n_2 with $n_1 + n_2 \le N$. Show that the conditional probability function of X_1 given $X_2 = x_2$ is hypergeometric. Find the mean and variance of the conditional distribution.

14*. In the World Chess Title Series two chess players A_1 and A_2 play against one another until A_1 or A_2 wins six games and the title. Some games end in a draw. Let $p_i = P(A_i \text{ wins})$, $i = 1, 2$, and $P(\text{draw}) = 1 - p_1 - p_2$. Assume $0 < p_1, p_2 < 1$,

$1 - p_1 - p_2 > 0$ and that the games are independent. Find the probability that:
  (i) A_1 wins the title on the kth game of the series.
  (ii) A_1 wins the title.
  (iii) A_1 or A_2 will eventually win the title.
  (iv) A_1 wins his or her first game on the kth game played.

15. Heritable characteristics depend on *genes*, which are passed on from parents to offsprings. Genes appear in pairs and, in the simplest case, each gene of a particular pair can assume one of two forms called alleles, A and a. Therefore, there are three possible *genotypes*: AA, Aa, and aa. (Biologically there is no difference between aA and Aa.) Assume random mating so that one gene is selected at random from each parent and the selections are independent. Moreover, the selection of genotypes for different offspring are also independent.
  (i) Suppose an individual of genotype Aa mates with another individual of genotype Aa. Find the probability that the offspring is of genotype:
      (a) aa,   (b) Aa,   (c) AA.
  (ii) If an individual of genotype AA mates with an individual of genotype Aa, find the probability that the offspring is of genotype:
      (a) aa,   (b) Aa,   (c) AA.
  (iii) If two parents of genotype Aa mate and have six offspring, what is the probability that:
      (a) Exactly three offspring will be genotype Aa?
      (b) Exactly two offspring will be of each genotype?
  (iv) Answer (iii) when the genotypes of parents are:
           (a) Aa and aa.   (b) AA and aa.   (c) Aa and AA.

16. Refer to Problem 15. If two parents of genotype Aa mate and have 10 offspring what is the probability that:
  (i) Two are of type Aa and two of type aa?
  (ii) Four are of type AA and four of type aa?
  (iii) At least two are of type aa and at least two of type AA?

6.8 REVIEW PROBLEMS

1. Ten balls are randomly distributed among five cells. Let X be the number of balls that land in cell 1. What is the probability function of X? What is the mean number of balls in cell 1? Find $P(X \geq 6)$. Find the smallest x such that $P(X \geq x) \leq .15$.

2. A small telephone exchange has a capacity of 15 calls. The switchboard gets jammed about 6 percent of the time. If the number of calls X coming in at any time is a Poisson random variable, what is its distribution? Find $P(X \leq \mathscr{E}X)$.

3. At a particular location on a river the number of fish caught per hour of fishing is approximately a Poisson random variable with mean $\lambda = 1.5$. If a man fishes there for one hour, what is the probability that he will catch:
  (i) At least two fish?   (ii) Exactly two fish?
  (iii) No more than three fish?

4. An urn contains 5 red and 20 white marbles. Four marbles are selected at random without replacement. Find the probability that:
  (i) The first two marbles drawn will be white.
  (ii) The second marble drawn is red.
  (iii) At least one marble drawn is red.

5. A lot of 30 light bulbs contains two defectives. A random sample of four light bulbs is examined.
   (i) What is the probability of at least one defective bulb in the sample?
   (ii) How large should the sample size be in order that the probability of finding at least one defective is greater than $\frac{1}{2}$?

6. Let Y_1, Y_2 respectively be the number of trials necessary to draw the first and the second defective from a large lot containing a proportion of p defectives. Find $P(Y_1 > t_1, Y_2 > t_2)$ where t_1, t_2 are fixed positive integers. Assume that Y_1, Y_2 are independent random variables.

7. Suppose X has a Poisson distribution such that

$$P(X = k) = P(X = k + 1)$$

   where $k \geq 0$ is an integer.
   (i) What is the distribution of X?   (ii) If $k = 3$, find $P(X \geq 2)$.

8. The green light at an intersection is on for 20 seconds, the yellow for 10 seconds, and the red for 20 seconds. Suppose the arrival times at the intersection are random due to traffic conditions so that making a green light is a chance event.
   (i) What is the probability of making a green light?
   (ii) What is the probability that of 20 cars arriving at the intersection, more than half will make the green light?

9. Three toothpaste-manufacturing competitors A, B, and C control respectively 40 percent, 20 percent, and 15 percent of the market. In a random sample of five independent buyers of toothpaste, what is the joint distribution of the numbers who buy the three toothpastes?
   Suppose in a random sample of 20 independent buyers, two buy toothpaste A, eight buy toothpaste B, and eight buy toothpaste C. How likely is this event?

10. Suppose Y has a binomial distribution with parameters n and p_1 and X given $Y = y$ has a binomial distribution with parameters y and p_2.
    (i) What is the unconditional distribution of X?   (ii) Find $\text{cov}(X, Y)$.

11. Given Y has a Poisson distribution with mean λ and the conditional distribution of X given $Y = y$ is binomial (y, p):
    (i) What is the unconditional distribution of X?   (ii) Find $\text{cov}(X, Y)$.

12. An ordered sample of size 2 is taken from N elements numbered $1, 2, \ldots, N$. Let X be the number on the first element drawn, and let Y be that on the second.
    (i) What is the joint distribution of (X, Y)? What are the marginal probability functions of X and Y?
    (ii) Find the conditional probability function of X given $Y = k$, $1 \leq k \leq N$.
    (iii) Find $P\{X \geq N - 1 | Y = k\}$

13. Let A and B be mutually exclusive events. Suppose the experiment is repeated until either A or B is observed. Show that the conditional probability that A happens before B does is given by $P(A)/[P(A) + P(B)]$.

14. It is estimated that the risk of going into coma with surgical anesthesia is 6 in 100,000. In the movie *Coma*, two patients out of 10 go into coma during surgery. How likely is this event?

15. The condition of low blood sugar is known as hypoglycemia. In order to compare two hypoglycemic compounds, A and B, each is experimentally applied to half of the diaphragm of each of 15 rats. Blood glucose uptake in milligrams per gram of tissue is then measured for each half. The following results were recorded:

Compound	Rat							
	1	2	3	4	5	6	7	8
A	4.7	5.2	8.5	9.4	3.9	6.3	6.8	3.3
B	5.1	5.4	8.7	8.4	3.6	6.2	6.8	4.6

Compound	Rat						
	9	10	11	12	13	14	15
A	10.2	4.7	7.0	6.3	7.0	9.4	3.6
B	10.0	4.1	7.0	6.3	6.1	9.4	3.6

The experimenter is interested in determining whether there is any difference between the two compounds. Is there evidence? You may choose α between .10 and .15.

16. Benzedrine and caffeine are widely believed to counteract the intoxicating effects of alcohol. In an experiment, 12 subjects were used to test the relative merits of the two drugs. It was found that benzedrine was more effective than caffeine in 10 cases. Is there enough evidence to support the hypothesis that the two drugs are equally effective?

17. The median time in minutes for a tune-up of a four cylinder car is stated to be 45 minutes. A random sample of 15 mechanics showed tune-up times in minutes as:

    42 45 32 47 55 36 45 37 38 40 34 38 49 36 38

What conclusion do you draw from the data?

18. A random sample of five boys and seven girls (aged 15–21) are asked their opinions on legalization of marijuana. Let p_1, p_2 be the proportions of boys and girls in the population who favor legalization of marijuana. It is believed that girls are more favorably disposed to legalization than boys. State the null and alternative hypotheses to test this belief. What is a reasonable critical region? Suppose two boys and four girls in the sample are in favor of the proposition. What is the P-value? State your conclusion.

CHAPTER 7

Some Continuous Models

7.1 INTRODUCTION

This chapter is devoted to some examples of continuous models. As remarked earlier, it is sufficient to concentrate on the characteristic properties of each model. The reader would do well to become familiar with the manipulative techniques involving each density function.

Section 7.2 deals with the uniform distribution, which is the continuous analogue of the discrete uniform model. In Section 7.3 we summarize the basic properties of gamma and beta functions. Gamma and beta integrals arise often in probability and statistics. The reader should at least familiarize himself or herself with the definitions. It will simplify many computations.

Section 7.4 considers a wide variety of models that are often used to characterize experiments that involve lifetimes of components, systems, or biometrical populations. Section 7.5 considers beta model which may be used as an alternative to the uniform model on $(0, 1)$ of Section 7.2.

In classical statistics normal model occupies the central position in that it is the most widely used model. This is partly due to the fact that the sample mean of a random sample from any population with finite variance has approximately a normal distribution provided the sample size is large enough. In Section 7.6 we study the normal model and its properties. It is important to be able to work with Table A5, which tabulates cumulative probabilities under the normal curve. In Section 7.7 we consider the bivariate normal model.

7.2 UNIFORM DISTRIBUTION

We recall that the discrete uniform distribution on n points assigns the same probability to each of the n points. In this section we consider a distribution that extends this idea to the continuous case. In that case there are uncountably many points in the sample space, each of which must have zero probability. The events of interest, however, are subsets of the outcome space that have a well defined measure (length in one dimension, area in the plane, volume in three dimensions, and so on). A natural extension of the discrete case is to assign, to any subset of the outcome space, a probability proportional to the size (measure) of the set irrespective of its shape or position.

DEFINITION 1.   Let X be any random variable (or random vector) that takes values in S with probability one. That is, $P(X \in S) = 1$. We say that X has uniform distribution over S if its density function is given by

(1)
$$f(x) = \begin{cases} \dfrac{1}{\text{measure}(S)}, & x \in S, \\ 0, & x \notin S \end{cases}$$

where measure(S) is length, area or volume depending on the dimensionality of S and $0 < \text{measure}(S) < \infty$.

According to this definition, for any subset A of S that has a well-defined measure:

$$P(A) = \int \cdots \int_A \frac{1}{\text{measure}(S)} \, dx = \frac{\text{measure}(A)}{\text{measure}(S)} .$$

We shall be mostly interested in one or two dimensions.

Uniform Distribution on (a, b), $-\infty < a < b < \infty$

A random variable X has uniform distribution on (a, b) if its density function is given by

$$f(x) = \begin{cases} \dfrac{1}{(b - a)} & \text{if } a < x < b \\ 0 & \text{otherwise.} \end{cases}$$

One may include either or both endpoints a and b.

Remark 1.   We can now give yet another definition of "random." *A point is randomly chosen from S simply means that an observation is taken on a random variable X with density given by* (1).

The uniform distribution is frequently used to model rounding-off errors in tabulating values to the nearest n decimal places. It is assumed that the error has uniform distribution on the interval $[-10^{-n}/2, 10^{-n}/2]$. Similarly, the error in rounding off figures expressed in binary scale is assumed to have uniform distribution on $[-2^{-n}, 2^{-n}]$ or $[0, 2^{-n}]$, depending on how the rounding-off is done.

Example 1.   Rounding-Off Error.   Sales tax is rounded off to the nearest cent. Assuming that the actual error—the true tax minus the rounded-off value—is random, that is, unsystematic, we can either use the discrete model

$$P(X = j) = \frac{1}{11}$$

for $j = -.005, -.004, \ldots, 0, .001, \ldots, .005$, or the simpler continuous model that the error X has density function

$$f(x) = \begin{cases} 100 & \text{if } -.005 \le x \le .005 \\ 0 & \text{otherwise.} \end{cases} \qquad \square$$

Example 2. *Uniform Distribution in Plane.* Suppose raindrops fall at random on a plane region S with area $A(S)$. Let (X, Y) denote the coordinates of the point in S where a particular raindrop strikes. Then (X, Y) has uniform distribution on S if the joint density is given by:

$$f(x, y) = \begin{cases} 1/A(S), & (x, y) \in S \\ 0, & (x, y) \notin S. \end{cases}$$

In particular, if S is the unit square bounded by $0 \le x \le 1, 0 \le y \le 1$, then $A(S) = 1$. If S is the triangle bounded by $0 \le y \le x \le 1$ (or $0 \le x \le y \le 1$), then $A(S) = \frac{1}{2}$, and so on (see Figure 1.). Since the probability that $(X, Y) \in B \subseteq S$ is simply the volume of the solid with base B and height $1/A(S)$, we need only compute the area of B either by integration or by using plane geometry. This observation greatly simplifies computation of probabilities under uniform distribution. $\qquad \square$

Let us now consider the uniform distribution on (a, b) with density function

(2)
$$f(x) = \begin{cases} 1/(b-a), & a < x < b \\ 0 & \text{otherwise.} \end{cases}$$

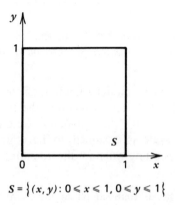

$S = \{(x, y) : 0 \le x \le 1, 0 \le y \le 1\}$

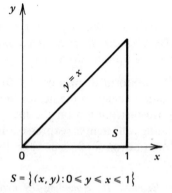

$S = \{(x, y) : 0 \le y \le x \le 1\}$

Figure 1

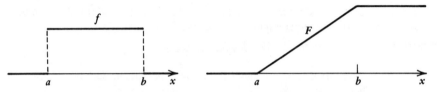

Figure 2.   Graphs of f and F given by (2) and (3).

Note that the distribution function corresponding to (2) is given by (see Figure 2):

$$(3) \qquad F(x) = \begin{cases} 0, & x \le a \\ (x - a)/(b - a), & a < x < b \\ 1, & b \le x. \end{cases}$$

Moreover,

$$\mathscr{E}X = \int_a^b x \frac{1}{b - a}\,dx = \frac{a + b}{2},$$

$$\mathscr{E}X^k = \int_a^b \frac{x^k}{b - a}\,dx = \frac{b^{k+1} - a^{k+1}}{(b - a)(k + 1)}.$$

In particular,

$$\mathrm{var}(X) = \frac{b^2 + a^2 + ab}{3} - \left(\frac{a + b}{2}\right)^2 = \frac{(b - a)^2}{12}$$

Mean and Variance of Uniform Distribution on (a, b)

$$\mathscr{E}X = \frac{a + b}{2}, \ \mathrm{var}(X) = \frac{(b - a)^2}{12}.$$

We note that

$$(4) \qquad\qquad P(c \le X \le d) = (d - c)/(b - a)$$

where $a \le c < d \le b$ so that the probability depends only on the length of (c, d) and not c and d. This property actually characterizes the uniform distribution.

    Example 3*.   A Characterizing Property of Uniform Distribution on (a, b).   Let us rewrite (4) as follows:

$$(5) \qquad\qquad F(x + t) - F(x) = F(t), \qquad a < x < x + t < b$$

and suppose that

$$(6) \qquad\qquad F(a) = 0 \quad \text{and} \quad F(b) = 1.$$

If F has a continuous derivative f on (a, b) and satisfies (5) and (6), then F must be of the form (3). In fact, differentiating (5) we respect to x, we get

$$f(x + t) - f(x) = 0.$$

Since t is arbitrary it follows that f is constant on (a, b). Suppose $f(x) = c$ for $x \in (a, b)$. Then

$$1 = \int_a^b f(x)\, dx = \int_a^b c\, dx = c(b - a)$$

and $c = \dfrac{1}{b - a}$.   □

Problems of geometric probability can frequently be solved with ease by using uniform distribution.

Example 4. *Choosing Two Points on (0, 1).* Let X and Y be two points chosen (independently) at random from the interval $(0, 1)$. Find the probability that the three line-segments so formed form a triangle.

Since X and Y are independent, each with uniform distribution on $(0, 1)$, their joint density is given by

$$f(x, y) = \begin{cases} 1 & \text{if } 0 < x < 1, 0 < y < 1 \\ 0 & \text{otherwise.} \end{cases}$$

Now a necessary and sufficient condition for any three line-segments to form a triangle is that the sum of lengths of any two of them is larger than the length of the third. Hence we must have

$$\text{either} \quad 0 < X < 1/2 < Y < 1 \quad \text{and} \quad Y - X < 1/2$$
$$\text{or} \quad 0 < Y < 1/2 < X < 1 \quad \text{and} \quad X - Y < 1/2.$$

The required probability can be obtained by integrating f over the two triangles. Since f is uniform on the unit square it is much easier to compute the volume of the solid of unit height bounded by the two shaded triangles in Figure 3. It is sufficient to compute the area of the shaded triangles, and the required probability is $\frac{1}{4}$.   □

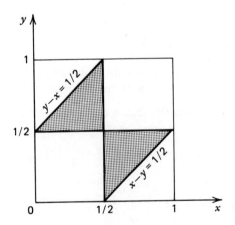

Figure 3

Solution of any problem in geometric probability, however, depends on the definition of randomness being used as illustrated in the following example.

Example 5.   Bertrand's Paradox.   A chord is drawn at random in the unit circle. What is the probability that the chord is longer than the side of the equilateral triangle inscribed in the circle?

Let A be the event of interest. Then $P(A)$ depends on the definition of randomness used.

Solution 1.   The length of the chord is uniquely determined by the position of its midpoint. Choose a point C at random in the circle and draw a line through C passing through the center of the circle O. Now draw the chord through C perpendicular to OC (see Figure 4a). Let l_1 be the length of the chord through C. Then $l_1 > \sqrt{3}$ if and only if the point C lies inside the circle with center O and radius $\frac{1}{2}$. It follows that $P(A) = \pi(1/2)^2/\pi = \frac{1}{4}$. Here Ω is the circle with center O and a radius of one, so that

$$f(x, y) = \frac{1}{\pi}, \qquad x^2 + y^2 \leq 1 \text{ and zero elsewhere}$$

where C is the point (x, y).

Solution 2.   In view of the symmetry of the circle, we choose and fix a point P on the circumference of the unit circle and take it to be one end point of the chord. The other end point P_1 of the chord is now chosen at random. Thus $\Omega = [0, 2\pi)$ and

$$f(x) = \frac{1}{2\pi}, \qquad 0 \leq x < 2\pi, \text{ and zero elsewhere,}$$

is the density of the length of the arc PP_1. The inscribed equilateral triangle with P as one of its vertices divides the circumference of the unit circle into three equal parts. A chord through P will be longer than the side of this triangle if and only if P_1 lies in that one-third part of the circumference which is opposite to P (see Figure 4b). It follows that $P(A) = 1/3$.

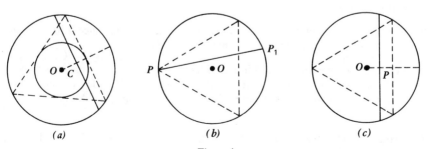

(a)   (b)   (c)

Figure 4

Solution 3. The length of the chord is uniquely determined by the distance of the midpoint of the chord from the center of the unit circle. In view of the symmetry of the circle, assume that the midpoint P of the chord (see Figure 4c) lies on a fixed radius of the circle. Choose P at random so that the probability that P lies in any given segment equals the length of that segment. Thus $\Omega = [0, 1]$ and

$$f(x) = 1, \qquad 0 \leq x \leq 1, \text{ and zero otherwise}$$

where x denotes the distance of P from center O. The length of the chord will be longer than the length of the side of the inscribed equilateral triangle, if and only if $x < 1/2$. It follows that $P(A) = 1/2$.   □

Uniform distribution has important applications in nonparametric statistics and in simulation where it is used to generate random samples from a continuous distribution function F. These are taken up in Section 8.2 and later sections.

Problems for Section 7.2

1. The radius X of a ball bearing has uniform distribution over the interval $(0, 1.5)$. Find:
   (i) $P(X > .5), P(X < .4), P(.3 < X < 1.2)$.   (ii) $P\{X > 1.2 | X > .5\}$.

2. The opening price of a particular stock on the New York Stock Exchange has uniform distribution on $[35\ 1/8,\ 45\ 1/8]$.
   (i) Find the probability that the opening price is less than \$40.
   (ii) What is the probability that the opening price is less than \$44 given that it opened at a price greater than \$40?

3. Consider a computer storage system where the information is retrieved by entering the system at a random point and systematically checking all the items in the system. Suppose it takes a seconds to begin the search and b seconds to run through the entire system. Let X be the time it takes to search a particular piece of information. Suppose $b = ka$, $k > 1$ fixed and $P(X > 1) = lP(X < 1)$ for some $l > 0$. Find a.

4. A point X is chosen at random from the unit interval $(0, 1)$. What is the probability that the area of the rectangle formed by the two line-segments (from X to the two ends) is $> 1/4$? $< 1/8$?

5. Given that X has uniform distribution with mean c and variance d, $0 < d$, find the density of X. Find the probability that $X > c + 1/2$.

6. Let X be the time (in minutes) it takes Jones to drive from his suburban home to his office. Suppose X has uniform distribution on $[25, 40]$. If he leaves his house every day at 7:25 a.m. and has to be in by 8:00 a.m., what is the probability that he will be late for work?

7. Express buses leave for downtown every 10 minutes between 7:00 and 8:00 a.m., starting at 7:00 a.m. A person who does not know the times of departure arrives at

the bus stop at X minutes past 7:00 a.m. where X has uniform distribution over
(i) 7:00 to 7:30 a.m.   (ii) 7:00 to 8:00 a.m.   (iii) 7:15 to 7:30 a.m.
Find in each case the probability that the traveler will have to wait for
(a) More than five minutes.   (b) Less than two minutes.

8. Let X have uniform distribution on $[-\theta, \theta]$, $\theta > 0$.
(i) Find θ when $P(X > 1) = 1/4$ and when $P(|X| > 1) = 1/6$.
(ii) Find $P\{X \in (-\theta/2, \theta/2)|X > \theta/4\}$.

9. Let X be uniformly distributed over the interval $(1 - 1/\sqrt{3}, 1 + 1/\sqrt{3})$. Find:
(i) $\mathscr{E}X$ and var(X).   (ii) $P(|X - \mathscr{E}X| \geq \frac{3}{2}\sqrt{\text{var }X})$ exactly.
(iii) A bound for $P(|X - \mathscr{E}X| \geq 3\sqrt{\text{var }X}/2)$ using Chebyshev's inequality.

10. A point X is picked at random from the interval $(0, \theta)$. Suppose θ equals $1, 2, \ldots,$ or N, each with probability $1/N$. Find $P\{\theta = k|X < j\}$ where $k = 1, 2, \ldots, N$ and $j \in (0, N]$ is fixed.

11. A point X is chosen at random from $(0, 1/2)$ and a point Y is chosen (independently) at random from $(1/2, 1)$. Find the probability that the three line-segments form a triangle.

12. Two points are chosen independently at random on a line of unit length. Find the probability that each of the three line-segments so formed will have length larger than $1/4$.

13. Two numbers are chosen independently at random from the unit interval. Find the probability that their sum is less than 1 and the product is less than $3/16$.

14. Suppose (X, Y) have uniform distribution over a circle of radius 2 and center $(0, 0)$.
(i) Find the joint density of (X, Y) and the two marginal densities.
(ii) Find the conditional density of X given $Y = y$.
(iii) Find $P(X^2 + Y^2 \leq 1)$.

15. A point (X, Y) is chosen at random from the triangle bounded by $x = 0$, $x = y$, and $y = 1$.
(i) Find the joint distribution of (X, Y) and the two marginal densities.
(ii) Find the conditional density of X given $Y = y$.

16. A point X is chosen at random from the interval $(0, 1)$ and then a point Y is chosen randomly from the interval $(X, 1)$.
(i) Find the joint density of (X, Y).   (ii) Find the marginal density of Y.
(iii) Find the conditional density of X given $Y = y$.
(See Problem 3.5.9.)

17. Two points are randomly selected from the interval $(0, a)$ so as to be on the opposite sides of the midpoint $a/2$. Find the probability that the distance between the two points is greater than $a/3$.

18. Let X, Y, Z be independent random variables each of which has a common uniform distribution on $(0, 1)$.
(i) Find $P\{0 \leq X \leq Y \leq Z \leq 1\}$.   (ii) Find $P\{0 \leq 3Z \leq 2Y \leq X \leq 1\}$.

19. Let (X, Y, Z) have uniform distribution over $S = \{(x, y, z): 0 \le x \le y \le z \le 1\}$. Find
    (i)   The joint density of (X, Y, Z).
    (ii) The marginal densities of X, Y and Z.
    (iii) The joint marginal densities of (X, Y), (Y, Z) and (X, Z).

20. Let X have uniform distribution on $(0, 4)$. Find the probability that the roots of the quadratic equation $x^2 + 4xX + X + 1 = 0$ are real.

21. Let X, Y, Z be independent random variables having common uniform distribution on $(0, 1)$. Find the probability that $\alpha^2 X + \alpha Y + Z = 0$ has real roots.

7.3*   GAMMA AND BETA FUNCTIONS

In many problems, the integrals

(1) $$\int_{0^+}^{\infty} x^{\alpha-1} e^{-x} \, dx$$

and

(2) $$\int_{0^+}^{1^-} x^{\alpha-1} (1-x)^{\beta-1} \, dx$$

will be encountered. Let us first consider the integral in (1). The following result holds.

PROPOSITION 1. *The integral $\int_{0^+}^{\infty} x^{\alpha-1} e^{-x} \, dx$ converges or diverges according as $\alpha > 0$ or ≤ 0. [For $\alpha > 0$ the integral in (1) is called the gamma function and is denoted by $\Gamma(\alpha)$.]*

Proof. We note that for $\alpha > 0$, $x^{\alpha+1} e^{-x} \to 0$ as $x \to \infty$. Hence for every $\varepsilon > 0$ there exists an x_0 such

$$x^{\alpha+1} e^{-x} < \varepsilon \qquad \text{for } x > x_0.$$

Let $a > x_0$. Then for $\alpha > 0$

$$\int_{0^+}^{a} x^{\alpha-1} e^{-x} \, dx \le \int_{0^+}^{x_0} x^{\alpha-1} \, dx + \int_{x_0}^{a} \frac{\varepsilon}{x^2} \, dx$$

$$= \frac{x_0^{\alpha}}{\alpha} + \varepsilon \left(\frac{1}{x_0} - \frac{1}{a} \right) \le \frac{x_0^{\alpha}}{\alpha} + \frac{\varepsilon}{x_0}.$$

Letting $a \to \infty$ we have:

$$\lim_{a \to \infty} \int_{0^+}^{a} x^{\alpha-1} e^{-x} \, dx \le \frac{x_0^{\alpha}}{\alpha} + \frac{\varepsilon}{x_0} < \infty.$$

If, on the other hand, $\alpha \le 0$, then for $x < 1$, $x^{\alpha} \ge 1$ so that $x^{\alpha-1}e^{-x} \ge e^{-x}/x \ge (xe)^{-1}$. It follows that for $0 < a < 1$:

$$\int_a^{\infty} x^{\alpha-1}e^{-x}\,dx \ge \int_a^1 x^{\alpha-1}e^{-x}\,dx \ge \int_a^1 (xe)^{-1}\,dx$$

$$= e^{-1}\ln\frac{1}{a}.$$

Letting $a \to 0$ we see that $\int_0^{\infty} x^{\alpha-1}e^{-x}\,dx$ cannot be finite for $\alpha \le 0$. $\square$

A greater flexibility in choice of underlying distributions is achieved by introducing yet another parameter. We write $x = y/\beta$, $\beta > 0$ in

$$\Gamma(\alpha) = \int_{0^+}^{\infty} x^{\alpha-1}e^{-x}\,dx$$

to get

(3) $$\beta^{\alpha}\Gamma(\alpha) = \int_{0^+}^{\infty} y^{\alpha-1}e^{-y/\beta}\,dy.$$

Some simple properties of the gamma function are given below.

PROPOSITION 2. *We have*:

(i) $\Gamma(\alpha + 1) = \alpha\Gamma(\alpha)$ for $\alpha > 0$,
(ii) $\Gamma(\alpha + 1) = \alpha!$ if $\alpha \ge 0$ is an integer, and
(iii) $\Gamma(1/2) = \sqrt{\pi}$.

Proof. On integration by parts we have:

$$\Gamma(\alpha + 1) = -x^{\alpha}e^{-x}\big|_0^{\infty} + \alpha\int_0^{\infty} x^{\alpha-1}e^{-x}\,dx$$

$$= \alpha\int_0^{\infty} x^{\alpha-1}e^{-x}\,dx = \alpha\Gamma(\alpha).$$

In particular, if α is integral then

$$\Gamma(\alpha + 1) = \alpha\Gamma(\alpha) = \alpha(\alpha - 1)\Gamma(\alpha - 1) = \cdots = \alpha(\alpha - 1)\ldots\cdot 2 \cdot 1 = \alpha!.$$

To prove (iii) we change the variable by writing $x = y^2/2$ in

$$\Gamma\left(\frac{1}{2}\right) = \int_0^{\infty} x^{-1/2}e^{-x}\,dx$$

to obtain

$$\Gamma\left(\frac{1}{2}\right) = \sqrt{2}\int_0^{\infty} e^{-y^2/2}\,dy = \frac{\sqrt{2}}{2}\int_{-\infty}^{\infty} e^{-y^2/2}\,dy$$

and it is sufficient to show that $I = \int_{-\infty}^{\infty} e^{-y^2/2} \, dy = \sqrt{2\pi}$. Since

$$0 < e^{-y^2/2} < e^{-|y|+1} \quad \text{for} \quad -\infty < y < \infty$$

the integral is finite. Moreover,

$$I^2 = \int_{-\infty}^{\infty} \int_{-\infty}^{\infty} \exp\{-(x^2 + y^2)/2\} \, dx \, dy.$$

Changing to polar coordinates by setting $x = r\cos\theta$ and $y = r\sin\theta$, we have:

$$I^2 = \int_0^{2\pi} \left(\int_0^{\infty} r \exp\left(-\frac{r^2}{2}\right) dr \right) d\theta$$

$$= \int_0^{2\pi} \left(\int_0^{\infty} \exp(-s) \, ds \right) d\theta \qquad \left(s = \frac{r^2}{2} \right)$$

$$= 2\pi.$$

Consequently, $I = \sqrt{2\pi}$ and (iii) follows. We have also shown, in particular, that

$$f(x) = \frac{1}{\sqrt{2\pi}} \exp(-x^2/2), \qquad x \in (-\infty, \infty)$$

is a density function.   □

Let us now consider the integral (2).

PROPOSITION 3. *The integral $\int_{0^+}^{1^-} x^{\alpha-1}(1-x)^{\beta-1} \, dx$ converges for $\alpha > 0, \beta > 0$ and is called the beta function. For $\alpha > 0, \beta > 0$ we write*

(4) $$B(\alpha, \beta) = \int_{0^+}^{1^-} x^{\alpha-1}(1-x)^{\beta-1} \, dx.$$

(*For $\alpha \leq 0$ or $\beta \leq 0$ the integral in (2) diverges.*)

Proof. Let us write

(5) $$B(\alpha, \beta) = \int_{0^+}^{1/2} x^{\alpha-1}(1-x)^{\beta-1} \, dx + \int_{1/2}^{1^-} x^{\alpha-1}(1-x)^{\beta-1} \, dx$$

and note that

$$\int_{1/2}^{1^-} x^{\alpha-1}(1-x)^{\beta-1} \, dx = \int_{0^+}^{1/2} y^{\alpha-1}(1-y)^{\beta-1} \, dy \qquad (y = 1 - x).$$

It is sufficient, therefore, to consider the first integral on the right side of (5). It clearly diverges for $\alpha \leq 0$, converges for $0 < \alpha < 1$, and for $\alpha \geq 1$ it is a proper integral. □

Some properties of the beta function are given below.

PROPOSITION 4. *For $\alpha > 0$, $\beta > 0$ we have*:

(i) $B(\alpha, \beta) = B(\beta, \alpha)$,

(ii) $B(\alpha, \beta) = \displaystyle\int_{0^+}^{\infty} x^{\alpha-1}(1 + x)^{-\alpha-\beta}\, dx$,

(iii) $B(\alpha, \beta) = \dfrac{\Gamma(\alpha)\Gamma(\beta)}{\Gamma(\alpha + \beta)}$.

Proof. The proof of (i) is obtained by the change of variable $y = 1 - x$ in (4) and that of (ii) by the change of variable $y = x(1 + x)^{-1}$ in (4). For the proof of (iii) we let $x = y^2$ in

$$\Gamma(\alpha) = \int_0^{\infty} x^{\alpha-1}e^{-x}\, dx$$

to obtain:

$$\Gamma(\alpha) = 2\int_0^{\infty} y^{2\alpha-1}e^{-y^2}\, dy.$$

Then

$$\Gamma(\alpha)\Gamma(\beta) = 4\int_0^{\infty}\int_0^{\infty} x^{2\beta-1}y^{2\alpha-1}e^{-(x^2+y^2)}\, dx\, dy.$$

Changing to polar coordinates with $x = r\cos\theta$ and $y = r\sin\theta$, we obtain:

$$\Gamma(\alpha)\Gamma(\beta) = 4\int_{\theta=0}^{\pi/2}\int_{r=0}^{\infty} r^{2(\alpha+\beta)-1}e^{-r^2}(\sin\theta)^{2\alpha-1}(\cos\theta)^{2\beta-1}\, dr\, d\theta$$

$$= 4\left(\int_0^{\infty} r^{2(\alpha+\beta)-1}e^{-r^2}\, dr\right)\left(\int_0^{\pi/2}(\sin\theta)^{2\alpha-1}(\cos\theta)^{2\beta-1}\, d\theta\right)$$

$$= 2\Gamma(\alpha + \beta)\int_0^{\pi/2}(\sin\theta)^{2\alpha-1}(\cos\theta)^{2\beta-1}\, d\theta$$

$$= \Gamma(\alpha + \beta)B(\alpha, \beta) \qquad \text{(Problem 7)}.$$

A simpler probabilistic proof of (iii) is given later (see Example 8.4.3). ☐

For convenience, we write:

$$\Gamma(\alpha) = \int_0^{\infty} x^{\alpha-1}e^{-x}\, dx \quad \text{and} \quad B(\alpha, \beta) = \int_0^1 x^{\alpha-1}(1 - x)^{\beta-1}\, dx.$$

It will always be clear from the context what the limits of integration are. We conclude this section with the following application of gamma functions.

PROPOSITION 5. (STIRLING'S APPROXIMATION). *For sufficiently large n*

(6) $$n! \doteq \sqrt{2\pi n}\, n^n e^{-n}.$$

Proof. We will sacrifice rigor by proceeding informally. We have

(7) $$\Gamma(n+1) = \int_0^\infty x^n e^{-x}\,dx = \int_0^\infty \exp\{n \ln x - x\}\,dx.$$

The function $n \ln x - x$ has a relative maximum at $x = n$. We therefore substitute $x = n + y$ on the right side of (7) to obtain

$$\Gamma(n+1) = e^{-n} \int_{-n}^\infty \exp\{n \ln(n+y) - y\}\,dy$$

$$= e^{-n} \int_{-n}^\infty \exp\left\{n \ln n + n \ln\left(1 + \frac{y}{n}\right) - y\right\}\,dy$$

$$= n^n e^{-n} \int_{-n}^\infty \exp\left\{n \ln\left(1 + \frac{y}{n}\right) - y\right\}\,dy.$$

Let us now use Taylor expansion for $\ln(1 + y/n)$ to obtain

$$\Gamma(n+1) = n^n e^{-n} \int_{-n}^\infty \exp\left\{n\left(\frac{y}{n} - \frac{y^2}{2n^2} - \cdots\right) - y\right\}\,dy$$

$$= n^n e^{-n} \int_{-n}^\infty \exp\left\{-\frac{y^2}{2n} + \frac{y^3}{3n^2} - \cdots\right\}\,dy$$

$$= n^n e^{-n} \sqrt{n} \int_{-\sqrt{n}}^\infty \exp\left\{-\frac{u^2}{2} + \frac{u^3}{3\sqrt{n}} - \cdots\right\}\,du \qquad (y = \sqrt{n}\,u).$$

Hence for large n, we get:

$$\Gamma(n+1) \doteq n^n e^{-n} \sqrt{n} \int_{-\infty}^\infty \exp\left(-\frac{u^2}{2}\right)\,du = \sqrt{2\pi n}\,n^n e^{-n}$$

in view of the fact that

$$\int_{-\infty}^\infty \exp(-u^2/2)\,du = \sqrt{2\pi}$$

[see proof of Proposition 2(iii)].   □

Problems for Section 7.3

1. Compute $\Gamma(3/2)$, $\Gamma(9/2)$, $\Gamma(4)$, $\Gamma(10)$.

2. Compute $\displaystyle\int_0^\infty \exp(-x\sqrt{x})\,dx$.

3. Compute $\displaystyle\int_0^\infty x^9 e^{-3x}\,dx$.

4. Compute $\displaystyle\int_0^\infty a^{-4x^2}\,dx$ for $a > 0$.

5. Compute

   (i) $\int_0^1 x^2(1-x)^3\,dx.$ (ii) $\int_0^1 x^3(1-x)^2\,dx.$ (iii) $\int_0^\infty x^m e^{-ax^n}\,dx,$

   $a, m, n > 0.$

6. Compute $\int_0^\infty \dfrac{x^4}{(1+x)^6}\,dx.$

7. Show that $B(\alpha, \beta) = 2\int_{0^+}^{\pi/2^-} (\sin x)^{2\alpha-1}(\cos x)^{2\beta-1}\,dx.$ [Hint: Substitute $x = \sin^2 y$ in (4).]

8. Show that $B(\alpha, \alpha) = 2^{1-2\alpha}B(\alpha, \tfrac{1}{2})$ for $\alpha > 0.$

9. Compute

   (i) $\int_0^a x^4(a^2-x^2)^{1/2}\,dx.$ (ii) $\int_0^3 x^3(3-x)^{-1/2}\,dx.$

10. Let n be an odd positive integer. Show that

$$\Gamma\left(\frac{n}{2}\right) = \frac{\sqrt{\pi}\,(n-1)!}{2^{(n-1)}\left(\dfrac{n-1}{2}\right)!}.$$

11. Show that for $\alpha > 0$:

   (i) $\Gamma(\alpha) = \int_0^1 \left(\ln\dfrac{1}{x}\right)^{\alpha-1}\,dx.$ (ii) $\Gamma\left(1 + \dfrac{1}{\alpha}\right) = \int_0^\infty e^{-x^\alpha}\,dx.$

   Use (ii) to derive $\int_0^\infty e^{-x^2}\,dx = \dfrac{1}{2}\sqrt{\pi}.$

   (iii) $\Gamma(\alpha + 1) = 2^{-\alpha}\int_0^\infty x^{2\alpha+1}e^{-x^2/2}\,dx\ (\alpha > -1).$

   Use (iii) to show that for nonnegative integral n and $\sigma > 0$

$$\int_0^\infty x^n \exp\left\{-\frac{x^2}{2\sigma^2}\right\}\,dx = 2^{(n-1)/2}\sigma^{n+1}\Gamma\left(\frac{n+1}{2}\right).$$

   (This last result essentially gives the moments of normal distribution with mean 0 and variance σ^2. See Section 7.6.)

12. Use Proposition 4 parts (ii) and (iii) for $\alpha = \beta = \tfrac{1}{2}$ to prove Proposition 2 (iii).

7.4 EXPONENTIAL, GAMMA, AND WEIBULL DISTRIBUTIONS

We recall that the distribution of the number of trials needed for the first success in a sequence of Bernoulli trials is geometric. A continuous analogue of this result is very useful in modeling life length (of organisms or pieces of equipment). Consider a sequence of events that occur randomly in time according to the Poisson distribution at rate $\lambda > 0$. In Section 6.6 we derived the distribution of the number of events $X(t)$ in the interval $[0, t]$. Suppose we now ask for the distribution of the *waiting time* for the first (Poissonian) event. Let T denote this random variable. Then

$$P(T > t) = P(\text{no event in } [0, t]) = P(X(t) = 0) = e^{-\lambda t},$$

and the density of T is given by

$$(1) \qquad f(t) = -\frac{d}{dt} P(T > t) = \begin{cases} \lambda e^{-\lambda t} & \text{for } t \geq 0 \\ 0 & \text{for } t < 0. \end{cases}$$

The density function defined in (1) is called an *exponential density function*. The distribution function is given by

$$F(t) = \begin{cases} 0, & \text{if } t < 0 \\ 1 - e^{-\lambda t}, & \text{if } t \geq 0. \end{cases}$$

We note that f is strictly decreasing with unique mode at $t = 0$ where f takes the value λ. The median of f is at $t = (\ln 2)/\lambda$ and t-axis is in asymptote of f. The distribution function F rises steadily from zero at $t = 0$ to one at $t = \infty$ $[F(t) = 1$ is an asymptote]. Figure 1 gives sketches of f and F for $\lambda = 1$.

The moments of T are easily computed. Indeed, for $\alpha > 0$:

$$\mathscr{E}T^\alpha = \int_0^\infty t^\alpha \lambda e^{-\lambda t}\, dt = \frac{\Gamma(\alpha + 1)}{\lambda^\alpha} \qquad [(7.3.3)].$$

Hence

$$(2) \qquad \mathscr{E}T = \mu = \frac{1}{\lambda},\ \mathscr{E}T^2 = \frac{2}{\lambda^2},\ \sigma^2 = \text{var}(T) = \frac{1}{\lambda^2}.$$

It will be convenient to write λ for $1/\lambda$.

Exponential Density Function

$$f(t) = \lambda^{-1} e^{-t/\lambda} \qquad \text{for } t \geq 0 \text{ and } f(t) = 0 \text{ if } t < 0.$$

$$\mu = \mathscr{E}T = \lambda \text{ and } \sigma^2 = \text{var}(T) = \lambda^2.$$

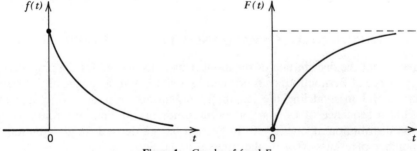

Figure 1. Graphs of f and F.

Example 1. Memoryless Property of Exponential Distributions.   An interesting property is characteristic of the exponential distribution. Let $s, t \geq 0$. Then

$$(3) \quad P\{T > t + s | T > s\} = \frac{P(T > t + s)}{P(T > s)} = \frac{e^{-(t+s)}}{e^{-s}} = e^{-t} = P(T > t).$$

If we think of T as lifetime of a piece of equipment, then (3) states that if the equipment has been working for time s then the probability that it will survive an additional time t depends only on t and equals the probability of survival for time t of a new piece of equipment. Thus the equipment does not "remember" that it has been in use for time s. In this sense there has been no deterioration. We recall that property (3) holds also for geometric random variables and just as (3) characterizes geometric distribution it also characterizes the exponential distribution in the sense that if, for a *continuous nonnegative* random variable T

$$(3') \qquad\qquad P\{T > t + s | T > s\} = P\{T > t\}$$

holds for all $s, t \geq 0$ then T must have an exponential distribution.

The lack of memory property simplifies many calculations and is precisely the reason for wide applicability of the exponential model. It should be understood that under this model an item that has not failed is as good as new, so that the operating time of the item may be identified with operating age. This is certainly not the case for any other (nonnegative continuous type) random variable. The conditional probability

$$P\{T > t + s | T > s\} = \frac{1 - P(T \leq t + s)}{1 - P(T \leq s)} = \frac{1 - F(t + s)}{1 - F(s)}$$

depends on s (except when T is exponentially distributed) so that T here is the total length of life rather than the time in service (as in the exponential case).   □

Example 2. Waiting Time at a Restaurant.   Suppose that the amount of time a customer spends at a fast food restaurant has an exponential distribution with a mean of six minutes. Then the probability that a randomly selected customer will spend more than 12 minutes in the restaurant is given by

$$P(T > 12) = e^{-12/\lambda} = e^{-12/6} = e^{-2} = .1353$$

whereas the (conditional) probability that the customer will spend more than 12 minutes in the restaurant given that she has been there for more than six minutes is

$$P\{T > 12 | T > 6\} = P(T > 6) = e^{-1} = .3679.$$ □

Example 3. Life Length of a Water Heater.   Suppose that the life length of a water heater produced by a certain company has an exponential distribution with mean 12 years. Mrs. Tennenbaum buys an old house with this brand of water heater. She would like to sell the house in four years. What is the probability that she will not have to replace the water heater before she sells?

In view of the lack of memory property, the remaining lifetime of the water heater has an exponential distribution with mean 12 years. Indeed, if T is the life length of the water heater then

$$P\{T > t + 4 \mid T > t\} = P(\text{water heater survives 4 more years})$$

$$= P\{T - t > 4 \mid T - t > 0\}$$

$$= P(\text{remaining lifetime} > 4)$$

$$= e^{-4/12} = e^{-1/3} = .7166.$$

This computation would not be possible for any other lifetime distribution F since in that case we need to know time t, the prior use of the water heater. $\qquad\square$

Weibull and Gamma Distributions*

Frequently in life testing the representation by an exponential model is not adequate. An electric component subjected to continuous vibrations or a steel beam under continuous stress will deteriorate over time so that an exponential model to describe the distribution of length of life in such cases will not be adequate. We now consider some modifications of the exponential model that allow greater flexibility. The following examples will help motivate our discussion.

> ***Example 4. Waiting Time for the rth Customer.***   In the beginning of Section 7.4 we showed that the distribution of the waiting time of the first event in a sequence of events occurring randomly in time according to the Poisson law is exponential. Frequently we are interested in the distribution of the waiting time for the rth occurrence.
>
> Suppose customers arrive at a service desk (automobiles at a toll booth or calls into an exchange) according to the Poisson law at mean rate λ per unit time. Let T_r be the time of occurrence of the rth event and let

$$F_r(t) = P(T_r \le t), \qquad \text{for } t \ge 0.$$

Then

$$P(T_r > t) = 1 - F_r(t) = P(r\text{th event occurs after time } t)$$

$$= P(\text{number of events in } (0, t] \text{ is less than } r)$$

$$= P(X(t) < r)$$

where $X(t) =$ number of events in time $[0, t]$. Since $X(t)$ has Poisson distribution with $\mathscr{E} X(t) = t/\lambda$ (Why?), we have

$$P(T_r > t) = P(X(t) \le r - 1) = \sum_{k=0}^{r-1} e^{-t/\lambda} \frac{(t/\lambda)^k}{k!}$$

so that for $t \ge 0$

$$f_r(t) = -\frac{d}{dt} P(T_r > t) = \frac{t^{r-1} e^{-t/\lambda}}{\lambda^r (r - 1)!}$$

and

$$f_r(t) = 0 \qquad \text{for } t < 0$$

gives the density function of T_r. We note that $f_r(t)$ is a gamma function with $\alpha = r$ and $\beta = \lambda$ except that r is an integer, $r \geq 1$, and $r = 1$ gives the exponential distribution (with mean λ). □

Example 5. Survival Analysis.   To model the length of life T of a living organism or a system or a component, it is a more convenient to consider the *survival function*:

$$(4) \qquad S(t) = P(T > t) = 1 - P(T \leq t) = 1 - F(t)$$

where F is the distribution function of T with density function f. Yet another function plays a key role in survival analysis. This is the *hazard (failure rate) function* $\lambda(t)$, defined by

$$(5) \qquad \lambda(t) = \frac{f(t)}{S(t)}$$

for $S(t) > 0$. The function $\lambda(t)$ may be interpreted as the conditional probability density that a t-year old component will fail. Indeed, for Δt small

$$P\{T \in [t, t + \Delta t] | T > t\} = \frac{P\{T \in (t, t + \Delta t]\}}{P(T > t)} \doteq \frac{f(t)\,\Delta t}{S(t)} = \lambda(t)\,\Delta t.$$

We note that $\lambda(t)$ uniquely determines f. For,

$$\lambda(t) = \frac{f(t)}{S(t)} = \frac{-S'(t)}{S(t)} \qquad (S'(t) = -f(t))$$

so that if $F(0) = 0$, then

$$\int_0^t \lambda(s)\,ds = -\int_0^t \frac{S'(p)}{S(p)}\,dp = -\ln S(t).$$

Hence

$$(6) \qquad S(t) = \exp\left\{ -\int_0^t \lambda(s)\,ds \right\}.$$

Suppose for fixed $\lambda > 0$,

$$\lambda(t) = 1/\lambda \qquad \text{for all } t > 0.$$

Then

$$S(t) = \exp\{-t/\lambda\}$$

and

$$f(t) = -S'(t) = \begin{cases} (1/\lambda)e^{-t/\lambda} & \text{for } t > 0 \\ 0 & \text{elsewhere.} \end{cases}$$

That is, an exponential density function is the only distribution with constant failure rate. In some situations $\lambda(t) =$ constant is not a reasonable assumption. A modification is obtained if we assume instead that the failure rate associated with T is of the form

$$(7) \qquad \lambda(t) = \left(\frac{\alpha}{\beta}\right)t^{\alpha-1}, \qquad \alpha > 0, \beta > 0.$$

Then (6) yields the density

$$(8) \qquad f(t) = -S'(t) = \lambda(t)\exp\left\{-\int_0^t \lambda(s)\,ds\right\}$$

$$= \begin{cases} (\alpha/\beta)t^{\alpha-1}e^{-t^{\alpha}/\beta} & \text{for } t > 0 \\ 0 & \text{elsewhere.} \end{cases}$$

A random variable with density function (8) is said to have a *Weibull distribution*. We note that $\alpha = 1$, $\beta = \lambda$ in (8) yields the exponential law with mean λ. A special case of (8) with $\alpha = 2$ is the *Rayleigh density function*

$$(9) \qquad f(t) = \begin{cases} \{2te^{-t^2/\beta}\}/\beta, & t > 0 \\ 0 & \text{elsewhere} \end{cases}$$

which has linear hazard rate $\lambda(t) = 2t/\beta$. Actually, Weibull law is closely related to the exponential law. Indeed

$$\int_t^\infty \frac{\alpha}{\beta} s^{\alpha-1} e^{-s^{\alpha}/\beta}\,ds = \int_{t^\alpha}^\infty \frac{1}{\beta} e^{-y/\beta}\,dy \qquad (y = s^\alpha)$$

so that if T_1 is Weibull with density (8) and T_2 is exponential with mean β, then

$$P(T_1 > t) = P(T_2 > t^\alpha). \qquad \qquad \square$$

The distributions encountered in Examples 4 and 5 and many more are simply related to (or are special cases of) the *gamma probability density function* with parameters $\alpha > 0$, $\beta > 0$, given by

$$(10) \qquad f(x) = \begin{cases} [x^{\alpha-1}\exp(-x/\beta)]/(\Gamma(\alpha)\beta^\alpha), & x > 0 \\ 0, & x \le 0. \end{cases}$$

In view of (7.3.3), we note that (10) does indeed define a density function.

The gamma density function takes on a wide variety of shapes depending on the values of α and β. For $\alpha < 1$, f is strictly decreasing and $f(x) \to 0$ as $x \to \infty$, $f(x) \to \infty$ as $x \to 0$. For $\alpha > 1$, the density f has a unique mode at $x = (\alpha - 1)\beta$ with maximum value $[(\alpha - 1)e^{-1}]^{\alpha-1}/(\beta\Gamma(\alpha))$. Figure 2 gives graphs of some typical gamma probability density functions.

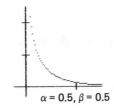

$\alpha = 0.5, \beta = 0.5$

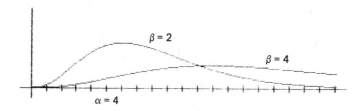

$\beta = 0.5$

$\beta = 1$

$\beta = 2$

$\alpha = 2$

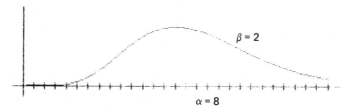

$\beta = 2$

$\beta = 4$

$\alpha = 4$

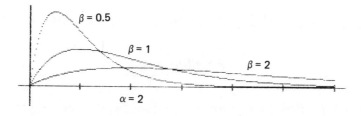

$\beta = 2$

$\alpha = 8$

Figure 2. Gamma density functions.

We now compute the moments of (10). Indeed, for any $s > 0$

$$\mathscr{E}X^s = \int_0^\infty x^s \frac{x^{\alpha-1}e^{-x/\beta}}{\Gamma(\alpha)\beta^\alpha}\,dx = \frac{1}{\Gamma(\alpha)\beta^\alpha}\int_0^\infty x^{s+\alpha-1}e^{-x/\beta}\,dx$$

$$= \frac{\Gamma(s+\alpha)\beta^{s+\alpha}}{\Gamma(\alpha)\beta^\alpha} = \beta^s\Gamma(s+\alpha)/\Gamma(\alpha)$$

$$= \beta^s(\alpha+s-1)(\alpha+s-2)\ldots\alpha.$$

In particular

(11) $\mathcal{E}X = \alpha\beta, \mathcal{E}X^2 = \alpha(\alpha + 1)\beta^2, \text{var}(X) = \alpha\beta^2.$

Mean and Variance of Gamma Density

$$f(x) = \begin{cases} \dfrac{x^{\alpha-1}e^{-x/\beta}}{\Gamma(\alpha)\beta^\alpha}, & x > 0, \alpha, \beta > 0 \\ 0, & x \le 0 \end{cases}$$

$$\mu = \alpha\beta, \sigma^2 = \alpha\beta^2$$

Example 6. Peak Signal Strength. The peak signal X (in volts) arriving at an antenna has a Weibull (or Rayleigh) distribution with parameters $\alpha = 2$ and β so that the density function of X is given by

$$f(x) = \begin{cases} \dfrac{2x}{\beta}e^{-x^2/\beta}, & x > 0 \\ 0, & x \le 0. \end{cases}$$

Then for $t > 0$

$$S(t) = P(X > t) = 1 - \int_0^t \frac{2x}{\beta}e^{-x^2/\beta}\, dx = e^{-t^2/\beta}$$

so that for $s, t > 0$

$$P\{X > s + t | X > t\} = \frac{P(X > s + t)}{P(X > t)}$$

$$= \exp\left\{-\frac{(s+t)^2 - t^2}{\beta}\right\} = \exp\left\{-\frac{s(s + 2t)}{\beta}\right\}$$

which depends on t as well as s. In particular, if s, t are known and $P\{X > s + t | X > t\} = \gamma, 0 < \gamma < 1$, is also known, then β is given by

$$\frac{s(s + 2t)}{\beta} = \ln\left(\frac{1}{\gamma}\right)$$

so that

$$\beta = \frac{s(s + 2t)}{\ln(1/\gamma)}.\qquad\qquad\square$$

Some important special cases of gamma distribution are widely used and have been given special names. Letting $\alpha = 1$ in (10) gives an exponential density with mean β. Letting $\alpha = n/2$ and $\beta = 2\sigma^2$ in (10) yields a $\sigma^2\chi^2$ (σ^2 *chi-square*) *density with n degrees of freedom* (d.f.) given by

$$(12) \quad f(x) = \frac{1}{2^{n/2}\sigma^n\Gamma\left(\dfrac{n}{2}\right)} x^{n/2-1}e^{-x/2\sigma^2} \text{ for } x > 0, \text{ and zero elsewhere.}$$

Here the parameter n takes positive integral values and parameter $\sigma > 0$. If we take $\sigma = 1$ in (12) we get the well known *chi-square* (χ^2) *density*

$$(13) \quad f(x) = \frac{1}{2^{n/2}\Gamma\left(\dfrac{n}{2}\right)} x^{n/2-1}e^{-x/2} \text{ for } x > 0, \text{ and zero otherwise}$$

with n d.f. Since f in (12) is a density function, we have

$$1 = \frac{1}{2^{n/2}\sigma^n\Gamma(n/2)} \int_0^\infty x^{n/2-1}e^{-x/2\sigma^2}\,dx$$

$$= \frac{1}{2^{n/2}\sigma^n\Gamma(n/2)} \int_0^\infty \left(y^2\sqrt{n}\right)^{n/2-1}e^{-ny^2/2\sigma^2}\left(2y\sqrt{n}\right)\,dy \qquad (x = y^2\sqrt{n})$$

$$= \frac{2(n/2)^{n/2}}{\sigma^n\Gamma(n/2)} \int_0^\infty y^{n-1}e^{-ny^2/2\sigma^2}\,dy$$

and it follows that the function

$$(14) \qquad f(x) = \begin{cases} \dfrac{2(n/2)^{n/2}}{\sigma^n\Gamma(n/2)} x^{n-1}e^{-nx^2/2\sigma^2}, & x > 0 \\ 0, & x \le 0 \end{cases}$$

also defines a density function. It is called a $\sigma\chi$ *density* with parameters $n = 1, 2, \ldots$ and $\sigma > 0$. The special case $\sigma = 1$ is known as a χ *distribution with n degree of freedom*. Some special cases of (14) are widely in use.

Rayleigh distribution with density function

$$f(x) = \begin{cases} (x/\alpha^2)e^{-x^2/2\alpha^2}, & x > 0 \\ 0, & x \le 0 \end{cases}$$

is the special case $n = 2$ and $\sigma = \alpha\sqrt{2}$. (This is the density (9) with $\beta = 2\alpha^2$.)

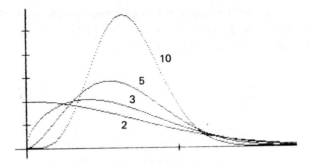

Figure 3. χ density for $n = 2, 3, 5, 10$.

Maxwell density function

$$f(x) = \begin{cases} (4/\alpha^3\sqrt{\pi})x^2 e^{-x^2/\alpha^2}, & x > 0 \\ 0, & x \le 0 \end{cases}$$

is the special case $n = 3$ and $\sigma = \alpha\sqrt{3/2}$.

Weibull distribution (8) is also obtainable from (12) by substituting $n = 2$, $\sigma^2 = \beta$, and the change of variable $x = 2t^\alpha$. Figures 3 and 4 give graphs of chi and Weibull densities.

Example 7. Weekly Consumer Demand for Bread. Suppose the weekly consumer demand X for bread in a certain town has a gamma distribution with parameters α and β. What is the probability that the demand for bread in a certain week in this town will be greater than t? What we need to do is to compute the survival function $P(X > t)$ of X. We have:

$$P(X > t) = \frac{1}{\Gamma(\alpha)\beta^\alpha} \int_t^\infty x^{\alpha-1} e^{-x/\beta} \, dx = 1 - \int_0^t \frac{x^{\alpha-1} e^{-x/\beta}}{\Gamma(\alpha)\beta^\alpha} \, dx.$$

Given t, α, β one can look up the probability in a table of incomplete gamma function.

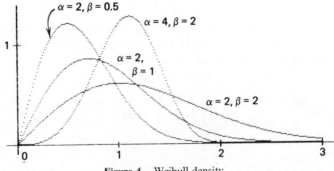

Figure 4. Weibull density.

There is yet another way to look at this probability. In Example 4 we showed that for a gamma random variable with parameters $\alpha = r$ and $\beta > 0$ where $r \geq 1$ is an integer

(15)
$$P(X > t) = \sum_{k=0}^{r-1} \frac{e^{-t/\beta}(t/\beta)^k}{k!}.$$

Hence in the special case when α is a positive integer, we can find $P(X > t)$ from a table of cumulative Poisson distribution function (such as Table A4) with parameter $\lambda = t/\beta$.

In particular, if $\alpha = 12$, $\beta = 3$ and $t = 48$ then

$$P(X > 48) = \sum_{k=0}^{11} e^{-16}(16)^k/k! = .167. \qquad \square$$

Problems for Section 7.4

Problems 11 to 25 refer to the Subsection on gamma distribution.

1. The times between successive arrivals of customers in a store are governed by the density function

$$f(x) = \begin{cases} \lambda e^{-\lambda x}, & x \geq 0 \\ 0 & \text{elsewhere.} \end{cases}$$

(i) Find the probability that a customer arrives after time t.
(ii) Find the probability that a customer arrives after time t_1 but not after time t_2, $0 < t_1 < t_2$.

2. On the Boston–Chicago route, the number of minutes a plane is late is an exponential random variable with a mean of 10 minutes. Find the probability that the flight will be late by more than 10 minutes on a randomly selected day. Show that your result is independent of the mean of the exponential distribution. That is, show that $P(X > \mathscr{E}X)$ is independent of $\mathscr{E}X = \lambda$.

3. Let X measure the time to failure (in hours) of a car battery of a certain brand. Suppose X has exponential distribution with mean λ. If n such batteries are installed and operate independently, what is the probability that one-half or more of them are still functioning at the end of t hours? In particular, if $\mathscr{E}X = 200$ and $t = 100$ hours, find the probability that 16 or more of the batteries are still functioning out of the 25 batteries installed.

4. (i) If X has exponential distribution with mean, λ, find c such that $P(X > c) = \alpha$, $0 < \alpha < 1$.
   (ii) Find c_1 and c_2 such that $P(c_1 < X < c_2) = \beta$ and $P(X < c_1) = \gamma$; $\beta, \gamma > 0$, $0 < \beta + \gamma < 1$.
   (iii) Find $P(X < -\lambda \ln(1 - p))$ for $0 < p < 1$.

5. Three processes are used to produce fuses for electrical circuits. Process 1 yields a mean life length of 100 hours, process 2 has an expected life length of 150 hours, and Process 3 yields an expected life of 200 hours. Suppose Process 1 costs $\$c$ per unit,

Process 2 costs $(3/2)$ c per unit, and Process 3 costs $2c$ per unit. Moreover, if a fuse lasts for less than 200 hours there is an assessment of d against the manufacturer. Find the expected cost of producing a fuse in each case. Assume that the life length has an exponential distribution.

6. An electronic device has a life length X (in 1000 hour units). Suppose X has an exponential density function with mean 1. Suppose further that the cost of manufacturing an item is $2.50 and the proceeds from the sale of one unit is $3. If the item is guaranteed a refund, if it does not survive 100 hours, what is the manufacturer's expected profit per item?

7. Let X have an exponential distribution with parameter $\lambda > 0$. Find $P(k < X \le k + 1)$, $k \ge 0$ integral, and show that it is of the form $p(1 - p)^k$. Find p.

8. Suppose that a car battery has life length X (hours) which has an exponential distribution with mean λ. A machine using this battery costs k_1 per hour to run, and while the machine is functioning a profit of k_2 per hour is realized. Suppose that this machine needs an operator to run it and the operator is paid T per hour. If the operator is hired for t hours, find:
   (i) The expected profit.
   (ii) The minimum number of hours t for which the expected profit is maximized.

9. The compressor of a new heat pump is guaranteed for 5 years. The mean life is estimated to be 6 years and the time to failure can be assumed to be exponential. If each failure of the compressor costs $50 and the realized profit on each heat pump is $700, what is the expected profit per heat pump? (Hint: The number of failures in $[0, t]$ has a Poisson distribution with $\lambda = t/6$.)

10. Suppose X has an exponential distribution with a mean of $\lambda > 0$. Compare the upper bound on the probability $P(|X - \mu| \ge 2\sigma)$ obtained from Chebyshev's inequality with the exact probability. [Here $\mu = \mathscr{E}X = \lambda$ and $\sigma^2 = \text{var}(X) = \lambda^2$.]

11. The demand X for a new product in a certain community can be assumed to be a random variable with Rayleigh density function

$$f(x) = \begin{cases} \dfrac{x}{25} e^{-x^2/50}, & x > 0 \\ 0 & \text{elsewhere.} \end{cases}$$

Find $P(X \le 10)$, $P(5 \le X \le 15)$, and $P(X > 5)$.

12. The percentage X of impurity in a high-grade alloy can be modeled by the Maxwell distribution with density function

$$f(x) = \begin{cases} \dfrac{4}{\sqrt{\pi}} x^2 e^{-x^2}, & x > 0 \\ 0 & \text{elsewhere.} \end{cases}$$

Find $P(X > 5)$ and $P(X < 25)$. (No closed form.)

13. The actual time required to repair an article at a large appliance repair shop can be approximately modeled by a gamma distribution with parameters $\alpha = 4$ hours and

$\beta = 1$ hour. What is the probability that the time it takes to repair a defective appliance will be
(i) More than one hour?   (ii) Between 2 and 4 hours?

14. Suppose the length of life X of a battery (in months) has a gamma distribution with parameters $\alpha = 3$ anc. $\beta = 15$. Find the probability that:
(i) $P(X > 45)$.   (ii) $P(30 < X < 60)$.   (iii) $P(X < 20)$.

15. Customers arrive at a supermarket according to Poisson distribution with intensity $\lambda = 1$ per minute. Find the probability that the first three customers do not arrive in the first 10 minutes. What is the probability that the shopkeeper will have to wait longer than 10 minutes for the first 10 customers?

16. The duration, in days, of an epidemic can be approximated by a Weibull distribution with parameters $\alpha = 2$ and $\beta = 81$. Find the probability that:
(i) An epidemic will last no more than 7 days.
(ii) An epidemic will last at least 20 days.
Find also the expected length of duration of an epidemic.

17. Let X be an observation from the density

$$f(x) = \frac{x^{\alpha-1}e^{-x/\beta}}{\Gamma(\alpha)\beta^{\alpha}}, \qquad x > 0, \text{ and } 0 \text{ otherwise.}$$

Let Y be a randomly chosen point from the interval $(0, X)$.
(i) Find the unconditional density of Y.
(ii) Find the conditional density of X given $Y = y$.
(iii) Find $P(X + Y \le 2)$.

18. Customers arrive at a supermarket according to Poisson distribution with intensity λ. However, λ itself varies and has a gamma distribution with parameters α and β.
(i) Find the unconditional distribution of the number of customers $X(t)$ who arrive in time t.
(ii) Find the conditional distribution of λ given $X(t) = x$.

19. The lead time for orders of a certain component from a manufacturer can be assumed to have a gamma distribution with mean μ and variance σ^2. Find the probability that the lead time for an order is more than time t.

20. For each of the following density functions find the survival function $S(t) = P(T > t)$ and the hazard function $\lambda(t) = f(t)/S(t)$.
(i) Gamma: $f(t) = (t^{\alpha-1}e^{-t/\beta})/(\Gamma(\alpha)\beta^{\alpha})$ if $t > 0$, $f(t) = 0$ if $t \le 0$.
(ii) Rayleigh:

$$f(t) = \begin{cases} (t/\alpha^2)\exp\{-t^2/2\alpha^2\}, & t > 0 \\ 0 & \text{elsewhere.} \end{cases}$$

(iii) Lognormal:

$$f(t) = \begin{cases} (1/(t\sigma\sqrt{2\pi}))\exp\{-(\ln t - \mu)^2/2\sigma^2\}, & \text{for } t > 0 \\ 0 & \text{elsewhere.} \end{cases}$$

(iv) Pareto: $f(t) = \alpha a^{\alpha}/t^{\alpha+1}$ for $t > a$, and 0 elsewhere.

21. If the hazard rate $\lambda(t)$ for a density f is increasing, we say that f has an *increasing failure rate*. Similarly we can define *decreasing failure rate*. Show that the results in the following table hold.

Failure Rate

Constant	Increasing	Decreasing
Exponential	Weibull ($\alpha > 1$)	Weibull ($\alpha < 1$)
	Gamma ($\alpha > 1$)	Gamma ($\alpha < 1$)
	Rayleigh ($\alpha \neq 0$)	Pareto ($t > a$)

22. For the gamma density function show that $\lim_{t \to \infty} \lambda(t) = 1/\beta$.

23. Show that for the Weibull density

$$f(t) = \begin{cases} (\alpha/\beta)t^{\alpha-1}e^{-t^\alpha/\beta}, & t > 0 \\ 0 & t \leq 0 \end{cases}$$

the mean and the variance are given by

$$\mu = \beta^{1/\alpha}\Gamma\left(1 + \frac{1}{\alpha}\right), \quad \sigma^2 = \beta^{2/\alpha}\left\{\Gamma\left(1 + \frac{2}{\alpha}\right) - \left[\Gamma\left(1 + \frac{1}{\alpha}\right)\right]^2\right\}.$$

24. Find the mean and the variance of a σ chi distribution with density

$$f(x) = \begin{cases} \dfrac{2(n/2)^{n/2}}{\sigma^n\Gamma(n/2)}x^{n-1}e^{-nx^2/2\sigma^2}, & x > 0 \\ 0, & x \leq 0 \end{cases}$$

and hence also for Rayleigh and Maxwell densities.

25. For the Pareto density function $f(t) = \alpha a^\alpha/t^{\alpha+1}$, for $t > a$ and $f(t) = 0$ if $t \leq a$, find the mean and the variance whenever they exist.

7.5* BETA DISTRIBUTION

We now introduce another distribution on $(0, 1)$ which can be used to model distribution of proportions and which provides greater flexibility than the uniform distribution on $(0, 1)$. In Section 7.3 we defined a beta function by the integral

(1)
$$B(\alpha, \beta) = \int_0^1 x^{\alpha-1}(1 - x)^{\beta-1}\, dx$$

where $\alpha, \beta > 0$. The integrand in (1) is nonnegative on $(0, 1)$, and in view of (1) the function

(2) $$f(x) = \begin{cases} [B(\alpha, \beta)]^{-1} x^{\alpha-1}(1-x)^{\beta-1}, & 0 < x < 1 \\ 0 & \text{elsewhere} \end{cases}$$

defines a density function. The density function in (2) is known as a *beta density function with parameters α and β*. The special case $\alpha = \beta = 1$ yields the uniform distribution on $(0, 1)$.

In order to graph a beta probability density function, we note that if $\alpha > 1$, $\beta > 1$ then $f(x) \to 0$ as $x \to 0$ or $x \to 1$. If $0 < \alpha < 1$ then $f(x) \to \infty$ as $x \to 0$, and if $0 < \beta < 1$ then $f(x) \to \infty$ as $x \to 1$. For $\alpha > 1$ and $\beta > 1$, f has a unique mode at $x = (\alpha - 1)/(\alpha + \beta - 2)$; for $\alpha < 1$ and $\beta < 1$, f has a unique minimum at $x = (\alpha - 1)/(\alpha + \beta - 2)$, so that the density is U-shaped. In the special case when $\alpha = \beta$, the density is symmetric about the median $x = 1/2$. Figure 1 gives graphs of several beta probability density functions.

For any $r > 0$ we have

$$\mathcal{E} X^r = [B(\alpha, \beta)]^{-1} \int_0^1 x^r x^{\alpha-1}(1-x)^{\beta-1} \, dx$$

$$= [B(\alpha, \beta)]^{-1} \int_0^1 x^{r+\alpha-1}(1-x)^{\beta-1} \, dx = \frac{B(r+\alpha, \beta)}{B(\alpha, \beta)}.$$

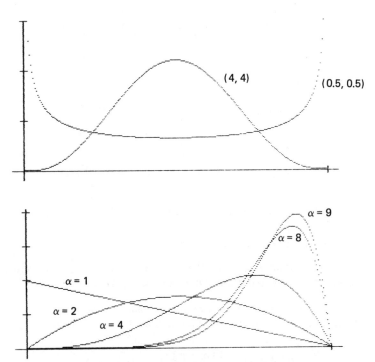

Figure 1.　Beta density for $(\alpha, \beta) = (1, 2), (2, 2), (4, 2), (8, 2)$ and $(9, 2), (4, 4)$ and $(.5, .5)$.

Now recall that $B(\alpha, \beta) = \Gamma(\alpha)\Gamma(\beta)/\Gamma(\alpha + \beta)$.
Hence:

$$(3) \qquad \mathscr{E} X^r = \frac{\Gamma(\alpha + \beta)\Gamma(\alpha + r)}{\Gamma(\alpha + \beta + r)\Gamma(\alpha)}.$$

In particular,

$$(4) \qquad \mathscr{E} X = \frac{\alpha}{\alpha + \beta} \quad \text{and} \quad \text{var}(X) = \frac{\alpha\beta}{(\alpha + \beta)^2(\alpha + \beta + 1)}.$$

Beta density function arises naturally in sampling from a normal population and also as the distribution of order statistics in sampling from a uniform distribution. In Bayesian statistics (see Sections 10.6 and 10.7) it is a preferred prior for the parameter p of a binomial distribution.

Example 1. Proportion of Erroneous Income Tax Returns. Suppose that the proportion X of erroneous income tax returns filed with the Internal Revenue Service each year can be viewed as a beta random variable with parameters α and β.

The probability that in a randomly chosen year there will be at most 100γ percent erroneous returns is

$$P(X \leq \gamma) = \frac{1}{B(\alpha, \beta)} \int_0^\gamma x^{\alpha - 1}(1 - x)^{\beta - 1} \, dx.$$

In particular, if $\alpha = 2$ and $\beta = 12$ then

$$P(X \leq \gamma) = \frac{\Gamma(14)}{\Gamma(2)\Gamma(12)} \int_0^\gamma x(1 - x)^{11} \, dx$$

$$= \frac{13!}{11!} \left\{ -\frac{\gamma(1 - \gamma)^{12}}{12} + \frac{1 - (1 - \gamma)^{13}}{12(13)} \right\}$$

$$= 1 - (1 + 12\gamma)(1 - \gamma)^{12}.$$

If $\gamma = .1$ then $P(X \leq \gamma) = 1 - 2.2(.9)^{12} = .3787$.   □

In some cases, there is a simpler way to compute probabilities such as $P(X < \gamma)$ under beta distribution. One can use the binomial tables.

Example 2. Relation Between Beta and Binomial Distributions. Suppose X has a beta distribution with parameters α and β and let $0 < p < 1$. Then

$$(5) \qquad P(X \leq p) = \frac{\Gamma(\alpha + \beta)}{\Gamma(\alpha)\Gamma(\beta)} \int_0^p x^{\alpha - 1}(1 - x)^{\beta - 1} \, dx.$$

Suppose both α and β are positive integers. Then we can rewrite (5) as

$$P(X \le p) = \frac{(\alpha + \beta - 1)!}{(\alpha - 1)!(\beta - 1)!} \int_0^p x^{\alpha-1}(1 - x)^{\beta-1}\, dx$$

$$= (\alpha + \beta - 1)\binom{\alpha + \beta - 2}{\alpha - 1} \int_0^p x^{\alpha-1}(1 - x)^{\beta-1}\, dx.$$

Integrating by parts $(\alpha - 1)$ times we see that

$$P(X \le p) = \sum_{j=\alpha}^{\alpha+\beta-1} \binom{\alpha + \beta - 1}{j} p^j (1 - p)^{\alpha+\beta-1-j}$$

$$= P(Y \ge \alpha)$$

where Y is a binomial random variable with parameters $\alpha + \beta - 1$ and p. For convenience write $\alpha + \beta = n + 1$ and $\alpha = k$. Then we have (see also Problem 6.3.22)

Relation Between Beta and Binomial

$$n\binom{n - 1}{k - 1} \int_0^p x^{k-1}(1 - x)^{n-k}\, dx = \sum_{j=k}^{n} \binom{n}{j} p^j (1 - p)^{n-j}$$

In Example 1, $p = .1$ and $\alpha = 2$, $\beta = 12$ so that $n = 13$, $k = 2$ and

$$P(X < .1) = (Y \ge 2)$$

where Y is binomial with $n = 13$, $p = .1$. From Table A3:

$$P(X < .1) = 1 - P(Y \le 1) = 1 - .6213 = .3787. \qquad \square$$

Yet another application is the following. Let Y_1, Y_2 be binomial with parameters n, p_1, and n, p_2 respectively. Then $p_2 > p_1$ implies

$$\int_0^{p_2} x^{k-1}(1 - x)^{n-k}\, dx \ge \int_0^{p_1} x^{k-1}(1 - x)^{n-k}\, dx$$

and it follows that

(6) $$P(Y_2 \ge k) \ge P(Y_1 \ge k).$$

Alternatively, we can write

(6') $$\sum_{j=0}^{k-1} \binom{n}{j} p_2^j(1 - p_2)^{n-j} \le \sum_{j=0}^{k-1} \binom{n}{j} p_1^j(1 - p_1)^{n-j}.$$

Thus if $p_2 > p_1$ then Y_2 has more probability in the right tail than Y_1 does; that is, the larger the p the more shifted the distribution to the right.

Problems for Section 7.5

1. Let X be a random variable with beta density function with parameters α and β. In each of the following cases verify that the beta density integrates to one:
   (i) $\alpha = 1$ and $\beta = 8$.   (ii) $\alpha = 2$ and $\beta = 4$.   (iii) $\alpha = 3$ and $\beta = 4$.

2. The proportion of defective television tubes made by some company has a beta distribution with $\alpha = 2$ and $\beta = 3$.
   (i) What is the average proportion of defective tubes that this company makes?
   (ii) Find the probability that the proportion of defective tubes will be between .2 and .5.

3. The annual proportion of new gas stations that fail in a city may be considered as a beta random variable with $\alpha = 2$ and $\beta = 10$.
   (i) Find the average annual proportion of new gas stations that fail in this city.
   (ii) Find the probability that 20 percent or more of all new gas stations will fail in any one year.
   (iii) Find the probability that between 5 and 10 percent of new gas stations will fail in a year.

4. The relative humidity measured at a certain location can be thought of as a random variable with density function

$$f(y) = \begin{cases} cy^4(1-y)^3 & 0 < y < 1 \\ 0 & \text{elsewhere.} \end{cases}$$

   (i) Find c.
   (ii) Find the probability that the relative humidity at this location will be at least 35 percent.

5. The percentage of tires per lot manufactured by a certain company has a beta distribution with $\alpha = 2$ and $\beta = 6$. It is known that any lot containing more than 25 percent defective tires will have to be sold as blemished. What is the probability that a randomly selected lot will be sold as blemished?

6. The proportion of blood-spotted eggs laid by hens in a flock may be considered a beta random variable with parameters $\alpha = 3$ and $\beta = 11$. Find the probability that
   (i) The proportion of blood-spotted eggs on a given day is between .05 and .15.
   (ii) The proportion is larger than the mean proportion.

7. Let X have a beta distribution with mean μ and variance σ^2. Find α and β. In particular, find α and β when $\mu = .1$, $\sigma^2 = .05$.

8. It is known that for a beta random variable X with parameters α and β $P(X < .2) = .22$. If $\alpha + \beta = 26$, find α and β. (Hint: Use Binomial Table A3.)

9. The proportion of patients who respond favorably to a new drug is known to have a beta distribution with parameters α and β. Let X be the number of patients who show improvement out of a randomly selected n patients all of whom were adminis-

tered the new drug. Find the unconditional distribution of X. Find the conditional distribution of the proportion of patients who respond favorably to the new drug given $X = k$.

10. The proportion of patients who respond favorably to a new drug is known to have a beta distribution with parameters α and β. Randomly selected patients are administered the new drug and are tested sequentially. Let X be the number of patients needed to get r patients who respond favorably. Find the unconditional distribution of X, its mean and variance. Find also the conditional distribution of the proportion of favorable responses given $X = k + r$, $k = 0, 1, 2 \ldots$.

11. Let X have beta distribution with parameters α and β. Find $\mathscr{E}\{X^p(1 - X)^q\}$.

7.6 NORMAL DISTRIBUTION

The normal distribution has a well deserved claim to be the most commonly used distribution in statistics. Part of the reason is the central limit theorem (Section 9.3) which essentially says that in random sampling from a population with finite variance the sample mean has a distribution that looks more and more like normal distribution as the sample size increases. In this section we consider some useful properties of the normal distribution.

A random variable X is said to have a *normal distribution with parameters μ and σ^2* if its density function is given by

$$(1) \qquad f(x) = \frac{1}{\sigma\sqrt{2\pi}} \exp\left\{ -\frac{(x - \mu)^2}{2\sigma^2} \right\}, \qquad x \in \mathbb{R}; \mu \in \mathbb{R}, \sigma > 0.$$

We will show that $\mu = \mathscr{E}X$ and $\sigma^2 = \mathrm{var}(X)$. In particular, if $\mu = 0$ and $\sigma = 1$, we say that X has a *standard normal distribution*. (We write Z for a standard normal random variable.)

We first check that (1) does indeed define a density function. In fact, $f(x) > 0$ for $x \in \mathbb{R}$ and

$$\int_{-\infty}^{\infty} f(x)\, dx = \int_{-\infty}^{\infty} \frac{1}{\sigma\sqrt{2\pi}} \exp\left\{ -\frac{(x - \mu)^2}{2\sigma^2} \right\} dx$$

$$= \int_{-\infty}^{\infty} \frac{1}{\sqrt{2\pi}} \exp\left\{ -\frac{y^2}{2} \right\} dy \qquad (y = (x - \mu)/\sigma)$$

$$= 1$$

in view of Proposition 7.3.2(iii). We will use this fact often, and we restate it for emphasis. Let $a \in \mathbb{R}$ and $b > 0$. Then:

$$(2) \qquad b\sqrt{2\pi} = \int_{-\infty}^{\infty} \exp\left\{ -\frac{(x - a)^2}{2b^2} \right\} dx.$$

Let us next compute the mean and the variance of X. We have

$$\mathscr{E}X = \int_{-\infty}^{\infty} x \frac{1}{\sigma\sqrt{2\pi}} \exp\left\{-\frac{(x-\mu)^2}{2\sigma^2}\right\} dx$$

$$= \int_{-\infty}^{\infty} (\sigma y + \mu) \frac{1}{\sqrt{2\pi}} \exp(-y^2/2)\, dy \qquad (y = (x-\mu)/\sigma)$$

$$(3) \qquad = \sigma \int_{-\infty}^{\infty} y e^{-y^2/2}\, dy + \mu \int_{-\infty}^{\infty} \frac{1}{\sqrt{2\pi}} e^{-y^2/2}\, dy = \mu.$$

Here the first integral vanishes since the integrand is an odd function and the second integral equals one in view of (2). Next we have

$$\text{var}(X) = \mathscr{E}(X-\mu)^2 = \int_{-\infty}^{\infty} (x-\mu)^2 \frac{1}{\sigma\sqrt{2\pi}} \exp\left\{-\frac{(x-\mu)^2}{2\sigma^2}\right\} dx$$

$$= \sigma^2 \int_{-\infty}^{\infty} \frac{y^2}{\sqrt{2\pi}} e^{-y^2/2}\, dy \qquad (y = (x-\mu)/\sigma)$$

$$= 2\sigma^2 \int_{0}^{\infty} \frac{z}{\sqrt{2\pi}} e^{-z/2} \frac{dz}{2\sqrt{z}} \qquad (z = y^2)$$

$$(4) \qquad = \frac{\sigma^2}{\sqrt{2\pi}} \int_{0}^{\infty} z^{1/2} e^{-z/2}\, dz = \sigma^2 \frac{\Gamma(3/2)2^{3/2}}{\sqrt{2\pi}} = \sigma^2.$$

Normal Probability Density Function

$$f(x) = \frac{1}{\sigma\sqrt{2\pi}} \exp\left\{-\frac{(x-\mu)^2}{2\sigma^2}\right\}, \quad x \in \mathbb{R}; \qquad \mu \in \mathbb{R}, \quad \sigma > 0.$$

$$\mathscr{E}X = \mu, \text{var}(X) = \sigma^2.$$

Special case $\mu = 0$, $\sigma = 1$: standard normal.

To graph a normal density with mean μ and variable σ^2, we note that the density is symmetric about μ. That is,

$$f(\mu + x) = f(\mu - x) \qquad \text{for all } x \in \mathbb{R}.$$

Moreover, f has a maximum at $x = \mu$ with maximum value $f(\mu) = 1/(\sigma\sqrt{2\pi})$. As $|x| \to \infty$, $f(x) \to 0$. Thus f is unimodal, and mean, median, and mode coincide at $x = \mu$. Graphs of several normal distributions are given in Figure 1.

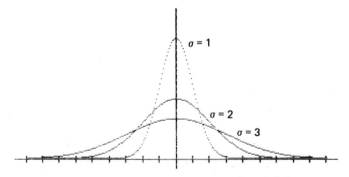

Figure 1. Normal density for $\mu = 0$ and $\sigma = 1, 2, 3$.

In view of the great importance of normal distribution, the distribution function of a standard normal random variable Z has been tabulated extensively. Table A5 in the Appendix gives the probability $P(0 < Z < z)$ (see Figure 2) for values of $z = .00(.01)3.09$. Probabilities corresponding to negative values of z can be obtained by symmetry. Thus, for $z > 0$:

$$P(-\infty < Z < z) = .5 + P(0 < Z < z)$$

$$P(0 < Z < z) = P(-z < Z < 0)$$

$$P(|Z| < z) = P(-z < Z < z) = 2P(0 < Z < z)$$

$$P(|Z| > z) = 2P(Z > z) = 2[\cdot 5 - P(0 < Z < z)]$$

$$P(z_1 < Z < z_2) = P(Z < z_2) - P(Z \leq z_1).$$

How do we use Table A5 when $\mu \neq 0$ and/or $\sigma \neq 1$? We standardize X. That is, suppose we want $P(X \leq c)$ for a normal random variable with mean μ and

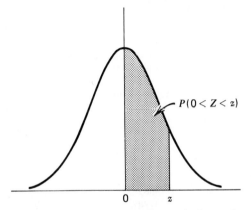

Figure 2. $P(0 < Z < z)$ under standard normal.

variance σ^2. Then on standardizing X we get

(5) $$P(X \leq c) = P\left(\frac{X - \mu}{\sigma} \leq \frac{c - \mu}{\sigma}\right) = P\left(Z \leq \frac{c - \mu}{\sigma}\right)$$

where Z is standard normal. This is so because

$$P(X \leq c) = \int_{-\infty}^{c} \frac{1}{\sigma\sqrt{2\pi}} \exp\left\{-\frac{(x - \mu)^2}{2\sigma^2}\right\} dx$$

$$= \int_{-\infty}^{(c-\mu)/\sigma} \frac{1}{\sqrt{2\pi}} \exp\left\{-\frac{y^2}{2}\right\} dy \qquad \left(y = \frac{x - \mu}{\sigma}\right).$$

An alternative way to look at $P(X \leq c)$ is as follows. Table A5 gives the probability that the (standard) *normal variable Z will be within z* standard deviation units of the mean. To find $P(X \leq c)$ we need to find the number of standard deviation units that c is from the mean μ. Since

$$c = \frac{c - \mu}{\sigma} \cdot \sigma + \mu,$$

we see that c is $(c - \mu)/\sigma$ standard deviation units from μ. Hence $P(X \leq c) = P(Z \leq (c - \mu)/\sigma)$.

Example 1. How Good Is Chebyshev's Bound? We recall that for a random variable X with mean μ and variance σ^2

$$P(|X - \mu| \leq k\sigma) \geq 1 - \frac{1}{k^2}$$

for $k > 1$. If, however, X is normal, then

$$P(|X - \mu| \leq k\sigma) = P\left(\frac{|X - \mu|}{\sigma} \leq k\right) = P(|Z| \leq k).$$

Table 1 shows that almost all the probability under a normal distribution is assigned to the interval $(\mu - 3\sigma, \mu + 3\sigma)$ even though $f(x)$ never touches the x-axis. Chebyshev's inequality always underestimates the exact probability.   □

TABLE 1.   Comparison of Chebyshev Bounds and Exact Probabilities, $P(|X - \mu| \leq k\sigma)$.

k	Chebyshev Bound ($\geq$)	Exact Probability
1.0	0	.6826
1.5	.5556	.7062
2.0	.75	.9544
2.5	.84	.9876
3.0	.8889	.9974
3.5	.9184	.9978

Example 2. Finding Probabilities Under Normal Curve. Let Z be a standard normal random variable. Then, from Figure 3 and Table A5:

$$P(Z \leq 1.75) = .5 + P(0 < Z \leq 1.75) = .9599,$$

$$P(Z \leq -.75) = P(Z \geq .75) = .5 - .2734 = .2266,$$

and

$$P(-.38 \leq Z \leq 1.42) = P(0 \leq Z \leq 1.42) + P(0 \leq Z \leq .38)$$

$$= .4222 + .1480 = .5702.$$

Suppose we want to find z such that $P(Z > z) = .05$ or $P(|Z| > z) = .02$. To find $P(Z > z) = .05$ we note from Figure 4 that z must satisfy $P(0 < Z < z) = .45$. From Table A5 we see that $z = 1.645$. To find z with $P(|Z| > z) = .02$ we note from Figure

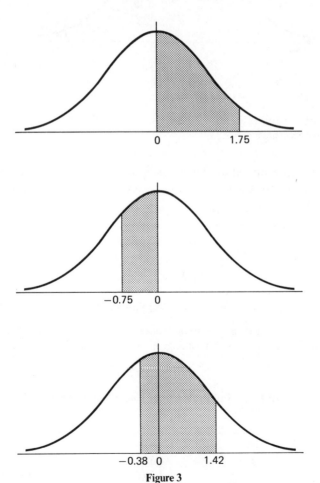

Figure 3

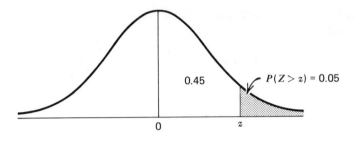

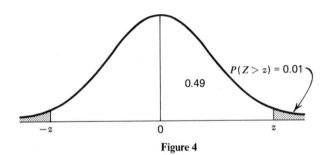

Figure 4

4 that z must satisfy $P(Z > z) = .01$ by symmetry so that $P(0 < Z < z) = .49$. Hence from Table A5 we have $z = 2.33$.   □

In statistics, tail probabilities under the normal curve have special significance. For that reason we will use the notation z_α, for $0 < \alpha < 1$, to denote the quantile of order $1 - \alpha$. That is

$$P(Z > z_\alpha) = \alpha$$

$$P(|Z| > z_{\alpha/2}) = \alpha$$

In Table 2 we list some commonly used values of α and z_α.

Example 3.   Distribution of Ages of Workers.   The ages of workers in a large plant have a normal distribution with mean age 40 years and standard deviation 7 years. Let

TABLE 2.   Tail Probabilities Under Normal Distribution

α	$z_{\alpha/2}$	z_α
.01	2.575	2.33
.02	2.33	2.05
.05	1.96	1.645
.10	1.645	1.28

X be the age of a randomly selected worker. Then X has a normal distribution with $\mu = 40$, $\sigma = 7$. The reader may wonder how X can have a normal distribution. Does not a normal random variable take negative values? The answer is that almost all of the probability is assigned to the interval $(\mu - 3\sigma, \mu + 3\sigma)$, as seen in Example 1. In fact,

$$P(X < 0) = P\left(\frac{X - \mu}{\sigma} < 0 - \frac{\mu}{\sigma}\right) = P\left(Z < -\frac{40}{7}\right)$$

$$= P(Z < -5.71) \doteq 0.$$

The proportion of workers aged between 50 and 60 is given by

$$P(50 < X < 60) = P\left(\frac{50 - 40}{7} < Z < \frac{60 - 40}{7}\right)$$

$$= P(0 < Z < 2.86) - P(0 < Z < 1.43)$$

$$= .4979 - .4236 = .0743.$$

The proportion of workers aged 35 years or younger is given by

$$P(X \leq 35) = P\left(\frac{X - \mu}{\sigma} \leq \frac{35 - 40}{7}\right) = P(Z < -.71)$$

$$= .5 - .2611 = .2389.$$

Let us find the age x below which are 20 percent of the workers. We want to find x such that

$$P(X \leq x) = .20.$$

Clearly, $x < \mu$ (see Figure 5). Also:

$$P\left(\frac{X - \mu}{\sigma} \leq \frac{x - 40}{7}\right) = P\left(Z \leq \frac{x - 40}{7}\right) = .20.$$

From Table A5, $(x - 40)/7 = -.84$ so that $x = 34.12$. Alternatively, we note from

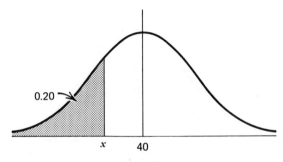

0.20

x 40

Figure 5

Table A5 that .2995 ($\doteq$ .30) of the area under the normal curve is between μ and $\mu - .84\sigma$. It follows that if x is located at $-.84\sigma$ distance from μ, then about .20 proportion of the workers will be younger than age x. Hence

$$x = \mu - .84\sigma = 40 - .84(7) = 34.12 \text{ years.}$$ $\square$

Example 4. Grades and Normal Distribution.   The grades in a large section of a calculus course are approximately normally distributed with mean 75 and standard deviation 6. The lowest D is 60, the lowest C is 68, the lowest B is 83, and the lowest A is 90. What proportion of the class will get A's, B's, C's, D's and F's?
We have (see Figure 6)

$$\text{proportion of A's} = P(X \ge 90) = P\left(Z \ge \frac{15}{6}\right) = .0062,$$

$$\text{proportion of B's} = P(83 \le X < 90) = P(Z < 2.5) - P(Z < 1.33)$$

$$= .9938 - .9082 = .0856,$$

$$\text{proportion of C's} = P(68 \le X < 83) = P(Z < 1.33) - P(Z < -1.17)$$

$$= .9082 - .1210 = .7872,$$

$$\text{proportion of D's} = P(60 \le X < 68) = P(Z < -1.17) - P(Z < -2.5)$$

$$= .1210 - .0062 = .1148,$$

$$\text{proportion of F's} = 1 - (.0062 + .0856 + .7872 + .1148) = .0062.$$

If, on the other hand, the instructor, after looking at the lopsided nature of the grade distribution, decides to change the grading scale so as to have 5 percent A's, 9.01 percent B's, 71.98 percent C's, and 11.68 percent D's, then what is the new grading scale? Now we are given probabilities under normal curve (Figure 7) and we wish to find the corresponding scores. We first find z scores from Table A5 and then the corresponding x scores as $x = \mu + z\sigma = 75 + 6z$. Thus the lowest A score is $75 + 6(1.645) = 84.87$, the lowest B corresponds to a score of $75 + 6(1.08) = 81.48$, the lowest C score is $75 - 6(1.08) = 68.52$, and the lowest D score is $75 - 6(1.99) = 63.06$.
                                                                                        $\square$

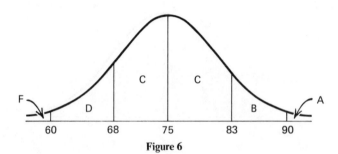

Figure 6

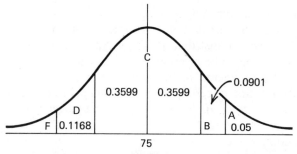

Figure 7

Example 5. Standard Deviation Setting of a Coffee Vending Machine. A coffee vending machine dispenses 6 ounces of coffee per cup, on the average. Assuming that the amount dispensed is normally distributed, at what level should the standard deviation be set so that 98 percent of the cups will have at least 5.85 ounces of coffee?

Since 98 percent of the cups will have at least 5.85 ounces of coffee (Figure 8), it follows from Table A5 that 5.85 is 2.05σ smaller than μ. Thus,

$$5.85 = \mu - 2.05\sigma = 6 - 2.05\sigma$$

and it follows that $\sigma = .15/2.05 = 0.0732$.

If the cups can hold at most 6.1 ounces of coffee, the proportion of those that will spill over is given by

$$P(X > 6.1) = P\left(\frac{X - \mu}{\sigma} > \frac{6.1 - 6}{.0732}\right) = P(Z > 1.37) = .0853. \qquad \square$$

As mentioned earlier, normal distribution arises often in practice. For example, according to Maxwell's law in physics the components of velocity of a molecule of mass m in a gas at absolute temperature T has a normal distribution with mean 0 and variance $\sigma^2 = m/(kT)$ where k is a physical constant. Frequently, though, normal distribution is only an approximation to the random phenomenon under

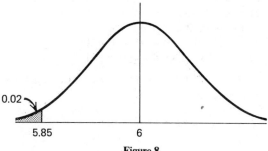

Figure 8

study. Thus when we say that height of a group of individuals obeys a normal law it simply means that the distribution is symmetric and bell-shaped and may be approximated by a normal distribution. Other typical examples include distribution of weights of a biometrical population, distribution of errors of measurements in physical experiments, distribution of scores on the Graduate Record Examination, distribution of certain lengths of life, distribution of output from a production line, and so on. If the population is heterogeneous, one needs to be careful. Thus the distribution of heights of teenagers aged 15–18 may be bimodal and hence not normal since it includes both boys and girls. The two distributions separately are approximately normal. Similarly the distribution of incomes of a large heterogeneous group is not likely to be normal. Later on (Section 9.3), we will study the central limit theorem and see that the distribution of sample averages is frequently approximately normal when the sample size is large.

Problems for Section 7.6

1. For a standard normal random variable Z find
    (i) $P(Z \le .54)$.   (ii) $P(Z > .27)$.   (iii) $P(|Z| \le 1.63)$.
    (iv) $P(-.64 < Z < 1.29)$.   (v) $P(.69 < Z < 2.37)$.
    (vi) $P(.45 < |Z| < 1.96)$.   (vii) $P(Z < -.86)$.

2. Let Z be a standard normal random variable. Find c such that:
    (i) $P(Z \le c) = .1772$.   (ii) $P(Z \ge c) = .0068$.
    (iii) $P(Z \ge -c) = .0044$.   (iv) $P(|Z| \ge c) = .05$.
    (v) $P(Z \le c) = .9564$.   (vi) $P(|Z| \le c) = .90$.

3. Let X have a normal distribution with mean 7 and variance 9. Find:
    (i) $P(X > 8)$.   (ii) $P(X > 5)$.   (iii) $P(4 < X < 9)$.
    (iv) $P(X < 3 \text{ or } X > 7.5)$.   (v) $P(|X| > 5)$.   (vi) $P(|X| < 8)$.

4. Suppose X has a normal distribution with mean 1 and variance 4. Find:
    (i) $P(X = 1)$.   (ii) $P(X \ge 0)$.   (iii) $P(-1 < X < 0.5)$.
    (iv) $P(|X| < 2)$.   (v) $P(|X| > 1.5)$.   (vi) $P(X > 6.5)$.

5. Let X have a normal distribution with mean μ and standard deviation σ. Find x such that
    (i) $P(X \ge x) = .95$, if $\mu = 1$, $\sigma = 2$.
    (ii) $P(X \ge x) = .0068$, if $\mu = 6$, $\sigma = 3$.
    (iii) $P(x < X < 1.40) = .2712$, if $\mu = 0$, $\sigma = 1$.
    (iv) $P(X > x) = 3P(X \le x)$, if $\mu = 2$, $\sigma = 3$.
    (v) $P(|X| > x) = .0076$, if $\mu = 0$, $\sigma = 1$.
    (vi) $P(|X - \mu| \le x) = .9660$, if $\mu = 50$, $\sigma = 10$.
    (vii) $P(|X| > x) = P(|X| \le x)$, if $\mu = 0$, $\sigma = 1$.

6. For what values of c are the following functions density functions? In each case find the corresponding mean and the variance.
    (i) $f(x) = (c/\sqrt{\pi})\exp(-x^2 + x)$, $x \in \mathbb{R}$.
    (ii) $f(x) = c\exp(-5x^2 + 9x - 11)$, $x \in \mathbb{R}$.
    (iii) $f(x) = (1/\sqrt{\pi})\exp(-x^2 + x - c)$, $x \in \mathbb{R}$.
    (iv) $f(x) = \sqrt{c}\exp\{-cx^2 - x - 4\}$, $x \in \mathbb{R}$.

7. The final examination scores in a large class on Principles of Economics are normally distributed with a mean score of 60 and a standard deviation of 10.
   (i) If the lowest passing score is 48, what proportion of the class is failing?
   (ii) If the highest 80 percent of the class is to pass, what is the lowest passing score?

8. The diameter of a large shipment of ball bearings has a normal distribution with mean 2 cm and standard deviation 0.01 cm.
   (i) What is the proportion of ball bearings from this shipment that have a diameter larger than 2.02 cm?
   (ii) Suppose ball bearings with diameter larger than 2.02 cm or smaller than 1.98 cm are unacceptable. If a random sample of three ball bearings is selected, what is the probability that exactly two of them will be unacceptable?

9. A certain machine makes iron rods with a diameter of 0.5 cm. Because various factors influence a production process, however, the diameter is $.505 + .01Z$ where Z is a standard normal random variable. What is the probability that a given iron rod produced by this machine will have a diameter that will be within .01 cm of the nominal diameter?

10. Optimum mean weight for a particular height and body frame for males can be assumed to have a normal distribution with a mean of 126 pounds and a standard deviation of 5 pounds. Suppose a male with weight 2.5 standard deviation above the mean is considered to be obese.

    What is the proportion of males of this particular height and body frame who are obese? Is a weight of 138 pounds obese? In a sample of 15 such males what is the probability that at least one will be obese?

11. Suppose the systolic blood pressure for normal males is a normal random variable with mean 120 mm Hg and standard deviation 20 mm Hg. If a person with blood pressure 2.5σ above the mean is considered hypertensive and to be referred for further diagnosis, what proportion of males fall in this category? How many males have a blood pressure below 100 mm Hg. and above 160 mm Hg? How would you classify a reading of 180 mm Hg?

12. The birth weight of girls in the United States is a normal random variable with a mean of 7.41 pounds and a standard deviation of 2 pounds. How probable is a birth weight of 3 pounds or less or 11 pounds or more? If a birth weight of 12 pounds is considered overweight, what proportion of newly born infant girls are overweight?

13. The scores of students taking the verbal portion of the Graduate Record Examination are artificially standardized to have mean score 500 and standard deviation 100.
    (i) If your sister scored 654, what percentage of students taking the test ranked higher?
    (ii) If your brother scored 540, what is his percentile rank?
    (iii) What score qualifies a student for the top 10 percent of all candidates?
    (iv) The top 5 percent?

14. The heights of male freshmen entering a large state university are normally distributed with a mean of 68 inches. It is known that about 5 percent of the

freshmen are taller than 72 inches.
(i)   What is the standard deviation of the distribution of heights?
(ii) What is the proportion of the freshmen males at this college who have a height of 63 inches or less?

15. The weights of instant coffee jars packed by a food processor are normally distributed with a standard deviation of .2 ounce. The processor has set the mean such that about two percent of the jars weigh more than 8.41 ounces. What is the mean setting?

16. The pebbles at a concrete mixing company can be modeled approximately by using a normal distribution.
(i)   If 10 percent of the pebbles are 4 inches or more in diameter and 15 percent are less than 1.5 inches or less in diameter, what is the mean diameter and what is the standard deviation?
(ii) In a random sample of 15 pebbles from this company's yard, what is the probability that five pebbles will have a diameter of less than 2 inches, and six will have a diameter of between 2 and 4 inches?

17. If the filling process in a food processing plant dispenses flour at a mean rate of 16 ounces and if the fill is normally distributed, what is the standard deviation if it is known that 6.68 percent of the packages weigh more than 16.1 ounces?

18. The pulse rate of infants one month or older is approximately a normal random variable with a mean of 120 beats per minute and standard deviation of 15 beats per minute. What is the probability that an infant aged 1–11 months will have a pulse rate of 80 beats per minute or lower? 160 beats per minute or higher? What is the probability that in a randomly selected sample of five infants exactly two will have a pulse rate of 160 beats per minute or higher?

19. In a large calculus course the final examination scores are normally distributed with a mean of 65 and a standard deviation of 10. What percentage of the class is failing if the lowest passing score is 45? If only 6.28 percent of the class is to be awarded A, what is the lowest A score?

20. A machine produces ball bearings. The diameters of the ball bearings are normally distributed with a standard deviation of 0.01 mm. What should be the "nominal" diameter at which the machine is set so that at most two percent of the ball bearings it produces will have diameters larger than 3 mm?

21. The breaking strength X of the rope produced by a certain manufacturer is approximately normal with parameters μ and σ. If $X > 100$ pounds, the net profit is $\$ p_1$ per 100 feet, if $90 \leq X \leq 100$ then the profit is $\$ p_2 (> p_1)$ per 100 feet and if $X < 90$ the rope is to be destroyed with a loss of $\$ p_3$. What is the expected profit per 100 feet? What should be μ to maximize the expected profit?

22. Suppose X has a normal distribution with mean μ and variance σ^2.
(i)   For what value of x is each of the following probabilities maximum?
(a) $P(x - c < X < x + c)$, $c > 0$ a fixed constant.
(b) $P(x < X < x + c)$, $c > 0$ a fixed constant.
(ii) Suppose $\mu = 0$ and $\sigma = 1$ and we wish to find $a > 0$ and $b > 0$ such that $P(-a < X < b) = 1 - \alpha$ and $L = b + a$, the length of $(-a, b)$ is minimum.

Show that $a = b$. (This result is true for all symmetric densities and is extensively used in constructing confidence intervals. See also Problem 4.3.12.)

23. Let

$$\phi(x) = \frac{1}{\sqrt{2\pi}} \exp\left(-\frac{x^2}{2}\right), \quad x \in \mathbb{R}$$

be the normal density function.
   (i) Show that $\phi^k(x)$ is integrable for every $k > 0$. Find $\int_{-\infty}^{\infty} \phi^k(x)\, dx$.
   (ii) For what value of c is $c\phi^k(x)$ a density function?
   (iii) Find the mean and the variance of the density function in (ii).

24*. Let Z be a standard normal random variable with distribution function Φ and density function ϕ. Show that

$$P(Z > z) < \frac{\phi(z)}{z} \quad \text{for } z > 0$$

and as $z \to \infty$

$$P(Z > z) \doteq \phi(z)/z.$$

[Hint: $\phi'(z) = -z\phi(z)$ so $\phi(z) = \int_z^{\infty} x\phi(x)\, dx$. Integrate by parts.]

25*. Let X be a standard normal random variable with density ϕ and distribution function Φ. Show that:
   (i) $\mathscr{E}\{Z\Phi(Z)\} = 1/(2\sqrt{\pi})$.
   (ii) $\mathscr{E}\{Z^2\Phi(Z)\} = 1/2$.
   (iii) $\mathscr{E}\{Z^3\Phi(Z)\} = 3/(2\sqrt{\pi})$.
   [Hint: $\Phi'(x) = \phi(x)$, $\phi'(x) = -x\phi(x)$, $\phi''(x) = (x^2 - 1)\phi(x)$, $\phi'''(x) = (3x - x^3)\phi(x)$.]

26*. Let X be a normal random variable with mean μ and variance σ^2. Let Φ be the standard normal distribution function. Show that $\mathscr{E}\Phi(X) = \Phi(\mu/\sqrt{1 + \sigma^2})$. Note that $\mathscr{E}\Phi((X - \mu)/\sigma) = 1/2$, since $\Phi((X - \mu)/\sigma)$ has a uniform distribution on $(0, 1)$ (Section 8.2). [Hint: Let Z and X be independent random variables where Z is a standard normal random variable. Then

$$\Phi(X) = P\{Z \le X | X\}$$

so that $\mathscr{E}\Phi(X) = P(Z \le X)$. Integrate.]

7.7* BIVARIATE NORMAL DISTRIBUTION

The bivariate distribution that is most commonly used in analyzing covariance or correlation is the *bivariate normal distribution*. Let $\mu_1 \in \mathbb{R}$, $\mu_2 \in \mathbb{R}$, $\sigma_1 > 0$,

$\sigma_2 > 0$, and $|\rho| < 1$, and write

(1)
$$Q(x, y) = \frac{1}{(1 - \rho^2)} \left\{ \frac{(x - \mu_1)^2}{\sigma_1^2} - 2\rho\left(\frac{x - \mu_1}{\sigma_1}\right)\left(\frac{y - \mu_2}{\sigma_2}\right) \right.$$
$$\left. + \frac{(y - \mu_2)^2}{\sigma_2^2} \right\}.$$

Then the function f defined on $\mathbb{R}_2$ by

(2)
$$f(x, y) = \frac{\exp\{-Q/2\}}{2\pi\sigma_1\sigma_2\sqrt{1 - \rho^2}}, \qquad \begin{matrix} -\infty < x < \infty, \\ -\infty < y < \infty, \end{matrix}$$

is known as a bivariative normal distribution with parameters μ_1, μ_2, σ_1, σ_2 and ρ. We first show that (2) defines the density function of a pair (X, Y) such that $\mathscr{E}X = \mu_1$, $\mathscr{E}Y = \mu_2$, $\text{var}(X) = \sigma_1^2$, and $\text{var}(Y) = \sigma_2^2$. It suffices to integrate $f(x, y)$ on (x, y) over $\mathbb{R}_2$. Let

$$g(x) = \int_{-\infty}^{\infty} f(x, y)\, dy = \int_{-\infty}^{\infty} \frac{\exp(-Q/2)}{2\pi\sigma_1\sigma_2\sqrt{(1 - \rho^2)}}\, dy.$$

The trick is to write the exponent in the integrand on the right side in the form $\exp\{-(y - \mu)^2/2\sigma^2\}$ and then use the definition of univariate normal distribution. We have

$$Q = \frac{1}{1 - \rho^2} \left\{ \left[\frac{y - \mu_2}{\sigma_2} - \rho\left(\frac{x - \mu_1}{\sigma_1}\right) \right]^2 + (1 - \rho^2)\left(\frac{x - \mu_1}{\sigma_1}\right)^2 \right\}.$$

Consequently

(3)
$$g(x) = \frac{\exp\left\{ -\frac{1}{2}\left(\frac{x - \mu_1}{\sigma_1}\right)^2 \right\}}{\sqrt{2\pi}\,\sigma_1}$$

$$\times \int_{-\infty}^{\infty} \frac{\exp\left\{ -\frac{1}{2(1 - \rho^2)}\left(\frac{y - \mu_2}{\sigma_2} - \rho\frac{x - \mu_1}{\sigma_1}\right)^2 \right\}\, dy}{\sigma_2\sqrt{2\pi(1 - \rho^2)}}$$

and for $|\rho| < 1$ we note that the integrand is a normal density with mean $\mu_2 + \rho(\sigma_2/\sigma_1)(x - \mu_1)$ and variance $\sigma_2^2(1 - \rho^2)$. It follows that

(4)
$$g(x) = \frac{1}{\sigma_1\sqrt{2\pi}} \exp\left\{ -\frac{(x - \mu_1)^2}{2\sigma_1^2} \right\}, \qquad x \in \mathbb{R}.$$

Since g is the density of a normal random variable with mean μ_1 and variance σ_1^2, it follows that $\mathscr{E}X = \mu_1$, var$(X) = \sigma_1^2$ and, moreover, that f is indeed a bivariate density with marginal density $f_1(x) = g(x)$ of X given by (4). By symmetry, $\mathscr{E}Y = \mu_2$, var$(Y) = \sigma_2^2$ and the marginal density of Y is also normal.

From (3) we see that the conditional density of Y, given $Y = x$, is given by

$$(5) \quad h(y|x) = \frac{\exp\left\{-\dfrac{1}{2(1-\rho^2)\sigma_2^2}\left[y - \mu_2 - \rho\dfrac{\sigma_2}{\sigma_1}(x - \mu_1)\right]^2\right\}}{\sigma_2\sqrt{2\pi(1-\rho^2)}}, \quad y \in \mathbb{R}.$$

It follows that

$$(6) \qquad \mathscr{E}\{Y|x\} = \mu_2 + \left(\frac{\rho\sigma_2}{\sigma_1}\right)(x - \mu_1)$$

and

$$(7) \qquad \text{var}\{Y|x\} = \sigma_2^2(1 - \rho^2).$$

In a similar manner we show that the conditional density of X given $Y = y$ is also normal with

$$(8) \qquad \mathscr{E}\{X|y\} = \mu_1 + \left(\frac{\rho\sigma_1}{\sigma_2}\right)(y - \mu_2)$$

and

$$(9) \qquad \text{var}\{X|y\} = \sigma_1^2(1 - \rho^2).$$

Finally, we show that cov$(X, Y) = \rho\sigma_1\sigma_2$ so that ρ is the correlation coefficient between X and Y. In fact,

$$\mathscr{E}(XY) = \mathscr{E}(\mathscr{E}\{XY|X\})$$

where

$$\mathscr{E}\{XY|x\} = x\mathscr{E}\{Y|x\} = x\mu_2 + \left(\frac{\rho\sigma_2}{\sigma_1}\right)x(x - \mu_1)$$

in view of (6). Hence

$$\mathscr{E}(\mathscr{E}\{XY|X\}) = \mathscr{E}\left\{X\mu_2 + \left(\frac{\rho\sigma_2}{\sigma_1}\right)X(X - \mu_1)\right\}$$

$$= \mu_1\mu_2 + \left(\frac{\rho\sigma_2}{\sigma_1}\right)(\mathscr{E}X^2 - \mu_1^2)$$

$$= \mu_1\mu_2 + \rho\sigma_1\sigma_2.$$

It follows that

$$\text{cov}(X, Y) = \mathscr{E}(XY) - \mathscr{E}X\mathscr{E}Y = \rho\sigma_1\sigma_2$$

as asserted.

In the special case when $\mu_1 = \mu_2 = 0$, $\sigma_1 = \sigma_2 = 1$, and $\rho = 0$, we say that (X, Y) has a standard bivariate normal distribution, and when $\rho = 0$ and $\sigma_1 = \sigma_2$ the joint density is called *circular normal density*.

Bivariate Normal Distribution

$$f(x, y) = \frac{1}{2\pi\sigma_1\sigma_2\sqrt{1 - \rho^2}} \exp\left\{ -\frac{1}{2(1 - \rho^2)} \left[\left(\frac{x - \mu_1}{\sigma_1}\right)^2 \right.\right.$$

$$\left.\left. - 2\rho\left(\frac{x - \mu_1}{\sigma_1}\right)\left(\frac{y - \mu_2}{\sigma_2}\right) + \left(\frac{y - \mu_2}{\sigma_2}\right)^2 \right] \right\},$$

for $(x, y) \in \mathbb{R}_2$ where

$$\mathscr{E}X = \mu_1, \mathscr{E}Y = \mu_2, \text{var}(X) = \sigma_1^2, \text{var}(Y) = \sigma_2^2, \text{ and cov}(X, Y) = \rho\sigma_1\sigma_2.$$

Marginal densities of X and Y are normal. Conditional density of X given y is normal with

$$\mathscr{E}\{X|y\} = \mu_1 + \left(\frac{\rho\sigma_1}{\sigma_2}\right)(y - \mu_2)$$

$$\text{var}\{X|y\} = \sigma_1^2(1 - \rho^2)$$

Conditional density of Y given $X = x$ is also normal.

We recall that X and Y are uncorrelated if and only if $\rho = 0$ [or $\text{cov}(X, Y) = 0$] but $\rho = 0$ does not in general imply independence. In the case of bivariate normal distribution X and Y uncorrelated *does* imply independence of X and Y.

PROPOSITION 1. *If X and Y have a bivariate normal distribution, then X and Y are independent if and only if $\rho = 0$.*

Proof. It is sufficient to show that $\rho = 0$ implies independence since the converse holds trivially. This follows by substituting $\rho = 0$ in $f(x, y)$. □

We note that if X and Y are independent normal random variables, then (X, Y) has a bivariate normal distribution. Moreover, a bivariate normal density has marginal normal densities. Given that a density f has marginal densities that are normal, it does not follow that f has to be a bivariate normal density.

In Figure 1 we plot the bivariate normal distribution with $\mu_1 = \mu_2 = 0$, $\sigma_1 = \sigma_2 = 1$ for some selected values of ρ.

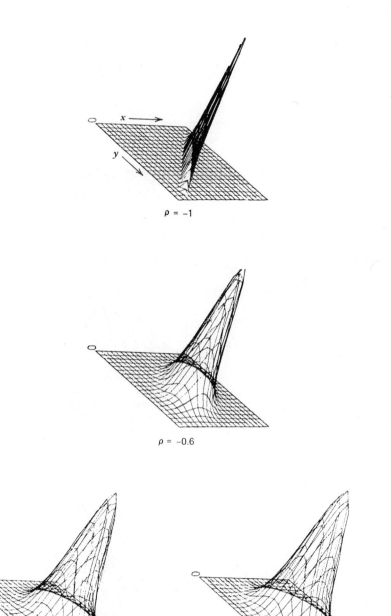

Figure 1. Bivariate normal distribution with $\mu_1 = \mu_2 = 0$, $\sigma_1 = \sigma_2 = 1$. Reproduced from N. L. Johnson and S. Kotz, *Distributions in Statistics: Continuous Multivariate Distributions*, Wiley, New York, 1972, pp. 89–90, with permission.

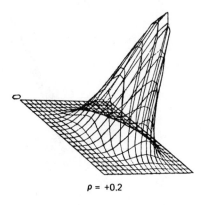

$\rho = +0.2$

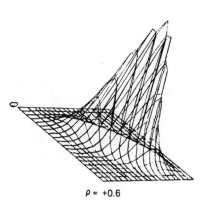

$\rho = +0.6$

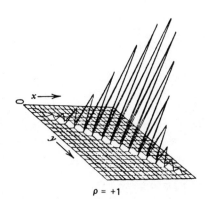

$\rho = +1$

Figure 1. (*continued*)

Problems for Section 7.7

1. For what values of c are the following functions bivariate normal densities? In each case give the five parameters of the bivariate normal distribution.

   (i) $f(x, y) = c \exp\{-\frac{1}{2}(x^2 - xy + y^2)\}, (x, y) \in \mathbb{R}_2$.

   (ii) $f(x, y) = c \exp\{-(2x^2 - xy + 4y^2)/2\}, (x, y) \in \mathbb{R}_2$.

   (iii) $f(x, y) = c \exp\{-\frac{1}{2}[y^2 + (xy/2) + (x^2/4) + (5/2)x + (5/2)y + 25/4]\}$.

2. Let X be the height and Y the weight at birth of boys born in the United States. Suppose (X, Y) has a bivariate normal distribution with $\mu_1 = 20$ inches, $\mu_2 = 7.5$ pounds, $\sigma_1 = 1.5$ inches, $\sigma_2 = 2$ pounds, and $\rho = 0.8$. Find

   (i) The expected height at birth of a child weighing 12 pounds.

   (ii) The expected weight at birth of a child whose height is 22 inches.

3. Let (X, Y) have a bivariate density given by

$$f(x, y) = \frac{1}{6\pi\sqrt{7}} \exp\left\{-\frac{8}{7}\left(\frac{x^2}{16} - \frac{31}{32}x + \frac{xy}{8} + \frac{y^2}{9} - \frac{4}{3}y + \frac{71}{16}\right)\right\}.$$

   for $-\infty < x < \infty, -\infty < y < \infty$.

   (i) Find the means and variances of X and Y and find the correlation coefficient between X and Y.

   (ii) Find the conditional density of Y given $X = x$ and find the mean and the variance of this density.

   (iii) Find $P\{6 \geq Y \geq 4 | X = 4\}$.

4*. Suppose the joint distribution of average scores obtained by students in their freshman and senior year, respectively X and Y, have a bivariate normal distribution with $\mu_1 = 65, \mu_2 = 80, \sigma_1 = 4, \sigma_2 = 2,$ and $\rho = .8$. Find $P(Y > X)$, the probability that a student is improving with maturity. (Hint: Find $P\{Y > X | X = x\}$ and integrate. Use Problem 7.6.26.)

5. Let X, Y have a circular normal density with $\sigma_1 = \sigma_2 = 16$ and $\mu_1 = \mu_2 = 0$. Find:

   (i) The probability $P(X \leq 8, Y \leq 8)$.

   (ii) The probability that $(X, Y) \in \{(x, y): x^2 + y^2 \leq 64\}$.

   (iii) k such that $P\{(X, Y) \in [(x, y): x^2 + y^2 \leq k^2]\} = .90$.

   [Hint: For (ii) and (iii) use polar coordinates.]

6. Let (X, Y) have a circular normal density with $\mu_1 = \mu_2 = 0$ and $\sigma_1 = \sigma_2 = 12$. Find:

   (i) $P(X < Y)$.   (ii) $P(X > 0, Y < 0)$.   (iii) $P(|X| < 5, |Y| < 3)$.

7. Let $A > 0, C > 0$ be constants. Suppose (X, Y) has joint density given by

$$f(x, y) = c \exp\{-Ax^2 + 2Bxy - Cy^2\}, (x, y) \in \mathbb{R}_2$$

   where $B \in \mathbb{R}$ such that $B^2 < AC$ and $c > 0$ are also constants.

   (i) Find the two marginal densities.

   (ii) When are X and Y independent? What is c in that case?

8. Let (X, Y) be a bivariate normal random vector with $\mu_1 = \mu_2 = 0$. Show that
    (i) $\mathscr{E}(X^2 Y^2) = \sigma_1^2 \sigma_2^2 (1 + 2\rho^2)$.
    (ii) $\mathrm{var}(XY) = \sigma_1^2 \sigma_2^2 (1 + \rho^2)$.
    [Hint: Condition on X (or Y). (In Example 8.6.6 we show that $\mathscr{E} X^4 = 3\sigma^4$.) Follow the proof of $\mathrm{cov}(X, Y) = \rho \sigma_1 \sigma_2$ in text.]

9*. For a bivariate normal random variable with parameters $\mu_1, \mu_2, \sigma_1, \sigma_2$, and ρ show that

$$P(X > \mu_1, Y > \mu_2) = \frac{1}{4} + \frac{1}{2\pi} \tan^{-1} \frac{\rho}{\sqrt{1 - \rho^2}}.$$

[Hint: $P(X > \mu_1, Y > \mu_2) = P((X - \mu_1)/\sigma_1 > 0, (Y - \mu_2)/\sigma_2 > 0)$. Integrate by changing to polar coordinates.]

7.8. REVIEW PROBLEMS

1. Two points are chosen at random from the interval $(0, 1)$. Consider the distance functions: (i) $Y_1 = |X_1 - X_2|$, (ii) $Y_2 = (X_1 - X_2)^2$. Find $\mathscr{E} Y_1, \mathscr{E} Y_2$, $\mathrm{var}(Y_1), \mathrm{var}(Y_2)$, and $P(Y_i \geq 1/2)$, $i = 1, 2$.

2. Let X be a nonnegative continuous type random variable with density function f. Find the probability that the roots of the quadratic equation $ax^2 + bxX + X = 0$ are real $(a, b \neq 0)$ when
    (i) X is uniform on $(0, 1)$.
    (ii) X is exponential with mean θ.

3. A point X is randomly chosen from the interval $(0, 1)$. Suppose $X = x$ is observed. Then a coin with $P(\text{heads}) = x$ is tossed independently n times. Let Y be the number of heads in n tosses. Find the unconditional distribution of Y.

4*. The proportion of defectives shipped by a wholesaler varies from shipment to shipment. If beta distribution with $\alpha = 3$ and $\beta = 2$ adequately approximates the distribution of the proportion of defectives, what is the probability that in a large shipment there will be at most 20 percent defectives?

5*. Show that if X has beta distribution with parameters (α, β), then $1 - X$ has beta distribution with parameters (β, α).

6*. Suppose X has normal distribution with mean 0 and variance 1. Show that

$$\mathscr{E}|X|^\alpha < \infty \quad \text{for } \alpha > -1 \quad \text{and} \quad \mathscr{E}|X|^\alpha = \infty \quad \text{for } \alpha \leq -1.$$

7*. Suppose X has density

$$f(x) = \frac{1}{\pi} \frac{1}{1 + x^2}, x \in \mathbb{R}.$$

Show that

$$\mathscr{E}|X|^\alpha < \infty \quad \text{for } 0 < \alpha < 1 \quad \text{and} \quad \mathscr{E}|X| = \infty.$$

8. Suppose the length of pregnancy of a certain population of females has a normal distribution with mean 270 days and a standard deviation of 12 days. What is the probability that a child was conceived 245 days or less before birth? In particular, suppose in a paternity suit the male defendant is able to prove that he was away for a period starting 300 days prior to the birth in question and ending 245 days prior to the birth. Is there sufficient evidence to conclude that he is not the father of the child?

9. Let $\alpha_i > 0$ for $i = 1, 2, \ldots, k$ and $c = \Gamma(\alpha_1 + \alpha_2 + \cdots + \alpha_k)/\{\Gamma(\alpha_1)\ldots\Gamma(\alpha_k)\}$. For $x_i > 0$, $i = 1, 2, \ldots, k - 1$ such that $x_1 + \cdots + x_{k-1} < 1$, define f by:

$$f(x_1, x_2, \ldots, x_{k-1}) = c(1 - x_1 - \cdots - x_{k-1})^{\alpha_k - 1} \prod_{j=1}^{k-1} x_j^{\alpha_j - 1}.$$

Show that f is a joint density function. [A random vector $(X_1, X_2, \ldots, X_k)$ is said to have a k-variate *Dirichlet distribution* with parameters $\alpha_1, \alpha_2, \ldots, \alpha_k$ if and only if $(X_1, X_2, \ldots, X_{k-1})$ has joint density function f and $X_k = 1 - \sum_{j=1}^{k-1} X_j$. The special case $k = 2$ gives the beta density function.]

10. The length of life of a 9-volt battery used in a radio has an exponential distribution with a mean of 20 hours. What is the probability that a randomly selected battery will last more than 30 hours? Suppose you are given a radio that has been running on a battery for 10 hours. What is the probability that it will continue to run for 10 more hours (on the same battery)?

11*. Suppose customers arrive at a store according to Poisson distribution with mean $\lambda = 4$ per hour. What is the expected waiting time until the 10th customer arrives? What is the probability that the waiting time between the arrival of the 10th and 11th customers exceeds 1 hour?

12. The amount of time (in minutes) that patients have to wait in the waiting room of a certain physician's office can be modeled by an exponential distribution with a mean of 30 minutes. Find the probability that a patient will have to wait:
(i)   More than an hour.   (ii)   At most 10 minutes.
(iii)   Between 15 and 45 minutes.

13. The joint distribution of temperatures (in degrees centigrade) at two locations at a particular time on a particular day is bivariate normal with parameters $\mu_1 = \mu_2 = 25$, $\sigma_1 = 1.5$, $\sigma_2 = 2$, and $\rho = .6$. What is the conditional density of Y given $X = x$? If $X = 22°C$, what is your best predictor of Y? Find $P\{22.5 \le Y \le 27 | X = 22\}$.

14. The length of guinea pigs X from Supplier A has a normal distribution with mean μ_1 (inches) and standard deviation σ_1. The length of guinea pigs Y from Supplier B has also a normal distribution with mean μ_2 (inches) and standard deviation σ_2. Find $P(X > \text{median}(Y))$ and $P(Y > \text{median}(X))$. When are the two probabilities equal?

15*. The number of defects per roll on a wallpaper produced by a company has a Poisson distribution with mean λ. The mean λ, however, changes from day to day and has an exponential distribution with mean 1. Find the unconditional distribution of the number of defects per roll. If a roll has one defect, what is the probability that it was produced on days for which $1 < \lambda < 3$?

16*. The number of trees per acre in a forest that have a certain disease has a Poisson
distribution with mean λ. The mean number of diseased trees, however, changes
from area to area and has a gamma distribution with parameters α and β. What is
the unconditional distribution of the number of diseased trees per acre? Find also
the conditional density of λ given that the number of diseased trees in an acre is k.

17. Suppose X has normal distribution with mean 25 and variance 64. Find
   (i) c such that $P(X \le c) = .4207$.
   (ii) a and b that are equidistant from the mean and such that $P(a < X < b) =
.9660$.
   (iii) $P(|X - 25| < 12)$ exactly and compare your result with the lower bound
obtained from Chebyshev's inequality.

18. Brightness of light bulbs (measured in foot-candles) manufactured by a certain
manufacturer has a normal distribution with mean 2600 foot-candles and standard
deviation 75 foot-candles. Bulbs are tested. Those brighter than 2750 foot-candles
are labeled high quality and sold at a premium, whereas those which show
brightness of 2500 foot-candles or less are heavily discounted. A random sample of
25 light bulbs is taken. What is the probability that five bulbs will be of high quality
and two bulbs will have to be discounted?

19. Let X have uniform distribution on the union of intervals $(0, 1)$, $(2, 4)$, and $(5, 8)$.
Find the distribution function of X. In a random sample of 15 observations, how
many do you expect to fall in each interval? Find the probability that six of the
observations will be less than 1 and four will be larger than 7.

20. The proportion of defectives X produced by a certain manufacturer has density
function:

$$f(x) = 6x(1 - x), \qquad 0 \le x \le 1 \text{ and zero otherwise.}$$

   (i)   Find a such that $P(X < a) = 2P(X > a)$.
   (ii) Find $P\{ X \le 1/2 | 1/3 < X < 2/3 \}$.

21. An electronic device has a life length X (in 1000 hours) that can be assumed to have
an exponential distribution with mean 1 (that is, 1000 hours). The cost of manufacturing an item is \$1.50 and the selling price is \$5 per item. However, a full refund is
guaranteed if $X < .9$. Find the manufacturer's expected profit per item.

22*. Recall that $\lambda(t) = f(t)/S(t)$ and $S(t) = \exp(-\int_0^t \lambda(s)\, ds)$. (See Example 7.4.5.)
Define $\Lambda(t) = \int_0^t \lambda(s)\, ds$. Then $S(t) = \exp(-\Lambda(t))$. A distribution function F is
said to have *increasing failure rate on the average* (IFRA) if $\Lambda(t)/t$ is an increasing
function of t for $t \ge 0$.
   (i)   Show that if F has an increasing failure rate (see Problem 7.4.21) then F has
IFRA.
   (ii) Show that F has IFRA if and only if for $0 \le \alpha \le 1$ and $t \ge 0$,

$$S(\alpha t) \ge [S(t)]^{\alpha}.$$

CHAPTER 8

Functions of Random Variables and Random Vectors

8.1 INTRODUCTION

In many practical problems we observe some real valued function g of a random variable X [or a random vector $(X_1, X_2, \ldots, X_n)$]. It is therefore of interest to determine the distribution of $g(X)$ [or $g(X_1, X_2, \ldots, X_n)$] given the distribution of X [or $(X_1, X_2, \ldots, X_n)$]. {All functions g considered here will be such that $g(X)$ [or $g(X_1, X_2, \ldots, X_n)$] is itself a random variable.} Consider, for example, a projectile fired with initial velocity u at an angle Θ, where Θ is uniformly distributed over the interval $(0, \pi/2)$. Let X be the distance from the launch site to the point where the projectile returns to earth, and assume that the projectile travels only under the effect of gravity. Then what is the density of X? What is the probability that X is greater than 10? From calculus we know that $X = g^{-1}u^2 \sin 2\Theta$ where g is the gravitational constant. It follows that X has distribution function given by

$$P\left(g^{-1}u^2 \sin 2\Theta \leq x\right) = \int_{g^{-1}u^2 \sin 2\theta \leq x} f_\Theta(\theta)\, d\theta$$

where f_Θ is the density function of Θ. One can obtain the density function of X by differentiating the distribution function. Similarly,

$$P(X > 10) = \int_{g^{-1}u^2 \sin 2\theta > 10} f_\Theta(\theta)\, d\theta$$

or

$$P(X > 10) = \int_{x > 10} f_X(x)\, dx$$

where f_X is the density function of X. The reader will notice that problems of this type were encountered in Chapter 3. All we do now is go one step further and find the density of X. Moreover, we note that to find $P(X > 10)$ we do not really need the density function of X as long as we know f_Θ. See Example 8.2.7.

In Section 8.2 we compute the distribution function of g from first principles in order to compute its density function. We use this so-called method of distribution functions to show that if X has a continuous distribution function then $F(X)$ has a uniform distribution on $(0, 1)$. This simple transformation allows us to transform any continuous type random variable into a random variable that has any desired continuous distribution. This result has applications in simulation and nonparametric inference.

In Section 8.3 we consider the special case when g is either strictly increasing or strictly decreasing and derive an expression for the density function of $Y = g(X)$. This so-called method of transformation bypasses the computation of distribution functions. We use here our knowledge of change of variable technique from calculus. We also consider the multivariate case.

In Section 8.4 we consider some special transformations that arise frequently. These are $g(x_1, x_2) = x_1 + x_2$, $g(x_1, x_2) = x_1 - x_2$, $g(x_1, x_2) = x_1 x_2$ and $g(x_1, x_2) = x_1/x_2$. We compute the density functions of $g(X_1, X_2)$ in terms of the joint distribution of (X_1, X_2). Several important special cases are considered, leading to t-, and F-distributions.

In Section 8.5 we derive the distribution of order statistics such as $\max(X_1, X_2, \ldots, X_n)$ or $\min(X_1, X_2, \ldots, X_n)$. Applications to inference concerning the quantile of order p are also considered.

Section 8.6 deals with the computation of $\mathscr{E}g(X)$ or $\mathscr{E}g(X_1, X_2, \ldots, X_n)$ for special classes of functions g. The idea here is to exploit the following property of exponential functions:

$$a^{x+y} = a^x \cdot a^y.$$

If X is integer-valued we consider the function $g(x) = s^x$. Then $P(s) = \mathscr{E}g(X) = \mathscr{E}s^X$ defines the probability generating function of X. The terms in the expansion of $P(s)$ give the probability function of X. Probability generating functions have important analytic properties that make them a useful tool in statistics. In the general case we consider the function $g(x) = \exp(tx)$, t real, and hence the function $M(t) = \mathscr{E}g(X) = \mathscr{E}\exp(tX)$. The function M, when it exists in an open neighborhood of the origin, also has important analytic properties that make it an indispensible tool in statistics. Section 8.6 supplements Sections 8.2 and 8.3, providing yet another method for determining the distribution of certain statistics.

Section 8.7 deals exclusively with the case of random sampling from a normal population. Section 8.7 together with Sections 9.6 and 9.7 form the basis of classical statistics. Here we compute the distribution of statistics such as $\overline{X}$, S^2, $(\overline{X} - \mu)/S$. These results are then used to consider inference problems concerning the mean μ and/or the variance σ^2.

Sections 8.3, 8.4, and 8.6 are of a technical nature and may not be of interest to every student. Nevertheless, the methods developed in these sections (and Chapter

9) are very useful in the calculation of the numerical limits of a confidence interval or in the calculation of *P*-value in testing a hypothesis or in the numerical assessment of the precision of an estimate. Every student of statistics at this level should be familiar with the statistical methods developed in Sections 8.5 and 8.7.

8.2   THE METHOD OF DISTRIBUTION FUNCTIONS

Let (Ω, ζ, P) be a probability space and let $\mathbf{X}$ be a random vector defined on Ω with (induced) distribution function F. In many applications the outcome of interest is not $\mathbf{X}$ but a certain function of $\mathbf{X}$. For example, if a projectile is fired with an initial velocity u at an angle X then one is interested in the point at which the projectile returns to earth (traveling only under the effect of gravity). If X_1 is the total time spent by a customer at a bank and X_2 is the time spent by the customer waiting in line, then it is of interest to the manager to know the distribution of the service time $X_1 - X_2$. When we toss a coin n times we do not usually keep record of each X_i ($X_i = 1$ if ith trial is a head and $X_i = 0$ otherwise). Rather we keep count of $\sum_{i=1}^{n} X_i$, the number of heads in n trials. What is the distribution of $\sum_{i=1}^{n} X_i$?

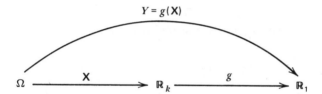

Let $\mathbf{X} = (X_1, X_2, \ldots, X_k)$ be defined on (Ω, ζ, P) and let g be a real valued function defined on $\mathbb{R}_k$. We assume that $Y = g(\mathbf{X})$ defined by

$$g(\mathbf{X})(\omega) = g(X_1(\omega), X_2(\omega), \ldots, X_k(\omega)), \qquad \omega \in \Omega$$

itself is a random variable. Given the distribution of $\mathbf{X}$ and the function g, we wish to find the distribution of $Y = g(\mathbf{X})$.

Let $y \in \mathbb{R}$. Then by definition the distribution function of Y is given by

$$(1) \qquad P(Y \le y) = P(\omega : g(\mathbf{X})(\omega) \le y)$$

$$= \begin{cases} \displaystyle\sum_{\{(x_1,\ldots,x_k):\ g(x_1,\ldots,x_k)\le y\}} \cdots \sum P(X_1 = x_1, \ldots, X_k = x_k) \\[4pt] \text{or} \\[4pt] \displaystyle\int_{\{(x_1,\ldots,x_k):\ g(x_1,\ldots,x_k)\le y\}} \cdots \int f(x_1,\ldots,x_k)\, dx_1 \ldots dx_k \end{cases}$$

according as $\mathbf{X}$ is the discrete or continuous type with joint density function f. If

X is the continuous type we can obtain the density of Y by differentiating the distribution function $P(Y \le y)$ with respect to y *provided that Y is also of the continuous type*. We will assume not only that Y is a random variable but also that it is of the continuous type if **X** is.

If, on the other hand, **X** is the discrete type then it is easier to compute the joint probability function of Y. Indeed,

$$(2) \quad P(Y = y) = \sum_{\{(x_1,\ldots,x_k):\ g(x_1,\ldots,x_k)=y\}} \cdots \sum P(X_1 = x_1,\ldots,X_k = x_k).$$

    Example 1.   *Function of Poisson Random Variable.*   Let X have probability function

$$P(X = x) = \begin{cases} e^{-\lambda}\lambda^x/x!, & x = 0,1,2,\ldots . \\ 0 & \text{elsewhere.} \end{cases}$$

Let $Y = g(X)$ where $g(x) = 2x + 3$. Then

$$P(Y = y) = P(2X + 3 = y) = P(X = (y - 3)/2)$$

$$= \begin{cases} \dfrac{e^{-\lambda}\lambda^{(y-3)/2}}{[(y - 3)/2]!}, & y = 3,5,7,\ldots \\ 0 & \text{elsewhere.} \end{cases} \qquad \square$$

    Example 2.   *Function of Discrete Uniform.*   Let X have uniform distribution on $\{-2, -1, 0, 1, 2\}$. Then

$$P(X = x) = 1/5 \qquad \text{for } x = \pm 2, \pm 1, 0$$

and

$$P(X = x) = 0 \qquad \text{elsewhere.}$$

Let $Y = X^2 - X$. Then Y takes values $0, 2, 6$ with probability function

$$P(Y = y) = P(X^2 - X = y)$$

$$= \begin{cases} P(X = 0 \text{ or } 1) = 2/5, & y = 0 \\ P(X = -1 \text{ or } 2) = 2/5, & y = 2 \\ P(X = -2) = 1/5, & y = 6 \\ 0 & \text{elsewhere.} \end{cases} \qquad \square$$

    Example 3.   *Distribution of $|X_1 - X_2|$ Given that of (X_1, X_2).*   Let (X_1, X_2) have joint probability function given by

X_2 \ X_1	0	1	2	$P(X_2 = x_2)$
1	0	1/6	1/6	1/3
2	1/6	0	1/6	1/3
3	1/12	1/12	1/6	1/3
$P(X_1 = x_1)$	1/4	1/4	1/2	1

Let $Y = |X_1 - X_2|$. Then Y takes values 0, 1, 2, and 3 with probability function

$$P(Y = y) = \begin{cases} P(X_1 = 1, X_2 = 1) + P(X_1 = 2, X_2 = 2) = 1/3, & y = 0 \\ P(X_1 = 0, X_2 = 1) + P(X_1 = 2, X_2 = 1) \\ \quad + P(X_1 = 2, X_2 = 3) + P(X_1 = 1, X_2 = 2) = 1/3, & y = 1 \\ P(X_1 = 0, X_2 = 2) + P(X_1 = 1, X_2 = 3) = 1/4, & y = 2 \\ P(X_1 = 0, X_2 = 3) = 1/12, & y = 3 \\ 0 & \text{elsewhere.} \end{cases}$$

$\square$

Example 4. Sum of Independent Poisson Random Variables. Let X_1 and X_2 be independent Poisson random variables with probability functions

$$P(X_i = x) = \begin{cases} e^{-\lambda}\lambda_i^x/x!, & x = 0,1,2,\ldots; i = 1,2 \\ 0 & \text{elsewhere.} \end{cases}$$

Let $Y = g(X_1, X_2) = X_1 + X_2$. Clearly, Y takes value $0,1,2,\ldots$. For nonnegative integer y,

$$P(Y = y) = P(X_1 + X_2 = y)$$

$$= \sum_{x=0}^{\infty} P(X_1 = x, X_1 + X_2 = y) \qquad \text{(total probability rule)}$$

$$= \sum_{x=0}^{y} P(X_1 = x, X_2 = y - x)$$

$$= \sum_{x=0}^{y} P(X_1 = x)P(X_2 = y - x) \qquad \text{(independence)}$$

$$= \sum_{x=0}^{y} \left(\frac{e^{-\lambda_1}\lambda_1^x}{x!}\right)\left(\frac{e^{-\lambda_2}\lambda_2^{y-x}}{(y-x)!}\right)$$

$$= e^{-(\lambda_1+\lambda_2)} \sum_{x=0}^{y} \frac{1}{y!}\frac{y!}{x!(y-x)!}\lambda_1^x\lambda_2^{y-x}$$

$$= \frac{e^{-(\lambda_1+\lambda_2)}}{y!} \sum_{x=0}^{y} \binom{y}{x}\lambda_1^x\lambda_2^{y-x}$$

$$= \frac{e^{-(\lambda_1+\lambda_2)}}{y!}(\lambda_1 + \lambda_2)^y \qquad \text{(binomial theorem)}$$

so that Y itself has a Poisson distribution with parameter $\lambda_1 + \lambda_2$. $\square$

Next we consider some examples where Y is of the continuous type.

Example 5.   Chi-Square Distribution As Square of Standard Normal.   Let X have standard normal density function

$$f(x) = e^{-x^2/2}/\sqrt{2\pi}, \quad x \in \mathbb{R}.$$

Then the distribution of $Y = X^2$ is given by

$$P(Y \le y) = 0, \quad \text{if } y \le 0,$$

and for $y > 0$ (see Figure 1)

$$P(Y \le y) = P(X^2 \le y) = P(|X| \le y^{1/2}) = 2P(0 < X \le y^{1/2})$$

$$= 2P(X \le y^{1/2}) - 1.$$

Differentiating with respect to y we get the density function of Y as

$$f_Y(y) = 0 \quad \text{if } y \le 0$$

and

$$f_Y(y) = 2f(y^{1/2}) \cdot \frac{1}{2}y^{-1/2} = \frac{1}{\sqrt{2\pi}}y^{-1/2}e^{-y/2} \quad \text{if } y > 0$$

which is the density function of a chi-square random variable with one d.f.   □

Example 6.   Rayleigh Distribution As Square Root of Exponential.   Suppose X has exponential distribution with mean $\lambda > 0$. Then the density function of X is given by

$$f(x) = \begin{cases} \lambda^{-1}e^{-x/\lambda}, & x > 0 \\ 0, & x \le 0. \end{cases} \quad (\lambda > 0)$$

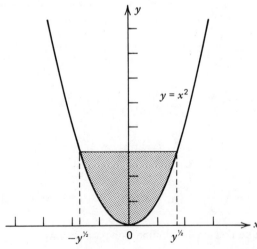

Figure 1.　$\{y \ge x^2\}$.

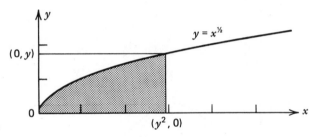

Figure 2. $\{x^{1/2} \le y\}$.

Let $Y = X^{1/2}$. Then for $y > 0$ (see Figure 2)

$$P(Y \le y) = P(X^{1/2} \le y) = P(X \le y^2)$$

and for $y \le 0$

$$P(Y \le y) = 0.$$

The density function of Y is obtained by differentiating $P(Y \le y)$ with respect to y, and we have $f_Y(y) = 2yf(y^2)$ for $y > 0$. Hence

$$f_Y(y) = \begin{cases} 0, & \text{if } y \le 0 \\ (2y/\lambda)e^{-y^2/\lambda}, & \text{if } y > 0. \end{cases}$$

This is the density function of a Rayleigh distribution. $\square$

Example 7.   *Range of a Projectile.*   Let a projectile be fired with initial velocity u at an angle X where X has uniform distribution on $(0, \pi/2)$. Let Y be the range of the projectile, that is, the distance between the point it was fired from and the point at which it returns to earth (Figure 3). We assume that the projectile is only under the effect of gravitational force. From calculus we know that $Y = g^{-1}u^2 \sin 2X$ where g is the gravitational constant assuming without loss of generality that the projectile was fired from the origin. Then the density function of X is given by

$$f(x) = \begin{cases} 2/\pi & \text{if } 0 < x < \pi/2 \\ 0 & \text{elsewhere.} \end{cases}$$

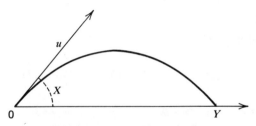

Figure 3.   Range of the projectile.

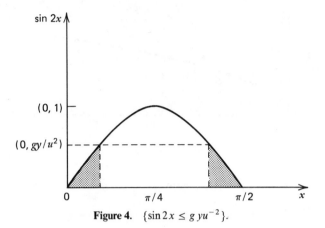

Figure 4. $\{\sin 2x \le g\, yu^{-2}\}$.

Now for $0 < y < g^{-1}u^2$ we have (see Figure 4)

$$P(Y \le y) = P(g^{-1}u^2 \sin 2X \le y) = P(\sin 2X \le gyu^{-2})$$

$$= 2P(X \le \tfrac{1}{2}\arcsin(gyu^{-2})) \qquad \text{(by symmetry)}$$

$$= 2 \cdot \frac{(\tfrac{1}{2})\arcsin(gyu^{-2})}{\pi/2} = (2/\pi)\arcsin(gyu^{-2}).$$

Differentiating with respect to y we get the density function of Y as

$$f_Y(y) = \begin{cases} \dfrac{2}{\pi}\dfrac{gu^{-2}}{(1 - g^2y^2u^{-4})^{1/2}}, & 0 < y < \dfrac{u^2}{g}, \\ 0 & \text{elsewhere.} \end{cases}$$

In particular,

$$P(Y > 10) = \begin{cases} 0 & \text{if } \dfrac{u^2}{g} < 10 \\ \int_{10}^{u^2/g} f_Y(y)\, dy & \text{if } \dfrac{u^2}{g} > 10. \end{cases}$$

Indeed, if $u^2 > 10g$, then

$$P(Y > 10) = 1 - P(Y \le 10) = 1 - \frac{2}{\pi}\arcsin(10\, gu^{-2}). \qquad \square$$

Example 8. *Exponential Distribution As a Function of Uniform Distribution.* Let X have uniform distribution on $(0,1)$. Then the distribution of $Y = \ln(1/X) = -\ln X$ is given by

$$P(Y \le y) = \begin{cases} 0, & y \le 0 \\ P(-\ln X \le y), & y > 0. \end{cases}$$

Hence, for $y > 0$ (Figure 5)

$$P(Y \le y) = P(X \ge e^{-y}) = \int_{e^{-y}}^{1} dx = 1 - e^{-y}$$

and

$$f_Y(y) = \begin{cases} 0 & \text{if } y \le 0 \\ e^{-y} & \text{if } y > 0 \end{cases}$$

so that Y has an exponential distribution with mean one. $\square$

Example 9. X Is Continuous, Y = g(X) Is Discrete. Let X have density function given by

$$f(x) = \begin{cases} 0, & x \le 0 \\ e^{-x}, & x > 0. \end{cases}$$

Let $Y = g(X)$ where $g(x) = [x]$ is the integral part of x, $x > 0$ (see Figure 6). Then for $y \ge 0$, y integral,

$$P(Y = y) = P([X] = y) = \int_{y}^{y+1} e^{-x} dx = e^{-y} - e^{-(y+1)},$$

and

$$P(Y = y) = 0 \qquad \text{otherwise.} \qquad \square$$

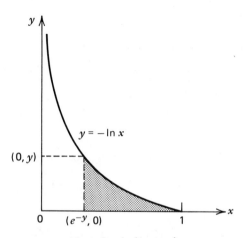

$y = -\ln x$

$(0, y)$

0

$(e^{-y}, 0)$

1

x

Figure 5. $\{-\ln x \le y\}$.

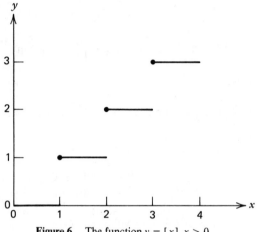

Figure 6. The function $y = [x]$, $x > 0$.

Next we consider some examples when $\mathbf{X}$ is a random vector and $Y = g(\mathbf{X})$ is a continuous type random variable.

    Example 10. ***Sum of Jointly Distributed Random Variables.*** Let (X_1, X_2) have uniform distribution on the triangle $\{0 \le x_1 \le x_2 \le 1\}$. Then (X_1, X_2) has joint density function

$$f(x_1, x_2) = \begin{cases} 2, & 0 \le x_1 \le x_2 \le 1 \\ 0, & \text{elsewhere.} \end{cases}$$

Let $Y = X_1 + X_2$. Then for $y < 0$, $P(Y \le y) = 0$, and for $y > 2$, $P(Y \le y) = 1$. For $0 \le y \le 2$, we have

$$P(Y \le y) = P(X_1 + X_2 \le y) = \iint\limits_{\substack{0 \le x_1 \le x_2 \le 1 \\ x_1 + x_2 \le y}} f(x_1, x_2)\, dx_1\, dx_2.$$

There are two cases to consider according to whether $0 \le y \le 1$ or $1 \le y \le 2$ (Figures 7*a* and 7*b*). In the former case,

$$P(Y \le y) = \int_{x_1=0}^{y/2}\left(\int_{x_2=x_1}^{y-x_1} 2\, dx_2\right) dx_1 = 2\int_0^{y/2}(y - 2x_1)\, dx_1 = y^2/2$$

and in the latter case,

$$P(Y \le y) = 1 - P(Y > y) = 1 - \int_{x_2=y/2}^{1}\left(\int_{x_1=y-x_2}^{x_2} 2\, dx_1\right) dx_2$$

$$= 1 - 2\int_{y/2}^{1}(2x_2 - y)\, dx_2 = 1 - \frac{(y - 2)^2}{2}.$$

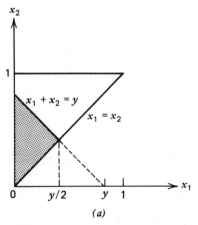

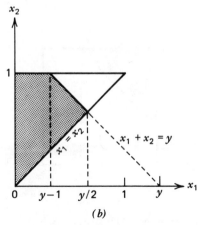

Figure 7.   (a) $\{x_1 + x_2 \leq y, 0 \leq x_1 \leq x_2 \leq 1, 0 \leq y \leq 1\}$. (b) $\{x_1 + x_2 \leq y, 0 \leq x_1 \leq x_2 \leq 1 \leq y \leq 2\}$.

Hence the density function of Y is given by

$$f_Y(y) = \begin{cases} y, & 0 \leq y \leq 1 \\ 2 - y, & 1 \leq y \leq 2 \\ 0, & \text{elsewhere.} \end{cases}$$ $\square$

*Example 11.   **Difference of Independent and Uniformly Distributed Random Variables.*** Let (X_1, X_2) have a joint uniform distribution on the unit square $\{(x_1, x_2): 0 < x_1 < 1, 0 < x_2 < 1\}$. Let $Y = X_1 - X_2$. Then

$$P(Y \leq y) = \iint\limits_{x_1 - x_2 \leq y} f(x_1, x_2) \, dx_1 \, dx_2.$$

Clearly, $P(Y \leq y) = 0$ if $y \leq -1$, and $P(Y \leq y) = 1$ if $y \geq 1$. For $|y| < 1$ we need to consider two cases according to whether $-1 < y \leq 0$ or $0 < y < 1$ (see Figures 8a and 8b). For $-1 < y \leq 0$, we have:

$$P(Y \leq y) = \int_{x_1=0}^{y+1} \left(\int_{x_2 = x_1 - y}^{1} 1 \, dx_2 \right) dx_1 = \int_0^{y+1} (1 - x_1 + y) \, dx_1$$

$$= \frac{(y + 1)^2}{2},$$

and for $0 < y < 1$,

$$P(Y \leq y) = 1 - P(Y > y) = 1 - \int_{x_1=y}^{1} \left(\int_{x_2=0}^{x_1 - y} 1 \, dx_2 \right) dx_1$$

$$= 1 - \int_y^1 (x_1 - y) \, dx_1 = 1 - \frac{(y - 1)^2}{2}.$$

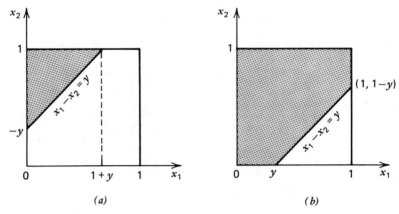

Figure 8. (a) $\{(x_1, x_2): -1 < x_1 - x_2 \le y \le 0\}$. (b) $\{(x_1, x_2): 0 < x_1 - x_2 \le y < 1\}$.

Hence

$$f_Y(y) = \begin{cases} 1 + y, & -1 < y \le 0 \\ 1 - y, & 0 < y < 1 \\ 0 & \text{elsewhere} \end{cases} = \begin{cases} 1 - |y|, & |y| < 1 \\ 0 & \text{elsewhere.} \end{cases}$$

In this particular case (and typically in cases where the distribution is uniform over a region) it is easy to compute $P(Y \le y)$ by simply computing the volume of the solid with height $f(x_1, x_2) = 1$ and triangular cross section as shown in Figure 8a or 8b. Indeed, for $-1 < y \le 0$,

$$P(Y \le y) = (\text{Area of the shaded triangle in Figure 8a}) \times \text{Altitude}$$

$$= \frac{(y + 1)^2}{2}(1) = \frac{(y + 1)^2}{2}$$

and for $0 < y < 1$

$$P(Y \le y) = 1 - \frac{(1 - y)^2}{2}.$$

We leave the reader to check that the density function of the random variable Y in Example 10 can also be obtained by a similar computation. □

Example 12. Product of Two Uniformly Distributed Random Variables. Let X_1, X_2 be a random sample of size 2 from a uniform distribution on $(0, 1)$. Let $Y = X_1 X_2$. Then for $0 < y < 1$ we have (Figure 9)

$$P(Y \le y) = P(X_1 X_2 \le y) = \iint\limits_{x_1 x_2 \le y} f(x_1, x_2) \, dx_1 \, dx_2$$

$$= 1 - \iint\limits_{x_1 x_2 > y} f(x_1, x_2) \, dx_1 \, dx_2 = 1 - \int_{x_1 = y}^{1} \left(\int_{x_2 = y/x_1}^{1} (1) \, dx_2 \right) dx_1$$

$$= 1 - \int_{y}^{1} [1 - y/x_1] \, dx_1 = y(1 - \ln y)$$

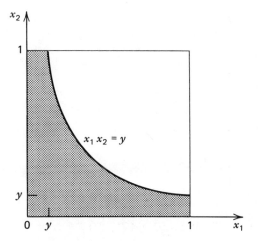

Figure 9. $\{(x_1, x_2): x_1 x_2 \leq y, 0 < x_1, x_2 < 1\}$.

and

$$P(Y \leq y) = \begin{cases} 0 & \text{for } y \leq 0 \\ 1 & \text{for } y \geq 1. \end{cases}$$

It follows that the density function of Y is given by

$$f_Y(y) = \begin{cases} -\ln y, & 0 < y < 1 \\ 0 & \text{elsewhere.} \end{cases} \qquad \square$$

Example 13. ***Distribution of Proportion of Times a Machine Is in Operation.*** Let X_1 be the time to failure of a machine and X_2 the repair time. Suppose X_1 and X_2 are independent exponentially distributed random variables with means λ_1 and λ_2, respectively. Then the proportion of time that the machine is in operation during any cycle is a random variable Y given by $Y = X_1/(X_1 + X_2)$. The distribution function of Y is given by (Figure 10)

$$P(Y \leq y) = P\big(X_1 \leq y(X_1 + X_2)\big) = P\big(X_1(1 - y) \leq yX_2\big)$$

$$= \iint\limits_{x_2 \geq \frac{1-y}{y} x_1} f(x_1, x_2)\, dx_1\, dx_2$$

$$= \frac{1}{\lambda_1 \lambda_2} \iint\limits_{x_2 \geq \frac{1-y}{y} x_1} e^{-x_1/\lambda_1} e^{-x_2/\lambda_2}\, dx_1\, dx_2, \qquad 0 < y < 1.$$

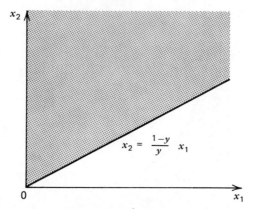

Figure 10. $\{(x_1, x_2): x_2 \geq \dfrac{1-y}{y} x_1, 0 < x_1, x_2 < \infty\}.$

Hence for $0 < y < 1$,

$$P(Y \leq y) = \frac{1}{\lambda_1 \lambda_2} \int_{x_1=0}^{\infty} e^{-x_1/\lambda_1} \left(\int_{x_2 = \frac{1-y}{y} x_1}^{\infty} e^{-x_2/\lambda_2} \, dx_2 \right) dx_1$$

$$= \frac{1}{\lambda_1} \int_0^{\infty} e^{-x_1/\lambda_1} e^{-[(1-y)/y]x_1/\lambda_2} \, dx_1$$

$$= \frac{1}{\lambda_1} \left[\frac{1}{\lambda_1} + \frac{1-y}{y} \frac{1}{\lambda_2} \right]^{-1} = \frac{\lambda_2 y}{\lambda_2 y + \lambda_1 (1-y)}$$

and the density of Y is given by

$$f_Y(y) = \begin{cases} 0 & \text{if } y \leq 0 \text{ or } y \geq 1, \\ \lambda_1 \lambda_2 [\lambda_2 y + (1-y)\lambda_1]^{-2} & \text{if } 0 < y < 1. \end{cases}$$

In particular, if $\lambda_1 = \lambda_2$ then Y has uniform distribution on $(0, 1)$. $\qquad \square$

The method of distribution functions can also be applied to compute the joint distribution of two or more functions of a random vector $(X_1, X_2, \ldots, X_k)$. The integration, however, becomes very complicated. We content ourselves here with the following example.

Example 14. Joint Distribution of $X_1 + X_2$ and $X_1 - X_2$ (Service Time at a Service Desk). Let X_1 be the time that a customer takes from getting on line at a service desk in a bank to completion of service, and let X_2 be the time she waits in line before she reaches the service desk. Then $X_1 \geq X_2$ and $X_1 - X_2$ is the service time of the customer. Suppose the joint density of (X_1, X_2) is given by

$$f(x_1, x_2) = \begin{cases} e^{-x_1}, & 0 \leq x_2 \leq x_1 < \infty \\ 0 & \text{elsewhere.} \end{cases}$$

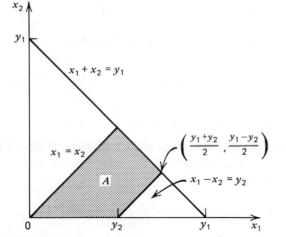

Figure 11. $\{(x_1, x_2): x_1 + x_2 \le y_1, x_1 - x_2 \le y_2, 0 \le x_2 \le x_1 < \infty\}.$

Let $Y_1 = X_1 + X_2$ and $Y_2 = X_1 - X_2$. Then the joint distribution of (Y_1, Y_2) is given by

$$P(Y_1 \le y_1, Y_2 \le y_2) = \iint_A f(x_1, x_2)\, dx_1\, dx_2$$

where $A = \{(x_1, x_2): x_1 + x_2 \le y_1, x_1 - x_2 \le y_2, 0 \le x_2 \le x_1 < \infty\}$. Clearly, $x_1 + x_2 \ge x_1 - x_2$ so that the set A is as shown in Figure 11. It follows that

$$P(Y_1 \le y_1, Y_2 \le y_2) = \int_{x_2=0}^{(y_1-y_2)/2} \left(\int_{x_1=x_2}^{x_2+y_2} e^{-x_1}\, dx_1 \right) dx_2$$

$$+ \int_{x_2=(y_1-y_2)/2}^{y_1/2} \left(\int_{x_1=x_2}^{y_1-x_2} e^{-x_1}\, dx_1 \right) dx_2$$

$$= \int_0^{(y_1-y_2)/2} e^{-x_2}(1 - e^{-y_2})\, dx_2$$

$$+ \int_{(y_1-y_2)/2}^{y_1/2} (e^{-x_2} - e^{-y_1+x_2})\, dx_2$$

$$= (1 - e^{-y_2})(1 - e^{-(y_1-y_2)/2})$$

$$+ (e^{-(y_1-y_2)/2} - e^{-y_1/2}) - e^{-y_1}(e^{y_1/2} - e^{(y_1-y_2)/2})$$

$$= 1 - e^{-y_2} - 2e^{-y_1/2} + 2e^{-(y_1+y_2)/2}.$$

Hence the joint density of Y_1, Y_2 is given by

$$f_{Y_1, Y_2}(y_1, y_2) = \begin{cases} \frac{1}{2} e^{-(y_1+y_2)/2}, & 0 \le y_2 \le y_1 < \infty \\ 0 & \text{elsewhere.} \end{cases}$$

The marginal densities of Y_1, Y_2 are easily obtained as

$$f_{Y_1}(y_1) = e^{-y_1} \quad \text{for } y_1 \geq 0, \text{ and } 0 \text{ elsewhere;}$$

$$f_{Y_2}(y_2) = e^{-y_2/2}(1 - e^{-y_2/2}), \quad \text{for } y_2 \geq 0, \text{ and } 0 \text{ elsewhere.} \qquad \square$$

The Probability Integral Transformation and Its Applications

Let X be a random variable with distribution function F so that $F(x) = P(X \leq x)$ for $x \in \mathbb{R}$. What is the distribution of the random variable $Y = F(X)$? Clearly, $0 \leq Y \leq 1$, and for $0 < y < 1$

$$P(Y \leq y) = P(F(X) \leq y).$$

If we can write $F(X) \leq y$ as $X \leq F^{-1}(y)$, then

$$(3) \qquad P(Y \leq y) = P(X \leq F^{-1}(y)) = F(F^{-1}(y)) = y$$

and Y will have uniform distribution on $(0, 1)$. Unfortunately, F may not be strictly increasing or even continuous. If F is not continuous, then there is a jump at, say, $X = a$. That is, $P(X = a) > 0$ for some a. In that case, $Y = F(X)$ takes the value $F(a)$ with positive probability and Y itself is not of the continuous type. If, on the other hand, F is not strictly increasing, then F is "flat" on some interval, say (a, b) and $F(x) = y_0$ for all $x \in (a, b)$. In that case, F does not have a unique inverse at y_0 (Figure 12). If, however, F is continuous and strictly increasing, then F has a continuous and strictly increasing inverse F^{-1} over $0 < y < 1$ so that for $0 < y < 1$

$$P(Y \leq y) = P(X \leq F^{-1}(y)) = y,$$

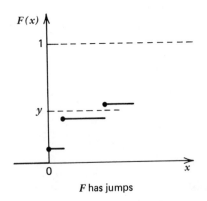

F has jumps

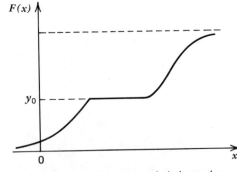

F is continuous but not strictly increasing

Figure 12

and

$$P(Y \leq y) = \begin{cases} 0 & \text{for } y < 0 \\ 1 & \text{for } y \geq 1. \end{cases}$$

Consequently Y has uniform distribution on $(0,1)$.

Suppose F is continuous. Then for $0 < y < 1$ there exists a unique smallest $x(y)$ such that $F(x(y)) = y$. Formally:

(4) $$x(y) = F^{-1}(y) = \inf\{x : F(x) \geq y\}.$$

Then

$$F(x) \leq y \quad \text{if and only if} \quad x \leq F^{-1}(y).$$

It follows that for $0 < y < 1$

$$P(F(X) \leq y) = P(X \leq F^{-1}(y)) = F(F^{-1}(y)) = F(x(y)) = y$$

and Y has a uniform distribution on $(0,1)$. The transformation $Y = F(X)$ is known as the *probability integral transformation*.

Probability Integral Transformation

If X has a continuous distribution function F, then $F(X)$ has uniform distribution on $(0,1)$.

In the case when X is discrete, F has jumps and it is clear from Figure 12 that for $0 < y < 1$

(5) $$P(F(X) \leq y) \leq y.$$

In fact, if X takes values $x_1 < x_2 < \cdots$ with probabilities $p_1, p_2,\ldots$ then

$$F(x) = P(X \leq x) = \sum_{j \leq i} p_j, \qquad x_i \leq x < x_{i+1}.$$

Thus $Y = F(X)$ is a discrete random variable which takes values $p_1, p_1 + p_2, p_1 + p_2 + p_3,\ldots$ with probabilities $p_1, p_2, p_3,\ldots$. Moreover,

(6) $$P(Y \leq y) = \sum_{j \leq i} p_j, \qquad \sum_{j \leq i} p_j \leq y < \sum_{j \leq i+1} p_j.$$

Probability integral transformation has important consequences in nonparametric statistics and in computer simulation when one needs a random sample from a given distribution G.

PROPOSITION 1. *Let G be any distribution function and X have uniform distribution on* $(0,1)$. *Then there exists a function h such that*

$$P(h(X) \le x) = G(x), \qquad x \in \mathbb{R}.$$

Proof. If G is continuous, we choose $h(X) = G^{-1}(X)$. Then

$$P(h(X) \le x) = P(X \le G(x)) = G(x).$$

If G has jumps of size p_k at y_k, $k = 1, 2, \dots$ then we choose h as follows.

$$h(x) = \begin{cases} y_1 & 0 < x \le p_1 \\ y_2 & p_1 < x \le p_1 + p_2 \\ \vdots \end{cases}$$

Then

$$P(h(X) = y_1) = P(0 \le X \le p_1) = p_1,$$

$$P(h(X) = y_2) = P(p_1 < X \le p_1 + p_2) = p_2,$$

and, in general,

$$P(h(X) = y_k) = p_k, \qquad k = 1, 2, \dots.$$

Thus $h(X)$ is a discrete random variable with distribution function G. $\square$

According to probability integral transformation, if X has density f and distribution function F, then $F(X)$ has uniform distribution on $(0,1)$. According to Proposition 1, if Y has density g and distribution function G, then $Y = G^{-1}(F(X))$ transforms the density f into g. Thus any continuous type random variable can be transformed into a random variable that has any desired continuous distribution by using the probability integral transformation.

Example 15. **Finding h Such That Y and h(X) Have the Same Distribution.**   Let X have density

$$f(x) = \begin{cases} (1/\lambda)e^{-x/\lambda}, & x \ge 0 \\ 0, & x < 0. \end{cases}$$

Then

$$F(x) = P(X \le x) = \begin{cases} 1 - e^{-x/\lambda}, & x \ge 0 \\ 0, & x < 0. \end{cases}$$

It follows that $Z = F(X) = 1 - e^{-X/\lambda}$ has uniform distribution on $(0,1)$.
Suppose Y has distribution function

$$G(y) = \begin{cases} 1 - \dfrac{1}{y}, & y \ge 1 \\ 0, & y < 0. \end{cases}$$

Then $U = G(Y) = 1 - 1/Y$ has uniform distribution on $(0,1)$. It also follows that $Y = 1/(1 - U) = 1/e^{-X/\lambda} = e^{X/\lambda}$ has density g where g is the density corresponding to G.

Also, $-\lambda \ln(1 - Z)$ has density f and $1/(1 - U)$ has density g in view of Proposition 1. Given a table of random numbers from uniform distribution we can use it to obtain random samples from f and g.

Suppose, for example, that .41, .13, .90, .43, and .52 is a random sample of size 5 from the uniform distribution on $(0,1)$. Then 1.06, 0.28, 4.61, 1.12, and 1.47 is a random sample from the exponential density f with mean $\lambda = 2$ and 1.69, 1.15, 10.00, 1.75, and 2.08 is a random sample from the density function g. □

Problems for Section 8.2

1. Given X with probability function

$$P(X = x) = \binom{n}{x} p^x (1 - p)^{n-x}, \qquad x = 0,1,2,\ldots,n; \quad 0 < p < 1.$$

Find the probability functions of:
(i) $Y = aX + b$ where a, b are nonnegative integers. (ii) $Y = X^2$.

2. Suppose X has probability function

$$P(X = x) = p(1 - p)^{x-1}, \qquad x = 1,2,\ldots; \quad 0 < p < 1.$$

Let $Y = g(X) = 1$ if X is an even integer and $Y = -1$ if X is an odd integer. Find the probability function of Y.

3. Suppose X has probability function

$$P(X = x) = e^{-\lambda} \frac{\lambda^x}{x!}, \qquad x = 0,1,2,\ldots.$$

Find the probability function of $Y = X^2 + 4$.

4. Suppose X has probability function

$$P(X = x) = \frac{1}{N}, \qquad x = 1,2,\ldots,N.$$

Find the probability function $Y = g(X) = X(\mathrm{mod}\,3)$. That is, $Y = 1$ if X is of the form $3m + 1$, $Y = 2$ if X is of the form $3m + 2$, and $Y = 0$ if X is of the form $3m + 3$; $m \geq 0$ is an integer.

5. Let Y be the output and X the input of an electrical device. The device is called a "full-wave" rectifier if $Y = |X|$. For a full-wave rectifier, find the density of the output if the density of the input is standard normal.

6. Let X be the (random) number of taxis that arrive at a taxi stand at O'Hare Airport (Chicago) during an interval of one hour. Let Y be the number of passengers who arrive at O'Hare and need taxi service during this time interval. Assume that X and Y are independent Poisson random variables; find the probability function of the excess of users over taxis, $Y - X$.

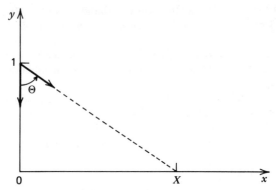

Figure 13. Position of the pointer after it comes to rest.

7. The temperature F, recorded in Fahrenheit, is related to temperature C measured in Centigrade by the relation C = 5(F − 32)/9. Suppose the temperature in Fahrenheit of a certain object has a normal distribution with mean μ and variance σ^2. Find the distribution of C.

8. According to the kinetic theory of gases, the velocity V of a molecule of mass m in a gas at absolute temperature T has a Maxwell distribution with density

$$f(x) = 4(\alpha^3\sqrt{\pi})^{-1}x^2\exp\left\{-\frac{x^2}{\alpha^2}\right\} \qquad \text{for } x > 0, \quad \text{and 0 elsewhere,}$$

where $\alpha = (2kT/m)^{1/2}$, k being the Boltzmann's constant. Find the density function of the kinetic energy $E = mV^2/2$ of a molecule.

9. A pointer supported vertically one unit above the point of origin is spun and comes to rest at an angle Θ with the y-axis (Figure 13). Suppose Θ has uniform distribution over $[-\pi/2, \pi/2]$. Let X be the distance (from the origin) of the point on the x-axis towards which the pointer is pointing. Find the distribution and density function of X.

10. The tensile strength T of a wire is assumed to be proportional to the square of its diameter X. If X has uniform distribution over (a, b), $0 < a < b$, find the distribution of T.

11. Let (X_1, X_2) have the joint probability function given in the following table

X_2 \ X_1	0	1	2	
0	25/144	10/144	1/144	36/144
1	50/144	20/144	2/144	72/144
2	25/144	10/144	1/144	36/144
	100/144	40/144	4/144	1

Find the probability functions of:
(i) $X_1 + X_2$.   (ii) $X_1 - X_2$.   (iii) $|X_1 - X_2|$.   (iv) $X_1 X_2$.

12. Let X_1, X_2 be independent random variables such that X_i has a negative binomial distribution with parameters r_i and p, $i = 1, 2$. Find the probability function of $X_1 + X_2$. [Hint: $P(X_1 = x) = \binom{x + r_1 - 1}{r_1 - 1} p^{r_1}(1 - p)^x$, $x = 0, 1, \ldots$. Use the identity $\sum_{y=0}^{k} \binom{a + k - y - 1}{a - 1}\binom{b + y - 1}{b - 1} = \binom{a + b + k - 1}{k}$.]

13. Let X_1, X_2 be independent random variables such that X_i has a binomial distribution with parameters n_i and p, $i = 1, 2$. Find the distribution of $X_1 + X_2$.

14. An urn contains four marbles marked 1 through 4. One marble is drawn at random and then all marbles with face values less than the one drawn are removed from the urn. The first marble drawn is then replaced in the urn and a second marble is drawn. Let X_1 be the number on the first marble drawn, and X_2 the number on the second marble drawn. Find the distributions of:
(i) $X_1 + X_2$.   (ii) $X_2 - X_1$.

15. A soft drink vending machine has a random amount X_1 gallons in supply at the beginning of a given day. It dispenses an amount X_2 gallons during the day. Assuming that the machine is not resupplied during the day and that X_1, X_2 have joint density:

$$f(x_1, x_2) = \begin{cases} 1/2, & 0 \le x_2 \le x_1 \le 2 \\ 0 & \text{otherwise} \end{cases}$$

find the density of $X_1 - X_2$, the amount sold during the day.

16. Let X_1, X_2 have joint density

$$f(x_1, x_2) = \begin{cases} 2(1 - x_1), & 0 \le x_1 \le 1, 0 \le x_2 \le 1, \\ 0 & \text{elsewhere.} \end{cases}$$

Find the density of $X_1 X_2$.

17. Let R be the radius and T the thickness of hockey pucks manufactured by a certain manufacturer. Suppose R and T are independently distributed with uniform distributions over the intervals $(5.0, 5.2)$ and $(2, 2.05)$ centimeters respectively. What is the density function of the volume of the puck?

18. Suppose X has a gamma density with parameters $\alpha > 0$ and $\beta = 1$. Let Y be independent of X and have the same distribution as X. Find the distributions of $X + Y$, X/Y, and $X/(X + Y)$.

19. Let (X, Y) have joint density f which is uniform on the triangle bounded by $0 \le x \le 3, 0 \le y \le 1$, and $x \ge 3y$. Show that $U = X - Y$ has density

$$f_U(u) = \begin{cases} u/3, & 0 < u < 2 \\ \tfrac{2}{3}(3 - u), & 2 < u < 3 \\ 0 & \text{elsewhere.} \end{cases}$$

20. Suppose X has exponential density with mean λ and Y has density

$$g(y) = \begin{cases} 1/(2\sqrt{y}), & 0 \le y \le 1 \\ 0 & \text{elsewhere.} \end{cases}$$

Find a transformation $y = h(x)$ such that if X has exponential density then $h(X)$ has density g.

21. Let (X, Y) have uniform density on the triangle $0 \le y \le x \le 1$. Show that:
    (i) $W = X/Y$ has density $f_W(w) = 1/w^2$ for $w \ge 1$ and $f_W(w) = 0$ elsewhere.
    (ii) $U = XY$ has density $f_U(u) = 2 + \ln u$ for $0 < u < 1$ and $f_U(u) = 0$ elsewhere.

22. Let X have uniform distribution on $(0, 1)$. Show that $Y = \cos 3x$ has density function given by $f_Y(y) = 1/(3\sqrt{1 - y^2})$, $\cos 3 < y < 1$ and 0 elsewhere.

23. Suppose X has density

$$f(x) = \begin{cases} 6x(1 - x), & 0 \le x \le 1 \\ 0 & \text{elsewhere} \end{cases}$$

and Y has density

$$g(y) = \begin{cases} 2(1 - y), & 0 \le y \le 1 \\ 0 & \text{elsewhere.} \end{cases}$$

Find the transformation $Y = h(X)$.

24. The following random sample was taken from the uniform distribution on $(0, 1)$:

$$.211 \quad .023 \quad .533 \quad .144 \quad .730 \quad .641 \quad .963.$$

   (i) Generate a random sample of size 7 from density function g of Problem 23.
   (ii) Generate a random sample from density function g of Problem 20.

8.3*   THE METHOD OF TRANSFORMATIONS

In this section we restrict our attention to the case when X has a density function f and $Y = g(X)$ itself is a continuous type random variable. Frequently the function g is either (strictly) increasing or (strictly) decreasing. In such cases it is possible to compute the density of $Y = g(X)$ directly without computing the distribution function. Recall that for $y \in \mathbb{R}$

(1) $$P(Y \le y) = P(\omega : g(X(\omega)) \le y) = \int_{g(x) \le y} f(x) \, dx.$$

We should like to be able to write $P(g(X) \le y)$ as $P(X \in A)$ where A is a subset of the real line (and is a function of y and g). Whenever this is possible, we can obtain the density of Y by differentiating $P(X \in A)$ with respect to y. This is certainly the case when g is (strictly) increasing or (strictly) decreasing over the set where $f > 0$. In either case, the mapping $x \to g(x)$ has a unique inverse g^{-1}. That is, for every y there is one and only one x given by $x = g^{-1}(y)$ and, moreover, $\{g(X) \le y\} = \{X \le g^{-1}(y)\}$ or $\{X \ge g^{-1}(y)\}$ according to whether g is increasing or decreasing. See Figures 1a and 1b.

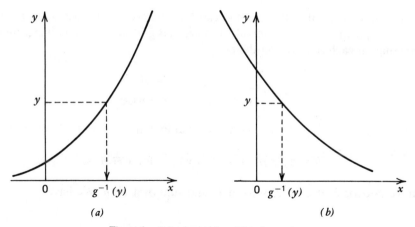

Figure 1.   (*a*) *g* increasing. (*b*) *g* decreasing.

Suppose *g* is increasing. In that case:

(2) $$P(Y \le y) = P(X \le g^{-1}(y)) = F(g^{-1}(y))$$

where F is the distribution function of X. Differentiating with respect to y we get the density of Y as

(3) $$f_Y(y) = \frac{d}{dy} F(g^{-1}(y)) = f(g^{-1}(y)) \cdot \frac{dx}{dy}$$

where $x = g^{-1}(y)$. In the case when g is decreasing, on the other hand,

(4) $$P(Y \le y) = P(X \ge g^{-1}(y)) = 1 - P(X \le g^{-1}(y))$$

since X is a continuous type random variable. Hence

(5) $$f_Y(y) = -f(g^{-1}(y)) \frac{dx}{dy}, \qquad x = g^{-1}(y).$$

Since, however, g is decreasing, so is g^{-1} and dx/dy is negative. It follows that in either case

(6) $$f_Y(y) = f(g^{-1}(y)) \left| \frac{dx}{dy} \right|, \qquad x = g^{-1}(y).$$

The Density Function of $Y = g(X)$ When g Is Increasing or Decreasing

$$f_Y(y) = f(g^{-1}(y)) \left| \frac{dx}{dy} \right|, \qquad x = g^{-1}(y)$$

It should be clear that this method can be extended to the case when the real line can be partitioned into intervals A_n such that g is either strictly increasing or decreasing on each A_n. We can define

$$g_n(x) = \begin{cases} g(x), & x \in A_n \\ 0 & \text{otherwise} \end{cases}$$

and write $g(x) = \sum_{n=1}^{\infty} g_n(x)$. Then g_n has a unique inverse in A_n. If we write

$$y = g_n(x), \quad x = g_n^{-1}(y) \qquad \text{for } x \in A_n$$

then we can consider each g_n separately and sum over all n. We have

(7) $$f_Y(y) = \sum_{n=1}^{\infty} f\big(g_n^{-1}(y)\big)\left|\frac{d}{dy} g_n^{-1}(y)\right|.$$

A few examples of this type will also be given in this section.

Example 1. *A Linear Function of Uniformly Distributed Random Variable.* Let X have uniform distribution on $(0,1)$ and let $y = g(x) = 2x + 4$. Since $Y = 2X + 4$ simply changes the origin and scale, we expect the distribution of Y to be uniform. Note that g is increasing on $(0,1)$ and the inverse function is obtained by solving $y = 2x + 4$ for x. Indeed,

$$x = g^{-1}(y) = \frac{(y - 4)}{2},$$

so that

$$\frac{dx}{dy} = \frac{1}{2}.$$

It follows that

$$f_Y(y) = f\big(g^{-1}(y)\big)\left|\frac{dx}{dy}\right| = f\left(\frac{y-4}{2}\right) \cdot \frac{1}{2}$$

$$= \begin{cases} \frac{1}{2} & \text{if } 0 < \dfrac{y-4}{2} < 1 \\ 0 & \text{elsewhere.} \end{cases}$$

We note that Y has uniform distribution on $(4, 6)$. $\qquad\qquad\square$

Example 2. *Square of Exponential Random Variable.* Let X have exponential density

$$f(x) = \begin{cases} e^x & \text{if } x < 0 \\ 0 & \text{elsewhere.} \end{cases}$$

The function $y = g(x) = x^2$ is strictly decreasing on the set $\{f(x) > 0\} = (-\infty, 0)$ so that the inverse function is $x = g^{-1}(y) = -y^{1/2}$. Thus

$$\frac{dx}{dy} = -\frac{1}{2\sqrt{y}}$$

and

$$f_Y(y) = \left|\frac{dx}{dy}\right| f(x) = \frac{1}{2\sqrt{y}} e^{-y^{1/2}}$$

for $y > 0$. For $y \le 0$, however,

$$f_Y(y) = 0. \qquad \square$$

Example 3.   Square of Normal (0, 1).   We consider Example 8.2.5 again, where X has a standard normal distribution and $g(X) = X^2$. The function g is decreasing over $(-\infty, 0)$ and increasing over $(0, \infty)$ and we can no longer apply (3) or (6) directly. The function $y = x^2$, for $y > 0$, has a unique inverse $x_1 = -\sqrt{y}$ on $(-\infty, 0)$ and a unique inverse $x_2 = \sqrt{y}$ on $(0, \infty)$ (Figure 2). We apply (7) to get

$$f_Y(y) = f(-\sqrt{y}) \cdot \frac{1}{2\sqrt{y}} + f(\sqrt{y}) \frac{1}{2\sqrt{y}} = \frac{1}{\sqrt{2\pi y}} e^{-y/2}$$

for $y > 0$ and $f_Y(y) = 0$ for $y \le 0$. $\qquad \square$

Example 4.   Distribution of $Y = \sin X$ for X Uniform.   Let X have uniform distribution on the circumference of unit circle with density function

$$f(x) = \begin{cases} 1/(2\pi), & 0 < x < 2\pi \\ 0 & \text{otherwise.} \end{cases}$$

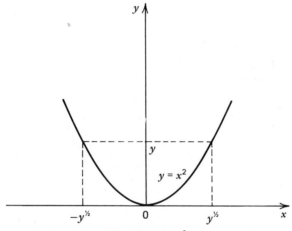

Figure 2. $y = x^2$.

Let $Y = g(X) = \sin X$. Clearly $-1 < Y < 1$ (with probability one). We note that $g'(x) = \cos x > 0$ for $0 < x < \pi/2$ or $3\pi/2 < x < 2\pi$ and $g'(x) < 0$ for $x \in (\pi/2, 3\pi/2)$, so that g is increasing over $(0, \pi/2) \cup (3\pi/2, 2\pi)$ and decreasing over $(\pi/2, 3\pi/2)$. (See Figure 3.) We need to separate the two cases according to whether $0 < y < 1$ (equivalently, $0 < x < \pi$) or $-1 < y < 0$ (equivalently, $\pi < x < 2\pi$). In the former case there are two inverses $x_1 = \sin^{-1}y$, $x_2 = \pi - \sin^{-1}y$; in the latter case the inverses are

$$x_3 = \pi - \sin^{-1}y \quad \text{and} \quad x_4 = 2\pi + \sin^{-1}y.$$

Hence for $0 < y < 1$:

$$f_Y(y) = f(x_1)\left|\frac{dx_1}{dy}\right| + f(x_2)\left|\frac{dx_2}{dy}\right|$$

$$= f(\sin^{-1}y) \cdot \frac{1}{\sqrt{1-y^2}} + f(\pi - \sin^{-1}y)\frac{1}{\sqrt{1-y^2}}$$

$$= \frac{1}{\pi\sqrt{1-y^2}},$$

and for $-1 < y < 0$:

$$f_Y(y) = f(\pi - \sin^{-1}y)\frac{1}{\sqrt{1-y^2}} + f(2\pi + \sin^{-1}y)\frac{1}{\sqrt{1-y^2}}$$

$$= \frac{1}{\pi\sqrt{1-y^2}}.$$

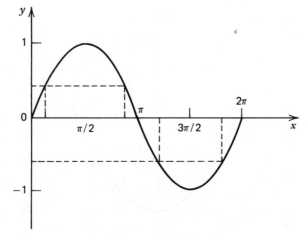

Figure 3. $y = \sin x$ on $(0, 2\pi)$.

Clearly, $f_Y(y) = 0$ for $|y| > 1$. Hence

$$f_Y(y) = \begin{cases} 1/\left[\pi\sqrt{1 - y^2}\right] & \text{for } |y| < 1 \\ 0 & \text{elsewhere.} \end{cases} \qquad \Box$$

The method of transformations can also be applied in the multivariate case. Rather than develop a general method, we will content ourselves with some examples to illustrate how the above method for the univariate case can be used in the multivariate case as well. In the bivariate case, the trick is to hold one of the variables, say X_2, fixed and then find the joint distribution of $Y = g(X_1, X_2)$ and X_1 (or X_2 if X_1 is held fixed). The distribution of Y is then obtained by computing the marginal distribution of Y from the joint distribution of (Y, X_1).

Example 5.   Sum of Poisson Random Variables.   We consider Example 8.2.4 again, where we computed the distribution of $Y = g(X_1, X_2) = X_1 + X_2$ in the case when X_1, X_2 are independent random variables with Poisson distributions. Assume $\lambda_1 = \lambda_2$. Then

$$P(X_1 = x_1, X_2 = x_2) = \frac{e^{-2\lambda}\lambda^{x_1 + x_2}}{x_1! x_2!}, \qquad x_1 = 0, 1, \ldots; \quad x_2 = 0, 1, \ldots$$

and the joint probability function of Y and X_1 is given by

$$P(X_1 = x_1, Y = y) = \frac{e^{-2\lambda}\lambda^y}{x_1!(y - x_1)!}, \qquad x_1 = 0, 1, 2, \ldots, y; \quad y = 0, 1, \ldots$$

Summing over $x_1 = 0, 1, 2, \ldots$ we get

$$P(Y = y) = \sum_{x_1=0}^{y} P(X_1 = x_1, Y = y) = e^{-2\lambda}\lambda^y \sum_{x_1=0}^{y} \frac{1}{x_1!(y - x_1)!}$$

$$= \frac{e^{-2\lambda}\lambda^y}{y!} \sum_{x_1=0}^{y} \binom{y}{x_1} = \frac{e^{-2\lambda}(2\lambda)^y}{y!}, \qquad y = 0, 1, \ldots. \qquad \Box$$

Example 6.   Distribution of $X_1/(X_1 + X_2)$ for Independent Exponentially Distributed Random Variables.   We return to Example 8.2.13, where we computed the distribution of $Y = X_1/(X_1 + X_2)$ where X_1, X_2 were independent exponentially distributed random variables with means λ_1 and λ_2 respectively. For convenience, we consider only the $\lambda_1 = \lambda_2 = \lambda$ case. Then X_1 and X_2 have joint density function

$$f(x_1, x_2) = \begin{cases} \lambda^{-2}e^{-(x_1 + x_2)/\lambda}, & \text{if } x_1 > 0, x_2 > 0 \\ 0 & \text{otherwise.} \end{cases}$$

Let $y = x_1/(x_1 + x_2)$ and suppose we hold x_1 constant. Then y decreases as x_2 increases and we can apply (6) to find the joint density of Y and X_1. Now $y = x_1/(x_1 + x_2)$ implies $x_2 = x_1(1 - y)/y$ so that $\partial x_2/\partial y = -x_1/y^2$. It follows from (6) that X_1 and Y have joint density

$$u(x_1, y) = f(x_1, x_1(1 - y)/y)|\partial x_2/\partial y| = \lambda^{-2}\exp\{-x_1/\lambda y\}(x_1/y^2)$$

for $x_1 > 0, 0 < y < 1$, and $u(x_1, y) = 0$ elsewhere. Hence the marginal of Y is given by

$$f_Y(y) = \int_{-\infty}^{\infty} u(x_1, y)\, dx_1 = \lambda^{-2} y^{-2} \int_0^{\infty} x_1 \exp\left\{ -\frac{x_1}{\lambda y} \right\} dx$$

$$= (\lambda y)^{-2} \Gamma(2) \cdot (\lambda y)^2 = 1 \qquad \text{for } 0 < y < 1$$

and $f_Y(y) = 0$ elsewhere. □

Example 7. Distribution of $X_1 + X_2 + X_3$. Let X_1, X_2, X_3 have joint density

$$f(x_1, x_2, x_3) = \begin{cases} 6(1 + x_1 + x_2 + x_3)^{-4}, & x_1, x_2, x_3 > 0 \\ 0 & \text{elsewhere.} \end{cases}$$

Let $y = g(x_1, x_2, x_3) = x_1 + x_2 + x_3$. We first find the density of Y, X_2, X_3. Note that g increases as x_1 increases (for fixed x_2, x_3), so that from (6)

$$u(y, x_2, x_3) = f(x_1, x_2, x_3) \left| \frac{\partial x_1}{\partial y} \right|.$$

Now $x_1 = y - x_2 - x_3$ and $\partial x_1 / \partial y = 1$ and it follows that

$$u(y, x_2, x_3) = f(y - x_2 - x_3, x_2, x_3)(1) = 6(1 + y)^{-4}$$

for $x_2, x_3 > 0$ and $y > x_2 + x_3$, and $u(y, x_2, x_3) = 0$ otherwise. Finally, the marginal of Y is given by

$$f_Y(y) = \int_0^y \left(\int_0^{y - x_3} \frac{6}{(1 + y)^4}\, dx_2 \right) dx_3$$

$$= \frac{6}{(1 + y)^4} \int_0^y (y - x_3)\, dx_3 = \frac{3y^2}{(1 + y)^4}, \qquad \text{for } y > 0$$

and $f_Y(y) = 0$ elsewhere. □

In Section 8.6 we will consider yet another method of obtaining distribution of $g(X_1, X_2, \ldots, X_n)$ given that of $(X_1, X_2, \ldots, X_n)$. This method is frequently useful when $X_1, X_2, \ldots, X_n$ are independently distributed and $g(x_1, \ldots, x_n) = \sum_{i=1}^{n} x_i$.

Problems for Section 8.3

1. Let X have density function f and $Y = aX + b$ where $a (\neq 0)$ and b are constants. Find the density of Y.

2. Let X have density $f(x) = 2x$ for $0 < x < 1$ and $f(x) = 0$ otherwise. Find the density function of $Y = e^{-X}$.

3. Let X be uniformly distributed over $(0,1)$. Find the density function of:
    (i) $Y = X^2$.   (ii) $Y = 1/(X + 1)$.   (iii) $Y = 1/X$.
    (iv) $Y = X^{1/\beta}, \beta \neq 0$.   (v) $Y = -\lambda \ln X, \lambda > 0$.   (vi) $Y = X/(1 + X)$.

4. A straight line is drawn at random through the point $(0,1)$ in the sense that the angle Θ between the line and $y = 1$ is uniformly distributed over the interval $(0, \pi)$. Let X be the length of the perpendicular from the origin to this line (Figure 4). What is the probability density function of X?

5. Let X have standard normal distribution. Find the density of $Y = \{|X|\}^{1/2}$.

6. Let X have density f. Find the density of
    (i) $Y = X^n$, when n is an odd positive integer > 1.
    (ii) $Y = X^n$, when n is an even positive integer.
    (iii) $Y = \exp\{aX + b\}, a \neq 0$.

7. Let X be normal with mean μ and variance σ^2. Show that $Y = e^X$ has *lognormal density function*

    $$f_Y(y) = \begin{cases} (y\sigma\sqrt{2\pi})^{-1}\exp\{-(\ln y - \mu)^2/2\sigma^2\}, & y > 0 \\ 0 & \text{elsewhere.} \end{cases}$$

8. Let X be a positive random variable having density function f. Find the density functions of:
    (i) $Y = (X)^{-1}$.   (ii) $Y = 1/(X + 1)$.
    (iii) $Y = X/(X + 1)$.   (iv) $Y = e^{-X}$.

9. Let X have density $f(x) = (1/\lambda)e^{-x/\lambda}, x > 0$, and $f(x) = 0$ otherwise. Find the density functions of:
    (i) $Y = -X$.   (ii) $Y = X/(1 + X)$.

10. Let X have Cauchy density $f(x) = [\pi(1 + x^2)]^{-1}, x \in \mathbb{R}$. Show that $Y = 1/X$ has the same density as X.

11. Suppose X has density $f(x) = (1 + x)^{-2}$ for $x > 0$ and $f(x) = 0$ elsewhere. Show that $Y = 1/X$ also has density f.

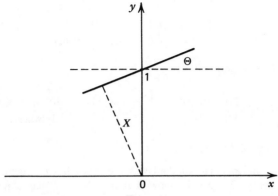

Figure 4

12. Suppose X has density function

$$f(x) = 1/x^2 \quad \text{if } x \geq 1 \qquad \text{and} \qquad f(x) = 0 \quad \text{if } x < 1.$$

Show that $Y = e^{-X}$ has density function

$$f_Y(y) = \left[y(\ln y)^2 \right]^{-1} \quad \text{for } 0 \leq y \leq e^{-1} \qquad \text{and} \qquad f_Y(y) = 0 \text{ elsewhere.}$$

13. Let X have density function

$$f(x) = (\tfrac{1}{2})\exp(-|x|), \qquad x \in \mathbb{R}.$$

Show that $Y = X^2$ has density

$$f_Y(y) = (2\sqrt{y})^{-1}\exp(-\sqrt{y}), \qquad \text{for } y > 0, \text{ and } 0 \text{ elsewhere.}$$

14. Let X_1, X_2 be the lengths of life of two different types of components operating in a system. Then $Y = X_1/X_2$ is a measure of the relative efficiency of component 1 with respect to component 2. Suppose (X_1, X_2) has joint density given by

$$f(x_1, x_2) = \begin{cases} \dfrac{1}{\Gamma(\alpha)\beta^{\alpha+1}} x_1^{\alpha-1}\exp\{-(x_1 + x_2)/\beta\}, & x_1 > 0, x_2 > 0, \\ 0 & \text{elsewhere.} \end{cases}$$

Find the density of Y.

15. Let X_1, X_2 be independent random variables with common uniform density

$$f(x) = \begin{cases} 1, & \text{for } 0 < x < 1 \\ 0 & \text{elsewhere.} \end{cases}$$

Show by transformation technique that the density of $Y = X_1 + X_2$ is given by

$$f_Y(y) = \begin{cases} y, & 0 < y < 1 \\ 2 - y, & 1 < y < 2 \\ 0 & \text{elsewhere.} \end{cases}$$

16*. Let X_1, X_2, X_3 have joint density given by

$$f(x_1, x_2, x_3) = \begin{cases} \lambda^3 \exp\{-\lambda(x_1 + x_2 + x_3)\}, & \text{for } x_1 > 0, x_2 > 0, x_3 > 0 \\ 0 & \text{elsewhere.} \end{cases}$$

Find the density of $Y = X_1 + X_2 + X_3$.

17. Let X_1, X_2 be independent random variables each having geometric distribution with parameter p, $0 < p < 1$. Show that $Y = X_1 + X_2$ has a negative binomial distribution.

18. Let X be a uniformly distributed random variable on $(0,1)$ and let Y be independent of X and have density $f_Y(y) = 1 - |y|$ for $|y| < 1$ and $f_Y(y) = 0$ otherwise. The random variable $X + Y$ is observed. Find its density.

19. Let X and Y be independent random variables having a common exponential density with mean one. Find the density of $Z = \sqrt{X + Y}$.

20. A missile is fired and lands at (X, Y) where X is the latitudinal distance in meters and Y the longitudinal distance in meters from the target. If X and Y are independent with common normal distribution having mean zero and variance σ^2, show that $Z = \sqrt{X^2 + Y^2}$ has a Rayleigh density given by $f_Z(z) = (z/\sigma^2)\exp(-z^2/2\sigma^2)$, $z > 0$, and $= 0$ elsewhere.

21*. If X and Y are independent with common uniform distribution on $(0,1)$, find the joint density of $X + Y$ and $X - Y$.

22. Suppose a simple electrical circuit contains a constant voltage source and a resistor. Let E be the voltage assumed to be constant. Suppose the resistor is chosen randomly from a stack where the resistance R has density function $f(r) = (r/\lambda^2)\exp\{-r/\lambda\}$ for $r > 0$ and $f(r) = 0$ for $r \le 0$. According to Ohm's law the current I passing through the circuit is determined by $I = E/R$.
   (i) Find the density function of I.
   (ii) Find the density function of W, the power dissipated in the resistor, where $W = I^2 R = E^2/R = EI$.

8.4 DISTRIBUTIONS OF SUM, PRODUCT, AND QUOTIENT OF TWO RANDOM VARIABLES

Let (X_1, X_2) have a joint distribution given by its joint probability or density function. In many applications we shall be interested in the distribution of random variables $X_1 + X_2$, $X_1 X_2$, and X_1/X_2. For example, if X_1, X_2 are two sides of a rectangle, then the area is given by $X_1 X_2$, the perimeter by $2(X_1 + X_2)$ and, given the area and one of the sides X_1, the remaining side is given by area $/X_1$. If proportion X_2 of a compound is sampled and X_1 is the proportion of chemical A in the sample, then the proportion of chemical A in the compound is $X_1 X_2$, and so on.

In this section we compute the distributions of $X_1 + X_2$, $X_1 X_2$, and X_1/X_2 in terms of the joint distribution of (X_1, X_2) by using methods of Section 8.2. Methods of Section 8.3 can also be applied to yield the same results. Some special cases have already been considered in several examples in Sections 8.2 and 8.3. See also Problem 5.5.13.

Let us first dispense with the discrete case. Let

$$p_{ij} = P(X_1 = x_i, X_2 = y_j), \qquad i, j = 1, 2, \ldots$$

be the joint probability function of (X_1, X_2). Then

$$P(X_1 + X_2 = y) = \sum_{j=1}^{\infty} P(X_1 + X_2 = y, X_1 = x_j)$$

$$= \sum_{j=1}^{\infty} P(X_1 = x_j, X_2 = y - x_j)$$

for those y for which $y = x_i + x_j$ for some x_i and some x_j and

$$P(X_1 + X_2 = y) = 0 \qquad \text{otherwise.}$$

The computations for $X_1 - X_2$, $X_1 X_2$ and X_1/X_2 (whenever well defined) are similar.

Example 1. Sum of $X_1 + X_2$ when (X_1, X_2) has Trinomial Distribution. Let (X_1, X_2) have joint probability function

$$p_{ij} = \begin{cases} \dfrac{n!}{i!j!(n-i-j)!} p_1^i p_2^j (1 - p_1 - p_2)^{n-i-j}, & i, j = 0,1,\ldots n, \quad i+j \le n, \\ 0 & \text{elsewhere,} \end{cases}$$

where $0 < p_1 < 1$, $0 < p_2 < 1$, $0 < p_1 + p_2 < 1$. Then

$$P(X_1 + X_2 = k) = \sum_{i=0}^{n} P(X_1 = i, X_1 + X_2 = k)$$

$$= \sum_{i=0}^{k} P(X_1 = i, X_2 = k - i) = \sum_{i=0}^{k} p_{i, k-i}$$

$$= \sum_{i=0}^{k} \frac{n!}{i!(k-i)!(n-k)!} p_1^i p_2^{k-i} (1 - p_1 - p_2)^{n-k}$$

$$= \binom{n}{k} (1 - p_1 - p_2)^{n-k} \sum_{i=0}^{k} \binom{k}{i} p_1^i p_2^{k-i}$$

$$= \binom{n}{k} (p_1 + p_2)^k (1 - p_1 - p_2)^{n-k} \qquad \text{(binomial theorem)}$$

for $k = 0, 1, \ldots, n$. We note that $X_1 + X_2$ has a binomial distribution with parameters n and $p_1 + p_2$. An informal derivation of this result was obtained in Section 6.7.

Similar computations can be made for the distributions of $X_1 X_2$ and $X_1 - X_2$ although it is not possible to give simple explicit expressions. Since $P(X_1 = 0) > 0$, X_2/X_1 is not a random variable. Let us briefly outline the computations for the probability function of $X_1 X_2$. Clearly $X_1 X_2$ takes values of the form $k = ij$ where

$i, j \in \{0, 1, 2, \ldots, n\}$. For $k = 0$

$$P(X_1 X_2 = k) = P(X_1 = 0, X_2 \geq 1) + P(X_1 \geq 1, X_2 = 0) + P(X_1 = X_2 = 0)$$

$$= \sum_{k=1}^{n} \binom{n}{k} p_2^k (1 - p_1 - p_2)^{n-k}$$

$$+ \sum_{k=1}^{n} \binom{n}{k} p_1^k (1 - p_1 - p_2)^{n-k} + (1 - p_1 - p_2)^n$$

$$= (1 - p_1)^n + (1 - p_2)^n - (1 - p_1 - p_2)^n.$$

For $k \neq 0$ and $k = ij$, $i, j \in \{1, 2, \ldots, n\}$

$$P(X_1 X_2 = k) = \sum_j P(X_1 X_2 = k, X_2 = j) = \sum_{\{(i, j): ij = k\}} p_{ij}.$$

Thus

$$P(X_1 X_2 = 1) = P(X_1 = 1, X_2 = 1) = p_{11},$$

$$P(X_1 X_2 = 2) = P(X_1 = 1, X_2 = 2) + P(X_1 = 2, X_2 = 1) = p_{12} + p_{21},$$

and so on. For $n \geq 8$, for example,

$$P(X_1 X_2 = 8) = P(X_1 = 1, X_2 = 8) + P(X_1 = 8, X_2 = 1) + P(X_1 = 2, X_2 = 4)$$

$$+ P(X_1 = 4, X_2 = 2)$$

$$= p_{18} + p_{81} + p_{24} + p_{42}. \qquad \square$$

In what follows we assume that (X_1, X_2) have a joint density function f and that f_i, $i = 1, 2$, is the marginal density of X_i. The following result is of great importance in statistics.

THEOREM 1.   *Let $Z_1 = X_1 + X_2$, $Z_2 = X_2 - X_1$, $Z_3 = X_1 X_2$, and $Z_4 = X_2/X_1$. Then the densities of Z_1, Z_2, Z_3 and Z_4 are given by*

(1) $$f_{Z_1}(z) = \int_{-\infty}^{\infty} f(x, z - x) \, dx = \int_{-\infty}^{\infty} f(z - y, y) \, dy, \qquad z \in \mathbb{R},$$

(2) $$f_{Z_2}(z) = \int_{-\infty}^{\infty} f(x, z + x) \, dx = \int_{-\infty}^{\infty} f(z + y, y) \, dy, \qquad z \in \mathbb{R},$$

(3) $$f_{Z_3}(z) = \int_{-\infty}^{\infty} \frac{1}{|x|} f\left(x, \frac{z}{x}\right) dx = \int_{-\infty}^{\infty} \frac{1}{|y|} f\left(\frac{z}{y}, y\right) dy, \qquad z \in \mathbb{R},$$

(4) $$f_{Z_4}(z) = \int_{-\infty}^{\infty} |x| f(x, zx) \, dx = \int_{-\infty}^{\infty} \frac{|y|}{z^2} f\left(\frac{y}{z}, y\right) dy, \qquad z \in \mathbb{R}.$$

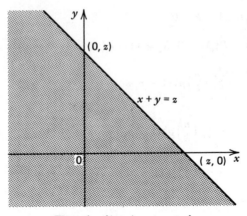

Figure 1. $\{(x, y): x + y \le z\}$.

Proof.   From Figure 1

$$P(Z_1 \le z) = \iint\limits_{x+y \le z} f(x, y) \, dx \, dy = \int_{-\infty}^{\infty} \left(\int_{-\infty}^{z-x} f(x, y) \, dy \right) dx$$

$$= \int_{-\infty}^{\infty} \left(\int_{-\infty}^{z} f(x, u - x) \, du \right) dx \qquad (y = u - x)$$

$$= \int_{-\infty}^{z} \left(\int_{-\infty}^{\infty} f(x, u - x) \, dx \right) du \qquad \text{(interchanging integrals)}$$

so that on differentiating with respect to z we have

$$f_{Z_1}(z) = \int_{-\infty}^{\infty} f(x, z - x) \, dx, \qquad z \in \mathbb{R}.$$

Interchanging x and y we get the second part of (1).

As for Z_2, we have from Figure 2:

$$P(Z_2 \le z) = \iint\limits_{y-x \le z} f(x, y) \, dx \, dy = \int_{-\infty}^{\infty} \left(\int_{-\infty}^{x+z} f(x, y) \, dy \right) dx$$

$$= \int_{-\infty}^{\infty} \left(\int_{-\infty}^{z} f(x, u + x) \, du \right) dx \qquad (y = u + x)$$

$$= \int_{-\infty}^{z} \left(\int_{-\infty}^{\infty} f(x, u + x) \, dx \right) du \qquad \text{(interchanging integrals)}$$

and on differentiation with respect to z we get:

$$f_{Z_2}(z) = \int_{-\infty}^{\infty} f(x, x + z) \, dx, \qquad z \in \mathbb{R}.$$

Interchanging the role of x and y we get the second part in (2).

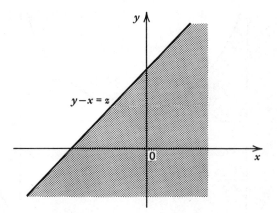

Figure 2. $\{(x, y): y - x \le z\}$.

Next we consider Z_3. In this case we note that

$$\{(x, y): xy \le z\} = \left\{(x, y): x < 0, y \ge \frac{z}{x}\right\} \cup \left\{(x, y): x > 0, y \le \frac{z}{x}\right\}.$$

Consequently (see Figures 3a and 3b) we have

$$P(Z_3 \le z) = \int_{-\infty}^{0} \left(\int_{z/x}^{\infty} f(x, y) \, dy\right) dx + \int_{0}^{\infty} \left(\int_{-\infty}^{z/x} f(x, y) \, dy\right) dx$$

$$= \int_{-\infty}^{0} \left(\int_{z}^{-\infty} x^{-1} f\left(x, \frac{u}{x}\right) du\right) dx + \int_{0}^{\infty} \left(\int_{-\infty}^{z} x^{-1} f\left(x, \frac{u}{x}\right) du\right) dx \qquad \left(y = \frac{u}{x}\right)$$

$$= \int_{-\infty}^{\infty} \left(\int_{-\infty}^{z} \frac{1}{|x|} f\left(x, \frac{u}{x}\right) du\right) dx$$

$$= \int_{-\infty}^{z} \left(\int_{-\infty}^{\infty} \frac{1}{|x|} f\left(x, \frac{u}{x}\right) dx\right) du \qquad \text{(interchanging integrals)}.$$

It follows that

$$f_{Z_3}(z) = \int_{-\infty}^{\infty} \frac{1}{|x|} f\left(x, \frac{z}{x}\right) dx, z \in \mathbb{R}.$$

Interchanging the role of x and y we get the second equality in (3).

Finally we consider Z_4. In this case,

$$P(Z_4 \le z) = \iint_{y/x \le z} f(x, y) \, dx \, dy.$$

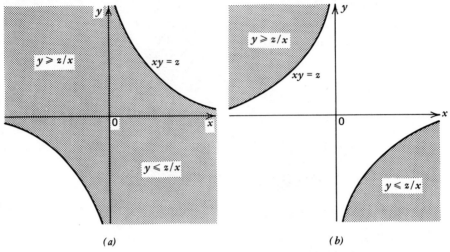

(a)   (b)

Figure 3.   (a) $\{xy \le z\}$ for $z > 0$. (b) $\{xy \le z\}$ for $z < 0$.

Note, however, that if $x < 0$, then

$$\frac{y}{x} \le z \quad \text{if and only if} \quad y \ge xz$$

so that

$$\left\{(x, y): \frac{y}{x} \le z\right\} = \{(x, y): x < 0, y \ge xz\} \cup \{(x, y): x > 0, y \le xz\}.$$

In view of Figures 4a and 4b, we have

$$P(Z_4 \le z) = \int_{-\infty}^{0} \left(\int_{xz}^{\infty} f(x, y)\, dy\right) dx + \int_{0}^{\infty} \left(\int_{-\infty}^{xz} f(x, y)\, dy\right) dx$$

$$= \int_{-\infty}^{0} \left(\int_{z}^{-\infty} xf(x, ux)\, du\right) dx + \int_{0}^{\infty} \left(\int_{-\infty}^{z} xf(x, ux)\, du\right) dx \ (y = ux)$$

$$= \int_{-\infty}^{\infty} \left(\int_{-\infty}^{z} |x| f(x, ux)\, du\right) dx$$

$$= \int_{-\infty}^{z} \left(\int_{-\infty}^{\infty} |x| f(x, ux)\, dx\right) du \quad \text{(interchanging integrals)}.$$

Differentiating with respect to z we get

$$f_{Z_4}(z) = \int_{-\infty}^{\infty} |x| f(x, zx)\, dx, \qquad z \in \mathbb{R}$$

and interchanging the roles of x and y we get the second part of (4).   □

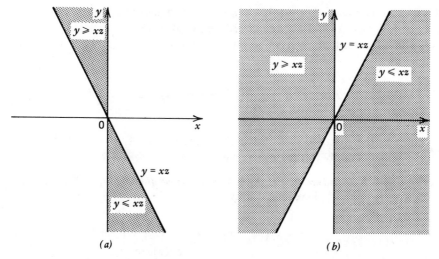

(a)   (b)

Figure 4.   (a) $\{y/x \le z\}$ for $z > 0$. (b) $\{y/x \le z\}$ for $z < 0$.

The case of most interest in applications is the case when X_1 and X_2 are independent. In this case we can rewrite Theorem 1 in terms of the marginal densities of X_1 and X_2 as follows.

THEOREM 2.   *Let X_1, X_2 be independent random variables of the continuous type with densities f_1 and f_2 respectively. Then, for $z \in \mathbf{R}$:*

$$(5) \qquad f_{Z_1}(z) = \int_{-\infty}^{\infty} f_1(x) f_2(z - x)\, dx = \int_{-\infty}^{\infty} f_1(z - y) f_2(y)\, dy,$$

$$(6) \qquad f_{Z_2}(z) = \int_{-\infty}^{\infty} f_1(x) f_2(z + x)\, dx = \int_{-\infty}^{\infty} f_1(z + y) f_2(y)\, dy,$$

$$(7) \qquad f_{Z_3}(z) = \int_{-\infty}^{\infty} \frac{1}{|x|} f_1(x) f_2\left(\frac{z}{x}\right) dx = \int_{-\infty}^{\infty} \frac{1}{|y|} f_1\left(\frac{z}{y}\right) f_2(y)\, dy,$$

$$(8) \qquad f_{Z_4}(z) = \int_{-\infty}^{\infty} |x| f_1(x) f_2(zx)\, dx = \int_{-\infty}^{\infty} \frac{|y|}{z^2} f_1\left(\frac{y}{z}\right) f_2(y)\, dy.$$

The two right sides of (5) suggest a simple method of obtaining densities. Given two (one-dimensional) densities f and g, the function h defined by

$$(9) \qquad h(z) = \int_{-\infty}^{\infty} f(x) g(z - x)\, dx = \int_{-\infty}^{\infty} f(z - y) g(y)\, dy$$

itself is a one-dimensional density function which is known as the *convolution of f and g*. In view of Theorem 2, the density of the sum of two independent continuous type random variables is the convolution of their individual densities.

However, *it does not follow that if h is a convolution density obtained from densities of X_1 and X_2 then X_1 and X_2 are independent.* (See Example 2 below.)

A similar definition is given in the discrete case. The probability function defined by

$$P(Z = z) = \sum_x P(X_1 = x)P(X_2 = z - x)$$

is known as the *convolution* of probability functions $P(X_1 = x)$ and $P(X_2 = y)$. Similar remarks apply to the discrete case.

Example 2. **Random Variable $X_1 + X_2$ with Convolution Density of Non-independent X_1, X_2.** Let (X_1, X_2) have joint density

$$f(x_1, x_2) = \begin{cases} \left[1 + x_1 x_2(x_1^2 - x_2^2)\right]/4 & \text{for } |x_1| < 1 \text{ and } |x_2| < 1 \\ 0 & \text{elsewhere.} \end{cases}$$

Then $f_1(x_1) = f_2(x_1) = 1/2$ for $|x_1| < 1$ and 0 elsewhere. It follows that X_1 and X_2 are dependent. Let $Z = X_1 + X_2$. We show that the density of Z is the convolution of f_1 and f_2. In fact, from (1):

$$f_Z(z) = \int_{-\infty}^{\infty} f(x_1, z - x_1) \, dx_1.$$

Now $f(x_1, z - x_1) > 0$ over $|x_1| < 1$ and $|z - x_1| < 1$, that is, over $|x_1| < 1$ and $x_1 - 1 < z < x_1 + 1$ (Figure 5). Rewriting the domain of positivity of $f(x_1, z - x_1)$ as $-1 < x_1 < 1$ and $z - 1 < x_1 < z + 1$, and noting that for $z < 0$, $z - 1 < 1$ and

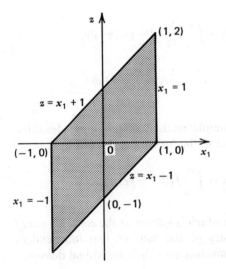

Figure 5. $\{|x_1| < 1, |z - x_1| < 1\}$.

$z + 1 < 1$, and for $z > 0$, $z - 1 > -1$ and $z + 1 > 1$, we have

$$
f_Z(z) = \begin{cases} \int_{-1}^{z+1}[1 + 3z^2x_1^2 - 2zx_1^3 - z^3x_1]/4 \, dx_1 = (2 + z)/4, & -2 < z < 0 \\ \int_{z-1}^{1}[1 + 3z^2x_1^2 - 2zx_1^3 - z^3x_1]/4 \, dx_1 = (2 - z)/4, & 0 < z < 2 \\ 0 & \text{elsewhere.} \end{cases}
$$

Finally, the convolution of f_1 and f_2 is given by

$$
h(z) = \int_{-\infty}^{\infty} f_1(x_1)f_2(z - x_1) \, dx_1
$$

$$
= \int_{|x_1| < 1, \, |z-x_1| < 1} (1/2)(1/2) \, dx_1
$$

$$
= \begin{cases} \int_{-1}^{z+1}(1/4) \, dx_1 = (2 + z)/4 & \text{for } -2 < z < 0, \\ \int_{z-1}^{1}(1/4) \, dx_1 = (2 - z)/4 & \text{for } 0 < z < 2, \\ 0 & \text{elsewhere.} \end{cases}
$$

It follows that even though the density of $X_1 + X_2$ is the convolution of the densities of X_1 and X_2, the random variables X_1 and X_2 are not independent.   □

Example 3*.   An Application to Beta Functions.   Let $\alpha_1, \alpha_2 > 0$. As an application of (5) we show that

$$
B(\alpha_1, \alpha_2) = \frac{\Gamma(\alpha_1)\Gamma(\alpha_2)}{\Gamma(\alpha_1 + \alpha_2)}
$$

which was proved in Proposition 7.3.4 (iii) by a somewhat complicated method. Let X_1, X_2 be independent random variables such that X_i has a gamma density with parameters $\alpha = \alpha_i$ and $\beta = 1$, $i = 1, 2$. Let $Z = X_1 + X_2$. Then for $z \leq 0$, $f_Z(z) = 0$ and for $z > 0$ in view of (5)

$$
f_Z(z) = \frac{e^{-z}}{\Gamma(\alpha_1)\Gamma(\alpha_2)} \int_0^z x_1^{\alpha_1 - 1}(z - x_1)^{\alpha_2 - 1} \, dx_1
$$

$$
= \frac{e^{-z}}{\Gamma(\alpha_1)\Gamma(\alpha_2)} \int_0^1 z^{\alpha_1 + \alpha_2 - 1}y^{\alpha_1 - 1}(1 - y)^{\alpha_2 - 1} \, dy \qquad (x_1 = zy)
$$

$$
= \frac{z^{\alpha_1 + \alpha_2 - 1}e^{-z}}{\Gamma(\alpha_1)\Gamma(\alpha_2)} \cdot B(\alpha_1, \alpha_2) \qquad \text{(definition of beta function).}
$$

Since f_Z is a density function we must have

$$\int_0^\infty f_Z(z)\, dz = 1 = \frac{B(\alpha_1, \alpha_2)}{\Gamma(\alpha_1)\Gamma(\alpha_2)} \int_0^\infty z^{\alpha_1 + \alpha_2 - 1} e^{-z}\, dz$$

$$= \frac{B(\alpha_1, \alpha_2)}{\Gamma(\alpha_1)\Gamma(\alpha_2)} \Gamma(\alpha_1 + \alpha_2) \qquad \text{(definition of gamma function)}.$$

It follows that

$$B(\alpha_1, \alpha_2) = \frac{\Gamma(\alpha_1)\Gamma(\alpha_2)}{\Gamma(\alpha_1 + \alpha_2)}$$

and, moreover, that Z itself has a gamma distribution with parameters $\alpha = \alpha_1 + \alpha_2$, $\beta = 1$. $\square$

Example 4.   Lifetime of a System with Two Components.   Let X_1, X_2 represent the lifetimes of two components, an original and its replacement. Then $Y = X_1 + X_2$ represents the lifetime of the system consisting of this pair. It is of interest to find the survival function $P(X_1 + X_2 > a)$ for $a > 0$. Suppose X_1 and X_2 are independent exponential random variables with common mean $\lambda > 0$. Then the joint density of (X_1, X_2) is given by

$$f(x_1, x_2) = \begin{cases} \lambda^{-2} \exp\{-(x_1 + x_2)/\lambda\} & \text{for } x_1 > 0, x_2 > 0, \\ 0 & \text{elsewhere.} \end{cases}$$

The density of $Y = X_1 + X_2$ is given by

$$f_Y(y) = \int_0^y (\lambda^{-1} e^{-x_1/\lambda})(\lambda^{-1} e^{-(y - x_1)/\lambda})\, dx_1$$

$$= \lambda^{-2} e^{-y/\lambda} \int_0^y dx = \lambda^{-2} y e^{-y/\lambda}$$

for $y > 0$ and

$$f_Y(y) = 0 \qquad \text{for } y \le 0.$$

Thus Y has a gamma distribution with parameters $\alpha = 2$ and $\beta = \lambda$. Moreover, for $a > 0$

$$P(X_1 + X_2 > a) = \int_a^\infty \lambda^{-2} y e^{-y/\lambda}\, dy = e^{-a/\lambda}\left(1 + \frac{a}{\lambda}\right).$$

Suppose next we wish to find the probability that the replacement outlasts the original component. Since $X_2 - X_1$ is a symmetric random variable (see Problem

3.8.16),

$$P(X_2 > X_1) = P(X_2 - X_1 > 0) = 1/2.$$

Let us compute the density of $Z = X_2 - X_1$. In view of (6), we have for $z > 0$

$$f_Z(z) = \int_0^\infty (\lambda^{-1} e^{-x_1/\lambda})(\lambda^{-1} e^{-(z+x_1)/\lambda}) \, dx_1$$

$$= \lambda^{-2} e^{-z/\lambda} \int_0^\infty e^{-2x_1/\lambda} \, dx_1 = (2\lambda)^{-1} e^{-z/\lambda},$$

and for $z < 0$

$$f_Z(z) = \int_{-z}^\infty (\lambda^{-1} e^{-x_1/\lambda})(\lambda^{-1} e^{-(z+x_1)/\lambda}) \, dx_1$$

$$= \lambda^{-2} e^{-z/\lambda} \int_{-z}^\infty e^{-2x_1/\lambda} \, dx_1 = (2\lambda)^{-1} e^{z/\lambda}.$$

It follows that

$$f_Z(z) = (2\lambda)^{-1} \exp\left\{ -\frac{|z|}{\lambda} \right\}, \quad z \in \mathbb{R}$$

which is known as a *double exponential* or *Laplace density function*. $\quad\square$

In Examples 5 and 6 we consider some important distributions in Statistics which arise as ratios of certain independent random variables.

Example 5. Student's t-distribution. Let X have a standard normal distribution and let Y be independent of X and have a chi-square distribution with n degrees of freedom. Then the random variable $T = X/\sqrt{Y/n}$ is said to have *Student's t-distribution* with n degrees of freedom. Let us compute the density function of T. We note that Y has density

$$f_2(y) = \begin{cases} \dfrac{y^{n/2-1} e^{-y/2}}{\Gamma(n/2) 2^{n/2}}, & y > 0 \\ 0 & \text{elsewhere} \end{cases}$$

since Y is a gamma variable with $\alpha = n/2$ and $\beta = 2$ (see Section 7.4). The density of $U = \sqrt{Y/n}$, therefore, is given by

$$f_U(u) = \left| \frac{\partial y}{\partial u} \right| f_2(u) = \frac{2un}{\Gamma(n/2) 2^{n/2}} (nu^2)^{n/2-1} e^{-nu^2/2}$$

$$= \frac{\sqrt{2n}}{\Gamma(n/2)} \left(\frac{u\sqrt{n}}{\sqrt{2}} \right)^{n-1} e^{-nu^2/2} \qquad \text{for } u > 0$$

and

$$f_U(u) = 0 \qquad \text{for } u \le 0.$$

Since X and Y are independent, so are X and U, and from Theorem 2 [equation (8)] the density of T is given by

$$f_T(t) = \int_0^\infty u \left\{ \frac{\sqrt{2n}}{\Gamma(n/2)} \left(\frac{u\sqrt{n}}{\sqrt{2}} \right)^{n-1} e^{-nu^2/2} \right\} \left\{ \frac{1}{\sqrt{2\pi}} e^{-t^2u^2/2} \right\} du$$

$$= \frac{\sqrt{n}}{\sqrt{\pi}\,\Gamma(n/2)} \int_0^\infty u \left(\frac{u\sqrt{n}}{\sqrt{2}} \right)^{n-1} \exp\left\{ -\frac{u^2}{2}(n + t^2) \right\} du, \qquad t \in \mathbb{R}.$$

Setting $z = u^2(n + t^2)/2$ so that $u = (2z/(n + t^2))^{1/2}$ and $dz = (n + t^2)u\,du$, we have

$$f_T(t) = \frac{\sqrt{n}}{\sqrt{\pi}\,\sqrt{(n/2)}} \int_0^\infty \left(\frac{zn}{n + t^2} \right)^{(n-1)/2} e^{-z} \frac{dz}{(n + t^2)}$$

$$= \frac{\Gamma((n + 1)/2)}{\sqrt{\pi n}\,\Gamma(n/2)} \left(1 + \frac{t^2}{n} \right)^{-(n+1)/2}, \qquad t \in \mathbb{R}. \qquad \square$$

Example 6. The F-Distribution. Let X and Y be independent random variables such that X has a chi-square distribution with m d.f. and Y has a chi-square distribution with n d.f. Then the random variable

$$F = \frac{X/m}{Y/n}$$

is said to have an *F-distribution with* (m, n) d.f. We compute the density of F. We note that the density of X/m is given by

$$f_1(x) = \begin{cases} \dfrac{m(mx)^{m/2-1}e^{-mx/2}}{\Gamma(m/2)2^{m/2}}, & x > 0 \\ 0 & \text{elsewhere} \end{cases}$$

and that of Y/n by

$$f_2(y) = \begin{cases} \dfrac{n(ny)^{n/2-1}e^{-ny/2}}{\Gamma(n/2)2^{n/2}}, & y > 0 \\ 0 & \text{elsewhere.} \end{cases}$$

In view of (8), the density of F is given by

$$f(z) = \int_0^\infty y \left(\frac{n(ny)^{n/2-1} e^{-ny/2}}{\Gamma(n/2)2^{n/2}} \right) \left(\frac{m(mzy)^{m/2-1} e^{-mzy/2}}{\Gamma(m/2)2^{m/2}} \right) dy$$

$$= \frac{(n/2)^{n/2}(m/2)^{m/2}}{\Gamma(n/2)\Gamma(m/2)} z^{m/2-1} \int_0^\infty y^{(m+n)/2-1} e^{-y(n+mz)/2} \, dy$$

$$= \frac{(n/2)^{n/2}(m/2)^{m/2}}{\Gamma(n/2)\Gamma(m/2)} z^{m/2-1} \Gamma\left(\frac{m+n}{2}\right) \left(\frac{2}{n+mz}\right)^{(m+n)/2}$$

$$= \frac{\Gamma((m+n)/2) m^{m/2} n^{n/2}}{\Gamma(m/2)\Gamma(n/2)} \frac{z^{m/2-1}}{(n+mz)^{(m+n)/2}} \qquad \text{for } z > 0$$

and

$$f(z) = 0 \qquad \text{for } z \le 0. \qquad \qquad \square$$

Problems for Section 8.4

1. The random variables X_1 and X_2 measure the errors in two measurements. Assume that X_1, X_2 are independent and each is uniformly distributed over $(0,1)$. Find the density functions of
   (i) $Y_1 = X_1 - X_2$.   (ii) $Y_2 = X_1 X_2$.   (iii) $Y_3 = X_1/X_2$.

2. Let X_1, X_2 measure the lifetimes of two components operating independently. Suppose each has density (in units of 100 hours)

$$f(x) = \begin{cases} 1/x^2, & x \ge 1 \\ 0 & \text{elsewhere.} \end{cases}$$

If $Y = \sqrt{X_1 X_2}$ measures the quality of the system, show that Y has density function

$$f_Y(y) = 4\frac{\ln y}{y^3} \quad \text{for } y \ge 1 \qquad \text{and} \quad f_Y(y) = 0 \text{ for } y < 1.$$

3. Let X_1 and X_2 be independent standard normal random variables. Show that $Y = X_1/X_2$ is a Cauchy random variable with density $f_Y(y) = [\pi(1+y^2)]^{-1}$, $y \in \mathbb{R}$.

4. Let X_1 and X_2 be independent with common density

$$f_1(x) = f_2(x) = \begin{cases} (x\sqrt{2\pi})^{-1} \exp\left\{ -\frac{(\ln x)^2}{2} \right\}, & \text{for } x > 0 \\ 0 & \text{elsewhere.} \end{cases}$$

Find the density of $X_1 X_2$.

5. Suppose X_1, X_2 are independent random variables with common density

$$f_1(x) = f_2(x) = \begin{cases} e^{-x/\beta}/\beta, & \text{for } x > 0 \\ 0 & \text{elsewhere.} \end{cases}$$

Find the density of X_1/X_2.

6. Find the density of X_1/X_2 if X_1 and X_2 have joint density

$$f(x_1, x_2) = \begin{cases} x_1 x_2 \exp\{-(x_1^2 + x_2^2)/2\}, & x_1 > 0, x_2 > 0 \\ 0 & \text{elsewhere.} \end{cases}$$

7. Let X_1, X_2 be independent random variables with common probability function $P(X_i = x) = 1/N$, $x = 1, 2, \ldots, N$; $i = 1, 2$. Find the probability functions of $X_1 + X_2$ and $X_1 - X_2$.

8. Let (X_1, X_2) have joint probability function given by

$$p_{ij} = \frac{\binom{n_1}{i}\binom{n_2}{j}\binom{N - n_1 - n_2}{n - i - j}}{\binom{N}{n}}, \qquad \begin{array}{l} i = 1, 2, \ldots, n_1; \quad j = 1, 2, \ldots, n_2; \\ i + j \le n. \end{array}$$

Find the probability function of $X_1 + X_2$.

9*. Suppose X_1 is uniform on $(-2, 2)$ and X_2 has density $f_2(x_2) = 1 - |x_2|$ for $|x_2| < 1$ and $f_2(x_2) = 0$ elsewhere. Show that their convolution density is given by

$$h(y) = \begin{cases} (y + 3)^2/8 & -3 < y < -2 \\ \left[1 - \dfrac{(y + 1)^2}{2}\right]\Big/4 & -2 < y < -1 \\ 1/4 & -1 < y < 1 \\ \left[1 - \dfrac{(y - 1)^2}{2}\right]\Big/4 & 1 < y < 2 \\ (y - 3)^2/8 & 2 < y < 3 \\ 0 & \text{elsewhere.} \end{cases}$$

10. Suppose (X, Y) has joint density given by

$$f(x, y) = \begin{cases} x + y, & 0 \le x \le 1, 0 \le y \le 1 \\ 0 & \text{elsewhere.} \end{cases}$$

Show that
(i) $X + Y$ has density $f_1(z) = z^2$, $0 < z < 1$, $= z(2 - z)$, $1 < z < 2$, and 0 elsewhere.
(ii) XY has density $f_2(z) = 2(1 - z)$, $0 < z < 1$, and 0 elsewhere.
(iii) Y/X has density $f_3(z) = (1 + z)/3$, $0 < z < 1$, $= (1 + z)/3z^3$, $z > 1$, and 0 elsewhere.
(iv) $Y - X$ has density $f_4(z) = 1 - |z|$, $|z| < 1$, and 0 elsewhere.

11*.   Suppose (X, Y) has joint density

$$f(x, y) = \begin{cases} 1 & 0 \le x \le 2, 0 \le y \le 1, 2y \le x \\ 0 & \text{elsewhere.} \end{cases}$$

Show that $Z = X + Y$ has density

$$f_Z(z) = \begin{cases} (1/3)z, & 0 < z < 2 \\ 2 - (2/3)z, & 2 < z < 3 \\ 0 & \text{elsewhere.} \end{cases}$$

12.   Suppose (X, Y) has joint density

$$f(x, y) = \begin{cases} xy \exp\{-(x + y)\}, & x, y > 0 \\ 0 & \text{elsewhere.} \end{cases}$$

Find the density of $X + Y$.

13*.   Let F be any distribution function and $\varepsilon > 0$. Show that

$$F_1(x) = \varepsilon^{-1} \int_x^{x+\varepsilon} F(y) \, dy$$

and

$$F_2(x) = (2\varepsilon)^{-1} \int_{x-\varepsilon}^{x+\varepsilon} F(y) \, dy$$

define distribution functions. (Hint: Use convolution result.)

14*.   Let X, Y, Z be independent random variables with common uniform distribution on $[0, 1]$. Show that $X + Y + Z$ has density

$$g(u) = \begin{cases} u^2/2 & 0 < u < 1, \\ 3u - u^2 - 3/2 & 1 \le u < 2, \\ (u - 3)^2/2 & 2 \le u \le 3, \\ 0 & \text{elsewhere.} \end{cases}$$

(Hint: Convolute $X + Y$ with Z and use the result of Problem 8.3.15.)

15.   For the bivariate distribution given by

x y	-1	0	1	
-1	1/9	1/9	1/9	1/3
0	1/9	1/9	1/9	1/3
1	1/9	1/9	1/9	1/3
	1/3	1/3	1/3	1

show that the probability function of $X + Y$ is the convolution of probability functions of X and Y.

16*. Let X, Y have a bivariate normal distribution with $\mu_1 = \mu_2 = 0$, $\sigma_1 = \sigma_2 = 1$, and correlation coefficient ρ. Show that X/Y has density

$$f_Z(z) = \frac{\sqrt{1 - \rho^2}}{\pi[1 - 2\rho z + z^2]}, z \in \mathbb{R}.$$

17. Let (X, Y) have joint density

$$f(x, y) = \begin{cases} 2 & 0 < y < x < 1 \\ 0 & \text{otherwise.} \end{cases}$$

Find the convolution density of X and Y. Is it the same as the density of $X + Y$?

18*. Let X_1, X_2, X_3 be independent random variables each having exponential density with mean λ. Find the density of $X_1 + X_2 + X_3$.

19. If F has an F distribution with (m, n) d.f. show that $1/F$ has an F distribution with (n, m) d.f.

20. Suppose X has an F distribution with (m, n) d.f. Show that $Y = mX/(mX + n)$ has a beta distribution.

8.5 ORDER STATISTICS

In many applications the relative magnitude of the observations is of great interest. If a manufacturing firm wants to estimate the average time to failure of its bulbs based on a sample of size n, the sample mean $\bar{X}$ can be used as an estimate of μ. In Section 5.5 we saw that $\bar{X}$ is an unbiased estimate of μ. In fact, $\bar{X}$ is the best estimate of μ in many models when the object is to minimize the variance of the estimate (Section 10.3). In order to compute $\bar{X}$, however, the manufacturer needs to wait until the last bulb has fused, and it may be a very long wait indeed. On the other hand, an estimate can be used based on the least time to failure. This estimate may not be as good as $\bar{X}$ in that its variance may be larger than that of $\bar{X}$, but it can be computed as soon as the first bulb fuses. (See Example 2.) In this section we consider distributions of certain statistics obtained by ordering the observations.

Let $X_1, X_2, \ldots, X_n$ be a random sample from a distribution function F with density function f. Let us order the observations according to increasing order of their magnitude and write $X_{(1)}, X_{(2)}, \ldots, X_{(n)}$ for the ordered random variables. Then

$$X_{(1)} \leq X_{(2)} \leq \cdots \leq X_{(n)},$$

and $X_{(j)}, j = 1, 2, \ldots, n$ are called *order statistics*. Note that equality holds with probability zero (why?). Thus

(1) $$X_{(1)}(\omega) = \min(X_1(\omega), X_2(\omega),\ldots,X_n(\omega)), \qquad \omega \in \Omega$$

is the minimum of the X's, and

(2) $$X_{(n)}(\omega) = \max(X_1(\omega), X_2(\omega),\ldots,X_n(\omega)), \qquad \omega \in \Omega$$

is the maximum of the X's. The $X_{(j)}$'s are also random variables, and given F and f we can find the joint and marginal distributions of $X_{(1)}, X_{(2)},\ldots,X_{(n)}$.

In this section we will restrict ourselves to finding the joint distribution function (and joint density function) of $X_{(r)}$ and $X_{(s)}$ for $r < s$. The argument given here is valid also for the discrete case and can be extended to more than two order statistics.

First note that if $x \geq y$ then $\{X_{(s)} \leq y\} \subseteq \{X_{(r)} \leq x\}$ and it follows that

(3) $$P(X_{(r)} \leq x, X_{(s)} \leq y) = P(X_{(s)} \leq y).$$

Next let $x < y$. Then

$$P(X_{(r)} \leq x, X_{(s)} \leq y) = P(\text{at least } r \ X\text{'s} \leq x, \text{ at least } s \ X\text{'s} \leq y)$$

$$= \sum_{i=r}^{n} \sum_{j=s-i}^{n-i} P(\text{exactly } i \ X\text{'s} \leq x, \text{exactly } j \ X\text{'s with } x < X \leq y)$$

unless $i > s$, in which case j starts at 0. Hence, for $x < y$:

(4) $$P(X_{(r)} \leq x, X_{(s)} \leq y)$$

$$= \sum_{i=r}^{n} \sum_{j=\max(0,\,s-i)}^{n-i} \frac{n!}{i!j!(n-i-j)!} F^i(x)[F(y) - F(x)]^j[1 - F(y)]^{n-i-j}.$$

In the continuous case, we differentiate partially with respect to y and then with respect to x to get

(5) $$f_{rs}(x, y) = \frac{\partial^2}{\partial x \, \partial y} P(X_{(r)} \leq x, X_{(s)} \leq y)$$

$$= \frac{n!}{(r-1)!(s-r-1)!(n-s)!} F^{r-1}(x)f(x)$$

$$\times [F(y) - F(x)]^{s-r-1} f(y)[1 - F(y)]^{n-s},$$

for $x < y$, and $f_{rs}(x, y) = 0$ otherwise. In particular, if $r = 1$ and $s = n$ we get:

$$(6) \quad f_{1n}(x, y) = \begin{cases} n(n-1)[F(y) - F(x)]^{n-2}f(x)f(y), & x < y \\ 0 & \text{otherwise.} \end{cases}$$

The marginal density of $X_{(r)}$, $1 \le r \le n$ can be obtained by integrating $f_{rs}(x, y)$ with respect to y or directly by computing the distribution function of $X_{(r)}$. In fact,

$$P(X_{(r)} \le x) = P(\text{at least } r \text{ of the } X\text{'s} \le x)$$

$$= \sum_{i=r}^{n} P(\text{exactly } i \ X\text{'s} \le x, \text{ and } (n-i) X\text{'s} > x)$$

$$(7) \qquad = \sum_{i=r}^{n} \binom{n}{i} F^i(x)[1 - F(x)]^{n-i}.$$

In the continuous case, we differentiate[†] (7) with respect to x to obtain

$$f_r(x) = \frac{d}{dx} P(X_{(r)} \le x)$$

$$= \sum_{i=r}^{n} \binom{n}{i} i F^{i-1}(x) f(x)[1 - F(x)]^{n-i}$$

$$- \sum_{i=r}^{n-1} \binom{n}{i} F^i(x)(n-i)[1 - F(x)]^{n-i-1} f(x)$$

$$= \sum_{i=r}^{n} \frac{n!}{(i-1)!(n-i)!} F^{i-1}(x)[1 - F(x)]^{n-i} f(x)$$

$$- \sum_{i=r+1}^{n} \frac{n!}{(i-1)!(n-i)!} F^{i-1}(x)[1 - F(x)]^{n-i} f(x),$$

so that for $x \in \mathbb{R}$

$$(8) \qquad f_r(x) = \frac{n!}{(r-1)!(n-r)!} F^{r-1}(x)[1 - F(x)]^{n-r} f(x).$$

[†]Alternatively, one could use the relation between beta and binomial distributions (Example 7.5.2) to obtain

$$P(X_{(r)} \le x) = n \binom{n-1}{r-1} \int_0^{F(x)} t^{r-1} (1-t)^{n-r} \, dt.$$

Differentiate with respect to x to get (8).

In particular,

$$f_1(x) = n[1 - F(x)]^{n-1}f(x), \qquad x \in \mathbb{R},$$

and

$$f_n(x) = nF^{n-1}(x)f(x), \qquad x \in \mathbb{R}.$$

In general, it can be shown (in the continuous case) that

$$(9) \qquad f_{12\ldots n}(x_1, x_2, \ldots, x_n) = \begin{cases} n! \displaystyle\prod_{j=1}^{n} f(x_j), & x_1 < x_2 < \cdots < x_n \\ 0 & \text{otherwise.} \end{cases}$$

Distribution of Order Statistics in Sampling from a Density Function f with Distribution Function F

$$f_r(x) = \frac{n!}{(r-1)!(n-r)!} F^{r-1}(x)[1 - F(x)]^{n-r}f(x), \qquad x \in \mathbb{R}$$

$$f_{rs}(x, y) = \begin{cases} \dfrac{n!}{(r-1)!(s-r-1)!(n-s)!} F^{r-1}(x)f(x)[F(y) - F(x)]^{s-r-1} \\ \qquad\qquad\qquad\qquad f(y)[1 - F(y)]^{n-s}, \qquad x < y \\ 0 \qquad\qquad\qquad\qquad\qquad\qquad\qquad\qquad\qquad x \geq y. \end{cases}$$

$$f_{12\ldots n}(x_1, \ldots, x_n) = \begin{cases} n! \displaystyle\prod_{j=1}^{n} f(x_j), & x_1 < x_2 < \cdots < x_n, \\ 0 & \text{elsewhere.} \end{cases}$$

Remark 1. Since $X_{(1)} \leq X_{(2)} \leq \cdots \leq X_{(n)}$ we cannot expect the $X_{(i)}$'s to be independent. Indeed they are dependent unless the common distribution of the X's is degenerate at some constant c. We leave the reader to supply a proof in Problem 27.

Example 1. **Distribution of Maximum in Binomial Sampling.** Let $X_1, X_2, \ldots, X_k$ be a sample from a binomial population with parameters n and p. Let $X_{(r)}, 1 \leq r \leq k$ be the rth order statistic. Then, from (7):

$$P(X_{(r)} \leq l) = \sum_{i=r}^{k} \binom{k}{i} F^i(l)[1 - F(l)]^{k-i}$$

for $l = 0, 1, 2, \ldots, n$ where $F(l) = P(X_i \le l) = \sum_{j=0}^{l} \binom{n}{j} p^j (1 - p)^{n-j}$ Hence, for $l \ge 1$,

$$P(X_{(r)} = l) = P(X_{(r)} \le l) - P(X_{(r)} \le l - 1)$$

$$= \sum_{i=r}^{k} \binom{k}{i} F^i(l)[1 - F(l)]^{k-i} - \sum_{i=r}^{k} \binom{k}{i} F^i(l-1)[1 - F(l-1)]^{k-i}$$

and

$$P(X_{(r)} = 0) = \sum_{i=r}^{k} \binom{k}{i} ((1-p)^n)^i [1 - (1-p)^n]^{k-i}. \qquad \square$$

Example 2. Length of Life of a System with n Components Operating in Series or Parallel. Consider a system of n identical batteries operating independently in a system. Suppose the length of life has common density function

$$f(x) = \begin{cases} \lambda^{-1} e^{-x/\lambda}, & x > 0 \\ 0 & \text{elsewhere.} \end{cases}$$

We first consider the case when the batteries operate in series (Figure 1) so that the system fails at the first battery failure.

Since the system fails at the first failure, the length of life Y of the system is given by

$$Y = \min(X_1, X_2, \ldots, X_n).$$

In view of (8) with $r = 1$, the density function of Y is given by

$$f_Y(y) = f_1(y) = \begin{cases} n[1 - F(y)]^{n-1} f(y), & \text{for } y > 0 \\ 0 & y \le 0. \end{cases}$$

Since $F(y) = 1 - e^{-y/\lambda}$ for $y > 0$ we have

$$f_Y(y) = \begin{cases} (n/\lambda) e^{-ny/\lambda}, & y > 0 \\ 0 & y \le 0. \end{cases}$$

It follows that Y itself has an exponential distribution with mean λ/n.

If the object is to estimate λ, then

$$\mathscr{E}Y = \lambda/n \Rightarrow \mathscr{E}(nY) = \lambda$$

so that nY is an unbiased estimate of λ. Note that $\overline{X}$ is also unbiased. Moreover,

$$\text{var}(\overline{X}) = \lambda^2/n.$$

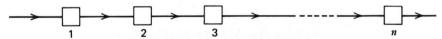

Figure 1. n Batteries in series.

Also,

$$\mathrm{var}(nY) = n^2\,\mathrm{var}(Y) = n^2\frac{\lambda^2}{n^2} = \lambda^2$$

so that

$$\mathrm{var}(nY) > \mathrm{var}(\overline{X}) \qquad \text{for } n > 1.$$

Yet nY may be preferred over $\overline{X}$ as an estimate of λ.

On the other hand, if the batteries operate in parallel (Figure 2), the system fails if all the n batteries fail. In this case, the length of life of the system equals

$$Z = \max(X_1, X_2, \ldots, X_n)$$

with density function

$$f_Z(z) = f_n(z) = n[F(z)]^{n-1}f(z)$$

$$= \begin{cases} (n/\lambda)e^{-z/\lambda}[1 - e^{-z/\lambda}]^{n-1} & \text{for } z > 0 \\ 0 & \text{elsewhere.} \end{cases} \qquad \square$$

Example 3. The Distribution of Range of a Random Sample. Let $X_1, X_2, \ldots, X_n$ be a random sample from a distribution function F with density function f. A somewhat crude measure of the variation in the sample is the range R defined by $R = X_{(n)} - X_{(1)}$. We compute the distribution of R by using (6) and (8.4.2). In view of (8.4.2), we have

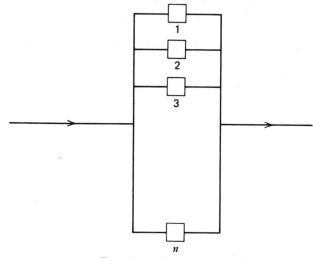

Figure 2. n Batteries in parallel.

for $r > 0$

$$f_R(r) = \int_{-\infty}^{\infty} f_{1n}(x, r + x)\, dx$$

(10)
$$= n(n - 1)\int_{-\infty}^{\infty} [F(r + x) - F(x)]^{n-2} f(x) f(x + r)\, dx$$

and for $r \le 0$,

$$f_R(r) = 0.$$

If in particular $X_1, X_2, \ldots, X_n$ are n points chosen at random and independently from $[0, 1]$, then

$$f(x) = \begin{cases} 1, & 0 \le x \le 1 \\ 0 & \text{elsewhere} \end{cases} \quad \text{and} \quad F(x) = \begin{cases} 0, & x < 0 \\ x, & 0 \le x < 1 \\ 1, & x \ge 1. \end{cases}$$

In this case $f(x)f(x + r) > 0$ if and only if $0 \le x < x + r \le 1$, and it follows (Figure 3) from (10) that for $0 < r < 1$

$$f_R(r) = n(n - 1)\int_0^{1-r} r^{n-2}\, dx = n(n - 1) r^{n-2}(1 - r)$$

and

$$f_R(r) = 0 \quad \text{for } r \le 0 \ \text{ or } r \ge 1.$$

We note that R has a beta density with $\alpha = n - 1$ and $\beta = 2$. For $n = 2$, the density of the range is given by

$$f_R(r) = \begin{cases} 2(1 - r), & 0 < r < 1 \\ 0 & \text{otherwise,} \end{cases}$$

which is *not* the density of $X_1 - X_2$ as computed in Example 8.2.11. The range in this case is in fact

$$R = X_{(2)} - X_{(1)} = |X_1 - X_2|.$$

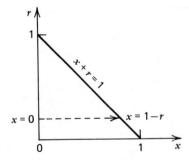

Figure 3. $\{(x, r): f(x)f(x + r) > 0\}$.

We leave the reader to check that the density of $|X_1 - X_2|$ as computed from Example 8.2.11 coincides with the one obtained above. $\square$

Example 4.   Estimating θ for Uniform Distribution on $(0, \theta)$.   Let $X_1, X_2, \ldots, X_n$, $n \geq 2$, be a sample from a uniform distribution on $(0, \theta)$. We consider the performance of several estimates of θ some based on order statistics. First we note that

$$\hat{\theta}_1 = 2\overline{X}$$

is unbiased for θ with variance $\sigma_1^2 = 4\,\theta^2/12n = \theta^2/3n$. Next we note that $M_n = \max(X_1, \ldots, X_n)$ has density function

$$f_n(x) = nF^{n-1}(x)f(x) = \begin{cases} nx^{n-1}/\theta^n & \text{for } 0 < x < \theta \\ 0 & \text{otherwise} \end{cases}$$

so that

$$\mathscr{E}M_n = (n/\theta^n)\int_0^\theta xx^{n-1}\,dx = \frac{n}{n+1}\theta$$

and

$$\mathscr{E}M_n^2 = (n/\theta^n)\int_0^\theta x^2x^{n-1}\,dx = \frac{n}{n+2}\theta^2.$$

It follows that

$$\hat{\theta}_2 = \frac{n+1}{n}M_n$$

is unbiased for θ with variance

$$\sigma_2^2 = \left(\frac{n+1}{n}\right)^2 \text{var}(M_n) = \left(\frac{n+1}{n}\right)^2\left[\frac{n}{n+2}\theta^2 - \left(\frac{n}{n+1}\right)^2\theta^2\right] = \frac{\theta^2}{n(n+2)}.$$

Finally, we consider the range $R = \max\{X_1, \ldots, X_n\} - \min\{X_1, \ldots, X_n\}$ and note that R has density

$$f_R(r) = \begin{cases} n(n-1)r^{n-2}(\theta - r)/\theta^n & \text{for } 0 < r < \theta \\ 0 & \text{otherwise.} \end{cases}$$

Hence

$$\mathscr{E}R = \frac{n(n-1)}{\theta^n}\int_0^\theta rr^{n-2}(\theta - r)\,dr$$

$$= \frac{n(n-1)}{\theta^n}\left[\theta\frac{\theta^n}{n} - \frac{\theta^{n+1}}{n+1}\right] = \frac{n-1}{n+1}\theta$$

and it follows that

$$\hat{\theta}_3 = \frac{n+1}{n-1} R$$

is also unbiased for θ. Moreover,

$$\mathscr{E}R^2 = \frac{n(n-1)}{\theta^n} \int_0^\theta r^2 r^{n-2}(\theta - r)\, dr = \frac{n(n-1)}{\theta^n}\left[\theta \frac{\theta^{n+1}}{n+1} - \frac{\theta^{n+2}}{n+2}\right]$$

$$= \frac{n(n-1)}{(n+1)(n+2)}\theta^2.$$

Hence

$$\sigma_3^2 = \mathrm{var}(\hat{\theta}_3) = \left(\frac{n+1}{n-1}\right)^2 \mathrm{var}(R) = \left(\frac{n+1}{n-1}\right)^2 \theta^2\left[\frac{n(n-1)}{(n+1)(n+2)} - \left(\frac{n-1}{n+1}\right)^2\right]$$

$$= \frac{n+1}{n-1}\theta^2\left(\frac{2}{(n+1)(n+2)}\right) = \frac{2\theta^2}{(n-1)(n+2)}.$$

It is easily checked that $\sigma_3^2 > \sigma_2^2$ for $n \geq 2$ so that $\hat{\theta}_2$ is a better estimate of θ than $\hat{\theta}_3$ if the criteria used is to choose an estimate with a smaller variance. Moreover, $\sigma_1^2 > \sigma_2^2$ so that $\hat{\theta}_2$ is the best of the three estimates considered here in terms of minimum variance.

This is essentially a continuous version of the taxi problem considered in Example 6.2.1.   □

8.5.1 Estimation of Quantiles of a Distribution

The quantiles provide important information about the shape of a distribution. Often one knows little about the distribution being sampled. Knowledge of a few quantiles provides important information about how the distribution assigns probabilities.

We recall (Section 3.7) that a quantile of order p of a random variable with distribution function F is a number $\zeta_p(F)$ that satisfies

$$P\big(X \leq \zeta_p(F)\big) \geq p \quad \text{and} \quad P\big(X \geq \zeta_p(F)\big) \geq 1 - p, \quad 0 < p < 1$$

so that $\zeta_p(F)$ is a solution of

$$(11) \qquad\qquad p \leq F(x) \leq p + P\{X = x\}.$$

If $P(X = x) = 0$, as is the case, in particular, if X is of the continuous type, then ζ_p is a solution of the equation

$$(12) \qquad\qquad\qquad F(x) = p.$$

If F is strictly increasing, ζ_p is uniquely determined by (12). We recall also that for $p = 1/2$ we call $\zeta_p(F)$ a median of F. Quantiles have the advantage that they can be estimated from the sample without the knowledge of the form of F, that is, nonparametrically. Moreover these estimates are easy to compute.

In what follows we will assume, unless otherwise specified, *that F is continuous and strictly increasing* so that (12) has a unique solution. Let $X_1, X_2, \ldots, X_n$ be a random sample from F. How do we estimate $\zeta_p(F)$, given p, $0 < p < 1$? An intuitive answer is to use the sample quantile of order p to estimate ζ_p. (This is the so-called *substitution principle*, which is discussed in detail in Section 10.4.) Since F is continuous, we can arrange the observations in an increasing order of magnitude as

$$X_{(1)} < X_{(2)} < \cdots < X_{(n)}.$$

Recall (Remark 3.8.5) that when $n = 2m + 1$, there is a unique middle value $X_{(m+1)}$ which is the sample median. If $n = 2m$, then $X_{(m)}$ and $X_{(m+1)}$ are two middle values and any number between $X_{(m)}$ and $X_{(m+1)}$ may be taken as a median. We adopted the convention that in this case we will take $(X_{(m)} + X_{(m+1)})/2$ as the sample median. Similar problems arise when we try to define a sample quantile of order p, $0 < p < 1$. If n is large, however, the range of possible values of the sample quantile of order p will be small, and it does not really matter which possible value is taken to define the sample quantile. For this reason, we define the *sample quantile of order p* by

$$\hat{\zeta}_p = \begin{cases} X_{(np)} & \text{if } np \text{ is integral} \\ X_{([np]+1)} & \text{if } np \text{ is not an integer.} \end{cases}$$

where $[np]$ is the largest integer[†] less than or equal to np. For $n = 5$, $p = .3$, for example, $np = 5(.3) = 1.5$ so that $\hat{\zeta}_{.3} = X_{(2)}$. If $n = 30$, $p = 1/4$, then $np = 7.5$, $[np] = 7$, and $\hat{\zeta}_{.25} = X_{(8)}$. We use $\hat{\zeta}_p$ to estimate $\zeta_p(F)$.

Example 5. Estimating Quantiles of a Life Length Distribution. The length of life X of a component used in a complex system has a continuous distribution assumed to be strictly increasing. In this case the mean and variance of X may not provide information as useful as some of the quantiles provide. It may be more pertinent to know, for example, that "15 percent of the components fail before time t_1 and 25 percent fail before time t_2," and so on. Suppose nine observations on X (in 1000 hours) were 3.2, 5.8, 7.3, 2.9, 4.1, 6.5, 4.7, 5.9, 6.3. The order statistic is 2.9, 3.2, 4.1, 4.7, 5.8, 5.9, 6.3, 6.5, 7.3. The .25 quantile corresponds to $[np] + 1 = [9(.25)] + 1 =$ third order statistic 4.1. The median corresponds to fifth order statistic 5.8, and the quantile of order .75 corresponds to seventh order statistic 6.3. Thus estimates of $\hat{\zeta}_{.25}, \hat{\zeta}_{.5}, \hat{\zeta}_{.75}$ are, respectively, 4.1, 5.8, and 6.3. □

[†]One could instead take $[x]$ to be the smallest integer greater than or equal to x. In that case $\hat{\zeta}_p = X_{([np])}$.

Example 6. Quantiles of Uniform Distribution on (0, θ). If we know the form of the underlying distribution F then we cannot expect sample quantiles to be good estimates of population quantiles. Let $X_1, X_2, \ldots, X_n$ be a sample from a uniform distribution on $(0, \theta)$. Suppose $0 < p < 1$ and we wish to estimate ζ_p. Clearly,

$$F(\zeta_p) = p = \frac{\zeta_p}{\theta}$$

so that $\zeta_p(F) = \theta p$. If we ignore the knowledge that F is uniform on $(0, \theta)$, our estimate of ζ_p will be $\hat{\zeta}_p$. Let $r = np$ if np is an integer, and $r = [np] + 1$ if np is not an integer. Then $\hat{\zeta}_p = X_{(r)}$ has density

$$f_r(x) = \begin{cases} n\binom{n-1}{r-1}(x/\theta)^{r-1}(1 - x/\theta)^{n-r}(1/\theta), & 0 < x < \theta \\ 0 & \text{otherwise} \end{cases}$$

so that

$$\mathscr{E}X_{(r)} = n\binom{n-1}{r-1}\frac{1}{\theta^r}\int_0^\theta x^r\left(1 - \frac{x}{\theta}\right)^{n-r} dx$$

$$= n\binom{n-1}{r-1}\theta\int_0^1 y^r(1 - y)^{n-r} dy \qquad \left(y = \frac{x}{\theta}\right)$$

$$= n\binom{n-1}{r-1}\theta B(r + 1, n - r + 1) = \frac{r\theta}{n+1}.$$

Also,

$$\mathscr{E}X_{(r)}^2 = n\binom{n-1}{r-1}\frac{1}{\theta^r}\int_0^\theta x^{r+1}\left(1 - \frac{x}{\theta}\right)^{n-r} dx$$

$$= n\binom{n-1}{r-1}\theta^2\int_0^1 y^{r+1}(1 - y)^{n-r} dy \qquad \left(y = \frac{x}{\theta}\right)$$

$$= n\binom{n-1}{r-1}\theta^2 B(r + 2, n - r + 1) = \frac{r(r+1)}{(n+1)(n+2)}\theta^2.$$

Thus

$$\text{MSE}(X_{(r)}) = \mathscr{E}\left(X_{(r)} - \zeta_p\right)^2 = \mathscr{E}X_{(r)}^2 + \zeta_p^2 - 2\zeta_p\mathscr{E}X_{(r)}$$

$$= \theta^2\left[\frac{r(r+1)}{(n+1)(n+2)} + p^2 - 2p\frac{r}{n+1}\right].$$

On the other hand, the knowledge of F allows us to choose the estimate

$$\hat{\theta}_1 = \frac{n+1}{n}M_n p = \frac{n+1}{n}pX_{(n)}$$

from Example 4. It is clear that $\mathscr{E}\theta_1 = p\mathscr{E}\left(\dfrac{n+1}{n}M_n\right) = \theta p$ so that θ_1 is unbiased and

$$\text{MSE}(\theta_1) = \text{var}(\theta_1) = p^2 \text{var}\left(\frac{n+1}{n}M_n\right) = \frac{p^2\theta^2}{n(n+2)}$$

again from Example 4. It should not be surprising if $\text{MSE}(\theta_1) \leq \text{MSE}(X_{(r)})$ since θ_1 uses the knowledge that F is uniform on $(0, \theta)$ whereas $X_{(r)}$ is the estimate of ζ_p no matter what F is.

We leave the reader to check that $\text{MSE}(\theta_1) < \text{MSE}(X_{(r)})$.   □

Some large sample properties of order statistics as estimates of quantiles will be taken up in Section 9.8. We now consider the problem of constructing confidence intervals for quantiles. This problem has an easy solution, as we show. Let $X_1, X_2, \ldots, X_n$ be a random sample from a distribution function F which may be discrete or continuous. Let $X_{(1)} \leq X_{(2)} \leq \cdots \leq X_{(n)}$ be the order statistic. Suppose we want a confidence interval for $\zeta_p(F) = \zeta_p$ for $0 < p < 1$ (p known). Recall that ζ_p is a solution of

$$F(\zeta_p) - P(X = \zeta_p) \leq p \leq F(\zeta_p).$$

Let Y be the number of X_i's that are less than or equal to ζ_p. Then Y has a binomial distribution with parameters n and $P(X_i \leq \zeta_p) = F(\zeta_p)$. Hence

$$(13) \qquad P(X_{(r)} \leq \zeta_p) = P(Y \geq r) = \sum_{j=r}^{n}\binom{n}{j}F^j(\zeta_p)\left[1 - F(\zeta_p)\right]^{n-j}.$$

Similarly,

$$(14) \quad P(X_{(s)} \geq \zeta_p) = 1 - P(X_{(s)} < \zeta_p) = 1 - \sum_{j=s}^{n}\binom{n}{j}F^j(\zeta_{p^-})\left[1 - F(\zeta_{p^-})\right]^{n-j}$$

$$= \sum_{j=0}^{s-1}\binom{n}{j}F^j(\zeta_{p^-})\left[1 - F(\zeta_{p^-})\right]^{n-j}.$$

In view of (13) and (14), we have for $r < s$

$$P(X_{(r)} \leq \zeta_p \leq X_{(s)}) = P(X_{(s)} \geq \zeta_p) - P(X_{(r)} > \zeta_p)$$

$$= \sum_{j=0}^{s-1}\binom{n}{j}F^j(\zeta_{p^-})\left[1 - F(\zeta_{p^-})\right]^{n-j} - \sum_{j=0}^{r-1}\binom{n}{j}F^j(\zeta_p)\left[1 - F(\zeta_p)\right]^{n-j}.$$

$$(15) \qquad \geq \sum_{j=r}^{s-1}\binom{n}{j}p^j(1-p)^{n-j},$$

since $F(\zeta_p-) \leq p \leq F(\zeta_p)$ with equality if F is continuous. [We have used inequality (7.5.6′).] Suppose we want a confidence interval for ζ_p at level at least $1 - \alpha$. Then we choose r and s such that

(16)
$$\sum_{j=r}^{s-1} \binom{n}{j} p^j (1 - p)^{n-j} \geq 1 - \alpha$$

so that

$$P\left(X_{(r)} \leq \zeta_p \leq X_{(s)} \right) \geq 1 - \alpha.$$

This is done easily with the help of a binomial distribution table such as Table A3. In practice we choose r and s to minimize $s - r$ subject to (16).

We emphasize that since the binomial distribution function [used on the left side of (16)] is discrete given α, a $1 - \alpha$ level confidence interval $[X_{(r)}, X_{(s)}]$ for ζ_p may or may not exist. If F is continuous, then there is equality in (15). Nevertheless, not every level $1 - \alpha$ is attainable. (See Problems 18 and 28.)

Example 7. Confidence Interval for "Half-Lifetime" or "Median Lifetime" for Example 5. Let us compute a 95 percent confidence interval for the median lifetime $\zeta_{.5}$ of the component. Since $n = 9$, $p = .5$, and $\alpha = .05$, we need to choose r and s such that $s - r$ is the least from

$$\sum_{j=r}^{s-1} \binom{9}{j} \left(\frac{1}{2}\right)^9 \geq .95.$$

From Table A3 we see that $s = 8$ and $r = 2$ gives a confidence interval of level $.9806 - .0196 = .9610$. Hence a 95 percent confidence interval is $[X_{(2)}, X_{(8)}] = [3.2, 6.5]$.

Often the problem is to find the sample size n such that $[X_{(r)}, X_{(s)}]$ is a confidence interval for ζ_p at level $\geq 1 - \alpha$. Given r, s, p and α this is done by solving (16) for n. In particular, suppose we want the sample size n such that $[X_{(1)}, X_{(n)}]$ is a 99 percent confidence interval for the median. Then we need to solve

$$\sum_{j=1}^{n-1} \binom{n}{j} \left(\frac{1}{2}\right)^j \geq .99$$

for n. It follows from Table A3 that $n = 8$.   □

8.5.2 Testing Hypothesis Concerning Quantiles

The problem of testing $H_0 : \zeta_p(F) = \zeta_0$ against one-sided or two-sided alternatives when F is continuous is also simple. For example, if the alternative is $H_1 : \zeta_p(F) > \zeta_0$, then we can use some order statistic $X_{(r)}$ to test H_0. Under H_0,

$$P(X \leq \zeta_0) = F(\zeta_0) = p.$$

If the level of significance α is specified, then we choose the largest r such that

$$\alpha \geq P\{X_{(r)} > \zeta_0 | H_0\} = 1 - P\{X_{(r)} \leq \zeta_0 | H_0\}$$

$$= 1 - \sum_{j=r}^{n} \binom{n}{j} p^j (1-p)^{n-j} = \sum_{j=0}^{r-1} \binom{n}{j} p^j (1-p)^{n-j},$$

in view of (13). We reject H_0 if the observed value of $X_{(r)}$ is greater than ζ_0.

A simpler procedure is to consider $Y_i = X_i - \zeta_0$, $i = 1, 2, \ldots, n$. Then count the number S of negative signs among the Y's. A small value of S tends to support H_1 so that we reject H_0 if S is small. This is precisely the sign test of Section 6.3.2. In fact, under H_0, $P(Y_i < 0) = P(X_i < \zeta_0) = p$ and S has a binomial distribution with parameters n and p. We reject H_0 if $S \leq c$, say, where c is chosen from

$$P\{S \leq c | H_0\} = \sum_{j=0}^{c} \binom{n}{j} p^j (1-p)^{n-j} \leq \alpha.$$

Alternatively, we compute the P-value

$$P\{S \leq s_0 | H_0\} = \sum_{j=0}^{s_0} \binom{n}{j} p^j (1-p)^{n-j}$$

where s_0 is the observed number of negative Y's. The magnitude of the P-value determines whether or not we reject H_0 at level α.

Note that

$X_{(r)} > \zeta_0$ if and only if there are at least $n - r + 1$ plus signs in the Y's or equivalently at most $r - 1$ negative signs among the Y's.

Hence the test with critical region $X_{(r)} > \zeta_0$ is equivalent to the test with critical region $S \leq r - 1$. The procedure is similar for testing $\zeta_p(F) = \zeta_0$ against $\zeta_p(F) < \zeta_0$ or $\zeta_p(F) \neq \zeta_0$. We refer to Section 6.3.

We conclude our discussion with the following example.

Example 8. I.Q. Scores. A company claims that 75 percent of its employees have an I.Q. score in excess of 115. A random sample of 10 employees showed I.Q. scores of

$$96, \quad 110, \quad 98, \quad 129, \quad 142, \quad 131, \quad 137, \quad 115, \quad 113, \quad 153.$$

Do the data support the company's claim?

The problem is to test H_0: $\zeta_{.25} = 115$ against H_1: $\zeta_{.25} < 115$. Since one of the observations is 115, it has to be discarded. Of the remaining 9 observations 4 are below 115 (that is, $X_i - 115$ is negative for four values of i). Since under H_0

$$P(X_i \leq 115) = .25 \quad \text{and} \quad P(X_i \geq 115) = .75,$$

the probability of 4 or more negative signs is

$$\sum_{j=4}^{9} \binom{9}{j}(.25)^{j}(.75)^{9-j} = 1 - .8343 = .1657.$$

The descriptive level of our test, or the P-value, is .1657. That is, if we reject H_0 whenever four or more observations in samples of size 9 are below 115, then we will be rejecting a true null hypothesis in about 16.57 percent of the samples. □

Problems for Section 8.5

1. Let $X_1, X_2, \ldots, X_k$ be a random sample from a discrete distribution. In each of the following cases find the probability function of $X_{(r)}, 1 \le r \le k$.
   (i) The common distribution of the X's is Poisson (λ).
   (ii) The common distribution of the X's is geometric (p). [Hint: Use (7) and the relation $P(X_{(r)} = l) = P(X_{(r)} \le l) - P(X_{(r)} \le l - 1)$ where $l \ge 1$ is an integer.]
   Specialize the results to obtain the probability function of the maximum of a sample of size $k = 2$ in each case.

2. Let (X_1, X_2) have joint distribution given by

X_2 \ X_1	-1	0	1	
0	1/6	1/3	1/6	2/3
1	1/6	0	1/6	1/3
	1/3	1/3	1/3	1

   Find the joint probability function of $(\min(X_1, X_2), \max(X_1, X_2))$ and find the two marginal distributions.

3. Let X_1, X_2 be independent random variables with joint distribution given by

X_2 \ X_1	1	2	3	
1	1/12	1/6	1/12	1/3
2	1/12	0	1/12	1/6
3	1/6	0	1/3	1/2
	1/3	1/6	1/2	1

   Find the joint distribution of $\min(X_1, X_2)$ and $\max(X_1, X_2)$. Also find the marginal probability functions.

4. Let $X_{(1)} < X_{(2)} < X_{(3)} < X_{(4)}$ be the order statistics of a random sample of size 4 from density function

$$f(x) = \begin{cases} 2(1 - x) & 0 \leq x \leq 1 \\ 0 & \text{elsewhere.} \end{cases}$$

   Find:
   (i) $P(.3 < X_{(3)} < .5)$.   (ii) $P(1/3 < X_{(2)} < 3/4)$.

5. A sample of size 9 is taken from the uniform distribution on $(0, 1)$. Find the probability that the sample median is less than .3.

6. A sample of size 9 is taken from a continuous type distribution. Suppose $\xi_{.75} = 16.4$.
   (i) Find $P(X_{(6)} < 16.4)$.   (ii) Find $P(X_{(4)} < 16.4 < X_{(6)})$.

7. Let X and Y be independent random variables with common probability function $P(X = x) = P(Y = x) = p(1 - p)^{x-1}, x = 1, 2, \ldots$. Show that $X_{(1)}$ and $X_{(2)} - X_{(1)}$ are independent.

8. Let $X_{(1)}, X_{(2)}$ be the order statistics in a sample of size 2 from density

$$f(x) = \begin{cases} (1/\lambda)^{-x/\lambda}, & x \geq 0 \\ 0 & x < 0. \end{cases}$$

   (i) Show that $X_{(1)}$ and $X_{(2)} - X_{(1)}$ are independent.
   (ii) Find the density of $X_{(2)} - X_{(1)}$.

9. Let $X_{(1)}, X_{(2)}, \ldots, X_{(n)}$ be the order statistics in a random sample of size n from exponential density with mean λ. Find the density function of the range $R = X_{(n)} - X_{(1)}$.

10. A manufacturer produces wire for screen doors on two machines operating independently. The tensile strength of the wire produced by each machine (when the machines are operating correctly) is a random variable with density

$$f(x) = \begin{cases} (1/\sigma)\exp\{-(x - \mu)/\sigma\}, & x \geq \mu \\ 0 & x < \mu. \end{cases}$$

   What is the distribution of the maximum tensile strength of wire produced by this manufacturer?

11. Let $X_1, X_2, \ldots, X_n$ be independent random variables and suppose that X_i has density $f_i, i = 1, 2, \ldots, n$. Find the density of:
    (i) $X_{(1)} = \min(X_1, \ldots, X_n)$.   (ii) $X_{(n)} = \max(X_1, \ldots, X_n)$.

12. Consider a system with n components $C_1, C_2, \ldots, C_n$ which are connected in series. If the component C_i has failure density that is exponential with mean $\lambda_i, i = 1, 2, \ldots, n$,
    (i) What is the reliability of the system? That is, find the survival function.
    (ii) What is the mean failure time for this system?

13. In Problem 12, suppose the n components are connected in parallel. Find the reliability of the system and an expression for its mean failure time.

14. Consider a system with n components each of which has an exponential failure density with parameter λ. Suppose the system does not fail as long as at least r components are operating. Find an expression for the mean system failure time.

15. A random sample of size $n = 2m + 1$ is taken from exponential density with mean λ. Find the density of the sample median $X_{(m+1)}$.

16. Let X and Y be independent and identically distributed with density f. Find the joint density of $(\min(X, Y), \max(X, Y))$ and the density of $\min(X, Y)/\max(X, Y)$ in each of the following cases.
    (i) $f(x) = 1, 0 \le x \le 1$ and 0 otherwise.
    (ii) $f(x) = e^{-x}$ for $x > 0$ and 0 elsewhere.

17. Find a 90 percent confidence interval for the quantile of order .70 based on a sample of size 10 so as to minimize $s - r$, where $[X_{(r)}, X_{(s)}]$ is the confidence interval used.

18. Consider the confidence interval $[X_{(1)}, X_{(n)}]$ for the median of a continuous distribution based on a sample of size n. Find the minimum confidence level of the interval for $n = 2, 3, \ldots, 10$. (See also Problem 28.)

19. To estimate the gas mileage that she is getting from her new car, Francine recorded the following number of miles per gallon on 12 tankfuls:

    27.2, 29.8, 30.4, 26.5, 29.4, 30.8, 30.9, 31.0, 28.6, 27.9, 29.6, 30.3.

    Find a point estimate for the median gasoline mileage and find a 90 percent confidence interval for the median.

20. Eight randomly selected families in a small town have the following incomes (in thousands of dollars): 24.5, 36.8, 19.4, 27.9, 40.5, 32.7, 28.9, 38.3. Find a point estimate for the median family income in the town and find a 98 percent confidence interval for the median income.

21. Find the smallest n such that $[X_{(1)}, X_{(n)}]$ is a 90 percent confidence interval for:
    (i) The quantile of order .25. (ii) The quantile of order .70.
    (iii) The quantile of order .80.

22. A light bulb manufacturer advertises that 90 percent of its 60-watt bulbs last more than 1000 hours. To check the company's claim, lifetimes in hours of 15 bulbs were recorded:

    1500, 2010, 1141, 233, 119, 495, 2400, 2200,

    1100, 1901, 1845, 1368, 1479, 3284, 2107.

    Do the data support the manufacturer's claim that 90 percent of its bulbs last more than 1000 hours? Find the P-value of your test of H_0: $\zeta_{.1} \ge 1000$.

23. A tire manufacturer claims that 75 percent of its tires last 40,000 miles or more. To test this claim a consumer group recorded the number of miles each of eight drivers

obtained from identical sets of tires:

32,100, 38,000, 46,500, 49,000, 52,700, 39,300, 47,600, 48,500.

Is there evidence to support the manufacturer's claim? What is the P-value of your test for testing H_0: $\zeta_{.25} \geq 40{,}000$? What is the P-value for testing H_0: $\zeta_{.25} = 40{,}000$ against H_1: $\zeta_{.25} \neq 40{,}000$?

24. The median time in minutes for a tune-up of a four cylinder car is known to be 45 minutes. A random sample of nine mechanics showed tune-up times of 32, 35, 47, 45, 57, 38, 34, 30, 36. What conclusion do you draw from the data?

25. The systolic blood pressures of 11 men were recorded:

160, 120, 113, 114, 135, 116, 126, 125, 119, 142, 134.

Use the sign test to test the hypothesis that the median systolic blood pressure in men is 120.

26*. In Example 6 we computed the mean square errors of $X_{(r)}$ and $\hat{\theta}_1 = (n+1)pX_{(n)}/n$ as estimators of θp in sampling from a uniform distribution on $(0, \theta)$.
   (i) Show that $\text{MSE}(\hat{\theta}_1) < \text{MSE}(X_{(r)})$ by considering the two cases when $r = np$ is integral or when $r = [np] + 1$.
   (ii) Note that $(n+1)X_{(r)}/r$ is unbiased for θ. Compute the variance of $(n+1)X_{(r)}/r$ and show that $\text{var}((n+1)X_{(r)}/r) > \text{var}[(n+1)X_{(n)}/n]$ for $r < n$. (See Example 4.)
   (iii) The estimate $p(n+1)X_{(r)}/r$ is unbiased for ζ_p but has larger variance than that of $\hat{\theta}_1$ for $r < n$.

27*. Let $X_{(1)} \leq X_{(2)} \leq \cdots \leq X_{(n)}$ be the order statistics of random variables $X_1, X_2, \ldots, X_n$. Show that $X_{(1)}, X_{(2)}, \ldots, X_{(n)}$ are dependent random variables unless they are all degenerate.

28. Find α for which no level $1 - \alpha$ confidence interval $[X_{(1)}, X_{(n)}]$ for ζ_p exists. In particular, let $p = .5$. Find n for which no level $1 - \alpha$ confidence interval for ζ_p exists for values of $\alpha = .01, .02, .05$. (See also Problem 18.)

8.6* GENERATING FUNCTIONS

In this section we consider some special functions that generate moments and probability functions. Often these functions simplify computations of induced distributions.

Let $\{a_n, n = 0, 1, \ldots\}$ be a sequence of real numbers. Then the generating function of $\{a_n\}$ is defined by

$$(1) \qquad\qquad A(s) = \sum_{k=0}^{\infty} a_k s^k$$

provided that the (power) series on the right side of (1) converges for all s in some open neighborhood $|s| < h$ of zero. In particular, let $a_k = P(X = k)$, $k = 0, 1, \ldots$ be the probability function of a nonnegative integer valued random variable. Let

$$(2) \qquad P(s) = \mathscr{E} s^X = \sum_{k=0}^{\infty} a_k s^k = \sum_{k=0}^{\infty} s^k P(X = k).$$

Then $P(s)$ is defined at least for $|s| \leq 1$ and is known as the *probability generating function* of X (or of the probability function $\{a_k\}$). Since the right side of (1) or (2) is a power series in s we can use properties of power series to great advantage.

PROPOSITION 1. *The generating function P is infinitely differentiable for $|s| < 1$ and the kth derivative of P is given by*

$$(3) \qquad P^{(k)}(s) = \sum_{n=k}^{\infty} n(n-1)\cdots(n-k+1)P(X=n)s^{n-k}.$$

In particular,

$$(4) \qquad P^{(k)}(0) = k! P(X = k).$$

Proof. We use the fact that a power series with radius of convergence r can be differentiated any number of times termwise in $(-r, r)$.[†] Equation (3) follows on successive differentiation of (2) for $|s| < 1$. $\square$

We note that $P^{(k)}$ is the generating function of the sequence $\{(k+l)(k+l-1)\ldots(l+1)P(X=k+l)\}_{l=0}^{\infty}$ since we can rewrite (3) as follows:

$$P^{(k)}(s) = \sum_{l=0}^{\infty} (k+l)(k+l-1)\ldots(l+1)P(X=k+l)s^l \qquad (n = k+l).$$

Moreover, it follows from (4) that

$$(5) \qquad P(X = k) = (k!)^{-1} P^{(k)}(0)$$

so that $P(X = k)$ for $k = 0, 1, \ldots$ are determined from the values of P and its derivatives at 0.

Proposition 1 allows us to compute moments of X. For example, we note that $P^{(k)}(s)$ converges at least for $|s| < 1$, and for $s = 1$ the right side of (3) reduces to

$$(6) \qquad \sum_{n=k}^{\infty} n(n-1)\ldots(n-k+1)P(X=n) = \mathscr{E}\{X(X-1)\ldots(X-k+1)\}.$$

[†]R. E. Johnson and E. S. Kiokemeister, *Calculus with Analytic Geometry*, 5th ed., Allyn and Bacon, Boston, 1974, p. 450.

Whenever $\mathscr{E}\{X(X-1)\ldots(X-k+1)\} < \infty$ the derivative $P^{(k)}(s)$ will be continuous in $|s| \leq 1$. If the series on the left side of (6) diverges (to ∞), then $P^{(k)}(s) \to \infty$ as $s \to 1$ and we say that

$$\mathscr{E}\{X(X-1)\ldots(X-k+1)\} = \infty.$$

In particular, we have the following result.

PROPOSITION 2. *If $\mathscr{E}X < \infty$ then*

$$(7) \qquad\qquad \mathscr{E}X = P'(1) \qquad (P' \equiv P^{(1)}).$$

If $\mathscr{E}X^2 < \infty$ then

$$(8) \qquad\qquad \mathrm{var}(X) = P''(1) + P'(1) - \big[P'(1)\big]^2 \qquad (P'' \equiv P^{(2)}).$$

In general, if $\mathscr{E}\{X(X-1)\ldots(X-k+1)\} < \infty$ then

$$(9) \qquad\qquad \mathscr{E}\{X(X-1)\ldots(X-k+1)\} = P^{(k)}(1).$$

Proof. It is only necessary to check (8). We have

$$\mathrm{var}(X) = \mathscr{E}X^2 - (\mathscr{E}X)^2 = \mathscr{E}\{X(X-1)\} + \mathscr{E}X - (\mathscr{E}X)^2$$

$$= P''(1) + P'(1) - \big[P'(1)\big]^2$$

as asserted.   □

Example 1. *Probability Generating Function and Moments of Poisson Distribution.*
Suppose X has probability function

$$P(X = k) = e^{-\lambda}\lambda^k/k!, \qquad k = 0,1,\ldots.$$

Then

$$P(s) = \sum_{k=0}^{\infty} s^k P(X=k) = e^{-\lambda}\sum_{k=0}^{\infty} (s\lambda)^k/k! = \exp\{-\lambda(1-s)\}$$

for $s \in \mathbb{R}$. Moreover,

$$P'(s) = \lambda e^{-\lambda(1-s)}, \qquad P''(s) = \lambda^2 e^{-\lambda(1-s)},$$

$$P'(1) = \lambda, \qquad P''(1) = \lambda^2,$$

so that $\mathscr{E}X = \lambda$ and $\mathrm{var}(X) = \lambda$.   □

Before we take some more examples, let us obtain an important property of probability generating functions that makes them an important tool in probability theory.

THEOREM 1. *A probability generating function determines the distribution function uniquely.*

Proof. It suffices to show that if P and Q are two generating functions such that $P(s) = Q(s)$ for $|s| < h$, $h > 0$, then P and Q generate the same distribution. Let

$$P(s) = \sum_{k=0}^{\infty} s^k P(X = k) \quad \text{and} \quad Q(s) = \sum_{k=0}^{\infty} s^k P(Y = k).$$

Then by (5):

$$P(X = k) = \frac{1}{k!} P^{(k)}(0) = \frac{1}{k!} Q^{(k)}(0) = P(Y = k)$$

for all $k = 0, 1, 2, \ldots$. That is, X and Y have the same distribution. $\square$

Yet another useful property of generating functions is as follows.

PROPOSITION 3. *Let $X_1, X_2, \ldots, X_n$ be independent random variables. Then (at least for $|s| \le 1$):*

$$(10) \qquad\qquad P_{\sum_{i=1}^n X_i}(s) = \prod_{i=1}^{n} P_{X_i}(s)$$

where $P_{\sum_{i=1}^n X_i}$ is the generating function of $\sum_{i=1}^n X_i$ and P_{X_i} is the generating function of X_i, $i = 1, 2, \ldots, n$. In particular, if X_i are, in addition, identically distributed, then:

$$(11) \qquad\qquad P_{\sum_1^n X_i}(s) = \left[P_{X_1}(s) \right]^n,$$

at least for $|s| \le 1$.

Proof. Since the X_i's are independent so are s^{X_i}, $i = 1, 2, \ldots, n$. Hence

$$P_{\sum_{i=1}^n X_i}(s) = \mathscr{E} s^{\sum_{i=1}^n X_i} = \mathscr{E} \prod_{i=1}^{n} s^{X_i} = \prod_{i=1}^{n} \mathscr{E}(s^{X_i}) = \prod_{i=1}^{n} P_{X_i}(s).$$

In the special case when $P_{X_i}(s) = P_{X_1}(s)$ for all i, (11) follows immediately from (10). $\square$

Example 2*. Probability Generating Function for the Binomial Distribution. In Problem 6.3.21 the reader was asked to show that

$$(12) \quad f(n, k) = p f(n - 1, k - 1) + (1 - p) f(n - 1, k), \qquad k = 1, 2, \ldots, n$$

where $f(n, k) = P(X = k)$ and X is the number of successes in n independent Bernoulli trials, each with constant probability p of success. Moreover, $f(n, 0) = (1 - p)^n$ by direct computation. Let us now use (12) and Theorem 1 to explicitly compute

$f(n, k)$. The generating function $P_n(s)$ of $f(n, k)$ satisfies

$$P_n(s) = \sum_{k=0}^{\infty} s^k f(n, k)$$

$$= (1-p)^n + \sum_{k=1}^{n} s^k [\, pf(n-1, k-1) + (1-p)f(n-1, k)]$$

$$= (1-p)^n + p \sum_{k=1}^{n} s(s)^{k-1} f(n-1, k-1) + (1-p) \sum_{k=1}^{n} s^k f(n-1, k)$$

$$= (1-p)^n + spP_{n-1}(s) + (1-p)\left[P_{n-1}(s) - (1-p)^{n-1}\right], \qquad s \in \mathbb{R},$$

where we have used the fact that $f(n, k) = 0$ for $k > n$. Hence

$$P_n(s) = P_{n-1}(s)(1-p+sp) = P_{n-2}(s)(1-p+sp)^2 = \cdots$$

$$= P_0(s)(1-p+sp)^n$$

where

$$P_0(s) = \sum_{k=0}^{\infty} s^k f(0, k) = f(0,0) = 1.$$

Hence

(13) $$P_n(s) = (1-p+sp)^n, \qquad s \in \mathbb{R}.$$

In view of Theorem 1, (13) uniquely determines the distribution $f(n, k)$ of X. However, by binomial theorem,

$$P_n(s) = \sum_{k=0}^{n} \binom{n}{k} p^k (1-p)^{n-k} s^k$$

and it follows that

$$f(n, k) = \text{coefficient of } s^k \text{ in } P_n(s) = \binom{n}{k} p^k (1-p)^{n-k}, \quad k = 0, 1, \ldots, n.$$

From (13) we have

$$P_n'(s) = np + (sp + 1 - p)^{n-1}, \quad P_n''(s) = n(n-1)p^2(sp + 1 - p)^{n-2},$$

and, in general,

$$P_n^{(k)}(s) = n(n-1)\ldots(n-k+1)p^k(sp+1-p)^{n-k}, \qquad 1 \le k \le n.$$

It follows that

$$\mathscr{E}X = P_n'(1) = np, \quad \mathrm{var}(X) = P_n''(1) + P_n'(1) - \left[P_n'(1)\right]^2 = np(1-p),$$

and

$$\mathscr{E}\{X(X-1)\dots(X-k+1)\} = P_n^{(k)}(1) = n(n-1)\dots(n-k+1)p^k, \text{ for } k \geq 1.$$

Thus

$$\mathscr{E}X^3 = \mathscr{E}\{X(X-1)(X-2) + 3X(X-1) + X\}$$

$$= n(n-1)(n-2)p^3 + 3n(n-1)p^2 + np,$$

and

$$\mathscr{E}X^4 = n(n-1)(n-2)(n-3)p^4 + 6n(n-1)(n-2)p^3 + 7n(n-1)p^2 + np.$$

$\square$

Example 3. Sum of Independent Binomial Random Variables. Let X_i have a binomial distribution with parameters n_i and p, $i = 1, 2, \dots, k$. Suppose $X_1, X_2, \dots, X_n$ are independent. Then the probability generating function of $S_k = \sum_{j=1}^k X_j$ is given by

$$P(s) = \prod_{j=1}^k P_{X_j}(s) = (sp + 1 - p)^{n_1 + \dots + n_k}, \quad s \in \mathbb{R}$$

in view of Proposition 3 and Example 2. It follows from Theorem 1 that S_k has a binomial distribution with parameters $n_1 + n_2 + \dots + n_k$ and p. In particular, if $n_i = n$ for $i = 1, \dots, k$ then S_k is binomial with parameters nk and p. $\square$

Example 4. Sum of a Random Number of Random Variables. Let $X_1, X_2 \dots$ be independent and identically distributed nonnegative integer valued random variables with common generating function P. Suppose N is also a nonnegative integer valued random variable that is independent of the X's, and let $S_N = \sum_{j=1}^N X_j$. Then the generating function of S_N is given by

$$P_1(s) = \mathscr{E}s^{S_N} = \mathscr{E}s^{X_1 + \dots + X_N} = \mathscr{E}\left\{\mathscr{E}\left\{s^{X_1 + \dots + X_N} | N\right\}\right\}$$

$$= \sum_{n=0}^\infty \mathscr{E}\left\{s^{X_1 + \dots + X_n} | N = n\right\} P(N = n)$$

$$= \sum_{n=0}^\infty \mathscr{E}\left(s^{X_1 + \dots + X_n}\right) P(N = n) \qquad (N \text{ and } X\text{'s are independent})$$

$$= \sum_{n=0}^\infty \left[P(s)\right]^n P(N = n) = \mathscr{E}\left[P(s)\right]^N = P_N\left[P(s)\right]$$

where P_N is the generating function of N. Clearly, P_1 converges at least for $|s| \le 1$. Differentiating with respect to s, we get

$$P_1'(s) = P_N'[P(s)]P'(s),$$

$$P_1''(s) = P_N''[P(s)][P'(s)]^2 + P''(s)P_N'[P(s)],$$

so that

(14)
$$P_1'(1) = \mathscr{E}S_N = P_N'[P(1)]P'(1) = P_N'(1)P'(1) = \mathscr{E}N.\mathscr{E}X,$$

and

$$P_1''(1) = P_N''(1)[\mathscr{E}X]^2 + P''(1)P_N'(1).$$

It follows from Proposition 2 that

$$\text{var}(S_N) = P_1''(1) + P_1'(1) - [P_1'(1)]^2$$

$$= (\mathscr{E}X)^2\text{var}(N) + \mathscr{E}N\,\text{var}(X).$$

The equation (14) is known as *Wald's equation* and was derived in Example 5.5.3 by a different method. □

Some Common Generating Functions

Distribution	Generating Function
Uniform on $\{1, 2, \ldots, N\}$	$(1 - s^{N+1})/[N(1 - s)]$, $s \in \mathbb{R}$.
Binomial	$(ps + 1 - p)^n$, $0 < p < 1$, $s \in \mathbb{R}$.
Poisson	$\exp\{-\lambda(1 - s)\}$, $\lambda > 0$, $s \in \mathbb{R}$.
Negative binomial	$(ps/[1 - s(1 - p)])^r$, $0 < p < 1$, $r \ge 1, s < 1/(1 - p)$.

Yet another generating function that generates moments has a wide variety of applications in probability and statistics.

DEFINITION 1. Let X be any random variable with distribution function F. Suppose $\mathscr{E}\exp(sX)$ is finite for s in some open interval containing zero. Then the *moment generating function* of X (or F) is defined by

(15)
$$M(s) = \mathscr{E}\exp(sX).$$

In the discrete case,

$$M(s) = \sum_{j=1}^{\infty} \exp(sx_j)P(X = x_j)$$

and in the case when X has density f,

$$M(s) = \int_{-\infty}^{\infty} e^{sx} f(x)\, dx.$$

In particular, if X is nonnegative integer valued with generating function P, then

$$M(s) = \mathscr{E} \exp(sX) = P(e^s).$$

We note that $M(0) \equiv 1$.

The name "moment generating function" derives from the fact that moments of X can be computed by differentiation of $M(s)$ at $s = 0$.

PROPOSITION 4. *Let X have moment generating function M, which is finite in* $|s| < s_0, s_0 > 0$. *Then,*

(i) *X has moments of all orders, and*

(ii) *M can be expanded in a Taylor series about zero.*

Moreover, the moments of X can be obtained by differentiating M under the integral sign and evaluating at $s = 0$. That is,

(16) $\mathscr{E} X^k = M^{(k)}(0), k = 0, 1, \dots$.

*Proof.** Clearly,

$$\exp(|sx|) \le \exp(sx) + \exp(-sx).$$

Since M is finite for s in $(-s_0, s_0)$ it follows that $\mathscr{E} \exp(|sX|) < \infty$ for $s < |s_0|$. Let $s \in (0, s_0)$. Then for every $n \ge 0$

$$\exp(|sx|) = \exp(s|x|) > \frac{s^n |x|^n}{n!}$$

and it follows that

$$\mathscr{E} |X|^n < n! s^{-n} \mathscr{E} \exp(|sX|) < \infty.$$

Also, for $|s| < s_0$ and any $n \ge 0$

$$\left| \sum_{k=0}^{n} \frac{s^k}{k!} x^k \right| \le \sum_{k=0}^{\infty} \frac{|sx|^k}{k!} = \exp(|sx|)$$

so that

$$\left| \sum_{k=0}^{\infty} \frac{s^k}{k!} \mathscr{E} X^k \right| < \infty.$$

It follows that we can interchange the expectation and summation signs. That is,

$$(17) \qquad M(s) = \mathscr{E}\exp(sX) = \mathscr{E}\left(\sum_{k=0}^{\infty} \frac{(sX)^k}{k!} \right) = \sum_{k=0}^{\infty} \frac{s^k \mathscr{E} X^k}{k!}$$

for $|s| < s_0$. Thus M has a Taylor series expansion about 0. Since M has a power series representation, we can differentiate it term by term in $|s| < s_0$, and we have

$$M^{(k)}(s)|_{s=0} = M^{(k)}(0) = \mathscr{E} X^k \qquad \text{for every } k \geq 1. \qquad \square$$

COROLLARY. *Let M be the moment generating function of X and set $\phi(s) = \ln M(s)$. Then*

$$(18) \qquad \mathscr{E} X = \phi'(0) \quad \text{and} \quad \text{var}(X) = \phi''(0).$$

Proof. We have

$$\phi'(s) = \frac{M'(s)}{M(s)}$$

so that

$$\phi'(0) = \frac{M'(0)}{M(0)} = M'(0) = \mathscr{E} X.$$

Moreover,

$$\phi''(s) = \frac{M''(s) M(s) - [M'(s)]^2}{[M(s)]^2}$$

so that

$$\phi''(0) = M''(0) - [M'(0)]^2 = \mathscr{E} X^2 - (\mathscr{E} X)^2 = \text{var}(X). \qquad \square$$

Example 5. Moment Generating Function and Moments of Binomial Distribution.
Let X be binomial with parameters n and p. Then, for $s \in \mathbb{R}$:

$$M(s) = \mathscr{E} e^{sX} = \sum_{k=0}^{n} e^{sk}\binom{n}{k}p^k(1-p)^{n-k} = (1 - p + pe^s)^n$$

by binomial theorem. Hence

$$M'(s) = np(1 - p + pe^s)^{n-1}e^s,$$

$$M''(s) = n(n-1)p^2(1 - p + pe^s)^{n-2}e^{2s} + np(1 - p + pe^s)^{n-1}e^s.$$

We have

$$M'(0) = np = \mathscr{E} X, \qquad M''(0) = \mathscr{E} X^2 = n(n-1)p^2 + np. \qquad \square$$

Example 6. Moment Generating Function and Moments of Normal Distribution.
Let X have standard normal distribution with density function

$$f(x) = (\sqrt{2\pi})^{-1}\exp\left(-\frac{x^2}{2}\right), \qquad x \in \mathbb{R}.$$

For $s \in \mathbb{R}$:

$$M(s) = (\sqrt{2\pi})^{-1}\int_{-\infty}^{\infty} \exp(sx)\exp\left(-\frac{x^2}{2}\right) dx$$

$$= (\sqrt{2\pi})^{-1}\int_{-\infty}^{\infty} \exp\left\{-\frac{1}{2}(x^2 - 2sx)\right\} dx$$

$$= (\sqrt{2\pi})^{-1}\int_{-\infty}^{\infty} \exp\left\{-\frac{1}{2}\left[(x - s)^2 - s^2\right]\right\} dx$$

$$= \exp(s^2/2)\int_{-\infty}^{\infty} \frac{\exp\left\{-(x - s)^2/2\right\}}{\sqrt{2\pi}} dx$$

$$= \exp\left(\frac{s^2}{2}\right)$$

since the integrand in the last but one equality is the density function of a normal random variable with mean s and variance 1. Hence

$$\phi(s) = \ln M(s) = \frac{s^2}{2}$$

and

$$\phi'(s) = s, \qquad \phi''(s) = 1.$$

It follows from (18) that $\mathscr{E}X = \phi'(0) = 0$ and $\text{var}(X) = \phi''(0) = 1$. In fact, we can expand M into the Taylor series:

$$M(s) = \exp\left(\frac{s^2}{2}\right) = \sum_{k=0}^{\infty} \frac{(s^2/2)^k}{k!} = \sum_{l=0}^{\infty} a_l s^l$$

where

$$a_l = \begin{cases} 0 & \text{if } l \text{ is odd} \\ \left[2^{l/2}\left(\frac{l}{2}\right)!\right]^{-1} & \text{if } l \text{ is even.} \end{cases}$$

Consequently,

$$\mathscr{E}X^l = \text{coefficient of } \frac{s^l}{l!} \text{ in } M(s)$$

$$= \begin{cases} 0 & \text{if } l \text{ is odd} \\ \dfrac{l!}{2^{l/2}(l/2)!} & \text{if } l \text{ is even.} \end{cases} \qquad \qquad \square$$

We now state without proof an important property of a moment generating function.

THEOREM 2. *If X has moment generating function M that is finite in $(-s_0, s_0)$, $s_0 > 0$, then M determines the distribution of X uniquely.*

Thus if two random variables X and Y have the same moment generating function, then X and Y must have the same distribution (and the same set of moments).

Remark 1. The requirement that M be finite in some open neighborhood $(-s_0, s_0)$ of 0 is a very strong requirement. It implies, in particular, that moments of all order exist (Proposition 4). This restricts the use of moment generating function as a tool. For most of the important probability models considered in this text, however, this requirement is satisfied. We note that the existence of moments of all orders is a necessary condition for the finiteness of M but *it is not a sufficient condition*. That is, the moments of all order may exist without the moment generating function being finite in some open neighborhood of zero. It also means that it is possible for two random variables to have the same set of moments so that moments do not necessarily determine a distribution. Clearly, if X has finite moments of order k and infinite higher order moments, then its moment generating function cannot be finite in an open neighborhood of zero.

Example 7. Probability Integral Transformation. Let X have an exponential density function with mean λ^{-1}. Then

$$f(x) = \begin{cases} \lambda e^{-\lambda x} & \text{for } x > 0 \\ 0 & \text{otherwise} \end{cases}$$

and the distribution function of X is given by

$$F(x) = \begin{cases} 0 & x \le 0 \\ 1 - e^{-\lambda x}, & x > 0 \end{cases}$$

Consider the random variable $Y = F(X) = 1 - e^{-\lambda X}$. The moment generating function of Y is given by

$$M_Y(s) = \mathscr{E}\exp(sY) = \int_0^\infty \exp\{s(1 - e^{-\lambda x})\} \lambda e^{-\lambda x}\, dx$$

$$= \lambda e^s \int_0^\infty \exp\{-\lambda x - se^{-\lambda x}\}\, dx = e^s \int_0^1 e^{-sy}\, dy \qquad (y = e^{-\lambda x})$$

$$= \frac{(e^s - 1)}{s} \qquad \text{for } s \in \mathbb{R}.$$

Since M_Y is the moment generating function of the uniform distribution on $(0, 1)$ (as is easily shown) it follows from Theorem 2 that Y has a uniform distribution on $(0, 1)$. This result, as we know (Section 8.2), is true for any continuous distribution function

F. In particular if f is the density function corresponding to a distribution function F then

$$M_Y(s) = \mathscr{E}\exp(sF(X)) = \int_{-\infty}^{\infty} e^{sF(x)}f(x)\, dx$$

$$= \frac{e^{sF(x)}}{s}\bigg|_{-\infty}^{\infty} = \frac{e^s - 1}{s}$$

so that $Y = F(X)$ has a uniform distribution on $(0,1)$.   □

Theorem 2 is specially useful in computing the distribution of a sum of independent random variables. The following result is easy to prove.

PROPOSITION 5. *Let* $X_1, X_2, \ldots, X_n$ *be independent random variables with corresponding moment generating functions* $M_1, M_2, \ldots, M_n$ *respectively. Suppose* $M_1, \ldots, M_n$ *are finite in* $(-s_0, s_0)$, $s_0 > 0$. *Then* $S_n = \sum_{i=1}^{n} X_i$ *has the moment generating function given by*

$$M(s) = \prod_{i=1}^{n} M_i(s), \qquad s \in (-s_0, s_0).$$

In particular, if X_i *are identically distributed, then*

$$M(s) = [M_1(s)]^n, \qquad s \in (-s_0, s_0).$$

Example 8. Sum of Independent Normal Random Variables. Let X and Y be independent standard normal random variables. Then for $s \in \mathbb{R}$,

$$M_{X+Y}(s) = M_X(s)M_Y(s) = e^{s^2/2}e^{s^2/2} = e^{s^2}$$

in view of Example 6. It follows that $X + Y$ itself is a normal random variable. In particular, if $Z = (X + Y)/\sqrt{2}$ then

$$M_Z(s) = M_{X+Y}\left(\frac{s}{\sqrt{2}}\right) = e^{s^2/2}$$

and Z itself is standard normal.   □

Example 9. Sum of Uniformly Distributed Random Variables. Let Z have a triangular density function

$$f(z) = \begin{cases} z, & 0 \le z \le 1 \\ 2 - z, & 1 \le z \le 2 \\ 0 & \text{otherwise.} \end{cases}$$

Then for $s \in \mathbb{R}$,

$$M_Z(s) = \int_0^1 e^{zs}z\, dz + \int_1^2 e^{zs}(2 - z)\, dz = \left(\frac{e^s - 1}{s}\right)^2.$$

Since $(e^s - 1)/s$ is the moment generating function of a uniformly distributed random variable on $(0,1)$ (Example 7), it follows from Proposition 5 and Theorem 2 that f is the density of the sum of two independent uniformly distributed random variables. (See also Example 8.2.10.)   □

It is easy to extend the two generating functions considered here to the multivariate case. We will content ourselves with the definitions. Results analogous to Theorem 1 and 2 and Propositions 1 through 4 hold also in the multivariate case.

Let $(X_1, X_2, \ldots, X_n)$ be a random vector. If all the X's are nonnegative integer valued we define the joint probability generating function of $(X_1, X_2, \ldots, X_n)$ by the power series

$$P(s_1, s_2, \ldots, s_n) = \mathscr{E}\left(s_1^{X_1} \ldots s_n^{X_n}\right)$$

$$= \sum_{k_1=0}^{\infty} \cdots \sum_{k_n=0}^{\infty} s_1^{k_1} \ldots s_n^{k_n} P(X_1 = k_1, \ldots, X_n = k_n)$$

which converges at least for $\{(s_1, \ldots, s_n) : |s_1| \le 1, \ldots, |s_n| \le 1\}$. In the general case when the X_i's are not necessarily integer valued, we define the *joint moment generating function* of $(X_1, X_2, \ldots, X_n)$ by

$$M(s_1, \ldots, s_n) = \mathscr{E}\exp(s_1 X_1 + \cdots + s_n X_n)$$

provided the expected value is finite for $|s_1| < h_1, \ldots, |s_n| < h_n$, where $h_i > 0$, $i = 1, 2, \ldots, n$.

Moment Generating Functions of Some Common Distributions

Distribution	Moment Generating Function
Uniform on $\{1, 2, \ldots, N\}$	$[1 - e^{s(N+1)}]/[N(1 - e^s)]$; $s \in \mathbb{R}$.
Binomial	$(pe^s + 1 - p)^n$, $0 < p < 1$; $s \in \mathbb{R}$.
Poisson	$\exp\{-\lambda(1 - e^s)\}$, $\lambda > 0$; $s \in \mathbb{R}$.
Negative binomial	$(pe^s/[1 - (1 - p)e^s])^r$, $0 < p < 1$, $r \ge 1$; $s < -\ln(1 - p)$.
Uniform on $(0, 1)$	$(e^s - 1)/s$; $s \in \mathbb{R}$.
Gamma	$(1 - \beta s)^{-\alpha}$, $\alpha, \beta > 0$; $s < 1/\beta$.
Normal	$\exp(\mu s + s^2\sigma^2/2)$, $\mu \in \mathbb{R}$, $\sigma > 0$; $s \in \mathbb{R}$.
Multinomial	$(p_1 e^{s_1} + \cdots + p_{k-1}e^{s_{k-1}} + p_k)^n$, $p_i > 0$, $\sum_{i=1}^{k} p_i = 1$; $s_i \in \mathbb{R}$, $i = 1, 2, \ldots, k - 1$.
Bivariate normal	$\exp\left(s_1\mu_1 + s_2\mu_2 + \dfrac{\sigma_1^2 s_1^2 + \sigma_2^2 s_2^2 + 2\rho\sigma_1\sigma_2 s_1 s_2}{2}\right)$, $\mu_1 \in \mathbb{R}$, $\mu_2 \in \mathbb{R}$, $\sigma_1 > 0, \sigma_2 > 0, |\rho| < 1$; $s_1 \in \mathbb{R}$, $s_2 \in \mathbb{R}$.

Problems for Section 8.6

1. Let P be the probability generating function of a random variable X. Find the generating functions of:
   (i) $X + k$ ($k \geq 0$ integral).   (ii) mX ($m \geq 1$ integral).   (iii) $mX + k$.

2. Let X be the face value in a throw with a fair die. Find its probability generating function. Find the distribution of the sum of face values in two independent throws with a fair die.

3. Find the probability generating function of X in the following cases:
   (i) Geometric: $P(X = k) = p(1 - p)^{k-1}, k = 1, 2, \ldots$.
   (ii) Negative binomial: $P(X = k) = \binom{k-1}{r-1} p^r (1 - p)^{k-r}, k = r, r + 1, \ldots, r \geq 1$.
   (iii) $P(X = k) = \{e^{-\lambda}/(1 - e^{-\lambda})\}(\lambda^k/k!), k = 1, 2, \ldots, \lambda > 0$.
   (iv) $P(X = k) = p(1 - p)^k[1 - (1 - p)^{N+1}]^{-1}, k = 0, 1, 2, \ldots N: 0 < p < 1$.
   (v) $P(X = k) = a_k \theta^k [f(\theta)]^{-1}, k = 0, 1, 2, \ldots, \theta > 0$, where $a_k \geq 0$ for all k and $f(\theta) = \sum_{k=0}^{\infty} a_k \theta^k$.
   (vi) $P(X = k) = 1/N, k = 1, 2, \ldots, N$.

4. Let P be the probability generating function of X. Find the generating functions of:
   (i) $P(X < n)$.   (ii) $P(X \leq n)$.   (iii) $P(X = 2n)$.
   (Note that these do not define probability functions for $n = 0, 1, 2, \ldots$.)

5. Let $P(X = k) = p_k, k = 0, 1, 2, \ldots$. Let $P(X > k) = q_k, k = 0, 1, 2, \ldots$. Then $q_k = p_{k+1} + p_{k+2} + \cdots, k \geq 0$. Show that the generating function of the q_k's in terms of the p_k's is given by

$$Q(s) = (1 - s)^{-1}[1 - P(s)], \quad |s| < 1,$$

where $P(s) = \sum_{k=0}^{\infty} s^k p_k$. Moreover, $\mathscr{E}X = \sum_{k=0}^{\infty} q_k = Q(1)$ and $\text{var}(X) = 2Q'(1) + Q(1) - Q^2(1)$. (Note that Q is not a probability generating function.)

6. Find the first two moments of each X in Problem 3 by differentiating the generating function.

7. Let $X_1, X_2, \ldots, X_n$ be independent random variables and let $S_n = \sum_{k=0}^{n} X_k, n \geq 1$.
   (i) If X_i has a geometric distribution with parameter $p, 0 < p < 1$, show that S_n has a negative binomial distribution with parameters n and p.
   (ii) If X_i has a negative binomial distribution with parameters r_i and p, $i = 1, 2, \ldots, n$ show that S_n has a negative binomial distribution with parameters $\sum_{i=1}^{n} r_i$ and p.
   (iii) Let X_i have a Poisson distribution with parameter $\lambda_i, i = 1, 2, \ldots, n$. Show that S_n has a Poisson distribution with parameter $\sum_{i=1}^{n} \lambda_i$.

8. Let $X_1, X_2, \ldots, X_k$ be independent and identically distributed with probability function $P(X_i = x) = 1/N, x = 1, 2, \ldots, N$. Find the probability generating function of $\sum_{j=1}^{k} X_j$.

9. Given the following probability generating functions, find the corresponding probability functions:
   (i) $P(s) = [(1 + s)/2]^n, s \in \mathbb{R}$.   (ii) $P(s) = s/(2 - s), |s| < 2$.
   (iii) $P(s) = \ln(1 - sp)/\ln(1 - p), |s| < 1/p$.

10. Let $X_1, X_2, \ldots$ be independent random variables with probability function

$$P(X_n = 0) = 1 - \frac{1}{n}, \quad P(X_n = 1) = \frac{1}{n}, \quad n = 1, 2, \ldots.$$

Find the probability generating function of $X_1 + X_2 + \cdots + X_n$ and hence find the probability function of $X_1 + X_2 + \cdots + X_n$, for $n = 2, 3, 4$.

11*. Let X be an integer valued random variable with

$$\mathscr{E}\{X(X-1)\ldots(X-k+1)\} = \begin{cases} k!\dbinom{n}{k} & \text{if } k = 0, 1, 2, \ldots, n \\ 0 & \text{if } k > n. \end{cases}$$

Show that X must be degenerate at n.

[Hint: Prove and use the fact that if $\mathscr{E}X^k < \infty$ for all k, then

$$P(s) = \sum_{k=0}^{\infty} \frac{(s-1)^k}{k!} \mathscr{E}\{X(X-1)\ldots(X-k+1)\}.$$

Write $P(s)$ as

$$P(s) = \sum_{k=0}^{\infty} P(X=k)s^k = \sum_{k=0}^{\infty} P(X=k) \sum_{i=0}^{k} \binom{k}{i}(s-1)^i$$

$$= \sum_{i=0}^{\infty} (s-1)^i \sum_{k=i}^{\infty} \binom{k}{i} P(X=k).]$$

12*. Let $p(n, k) = f(n, k)/n!$ where $f(n, k)$ is given by

$$f(n+1, k) = f(n, k) + f(n, k-1) + \cdots + f(n, k-n),$$

for $k = 0, 1, \ldots, \dbinom{n}{2}$ and

$$f(n, k) = 0 \quad \text{for } k < 0, f(1, 0) = 1, f(1, k) = 0 \text{ otherwise.}$$

Let

$$P_n(s) = \frac{1}{n!} \sum_{k=0}^{\infty} s^k f(n, k)$$

be the probability generating function of $p(n, k)$. Show that

$$P_n(s) = \prod_{k=2}^{n} \frac{1 - s^k}{1 - s} \quad |s| < 1.$$

(P_n is the generating function of Kendall's τ-statistic. See Section 11.9.)

13*. For $k = 0, 1, \ldots, \dbinom{n}{2}$ let $u_n(k)$ be defined recursively by

$$u_n(k) = u_{n-1}(k-n) + u_{n-1}(k)$$

with $u_0(0) = 1$, $u_0(k) = 0$ otherwise and $u_n(k) = 0$ for $k < 0$. Let $P_n(s) = \sum_{k=0}^{\infty} s^k u_n(k)$ be the generating function of $\{u_n\}$. Show that

$$P_n(s) = \prod_{j=1}^{n} (1 + s^k) \qquad \text{for } |s| < 1.$$

If $p_n(k) = u_n(k)/2^n$, find $\{p_n(k)\}$ for $n = 2, 3, 4$. (P_n is the generating function of one sample Wilcoxon test statistic. See Section 11.5.)

14. Find the moment generating functions of the following distributions.
    (i)   Poisson: $P(X = k) = e^{-\lambda}\lambda^k/k!$, $k = 0, 1, 2, \ldots$
    (ii)* Negative binomial: $P(X = k) = \binom{k-1}{r-1}p^r(1-p)^{k-r}$, $k = r, r+1, \ldots$
    (iii)* Gamma:

$$f(x) = \begin{cases} \dfrac{x^{\alpha-1}e^{-x/\beta}}{\Gamma(\alpha)\beta^\alpha}, & x > 0 \\ 0, & x \le 0. \end{cases}$$

    (iv) Laplace: $f(x) = (1/2)\exp\{-|x|\}$, $x \in \mathbb{R}$.

15. Given that $M(s) = (1-s)^{-2}$, $s < 1$ is a moment generating function, find the corresponding moments.

16. Find the moment generating function, the mean, and the variance in each of the following cases.
    (i) $f(x) = 2(1-x)$, $0 < x < 1$; $= 0$ otherwise.
    (ii) $f(x) = 6x(1-x)$, $0 \le x \le 1$, zero otherwise.
    (iii) $f(x) = 1/6$, $-2 < x < 4$, zero elsewhere.

17*. Let X be a random variable such that $\ln X$ has a normal distribution with mean μ and variance σ^2 so that X has lognormal density

$$f(x) = \begin{cases} (1/\sigma x\sqrt{2\pi})\exp\{-(\ln x - \mu)^2/2\sigma^2\}, & x > 0 \\ 0, & x \le 0. \end{cases}$$

Show that $\mathscr{E}X^s = \exp(\mu s + \sigma^2 s^2/2)$ and find the mean and the variance of X. [Hint: Find the moment generating function of $\ln X$].

18. Suppose X has density

$$f(x) = 1/x^2, \qquad x \ge 1 \text{ and } 0 \text{ elsewhere.}$$

Is the moment generating function finite in an open neighborhood of zero?

19. Let $X_1, X_2, \ldots, X_n$ be a random sample from density function f. Find the distribution of $S_n = \sum_{k=1}^{n} X_k$ in each of the following cases.
    (i) f is exponential with mean λ.
    (ii)* f is gamma with parameters α and β.
    (iii)* Do part (ii) when each X_i is gamma with parameters α_i and β, $i = 1, 2, \ldots$.

20. Let $X_1, X_2, \ldots, X_n$ be independent random variables. Suppose X_i has a normal distribution with mean μ_i and variance σ_i^2, $i = 1, 2, \ldots, n$. Show that $S_n = \sum_{i=1}^{n} a_i X_i$ has a normal distribution with mean $\sum_{i=1}^{n} a_i \mu_i$ and variance $\sum_{i=1}^{n} a_i^2 \sigma_i^2$. [Hint: First use Example 6 to show that X_i has moment generating function $\exp(\mu_i s + \sigma_i^2 s^2/2)$ by writing $Z = (X_i - \mu)/\sigma$ where Z has generating function $\exp(s^2/2)$.]

21*. Obtain the joint generating functions of multinomial and bivariate normal distributions.

22*. Let (X, Y) have moment generating function $M(s_1, s_2)$, which is finite for $|s_1| < h_1$ and $|s_2| < h_2$. Show that
   (i) $M(s_1, 0) = M_X(s_1)$, $M(0, s_2) = M_Y(s_2)$.
   (ii) X and Y are independent if and only if $M(s_1, s_2) = M(s_1, 0) \cdot M(0, s_2)$.
   (iii) $(\partial^{m+n})/(\partial s_1^m \, \partial s_2^n) M(s_1, s_2)|_{s_1 = 0, \, s_2 = 0} = \mathscr{E}(X^m Y^n)$.

23*. Let (X, Y) have joint density

$$f(x, y) = \begin{cases} x + y, & 0 < x < 1, 0 < y < 1 \\ 0, & \text{elsewhere.} \end{cases}$$

   (i) Find the moment generating function of (X, Y) and the moment generating functions of X and Y.
   (ii) Use (i) to show that $\mathscr{E}X = \mathscr{E}Y = 7/12$, $\text{var}(X) = \text{var}(Y) = 11/144$ and $\text{cov}(X, Y) = -1/144$.

24*. Let (X, Y) have joint density

$$f(x, y) = \begin{cases} \left[1 + xy(x^2 - y^2)\right]/4, & |x| \le 1, |y| \le 1 \\ 0 & \text{elsewhere.} \end{cases}$$

Find the moment generating function of (X, Y). Are X and Y independent? Are they uncorrelated?

25*. Let (X, Y) have uniform distribution on $\{0 < y < x < 1\}$. Find the moment generating function of (X, Y).

26*. Let (X, Y) have probability function

$$P(X = x, Y = y) = \frac{(x + y + k - 1)!}{x! \, y! \, (k - 1)!} p_1^x p_2^y (1 - p_1 - p_2)^k,$$

where $x, y = 0, 1, 2, \ldots, k \ge 1$ is an integer, $p_1, p_2 \in (0, 1)$ and $p_1 + p_2 < 1$. Show that $X + Y$ has a negative binomial distribution. [Hint: Find the moment generating function of (X, Y). Then

$$M_{X+Y}(s) = \mathscr{E} \exp(sX + sY) = M_{X, Y}(s, s)$$

in an obvious notation.]

27*. Let (X, Y) have a bivariate normal distribution with parameters $\mu_1, \mu_2, \sigma_1, \sigma_2$ and ρ. Let $U = a_1 X + a_2 Y$, $V = b_1 X + b_2 Y$.
   (i) Show that U and V have a bivariate normal distribution with parameters:

$$\mu_1 a_1 + \mu_2 a_2, \mu_1 b_1 + \mu_2 b_2, \sigma_1^2 a_1^2 + 2\rho a_1 a_2 \sigma_1 \sigma_2 + \sigma_2^2 a_2^2, \sigma_1^2 b_1^2$$
$$+ 2\rho b_1 b_2 \sigma_1 \sigma_2 + \sigma_2^2 b_2^2 \quad \text{and} \quad \text{cov}(U, V) = a_1 b_1 \sigma_1^2 + a_2 b_2 \sigma_2^2$$
$$+ (a_2 b_1 + a_1 b_2) \rho \sigma_1 \sigma_2.$$

   (ii) Show that the marginal distributions of U and V are also normal.

8.7 SAMPLING FROM A NORMAL POPULATION

In this section we consider the special case when $X_1, X_2, \ldots, X_n$ is a random sample from a normal distribution with mean μ and variance σ^2. Our object is to derive the distributions of some statistics that are functions of $\overline{X}$ and S^2. These results form the basis for small (or large) sample inference in sampling from a normal population. Let

$$\overline{X} = \sum_{i=1}^{n} \frac{X_i}{n} \quad \text{and} \quad S^2 = \frac{\sum_{i=1}^{n}(X_i - \overline{X})^2}{(n-1)}$$

be the sample mean and the sample variance. We know that

$$\mathcal{E}\overline{X} = \mu, \quad \operatorname{var}(\overline{X}) = \frac{\sigma^2}{n}, \quad \text{and} \quad \mathcal{E}S^2 = \sigma^2.$$

We first note that $\overline{X}$ has an exact normal distribution with mean μ and variance σ^2/n no matter what n is. In fact, for $s \in \mathbb{R}$,

$$M_{\overline{X}}(s) = \mathcal{E}e^{s\overline{X}} = \prod_{i=1}^{n} \mathcal{E}e^{sX_i/n}$$

$$= \prod_{i=1}^{n} \exp\left\{ \frac{s\mu}{n} + \frac{s^2\sigma^2}{2n^2} \right\}$$

$$= \exp\left\{ s\mu + \frac{s^2\sigma^2}{2n} \right\}$$

which is the moment generating function of a normal random variable with mean μ and variance σ^2/n. Thus

$$Z = \frac{\sqrt{n}\,(\overline{X} - \mu)}{\sigma}$$

has a standard normal distribution. From a practical point of view σ is usually unknown and it would be desirable to replace σ by its estimate s. Is S unbiased for σ? What is the distribution of

(1)
$$T = \frac{\sqrt{n}\,(\overline{X} - \mu)}{S}?$$

To answer these questions we prove the following fundamental result in statistics. The reader may skip the proof.

THEOREM 1. *Let $X_1, X_2, \ldots, X_n$, $n \geq 2$, be a random sample from a normal distribution with mean μ and variance σ^2. Then*

1. $\overline{X}$ *and* S^2 *are independent.*
2. $(n-1)S^2/\sigma^2$ *has a chi-square distribution with* $(n-1)$ *degrees of freedom* (d.f.).
3. $\sqrt{n}\,(\overline{X} - \mu)/S$ *has a t-distribution with* n *d.f.*

*Proof.** 1. We show that $\bar{X}$ is independent of $X_i - \bar{X}$ for each i. If we know that $\bar{X}$ and $X_i - \bar{X}$ have a joint bivariate normal distribution, then the solution is easy, since

$$\mathrm{cov}(\bar{X}, X_i - \bar{X}) = \mathscr{E}\bar{X}(X_i - \bar{X}) - \mathscr{E}\bar{X}\mathscr{E}(X_i - \bar{X})$$

$$= \mathscr{E}(X_i\bar{X}) - \mathscr{E}\bar{X}^2 \qquad (\mathscr{E}(X_i - \bar{X}) = 0)$$

$$= \mathscr{E}X_i^2/n + \mathscr{E}\left(X_i \sum_{\substack{j=1 \\ j \neq i}}^{n} \frac{X_j}{n}\right) - \left(\frac{\sigma^2}{n} + \mu^2\right)$$

$$= \frac{(\sigma^2 + \mu^2)}{n} + \frac{n-1}{n}\mu^2 - \left(\frac{\sigma^2}{n} + \mu^2\right) = 0.$$

The result would then follow from Proposition 7.7.1.

Since, however, we do not know that $\bar{X}$ and $X_i - \bar{X}$ have a bivariate normal distribution (although it is possible to show it by using methods of Sections 8.2, 8.3, and 8.6), we proceed directly to compute the joint moment generating function of $\bar{X}$ and $X_i - \bar{X}$. We have:

$$M(s_1, s_2) = \mathscr{E}\exp\{s_1\bar{X} + s_2(X_i - \bar{X})\} = \mathscr{E}\exp\{s_2 X_i + (s_1 - s_2)\bar{X}\}$$

$$= \mathscr{E}\exp\left\{X_i\left(s_2 + \frac{s_1 - s_2}{n}\right) + \frac{(s_1 - s_2)}{n}\sum_{\substack{j=1 \\ j \neq i}}^{n} X_j\right\}$$

$$= \mathscr{E}\exp\left\{X_i\left(s_2 + \frac{s_1 - s_2}{n}\right)\right\}\mathscr{E}\exp\left\{\frac{s_1 - s_2}{n}\sum_{\substack{j=1 \\ j \neq i}}^{n} X_j\right\} \qquad \text{(by independence)}$$

$$= \exp\left\{\mu\left(s_2 + \frac{s_1 - s_2}{n}\right) + \frac{\sigma^2}{2}\left(s_2 + \frac{s_1 - s_2}{2}\right)^2\right\}$$

$$\times\exp\left\{\frac{n-1}{n}(s_1 - s_2)\mu + \left(\frac{s_1 - s_2}{n}\right)^2(n-1)\frac{\sigma^2}{2}\right\}$$

$$\left(X_i \text{ as well as } \sum_{j \neq i} X_j \text{ are normal}\right)$$

$$= \exp\left\{\mu s_1 + \frac{\sigma^2 s_1^2}{2n}\right\}\exp\left\{(n-1)\frac{s_2^2\sigma^2}{2n}\right\}.$$

Since the right side is the product of moment generating functions of normal $(\mu, \sigma^2/n)$ and normal $(0, \sigma^2(n-1)/n)$ random variables, we have shown[†] that

(i) $\bar{X}$ and $X_i - \bar{X}$ are independent.

(ii) $\bar{X}$ is normal $(\mu, \sigma^2/n)$ whereas $X_i - \bar{X}$ is normal $(0, (n-1)\sigma^2/n)$. Consequently, $\bar{X}$ and $\sum_{i=1}^{n}(X_i - \bar{X})^2$ are independent and so also are $\bar{X}$ and S^2.

[†]See Problem 8.6.22(ii).

2. We have

$$\sum_{i=1}^{n} \left(X_i - \bar{X} \right)^2 = \sum_{i=1}^{n} \left(X_i - \mu + \mu - \bar{X} \right)^2 = \sum_{i=1}^{n} \left(X_i - \mu \right)^2 - n\left(\bar{X} - \mu \right)^2$$

so that

$$\sum_{i=1}^{n} \frac{\left(X_i - \mu \right)^2}{\sigma^2} = \sum_{i=1}^{n} \frac{\left(X_i - \bar{X} \right)^2}{\sigma^2} + n\frac{\left(\bar{X} - \mu \right)^2}{\sigma^2}.$$

We note that $(X_i - \mu)/\sigma$, for each i, is a normal $(0,1)$ random variable and $\sqrt{n}\,(\bar{X} - \mu)/\sigma$ is also normal $(0,1)$. In view of Example 8.2.5, $(X_i - \mu)^2/\sigma^2$ is a chi-square random variable with one d.f. and so also is $n(\bar{X} - \mu)^2/\sigma^2$. Since, however, the X_i's are independent and the sum of n independent chi-square random variables with one d.f. is itself a chi-square random variable with n d.f., $\sum_{i=1}^{n}(X_i - \mu)^2/\sigma^2$ is a chi-square random variable with n d.f. In view of the independence of $\bar{X}$ and S^2, it follows (why?) that $(n-1)S^2/\sigma^2$ must have a chi-square distribution with $n - 1$ d.f.

3. We write

$$T = \frac{(\bar{X} - \mu)\sqrt{n}\,/\sigma}{\sqrt{(n-1)S^2/\sigma^2 \cdot 1/(n-1)}}$$

Since $(\bar{X} - \mu)\sqrt{n}\,/\sigma$ is normal $(0,1)$ and $(n-1)S^2/\sigma^2$ is chi-square with $(n-1)$ d.f. it follows from part 1 and the definition of a t-statistic (Example 8.4.5) that T has a t-distribution with $n - 1$ d.f. $\square$

We can use part 2 to show that S is not an unbiased estimate of σ. In fact,

$$\mathscr{E}S = \left(\sqrt{\frac{(n-1)S^2}{\sigma^2}} \right) \frac{\sigma}{\sqrt{n-1}}$$

$$= \frac{\sigma}{\sqrt{n-1}} \int_0^\infty \sqrt{y} \left(\frac{1}{2^{(n-1)/2}\Gamma((n-1)/2)} y^{(n-1)/2-1} e^{-y/2} \, dy \right)$$

$$= \frac{\sigma}{2^{(n-1)/2}\sqrt{n-1}\,\Gamma((n-1)/2)} \int_0^\infty y^{n/2-1} e^{-y/2} \, dy$$

$$= \frac{\sigma}{\sqrt{n-1}} \frac{\Gamma(n/2)}{\Gamma((n-1)/2)} \cdot \frac{2^{n/2}}{2^{(n-1)/2}}$$

$$= \sigma\sqrt{\frac{2}{n-1}} \frac{\Gamma(n/2)}{\Gamma((n-1)/2)}$$

so that S is a biased estimate of σ. However, the estimate

(2)
$$\hat{\sigma} = \sqrt{\frac{n-1}{2}} \frac{\Gamma((n-1)/2)}{\Gamma(n/2)} S$$

is unbiased for σ.

Let us now consider some properties of t and chi-square densities. We recall (Example 8.4.5) that the density of t-distribution with n d.f. is given by

(3)
$$f_n(t) = \frac{\Gamma((n+1)/2)}{\Gamma(n/2)\sqrt{n\pi}} (1 + t^2/n)^{-(n+1)/2}, \qquad t \in \mathbb{R}.$$

Clearly,

$$f_n(t) = f_n(-t)$$

so that f_n is symmetric about zero. It is unimodal with mode at $t = 0$. By symmetry, the median is also at $t = 0$. The existence of mean of f_n depends on n. Thus for $n = 1$,

$$f_1(t) = \frac{1}{\pi} \frac{1}{1 + t^2}, \qquad t \in \mathbb{R}$$

and f_1 does not have finite mean. One can show that $\mathscr{E}|X|^r < \infty$ for $r < n$, and in that case all odd-order moments vanish. For $n > 2$, $\mathscr{E}X^2 = \text{var}(X) = n/(n-2)$. The largest value of f_n at $t = 0$ is $f_n(0) = [\Gamma((n+1)/2)/\Gamma(n/2)](1/\sqrt{n\pi})$. Clearly, $f_n(t) \to 0$ as $t \to \infty$ and one can show that $f_n(t) \to (1/\sqrt{2\pi})e^{-t^2/2}$ as $n \to \infty$ so that f_n can be approximated by a standard normal density for large n. For small n, however, t-distribution assigns more probability to its tails compared with standard normal distribution.

Table A6 in the Appendix gives the values of $t_{n,\alpha}$ where

(4)
$$P(t(n) > t_{n,\alpha}) = \alpha$$

for selected values of α and $n = 1, 2, \ldots, 30, 40, 60$, and 120. Here $t(n)$ has density f_n. Thus $t_{n,\alpha}$ is the quantile of order $1 - \alpha$. Left tail t values can be obtained by symmetry (Figure 1a). Indeed,

$$t_{n,1-\alpha} = -t_{n,\alpha}$$

Graphs of f_n for some values of n are given in Figure 2.

In Tables 1 and 2 we compare t-distribution and standard normal distribution (see also Figure 1b).

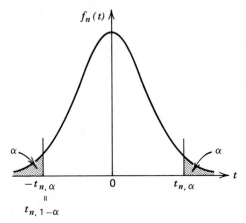

Figure 1a

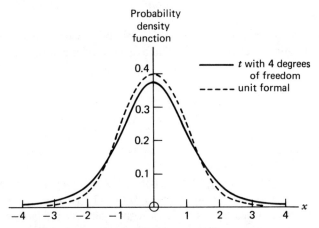

Figure 1b. Comparison of unit normal and $t(4)$ density functions. Reproduced from N. L. Johnson and S. Kotz, *Distributions in Statistics*: *Continuous Univariate Distributions*, Vol. 2, Wiley, New York, 1970, p. 77, with permission.

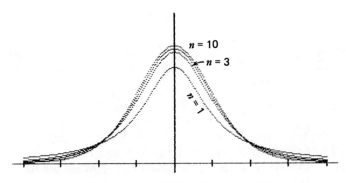

Figure 2. t-distribution for $n = 1, 3, 5, 10$.

TABLE 1.   Comparison of Tail Probabilities, $P(|X| > t)$

		X, t-distribution	
		with $n = 10$ d.f.[a]	with $n = 5$ d.f.
t	X, Normal $(0,1)$	probability greater than[b]	
2	.0456	.05	.1
2.5	.0124	.02	.05
3.0	.0026	.01	.02

[a]d.f., degree of freedom.
[b]For $n = 10$, the exact probabilities are, respectively, .0734, .0314, .0133; for $n = 5$, they are .1019, .0545, .0301.

TABLE 2.   Comparison of z_α and $t_{n,\alpha}$

		$t_{n,\alpha}$	
α	z_α	$n = 5$	$n = 15$
.10	1.28	1.476	1.341
.05	1.645	2.015	1.753
.01	2.33	3.365	2.602

Next we recall that a chi-square random variable with n d.f. is a gamma random variable with $\alpha = n/2$ and $\beta = 2$, so that the density is given by

$$(5) \qquad g_n(x) = \begin{cases} \dfrac{1}{\Gamma(n/2)2^{n/2}} x^{n/2-1} e^{-x/2}, & x > 0 \\ 0, & x \le 0. \end{cases}$$

We note that g_n is unimodal for $n > 2$ and has its maximum (or mode) at $x = n - 2$. For $n = 1$ it is strictly decreasing (with both axes as asymptotes). For $n = 2$, it is the exponential density with mean 2, and g_n is decreasing with mode[†] at $x = 0$. Figure 3 gives graphs of g_n for some selected values of n.

We note that for a chi-square (χ^2) random variable with n d.f.

$$\mu = n \quad \text{and} \quad \sigma^2 = 2n.$$

Table A7 in the Appendix gives the values of $\chi^2_{n,\alpha}$ where (Figure 4)

$$P\left(\chi^2(n) > \chi^2_{n,\alpha}\right) = \alpha$$

for selected values of α and $n = 1, 2, \ldots, 30$. Here $\chi^2(n)$ is a random variable with

[†]Mode at $x = 0$ if we define $g_n(x)$ at $x = 0$ by the first equality in (5).

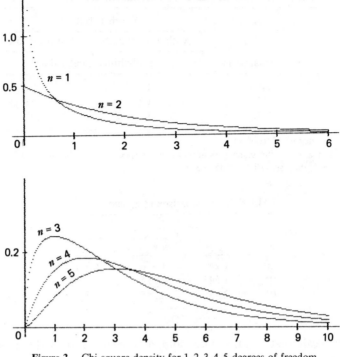

Figure 3. Chi-square density for 1, 2, 3, 4, 5 degrees of freedom.

density g_n. For larger values of n we use the result that $\sqrt{2\chi^2(n)} - \sqrt{2n-1}$ has approximately a normal distribution with mean 0, variance 1 (Problem 9.2.7).

Example 1. **Urine Chloride Level in Infants.** The mean chloride level in normal infants is assumed to have a normal distribution with a mean of 210 mEq per 24-hour period. A random sample of six premature infants showed a mean of 203 mEq per 24-hour period with a standard deviation of 15 mEq. Is there evidence to suggest that premature infants have a lower urine chloride level?

If X is the mean chloride level in full-term newborn infants, then X has a normal distribution with mean $\mu = 210$ mEq per 24-hour period. For the sample $n = 6$,

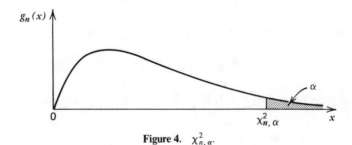

Figure 4. $\chi^2_{n,\alpha}$.

$\bar{x} = 203$ and $s = 15$. We wish to find $P\{\bar{X} \le 203 | \mu = 210\}$. We have

$$P\{\bar{X} \le 203 | \mu = 210\} = P\left\{\frac{\bar{X} - \mu}{s/\sqrt{n}} \le \frac{203 - 210}{15/\sqrt{6}}\right\}$$

$$= P\left(T \le \frac{7\sqrt{6}}{15}\right) = P(T \le -1.1431).$$

For $n - 1 = 6 - 1 = 5$ d.f. we see from Table A6 that

$$.20 > P(T \le -1.1431) > .15.$$

Therefore, there does not seem to be enough evidence to conclude that premature infants have a lower urine chloride level.

What is the probability that the sample comes from a population with standard deviation $\sigma = 10$? We need to compute

$$P\{S \ge 15 | \sigma = 10\} = P\{S^2 \ge 225 | \sigma^2 = 100\}$$

$$= P\left\{\frac{(n-1)S^2}{\sigma^2} \ge \frac{(5)(225)}{100}\right\} = P(\chi^2 \ge 11.25).$$

Here d.f. $= n - 1 = 5$ so that from Table A7

$$.05 > P(\chi^2 \ge 11.25) > .02.$$

That is, it is unlikely (the probability is between .02 and .05) that a random sample of size 6 from a normal population with variance $\sigma^2 = 100$ will have a sample variance as large as 225. $\quad\square$

In summary, therefore, we have:

Let $X_1, X_2, \ldots, X_n$ be a sample from a normal population with mean μ and variance σ^2. Let $\bar{X} = n^{-1}\sum_{i=1}^{n} X_i$, $S^2 = \sum_{i=1}^{n}(X_i - \bar{X})^2/(n-1)$. Then,

(i)   The statistic $Z = \sqrt{n}(\bar{X} - \mu)/\sigma$ has a standard normal distribution,
(ii) $\bar{X}$ and S^2 are independent,
(iii) $(n-1)S^2/\sigma^2$ has a chi-square distribution with $(n-1)$ d.f., and
(iv) $T = \sqrt{n}(\bar{X} - \mu)/S$ has a t-distribution with $(n-1)$ d.f.

These results can be extended to the case when we have two independent samples. Let $X_1, X_2, \ldots, X_m$ and $Y_1, Y_2, \ldots, Y_n$ be two independent random samples from normal populations with parameters μ_1, σ_1^2, and μ_2, σ_2^2 respectively. Let $\bar{X}, \bar{Y}$ and S_1^2, S_2^2 be the corresponding means and variances for the samples. By the methods of Section 8.6 we can show that

$$(6) \qquad Z = \frac{\bar{X} - \bar{Y} - (\mu_1 - \mu_2)}{\sqrt{\sigma_1^2/m + \sigma_2^2/n}}$$

has a standard normal distribution. If σ_1^2, σ_2^2 are unknown *assume that* $\sigma_1^2 = \sigma_2^2 = \sigma^2$ (unknown). Then

$$(7) \qquad S^2 = \frac{(m-1)S_1^2 + (n-1)S_2^2}{m+n-2}$$

is an unbiased estimate of σ^2 since

$$\mathscr{E}S^2 = \frac{(m-1)\mathscr{E}S_1^2 + (n-1)\mathscr{E}S_2^2}{m+n-2} = \frac{(m-1)\sigma^2 + (n-1)\sigma^2}{m+n-2} = \sigma^2.$$

Moreover, $(m + n - 2)S^2/\sigma^2$ has a chi-square distribution with $(m + n - 2)$ d.f. This follows (Problem 8.6.19) since $(m - 1)S_1^2/\sigma^2$ and $(n - 1)S_2^2/\sigma^2$ are *independent* random variables each having a chi-square distribution. Consequently,

$$(8) \qquad \frac{\left\{ [\bar{X} - \bar{Y} - (\mu_1 - \mu_2)]/(\sigma\sqrt{1/m + 1/n}) \right\}}{\sqrt{\dfrac{(m+n-2)S^2}{\sigma^2} \cdot \dfrac{1}{m+n-2}}}$$

$$= \frac{\bar{X} - \bar{Y} - (\mu_1 - \mu_2)}{S} \sqrt{\frac{mn}{m+n}}$$

has a *t*-distribution with $m + n - 2$ d.f. Moreover, in view of the definition of an *F* statistic (Example 8.4.6), it also follows that:

$$(9) \qquad \frac{((m-1)S_1^2/\sigma_1^2)/(m-1)}{((n-1)S_2^2/\sigma_2^2)/(n-1)} = \frac{\sigma_2^2}{\sigma_1^2} \frac{S_1^2}{S_2^2}$$

has an *F* distribution with $(m - 1, n - 1)$ d.f.

We recall that an *F* random variable with (m, n) d.f. has density function

$$(10) \quad h_{m,n}(f) = \begin{cases} \dfrac{\Gamma((m+n)/2)m^{m/2}n^{n/2}}{\Gamma(m/2)\Gamma(n/2)} \dfrac{f^{m/2-1}}{(n+mf)^{(m+n)/2}}, & f > 0 \\ 0, & f \le 0. \end{cases}$$

As $f \to \infty$, $h_{m,n}(f) \to 0$ and if $m > 2$, $h_{m,n}(f) \to 0$ as $f \to 0$. For $m > 2$, the density has a unique mode at $f = n(m - 2)/[m(n + 2)]$. If $m = 2$, the mode is at

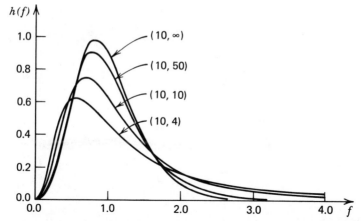

Figure 5. *F* density functions. Reproduced from N. L. Johnson and S. Kotz, *Distributions in Statistics: Continuous Univariate Distributions*, Vol. 2, Wiley, New York, 1970, p. 77, with permission.

$f = 0$ and if $m = 1$, $h_{m,n}(f) \to \infty$ as $f \to 0$. Figure 5 gives graphs of $h_{m,n}$ for selected pairs of values of (m, n).

It can be shown (Problem 8.8.26) that

$$\mathscr{E}F(m, n) = \frac{n}{n - 2} \qquad \text{for } n > 2$$

and

$$\text{var}(F(m, n)) = \frac{2n^2(m + n - 2)}{m(n - 2)^2(n - 4)} \qquad \text{for } n > 4.$$

Table A8 in the Appendix gives values of $F_{m, n, \alpha}$ where (Figure 6)

$$P(F(m, n) > F_{m, n, \alpha}) = \alpha$$

for selected pairs of values of (m, n) and $\alpha = .01$ and $\alpha = .05$. Note, however, that

$$F(m, n) = \frac{\chi^2(m)/m}{\chi^2(n)/n} = \frac{1}{\{\chi^2(n)/n\}/\{\chi^2(m)/m\}} = \{F(n, m)\}^{-1}$$

so that

$$\alpha = P(F(m, n) > F_{m, n, \alpha}) = P(F_{m, n, \alpha}^{-1} > F(n, m))$$

$$= 1 - P(F(n, m) > F_{m, n, \alpha}^{-1}).$$

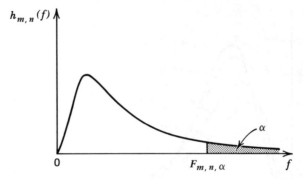

Figure 6. $F_{m, n, \alpha}$

Consequently,

(11) $$F_{n, m, 1-\alpha} = F_{m, n, \alpha}^{-1}.$$

8.7.1   Confidence Intervals for μ and σ^2

We now consider the problem of constructing confidence intervals for μ and σ^2 in sampling from a normal population. The results of this section give minimum level $1 - \alpha$ confidence intervals irrespective of the size of the sample.

If we want a confidence interval for μ at minimum level $1 - \alpha$ we have

$$P\left(\left|\frac{\overline{X} - \mu}{\sigma/\sqrt{n}}\right| \le z_{\alpha/2}\right) = 1 - \alpha$$

so that if σ is known, then a confidence interval for μ is

$$\left[\overline{X} - \frac{\sigma}{\sqrt{n}} z_{\alpha/2}, \overline{X} + \frac{\sigma}{\sqrt{n}} z_{\alpha/2}\right].$$

If, on the other hand, σ is unknown, then we replace it by S and we have

$$P\left(\left|\frac{\overline{X} - \mu}{S/\sqrt{n}}\right| \le t_{n-1, \alpha/2}\right) = 1 - \alpha$$

since $(\overline{X} - \mu)\sqrt{n}/S$ has a $t(n - 1)$ distribution. Hence a $(1 - \alpha)$ level confidence interval for μ is given by

$$\left[\overline{X} - \frac{S}{\sqrt{n}} t_{n-1, \alpha/2}, \quad \overline{X} + \frac{S}{\sqrt{n}} t_{n-1, \alpha/2}\right].$$

Confidence Interval for μ

Let $X_1, X_2, \ldots, X_n$ be a sample of size n from a normal distribution with mean μ and variance σ^2.

If σ is known

$$\left[\overline{X} - \frac{\sigma}{\sqrt{n}} z_{\alpha/2}, \overline{X} + \frac{\sigma}{\sqrt{n}} z_{\alpha/2} \right]$$

is a minimum level $1 - \alpha$ confidence interval for μ. If σ is unknown

$$\left[\overline{X} - \frac{S}{\sqrt{n}} t_{n-1, \alpha/2}, \overline{X} + \frac{S}{\sqrt{n}} t_{n-1, \alpha/2} \right]$$

is a minimum level $1 - \alpha$ confidence interval for μ.

If μ and σ^2 are unknown and a confidence interval for σ^2 is desired, we use the fact that $(n - 1)S^2/\sigma^2$ has a χ^2 distribution with $(n - 1)$ d.f. Hence given α and n we can find a, b such that

$$P\left(a \le (n - 1) \frac{S^2}{\sigma^2} \le b \right) = 1 - \alpha.$$

Each such pair (a, b) gives a $(1 - \alpha)$ level confidence interval for σ^2, namely,

$$\left[(n - 1)S^2/b, (n - 1)S^2/a \right].$$

Since the χ^2 distribution is not symmetric we cannot use symmetry to choose a and b (as we did in the case of computing confidence interval for μ). Although it is possible[†] to find a and b, numerically, such that the length of the confidence interval is minimum, it is simpler to assign equal probability to each tail and choose

$$a = \chi^2_{n-1, 1-\alpha/2}, \quad \text{and} \quad b = \chi^2_{n-1, \alpha/2}$$

Then the equal tails minimum level $1 - \alpha$ confidence interval for σ^2 is given by

---

Equal Tails Confidence Interval for σ^2

$$\left[(n - 1)S^2/\chi^2_{n-1, \alpha/2}, (n - 1)S^2/\chi^2_{n-1, 1-\alpha/2} \right]$$

(μ unknown).

[†] See Problem 8.8.24.

If μ is known, then one uses

$$\left[\frac{\sum_{i=1}^{n}(X_i - \mu)^2}{\chi_{n,\,\alpha/2}^2}, \frac{\sum_{i=1}^{n}(X_i - \mu)^2}{\chi_{n,\,1-\alpha/2}^2} \right]$$

as a $(1 - \alpha)$ level confidence interval for σ^2. This follows since $(X_i - \mu)/\sigma$, for $i = 1, 2, \ldots, n$ are independent standard normal random variables so that $(X_i - \mu)^2/\sigma^2$ is a $\chi^2(1)$ random variable for each i (Example 8.2.5). By independence their sum is a $\chi^2(n)$ random variable.

Example 2. *Confidence Interval for Average Length of a Desert Snake.* A herpetologist wishes to estimate the length of a rare species of desert snake in Nevada. She is able to trap a sample of nine snakes of the species and finds that the mean length is 3.7 feet with a standard deviation of 1 foot. Find a 90 percent confidence interval for the true mean length of this species of snake assuming that the distribution of lengths is normal.

Here $\bar{x} = 3.7$, $s = 1$, and $n = 9$. Since σ is not known, we use a confidence interval based on t-statistic with $n - 1 = 8$ d.f. Since $\alpha = .10$, $t_{8,.05} = 1.860$ and a 90 percent confidence interval is given by

$$\left[\bar{x} - \frac{s}{\sqrt{n}} t_{n-1,\,\alpha/2}, \bar{x} + \frac{s}{\sqrt{n}} t_{n-1,\,\alpha/2} \right] = \left[3.7 - \frac{1}{\sqrt{9}}(1.86), 3.7 + \frac{1}{\sqrt{9}}(1.86) \right]$$

$$= [3.08, 4.32] \text{ feet.}$$

We are therefore 90 percent confident that $3.08 \le \mu \le 4.32$ in the sense that for about 90 percent random samples of nine snakes of this species the random interval $[x - 1.86s/3, \bar{x} + 1.86s/3]$ will contain μ. □

Example 3. *Nicotine Content of a Certain Brand of Cigarettes.* The nicotine content of a certain brand of cigarettes measured in milligrams may be assumed to have a normal distribution. A sample of size 5 produces nicotine contents of 16, 16.5, 19, 15.4, and 15.6. Let us find a 98 percent confidence interval for the average nicotine content of this brand.

Here $n = 5$, $\alpha = .02$ so that $t_{n-1,\,\alpha/2} = t_{4,.01} = 3.747$. Moreover,

$$\bar{x} = \frac{16 + 16.5 + 19 + 15.4 + 15.6}{5} = 16.5,$$

$$s^2 = \frac{\sum_{i=1}^{n}(x_i - \bar{x})^2}{(n - 1)} = 2.13,$$

and $s = 1.459$. Hence a 98 percent confidence interval for μ is given by

$$\left[16.5 - \frac{1.459}{\sqrt{5}} 3.747, 16.5 + \frac{1.459}{\sqrt{5}} 3.747 \right] = [14.06, 18.94].$$

We are 98 percent confident that $14.06 \leq \mu \leq 18.94$. If the manufacturer asserts, for example, that the mean nicotine content is 13.5 milligram, we will have to be suspicious of his claim, whereas if he claims that $\mu = 14.5$, then the sample does support his claim at level .02. □

Example 4. Variation in Manufacture of Fuses. A random sample of 15 fuses for the 25 amperage value manufactured by a certain company shows a standard deviation of .85 amperage. If the amperage is assumed to be normally distributed, then a 95 percent confidence interval for σ^2 is given by

$$\left[\frac{(n-1)s^2}{\chi^2_{n-1,\alpha/2}}, \frac{(n-1)s^2}{\chi^2_{n-1,1-\alpha/2}} \right]$$

where $n = 15$, $s^2 = (.85)^2$, $\alpha = .05$ so that $\chi^2_{14,.025} = 26.119$ and $\chi^2_{14,.975} = 5.629$. The confidence interval is $[.39, 1.80]$. If the manufacturer claims that the variance in amperage of his fuses is .8, we would have to conclude that the sample supports his claim. If a confidence interval for σ is desired, we take the square root of the above interval. In the example under consideration, a 95 percent confidence interval for σ is given by $[.62, 1.34]$. □

Example 5. Sample Size Determination. Suppose we wish to find the sample size n in such a way that the length of the confidence interval for μ is fixed in advance at $2d$ and minimum level at $1 - \alpha$. Suppose σ is known. Then

$$\left[\overline{X} - \frac{\sigma}{\sqrt{n}} z_{\alpha/2}, \overline{X} + \frac{\sigma}{\sqrt{n}} z_{\alpha/2} \right]$$

is a $1 - \alpha$ level confidence interval for μ with length $l = (2\sigma/\sqrt{n}) z_{\alpha/2}$. It follows that

$$2d = \frac{2\sigma}{\sqrt{n}} z_{\alpha/2} \quad \text{or} \quad \boxed{n \doteq \frac{\sigma^2 z^2_{\alpha/2}}{d^2}}.$$

If n computed from this formula is not an integer we choose the smallest integer greater than $(\sigma^2 z^2_{\alpha/2}/d^2)$. Often σ is not known. In that case, we use any prior information on σ that is available, such as the standard deviation from a previous study.

In particular, suppose it is desired to find a 95 percent confidence interval for the mean systolic blood pressure of males in such a way that the length of the confidence interval is 6 mm Hg. Suppose it is known that $\sigma = 10$ mm Hg. Then $z_{\alpha/2} = z_{.025} = 1.96$, $2d = 6$, and

$$n = \frac{(10)^2 (1.96)^2}{3^2} = 42.68$$

so that we need a sample of size 43.

This problem can be equivalently phrased as follows. It is desired to estimate μ by the sample mean $\overline{X}$ such that the maximum error is ε with probability $1 - \alpha$. What should be n? In that case we want n to satisfy

$$P(|\bar{X} - \mu| \leq \varepsilon) = 1 - \alpha$$

so that

$$1 - \alpha = P\left(\frac{|\bar{X} - \mu|}{\sigma/\sqrt{n}} \leq \frac{\varepsilon}{\sigma}\sqrt{n}\right) = P\left(|Z| \leq \frac{\varepsilon\sqrt{n}}{\sigma}\right)$$

where Z is standard normal. It follows that

$$\varepsilon\sqrt{n}/\sigma = z_{\alpha/2} \quad \text{or} \quad \boxed{n \doteq \frac{\sigma^2 z_{\alpha/2}^2}{\varepsilon^2}}.$$

Thus the maximum error in estimating μ by $\bar{X}$ with probability $1 - \alpha$ equals (length of confidence interval)/2. In our example, if we choose a sample of size 43, then $\bar{X}$ will be within 3 mm Hg of μ with probability .95.   □

8.7.2 Confidence Intervals for $\mu_1 - \mu_2$ and σ_2^2 / σ_1^2.

The procedure for finding a confidence interval for $\mu_1 - \mu_2$ in sampling from two normal populations closely parallels the procedure for the one sample case described in Section 8.7.1. We summarize the results.

Confidence Interval for $\mu_1 - \mu_2$

Let $X_1, X_2, \ldots, X_m$ and $Y_1, Y_2, \ldots, Y_n$ be independent random samples from normal distributions with means μ_1, μ_2 and variances σ_1^2, σ_2^2 respectively. If σ_1, σ_2 are known, then

$$\left[\bar{X} - \bar{Y} - \sqrt{\frac{\sigma_1^2}{m} + \frac{\sigma_2^2}{n}}\, z_{\alpha/2}, \quad \bar{X} + \bar{Y} + \sqrt{\frac{\sigma_1^2}{m} + \frac{\sigma_1^2}{n}}\, z_{\alpha/2}\right]$$

is a $(1 - \alpha)$ level confidence interval for $\mu_1 - \mu_2$. If σ_1, σ_2 are unknown and $\sigma_1 = \sigma_2$ (unknown), then

$$\left[\bar{X} - \bar{Y} - S\sqrt{\frac{1}{m} + \frac{1}{n}}\, t_{m+n-2,\,\alpha/2}, \quad \bar{X} - \bar{Y} + S\sqrt{\frac{1}{m} + \frac{1}{n}}\, t_{m+n-2,\,\alpha/2}\right]$$

is a $(1 - \alpha)$ level confidence interval for $\mu_1 - \mu_2$ where S^2 is defined in (7).

Often, however, we have n pairs of independent observations from a bivariate distribution where the observations in each pair (X_i, Y_i) are dependent. Suppose we want a confidence interval for $\mu_1 - \mu_2$. If $X_i - Y_i$ is normally distributed, we can consider $X_1 - Y_1, X_2 - Y_2, \ldots, X_n - Y_n$ as a random sample from a normal population with, say, mean $\mu_1 - \mu_2$ and variance σ^2. To get a confidence interval for $\mu_1 - \mu_2$ we simply apply the methods of Section 8.7.1.

Confidence Interval for $\mu_1 - \mu_2$ for Dependent Pairs

Suppose $X_1 - Y_1, \ldots, X_n - Y_n$ are normally distributed with unknown mean $\mu_1 - \mu_2$ and unknown variance σ^2. Let $D_i = X_i - Y_i$, $\overline{D} = \Sigma_1^n D_i/n$, and S_D^2 be the sample variance of D_i's. Then

$$\left[\overline{X} - \overline{Y} - \frac{S_D}{\sqrt{n}} t_{n-1,\,\alpha/2}, \; \overline{X} - \overline{Y} + \frac{S_D}{\sqrt{n}} t_{n-1,\,\alpha/2} \right]$$

is a $(1 - \alpha)$ level confidence interval for $\mu_1 - \mu_2$.

To construct a confidence interval for the ratio σ_2^2/σ_1^2 in sampling from normal populations (when μ_1, μ_2 are unknown), we use (9) to find a and b such that

$$P\left(a \le \frac{\sigma_2^2}{\sigma_1^2} \frac{S_1^2}{S_2^2} \le b \right) = 1 - \alpha.$$

Since the distribution of an F statistic is asymmetric, we use an equal tails approach and choose $a = F_{m-1,\,n-1,\,1-\alpha/2}$ and $b = F_{m-1,\,n-1,\,\alpha/2}$. Then

$$P\left(F_{m-1,\,n-1,\,1-\alpha/2} \frac{S_2^2}{S_1^2} \le \frac{\sigma_2^2}{\sigma_1^2} \le F_{m-1,\,n-1,\,\alpha/2} \frac{S_2^2}{S_1^2} \right) = 1 - \alpha.$$

Confidence Interval for σ_2^2/σ_1^2

Let $X_1, X_2, \ldots, X_m$ and $Y_1, Y_2, \ldots, Y_n$ be independent samples from two normal populations. Then

$$\left[F_{m-1,\,n-1,\,1-\alpha/2} \frac{S_2^2}{S_1^2}, \quad F_{m-1,\,n-1,\,\alpha/2} \frac{S_2^2}{S_1^2} \right]$$

is a $(1 - \alpha)$ level confidence interval for σ_2^2/σ_1^2.

A confidence interval for σ_2/σ_1 is obtained by taking the square root of the above confidence interval for σ_2^2/σ_1^2.

Example 6. Confidence Interval for Difference in Performance of Two Groups of Students. A test for reading comprehension is given to an elementary school class consisting of 15 Anglo-American and 12 Mexican-American children. The results of the test are as follows:

| Anglo-American | $\bar{x} = 74, s_1 = 9$ |
| Mexican-American | $\bar{y} = 72, s_2 = 12$ |

Let us compute a 95 percent confidence interval for the difference in means of the two groups assuming that they form random samples from normal populations. In order to apply the confidence interval based on a t-statistic, we need to assume further that $\sigma_1^2 = \sigma_2^2$. In that case

$$s^2 = \frac{(m-1)s_1^2 + (n-1)s_2^2}{m+n-2} = \frac{14(9^2) + 11(12)^2}{25} = 108.72$$

so that $s = 10.43$. Also, for $\alpha = .05$, $t_{m+n-2,\alpha/2} = t_{25,.025} = 2.06$. Hence a 95 percent confidence interval for $\mu_1 - \mu_2$ is given by

$$\left[74 - 72 - 10.43\sqrt{\tfrac{1}{15} + \tfrac{1}{12}}\,(2.06), 74 - 72 + 10.43\sqrt{\tfrac{1}{15} + \tfrac{1}{12}}\,(2.06)\right]$$

$$= [-6.32, 10.32].$$

We are therefore 95 percent confident that $-6.32 \le \mu_1 - \mu_2 \le 10.32$. In particular, the evidence suggests that the difference in means is not significant at the 5 percent level.

What about the assumption $\sigma_1 = \sigma_2$? Let us compute a 90 percent confidence interval for σ_2^2/σ_1^2. We note that $F_{14,11,.05} = 2.74$ and $F_{14,11,.95} = 1/F_{11,14,.05} = 1/2.56$, so that a 90 percent confidence interval is

$$\left[\frac{1}{2.56}\frac{12^2}{9^2}, 2.74\frac{12^2}{9^2}\right] = [.69, 4.87].$$

Since this interval does contain $\sigma_2^2/\sigma_1^2 = 1$ our assumption that $\sigma_1^2 = \sigma_2^2$ is reasonable. □

Example 7. Difference in Birth Weight of Twins. A random sample of six sets of twins was taken, and it showed the following birth weights in pounds.

Birth Weight of Twins (in pounds)

Twins	1	2	3	4	5	6
First-born, X	7.4	6.2	3.4	4.8	5.3	5.1
Second-born, Y	6.5	5.8	3.6	4.6	5.5	4.8
Difference, $X - Y$	.9	.4	-.2	.2	-.2	.3

Assuming that the difference in birth weights is a normal random variable, we construct a 98 percent confidence interval for the true difference in weights of the first-born and the second-born in a set of twins. Here,

$$\bar{d} = \frac{.9 + .4 - .2 + .2 - .2 + .3}{6} = \frac{1.4}{6} = .23,$$

$$s_d^2 = \frac{1}{6-1}\left\{\sum_{i=1}^{6} d_i^2 - 6\bar{d}^2\right\} = .171, \qquad s_d = .413.$$

Since $\alpha = .02$, $t_{n-1,\alpha/2} = t_{5,.01} = 3.365$, a 98 percent confidence interval is given by $[.23 - (.413/\sqrt{6})3.365, .23 + (.413/\sqrt{6})3.365] = [-.34, .80]$. In particular, this means that the difference in mean birth weights of the first-born and second-born twins is not significant at level .02.   □

8.7.3 Testing Hypotheses Concerning Mean and Difference of Means

In order to test the hypothesis H_0: $\mu = \mu_0$ based on a random sample of size n from a normal population, the appropriate test statistic is

$$Z = \sqrt{n}\,\frac{\bar{X} - \mu_0}{\sigma}$$

if σ is known and

$$T = \sqrt{n}\,\frac{\bar{X} - \mu_0}{S}$$

if σ is unknown. Clearly, large values of Z (or T) support the alternative H_1: $\mu > \mu_0$ and lead to the rejection of H_0. Similar arguments apply in the case when H_1 is $\mu < \mu_0$ or $\mu \neq \mu_0$. Similar methods also apply in the two sample case. We state these results below. It will be shown in Section 11.3 that the same tests are valid when H_0 is $\mu \leq \mu_0$ or $\mu \geq \mu_0$.

Testing Hypotheses Concerning μ

Compute $t_0 = \sqrt{n}\,(\bar{x} - \mu_0)/s$.

H_0	H_1	Reject H_0 at level α if	P-value
$\mu \leq \mu_0$	$\mu > \mu_0$	$t_0 > t_{n-1,\alpha}$	$P(T \geq t_0)$
$\mu \geq \mu_0$	$\mu < \mu_0$	$t_0 < -t_{n-1,\alpha}$	$P(T \leq t_0)$
$\mu = \mu_0$	$\mu \neq \mu_0$	$|t_0| > t_{n-1,\alpha/2}$	$2P(T \geq |t_0|)$

Here T has t distribution with $n-1$ d.f. If σ is known, compute instead $z_0 = \sqrt{n}\,(\bar{x} - \mu_0)/\sigma$ and use z_α instead of $t_{n-1,\alpha}$.

Due to limitations of the t-tables, the P-value cannot be exactly computed. However, one can get bounds on the P-value.

Testing Hypotheses Concerning $\mu_1 - \mu_2$ when $\sigma_1 = \sigma_2$ (unknown)

Compute:

$$t_0 = (\bar{x} - \bar{y}) / \left\{ s\sqrt{\frac{1}{m} + \frac{1}{n}} \right\}, \qquad s^2 = \frac{(m-1)s_1^2 + (n-1)s_2^2}{(m+n-2)}.$$

H_0	H_1	Reject H_0 at level α if	P-value
$\mu_1 \le \mu_2$	$\mu_1 > \mu_2$	$t_0 > t_{m+n-2,\,\alpha}$	$P(T \ge t_0)$
$\mu_1 \ge \mu_2$	$\mu_1 < \mu_2$	$t_0 < -t_{m+n-2,\,\alpha}$	$P(T \le t_0)$
$\mu_1 = \mu_2$	$\mu_1 \ne \mu_2$	$|t_0| > t_{m+n-2,\,\alpha/2}$	$2P(T \ge |t_0|)$

Here T is a t-statistic with $m + n - 2$ d.f.

If σ_1, σ_2 are known, then one computes

$$z_0 = \frac{(\bar{x} - \bar{y})}{\left\{ \sigma_1^2/m + \sigma_2^2/n \right\}^{1/2}}$$

and uses z_α instead of $t_{m+n-2,\,\alpha}$.

We recall that there is a one-to-one correspondence between a size α test of $\theta = \theta_0$ against $\theta \ne \theta_0$ and a minimum level $1 - \alpha$ confidence interval for θ. (See Section 4.4.) Hence the confidence intervals of Section 8.7.1 can be used to accomplish the same purpose and conversely. Similar correspondence exists between one-sided tests and confidence bounds.

Example 8. Average Handling Costs Per Roll of Film. A mail-order photofinishing company charges 50¢ per roll as handling costs. The manager suspects that handling costs have gone up. In order to verify his hunch he takes a random sample of 16 orders and finds that the average cost is 55¢ with a standard deviation of 8¢. Suppose the handling cost for an order has a normal distribution. Then we wish to test $H_0: \mu = 50$ against $H_1: \mu > 50$. We have

$$t_0 = \frac{55 - 50}{8}\sqrt{16} = \frac{20}{8} = 2.5$$

and d.f. $= 16 - 1 = 15$. From Table A6, $t_{15,\,.01} = 2.602$ and $t_{15,\,.025} = 2.131$, so that the P-value satisfies

$$.01 \le P(T \ge t_0) \le .025.$$

At level $\alpha = .025$, we reject H_0 and conclude that the observed difference between the sample mean and $\mu = 50$ is significant. That is, the average handling costs are indeed higher than 50¢ per roll. $\qquad\square$

Example 9. Difference in Performance of Two Groups of Students (Example 6). In Example 6 we constructed a 95 percent confidence interval for the difference in mean scores of Anglo-American students and Mexican-American students. If we wish to test $H_0: \mu_1 = \mu_2$ against $H_1: \mu_1 \neq \mu_2$, then under $H_0: \mu_1 - \mu_2 = 0$, and since 0 is included in the 95 percent confidence interval $[-6.32, 10.32]$ we cannot reject H_0 at level .05.

Alternatively, let us compute the value of t_0 to see how significant it is. We have

$$t_0 = \frac{74 - 72}{10.43\sqrt{(1/15) + (1/12)}} = \frac{2}{4.04} = .495.$$

The associated P-value is $2P(T > .495)$ for 25 d.f. From Table A6 we note that $.3 < P(T > .495) < .35$ so that t_0 is not significant even at $\alpha = .60$ level. $\qquad\square$

Example 10. Sample Size Determination. In Example 8, suppose the manager wants to find the size of the sample required to ensure that if $\mu = 50¢$ then he wants no more than 5 in 100 chance of increasing the rate, but if the true average cost is $55¢$, then he is willing to raise the charge but is willing to allow a chance of one in 10 of not doing so. Suppose $\sigma = 8$.

Here the problem is to test $H_0: \mu = \mu_0$ against $H_1: \mu > \mu_0$ at, say, level α such that the probability of type II error when μ is the true value is β. We have, for $\mu > \mu_0$:

$$\beta = P\{\text{accept } H_0|\mu\} = P\{Z \leq z_\alpha|\mu\}$$

$$= P\left\{\overline{X} \leq \mu_0 + \frac{\sigma}{\sqrt{n}} z_\alpha \,\Big|\, \mu\right\} = P\left\{(\overline{X} - \mu)\frac{\sqrt{n}}{\sigma} \leq (\mu_0 - \mu)\frac{\sqrt{n}}{\sigma} + z_\alpha\right\}.$$

Hence

$$z_{1-\beta} = -z_\beta = z_\alpha - \frac{\sqrt{n}(\mu - \mu_0)}{\sigma}.$$

It follows that

$$n = \left\{(z_\alpha + z_\beta)\frac{\sigma}{\mu - \mu_0}\right\}^2.$$

In our case, $\alpha = .05$, $\beta = .10$, $\sigma = 8$, $\mu_0 = 50$, and $\mu = 55$, so that

$$n = \left\{(1.645 + 1.282)\frac{8}{5}\right\}^2 = 21.9.$$

Thus the minimum sample size needed is 22. $\qquad\square$

8.7.4 Testing Hypotheses Concerning Variances

In order to test $H_0: \sigma = \sigma_0$ against one-sided or two-sided alternatives, the appropriate test statistic is

$$V = \frac{(n - 1)S^2}{\sigma^2}$$

when μ is unknown, and

$$V' = \frac{\sum_{i=1}^n (X_i - \mu)^2}{\sigma^2}$$

when μ is known. We note that V has a chi-square distribution with $n - 1$ d.f. whereas V' has a chi-square distribution with n d.f. Large values of V (or V') support $\sigma > \sigma_0$ and result in rejection of H_0 in favor of H_1: $\sigma > \sigma_0$. A similar test is valid for H_0: $\sigma \leq \sigma_0$. We summarize the results below.

Testing Hypotheses Concerning σ^2

Compute $v_0 = (n - 1)s^2/\sigma_0^2$.

H_0	H_1	Reject H_0 at Level α if	P-value
$\sigma \leq \sigma_1$	$\sigma > \sigma_0$	$v_0 > \chi^2_{n-1,\,\alpha}$	$P(V \geq v_0)$
$\sigma \geq \sigma_0$	$\sigma < \sigma_0$	$v_0 < \chi^2_{n-1,\,1-\alpha}$	$P(V \leq v_0)$
$\sigma = \sigma_0$	$\sigma \neq \sigma_0$	$v_0 > \chi^2_{n-1,\,\alpha/2}$	2 (smaller tail
		or	probability)
		$v_0 < \chi^2_{n-1,\,1-\alpha/2}$	

Here V is a chi-square random variable with $n - 1$ d.f. If μ is known, compute $v_0' = \sum_{i=1}^{n}(x_i - \mu)^2/\sigma_0^2$ and use $\chi^2_{n,\,\alpha}$ instead of $\chi^2_{n-1,\,\alpha}$.

Finally we consider testing H_0: $\sigma_1 = \sigma_2$. In this case the appropriate test statistic is

$$F = \frac{\sigma_2^2 S_1^2}{\sigma_1^2 S_2^2},$$

which, under H_0: $\sigma_1 = \sigma_2$, reduces to $F = S_1^2/S_2^2$ and has an F-distribution with $(m - 1, n - 1)$ d.f. Clearly, large values of F support the claim that $\sigma_1 > \sigma_2$ and small values of F support the claim that $\sigma_1 < \sigma_2$. The same tests apply when H_0 is $\sigma_1 \leq \sigma_2$ or $\sigma_1 \geq \sigma_2$.

Testing Equality of Variances

Compute $f_0 = s_1^2/s_2^2$.

H_0	H_1	Reject H_0 at Level α if	P-value
$\sigma_1 \leq \sigma_2$	$\sigma_1 > \sigma_2$	$f_0 > F_{m-1,\,n-1,\,\alpha}$	$P(F \geq f_0)$
$\sigma_1 \geq \sigma_2$	$\sigma_1 < \sigma_2$	$f_0 < F_{m-1,\,n-1,\,1-\alpha}$	$P(F \leq f_0)$
$\sigma_1 = \sigma_2$	$\sigma_1 \neq \sigma_2$	$f_0 > F_{m-1,\,n-1,\,\alpha/2}$	2 (smaller tail
		or	probability)
		$f_0 < F_{m-1,\,n-1,\,1-\alpha/2}$	

Here F has an F distribution with $(m - 1, n - 1)$ d.f.

The tests can be modified when either or both μ_1 or μ_2 is known. We leave the reader to furnish the details. We emphasize that Table A8 gives only values of $F_{m,n,\alpha}$ for $\alpha = .01$ and $.05$ but does not give left tail critical values. For this purpose we use the relation (11).

Example 11.   Testing Variability in GPA's of Male and Female Students.   In a study of the grade point averages of seniors, a researcher wishes to use both female and male students. In order to do so she would like to know if the GPAs of male and female students have the same amount of variability. Independent random samples of male and female students gave the following results.

Sex	Size	Sample Standard Deviation
Female	12	3.7
Male	17	4.1

Here she wishes to test H_0: $\sigma_1 = \sigma_2$ against H_1: $\sigma_1 \neq \sigma_2$. We note that

$$f_0 = \frac{(3.7)^2}{(4.1)^2} = .814,$$

which is in the left tail, so we need $F_{m-1,n-1,1-\alpha/2}$. We have from (11):

$$F_{m-1,n-1,1-\alpha/2} = \frac{1}{F_{n-1,m-1,\alpha/2}} = \begin{cases} 1/2.7 = .37 & \text{if } \alpha = .10 \\ 1/4.21 = .24 & \text{if } \alpha = .02. \end{cases}$$

Since $f_0 = .814 \not< .37$ we do not reject H_0 at level $.10$.

A 90 percent confidence interval for σ_2^2/σ_1^2 is easily constructed. It is given by

$$\left[F_{m-1,n-1,1-\alpha/2}\frac{s_2^2}{s_1^2}, F_{m-1,n-1,\alpha/2}\frac{s_2^2}{s_1^2} \right] = [.37(1.23), 2.45(1.23)]$$

$$= [.46, 3.01]$$

and that for σ_1^2/σ_2^2 is given by $[.33, 2.20]$. Since 1 is included in the interval $[.46, 3.01]$ (or in $[.33, 2.20]$), we cannot reject H_0 at 10 percent level of significance. $\square$

Remark 1.   In the two sample t-test for testing H_0: $\mu_1 = \mu_2$ against one-sided or two-sided alternatives, we made the assumption that $\sigma_1 = \sigma_2 = \sigma$ (unknown). It is therefore tempting to use a pre-test to determine whether this assumption holds. There are, however, two reasons for not doing so. First, the combined level of significance of an F-test at level α followed by a t-test at level α is no longer α but $1 - (1 - \alpha)^2$, which is $> \alpha$. In general, if two tests are performed sequentially, with respective significance levels α_1 and α_2, then the significance level of the combined test is larger than $\max(\alpha_1, \alpha_2)$. Secondly, the F test may reject H_0

not only when H_0 is false but also when the underlying distribution is nonnormal. The t-test, on the other hand, is a good test even when the underlying distribution is only approximately normal.

Remark 2.   In Sections 6.3 and 8.5 we studied the sign test for testing hypotheses concerning the quantiles of a distribution or the mean of a continuous symmetric distribution. Thus t-test and sign test are competitors for testing hypotheses concerning the mean of a normal distribution. Which of the two tests should we use? If the underlying distribution is normal, it is intuitively clear that the t-test should be preferred over the sign test. A more difficult question to answer is, Can we use the t-test even when F is not normal but unimodal and symmetric? If so, which of the two tests performs better in terms of its power function? Power comparisons are often very difficult to make, especially when the form of F is unknown. In Section 11.5 we consider yet another test for the location parameter of a symmetric distribution. We also compare the performance of three tests there. Similar comparisons are made in the two-sample case in Section 11.6.

Problems for Section 8.7

1. Let T have a t-distribution with n d.f. Find:
   (i) $P(|T| < 3.365)$ if $n = 5$.
   (ii) $P(|T| > 1.812)$ if $n = 10$.
   (iii) Constant c, if $P(|T| \le c) = .95$, and $n = 20$.
   (iv) Constant c, if $P(|T| > c) = .10$, and $n = 12$.
   (v) Constant c, if $P(-c < T < 0) = .45$, and $n = 17$.

2. Consider a random sample of size 7 from a normal distribution with mean $\mu = 18$. Let $\bar{X}$ and S^2 be the sample mean and the sample variance. Find:
   (i) $P(|\bar{X} - 18| > (1.943/\sqrt{7})S)$.
   (ii) $P(|\bar{X} - 18| \le (1.44/\sqrt{7})S)$.
   Find the constant c such that:
   (iii) $P(|\bar{X} - 18| < cS) = .98$.
   (iv) $P(|\bar{X} - 18| > cS) = .20$.
   (v) Estimate $P(\bar{X} > 24)$, $P(\bar{X} < 13)$, and $P(12 < \bar{X} < 28)$ when $s = 10$.

3. Let X be a chi-square random variable with n d.f. Find:
   (i) $P(X \le 2.558)$ if $n = 10$.
   (ii) $P(10.600 \le X \le 21.337)$ if $n = 22$.
   (iii) Constant c, if $P(X > c) = .05$ and $n = 7$.
   (iv) Constant c, if $P(X < c) = .01$ and $n = 9$.
   (v) Constants c_1 and c_2, if $P(X < c_1) = P(X > c_2) = .05$ and $n = 10$.
   (vi) Constants c_1 and c_2, if $P(X < c_1) = P(X > c_2) = .01$ and $n = 15$.

4. Let S^2 be the sample variance of a random sample of size n from a normal population. Find
   (i) c such that $P(S^2 > c) = .025$ if $n = 15$ and $\sigma = .05$.
   (ii) c such that $P(S^2 > c) = .95$ if $n = 17$ and $\sigma = 8$.

(iii) c such that $P(S^2 > c\sigma^2) = .05$ if $n = 12$.
(iv) c such that $P(S^2 < c\sigma^2) = .975$ if $n = 11$.
Estimate the probabilities
(v) $P(S^2 > \sigma^2/4)$ if $n = 8$.
(vi) $P(S^2 < \sigma^2/2)$ if $n = 16$.

5. For an F random variable with (m, n) d.f., find c such that
(i) $P(F > c) = .01$, $m = 3$, $n = 7$.
(ii) $P(F > c) = .05$, $m = 8$, $n = 4$.
(iii) $P(F > c) = .01$, $m = 11$, $n = 14$.
(iv) $P(F < c) = .05$, $m = 16$, $n = 10$.
(v) $P(F < c) = .01$, $m = 7$, $n = 11$.

6. Independent random samples of sizes m and n are drawn from two normal populations with the same variance. Let S_1^2, S_2^2 be the two-sample variances. Find c such
that
(i) $P(S_1^2 > cS_2^2) = .05$, $m = 8$, $n = 4$.
(ii) $P(S_1^2 < cS_2^2) = .95$, $m = 11$, $n = 7$.
(iii) $P(S_1 > cS_2) = .01$, $m = 6$, $n = 14$.
(iv) $P(S_1 < cS_2) = .05$, $m = 12$, $n = 25$.

7. The reaction time of a certain animal to a certain stimulus may be assumed to be
normally distributed. In order to estimate the mean reaction time with 95 percent
confidence, a psychologist took a sample of eight animals and found $\bar{x} = .095$ second
with a standard deviation .015 second. What confidence interval should he report for
the mean reaction time?

8. In order to estimate the mean effect on pulse rate of a particular drug with a 98
percent confidence, a researcher selected 16 healthy males aged 22–35 years. Each
subject is administered one oral dose of 100 milligram of the drug and his pulse rate
in beats per minute is recorded after 90 minutes. The sample mean is found to be
$\bar{x} = 70$ beats per minute. If a clinically accepted standard deviation of pulse rate is
10 beats per minute, find a 98 percent confidence interval for the mean pulse rate for
patients on this drug assuming that the pulse rate has a normal distribution.

9. The systolic blood pressure of people undergoing a certain controlled exercise
program is normally distributed. A random sample of nine subjects in the program
showed a mean systolic blood pressure of 212.4 mm Hg with a standard deviation of
10.1 mm Hg. Find a 90 percent confidence interval for the mean systolic blood
pressure of persons in the program.

10. In order to estimate the average temperature of Indians living on a reservation, a
nurse randomly selected 11 people on the reservation and recorded their body
temperatures in degrees Fahrenheit with the following results.

$$\bar{x} = 98.36\,^\circ F, s = .15\,^\circ F.$$

Find a 95 percent confidence interval for the mean temperature of Indians on this
reservation.

11. In Problem 8, suppose the pulse rates of all 16 subjects were recorded before the administration of the drug and 90 minutes after the drug, with following results.

Subject	1	2	3	4	5	6	7	8
Pulse rate before	74	80	86	95	92	98	74	77
Pulse rate after	65	74	71	73	74	68	75	65
Subject	9	10	11	12	13	14	15	16
Pulse rate before	89	87	95	97	85	83	73	84
Pulse rate after	68	69	67	70	71	70	74	66

Find a 98 percent confidence interval for the mean difference in pulse rates.

12. The copper content of an alloy is normally distributed with variance 7 $(percent)^2$. What is the sample size required in order to estimate the average copper content by $\overline{X}$ to within 2 percent with 95 percent confidence?

13. A random sample of scores on a standardized college entrance examination is to be taken to test $H_0: \mu = 100$ against $H_1: \mu < 100$ at level $\alpha = .10$. How large should the sample size be to ensure that the probability of accepting H_0 when $\mu = 90$ is .20? Take $\sigma = 20$.

14. The resistance of a wire (in ohms) produced by a manufacturer is normally distributed. In order to check if the wires produced on two different machines have the same average resistance (when both machines have been set to have same variability), two samples were taken with the following results.

Resistance (Ohms)

Machine 1:	0.136	0.141	0.140	0.142	0.139	0.145	0.141
Machine 2:	0.136	0.135	0.143	0.138	0.140		

Find a 98 percent confidence interval for the true difference in mean resistances.

15. A random sample of 25 participants in a diet program showed an average weight loss of 6.8 pounds after the completion of the program. Find a 90 percent confidence interval for the mean decrease in weight of participants in this program, if the sample variance is 3 $(pounds)^2$ assuming that the difference in weights are normally distributed.

16. In order to estimate with 90 percent confidence the difference in mean monthly income of male and female graduates of a university, the Placement Office took two independent random samples of graduates (within two months of graduation) with

the following results.

Sex	Size	Sample Mean	Standard Deviation
Male	20	$1,205	$152
Female	17	$1,092	$136

What is the confidence interval? Assume population variances are equal.

17. The tar content of two brands of cigarettes is to be compared based on the following sample results:

Brand 1:	18	16	14	16	19	
Brand 2:	16	13	15	12	14	13

Is the difference in mean tar content of the two brands significant? At what level?

18. The standard deviation of the fill in a soup-vending machine is important to the vending machine operator. He takes a random sample of 15 cups and finds that the sample standard deviation is 0.53 ounce. Find a 95 percent confidence interval for the true standard deviation of the fill on this machine.

19. A sample of 16 cigarettes showed a sample mean tar content of 11.2 with a standard deviation of .9. Find a 90 percent confidence interval for the true standard deviation.

20. A phosphate fertilizer was applied to seven plots and a nitrogen fertilizer to five plots. The yield in bushels for each plot was recorded:

Phosphate:	46	39	47	36	50	38	45
Nitrogen:	46	49	50	41	48		

Find a 98 percent confidence interval for the ratio σ_1^2/σ_2^2 of variances in yields of the two fertilizers.

21. A random sample of 16 half-hour intervals of daytime television programs showed that the mean time devoted to commercials was 8.5 minutes with a standard deviation of 1 minute. A random sample of 10 half-hour prime-time programs showed a mean time of 7.8 minutes with a standard deviation of 1.2 minutes. Find a 90 percent confidence interval for the ratio of variances of the two times devoted to commercials.

22. In Problem 7, suppose the psychologist wishes to test that the mean reaction time is no more than .085 second, how significant is the observed value of $\bar{x} = .095$ second? What would be your conclusion at level $\alpha = .05$?

23. In Problem 21, if it is desired to check if there is any difference between the mean time devoted to commercials during day and prime time programming, how significant is the observed difference between the sample means, assuming that $\sigma_1 = \sigma_2$? If $\alpha = .10$, is the difference significant?

24. A manufacturer claims that the average mileage of a particular model it makes equals or exceeds 42 miles per gallon. A consumer protection agency conducts a field test with 10 cars of the model, each of which is run on one gallon of gasoline. The following results were obtained:

$$41 \quad 40 \quad 38 \quad 43 \quad 44 \quad 39 \quad 37 \quad 38 \quad 42 \quad 40.$$

Do the data corroborate the manufacturer's claim at level $\alpha = .05$? You may assume that the mileage per gallon for this model has a normal distribution. How significant is the observed sample mean?

25. A filling machine is set to fill 2 kg of sugar (when under control). In order to check that the machine is under control, a random sample of seven bags is selected and their weights recorded:

$$2.20 \quad 2.18 \quad 2.22 \quad 2.05 \quad 2.28 \quad 2.15 \quad 2.14.$$

How strong is the evidence that the machine is out of control?

26. In Problem 11, how strong is the evidence that the administration of the drug reduces the pulse rate by 20 beats per minute or more?

27. In Problem 14, how strong is the evidence that the resistance of wires produced on the two machines are different? What will your conclusion be if $\alpha = .05$?

28. In Problem 15, is there strong evidence to conclude that the diet program results in a weight reduction of 7.5 pounds or more?

29. In Problem 16, is there evidence to suggest that the mean salary of male graduates exceeds that of female graduates? Use $\alpha = .05$.

30. A random sample of 20 American adults showed a mean weight of 162.8 pounds with a standard deviation of 10.4 pounds. A random sample of 16 Australian adults showed a mean weight of 156.5 pounds with a standard deviation of 9.2 pounds. How significant is the difference between the two sample means assuming that $\sigma_1 = \sigma_2$? Is the difference significant at 5 percent level?

31. Brand A 40-watt bulbs sell for 75¢ each and Brand B bulbs sell for 89¢ each. Brand B bulbs claim to have a higher mean length of life than Brand A bulbs. A random sample of 12 Brand A bulbs showed a mean life length of 2010 hours with a standard deviation of 150 hours. A random sample of 15 Brand B bulbs showed a mean life length of 2150 hours with a standard deviation of 165 hours. Do the data support the claim at level $\alpha = .05$? Find bounds for the P-value of the observed difference.

32. In Problem 18, the vending machine is considered to be under control if the standard deviation of fill is 0.40 ounces. How strong is the evidence that the machine is not under control? What will be your conclusion if $\alpha = .05$?

33. A machine is set to produce flathead wood screws with a variance of .0001 (inch)^2. A random sample of seven screws showed a sample standard deviation of .016 inch. Would you conclude that the machine is not under control?

34. A jar of peanut butter is supposed to contain an average of 16 ounces with a standard deviation of 1.5 ounces. A random sample of 16 jars showed a sample standard deviation of 2.8 ounces. Is it reasonable to conclude that $\sigma > 1.5$ ounces?

35. In Problem 14, is it reasonable to conclude that the two machines produce wires whose resistance have equal variance?

36. In Problem 20, is it reasonable to assume that the yield with nitrogen fertilizer has the same variance as that with phosphate fertilizer?

37. In Problem 16, is it reasonable to conclude that the monthly income of male graduates is more variable than that of female graduates?

38. Let $X_1, X_2, \ldots, X_n$ and $Y_1, Y_2, \ldots, Y_n$ be independent random samples from normal populations with means μ_1, μ_2 and variances σ_1^2, σ_2^2 respectively where σ_i^2, $i = 1, 2$, are known. It is desired to test $H_0: \mu_1 = \mu_2$ against $H_1: \mu_1 \neq \mu_2$ at level α such that the probability of accepting H_0 when $|\mu_1 - \mu_2| = \delta$ is β. What should the minimum sample size be? In particular, if $\alpha = \beta = .05$, $\sigma_1 = \sigma_2 = 4$, and $\delta = 1$, find n.

8.8   REVIEW PROBLEMS

1. Suppose X is a random variable with distribution function F. Define:

$$X^+ = X \text{ if } X \geq 0 \qquad \text{and zero otherwise,}$$

and

$$X^- = X \text{ if } X \leq 0 \qquad \text{and zero otherwise.}$$

(i) Show that the distribution function of X^+ is given by:

$$F^+(x) = 0 \text{ if } x < 0, \quad = F(0) \text{ if } x = 0, \quad \text{and } = F(x) \text{ if } x > 0,$$

and that of X^- by

$$F^-(x) = 1 \quad \text{if } x \geq 0, \qquad \text{and } = F(x) \quad \text{if } x < 0.$$

(ii) In particular, if X is a standard normal random variable show that X^+ is a mixed type random variable with distribution given by

$$P(X^+ = 0) = \frac{1}{2}, \qquad f(x) = \frac{1}{\sqrt{2\pi}} e^{-x^2/2} \text{ for } x > 0, \quad \text{and } f(x) = 0 \text{ otherwise.}$$

(iii) If X has a uniform distribution in $[-1, 1]$, show that X^+ has distribution given by

$$P(X^+ = 0) = 1/2, \qquad f(x) = 1/2 \text{ if } 0 < x < 1, \quad \text{and } f(x) = 0 \text{ otherwise.}$$

2. Suppose X has density function $f(x) = 2x/\pi^2$ for $0 < x < \pi$, and zero elsewhere. Show that $Y = \sin X$ has density given by

$$f_Y(y) = \frac{2}{\pi \left(1 - y^2\right)^{1/2}}, \qquad 0 < y < 1, \text{ and zero otherwise.}$$

3. Suppose X has density function f and distribution function F. Let $Y = g(X)$ be defined as follows:
   (i) $g(x) = 1$ if $x > 0$, and $= -1$ if $x \leq 0$,
   (ii) $g(x) = x$ if $|x| \geq c$, and $= 0$ if $|x| < c$, where c is a constant, $c > 0$,
   (iii) $g(x) = c$ if $x \geq c$, $= x$ if $|x| < c$, and $= -c$ if $x \leq -c$, where $c > 0$ is a constant.
   In each case find the distribution of Y.

4. Let X_1, X_2 be a random sample from a standard normal distribution. Let S^2 be the sample variance.
   (i) Show that S^2 has a chi-square distribution with 1 d.f.
   [Hint: $S^2 = (X_1 - X_2)^2/2$. Use Problem 8.6.20 or prove directly that $(X_1 - X_2)/\sqrt{2}$ has a standard normal distribution. Then use Example 8.2.5 or 8.3.3. For the general case, see Section 8.7.]
   (ii) Show that $X_1 + X_2$ and $X_1 - X_2$ are independent. Hence conclude that $\bar{X}$ and S^2 are independent.

5. Suppose X has uniform distribution on $(0, 1)$. Find the function h such $Y = h(X)$ has the distribution given by
   (i) $P(Y = y) = p(1 - p)^{y-1}, y = 1, 2, \ldots$
   (ii) $P(Y = y) = \binom{n}{y} p^y (1 - p)^{n-y}, y = 0, 1, \ldots, n$.
   (iii) $f_Y(y) = 1/y^2, y \geq 1$, and zero elsewhere.
   (iv) $f_Y(y) = 3y^2, 0 < y < 1$, and zero elsewhere.
   (v) $f_Y(y) = y$ if $0 < y < 1$, $= 2 - y$ if $1 \leq y < 2$, and zero otherwise.

6. Suppose X_1, X_2 is a random sample from an exponential distribution with mean λ. Show that $X_1 + X_2$ and X_1/X_2 are independent random variables. Hence conclude that $X_1 + X_2$ and $X_1/(X_1 + X_2)$ are also independent. [Hint: Show that $P(X_1 + X_2 \leq u, X_1/X_2 \leq v) = P(X_1 + X_2 \leq u) \cdot P(X_1/X_2 \leq v)$ for all u, v.]

7. Let X_1 be the total time between a customer's arrival at a service station and his departure from the service desk. Let X_2 be the time spent in line before reaching the service desk. Suppose (X_1, X_2) have a joint density function given by

$$f(x_1, x_2) = (x_1/2)e^{-x_1}, \qquad 0 \leq x_2 \leq x_1 < \infty, \text{ and zero otherwise.}$$

Show that the service time $Y = X_1 - X_2 (\geq 0)$ has density function given by

$$f_Y(y) = \left(\frac{1 + y}{2}\right)e^{-y}, \qquad \text{for } y \geq 0, \text{ and zero elsewhere.}$$

8. Suppose X and Y are independent random variables and $k \geq 0$ is a fixed nonnegative integer. Find the distribution of $Z = X + Y + k$ for the following cases:
   (i) When X is binomial (m, p) and Y is binomial (n, p).
   (ii) When X is Poisson with mean λ and Y is Poisson with mean μ.
   In each case find the mean and the variance of Z.

9. A foot ruler is broken into two pieces at a point chosen at random. Let Z be the ratio of the length of the left piece to the length of the right piece. Show that Z has density function

$$f(z) = 1/(1 + z)^2, \qquad z > 0, \text{ and zero elsewhere.}$$

What is the probability that the length of the longer piece is larger than twice the length of the shorter piece?

10. In Problem 9 find the density of the ratio (length of the longer piece/length of the shorter piece).

11. Let X and Y be tensile strengths of two kinds of nylon fibers and suppose that X and Y are independent with common density function

$$f(x) = (x/\lambda^2)e^{-x/\lambda}, \qquad x > 0, \text{ and } 0 \text{ otherwise.}$$

Find the density function of $Z = X + Y$. (See also Example 3.4.3.)

12. Suppose X, Y, Z, U have joint density function

$$f(x, y, z, u) = \frac{24}{(1 + x + y + z + u)^5}, \qquad x > 0, y > 0, z > 0, u > 0$$

and zero elsewhere. Find the density function of $X + Y + Z + U$. (See also Problem 3.4.15.)

13. Let X_1, X_2 be a random sample from density function f given below. Find the density function of $Z = \min(X, Y)/\max(X, Y)$.
    (i) $f(x) = 6x(1 - x), 0 < x < 1$, and zero otherwise.
    (ii) $f(x) = xe^{-x}, x > 0$, and zero otherwise.

14. Let $X_1, X_2, \ldots, X_n$ be a random sample from the uniform distribution on $(0, 1)$ and let $X_{(1)} < \cdots < X_{(n)}$ be the set of order statistics. Show that

$$\text{var}(X_{(1)}) = \text{var}(X_{(n)}) = \frac{n}{(n + 1)^2(n + 2)} \quad \text{and } \rho(X_{(1)}, X_{(n)}) = \frac{1}{n}.$$

15. Suppose X_1, X_2 is a random sample from a normal distribution with mean μ and variance σ^2.
    (i) Show that

$$\mathscr{E}R = \mathscr{E}(X_{(2)} - X_{(1)}) = \frac{2\sigma}{\sqrt{\pi}}.$$

Thus $\hat{\sigma} = R\sqrt{\pi}/2$ is an unbiased estimate of σ. Find $\text{var}(\hat{\sigma})$.
    (ii) Show that $\mathscr{E}X_{(2)} = \sigma/\sqrt{\pi}$ so that $X_{(2)}\sqrt{\pi}$ is an unbiased estimate of σ.

16*. Suppose X has a normal distribution with mean μ and variance σ^2. Show that $Y = X^2$ has the density function given by

$$f_Y(y) = \frac{1}{\sigma\sqrt{2\pi y}} \cosh\left(\frac{\mu\sqrt{y}}{\sigma^2}\right) \exp\left\{-\frac{1}{2\sigma^2}(y + \mu^2)\right\}, y > 0,$$

and zero otherwise.

17. Suppose $X_1, X_2, \ldots, X_n$ is a random sample from a Poisson distribution with mean λ.
    (i)   Find the conditional distribution of X_k given $\sum_{i=1}^{n} X_i = t$.
    (ii) Show that the joint conditional distribution of $X_1, X_2, \ldots, X_n$ given $X_1 + \cdots + X_n = t$ is multinomial.

18. Suppose X has a Poisson distribution with mean λ and c is a constant. Show that
    (i) $\mathscr{E}\{cX\exp(-cX)\} = \lambda c \exp\{\lambda(e^{-c} - 1) - c\}$.
    (ii) $\mathscr{E}\exp(-cX) = \exp\{-\lambda(1 - e^{-c})\}$.
    (iii) If $X_1, X_2, \ldots, X_n$ are n observations on X, show that $\mathscr{E}\exp(-\bar{X}) = \exp\{-n\lambda(1 - e^{-1/n})\}$.

19. Consider the confidence interval $[X_{(1)}, X_{(n)}]$ for $\zeta_{.25}$ when X_i's have a continuous distribution. Find the minimum confidence level of $[X_{(1)}, X_{(n)}]$ for $n = 2, 3, \ldots, 10$.

20. In Example 8.5.8 the random sample of 10 I.Q. scores was as follows:

$$96, \quad 110, \quad 98, \quad 129, \quad 142, \quad 131, \quad 137, \quad 115, \quad 113, \quad 153$$

    (i)   Find an estimate of the quantile of order .25.
    (ii) Does there exist a confidence interval for $\zeta_{.25}$ at level at least .95?
    (iii) Find a level .90 confidence interval for $\zeta_{.25}$. Does $\zeta_{.25} = 115$ lie in this interval? What do you conclude?
    (iv) Find a level .90 confidence interval for $\zeta_{.75}$. Does the value $\zeta_{.75} = 140$ lie in this confidence interval? What is your conclusion?
    (v)   Test $H_0: \zeta_{.75} = 140$ against $H_1: \zeta_{.75} < 140$ directly.

21. In Problem 8.5.23 the number of miles that each of eight drivers obtained from identical tire sets is:

$$32{,}100, \quad 38{,}000, \quad 46{,}500, \quad 49{,}000, \quad 52{,}700, \quad 39{,}300, \quad 47{,}600, \quad 48{,}500.$$

    (i)   Find a 90 percent confidence interval for the quantile of order .25 if it exists.
    (ii) Find an 80 percent confidence interval for $\zeta_{.25}$.

22. Find the relation between n and p such that $[X_{(1)}, X_{(n)}]$ is a confidence interval for ζ_p at a level at least $1 - p$. If $p = .1$, find the smallest n such that $[X_{(1)}, X_{(n)}]$ is a .9 level confidence interval for p.

23. The survival times (weeks) of a random sample of 12 patients who died of acute myelogenous leukemia are as follows:

$$6, \quad 2, \quad 1, \quad 19, \quad 100, \quad 124, \quad 16, \quad 65, \quad 12, \quad 10, \quad 28, \quad 33$$

    (i)   Find a 98 percent confidence interval for the quantile of order .40.
    (ii) It is claimed that two-thirds of the patients survive 12 weeks or more. Is there evidence to support this claim?

24. (i) Suppose X has a chi-square distribution with $n = 20$ d.f. Find two pairs (a, b) such that $P(a \leq X \leq b) = .9$. (Note that there are infinitely many such pairs.)
    (ii) Suppose we wish to find $a < b$ such that $P(a < X < b) = 1 - \alpha$, $0 < \alpha < 1$, where X has a chi-square distribution with n d.f. The equal tails method is to choose a and b such that $P(X < a) = P(X > b) = \alpha/2$. The shortest length

method is to choose a and b such that the length of (a, b) is least. Show how you will find a and b (numerically) to minimize $b - a$ subject to $P(a < X < b) = 1 - \alpha$.

25*. Suppose T has Student's t-distribution with n d.f. Let $n > 2$. Show that $\mathcal{E}T = 0$ and var$(T) = n/(n - 2)$. [Hint: $T = X/\sqrt{(Y/n)}$ where X has standard normal and Y has $\chi^2(n)$ distribution, and X and Y are independent. Thus $\mathcal{E}T = \mathcal{E}X . \mathcal{E}(\sqrt{(n/Y)})$, $\mathcal{E}T^2 = \mathcal{E}X^2\mathcal{E}(n/Y)$.]

26*. Suppose F has an $F(m, n)$ distribution. Show that $\mathcal{E}F = n/(n - 2)$ for $n > 2$ and var$(F) = n^2(2m + 2n - 4)/[m(n - 2)^2(n - 4)]$ for $n > 4$. (Hint: $\mathcal{E}F = \mathcal{E}(X/m)\mathcal{E}(n/Y)$ where X and Y are independent chi-square random variables, etc.)

27. If T has a t-distribution with n d.f., show that T^2 has an F distribution with $(1, n)$ d.f. (No computations are required.)

28. Let $X_1, X_2, \ldots, X_{12}$ be a random sample from a normal population with mean 0 and variance 9.
(i)   Find $P(\sum_{i=1}^{12} X_i^2 \geq 142.31)$.
(ii) Find $P(\sum_{i=1}^{12}(X_i - \bar{X})^2 \geq 142.31)$.

29. The average birth weight of all infants in the United States is 7 pounds. In order to study the effect of the mother's diet on the birth weight of her infant, a random sample of 12 pregnant mothers was taken and the mothers were put on a diet beginning the 12th week of pregnancy. The sample mean birth weight of their infants was recorded to be 6.4 pounds with a standard deviation of 1.12 pounds. Did the diet have an effect on infants' birth weight?

30. Suppose the systolic blood pressure of all males is normally distributed. A random sample of seven males showed an average systolic blood pressure of 122.7 mm Hg with a standard deviation of 21.8 mm Hg.
(i)   Find a 90 percent confidence interval for the true mean systolic blood pressure.
(ii) Find a 95 percent confidence interval for the standard deviation of systolic blood pressure.
(iii) Is the observed average systolic blood pressure in the sample significantly different from the national average of 120 mm Hg?
(iv) Is the observed sample standard deviation significantly different from the standard deviation $\sigma = 20$ mm Hg for the population of all males?

31. Two groups of experimental animals were run through a maze independently under different conditions. The respective sample means and sample standard deviations of the times it took to run the maze were recorded as follows:

Animals	Sample		
	Size	Mean	Standard Deviation
Group 1	9	11.4	6.2
Group 2	6	8.3	6.6

(i)   Is the difference between the two group means significant?
(ii) Find a 98 percent confidence interval for the difference in means.
(iii) Is the difference between the two standard deviations significant?

32. In order to test the difference in grades in mathematics on a standardized test for two high schools, independent random samples of students were selected from each school and the test was administered with the following results:

High School	Sample		
	Size	Mean	Standard Deviation
A	12	94.1	1.4
B	15	91.8	2.3

(i) Is the difference between the two standard deviations significant?
(ii) Find a 90 percent confidence interval for σ_B/σ_A, the ratio of two standard deviations. Also find a 98 percent confidence interval for σ_B/σ_A.
(iii) Is the difference between the two grades significant?

CHAPTER 9

Large-Sample Theory

9.1 INTRODUCTION

We have seen that in any problem of statistical inference the choice of a statistic, though very important, is only the first step. In actual application of a test of hypothesis (or in the construction of a confidence interval) we also need to compute the P-value (or the numerical limits of the confidence interval). In order to do so we need to know the exact distribution of the test statistic.

Suppose, for example, that we wish to estimate the mean age at which married women have their first child in a large city. A random sample of n such women shows a mean age of 24.8 years. We have seen that $\bar{x} = 24.8$ is a good estimate of the mean age. How good is this estimate? Its precision as measured by its variance is given by σ^2/n where σ^2 can be estimated from the sample. What if we want a 95 percent confidence interval for μ, the unknown mean? We will need to know the distribution of $\bar{X}$. If the underlying distribution of the X_i's is known, we may be able to compute the distribution of $\bar{X}$ by methods of Sections 8.3, 8.4, or 8.6. Given the distribution of $\bar{X}$ it is at least theoretically possible to find a and b such that

$$P(a \leq \bar{X} \leq b) \geq .95.$$

In order to find a confidence interval for μ we then have to be able to invert the inequality $a \leq \bar{X} \leq b$ in the form $l(\bar{X}) \leq \mu \leq u(\bar{X})$.

In practice it may be either difficult or inconvenient to find the distribution of $\bar{X}$. Even if a and b are known it may be inconvenient to invert $a \leq \bar{X} \leq b$ in the form $l(\bar{X}) \leq \mu \leq u(\bar{X})$ to give a $1 - \alpha$ level confidence interval for μ.

We need therefore to investigate some alternative approach to find the numerical limits of the confidence interval. One possible approach is to approximate the distribution of $\bar{X}$. If the approximating distribution is nice enough and the approximation is good enough, then it may be possible to compute a and b with the help of a table of probabilities such as normal or t-distribution.

In this chapter we consider the special case when the sample size n is large. By the law of large numbers we know that $\bar{X}$ is likely to be near μ (that is, consistent for μ) when n is large. We will show that when $\text{var}(X) = \sigma^2$ is finite we can closely approximate the distribution of $\overline{X}$ by a normal distribution with mean μ and variance σ^2/n. This so-called central limit theorem is proved in Section 9.3. It is basic to all large sample theory of statistical inference. In Section 9.4 we improve upon the normal approximation in the special case when the population being sampled is (discrete) integer valued. In Section 9.5 we consider the concept of consistency of an estimate (first introduced in Section 4.3) in some detail. Sections 9.6, 9.7, and 9.8 deal with the estimation and testing of hypothesis for means, proportions and quantiles. In Section 9.9 we return to the problem of testing whether the data can be fit by a specific (known) distribution.

Section 9.2 deals with the general problem of approximating the distribution of a statistic. The reader may skip this section and the proof of the central limit theorem given in Section 9.3, at least on a first reading. Neither is crucial to the understanding of other sections or subsequent chapters. It is important, however, to learn the application of the central limit theorem and its ramifications as explained in Sections 9.3, 9.6, 9.7, and 9.8.

9.2 APPROXIMATING DISTRIBUTIONS: LIMITING MOMENT GENERATING FUNCTION

Approximating the probability of an event is an important part of statistics. Indeed, all the stochastic models studied in this book are approximations in that the assumptions underlying the model may not be strictly correct and yet it may often give reasonably satisfactory representation. In any event, it is a starting point from which departures may be measured.

Yet another type of approximation is needed when we wish to compute probabilities involving a complicated statistic. Even if we know the distribution of the statistic, the computation involved may be tedious. For example, if X has a binomial distribution with parameters n and p, it is often difficult to compute

$$P(X \geq k) = \sum_{j=k}^{n} \binom{n}{j} p^j (1-p)^{n-j}$$

when n is very large, say 500. Similarly, if $X_1, X_2, \ldots, X_n$ are independent round-off errors in n numbers, then we can assume that each X_i has uniform distribution on $(-1/2, 1/2)$. What is the probability that the total error will exceed $n/4$? In this case, the distribution of the total error $S_n = \sum_{k=1}^{n} X_k$ can be computed (although the computation is not easy—see Example 9.3.2), but it is a formidable task to compute $P(S_n > n/4)$ exactly since the density of S_n is quite complicated. Often one simply does not know the exact distribution of the statistic.

Let $X_1, X_2, \ldots, X_n$ be a sample from a distribution function F. Suppose we wish to find $p_n = P(T(X_1, X_2, \ldots, X_n) \in A)$ for some set A on the real line. If this cannot be done exactly and simply for any of the reasons outlined above, we should like to approximate p_n. This can be done if we can approximate the distribution of $T(X_1, X_2,\ldots,X_n)$ by some known and simpler distribution F. In that case we can simply approximate p_n by the probability that F assigns to the set A.

Examples of this kind have already appeared in Chapter 6, where we approximated (under certain conditions) binomial by Poisson, hypergeometric by binomial, and so on.

In simple mathematical terms the problem is as follows: Let $T_n = T(X_1, X_2,\ldots,X_n)$, and $F_n(x) = P(T_n \le x)$ for $n \ge 1$. Does there exist a distribution function F such that $F_n \to F$ as $n \to \infty$? If so, then for (sufficiently) large n we can replace F_n by F. The following property of moment generating functions is frequently useful in finding F the limiting distribution function of F_n.

THEOREM 1. (CONTINUITY THEOREM). *Let T_n be a random variable with moment generating function M_n, $n = 1, 2, \ldots$. Let T be a random variable with moment generating function M and suppose that M_n, M are finite for $|s| < s_0, s_0 > 0$ and all n. If for every $|s| < s_0$*

$$\lim_{n \to \infty} M_n(s) = M(s)$$

then

$$\lim_{n \to \infty} F_n(x) = F(x)$$

for each $x \in \mathbb{R}$ at which F is continuous. (Here F_n is the distribution function of T_n.)

The proof of Theorem 1 is beyond the scope of this text and is omitted. We consider some examples.

Example 1. **Two-Point Distribution.** Let X_n have probability function

$$P(X_n = 1) = \frac{1}{n}, \quad P(X_n = 0) = 1 - \frac{1}{n}, \quad n = 1, 2, \ldots .$$

Then for $s \in \mathbb{R}$

$$M_n(s) = \mathscr{E}\exp(sX_n) = \left(\frac{1}{n}\right)e^s + \left(1 - \frac{1}{n}\right)(1) \to M(s) \equiv 1$$

for all $s \in \mathbb{R}$. Since $M(s) \equiv 1$ is the moment generating function of random variable X degenerate at 0, we conclude from Theorem 1 that

$$P(X_n \le x) \to P(X \le x) \qquad \text{for } x \ne 0.$$

(The point $x = 0$ is *not* a continuity point of $F(x) = P(X \le x) = 0$ if $x < 0$, $= 1$ if $x \ge 0$.) The result also holds at $x = 0$ since $P(X_n \le 0) = 1 - 1/n \to 1 = P(X \le 0)$.

$\square$

Example 2. **Poisson Approximation to Negative Binomial.** Let X have probability function

$$P(X = k + r) = \binom{k + r - 1}{r - 1} p^r (1 - p)^k, \qquad k = 0, 1, 2, \ldots, 0 < p < 1;$$

then X has a negative binomial distribution. The probability $P(X = k + r)$, is the probability of the rth success on the $(k + r)$th trial in a sequence of independent Bernoulli trials with constant probability p of success on each trial. The moment generating function of X is given by

$$M(s) = \frac{p^r}{[1 - (1 - p)e^s]^r},$$

which is finite for $s < -\ln(1 - p)$. Taking logarithms and expanding into a Taylor series, we have for $s < -\ln(1 - p)$:

$$\ln M(s) = r\{\ln p - \ln[1 - (1 - p)e^s]\}$$

$$= r\{\ln(1 - (1 - p)) - \ln(1 - (1 - p)e^s)\}$$

$$= r\left\{-\left[(1 - p) + (1/2)(1 - p)^2 + (1/3)(1 - p)^3 + \cdots\right]\right.$$

$$\left. + \left[(1 - p)e^s + (1/2)(1 - p)^2 e^{2s} + \cdots\right]\right\}$$

$$= r\left\{(1 - p)(e^s - 1) + (1/2)(1 - p)^2(e^{2s} - 1) + \cdots\right\}$$

$$= \left\{r(1 - p)(e^s - 1) + (1/2)r(1 - p)\cdot(1 - p)(e^{2s} - 1) + \cdots\right\}.$$

Let $p \to 1$ and $r \to \infty$ such that $r(1 - p) \to \lambda > 0$. Then

$$\ln M(s) \to \lambda(e^s - 1)$$

and it follows that

$$M(s) \to \exp\{\lambda(e^s - 1)\}.$$

Since $e^{\lambda(e^s - 1)}$ is the moment generating function of the Poisson distribution with mean λ, it follows from Theorem 1 that the negative binomial distribution can be approximated by the Poisson distribution for large r, large p, and $\lambda = r(1 - p)$. $\quad\square$

Example 3. *Normal Approximation to Poisson.* Let X have probability function

$$P(X = k) = \begin{cases} e^{-\lambda}\lambda^k/k!, & k = 0, 1, \dots \\ 0, & \text{elsewhere,} \end{cases}$$

where $\lambda > 0$ and $\mathscr{E}X = \text{var}(X) = \lambda$. Consider the standardized X, defined by

$$Z = \frac{X - \mathscr{E}X}{\sqrt{\text{var } X}} = \frac{X - \lambda}{\sqrt{\lambda}}.$$

We have for $s \in \mathbb{R}$

$$M_Z(s) = \mathscr{E}\exp\left(\frac{s(X - \lambda)}{\sqrt{\lambda}}\right) = e^{-s\sqrt{\lambda}}\exp\{\lambda(e^{s/\sqrt{\lambda}} - 1)\}$$

so that

$$\ln M_Z(s) = -s\sqrt{\lambda} + \lambda\left(\frac{s}{\sqrt{\lambda}} + \frac{1}{2}\frac{s^2}{\lambda} + \frac{1}{3!}\frac{s^3}{\lambda^{3/2}} + \cdots\right)$$

$$= \frac{1}{2}s^2 + \frac{1}{6}\frac{s^3}{\lambda^{1/2}} + \cdots.$$

Letting $\lambda \to \infty$ we see that

$$M_Z(s) \to \exp\left(\frac{s^2}{2}\right) \qquad \text{for } s \in \mathbb{R},$$

and it follows from Theorem 1 that the distribution of Z can be approximated by a standard normal distribution for large λ. That is, for large λ, X has approximately a normal distribution with mean λ and variance λ. $\square$

Example 4.   Borel Weak Law of Large Numbers.   Let $X_1, X_2, \ldots$ be independent and identically distributed Bernoulli random variables with parameter p, $0 < p < 1$. Let $S_n = \sum_{k=1}^n X_k$ and consider the sequence $n^{-1}S_n$. We have

$$M_n(s) = \mathscr{E}\exp(sS_n/n) = \prod_{j=1}^n \mathscr{E}\exp(sX_j/n)$$

$$= \prod_{j=1}^n \{pe^{s/n} + 1 - p\} = (1 - p + pe^{s/n})^n, \qquad s \in \mathbb{R}.$$

Hence

$$\ln M_n(s) = n\ln(1 - p + pe^{s/n})$$

$$= n\ln\{1 + p(e^{s/n} - 1)\}$$

$$= n\left\{p(e^{s/n} - 1) - \frac{1}{2}p^2(e^{s/n} - 1)^2 + \frac{1}{3}p^3(e^{s/n} - 1)^3 + \cdots\right\}$$

$$= n\left\{p\left(\frac{s}{n} + \frac{1}{2}\frac{s^2}{n^2} + \cdots\right) - \frac{1}{2}p^2\left(\frac{s}{n} + \frac{1}{2}\frac{s^2}{n^2} + \cdots\right)^2 + \cdots\right\}$$

$$\to ps$$

as $n \to \infty$ for all $s \in \mathbb{R}$. It follows that $M_n(s) \to \exp(ps)$, which is the moment generating function of the random variable degenerate at p. Hence as $n \to \infty$

$$P(n^{-1}S_n \leq x) \to F(x) = \begin{cases} 0 & x < p \\ 1 & x \geq p, \end{cases}$$

which is Borel's weak law of large numbers obtained in Section 6.3.

We can also prove the weak law of large numbers by the same argument provided we assume that the moment generating function of X_i is finite for $|s| < s_0$, for some $s_0 > 0$. This is a much stronger assumption than that used in Section 5.4.   □

Problems for Section 9.2

1. Let $X_1, X_2, \ldots$ be a sequence of independent Bernoulli random variables. Show that the distribution of S_n can be approximated by Poisson distribution with parameter λ as $n \to \infty$ in such a way that $np \to \lambda$.

2*. Suppose X has a negative binomial distribution with parameters r and p. Show that $2pX$ has a limiting chi-square distribution with $2r$ d.f. as $p \to 0$.

3*. Suppose X_n has a gamma distribution with parameters $\alpha = n$ and β. Find the limiting distribution of X_n/n as $n \to \infty$.

4. Suppose X_n has a normal distribution with mean μ_n and variance σ_n^2. Let $\mu_n \to \mu$ and $\sigma_n \to \sigma$ as $n \to \infty$. What is the limiting distribution of X_n?

5. Let $S_n = \sum_{i=1}^n X_i$, where $X_1, X_2, \ldots, X_n$ is a random sample from a Poisson distribution with mean λ. Show that the sequence $\{(S_n - n\lambda)/\sqrt{n\lambda}\}$ has a limiting standard normal distribution as $n \to \infty$.

6. Let X_n be a geometric random variable with parameter λ/n, $n > \lambda > 0$. Show that $\{X_n/n\}$ has a limiting exponential distribution with mean $1/\lambda$ as $n \to \infty$.

7*. Let X_n have a chi-square distribution with n d.f. Show that as $n \to \infty$ $\{\sqrt{2X_n} - \sqrt{2n-1}\}$ has a limiting standard normal distribution.

8. Suppose $X_1, X_2, \ldots, X_n, \ldots$ is a sequence of independent Bernoulli random variables with parameter p. Show that $S_n = \sum_{k=1}^n X_k$ has an approximate normal distribution with mean np and variance $np(1-p)$ for sufficiently large n. [Hint: Consider $(S_n - np)/\sqrt{np(1-p)}$.]

9. An estimate is required of the difference in proportions of male and female students who have a grade point average of 3.5 or better at a large university. A sample of n students is taken from the population of students who have a grade point average of 3.5 or better. If Y is the number of male students in the sample, then $(2Y/n) - 1$ is a good estimate of $p_1 - p_2$ where p_1 is the true proportion of male students with 3.5 or better grade point average. Show that $(2Y/n) - 1$ has an approximate normal distribution with mean $p_1 - p_2 = 2p_1 - 1$ and variance $4p_1(1 - p_1)/n = 4p_1 p_2/n$ for large n. [Hint: Use Problem 8.]

9.3 THE CENTRAL LIMIT THEOREM OF LÉVY

We now demonstrate the key role played by the normal distribution in probability and statistics. The main result of this section is the central limit theorem due to Lévy (Theorem 1) which says that the distribution of the sum of n independent identically distributed (nondegenerate) random variables with finite variance can be approximated by the normal distribution provided the sample size n is sufficiently large. To motivate the central limit theorem, we first consider some examples.

Example 1.   Sample Size Required to Estimate μ with Prescribed Error.   Let $X_1, X_2, \ldots, X_n$ be a random sample from a population with mean and (finite) variance σ^2. Let $S_n = \sum_{k=1}^{n} X_k$. We know that $\bar{X} = n^{-1} S_n$ is unbiased for μ and has variance σ^2/n. Moreover, the weak law of large numbers tells us that

$$(1) \qquad\qquad P(|\bar{X} - \mu| \le \varepsilon) \to 1 \qquad \text{as } n \to \infty$$

so that for large n, $\bar{X}$ is likely to be close to μ. Hence $\bar{X}$ is a candidate as an estimate for μ. Suppose we estimate μ by $\bar{X}$ and want to ensure that the error $|\bar{X} - \mu|$ in estimation is no more than $\varepsilon > 0$ with a probability of, say, at least $1/2$. How large should the sample size be? In other words, we want to find n such that

$$(2) \qquad\qquad P(|\bar{X} - \mu| \le \varepsilon) \ge \tfrac{1}{2}.$$

In the particular case when the X_i's have a normal distribution, the solution is simple. It was given in Example 8.7.5. We choose n to satisfy

$$(3) \qquad\qquad n = \frac{\sigma^2 (.67)^2}{\varepsilon^2} \qquad \left(z_{\alpha/2} = z_{.25} = .67 \right).$$

In the nonnormal case, however, we can apply the Chebyshev's inequality to choose n. Indeed,

$$(4) \qquad P(|\bar{X} - \mu| \le \varepsilon) = P\left(\frac{|\bar{X} - \mu|}{\sigma/\sqrt{n}} \le \sqrt{n}\,\frac{\varepsilon}{\sigma} \right) \ge 1 - \frac{\sigma^2}{n \varepsilon^2}$$

(with $k = \varepsilon \sqrt{n}/\sigma$); on comparing (2) and (4) we see that

$$(5) \qquad\qquad \frac{1}{2} = \frac{\sigma^2}{n \varepsilon^2} \quad \text{or} \quad n = \frac{2\sigma^2}{\varepsilon^2}.$$

It is clear that n given by (5) is much larger than that given by (3). Is (5) the best we can do in the nonnormal case? We will see that even in this case n given by (3) suffices.   □

Example 2*.   Magnitude of Round-off Error.   In Example 7.2.1 we saw that the uniform distribution may be used to model round-off error. Suppose n numbers are rounded off to the nearest integer. Let $X_1, X_2, \ldots, X_n$ be the round-off errors. Then $X_1, X_2, \ldots, X_n$ are independent identically distributed random variables with uniform distribution on $[-1/2, 1/2]$. The total round-off error is given by $S_n = \sum_{k=1}^{n} X_k$. What is the probability that the round-off error is more than $n/4$? That is, what is $P(S_n > n/4)$? If f_n is the density of S_n, then

$$P(S_n > n/4) = \int_{n/4}^{\infty} f_n(x)\, dx.$$

Let us compute f_n. Since $S_n = X_n + S_{n-1}$, $n \geq 2$, we have from (8.4.5)

(6) $f_n(x) = \int_{-\infty}^{\infty} f_{n-1}(x - y) f_1(y)\, dy$

$$= \int_{-1/2}^{1/2} f_{n-1}(x - y)\, dy = \int_{x-1/2}^{x+1/2} f_{n-1}(u)\, du \qquad (u = x - y)$$

for $x \in [-n/2, n/2]$ where we have used the fact that $f_1(y) = 1$ for $y \in [-1/2, 1/2]$ and $f_1(y) = 0$ otherwise. Let us write

$$\delta(x) = \begin{cases} 0, & x \leq 0 \\ x, & x \geq 0. \end{cases}$$

Then we can rewrite

(7) $f_1(x) = [\delta(-x + 1/2)]^0 - [\delta(-x - 1/2)]^0$

using the convention $0^0 = 0$. Using (6) with $n = 2$ we have (see Figure 1)

$$f_2(x) = \int_{x-1/2}^{x+1/2} f_1(u)\, du = \begin{cases} \int_{-1/2}^{x+1/2} 1\, du = 1 + x, & -1 \leq x \leq 0 \\ \int_{x-1/2}^{1/2} 1\, du = 1 - x, & 0 \leq x \leq 1 \\ 0, & \text{elsewhere.} \end{cases}$$

We can rewrite f_2 in terms of δ as follows

(8) $f_2(x) = \delta(-1 - x) - 2\delta(-x) + \delta(1 - x).$

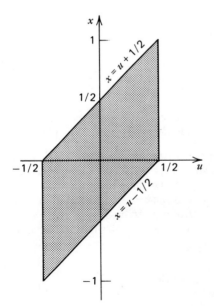

Figure 1

We show inductively that

(9) $$f_n(x) = \frac{1}{(n-1)!} \sum_{j=0}^{n} (-1)^{n-j} \binom{n}{j} \left[\delta\left(\frac{2j-n}{2} - x\right) \right]^{n-1}.$$

For this purpose we need the following integration formula.

(10) $$\int_a^b [\delta(c-x)]^{k-1}\, dx = \frac{1}{k} \left\{ [\delta(c-a)]^k - [\delta(c-b)]^k \right\}.$$

The proof of (10) is straightforward. [We can use (10) and (7) in (6) to get an alternative derivation of (9).]

   In view of (7) and (8) we see that (9) holds for $m = 1, 2$. Assume that (9) holds for $1 \le m \le n - 1$; we show that it also holds for $m = n$. In fact, by inductive hypothesis and (10):

$$f_n(x) = \int_{x-1/2}^{x+1/2} f_{n-1}(u)\, du$$

$$= \int_{x-1/2}^{x+1/2} \frac{1}{(n-2)!} \sum_{j=0}^{n-1} (-1)^{n-1-j} \binom{n-1}{j} \left[\delta\left(\frac{2j-n+1}{2} - x\right) \right]^{n-2}$$

$$= \frac{1}{(n-2)!} \sum_{j=0}^{n-1} (-1)^{n-1-j} \binom{n-1}{j} \left\{ \frac{1}{n-1} \left(\left[\delta\left(\frac{2j-n+1}{2} - x + \frac{1}{2}\right) \right]^{n-1} \right. \right.$$

$$\left. \left. - \left[\delta\left(\frac{2j-n+1}{2} - x - \frac{1}{2}\right) \right]^{n-1} \right) \right\}$$

$$= \frac{1}{(n-1)!} \sum_{j=0}^{n-1} (-1)^{n-1-j} \left\{ \binom{n-1}{j} \left[\delta\left(\frac{2j-n+2}{2} - x\right) \right]^{n-1} \right.$$

$$\left. - \binom{n-1}{j} \left[\delta\left(\frac{2j-n}{2} - x\right) \right]^{n-1} \right\}$$

$$= \frac{1}{(n-1)!} \left\{ \sum_{j=1}^{n} (-1)^{n-j} \binom{n-1}{j-1} \left[\delta\left(\frac{2j-n}{2} - x\right) \right]^{n-1} \right.$$

$$\left. - \sum_{j=0}^{n-1} (-1)^{n-1-j} \binom{n-1}{j} \left[\delta\left(\frac{2j-n}{2} - x\right) \right]^{n-1} \right\}$$

$$= \frac{1}{(n-1)!} \sum_{j=0}^{n} (-1)^{n-j} \left[\delta\left(\frac{2j-n}{2} - x\right) \right]^{n-1} \left\{ \binom{n-1}{j-1} + \binom{n-1}{j} \right\}$$

$$= \frac{1}{(n-1)!} \sum_{j=0}^{n} (-1)^{n-j} \binom{n}{j} \left[\delta\left(\frac{2j-n}{2} - x\right) \right]^{n-1}$$

where we interpret $\binom{n}{r} = 0$ for $r < 0$ or $r > n$ and have used the simple fact that

$$\binom{n-1}{k-1} + \binom{n-1}{k} = \binom{n}{k}.$$

In Figure 2 we have plotted f_n for $n = 1, 2, 3, 4$. We observe that f_3 and f_4 can hardly be distinguished from the normal density with a naked eye.

Returning to the computation of $P(S_n > n/4)$, we note that even though we have been able to come up with an explicit expression for f_n (with a lot of work), it is practically useless in computing $P(S_n > n/4)$. Shapes of f_3 and f_4 suggest that we may be able to approximate $P(S_n > n/4)$ by using normal distribution. $\square$

Example 2 is typical in that in many applications we need to compute probabilities under the distribution of S_n or $\overline{X} = n^{-1}S_n$. It suggests that these probabilities may be approximated by normal distribution.

In Figure 3 we graph the distribution of $\overline{X}$ for several random variables for selected values of n. In each case we have also superimposed a graph of the

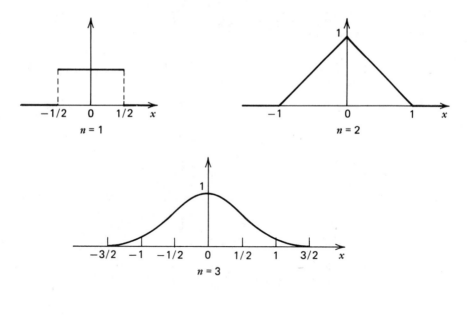

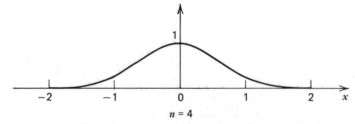

Figure 2.   Graphs of f_n [given by (9)] for $n = 1, 2, 3, 4$.

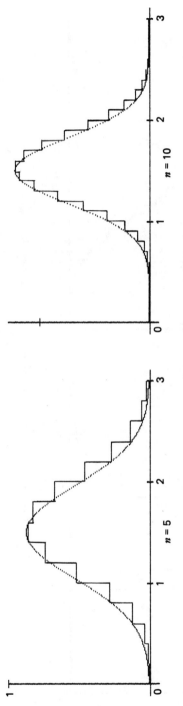

Figure 3a. Distribution of $\bar{X}$ for binomial $(5, p = .3)$ random variables.

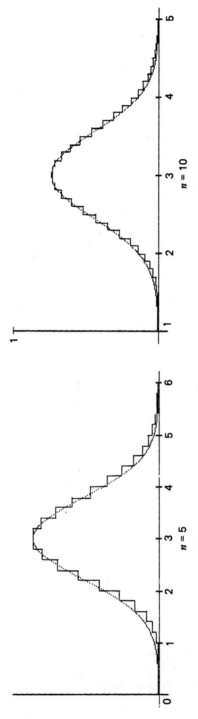

Figure 3b. Distribution of $\bar{X}$ for Poisson random variables with mean 3.

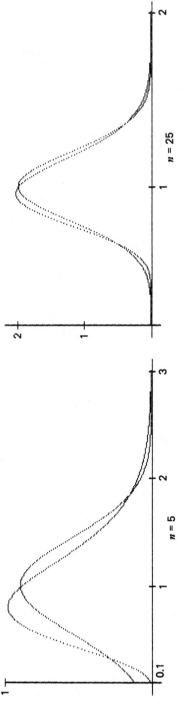

Figure 3c. Distribution of $\overline{X}$ for exponential random variables with mean 1.

normal curve with the same mean ($= \mu$) and same variance ($= \sigma^2/n$). In each case the two curves are rather similar for larger values of n, and agreement improves as n gets larger.

Let us now explore the possibility of approximating the distribution of $\overline{X}$ by normal distribution. Note that if $P(X_i > 0) = 1$, then $\mathscr{E}X_i = \mu > 0$, $P(S_n > 0) = 1$ so that $P(S_n \leq x) = 0$ for $x \leq 0$. Moreover, $P(|\overline{X} - \mu| > \varepsilon) \to 0$ so that for large n, S_n "behaves like" $n\mu$. In particular, $S_n \to \infty$.[†] We cannot, therefore, expect $P(S_n \leq x)$ to converge to the normal distribution (see Section 9.2). The case when X_i's have a normal distribution with mean μ and variance σ^2 suggests that we need to "adjust" S_n suitably. In Section 8.7 we saw that $Z = [\sqrt{n}(\overline{X} - \mu)]/\sigma = (S_n - n\mu)/\sqrt{n}\,\sigma$ has an exact normal distribution. This is simply the standardized $\overline{X}$ or S_n.

Following the normal case, let us consider the standardized sums

$$Z_n = \frac{S_n - \mathscr{E}S_n}{\sqrt{\mathrm{var}(S_n)}} = \frac{S_n - n\mu}{\sigma\sqrt{n}}.$$

Clearly $\mathscr{E}Z_n = 0$ and $\mathrm{var}(Z_n) = 1$. We show that $P(Z_n \leq x) \to \int_{-\infty}^{x} (e^{-t^2/2}/\sqrt{2\pi})\, dt$ as $n \to \infty$. That is, we can approximate the distribution of Z_n by standard normal distribution.

THEOREM 1. (LÉVY'S CENTRAL LIMIT THEOREM). *Let $\{X_n\}$ be a sequence of nondegenerate independent identically distributed random variables with common mean μ and common variance σ^2, $0 < \sigma^2 < \infty$. Let $S_n = \sum_{j=1}^{n} X_j$, $n = 1, 2, \ldots$, and $Z_n = (\sigma\sqrt{n})^{-1}(S_n - n\mu)$. Then for all $z \in \mathbb{R}$*

(11)
$$\lim_{n \to \infty} P(Z_n \leq z) = (1/\sqrt{2\pi})\exp(-z^2/2).$$

Proof.[*] A complete proof of Theorem 1 is beyond the scope of this text. We restrict ourselves to the case when X_n's have a moment generating function $M(s)$ which is finite in $|s| < s_0$ for some $s_0 > 0$. (This, in particular, implies that $\mathscr{E}|X_j|^k < \infty$ for all k, which is considerably more than what we are given in the statement of the theorem.) For ease in writing and for computational convenience, we assume without loss of generality[‡] that $\mu = 0$ and $\sigma = 1$. Then

$$M_n(s) = \mathscr{E}\exp(sZ_n) = \mathscr{E}\exp(sn^{-1/2}S_n)$$

$$= \prod_{j=1}^{n} \exp(sX_j/\sqrt{n}) = [M(s/\sqrt{n})]^n.$$

We need to show (in view of the continuity theorem of Section 9.2) that

$$M_n(s) \to \exp(s^2/2) \qquad \text{as } n \to \infty.$$

[†]Since S_n is a random variable, we mean $P(S_n \to \infty$ as $n \to \infty) = 1$.
[‡]For the general case when $\mu \neq 0$, $\sigma \neq 1$ define $Y_i = (X_i - \mu)/\sigma$, $i = 1, 2, \ldots, n$. Then $\mathscr{E}Y_i = 0$ and $\mathrm{var}(Y_i) = 1$.

Taking logarithms, we have

$$\ln M_n(s) = n \ln M(s/\sqrt{n}).$$

Set $\phi(x) = \ln M(x)$. Then

$$\ln M_n(s) = n\phi(s/\sqrt{n}) = \phi(s/\sqrt{n})/(1/n).$$

We note that $\phi(0) = \ln M(0) = \ln 1 = 0$. Clearly ϕ is differentiable (Section 8.6). By L'Hôpital's rule

$$\lim_{n \to \infty} \ln M_n(s) = \lim_{n \to \infty} \frac{\phi(s/\sqrt{n})}{1/n} = \lim_{n \to \infty} \frac{\phi'(s/\sqrt{n})\left(-\dfrac{s/2}{n^{3/2}}\right)}{(-1/n^2)}$$

$$= \lim_{n \to \infty} (1/2) \frac{s\phi'(s/\sqrt{n})}{1/\sqrt{n}}.$$

Again $\phi'(0) = \mathscr{E} X_i = \mu = 0$ (by (8.6.18)) so that the conditions for L'Hôpital's rule are again satisfied. We have

$$\lim_{n \to \infty} \ln M_n(s) = (1/2) \lim_{n \to \infty} \frac{s^2\phi''(s/\sqrt{n})(-1/2n^{3/2})}{(-1/2n^{3/2})}$$

$$= (s^2/2) \lim_{n \to \infty} \phi''(s/\sqrt{n})$$

$$= (s^2/2)\phi''(0)$$

by continuity of ϕ''. In view of (8.6.18), $\phi''(0) = \text{var}(X_i) = \sigma^2 = 1$ and it follows that

$$M_n(s) \to \exp(s^2/2) \qquad \text{as } n \to \infty.$$

Since $\exp(s^2/2)$ is the moment generating function of a standard normal random variable, it follows from the continuity theorem (Theorem 9.2.1) that

$$P(Z_n \le z) \to (1/\sqrt{2\pi}) \int_{-\infty}^{z} e^{-x^2/2} \, dx$$

which is (11). $\qquad\qquad\qquad\qquad\qquad\qquad\qquad\qquad\qquad\qquad\qquad\qquad\qquad\qquad\square$

Remark 1. Theorem 1 is a remarkable result. It says that the limiting distribution of the standardized partial sums is standard normal irrespective of the distribution of the random variables being summed provided the variance is finite.

Remark 2. We can write (11) in terms of $\overline{X}$. Indeed,

$$P(Z_n \le x) = P\left(\frac{\overline{X} - \mu}{\sigma/\sqrt{n}} \le x\right) \to \Phi(x).$$

where

$$\Phi(x) = \left(1/\sqrt{2\pi}\right)\int_{-\infty}^{x} e^{-t^2/2}\, dt.$$

Remark 3.   According to (11) for large n,

$$P(a \le S_n \le b) = P\left(\frac{a - n\mu}{\sigma\sqrt{n}} \le Z_n \le \frac{b - n\mu}{\sigma\sqrt{n}}\right)$$

$$\doteq \Phi\left(\frac{b - n\mu}{\sigma\sqrt{n}}\right) - \Phi\left(\frac{a - n\mu}{\sigma\sqrt{n}}\right)$$

$$= \left(1/\sigma\sqrt{2\pi n}\right)\int_{a}^{b}\exp\left\{-(y - n\mu)^2/(2n\sigma^2)\right\}\, dy.$$

That is S_n is *approximately* (or asymptotically) normally distributed with mean $n\mu$ and variance $n\sigma^2$. Similarly, $\overline{X}$ is approximately normally distributed with mean μ and variance σ^2/n. Often we will say that X_n is asymptotically normal to mean that, when standardized, X_n has a limiting standard normal distribution.

Remark 4.   How large must n be before we can apply the approximation? What is the error in this approximation? Both these questions concern the speed of convergence of $P(Z_n \le x)$ to $\Phi(x)$. That is, the rate of convergence of $|P((S_n - n\mu)/\sigma\sqrt{n} \le x) - \Phi(x)|$ to zero. Clearly, the rate of convergence should depend on the distribution of the X_i's. If X_i's have a symmetric distribution to begin with, one would expect the approximation to be good even for moderately large n. The following estimate of error in normal approximation appears to be the best result known so far. If $\mathscr{E}|(X_i - \mu)/\sigma|^3 < \infty$ then

$$(12) \qquad \left|P\left(\frac{S_n - n\mu}{\sigma\sqrt{n}} \le x\right) - \Phi(x)\right| = \left|P\left(\frac{\overline{X} - \mu}{\sigma}\sqrt{n} \le x\right) - \Phi(x)\right|$$

$$\le .7975\,\mathscr{E}|X_i - \mu|^3/\sigma^3.$$

Table 1 gives the error bound for some common distributions.

It is wise to use caution in using normal approximation. Although the error estimate in (12) can frequently be used to show that the error in approximation is not too large, it should be emphasized that this estimate gives a bound for *any* distribution F for which μ, σ and $\mathscr{E}|X_k - \mu|^3$ is all that is known. Needless to say that (12) can be improved in special cases when F is completely specified.

Remark 5.   The weak law of large numbers tells us that $\overline{X}$ converges to μ (in the sense that $P(|\overline{X} - \mu| > \varepsilon) \to 0$ as $n \to \infty$). By Chebyshev's inequality

$$P(|\overline{X} - \mu| > \varepsilon) \le \sigma^2/(n\varepsilon^2)$$

so that the rate of convergence is that of n^{-1}. The central limit theorem, on the other hand, gives a more precise estimate of the rate of convergence. Indeed, for large n

$$
(13) \qquad P\big(|\overline{X} - \mu| > \varepsilon\big) = P\bigg(|\overline{X} - \mu|\frac{\sqrt{n}}{\sigma} > \frac{\varepsilon\sqrt{n}}{\sigma}\bigg)
$$

$$
\doteq P\bigg(|Z| > \frac{\varepsilon\sqrt{n}}{\sigma}\bigg) = \frac{2}{\sqrt{2\pi}}\int_{\varepsilon\sqrt{n}/\sigma}^{\infty} e^{-x^2/2}\, dx
$$

$$
\doteq \frac{2\sigma}{\sqrt{2\pi}\,\varepsilon\sqrt{n}}\exp\bigg(-\frac{\varepsilon^2 n}{2\sigma^2}\bigg)
$$

in view of Problem 7.6.24.

Remark 6.   According to Chebyshev's inequality

$$
(14) \qquad P\bigg(\frac{|\overline{X} - \mu|}{\sigma}\sqrt{n} \le k\bigg) \ge 1 - \frac{1}{k^2} \qquad \text{for } k > 1
$$

whereas according to central limit theorem

$$
(15) \qquad P\bigg(\frac{|\overline{X} - \mu|}{\sigma}\sqrt{n} \le k\bigg) \doteq \int_{-k}^{k}\frac{1}{\sqrt{2\pi}}e^{-x^2/2}\, dx
$$

for large n, which may be thought of as a refinement of (14) in that (15) provides a much more precise estimate. (See Table 7.6.1 for a comparison.)

Example 3.   Breakdown of a Machine: Normal Approximation to Gamma Distribution.   The times between breakdowns of a machine can be modeled as an exponential random variable with mean λ with density

$$
f(x) = \begin{cases} \lambda^{-1}e^{-x/\lambda}, & x > 0 \\ 0, & x \le 0. \end{cases}
$$

TABLE 1.   Error Bound for Some Common Distributions

Distribution	Error Bound
Bernoulli, p	$.7975[1 - 2p(1 - p)]/\sqrt{np(1 - p)}$
Poisson, λ	$.7975(8\lambda^2 + 6\lambda + 1)/\sqrt{n\lambda}$
Uniform on $(-1/2, 1/2)$	$1.036/\sqrt{n}$
Exponential (mean λ)	$.1653/\sqrt{n}$

Let $X_1, X_2, \ldots, X_n$ be a random sample from f so that X_1 is the waiting time for the first breakdown, X_2 for the second breakdown, and so on. Then $S_n = \sum_{k=1}^{n} X_k$ is the total time to the nth breakdown. What is $P(S_n > t)$, the probability that the nth breakdown does not occur before time t? We recall (Problem 8.6.19) that S_n has a gamma distribution with parameters $\alpha = n$ and $\beta = \lambda$. Hence

$$P(S_n > t) = \int_t^\infty \frac{1}{\Gamma(n)\lambda^n} x^{n-1} e^{-x/\lambda} \, dx$$

and in view of Example 7.4.4 we have

$$P(S_n > t) = \sum_{j=0}^{n-1} e^{-t/\lambda} \frac{(t/\lambda)^j}{j!}.$$

Given n, t, and λ we can compute this probability by using a table of Poisson probabilities.

Note that $\mathscr{E} X_1 = \lambda$ and $\text{var}(X_1) = \lambda^2$ so that for large n we can apply the central limit theorem. We have

$$P(S_n > t) = P\left(\frac{S_n - n\lambda}{\sqrt{n}\,\lambda} > \frac{t - n\lambda}{\sqrt{n}\,\lambda} \right)$$

$$\doteq P\big(Z > (t - n\lambda)/\lambda\sqrt{n} \big)$$

which can be read off from a table of normal distribution.

In particular let $\lambda = 2$ months, $n = 30$, and $t = 36$. Then the exact probability from a table of cumulative Poisson probabilities with $\lambda = 18$ is given by

$$P(S_n > 36) = \sum_{j=0}^{29} e^{-18} \frac{(18)^j}{j!} = .9941.$$

Normal approximation gives

$$P(S_n > 36) \doteq P\left(Z > \frac{36 - 60}{2\sqrt{30}} \right) = P(Z > -2.19)$$

$$= .9857$$

with error $|.9941 - .9857| = .0016$, which is very small even though n is not very large. (The error bound from Table 1 gives .0301 but the actual error .0016 is much less.)

How large should n be such that

$$P(S_n > t) = 1 - \alpha?$$

From normal approximation

$$P(S_n > t) \doteq P\left(Z > \frac{t - n\lambda}{\lambda\sqrt{n}} \right) = 1 - \alpha$$

so that

$$\frac{t - n\lambda}{\lambda\sqrt{n}} = z_{1-\alpha}.$$

Solving for n, we get (choosing the larger root):

$$n = \frac{2\lambda t + \lambda^2 z_{1-\alpha}^2 + \sqrt{4t\lambda^3 z_{1-\alpha}^2 + \lambda^4 z_{1-\alpha}^4}}{2\lambda^2}.$$

In the particular case when $\lambda = 1$, $t = 36$, and $\alpha = .95$, $z_{.05} = 1.645$ and we have

$$n = 47.3 \quad \text{or} \quad n = 48.$$

The exact probability for $n = 48$ is given by $P(S_n > 36) = .968$ so that the approximation is quite satisfactory. $\qquad\qquad\qquad\qquad\qquad\qquad\qquad\qquad\qquad\qquad\qquad\square$

Example 4.   Sample Size Determination (Example 1).   In Example 1 we wanted to choose n such that the estimate $\overline{X}$ of μ has a maximum error of ε with probability $\geq 1/2$. Thus we choose n to satisfy

$$P(|\overline{X} - \mu| \leq \varepsilon) \geq \tfrac{1}{2}.$$

By central limit theorem

$$P(|\overline{X} - \mu| \leq \varepsilon) \doteq P\left(|Z| \leq \frac{\varepsilon\sqrt{n}}{\sigma}\right) \geq \frac{1}{2}$$

so that from Table A5 we choose n satisfying

$$.67 = \frac{\varepsilon\sqrt{n}}{\sigma} \quad \text{or} \quad n = \frac{\sigma^2(.67)^2}{\varepsilon^2}$$

which is much smaller than the one suggested by Chebyshev's inequality. In particular, let X_i be Bernoulli random variables with $p = .30$ and let $\varepsilon = .05$. In that case Chebyshev's inequality requires that we choose n according to (5), that is,

$$n = \frac{2p(1-p)}{\varepsilon^2} = \frac{2(.3).7}{(.05)^2} = 168.$$

The normal approximation requires only that we choose

$$n = \frac{(.67)^2(.3).7}{(.05)^2} = 37.7 \quad \text{or} \quad n = 38$$

which is considerably smaller. For comparison purposes let us compute the exact value of n from binomial tables. We have

$$P(|\overline{X} - \mu| \le \varepsilon) = P(|S_n - np| \le n\varepsilon)$$

$$= P(.3n - .05n \le S_n \le .3n + .05n)$$

$$= P(.25n \le S_n \le .35n).$$

Here S_n has a binomial distribution with parameters n (unknown) and $p = .3$. By trial and error and using a more comprehensive binomial table than Table A3 (such as the Harvard University Table), we get $n = 32$. $\square$

Example 5.   Round-Off Error (Example 2).   We return to the problem of Example 2 where we wish to compute the probability that total round-off error in rounding off n numbers exceeds $n/4$. Since each X_i is uniform on $[-1/2, 1/2]$, $\mu = \mathscr{E}X_i = 0$ and $\sigma^2 = \text{var}(X_i) = 1/12$. By central limit theorem S_n is asymptotically normally distributed with mean 0 and variance $n/12$. Consequently,

$$P\left(S_n > \frac{n}{4}\right) = P\left(\frac{S_n - n\mu}{\sigma\sqrt{n}} > \frac{n/4}{(1/\sqrt{12})\sqrt{n}}\right)$$

$$\doteq P\left(Z > \frac{\sqrt{3n}}{2}\right)$$

which can be computed once n is given.

Suppose $n = 1000$ and we want $P(S_n > 25)$. Then

$$P(S_n > 25) \doteq P\left(Z > \frac{25\sqrt{12}}{\sqrt{1000}}\right) = P(Z > 2.7386)$$

$$= .0031.$$

In this case we cannot compare our result with the exact probability since we cannot integrate f_{1000} as given in (9). Table 1 gives the error in normal approximation to be at most $1.036/\sqrt{n} = 1.036/\sqrt{1000} = .0328$, quite large in relation to the magnitude of the probability being estimated. The actual error in normal approximation is in fact much smaller in this case, but we do not propose to investigate it any further. $\square$

Example 6.   Conviction Rate.   In a very large city it is known that about 0.1 percent of the population has been convicted of a crime. What is the probability that in a random sample of 100,000 from this population, at most 120 have been convicted of a crime?

Let X be the number in the sample who have been convicted of a crime. Then X has a binomial distribution with $n = 100,000$ and $p = .001$ so that

$$P(X \le 120) = \sum_{k=0}^{120} \binom{100,000}{k}(.001)^k(.999)^{100,000-k}.$$

We consider X as the sum of $n = 100{,}000$ independent identically distributed random variables X_i, each with parameter $p = .001$. Then

$$\mathscr{E} X = n \mathscr{E} X_i = 100{,}000(.001) = 100$$

and

$$\mathrm{var}(X) = n \, \mathrm{var}(X_i) = np(1 - p) = 99.9.$$

Since $\mathscr{E} X = 100$ is large, it is inconvenient to apply the Poisson approximation (6.6.3). By the central limit theorem, however,

$$P(X \le 120) = P\left(\frac{X - np}{\sqrt{np(1 - p)}} \le \frac{120 - 100}{\sqrt{99.99}}\right) \doteq P(Z \le 2.001)$$

$$= .9772.$$

The error bound from Table 1 is

$$\frac{.7975[1 - 2p(1 - p)]}{\sqrt{np(1 - p)}} = .0796$$

which is not very precise. What if we have to estimate a small probability, such as $P(X = 120)$? First note that we cannot apply the central limit approximation directly to $P(X = 120)$ since it will lead to the approximation $P(X = 120) \doteq 0$ (Why?). Let us write

$$P(X = 120) = P(X \le 120) - P(X \le 119).$$

Then

$$P(X = 120) \doteq P(Z \le -2.001) - P(Z \le 1.901)$$

$$= .9772 - .9713 = .0059$$

Now the error bound is practically meaningless.   □

In Section 9.4 we consider some further approximations and also a correction that leads to an improvement in normal approximation when applied in the special case when the X_i's are integer valued.

The central limit theorem holds also in the case when the X_i's are independent but not necessarily identically distributed. Extensions to random vectors is also known, but we will not pursue the subject any further. We conclude this section with the following simple application of the central limit theorem.

Example 7. Normal Approximation to Poisson and an Application. Let $X_1, X_2, \ldots, X_n$ be a random sample from a Poisson distribution with mean λ. Then $S_n = \sum_{i=1}^{n} X_i$ has a Poisson distribution with parameter $n\lambda$. Since $\mathscr{E} S_n = \mathrm{var}(S_n) = n\lambda$ it follows that

$$P\left(\frac{S_n - n\lambda}{\sqrt{n\lambda}} \le x\right) \to \int_{-\infty}^{x} \frac{e^{-t^2/2}}{\sqrt{2\pi}} \, dt.$$

In particular, let $\lambda = 1$ and $x = 0$. Then

$$P(S_n \leq n) \to \int_{-\infty}^{0} \frac{e^{-t^2/2} \, dt}{\sqrt{2\pi}} = \frac{1}{2}.$$

Since S_n has a Poisson distribution with parameter n, however,

$$P(S_n \leq n) = e^{-n} \sum_{k=0}^{n} \frac{n^k}{k!}.$$

It follows that

$$\lim_{n \to \infty} e^{-n} \sum_{k=0}^{n} \frac{n^k}{k!} = \frac{1}{2}. \qquad \square$$

Problems for Section 9.3

1. Check to see if the sufficient conditions for the central limit theorem hold in sampling from the following distributions.
   (i) $f(x) = 1/x^2$, $x \geq 1$, and zero elsewhere.
   (ii) $f(x) = x^{\alpha-1}(1 - x)^{\beta-1}/B(\alpha, \beta)$, $0 < x < 1$, and zero elsewhere.
   (iii) $f(x) = 1/[\pi(1 + x^2)]$, $x \in \mathbb{R}$.
   (iv) $P(X = x) = \binom{k-1}{r-1} p^r (1 - p)^{k-r}$, $k = r, r + 1, \dots$.
   (v) $f(x) = (1/x\sqrt{2\pi})\exp\{-(\ln x - \mu)^2/2\}$, $x > 0$, and zero elsewhere.
   (vi)* $f(x) = c\exp\{-|x|^\alpha\}$, $x \in \mathbb{R}$, $\alpha \in (0, 1)$.

2. Customers arrive at a store at the Poisson rate of one per every five minute interval. What is the probability that the number of arrivals during a 12-hour period is between 114 and 176?

3. Interest is calculated at a bank on randomly selected 100 accounts and rounded up or down to the nearest dollar. Assuming that the error in round-off has uniform distribution on $(-1/2, 1/2)$, find the probability that the total error is at most $2.

4. A new drug is claimed to be 80 percent effective in curing a certain disease. A random sample of 400 people with this disease is given the drug and it is found that 306 were cured. Is there sufficient evidence to conclude that the drug is effective?

5. In order to estimate the mean duration of telephone conversations on residential phones, a telephone company monitored the durations of a random sample of 100 phone conversations. A preliminary study indicated that the mean duration of such conversations is 3 minutes with a standard deviation of 3.5 minutes. What is the probability that the sample mean is
   (i) Larger than 4 minutes?
   (ii) Between 2 and 4 minutes?
   (iii) Smaller than 3.5 minutes?

6. A psychologist wishes to estimate the mean I.Q. of a population of students at a large high school. She takes a sample of n students and would like to use the sample

mean $\overline{X}$ as an estimate of μ. If she wants to ensure that the error in her estimation is at most 2 with probability .95, how large a sample need she take? It is known that $\sigma = 14$. What if $\sigma = 28$?

7. An economist would like to estimate the mean income (μ) in a rich suburb of a large eastern city. She decides to use the sample mean as an estimate of μ and would like to ensure that the error in estimation is no more than $400 with probability .90. How large a sample should she choose if the standard deviation is known to be $4000?

8. What is the probability that the sample mean living area in a sample of 49 houses in a city is > 1900 square feet when $\mu = 1800$ square feet and $\sigma = 300$ square feet?

9. In a large automobile plant the average hourly wage is $10.00, with a standard deviation of $2.50. A women's group at this plant sampled 36 female workers and found their average wage to be $8.50 per hour. Is there evidence to suggest that the average hourly salary of female workers is lower than the plant average?

10. A food processing plant packages instant coffee in eight-ounce jars. In order to see if the filling process is operating properly, it is decided to check a sample of 100 jars every two hours. If the sample mean is smaller than a critical value c, the process is stopped, otherwise filling is continued. What should the value of c be in order that there is only 2 percent chance of stopping the filling process which has been set to produce a mean of 8.10 ounces with a standard deviation of .05 ounce?

11. Cigarette manufacturers frequently make claims about the average amount of tar per cigarette that their brand contains. Suppose a manufacturer of low-tar cigarettes claims that its cigarettes contain an average of 16.3 milligrams of tar per cigarette with a standard deviation of 3.1 milligrams per cigarette. What is the probability that in a sample of 225 cigarettes, the sample mean (of tar per cigarette in milligrams) is larger than 16.8? If the sample mean is in fact $\overline{x} = 16.8$ what would you conclude?

12. An automobile manufacturer claims that its XYZ compact models give on the average 54 miles per gallon on the highway, with a standard deviation of 5 miles per gallon. A random sample of 100 such cars were driven under identical conditions on open road. What is the probability that the sample mean will be between 53 and 56 miles per gallon? If $\overline{x} = 52.5$, what would you conclude?

13. All weighing machines have errors in their calibrations. In order to find the true weight μ of an object, it is weighed 64 times on a given set of scales. Find the probability that the sample average will be within .1 pound of the true weight μ? Assume that $\sigma = 1$. If an estimate of μ to within .1 pound with 95 percent probability is required, how large a sample size n must be chosen?

14. A psychologist would like to find the average time required for a two-year-old to complete a simple maze. He knows from experience that about 2.5 percent samples of size 36 show a sample mean larger than 3.65 minutes, and 5 percent samples show the sample mean to be smaller than 3.35 minutes. What are μ and σ?

15. To estimate the average amount owed by a customer, a department store with a large number of charge accounts selects a random sample of 100 accounts. If it wants to be 90 percent sure that the true mean is within $10 of the sample mean of 100 accounts, what should the standard deviation σ be?

16. A system consists of n components $C_1, \ldots, C_n$ where only one component is necessary for the system to be in operation. If a component C_1 fails, component C_2 takes over, and so on. Suppose the times to failure of the components are independent and identically distributed exponential random variables with mean λ. How large should n be such that the reliability of the system at time t is at least .95? That is, find n such that

$$P\left(\sum_{j=1}^{n} X_j > t\right) \doteq .95.$$

Find n if $\lambda = 1$ and $t = 60$.

17. From a table of random numbers, one-digit numbers are selected from $0, 1, \ldots, 9$ (repetitions allowed). Let S_n be the sum of numbers in n selections. Approximate the probabilities for large n:
   (i) $P(S_n \geq 4.55n)$.   (ii) $P(S_n \leq 4.52n)$.
   Take $n = 625$, 2500, 10,000, 40,000, 62,500. What is the limit in each case as $n \to \infty$?

18*. Extend the result in Example 7 and show that for $\lambda > 0$

$$\lim_{n \to \infty} e^{-n\lambda} \sum_{k=0}^{n} (n\lambda)^k / k! = \begin{cases} 1 & \text{if } \lambda < 1 \\ 1/2 & \text{if } \lambda = 1 \\ 0 & \text{if } \lambda > 1. \end{cases}$$

(The case $\lambda = 1$ is proved in Example 7.)

19*. Use the central limit theorem to show that

$$\lim_{n \to \infty} \int_0^n e^{-t} \frac{t^{n-1}}{(n-1)!} \, dt = \frac{1}{2}.$$

[Hint: Consider the sum of independent exponential random variables with mean 1.]

20*. Let $X_1, X_2, \ldots, X_n$ be independent identically distributed random variables with common uniform distribution on $(0, 1)$. Imitate the proof in Example 2 to show that $\sum_{j=1}^{n} X_j$ has density function

$$f_n(x) = \frac{1}{(n-1)!} \sum_{k=0}^{n} (-1)^{n-k} \binom{n}{k} [\varepsilon(k - x)]^{n-1},$$

where $\varepsilon(x) = 0$, if $x \leq 0$, and $= x$ if $x \geq 0$.

21. Let $X_1, X_2, \ldots$ be a sequence of independent random variables with common mean μ_1 and variance σ_1^2. Let $\{Y_n\}$ be a sequence of independent random variables with common mean μ_2 and variance σ_2^2 and suppose that $\{X_n\}$ and $\{Y_n\}$ are independent. Show that $\bar{X} - \bar{Y}$ has an asymptotic normal distribution with mean $\mu_1 - \mu_2$ and variance $(\sigma_1^2 + \sigma_2^2)/n$. [Hint: Consider the sequence $Z_n = X_n - Y_n$, $n \geq 1$.]

22. In Problem 21 suppose that the sequences $\{X_n\}$ and $\{Y_n\}$ are dependent and $\mathrm{cov}(X_n, Y_n) = \rho \sigma_1 \sigma_2$ for all n. What is the asymptotic distribution of $\overline{X} - \overline{Y}$?

23. (i) Specialize Problem 21 to the case when $\{X_n\}$ is a sequence of independent Bernoulli random variables with parameter p_1 and $\{Y_n\}$ is an independent sequence of Bernoulli random variables with parameter p_2, $\{X_n\}$ and $\{Y_n\}$ are independent.
    (ii) Specialize part (i) to the case when (X_n, Y_n) have a multinomial distribution with parameters n, p_1, and p_2.

24. Let $\{X_n\}$ be a sequence of independent identically distributed random variables such that $\mathscr{E}|X_n|^{2k} < \infty$. Write $\overline{X^k} = \sum_{i=1}^n X_i^k / n$. Show that $\overline{X^k}$ is asymptotically normally distributed with mean $\mathscr{E}X_i^k$ and variance $[\mathscr{E}X_i^{2k} - (\mathscr{E}X_i^k)^2]/n$.

25*. Suppose $X_1, X_2, \ldots$ are independent identically distributed with $\mathscr{E}X_i = 0$ and $0 < \mathscr{E}X_i^2 = \sigma^2 < \infty$. Show that

$$P(\overline{X} \geq \varepsilon) \doteq \frac{\sigma}{\varepsilon\sqrt{n}} \frac{1}{\sqrt{2\pi}} \exp(-n\varepsilon^2/2\sigma^2)$$

for large n and any $\varepsilon > 0$. [Hint: Use Problem 7.6.24.]

9.4† THE NORMAL APPROXIMATION TO BINOMIAL, POISSON, AND OTHER INTEGER-VALUED RANDOM VARIABLES

We restrict attention to the case when the random variables X_n take only integer values $0, 1, 2, \ldots$. This includes, for example, binomial, Poisson, hypergeometric, and negative binomial random variables. According to the central limit theorem, $S_n = \sum_{k=1}^n X_k$ has approximately a normal distribution with mean $n\mathscr{E}X_i = n\mu$ and variance $n\,\mathrm{var}(X_i) = n\sigma^2$ for large n. In Section 9.3 we gave a bound for the error in this approximation and showed that the error is sometimes large. We suggest some alternative approximations and a correction that leads to a somewhat better approximation.

We note that the central limit theorem concerns the convergence of distribution functions (of S_n's) and not probability or density functions (to the normal density function). Suppose first that the X_n's are of the continuous type. Then S_n is also a continuous type random variable [this follows, for example, from (8.2.5)] with density function f_n. By the central limit theorem for large n

(1)
$$P(S_n \leq x) \doteq \Phi\left(\frac{x - n\mu}{\sigma\sqrt{n}}\right), \qquad x \in \mathbb{R}$$

where $\mu = \mathscr{E}X_i$, $\sigma^2 = \mathrm{var}(X_i)$ and Φ is the standard normal distribution function.

---

†This section deals with a special topic. For most readers it will be sufficient to learn only the continuity correction given in (7) or (8).

If we differentiate both sides of (1) with respect to x we get

$$(2) \qquad f_n(x) \doteq \frac{1}{\sigma\sqrt{n}} \phi\left(\frac{x - n\mu}{\sigma\sqrt{n}}\right) = \frac{1}{\sigma\sqrt{2n\pi}} \exp\left\{-\frac{(x - n\mu)^2}{2\sigma^2 n}\right\}.$$

A formal justification of (2) will not be given here but (2) provides a good approximation for large n. (We also need the condition that f_n is bounded for some n.) A result such as (2) is called a *local central limit theorem* as against (1), which is a *global* result. A similar result holds also in the discrete case. We assume that the X_i's have positive probability only over the set of integers $k = 0, \pm 1, \pm 2, \ldots$. Then the following result holds.

$$(3) \qquad \sigma\sqrt{n}\, P\big(S_n = n\mu + x\sigma\sqrt{n}\big) \to \frac{1}{\sqrt{2\pi}} \exp(-x^2/2)$$

as $n \to \infty$ where x is of the form $(k - n\mu)/\sigma\sqrt{n}$ for $k = 0, \pm 1, \pm 2, \ldots$. That is, for sufficiently large n

$$(4) \quad P(S_n = k) \doteq \frac{1}{\sigma\sqrt{2\pi n}} \exp\left\{-\frac{(k - n\mu)^2}{2\sigma^2 n}\right\}, \qquad k = 0, \pm 1, \pm 2, \ldots.$$

  Example 1. Early Births: Another Approximation for Binomial Probability. Suppose the X_n's are Bernoulli random variables so that S_n is a binomial random variable with parameter p, $0 < p < 1$. Since $\mu = \mathscr{E}X_i = p$, $\sigma^2 = \mathrm{var}(X_i) = p(1 - p)$, we have, for large n

$$(5) \quad P(S_n = k) = \binom{n}{k} p^k (1 - p)^{n-k} \doteq \frac{1}{\{2\pi np(1 - p)\}^{1/2}} \exp\left\{-\frac{(k - np)^2}{2np(1 - p)}\right\},$$

for $k = 0, 1, \ldots, n$.

  In a recent study of births it was found that about $1/3$ of the births arrived earlier than the doctor had predicted. What is the probability that in 100 randomly selected births exactly 35 will arrive earlier than predicted? Let X be the number of early arrivals. According to (5), $P(X = 35) \doteq .0795$. The exact probability from Table A3 is .0763.

  If we want $P(X \le l)$ we simply add the right side of (5) for $k = 0, 1, \ldots, l$. $\square$

  Example 2. Breakdowns of a Machine (Example 9.3.3). We consider once again the times $X_1, X_2, \ldots, X_n$ between breakdowns of a machine where each X_i has an exponential distribution with mean $\lambda > 0$. Then S_n has a gamma density given by

$$f_n(x) = \begin{cases} \dfrac{1}{(n - 1)!\,\lambda^n} x^{n-1} e^{-x/\lambda}, & x > 0 \\ 0 & \text{otherwise.} \end{cases}$$

According to (2) we can approximate f_n, for large n, by

$$(6) \qquad f_n(x) \doteq \frac{1}{\sqrt{2\pi n\lambda^2}} \exp\left\{-\frac{(x - n\lambda)^2}{2n\lambda^2}\right\}, \qquad x \in \mathbb{R}. \qquad \square$$

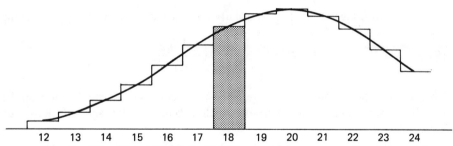

Figure 1. Binomial probabilities for $n = 50$, $p = .4$ and normal approximation (5).

The true probability or density functions of Examples 1 and 2 and the corresponding normal approximations (5) and (6) are plotted in Figures 1 and 2.

From Figure 1 we arrive at yet another method for approximating $P(S_n = k)$ in the discrete (integer valued) case which is usually more accurate than the original normal approximation formula (9.3.11) given by the central limit theorem or even (4). Indeed, from (4),

$$(7) \qquad P(S_n = k) \doteq \int_{k-1/2}^{k+1/2} \frac{1}{\sigma\sqrt{2\pi n}} \exp\left\{-\frac{(x-n\mu)^2}{2n\sigma^2}\right\} dx$$

$$= \Phi\left(\frac{k + \frac{1}{2} - n\mu}{\sigma\sqrt{n}}\right) - \Phi\left(\frac{k - \frac{1}{2} - n\mu}{\sigma\sqrt{n}}\right),$$

$k = 0, 1, 2, \ldots$. The area of the shaded region in Figure 1 is an approximation to $P(S_n = k)$. The idea is that we are approximating the area under the histogram for the distribution of S_n by the normal curve. Summing over k from $k = 0$ to $k = x$, say, we are led to the approximation

$$(8) \qquad P(S_n \le x) \doteq \Phi\left(\frac{x + 1/2 - n\mu}{\sigma\sqrt{n}}\right)$$

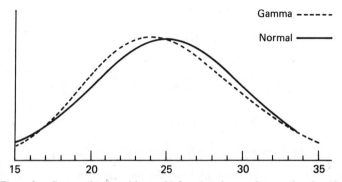

Figure 2. Gamma density with $n = 25$, $\lambda = 3$ and normal approximation (6).

where x is an integer ≥ 0. This is usually more accurate than the approximation

$$(9) \qquad P(S_n \leq x) \doteq \Phi\left(\frac{x - n\mu}{\sigma\sqrt{n}}\right).$$

This approximation procedure is referred to as *continuity correction* and applies for all integer valued random variables such as binomial, Poisson, geometric, and hypergeometric.

Remark 1.   The continuity correction helps to improve the accuracy of the central limit theorem approximation in all cases where the random variable X being sampled has a discrete distribution on equally spaced points. A random variable that takes only values $k + lh$, $l = 0, \pm 1, \pm 2, \ldots$ with positive probability is known as a *lattice random variable*. Continuity correction should be applied for all lattice random variables by writing, for example,

$$P\left(\sum_{j=1}^{n} X_j = t\right) = P\left(t - \frac{h}{2} < \sum_{j=1}^{n} X_j < t + \frac{h}{2}\right)$$

where t is a possible value of $\sum_{j=1}^{n} X_j$. The approximation (3) [or (4)] also holds for lattice random variables.

Example 3.   Early Births: Normal Approximation to Binomial with Continuity Correction (Example 1).   Suppose we wish to find the probability of at most 35 early births in a sample of 100 births. Then

$$P(X \leq 35) \doteq \Phi\left(\frac{35 + 1/2 - 100(1/3)}{\sqrt{100(1/3)(2/3)}}\right) = \Phi(.46) = .6772.$$

The exact probability from Table A4 is .7054, so the error is about 4 percent. If the continuity correction is not applied, then

$$P(X \leq 35) \doteq \Phi\left(\frac{35 - 100/3}{\sqrt{200/9}}\right) = \Phi(.35) = .6368$$

and the error is about 10 percent. Also,

$$P(X = 35) \doteq \Phi\left(\frac{35 + 1/2 - 100/3}{\sqrt{200/9}}\right) - \Phi\left(\frac{35 - 1/2 - 100/3}{\sqrt{200/9}}\right)$$

$$\doteq \Phi(.46) - \Phi(.25) = .6772 - .5987 = .0785$$

which is a better approximation than the one provided by (5). $\square$

Example 4*.   Reversal in an Election: Normal Approximation to Hypergeometric.
In Example 6.4.6 we computed the probability of reversal in an election with two

candidates A and B by applying the binomial approximation to hypergeometric distribution. We recall that if A gets a votes and B get b votes, $a > b$, and n votes are invalid, then the number of invalid votes for A is a hypergeometric random variable with distribution

$$P(X = x) = \binom{a}{x}\binom{b}{n-x} \Big/ \binom{a+b}{n}, \qquad x = 0,1,\ldots,a.$$

The probability of reversal is given by $P(X \geq (n + a - b)/2)$.

Since hypergeometric distribution can be approximated by binomial and binomial by normal, we will approximate the probability of reversal by normal distribution. (A formal proof is too complicated to bother with here.)

We recall that

$$\mu = \mathscr{E}X = \frac{na}{(a+b)} \quad \text{and} \quad \sigma^2 = \text{var}(X) = n\frac{ab(a+b-n)}{(a+b)^2(a+b-1)}.$$

Hence

$$P\left(X \geq \frac{n+a-b}{2}\right) = P\left(\frac{X-\mu}{\sigma} \geq \frac{\dfrac{n+a-b}{2} - \dfrac{na}{a+b}}{\sqrt{\dfrac{nab(a+b-n)}{(a+b)^2(a+b-1)}}}\right)$$

$$\doteq P(Z \geq z),$$

where Z is a standard normal random variable. It is possible to simplify the expression for z considerably. First we note that

$$\frac{n+a-b}{2} - \frac{na}{a+b} = \frac{(a+b-n)(a-b)}{2(a+b)}$$

so that

$$z = \frac{\dfrac{(a+b-n)(a-b)}{2(a+b)}}{\sqrt{\dfrac{nab(a+b-n)}{(a+b)^2(a+b-1)}}} \doteq \frac{(a-b)\sqrt{a+b-n}}{2\sqrt{\dfrac{nab}{a+b}}}$$

since when $a + b$ is large we can replace $a + b - 1$ by $a + b$. Now

$$4ab = (a+b)^2 - (a-b)^2 \leq (a+b)^2$$

so that

$$ab \leq ((a+b)/2)^2.$$

It follows that

$$z \gtrsim \frac{(a - b)\sqrt{a + b - n}}{\sqrt{n(a + b)}} = (a - b)\sqrt{\frac{1}{n} - \frac{1}{a + b}} \; .$$

Hence

$$P(\text{reversal}) \doteq P(Z \geq z) \lesssim P\left(Z > (a - b)\sqrt{\frac{1}{n} - \frac{1}{a + b}} \right)$$

for large $a + b$.

In the particular case of Example 6.4.6 we have $n = 75$, $a = 510$, $b = 490$, so that

$$P(\text{reversal}) \lesssim P\left(Z \geq 20\sqrt{\frac{1}{75} - \frac{1}{1000}} \right) = P(Z \geq 2.22) = .0132$$

as compared with .0158 obtained by applying binomial approximation.

The reader may justifiably wonder why we went through so much trouble to simplify the probability of reversal when a straightforward standardization and application of the central limit theorem (or binomial approximation) leads to practically the same result. Apart from the fact that we may not have the binomial probabilities available for the desired n and p, the approximation

$$P(\text{reversal}) \lesssim P\left(Z \geq (a - b)\sqrt{\frac{1}{n} - \frac{1}{a + b}} \right)$$

has a slight advantage. It tells us that the desired probability depends primarily on the plurality $a - b$ and the number of invalid votes n. The total number of votes cast is much less important since $1/(a + b)$ is nearly zero for large $a + b$. Thus given α, $0 < \alpha < 1$, where

$$\alpha \doteq \text{probability of reversal}$$

$a - b$ and n are related according to $(a - b)^2 \doteq nz_\alpha^2$.   □

Example 5. Salk Polio Vaccine. In 1954 a large field trial was conducted in the United States to test the effectiveness of the Salk vaccine as a protection against polio. The following table summarizes the results.

Children	Contracted Polio	Polio-Free	Total
Vaccinated	33	200,712	200,745
Control (not vaccinated)	115	201,114	201,229
Total	148	401,826	401,974

The null hypothesis to be tested was that treatment has no effect, that is, the vaccine offered no protection. Under the null hypothesis, therefore, we wish to test that the proportions of children who contracted polio is the same in the two groups, the treatment group and the control group. Under H_0 we can think of a lot of $N = 401,974$ children of which $D = 148$ are marked polio and the remaining $N - D = 401,826$ are marked "no polio." A random sample of size $n = 200,745$ is selected as the treatment group. Then the number X of these who have polio is a random variable with a hypergeometric distribution. According to the Fisher–Irwin test of Section 6.4, we reject H_0 if X is small. The P-value associated with the observation 33 is $P_{H_0}(X \leq 33)$. Since n is large and $D/n = 148/401,974 = 3.68 \times 10^{-4}$ is very small, we use the normal approximation. We have

$$\mathscr{E}X = n\frac{D}{N} = 73.91$$

and

$$\mathrm{var}(X) = \frac{N-n}{N-1}n\frac{D(N-D)}{N^2} = 36.9864, \qquad \sqrt{\mathrm{var}(X)} = 6.0816,$$

so that

$$P(X \leq 33) = P(X < 33.5) = P\left(\frac{X - \mathscr{E}X}{\sqrt{\mathrm{var}\,X}} < \frac{33.5 - 73.91}{6.08}\right)$$

$$\doteq \Phi(-6.64) < 10^{-8}$$

where Φ is the standard normal distribution function. Thus, the probability of observing as small as 33 cases of polio in random sampling under the null hypothesis is less than one in 100 million. There is little or no support for H_0 and we have every reason to conclude that the Salk vaccine does indeed offer protection from polio. □

Remark 2.   Examples 4 and 5 indicate that normal approximation can also be used to evaluate probabilities in sampling without replacement from a finite population. The following result holds. Let $\{X_1, X_2, \ldots, X_n\}$ be a random sample from a finite population of size N with mean μ and variance σ^2. Then, as $n \to \infty$ and $N/n \to \infty$ the sequence

$$\frac{(\overline{X} - \mu)}{\left\{\frac{\sigma}{\sqrt{n}}\left[\frac{(N-n)}{(N-1)}\right]^{1/2}\right\}}$$

has a limiting standard normal distribution.

We recall (Example 5.3.1) that $\mathscr{E}\overline{X} = \mu$, $\mathrm{var}(\overline{X}) = \sigma^2(N-n)/[n(N-1)]$ where $(N-n)/(N-1)$ is the finite population correction. When n/N is small $(< .05)$ we can use $\mathrm{var}(\overline{X}) \doteq \sigma^2/n$. Hypergeometric distribution is the special case when $P(X_i = 1) = D/N$ and $P(X_i = 0) = (N-D)/N$.

Remark 3.   According to (8)

$$P(S_n \leq x) \doteq \Phi\left(\frac{x + 1/2 - n\mu}{\sqrt{n}}\right), \qquad x \text{ integral.}$$

If n is so large that

$$\frac{x + 1/2 - n\mu}{\sigma\sqrt{n}} - \frac{x - n\mu}{\sigma\sqrt{n}} < .005$$

we can drop the continuity correction since Table A5 gives probabilities under the normal curve only at jumps of .01. (We are assuming that a number x, $N.005 \le x < N.01$ will be rounded off to $N.01$.) If $\sigma = 1$, for example, this means that for $n > 10,000$ we can drop the continuity correction.

Remark 4. Often the approximation is applied to $\overline{X}$. If t is a possible value of $\overline{X}$, then the continuity correction approximation for $P(\overline{X} \le t)$ is given by

$$P(\overline{X} \le t) \doteq \Phi\left(\sqrt{n}\,\frac{(t - \mu + 1/2n)}{\sigma}\right).$$

Alternatively, we can translate the statement in terms of $\sum_{i=1}^{n} X_i$ and apply the continuity correction. For example, let $\hat{p} = \overline{X} = Y/n$ be the proportion of successes in n trials. Let $n = 100$ and suppose we want $P(\hat{p} \le .45)$ when $p = .5$. Then

$$P(\hat{p} \le .45) \doteq \Phi\left(\sqrt{100}\left(.45 - .5 + \frac{1}{200}\right)\Big/\sqrt{.5(.5)}\right) = \Phi(-.9)$$

since .45 is a possible value of $\hat{p}$. If $n = 84$, say, and we want $P(\hat{p} \le .45)$ when $p = .5$, then we note that .45 is not a possible value of $\hat{p}$. In this case we write

$$P(\hat{p} \le .45) = P(Y \le 84(.45)) = P(Y \le 37.8)$$

$$= P(Y \le 37)$$

and use (8) to obtain

$$P(\hat{p} \le .45) = \Phi\left(\frac{37.5 - 42}{.5\sqrt{84}}\right) = \Phi(-.98).$$

We have seen that frequently there are several approximations to choose from. When should we approximate binomial by Poisson and when should we apply normal approximation to binomial? In Table 1 we summarize various approximations and provide guidelines to help in this choice. In applications of the central limit theorem we will often be required to compute probabilities such as $P(\overline{X} \ge \varepsilon)$ without any knowledge of the underlying distribution (other than knowledge of σ^2 or s^2 and the sample size n). In that case *we will approximate the required probability by using normal approximation as long as n is greater than or equal to* 30. Thus the phrase "large n" in this context means $n \ge 30$.

TABLE 1. Approximations to Distributions

A: Nonnormal Approximation

Underlying Distribution of X	Probability to be Approximated	Approximate by	When
Binomial (n, p)	$P(X = k)$	$e^{-\lambda}\lambda^k/k!, \lambda = np$	$p < .1$, np moderate (≤ 10)
Hypergeometric $\dbinom{D}{K}\dbinom{N-D}{n-k}\big/\dbinom{N}{n}$	$P(X = k)$	(i) $\dbinom{n}{k}\left(\dfrac{D}{N}\right)^k\left(1 - \dfrac{D}{N}\right)^{n-k}$	$n/N < .1$ (N large)
		(ii) $e^{-\lambda}\lambda^k/k!, \lambda = \dfrac{nD}{N}$	$D/N < .1$ and $\dfrac{nD}{N} \leq 10$ $(n, N$ large)
Negative binomial $\dbinom{k+r-1}{r-1}$	$P(X = k + r)$	$e^{-\lambda}\lambda^k/k!, \lambda = r(1-p)$	$p > .9, r(1-p) \leq 10$ $(k/r$ moderate, r large)

B: Normal Approximation

$X, X_1, X_2, \ldots, X_n$ independent and identically distributed, $S_n = \sum_{i=1}^n X_i$, $\mathscr{E}X_i = \mu$, $\mathrm{var}(X_i) = \sigma^2$, and X nondegenerate.

Distribution of X	Distribution of S_n	Probability to be Approximated	Approximate by	When
1. Any noninteger valued X	—	$P(a \leq S_n \leq b)$	$\Phi\left(\dfrac{b - n\mu}{\sigma\sqrt{n}}\right) - \Phi\left(\dfrac{a - n\mu}{\sigma\sqrt{n}}\right)$	n large
2. Integer valued X				
(i) Bernoulli, p	Binomial (n, p)	$P(k \leq S_n \leq l)$	$\Phi\left(\dfrac{l + 1/2 - np}{\sqrt{np(1-p)}}\right) - \Phi\left(\dfrac{k - 1/2 - np}{\sqrt{np(1-p)}}\right)$	$np(1 - p) > 10$

Table 1 *Continued*

Distribution of X	Distribution of S_n	Probability to be Approximated	Approximate by	When
		$P(S_n = k)$	$\Phi\left(\dfrac{k+1/2-np}{\sqrt{np(1-p)}}\right) - \Phi\left(\dfrac{k-1/2-np}{\sqrt{np(1-p)}}\right)$ or $\dfrac{1}{\sqrt{2\pi np(1-p)}}\exp\left\{-\dfrac{(k-np)^2}{2np(1-p)}\right\}$	Good near $p = 1/2$, unsatisfactory for small or large p
(ii) Poisson, λ	Poisson, $n\lambda$	$P(k \leq S_n \leq l)$	$\Phi\left(\dfrac{l+1/2-n\lambda}{\sqrt{n\lambda}}\right) - \Phi\left(\dfrac{k-1/2-n\lambda}{\sqrt{n\lambda}}\right)$	Large $n\lambda$
		$P(S_n = k)$	$\Phi\left(\dfrac{k+1/2-n\lambda}{\sqrt{n\lambda}}\right) - \Phi\left(\dfrac{k-1/2-n\lambda}{\sqrt{n\lambda}}\right)$ or $\dfrac{1}{\sqrt{2n\pi\lambda}}\exp\left\{-\dfrac{(k-n\lambda)^2}{2n\lambda}\right\}$	
(iii) Poisson, λ	—	$P(k \leq X \leq l)$	$\Phi\left(\dfrac{l+1/2-\lambda}{\sqrt{\lambda}}\right) - \Phi\left(\dfrac{k-1/2-\lambda}{\sqrt{\lambda}}\right)$	Large λ
		$P(X = k)$	$\Phi\left(\dfrac{k+1/2-\lambda}{\sqrt{\lambda}}\right) - \Phi\left(\dfrac{k-1/2-\lambda}{\sqrt{\lambda}}\right)$	
(iv) Hypergeometric		$P(k \leq X \leq l)$	$\Phi\left(\dfrac{l+1/2-nD/N}{\sigma}\right) - \Phi\left(\dfrac{k-1/2-nD/N}{\sigma}\right)$, $\sigma = \sqrt{\dfrac{N-n}{N-1}n\dfrac{D}{N}\left(1-\dfrac{D}{N}\right)}$	Large n, D/N not small

Problems for Section 9.4

1. A Congressman sampled 225 voters in his district to determine if his constituency favors military aid to a South American country. He believed that voters in his district were equally divided on the issue. The sample produced 95 in favor of military aid and 130 against. What do you conclude?

2. A company produces golf balls. From daily production a sample of 80 golf balls is inspected, and if the sample contains 10 percent or more defectives the process is stopped and readjusted. If the machine is actually producing 15 percent defective golf balls on a certain day, what is the probability that it will have to be stopped and readjusted?

3. It is estimated that about 60 percent of patients with lung disorders are heavy smokers.
    (i)   What is the probability that in a sample of 25 such patients more than half are heavy smokers?
    (ii) What is the probability that in a sample of 225 such patients more than half are heavy smokers?

4. A bag of 500 half-dollars is dumped at the counter of a local utility company by an irate consumer to protest the large bill she got from the company for the month of February. What is the probability that between 225 and 250 (both inclusive) half-dollars showed heads? Find the probability that fewer than 100 showed heads.

5. According to genetic theory, when two right varieties of peas are crossed, the probability of obtaining a smooth seed is $1/2$. An experimenter would like to be 95 percent certain of having at least 50 seeds that are smooth. How many seeds must she produce?

6. A multiple-choice test consists of 75 questions, with five choices for each answer, one of which is correct. Assume that a student answers all questions by pure guessing. What is the expected number of correct answers? What is the probability that the first four answers are correct? What is the probability that at least 20 answers are correct? Suppose each correct answer carries one point and each incorrect answer carries zero points. In order to eliminate the effect of guessing, how many points should be taken off for an incorrect answer so that the long-run score per question will average 0 points?

7. Candidate X believes that he can win a countywide election if he can poll at least 53 percent of the votes in the city which is the seat of the county. He believes that about 50 percent of the county's voters favor him. If 6150 voters show up to vote in the city, what is his chance of winning? What if 2500 vote in the city?

8. A state legislator in a certain state wishes to assess voters' attitudes in her district to determine if the voters favor repeal of the Equal Rights Amendment that the state legislature had passed a year ago. Polls show that the sentiment in the state is about 60 percent against repeal. If she were to select a random sample of 80 voters from her district, what is the probability that a majority of the sample would be against repeal? If it is found that only 45 percent in the sample are against repeal, what would you conclude?

9. A fair coin is tossed until the first head is observed. The number of trials is recorded. This experiment is repeated n times. Let S_n be the total number of trials required for n successes. What should n be so that the probability is .05 that the total number of trials needed is at least t? Find n when $t = 500$.

(Note that S_n has a negative binomial distribution. Use the relation in Section 6.5 between a binomial and a negative binomial distribution to find n.)

10. In the peak period of an outbreak of flu it is assumed that about .1 percent of the people in a city have the bug. In a sample of 10,000 people, what is the probability that 15 will have the flu? Estimate the same probability for a sample of 62,500.

11. It is estimated that the probability that a baby will be born on the date the obstetrician predicts is $1/40$. What is the probability that of 400 babies born 15 will be born on the date the doctor predicts? Use Poisson as well as normal approximations.

12. Let $X(t)$ be the number of customers that enter a store in interval $(0, t)$ and suppose $X(t)$ has a Poisson distribution with parameter λt where $\lambda = 50$ per hour. Find the smallest t such that

$$P(X(t) > 150) \doteq .95$$

13. A random sample of 50 swimmers is taken at a swim meet with 450 participants consisting of 200 girls and 250 boys. Estimate the probability that the number of girls in the sample is:
   (i) Greater than or equal to 30. (ii) Between 20 and 25 (inclusive).
   (iii) Larger than 30.

14. There are 500 participants at a convention of physicians. A random sample of size 60 is taken. Let $\overline{X}$ be the sample average income in thousands of dollars. Find the probability that:
   (i) $\overline{X} \geq 67$, (ii) $\overline{X} < 68$, (iii) $62 < \overline{X} < 66$,
   if it is known that $\mu = 65$, $\sigma = 10$.

15. Consider a finite population of N weights of pigs at a farm. To estimate the mean weight of pigs on the farm to within ε pounds with probability at least $1 - \alpha$, $0 < \alpha < 1$, a sample of size n is to be taken and the sample mean $\overline{X}$ is to be used to estimate μ. How large should n be if the standard deviation of weights of pigs is known to be σ pounds?

16. Let $X(t)$ be the number of telephone calls that come in to a service counter switchboard in t minutes. If $X(t)$ has a Poisson distribution with parameter $5t$, find the probability that in a five-hour period there will be:
   (i) Less than 1500 calls. (ii) More than 1600 calls.
   (iii) Between 1500 and 1550 calls.

17. A machine produces, under normal operations, on the average 1 percent defectives. Approximate the probability that in 100 items produced by the machine there will be 2 or more defectives. Compare the Poisson and normal approximations to the exact probability from Table A3 under appropriate independence assumption.

18. The number of automobile accidents on a two-day weekend at a particular traffic intersection can be modeled by a Poisson distribution with mean 3. The number of accidents at this intersection is recorded for all the 52 weeks of the year. Find the probability that the average number of accidents on nonholiday weekends (assume there are 45 nonholiday weekends in a year) at this intersection will be:
    (i) Greater than 4.   (ii) Less than 2.5.   (iii) Between 2.8 and 3.5 (inclusive).

19. A town has 5000 voters of which 60 percent are Republicans and 40 percent Democrats. A random sample of 100 voters is taken. Approximate the probability that the sample contains exactly 50 Democrats. Use binomial as well as normal approximations.

9.5 CONSISTENCY

We consider once again a desirable property of estimates first introduced in Section 4.3. Suppose we wish to estimate a parameter θ by an estimate $\hat{\theta}_n = \theta(X_1, \ldots, X_n)$ based on a sample of size n. If n is allowed to increase indefinitely, we are practically sampling the whole population. In that case we should expect $\hat{\theta}_n$ to be practically equal to θ. That is, if $n \to \infty$ it would be desirable to have $\hat{\theta}_n$ converge to θ in some sense.

DEFINITION 1. Let $X_1, X_2, \ldots, X_n$ be a sample of size n from a distribution function F and let $\theta(F) = \theta$ be some parameter determined by F. The sequence of estimates $\{\hat{\theta}_n\}$ where

$$\hat{\theta}_n = \theta(X_1, X_2, \ldots, X_n), \, n = 1, 2, \ldots$$

is said to be consistent for θ if for each $\varepsilon > 0$

$$P(|\hat{\theta}_n - \theta| > \varepsilon) \to 0 \text{ as } n \to \infty.$$

If $P(|\hat{\theta}_n - \theta| > \varepsilon) \to 0$ as $n \to \infty$ we say that $\hat{\theta}_n$ *converges in probability* to θ and write $\hat{\theta}_n \overset{P}{\to} \theta$. Thus $\{\hat{\theta}_n\}$ is consistent for θ simply means $\hat{\theta}_n \overset{P}{\to} \theta$. As pointed out earlier (in Sections 5.4 and 6.3.1), $\hat{\theta}_n \overset{P}{\to} \theta$ does not mean that $\hat{\theta}_n$ converges to θ pointwise. It only means that for large n, $\hat{\theta}_n$ is likely to be near θ.

Frequently the following simple criterion helps in determining the consistency of $\hat{\theta}_n$ for θ.

PROPOSITION 1. *If the sequence of real numbers $\{\mathscr{E}\hat{\theta}_n\}$ converges to θ and $var(\hat{\theta}_n) \to 0$ as $n \to \infty$ then $\hat{\theta}_n$ is consistent for θ.*

In particular, if $\mathscr{E}\hat{\theta}_n = \theta$ (that is, $\hat{\theta}_n$ is unbiased) and $var(\hat{\theta}_n) \to 0$ as $n \to \infty$ then $\hat{\theta}_n \overset{P}{\to} \theta$.

Proof.   We use Chebyshev's inequality to see that

$$P(|\hat{\theta}_n - \theta| > \varepsilon) \leq \frac{\mathscr{E}(\hat{\theta}_n - \theta)^2}{\varepsilon^2} = \frac{1}{\varepsilon^2}\mathscr{E}(\hat{\theta}_n - \mathscr{E}\hat{\theta}_n + \mathscr{E}\hat{\theta}_n - \theta)^2$$

$$= \frac{1}{\varepsilon^2}\left[\text{var}(\hat{\theta}_n) + (\mathscr{E}\hat{\theta}_n - \theta)^2\right].$$

Both the assertions now follow easily.   □

> **Example 1.   Sample Mean is Consistent for Population Mean.**   Let $X_1, X_2, \ldots, X_n$ be a random sample from a distribution with mean μ and variance σ^2, $0 < \sigma^2 < \infty$. Let $\bar{X} = n^{-1}\Sigma_{k=1}^n X_k$. Then $\mathscr{E}\bar{X} = \mu$ and $\text{var}(\bar{X}) = \sigma^2/n \to 0$ and it follows that $\bar{X}$ is consistent for μ. (This is simply the law of large numbers of Section 5.4.)
>
> If $\theta = \mathscr{E}X_i^k$ is to be estimated, then $\hat{\theta}_n = \Sigma_{i=1}^n X_i^k/n$ is both consistent and unbiased for θ.   □

The following result is often useful and is stated without proof.

PROPOSITION 2.

1.   *Let g be a continuous function. If $\hat{\theta}_n \overset{P}{\to} \theta$ then $g(\hat{\theta}_n) \overset{P}{\to} g(\theta)$ as $n \to \infty$.*

2.   *If $\hat{\theta}_n \overset{P}{\to} \theta_1$ and $\hat{\theta}_n' \overset{P}{\to} \theta_2$ then:*

   (i) $\hat{\theta}_n \pm \hat{\theta}_n' \overset{P}{\to} \theta_1 \pm \theta_2$.   (ii) $\hat{\theta}_n\hat{\theta}_n' \overset{P}{\to} \theta_1\theta_2$.   (iii) $\hat{\theta}_n/\hat{\theta}_n' \overset{P}{\to} \theta_1/\theta_2$ ($\theta_2 \neq 0$).

3.   *If $\hat{\theta}_n$ has a limiting distribution F (that is, $P(\hat{\theta}_n \leq x) \to F(x)$ as $n \to \infty$) and $\hat{\theta}_n' \overset{P}{\to} \theta$, ($\neq 0$), then $\hat{\theta}_n/\hat{\theta}_n'$ has the limiting distribution $F(x/\theta)$. That is,*

$$P(\hat{\theta}_n/\hat{\theta}_n' \leq x) \to F(x/\theta) \qquad \text{as } n \to \infty$$

   for all x at which F is continuous.

Proposition 2, part 1 tells us that if $\hat{\theta}_n$ is consistent for θ then so is $\hat{\theta}_n + c_n$ where c_n is a sequence of constants such that $c_n \to 0$. Thus consistency by itself cannot be a determining criterion.

> **Example 2.   Estimating Population Variance.**   We recall that if $X_1, X_2, \ldots, X_n$ is a random sample from a population with variance σ^2 then the sample variance $S^2 = (n - 1)^{-1}\Sigma_{i=1}^n(X_i - \bar{X})^2$ is unbiased for σ^2. Suppose $\mathscr{E}X_i^4 < \infty$. Then

$$S^2 = (n - 1)^{-1}\left\{ \sum_{i=1}^n X_i^2 - n\bar{X}^2 \right\}$$

$$= \frac{n}{n - 1}\left(\sum_{i=1}^n X_i^2/n \right) - \frac{n}{n - 1}\bar{X}^2 \overset{P}{\to} \mathscr{E}X^2 - \mu^2 = \sigma^2$$

by law of large numbers and Proposition 2 so that S^2 is consistent for σ^2. By part 1 of Proposition 2 we note that S is consistent for σ although S is not unbiased for σ. Note that $\sum_{i=1}^{n}(X_i - \bar{X})^2/n$ is also consistent for σ^2. □

Example 3. *t-Statistic Converges to Normal.* In sampling from a normal population we know that

$$T = \frac{\bar{X} - \mu}{S/\sqrt{n}}$$

has a t-distribution with $(n-1)$ d.f. (Section 8.7). Writing T as follows

$$T = \frac{(\bar{X} - \mu)\sqrt{n}/\sigma}{S/\sigma}$$

we note that $(\bar{X} - \mu)\sqrt{n}/\sigma$ has an exact normal distribution whereas $S/\sigma \overset{P}{\to} 1$ (by Example 2). Hence Proposition 2, part 3 implies that

$$P(T \le t) \to \frac{1}{\sqrt{2\pi}} \int_{-\infty}^{t} e^{-x^2/2}\, dx \qquad \text{for } t \in \mathbb{R}$$

as $n \to \infty$.

If X_i's are not necessarily normally distributed but have a finite variance σ^2 and $\mathscr{E}X_i^4 < \infty$ then the central limit theorem implies the same result. □

Example 4. *Empirical Distribution Function As a Consistent Estimate.* Let $X_1, X_2, \ldots X_n$ be a random sample from a distribution function F. We recall that the empirical distribution function of the sample is defined by

$$\hat{F}_n(x) = \frac{\text{number of } X_i\text{'s} \le x}{n}.$$

Let $Y = Y(x) = $ number of X_i's $\le x$. Then Y has a binomial distribution with parameters n and $p = p(x) = P(X_i \le x) = F(x)$. It follows from Example 1 that

$$\hat{F}_n(x) \to F(x) \qquad \text{as } n \to \infty.$$

for each $x \in \mathbb{R}$. □

Example 5. *The Taxi Problem.* In Example 6.2.5 we considered $X_{(n)} = M_n = \max(X_1, \ldots, X_n)$ as an estimate of N, the total number of taxis in a town. It is intuitively clear that M_n is consistent for N. In fact, for $\varepsilon > 0$,

$$P(|M_n - N| > \varepsilon) = P(M_n > N + \varepsilon) + P(M_n < N - \varepsilon)$$

$$= P(M_n < N - \varepsilon) \qquad (M_n \le N)$$

$$= \begin{cases} \left(\dfrac{[N-\varepsilon]}{N}\right)^n & \text{if } \varepsilon \text{ is not an integer} \\[4mm] \left(\dfrac{N-\varepsilon-1}{N}\right)^n & \text{if } \varepsilon \text{ is an integer.} \end{cases}$$

Since $\varepsilon > 0$, $[N - \varepsilon]/N < 1$ as well as $(N - \varepsilon - 1)/N < 1$ so that

$$P(|M_n - N| > \varepsilon) \to 0 \qquad \text{as } n \to \infty.$$

Hence M_n is consistent for N. Moreover the estimate $\hat{N}_1 = [(n + 1)/(n)]M_n$ is also consistent for N. $\qquad\square$

Problems for Section 9.5

1. (i) Let Y be the number of successes in n Bernoulli trials with parameter p. Show that Y/n, the proportion of successes is consistent for p.
   (ii) Show, with $\hat{p} = Y/n$, that $(\hat{p} - p)/\sqrt{\hat{p}(1 - \hat{p})/n}$ has approximately a standard normal distribution for large n.

2. Suppose X_1 and X_2 are the number of successes in n_1 and n_2 trials respectively. Show that $\hat{p}_1 - \hat{p}_2$ is consistent for $p_1 - p_2$ where $\hat{p}_i = X_i/n_i$, $i = 1, 2$.

3. Let p_1 be the proportion of male students at a state university and p_2 that of female students. An estimate of $p_1 - p_2$, the excess proportion of males over females, is required. Find a consistent estimate for $p_1 - p_2$ and find the limiting distribution of your estimate. [Hint: $p_1 = 1 - p_2$.]

4. Show that $(X/n)^2$ and $X(X - 1)/n(n - 1)$ are both consistent estimates of p^2 where X is the number of successes in n trials with constant probability p of success.

5. (i) Find a consistent estimate of λ in sampling from a Poisson distribution with parameter λ.
   (ii) Find a consistent estimate of $P(X = 0) = e^{-\lambda}$.

6. Let $X_1, X_2, \ldots, X_n$ and $Y_1, Y_2, \ldots, Y_n$ be independent random samples from two populations with means μ_1 and μ_2 and variances σ_1^2 and σ_2^2 respectively.
   (i) Show that $\bar{X} - \bar{Y}$ is consistent for $\mu_1 - \mu_2$.
   (ii)* If $\sigma_1 = \sigma_2 = \sigma$, and $\mathscr{E}X_i^4 < \infty$, $\mathscr{E}Y_i^4 < \infty$, show that

$$S^2 = \frac{\displaystyle\sum_{i=1}^{n} (X_i - \bar{X})^2 + \sum_{i=1}^{n} (Y_i - \bar{Y})^2}{2n - 2}$$

is consistent for σ^2.
   (iii) Show that $\sqrt{n}\,[\bar{X} - \bar{Y} - (\mu_1 - \mu_2)]/2S$ is asymptotically normal with mean 0 and variance 1.

7. Let $X_1, X_2, \ldots, X_n$ be a random sample from a geometric distribution with parameter p. Find a consistent estimate for p.

8. Let $X_1, X_2, \ldots, X_n$ be a random sample from a population with mean μ and variance σ^2. Which of the following estimates are consistent for μ?
   (i) $\hat{\mu}_1 = (X_{1_n} + X_2 + \cdots + X_n)/(n/2)$.
   (ii) $\hat{\mu}_2 = 2\Sigma_{i=1}^{n} iX_i/(n(n + 1))$.
   (iii) $\hat{\mu}_3 = \{(X_1 + X_2 + \cdots + X_{[n/2]})/[n/2]\}$, ($[x]$ is the largest integer $\leq x$).
   (iv) $\hat{\mu}_4 = X_1$.

9. Which of the following estimates are consistent for variance?
   (i) $\hat{\sigma}_1^2 = (X_1 - X_2)^2/2$.
   (ii) $\hat{\sigma}_2^2 = \sum_{i=1}^{n}(X_i - \mu)^2/n$ (μ known).
   (iii) $\hat{\sigma}_3^2 = \sum_{i=1}^{n}(X_i - \hat{\mu}_2)^2/n$, where $\hat{\mu}_2$ is as defined in Problem 8(ii).

10. Let $X_1, X_2, \ldots, X_n$ be a random sample from the uniform distribution on $(0, \theta)$. Which of the following estimates are consistent for θ?
    (i) $\hat{\theta}_1 = 2\bar{X}$.   (ii) $\hat{\theta}_2 = [(n+1)/n]\max(X_1, \ldots, X_n)$.
    (iii) $\hat{\theta}_3 = \max(X_1, \ldots, X_n)$.
    (iv) $\hat{\theta}_4 = (n+1)R/(n-1)$ where $R = \max(X_1, X_2, \ldots, X_n) - \min(X_1, X_2, \ldots, X_n)$.
    (v) $\hat{\theta}_5 = X_1 + (X_2 + \cdots + X_n)/(n-1)$.

11. Let $\{\hat{\theta}_n\}$ be any sequence of estimates for θ. We say that $\hat{\theta}_n$ is *mean square consistent* for θ, if $\mathscr{E}\hat{\theta}_n^2 < \infty$ and $\mathscr{E}|\hat{\theta}_n - \theta|^2 \to 0$ as $n \to \infty$.
    (i) Show that mean square consistency implies consistency.
    (ii) Show that $\{\hat{\theta}_n\}$ is mean square consistent if and only if $\mathrm{var}(\hat{\theta}_n) \to 0$ and $\mathscr{E}\hat{\theta}_n \to \theta$ as $n \to \infty$. ($\mathscr{E}\hat{\theta}_n \to \theta$ is known as *asymptotic unbiasedness* of $\hat{\theta}_n$. $\mathscr{E}\hat{\theta}_n - \theta \to 0$ can also be written as bias of $\hat{\theta}_n \to 0$.) [Hint: Use Chebyshev's inequality and the fact that $\mathscr{E}(\hat{\theta}_n - \theta)^2 = \mathrm{var}(\hat{\theta}_n) + (\mathscr{E}\hat{\theta}_n - \theta)^2$.]

12. Let $X_1, X_2, \ldots, X_n$ be a random sample from the uniform distribution on $(\theta, \theta+1)$. Show that $\hat{\theta}_1 = \bar{X} - 1/2$ and $\hat{\theta}_2 = \max(X_1, \ldots, X_n) - n/(n+1)$ are both consistent for θ.

13. Let $X_1, X_2, \ldots, X_n$ be a random sample from density function

$$f(x) = \begin{cases} \theta(1-x)^{\theta-1}, & 0 < x < 1 \\ 0 & \text{elsewhere}, \end{cases} \quad \theta > 0$$

    Is $\bar{X}$ consistent for θ?

14. Let $X_1, X_2, \ldots, X_n$ and $Y_1, Y_2, \ldots, Y_n$ be independent random samples from populations with variances σ_1^2 and σ_2^2 respectively. Find consistent estimates of
    (i) σ_1^2/σ_2^2.   (ii) $\sigma_1^2/(\sigma_1^2 + \sigma_2^2)$.   (iii) $\sigma_1^2\sigma_2^2$.
    (You may assume that $\mathscr{E}X_1^4 < \infty, \mathscr{E}Y_1^4 < \infty$.)

15. Let $X_1, X_2, \ldots, X_n$ be a random sample from the uniform distribution on $(0, \theta)$. Show that $(\prod_{i=1}^{n}X_i)^{1/n}$ is consistent for θ/e. [Hint: $\ln(\prod_{i=1}^{n}X_i)^{1/n} = (1/n)\sum_{i=1}^{n}\ln X_i$.]

9.6 LARGE-SAMPLE POINT AND INTERVAL ESTIMATION

Let $X_1, X_2, \ldots, X_n$ be a random sample from a distribution with mean μ and variance σ^2. We have seen that the sample mean $\bar{X}$ is an unbiased estimate of μ and the sample variance S^2 is an unbiased estimate of σ^2 no matter how large or how small n is. If n is small we have to make further assumptions about the functional form of the distribution of X_i's in order to construct a sharp (minimum level) confidence interval for μ (or σ) or to make statements concerning the error in estimation. The Chebyshev's inequality can be used to give a confidence interval or to provide an estimate of error but it leads to unnecessarily large intervals and poor estimates of error. Indeed, by Chebyshev's inequality

$$(1) \qquad P\left(|\bar{X} - \mu| \le k\frac{\sigma}{\sqrt{n}}\right) \ge 1 - \frac{1}{k^2}$$

and choosing $k = (1/\sqrt{\alpha})$ we get a level at least $1 - \alpha$ confidence interval

$$X - \frac{\sigma}{\sqrt{\alpha n}} \le \mu \le \bar{X} + \frac{\sigma}{\sqrt{\alpha n}}$$

for μ with length $(2\sigma/\sqrt{\alpha n})$. Similarly, an estimate of error in estimating μ by $\bar{X}$ is given by $\sigma/\sqrt{\alpha n}$ with probability at least $1 - \alpha$.

In the special case when X_i's have a normal distribution, this problem was considered in Section 8.7; for the case when X_i's are Bernoulli random variables, the problem was considered in Section 6.3.

We note that the precision of $\bar{X}$ as measured by its variance is σ^2/n, which gets smaller and smaller as n gets larger and larger. Indeed, by the law of large numbers $\bar{X}$ is consistent for μ (and S^2 for σ^2) so that $\bar{X}$ is likely to be near μ for large n. By the central limit theorem the distribution of $\bar{X}$ is highly concentrated near μ for large n and is approximately normal. This information provides much sharper confidence intervals for μ. We have

$$P\left(|\bar{X} - \mu| \le k\frac{\sigma}{\sqrt{n}}\right) \doteq P(|Z| \le k)$$

where Z is a standard normal random variable. Choosing $k = z_{\alpha/2} > 0$ we see that

$$(2) \qquad \bar{X} - z_{\alpha/2}\frac{\sigma}{\sqrt{n}} \le \mu \le \bar{X} + z_{\alpha/2}\frac{\sigma}{\sqrt{n}}$$

is an approximate $1 - \alpha$ level confidence interval for μ with length $(2\sigma/\sqrt{n})z_{\alpha/2}$. Writing $z_{\alpha/2} = z$ and using Problem 7.6.24, it is easy to see that

$$\frac{2\sigma}{\sqrt{\alpha n}} > \frac{2\sigma z}{\sqrt{n}}.$$

In fact,

$$\frac{\alpha}{2} = P\left(Z > z_{\alpha/2}\right) \le \frac{1}{z}\frac{e^{-z^2/2}}{\sqrt{2\pi}}$$

so that

$$\frac{1}{\alpha} \geq \sqrt{\frac{\pi}{2}}\, ze^{z^2/2} > z\left(1 + \frac{z^2}{2}\right) > z^2,$$

or

$$\frac{1}{\sqrt{\alpha}} > z.$$

If σ is unknown we can substitute S for σ since S is likely to be near σ for large n (by consistency and Example 9.5.3).

The method used above is very general and applies whenever we can find an estimate $\hat{\theta}$ of θ which has an asymptotic normal distribution with mean θ and variance $\sigma_{\hat{\theta}}^2$, that is, whenever

$$P\{(\hat{\theta} - \theta)/\sigma_{\hat{\theta}} \leq x\} \to (1/\sqrt{2\pi})\int_{-\infty}^{x} \exp(-y^2/2)\, dy \text{ as } n \to \infty,$$

where n is the sample size. In that case, for sufficiently large n we have

$$P\{|\hat{\theta} - \theta| \leq z_{\alpha/2}\sigma_{\hat{\theta}}\} \doteq 1 - \alpha$$

and it follows that $[\hat{\theta} - z_{\alpha/2}\sigma_{\hat{\theta}}, \hat{\theta} + z_{\alpha/2}\sigma_{\hat{\theta}}]$ is an approximate $1 - \alpha$ level confidence interval for θ. In practice, however, $\sigma_{\hat{\theta}}$ will hardly if ever be known and will have to be estimated from the sample. This is done by replacing in $\sigma_{\hat{\theta}}$ any unknown parameter or parameters by consistent estimates. In the following table we give some typical parameters of interest, their point estimates, and $\sigma_{\hat{\theta}}^2$. The confidence interval is easily obtained by an appropriate substitution in $[\hat{\theta} - z_{\alpha/2}\hat{\sigma}_{\hat{\theta}}, \hat{\theta} + z_{\alpha/2}\hat{\sigma}_{\hat{\theta}}]$, where $\hat{\sigma}_{\hat{\theta}}$ is an estimate of $\sigma_{\hat{\theta}}$.

The notation used in Table 1 is the same as that used earlier. Thus

$$\overline{X} = \sum_{i=1}^{m} \frac{X_i}{m},\ Y = \sum_{i=1}^{n} \frac{Y_i}{n},\ \hat{p} = \text{proportion of successes},$$

$$S_p^2 = \frac{(m-1)S_1^2 + (n-1)S_2^2}{(m+n-2)},\ S_D^2 = \frac{\sum_{i=1}^{n}(D_i - \overline{D})^2}{(n-1)},$$

where $D_i = (X_i - Y_i)$, and so on.

Remark 1. How large should n be before we can apply the methods of this section? As we saw in Sections 9.3 and 9.4, the answer very much depends on the distribution being sampled. In general, we will be guided by considerations of Sections 9.3 and 9.4, especially Table 1 of Section 9.4. Whenever applicable, it is also wise to use the continuity correction. If nothing is known about the

TABLE 1. Some Common Point Estimates with Their Asymptotic Variance

Parameter, θ	Sample Size(s)	Point Estimate, θ	σ_θ^2	$\widehat{\sigma_\theta^2}$
1. Single sample				
Mean, μ	n	$\bar{X}$	σ^2/n	S^2/n
Proportion, p	n	$\hat{p}$	$p(1-p)/n$	$\hat{p}(1-\hat{p})/n$
Standard deviation[a], σ	n	S	$\sigma^2/(2n)$	$S^2/(2n)$
2. Two independent samples				
Difference in means, $\mu_1 - \mu_2$	m and n	$\bar{X} - \bar{Y}$	$(\sigma_1^2/m) + (\sigma_2^2/n)$	$\begin{cases} (S_1^2/m) + (S_2^2/n), \text{ if } \sigma_1 \neq \sigma_2 \\ S_p^2[(m+n)/mn] \text{ if } \sigma_1 = \sigma_2 \end{cases}$
Difference in proportions, $p_1 - p_2$	m and n	$\hat{p}_1 - \hat{p}_2$	$p_1(1-p_1)/m + p_2(1-p_2)/n$	$\hat{p}_1(1-\hat{p}_1)/m + \hat{p}_2(1-\hat{p}_2)/n$
3. Paired observations				
Difference in means, $\mu_D = \mu_1 - \mu_2$	n	$\bar{D} = \bar{X} - \bar{Y}$	$\text{var}(X - Y)/n$	S_D^2/n

[a]We assume that the sample comes from a nearly normal population.

underlying distribution, a sample of size 30 or more is usually considered adequate for using large-sample results. See also Remark 3.

Example 1. *Mean Television Viewing Time per Family per Week.* In order to estimate the average television viewing time per family in a large southern city, a sociologist took a random sample of 500 families. The sample yielded a mean of 28.4 hours per week with a standard deviation of 8.3 hours per week. Then an approximate 95 percent confidence interval for the true unknown mean viewing time per family per week is given by

$$\left[28.4 - 1.96 \frac{8.3}{\sqrt{500}}, \; 28.4 + 1.96 \frac{8.3}{\sqrt{500}} \right] = [27.67, 29.13]. \qquad \Box$$

Example 2. *Percentage of Allergic Reactions.* A new soaking and wetting solution is being tested for contact lenses. A random sample of 484 contact-lens wearers tested the solution and only 27 reported an allergic reaction. Let p be the true proportion of contact-lens wearers who would suffer an allergic reaction to this solution. Then in Table 1, $\theta = p$ and a 96 percent confidence interval for p is given by

$$\left[\hat{p} - z_{.02} \sqrt{\frac{\hat{p}(1 - \hat{p})}{n}}, \; \hat{p} + z_{.02} \sqrt{\frac{\hat{p}(1 - \hat{p})}{n}} \right].$$

In our case $\hat{p} = 27/484 = .056$, $z_{\alpha/2} = z_{.02} = 2.06$ and

$$z_{.02} \sqrt{\frac{\hat{p}(1 - \hat{p})}{n}} = .022$$

so that a 96 percent (approximate) confidence interval for p is [.034, .078]. Thus we are 96 percent confident that between 3.4 and 7.8 percent of the users will have allergic reaction to the soaking and wetting solution. $\qquad \Box$

Remark 2. *(Sampling From a Finite Population).* The confidence intervals for μ are also valid (with minor modifications) when sampling without replacement from a finite population. Suppose a sample of size n is taken from a population of N units. Then we know (Example 5.3.1) that

$$\mathscr{E} \overline{X} = \mu \text{ and } \text{var}(\overline{X}) = \frac{N - n}{N - 1} \frac{\sigma^2}{n}$$

where μ is the population mean and σ^2 is the population variance. From Remark 9.4.2 the random variable

$$Z = \frac{\overline{X} - \mu}{\sigma \sqrt{(N - n)/(N - 1)}} \sqrt{n}$$

has approximately a standard normal distribution for large n. If σ is not known, which is usually the case, we replace σ by S (Example 9.5.3). Then an approximate $1 - \alpha$ level confidence interval for μ is given by

$$\left[\overline{X} - z_{\alpha/2}\frac{S\sqrt{N - n}}{\sqrt{n(N - 1)}}, \quad \overline{X} + z_{\alpha/2}\frac{S\sqrt{N - n}}{\sqrt{n(n - 1)}}\right].$$

Since N is large, $N \doteq N - 1$. Moreover, if n/N is small, say, $\leq .05$, then $(N - n)/(N - 1) \doteq 1$ and we can use the approximation

$$\text{var}(\overline{X}) = \sigma^2/n.$$

For example, if $N = 100$ and $n = 5$, then

$$\frac{N - n}{N - 1} = \frac{95}{99} = .96.$$

If $N = 10,000$ and $n = 250$, then $n/N = .025 < .05$ and

$$\frac{N - n}{N - 1} = \frac{9750}{9999} = .98.$$

In either case the finite population correction $(N - n)/(N - 1)$ can safely be ignored.

Remark 3. (Sample Size Determination). How large should the sample size n be such that the error in estimating μ by $\overline{X}$ (or p by $\hat{p}$, the sample proportion) is at most ε with a specified level of confidence? This question was essentially answered in Section 9.3 (see Examples 9.3.3 and 9.3.4). We want n to satisfy

$$P(|\overline{X} - \mu| < \varepsilon) \doteq 1 - \alpha.$$

If σ is known, we choose $n \doteq (z_{\alpha/2}\sigma/\varepsilon)^2$. Sometimes we want a prescribed width d for the confidence interval. In that case we choose n to satisfy

$$(2\sigma/\sqrt{n})z_{\alpha/2} = d,$$

or

$$n \doteq \left(\frac{2\sigma z_{\alpha/2}}{d}\right)^2.$$

In fact, $d = 2\varepsilon$ so it does not much matter whether d is given or ε is.

In the case of proportions, $\sigma^2 = p(1 - p)$ is unknown. We either use the fact that $\sigma^2 \leq 1/4$ so that

$$n \doteq (z_{\alpha/2}/2\varepsilon)^2$$

or use some prior information about p if such information is available.

In sampling from a finite population, if σ is known and we want the error in estimation to be at most ε with confidence level $1 - \alpha$, then n is chosen from

$$z_{\alpha/2}\sigma\sqrt{\frac{N - n}{Nn}} \le \varepsilon$$

so that

$$n \doteq \frac{Nz_{\alpha/2}^2\sigma^2}{N\varepsilon^2 + z_{\alpha/2}^2\sigma^2}.$$

See also Problems 26 and 27.

Example 3. Proportion of Voters in Favor of Proposition A. Suppose an estimate is required of the proportion p of voters who are in favor of a certain Proposition A on the ballot with 95 percent confidence to within 3 percent of p. How large a sample should be taken?

In this case $\varepsilon = .03$, and $1 - \alpha = .95$ so that $z_{\alpha/2} = z_{.025} = 1.96$. It follows that

$$n \doteq \left(\frac{1.96}{2(.03)}\right)^2 = 1067.11$$

so that we need a sample of minimum size 1068. If, on the other hand, we feel that the true p is near .6, then we choose n from

$$n \doteq \left(z_{\alpha/2}\sigma/\varepsilon\right)^2 = \left(1.96\sqrt{(.6)(.4)}/.03\right)^2 = 1024.43$$

so that the minimum sample size required is 1025. Table 1 in Section 2.8 gives the minimum sample size required for prescribed levels of confidence and margins of error in simple random sampling. □

Example 4. Difference in Mean Life of Nonsmokers and Smokers. In order to estimate the difference in mean life of nonsmokers and smokers in a large city, independent random samples of 36 nonsmokers and 49 smokers were taken with the following results.

		Sample	
	Size	Mean	Standard Deviation
Nonsmokers	$m = 36$	$\bar{x} = 72$	$s_1 = 9$
Smokers	$n = 44$	$\bar{y} = 62$	$s_2 = 11$

Let us find a 90 percent confidence interval for the difference in means $\mu_1 - \mu_2$. In Table 1, we take $\theta = \mu_1 - \mu_2$. Since $\alpha = 1 - .9 = .1$, $z_{\alpha/2} = z_{.05} = 1.645$ so that the confidence interval is given by

$$\left[72 - 62 - 1.645\sqrt{\frac{9^2}{36} + \frac{11^2}{44}}, \quad 72 - 62 + 1.645\sqrt{\frac{9^2}{36} + \frac{11^2}{44}}\right]$$

$$= [10 - 3.68, 10 + 3.68] = [6.32, 13.68].$$

Thus we are about 90 percent confident that the nonsmokers have a mean lifetime that is larger than the mean lifetime of smokers by a margin of between 6.3 to 13.7 years. □

Example 5. Difference in Proportions of People in Favor of a Proposal in Two Neighborhoods. In order to estimate the difference in proportions of people who favor a certain proposal in two neighborhoods, independent random samples were taken from both neighborhoods. In the sample of 60 from the first neighborhood, 48 were in favor of the proposal. In the sample of 100 from the second neighborhood, 50 were in favor of the proposal. Then $\theta = p_1 - p_2$ in Table 1 and

$$\hat{p}_1 = 48/60 = .8, \hat{p}_2 = 50/100 = .5$$

so that

$$\sqrt{\frac{\hat{p}_1(1 - \hat{p}_1)}{m} + \frac{\hat{p}_2(1 - \hat{p}_2)}{n}} = \sqrt{\frac{.8(.2)}{60} + \frac{.5(.5)}{100}} = .0719.$$

If $1 - \alpha = .95$, then $\alpha/2 = .025$ and $z_{\alpha/2} = 1.96$ so that an approximate 95 percent confidence interval for the difference in proportions is given by

$$[.8 - .5 - 1.96(.0719), .8 - .5 + 1.96(.0719)] = [.16, .44]. \qquad □$$

Problems for Section 9.6

1. A coffee packing plant produces 10-ounce coffee jars. The packing line has been set to pack an average of $\mu = 10.03$ ounces. In order to check the line a sample of 36 jars is taken and shows a mean of 10.01 ounces with a standard deviation of .02 ounces. Construct a 95 percent confidence interval for the mean and check to see if the packing line is working according to the specification.

2. A random sample of 100 marriage records is taken from the official records of the last 12 months in a certain large town and shows that the mean marriage age of women is 22.8 years with a standard deviation of 5.1 years. Find a 98 percent confidence interval for the true mean age at which women marry in this town.

3. A random sample of 49 entering male freshmen at a state university showed an average height of 68.2 inches with a standard deviation of 6 inches. Find a 90 percent confidence interval for the mean height of entering male freshmen at this university.

4. In order to estimate the mean monthly rental of two-bedroom apartments in a certain midwestern university town, a sample of 100 such units is taken. If the sample mean is $275 and the sample standard deviation is $60, find a 99 percent confidence interval for the true mean.

5. A new shipment of frozen apple juice cans has been received at a certain supermarket. A random sample of 64 cans yields a mean net weight of 12.03 ounces with a standard deviation of .10 ounce. Find a 98 percent confidence interval for the true mean net weight of apple juice (per can) in this shipment.

6. The grade point averages of 96 randomly selected students at a certain four-year college showed an average GPA of 2.37 with a standard deviation of 0.31. Find a 94 percent confidence interval for the true average GPA at this college.

7. A random sample of 200 voters in a county is taken and each voter is interviewed. It is found that 80 voters are in favor of the incumbent senator who is up for reelection. Find a 98 percent confidence interval estimate for the fraction of voters who favor the incumbent. Does the incumbent have reason for optimism?

8. A well-known chewing gum manufacturer advertises a new squirt gum with a flavored syrup in the center. In order to estimate the proportion of defective gum produced by the manufacturer, a sample of 800 gums was taken and it was found that 69 of these were without syrup. Find a 92 percent confidence interval for the true proportion of defectives.

9. A random sample of 400 Bowling Green wage earners showed that 175 have an income of more than $18,000 a year. Find a 90 percent confidence interval for the proportion of Bowling Green wage earners with incomes over $18,000 a year.

10. In a research study involving a new drug as a cure for leukemia, a random sample of 64 patients was selected and given the new drug. After five years, 50 patients were found to be alive. Find a 98 percent confidence interval estimate for the proportion of leukemia patients who will be alive five years after treatment with the new drug.

11. To estimate the five-year survival rate of the new drug in Problem 10 to within 10 percent of the true proportion with 95 percent confidence:
    (i)   How large a sample size is needed?
    (ii) If the success rate of the current treatment is known to be about 70 percent, how large should the sample be?

12. In order to estimate the proportion of families in Springfield who prefer female baby-sitters to male baby-sitters with an error of at most 8 percent and with 98 percent confidence, how large a sample should be taken?

13. A large tire outlet in a metropolitan area wishes to estimate with 90 percent confidence the average amount of time it takes its mechanics to balance and mount new tires and to do an "oil-and-lube" job. If it is desired that the estimate be within 10 minutes of the true value, how large a sample should be taken? It is known that a similar study a few years earlier had shown a standard deviation of 35 minutes.

14. An estimate is desired of the mean weight of a certain type of bird at a certain age. Suppose with 95 percent confidence we want to be within 12 grams of the true mean weight. How large a sample size is required if it is known from a pilot study that $\sigma = 100$ grams?

15. In order to estimate with 95 percent confidence the difference in mean weights of entering male and female students at a university, independent random samples were taken with following results.

Sex	Size	Sample Mean	Standard Deviation
Male	36	156 lb.	21 lb.
Female	48	118 lb.	16 lb.

Find a 95 percent confidence interval for $\mu_1 - \mu_2$:
(i)   When σ_1, σ_2 are assumed to be different.
(ii) When it is assumed that $\sigma_1 = \sigma_2$.
(μ_1 and σ_1 correspond to the male population.)

16. A confidence interval estimate is desired for the difference in average tread wear of steel-belted radial tires and bias ply tires manufactured by a certain company. Independent random samples of 30 radial and 42 bias ply tires are taken and driven 30,000 miles each on a machine. The tread wear is given below.

Type	Size	Sample Mean	Standard Deviation
Belted	30	0.6 cm	.27 cm
Bias Ply	42	0.9 cm	.33 cm.

(i)   Find a 96 percent confidence interval for the difference in means between the tread wear in 30,000 miles.
(ii) What is the confidence interval if it is assumed that $\sigma_1 = \sigma_2$?

17. Independent random samples of 800 voters in Detroit and 600 voters in Kansas City show 232 and 294 voters, respectively, in favor of the proposed constitutional amendment on a balanced budget. Find a 90 percent confidence interval for the difference $p_1 - p_2$ in proportions of voters in favor of the amendment in the two cities.

18. Suppose 90 randomly selected patients were given aspirin and 100 randomly selected patients were given an aspirin-free pain reliever. After one hour the patients were asked if the drug had desired effect. The following results were obtained.

Pain Reliever	Sample Size	Patients Obtaining Pain Relief
Aspirin	90	50
Aspirin-free	100	60

Find a 98 percent confidence interval for the difference in two proportions. Is there evidence to suggest that the aspirin-free pain reliever results in a higher proportion of patients with pain relief than aspirin?

19. In order to estimate the difference in proportions of defective parts produced by two assembly lines, independent random samples of 200 parts are selected from each assembly line. If Line I yields 15 defectives and Line II yields 27 defectives, find a 96 percent confidence interval for the true difference in proportions of defectives. Would you conclude that Line II is producing a larger proportion of defectives than Line I?

20. The weights of 64 randomly selected people were measured before they went on a diet program and after they completed the program. Let d_i be the difference (= weight before minus weight after) in weights for the ith individual, $i = 1, 2, \ldots, 64$.

The following results were obtained.

$$\sum_{i=1}^{64} d_i = 512 \text{ pounds}, \qquad \sum_{i=1}^{64} \left(d_i - \bar{d} \right)^2 = 1010 \text{ (pounds)}^2.$$

Find a 96 percent confidence interval for the weight loss as a result of the diet program.

21. The systolic blood pressures of 36 randomly selected nursing home patients were measured before and after a 15-minute program of back rub, with the following results:

$$\sum_{i=1}^{36} d_i = 414 \text{ mm Hg}, \qquad \sum_{i=1}^{36} \left(d_i - \bar{d} \right)^2 = 5915 \text{ (mm Hg)}^2,$$

where d_i is the reduction in systolic blood pressure of the ith patient. Find a 95 percent confidence interval for the average reduction in systolic blood pressure as a result of the back rub. Would you conclude that the program of back rub is effective in lowering systolic blood pressure?

22. (i) Suppose $(X_1, X_2, \ldots, X_k)$ has a multinomial distribution with parameters $n, p_1, p_2, \ldots, p_k, \sum_{j=1}^{k} p_j = 1$. Find an approximate large-sample confidence interval for $p_i - p_j$, $i \neq j$. [Hint: $(X_i - X_j)/n$ has approximately a normal distribution with mean $p_i - p_j$ and variance $\{ p_i(1 - p_i) + p_j(1 - p_j) + 2p_i p_j \}/n$. Indeed any linear combination of the X_i's has approximately a normal distribution.]

(ii) There are five candidates in an election, and it is believed that candidates A and B are top contenders. A random sample of 1000 voters showed that 370 favor Candidate A, 320 favor Candidate B, 140 favor C, 132 favor D, and 38 favor E. Find a 98 percent confidence interval for the difference in proportions of voters who favor candidates A and B. Is there evidence to suggest that Candidate A is ahead in the race?

23. (i) In Problem 22(i) take $k = 2$, and $p_1 = p$, $p_2 = 1 - p$. Find a $1 - \alpha$ level confidence interval for $p_1 - p_2$ either by specializing the result of Problem 22(i) or directly by noting that $p_1 - p_2 = 2p - 1$ and that X_1 has a binomial distribution.

(ii) A random sample of 600 people was polled for their opinion on the Equal Rights Amendment. If 400 respondents were found to be in favor of ERA, find a 95 percent confidence interval for the difference in proportions in the population of those in favor and those against ERA.

24[†]. Using the fact that S has an approximate normal distribution with mean σ and variance $\sigma^2/(2n)$ for large n, and manipulating the inequality in

$$P\left\{ |S - \sigma| \le z_{\alpha/2}\sigma/\sqrt{2n} \right\} \doteq 1 - \alpha,$$

[†]In Problems 24 and 25 we have assumed that the sample comes from a nearly normal population.

show that

$$\left[S \Big/ \left(1 + \frac{z_{\alpha/2}}{\sqrt{2n}} \right), S \Big/ \left(1 - \frac{z_{\alpha/2}}{\sqrt{2n}} \right) \right]$$

is also a $1 - \alpha$ level confidence interval for σ.

25[†]. Let $X_1, X_2, \ldots, X_m$ and $Y_1, Y_2, \ldots, Y_n$ be independent random samples from two populations with unknown variances σ_1^2 and σ_2^2. Find a large sample confidence interval for $\sigma_1 - \sigma_2$ with approximate confidence level $1 - \alpha$. {Hint: $S_1 - S_2$ has an asymptotic normal distribution with mean $\sigma_1 - \sigma_2$ and variance $[\sigma_1^2/(2m) + \sigma_2^2/(2n)]$.}

26. Sample size for difference of means: Suppose $X_1, X_2, \ldots, X_n$ and $Y_1, Y_2, \ldots, Y_n$ are independent random samples from two populations with means μ_1, μ_2 and variances σ_1^2, σ_2^2 respectively. Show that the sample size required in order to estimate $\mu_1 - \mu_2$ by $\overline{X} - \overline{Y}$ to within ε of $\mu_1 - \mu_2$ with $1 - \alpha$ level confidence is given by

$$n = z_{\alpha/2}^2 \left(\sigma_1^2 + \sigma_2^2 \right) / \varepsilon^2.$$

27. Sample size for difference of proportions: Let $\hat{p}_1, \hat{p}_2$ be two sample proportions (both based on samples of the same size n) from populations with proportions p_1 and p_2 respectively. Suppose $\hat{p}_1, \hat{p}_2$ are independent. Show that the smallest n required in order that

$$P\{|\hat{p}_1 - \hat{p}_2 - (p_1 - p_2)| \le \varepsilon\} \doteq 1 - \alpha$$

is given by

$$n = z_{\alpha/2}^2 (p_1(1 - p_1) + p_2(1 - p_2)) / \varepsilon^2.$$

If no prior knowledge of p_1 and p_2 is available, then we choose

$$n \doteq z_{\alpha/2}^2 / 2\varepsilon^2.$$

9.7   LARGE-SAMPLE HYPOTHESIS TESTING

In many problems the interest is not in the estimation of an unknown parameter θ but in checking the validity of a certain statement about θ. Do the coffee jars contain 10 ounces of coffee per jar as the label claims? Do the men and women students at a certain state university have the same grade point averages? This problem of hypothesis testing has been considered for small samples in Sections 6.5 and 8.7 for specific distributions. We now consider the case when the sample size is large.

[†] In Problems 24 and 25 we have assumed that the sample comes from a nearly normal population.

Let $\hat{\theta} = \hat{\theta}(X_1, X_2, \ldots, X_n)$ be a point estimate of θ based on a random sample $X_1, X_2, \ldots, X_n$. Suppose $\hat{\theta}$ has an asymptotic normal distribution with mean θ and variance $\sigma_{\hat{\theta}}^2$. This fact can be used to test hypotheses such as $H_0: \theta = \theta_0$ against one-sided or two-sided alternatives. Under H_0, the statistic

$$(1) \qquad\qquad Z = \frac{\hat{\theta} - \theta_0}{\sigma_{\hat{\theta}}(\theta_0)}$$

has approximately a standard normal distribution for sufficiently large n. By $\sigma_{\hat{\theta}}(\theta_0)$ in the denominator, we mean the standard deviation of $\hat{\theta}$ computed under the assumption that $\theta = \theta_0$. Thus if $\hat{\theta} = \overline{X}$, $\theta_0 = \mu_0$, then $\sigma_{\hat{\theta}}(\theta_0) = \sigma/\sqrt{n}$ irrespective of what θ_0 is. However, if $\hat{\theta} = \hat{p}$, $\theta_0 = p_0$, then $\sigma_{\hat{\theta}}(\theta_0) = \sqrt{p_0(1 - p_0)/n}$ and we do not need to use $\hat{\sigma}_{\hat{\theta}} = \sqrt{\hat{p}(1 - \hat{p})/n}$ in computing z. Similarly, if $\theta = p_1 - p_2$, $\hat{\theta} = \hat{p}_1 - \hat{p}_2$ and $\theta_0 = 0$, then under $H_0: p_1 = p_2 = p$ (unknown) so that

$$\sigma_{\hat{p}_1 - \hat{p}_2} = \left(\frac{p(1 - p)}{m} + \frac{p(1 - p)}{n} \right)^{1/2}.$$

Since p is unknown, however, we estimate it by

$$(2) \qquad\qquad \hat{p} = \frac{X + Y}{m + n}$$

where X and Y are the respective number of "successes" in the first and second samples. Then

$$(3) \qquad\qquad \hat{\sigma}_{\hat{p}_1 - \hat{p}_2} = \sqrt{\hat{p}(1 - \hat{p})}\,(1/m + 1/n)^{1/2}$$

and

$$(4) \qquad\qquad Z = \frac{\hat{p}_1 - \hat{p}_2}{\hat{\sigma}_{\hat{p}_1 - \hat{p}_2}}$$

is the appropriate test statistic.

If $\theta = \sigma$, then $\hat{\theta} = S$ and[†] $\sigma_{\hat{\theta}} = \sigma/\sqrt{2n}$ and under $H_0: \sigma = \sigma_0$, $\sigma_{\hat{\theta}}(\sigma_0) = \sigma_0/\sqrt{2n}$ so that the appropriate test statistic is

$$Z = \sqrt{2n}\,\frac{(S - \sigma_0)}{\sigma_0}.$$

Clearly, large values of $\hat{\theta}$ support $\theta > \theta_0$ so that large values of Z tend to discredit $H_0: \theta = \theta_0$ against $H_1: \theta > \theta_0$. Similar arguments apply in the remaining cases.

[†] Here we have assumed that the sample comes from a nearly normal population.

A Large Sample Test of H_0: $\theta = \theta_0$

Calculate

$$z = \frac{\hat{\theta} - \theta_0}{\sigma_{\hat{\theta}}(\theta_0)} \quad \left(\text{or } \frac{\hat{\theta} - \theta_0}{\hat{\sigma}_{\hat{\theta}}(\theta_0)} \text{ of } \sigma_{\hat{\theta}} \text{ unknown} \right)$$

from the data.

H_1	Reject H_0 at Level α if	P-value
$\theta > \theta_0$	$z > z_\alpha$	$P(Z \geq z)$
$\theta < \theta_0$	$z < -z_\alpha$	$P(Z \leq z)$
$\theta \neq \theta_0$	$|z| > z_{\alpha/2}$	$2P(Z > |z|)$

Here Z is a standard normal random variable.

Table 1 of Section 9.6 gives some common point estimates and their (estimated) asymptotic variances $\hat{\sigma}_{\hat{\theta}}^2$.

How large should the sample size be in order to apply this procedure? This question has been answered in Remark 9.6.1. See also Remark 1 and Problems 19 and 20 in the section.

Example 1. Net Weight of Beer in a Can. A new filling machine has been put in use to fill 12-ounce beer cans. In order to check if the machine is operating properly, a random sample of 49 beer cans is taken. The cans are found to contain an average of 11.89 ounces of beer with a standard deviation of .36 ounce. Is the difference between the stated average and the observed average due to random variation? Here the null hypothesis of interest is H_0: $\mu = 12$ ounces to be tested against H_1: $\mu \neq 12$. The observed value of the test statistics is

$$z = \frac{\bar{x} - \mu_0}{s/\sqrt{n}} = \frac{11.89 - 12}{.36/\sqrt{49}} = -2.14.$$

Under H_0,

$$P(|Z| \geq |z|) = P(|Z| \geq 2.14) = 2(.0162) = .0324.$$

Thus in random sampling, if the process were operating according to specifications, there is only about a 3.24 percent chance of obtaining such an extreme value of sample mean weight. Hence, at significance level $\alpha \geq .0324$ we conclude that the difference in the sample mean weight and stated population mean weight is not due to chance fluctuation.   □

When testing for the proportion p, the appropriate test statistic is

$$Z = \frac{\hat{p} - p_0}{\sqrt{p_0(1 - p_0)/n}} = \frac{n\hat{p} - np_0}{\sqrt{np_0(1 - p_0)}} = \frac{X - np_0}{\sqrt{np_0(1 - p_0)}},$$

where X, the number of successes, is integer valued. It would be wise to apply the continuity correction to improve normal approximation. In the following example it does not much matter whether or not the continuity correction is applied. It is given here to illustrate the procedure and an alternative analysis of the Salk vaccine data of Example 9.4.5.

Example 2. Salk Vaccine Field Trial. In Example 9.4.5 we considered the following data from Salk polio vaccine field trial of 1954.

Children	Contracted Polio	Polio-Free	Total
Vaccinated	33	200, 712	200,745
Control	115	201,114	201,229
Total	148	401, 826	401,974

We applied an approximation to the Fisher–Irwin test of Section 6.4 to test the null hypothesis that the treatment has no effect. That is, under H_0 the proportions of children who contracted polio is the same in the treatment as well as the control group.

There is yet another way to look at the data. There were about 402,000 children involved in the experiment. Each child had equal chance of being assigned to a treatment (vaccinated) or control (placebo) group. In all there were 148 cases of polio. Assuming that the assignment to the two groups had no effect on the outcome, could not the difference between 33 cases of polio in the treatment group and 115 in the control group be due to random variation?

Here $n = 148$, $H_0: p = 1/2$, $H_1: p \neq 1/2$, $\hat{p} = 33/148 = .22$. Rather than use

$$z = \frac{\hat{p} - p_0}{\sqrt{p_0(1 - p_0)/n}} = \frac{.22 - .5}{\sqrt{.5(.5)/148}} = -6.81$$

we note that X, the number of polio cases in the treatment group is 33 and the corresponding P-value is $P(X \leq 33)$. Since the critical region is in the left tail

$$P(X \leq 33) = P(X < 33.5)$$

and

$$z = \frac{33.5 - 148(1/2)}{\sqrt{148(1/2)(1/2)}} = -6.66.$$

The associated P-value $P(X \leq -6.66)$ being less than 1 in 100 million, we have to conclude that the difference cannot be explained by chance variation. The only reasonable explanation is that the vaccine is effective. This is the same conclusion we arrived at in Example 9.4.5. □

Remark 1. *(Sample Size Determination).* Suppose σ is known and we wish to find the smallest sample size required to test $H_0: \mu \leq \mu_0$ against $H_1: \mu > \mu_0$ such that the level is approximately α and the probability of a type II error at some specified alternative $\mu(> \mu_0)$ is β. (Here μ is the population mean.) Then $\theta = \mu$, $\hat{\theta} = \bar{X}$ and we reject H_0 at level α if

$$\sqrt{n}\left(\bar{X} - \mu_0\right) > \sigma z_\alpha, \quad \text{or if } \bar{X} > \mu_0 + \frac{\sigma z_\alpha}{\sqrt{n}}.$$

We need to choose n such that

$$\beta \doteq P(\text{type II error at } \mu) = P\{\text{accept } H_0 | \mu\}$$

$$= P\left\{\bar{X} \le \mu_0 + \frac{\sigma z_\alpha}{\sqrt{n}} \Big| \mu\right\} = P\left\{\frac{\sqrt{n}\left(\bar{X} - \mu\right)}{\sigma} \le z_\alpha + \frac{(\mu_0 + \mu)\sqrt{n}}{\sigma}\right\}.$$

It follows that

$$z_{1-\beta} = -z_\beta \doteq z_\alpha + \frac{(\mu_0 - \mu)\sqrt{n}}{\sigma}$$

so that

(5)
$$n \doteq \left\{\frac{(z_\alpha + z_\beta)\sigma}{(\mu - \mu_0)}\right\}^2.$$

The same argument applies to the two-sided case and we get

(6)
$$n \doteq \left\{\frac{(z_{\alpha/2} + z_{\beta/2})\sigma}{(\mu - \mu_0)}\right\}^2.$$

In practice, σ is not likely to be known. In that case, one uses any prior information on σ. For example, one can use the value of the sample standard deviation obtained from a prior sample if available.

Similar argument applies to the case when we wish to find n in order to test $p \ge p_0$ against $p < p_0$ at a level α such that at some specified alternative $p(< p_0)$ the type II error is β. We choose n from the relation

(7)
$$n \doteq \frac{\left\{z_\alpha\sqrt{p_0(1 - p_0)} + z_\beta\sqrt{p(1 - p)}\right\}^2}{(p_0 - p)^2}.$$

If the alternative is two-sided, we replace z_α and z_β by $z_{\alpha/2}$ and $z_{\beta/2}$ respectively in (7). See also Problems 19 and 20.

> ***Example 3. Custom Shirts.*** A mail-order company that sells custom-made shirts has 10 percent returns over the years it has been in business. A prospective buyer would like to test this assertion before it makes an offer to buy out the company. How large a

sample should it take if it is willing to tolerate errors of $\alpha = .05$ and $\beta = .05$ if true $p = .20$?

Since $z_\alpha = z_\beta = 1.65$ and $p_0 = .10$, $p = .20$ we have from (7)

$$n \doteq \frac{\left\{1.65\sqrt{.1(.9)} + 1.65\sqrt{.2(.8)}\,\right\}^2}{(.1)^2} = 133.4.$$

Thus he needs to take a random sample of 134 orders to test H_0: $p = .10$ against H_1: $p > .10$ at level $\alpha = .05$ to ensure that the type II error probability if true $p = .20$ is also approximately .05. ☐

Example 4.   Sex Differences in Opinion on Abortion.   In order to test if there is a significant difference between opinions of males and females on abortion, independent random samples of 100 males and 150 females were taken with the following results.

| | | Opinion on Abortion | |
Sex	Sample Size	Favor	Oppose
Male	100	52	48
Female	150	95	55

We wish to test H_0: $p_1 = p_2$ against H_1: $p_1 \neq p_2$. We have

$$\hat{p}_1 = .52, \hat{p}_2 = .63 \text{ and } \hat{p} = \frac{52 + 95}{100 + 150} = .59.$$

Hence

$$z = \frac{.52 - .63}{\sqrt{.59((1/100) + (1/150))}} = -\frac{.11}{.099} = -1.1.$$

The corresponding P-value is given by

$$2P(Z \geq 1.1) = 2(.1357) = .2714.$$

Thus the observed difference between male and female respondents could very well be due to chance variation. ☐

Example 5.   Mean Ages of Females Using IUD's.   It is suspected that the average age of females who retain interuterine devices as contraceptives is higher than those who expel the devices. Independent random samples of females who had retained IUD's and those who had expelled IUD's were taken with the following results.

| | | Sample | |
	Size	Mean Age (Years)	Standard Deviation
Expelled	64	26	7
Retained	36	31	13

We wish to test $H_0: \mu_1 - \mu_2 = 0$ against the alternative $H_1: \mu_1 < \mu_2$. We have

$$z = \frac{26 - 31}{\sqrt{7^2/64 + 13^2/36}} = -\frac{5}{2.34} = -2.14,$$

with associated P-value

$$P(Z < -2.14) = .0162.$$

Thus under H_0 there is only 1.6 percent chance of obtaining such an extreme value of z, and we have to conclude that the difference in ages is significant at levels $\alpha \geq .0162$. That is, we reject H_0 and conclude that the mean age of females who retain the IUD's is higher than those who expel it. $\square$

Remark 2. If the level of significance α is preassigned, then the methods of Section 9.6 and 9.7 are essentially the same. The difference lies in the objective. The objective of a test is to reach a decision whether or not to reject a null hypothesis. The confidence interval, on the other hand, gives a range of possible values for the unknown parameter. The objective is to state, with a high degree of confidence, an interval for the true value of the unknown parameter.

Remark 3. In view of Remark 9.6.2 we can test hypotheses concerning the mean of a finite population (or the difference between means) in sampling without replacement. The appropriate test statistic for testing $\mu = \mu_0$ is

$$Z = \frac{\sqrt{n}\,(\overline{X} - \mu_0)}{\left\{ S\sqrt{(N - n)/(N - 1)} \right\}}$$

where S is the sample variance, N is the size of the population and n is the sample size.

Problems for Section 9.7

In each problem, state clearly the null and alternative hypotheses, compute the P-value, and state your conclusion.

1. The average mortality rate for a certain disease is 24 out of 100 attacks. A new treatment for the disease is tested on 300 patients. There were 54 deaths. What can you say about the efficacy of this treatment?

2. A random sample of 400 registered voters showed that if the election were held on the day the poll was taken 216 of the respondents will vote for Candidate A. Does Candidate A have any reason for optimism?

3. A filling machine is set up to fill cans with orange juice with an average of 48.05 ounces. The standard deviation in fill is set at $\sigma = .20$ ounce. A random sample of 36 filled cans from this machine is taken and it is found that the sample mean is 47.82 ounces with a standard deviation of .28 ounce.

(i) Is the sample standard deviation of .28 ounce consistent with the setting of $\sigma = .20$ ounce?

(ii) Is the sample mean of 47.82 ounces consistent with the setting of $\mu = 48.05$ ounces?

4. A branch manager of a bank will hire a new teller if the average service time of customers is more than 4 minutes. She takes a random sample of 50 customers and finds that the sample mean is 4.5 minutes with a standard deviation of 1.5 minutes. Should she hire a new teller?

5. A circuit fuse is designed to burn out as the current reaches 15 amperes. A random sample of 49 fuses is selected from a large lot and each fuse is tested for its breaking point. The following results are obtained:

   Sample mean = 15.7 amperes; sample standard deviation = 1.6 amperes.

   Does the lot confirm to the specification?

6. The average miles per gallon (mpg) of lead-free gasoline for a new model is claimed to be 42. An independent consumer protection group, believing that the claim is exaggerated, tests 40 cars of this particular model and finds that the sample mean is 40.8 mpg with a standard deviation of 4.2 mpg. At what level should the agency reject the manufacturer's claim?

7. A leading manufacturer of dishwashers claims that 90 percent of its dishwashers do not require any repairs for the first four years of operation. A random sample of 164 dishwashers showed that 28 dishwashers required some repair during the first four years of operation. Is there evidence to doubt the manufacturer's claim?

8. The advertisements for a gasoline additive claim that it results in an increase in mileage. In order to check this claim a certain car model is selected with an average mpg of 18.7. A random sample of 49 runs with the additive showed a sample mean of 19.6 mpg with a standard deviation of 3.6 mpg. Is there evidence to support the advertised claim?

9. A new filtering device is under consideration in a chemical plant. Prior to installation of the new device a random sample of size 32 yields an average of 11.2 percent impurity with a standard deviation of 8.6 percent. After the device was installed a random sample of size 42 yielded an average of 9.4 percent impurity with a standard deviation of 8.1 percent. Is the reduction in impurity significant? At what level?

10. Two machines fill milk in one-pint cartons. Independent random samples from the two machines produced the following results on net fill weights.

	Sample		
	Size	Mean	Variance
Machine 1	46	15.8 oz.	$.49 \, (oz.)^2$
Machine 2	54	16.2 oz.	$.76 \, (oz.)^2$

Is the difference in sample averages significant? At what level?

11. Independent random samples of fraternity and nonfraternity students at a state university yielded the following grade point averages:

	Sample		
Students	Size	Mean	Standard Deviation
Fraternity	41	2.54	.43
Nonfraternity	69	2.76	.88

Is there evidence to suggest that the mean GPA of fraternity and nonfraternity students are different?

12. In an attempt to verify the claim that the absentee rate in Friday morning classes is higher than that in Monday to Thursday morning classes, independent random samples of students were taken. They yielded the following results.

	Sample Size	Percentage Absent
Friday	50	14
Monday–Thursday	175	6

Do the data support the claim?

13. Suppose that patients suffering from headaches are randomly assigned to one of two treatments. In the first treatment patients are given a placebo while in the second treatment patients are given a pain reliever. After four hours it is found that 10 out 36 patients in the first group and 20 out of 42 patients in the second group report no headache. At what level is the difference between the two proportions significant?

14. Independent random samples of voters are taken in the South and in the North, and they are asked if they favor a bill to ban handguns. The following results are obtained.

Region	Sample Size	Favoring Ban
North	1842	1607
South	694	502

What conclusion do you draw from this data?

15. A study is conducted to see if the average length of stay in a hospital of insured patients is more than that of uninsured patients. The following results are obtained.

	Sample		
Patients	Size	Mean (days)	Standard Deviation (days)
Insured	350	4.8	2.9
Uninsured	76	3.1	2.6

Do the data support the belief that the average length of stay in a hospital of insured patients is more than that of uninsured patients?

16. (i) Let $X_1, X_2, \ldots, X_m$ and $Y_1, Y_2, \ldots, Y_n$ be independent random samples from population distributions with respective means μ_1 and μ_2, and variances σ_1^2 and σ_2^2. How will you test H_0: $\sigma_1 = \sigma_2$ against the usual one-sided or two-sided alternatives in the case when m and n are large?

   (ii) Apply the result in (i) to test H_0: $\sigma_1 = \sigma_2$ against H_1: $\sigma_1 \neq \sigma_2$ for the data in Problem 15.

   (iii) For the data of Problem 10, is there evidence to suggest that the fill variance for machine 1 is less than that for the machine 2?

17. Suppose it is required to test H_0: $\mu = 5$ against H_1: $\mu > 5$ at level $\alpha = .05$ such that the probability of type II error when true $\mu = 6$ is $\beta = .05$. What should be the minimum sample size if σ is approximately 3.5?

18. How many tosses of a coin should be made in order to test H_0: $p = 1/2$ (coin fair) against H_1: $p > 1/2$ (loaded in favor of heads) at level $\alpha = .05$ such that if the true p is .6 then β, the probability of type II error is .1?

19. (i) Find the minimum sample sizes required in order to test H_0: $\mu_1 - \mu_2 = 0$ against H_1: $\mu_1 \neq \mu_2$ at level α such that the probability of type II error at (μ_1, μ_2), $\mu_1 \neq \mu_2$ is approximately β. Assume that independent samples of the same size are to be taken. Also find an expression for the common sample size in the case of one-sided alternatives. (Assume that the variances σ_1^2, σ_2^2 are known.)

   (ii) Find the common sample size n in order to test H_0: $\mu_1 = \mu_2$ against $\mu_1 > \mu_2$ at level $\alpha = .01$ such that $\beta = .05$ when $\mu_1 - \mu_2 = 5$.

   (iii) Do the same for testing H_0: $\mu_1 = \mu_2$ against H_1: $\mu_1 \neq \mu_2$ when $\alpha = .1$ and $\beta = .1$ when $\mu_1 - \mu_2 = 2$.

20. (i) In order to test H_0: $p_1 = p_2$ against $p_1 \neq p_2$ at level α, independent random samples of sizes n and n are to be taken to ensure that the probability of type II error at specified alternatives (p_1, p_2), $p_1 \neq p_2$, is approximately β. Show that the minimum common sample size required is given by

$$n \doteq \left\{ \frac{z_{\alpha/2}\sqrt{(p_1 + p_2)(q_1 + q_2)/2} + z_{\beta/2}\sqrt{p_1 q_1 + p_2 q_2}}{p_1 - p_2} \right\}^2$$

   where $q_i = 1 - p_i$, $i = 1, 2$. Find the minimum sample size when the alternative is one-sided.

   (ii) For the Salk vaccine field trial (Example 2), find the common sample size required for testing H_0: $p_1 = p_2$ against H_1: $p_1 > p_2$ ($p_1 =$ probability of a child from the control group getting polio, $p_2 =$ probability of a child from the treatment group getting polio) with $\alpha = .05$ and $\beta = .15$ when $p_1 = .0001$ and $p_2 = .00005$.

   (iii) Suppose H_0: $p_1 = p_2$ is to be tested against H_1: $p_1 \neq p_2$. Find the common sample size n if $\alpha = .01$ and $\beta = .05$ when $p_1 = .5$ and $p_2 = .6$.

21. Modify the Fisher–Irwin test of Section 6.4 for testing $p_1 = p_2$ against $p_1 < p_2$ (or $p_1 > p_2$) when the sample sizes m and n are large.

9.8 INFERENCE CONCERNING QUANTILES

Consider a random sample $X_1, \ldots, X_n$ from a continuous distribution function $F \in \mathscr{F}$. Nothing else is known about $\mathscr{F}$. Then $\mathscr{F}$ is a nonparametric family. Let $X_{(1)} < X_{(2)} < \cdots < X_{(n)}$ be the sample order statistic. In Section 8.5 we considered the problem of estimating the quantile of order p of F by the corresponding sample quantile. We now consider some large sample properties of sample quantiles as estimates of population quantiles.

Assume that the quantile of order p, ζ_p is uniquely determined by

$$F(\zeta_p) = p, \qquad 0 < p < 1.$$

Let $\hat{\zeta}_p$ be the sample quantile of order p defined by

$$\hat{\zeta}_p = \begin{cases} X_{(np)} & \text{if } np \text{ is an integer} \\ X_{([np]+1)} & \text{if } np \text{ is not an integer} \end{cases}$$

where $[x]$ is the integral part of np.

The following result is useful in statistical inference concerning quantiles. The proof is beyond our scope and is omitted.

THEOREM 1. *Let $X_{(r)}$ be the rth order statistic in sampling from a continuous distribution with density f. Then, as $n \to \infty$ such that $r/n \to p$ (fixed), the distribution of*

$$\left(\frac{n}{p(1-p)} \right)^{1/2} f(\mu)(X_{(r)} - \mu)$$

tends to that of the standard normal distribution, where μ satisfies $F(\mu) = p$ for $p = r/n$.

According to Theorem 1, for large n, the sample order statistic $\hat{\zeta}_p$ is approximately normally distributed with mean ζ_p, and variance $p(1 - p)/(n[f(\zeta_p)]^2)$. In particular, $\hat{\zeta}_p$ is a consistent estimate of ζ_p.

Theorem 1, however, is practically useless for constructing confidence intervals for ζ_p (unless, of course, f is known). The case of most interest is when f is not known. In that case we use normal approximation to binomial distribution.

From (8.5.15) we recall that for any distribution function, discrete or continuous,

$$(1) \qquad P\left(X_{(r)} \le \zeta_p \le X_{(s)} \right) \ge \sum_{j=r}^{s-1} \binom{n}{j} p^j (1 - p)^{n-j}$$

so that $[X_{(r)}, X_{(s)}]$ is a confidence interval for ζ_p at level $\ge 1 - \alpha$ if we choose r and s such that

$$(2) \qquad \sum_{j=r}^{s-1} \binom{n}{j} p^j (1 - p)^{n-j} \ge 1 - \alpha.$$

Let Y have a binomial distribution with parameters n and p. Then (2) can be rewritten as

$$(3) \qquad\qquad P(r \leq Y \leq s - 1) \geq 1 - \alpha$$

and using normal approximation with continuity correction we have, for large n,

$$(4) \qquad P(r \leq Y \leq s - 1) = P(r - 1/2 < Y < s - 1/2)$$

$$\doteq P\left(\frac{r - 1/2 - np}{\sqrt{np(1 - p)}} < Z < \frac{s - 1/2 - np}{\sqrt{np(1 - p)}} \right)$$

where Z has a standard normal distribution. We choose r and s by assigning equal probability to each tail. In that case

$$(5) \quad r = np + \frac{1}{2} - z_{\alpha/2}\sqrt{np(1 - p)}, \text{ and } s = np + 1/2 + z_{\alpha/2}\sqrt{np(1 - p)}.$$

However, r and s need not be integers as required. So we choose r to be the largest integer $\leq np + 1/2 - z_{\alpha/2}\sqrt{np(1 - p)}$ and s to be the smallest integer $\geq np + 1/2 + z_{\alpha/2}\sqrt{np(1 - p)}$. In the special case when $p = 1/2$ and $\zeta_{.5}$ is the median, then an approximate $1 - \alpha$ level confidence interval for large n is $[X_{(r)}, X_{(n-r+1)}]$ where r is the largest integer less than or equal to

$$\frac{(n + 1)}{2} - z_{\alpha/2}\frac{\sqrt{n}}{2}.$$

Large Sample Equal Tails Confidence Interval for ζ_p

An approximate $1 - \alpha$ level confidence interval for ζ_p for large n is given by $[X_{(r)}, X_{(s)}]$ where r is the largest integer $\leq np + 1/2 - z_{\alpha/2}\sqrt{np(1 - p)}$ and s is the smallest integer $\geq np + 1/2 + z_{\alpha/2}\sqrt{np(1 - p)}$.

We emphasize that since (1)–(3) are valid for any distribution, discrete or continuous, the above large sample confidence interval is also valid for any distribution.

Example 1. Median Family Income. A random sample of 36 family incomes in suburban Detroit yielded the following incomes (in thousands of dollars per year):

82, 112, 62, 35, 28, 26, 34, 38, 36, 92, 101, 34, 37, 36, 26, 22, 21, 28,

26, 63, 54, 50, 49, 48, 36, 28, 84, 72, 63, 54, 46, 42, 49, 38, 32, 34

Suppose we wish to estimate the median family income for this suburb by a 95 percent confidence interval. Since $\alpha = .05$, $z_{\alpha/2} = 1.96$, and

$$\frac{n+1}{2} - z_{\alpha/2}\frac{\sqrt{n}}{2} = \frac{37}{2} - 1.96\frac{6}{2} = 12.62,$$

so that $r = 12$ and $s = 25$. Thus an approximate 95 percent confidence interval is given by

$$X_{(12)} \leq \zeta_{.5} \leq X_{(25)}.$$

Ordering the incomes in increasing order, we get

21	28	35	38	50	72
22	28	36	42	54	82
26	32	36	46	54	84
26	34	36	48	62	92
26	34	37	49	63	101
28	34	38	49	63	112

so that $x_{(12)} = 34$ and $x_{(25)} = 50$ and the 95 percent confidence interval for median income is $34 \leq \zeta_{.5} \leq 50$ (in thousands of dollars).

If we want a 90 percent confidence interval for the .75 quantile, then $z_{\alpha/2} = 1.645$, $np = 36(.75) = 27$ and

$$np + 1/2 - z_{\alpha/2}\sqrt{np(1-p)} = 27.5 - 1.645\sqrt{27(1/4)} = 23.23,$$

$$np + 1/2 + z_{\alpha/2}\sqrt{np(1-p)} = 27.5 + 1.645\sqrt{27(1/4)} = 31.77.$$

It follows that $r = 23$ and $s = 32$ and the 90 percent confidence interval is $49 \leq \zeta_{.75} \leq 82$, in thousands of dollars.   □

Example 2. Sample Quantile as Estimate of Population Quantile. Suppose a sample of size 100 is taken from a normal population with mean μ and variance 1. What is the probability of observing a sample quantile of order .75 larger than .25?

We note that $\zeta_{.75}$ is given by

$$P(X \leq \zeta_{.75}) = .75$$

where X is normal with mean μ and variance 1. Hence

$$P(Z \leq \zeta_{.75} - \mu) = .75$$

so that $\zeta_{.75} = \mu + .67$. Also

$$f(\zeta_{.75}) = \frac{1}{\sqrt{2\pi}}\exp\left\{-\frac{(.67)^2}{2}\right\} = .3187.$$

In view of Theorem 1, the sample quantile of order .75, that is, $X_{(75)}$, has approximately a normal distribution with mean $\zeta_{.75}$ and variance

$$\frac{(.75)(.25)}{100[.3187]^2} = .0185.$$

Hence

$$P(X_{(75)} > .25) = P\left(\frac{X_{(75)} - \zeta_{.75}}{.1359} > \frac{.25 - \zeta_{.75}}{.1359}\right)$$

$$\doteq P\left(Z > \frac{.25 - \zeta_{.75}}{.1359}\right),$$

where Z is a standard normal random variable. If, for example, $\mu = -.50$, then $\zeta_{.75} = .17$ and

$$P(X_{(75)} > .25) \doteq P\left(Z > \frac{.08}{.1359}\right)$$

$$= P(Z > .59) = .2776.$$

In this case we can find a $1 - \alpha$ level confidence interval for $\zeta_{.75}$ from Theorem 1. In view of the approximate normality of $X_{(75)}$, a $1 - \alpha$ level confidence interval is given by

$$\left[X_{(75)} - z_{\alpha/2}(.1359), X_{(75)} + z_{\alpha/2}(.1359)\right].$$

We can also use Theorem 1 to see how the nonparametric estimate $X_{(75)}$ of $\zeta_{.75}$ performs compared with the parametric estimate. Because $\zeta_{.75} = .67 + \mu$, a good estimate (consistent, unbiased) of $\zeta_{.75}$ is given by

$$\hat{\zeta}_{.75} = .67 + \hat{\mu} = .67 + \overline{X}$$

with MSE($=$ variance) given by

$$\text{MSE}(\hat{\zeta}_{.75}) = \mathcal{E}_\mu(\hat{\zeta}_{.75} - \zeta_{.75})^2 = \frac{\sigma^2}{n} = \frac{1}{100}.$$

According to Theorem 1, however,

$$\text{MSE}(X_{(75)}) = .0185$$

so that the relative efficiency of $\hat{\zeta}_{.75}$ with respect to $X_{(75)}$ is given by

$$\text{RE}(\hat{\zeta}_{.75}, X_{(75)}) = \frac{\text{MSE}(X_{(75)})}{\text{MSE}(\hat{\zeta}_{.75})} = \frac{.0185}{.01} = 1.85.$$

This means that we are losing about half the sample size by using $X_{(75)}$ instead of $\hat{\zeta}_{.75}$. However, what we lose in terms of sample size is balanced by the assurance that $X_{(75)}$ is a good estimate of $\zeta_{.75}$ even if the underlying distribution is not normal (as long as it is continuous and has a unique quantile of order .75). □

Remark 1. How large should n be in order to apply results of this section? In general, we are guided by Table 1 of Section 9.4. As long as $np(1 - p) > 10$, the approximation is quite good. For $p = 1/2$ approximation is fairly good for $n > 20$. For values of p near zero or 1, one should check to see if n is large enough that $np(1 - p) > 10$. For intermediate values of p we will use the approximation even for values of n as small as 30 but it should be understood that the approximation may not be that good.

Remark 2. As pointed out in Section 8.5, given α and n there may not exist a confidence interval $[X_{(r)}, X_{(s)}]$ for ζ_p at level $1 - \alpha$. According to Problem 8.5.28, such is the case when $\alpha < p^n + (1 - p)^n$. Thus, for $n = 40$, $\alpha = .01$ there does not exist a confidence interval of the form $[X_{(r)}, X_{(s)}]$ for $\zeta_{.1}$. In such cases, either the value of r selected according to (5) will be necessarily less than 1 or the value of s selected according to (5) will be greater than n. If, however, $\alpha \geq p^n + (1 - p)^n$ and s selected according to (5) is $> n$, then one should find the largest r such that $[X_{(r)}, X_{(n)}]$ is a $1 - \alpha$ level confidence interval for ζ_p. Similarly, if $\alpha \geq p^n + (1 - p)^n$ and r selected according to (5) is < 1, then one chooses the smallest s such that $[X_{(1)}, X_{(s)}]$ is a $1 - \alpha$ level confidence interval for ζ_p. Thus for $\alpha = .01$, $n = 60$ and $p = .9$, $\alpha = .01 \geq p^n + (1 - p)^n = .002$, but s according to (5) is 61. Hence we find the largest r such that $[X_{(r)}, X_{(60)}]$ includes $\zeta_{.9}$. We have

$$.99 \leq \sum_{j=r}^{59} \binom{60}{j}(.9)^j(.1)^{59} = P(r \leq Y \leq 59)$$

where Y is binomial with $n = 60$ and $p = .9$. Applying normal approximation, we choose the largest r such that

$$.99 \leq P\left(r - \frac{1}{2} < Y < 59.5\right) \doteq P\left(\frac{r - \frac{1}{2} - 54}{\sqrt{5.4}} < Z < \frac{59.5 - 54}{\sqrt{5.4}}\right)$$

or such that

$$P(Z < 2.37) - .99 \geq P\left(Z \leq \frac{r - 1/2 - 54}{\sqrt{5.4}}\right).$$

Thus

$$.0011 \geq P\left(Z \leq \frac{r - 1/2 - 54}{\sqrt{5.4}}\right)$$

so that

$$r = 54.5 - 3.05\sqrt{5.4} = 47.4$$

and we choose $r = 48$. The level of $[X_{(48)}, X_{(60)}]$ is in fact .9915.

Remark 3. If $\alpha < p^n + (1 - p)^n$ so that no confidence interval $[X_{(r)}, X_{(s)}]$ at level $1 - \alpha$ exists, then we have two choices: either take a larger sample size (such that $\alpha \geq p^n + (1 - p)^n$) or increase α and hence decrease the confidence level.

Remark 4.   Testing Hypotheses Concerning Quantiles.   In Section 8.5.2 we saw that hypotheses concerning ζ_p of the form $\zeta_p = \zeta_0$ against one-sided or two-sided alternatives can be tested by using a sign test based on binomial distribution. If n is large (see Remark 1) we can use normal approximation of Theorem 1 if f is known and normal approximation to binomial if f is unknown. Since tests based on Z, a standard normal random variable, have been described in detail in Section 8.7.3, we omit these details. Similar remarks also apply to the median and the Fisher–Irwin tests of Section 6.4.

Problems for Section 9.8

1. The birth weights (in pounds) of 40 infants at a county hospital are found to be:

7.8	11.8	9.3	8.4	7.3
5.6	7.2	9.1	8.3	7.6
7.5	9.2	9.4	8.6	7.8
6.0	4.8	7.3	8.7	7.1
4.2	5.0	7.6	9.4	7.0
6.7	5.4	8.4	10.2	4.9
7.8	8.3	7.6	8.9	5.9
10.3	7.4	7.8	6.7	8.6

(i) Find an approximate $100(1 - \alpha)$ percent confidence interval for the median birth weight if (a) $\alpha = .10$; (b) $\alpha = .01$.

(ii) Find an approximate 95 percent confidence interval for the quantile of order .6.

2. The ages of patients in a coronary care unit of a hospital were recorded for a sample of 49 patients as follows.

66	63	46	55	39	37	75
39	35	37	56	37	39	73
28	50	40	60	38	48	74
54	40	41	63	46	55	69
56	38	43	61	48	66	65
55	39	42	64	39	68	62
60	42	54	38	40	72	54

Find an approximate $100(1 - \alpha)$ percent confidence interval for the median and the quantile of order .25 for the age distribution of coronary patients when:
(i) $\alpha = .02$.   (ii) $\alpha = .05$.

3. A random sample of 50 ranches in western Texas gave the following acreages:

532	20473	1746	1806	10400
1438	30834	1384	695	38070
395	15225	1239	478	987
874	1648	2143	2836	1748
966	2536	2586	997	5274
1004	3849	2795	1213	5136
1546	639	613	1346	3948
3248	647	889	2937	11200
873	894	1473	3246	200
10800	984	9300	4800	5640

Find an approximate $100(1 - \alpha)$ percent confidence interval for the median acreage of a ranch in western Texas if (a) $\alpha = .05$, (b) $\alpha = .10$.

4. The lifetimes of 121 light bulbs manufactured by a certain company shows a sample median of 1207 hours. The lifetimes may be assumed to have an exponential distribution.
   (i) Use Theorem 1 to find the probability that the sample median is less than 1207 hours. (Take the population mean to be 1250 hours.)
   (ii) Find a 95 percent confidence interval for the median lifetime of bulbs made by this company.
   (iii) Compare the relative efficiency of the parametric estimate $\hat{\mu} \ln 2$ of the median with the nonparametric estimate of the median. (Here μ is the mean lifetime of the exponential distribution.)

5. Suppose a sample of 75 observations is taken from the distribution

$$f(x) = 2x, \qquad 0 \le x \le 1, \text{ and zero elsewhere.}$$

Find the probability that the sample quantile of order .2 will be larger than .60.

6. A large company uses small foreign-made cars for its sales force. A random sample of 48 repair bills yielded the following results (in dollars):

25.00	132.17	242.16	542.48	82.17	46.17
186.10	38.16	102.11	102.39	103.36	39.48
295.14	28.17	86.95	136.68	37.18	103.00
238.17	124.16	94.36	16.39	36.32	109.96
342.14	28.95	36.39	48.84	48.56	39.87
6.17	6.38	48.15	39.96	94.65	28.63
14.68	17.32	24.12	87.56	201.38	7.14
86.32	63.39	104.65	88.06	149.68	16.89

Find an approximate 90 percent confidence interval for the median repair bill for this company's fleet of cars.

7. Suppose $X_1, X_2, \ldots, X_n$ is a random sample from density function $f(x|\theta)$ as given below. An estimate of the median of f is desired. Compute the relative efficiency of the parametric estimate (choose the best unbiased estimate) relative to the sample median.

   (i) $f(x|\theta) = (1/\sigma\sqrt{2\pi})\exp\{-(x-\mu)^2/2\sigma^2\}, \theta = (\mu, \sigma^2)$.
   (ii) $f(x|\theta) = 1/\theta, 0 < x < \theta$, and zero elsewhere (see Example 8.5.4).

   In each case find a $1 - \alpha$ level confidence interval for the median based on the sample median.

8. A random sample of 40 observations is taken from a uniform distribution on $[0, \theta]$ with the following results:

   0.24, 0.28, 0.38, 1.59, 1.73, 1.78, 1.30, 1.25, 0.57, 0.54, 1.73,
   0.06, 1.42, 0.93, 0.53, 0.64, 1.48, 1.69, 0.24, 0.14, 1.63, 0.51,
   0.48, 0.62, 0.14, 1.38, 1.21, 1.14, 0.59, 0.72, 0.84, 1.42, 1.25,
   0.63, 0.74, 0.53, 0.69, 0.98, 1.01, 1.03

   (i) Construct a 95 percent confidence interval for the median by ignoring the fact that the sample comes from a uniform distribution.
   (ii) Construct a 95 percent confidence interval for the median using the fact that the sample comes from a uniform distribution on $[0, \theta]$.

9. For the data of Problem 1 test the hypothesis that the median birth weight is less than or equal to 7.3 pounds at a 5 percent level of significance.

10. For the data on ages of patients in coronary care (Problem 2), test the hypothesis that the quantile of order .25 is 38 years against the two-sided alternative.

11. The diastolic blood pressures of 38 randomly selected men were found to be:

    65, 80, 68, 81, 82, 84, 83, 85, 92, 91, 93, 77, 86, 87, 90,
    94, 86, 97, 88, 93, 89, 95, 72, 73, 71, 76, 78, 80, 79, 64,
    83, 65, 82, 79, 88, 89, 70, 75.

    Test the hypothesis that the median diastolic blood pressure of men is 80.

12. In order to check the gasoline mileage that she is getting on her new car, a student recorded the miles per gallon for 35 tankfuls:

    18.1, 19.2, 24.7, 24.2, 25.6, 18.7, 18.5, 19.8, 23.8, 22.4, 27.6,
    24.3, 22.9, 23.2, 18.6, 18.2, 18.8, 19.5, 20.4, 20.6, 20.8, 22.3, 26.4,
    25.4, 19.8, 20.3, 20.8, 22.8, 23.7, 25.9, 26.3, 22.7, 24.6, 18.3, 19.3.

    Test the hypothesis that the median gasoline mileage that this student gets from her car is 24 miles per gallon.

13. For the data of Problem 11, construct a 90 percent confidence interval for the quantile of order .4.

14. (i) Does there exist a confidence interval of the form $[X_{(r)}, X_{(s)}]$ for $\zeta_{.05}$ at level .90 based on a sample of size 40? Answer the same question when the level is decreased to .85.

    (ii) Answer part (i) when n is increased to 60.

    (iii) Find a .95 level confidence interval for $\zeta_{.05}$ when $n = 60$.

15. For $n = 40$, $p = .80$ find an upper confidence bound for ζ_p at a level of approximately .98. [Hint: Find s such that $P(X_{(s)} \geq \zeta_p) \doteq .98$. Use relation (8.5.14).]

16. For $n = 60$, $p = .9$ find a lower confidence bound for ζ_p at a level of approximately .95. [Hint: Find r such that $P(X_{(r)} \leq \zeta_p) \doteq .95$. Use (8.5.13).]

9.9 GOODNESS OF FIT FOR MULTINOMIAL DISTRIBUTION: PRESPECIFIED CELL PROBABILITIES

Suppose we have the following record of 340 fatal automobile accidents, per hour, on a long holiday weekend in the United States.

Number of Fatal Accidents per Hour	Number of Hours
0 or 1	5
2	8
3	10
4	11
5	11
6	9
7	8
8 or more	10
Total	72

The number X of fatal auto accidents per hour is clearly a discrete random variable. We should like to know the distribution of X. In Section 4.5 we studied some graphical methods of looking at the data to determine if the data could have come from one of a particular family of distributions. One method that we had suggested was to draw a graph of the relative frequency histogram and check visually to see if the histogram looks much like the histogram of one of the known discrete distributions (studied in Chapter 6). We did not, however, provide an overall measure of the goodness of fit. How well does an assumed distribution fit the histogram data? For example, suppose we feel that the Poisson distribution with $\lambda = 5$ provides an adequate representation for the distribution of X. We can now draw the (observed) relative frequency histogram and the postulated (or expected) relative frequency histogram on the same graph to see how close the agreement is. The following table compares the observed and the postulated relative frequencies: (See Example 1 below for details of computations.)

X	Relative Frequency	Postulated Relative Frequency
0 or 1	.07	.04
2	.11	.08
3	.14	.14
4	.15	.18
5	.15	.18
6	.13	.15
7	.11	.10
8 or more	.14	.13

In Figure 1 below we have plotted the two relative frequency histograms.

The two histograms appear close. The question we have to answer, however, is the following: Is the difference between the two histograms due to chance fluctuation, or is there something wrong with our assumption about the model? We now study the chi-square test of goodness of fit which is specially designed to answer this question.

Consider a random sample $X_1, X_2, \ldots, X_n$ from a discrete distribution given by

$$P(X = x_j) = p_j, \, j = 1, 2, \ldots, k, \quad \text{and zero elsewhere,}$$

where $p_j > 0$ and $\sum_{j=1}^k p_j = 1$. How do we use the observations to test the null hypothesis

$$H_0: p_j = p_j^0, \quad j = 1, 2, \ldots k,$$

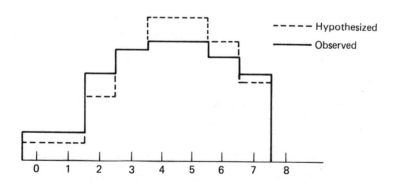

Figure 1. Relative frequency histograms. The open-ended class 8 or more is not included.

where p_j^0 are known numbers? Let

$$n_j = \text{number of times } x_j \text{ appears in the sample}$$

$$= \text{number of } X\text{'s that equal } x_j.$$

Then n_j, $j = 1, 2, \ldots, k$, are random variables such that $\sum_{j=1}^{k} n_j = n$ and, moreover, $(n_1, n_2, \ldots, n_{k-1})$ has, under H_0, a multinomial distribution with parameters $(p_1^0, p_2^0, \ldots, p_{k-1}^0)$. Since each n_j has a marginal binomial distribution with parameter p_j, we have

$$\mathscr{E} n_j = n p_j, \text{ so that } \mathscr{E}(n_j/n) = p_j$$

and we can use n_j/n, $j = 1, 2, \ldots, k$ respectively as estimates of $p_1, p_2, \ldots, p_k$. It is natural to compare the observed proportions n_j/n and the postulated proportions p_j^0. It is tempting to use a distance function such as

$$\sum_{j=1}^{k} \left| \frac{n_j}{n} - p_j^0 \right| \quad \text{or} \quad \sum_{j=1}^{k} \left(\frac{n_j}{n} - p_j^0 \right)^2$$

as a measure of the agreement between H_0 and the observations. The smaller this distance the greater the agreement. The measure most commonly used in practice, however, is the weighted sum of squares.

$$(1) \qquad Q = \sum_{j=1}^{k} \left(\frac{n}{p_j^0} \right) \left(\frac{n_j}{n} - p_j^0 \right)^2 = \sum_{j=1}^{k} \frac{\left(n_j - n p_j^0 \right)^2}{n p_j^0}.$$

Since large values of Q correspond to a large discrepancy between the data and H_0, large values of Q lead to the rejection of H_0.

Chi-Square Test of Goodness of Fit

Under H_0: $P(X = x_j) = p_j^0$, $j = 1, \ldots, k$, for large n the statistic

$$Q = \sum_{j=1}^{k} \frac{\left(n_j - n p_j^0 \right)^2}{n p_j^0}$$

has approximately a χ^2 distribution with $k - 1$ d.f.

Test: Reject H_0 at level α, $0 < \alpha < 1$, if the observed value of Q is larger than $\chi^2_{k-1, \alpha}$. P-value $= P_{H_0}(Q \geq q_0)$, where q_0 is the observed value of Q.

The derivation of the asymptotic distribution of Q is beyond our scope. For $k = 2$, however, the result easily follows from the central limit theorem. In fact,

$$Q = \frac{\left(n_1 - np_1^0\right)^2}{np_1^0} + \frac{\left(n - n_1 - n\left(1 - p_1^0\right)\right)^2}{n\left(1 - p_1^0\right)} = \frac{\left(n_1 - np_1^0\right)^2}{np_1^0} + \frac{\left(n_1 - np_1^0\right)^2}{n\left(1 - p_1^0\right)}$$

$$= \frac{\left(n_1 - np_1^0\right)^2}{np_1^0\left(1 - p_1^0\right)} = \left\{\frac{n_1 - np_1^0}{\sqrt{np_1^0\left(1 - p_1^0\right)}}\right\}^2 .$$

Since n_1 has a binomial distribution with parameters n and p_1^0 (under H_0) it follows from the central limit theorem that

$$\frac{n_1 - np_1^0}{\sqrt{np_1^0\left(1 - p_1^0\right)}}$$

has a limiting standard normal distribution. Now the square of a standard normal random variable has a chi-square distribution with 1 d.f. Consequently, Q has approximately a chi-square distribution with 1 d.f.

Remark 1. In the case $k = 2$, the problem reduces to testing H_0: $p = p_0$ against H_1: $p \neq p_0$ in sampling from a binomial population. In Section 9.6 we saw that under H_0 the statistic $(X - np_0)/\sqrt{np_0(1 - p_0)}$ has approximately a standard normal distribution for large n so that $(X - np_0)^2/(np_0(1 - p_0))$ has approximately a chi-square distribution with 1 d.f. under H_0. With $n_1 = X$ this is precisely Q.

Remark 2. Under H_0

$$\mathscr{E}Q = \sum_{j=1}^{k} \mathscr{E}\left\{\frac{\left(n_j - np_j^0\right)^2}{np_j^0}\right\} = \sum_{j=1}^{k} \frac{\text{var}(n_j)}{np_j^0}$$

$$= \sum_{j=1}^{k} \frac{np_j^0\left(1 - p_j^0\right)}{np_j^0} = k - 1.$$

How do we use Q to test that X has a specified probability density function f? The answer is surprisingly simple. Divide the real line into k intervals $I_1, \ldots, I_k$ and let $p_j^0 = P_f(X \in I_j)$, $j = 1, \ldots, k$. Let $\hat{p}_j$ be the observed proportion of observations that fall in the interval I_j. Then the statistic

$$Q = \sum_{j=1}^{k} \frac{\left(n\hat{p}_j - np_j^0\right)^2}{np_j^0}$$

has approximately a chi-square distribution with $(k - 1)$ d.f. under H_0.

How do we choose the intervals $I_1, I_2, \ldots, I_k$? The first point to remember is that the chi-square distribution is only an approximation to the true distribution of Q. So the choice has to be made in such a way that this approximation is good. The second point to remember is that the underlying assumption to the chi-square approximation is that each random variable $(n\hat{p}_j - np_j^0)/\sqrt{np_j^0}$ has approximately a normal distribution (with mean 0 and variance $1 - p_j^0$). This holds provided n is large. Suppose p_j^0 is small for some cell. Then for the approximation to be good we would want either n to be very large or np_j^0 to be large. (Recall that if the mean $\lambda \to \infty$ then the Poisson distribution can be approximated by the normal distribution.) If np_j^0 is small then the term $(n_j - np_j^0)^2/np_j^0$ may dominate the other terms of Q due to a small denominator. Although the approximation is often good even when np_j^0 is as small as 1, the following rules of thumb are often used in choosing $I_1, I_2, \ldots, I_k$.

(i) Choose $I_1, I_2, \ldots, I_k$ such that under H_0, $p_j^0 = P(X \in I_j)$ is approximately equal to $1/k$ and each $np_j^0 \doteq n/k \geq 5$.

(ii) If any of the np_j^0's is less than 5, pool the corresponding interval with one or more intervals to make the cell frequency at least 5. The decision which intervals (or cells) to pool is arbitrary but we restrict pooling to a minimum. The degrees of freedom associated with the chi-square approximation, after pooling, are reduced by the number of classes pooled.

Example 1. *Fatal Auto Accidents on a Long Weekend.* We return to the data on fatal auto accidents given at the beginning of this section. Suppose we wish to test the hypothesis H_0 that the number of fatal auto accidents X, per hour, has a Poisson distribution with mean $\lambda = 5$. Under H_0,

$$p_j^0 = P(X = j) = e^{-5}\frac{5^j}{j!}, \quad j = 0, 1, 2, \ldots.$$

From Table A4 we have

j	0	1	2	3	4	5	6	7
p_j^0	.007	.033	.085	.140	.175	.176	.146	.105

Moreover,

$$P_{H_0}(X \geq 8) = 1 - P_{H_0}(X \leq 7) = 1 - \sum_{j=0}^{7} p_j^0 = .133.$$

Under H_0 we expect $np_j^0 = 72\, p_j^0$ hours with j accidents. The following table gives the comparison between the observed and expected frequencies.

j	0 or 1	2	3	4	5	6	7	8 or more
Observed number of hours, n_j	5	8	10	11	11	9	8	10
Expected number of hours, np_j^0	2.88	6.12	10.08	12.60	12.67	10.51	7.56	9.58

Since for the first cell $np_j^0 < 5$, we pool it with the second cell to get the following table.

j	0, 1, or 2	3	4	5	6	7	8 or more
n_j	13	10	11	11	9	8	10
np_j^0	9.0	10.08	12.60	12.67	10.51	7.56	9.58

The observed value of Q is then given by

$$q_0 = \sum_j \left[\left(n_j - np_j^0 \right)^2 / np_j^0 \right] = 2.46.$$

The associated degrees of freedom equal the number of classes after pooling $-1 = 7 - 1 = 6$. Hence the P-value is given (approximately) by

$$.80 < P(Q \geq 2.46) < .90,$$

and we have to conclude that the Poisson model with $\lambda = 5$ does provide a reasonable fit for the data. (Without pooling, $q_0 = 2.82$ with P-value $\doteq .90$.) $\square$

Example 2. *Grading on a Curve.* In practically every course I teach, the students ask if I grade on the curve. First, what do we mean by grading on the curve? We are making the assumption that the numerical grades follow a normal distribution. Let μ be the mean and σ the standard deviation of the numerical grades. Then grades above $\mu + k_1\sigma$ are taken to be A, those between $\mu + k_2\sigma$ and $\mu + k_1\sigma$, $(k_1 > k_2)$ are taken to be B, and so on. The numbers k_1, k_2, k_3, and k_4 are determined by the instructor. Suppose we take $k_1 = 3/2$, $k_2 = 1/2$, $k_3 = -1/2$, and $k_4 = -3/2$. Then with these cut-off points, the proportion of A's (and also of F's) is easily seen to be .0668, that of B's (and also of D's) is .2417, and that of C's is $1 - 2(.0668) - 2(.2417) = .3830$. It is clear that these proportions depend only on the choice of constants k_1, k_2, k_3, k_4 but not on the values of μ and σ.

Consider now a large calculus class of 487 students with the following grade distribution:

A	B	C	D	F
48	110	219	85	25

Did the instructor grade on the curve? Here H_0: $P(A) = P(F) = .0668$, $P(B) = P(D) = .2417$, $P(C) = .3830$, is to be tested against the alternative hypothesis that the proportions are different than these. A comparison of the observed and the expected frequencies is given below:

Grade	A	B	C	D	F
Expected frequency, np_j^0	32.5	117.7	186.6	117.7	32.5
Observed frequency, n_j	48	110	219	85	25

The observed value of Q is given by

$$q_0 = \sum_j \frac{\left(n_j - np_j^0\right)^2}{np_j^0} = 24.33.$$

For 4 d.f. we note that under H_0

$$P(Q \geq 24.33) < .001$$

and one can hardly argue that such a large value of q_0 is due to chance variation. There is no agreement between the grades as given and grades that would have been given had the instructor been grading on the curve (with 6.68 percent each of A's and F's, 24.17 percent each of B's and D's and the rest C's). $\qquad\square$

Example 3. Sampling from the Uniform Distribution (Example 4.5.4 Continued). In Example 4.5.4 a random sample of size 80 was taken from the interval $(0, 1)$. In Table 4.5.2 we constructed the frequency distribution with 10 class intervals of equal width and computed the relative frequency histogram. After plotting the histogram and ogive estimates of the distribution along with the uniform distribution (in Figures 4.5.6 and 4.5.7), we found that a visual comparison was inconclusive. Let us now test the null hypothesis H_0, that the sample comes from the uniform distribution on $(0, 1)$. The following table gives the observed and expected frequencies.

Interval	(.005, .105)	(.105, .205)	(.205, .305)	(.305, .405)	(.405, .505)
n_j	5	11	6	3	11
np_j^0	8	8	8	8	8
Interval	(.505, .605)	(.605, .705)	(.705, .805)	(.805, .905)	(.905, 1.005)
n_j	11	9	11	6	7
np_j^0	8	8	8	8	8

We have

$$q_0 = \sum_j \frac{\left(n_j - np_j^0\right)^2}{np_j^0} = 10.0.$$

For 7 d.f. we note that

$$.20 > P\text{-value} = P(Q \geq 10.0) > .10.$$

Hence the observed value of Q is not significant at 10 percent level and we cannot reject H_0. $\qquad\square$

Remark 3. The procedure described here applies when the cell probabilities are completely specified; that is, when H_0 is a simple hypothesis. In many applications, however, this is not the case. Often the null hypothesis we wish to test is that the underlying distribution is normal or Poisson or gamma, and so on. The parameters are not specified and have to be estimated from the data. How do we modify the test procedure when parameters are estimated? This is done in Section 11.7.

Remark 4. The chi-square test of goodness of fit tests H_0: $p_j = p_j^0$, $j = 1, 2, \ldots, k$ against *all possible* alternatives. In a specific problem we may be more interested in testing H_0 against restricted alternatives such as H_1: $p_j < p_j^0$, $j = 1, 2, \ldots, k - 1$ and $p_k > p_k^0$. A test specifically designed to test H_0 against this particular H_1 will usually perform better than the chi-square test in the sense that the chi-square test will be more conservative. It is more difficult to reject H_0 using the chi-square test at a given level α then it would be if we were to use the test designed with the alternative H_1 in mind. Nevertheless, the chi-square test should be used as a first test to decide whether the data can be regarded as a random sample from one of a particular family of distributions. It should be used to answer questions such as: Do the data come from a normal distribution? Are the observations exponentially distributed?

Remark 5. We emphasize that the chi-square test is only a first step. If the test rejects a particular fit, we know that the postulated model is not a reasonable fit for the data. On the other hand, if the test accepts a certain fit, it does not necessarily follow that the underlying distribution is the hypothesized distribution. Often the test will accept two or more dissimilar distributions. The test will accept as normal almost all unimodal symmetric distributions (see, for example, Problem 6). In such cases, further analysis of the data is necessary. (See, for example, Sections 4.4, 11.2, and 11.3.).

Remark 6. We know that sufficiently large values of Q lead to a rejection of H_0 because these values of Q have small P-values. Very small values of Q should also be looked at with suspicion. Suppose, for example, that a student is asked to roll a fair die 120 times and comes up with the following data.

Face value	1	2	3	4	5	6
Frequency	18	21	19	20	20	22

Under H_0 we expect each face value to appear 20 times, so the data produced by the student look reasonable. In fact, they look too reasonable. We have

$$q_0 = .50$$

so that $$P_{H_0}(Q \geq .50) > .99 \quad \text{or} \quad P_{H_0}(Q \leq .50) < .01.$$

Thus there is less than 1 chance in 100 that such a small value of Q could have come from actual tossing of a fair die 120 times. It is possible that the student did actually perform 120 tosses and came up with these results but the data look suspicious. He may well have cooked up the data.

Problems for Section 9.9

1. An urn contains red, green, white, yellow, and blue marbles. In order to test the hypothesis

   $$H_0: P(\text{red}) = P(\text{green}) = P(\text{white}) = .10, \quad P(\text{yellow}) = .3, \quad P(\text{blue}) = .4$$

   a random sample of 200 marbles is drawn from the urn with replacement. If the sample contains 25, 15, 28, 45, 87 red, green, white, yellow, and blue marbles respectively, would you reject H_0 at level $\alpha = .10$?

2. In order to test whether a particular coin is fair, the following procedure was designed. Forty sequences of four tosses each with the coin were observed and the number of heads recorded in each sequence of four tosses as follows:

Number of heads	0	1	2	3	4
Number of sequences	1	15	12	7	5

   Would you accept the claim that the coin is fair?

3. The frequency distribution of 608 digits in the decimal expansion of π is as follows:

Digit	0	1	2	3	4	5	6	7	8	9
Frequency	60	62	67	68	64	56	62	44	58	67

   Is there sufficient evidence to conclude that each of the 10 digits $0, 1, \ldots, 9$ occurs with equal frequency?

4. In a well known experiment, the geneticist Gregor Mendel crossbred yellow round peas with green wrinkled peas. He observed the shape and color of a sample of 556 peas resulting from seeds of this crossbreeding with the following results:
   (i)   Round and yellow, 315.   (ii) Round and green, 108.
   (iii) Wrinkled and yellow, 101.   (iv) Wrinkled and green, 32.
   According to Mendel's theory, the corresponding frequencies should be in the ratio 9:3:3:1. Do the data support his theory? Is the fit too good to be reasonable?

5. A student is asked to draw a random sample of 50 digits from a random number table. He comes up with the following digits:

   $$0, 1, 5, 3, 6, 2, 5, 5, 9, 5, 2, 2, 5, 2, 7, 0, 6, 2, 4, 3, 6, 3, 6, 6, 1,$$

   $$8, 1, 8, 3, 7, 1, 1, 0, 0, 8, 5, 6, 4, 2, 0, 0, 5, 4, 6, 3, 4, 3, 3, 4, 2.$$

   Test the hypothesis that these digits are random:
   (i) Count the frequency of each digit and test $H_0: p_i = P(\text{digit } i) = 1/10$, $i = 0, 1, \ldots, 9$.
   (ii) Go through the list and count the number of times the next digit is the same as the preceding one or one away (that is, next smaller or next larger) from the preceding one (assume 0 is one away from 9). If the digits are random, $p_0 = P(\text{next digit same}) = 1/10$ and $p_1 = P(\text{next digit is one away}) = 2/10$. Test $H_0: p_0 = 1/10$, $p_1 = 2/10$, $p_2 = P(\text{neither}) = 7/10$, against all possible alternatives. (See also Problem 14.)

6. A random sample of size 50 from a population produced the following data:

.46,	.14,	2.46,	−.32,	−.07,	.29,	−.29,	1.30,	.24,	−.96
.06,	−2.53,	−.53,	−.19,	.54,	−1.56,	.19,	−1.19,	.02,	.53
1.49,	−.35,	−.63	.70	.93,	1.38,	.79,	−.96,	−.85,	−1.87
1.02,	−.47,	1.28,	3.52,	.57,	−1.85,	.19,	1.19,	−.50,	−.27
1.39,	−.56,	.05,	.32,	2.95,	1.97,	−.26,	.41,	.44	−.04

(i) Construct a frequency distribution with class intervals

$$(-\infty, -1.405), (-1.405, -1.005), (-1.005, -.605), (-.605, -.205),$$

$$(-.205, .205), (.205, .605), (.605, 1.005), (1.005, 1.405), (1.405, \infty)$$

(ii) Use the frequency distribution in part (i) to test if the data is normally distributed with mean 0 and variance 1.

(iii) Could the data have come from the double exponential density function

$$f(x) = (1/2)\exp(-|x|), \ x \in \mathbb{R}?$$

(See also Example 11.2.3.)

7. The failure times of 100 transistor batteries of a certain kind in hours have the following frequency distribution.

Hours	[0, 25)	[25, 50)	[50, 75)	[75, 100)	[100, ∞)
Frequency	33	28	20	12	7

(i) Could the data have come from an exponential distribution with mean 50 hours?

(ii) Could the data have come from an exponential distribution with mean 45 hours?

8. In order to test the hypothesis that a certain coin is fair, the coin is tossed until the first head shows up. The experiment is repeated 150 times with the following results:

Number of trials needed	1	2	3	4	5 or more
Frequency	60	48	22	11	9

Can we conclude that the coin is fair?

9. A computer program generates random numbers from the uniform distribution on [0, 10). A sample of 200 observations was found to have the following frequency distribution.

Interval	[0, 2)	[2, 4)	[4, 6)	[6, 8)	[8, 10)
Frequency	29	45	44	33	49

Is there strong evidence to conclude that the program is written properly?

10. The distribution of number of particles emitted from a radioactive source is thought to be Poisson with mean $\lambda = 2$ per 10 second. In order to test this hypothesis, the number of particles emitted in 150 consecutive 10-second intervals are recorded. They have the following frequency distribution:

Number of Particles	0	1	2	3	4	5 or more
Number of Intervals	15	48	33	30	14	10

How strong is the evidence against the claim?

11. The lifetime of a rotary motor is believed to have an exponential distribution with mean 2 years. A random sample of 80 motors was tested with the following results:

Time to failure (years)	$(0,1)$	$[1,2)$	$[2,3)$	$[3,4)$	4 or more
Frequency of failures	34	26	8	6	6

Is there sufficient evidence in support of the belief?

12. Show that

$$Q = \sum_{j=1}^{k} \left(\frac{n_j^2}{np_j^0} \right) - n.$$

13. Suppose a die is rolled 120 times with the following results:

Face value	1	2	3	4	5	6
Frequency	15	15	15	25	25	25

   (i) Test the hypothesis that the die is fair.
   (ii) Test the hypothesis that the die is fair by pooling the frequencies for face values 1 and 2 and those for face values 5 and 6.
   (iii) Test the same hypothesis by pooling the frequencies for the face values 1, 2, and 3 and those for the face values 4, 5, and 6. (Now one can apply either the χ^2-test or the (equivalent) test for equality of two proportions.)

14. Suppose a student produces the following sample of 50 random digits:

$$0, 0, 0, 0, 0, 0, 1, 1, 1, 1, 1, 2, 2, 2, 2, 2, 2, 2, 3, 3, 3, 3, 3, 3, 3, 3,$$

$$4, 4, 4, 4, 4, 5, 5, 5, 5, 5, 5, 5, 6, 6, 6, 6, 6, 6, 6, 7, 7, 8, 8, 8, 9$$

(See Problem 5.)
   (i) Apply the test of Problem 5(i) to check if the digits are random.
   (ii) Apply the test of Problem 5(ii) to check if the digits are random.

9.10   REVIEW PROBLEMS

1. Use the continuity theorem (Theorem 9.2.1) to show that the weak law of large numbers holds in sampling from the following distributions:
    (i) $f(x) = 1, 0 \leq x \leq 1$, and zero elsewhere.
    (ii) $f(x) = \lambda^{-1}\exp(-x/\lambda), x > 0$, and zero elsewhere.
    (iii) $f(x) = (1/2)\exp(-|x|), x \in \mathbb{R}$.
    (iv) $f(x) = 6x(1 - x), 0 \leq x \leq 1$, and zero elsewhere.
    (See Problem 5.4.8.)

2. Show that

$$[\Gamma(n)]^{-1}\int_0^{t\sqrt{n}+n} x^{n-1}e^{-x}\, dx \rightarrow \frac{1}{\sqrt{2\pi}}\int_{-\infty}^t e^{-x^2/2}\, dx$$

as $n \rightarrow \infty$. [Hint: Use the central limit theorem on exponentially distributed random variables.]

3. If X_n has a chi-square distribution with n d.f., what is the limiting distribution of X_n/n^2?

4*. Suppose $X_1, X_2,\ldots$ are independent random variables with probability function $P(X_n = 1) = P(X_n = -1) = 1/2$ for all $n = 1, 2,\ldots$. Let $Z_n = \sum_{j=1}^n X_j/2^j$. Show that the limiting distribution of Z_n is uniform on $[-1, 1]$.

5. Are the sufficient conditions for the Levy's central limit theorem satisfied in random sampling from the following distributions?
    (i) $f(x) = \dfrac{1}{1 + x^2}, x \in \mathbb{R}$.
    (ii) $f(x) = (1/2)\exp(-|x|), x \in \mathbb{R}$.
    (iii) $P(X = k) = \dfrac{c}{k^2}, k = 1, 2,\ldots$ where $c = \dfrac{6}{\pi^2} = \left[\sum_{k=1}^{\infty}(1/k^2)\right]^{-1}$.
    (iv) $f(x) = (2x/\beta)\exp(-x^2/\beta), x > 0$, and zero otherwise.

6. An urn contains eight identical marbles numbered 1 though 8. A random sample of n marbles is drawn *with replacement*.
    (i)   What does the law of large numbers say about the appearance of the marble numbered 1 in n draws?
    (ii) How many drawings are needed to ensure that with probability at least .90 the relative frequency of the occurrence of 1 will be between .12 and .13?
    (iii) Use the central limit theorem to approximate the probability that among n numbers drawn the number 4 will appear between $(n - \sqrt{n})/8$ and $(n + \sqrt{n})/8$ (inclusive).

7. Ever watchful of the elections every two years, an Ohio state senator bases his decision on whether to vote for or against politically sensitive issues on the results of private polls. A bill to increase the state income tax by 90 percent is before the Ohio Senate. The senator commissions a random sample of 200 registered voters in his district. He will vote for the bill only if 100 or more voters are in favor. What is the

probability that he will vote in favor of the bill if in fact only 40 percent of all voters in his district are in favor of it? What is the probability that he will not vote for the bill if in fact 55 percent of all voters in his district are in favor of it?

8. A random sample of 280 items is checked every hour from the production line of a certain machine. If the number of defectives in the sample is six or less, then the machine is considered in control. Otherwise the machine is stopped. Find the probability of:
   (i) Stopping the machine when it produces, on the average, 1.5 percent defective items.
   (ii) Not stopping the machine when it produces, on the average, 3 percent defective items.

9. From a very large shipment of 9-volt batteries, a random sample of 64 is checked. If the mean length of life of the sample is 75 hours or more the shipment is accepted; otherwise the shipment is rejected.
   (i) Find the probability of accepting a shipment that has a mean life length of 72 hours with a standard deviation of 10 hours.
   (ii) Find the probability of rejecting a shipment that has a mean life length of 80 hours with a standard deviation of 20 hours.
   (iii) Find the probability of accepting (or rejecting) a shipment that has a mean life of 75 hours.

10. There are 150 female and 350 male faculty members at a junior state college. A random sample of 50 faculty members is taken to fill 50 seats on the faculty senate. If the sample contains 12 female faculty members, is there sufficient evidence for the female faculty members to complain about underrepresentation?

11. In order to estimate the mean television viewing time (in hours) per American family per week, a social scientist takes a random sample of 600 families and finds that the sample mean is 24.8 hours with a standard deviation of 8.6 hours.
   (i) Find a 96 percent confidence interval estimate of the mean weekly viewing time.
   (ii) If the social scientist wishes to estimate the mean time to within 1.5 hours with 95 percent confidence, how large a sample should he take? Take $\sigma = 9$ hours.

12. Let $X_1, X_2, \ldots, X_n$ be a random sample from a distribution function F. Let c be a known constant and suppose $\theta(F) = \theta = P(X_i > c) = 1 - F(c)$.
   (i) Find a consistent estimate of θ.
   (ii) Find a consistent estimate of the odds ratio $\theta/(1 - \theta)$.

13. Let $X_1, X_2, \ldots, X_n$ be a random sample from a Poisson distribution. Find a consistent estimate of $P(X = 1)$.

14. Let $X_1, X_2, \ldots, X_n$ be the lifetimes of n batteries of the same type. Suppose $X_1, X_2, \ldots, X_n$ are independent and identically distributed with exponential distribution having mean θ. The estimate $\hat{\theta} = nX_{(1)}$ has certain advantages over the estimate $\bar{X}$. Is $\hat{\theta}$ a consistent estimate of θ? (Hint: See example 8.5.2.)

15. Let $(X_1, X_2, \ldots, X_{k-1})$ have a multinomial distribution with parameters n, $p_1, p_2, \ldots, p_{k-1}, p_i > 0$, $\sum_{i=1}^{k-1} p_i < 1$. Find a consistent estimate of $p_i - p_j, i \neq j$. (See also Problem 9.6.22.)

16. Let $(X_1, Y_1), (X_2, Y_2) \ldots, (X_n, Y_n)$ be a random sample from a bivariate distribution with

$$\mathscr{E} X_i = \mu_1, \mathscr{E} Y_i = \mu_2, \mathrm{var}(X_i) = \sigma_1^2, \mathrm{var}(Y_i) = \sigma_2^2, \rho(X_i, Y_i) = \rho.$$

   (i)   What is the limiting distribution of $\overline{X} - \overline{Y}$?
   (ii) Is $\overline{X} - \overline{Y}$ a consistent estimate of $\mu_1 - \mu_2$? Is it unbiased?
   (iii) How will you test the null hypothesis H_0: $\mu_1 - \mu_2 = \delta$ against the alternative hypothesis H_1: $\mu_1 - \mu_2 \neq \delta$ when σ_1, σ_2 are known and n is large? When σ_1, σ_2 are unknown and n is large?

17. Several studies have shown that twins have lower I.Q.'s in their early years than nontwins but they catch up with the nontwins by about age 10. A psychologist believes that the slower intellectual growth of twins may be due to parental neglect. She takes a random sample of 49 sets of twins aged two years and finds that the mean attention time given to each pair of twins in a particular week is 28 hours with a standard deviation of 10 hours.
   (i) Find a 92% confidence interval for the mean attention time given to all two-year-old twins.
   (ii) Suppose the national average mean attention time given two-year-old twins is 31 hours a week. Is the observed mean time of 28 hours a week for families in the sample significantly lower than the national average?

18. An experiment is to be designed to test the mean lifetime of a particular component. The experimenter is willing to take a chance of 1 in 100 of rejecting a true H_0: $\mu \leq \mu_0$, where μ is the mean lifetime. If the true mean, however, is $\mu = \mu_1$ ($> \mu_0$), the experimenter is willing to allow a chance of 5 in 100 of accepting H_0: $\mu \leq \mu_0$. How large should the sample size n be if the test statistic used is $\overline{X}$, the sample mean? Assume that the population standard deviation is σ hours. In particular, find n when $\mu_1 - \mu_0 = 5$ hours and $\sigma = 10$ hours.

19. The location of display of an item is important to the distributor. B. G. Vending Company has the option of putting soft drink vending machines in either (or both) of two different locations. The company decides to put one machine each in the two locations for a trial period of six weeks. The number of cans sold each day from each machine is recorded with the following results.

Location	Sample Mean	Sample Standard Deviation
1	42.5	7.1
2	37.3	5.9

   (i) Is the difference between the mean number of cans sold per day in the two locations significant?
   (ii) Find a 90 percent confidence interval for the true difference in means.

20. The mean shelf-life (in days) of various products in a supermarket is important to the manager. Independent random samples of 64 cans of peaches of Brand A and Brand B are tested with the following results.

	Sample	
Brand	Mean	Standard Deviation
A	4.8	1.7
B	4.1	1.5

(i) Is there sufficient evidence to conclude that the mean shelf-lives of the two brands are different?

(ii) Find a 96 percent confidence interval for the true mean difference in shelf-lives of the two brands.

21. Independent random samples of 100 cars in Washington, D. C., and 150 cars in Cleveland were inspected for five specific safety-related defects. An inspected car is considered unsafe if it does not pass at least one of the five tests. The following results were obtained.

City	Number of Unsafe Cars	Sample Size
Washington, D.C.	9	100
Cleveland	28	150

(i) Find a 90 percent confidence interval for the difference between the true proportions of unsafe cars in Washington, D.C. and Cleveland.

(ii) Is there sufficient evidence to conclude that the proportion of unsafe cars in Cleveland is larger than that in Washington, D. C.? (There is no mandatory city or state yearly inspection of cars in Cleveland. In Washington, D.C., every registered car is inspected for safety every year.)

22. Suppose a major automobile company wishes to allow enough headroom in its automobiles to accommodate all but the tallest 10 percent of the drivers. Previous studies show that the quantile of order .90 was 68.2 inches. In order to validate the results of the previous studies a random sample of 65 drivers is selected and it is found that the tallest 21 persons in the sample have the following heights: 68.1, 70.4, 72.6, 73.1, 70.2, 68.5, 68.0, 68.2, 69.4, 67.6, 69.8, 70.1, 70.2, 75.8, 72.3, 71.8, 69.1, 67.9, 68.4, 71.6, 74.2.

(i)   Find a 99 percent confidence interval estimate for the quantile of order .90 if it exists.

(ii) Find a 90 percent confidence interval estimate for $\zeta_{.9}$ if it exists.

(iii) Is it reasonable to use $\zeta_{.9} = 68.2$ inches?

23. The lengths of life in hours of 36 Brand X 9-volt batteries were recorded as follows:

25.1	43.5	31.2	34.0	67.5	51.4	9.3	21.3	5.4	14.3	60.3	24.9
20.6	151.5	12.1	48.8	8.0	53.7	150.7	8.4	45.5	17.1	24.2	21.5
21.7	59.0	40.3	27.2	22.1	11.2	3.1	11.3	14.2	38.7	77.5	31.6

(i) Find a 96 percent confidence interval for $\zeta_{.5}$.

(ii) Is it reasonable to use 31.2 hours as the median life length of the batteries? [The sample was generated from a table of random numbers from the uniform distribution on $(0, 1)$ by using the transformation $X = 45 \ln \dfrac{1}{1 - Y}$, where Y is uniform on $(0, 1)$. Thus the sample comes from an exponential distribution with mean 45 and median $45 \ln 2 = 31.2$.]

(iii) Test H_0: $\zeta_{.5} = 31.2$ against H_1: $\zeta_{.5} \neq 31.2$ when it is known that X has an exponential distribution so that under H_0, the mean of X is 45 hours. [The statistic $X_{(18)}$ in view of Theorem 9.8.1 has a normal distribution, under H_0, with mean 31.2 and variance $p(1 - p)/n[f(31.2)]^2$.]

24. Let $(X_1, Y_1), (X_2, Y_2), \ldots, (X_n, Y_n) \ldots$ be independent and identically distributed with $\mu_1 = \mathscr{E} X_i$, $\mu_2 = \mathscr{E} Y_i$, $\sigma_1^2 = \mathrm{var}(X_i)$, $\sigma_2^2 = \mathrm{var}(Y_i)$, and $\rho = \rho(X_i, Y_i)$. Show that $S_{11} = \displaystyle\sum_{i=1}^{n} (X_i - \bar{X})(Y_i - \bar{Y})/(n - 1)$ is a consistent estimate of $\rho \sigma_1 \sigma_2$ and $R = S_{11}/(S_1 S_2)$ is a consistent estimate of ρ under appropriate moment conditions. (See also Example 5.3.6.) Here S_1, S_2 are the sample standard diviations.

CHAPTER 10

General Methods of Point and Interval Estimation

10.1 INTRODUCTION

The problem of statistical estimation was first introduced in Section 4.3 where, for most part, we considered samples of size 1 or 2. This was mainly due to technical difficulties. We did not have enough tools in hand to evaluate the worth of point estimates or to compute the numerical values of the confidence interval estimates. Nevertheless, we were able to consider several special cases in Chapters 5 to 8. In Chapter 5 we showed that the sample mean is an unbiased and consistent estimate of the population mean. In Chapter 8 we considered the special case of sampling from a normal population and considered some point and interval estimates of the mean and the variance.

No general methods of obtaining point and interval estimates have yet been introduced. In this chapter we consider several methods of estimation. With the knowledge acquired in Chapter 8 (Sections 2–6), it is now possible to evaluate the performance of our procedures and to compute the numerical values of the confidence interval estimates.

Suppose $X_1, X_2, \ldots, X_n$ is a random sample from a (population) distribution function F, and suppose we wish to estimate (or make a guess at) some function $h(F)$. Thus $h(F)$ may be the mean of F or the quantile of order p, $0 < p < 1$, and so on. Let $\mathscr{F}$ be the family of possible distribution functions F of the X_i's. We recall that any statistic $T(X_1, X_2, \ldots, X_n)$ is said to be an estimate of $h(F)$ if T takes values in the set $\{h(F): F \in \mathscr{F}\}$. In particular, if $\mathscr{F}$ is parametric then we can write $\mathscr{F} = \{F_\theta : \theta \in \Theta\}$ for $\Theta \subseteq \mathbb{R}_k$ where θ is known as the parameter and Θ the parameter set. Here F_θ is known except for the numerical value of the parameter θ. The object of statistical estimation in this case is to find an estimate (statistic) $T(X_1, X_2, \ldots, X_n)$ for θ or for some parametric function $h(\theta)$ of θ.

In statistical inference a problem of great importance is to determine how to reduce the amount of data collection. Do we need to collect all the observations $x_1, x_2, \ldots, x_n$ in the sample? Is it possible instead to observe some function $T(x_1, x_2, \ldots, x_n)$ of the observations? For example, in a coin-tossing experiment if we write $x_i = 1$ if the ith toss results in a head and zero otherwise, then there are 2^n possible sequences of 0's and 1's. Do we need to record the specific sequence of 0's and 1's observed, or will some function of the x's suffice? If we can be sure that no information about $\theta = P(X_i = 1)$ is lost if we were to observe only $\sum_{i=1}^{n} x_i$ = number of heads, say, then there will be a great deal of reduction in the amount of data collected. We note that $\sum_{i=1}^{n} x_i$ takes only $n + 1$ values $0, 1, 2, \ldots, n$, considerably less than the 2^n possible values of $(x_1, x_2, \ldots, x_n)$. In Section 10.2 we introduce the concept of sufficiency, which tells us how to reduce data collection.

Section 10.3 deals with unbiased estimation. We explore, briefly, how sufficiency may be used to construct "better" unbiased estimates. In Section 10.4 we consider the first of the methods of estimation. This is the classical substitution principle according to which we estimate $h(F)$ by $h(\hat{F}_n)$ where $\hat{F}_n$ is the empirical distribution function estimate of F.

In Section 10.5 the method of maximum likelihood estimation is introduced. According to the maximum likelihood estimation principle, we choose that value of θ which maximizes the likelihood (or the joint density or probability) function as a function of θ. The reader has already seen many examples of this method in Chapter 6 where we often computed estimates of parameters to maximize $P(X = x)$.

Section 10.6 deals with a special topic, Bayes estimation. This method is based on Bayes's theorem of Section 2.6. In Section 10.7 we consider a general method of constructing confidence intervals.

10.2 SUFFICIENCY

Suppose you wish to play a coin-tossing game against an adversary who supplies the coin. If the coin is fair, and you win a dollar if you predict the outcome of a toss correctly and lose a dollar otherwise, then your expected net gain is zero. Since your adversary supplies the coin you may want to check if the coin is fair before you start playing the game. To test $H_0: p = 1/2$ (coin fair) you toss the coin n times. Should you record the outcome on each trial, or is it enough to know the total number of heads in n tosses to test H_0? Intuitively it seems clear that the number of heads in n trials contains all the information about unknown p and this is precisely the information we have used so far in problems of inference concerning p (in Sections 6.3, 9.6, and 9.7). Writing $X_i = 1$ if ith toss results in a head and zero otherwise and setting $T = T(X_1, \ldots, X_n) = \sum_{i=1}^{n} X_i$, we note that T is the number of heads in n trials. Clearly there is a substantial reduction in data collection and storage if we record values of $T = t$ rather than the observation vectors $(x_1, x_2, \ldots, x_n)$, there being only $n + 1$ values of t and 2^n values of $(x_1, x_2, \ldots, x_n)$. Whatever decision we make about H_0 should depend on the value of t and not on which of the $\binom{n}{t}$ n-length sequences of 0's and 1's with exactly t ones is observed. Similarly, if we wish to estimate the average height μ of entering freshmen based on a sample of size n, the record of n values $x_1, x_2, \ldots, x_n$ of heights of individuals in the sample is just too voluminous (especially when n is large) and involves large storage space and hence high cost. Once a decision is made about the underlying stochastic model for the distribution of heights under study, it is desirable to reduce the amount of data to be collected provided there is no loss of information.

To formalize and facilitate discussion, let $X_1, X_2, \ldots, X_n$ be a sample from a distribution function F. We will restrict ourselves to the parametric case where we know the functional form of F (or have made a decision about the model to use) through its probability or density function except for the specific value of the parameter $\theta \in \Theta$. Here θ may be a scalar or a vector. Consider a statistic $T = T(X_1, X_2, \ldots, X_n)$. If we can use T to extract all the information in the sample about θ, then it will be sufficient to observe only T and not the X's individually. What does it mean to say that T contains all the information in the sample about θ? Suppose the X_i's are discrete type random variables. Then a reasonable criterion would be to look at the (joint) conditional distribution of $(X_1, X_2, \ldots, X_n)$ given $T(X_1, X_2, \ldots, X_n) = t$. If

$$P\{ X_1 \le x_1, X_2 \le x_2, \ldots, X_n \le x_n | T = t \}$$

does not depend on θ (that is, does not vary with θ) for all possible values x_i of the random variables X_i, $i = 1, 2, \ldots, n$, $t \in \mathbb{R}$ and $\theta \in \Theta$, then we feel confident that T contains all the information about θ that is contained in the sample. Thus once the value of T is known, the sample $(X_1, X_2, \ldots, X_n)$ contains no further information about θ. Such a statistic is called a *sufficient statistic* for θ or for the family $\{ F_\theta : \theta \in \Theta \}$ of possible distributions of X. An equivalent definition in the discrete case is as follows.

Sufficient Statistic

A statistic $T = T(X_1, X_2, \ldots, X_n)$ is said to be sufficient for a parameter $\theta \in \Theta$ if the conditional probability function

$$P\{ X_1 = x_1, X_2 = x_2, X_n = x_n | T(X_1, X_2, \ldots, X_n) = t \}$$

does not depend on θ.

In the continuous case, some technical problems of a mathematical nature arise that are beyond our scope. We will, however, give a criterion to determine a sufficient statistic that is applicable in both discrete and continuous cases. We first consider some examples. We note that the trivial statistic $T(X_1, X_2, \ldots, X_n) = (X_1, X_2, \ldots, X_n)$ is always sufficient but does not provide any reduction in the data collection. Hence it will be excluded from consideration. Also, the definition does not require either T or θ to be scalars.

Example 1. Sufficient Statistic for Independent Bernoulli Random Variables. Let $X_1, X_2, \ldots, X_n$ be independent Bernoulli random variables with $P(X_i = 1) = p = 1 - P(X_i = 0)$, $i = 1, 2, \ldots, n$. We argued that $T = \sum_{i=1}^{n} X_i$ should be sufficient for p.

Indeed, for $x_i = 0$ or 1, $i = 1, \ldots, n$, and $0 \le t \le n$, we have

$$P\left\{ X_1 = x_1, X_2 = x_2, \ldots, X_n = x_n \middle| \sum_{i=1}^{n} X_i = t \right\}$$

$$= \begin{cases} 0, & \text{if } \sum_{i=1}^{n} x_i \neq t \\ \dfrac{P\left(X_1 = x_1, \ldots, X_n = x_n, \sum_{i=1}^{n} X_i = t \right)}{P\left(\sum_{i=1}^{n} X_i = t \right)}, & \text{if } \sum_{i=1}^{n} x_i = t. \end{cases}$$

For $\sum_{i=1}^{n} x_i \neq t$, there is nothing to prove. For $\sum_{i=1}^{n} x_i = t$,

$$X_1 = x_1, X_2 = x_2, \ldots, X_n = x_n \Rightarrow \sum_{i=1}^{n} X_i = t$$

so that

$$P\left\{ X_1 = x_1, \ldots, X_n = x_n \middle| \sum_{i=1}^{n} X_i = t \right\} = \frac{P(X_1 = x_1, \ldots, X_n = x_n)}{P\left(\sum_{i=1}^{n} X_i = t \right)}$$

$$= \frac{\left[p^{x_1}(1-p)^{1-x_1} \right] \cdots \left[p^{x_n}(1-p)^{1-x_n} \right]}{\binom{n}{t} p^t (1-p)^{n-t}} = \frac{1}{\binom{n}{t}},$$

which does not depend on p. Here we have used the fact that $\sum_{i=1}^{n} X_i$ is the number of 1's in n trials and has a binomial distribution with parameters n and p. Moreover, we have written

$$P(X_i = 1) = p = 1 - P(X_i = 0)$$

as $p^{x_i}(1-p)^{1-x_i}$, $x_i = 0$, or 1 for convenience. It follows that $T = \sum_{i=1}^{n} X_i$ is sufficient for p.

In this example we note that the statistics $(X_1, \sum_{i=2}^{n} X_i)$, $(X_1, X_2, \sum_{i=3}^{n} X_i)$, and so on are also sufficient for p, as is easily checked (or as follows from Theorem 1 below). However, it is clear that $\sum_{i=1}^{n} X_i$ reduces (or condenses) the data the most and is to be preferred over these competitors. Such a statistic is called a *minimal sufficient statistic*. □

Example 2. Example of a Statistic that Is Not Sufficient. Consider the model of Example 1 again with $n = 3$. Then $X_1 + X_2 + X_3$ is sufficient. Is $T = X_1 + 2X_2 + X_3$

sufficient? That T is not sufficient follows from the fact that

$$P\{X_1 = 1, X_2 = 0, X_3 = 1 | X_1 + 2X_2 + X_3 = 2\}$$

$$= \frac{P\{X_1 = 1, X_2 = 0, X_3 = 1\}}{P\{X_1 + 2X_2 + X_3 = 2\}}$$

$$= \frac{p^2(1 - p)}{P(X_1 = 1, X_2 = 0, X_3 = 1) + P(X_1 = X_3 = 0, X_2 = 1)}$$

$$= \frac{p^2(1 - p)}{p^2(1 - p) + p(1 - p)^2} = \frac{p}{p + (1 - p)} = p$$

which *does* depend on p. Hence $X_1 + 2X_2 + X_3$ is not sufficient for p. Interestingly, the statistic $X_1 + X_2 + X_3$ assumes value $0, 1, 2, 3$ without losing any information about p whereas the statistic $X_1 + 2X_2 + X_3$ assumes values $0, 1, 2, 3, 4$ and yet loses information on p. □

Example 3. Sufficient Statistic for Poisson Family. Let $X_1, X_2, \ldots, X_n$ be a random sample from a Poisson distribution with parameter λ. We show that $T(X_1, \ldots, X_n) = \sum_{i=1}^{n} X_i$ is sufficient for λ. We recall (Problem 8.6.7) that T has a Poisson distribution with parameter $n\lambda$. Hence for $x_1, \ldots, x_n = 0, 1, \ldots$ and $t = 0, 1, \ldots$ we have

$$P\{X_1 = x_1, \ldots, X_n = x_n | T = t\} = 0, \qquad \text{if } \sum_{i=1}^{n} x_i \neq t,$$

and if $\sum_{i=1}^{n} x_i = t$, then

$$P\{X_1 = x_1, \ldots, X_n = x_n | T = t\} = \frac{P(X_1 = x_1, \ldots, X_n = x_n)}{P(T = t)}$$

$$= \frac{\left(e^{-\lambda}\lambda^{x_1}/x_1!\right) \cdots \left(e^{-\lambda}\lambda^{x_n}/x_n!\right)}{e^{-n\lambda}\lambda^t/t!}$$

$$= \frac{t!}{x_1! x_2! \ldots x_n!},$$

which is independent of λ. Hence $\sum_{i=1}^{n} X_i$ is sufficient for λ. □

We note that the definition of sufficiency does not provide any clue to what the choice of T should be. Moreover, once such a choice is made, checking sufficiency by definition is a long and complicated process. If we are not able to choose the right T it does not follow that no sufficient statistic exists but merely that T is not sufficient and we have to start all over again. Fortunately, whenever a sufficient statistic exists it is often easy to obtain it by using the following criterion.

THEOREM 1. (THE NEYMAN FACTORIZATION CRITERION FOR SUF-FICIENCY). *Let $(X_1, X_2, \ldots, X_n)$ be a random vector governed by probability function $p_\theta(x_1, \ldots, x_n) = P_\theta\{X_1 = x_1, \ldots, X_n = x_n\}$ or probability density function $f_\theta(x_1, x_2, \ldots, x_n)$ where $\theta \in \Theta$ and the form of p_θ or f_θ is known except for the specific numerical value of θ in Θ. Then a statistic $T = T(X_1, X_2, \ldots, X_n)$ is sufficient for θ if and only if we can factor p_θ or f_θ in the following form:*

$$(1) \qquad p_\theta(x_1, \ldots, x_n) = h(x_1, \ldots, x_n) g_\theta(T(x_1, \ldots, x_n))$$

or

$$(2) \qquad f_\theta(x_1, \ldots, x_n) = h(x_1, \ldots, x_n) g_\theta(T(x_1, \ldots, x_n))$$

where h depends on the observation vector $(x_1, \ldots, x_n)$ and not on θ, and g_θ depends on θ and on $x_1, \ldots, x_n$ only through $T(x_1, \ldots, x_n)$. The statistic T or the parameter θ or both may be vectors.

Proof.* We will restrict ourselves only to the discrete case, and for convenience we assume that $n = 1$.

Suppose

$$P(X = x) = p_\theta(x) = h(x) g_\theta(T(x))$$

for all $x \in \mathbb{R}$ and $\theta \in \Theta$. Then

$$P\{X = x \mid T(X) = t\} = \begin{cases} 0 & \text{if } T(x) \neq t \\ \dfrac{P(X = x, T(x) = t)}{P(T(X) = t)}, & T(x) = t. \end{cases}$$

It is clear that if $T(x) \neq t$, $P\{X = x \mid T(X) = t\}$ does not vary with θ. If $T(x) = t$, then

$$P\{T(X) = t\} = \sum_{T(x)=t} P(X = x)$$

$$= \sum_{T(x)=t} p_\theta(x) = \sum_{T(x)=t} h(x) g_\theta(t) = g_\theta(t) \sum_{T(x)=t} h(x).$$

Hence

$$P\{X = x \mid T(X) = t\} = \frac{P(X = x, T(x) = t)}{P(T(X) = t)}$$

$$= \frac{P(X = x)}{g_\theta(t) \displaystyle\sum_{T(x)=t} h(x)} = \frac{h(x) g_\theta(t)}{g_\theta(t) \displaystyle\sum_{T(x)=t} h(x)}$$

$$= \frac{h(x)}{\displaystyle\sum_{T(x)=t} h(x)}$$

which does not depend on θ. It follows by definition that T is sufficient.

Next, suppose that T is sufficient. By multiplication rule

$$p_\theta(x) = P(X = x) = P(T(X) = t) P\{X = x \mid T(X) = t\}.$$

Since T is sufficient, $P\{X = x \mid T(X) = t\}$ does not depend on $\theta \in \Theta$. Hence by choosing $h(x) = P\{X = x \mid T(X) = t\}$ and

$$g_\theta(t) = P(T(X) = t) = \sum_{T(x)=t} p_\theta(x)$$

we see that (1) holds. □

Example 4. Sufficient Statistic for Poisson Family. We consider once again Example 3 and show how easy it is to obtain a sufficient statistic by using Theorem 1. We have

$$p_\lambda(x_1, \ldots, x_n) = \frac{e^{-\lambda}\lambda^{x_1}}{x_1!} \cdots \frac{e^{-\lambda}\lambda^{x_n}}{x_n!} = \frac{e^{-n\lambda}\lambda^{\sum_{i=1}^n x_i}}{\prod_{i=1}^n x_i!}$$

$$= \left(\frac{1}{\prod_{i=1}^n x_i!}\right)\left(e^{-n\lambda}\lambda^{\sum_{i=1}^n x_i}\right).$$

Taking $h(x_1, \ldots, x_n) = 1/\prod_{i=1}^n x_i!$ and $g_\lambda(T(x_1, \ldots, x_n)) = e^{-n\lambda}\lambda^{\sum_{i=1}^n x_i}$ with $T(x_1, \ldots, x_n) = \sum_{i=1}^n x_i$, we see that T is sufficient.

Note that the statistics $T_1 = (X_1, \sum_{i=2}^n X_i)$, $T_2 = (X_1, X_2, \sum_{i=3}^n X_i), \ldots, T_{n-2} = (X_1, X_2, \ldots, X_{n-2}, X_{n-1} + X_n)$ are also sufficient. It is clear, however, that $T = \sum_{i=1}^n X_i$ reduces the data the most and is to be preferred. The reader should always be looking for a sufficient statistic that results in the greatest reduction of the data. □

Example 5. Uniform Distribution. Let $X_1, X_2, \ldots, X_n$ be a random sample from a uniform distribution on $(0, \theta)$, $\theta \in \Theta = (0, \infty)$. Then

$$f_\theta(x_1, \ldots, x_n) = \begin{cases} 1/\theta^n & \text{if } 0 < x_i < \theta, i = 1, 2, \ldots, n \\ 0 & \text{elsewhere.} \end{cases}$$

In cases such as this, where the domain of definition of f_θ depends on θ, great care is needed. We use the following device to write f_θ in one piece without the "if" part.

Let A be any set. Define the *indicator function of A* by

(3) $$I_A(x) = \begin{cases} 1, & x \in A \\ 0, & x \notin A. \end{cases}$$

Indicator functions have simple properties, and the most useful property we need is the following:

(4) $$I_A(x) I_B(x) = I_{A \cap B}(x).$$

We rewrite $f_\theta(x_1,\ldots,x_n)$ as follows

$$f_\theta(x_1,\ldots,x_n) = \frac{1}{\theta^n} I_{[0<x_i<\theta,\,i=1,2,\ldots,n]}(x_1,\ldots,x_n).$$

Now $0 < x_i < \theta$, $i = 1,2,\ldots,n$ if and only if $0 < \max(x_1,\ldots,x_n) < \theta$. Hence

$$f_\theta(x_1,\ldots,x_n) = \frac{1}{\theta^n} I_{[0<\max(x_1,\ldots,x_n)<\theta]}(x_1,\ldots,x_n).$$

Choosing

$$h(x_1,\ldots,x_n) = 1, \, T(x_1,\ldots,x_n) = \max(x_1,\ldots,x_n)$$

and

$$g_\theta(T(x_1,\ldots,x_n)) = \frac{I_{[0<\max x_i<\theta]}(x_1,\ldots,x_n)}{\theta^n}$$

we see that $\max(X_1,\ldots,X_n)$ is sufficient for θ.

If, on the other hand, the common density of the X_i's is

$$f_\theta(x) = 1, \qquad \theta - \tfrac{1}{2} < x < \theta + \tfrac{1}{2}, \text{ and zero elsewhere}$$

then

$$f_\theta(x) = I_{[\theta-1/2<x<\theta+1/2]}(x)$$

and

$$f_\theta(x_1,\ldots,x_n) = I_{\cap_{i=1}^{n}[\theta-1/2<x_i<\theta+1/2]}.$$

Since

$$\bigcap_{i=1}^{n} \left[\theta - \tfrac{1}{2} < x_i < \theta + \tfrac{1}{2} \right] = \left\{ (x_1,\ldots,x_n) : \theta - \tfrac{1}{2} < \min x_i \le \max x_i < \theta + \tfrac{1}{2} \right\}$$

it follows that

$$f_\theta(x_1,\ldots,x_n) = I_{[\theta-1/2<\min x_i\le\max x_i<\theta+1/2]}(x_1,\ldots,x_n)$$

and choosing

$$h(x_1,\ldots,x_n) = 1, \, T(x_1,\ldots,x_n) = (\min x_i, \max x_i)$$

and

$$g_\theta(T(x_1,\ldots,x_n)) = I_{[\theta-1/2<\min x_i\le\max x_i<\theta+1/2]}(x_1,\ldots,x_n)$$

we see that T is sufficient for θ. Here the vector $(\min X_i, \max X_i)$ is jointly sufficient for the scalar parameter θ. We need to know both $\min X_i$ and $\max X_i$. The knowledge of only one of the two is not sufficient. $\quad\square$

Example 6. Sufficient Statistic for Normal Family. Let $X_1, X_2, \ldots, X_n$ be a random sample from a normal distribution with mean μ and variance σ^2. Suppose both μ and σ^2 are unknown. Then $\theta = (\mu, \sigma^2) \in \Theta = \{-\infty < \mu < \infty, \sigma^2 > 0\}$ and

$$f_\theta(x_1, \ldots, x_n) = \frac{1}{(2\pi\sigma^2)^{n/2}} \exp\left\{ -\frac{1}{2\sigma^2} \sum_{i=1}^n (x_i - \mu)^2 \right\}$$

$$= \frac{1}{(2\pi\sigma^2)^{n/2}} \exp\left\{ -\frac{1}{2\sigma^2} \left[\sum_{i=1}^n x_i^2 - 2\mu \sum_{i=1}^n x_i + n\mu^2 \right] \right\}.$$

Taking $T = (\sum_{i=1}^n X_i, \sum_{i=1}^n X_i^2)$, $h(x_1, \ldots, x_n) = 1$ and

$$g_\theta(T(x_1, \ldots, x_n)) = f_\theta(x_1, \ldots, x_n)$$

$$= \frac{1}{(2\pi\sigma^2)^{n/2}} e^{-n\mu^2/2\sigma^2} \exp\left\{ -\frac{1}{2\sigma^2} \left(\sum_{i=1}^n x_i^2 - 2\mu \sum_{i=1}^n x_i \right) \right\}$$

we see that T is sufficient. If σ^2 is not known, $\sum_{i=1}^n X_i$ *is not sufficient for* μ. Similarly, if μ is not known, $\sum_{i=1}^n X_i^2$ *is not sufficient for* σ^2. If σ^2 is known, the same factorization shows that $\sum_{i=1}^n X_i$ is sufficient for μ. If μ is known, we see easily from

$$f_{\sigma^2}(x_1, \ldots, x_n) = \frac{1}{(2\pi\sigma^2)^{n/2}} \exp\left\{ -\frac{1}{2\sigma^2} \sum_{i=1}^n (x_i - \mu)^2 \right\}$$

that $\sum_{i=1}^n (X_i - \mu)^2$ is sufficient for σ^2. We emphasize that $\sum_{i=1}^n (x_i - \mu)^2$ is computable from the observation $(x_1, \ldots, x_n)$ once μ is known.

It is possible to write $T(X_1, \ldots, X_n) = (\sum_{i=1}^n X_i, \sum_{i=1}^n X_i^2)$ in a more recognizable form. Knowledge of T is equivalent to that of $\overline{X} = \sum_{i=1}^n X_i/n$ and $S^2 = (\sum_{i=1}^n X_i^2 - n\overline{X}^2)/(n-1)$. That is, if we know T we know $(\overline{X}, S^2)$ and, conversely, if we know $(\overline{X}, S^2)$ we know T. It follows that $(\overline{X}, S^2)$ is sufficient for (μ, σ^2), which is reassuring. $\quad\square$

Example 7*. Sufficient Statistic for a Nonparametric Family. Let $\mathcal{F}$ be the class of all distribution functions on the real line that have densities. Let $X_1, X_2, \ldots, X_n$ be a random sample from a distribution function $F \in \mathcal{F}$. Then the order statistic $X_{(1)} < X_{(2)} < \cdots < X_{(n)}$ corresponding to the sample is sufficient for the family $\mathcal{F}$. A rigorous proof is beyond the scope of this text. Proceeding heuristically we note that each order statistic $x_{(1)} < x_{(2)} < \cdots < x_{(n)}$ corresponds to $n!$ n-tuples $x_1, x_2, \ldots, x_n$. That is, $n!$ n-tuples $x_1, \ldots, x_n \in \mathbb{R}$ yield the same order statistic $x_{(1)} < x_{(2)} < \cdots < x_{(n)}$. For example, the six triples $(1.1, 1, 2.4)$, $(1.1, 2.4, 1)$, $(1, 1.1, 2.4)$, $(1, 2.4, 1.1)$, $(2.4, 1, 1.1)$, $(2.4, 1.1, 1)$ all yield the same order statistic $1 < 1.1 < 2.4$. Since the observations are independent and identically distributed given $(X_{(1)}, X_{(2)}, \ldots, X_{(n)}) = (x_{(1)}, x_{(2)}, \ldots, x_{(n)})$, the conditional probability of observing any one of these $n!$ n-tuples is $1/n!$, which does not depend on the choice of F in $\mathcal{F}$. $\quad\square$

We conclude this section with a few remarks. Suppose $X_1, X_2, \ldots, X_n$ is a random sample from some distribution. Then any procedure that is based only on a part of the sample violates the criterion of sufficiency and cannot be as good as the one based on the whole sample. Indeed, if the X_i's are discrete, for example, then $X_1, X_2, \ldots, X_{n-1}$, say, is not sufficient. We have

$$P\{X_1 = x_1, \ldots, X_n = x_n | X_1 = x_1, \ldots, X_{n-1} = x_{n-1}\} = P(X_n = x_n)$$

which is not independent of the parameter.

Remark 1*. Since the object of looking for a sufficient statistic is to reduce the amount of data to be collected (without losing any information about the parameter that the sample contains) one should always be on the lookout for that sufficient statistic which results in the greatest reduction of data collection. Such a statistic is called a minimal sufficient statistic. We will not give a formal definition of a minimal sufficient statistic here, but the following informal discussion will be helpful.

We recall that every random variable (and hence also every statistic) partitions the sample space. Let $\mathcal{X}$ be the set of values of $(X_1, X_2, \ldots, X_n)$ and let $T(X_1, X_2, \ldots, X_n)$ be a statistic. Then $T = T(X_1, X_2, \ldots, X_n)$ partitions $\mathcal{X}$ into disjoint sets $A_t = \{(x_1, x_2, \ldots, x_n) : T(x_1, x_2, \ldots, x_n) = t\}$ where t is a point in the range of T. Thus $\mathcal{X} = {}_t U A_t$. A minimal sufficient statistic is one which induces on $\mathcal{X}$ the coarsest partition in the sense that if T_1 is any other sufficient statistic then the partition sets induced by T_1 are subsets of those induced by T. For example, in Example 1 we saw that $T = \sum_{i=1}^n X_i$ and $T_1 = (X_1, \sum_{i=2}^n X_i)$ are both sufficient for p. However, T is preferable over T_1 since it requires less data collection. For convenience, take $n = 4$. Then $\mathcal{X} = \{(x_1, x_2, x_3, x_4) : x_i = 0 \text{ or } 1, i = 1, 2, 3, 4\}$ has $2^4 = 16$ sample points. Partition sets induced by T are $A_t = \{(x_1, x_2, x_3, x_4) : \sum_{i=1}^4 x_i = t, x_i = 0 \text{ or } 1 \text{ for each } i\}$, $t = 0, 1, \ldots, 4$. On the other hand, T_1 induces a partition with sets

$$B_{x_1, t - x_1} = \{(y_1, y_2, y_3, y_4) \in \mathcal{X} : y_1 = x_1, y_2 + y_3 + y_4 = t - x_1\}$$

where $t = \sum_{i=1}^4 y_i$. Thus

$$A_0 = \{(0, 0, 0, 0)\} = B_{0,0},$$

$$A_1 = \{(1, 0, 0, 0), (0, 1, 0, 0), (0, 0, 1, 0), (0, 0, 0, 1)\}$$

$$= B_{0,1} \cup B_{1,0} = \{(0, 1, 0, 0), (0, 0, 1, 0), (0, 0, 0, 1)\} \cup \{1, 0, 0, 0\},$$

$$A_2 = \{(1, 1, 0, 0), (0, 1, 1, 0), (0, 0, 1, 1), (1, 0, 1, 0), (1, 0, 0, 1), (0, 1, 0, 1)\}$$

$$= B_{0,2} \cup B_{1,1}$$

$$= \{(0, 1, 1, 0), (0, 0, 1, 1), (0, 1, 0, 1)\} \cup \{(1, 1, 0, 0), (1, 0, 1, 0), (1, 0, 0, 1)\},$$

and so on. It is now clear that $\{A_t\}$ is coarser than $\{B_{x_1, t - x_1}\}$ (see Figure 1).

If $n(A)$ denotes the number of elements in set A, then clearly

$$n(\mathcal{X}) = 16, \ n(A_0) = n(A_4) = 1, \ n(A_1) = n(A_3) = 4, \ n(A_2) = 6$$

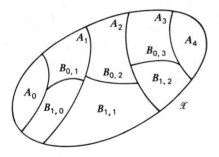

Figure 1

whereas

$$n(B_{0,0}) = n(B_{1,0}) = n(B_{0,3}) = n(B_{1,3}) = 1,$$

and

$$n(B_{0,1}) = n(B_{0,2}) = n(B_{1,1}) = n(B_{1,2}) = 3.$$

Problems for Section 10.2

1. Use the definition of sufficiency to show that the statistic $T = T(X_1, X_2, \ldots, X_n)$ is sufficient for θ.
   (i) $P(X = k) = \binom{m}{k} p^k (1 - p)^{m-k}$, $k = 0, 1, \ldots, m$; $\theta = p$, $0 \le p \le 1$, $T = \sum_{i=1}^{n} X_i$.
   (ii) $P(X = k) = p(1 - p)^{k-1}$, $k = 1, 2, \ldots$; $\theta = p \in \Theta = [0, 1]$, $T = \sum_{i=1}^{n} X_i$.
   (iii) $P(X = k) = \binom{k-1}{r-1} p^r (1 - p)^{k-r}$, $k = r, r + 1, \ldots$; $\theta = p$, $0 \le p \le 1$, $T = \sum_{i=1}^{n} X_i$.

2. Let X_1 and X_2 be independent random variables, each having Poisson distribution with mean λ. Show that $X_1 + 2X_2$ is not sufficient for λ.

3. Let X_1 and X_2 be independent random variables with $P(X_1 = 1) = p = 1 - P(X_1 = 0)$, and $P(X_2 = 1) = 5p = 1 - P(X_2 = 0)$, $0 \le p \le 1/5$. Is $X_1 + X_2$ sufficient for p? Is $X_1 + 3X_2$ sufficient for p?

4. Let X_1 and X_2 be independent Bernoulli random variables with $P(X_i = 1) = p = 1 - P(X_i = 0)$, $0 \le p \le 1$. Show that $X_1 + X_2$ and $X_1 + 5X_2$ are both sufficient for p. Which of the two statistics is preferred?

5. Let $X_1, X_2, \ldots, X_n$ be a random sample of size n from a distribution with density $f_\theta(x)$ given below. Find a sufficient statistic for θ by using factorization theorem.
   (i) $f_\theta(x) = 2(\theta - x)/\theta^2$, $0 < x < \theta$, and zero elsewhere; $\theta > 0$.
   (ii) $f_\theta(x) = \theta x^{\theta-1}$, $0 < x < 1$, and zero elsewhere; $\theta > 0$.
   (iii) $f_\theta(x) = (1/|\theta|\sqrt{2\pi})\exp\{-(x - \theta)^2/2\theta^2\}$, $x \in \mathbb{R}$; $\theta \in \mathbb{R} - \{0\}$.

6. Let $X_1, X_2, \ldots, X_n$ be a random sample from the following distributions. Find a sufficient statistic in each case.
   (i) $P(X = x) = 1/N$, $x = 1, 2, \ldots, N$; $\Theta = \{N \ge 1\}$.
   (ii) $P(X = x_1) = (1 - \theta)/2$, $P(X = x_2) = 1/2$ and $P(X = x_3) = \theta/2$, and zero elsewhere; $\Theta = \{0 < \theta < 1\}$.
   (iii) $f_\theta(x) = c(\theta)2^{-x/\theta}$, $x = \theta, \theta + 1, \theta + 2, \ldots$; $\Theta = \{\theta > 0\}$.

7. Use the factorization theorem to find a sufficient statistic based on a random sample of size n in each of the following cases.
   (i) Weibull: $f(x; \alpha, \beta) = (\alpha/\beta)x^{\alpha-1}e^{-x^{\alpha}/\beta}$ if $x > 0$ and zero otherwise when:
      (a) α, β are both unknown.   (b) α is known.   (c) β is known.
   (ii) Gamma: $f(x; \alpha, \beta) = x^{\alpha-1}e^{-x/\beta}/\Gamma(\alpha)\beta^{\alpha}$ if $x > 0$ and zero elsewhere, when:
      (a) α is unknown, β is known.   (b) β is unknown, α is known.
      (c) α and β are both unknown.
   (iii) Beta: $f(x; \alpha, \beta) = x^{\alpha-1}(1-x)^{\beta-1}/B(\alpha, \beta)$ if $0 < x < 1$ and zero elsewhere, when:
      (a) α is unknown.   (b) β is unknown.   (c) α and β are both unknown.
   (iv) Uniform: $f(x; \theta_1, \theta_2) = 1/(\theta_2 - \theta_1)$ if $\theta_1 < x < \theta_2$ and zero otherwise where both θ_1, θ_2 are unknown.
   (v) Uniform: $f(x; \theta) = 1$ if $\theta < x < \theta + 1$ and zero elsewhere.
   (vi) Exponential: $f(x; \mu, \lambda) = (1/\lambda)\exp\{-(x - \mu)/\lambda\}$ if $x > \mu$ and zero elsewhere, when:
      (a) λ is known.   (b) μ is known.   (c) both μ and λ are unknown.
   (vii) Double exponential: $f(x; \mu, \lambda) = (1/2\lambda)\exp\{-|x - \mu|/\lambda\}$, if $x \in \mathbb{R}$, where:
      (a) μ is known, $\lambda > 0$ unknown.   (b) μ is unknown.   (c) both μ and λ (> 0) are unknown.
   (viii) Pareto: $f(x) = \alpha a^{\alpha}/x^{\alpha+1}$ for $x > a$ and zero elsewhere when:
      (a) a is unknown,   (b) α is unknown, and   (c) a and α are both unknown.
   (ix) Log normal: $f(x) = 1/(x\sigma\sqrt{2\pi})\exp\{-(\ln x - \mu)^2/2\sigma^2\}$ if $x > 0$ and zero elsewhere when:
      (a) μ is unknown.   (b) σ is unknown.   (c) both μ and σ are unknown.

8*. Let $(X_1, Y_1), (X_2, Y_2), \ldots, (X_n, Y_n)$ be a random sample from a distribution given below. Find sufficient statistics for the indicated parameters in each case.
   (i) $P(X = x, Y = y) = [(x + y + k - 1)!/x!\, y!\, (k-1)!]\, p_1^x p_2^y (1 - p_1 - p_2)^k$, where $x, y = 0, 1, 2, \ldots, k \geq 1$ is an integer, $0 < p_1 < 1, 0 < p_2 < 1$ and $p_1 + p_2 < 1$; $\theta = (p_1, p_2)$.
   (ii) $f(x, y) = [\beta^{\alpha+\gamma}/\Gamma(\alpha)\Gamma(\gamma)]x^{\alpha-1}(y - x)^{\gamma-1}e^{-\beta y}$, $0 < x < y$, and zero elsewhere; $\alpha, \beta, \gamma > 0$; $\theta = (\alpha, \beta, \gamma)$.
   (iii)

$$f(x, y) = \frac{1}{2\pi\sigma_1\sigma_2\sqrt{1 - \rho^2}} \exp\left(-\frac{1}{2(1-\rho^2)}\left[\left(\frac{x - \mu_1}{\sigma_1}\right)^2 - 2\rho\left(\frac{x - \mu_1}{\sigma_1}\right)\right.\right.$$

$$\left.\left. \times \left(\frac{y - \mu_2}{\sigma_2}\right) + \left(\frac{y - \mu_2}{\sigma_2}\right)^2\right]\right\},$$

for $(x, y) \in \mathbb{R}_2$; $\theta = (\mu_1, \mu_2, \sigma_1, \sigma_2, \rho)$.

9. Let $T(X_1, \ldots, X_n)$ be sufficient for θ where both T and θ are scalars. Let f be a one-to-one function on $\mathbb{R}$ with inverse f^{-1}.
    (i) Show that $f(T)$ is sufficient for θ.
    (ii) Let d be a one-to-one function on $\mathbb{R}$. Show that T is sufficient for $d(\theta)$.
    [Hint: Use factorization theorem. The results hold when either or both T and θ are vectors.]

10. Let T_1 and T_2 be two distinct sufficient statistics for θ. Show that T_1 is a function of T_2. Does it follow that every function of a sufficient statistic is itself sufficient? [Hint: Use factorization theorem. Is $\overline{X}^2$ sufficient for μ in sampling from a normal population?]

11. If $T = \phi(U)$ and T is sufficient for θ then so is U.

12. If T is sufficient for $\theta \in \Theta$, then it is also sufficient for $\theta \in \omega \subseteq \Theta$.
    [Hint: Use definition of sufficiency.]

10.3 UNBIASED ESTIMATION

Let $X_1, X_2, \ldots, X_n$ be a random sample from a distribution function $F_\theta, \theta \in \Theta \subseteq \mathbb{R}$. We recall (Section 4.3) that an estimate $T(X_1, \ldots, X_n)$ of θ is said to be *unbiased* for θ if $\mathscr{E}|T| < \infty$ and

$$(1) \qquad \mathscr{E}_\theta T(X_1, \ldots, X_n) = \theta \qquad \text{for all } \theta \in \Theta.$$

We have subscripted $\mathscr{E}$ by θ to indicate that the distribution function to be used in computation of the expected value in (1) is F_θ. Thus, in the discrete case:

$$\mathscr{E}_\theta T(X_1, \ldots, X_n) = \sum_{x_1} \cdots \sum_{x_n} T(x_1, \ldots, x_n) \prod_{j=1}^{n} P_\theta(X = x_j).$$

Then (1) is simply a property of the estimate that may be used to restrict our search to only those estimates which satisfy (1). An estimate T that does not satisfy (1) is called a *biased estimate*. We defined bias of T by

$$(2) \qquad b(T, \theta) = \mathscr{E}_\theta T(X_1, \ldots, X_n) - \theta.$$

For an unbiased estimate T, $b(T, \theta) = 0$ for every $\theta \in \Theta$. Bias arises due to systematic error and measures, on the average, how far away and in what direction $\mathscr{E}_\theta T$ is from the parameter θ.

Example 1. Nonparametric Unbiased Estimates for Population Moments. Let $X_1, X_2, \ldots, X_n$ be a random sample from a distribution $F \in \mathscr{F}$ which has finite kth moment $\mathscr{E}_F X_i^k = m_k(F)$. Nothing else is known or assumed about F. Thus $\mathscr{F} = \{F: m_k(F) \text{ finite}\}$. Then the kth sample moment $\overline{X^k} = \sum_{i=1}^{n} X_i^k / n$ is unbiased for $m_k(F)$ since

$$\mathscr{E}\, \overline{X^k} = \sum_{i=1}^{n} \mathscr{E} \frac{X_i^k}{n} = \sum_{i=1}^{n} \frac{m_k(F)}{n} = m_k(F), \qquad F \in \mathscr{F}$$

by Theorem 5.3.1. In particular, $\overline{X}$ is unbiased for μ_F, the population mean. Moreover, we have seen in Section 5.3 that S^2 is unbiased for σ_F^2, the variance of F. See also Problem 5.4.4. $\qquad\qquad\qquad\qquad\qquad\qquad\qquad\qquad\qquad\qquad\qquad\qquad\qquad\qquad$ ☐

Frequently, however, there will be many unbiased estimates for a parameter. For example, if $\mathscr{E}X_i = \mu = \text{var}(X_i) = \sigma^2$ (which is the case when X_i's have a common Poisson distribution), then both $\overline{X}$ and S^2 are unbiased for μ as is also the estimate

$$T_\alpha(X_1,\ldots,X_n) = \alpha\overline{X} + (1-\alpha)S^2, \text{ for all } \alpha \in \mathbb{R}$$

since

$$\mathscr{E}T = \alpha\mathscr{E}\overline{X} + (1-\alpha)\mathscr{E}S^2$$
$$= \alpha\mu + (1-\alpha)\mu = \mu.$$

Which of these should we choose? Moreover, the property of unbiasedness by itself is not enough. We also need to have a procedure to measure the precision of an estimate. An unbiased estimate with a distribution highly concentrated near θ should be preferable to the one with a distribution that is very spread out. In Section 4.3 we said that the precision of any estimate T (biased or unbiased) for θ is measured by its mean square error (MSE) defined by

$$(3) \qquad\qquad\qquad \text{MSE}_\theta(T) = \mathscr{E}_\theta(T-\theta)^2.$$

We recall that

$$(4) \qquad\qquad\qquad \text{MSE}_\theta(T) = \text{var}_\theta(T) + (b(T,\theta))^2.$$

In particular, if $b(T,\theta) = 0$, then T is unbiased for θ and $\text{MSE}_\theta(T) = \text{var}_\theta(T)$. Recall also that an estimate T_1 for θ is preferable to an estimate T_2 (for θ) if

$$(5) \qquad\qquad\qquad \text{MSE}_\theta(T_1) \le \text{MSE}_\theta(T_2), \qquad \theta \in \Theta$$

with strict inequality holding for at least one θ in Θ. For unbiased estimates T_1 and T_2, (5) reduces to

$$(6) \qquad\qquad\qquad \text{var}_\theta(T_1) \le \text{var}_\theta(T_2) \qquad \text{for } \theta \in \Theta$$

with strict inequality for at least one θ in Θ. If there exists an unbiased estimate for θ which has the smallest variance amongst all unbiased estimates it is called a *uniformly minimum variance unbiased estimate* (UMVUE) for θ. Thus T is UMVUE of θ if

$$\mathscr{E}_\theta T = \theta \qquad \text{for all } \theta \in \Theta$$

and

(7) $\quad \text{var}_\theta(T) \le \text{var}_\theta(T')$ $\qquad$ for all $\theta \in \Theta$ and all T' such that $\mathscr{E}_\theta T' = \theta$.

How does sufficiency fit in here? Intuitively, if there are two unbiased estimates for θ the one based on a sufficient statistic should be preferable to the one that is not based on a sufficient statistic. It would appear therefore that we need only concentrate on estimates based on sufficient statistics. If so, does it mean that an estimate based on a sufficient statistic necessarily has a smaller variance than an estimate that is not based on a sufficient statistic? The answer is yes, but let us first consider some examples.

Example 2. Estimating Variance of a Normal Distribution. Let $X_1, X_2, \ldots, X_n$ be a random sample from a normal distribution with unknown mean μ and unknown variance σ^2. Suppose we wish to estimate σ^2. We know that S^2 is always unbiased for σ^2. Suppose, however, that biasedness of an estimate for σ^2 is not a drawback. Rather, we prefer an estimate that has a smaller MSE. Consider all estimates of the form $T_c = cS^2$ where $c > 0$ is a constant to be determined. We will find one estimate having the least MSE in the class $\{T_c : c > 0\}$. Since $(n-1)S^2/\sigma^2$ has a chi-square distribution with $(n-1)$ d.f. (Theorem 8.7.1), we note that

$$\text{var}\left(\frac{(n-1)S^2}{\sigma^2}\right) = 2(n-1)$$

so that

$$\text{var}(S^2) = 2\frac{\sigma^4}{n-1}.$$

Now

$$b_{\mu,\sigma^2}(T_c, \sigma^2) = \mathscr{E}_{\mu,\sigma^2}(T_c - \sigma^2) = c\sigma^2 - \sigma^2 = (c-1)\sigma^2$$

and from (4)

$$\begin{aligned}
\text{MSE}_{\mu,\sigma^2}(T_c, \sigma^2) &= \mathscr{E}_{\mu,\sigma^2}(T_c - \sigma^2)^2 \\
&= \text{var}_{\mu,\sigma^2}(T_c) + \left[b(T_c, \sigma^2)\right]^2 \\
&= \text{var}_{\mu,\sigma^2}(cS^2) + (c-1)^2\sigma^4 \\
&= \frac{c^2(2\sigma^4)}{n-1} + (c-1)^2\sigma^4.
\end{aligned}$$

Let

$$m(c) = \sigma^4\left(\frac{2c^2}{n-1} + (c-1)^2\right).$$

Then $m(c)$ is minimized for $c = c_0$ given by

$$m'(c) = \frac{\partial}{\partial c} m(c) = 0$$

provided that $m''(c)|_{c=c_0} > 0$. We have

$$m'(c) = \sigma^4\left\{\frac{4c}{n-1} + 2(c-1)\right\} = 0$$

which yields $c_0 = (n-1)/(n+1)$. Moreover,

$$m''(c) = \sigma^4\left\{\frac{4}{n-1} + 2\right\} > 0.$$

It follows that

$$T_{c_0} = c_0 S^2 = \left(\frac{n-1}{n+1}\right)\sum_{i=1}^{n}\frac{(X_i - \bar{X})^2}{n-1} = \sum_{i=1}^{n}\frac{(X_i - \bar{X})^2}{n+1}$$

has the smallest MSE in the class of all estimates $\{cS^2 : c > 0\}$. We note that

$$b\left(T_{c_0}, \sigma^2\right) = \left(\frac{n-1}{n+1} - 1\right)\sigma^2 = -\frac{2\sigma^2}{n+1}$$

so that T_{c_0} is negatively biased for σ^2 (tends to underestimate σ^2). Moreover, T_{c_0} is based on the sufficient statistic $(\bar{X}, S^2)$ for σ^2 (when μ is unknown). □

Example 3. Estimating p^2 for Bernoulli Distribution. Let $X_1, X_2, \ldots, X_n$ be a sample from a Bernoulli distribution with parameter p, $0 \le p \le 1$. First suppose that $n = 1$. Can we estimate p^2 unbiasedly based on one observation? The answer is no. To see this, let $T(X)$ be an estimate of p^2. If T is unbiased, then

(8) $p^2 = \mathscr{E}_p T(X) = pT(1) + (1-p)T(0)$ for all $p \in [0,1]$.

Since (8) is true for all p, the coefficient of various powers of p on both sides of (8) must be the same. This being impossible, no such T exists.

Suppose now that $n > 1$. Can we estimate p^2 unbiasedly and if so how? We note that $T = \sum_{i=1}^{n} X_i$ is sufficient for p and has a binomial distribution with

$$\mathscr{E}T = np \quad \text{and} \quad \text{var}(T) = np(1-p).$$

Since $\mathscr{E}(T/n) = p$, T/n is unbiased for p and since $p^2 = p \cdot p$, $(T/n)^2$ appears to be a candidate. But $(T/n)^2$ cannot be unbiased for p^2. This is typical of unbiased estimates. If T is unbiased for θ, $f(T)$ is not in general unbiased for $f(\theta)$ (except when f is a linear function of θ). Integrals and sums simply do not behave that nicely. In our example

$$\mathscr{E}_p\left(T^2/n^2\right) = \frac{1}{n^2}\mathscr{E}_p T^2 = \frac{1}{n^2}\left[\text{var}_p(T) + \left(\mathscr{E}_p T\right)^2\right]$$

$$= \frac{1}{n^2}\left[np(1-p) + n^2 p^2\right] \neq p^2.$$

However, this computation is not in vain. It does provide us an unbiased estimate for p^2. Indeed,

$$\mathscr{E}_p \frac{T^2}{n^2} = \frac{np(1-p)}{n^2} + p^2 = \frac{p(1-p)}{n} + p^2$$

so that all we need is an unbiased estimate T_1 for $p(1-p)/n = \text{var}_p(X_i)/n$. In that case,

$$\mathscr{E}_p \left(\frac{T^2}{n^2} - T_1 \right) = p^2$$

and $(T^2/n^2) - T_1$ is unbiased for p^2. We know that $S^2 = \sum_{i=1}^n (X_i - \bar{X})^2/(n-1)$ is always unbiased for $\text{var}(X_i)$. In our case, $\text{var}(X_i) = p(1-p)$ so that $T_1 = S^2/n$ is unbiased for $p(1-p)/n$. Note also that

$$S^2 = \frac{1}{n-1} \sum_{i=1}^n (X_i - \bar{X})^2 = \frac{1}{n-1} \left[\sum_{i=1}^n X_i^2 - n\bar{X}^2 \right]$$

$$= \frac{1}{n-1} \left[T - n\left(\frac{T}{n}\right)^2 \right] \quad \left(T = \sum_{i=1}^n X_i = \sum_{i=1}^n X_i^2 \right)$$

$$= \frac{1}{n-1} \left[T - \frac{T^2}{n} \right].$$

Hence

$$\frac{T^2}{n^2} - T_1 = \frac{T^2}{n^2} - \frac{1}{n(n-1)} \left[T - \frac{T^2}{n} \right] = \frac{T}{n} \cdot \frac{(T-1)}{(n-1)}$$

is unbiased for p^2. The point is that frequently an examination of the expression for the right order moment will lead to an unbiased estimate. In this example, since we want an unbiased estimate for p^2 and T has binomial distribution, it is natural to investigate $T(T-1)$. (See Section 6.3.1.) We have

$$\mathscr{E}T(T-1) = n(n-1)p^2$$

so that $T(T-1)/n(n-1)$ is unbiased for p^2. The same argument applies to finding an unbiased estimate for p^3. The proper choice here is $T(T-1)(T-2)$, and since

$$\mathscr{E}T(T-1)(T-2) = n(n-1)(n-2)p^3$$

it follows that $T(T-1)(T-2)/[n(n-1)(n-2)]$ is unbiased for p^3. Since the estimates for p^2 and p^3 constructed above are both based on a sufficient statistic, we expect them to be good. It turns out that they have the least variance in the class of all unbiased estimates. □

Example 4. Taxi Problem Revisited. Let $X_1, X_2, \ldots, X_n$ be a random sample from the uniform distribution on $\{1, 2, \ldots, N\}$. In Section 6.2 we considered several estimates of N (Example 6.2.1, Problem 6.2.7). Here the parameter is N and takes values in $\Theta = \{\text{all integers} \geq 2\}$. We note that

$$\mathscr{E} X_i = \frac{(N+1)}{2} = \mathscr{E} \bar{X}$$

so that $\hat{N}_1 = 2\bar{X} - 1$ is unbiased for N. We showed in Example 6.2.1 that $\hat{N}_2 = (n + 1) M_n/n$ is nearly unbiased for N but is not unbiased. In Problem 6.2.7 we showed that $\hat{N}_3 = M_n + N_n - 1$ is unbiased for N. [Here $M_n = \max(X_1, \ldots, X_n)$ and $N_n = \min(X_1, \ldots, X_n)$.] We note that M_n is sufficient for N so that $\hat{N}_2$ should be preferable over $\hat{N}_1$. Is $\mathrm{MSE}(\hat{N}_2) < \mathrm{MSE}(\hat{N}_1)$? We note that

$$\mathrm{MSE}(\hat{N}_1) = \mathrm{var}(\hat{N}_1) = \mathrm{var}(2\bar{X} - 1) = 4\,\mathrm{var}(X)/n$$

$$= (N^2 - 1)/(3n)$$

and it is possible to show by approximating the sums by integrals (as was done in Example 6.2.1) that

$$\mathrm{MSE}(\hat{N}_2) \doteq \frac{N^2}{(n+2)} - \frac{n+1}{n^2} N.$$

We have

$$\mathrm{MSE}(\hat{N}_2) - \mathrm{MSE}(\hat{N}_1) \doteq \frac{N^2}{n}\left[\frac{1}{n+2} - \frac{1}{3}\right] - \frac{n+1}{n^2} N + \frac{1}{3n}$$

$$< 0 \text{ for all } n \geq 2.$$

What about $\hat{N}_3$? Computation of MSE (or variance) of $\hat{N}_3$ is messy. So we argue as follows. The estimate $\hat{N}_3$ is not based on the "best" sufficient statistic M_n. Hence we should not expect it to perform better than $\hat{N}_2$. Since M_n is sufficient for N, an unbiased estimate based on M_n should perform well in terms of its variance. It turns out that the estimate

$$\hat{N}_4 = \frac{(M_n)^{n+1} - (M_n - 1)^{n+1}}{(M_n)^n - (M_n - 1)^n}$$

based on M_n is unbiased for N and has the least variance among all unbiased estimates for N. We will not prove this assertion. $\square$

Unbiasedness and Sufficiency

Let $T_1(X_1, \ldots, X_n)$ and $T_2(X_1, \ldots, X_n)$ be two unbiased estimates for a parameter θ. Let $T(X_1, \ldots, X_n)$ be sufficient for θ. Suppose $T_1 = f(T)$ for some function f. If sufficiency of T for θ is to have any meaning, we should expect T_1 to perform better than T_2 in the sense that $\mathrm{var}(T_1) \leq \mathrm{var}(T_2)$. Otherwise there would be no reason to study sufficiency. More generally, given an unbiased estimate h for θ, is it possible to improve upon h by using a sufficient statistic for θ? The answer is yes. Suppose T is sufficient for θ. Then the conditional distribution of $(X_1,\ldots,X_n)$ given T does not depend on θ. Consider $\mathscr{E}_\theta\{h(X_1,\ldots,X_n)|T(X_1,\ldots,X_n)\}$. Since the expected value is computed with respect to the conditional distribution of $(X_1,\ldots,X_n)$ given T it also *does not depend on* θ. Hence

$$T_1 = \mathscr{E}\{h(X_1,\ldots,X_n)|T(X_1,\ldots,X_n)\}$$

is itself an estimate of θ. Recall that $\mathscr{E}X = \mathscr{E}\{\mathscr{E}\{X|Y\}\}$. It follows that

$$\mathscr{E}_\theta T_1 = \mathscr{E}_\theta\{\mathscr{E}\{h(X_1,\ldots,X_n)|T(X_1,\ldots,X_n)\}\}$$

$$= \mathscr{E}_\theta h(X_1,\ldots,X_n) = \theta$$

since h is unbiased for θ. Thus we have found another unbiased estimate of θ that is a function of the sufficient statistic. What about the precision of T_1? We have, in view of (5.5.9),

$$\mathrm{var}_\theta(h) = \mathrm{var}_\theta(T_1) + \mathscr{E}_\theta \mathrm{var}\{h|T\}$$

$$> \mathrm{var}_\theta(T_1)$$

since $\mathrm{var}\{h|T\} > 0$ so that $\mathscr{E}_\theta \mathrm{var}\{h|T\} > 0$. If T happens to be the "best" sufficient statistic, T_1 should be the "best" estimate of θ. Unfortunately, computation of $\mathscr{E}\{h|T\}$ is usually quite involved. We will consider a few examples.

Example 5.   Estimation of p for Bernoulli Distribution.   Let $X_1, X_2,\ldots,X_n$ be a random sample from a Bernoulli distribution with parameter p. Then $T = \sum_{i=1}^n X_i$ is sufficient. Also $\mathscr{E}X_1 = p$ so that X_1 is unbiased. But X_1 is not based on T (that is, X_1 is not a function of T). According to our discussion $T_1 = \mathscr{E}\{X_1|T\}$ should be better than X_1 as an estimate of p. We have for $1 \le t \le n$

$$\mathscr{E}\{X_1|T = t\} = 1 \cdot P\{X_1 = 1|T = t\} + 0 \cdot P\{X_1 = 0|T = t\}$$

$$= \frac{P\left(X_1 = 1, \sum_{i=1}^n X_i = t\right)}{P(T = t)} = \frac{P\left(X_1 = 1, \sum_{i=2}^n X_i = t - 1\right)}{P(T = t)}$$

$$= \frac{p \cdot \binom{n-1}{t-1} p^{t-1}(1 - p)^{n-1-(t-1)}}{\binom{n}{t} p^t(1 - p)^{n-t}}$$

by independence of X_1 and $\sum_{i=2}^n X_i$ and the fact that $\sum_{i=1}^r X_i$ is a binomial random variable with parameters r and p. Hence

$$\mathscr{E}\{X_1|T = t\} = \frac{\binom{n-1}{t-1}}{\binom{n}{t}} = \frac{t}{n}.$$

Since $\mathscr{E}\{X_1|T=0\}=0$, it follows that $\mathscr{E}\{X_1|T\}=T/n=\sum_{i=1}^{n}X_i/n=\bar{X}$ is unbiased for p and has a smaller variance than X_1. Indeed, $\mathrm{var}(\bar{X})=p(1-p)/n$ and $\mathrm{var}(X_1)=p(1-p)$.

Note that $T_1=(X_n,\sum_{i=1}^{n-1}X_i)$ is also sufficient for p so that $\mathscr{E}\{X_1|T_1\}$ is also unbiased and has a smaller variance than X_1. Which of $\mathscr{E}\{X_1|T\}$ and $\mathscr{E}\{X_1|T_1\}$ should we choose? Since T condenses the data the most it would appear that $\mathscr{E}\{X_1|T\}$ is to be preferred. In fact, for $t_1=(x_n,t_0)$, $x_n=0$ or 1 and $t_0\geq 1$ we have

$$\mathscr{E}\{X_1|T_1=t_1\}=P\{X_1=1|T_1=t_1\}$$

$$=\frac{P\left\{X_1=1,\,X_n=x_n,\,\sum_{i=1}^{n-1}X_i=t_0\right\}}{P\left\{X_n=x_n,\,\sum_{i=1}^{n-1}X_i=t_0\right\}}$$

$$=\frac{P\left\{X_1=1,\,X_n=x_n,\,\sum_{i=2}^{n-1}X_i=t_0-1\right\}}{P\{X_n=x_n\}P\left\{\sum_{i=1}^{n-1}X_i=t_0\right\}}$$

$$=\frac{pp^{x_n}(1-p)^{1-x_n}\binom{n-2}{t_0-1}p^{t_0-1}(1-p)^{n-2-t_0+1}}{p^{x_n}(1-p)^{1-x_n}\binom{n-1}{t_0}p^{t_0}(1-p)^{n-1-t_0}}$$

$$=\frac{\binom{n-2}{t_0-1}}{\binom{n-1}{t_0}}=\frac{t_0}{n-1}$$

by independence. It follows that $\sum_{i=1}^{n-1}X_i/(n-1)$ is also unbiased for p and has smaller variance than X_1. Since

$$\mathrm{var}\left(\sum_{i=1}^{n-1}\frac{X_i}{n-1}\right)=\frac{p(1-p)}{n-1}>\mathrm{var}(\bar{X})$$

we see that $\bar{X}$ is to be preferred. Thus, even though every sufficient statistic will yield a better estimate for p than X_1, the estimate based on a sufficient statistic that reduces the data the most (minimal sufficient statistic) is the one to look for. This result is true in general but it is beyond our scope to prove it. $\square$

Example 6.   Estimation of Exponential Mean.   Let $X_1,X_2,\ldots,X_n$ be a random sample from an exponential density function with mean θ. Then $\sum_{i=1}^{n}X_i$ is sufficient for θ. To estimate θ, therefore, all we need to do is find an unbiased estimate based on $\sum_{i=1}^{n}X_i$. This is given by $\bar{X}$. In fact, $\bar{X}$ is the best unbiased estimate of θ, but we cannot prove it here. This is related to the fact that $\sum_{i=1}^{n}X_i$ is the "best" sufficient statistic.

To estimate θ^2, for example, we note that $\mathrm{var}(X)=\theta^2$ so that S^2, the sample variance, is an unbiased estimate for θ^2. But S^2 is not based on $\bar{X}$ (or $\sum_{i=1}^{n}X_i$) so that it should perform worse than an unbiased estimate based on $\bar{X}$. Is there such an estimate?

It is natural to compute $\mathscr{E}\overline{X}^2$. We have

$$\mathscr{E}\overline{X}^2 = \text{var}(\overline{X}) + (\mathscr{E}\overline{X})^2 = \frac{\theta^2}{n} + \theta^2 = \frac{(n+1)}{n}\theta^2.$$

It follows that $[n/(n+1)]\overline{X}^2$ is unbiased for θ, which is a function of the sufficient statistic $\overline{X}$. It turns out to be the unbiased estimate for θ^2 with the least variance. □

The procedure for finding unbiased estimates with smaller variance can now be summarized:

(i) Find, if possible, a sufficient statistic that reduces the data the most.
(ii) Find a function of this sufficient statistic that is unbiased by trial and error.
(iii) If your sufficient statistic is the best possible (minimal), then your unbiased estimate will have the least variance. If not, the unbiased estimate you construct will not be the best possible but you have the assurance that it is based on a sufficient statistic.

Example 7. Estimation of Mean and Variance of a Nonparameteric Family. In Example 1 we considered the unbiased estimation of μ and σ^2 without any restrictions on F. Let $\mathscr{F}$ be the family of all distribution functions on the real line which have densities and finite mean and finite variance. In Example 10.2.7 we argued that the ordered statistic $X_{(1)} < X_{(2)} < \cdots < X_{(n)}$ based on a random sample from $F \in \mathscr{F}$ is sufficient. In Example 1 of this section we saw that $\mathscr{E}_F\overline{X} = \mu_F$ and $\mathscr{E}_F S^2 = \sigma_F^2$ so that $\overline{X}$ and S^2 are unbiased estimates of μ and σ^2. Since

$$\overline{X} = \sum_{i=1}^{n} \frac{X_i}{n} = \sum_{i=1}^{n} \frac{X_{(i)}}{n}$$

and

$$S^2 = \sum_{i=1}^{n} \frac{\left(X_i - \overline{X}\right)^2}{n-1} = \sum_{i=1}^{n} \frac{\left(X_{(i)} - \overline{X}\right)^2}{n-1}$$

are both functions of the sufficient statistic $(X_{(1)},\ldots,X_{(n)})$ we should be confident that these are good estimates of μ and σ^2 respectively, certainly better than unbiased estimates not based on $(X_{(1)},\ldots,X_{(n)})$ since it is the best sufficient statistic. □

Example 8. Empirical Distribution Function. Let $X_1, X_2,\ldots,X_n$ be a random sample from a continuous population distribution function $F \in \mathscr{F}$ with density f. Then the order statistic $X_{(1)} < X_{(2)} < \cdots < X_{(n)}$ is sufficient (Example 10.2.7). We show that the empirical distribution function $\hat{F}_n$ defined by

$$\hat{F}_n(x) = \frac{\text{Number of } X_i\text{'s} \leq x}{n}$$

is a function of the order statistic and is unbiased for F. Indeed, we can rewrite

$$\hat{F}_n(x) = \begin{cases} 0, & X_{(1)} < x \\ k/n, & X_{(k)} \le x < X_{(k+1)} \\ 1, & \text{if } x \ge X_{(k)} \end{cases}$$

$$= \frac{1}{n} \sum_{j=1}^{n} \varepsilon(x - X_{(j)}),$$

where $\varepsilon(x) = 1$ if $x \ge 0$ and zero otherwise. Moreover,

$$\mathscr{E}_F \hat{F}_n(x) = \frac{1}{n} \sum_{j=1}^{n} \mathscr{E}_F \varepsilon(x - X_{(j)}) = \frac{1}{n} \sum_{j=1}^{n} P(X_{(j)} \le x)$$

$$= \frac{1}{n} \sum_{j=1}^{n} \sum_{k=j}^{n} \binom{n}{k}[F(x)]^k [1 - F(x)]^{n-k}$$

$$= \frac{1}{n} \sum_{k=1}^{n} \binom{n}{k}[F(x)]^k [1 - F(x)]^{n-k} \sum_{j=1}^{k} 1$$

$$= \frac{1}{n} \sum_{k=1}^{n} k\binom{n}{k}[F(x)]^k [1 - F(x)]^{n-k}$$

$$= \frac{1}{n}(nF(x)) = F(x).$$

Alternatively, we observe that $n\hat{F}(x)$ has a binomial distribution with mean $F(x)$ and variance $nF(x)[1 - F(x)]$, so that $\mathscr{E}_F(n\hat{F}_n(x)) = nF(x)$. It also follows by the central limit theorem that the sequence

$$\left\{ \frac{\hat{F}_n(x) - F(x)}{\sqrt{F(x)(1 - F(x))}} \sqrt{n} \right\}$$

has a limiting normal distribution for each $x \in \mathbb{R}$. We remark that the assumption that F be continuous (with density f) is not necessary. All the results in this example hold also in the discrete case.

In the same vein, suppose we wish to estimate $p = P(X \in I)$ for some fixed interval (or set) I. Let $X_1, X_2, \ldots, X_n$ be a random sample of size n from the distribution of X. Let Y be the number of observations in the sample that belong to I. Then Y/n is an unbiased estimate of p. This follows since Y has a binomial distribution with parameters p and n. It also follows that Y/n is consistent and has an approximate normal distribution with mean p and variance $p(1 - p)/n$. $\square$

Problems for Section 10.3

1. Let X have a binomial distribution with parameters n (known) and p. Find unbiased estimates of:
   (i) Probability of failure $= 1 - p$.
   (ii) Difference in success and failure probabilities $= p - (1 - p)$.

2. Let T_1 and T_2 be independent unbiased estimates of θ. Find an unbiased estimate of:
   (i) θ^2.   (ii) $\theta(1 - \theta)$.

3. Suppose:
   (i) $\mathscr{E}_\theta T = (\theta - 1)/(\theta + 1)$.   (ii) $\mathscr{E}_\theta T = a\theta + b, a \neq 0$.
   Find, if possible, an unbiased estimate of θ in each case.

4. Let T_1, T_2 be two unbiased estimates of θ with variances σ_1^2, σ_2^2 (both known) respectively. Consider the estimate $\hat{\theta} = \alpha T_1 + (1 - \alpha)T_2$. Find α such that $\hat{\theta}$ has minimum variance both
   (i) When T_1, T_2 have $\mathrm{cov}(T_1, T_2) = \rho\sigma_1\sigma_2$.   (ii) When T_1, T_2 are independent.

5. Let X have a hypergeometric distribution:

$$P(X = x) = \frac{\binom{D}{x}\binom{N - D}{n - x}}{\binom{N}{n}}, \qquad x = 0, 1, \ldots, D.$$

   (i) What does X^2 estimate (without bias)?
   (ii) Find an unbiased estimate of D^2 assuming n, N are known.
   [Hint: Use (6.4.6) and (6.4.7).]

6. Let X have a binomial distribution with parameters n and p where n is known. Is it possible to find an unbiased estimate of $1/p$? (See also Problem 6.5.29.)

7*. Let X have a negative binomial distribution with parameters r and p where p is known. Find an unbiased estimate of r. What is the variance of your estimate? Also find an unbiased estimate of r^2.

8. Let X have a binomial distribution where p is known but not n. Find an unbiased estimate of n and the variance of your estimate. Also find an unbiased estimate of n^2.

9. The number of defects in a roll of fabric produced by a manufacturer has a Poisson distribution with mean λ. If seven rolls are examined and found to have 7, 3, 9, 4, 8, 10, 6 defects,
   (i) Find an unbiased estimate of λ.
   (ii) Find an estimate of the precision (variance) of your estimate.

10. The complete remission times of nine adult patients with acute leukemia were recorded (in weeks) as follows: 17, 25, 21, 57, 38, 72, 59, 46, 45.
    (i) Find an unbiased estimate of the mean remission time assuming that the remission times have an exponential distribution.
    (ii) Find an estimate of the precision (variance) of your estimate.

11. An estimate is required of the true weight μ of a substance on a given set of scales. Due to errors in calibrations, the set of scales measures $X = \mu + \varepsilon$ where ε is a normal random variable with mean 0 and variance 1. A sample of 50 measurements showed a sample mean of 2.04 grams. Find an unbiased estimate of μ and its precision (variance).

12. Let $X_1, X_2, \ldots, X_n$ be a random sample from a uniform distribution on $(\theta, \theta + 1)$, $\theta \in \mathbb{R}$. Which of the following estimates are unbiased for θ?
    (i) $\hat{\theta}_1 = \overline{X}$.   (ii) $\hat{\theta}_2 = \overline{X} - 1/2$.   (iii) $\hat{\theta}_3 = (\min X_i + \max X_i - 1)/2$.
    (iv) $\hat{\theta}_4 = \min X_i - 1/(n + 1)$.   (v) $\hat{\theta}_5 = \max X_i - n/(n + 1)$.

13. Suppose $X_1, X_2, \ldots, X_n$ is a random sample from density function

$$f(x) = \frac{2}{(b - \theta)^2}(x - \theta), \qquad \theta < x < b, \text{ and zero elsewhere.}$$

Find an unbiased estimate of θ when b is known. [Hint: Find $\mathscr{E}X$.]

14. Let $X_1, X_2, \ldots, X_n$ be n independent and identically distributed random variables with common uniform distribution on $[0, \theta]$. Find an unbiased estimate of θ^α based on the sufficient statistics $X_{(n)} = \max(X_1, \ldots, X_n)$. (Here α is known.) [Hint: What is $\mathscr{E}X_{(n)}^\alpha$?]

15. Let $X_1, \ldots, X_n$ be a random sample from density function $f(x) = 2(\theta - x)/\theta^2$, $0 \le x \le \theta$, and zero elsewhere. Find an unbiased estimate of θ.

16. Suppose X has a Bernoulli distribution with parameter θ^2. That is,

$$P(X = 1) = \theta^2 = 1 - P(X = 0).$$

Show that there exists no unbiased estimate of θ.

17. Let X have a Poisson distribution with mean λ. Let $\theta = 1/\lambda$. Show that there does not exist an unbiased estimate of θ.

18. Let $X_1, X_2, \ldots, X_n$ be a random sample from a Bernoulli distribution with parameter p. Let $0 < r < n$ be an integer. Find unbiased estimates of:
    (i) $d(p) = p^r$.   (ii) $d(p) = p^r + (1 - p)^{n-r}$.

19. Let $X_1, X_2, \ldots, X_n$ be a random sample from a normal distribution with unknown mean μ and unknown variance σ^2. Find an unbiased estimate of ζ_p, the quantile of order p where

$$p = P(X \le \zeta_p)$$

based on the sufficient statistic $(\overline{X}, S^2)$.
[Hint: Standardize and find the relation between μ, σ and ζ_p. Use (8.7.2).]

20. An example where an unbiased estimate may not be desirable (see Example 8.5.2): Let $X_1, X_2, \ldots, X_n$ be the survival times (in months) of n patients with advanced lung cancer. Suppose X_i's are independent, having a common exponential distribution with mean $\lambda > 0$. The object of the study is to estimate λ and an unbiased estimate is desired. Show that
    (i) $\hat{\lambda}_1 = \overline{X}$,   (ii) $\hat{\lambda}_2 = n \min(X_1, \ldots, X_n)$,
    are both unbiased estimates of λ.
    If smaller variance is the criterion used, which of the two would you choose?
    If the time required to complete the study is the determining factor, which estimate would you choose?

In particular, consider the sample survival times (in months) 4, 77, 55, 4, 14, 23, 110, 10, 9, 62, and compare the two estimates and their precisions.

21. Let $X_1, X_2, \ldots, X_n$ be independent and identically distributed random variables with common Poisson distribution with mean λ. Find an unbiased estimate for $d(\lambda)$ where:

    (i) $d(\lambda) = \lambda^2$.   (ii) $d(\lambda) = \lambda^3$.   (iii) $d(\lambda) = \lambda^k$, $k \geq 1$ integral, $k \leq n$.
    [Hint: Consider product moments. See also Problem 10.8.9.]

22. Let $X_1, X_2, \ldots, X_n$ be independent Bernoulli random variables with common parameter p. Find an unbiased estimate of p^k ($k \geq 1$ integral $k < n$) based on a sufficient statistic. [Hint: Consider product moments. Follow Example 5.]

23*. The number of accidents X per year in a plant may be modeled as a Poisson random variable with mean λ. Assume that the accidents in successive years are independent random variables and suppose you have n observations. How will you find a good unbiased estimate for the probability $e^{-\lambda}$ that in any given year the plant has no accidents? [Hint: Define $Y_i = 1$ if $X_i = 0$ and $Y_i = 0$ otherwise. Then Y_i is unbiased. Find $\mathscr{E}\{Y_i | T\}$ where T is a sufficient statistic.]

24*. In Problem 23, suppose we wish to estimate λ.
    (i)   Show that $\bar{X}$ and S^2 are both unbiased for λ.
    (ii) Note that $\bar{X}$ is sufficient for λ. Which of the two unbiased estimates would you choose if your estimate is to have smaller variance? (Do not compute variance of S^2.)
    (iii) Show that $\mathscr{E}\{S^2 | \bar{X}\} = \bar{X}$ and hence conclude that $\text{var}(\bar{X}) \leq \text{var}(S^2)$. [Hint: $\mathscr{E}\{S^2 | \sum_{i=1}^n X_i = t\} = [1/(n-1)][\sum_{i=1}^n \mathscr{E}\{X_i^2 | \sum_{i=1}^n X_i = t\} - t^2/n]$. The conditional distribution of X_i given $\sum_{i=1}^n X_i = t$ is binomial, with parameters t and $1/n$ (Problem 8.8.17).]

25*. Let $X_1, X_2, \ldots, X_n$ be a random sample from a normal distribution with mean θ and variance θ. Then necessarily $\theta > 0$ and both $\bar{X}$ and S^2 are unbiased for θ.
    (i) Which of the two estimates is preferable?
    (ii) Is either $\bar{X}$ or S^2 a function of the "best" sufficient statistic?

26. Let $X_1, X_2, \ldots, X_n$ be a random sample from a population distribution with mean μ and variance σ^2. Let $S_n = \sum_{i=1}^n a_i X_i$.
    (i) Show that S_n is an unbiased estimate of μ if $\sum_{i=1}^n a_i = 1$.
    (ii) Consider the class of all unbiased estimates of μ of the form S_n where $a_1, \ldots, a_n \in \mathbb{R}$ are such that $\sum_{i=1}^n a_i = 1$. Show that the estimate with the least variance in this class is given by $a_i = 1/n, i = 1, \ldots, n$. [Hint: Note that $\sum_{i=1}^n a_i = 1$ implies that $\sum_{i=1}^n a_i^2 = \sum_{i=1}^n (a_i - 1/n)^2 + 1/n$.]

10.4   THE SUBSTITUTION PRINCIPLE (THE METHOD OF MOMENTS)

One of the simplest and oldest methods of estimation is the *substitution principle*: Let $d(\theta)$, $\theta \in \Theta$ be a parametric function to be estimated on the basis of a random sample $X_1, X_2, \ldots, X_n$ from a population distribution function F. Suppose d can be written as $d(\theta) = h(F)$ for some function h. Then the substitution principle estimate of $d(\theta)$ is $h(\hat{F}_n)$ where $\hat{F}_n$ is the empirical distribution function defined by

$$\hat{F}_n(x) = \text{proportion of } X_i\text{'s} \leq x.$$

(See Section 4.5.) This method leads to estimates that are quickly and easily computed. Often estimates have to be computed by some numerical approximation procedure. In that case this method provides a preliminary estimate.

> *Example 1.   **Estimating the Average Number of Blood-Spotted Eggs in a Flock of Chickens.***   The following problem has attracted wide attention in the literature. A poultry scientist wishes to estimate the frequency of occurrence of blood-spotted eggs for a flock by observing a small number of eggs laid early in the life of each hen.
>
> Let p be the probability that a given chicken will lay a blood-spotted egg and let m be the number of eggs laid. The number of blood-spotted eggs X is assumed to have a binomial distribution with parameters m and p. The probability p, however, varies from chicken to chicken and can be thought of as having a continuous distribution on $(0,1)$. This probability distribution is not observable. We assume that the distribution function of p is F (with density function f) and that the values of p for each bird in the flock are independently assigned according to the distribution F (although these values are not observed). Thus the conditional distribution of X is binomial and the unconditional distribution is given by
>
> $$p(x; m) = \int_0^1 \binom{m}{x} p^x (1-p)^{m-x} f(p)\, dp, \qquad x = 0, 1, \ldots, m.$$
>
> The problem is to estimate
>
> $$\mu_F = \int_0^1 p f(p)\, dp,$$
>
> the average probability of laying a blood-spotted egg for the flock.
>
> Let us digress a little to consider the problem of estimating the kth moment $\mathscr{E}_F X^k$ of a distribution function X. The substitution principle says that $\mathscr{E}_{\hat{F}_n}(X^k)$ is an estimate of $\mathscr{E}_F(X^k)$. Since $\hat{F}_n$ assigns probability $1/n$ to each of $X_1, X_2, \ldots, X_n$,
>
> $$\mathscr{E}_{\hat{F}_n}(X^k) = \sum_{j=1}^n \frac{X_j^k}{n}$$
>
> is the kth sample moment. Thus a moment of any order is estimated by the sample moment of the same order. Hence the name *method of moments*.
>
> Returning to the problem at hand, we note that
>
> $$\mathscr{E}X_i = \sum_{x=0}^m x p(x; m) = \sum_{x=0}^m x \int_0^1 \binom{m}{x} p^x (1-p)^{m-x} f(p)\, dp$$
>
> $$= \int_0^1 \left[\sum_{x=0}^m x \binom{m}{x} p^x (1-p)^{m-x} \right] f(p)\, dp$$
>
> $$= \int_0^1 m p f(p)\, dp = m \mu_F$$

so that

$$\mu_F = \mathscr{E}\frac{X_i}{m}.$$

By the substitution principal, therefore, we estimate μ_F by $\hat{\mu}$ where

$$\hat{\mu} = \frac{\overline{X}}{m} = \frac{\displaystyle\sum_{i=1}^{n} X_i}{mn}$$

where $X_1, X_2, \ldots, X_n$ is a random sample on X. We note that

$$\mathscr{E}\hat{\mu} = \mathscr{E}\frac{\overline{X}}{m} = \mu_F$$

and

$$\hat{\mu} \xrightarrow{P} \mu_F$$

as $n \to \infty$ so that $\hat{\mu}$ is unbiased and consistent for μ_F.   □

Example 2. *Randomized Response Survey Technique to Eliminate Evasive Answer Bias.* Certain sensitive questions elicit either wrong answers or noncooperation from the respondent. Answers to questions such as "Have you had an abortion?" or "Have you smoked marijuana?" or "Have you committed adultery?" may not be forthcoming or may be evasive simply because of the respondent's reluctance to confide secrets to a stranger or because of modesty or embarrassment. The procedure we describe here has received much attention. Suppose the population being surveyed can be partitioned into two groups A and B and an estimate is required of the proportion π of group A. A random sample of size n is drawn with replacement and each person is interviewed. Rather than asking the interviewee if he or she belongs to group A, the interviewee is asked to spin a spinner unobserved by the interviewer. The spinner is marked so that the probability that it points to A after a spin is p (known) and to B is $(1 - p)$. The interviewee is simply asked to report whether or not the spinner points to the letter representing the group to which he or she belongs. We assume that the respondents answer truthfully. Let

$$X_i = \begin{cases} 1 & \text{if } i\text{th respondent says yes} \\ 0 & \text{if } i\text{th respondent says no.} \end{cases}$$

Then

$$P(X_i = 1) = P(A) \cdot P\{X_i = 1|A\} + P(B)P\{X_i = 1|B\}$$

$$= \pi p + (1 - \pi)(1 - p)$$

and

$$P(X_i = 0) = P(A) \cdot P\{X_i = 0|A\} + P(B)P\{X_i = 0|B\}$$

$$= \pi(1 - p) + (1 - \pi)p.$$

Let Y be the number of yes responses. Then Y has a binomial distribution with parameters n and $P(X_i = 1)$ and $Y = \sum_{i=1}^{n} X_i$. By the substitution principle we estimate $P(X_i = 1)$ by Y/n, the proportion of yes answers and consequently $P(X_i = 0)$ by $1 - Y/n$. It follows again by the substitution principle that we estimate π by $\hat{\pi}$ where $\hat{\pi}$ is the solution of

$$\frac{Y}{n} = \pi p + (1 - \pi)(1 - p) = \pi[p - (1 - p)] + 1 - p$$

so that

$$\hat{\pi} = \frac{Y/n - (1 - p)}{2p - 1} \qquad \text{for } p \neq \frac{1}{2}.$$

Note that

$$\mathscr{E}\hat{\pi} = (2p - 1)^{-1}\left(\mathscr{E}\left(\frac{Y}{n}\right) - (1 - p)\right)$$

$$= (2p - 1)^{-1}[\pi p + (1 - \pi)(1 - p) - (1 - p)] = \pi$$

so that $\hat{\pi}$ is unbiased with variance

$$\text{var}(\hat{\pi}) = (2p - 1)^{-2}\text{var}\left(\frac{Y}{n}\right) = (2p - 1)^{-2}\left(n\frac{\text{var}(X_i)}{n^2}\right)$$

$$= \frac{[\pi p + (1 - \pi)(1 - p)][1 - \pi p - (1 - \pi)(1 - p)]}{n(2p - 1)^2}$$

$$= \frac{1}{n}\left[\frac{1}{16(p - \frac{1}{2})^2} - \left(\pi - \frac{1}{2}\right)^2\right], \qquad \text{for } p \neq \frac{1}{2}.$$

It should be noted that $\hat{\pi}$ may not be in $(0,1)$. In fact,

$$P(\hat{\pi} > 1) = P\left(\frac{Y}{n} > p\right) = P(Y > np) > 0.$$

However, by Borel's law of large numbers (Section 6.3), we have

$$\hat{\pi} = \frac{(Y/n) - (1 - p)}{2p - 1} \xrightarrow{P} \frac{P(X_i = 1) - (1 - p)}{2p - 1} = \pi$$

so that $\hat{\pi}$ is consistent for π and for large n the possibility that $\hat{\pi} \notin (0,1)$ is remote. $\square$

In general, suppose we wish to estimate a continuous function h of the first r moments $m_1, m_2, \ldots, m_r$ of a distribution F. Let

$$d(\theta) = h(m_1(\theta), m_2(\theta), \ldots, m_r(\theta)).$$

Then we estimate d by the statistic

$$T(X_1, X_2, \ldots, X_n) = h(\hat{m}_1, \hat{m}_2, \ldots, \hat{m}_r)$$

where $\hat{m}_j$ is the jth sample moment given by

$$\hat{m}_j = \sum_{i=1}^{n} \frac{X_i^j}{n}, \qquad j = 1, 2, \ldots, r.$$

Frequently there are several method of moments estimates for the same d. Since the method is used to provide a preliminary estimate, it is usual to choose that estimate of d which is based on moments of lowest possible order. However, one can also use other criteria to make this choice, such as minimum mean square error. To estimate the mean of a distribution where the mean and the variance coincide (example: Poisson), one has to choose between the sample mean and the sample variance. Since the sample mean is the lower order moment, it is usual to choose the sample mean to estimate the population mean unless some other considerations make the use of S^2 more desirable as an estimate. In the Poisson case, it can be shown that $\text{var}(\overline{X}) \leq \text{var}(S^2)$ so that $\overline{X}$ would be the choice to estimate $\mathscr{E}X_i$ if smaller variance was desired.

The method of moments estimates are consistent under mild conditions and have an asymptotic normal distribution.

Example 3. Estimating the Number of Taxis in a Town. In Example 6.2.1 we considered several estimates of N in sampling from the probability function

$$P(X = k) = \frac{1}{N}, \qquad k = 1, 2, \ldots, N.$$

To find the substitution principle estimate of N we note that

$$\mathscr{E}X = \frac{N + 1}{2}$$

so that $N = 2\mathscr{E}X - 1$. By the substitution principle we estimate N by

$$\hat{N} = 2\mathscr{E}_{\hat{F}_n}X - 1 = 2\overline{X} - 1$$

where $\overline{X}$ is the sample mean of a sample of size n. We note that $\hat{N}$ is unbiased and consistent and by the central limit theorem

$$P\left(\frac{\hat{N} - N}{\sigma_{\hat{N}}} \leq x\right) \to \frac{1}{\sqrt{2\pi}} \int_{-\infty}^{x} e^{-t^2/2}\, dt \qquad \text{as } n \to \infty$$

for all $x \in (-\infty, \infty)$. Here $\sigma_{\hat{N}} = \{(N^2 - 1)/12\}^{1/2}$ (Problem 6.2.1).

We note that $\hat{N}$ is not based on the sufficient statistic $\max(X_1, \ldots, X_n)$. In this case we said in Example 10.3.4 that the best unbiased estimate of N is a function of

$M_n = \max(X_1,\ldots,X_n)$ and is given by

$$\hat{N}_4 = \frac{(M_n)^{n+1} - (M_n - 1)^{n+1}}{(M_n)^n - (M_n - 1)^n}.$$ □

We note that the method of moments estimate of

$$\mathscr{E}_F g(x) = \begin{cases} \int_{-\infty}^{\infty} g(x)f(x)\,dx \\ \sum_x g(x)P(X = x), \end{cases}$$

given by

$$\mathscr{E}_{\hat{F}_n} g(x) = \sum_{i=1}^{n} \frac{g(X_i)}{n},$$

is a nonparametric estimate since it does not depend on the form of F. Similarly, if an estimate of a median of X is desired, it is the sample median. Indeed, by definition median of X is a number m such that

$$P(X < m) = F(m-) \le \tfrac{1}{2} \le P(X \le m) = F(m)$$

so that by substitution principle the estimate $\hat{m}$ of m satisfies

$$\hat{F}_n(\hat{m}-) \le \tfrac{1}{2} \le \hat{F}_n(\hat{m}).$$

Hence it is the sample median, given by

$$\hat{m} = \begin{cases} X_{(n+1)/2} & \text{if } n \text{ is an odd integer} \\ (X_{(n)} + X_{(n+1)})/2 & \text{if } n \text{ is an even integer.} \end{cases}$$

Problems for Section 10.4

1. Let $X_1, X_2,\ldots,X_n$ be a sample from a Bernoulli population with parameter p.
   (i) Show that $\bar{X} = n^{-1}\sum_{k=1}^{n} X_k$ is a method of moments estimator of p.
   (ii) Find a method of moments estimate of $\mathrm{var}(X_1) = p(1 - p)$ based only on $\bar{X}$.
   (iii) Find a method of moments estimate of $\mathrm{var}(X_1) = p(1 - p)$ based on the second sample moment.
   (iv) Show that the estimates in (ii) and (iii) are identical. [Hint: $P(X_i = 1) = p$, $P(X_i = 0) = 1 - p$ so $\sum_{i=1}^{n} X_i = \sum_{i=1}^{n} X_i^2$.]

2. In a clinical trial n patients with advanced cancer of the prostate survived times $t_1, t_2,\ldots,t_n$ (months). Suppose that the survival distribution is exponential with

mean λ:

(i) Find two different method of moments estimates of λ.

(ii) Find a method of moment estimate of the survival function

$$S(t) = P(T > t) = e^{-t/\lambda}.$$

(iii) If $n = 8$ and the survival times were recorded (in months) as 9, 15, 8, 25, 6, 17, 23, 5 find these estimates. In part (ii) take $t = 20$.

3. A certain blood disease occurs in a proportion p of white females. A random sample of n females was selected and it was found that X females in the sample have the blood disease. Unfortunately, the researcher lost track of the total number n of females surveyed. Find the method of moments estimate of n if p is known.

4. How will you estimate the proportions of freshmen, sophomores, juniors, and seniors at a large state university based on a sample of size n?

5. Heritable characteristics depend on genes, which are passed on from parents to offsprings. Genes appear in pairs, and in the simplest case each gene of a particular pair can assume one of two forms called alleles, A and a. Therefore, there are three possible genotypes: AA, Aa, and aa. (Biologically, there is no difference between aA and Aa.) According to the *Hardy–Weinberg stability principle* the genotype probabilities stabilize after one generation in a population with random mating (see Problems 6.7.15 and 6.7.16). We are therefore led to assume that after one generation there are three types of individuals with proportions

$$p_1 = p^2, \quad p_2 = 2p(1-p), \quad \text{and} \quad p_3 = (1-p)^2, \quad 0 < p < 1.$$

Let X_i be the number of individuals of type i in a sample of n individuals. How will you estimate p by method of moments?

6. The diameters of a large shipment of ball bearings have a normal distribution with mean μ and variance σ^2. A sample of size n is taken. Find estimates for μ, σ^2, and σ.

7*. The lead time for orders of diodes from a certain manufacturer has a gamma distribution with mean μ and standard deviation σ. Estimate the parameters of the gamma distribution based on a sample of size n.

8*. The proportion of defective tires manufactured by a certain company has a beta distribution with parameters α and β. Estimate α and β based on a sample of n proportions.

9*. Let $X_1, X_2, \ldots, X_n$ be a sample from a negative binomial distribution with parameters p and r, $0 < p < 1$, and $r \geq 1$ integral. Find estimates of r and p.

10. Let ζ_p be the pth percentile of a distribution function F so that ζ_p is a value satisfying

$$F(\zeta_p -) \leq p \leq F(\zeta_p).$$

Find an estimate of ζ_p based on a sample of size n. Find also an estimate of the survival function $S(t) = 1 - F(t)$.

11*. The lifetime of a mechanical part is a random variable with density function

$$f(x) = \frac{x}{\alpha^2} e^{-x^2/2\alpha^2} \quad \text{for } x > 0, \quad \text{and} \quad = 0 \quad \text{for } x < 0.$$

Find an estimate of α and of the survival function $S(t)$.

12. In Example 1 we considered $\hat{\mu}$ defined by

$$\hat{\mu} = \frac{\overline{X}}{m}$$

as an estimate of $\mu_F = \int_0^1 p f(p)\, dp$ in sampling from

$$p(x; m) = \int_0^1 \binom{m}{x} p^x (1 - p)^{m-x} f(p)\, dp.$$

Show that

$$\text{var}(\hat{\mu}) = \frac{\text{var}(X_1)}{m^2 n}$$

and that $\hat{\mu}$ is asymptotically normal with mean μ_F and variance $\text{var}(X_1)/m^2 n$ for large n.

10.5 MAXIMUM LIKELIHOOD ESTIMATION

Suppose you had two loaded coins, one with a probability $p = 1/10$ of heads and the other with $p = 9/10$ of heads. One of the coins is lost but you do not know which one. To find p for the coin you now have left, you do what comes naturally, you toss it. Since p is either $1/10$ or $9/10$ it is natural to choose $p = 9/10$ if a head shows up and choose $p = 1/10$ if a tail shows up. Intuitively you feel that you are most likely to get a head with a coin with $p = 9/10$ then with $p = 1/10$. This intuitive procedure turns out to be a good one, as we shall see. First let us examine what your procedure does. It chooses that value of p which maximizes the probability of what you have observed. Indeed, if we write X to be the number of heads, then

$$P_p(X = x) = p^x (1 - p)^{1-x} = \begin{cases} 9/10 & \text{if } x = 1 \text{ and you choose } p = 9/10 \\ 9/10 & \text{if } x = 0 \text{ and you choose } p = 1/10. \end{cases}$$

This is essentially *the method of maximum likelihood estimation*. Before we give a formal definition, let us examine the estimate $\hat{p}(X)$ given by

$$\hat{p}(x) = \begin{cases} 9/10, & x = 1 \\ 1/10, & x = 0. \end{cases}$$

We note that

$$\mathscr{E}_p \hat{p}(X) = p\hat{p}(1) + (1-p)\hat{p}(0) = \frac{(1+8p)}{10}$$

$$= \begin{cases} 9/50 & \text{if } p = 1/10 \\ 41/50 & \text{if } p = 9/10 \end{cases}$$

so that $\hat{p}$ is biased. Moreover,

$$\text{MSE}_p(\hat{p}(X)) = \mathscr{E}_p[\hat{p}(X) - p]^2 = p\left(\frac{9}{10} - p\right)^2 + (1-p)\left(\frac{1}{10} - p\right)^2$$

$$= \frac{1}{100} + \frac{3}{5}p(1-p),$$

and since $p(1-p) = 9/100$ whether $p = 1/10$ or $9/10$,

$$\text{MSE}(\hat{p}(X)) = \frac{8}{125}.$$

If you choose estimates $\hat{p}_1 = 1/10$ or $\hat{p}_2 = 9/10$ no matter what the outcome, then in either case the mean square error is zero if you are correct and $64/100$ if you are not correct.

Of course, you could toss the coin a few more times. For $n = 2$ or 3 tosses, the situation is as follows.

Probability of x Heads in n Tosses

	$n = 2$		$n = 3$	
x	$p = 1/10$	$p = 9/10$	$p = 1/10$	$p = 9/10$
0	81/100	1/100	729/1000	1/1000
1	18/100	18/100	243/1000	27/1000
2	1/100	81/100	27/1000	243/1000
3	—	—	1/1000	729/1000

In the case when $n = 2$ we have no difficulty in concluding that $p = 1/10$ if $x = 0$ and $p = 9/10$ if $x = 2$. But when $x = 1$, we cannot make a choice. The reason is that $1/10 = 1 - 9/10$ so that if the coin is tossed an even number of times, the probability of $n/2$ successes and $n/2$ failures with both coins is the same, namely

$$\binom{n}{n/2} p^{n/2}(1-p)^{n/2} = \binom{n}{n/2}(9/100)^{n/2}.$$

In the case when $n = 3$, we see that our choice of p that maximizes the probability of observing x is given by

$$\hat{p}(x) = \begin{cases} 1/10 & \text{if } x = 0 \text{ or } 1 \\ 9/10 & \text{if } x = 2 \text{ or } 3. \end{cases}$$

You can now see that the procedure we have used is the following: Given that the outcome is x, *estimate p by a value $\hat{p} \in \Theta$, which maximizes $P_p(X = x)$.*

This is all very well when X is the discrete type but what happens when X has a continuous distribution? In that case, for any outcome x

$$P_\theta(X = x) = 0 \qquad \text{for all } \theta \in \Theta.$$

We can argue as follows. If $X = x$ is observed, the probability that the outcome x falls in an interval of (small) length dx around x is given by

$$P_\theta(X \in dx) = f_\theta(x)\, dx$$

where f_θ is the density function of X. The above procedure can now be applied as follows: *Estimate θ by that value in Θ which maximizes $f_\theta(x)$* so that the probability of the outcome falling in a small neighborhood of x is maximum.

To formalize this idea, let $(X_1, X_2, \ldots, X_n)$ be a random vector with joint density or probability function $f_\theta(x_1, \ldots, x_n)$ where $\theta \in \Theta$ is the unknown parameter. Here θ may be a scalar or a vector.

DEFINITION 1. The likelihood function of the observation $(x_1, x_2, \ldots, x_n)$ is a function of θ defined on Θ by

$$L(\theta; x_1, \ldots, x_n) = f_\theta(x_1, x_2, \ldots, x_n).$$

In the discrete case $L(\theta; x_1, \ldots, x_n)$ is just the probability of observing $X_1 = x_1, \ldots, X_n = x_n$. In the continuous case, however, we need a constant multiple of $f_\theta(x_1, \ldots, x_n)$ to represent the volume of a small neighborhood of the observation $(x_1, \ldots, x_n)$ to give the same interpretation to the likelihood function. That is, technically

$$L(\theta; x_1, \ldots, x_n) = cf_\theta(x_1, \ldots, x_n)$$

for some arbitrary constant $c > 0$. But we will take $c = 1$ as above.

In the special case (of most interest) when $X_1, X_2, \ldots, X_n$ is a random sample

$$(1) \quad L(\theta; x_1, \ldots, x_n) = \begin{cases} \displaystyle\prod_{j=1}^{n} P_\theta(X_j = x_j) & \text{if } X_j\text{'s are discrete type,} \\ \displaystyle\prod_{j=1}^{n} f_\theta(x_j) & \text{if } X_j\text{'s are continuous type.} \end{cases}$$

Maximum Likelihood Principle

Choose as an estimate of θ a value $\hat{\theta}(X_1,\ldots,X_n) \in \Theta$ that maximizes the likelihood function, that is, a $\hat{\theta}$ satisfying

$$L(\hat{\theta}; x_1,\ldots,x_n) = \sup_{\theta \in \Theta} L(\theta; x_1,\ldots,x_n).$$

If such a $\hat{\theta}$ exists, it is known as a *maximum likelihood estimate* of θ. In many practical problems, maximum likelihood estimates are unique although this is not true in general. Sometimes an observed value of $(x_1,\ldots,x_n)$ will produce a value of $\hat{\theta}$ that is not in Θ. In that case we say that a maximum likelihood estimate does not exist.

Since logarithm is a monotone function we note that

$$\ln L(\hat{\theta}; x_1,\ldots,x_n) = \sup_{\theta \in \Theta} \ln L(\theta; x_1,\ldots,x_n).$$

In the case of random sampling when L is of the form (1) it is convenient to work with the log likelihood function. Often the likelihood function is differentiable in Θ. In that case we solve the *likelihood equation(s)*

$$\frac{\partial \ln L(\theta; x_1,\ldots,x_n)}{\partial \theta} = 0$$

to find $\hat{\theta}$, but then we need to check if $\hat{\theta}$ is indeed a maximum. (Check to see if $\partial^2 \ln L / \partial \theta^2 |_{\theta = \hat{\theta}} < 0$.)

Example 1. **Maximum Likelihood Estimate of Poisson Mean—Wrong Telephone Connections.** The number of connections to a wrong number at a telephone exchange is frequently modeled by a Poisson distribution. Let $x_1, x_2,\ldots,x_n$ be the observed number of wrong connections on n different days. If we assume that the mean number of wrong connections is the same, namely λ, on each day and that the events are independent, then

$$L(\lambda; x_1,\ldots,x_n) = \prod_{j=1}^{n} P_\lambda(X_j = x_j) = \prod_{j=1}^{n} \left(e^{-\lambda} \frac{\lambda^{x_j}}{x_j!} \right) = \frac{e^{-n\lambda} \lambda^{\sum_{j=1}^{n} x_j}}{\prod_{j=1}^{n} (x_j!)}$$

so that

$$\ln L(\lambda) = -n\lambda + \sum_{j=1}^{n} x_j \ln \lambda - \sum_{j=1}^{n} \ln(x_j!).$$

It follows that

$$\frac{d}{d\lambda} \ln L(\lambda) = -n + \frac{\left(\sum\limits_{j=1}^{n} x_j\right)}{\lambda}$$

and

$$\frac{d^2}{d\lambda^2} \ln L(\lambda) = -\sum_{j=1}^{n} \frac{x_j}{\lambda^2} < 0 \qquad \text{for all } \lambda > 0.$$

Hence

$$\frac{d}{d\lambda} \ln L(\lambda) = 0 = -n + \sum_{j=1}^{n} \frac{x_j}{\lambda}$$

gives the maximum likelihood estimate as

$$\lambda = \sum_{j=1}^{n} \frac{x_j}{n} = \bar{x},$$

provided that $\bar{x} \neq 0$ since $\lambda = 0 \notin \Theta = (0, \infty)$. This problem is sometimes avoided by taking the parameter set to be the closure of Θ.

We note that λ is unbiased and is a sufficient statistic for λ. Moreover λ is consistent for λ.

In particular, if we observe 6, 5, 11, 2, 9, 7, 7, 6, 10 wrong connections on nine randomly chosen days at this exchange, then $\lambda = \sum_{i=1}^{9} x_i/9 = 63/9 = 7$. It means that the likelihood function, given this set of observations, namely

$$L(\lambda; 6, 5, 11, 2, 9, 7, 7, 6, 10) = \frac{e^{-9\lambda}\lambda^{63}}{(6!)^2 5!11!2!9!(7!)^2 10!}$$

has a larger value for $\lambda = \lambda = 7$ than for any other $\lambda \in \Theta = (0, \infty)$. Moreover, no matter what the sample (of size 9), as long as $\bar{x} = 7$, the likelihood function is maximized at $\lambda = 7$. $\square$

Example 2. Estimating θ for Uniform Distribution. Suppose a bus arrives at a bus stop between time 0 and time θ (inclusive) and that the probability of arrival in any subinterval of time is proportional to the length of the subinterval. Then the time X that a person who arrives at time 0 has to wait for the bus has a uniform distribution on $[0, \theta]$. In order to estimate θ suppose we take n independent observations on X, say $x_1, x_2, \ldots, x_n$. What is the maximum likelihood estimate of θ? The likelihood function is given by

$$L(\theta; x_1, \ldots, x_n) = \begin{cases} 1/\theta^n, & 0 \le x_i \le \theta, i = 1, 2, \ldots, n. \\ 0 & \text{elsewhere.} \end{cases}$$

Here the domain of definition depends on θ so that $1/\theta^n$ is not the joint density. If we were to differentiate ln $L(\theta)$ blindly, we end up with

$$-n/\theta = 0$$

which gives $\hat{\theta} = \infty$, the minimum of $L(\theta)$. The likelihood function is in fact

$$L(\theta; x_1, \ldots, x_n) = \frac{1}{\theta^n} I_{[0 \leq \max x_i \leq \theta]}(x_1, \ldots, x_n)$$

where I_A is the indicator function of set A. If we plot the likelihood function we get Figure 1. The maximum is at $\theta = \max(x_1, \ldots, x_n)$ and the maximum likelihood estimate for θ is given by

$$\hat{\theta}(x_1, \ldots, x_n) = \max(x_1, x_2, \ldots, x_n) = x_{(n)}, \text{ provided } x_n \neq 0.$$

The result is obtained if we observe that $L(\theta)$ is a strictly decreasing function of θ so that the maximum is achieved at the smallest possible value of θ, namely, $\max(x_1, \ldots, x_n)$.

We note that $\hat{\theta}$ is consistent for θ but not unbiased (Example 8.5.4). Again $\hat{\theta}$ is a sufficient statistic (in fact the best) for θ. $\qquad \square$

Example 3. Restricted Bernoulli Parameter. Suppose we take one observation from a Bernoulli distribution where we know that $p \in [1/3, 2/3]$. Then the likelihood function is given by

$$L(p; x) = p^x (1 - p)^{1-x}, p \in [1/3, 2/3].$$

A plot of $L(p; x)$ is given in Figure 2. It is clear that the maximum of $L(p; x)$ is achieved at $p = 1/3$ when $x = 0$ and at $p = 2/3$ when $x = 1$. It follows that the maximum likelihood estimate is given by

$$\hat{p}(x) = \begin{cases} 1/3 & \text{if } x = 0 \\ 2/3 & \text{if } x = 1. \end{cases}$$

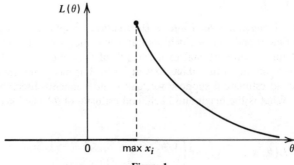

Figure 1

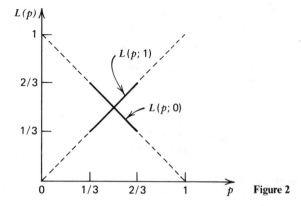

Figure 2

We can rewrite

$$\hat{p}(x) = \frac{(x+1)}{3}$$

and note that

$$\mathscr{E}_p \hat{p}(X) = \mathscr{E}\frac{(X+1)}{3} = \frac{(p+1)}{3} \neq p$$

so that $\hat{p}$ is biased. Moreover,

$$\text{MSE}_p(\hat{p}(X)) = \mathscr{E}_p(\hat{p}(X) - p)^2 = \mathscr{E}_p\left(\frac{X+1}{3} - p\right)^2 = \frac{1}{9}\mathscr{E}_p(X + 1 - 3p)^2$$

$$= \frac{1}{9}\left[\mathscr{E}_p(X - p)^2 + (1 - 2p)^2\right] = \frac{1}{9}\left[p(1-p) + (1 - 2p)^2\right]$$

$$= \frac{1 - 3p(1-p)}{9}.$$

Consider now the trivial estimate $\hat{p}_1 = 1/2$, which ignores the observation. The MSE of $\hat{p}_1$ is given by

$$\mathscr{E}_p(\hat{p}_1 - p)^2 = \left(\frac{1}{2} - p\right)^2 \leq \text{MSE}_p(\hat{p}(X)) = \frac{1 - 3p(1-p)}{9}$$

for $p \in [1/3, 2/3]$ and the maximum likelihood estimate is worse than the trivial estimate $\hat{p}_1 \equiv \frac{1}{2}$. □

Example 4. *Estimating Mean and Variance of a Normal Distribution.* Let $X_1, X_2, \ldots, X_n$ be a random sample from a normal distribution with mean μ and variance σ^2. We first suppose that both μ and σ^2 are unknown so that $\theta = (\mu, \sigma^2) \in$

$\{(\mu, \sigma^2): \mu \in \mathbb{R}, \sigma^2 > 0\}$. The likelihood function is given by

$$L(\theta; x_1, \ldots, x_n) = \frac{1}{(\sigma\sqrt{2\pi})^n} \exp\left\{ -\sum_{i=1}^{n} \frac{(x_i - \mu)^2}{2\sigma^2} \right\}$$

so that

$$\ln L(\theta) = -n \ln \sigma - n \ln\sqrt{2\pi} - \sum_{i=1}^{n} \frac{(x_i - \mu)^2}{2\sigma^2}.$$

Differentiating with respect to μ and σ^2 we get

$$\frac{\partial \ln L(\theta)}{\partial \mu} = \frac{1}{\sigma^2} \sum_{i=1}^{n} (x_i - \mu)$$

and

$$\frac{\partial \ln L(\theta)}{\partial \sigma^2} = -\frac{n}{2\sigma^2} + \frac{1}{2\sigma^4} \sum_{i=1}^{n} (x_i - \mu)^2.$$

Solving $\partial \ln L/\partial \mu = 0$ and $\partial \ln L/\partial \sigma^2 = 0$ simultaneously gives maximum likelihood estimates

$$\hat{\mu} = \bar{x} \text{ and } \hat{\sigma}^2 = \sum_{i=1}^{n} \frac{(x_i - \bar{x})^2}{n}.$$

We leave the reader to show that $\hat{\mu}$, $\hat{\sigma}^2$ do indeed give maxima for $L(\theta)$. We note that although $\hat{\mu}$ is unbiased for μ, $\hat{\sigma}^2$ is biased since

$$\mathscr{E}\hat{\sigma}^2 = \mathscr{E}\frac{n-1}{n} S^2 = \frac{n-1}{n} \sigma^2.$$

Moreover, both $\hat{\mu}$ and $\hat{\sigma}^2$ are consistent estimates for μ and σ^2 respectively, and $(\hat{\mu}, \hat{\sigma}^2)$ is sufficient for (μ, σ^2).

If μ is unknown and σ is known, the same argument leads to $\hat{\mu} = \bar{X}$ as the maximum likelihood estimate. If σ is unknown and μ is known, the maximum likelihood estimate for σ^2 is given by

$$\hat{\sigma}^2 = \sum_{i=1}^{n} \frac{(X_i - \mu)^2}{n}. \qquad \square$$

We have thus far seen that a maximum likelihood estimate may not in general be unique, may not be unbiased, and appears to be consistent and a function of a sufficient statistic. Let us consider the last two properties and also the invariance property of maximum likelihood estimates.

Under some reasonable conditions it can be shown that the method of maximum likelihood estimation leads to a consistent estimate which has an asymptotic normal distribution. A proof of this property is beyond our scope. However, this large sample property of maximum likelihood estimates is frequently very useful.

PROPOSITION 1. (SUFFICIENCY AND MAXIMUM LIKELIHOOD ESTIMATES). *Suppose $X_1, X_2, \ldots, X_n$ is a random sample from a discrete or a continuous type distribution. Let $T(X_1, \ldots, X_n)$ be a sufficient statistic for the parameter $\theta \in \Theta$. If there exists a unique maximum likelihood estimate $\hat{\theta}$ of θ, then $\hat{\theta}$ is a (nonconstant) function of T. If a maximum likelihood estimate of θ exists but is not unique, then one can find a maximum likelihood estimate $\hat{\theta}$ that is a function of T.*

Proof. By factorization theorem,

$$L(\theta; x_1, \ldots, x_n) = h(x_1, \ldots, x_n) g_\theta(T(x_1, \ldots, x_n))$$

for all $x_1, \ldots, x_n$, θ, and some h and g_θ. If a unique $\hat{\theta}$ exists that maximizes $L(\theta)$, it also maximizes $g_\theta(T(x_1, \ldots, x_n))$ and hence $\hat{\theta}$ is a function of T. If a maximum likelihood estimate of θ exists but is not unique, we choose a particular maximum likelihood estimate $\hat{\theta}$ which is a function of T from the set of all maximum likelihood estimates. $\square$

Example 5. Uniform Distribution on $[\theta - 1, \theta + 1]$. Let $X_1, X_2, \ldots, X_n$ be a random sample from the uniform distribution on $[\theta - 1, \theta + 1]$. Then the likelihood function is given by

$$L(\theta; x_1, \ldots, x_n) = \left(\tfrac{1}{2}\right)^n I_{[\theta - 1 \leq \min x_i \leq \max x_i \leq \theta + 1]}(x_1, \ldots, x_n).$$

We note that $T(X_1, \ldots, X_n) = (\min X_i, \max X_i)$ is jointly sufficient for θ. Now any θ satisfying

$$\theta - 1 \leq \min x_i \leq \max x_i \leq \theta + 1$$

or, equivalently,

$$\max x_i - 1 \leq \theta \leq \min x_i + 1$$

maximizes the likelihood and hence is a maximum likelihood estimate. In particular, we can choose

$$\hat{\theta} = \frac{(\min X_i + \max X_i)}{2}$$

which is a function of T. In fact, we can choose any $\hat{\theta}_\alpha$, $0 \leq \alpha \leq 1$

$$\hat{\theta}_\alpha = \alpha(\max X_i - 1) + (1 - \alpha)(\min X_i + 1),$$

as a maximum likelihood estimate of θ. If α is constant independent of the X's, then $\hat{\theta}_\alpha$ is a function of T. On the other hand, if α depends on the X's such that $0 \leq \alpha \leq 1$, then $\hat{\theta}_\alpha$ may not be a function of T. For example, we can choose $\alpha = \sin^2 X_1$. Then $0 \leq \sin^2 X_1 \leq 1$ and

$$\hat{\theta}_\alpha = \sin^2 X_1 (\max X_i - 1) + \cos^2 X_1 (\min X_i + 1)$$

is no longer a function of T alone.   □

We next consider another desirable property of maximum likelihood estimates. Let $d: \Theta \to \Theta$ be a one to one function. Then the inverse function d^{-1} exists and

$$L(\theta; x_1, \ldots, x_n) = L\big(d^{-1}(d(\theta)); x_1, \ldots, x_n\big)$$

and it follows that if $\hat{\theta}$ is a maximum likelihood estimate of θ, then $d(\hat{\theta})$ is a maximum likelihood estimate of $d(\theta)$. This follows since $L(\theta)$ and $L(d^{-1}(d(\theta))$ are both maximized by $\hat{\theta}$. Hence $\hat{\theta} = d^{-1}(\widehat{d(\theta)})$ so that $\widehat{d(\theta)} = d(\hat{\theta})$. If d is not one to one, we could still choose $d(\hat{\theta})$ as a maximum likelihood estimate of $d(\theta)$ (and this is frequently done) although it is not clear what $d(\hat{\theta})$ maximizes. Certainly if we can show that $d(\theta)$ also determines the distribution of the X's and $d(\hat{\theta})$ maximizes the likelihood, then $d(\hat{\theta})$ is a maximum likelihood estimate of $d(\theta)$.

Example 6. *Estimating p^2 in Bernoulli Distribution.*   Let $X_1, X_2, \ldots, X_n$ be a random sample from a Bernoulli distribution with parameter p. Suppose we wish to estimate $\theta = p^2$. We note $\theta = d(p) = p^2$ so that d is a one to one function of $[0, 1]$ to $[0, 1]$ with inverse function $p = \theta^{1/2}$. Since $\bar{X}$ is the maximum likelihood estimate of p, the maximum likelihood estimate of θ is $\bar{X}^2$. Formally, the likelihood function of $x_1, x_2, \ldots, x_n$ in terms of θ is given by

$$L(\theta; x_1, \ldots, x_n) = (\theta^{1/2})^{\sum_{i=1}^n x_i} (1 - \theta^{1/2})^{n - \sum_{i=1}^n x_i}$$

so that

$$\ln L(\theta) = \frac{1}{2} \sum_{i=1}^n x_i \ln \theta + \left(n - \sum_{i=1}^n x_i\right) \ln(1 - \theta^{1/2}).$$

Moreover,

$$\frac{d \ln L(\theta)}{d\theta} = \frac{1}{2\theta} \sum_{i=1}^n x_i - \frac{n - \sum_{i=1}^n x_i}{1 - \theta^{1/2}} \left(\frac{1}{2\theta^{1/2}}\right),$$

$$\frac{d^2 \ln L(\theta)}{d\theta^2} = -\frac{1}{2\theta^2} \sum_{i=1}^n x_i + \frac{n - \sum_{i=1}^n x_i}{(\theta^{1/2} - \theta)^2} \left(\frac{1}{2\sqrt{\theta}} - 1\right),$$

so that

$$\frac{d \ln L(\theta)}{d\theta} = 0$$

gives the maximum as

$$\hat{\theta} = \left(\sum_{i=1}^{n} \frac{x_i}{n} \right)^2.$$ $\qquad \square$

The following example shows that the maximum likelihood estimate may not be computable explicitly as a function of the observations.

Example 7. Estimating Median of Cauchy Distribution. Let $X_1, X_2, \ldots, X_n$ be a random sample from Cauchy density function

$$f(x) = \frac{1}{\pi} \frac{1}{1 + (x - \theta)^2}, \qquad x \in \mathbb{R}$$

where $\theta \in \Theta = \mathbb{R}$. Then

$$\ln L(\theta; x_1, \ldots, x_n) = -n \ln \pi - \sum_{i=1}^{n} \ln \left[1 + (x_i - \theta)^2 \right]$$

and

$$\frac{d \ln L}{d\theta} = \sum_{i=1}^{n} \frac{2(x_i - \theta)}{1 + (x_i - \theta)^2}.$$

For any given set of observations $x_1, \ldots, x_n$ we have to use numerical methods to solve for the maximum likelihood estimate of θ. $\qquad \square$

Example 8. Empirical Distribution Function Is a Maximum Likelihood Estimate. Let $X_1, X_2, \ldots, X_n$ be a random sample from a distribution function $F \in \mathcal{F}$, where $\mathcal{F}$ is the class of all distribution functions on the real line. Suppose we observe $x_1, x_2, \ldots, x_n$ which are all different. That is, $x_i \neq x_j, i \neq j$. What distribution function from $\mathcal{F}$ makes this outcome most likely? We show that $\hat{F}_n$ is the required distribution function. The object is to find F such that

$$P_F(X_1 = x_1) P_F(X_2 = x_2) \cdots P_F(X_n = x_n)$$
$$= P_F(X_1 = x_1) P_F(X_1 = x_2) \cdots P_F(X_1 = x_n)$$

is maximum. [Here P_F means that we compute $P(X_1 = x_1)$ with respect to distribution function F for X_1]. We have used the fact that X_i's are identically distributed so that $P_F(X_i = x_i) = P_F(X_1 = x_i)$ for all i. We first note that an $F \in \mathcal{F}$ that maximizes this likelihood must be concentrated only on $x_1, \ldots, x_n$ so that

$$P_F(X_1 = x_1) + \cdots + P_F(X_n = x_n) = 1.$$

Otherwise, there is some interval I for which $P_F(X_1 \in I) > 0$, and I contains none of $x_1, x_2, \ldots, x_n$. In that case we can make the likelihood function larger by putting $P(X_1 \in I)$ at one of $x_1, \ldots, x_n$, say x_1. Thus

$$P_F(X_1 = x_1) \cdots P_F(X_n = x_n) = p_1 p_2 \cdots p_n$$

is to be a maximized subject to $\sum_{i=1}^{n} p_i = 1$. Let

$$L(p_1, p_2, \ldots, p_{n-1}) = p_1 p_2 \cdots p_n = p_1 p_2 \cdots p_{n-1}(1 - p_1 - \cdots - p_{n-1}).$$

Taking logarithms, we have

$$\ln L = \sum_{i=1}^{n-1} \ln p_i + \ln(1 - p_1 - \cdots - p_{n-1})$$

so that

$$\frac{\partial \ln L}{\partial p_i} = \frac{1}{p_i} - \frac{1}{1 - p_1 - \cdots - p_{n-1}} \qquad \text{for } i = 1, 2, \ldots, n - 1.$$

Equating $\partial \ln L / \partial p_i$ to zero, we see that

$$p_1 = p_2 = \cdots = p_{n-1}$$

so that $p_1 = p_2 = \cdots = p_n = 1/n$ is the maximum likelihood estimate of F. This is precisely the empirical distribution function. □

Problems for Section 10.5

1. Let $X_1, X_2, \ldots, X_n$ be a random sample from a uniform distribution on $\{1, 2, \ldots, N\}$, $N \in \Theta = \{1, 2, \ldots\}$.
   (i) Find the maximum likelihood estimate of N.
   (ii) Show that the maximum likelihood estimate is consistent.

2. Let $X_1, X_2, \ldots, X_n$ be a random sample from the following distributions. Find the maximum likelihood estimate of θ in each case.
   (i) $f(x|\theta) = \theta(1 - x)^{\theta-1}, 0 \le x \le 1$, and zero elsewhere, $\theta > 0$.
   (ii) $f(x|\theta) = \theta x^{\theta-1}, 0 \le x \le 1$, and zero elsewhere, $\theta > 0$.
   (iii) $f(x|\theta) = \exp\{-(x - \theta)\}, x > \theta$, and zero elsewhere.
   (iv) $f(x|\theta) = \frac{1}{2}\exp\{-|x - \theta|\}, x \in \mathbb{R}$.
   (v) $f(x|\theta) = (\alpha/\theta)x^{\alpha-1}e^{-x^{\alpha}/\theta}, x > 0$, and zero elsewhere, and $\alpha > 0$ is known, $\theta > 0$.
   (vi) $P(X = x|\theta) = \theta(1 - \theta)^{x-1}, x = 1, 2, \ldots, 0 < \theta < 1$.
   (vii) $P(X = x|\theta) = \binom{x+r-1}{r-1}\theta^r(1 - \theta)^{x-r}, x = r, r + 1, \ldots, 0 < \theta < 1$.
   (viii) $f(x|\theta) = \frac{1}{\sqrt{2\pi\theta}}\exp\{-(x - \theta)^2/2\theta\}, x \in \mathbb{R}, \theta > 0$.
   (ix) $f(x|\theta) = 1/(\alpha\theta + \beta), 0 < x < \alpha\theta + \beta$, and zero elsewhere; α, β known, $\alpha \ne 0$.

3. Let $X_1, X_2, \ldots, X_n$ be a random sample from probability function

$$P(X = x_1) = \frac{(1 - \theta)}{2}, P(X = x_2) = \frac{1}{2} \quad \text{and} \quad P(X = x_3) = \frac{\theta}{2}, \qquad 0 \le \theta \le 1.$$

Find the maximum likelihood estimate of θ.

4. Let $X_1, \ldots, X_n$ be a random sample from an exponential distribution with mean $1/\theta$, $\theta > 0$. Find the maximum likelihood estimate of $1/\theta$, and show that it is consistent and asymptotically normal. What is the maximum likelihood estimate of θ?

5. Let $X_1, \ldots, X_n$ be a random sample from the uniform distribution on (θ_1, θ_2), θ_1, θ_2 both unknown. Find maximum likelihood estimates of θ_1 and θ_2.

6. Let $X_1, \ldots, X_n$ be a random sample from density function

$$f(x, \alpha, \theta) = \theta^{-1}\exp\{-(x - \alpha)/\theta\}, \qquad x > \alpha, \text{ and zero elsewhere}$$

   where $\alpha \in \mathbb{R}$ and $\theta > 0$ are both unknown.
   (i)   Find the maximum likelihood estimate of (α, θ).
   (ii) Find the maximum likelihood estimate of $P(X_1 \geq 1)$ where $\alpha < 1$.

7. Let $X_1, \ldots, X_n$ be a random sample from a normal population with mean μ and variance σ^2 (both unknown). Find the maximum likelihood estimate of the quantile of order p.

8. In the randomized response model of Example 10.4.2 where we wish to estimate π based on a sample $X_1, \ldots, X_n$ of independent Bernoulli random variables with

$$P(X_i = 1) = \pi p + (1 - \pi)(1 - p), \qquad p \text{ known}, 0 < p < 1, p \neq \tfrac{1}{2}$$

   where $\pi \in [0, 1]$, find the maximum likelihood estimate of π.

9. Let $X_1, X_2, \ldots, X_n$ be survival times of n patients with malignant melanoma with metastases. An estimate is required of the probability of a patient surviving time t, that is, $S(t) = P(X > t)$. Find the maximum likelihood estimates of $S(t)$ in the following cases.
   (i)   The death density function is exponential with mean $\theta > 0$.
   (ii) The death density function is Weibull density

$$f(x) = \frac{\alpha}{\beta}x^{\alpha-1}\exp(-x^\alpha/\beta), \qquad x > 0, \text{ and zero otherwise},$$

   where both α, β are unknown.
   (iii) The death density function is Rayleigh density

$$f(x) = \frac{2x}{\beta}\exp\left\{-\frac{x^2}{\beta}\right\}, \qquad x > 0, \text{ and zero elsewhere.}$$

   (iv) In each case also find the maximum likelihood estimate of the hazard rate $\lambda(t) = f(t)/S(t)$.
   [Hint: Use invariance of maximum likelihood estimates.]

10. Let $X_1, X_2, \ldots, X_n$ be a random sample from a normal population with mean μ and variance 1. It is known that $\mu \geq 0$. Find the maximum likelihood estimate of μ.

11. A sample of n observations is taken from a normal distribution with mean μ and variance 1. It was found that m observations ($m \leq n$) were less than 0 and no record of these observations was made. Observations greater than 0 were recorded. Find the maximum likelihood estimate of μ.

12. A sample of n observations was taken from an exponential density with mean θ as follows. Observations larger than t were not recorded; those smaller than t were recorded. It was found that k ($0 \le k \le n$) observations were larger than t. Find the maximum likelihood estimate of θ.

13*. Let $(X_1, Y_1), \ldots, (X_n, Y_n)$ be a random sample from a bivariate normal distribution with parameters $\mu_1, \mu_2, \sigma_1^2, \sigma_2^2$ and ρ. Find the maximum likelihood estimates of μ_1, $\mu_2, \sigma_1^2, \sigma_2^2$ and ρ.

14. Let $X_1, \ldots, X_n$ be a random sample from a Poisson distribution with parameter λ. Find the maximum likelihood estimate of $P(X = 0) = e^{-\lambda}$.

15. Let $(X_1, \ldots, X_{k-1})$ have a multinomial distribution with parameters $n, p_1, \ldots, p_{k-1}$, $0 \le p_1, \ldots, p_{k-1} \le 1$, $\sum_{j=1}^{k-1} p_j \le 1$ where n is known. Find the maximum likelihood estimate of $(p_1, \ldots, p_{k-1})$.

16. Consider the Hardy–Weinberg genetic equilibrium principle with respect to a single gene with two alleles (Problem 10.4.5). Suppose the three types of possible individuals have proportions

$$p_1 = \theta^2, \quad p_2 = 2\theta(1 - \theta), \quad p_3 = (1 - \theta)^2, \quad 0 < \theta < 1.$$

Let n_i be the number of individuals of type i, $i = 1, 2, 3$, in a sample of n individuals (labeled 1, 2, and 3), $n = n_1 + n_2 + n_3$. Find the maximum likelihood estimate of θ (whenever it exists).

10.6* BAYESIAN ESTIMATION

Up to now we have associated probabilities only with random experiments: experiments that can be repeated (under identical conditions) and have an element of uncertainty. The probability of a random event is an objective value (not depending on the observer) which can be approximated by the relative frequency of the event in a large number of replications of this experiment. The same frequentist interpretation was used to interpret inferential techniques. Accordingly, an unbiased estimate $\hat{\theta}$ of a parameter θ is desirable since the mean of the sampling distribution of $\hat{\theta}$ is θ so that in a large number of samples each of the same size the average value of $\hat{\theta}$ will be nearly θ. Similarly, according to this interpretation, a $(1 - \alpha)$ level confidence interval simply means that the relative frequency of intervals, based on a large number of samples each of the same size that trap the true value of the parameter, will be approximately $1 - \alpha$. No probability assignment was made on the parameter set Θ. We do not speak of the probability that a null hypothesis is correct, that is, the probability that $\theta \in \Theta_0 \subseteq \Theta$.

In everyday use of probability, however, we do make subjective judgments about the probability of an event. What we have done so far does not deal with subjective judgments. The Bayesian school of statisticians subscribes to the notion that our degree of belief in the true parameter value should be reflected by our (subjective) faith in various values of θ. It is argued that this subjective belief should be reflected by a probability distribution on Θ. This subjective probability is not empirically verifiable, since only X is observable and not θ.

The roots of Bayesian analysis lie in the Bayes rule (Proposition 2.6.3), according to which given that an event A has happened we should revise the prior probability $P(H_j)$ of event H_j according to the formula

$$(1) \qquad P\{H_j|A\} = \frac{P(H_j)P\{A|H_j\}}{\sum\limits_{j=1}^{\infty} P(H_j)P\{A|H_j\}}$$

where the H_j's are disjoint events with $\bigcup_{j=1}^{\infty} H_j = \Omega$ the sample space, and $P(H_j) > 0$ for each j. We recall that $P\{H_j|A\}$ is called the posterior probability of H_j. The same result can be translated to random variables as follows. If (X, Y) has joint probability function $p(x, y)$, then

$$(2) \qquad P\{X = x|Y = y\} = \frac{P(X = x)P\{Y = y|X = x\}}{\sum\limits_{x} P(X = x)P\{Y = y|x = x\}},$$

whenever the denominator is positive. Note that

$$\sum_{x} P(X = x)P\{Y = y|X = x\} = \sum_{x} p(x, y) = P(Y = y).$$

If (X, Y), on the other hand, has a joint (continuous) density function f, then an analog of (2) is

$$(3) \qquad h(x|y) = \frac{f_X(x)h_1(y|x)}{\int_{-\infty}^{\infty} f_X(x)h_1(y|x)\, dx} = \frac{f_X(x)h_1(y|x)}{f_Y(y)}$$

provided $f_Y(y) > 0$, in an obvious notation. The probability function $P(X = x)$ [or density function $f_X(x)$] is called the *prior probability function* (or *density function*) of X, and $P\{X = x|Y = y\}$ [or $h(x|y)$] is called the *posterior probability function* (or *density function*) of X.

Rather than give the arguments for and against the Bayesian viewpoint, let us consider how the Bayesian inference is made. Let us consider θ as a value of random variable Θ (the same notation is used for the parameter set) with prior probability or density function $\pi(\theta)$. The distribution π is regarded as known independently of the data that is being analyzed. Then π contains information about θ which, if correct, will often sharpen the inference concerning θ. Let $x_1, x_2, \ldots, x_n$ be a realization of $(X_1, X_2, \ldots, X_n)$ with common probability or density function $f_\theta(x)$. We write $f_\theta(x) = f(x|\theta)$ and interpret it as the probability (or density) of X given $\Theta = \theta$. Then the joint distribution of the X's and Θ is given by

(4) $$u(x_1, x_2, \ldots, x_n, \theta) = \pi(\theta) \prod_{j=1}^{n} f(x_j|\theta)$$

and the unconditional (or marginal) distribution of $X_1, \ldots, X_n$ is given by

(5) $$g(x_1, \ldots, x_n) = \begin{cases} \int_{-\infty}^{\infty} \pi(\theta) \prod_{j=1}^{n} f(x_j|\theta) \, d\theta & \text{if } \pi \text{ is a density function} \\ \sum_{\theta} \pi(\theta) \prod_{j=1}^{n} f(x_j|\theta) & \text{if } \pi \text{ is a probability function.} \end{cases}$$

It follows that the posterior distribution of Θ given the x's is given by

(6) $$h(\theta|x_1, \ldots, x_n) = \frac{u(x_1, x_2, \ldots, x_n, \theta)}{g(x_1, \ldots, x_n)}$$

provided the denominator is positive. This is (2) or (3). Rewriting (6) as

$$u(x_1, \ldots, x_n, \theta) = \pi(\theta) \prod_{j=1}^{n} f(x_j|\theta) = h(\theta|x_1, \ldots, x_n) g(x_1, \ldots, x_n)$$

and recognizing $\prod_{j=1}^{n} f(x_j|\theta)$ as the likelihood function of the x's, we have

(7) $$L(\theta; x_1, \ldots, x_n) = \prod_{j=1}^{n} f(x_j|\theta) = \frac{h(\theta|x_1, \ldots, x_n) g(x_1, \ldots, x_n)}{\pi(\theta)}.$$

which gives the likelihood function in terms of the posterior and the prior distribution of θ.

The Bayes result (6) says that the observation $x_1, \ldots, x_n$ changes our belief in the possible values of the parameter θ by changing the prior distribution π into a posterior distribution $h(\theta|x_1, \ldots, x_n)$.

Example 1.   Introducing a New Drug in the Market.   A drug company would like to introduce a drug to reduce acid indigestion. It is desirable to estimate θ, the proportion of the market share that this drug will capture. The company interviews n people and X of them say that they will buy the drug. In the non-Bayesian analysis $\theta \in [0, 1]$ and X has a binomial distribution with parameters n and θ. We know that $\hat{\theta} = X/n$ is a very good estimate of θ. It is unbiased, consistent, and asymptotically normal. Moreover, $\hat{\theta}$ is the maximum likelihood estimate of θ, which is also the minimum variance unbiased estimate.

A Bayesian may look at the past performance of new drugs of this type. If in the past new drugs tend to capture a proportion between say .05 and .15 of the market, and if all values in between are assumed equally likely, then Θ has uniform distribution on

[.05, .15] so that

$$\pi(\theta) = \begin{cases} 1/(.15 - .05) = 10, & .05 \le \theta \le .15 \\ 0 & \text{otherwise.} \end{cases}$$

Then

$$u(x, \theta) = \pi(\theta)f(x|\theta) = 10\binom{n}{x}\theta^x(1-\theta)^{n-x},$$

for $x = 0,1,\ldots,n$, and $\theta \in [.05,15]$,   and

$$g(x) = 10\binom{n}{x}\int_{.05}^{.15}\theta^x(1-\theta)^{n-x}\,d\theta.$$

It follows that the posterior distribution of θ is given by

$$h(\theta|x) = \frac{\theta^x(1-\theta)^{n-x}}{\int_{.05}^{.15}\theta^x(1-\theta)^{n-x}\,dx}. \qquad\qquad \square$$

Example 2.   Sampling Inspection.   Suppose a manufacturer produces a proportion θ of defective items. Let X be the number of defectives in a lot of N items from this production process. Then X has a binomial distribution with parameters N and θ. Suppose a random sample of size n is taken from the lot without replacement and let Y be the number of defectives in the sample. Then the conditional distribution of Y, given $X = x$, is hypergeometric. Hence

$$u(x, y) = \pi(x)\cdot f(y|x) = \binom{N}{x}\theta^x(1-\theta)^{N-x}\frac{\binom{x}{y}\binom{N-x}{n-y}}{\binom{N}{n}}$$

so that

$$u(x, y) = \theta^x(1-\theta)^{N-x}\frac{N!}{x!(N-x)!}\frac{x!}{y!(x-y)!}\frac{(N-x)!n!(N-n)!}{(N-x-n+y)!(n-y)!N!}$$

$$= \theta^x(1-\theta)^{N-x}\binom{n}{y}\binom{N-n}{x-y},$$

for $x = 0,1,\ldots,N$,   and $\max(0, x + n - N) \le y \le \min(x, n)$.   Consequently,

$$g(y) = P(Y = y) = \binom{n}{y}\sum_{x=y}^{N}\binom{N-n}{x-y}\theta^x(1-\theta)^{N-x}$$

$$= \binom{n}{y}\sum_{z=0}^{N-y}\binom{N-n}{z}\theta^{z+y}(1-\theta)^{N-z-y} \qquad (z = x - y)$$

$$= \binom{n}{y}\left(\frac{\theta}{1-\theta}\right)^y(1-\theta)^N\sum_{z=0}^{N-n}\binom{N-n}{z}\left(\frac{\theta}{1-\theta}\right)^z \qquad (N - y \ge N - n)$$

$$= \binom{n}{y}\left(\frac{\theta}{1-\theta}\right)^y(1-\theta)^N\left(1 + \frac{\theta}{1-\theta}\right)^{N-n}$$

$$= \binom{n}{y}\theta^y(1-\theta)^{n-y}, \qquad y = 0,1,\ldots,n.$$

It follows that for $x = y, y + 1, \dots, y + N - n$

$$h(x|y) = \frac{u(x, y)}{g(y)} = \binom{N - n}{x - y}\theta^{x-y}(1 - \theta)^{N-n-(x-y)}$$

which is binomial with parameters $N - n$ and θ (except that it is $x - y$ that takes values $0, 1, \dots, N - n$). $\qquad\square$

Example 3. Diseased White Pine Trees. White pine is one of the best known species of pines in the northeastern United States and Canada. White pine is suscepti-ble to blister rust, which develops cankers on the bark. These cankers swell, resulting in death of twigs and small trees. A forester wishes to estimate the average number of diseased pine trees per acre in a forest. The number of diseased trees per acre can be modeled by a Poisson distribution with parameter λ. Since λ changes from area to area, the forester believes that λ has an exponential distribution with mean θ so that

$$\pi(\lambda) = (1/\theta)e^{-\lambda/\theta}, \qquad \text{if } \lambda > 0, \text{ and zero elsewhere.}$$

The forester takes a random sample of size n from n different one-acre plots. Then

$$u(x_1, \dots, x_n, \lambda) = \{(1/\theta)e^{-\lambda/\theta}\}\left(\frac{\lambda^{\sum_{i=1}^{n} x_i}e^{-n\lambda}}{\prod_{i=1}^{n} x_i!}\right)$$

and

$$g(x_1, \dots, x_n) = \frac{1}{\theta\prod_{i=1}^{n} x_i!}\int_0^\infty e^{-\lambda(n+1/\theta)}\lambda^{\sum_{i=1}^{n} x_i}\, d\lambda$$

$$= \Gamma\left(\sum_{i=1}^{n} x_i + 1\right)\frac{1}{(n + 1/\theta)^{\sum_{i=1}^{n} x_i + 1}}\cdot\frac{1}{\theta\prod_{i=1}^{n} x_i!}.$$

in view of the fact that the integral on the right side of the middle equality is a gamma integral with $\alpha = \sum_{i=1}^{n} x_i + 1$ and $\beta = 1/(n + 1/\theta)$.

Consequently, the posterior distribution of λ is given by

$$h(\lambda|x_1, \dots, x_n) = \frac{(n + 1/\theta)^{\sum_{i=1}^{n} x_i + 1}}{\Gamma\left(\sum_{i=1}^{n} x_i + 1\right)}\cdot\lambda^{\sum_{i=1}^{n} x_i}e^{-\lambda(n+1/\theta)}$$

which is a gamma density with $\alpha = \sum_{i=1}^{n} x_i + 1$ and $\beta = \theta/(n\theta + 1)$. $\qquad\square$

Having determined the posterior distribution of Θ, how do we estimate θ? This is done by choosing that value of θ which minimizes some measure of precision of the estimate such as

$$\mathscr{E}_\pi\big(T(X_1,\ldots,X_n) - \theta\big)^2 \quad \text{or} \quad \mathscr{E}_\pi|T(X_1,\ldots,X_n) - \theta|$$

where we note, for example, that

(8) $\mathscr{E}_\pi\big(T(X_1,\ldots,X_n) - \theta\big)^2$

$$= \int\cdots\int_{x_1\cdots x_n}\int_\theta (T(x_1,\ldots,x_n) - \theta)^2 u(x_1,\ldots,x_n,\theta)\, d\theta \prod_{j=n}^1 dx_j.$$

The quantity in (8) is called the *risk* (of using T when π is the prior distribution of Θ) *function*, and we write

(9) $$R(\pi, T) = \mathscr{E}_\pi\big(T(X_1,\ldots,X_n) - \theta\big)^2.$$

If X or Θ is of the discrete type, we replace integrals by sums. The functions $(T - \theta)^2$ and $|T - \theta|$ are called *loss functions*.

THEOREM 1. *Let $X_1,\ldots,X_n$ be a random sample from a distribution with probability or density function $f(x|\theta)$, and suppose Θ has prior probability or density function $\pi(\theta)$. Then the Bayes estimate of θ, which minimizes $\mathscr{E}_\pi(T(X_1,\ldots,X_n) - \theta)^2$, is given by*

(10) $$T(x_1,\ldots,x_n) = \mathscr{E}_\pi\{\Theta | X_1 = x_1,\ldots,X_n = x_n\}$$

which is the mean of the posterior distribution.

Proof. To prove (10) we use (6) to rewrite

$\mathscr{E}_\pi\big(T(X_1,\ldots,X_n) - \theta\big)^2$

$$= \int\cdots\int_{x_1\cdots x_n} g(x_1,\ldots,x_n)\int_\theta\big((T(x_1\ldots x_n) - \theta)^2 h(\theta|x_1,\ldots,x_n)\, d\theta\big)\prod_{j=1}^n dx_j.$$

It follows that $\mathscr{E}_\pi(T - \theta)^2$ is minimized whenever $\int (T(x_1,\ldots,x_n) - \theta)^2 h(\theta|x_1,\ldots,x_n)\, d\theta$ is. We know, however, from Remark 3.7.4 that $\mathscr{E}(X - c)^2$ is minimized whenever $c = \mathscr{E}X$. It follows that the minimum of $\mathscr{E}_\pi(T - \theta)^2$ occurs for the value of θ given by

$$\hat\theta(x_1,\ldots,x_n) = \mathscr{E}\{\Theta | X_1 = x_1,\ldots,X_n = x_n\}. \qquad \square$$

Example 4. Bayes Estimate of Proportion. In Example 1 we wish to estimate the proportion θ of the market share that a new drug will capture given $X = x$ where X is the number of respondents who say that they will buy the drug. If we wish to minimize $\mathscr{E}_\pi(T - \theta)^2$, then the Bayesian estimate is given by

$$\hat\theta = \mathscr{E}\{\Theta | x\} = \int_{.05}^{.15}\theta h(\theta|x)\, d\theta$$

$$= \frac{\int_{.05}^{.15}\theta\cdot\theta^x(1 - \theta)^{n-x}\, d\theta}{\int_{.05}^{.15}\theta^x(1 - \theta)^{n-x}\, d\theta}$$

which can be evaluated numerically or from appropriate tables once x and n are known.

Suppose $x = 15$ and $n = 90$. Then the traditional (non-Bayesian) estimate is $\hat{\theta}_1 = 15/90 = .16667$, the sample proportion, whereas the Bayes estimate is given by

$$\hat{\theta} = \frac{\int_{.05}^{.15} \theta^{16} (1 - \theta)^{75} \, d\theta}{\int_{.05}^{.15} \theta^{15} (1 - \theta)^{75} \, d\theta}.$$

We now use the relation between beta and binomial distributions (Example 7.5.2). We have

$$N = \int_{.05}^{.15} \theta^{16} (1 - \theta)^{75} \, d\theta = B(17, 76) \{ P(Y \le .15) - P(Y \le .05) \}$$

where Y has a beta distribution with $\alpha = 17$, $\beta = 76$. Hence

$$N = B(17, 76) \{ P(U_1 \ge 17) - P(U_2 \ge 17) \}$$

where U_1 is binomial with parameters $n = \alpha + \beta - 1 = 92$ and $p = .15$, whereas U_2 is binomial with $n = 92$ and $p = .05$. Consequently,

$$N = B(17, 76)[.21137 - .00000] = .21137 \, B(17, 76)$$

where we have used the Harvard University Tables of Cumulative Binomial Distribution.[†] Similarly

$$D = \int_{.05}^{.15} \theta^{15} (1 - \theta)^{75} \, d\theta = B(16, 76) \{ P(Y \le .15) - P(Y \le .05) \}$$

$$= B(16, 76) \{ P(U_3 \ge 16) - P(U_4 \ge 16) \}$$

where U_3 is binomial with $n = 91$, $p = .15$ and U_4 with $n = 91$, $p = .05$ so that

$$D = B(16, 76) \{ .28548 - .00001 \} = .28547 \, B(16, 76).$$

It follows that

$$\hat{\theta} = \frac{.21137}{.28547} \cdot \frac{16! 75!}{92!} \cdot \frac{91!}{15! 75!} = .12877.$$

We note that $\hat{\theta}_1 = .16667 \notin [.05, .15]$ whereas $\hat{\theta} \in [.05, .15]$, as it must.   □

Example 5. Estimating the Mean Number of Diseased Trees per Acre. In Example 3 the problem is to estimate λ, the mean number of diseased pine trees per acre. In view of Theorem 1 we see that the mean of the posterior distribution minimizes

---

[†]*Tables of Cumulative Binomial Probability Distribution*, Harvard University Press, Cambridge, Massachusetts, 1955.

$\mathscr{E}_{\pi}(T - \lambda)^2$. Hence the Bayes estimate of λ is given by

$$\lambda_1 = \mathscr{E}\{\lambda | x_1, \dots, x_n\} = \int_0^\infty \frac{\lambda(n + 1/\theta)^{\Sigma_{i=1}^n x_i + 1}}{\Gamma\left(\sum x_i + 1\right)} \lambda^{\Sigma_{i=1}^n x_i} e^{-\lambda(n+1/\theta)} \, d\lambda$$

$$= \frac{\Gamma\left(\sum_{i=1}^n x_i + 2\right)(n + 1/\theta)^{\Sigma_{i=1}^n x_i + 1}}{\Gamma\left(\sum_{i=1}^n x_i + 1\right)(n + 1/\theta)^{\Sigma_{i=1}^n x_i + 2}} = \frac{\sum_{i=1}^n x_i + 1}{(n + 1/\theta)}.$$

We recall that the classical (non-Bayesian) estimate of λ is $\hat{\lambda} = \bar{x}$, which has nice properties. □

Problems for Section 10.6

1. Suppose the probability of success p in a sequence of independent Bernoulli trials has a uniform prior distribution on $[0, 1]$. (This is a so-called *noninformation* prior which favors no value of p or which contains no information about p.)
   (i) Show that the probability of k successes in n trials is uniformly distributed on $\{0, 1, 2, \dots, n\}$.
   (ii) If it is known that there were k successes in n trials, what is the probability $\hat{p}$ that the next trial will be a success?
   (iii) Find the Bayes risk $\mathscr{E}_{\pi}(\hat{p} - p)^2$.

2. Suppose X has uniform distribution on $[0, \theta]$ where Θ has prior density $\pi(\theta) = \theta e^{-\theta}$, if $\theta > 0$, and zero elsewhere.
   (i) Find the posterior distribution of Θ.   (ii) Find the Bayes estimate of θ.

3. Let X be the length of life (in thousands of hours) of a light bulb manufactured by a certain company. It is known that X can be modeled by density $f(x|\theta) = \theta e^{-\theta x}$, if $x > 0$, and zero elsewhere. Suppose Θ has improper prior $\pi(\theta) = 1/\theta, 0 < \theta < \infty$. Find
   (i) The posterior density of Θ.   (ii) The Bayes estimate of θ.

4. In Problem 1, show that the Bayes estimate of $\theta = p^2$ is given by

$$\hat{\theta} = \frac{(X + 1)(X + 2)}{(n + 2)(n + 3)}.$$

5. The distribution of lifetimes of bulbs made by a certain manufacturer is exponential with mean θ^{-1}. Suppose Θ has a priori exponential distribution with mean 1. Show that the Bayes estimate of θ based on a random sample $X_1, \dots, X_n$ is given by $\hat{\theta} = (n + 1)/(\Sigma_{i=1}^n X_i + 1)$.

6. Repeat Problem 5 with (improper) prior density

$$\pi(\theta) = \theta^{\alpha-1} e^{-\beta\theta}, \qquad \theta > 0 \text{ and zero elsewhere.}$$

7. In Example 5 we estimated λ, the mean number of diseased trees per acre when the number of diseased trees per acre has a Poisson distribution and λ has an exponential distribution with mean θ (thousand). Take $\theta = 1$. Suppose we wish to estimate the probability $P(X = 0) = e^{-\lambda}$ instead. Find the Bayes estimate. [Hint: Find $\mathscr{E}\{e^{-\lambda}|x_1,\ldots,x_n\}$.]

8. Suppose X has a Poisson distribution with parameter λ where λ has a prior distribution with gamma density having parameters α and β. A sample of size n is taken.
   (i) Show that the posterior distribution of λ is itself gamma with parameters $\alpha + \sum_{i=1}^{n} x_i$ and $\beta/(1 + n\beta)$.
   (ii) Show that the Bayes estimate of λ is given by $(\alpha + \sum_{i=1}^{n} x_i)\beta/(n\beta + 1)$.

9. Suppose X has a Bernoulli distribution with parameter θ where Θ has prior uniform distribution on $[0,1]$. A random sample of size n is taken.
   (i) Show that the posterior density of Θ is beta with parameters $\sum_{i=1}^{n} x_i + 1$ and $n + 1 - \sum_{i=1}^{n} x_i$.
   (ii) Find the Bayes estimate of θ.
   (iii) Find the Bayes risk, $\mathscr{E}_\pi\{\mathscr{E}\{\Theta|X_1,\ldots,X_n\} - \theta\}^2$. (See also Problem 1.)

10. Suppose $X_1, X_2,\ldots,X_n$ is a sample from a geometric distribution with parameter θ, $0 \le \theta \le 1$. If Θ has prior beta density with parameters α and β, find the Bayes estimate for θ.

11. Let $X_1, X_2,\ldots,X_n$ be a random sample from a Poisson distribution with parameter λ where λ has prior distribution which is exponential with mean 1. Find the Bayes estimate for λ^2. (See Problem 7.) [Hint: The Bayes estimate is $\mathscr{E}\{\lambda^2|x_1,\ldots,x_n\}$.]

12. Let $X_1, X_2,\ldots,X_n$ be a random sample from a normal distribution with mean μ and variance $\sigma^2 = 1$. It is known that $\mu > 0$. Find the Bayes estimates of μ using the following priors:
   (i) $\pi(\mu) = e^{-\mu}$ for $\mu > 0$ and zero elsewhere.
   (ii) $\pi(\mu) = 1$ for $0 < \mu < \infty$. [Here $\pi(\mu)$ is *not* a density function but is frequently used as a noninformative improper prior density. Caution: No closed form.]

13*. The brain weight of adult females (in grams) can be assumed to have a normal distribution with mean μ and variance σ^2 where σ^2 is known. In order to estimate μ, a random sample of size n is taken. The experimenter believes that μ itself has a normal distribution with parameters μ_0 and $\alpha(> 0)$.
   (i) Show that the posterior distribution of μ is normal with mean θ_n and variance σ_n^2 where

$$\theta_n = \frac{n\bar{x}/\sigma^2 + \mu_0/\alpha}{n/\sigma^2 + 1/\alpha} \quad \text{and} \quad \sigma_n^{-2} = (n/\sigma^2) + 1/\alpha = (\alpha n + \sigma^2)/\alpha\sigma^2.$$

   (ii) Find the Bayes estimate of μ.
   (iii) Show that the ratio of the variance of classical estimate $\bar{X}$ (which equals σ^2/n) to the Bayes risk $\mathscr{E}_\pi\{\theta_n - \theta\}^2$ is given by $(\alpha n + \sigma^2)/\alpha n$.
   (iv) Suppose $n = 100$, $\sum_{i=1}^{n} x_i = 130,271$ grams and $\sigma^2 = 4$ (grams)^2. Take $\mu_0 = 1400$ and $\alpha = 5$. Compare the two estimates.

14. Recall that the maximum likelihood estimate of a parameter maximizes the likelihood function $L(\theta) = f(x_1,\ldots,x_n|\theta)$. An analogous Bayesian procedure can be obtained by maximizing $h(\theta|x_1,\ldots,x_n)$. A value of θ that maximizes the posterior distribution $h(\theta|x_1,\ldots,x_n)$ is known as a *generalized maximum likelihood* estimate of θ.
    (i) Let X be an observation from an exponential density $f(x|\theta) = e^{-(x-\theta)}$, if $x > \theta$, and zero elsewhere. Suppose Θ has a prior distribution given by Cauchy density $\pi(\theta) = 1/(\pi[1 + \theta^2])$, $-\infty < \theta < \infty$. Show that the generalized maximum likelihood estimate of θ is given by $\hat{\theta} = X$.
    (ii) Let $X_1, X_2,\ldots,X_n$ be a sample from a normal distribution with mean μ and variance 1. Find the generalized maximum likelihood estimate of μ where it is known that $\mu > 0$ using prior density that is exponential with mean 1.

15. In all the problems so far, we have found the Bayes estimate by minimizing $\mathscr{E}_\pi\{T - \theta\}^2$. Suppose, instead, that we wish to minimize the risk $\mathscr{E}_\pi|T - \theta|$.
    (i) Show that an analog of Theorem 1 holds. That is, show that the Bayes estimate that minimizes

$$\mathscr{E}_\pi|T - \theta| = \int \cdots \int_{x_1\cdots x_n} \int_\theta |T(x_1,\ldots,x_n) - \theta|\pi(\theta)f(x_1,\ldots,x_n|\theta)\, d\theta\, dx_1 \ldots dx_n$$

    is given by any median of the posterior distribution $h(\theta|x_1,\ldots,x_n)$.
    (ii) Let $f(x|\theta) = 1/\theta$, $0 \le x \le \theta$, and zero elsewhere and $\pi(\theta)$ be uniform on $(0,1)$. Suppose $n = 1$. Find the Bayes estimate of θ that minimizes $\mathscr{E}_\pi|T - \theta|$.
    (iii) Suppose $X_1,\ldots,X_n$ is a random sample from a normal population with mean μ and variance 1. A Bayes estimate of μ is required that minimizes $\mathscr{E}_\pi|T(\overline{X}) - \mu|$. If the prior distribution of μ is normal with mean μ_0 and variance α, find the Bayes estimate of μ. (See Problem 13.)

16. The number of successes in n independent Bernoulli trials with constant probability p is observed to be k, $0 \le k \le n$. It is believed that p has a beta density with known parameters α and β. Find the Bayes estimate of p that minimizes $\mathscr{E}_\pi(T - p)^2$. (Problem 9 is a special case.)

17. An insurance company estimates that about 20 percent of the passengers purchase flight insurance. The proportion varies from terminal to terminal, and the company estimates that this variation is measured by a standard deviation of 10 percent. A random sample of 50 passengers at an air terminal showed that only five purchased flight insurance. Estimate the true proportion of passengers who buy flight insurance at this terminal assuming a beta prior density. [Hint: Use the result of Problem 16.]

18. The daily number of incoming calls at a small city telephone exchange has a Poisson distribution with mean λ. The mean λ, however, varies from day to day. This variation is represented by a gamma density with mean 100 and variance 200. If there were 90 incoming calls on a given day, what is your estimate of the average number of daily incoming calls? Compare it with your estimate of the average calls:
    (i) Based only on the prior information.
    (ii) Based only on the information that 90 calls came in that day.
    [Hint: Use the result of Problem 8.]

19. The average demand for a brand of lawn fertilizer in the months of March and April at lawn and garden stores is 5000 bags. The demand varies from store to store,

and it is felt that the distribution of mean demand is adequately represented by a normal distribution with a standard deviation of 500 bags. Suppose the number of bags of this fertilizer sold by a particular store during the months of March and April is a normal random variable with a standard deviation of 100 bags. Over nine years the store averaged sales of 4600 bags. What is the estimate of the average yearly sales at this store? Find the probability that for this store μ is between 4500 and 4800 bags. [Hint: Use the result of Problem 13.]

10.7 CONFIDENCE INTERVALS

In this section we consider a general method of constructing confidence intervals (or confidence bounds) for a real valued parameter. The reader has by now encountered confidence intervals for means, proportions, quantiles, and so on, for both large and small samples. The object here is to describe a general method of constructing confidence intervals.

Let $X_1, X_2, \ldots, X_n$ be a random sample from a distribution function F_θ where $\theta \in \Theta$ and Θ is an interval of the real line. The object is to find statistics $\underline{\theta}(X_1, \ldots, X_n)$ and $\bar{\theta}(X_1, \ldots, X_n)$ such that

(1) $\quad P_\theta \{ \underline{\theta}(X_1, \ldots, X_n) \leq \theta \leq \bar{\theta}(X_1, \ldots, X_n) \} \geq 1 - \alpha \qquad$ for all $\theta \in \Theta$

where the probability on the left side of (1) is computed with respect to $F_\theta^\dagger$. The following result holds:

THEOREM 1. *Let* $T(X_1, \ldots, X_n, \theta) = T$ *be a real valued function defined on* $\mathbb{R}_n \times \Theta$, *such that, for each fixed* θ *in* Θ, T *is a statistic and as a function of* θ, T *is either strictly increasing or strictly decreasing at every* $(x_1, \ldots, x_n) \in \mathbb{R}_n$. *Let* $\Lambda \subseteq \mathbb{R}$ *be the range of* T *and suppose that for every* $\lambda \in \Lambda$ *and* $(x_1, \ldots, x_n) \in \mathbb{R}_n$ *the equation* $\lambda = T(x_1, \ldots, x_n, \theta)$ *is solvable uniquely for* θ. *If the distribution of* $T(X_1, \ldots, X_n, \theta)$ *does not depend on* θ, *then one can construct a confidence interval for* θ *at any level* $1 - \alpha$, $0 < \alpha < 1$.

Proof. We need to show that there exists a pair of statistics $\underline{\theta}(X_1, \ldots, X_n)$ and $\bar{\theta}(X_1, \ldots, X_n)$ such that (1) holds for every α, $0 < \alpha < 1$. Since the distribution of T does not depend on θ, we can find real numbers $\lambda_1(\alpha), \lambda_2(\alpha) \in \Lambda$, not necessarily uniquely such that

(2) $\qquad P_\theta \{ \lambda_1(\alpha) \leq T(X_1, \ldots, X_n, \theta) \leq \lambda_2(\alpha) \} \geq 1 - \alpha$

for all $\theta \in \Theta$. The pair $(\lambda_1(\alpha), \lambda_2(\alpha))$ does not depend on θ. By monotonicity of T in θ we can solve the equations

$$T(x_1, \ldots, x_n, \theta) = \lambda_i(\alpha), \quad i = 1, 2$$

uniquely for θ for every $(x_1, \ldots, x_n)$. It follows that (1) holds. $\qquad \square$

Remark 1. The condition that $\lambda = T(x_1, \ldots, x_n, \theta)$ be solvable for θ will be satisfied if, for example, T is a continuous and monotonic function of θ on Θ for

---

†If the distributions of $\underline{\theta}$ and $\bar{\theta}$ are continuous, it is clear that either one or both end points $\underline{\theta}, \bar{\theta}$ may be excluded from the confidence interval. In the following we will choose $[\underline{\theta}, \bar{\theta}]$ as confidence interval.

every fixed $(x_1, \ldots, x_n)$. The relation (2) holds even if T is not monotonic in θ, but in that case the inversion of the inequalities within parentheses in (2) may yield a set of random intervals in Θ.

Remark 2. If the distribution of T is continuous, then we can find a minimum level $1 - \alpha$ confidence interval for θ [that is, with equality in (2) and hence also in (1)] but in the discrete case this is usually not possible.

DEFINITION 1. A random variable $T(X_1, \ldots, X_n, \theta)$ whose distribution is independent of θ is called a pivot.

In view of Theorem 1 the problem of finding a confidence interval simply reduces to finding a pivot. What statistics should we choose for a pivot? The choice of the pivot in some way will determine the precision of a confidence interval based on it. The obvious choice is to find a pivot based on a sufficient statistic (possibly the best sufficient statistic). We consider some examples.

Example 1. Waiting Time for Emergency Care. Suppose the waiting times at the emergency room of a large city hospital can be modeled by a normal distribution. A random sample of 15 patients showed a mean waiting time of 40 minutes with a standard deviation of eight minutes. Find a 95 percent confidence interval for the true mean waiting time at this emergency room.

We have already solved problems like this in Section 8.7. How do we use Theorem 1 to solve this problem? Let $X_1, \ldots, X_n$ be a random sample from a normal population with mean μ and variance σ^2, both unknown. The object is to find a confidence interval for μ. We note that $(\bar{X}, S^2)$ is sufficient for (μ, σ^2). Moreover, the random variable

$$T(X_1, \ldots, X_n, \mu) = \frac{\bar{X} - \mu}{S/\sqrt{n}}$$

which is a function of the sufficient statistic $(\bar{X}, S^2)$ and unknown parameter μ has a t distribution with $n - 1$ d.f. (independently of μ). Hence T is a reasonable choice for the pivot. We note that for fixed $(x_1, \ldots, x_n)$, T is a monotonic decreasing function of μ. Conditions of Theorem 1 are satisfied and we can find $t_1 = t_1(\alpha)$ and $t_2 = t_2(\alpha)$ such that

$$(3) \qquad P_\mu\left(t_1 \le \frac{\bar{X} - \mu}{S/\sqrt{n}} \le t_2 \right) = 1 - \alpha.$$

We note that equality is possible in (3) since t distribution is of the continuous type and, moreover, there are infinitely many pairs (t_1, t_2) for which (3) holds. Inverting the inequalities in (3) we get

$$P_\mu\left(\bar{X} - t_2 \frac{S}{\sqrt{n}} \le \mu \le \bar{X} - t_1 \frac{S}{\sqrt{n}} \right) = 1 - \alpha \qquad \text{for all } \mu$$

so that $[\bar{X} - t_2(S/\sqrt{n}), \bar{X} - t_1(S/\sqrt{n})]$ is a minimum level $1 - \alpha$ confidence interval for μ.

In our case $n = 15$, $\bar{x} = 40$ and $s = 8$ so that we get a minimum level $1 - \alpha$ class of confidence intervals.

$$\left[40 - \frac{8}{\sqrt{15}} t_2, 40 - \frac{8}{\sqrt{15}} t_1 \right] = [40 - 2.07t_2, 40 - 2.07t_1].$$

In view of the symmetry of the t distribution, the interval with the shortest length $L = 2.07(t_2 - t_1)$ is obtained if we choose $t_1 = -t_2$ (Problem 4.3.12). In that case t_2 is chosen from

$$P_\mu\left(\left| \frac{\bar{X} - \mu}{S/\sqrt{n}} \right| < t_2 \right) = 1 - \alpha$$

and hence $t_2 = t_{n-1, \alpha/2} = t_{14, .025} = 2.145$. Thus the required shortest length confidence interval for μ based on T is given by

$$[40 - 4.44, 40 + 4.44] \quad \text{or} \quad [35.66, 44.44].$$

If σ is known, then $\bar{X}$ is sufficient and an appropriate pivot is $(\bar{X} - \mu)\sqrt{n}/\sigma$. If a confidence interval for σ^2 is desired, then if μ is known we choose $\sum_{i=1}^n (X_i - \mu)^2/\sigma^2$ and if μ is unknown we choose $(n - 1)S^2/\sigma^2$. This is precisely what we did in Section 8.7.   □

Example 2. **Confidence Interval for θ in Sampling from Uniform Distribution.** Let $X_1, X_2, \ldots, X_n$ be a random sample from a uniform distribution on $(0, \theta)$, $\theta \in (0, \infty)$. We note that $X_{(n)} = \max(X_1, \ldots, X_n)$ is sufficient for θ. Moreover the random variable

$$T(X_1, \ldots, X_n, \theta) = \frac{X_{(n)}}{\theta}$$

has density function

$$h(t) = nt^{n-1}, \quad 0 < t < 1, \text{ and zero elsewhere}$$

which does not depend on θ. Hence T is a good choice for a pivot. Since T is a decreasing function of θ, conditions of Theorem 1 are satisfied and there exist numbers $\lambda_1(\alpha), \lambda_2(\alpha)$ such that

$$P_\theta\{\lambda_1 \le T(X_1, \ldots, X_n, \theta) \le \lambda_2\} = 1 - \alpha$$

for all $\theta \in \Theta = (0, \infty)$. In fact, λ_1, λ_2 satisfy

$$n \int_{\lambda_1}^{\lambda_2} t^{n-1} \, dt = \lambda_2^n - \lambda_1^n = 1 - \alpha, \quad 0 < \lambda_1, \lambda_2 < 1.$$

This equation gives infinitely many pairs (λ_1, λ_2) each of which gives a minimum level $1 - \alpha$ confidence interval for θ of the form $[X_{(n)}/\lambda_2, X_{(n)}/\lambda_1]$. The distribution of T is not symmetric and cannot be used to make a unique choice for (λ_1, λ_2).   □

We note that a confidence interval for a parameter crucially depends on the choice of the pivot. Once a pivot is chosen (usually based on the best possible sufficient statistic that can be found), there is still the problem of choosing λ_1, λ_2. This choice is based on the precision of the confidence interval as measured by its length or by the expected length. The smaller the length (or expected length) the higher the precision. Hence we choose λ_1, λ_2 such that the length (or the expected length) of the confidence interval is the least. If the length of the confidence interval is proportional to $\lambda_2 - \lambda_1$ and the distribution of T is symmetric (about zero) then we have seen (Problem 4.3.12) that the choice $\lambda_2 = -\lambda_1$ minimizes the length of the confidence interval.

Example 3.   *Shortest Length Confidence Interval for θ in Uniform Distribution.* In Example 2 we find that $[X_{(n)}/\lambda_2, X_{(n)}/\lambda_1]$ where λ_1, λ_2 are chosen from

$$(4) \qquad\qquad \lambda_2^n - \lambda_1^n = 1 - \alpha$$

is a minimum level $1 - \alpha$ confidence interval for θ. The length of this confidence interval is given by

$$(5) \qquad\qquad L(\lambda_1, \lambda_2) = X_{(n)}\left(\frac{1}{\lambda_1} - \frac{1}{\lambda_2} \right)$$

which is to be minimized subject to (4). We have

$$\frac{\partial L}{\partial \lambda_2} = X_{(n)}\left(\frac{1}{\lambda_2^2} - \frac{1}{\lambda_1^2}\frac{\partial \lambda_1}{\partial \lambda_2} \right)$$

and from (4)

$$n\lambda_2^{n-1} - n\lambda_1^{n-1}\frac{\partial \lambda_1}{\partial \lambda_2} = 0.$$

Thus

$$\frac{\partial \lambda_1}{\partial \lambda_2} = \left(\frac{\lambda_2}{\lambda_1} \right)^{n-1}$$

and

$$\frac{\partial L}{\partial \lambda_2} = X_{(n)}\left\{ \frac{1}{\lambda_2^2} - \frac{\lambda_2^{n-1}}{\lambda_1^{n+1}} \right\} = X_{(n)}\frac{\lambda_1^{n+1} - \lambda_2^{n+1}}{\lambda_2^2\lambda_1^{n+1}} < 0.$$

It follows that L is a decreasing function of λ_2 and the minimum occurs at $\lambda_2 = 1$. In that case, $\lambda_1 = \alpha^{1/n}$ from (4) and the shortest length confidence interval for θ based on $X_{(n)}$ is given by $[X_{(n)}, X_{(n)}\alpha^{-1/n}]$.

We note that

$$\mathscr{E}L = \left(\frac{1}{\lambda_1} - \frac{1}{\lambda_2} \right)\mathscr{E}X_{(n)} = \frac{n\theta}{n+1}\left(\frac{1}{\lambda_1} - \frac{1}{\lambda_2} \right)$$

which, when minimized subject to (4), also yields the same confidence interval.   □

Example 4.   The Role of the Pivot (Example 3).   We consider the problem of Examples 2 and 3 again and exhibit the role played by the choice of the pivot. For convenience, we take $n = 2$. The shortest length confidence interval based on $X_{(2)}$ as constructed in Example 3 is $[X_{(2)}, X_{(2)}/\sqrt{\alpha}]$.

We note that X/θ has a uniform distribution on $(0, 1)$ so that $(X_1 + X_2)/\theta$ has the triangular density

$$f(u) = u, \quad 0 < u < 1, \ = 2 - u, \quad 1 < u < 2, \text{ and zero elsewhere.}$$

We can therefore choose $(X_1 + X_2)/\theta$ also as a pivot. We emphasize that $X_1 + X_2$ is not sufficient for θ. Since the density f is symmetric about $u = 1$, it is clear that we choose $\lambda_2 = 2 - \lambda_1$ where

$$\int_{\lambda_1}^{\lambda_2} f(u) \, du = 1 - \alpha.$$

This yields $\lambda_1 = \sqrt{\alpha}$ so that $\lambda_2 = 2 - \sqrt{\alpha}$ and the minimum length confidence interval based on $X_1 + X_2$ is given by

$$\left[\frac{X_1 + X_2}{2 - \sqrt{\alpha}}, \frac{X_1 + X_2}{\sqrt{\alpha}} \right].$$

The length of the confidence interval based on $X_{(2)}$ is

$$L_1 = X_{(2)} \left(\frac{1}{\sqrt{\alpha}} - 1 \right)$$

and that based on $X_1 + X_2$ is

$$L_2 = (X_1 + X_2) \left(\frac{1}{\sqrt{\alpha}} - \frac{1}{2 - \sqrt{\alpha}} \right).$$

Clearly $L_2 > L_1$ (with probability one).

Yet another confidence interval can be based on the range $R = X_{(2)} - X_{(1)}$. In Example 8.5.4 we showed that R has density

$$f_R(r) = 2(\theta - r)/\theta^2, \quad 0 < r < \theta, \text{ and zero elsewhere}$$

so that R/θ has density

$$f(y) = 2(1 - y), \quad 0 < y < 1, \text{ and zero elsewhere.}$$

Conditions of Theorem 1 hold with $T = R/\theta$ and $[R/\lambda_2, R/\lambda_1]$ is a $1 - \alpha$ level confidence interval for θ provided we choose λ_1, λ_2 such that

$$\int_{\lambda_1}^{\lambda_2} 2(1 - y) \, dy = 1 - \alpha.$$

We choose λ_1, λ_2 to minimize $L(\lambda_1, \lambda_2) = R(1/\lambda_1 - 1/\lambda_2)$ subject to

$$1 - \alpha = \int_{\lambda_1}^{\lambda_2} 2(1 - y) \, dy = 2\left[\lambda_2 - \lambda_1 - \frac{\lambda_2^2}{2} + \frac{\lambda_1^2}{2}\right].$$

Differentiating partially with respect to λ_2 we get

$$\frac{\partial L}{\partial \lambda_2} = R\left(\frac{1}{\lambda_2^2} - \frac{1}{\lambda_1^2} \frac{\partial \lambda_1}{\partial \lambda_2}\right)$$

and

$$0 = 2\left[1 - \frac{\partial \lambda_1}{\partial \lambda_2} - \lambda_2 + \lambda_1 \frac{\partial \lambda_1}{\partial \lambda_2}\right].$$

Hence

$$\frac{\partial L}{\partial \lambda_2} = R\left(\frac{1}{\lambda_2^2} - \frac{1}{\lambda_1^2} \cdot \frac{1 - \lambda_2}{1 - \lambda_1}\right) = R \frac{(\lambda_1 - \lambda_2)\left[\lambda_1 + \lambda_2 + \lambda_1\lambda_2 - (\lambda_1 + \lambda_2)^2\right]}{\lambda_1^2 \lambda_2^2 (1 - \lambda_1)}$$

$$< 0$$

so that L is a decreasing function of λ_2. Hence minimum occurs at $\lambda_2 = 1$. In that case λ_1 is given by

$$1 - \alpha = 2\left(1 - \lambda_1 - \frac{1}{2} + \frac{\lambda_1^2}{2}\right) = 1 - 2\lambda_1 + \lambda_1^2 = (1 - \lambda_1)^2$$

so that $\lambda_1 = 1 - \sqrt{1 - \alpha}$. The shortest length confidence interval for θ based on R is then $[R, R/\lambda_1]$ with length

$$L_3 = R\left(\frac{1}{1 - \sqrt{1 - \alpha}} - 1\right) = R \frac{\sqrt{1 - \alpha}}{1 - \sqrt{1 - \alpha}}.$$

Then

$$L_3 - L_1 = X_{(2)}\left(\frac{\sqrt{1 - \alpha}}{1 - \sqrt{1 - \alpha}} - \frac{1}{\sqrt{\alpha}} + 1\right) - X_{(1)}\frac{\sqrt{1 - \alpha}}{1 - \sqrt{1 - \alpha}}$$

$$= X_{(2)}\left(\frac{\sqrt{\alpha} + \sqrt{1 - \alpha} - 1}{\sqrt{\alpha}\,(1 - \sqrt{1 - \alpha})}\right) + X_{(1)}\frac{\sqrt{1 - \alpha}}{1 - \sqrt{1 - \alpha}}$$

$$= \frac{X_{(2)}(\sqrt{\alpha} + \sqrt{1 - \alpha} - 1) - X_{(1)}\sqrt{\alpha(1 - \alpha)}}{\sqrt{\alpha}\,(1 - \sqrt{1 - \alpha})}.$$

In particular, $L_3 > L_1$ whenever

$$X_{(2)}(\sqrt{\alpha} + \sqrt{1 - \alpha} - 1) > X_{(1)}\sqrt{\alpha(1 - \alpha)}$$

which has to be checked for the particular sample. We note, however, that

$$\mathscr{E}L_3 - \mathscr{E}L_1 = \frac{\sqrt{1 - \alpha}}{1 - \sqrt{1 - \alpha}} \frac{\theta}{3} - \left(\frac{1}{\sqrt{\alpha}} - 1\right)\frac{2\theta}{3}$$

$$= \frac{2\theta}{3\sqrt{\alpha}(1 - \sqrt{\alpha})}\left\{\sqrt{\alpha} + \sqrt{1 - \alpha} - 1 - \frac{\sqrt{\alpha(1 - \alpha)}}{2}\right\}$$

$$> 0$$

so that on the average L_1 is smaller. We leave the reader to check that for $0 < \alpha < 1$

$$\sqrt{\alpha} + \sqrt{1 - \alpha} > 1 + (1/2)\sqrt{\alpha(1 - \alpha)}. \qquad \square$$

Bayesian Confidence Intervals*

The Bayesian approach to confidence intervals uses the prior information on θ through a distribution on Θ. Once the posterior distribution of Θ is computed as described in Section 10.6, it is easy to compute $\underline{\theta}(x_1, \dots, x_n)$ and $\bar{\theta}(x_1, \dots, x_n)$ such that

$$(6) \qquad P\{\underline{\theta}(X_1, \dots, X_n) \le \Theta \le \bar{\theta}(X_1, \dots, X_n)|x_1, \dots, x_n\} \ge 1 - \alpha.$$

Let $X_1, \dots, X_n$ be a random sample from a distribution with probability (or density) function $f(x|\theta)$ and let Θ have prior probability (or density) function $\pi(\theta)$. Let $h(\theta|x_1, \dots, x_n)$ be the posterior density of Θ given $X_1 = x_1, \dots, X_n = x_n$. (Here h may be a probability function if Θ is a discrete type.) Then any interval $[\underline{\theta}(x_1, \dots, x_n), \bar{\theta}(x_1, \dots, x_n)]$ satisfying

$$\int_{\underline{\theta}}^{\bar{\theta}} h(\theta|x_1, \dots, x_n)\, d\theta \ge 1 - \alpha$$

is called a $1 - \alpha$ level *Bayesian confidence interval* for θ. One can define lower and upper confidence bounds for θ in an analogous manner. It is clear that we choose $\underline{\theta}, \bar{\theta}$ in such a way that $\bar{\theta} - \underline{\theta}$ is least.

Example 5. Confidence Interval for the Mean of Normal Distribution. Let $X_1, \dots, X_n$ be a random sample from a normal population with mean θ and variance 1. Suppose Θ has a prior standard normal distribution. Then the posterior distribution of Θ is normal (Problem 10.6.13) with mean θ_n and variance σ_n^2 where

$$\theta_n = \frac{n\bar{x}}{n + 1} \quad \text{and} \quad \sigma_n^2 = \frac{1}{n + 1}.$$

It follows that the shortest minimum level $1 - \alpha$ Bayesian confidence interval for θ is given by

$$\left[\frac{n\bar{X}}{n+1} - \frac{z_{\alpha/2}}{\sqrt{n+1}}, \frac{n\bar{X}}{n+1} + \frac{z_{\alpha/2}}{\sqrt{n+1}}\right].$$

This interval has a smaller length than the non-Bayesian interval $[\bar{X} - z_{\alpha/2}/\sqrt{n}, \bar{X} + z_{\alpha/2}/\sqrt{n}]$, which is to be expected since we have assumed knowledge of more information in the Bayesian case. □

Example 6.   Confidence Interval for θ in Uniform Distribution.   Let $X_1, \ldots, X_n$ be a random sample from the uniform distribution on $(0, \theta)$. Then

$$f(x_1, \ldots, x_n | \theta) = \frac{1}{\theta^n}, \qquad 0 < \max x_i < \theta, \text{ and zero elsewhere.}$$

If the prior distribution of Θ is given by the density function

$$\pi(\theta) = \frac{\theta^n e^{-\theta}}{n!}, \qquad \theta > 0, \text{ and zero elsewhere}$$

then

$$g(x_1, \ldots, x_n) = \frac{1}{n!} \int_{\max x_i}^{\infty} \theta^n e^{-\theta} \frac{1}{\theta^n} \, d\theta = \frac{1}{n!} e^{-\max x_i}$$

so that

$$h(\theta | x_1, \ldots, x_n) = \begin{cases} e^{-\theta + \max x_i}, & \theta > \max x_i \\ 0 & \text{elsewhere.} \end{cases}$$

Hence a minimum level $1 - \alpha$ Bayesian confidence interval is given by $[\underline{\theta}, \bar{\theta}]$ where

$$1 - \alpha = \int_{\underline{\theta}}^{\bar{\theta}} h(\theta | x_1, \ldots, x_n) \, d\theta = e^{\max x_i}\left[e^{-\underline{\theta}} - e^{-\bar{\theta}}\right].$$

We leave the reader to show that the shortest length $\bar{\theta} - \underline{\theta}$ is achieved by choosing the least value of θ for $\underline{\theta}$, namely $\underline{\theta} = \max x_i$. In that case $\bar{\theta} = \max x_i + \ln(1/\alpha)$ so that $[\max X_i, \max X_i + \ln(1/\alpha)]$ is the shortest length Bayes confidence interval for θ at minimum level $1 - \alpha$. □

Problems for Section 10.7

Find minimum level $1 - \alpha$ confidence intervals whenever possible. Problems 13–18 refer to the Subsection on Bayesian confidence intervals.

1. Find a suitable pivot for each of the following densities based on a sample of size 1.
    (i) $f(x|\theta) = \dfrac{2}{\theta^2}(\theta - x), \qquad 0 < x < \theta$, and zero elsewhere.
    (ii) $f(x|\theta) = \exp\{-(x - \theta)\}, \qquad x > \theta$, and zero elsewhere.

(iii) $f(x|\theta) = \dfrac{1}{\theta} e^{-x/\theta}$, $x > 0$, and zero elsewhere, $\theta > 0$.

(iv) $f(x|\theta) = \dfrac{1}{\theta^2} x e^{-x/\theta}$, $x > 0$, and zero elsewhere, $\theta > 0$

(v) $f(x|\theta) = \dfrac{1}{\theta\sqrt{2\pi}} \exp\{-\dfrac{x^2}{2\theta^2}\}$, $x \in \mathbb{R}, \theta > 0$.

(vi) $f(x, y|\mu_1, \mu_2) = \dfrac{1}{2\pi} \exp\{-\dfrac{(x - \mu_1)^2}{2} - \dfrac{(y - \mu_2)^2}{2}\}$, $x \in \mathbb{R}, y \in \mathbb{R}$.

2. Let $X_1, X_2, \ldots, X_n$ be a random sample from a normal population with mean μ and variance σ^2.
   (i) If μ is known, find a minimum level $1 - \alpha$ confidence interval for σ^2 based on the sufficient statistic $\sum_{i=1}^{n}(X_i - \mu)^2$.
   (ii) When is the length of your confidence interval the least? (Just give the relation between λ_1 and λ_2; do not attempt to solve the equation.)
   (iii) Find the equal tails $1 - \alpha$ level confidence interval for σ^2. That is, choose λ_1, λ_2 such that the probability in each tail is $\alpha/2$.
   (iv) Repeat (i), (ii), and (iii) when μ is unknown, using the sufficient statistic $\sum_{i=1}^{n}(X_i - \bar{X})^2$.

3. Let $X_1, \ldots, X_n$ be a random sample from an exponential distribution with mean θ. Find a $1 - \alpha$ level minimum length confidence interval for θ based on the sufficient statistic $\sum_{i=1}^{n} X_i$. [Hint: $\sum_{i=1}^{n} X_i$ has a gamma distribution with $\alpha = n$, $\beta = \theta$. Use $\sum_{i=1}^{n} X_i/\theta$ or $2\sum_{i=1}^{n} X_i/\theta$ as pivot. Derive the relation between λ_1, λ_2. Do not attempt to solve the equation.]

4. Let $X_1, \ldots, X_n$ be a random sample from the shifted exponential density

$$f(x|\theta) = \exp\{-(x - \theta)\}, \quad x > \theta, \text{ and zero elsewhere.}$$

Find the shortest length $1 - \alpha$ level confidence interval based on the sufficient statistic $\min(X_1, \ldots, X_n)$.

5. Show how you will find a $1 - \alpha$ level shortest length confidence interval based on one observation from the density function

$$f(x|\theta) = \dfrac{2}{\theta^2}(\theta - x), \quad 0 < x < \theta, \text{ and zero elsewhere.}$$

6. An observation is taken from density function

$$f(x|\theta) = 2\dfrac{(x - \theta)}{(b - \theta)^2}, \quad \theta < x < b, \text{ and zero elsewhere,}$$

where b is known.
   (i) Show that $(X - \theta)/(b - \theta)$ is a pivot.
   (ii) Show that $[(X - b\lambda_2)/(1 - \lambda_2), (X - b\lambda_1)/(1 - \lambda_1)]$ is a $1 - \alpha$ level confidence interval with length $(b - X)(\lambda_2 - \lambda_1)/[(1 - \lambda_1)(1 - \lambda_2)]$ where λ_1, λ_2 are chosen from $1 - \alpha = \lambda_2^2 - \lambda_1^2$.

7. Let $X_1, \ldots, X_n$ be a random sample from the uniform distribution on $\{1, 2, \ldots, N\}$. Find an upper $1 - \alpha$ level confidence bound for N.

8. Let $X_1, X_2, \ldots, X_n$ be a random sample from uniform distribution on $(0, \theta)$. Then $[X_{(n)}, \alpha^{-1/n} X_{(n)}]$ is the shortest length confidence interval at level $1 - \alpha$ based on $X_{(n)}$ as shown in Example 3. If it is known that $\theta \leq a$ where a is a known constant, find the smallest sample size required such that the length of the shortest confidence interval is at most $d(> 0)$.

9. Let $X_1, \ldots, X_m$ and $Y_1, \ldots, Y_n$ be independent random samples from normal populations with parameters μ_1, σ^2 and μ_2, σ^2 respectively. Find the smallest length $1 - \alpha$ level confidence interval for $\mu_1 - \mu_2$ based on an appropriate sufficient statistic when σ is known.

10. In Problem 9, suppose the variances are σ_1^2 and σ_2^2 respectively and a confidence interval for σ_1^2/σ_2^2 is desired. Find $1 - \alpha$ level confidence intervals based on appropriate sufficient statistics when
   (i) μ_1, μ_2 are known.
   (ii) Only one of μ_1 or μ_2 is known.
   (iii) Both μ_1, μ_2 are unknown.

11. In Problem 10, find a $1 - \alpha$ level shortest length confidence interval for $\mu_1 - \mu_2$ when σ_1^2, σ_2^2 are both known.

12. Let $X_1, X_2, \ldots, X_n$ be a random sample from Laplace density function

$$f(x|\theta) = (1/2\theta) \exp\{-|x|/\theta\}, x \in \mathbb{R}; \theta > 0.$$

   (i) Show that $\sum_{j=1}^{n}|X_j|$ is sufficient and $2\sum_{j=1}^{n}|X_j|/\theta$ has a chi-square distribution with $2n$ d.f.
   (ii) Show how you will find the shortest length confidence interval based on the pivot $2\sum_{j=1}^{n}|X_j|/\theta$.

13. Let X be the number of successes in n independent Bernoulli trials with constant probability p of success. The prior distribution of p is uniform on $[0, 1]$. Show how you will find a $1 - \alpha$ level Bayesian confidence interval for p. [Hint: Use the result of Problem 10.6.1.]

14. Suppose the death density function of a system is given by $f(x|\theta) = \theta e^{-x\theta}$, if $x > 0$, and zero elsewhere. If Θ has improper prior density $\pi(\theta) = 1/\theta$, $\theta > 0$, find the shortest length $1 - \alpha$ level Bayesian confidence interval for θ. [Hint: Use the posterior density of Problem 10.6.3.]

15. Suppose X has a Bernoulli distribution with parameter θ where Θ has a prior uniform distribution on $(0, 1)$. Show how you will find a Bayesian $1 - \alpha$ level confidence interval for θ based on a sample of size n. [Hint: Use the posterior distribution from Problem 10.6.9.]

16. Let $X_1, \ldots, X_n$ be a random sample from a Poisson distribution with parameter λ where λ itself has a prior gamma density with known parameters α and β. Show how you will find a $1 - \alpha$ level Bayesian confidence interval for λ. [Hint: Use the posterior distribution computed in Problem 10.6.8.]

17. Suppose you have an observation from density function

$$f(x|\theta) = \frac{2}{\theta^2}(\theta - x), \quad 0 < x < \theta, \text{ and zero elsewhere.}$$

Let Θ have prior density given by

$$\pi(\theta) = \tfrac{1}{16}\theta^2 e^{-\theta/2}, \qquad \theta > 0, \text{ and zero elsewhere.}$$

Show how you will find a $1 - \alpha$ level Bayesian confidence interval for θ.

18. Let $X_1, X_2, \ldots, X_n$ be a random sample from a geometric distribution with parameter θ, $\theta \in [0,1]$. If the prior distribution of Θ is beta with known parameters α and β, how will you find a $1 - \alpha$ level Bayesian confidence interval for θ? [Hint: Use the posterior distribution computed in Problem 10.6.10.]

10.8 REVIEW PROBLEMS

1. Let $X_1, X_2, \ldots, X_n$ be a random sample from a distribution with density function $f(x, \theta)$ given below.
   (i) $f(x, \theta) = (x/\theta^2)\exp\{-x^2/\theta^2\}$ for $x > 0$ and zero elsewhere; $\theta > 0$.
   (ii) $f(x, \theta) = 1/2\theta$, $-\theta < x < \theta$, and zero otherwise; $\theta > 0$.
   (iii) $f(x, \theta) = 3x^2/\theta^3$, $0 < x < \theta$, and zero otherwise.
   (iv) $f(x, \theta) = [\ln \theta/(\theta - 1)]\theta^x$, $0 < x < 1$, and zero otherwise; $\theta > 1$.
   (a) Find a sufficient statistic for θ in each case. Find a (the) maximum likelihood estimate of θ.
   (b) In part (iii) find an unbiased estimate of θ based on your sufficient statistic. In part (iv) find an unbiased estimate of $\mathscr{E}_\theta X_1$ based on your sufficient statistic.

2. Let $X_1, X_2, \ldots, X_n$ be a random sample from the density function

$$f(x, \theta) = \frac{1}{\sigma}\exp\{-(x - \mu)/\sigma\} \qquad \text{for } x \geq \mu \text{ and zero otherwise.}$$

Let $\theta = (\mu, \sigma^2)$, $\mu \in \mathbb{R}$, $\sigma^2 > 0$ and let t be a known constant. Find the maximum likelihood estimate of $P_\theta(X_1 \geq t)$. (Assume that $t > \mu$.)

3. Let $X_1, X_2, \ldots, X_n$ be a random sample from the probability function $p(x, \theta)$ given below. Is the set of order statistics $X_{(1)}, X_{(2)}, \ldots, X_{(n)}$ sufficient for θ?
   (i) $p(x) = e^{-\theta}\theta^x/x!$, $x = 0, 1, 2, \ldots, n$.
   (ii) $p(x) = \binom{n}{x}p^x(1 - p)^{n-x}$, $x = 0, 1, 2, \ldots, n$.
   [Hint: Use the factorization theorem.]

4. Suppose $(X_1, X_2, \ldots, X_{k-1})$ has a multinomial distribution with parameters $n, p_1, p_2, \ldots, p_{k-1}$, where $p_i > 0$, $\sum_{i=1}^{k-1}p_i < 1$. Is the set of order statistics $X_{(1)}, X_{(2)}, \ldots, X_{(k)}$ sufficient for $p_1, p_2, \ldots, p_k$ when n is known?

5*. Let p be the proportion of males in a population. A random sample of size n is taken. Let $X_i = 1$ if the ith member of the sample is a male and $X_i = 0$ zero otherwise, $i = 1, 2, \ldots, n$. Does there exist an unbiased estimate of the sex ratio $p/(1 - p)$ based on the sufficient statistic $\sum_{i=1}^{n}X_i$?

6. Let $X_1, X_2, \ldots, X_n$ be the number of particles emitted from a radioactive source per 20-second interval. If $X_1, \ldots, X_n$ is assumed to be a random sample from a Poisson distribution with parameter $\lambda > 0$, find an unbiased estimate of $P(X = k)$ where k is a known nonnegative integer. Find an unbiased estimate for $P(X = k)$ based on

the sufficient statistic ΣX_i. (Hint: Let $Y_i = 1$ if $X_i = k$ and $Y_i = 0$ otherwise. Compute $\mathscr{E}\{Y_i | \Sigma_{j=1}^{n} X_j = t\}$.)

7. Let $X_1, X_2, \ldots, X_n$ be a random sample from the shifted exponential density function $f(x) = \exp(-x + \theta)$ if $x > \theta$, and zero otherwise.
   (i)   Show that $\hat{\theta}_1 = \bar{X} - 1$ and $\hat{\theta}_2 = \min X_i - \dfrac{1}{n}$ are both unbiased for θ.
   (ii) Find the relative efficiency of $\hat{\theta}_1$ with respect to $\hat{\theta}_2$.
   (iii) Explain why $\hat{\theta}_2$ turns out to be more efficient than $\hat{\theta}_1$ in part (ii).

8*. The geometric mean of a set of real numbers $x_1, x_2, \ldots, x_n$ is defined by $(\prod_{i=1}^{n} x_i)^{1/n}$. It is well known that

$$\left(\prod_{i=1}^{n} x_i \right)^{1/n} \leq \bar{x} = \Sigma_{i=1}^{n} \frac{x_i}{n}.$$

Use this property of a set of real numbers to show that the empirical distribution function is the maximum likelihood estimate of a distribution function. (See Example 10.5.8 for an alternative proof.) [Hint: For any positive real numbers $p_1, \ldots, p_n$, such that $\Sigma_{i=1}^{n} p_i \leq 1$, $(\prod_{i=1}^{n} p_i)^{1/n} \leq \bar{p} \leq 1/n$.]

9. Let $X_1, X_2, \ldots, X_n$ be a random sample from a Poisson distribution with parameter $\lambda > 0$. In Problem 10.3.21 we asked the reader to find an unbiased estimate of $d(\lambda) = \lambda^k$ for various values of the constant k (integral). Let $T = \Sigma_{i=1}^{n} X_i$. Then T is sufficient for λ.
   (i)   Show that $T(T-1)\ldots(T-k+1)/n^k$ is an unbiased estimate for $d(\lambda) = \lambda^k$ based on T.
   (ii) Let $k = 2$. Then $X_1 X_2$ is unbiased for $d(\lambda) = \lambda^2$. Show that $\mathscr{E}\{X_1 X_2 | T\} = T(T-1)/n^2$. [Hint: For part (ii) use properties of conditional expectation studied in Section 5.5 (Proposition 1). Show that $\mathscr{E}\{X_1 | T, X_2\} = (T - X_2)/(n - 1)$, and now use Problem 8.8.17 to show that $\mathscr{E}\{X_1 X_2 | T\} = T(T-1)/n^2$.]

10. Let X have Poisson distribution with parameter λ. Show that
    (i) $(-1)^X$ is an unbiased estimate of $d(\lambda) = e^{-2\lambda}$.
    (ii) $(-2)^X$ is an unbiased estimate of $d(\lambda) = e^{-3\lambda}$.
    [The estimate in parts (i) and (ii) are in fact the best unbiased estimates of $d(\lambda)$ in the sense of minimum variance. Yet each estimate takes negative values with positive probability whereas $d(\lambda) > 0$.]

11. Find the maximum likelihood estimate of θ based on a sample of size 1 from the density function

$$f(x, \theta) = \frac{2}{\theta^2}(\theta - x), \qquad 0 < x < \theta \text{ and zero elsewhere.}$$

Is your estimate unbiased for θ?

12. Let $X_1, X_2, \ldots, X_m$ and $Y_1, Y_2, \ldots, Y_n$ be independent random samples from exponential densities with means $1/\lambda_1$ and $1/\lambda_2$.
    (i)   Find maximum likelihood estimates of λ_1 and λ_2.
    (ii) Find the maximum likelihood estimate of λ if $\lambda_1 = \lambda_2 = \lambda$.

13. Let $X_1, X_2, \ldots, X_m$ be m independent observations on a binomial distribution with parameters n and p both unknown. Find method of moments estimates of n and p.

14.   (i)   Let $X_1, X_2, \ldots, X_m$ and $Y_1, Y_2, \ldots, Y_n$ be independent random samples from normal distributions with respective parameters μ_1 and σ^2, and μ_2 and σ^2. Find maximum likelihood estimates of μ_1, μ_2, and σ^2.

      (ii)   Suppose $(X_1, Y_1), \ldots, (X_n, Y_n)$ is a random sample from a bivariate normal population with parameters $\mu, \mu, \sigma^2, \sigma^2$, and ρ. Find maximum likelihood estimates of μ, σ^2 and ρ.

15.   Suppose $X_1, X_2, \ldots, X_n$ are n independent observations from the density function

$$f(x) = 1/\theta, \qquad 0 < x < \theta, \text{ and zero elsewhere.}$$

Let $R = X_{(n)} - X_{(1)}$ be the range of the sample. Find a minimum level $1 - \alpha$ confidence interval for θ of the form $[R, R/c]$ where $0 < c < 1$ is a constant. Compare the expected length of your confidence interval with the expected length of the minimum level $1 - \alpha$ shortest length confidence interval $[X_{(n)}, \alpha^{-1/n} X_{(n)}]$ based on the (best) sufficient statistic $X_{(n)}$ for θ (derived in Example 10.7.3). Use the results of Example 8.5.4. (See also Example 10.7.4.)

16.   Let $X_1, X_2, \ldots, X_n$ be a random sample from a Weibull density function $f(x)$ $= \dfrac{\gamma}{\beta} x^{\gamma-1} \exp(-x^\gamma/\beta)$, $x > 0$ and zero elsewhere, where $\gamma > 0$ is known and $\beta > 0$ is unknown. Find a $1 - \alpha$ level confidence interval for β based on the pivot $2\Sigma_{j=1}^{n} X_j^\gamma / \beta$.

17.   In the randomized response model of Example 10.4.2 we showed that

$$\sigma_{\hat{\pi}}^2 = \frac{1}{n}\left[\frac{1}{16(p - 1/2)^2} - (\pi - 1/2)^2\right].$$

   (i)   Find n to ensure the $|\hat{\pi} - \pi| < \varepsilon$ with probability $1 - \alpha$.

   (ii)   Find an approximate $1 - \alpha$ level confidence interval for π based on Y, the number of "yes" responses when there are n participants, where n is large.

      Suppose $n = 200$, $Y = 110$, and $p = .9$. Find a point estimate of π and also a 90 percent confidence interval estimate for π.

18.   Suppose $X_1, X_2, \ldots, X_m$ and $Y_1, Y_2, \ldots, Y_n$ are independent random samples from normal distributions with respective parameters (μ_1, σ_1^2) and (μ_2, σ_2^2). Suppose $\sigma_1 = \sigma_2 = \sigma$ is unknown. Show that $(\bar{X}, \bar{Y}, S_p^2)$ is sufficient for (μ_1, μ_2, σ^2) where S_p^2 is the pooled variance: $S_p^2 = [(m - 1)S_1^2 + (n - 1)S_2^2]/(m + n - 2)$. Show further that an appropriate pivot is the t-statistic

$$T = \frac{\bar{X} - \bar{Y} - (\mu_1 - \mu_2)}{S_p\sqrt{\dfrac{1}{m} + \dfrac{1}{n}}}$$

with $m + n - 2$ d.f.

19.   Let $X_1, X_2, \ldots, X_n$ be a random sample from a uniform distribution on $(0, \theta)$ and suppose that Θ has prior distribution given by the density function

$$f(\theta) = \frac{\alpha a^\alpha}{\theta^{\alpha+1}}, \qquad \theta > a \text{ and } = 0 \text{ if } \theta \le a.$$

(i)   Find the posterior distribution of Θ and the Bayes estimate of θ using squared error loss function.

(ii) Find the Bayes estimate δ of θ using loss function $|\delta - \theta|$.

(iii) Find the minimum length $1 - \alpha$ level Bayes confidence interval for θ in (i).

20. A sample of n observations is taken from a normal distribution with mean μ and variance 1. Suppose all one records is a count of the number of observations less than 0. If m is the number of X_i's that are negative show that the maximum likelihood estimate of μ is given by $\hat{\mu} = -\Phi^{-1}(m/n)$ where Φ is the standard normal distribution function. (See also Problem 10.5.11.)

21. Suppose X has a binomial distribution with parameter θ where θ itself has a prior uniform distribution on $(0, 1)$. Find the generalized maximum likelihood estimate of θ. That is, find the estimate of θ that maximizes the posterior density of θ. (See Problem 10.6.14.)

22. (i)   Suppose X has distribution given by

$$P(X = x|\theta) = p(1 - p)^{x - \theta}, \qquad x = \theta, \theta + 1, \ldots, p \text{ known.}$$

How will you find a $1 - \alpha$ level confidence interval for θ?

(ii) Do the same for the distribution

$$P(X = x|\theta) = \binom{n}{x - \theta} p^{x - \theta}(1 - p)^{n - x + \theta},$$

where $x = \theta, \theta + 1, \ldots, n + \theta$, and p is known.

23. Let $X_1, X_2, \ldots, X_n$ be a random sample from an exponential distribution with mean $\theta > 0$. Show that S^2, $\sum_{i=1}^{n} X_i^2/2n$ and $n\bar{X}^2/(n + 1)$ are all unbiased for θ^2. Which of these estimates has the least variance?

CHAPTER 11

Testing Hypotheses

11.1 INTRODUCTION

The basic ideas of hypothesis testing were first introduced in Section 4.4, where for the most part we considered samples of size 1 or 2 to avoid computational difficulties. In Chapters 6 and 8, however, we considered several important and widely applicable tests. In Chapter 9 we considered some large-sample tests that were based on the assumption that the test statistic has a limiting normal distribution. In developing these tests, however, we mostly relied on our intuition: if large (small) values of the test statistic supported the alternative hypothesis, we rejected H_0 if the observed value of the test statistic is large (small). Although we often made power computations we did not show (except in some examples of Section 4.4) that the resulting procedure maximized the power function. In Section 11.2 we prove the Neyman–Pearson lemma, which gives the general procedure of constructing best (most powerful) tests in the case when a simple null hypothesis is tested against a simple alternative hypothesis. In Section 11.3 we show how the methods of Section 11.2 can sometimes be extended to yield best tests in more complicated situations.

A general method of constructing tests, especially when two or more unknown parameters are involved even though we may only be interested in testing for one of them, is given in Section 11.4. This so-called likelihood ratio test often yields reasonable results.

In Sections 11.5 through 11.9 we consider some nonparametric alternatives of tests introduced in earlier chapters. In Section 11.5 we consider the signed rank test, which may be used as an alternative to the one sample t-test of Section 8.7. The two sample tests of Section 11.6 may be used as alternatives to the two sample t-tests of Section 8.7. In Sections 11.7 and 11.8 we return to the problem of goodness of fit considered earlier in Sections 4.5 and 9.9. In Section 11.9 we develop two procedures to test the null hypothesis that no association exists between two variables against the alternative that a direct or indirect association exists.

11.2   NEYMAN–PEARSON LEMMA

In this section we concentrate on the case when Θ, the parameter set, contains only two values θ_0 and θ_1. Let H_0, the null hypothesis, be that X has distribution P_{θ_0} and let H_1 be that X has distribution P_{θ_1}. Both null and alternative hypotheses are simple hypotheses so that the distribution of X is completely specified under H_0 as well as H_1. Our object is to find a critical region $C(X_1, \ldots, X_n)$ based on the sample $X_1, X_2, \ldots, X_n$ such that it has maximum power (or equivalently, the smallest type II error probability) at $\theta = \theta_1$. We recall that the power function of any critical region is defined by

$$\beta(\theta) = P_\theta(C) = \begin{cases} \beta(\theta_0) = P(\text{type I error}) & \text{if } \theta = \theta_0, \\ \beta(\theta_1) = 1 - P(\text{type II error}) & \text{if } \theta = \theta_1. \end{cases}$$

The following result due to Neyman and Pearson gives a complete solution to the problem.

THEOREM 1.   (NEYMAN–PEARSON LEMMA).   *Let $X_1, X_2, \ldots, X_n$ be a random sample from a distribution with probability or density function $f(x; \theta)$. Suppose $\Theta = \{\theta_0, \theta_1\}$ and $f(x; \theta_0) = f_0(x), f(x; \theta_1) = f_1(x)$. Let*

$$L(\theta) = \prod_{j=1}^{n} f(x_j; \theta)$$

be the likelihood function of the sample. Then any critical region C of the form

$$C = \{(x_1, x_2, \ldots, x_n) : L(\theta_1)/L(\theta_0) \geq k\}$$

for some $0 \leq k < \infty$ is best of its size for testing $H_0 : \theta = \theta_0$ against $H_1 : \theta = \theta_1$. For $k = \infty$, the critical region

$$C_0 = \{(x_1, \ldots, x_n) : L(\theta_0) = 0\}$$

is best of size zero for testing H_0 against H_1. ("Best" means the critical region has maximum power at $\theta = \theta_1$.)

Proof.   We note that for $0 \leq k < \infty$ the size of C in the continuous case is given by

$$P_{\theta_0}(C) = \int \cdots \int_C L(\theta_0)\, dx_n \ldots dx_1.$$

Let C' be any critical region such that its level is at most $P_{\theta_0}(C)$. That is, let

(1) $$P_{\theta_0}(C') \leq P_{\theta_0}(C).$$

Now (see Figure 1)

$$C = (C \cap C') \cup (C \cap \bar{C}')$$

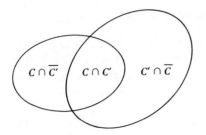

Figure 1

and

$$C' = (C' \cap C) \cup (C' \cap \bar{C})$$

so that

(2) $$P_{\theta_1}(C') = \int_{C'} L(\theta_1) = \int_{C' \cap C} L(\theta_1) + \int_{C' \cap \bar{C}} L(\theta_1)$$

and

(3) $$P_{\theta_1}(C) = \int_C L(\theta_1) = \int_{C \cap C'} L(\theta_1) + \int_{C \cap \bar{C'}} L(\theta_1).$$

Here we have written for convenience

$$P_{\theta}(C) = \int \cdots \int_C L(\theta)\, dx_n \ldots dx_1 = \int_C L(\theta).$$

Subtracting (3) from (2) we have

(4) $$P_{\theta_1}(C') - P_{\theta_1}(C) = \int_{C' \cap \bar{C}} L(\theta_1) - \int_{C \cap \bar{C'}} L(\theta_1).$$

Next we note that for $(x_1, \ldots, x_n) \in C' \cap \bar{C} \subseteq \bar{C}$, $L(\theta_1) < kL(\theta_0)$ and for $(x_1, \ldots, x_n) \in C \cap \bar{C'} \subseteq C$, $L(\theta_1) \geq kL(\theta_0)$. It follows from (4) that

$$P_{\theta_1}(C') - P_{\theta_1}(C) \leq k \int_{C' \cap \bar{C}} L(\theta_0) - k \int_{C \cap \bar{C'}} L(\theta_0)$$

$$= k \left\{ \int_{C' \cap \bar{C}} L(\theta_0) - \int_{C \cap \bar{C'}} L(\theta_0) \right\}$$

$$= k \left\{ \int_{C'} L(\theta_0) - \int_C L(\theta_0) \right\} \quad \text{(Why?)}$$

$$= k \left\{ P_{\theta_0}(C') - P_{\theta_0}(C) \right\}$$

$$\leq 0 \qquad\qquad\qquad \text{[from (1)]}.$$

A similar argument applies in the discrete case.

For the case $k = \infty$ we have $P_{\theta_0}(C_0) = 0$ so that C_0 has size zero. Moreover, if critical region C_0' has level zero then $P_{\theta_0}(C_0') = 0$ and $C_0' \subseteq C_0$. (If C_0' contains any points $(x_1, x_2, \ldots, x_n)$ for which $L(\theta_0) > 0$, then $P_{\theta_0}(C_0')$ cannot be zero.) Hence

$$P_{\theta_1}(C_0') \le P_{\theta_1}(C_0).$$ □

Example 1. Number of System Failures. Suppose the number of system failures each month has a Poisson distribution with parameter λ. The number of such failures was observed for 12 months in order to test $H_0 : \lambda = 2$ against $H_1 : \lambda = 3$. The likelihood function is given by

$$L(\lambda) = \prod_{j=1}^{12} P(X_j = x_j) = \prod_{j=1}^{12} \left\{ \frac{e^{-\lambda}\lambda^{x_j}}{x_j!} \right\} = \frac{e^{-12\lambda}\lambda^{\sum_{j=1}^{12} x_j}}{\prod_{j=1}^{12} x_j!}$$

and

$$\frac{L(\lambda_1)}{L(\lambda_0)} = \frac{L(3)}{L(2)} = \frac{e^{-12(3)}(3)^{\sum_{j=1}^{12} x_j}}{e^{-12(2)}(2)^{\sum_{j=1}^{12} x_j}} = e^{-12}\left(\frac{3}{2} \right)^{\sum_{j=1}^{12} x_j}.$$

The Neyman–Pearson critical region is of the form

$$C = \left\{ (x_1, \ldots, x_n) : \frac{L(3)}{L(2)} \ge k \right\}, \qquad \text{for } k \ge 0.$$

Note, however, that $L(3)/L(2)$ is an increasing function of $\sum_{i=1}^{12} x_i$, so that $L(3)/L(2) \ge k$ if and only if $\sum_{i=1}^{12} x_i \ge k'$ for some constant $k' \ge 0$ (see Figure 2). Thus the critical region $C_0 = \{(x_1, \ldots, x_n) : \sum_{i=1}^{12} x_i \ge k'\}$ is the best critical region of its size $P_{\lambda=2}(C_0)$ for testing $H_0 : \lambda = 2$ against $H_1 : \lambda = 3$. Since $\sum_{i=1}^{k} X_i$ has a Poisson distri-

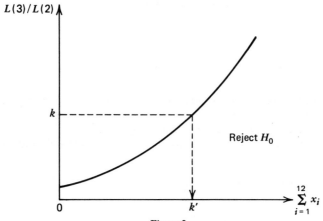

Figure 2

bution with parameter $12\lambda_0 = 24$ (under H_0), we can compute the size for any given k'. If, however, α, $0 < \alpha < 1$, is given as the level of significance we choose k' to be the smallest integer such that

$$\sum_{k=k'}^{\infty} e^{-24}\frac{(24)^k}{k!} \le \alpha. \qquad \square$$

Example 2. A Case Where X Has Continuous Density. Let X have density function

$$f(x; \theta) = 2\theta x + 2(1 - \theta)(1 - x), \qquad 0 < x < 1, \text{ and zero elsewhere,}$$

where $\theta \in [0, 1]$. Suppose it is known that θ is either θ_0 or θ_1. The object is to choose between $f(x; \theta_0)$ and $f(x; \theta_1)$ based on a single observation. We have

$$\frac{L(\theta_1)}{L(\theta_0)} = \frac{2\theta_1 x + 2(1 - \theta_1)(1 - x)}{2\theta_0 x + 2(1 - \theta_0)(1 - x)} = \frac{x[\theta_1 - (1 - \theta_1)] + 1 - \theta_1}{x[\theta_0 - (1 - \theta_0)] + 1 - \theta_0}.$$

For definiteness, let $\theta_0 < \theta_1$. Let us examine $L(\theta_1)/L(\theta_0)$ as a function of x. Set $\lambda(x) = L(\theta_1)/L(\theta_0)$. Then

$$\lambda'(x) = \frac{[x(2\theta_0 - 1) + 1 - \theta_0](2\theta_1 - 1) - (2\theta_0 - 1)[x(2\theta_1 - 1) + 1 - \theta_1]}{[x(2\theta_0 - 1) + 1 - \theta_0]^2}$$

$$= \frac{(\theta_1 - \theta_0)}{[x(2\theta_0 - 1) + 1 - \theta_0]^2} > 0$$

and it follows that $\lambda(x)$ is an increasing function of x. Hence

$$\lambda(x) \ge k \quad \text{if and only if} \quad x \ge k'$$

so that

$$C = \{x : 0 \le k' \le x < 1\}.$$

It is instructive to compare the graphs of $L(\theta_0)$ and $L(\theta_1)$ as done in Figure 3. Since large values of X are more probable under H_1 than under H_0, the method of analysis we used in Section 4.4 also leads to the critical region C.

Given k' it is easy to compute the size of C and its power at θ_1. Indeed,

$$\beta(\theta) = P_\theta(C) = 2\int_{k'}^{1}[x(2\theta - 1) + 1 - \theta]\, dx$$

$$= (2\theta - 1)(1 - k'^2) + 2(1 - \theta)(1 - k').$$

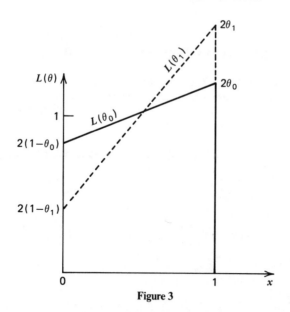

Figure 3

If the size is given to be α, then we choose k' from

$$\alpha = \beta(\theta_0) = (2\theta_0 - 1)(1 - k'^2) + 2(1 - \theta_0)(1 - k')$$

so that

$$k'^2(2\theta_0 - 1) + 2k'(1 - \theta_0) + \alpha - 1 = 0.$$

It follows that

$$k' = \frac{-2(1 - \theta_0) \pm \sqrt{4(1 - \theta_0)^2 + 4(2\theta_0 - 1)(1 - \alpha)}}{2(2\theta_0 - 1)}, \qquad \theta_0 \neq \frac{1}{2}.$$

Since $0 \le k' < 1$, we choose

$$k' = \frac{-(1 - \theta_0) + \sqrt{(1 - \theta_0)^2 + (2\theta_0 - 1)(1 - \alpha)}}{2\theta_0 - 1}$$

$$= \frac{-(1 - \theta_0) + \sqrt{\theta_0^2 - \alpha(2\theta_0 - 1)}}{2\theta_0 - 1},$$

for $\theta_0 \neq \frac{1}{2}$. If $\theta_0 = \frac{1}{2}$ then $k' = 1 - \alpha$. In either case we note that k' depends only on α and θ_0 but not on θ_1 (as long as $\theta_0 < \theta_1$).   □

Example 3.   Testing Normal Against Double Exponential.   Suppose X has density function

$$f(x) = \frac{1}{\sqrt{2\pi}} \exp\left\{-\frac{x^2}{2}\right\}, \qquad x \in \mathbb{R}$$

under H_0 and

$$g(x) = \tfrac{1}{2}\exp\{-|x|\}, \qquad x \in \mathbb{R}$$

under H_1. How do we choose between f and g on the basis of a single observation? Since both the densities are completely specified (that is, free of any unknown parameters) we apply the Neyman–Pearson lemma and conclude that the most powerful test has a critical region of the form

$$\frac{g(x)}{f(x)} \geq k.$$

Let

$$\lambda(x) = \frac{g(x)}{f(x)} = \frac{\sqrt{2\pi}}{2}\exp\left\{-|x| + \frac{x^2}{2}\right\}.$$

We note that

$$\frac{x^2}{2} - |x| = \left(\tfrac{1}{2}\right)\left(x^2 - 2|x| + 1 - 1\right) = \left(\tfrac{1}{2}\right)\left[\left(|x| - 1\right)^2 - 1\right]$$

so that

$$\lambda(x) = \sqrt{\frac{\pi}{2}}\, e^{-1/2}e^{(|x|-1)^2/2}$$

which is an increasing function of $\big||x| - 1\big|$. Hence

$$\lambda(x) \geq k \quad \text{if and only if} \quad \big||x| - 1\big| \geq k'$$

and it follows that C is of the form $C = \{x : |x| \geq k_1 \text{ or } |x| \leq k_2\}$ (see Figure 4). Thus if either a very large or a very small value of X is observed, we suspect that H_1 is

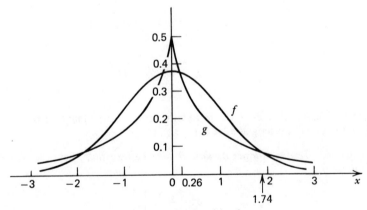

Figure 4. Graphs of f and g.

true rather than H_0. This is consistent with the fact that g has more probability in its tails and near zero than f has. The size of C is given by

$$P_f(C) = \int_{|x| \geq k_1} (1/\sqrt{2\pi}) e^{-x^2/2} \, dx + \int_{|x| \leq k_2} (1/\sqrt{2\pi}) e^{-x^2/2} \, dx$$

$$= 2\{P(Z \geq k_1) + P(0 < Z \leq k_2)\}$$

where Z has a standard normal distribution. The power of C is given by

$$P_g(C) = \int_{|x| \geq k_1} \left(\tfrac{1}{2}\right) e^{-|x|} \, dx + \int_{|x| \leq k_2} \left(\tfrac{1}{2}\right) e^{-|x|} \, dx$$

$$= e^{-k_1} + 1 - e^{-k_2}. \qquad \qquad \square$$

Problems for Section 11.2

Find the Neyman–Pearson most powerful test of its size in Problems 1–10. Assume that a random sample of size n is given. Simplify the test procedure as much as possible.

1. Let X have probability function $P(X = 1) = \theta = 1 - P(X = 0)$, $0 < \theta < 1$. Take $H_0 : \theta = \theta_0$ and $H_1 : \theta = \theta_1$ where $\theta_1 > \theta_0$.

2. Let X have probability function $P(X = x) = \theta(1 - \theta)^{x-1}$, $x = 1, 2, \ldots$, and zero elsewhere, $0 < \theta < 1$. Take $H_0 : \theta = \theta_0$ and $H_1 : \theta = \theta_1$ where $\theta_1 < \theta_0$.

3. Let X have density function $f(x; \theta) = 2(\theta - x)/\theta^2$, $0 < x < \theta$, and zero otherwise. Take $n = 1$, $H_0 : \theta = \theta_0$ and $H_1 : \theta = \theta_1$ $(\theta_1 > \theta_0)$.

4. Let $f(x; \theta) = \exp\{-(x - \theta)\}$, $x \geq \theta$, and zero elsewhere. Take $H_0 : \theta = \theta_0$ and $H_1 : \theta = \theta_1$ $(> \theta_0)$.

5. Let $f(x; \theta) = \theta x^{\theta - 1}$, $0 < x < 1$, and zero elsewhere. Take $n = 1$, $H_0 : \theta = 1$, and $H_1 : \theta = \theta_1$ (> 1).

6. Let $f(x; \theta) = \theta/x^2$, $0 < \theta \leq x < \infty$, and zero otherwise. Take $H_0 : \theta = \theta_0$ and $H_1 : \theta = \theta_1$ $(> \theta_0)$.

7. Let $n = 1$ and, under H_0, let X have density

$$f_0(x) = \begin{cases} 4x, & 0 < x < \tfrac{1}{2}, \\ 4 - 4x, & \tfrac{1}{2} \leq x < 1, \\ 0 & \text{elsewhere.} \end{cases}$$

Under H_1, let X have density $f_1(x) = 1$, $0 < x < 1$, and zero elsewhere.

8. Let $f(x; \theta) = (1/\sqrt{2\pi}) \exp\{-(x - \theta)^2/2\}$, $x \in \mathbb{R}$. Take $H_0 : \theta = \theta_0$ and $H_1 : \theta = \theta_1$ $(< \theta_0)$.

9. Let $f(x; \theta) = (1/(\theta\sqrt{2\pi})) \exp\{-x^2/2\theta^2\}$, $x \in \mathbb{R}$; $\theta > 0$. Take $H_0 : \theta = \theta_0$ and $H_1 : \theta = \theta_1$ $(> \theta_0)$.

10. Let $f(x; \theta) = 1/\theta, 0 < x < \theta$, and zero elsewhere. Take $H_0 : \theta = \theta_0$, and $H_1 : \theta = \theta_1$, $\theta_1 > \theta_0$.

11. (i) In Problems 4, 5, 6, 8, and 9 find the Neyman–Pearson size α test, $0 < \alpha < 1$.
    (ii) In each case in part (i) write down the power of the test.

12. Let $X_1, X_2, \ldots, X_n$ be a random sample from the uniform distribution on $(0, \theta)$. Find the Neyman–Pearson size α test of $H_0 : \theta = \theta_0$ against $H_1 : \theta = \theta_1, \theta_1 < \theta_0$.

13. Show that the Neyman–Pearson test is a function of every sufficient statistic.

11.3* COMPOSITE HYPOTHESES

We now consider briefly the case when both H_0 and H_1 are composite hypotheses. In a large class of problems where the null hypothesis is one-sided, such as $\theta \leq \theta_0$ or $\theta \geq \theta_0$, the power function is monotonic and, moreover, the critical region depends on θ_0 and the given level α but not on the alternative. In such cases it is not too difficult to use the Neyman–Pearson lemma to find the best critical region, that is, a critical region of level α that has maximum power uniformly in θ, $\theta \in \Theta_1$. We begin with some examples.

Example 1. Number of System Failures (Example 11.2.1). In Example 11.2.1 we considered the number of system failures X to be a random variable having a Poisson distribution with parameter λ. Suppose we wish to test the null hypothesis $H_0 : \lambda \leq 2$ against the alternative $H_1 : \lambda > 2$ based on a sample of size n.

Let $\lambda_1(> 2)$ be fixed and consider the problem of testing $H_0' : \lambda = 2$ against $H_1' : \lambda = \lambda_1(> 2)$. By the Neyman–Pearson lemma, proceeding exactly as in Example 11.2.1, we see that any critical region of the form

$$C = \left\{ (x_1, x_2, \ldots, x_n) : \sum_{i=1}^{n} x_i \geq k \right\}$$

is a most powerful region of its size. The power function of C is given by

$$P_\lambda(C) = \sum_{j=k}^{\infty} e^{-n\lambda} \frac{(n\lambda)^j}{j!} = 1 - \sum_{j=0}^{k-1} e^{-n\lambda} \frac{(n\lambda)^j}{j!}$$

$$= 1 - P(Y > n\lambda) = P(Y \leq n\lambda) \qquad \text{(Example 7.4.7)}$$

where Y has a gamma distribution with parameters $\alpha = k$ and $\beta = 1$. It follows that $P_\lambda(C)$ is an increasing function of λ and

$$\sup_{\lambda \leq 2} P_\lambda(C) = P_{\lambda=2}(C) = \sum_{j=k}^{\infty} e^{-2n} \frac{(2n)^j}{j!}.$$

Choose k such that C has level α; that is, choose k to be the smallest integer such that

$$\sum_{j=k}^{\infty} e^{-2n} \frac{(2n)^j}{j!} \leq \alpha, \qquad 0 < \alpha < 1.$$

Then C is a most powerful critical region of its size for testing $\lambda \leq 2$ against $\lambda = \lambda_1$.

Next we note that C depends only on α but not on λ_1 as long as $\lambda_1 > 2$. Thus, no matter what λ_1 we choose, as long as $\lambda_1 > 2$, the same critical region has maximum power against λ_1. It follows that C is uniformly most powerful for testing $H_0 : \lambda \leq 2$ against $H_1 : \lambda > 2$.   □

Example 2. Radiation Level of Microwave Ovens. Let $X_1, X_2, \ldots, X_n$ be the radiation readings in a random sample of n microwave ovens produced by a certain manufacturer. Suppose X_i has a normal distribution with mean μ and variance σ^2 (known). The recommended limit for this type of radiation exposure is μ_0 (milliroentgens per hour). Do the microwave ovens produced by this manufacturer meet the recommended limit?

We wish to test $H_0 : \mu \leq \mu_0$ against $H_1 : \mu > \mu_0$. Let $\mu_1 (> \mu_0)$ be fixed and consider the problem of testing $\mu = \mu_0$ against $\mu = \mu_1$. By the Neyman–Pearson lemma it is easy to see that a critical region of the form

$$C = \left\{ (x_1, \ldots, x_n) : \sum_{i=1}^{n} x_i \geq k \right\}$$

is a most powerful region of its size with power function

$$P_\mu(C) = P_\mu\left\{ \sum_{i=1}^{n} X_i \geq k \right\} = \int_k^\infty \frac{1}{\sigma\sqrt{2\pi n}} \exp\left\{ -\frac{(x - n\mu)^2}{2n\sigma^2} \right\} dx.$$

This follows since $\sum_{i=1}^n X_i$ has a normal distribution with mean $n\mu$ and variance $n\sigma^2$. Now

$$P_\mu(C) = \int_{(k-n\mu)/\sigma\sqrt{2n}}^\infty (1/\sqrt{\pi}) \exp(-y^2)\, dy$$

so that $P_\mu(C)$ is an increasing function of μ. Thus

$$\sup_{\mu \leq \mu_0} P_\mu(C) = P_{\mu_0}(C) = \int_{(k-n\mu_0)/\sigma\sqrt{2n}}^\infty (1/\sqrt{\pi}) \exp(-y^2)\, dy$$

which is the size of C (for testing $H_0' : \mu = \mu_0$ against $H_1' : \mu = \mu_1$). Hence the same test is a most powerful test of its size for testing $H_0 : \mu \leq \mu_0$ against $H_1' : \mu = \mu_1$. Next note that C is independent of μ_1 (as long as $\mu_1 > \mu_0$) so that no matter what $\mu_1 (> \mu_0)$ we choose C has maximum power against μ_1. Consequently, the test with critical region C is uniformly most powerful of size $P_{\mu_0}(C)$ for testing $\mu \leq \mu_0$ against $\mu > \mu_0$.   □

Examples 1 and 2 are typical special cases of a general result that holds for a large family of distributions. Rather than specify this general family and study its properties, we use the arguments used in these examples to prove the following two results.

THEOREM 1. *Suppose the Neyman–Pearson test of its size for the simple null hypothesis $H_0 : \theta = \theta_0$ against alternative $\theta = \theta_1$ does not depend on $\theta_1 \in \Theta_1$. Then this test is a uniformly most powerful test of its size for testing $H_0 : \theta = \theta_0$ against $H_1 : \theta \in \Theta_1$.*

Proof. Let C be the critical region of the Neyman–Pearson test of H_0 against $\theta = \theta_1$ of size $P_{\theta_0}((X_1, X_2, \ldots, X_n) \in C)$. Suppose C does not depend on $\theta_1 \in \Theta_1$. Then C is most powerful against every $\theta_1 \in \Theta_1$ and hence is a uniformly most powerful test of its size. $\qquad\square$

We can now extend Theorem 1 to the case when H_0 is also composite, as follows.

THEOREM 2. *Suppose we wish to test $H_0 : \theta \in \Theta_0$ against $H_1 : \theta \in \Theta_1$ where both Θ_0 and Θ_1 have at least two elements. Suppose further that for some $\theta_0 \in \Theta_0$ there is a uniformly most powerful test of its size for testing $\theta = \theta_0$ against $\theta \in \Theta_1$ of the type described in Theorem 1. Let C be the critical region of this test. Consider the test with critical region C as a test of H_0 vs. H_1. If the level of this test of H_0 is $P_{\theta_0}(C)$ (the size of C for testing $\theta = \theta_0$ against $\theta \in \Theta_1$) then C is a uniformly most powerful test of H_0 against H_1.*

Proof. By hypothesis C is a uniformly most powerful size $P_{\theta_0}(C)$ test of $\theta = \theta_0$ against $\theta \in \Theta_1$. Now any test of H_0 against H_1 is also a test of $\theta = \theta_0$ against $\theta \in \Theta_1$. Since C is a level $P_{\theta_0}(C)$ critical region of H_0 against H_1, it must be a uniformly most powerful level $P_{\theta_0}(C)$ test of H_0 versus H_1. $\qquad\square$

Remark 1. It follows from Theorem 1 that if the Neyman–Pearson test of $\theta = \theta_0$ against $\theta = \theta_1$ depends on θ_1, then there is no uniformly most powerful test of $\theta = \theta_0$ against $\theta \in \Theta_1$. In particular, this means that a uniformly most powerful test of $\theta = \theta_0$ against $\theta \neq \theta_0$ usually does not exist. For instance, in Example 1 there does not exist a uniformly most powerful test of $\lambda = 2$ vs. $\lambda \neq 2$. This is essentially because there are two distinct best tests, one for $\lambda = 2$ vs. $\lambda < 2$ and one for $\lambda = 2$ vs. $\lambda > 2$.

Example 3. Testing $\sigma^2 \geq \sigma_0^2$ vs. $\sigma^2 < \sigma_0^2$. Let $X_1, X_2, \ldots, X_n$ be a random sample from a normal population with mean 0 and variance σ^2. Suppose we wish to test $H_0 : \sigma^2 \geq \sigma_0^2$ against $H_1 : \sigma^2 < \sigma_0^2$. Let $\sigma_1 < \sigma_0$ and consider the Neyman–Pearson test of σ_0^2 vs. σ_1^2 with critical region

$$C = \left\{ (x_1, \ldots, x_n) : \sum_{i=1}^{n} x_i^2 < k \right\},$$

and size

$$P_{\sigma_0^2}(C) = P_{\sigma_0^2}\left(\sum_{i=1}^{n} X_i^2 < k \right).$$

We show that C is of level $P_{\sigma_0^2}(C)$ for testing H_0 against H_1. We note that $\sum_{i=1}^{n} X_i^2 / \sigma^2$ has a chi-square distribution with n d.f. when σ^2 is the true variance. Thus

$$P_{\sigma_0^2}(C) = P_{\sigma_0^2}\left(\sum_{i=1}^{n} \frac{X_i^2}{\sigma_0^2} < k/\sigma_0^2 \right) = P\{ Y_n < k/\sigma_0^2 \}$$

where Y_n is a chi-square random variable with n d.f. Moreover for $\sigma > \sigma_0$ we have

$$P_{\sigma^2}(C) = P\left\{ \sum_{i=1}^{n} X_i^2/\sigma^2 < k/\sigma^2 \right\}$$

$$= P\{ Y_n < k/\sigma^2 \} < P\{ Y_n < k/\sigma_0^2 \} = P_{\sigma_0^2}(C).$$

In view of Theorem 2 it follows that the test with critical region C is a size $P_{\sigma_0^2}(C)$ test of H_0 against H_1.   ☐

Problems for Section 11.3

1. (i)   A coin is tossed n times in order to test if it is loaded in favor of heads. Find a uniformly most powerful test.
    (ii) In particular, let $n = 25$ and suppose 17 heads are observed. Would you reject $H_0: p = 1/2$ in favor of $H_1: p > 1/2$ at level .10?

2. Let X be the lifetime of a light bulb manufactured by a certain company. Suppose X has density function

    $$f(x) = (1/\lambda)\exp(-x/\lambda), \qquad x > 0 \text{ and zero otherwise.}$$

    Find a uniformly most powerful test of $H_0: \lambda \leq \lambda_0$ against $H_1: \lambda > \lambda_0$.

3. The length of service of a component can be modeled by density function

    $$f(x) = (x/\beta^2)\exp(-x/\beta), \qquad x > 0 \text{ and zero elsewhere.}$$

    Find a uniformly most powerful critical region for testing $H_0: \beta \leq \beta_0$ against $\beta > \beta_0$ based on a sample of n observations.

4. The number of accidents per week in a plant may be modeled by a Poisson random variable with mean λ where $\lambda \geq 4$. A number of steps were taken to increase safety and hence reduce the average number of accidents. After the campaign, an eight-week interval produced an average of three accidents per week. Construct a uniformly most powerful test of $\lambda \geq 4$. Is there a significant improvement at level $\alpha = .05$?

5. Let $X_1, X_2, \dots, X_n$ be the number of trials necessary to obtain the first success in n independent sequences of tosses of the same coin. The object is to test if the coin is fair against the alternative that it is loaded in favor of heads. Find a uniformly most powerful test of $H_0: p \leq 1/2$ where p is the probability of a success.

6. From a group of N 18-year-old boys a sample of size n showed that x have registered for the draft. Find a uniformly most powerful test of $H_0: D \leq D_0$ against $H_1: D > D_0$ where D is the total number of boys in the group who have registered for the draft. If $N = 1000$, $n = 25$, and $D_0 = 500$, how significant is the observed value $x = 15$?

7. Let $X_1, X_2, \dots, X_n$ be a random sample from the following distributions. In each case find a uniformly most powerful test of its size for testing $H_0: \theta \leq \theta_0$ against $H_1: \theta > \theta_0$.
    (i) $f(x) = (1/\theta)x^{\theta-1}, 0 < x < 1$, and zero otherwise; $\theta > 0$.
    (ii) $f(x) = (1/\Gamma(\theta))x^{\theta-1}e^{-x}, x > 0$, and zero otherwise; $\theta > 0$.
    (iii)* $f(x) = \exp(-x + \theta), x > \theta$, and zero elsewhere.

8*. Let $X_1, X_2, \ldots, X_n$ be a random sample from a uniform distribution on $(0, \theta)$. Find a uniformly most powerful test of
  (i) $H_0: \theta \le \theta_0$ against $H_1: \theta > \theta_0$.
  (ii) $H_0: \theta \ge \theta_0$ against $H_1: \theta < \theta_0$.

11.4* LIKELIHOOD RATIO TESTS

We now consider a general procedure based on an analysis of the likelihood function, which frequently leads to good test procedures when Θ_1 has more than one point (that is, when H_1 is composite). Moreover, the resulting test statistic has some desirable large-sample properties.

Let $\Theta \subseteq \mathbb{R}_k$ and let $(X_1, X_2, \ldots, X_n)$ be a random vector with probability or density function f_θ, $\theta \in \Theta$. Consider the problem of testing the null hypothesis $H_0: (X_1, X_2, \ldots, X_n)$ has distribution f_θ, $\theta \in \Theta_0$ against the alternative $H_1: (X_1, X_2, \ldots, X_n)$ has distribution f_θ, $\theta \in \Theta_1$.

  DEFINITION 1. (LIKELIHOOD RATIO). The likelihood ratio for testing $H_0: \theta \in \Theta_0$ is defined to be the ratio

$$\lambda(x_1, x_2, \ldots, x_n) = \frac{\sup\limits_{\theta \in \Theta_0} f_\theta(x_1, x_2, \ldots, x_n)}{\sup\limits_{\theta \in \Theta} f_\theta(x_1, x_2, \ldots, x_n)}.$$

The numerator of the likelihood ratio is the best explanation of the observation $(x_1, x_2, \ldots, x_n)$ in the sense of maximum likelihood estimation that the null hypothesis provides. The denominator is the best possible explanation of the observation (under the model). Intuitively we should reject H_0 if there is a better explanation of $(x_1, x_2, \ldots, x_n)$ than the best one provided by H_0. Clearly $0 \le \lambda(x_1, x_2, \ldots, x_n) \le 1$.

  DEFINITION 2. (LIKELIHOOD RATIO TEST). The likelihood ratio test for testing $H_0: \theta \in \Theta_0$ against $H_1: \theta \in \Theta_1$ rejects H_0 if $\lambda(x_1, x_2, \ldots, x_n) < c$ for some constant c; $0 < c < 1$.

The constant c is determined from the size α, $0 < \alpha < 1$, according to the relation

$$\sup_{\theta \in \Theta_0} P_\theta \{\lambda(X_1, X_2, \ldots, X_n) < c\} = \alpha.$$

If no such c exists then we choose the largest c such that

$$P_\theta \{\lambda(X_1, X_2, \ldots, X_n) < c\} \le \alpha \qquad \text{for all } \theta \in \Theta_0.$$

  Example 1. Testing for p in Sampling from a Bernoulli Distribution. Let $X_1, X_2, \ldots, X_n$ be a random sample from a Bernoulli distribution with parameter p, $0 \le p \le 1$. We seek a level α likelihood ratio test of $H_0: p \le p_0$ against $H_1: p > p_0$.

The likelihood function is given by

$$L(p) = \prod_{j=1}^{n} P(X = x_j) = \prod_{j=1}^{n} \left[p^{x_j}(1 - p)^{1 - x_j} \right]$$

$$= p^{\sum_{j=1}^{n} x_j}(1 - p)^{n - \sum_{j=1}^{n} x_j},$$

where $x_1, x_2, \ldots, x_n \in \{0,1\}$. Let $s = \sum_{j=1}^{n} x_j$ be the number of 1's in the sample. Now

$$\sup_{0 \le p \le 1} p^s(1 - p)^{n - s} = \left(\frac{s}{n} \right)^s \left(1 - \frac{s}{n} \right)^{n - s}$$

since the maximum likelihood estimate of p is $\hat{p} = s/n$. We note that

$$L'(p) = p^{s-1}(1 - p)^{n - s - 1}(s - pn)$$

so that $L'(p) > 0$ for $p < s/n$ and $L'(p) < 0$ for $p > s/n$. Thus $L(p)$ first increases, achieves its maximum at s/n, and finally decreases. It follows (see Figure 1) that

$$\sup_{p \le p_0} L(p) = \begin{cases} p_0^s(1 - p_0)^{n - s} & \text{if } p_0 < s/n, \\ (s/n)^s (1 - s/n)^{n - s} & \text{if } p_0 \ge s/n. \end{cases}$$

Consequently,

$$\lambda(x_1, \ldots, x_n) = \lambda(s) = \begin{cases} \dfrac{p_0^s(1 - p_0)^{n - s}}{(s/n)^s (1 - s/n)^{n - s}} & \text{if } p_0 < s/n, \\ 1 & \text{if } p_0 \ge s/n. \end{cases}$$

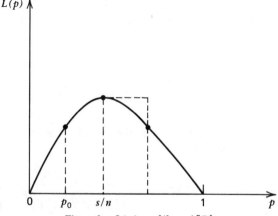

Figure 1. $L(p) = p^s(1 - p)^{n - s}$.

Since $\lambda(s) \leq 1$ for $np_0 < s$ and $\lambda(s) = 1$ for $s \leq np_0$, $\lambda(s)$ is a decreasing function of s. Thus

$$\lambda(s) < c \quad \text{if and only if} \quad s > c'$$

and the likelihood ratio test rejects H_0 if $s > c'$, that is, if the number of 1's in the sample is large. Here $c' > 0$ is some constant (integer). We note that S has a binomial distribution with parameters n and p. The power function of the likelihood ratio test, given by

$$P_p(S > c') = \sum_{j=c'+1}^{n} \binom{n}{j} p^j (1 - p)^{n-j},$$

is a nondecreasing function of p (Example 7.5.2) so that

$$\sup_{p \leq p_0} P(S > c') = \sum_{j=c'+1}^{n} \binom{n}{j} p_0^j (1 - p_0)^{n-j}.$$

If the level α, $0 < \alpha < 1$, is preassigned then we choose c' from

$$\alpha = P_{p_0}(S > c').$$

Often no such c' exists because S has a discrete distribution. In that case we choose an integer c' such that

$$P_{p_0}(S > c') \leq \alpha \quad \text{and} \quad P_{p_0}(S > c' - 1) > \alpha. \qquad \square$$

Example 2. *Student's t-test as a Likelihood Ratio Test.* Let $X_1, X_2, \ldots, X_n$ be a random sample from a normal distribution with mean μ and variance σ^2, both unknown. Suppose we wish to test $H_0 : \mu = \mu_0$ against $H_1 : \mu \neq \mu_0$. Here $\Theta_0 = \{(\mu_0, \sigma^2), \sigma^2 > 0\}$, $\Theta_1 = \{(\mu, \sigma^2) : \mu \in \mathbb{R}, \sigma^2 > 0\}$ are both composite. The likelihood function of a sample of size n is given by

$$L(\mu, \sigma^2) = \frac{1}{(\sigma\sqrt{2\pi})^n} \exp\left\{ -\frac{\sum\limits_{i=1}^{n} (x_i - \mu)^2}{2\sigma^2} \right\}.$$

Since

$$\hat{\mu} = \bar{x}, \qquad \hat{\sigma}^2 = \sum_{i=1}^{n} (x_i - \bar{x})^2 / n$$

are the unrestricted maximum likelihood estimates of μ and σ^2 (see Example 10.5.4) respectively, we have, after some simplification,

$$\sup_{(\mu, \sigma^2) \in \Theta} L(\mu, \sigma^2) = L(\hat{\mu}, \hat{\sigma}^2) = \frac{e^{-n/2}}{(2\pi/n)^{n/2} \left\{ \sum\limits_{i=1}^{n} (x_i - \bar{x})^2 \right\}^{n/2}}.$$

Under H_0, the maximum likelihood estimate of σ^2 is easily shown to be

$$\hat{\sigma}^2 = \sum_{i=1}^{n} \frac{(x_i - \mu_0)^2}{n}$$

so that (after some simplification)

$$\sup_{(\mu,\sigma^2)\in\Theta_0} L(\mu,\sigma^2) = L(\mu_0,\hat{\sigma}^2) = \frac{e^{-n/2}}{(2\pi/n)^{n/2}\left\{\sum_{i=1}^{n}(x_i - \mu_0)^2\right\}^{n/2}}.$$

Hence

$$\lambda(x_1,\ldots,x_n) = \left\{\frac{\sum_{i=1}^{n}(x_i - \bar{x})^2}{\sum_{i=1}^{n}(x_i - \mu_0)^2}\right\}^{n/2} = \left\{\frac{\sum_{i=1}^{n}(x_i - \bar{x})^2}{\sum_{i=1}^{n}(x_i - \bar{x})^2 + n(\bar{x} - \mu_0)^2}\right\}^{n/2}$$

$$= \left\{\frac{1}{1 + \left[n(\bar{x} - \mu_0)^2 \Big/ \sum_{i=1}^{n}(x_i - \bar{x})^2\right]}\right\}^{n/2}.$$

We note that $\lambda(x_1,\ldots,x_n)$ is a decreasing function of $n(\bar{x} - \mu_0)^2/\sum_{i=1}^{n}(x_i - \bar{x})^2$ or of $\sqrt{n}\,|\bar{x} - \mu_0|/\{\sum_{i=1}^{n}(x_i - \bar{x})^2\}^{1/2}$. Hence

$$\lambda(x_1,\ldots,x_n) < c \quad \text{if and only if} \quad \frac{\sqrt{n}\,|\bar{x} - \mu_0|}{\{(n-1)s^2\}^{1/2}} > c'$$

where s^2 is the sample variance. We can rewrite the likelihood ratio test as

$$\frac{\sqrt{n}\,|\bar{x} - \mu_0|}{s} > c'(n-1)^{1/2} = c''.$$

The statistic $\sqrt{n}\,(\bar{X} - \mu_0)/S$ has a t-distribution with $(n-1)$ d.f. under $H_0: \mu = \mu_0$ so that the likelihood ratio test is precisely the two-tailed t-test. Given α, we choose $c'' = t_{n-1,\alpha/2}$. $\square$

Example 3. Testing for Independence of X and Y. Let (X_1, Y_1), $(X_2, Y_2),\ldots,(X_n, Y_n)$ be a random sample from a bivariate normal distribution with $\mathscr{E}X_1 = \mu_1$, $\mathscr{E}Y_1 = \mu_2$, $\mathrm{var}(X_1) = \sigma_1^2$, $\mathrm{var}(Y_1) = \sigma_2^2$, and $\rho(X_1, Y_1) = \rho$. We recall that X and Y are independent if and only if $\rho = 0$. Hence a test of independence of X and Y can be based on ρ. More precisely, we wish to test $H_0: \rho = 0$ against $H_1: \rho \neq 0$, say. Then

$$\Theta_0 = \{(\mu_1,\mu_2,\sigma_1,\sigma_2,0): \mu_i \in \mathbb{R}, \sigma_i > 0, i = 1,2\}$$

and

$$\Theta = \left\{ (\mu_1, \mu_2, \sigma_1, \sigma_2, \rho) : \mu_i \in \mathbb{R}, \sigma_i > 0, i = 1, 2, |\rho| < 1 \right\}.$$

The likelihood function is given by

$$L(\theta) = \frac{1}{\left[2\pi\sigma_1\sigma_2 (1 - \rho^2)^{1/2} \right]^n} \exp\left\{ -\frac{1}{2(1 - \rho^2)} \sum_{i=1}^{n} \left[\left(\frac{x_i - \mu_1}{\sigma_1} \right)^2 - 2\rho \left(\frac{x_i - \mu_1}{\sigma_1} \right) \right. \right.$$

$$\left. \left. \times \left(\frac{y_i - \mu_2}{\sigma_2} \right) + \left(\frac{y_i - \mu_2}{\sigma_2} \right)^2 \right] \right\},$$

where $\theta = (\mu_1, \mu_2, \sigma_1, \sigma_2, \rho)$. Under H_0, $\rho = 0$ and the maximum likelihood estimates of $\mu_1, \mu_2, \sigma_1^2, \sigma_2^2$ are given respectively by

$$\bar{x}, \bar{y}, \hat{\sigma}_1^2 = \sum_{i=1}^{n} (x_i - \bar{x})^2 / n, \hat{\sigma}_2^2 = \sum_{i=1}^{n} (y_i - \bar{y})^2 / n.$$

It follows that

$$\sup_{\theta \in \Theta_0} L(\theta) = \frac{1}{(2\pi\hat{\sigma}_1\hat{\sigma}_2)^n} \exp(-n).$$

Under Θ, the (unrestricted) maximum likelihood estimates of $\mu_1, \mu_2, \sigma_1, \sigma_2$ are the same as under Θ_0, whereas the maximum likelihood estimate of ρ is the sample correlation coefficient $\hat{\rho} = r$ given (Problem 10.5.13) by

$$r = \frac{1}{n\hat{\sigma}_1\hat{\sigma}_2} \sum_{i=1}^{n} (x_i - \bar{x})(y_i - \bar{y})$$

$$= \frac{\displaystyle\sum_{i=1}^{n} (x_i - \bar{x})(y_i - \bar{y})}{\sqrt{\displaystyle\sum_{i=1}^{n} (x_i - \bar{x})^2 \sum_{i=1}^{n} (y_i - \bar{y})^2}}.$$

It follows that

$$\sup_{\theta \in \Theta} L(\theta) = \frac{1}{(2\pi\hat{\sigma}_1\hat{\sigma}_2)^n (1 - r^2)^{n/2}} \exp(-n).$$

The likelihood ratio test therefore rejects H_0 if

$$(1 - r^2)^{n/2} < c$$

or, equivalently, if

$$|r| > c'.$$

The exact distribution of the statistic R under H_0 is known but we will not derive it here. It should be mentioned that the statistic T defined by

$$T = \frac{R}{\sqrt{1 - R^2}} \sqrt{n - 2}$$

has, under H_0, Student's t-distribution with $n - 2$ d.f. This fact may be used to compute c' for a given α, or the P-value associated with an observed value of R.   □

Remark 1. *(Relation Between Likelihood Ratio Test and Neyman–Pearson Test).* Does the likelihood ratio test of a simple null hypothesis against a simple alternative hypothesis coincide with the Neyman–Pearson test? The following result holds. Suppose the Neyman–Pearson and the likelihood ratio tests exist for a given size α. Then they are equivalent. We will not prove this result here.

Remark 2. *(Large Sample Distribution of λ).* In the examples considered above, the test statistic turns out to have a well known distribution under H_0. This, however, is not always the case. How then do we compute the P-value or perform the test? The main advantage of the likelihood ratio test is that under certain regularity conditions on the underlying distribution, the statistic $-2 \ln \lambda(X_1, X_2, \ldots, X_n)$ has, for large n, approximately a chi-square distribution under H_0. The degrees of freedom equal the number of unknown parameters under Θ minus the number of unknown parameters under Θ_0. In Examples 1, 2, and 3 the degree of freedom is 1. In Example 2,

$$-2 \ln \lambda = n \ln \left\{ 1 + \frac{n(\overline{X} - \mu_0)^2}{\sum_{i=1}^{n} (X_i - \overline{X})^2} \right\}$$

has, for large n, approximately a chi-square distribution with 1 d.f. Hence the likelihood ratio test rejects H_0 if

$$\lambda < c, \quad \text{or if } -2 \ln \lambda > -2 \ln c = c_1,$$

or if

$$n \ln \left\{ 1 + n \frac{(\overline{X} - \mu_0)^2}{\sum_{i=1}^{n} (X_i - \overline{X})^2} \right\} > c_1.$$

The approximate P-value is therefore given by

$$p \doteq P\left(n \ln\left\{ 1 + n\frac{(\bar{X} - \mu_0)^2}{(n-1)S^2} \right\} > c_1 \right)$$

which is computed from $\chi^2(1)$ tables. However, for large n, a t-statistic T is approximately normal so that T^2 is approximately $\chi^2(1)$. Since

$$\frac{n(\bar{X} - \mu_0)^2}{(n-1)S^2} = \frac{T^2}{n-1},$$

we have

$$p \doteq P\left(n \ln\left(1 + \frac{T^2}{n-1} \right) > c_1 \right)$$

so that the P-value is (approximately) the same whether we compute it under $\chi^2(1)$ or under t-distribution.

Problems for Section 11.4

1. Let $X_1, X_2, \ldots, X_k$ be a sample from probability function

   $$P_N(X = j) = 1/N, \qquad j = 1, 2, \ldots, N, N \geq 1 \text{ is an integer, and zero otherwise.}$$

   (i) Find the likelihood ratio test of $H_0 : N \leq N_0$ against $H_1 : N > N_0$.
   (ii) Find the likelihood ratio test of $H_0 : N = N_0$ against $H_1 : N \neq N_0$.

2. Find the likelihood ratio test of $\theta = \theta_0$ against $\theta \neq \theta_0$ based on a sample of size 1 from density function

   $$f(x) = 2\frac{(\theta - x)}{\theta^2}, \qquad 0 < x < \theta, \text{ and zero otherwise.}$$

3. Let $X_1, X_2, \ldots, X_n$ be a sample from density function

   $$f(x) = \frac{1}{\theta}\exp(-x/\theta), \qquad x > 0 \text{ and zero elsewhere.}$$

   Find the likelihood ratio tests of
   (i) $H_0 : \theta \leq \theta_0$ against $H_1 : \theta > \theta_0$,   (ii) $H_0 : \theta = \theta_0$ against $H_1 : \theta \neq \theta_0$.

4. Let $X_1, X_2, \ldots, X_n$ be a random sample from a Poisson distribution with parameter $\lambda > 0$. Find the likelihood ratio tests of
   (i) $H_0 : \lambda = \lambda_0$ against $H_1 : \lambda \neq \lambda_0$.
   (ii) $H_0 : |\lambda - \lambda_0| \leq a$ against $H_1 : |\lambda - \lambda_0| > a$ where a is a known constant.
   (iii) $H_0 : \lambda \geq \lambda_0$ against $H_1 : \lambda < \lambda_0$.

5. Find the likelihood ratio test of $H_0: p = p_0$ against $p \neq p_0$ based on a sample of size 1 from probability function

$$P(X = k) = \binom{n}{k} p^k (1-p)^{n-k}, \qquad k = 0, 1, 2, \ldots, n,$$

and zero otherwise, $0 \leq p \leq 1$.

6. Let $X_1, X_2, \ldots, X_n$ be a sample from a normal distribution with unknown parameters μ and σ^2. Find the likelihood ratio tests of
   (i) $H_0: \sigma = \sigma_0$ against $H_1: \sigma \neq \sigma_0$,
   (ii) $H_0: \sigma \leq \sigma_0$ against $H_1: \sigma > \sigma_0$.

7*. Let $X_1, X_2, \ldots, X_m$ and $Y_1, Y_2, \ldots, Y_n$ be independent random samples from normal distributions with means μ_1 and μ_2, and variances σ_1^2 and σ_2^2 respectively. Find the likelihood ratio tests of
   (i) $H_0: \sigma_1 = \sigma_2$ against $H_1: \sigma_1 \neq \sigma_2$ (μ_1, μ_2 both unknown).
   (ii) $H_0: \mu_1 = \mu_2$ against $H_1: \mu_1 \neq \mu_2$ (σ_1, σ_2 unknown but equal).

8. Let $X_1, X_2, \ldots, X_m$ and $Y_1, Y_2, \ldots, Y_n$ be independent random samples. Suppose the X's have common density

$$f_1(x) = (1/\lambda_1) e^{-x/\lambda_1}, \qquad x > 0, \text{ and zero elsewhere,}$$

and the Y's have common density

$$f_2(y) = (1/\lambda_2) e^{-y/\lambda_2}, \qquad y > 0, \text{ and zero elsewhere.}$$

Find the likelihood ratio test of $H_0: \lambda_1 = \lambda_2$ against $H_1: \lambda_1 \neq \lambda_2$.

9. Let $(X_1, X_2, \ldots, X_{k-1})$ be one observation from multinomial distribution

$$P(X_1 = x_1, X_2 = x_2, \ldots, X_{k-1} = x_{k-1}) = \frac{n!}{x_1! x_2! \ldots x_k!} \prod_{j=1}^{k} p_j^{x_j},$$

where $x_k = n - \sum_{i=1}^{k-1} x_i$, $0 \leq x_i \leq n$, $0 \leq p_i \leq 1$, $\sum_{i=1}^{k} p_i = 1$. Find the likelihood ratio test of $H_0: p_1 = p_2 = \cdots = p_k = 1/k$ against H_1 that at least one inequality holds in $p_1 = p_2 = \cdots = p_k$.

10*. Let $X_{11}, X_{12}, \ldots, X_{1n_1}$, $X_{21}, X_{22}, \ldots, X_{2n_2}$, and $X_{31}, X_{32}, \ldots, X_{3n_3}$ be independent random samples from normal distributions with unknown means μ_1, μ_2, μ_3 and unknown variances $\sigma_1^2, \sigma_2^2, \sigma_3^2$ respectively. Find the likelihood ratio tests of
   (i) $H_0: \sigma_1 = \sigma_2 = \sigma_3$ against the alternative that there is at least one inequality.
   (ii) $H_0: \mu_1 = \mu_2 = \mu_3$ against the alternative that there is at least one inequality, assuming that $\sigma_1 = \sigma_2 = \sigma_3$ but the common value is unknown.

11. Show that the likelihood ratio test statistic is a function of every sufficient statistic.

11.5 THE WILCOXON SIGNED RANK TEST

In Section 6.3.3 we discussed the sign test as a test of median or for comparing two populations based on paired observations. We recall that the sign test is rather crude in that it uses only the sign of the difference between each observation and the postulated value of the median but ignores the magnitude of these differences. Often the magnitude of these differences are also available. The Wilcoxon signed rank test that we now study uses both the sign and the magnitude of the differences.

Let $X_1, X_2, \ldots, X_n$ be a random sample from a continuous symmetric distribution. Let $\tilde{\mu} = \text{med}(X)$ and suppose we wish to test the null hypothesis $H_0 : \tilde{\mu} = \tilde{\mu}_0$ against the alternative $H_1 : \tilde{\mu} > \tilde{\mu}_0$. Consider the differences $X_1 - \tilde{\mu}_0, X_2 - \tilde{\mu}_0, \ldots, X_n - \tilde{\mu}_0$. Clearly, under H_0, $X_i - \tilde{\mu}_0$ has a continuous symmetric distribution about zero. Consequently, under H_0, we expect negative and positive differences to be about evenly spread out. That is, the expected number of negative differences is $n/2$ and, moreover, negative and positive differences of equal absolute magnitude should occur with equal probability.

For convenience, set $D_i = X_i - \tilde{\mu}_0$. Since the X_i's have a continuous distribution $P(D_i = 0) = 0$ and we assume that $|D_i| > 0$ for $i = 1, 2, \ldots, n$ and $D_i \neq D_j$ for $i \neq j$. Consider the absolute values $|D_1|, |D_2|, \ldots, |D_n|$ and rank them from 1 (for the smallest) to n (for the largest). Let W_+ be the sum of ranks assigned to those D_i's that are positive and W_- be the sum of ranks assigned to D_i's that are negative. Clearly,

$$(1) \qquad W_+ + W_- = \sum_{k=1}^{n} k = \frac{n(n+1)}{2}$$

so that W_+ and W_- offer equivalent test statistics. Under H_0 we expect W_+ and W_- to be the same. Since a large of value of W_+ indicates that most of the larger ranks are assigned to positive D_i's, it follows that large values of W_+ support $H_1 : \tilde{\mu} > \tilde{\mu}_0$. In view of relation (1) it is sufficient to compute the smaller of W_+ and W_- and then compute the larger value by differencing. The same argument applies to the alternative $H_1 : \tilde{\mu} < \tilde{\mu}_0$ or $H_1 : \tilde{\mu} \neq \tilde{\mu}_0$. Let w_0 be the observed value of W_+ (or W_-).

The Wilcoxon Signed Rank Test

Null, H_0	Alternative, H_1	Reject H_0 if	P-value
$\tilde{\mu} = \tilde{\mu}_0$	$\tilde{\mu} > \tilde{\mu}_0$	W_+ is large	$P_{H_0}(W_+ \geq w_0)$
$\tilde{\mu} = \tilde{\mu}_0$	$\tilde{\mu} < \tilde{\mu}_0$	W_- is large	$P_{H_0}(W_- \geq w_0)$
$\tilde{\mu} = \tilde{\mu}_0$	$\tilde{\mu} \neq \tilde{\mu}_0$	W_+ or W_- is large	2 (smaller tail probability)

Even though we assume that the distribution is continuous, ties and zeros do occur in practice. As in the case of a sign test, the recommended procedure is to drop the zeros and use the reduced sample size. If there are ties, that is, if some of the D_i's are equal, assign to each such D_i a rank equal to the simple arithmetic average of the ranks these D_i's would have received if they were not equal. Thus the ordered observations 2, 2, 5, 5, 7, 8, 8, 8, 10, 11 are assigned, respectively, the ranks 1.5, 1.5, 3.5, 3.5, 5, 7, 7, 7, 9, 10.

Table A11 in the Appendix gives the cumulative right tail probabilities $P_{H_0}(W \geq w_0)$ for values of $w_0 \geq n(n + 1)/4$ and $n = 2, 3, \ldots, 15$. (Here W is interpreted as either W_+ or W_-.) For larger values of n we use the normal approximation. Under H_0, the common distribution of W_+ and W_- is symmetric about the mean

$$\mathscr{E}W_+ = n(n + 1)/4$$

with variance

$$\text{var}(W_+) = n(n + 1)(2n + 1)/24$$

and the standardized W_+ has approximately a standard normal distribution. Thus

$$P_{H_0}(W_+ \geq w_0) = P_{H_0}(W_+ > w_0 - 1/2) \qquad \text{(continuity correction)}$$

$$= P_{H_0}\left(\frac{W_+ - n(n + 1)/4}{\sqrt{n(n + 1)(2n + 1)/24}} > \frac{w_0 - .5 - n(n + 1)/4}{\sqrt{n(n + 1)(2n + 1)/24}} \right)$$

$$\doteq P\left(Z > \frac{w_0 - .5 - n(n + 1)/4}{\sqrt{n(n + 1)(2n + 1)/24}} \right).$$

Example 1. Pulse Rate of Infants Aged One Month or Over. A random sample of 15 infants one month or older shows the following pulse rates (beats per minute):

119, 120, 125, 122, 118, 117, 126, 114, 115, 123, 121, 120, 124, 127, 126.

Assuming that the distribution of pulse rates is symmetric and continuous, is there evidence to suggest that the median pulse rate of one month or older infants is different from 120 beats per minute?

We wish to test $H_0: \tilde{\mu} = 120$ against $H_1: \tilde{\mu} \neq 120$. Let us compute $D_i = X_i - \tilde{\mu}_0 = X_i - 120$, $|D_i|$ and rank of $|D_i|$.

$$\begin{array}{lrrrrrrrrrrrrrrr}
D_i: & -1, & 0, & 5, & 2, & -2, & -3, & 6, & -6, & -5, & 3, & 1, & 0, & 4, & 7, & 6 \\
|D_i|: & 1, & 0, & 5, & 2, & 2, & 3, & 6, & 6, & 5, & 3, & 1, & 0, & 4, & 7, & 6
\end{array}$$

Excluding the two 0's we arrange the observations in increasing order as

1, 1, 2, 2, 3, 3, 4, 5, 5, 6, 6, 6, 7

with ranks

$$1.5, \quad 1.5, \quad 3.5, \quad 3.5, \quad 5.5, \quad 5.5, \quad 7, \quad 8.5, \quad 8.5, \quad 11, \quad 11, \quad 11, \quad 13.$$

Hence

$$W_- = 1.5 + 3.5 + 5.5 + 8.5 + 11 = 30$$

so that

$$W_+ = 13(13 + 1)/2 - 30 = 61 = w_0.$$

For the reduced sample size $n = 13$, $P_{H_0}(W_+ \geq w_0) = .153$ so that the associated P-value is $2(.153) = .306$. Thus, under H_0 the chance of observing as large a value of W_+ as 61 is 30.6 percent and we can hardly reject H_0.

Let us analyze the data using the sign test. The number of negative signs is 5 so that the P-value under $H_0 : p = 1/2$ is

$$2P(S \leq 5) = 2(.295) = .59.$$

The sign test is more conservative.

If we assume that the distribution of pulse rates is normal, then we can test $H_0 : \mu = 120$ against $H_1 : \mu \neq 120$ using a two-tailed t-test. We have

$$\bar{x} = 1817/15 = 121.13,$$

$$s^2 = \frac{\sum_{i=1}^{n} x_i^2 - n\bar{x}^2}{n - 1} = \frac{220331 - 220099.3}{14} = 16.55, \qquad s = 4.07,$$

so that

$$t_0 = \frac{121.13 - 120}{4.07} \sqrt{15} = 1.08.$$

For 14 d.f.

$$P(T \geq 1.08) \doteq .15$$

so that the P-value is about .30, which is approximately the same as that obtained by the Wilcoxon signed rank test. In general, the sign test performs worse than the t- and signed rank tests, at least for large samples. For large samples, the signed-rank test performs relatively well in comparison with the t-test (even when the underlying distribution is normal). For small samples from nonnormal populations, either the sign test or signed rank test may be more powerful than the t-test. The sign-ranked test should be preferable to the sign test in situations when the underlying distribution is symmetric. $\square$

Remark 1. The Wilcoxon signed rank test can also be used as a test for symmetry. Suppose $x_1, x_2, \ldots, x_n$ is a random sample from a continuous population with median $\tilde{\mu}$. Then we can test either

$$H_0 : \text{distribution is symmetric about } \tilde{\mu} \text{ (known)}$$

or

$$H_0' : \text{distribution is symmetric and } \tilde{\mu} = \tilde{\mu}_0$$

by using W_+ or W_- exactly as described above.

Remark 2. The same test can be used for paired data $(X_1, Y_1), (X_2, Y_2), \ldots,$ (X_n, Y_n) to test $H_0 : \text{median}(X_i - Y_i) = \tilde{\mu}_0$ against one-sided or two-sided alternatives. In this case we assume that $X_i - Y_i$ have a continuous and symmetric distribution about their median $\tilde{\mu}$. The test is performed exactly as above by taking $D_i = |X_i - Y_i - \tilde{\mu}_0|$. The case of most interest is when $\tilde{\mu}_0 = 0$.

Example 2. Effect of Smoking on Pulse Rate. In order to determine if smoking results in increased heart activity a random sample of 20 people was taken. Their pulse rates (beats per minute) before smoking and after smoking a certain brand of cigarette are measured with the following results:

	Subject									
	1	2	3	4	5	6	7	8	9	10
Pulse rate before	70	69	72	74	66	68	69	70	71	69
Pulse rate after	69	72	71	74	68	67	72	72	72	70
	11	12	13	14	15	16	17	18	19	20
Pulse rate before	73	72	68	72	67	70	68	69	70	71
Pulse rate after	75	73	71	72	69	71	72	70	71	71

We wish to test $H_0 : \tilde{\mu}_D = 0$ against $H_1 : \tilde{\mu}_D < 0$ where $D_i = X_i - Y_i$, $i = 1, 2, \ldots, n$. Assuming that D_i have a continuous symmetric distribution we can test H_0 by using Wilcoxon signed rank test. Ignoring the three observations where $x_i = y_i$, and arranging $|d_i|$ in increasing order of magnitude, we have

d_i	1	-3	1	-2	1	-3	-2	-1	-1	-2	-1
$|d_i|$	1	3	1	2	1	3	2	1	1	2	1
Rank of $|d_i|$	5	15	5	11.5	5	15	11.5	5	5	11.5	5
d_i	-3	-2	-1	-4	-1	-1					
$|d_i|$	3	2	1	4	1	1					
Rank of $|d_i|$	15	11.5	5	17	5	5					

Thus

$$w_+ = 5 + 5 + 5 = 15$$

so that

$$w_- = 17\frac{(17 + 1)}{2} - 15 = 153 - 15 = 138.$$

Since $n > 15$, we use normal approximation. We have

$$\mathscr{E}_{H_0}W_- = \frac{17(17 + 1)}{4} = 76.5, \quad \text{var}_{H_0}(W_-) = \frac{17(18)(35)}{24} = 446.25.$$

Hence

$$P_{H_0}(W_- \geq 138) = P_{H_0}(W_- > 138 - .5) \doteq P\left(Z > \frac{137.5 - 76.5}{\sqrt{446.25}}\right)$$

$$= P(Z > 2.89) = .0019$$

so that $w_- = 138$ is highly significant and it is reasonable to conclude that smoking does increase the pulse rate.

For comparative purposes let us also apply the sign test. The number of negative signs S is 14. Under $H_0, p = 1/2$ and under $H_1 : p > 1/2$ where $p = P(X_i < Y_i)$. Thus the associated P-value is

$$P_{H_0}(S \geq 14) = 1 - .9936 = .0064$$

and we reject H_0 at level $\alpha \geq .0064$. We leave the reader to show that the t-test also results in rejection of H_0 with P-value $< .005$.   □

Problems for Section 11.5

1. A certain university's brochure claims that the average amount of money needed for boarding and lodging in the town for a single student is $75 per week. A random sample of nine single students from this university showed the following weekly expenditures:

$$75, \quad 92, \quad 80, \quad 84, \quad 73, \quad 60, \quad 84, \quad 91, \quad 78.$$

Is there evidence to suggest that the university's estimate is not correct? Assuming that the weekly expenditures are normally distributed, also apply the t-test to analyze the same data.

2. Assume that the average birth weight of infants is known to be 7 pounds. The birth weights of a random sample of 11 children born to migrant workers were recorded (in pounds) as follows:

$$4.8, \quad 6.0, \quad 5.4, \quad 7.6, \quad 6.1, \quad 6.9, \quad 7.2, \quad 7.0, \quad 5.8, \quad 6.8, \quad 6.4.$$

Is it reasonable to conclude that the average birth weight of children of migrant workers is smaller than the national average? Analyze the same data by performing the *t*-test and the sign test.

3. In order to keep track of inflation, a Network News Program visits 10 selected supermarkets on the first Monday of every month and buys a typical market basket of 30 preselected items. The total cost data for November 1, 1982, and November 7, 1983, were as follows:

Market Basket Costs
(dollars)

Date	Supermarket				
	1	2	3	4	5
November 1, 1982	50.16	52.84	49.38	49.85	55.47
November 7, 1983	53.18	56.94	52.42	49.82	54.86
	6	7	88	9	10
November 1, 1982	53.51	51.49	49.20	53.10	50.70
November 7, 1983	56.42	51.00	52.10	55.80	54.75

Is there evidence to suggest that the average market cost remained the same over the 12-month period? Analyze the same data by using a *t*-test.

4. In order to determine if children watch more TV in preteen years, a random sample of 20 children aged nine, ten, or eleven was selected and their daily average TV viewing times were recorded. The same children were then canvassed four years later and their daily average viewing time was recorded.

Children's Daily TV Viewing Time
(hours)

Age	Subject									
	1	2	3	4	5	6	7	8	9	10
Preteen	3.5	2.8	4.6	3.7	3.6	4.2	2.2	1.6	3.6	5.0
Teen	4.2	2.2	5.2	2.1	0.5	5.4	2.2	1.0	2.8	4.6
	11	12	13	14	15	16	17	18	19	20
Preteen	3.0	4.8	1.5	2.5	3.2	3.4	1.2	0.5	1.8	3.5
Teen	4.0	2.2	1.3	2.5	3.0	2.6	2.6	2.3	0.5	2.7

How strong is the evidence that the TV watching habits in preteen and teen years is the same?

5. A group of 10 students is given a task to perform while sober. Each student is then given three beers over a one-hour period and asked to repeat the task. The results are

as follows:

Task Performance Time (minutes)

Condition	Subject									
	1	2	3	4	5	6	7	8	9	10
Sober	5.0	3.5	6.5	8.5	4.5	5.0	2.5	3.0	4.5	3.5
After drinking	8.5	4.0	6.0	8.0	6.0	5.0	4.0	5.5	4.0	6.0

How strong is the evidence that alcohol dulls mental ability?

6. In order to compare the effectiveness of two sunburn lotions, a random sample of seven subjects is selected. Lotion A is applied to the left side of their faces and Lotion B to the right side. After the subjects have sat in the sun watching a three-hour tennis match, the degree of sunburn is measured on a scale.

Lotion	Subject						
	1	2	3	4	5	6	7
A	48	62	42	69	74	35	84
B	46	49	48	63	43	32	53

Do the data support the claim that the two lotions are equally effective?

7. The following observations were taken from a table of random numbers from a distribution F with median 0.

0.464	0.137	2.455	-0.323	-0.068
0.906	-0.513	-0.525	0.595	0.881
-0.482	1.678	-0.057	-1.229	-0.486
-1.787	-0.261	1.237	1.046	

Is it reasonable to conclude that F is a symmetric distribution?

8. The median time for a tune-up of a four-cylinder car is known to be 45 minutes. A random sample of 12 mechanics showed tune-up times, in minutes, of

$$32, \quad 56, \quad 48, \quad 45, \quad 47, \quad 57, \quad 38, \quad 33, \quad 32, \quad 35, \quad 40, \quad 50.$$

What conclusion do you draw from the sample?

9. Consider the density function

$$g(y) = (\theta - y)/\theta^2 \quad \text{for} \quad 0 < y < \theta, \; = (y - \theta)/\theta^2 \quad \text{for} \quad \theta \le y < 2\theta,$$

and 0 elsewhere.
   (i) Graph g.
   (ii) Show that h defined by $h(X) = \theta(1 - \sqrt{1 - 2X})$ if $X < 1/2$, $= \theta(1 + \sqrt{2X - 1})$ if $X > 1/2$, where X is the uniform distribution on $(0,1)$ has density $g(y)$.

(iii) Use part (ii) to generate a sample of size 20 from density g with $\theta = 1$ given a random sample of size 20 from a uniform distribution on $(0,1)$:

.277	.435	.130	.143	.853	.889	.294	.697	.940	.648
.324	.482	.540	.152	.477	.667	.741	.882	.885	.740

(iv) Use the sample in part (iii) to test H_0: median of $g = \zeta_{.5}(g) = 1$ against $H_1: \zeta_{.5}(g) \neq 1$. Compare the performance of sign, Wilcoxon signed rank, and t-tests.

11.6 SOME TWO-SAMPLE TESTS

The sign and the Wilcoxon signed rank tests are one-sample tests that are the nonparametric analogs of the one-sample t-test based on normal theory. We now consider some nonparametric analogs of the two-sample t-test. Recall that if $X_1, X_2, \ldots, X_m$ and $Y_1, Y_2, \ldots, Y_n$ are independent random samples from two normal populations with common (unknown) variance, then the t-statistic defined by

$$T = \frac{\bar{X} - \bar{Y}}{S_p \sqrt{(1/m) + (1/n)}}$$

where

$$\bar{X} = \sum_{i=1}^{m} X_i / m, \qquad \bar{Y} = \sum_{j=1}^{n} Y_j / n$$

and

$$S_p^2 = \frac{\sum_{i=1}^{m} (X_i - \bar{X})^2 + \sum_{j=1}^{n} (Y_j - \bar{Y})^2}{m + n - 2}$$

is used to test the null hypothesis $H_0 : \mathscr{E}X = \mathscr{E}Y$. Under H_0, both the X's and Y's have the same distribution.

We now consider the case when the normality assumption is dropped. Let $X_1, X_2, \ldots, X_m$ and $Y_1, Y_2, \ldots, Y_n$ be independent random samples from continuous distribution functions F and G respectively. How do we test the hypothesis that X's and Y's have the same distribution? The hypothesis of interest is $H_0 : F(x) = G(x)$ for all $x \in \mathbb{R}$ to be tested against the one-sided or two-sided alternatives $F(x) \geq G(x)$ or $F(x) \leq G(x)$ or $F(x) \neq G(x)$. [Here $F(x) \geq G(x)$ means that $F(x) > G(x)$ for at least one x, and so on.]

First note that

$$F(x) \geq G(x) \quad \text{if and only if} \quad P(X \leq x) \geq P(Y \leq x)$$

so that

$$F(x) \geq G(x) \quad \text{if and only if} \quad P(X > x) \leq P(Y > x).$$

Thus $F(x) \geq G(x)$ means that the Y's tend to be larger than the X's—and we say that Y is *stochastically larger* than X. Similarly $F(x) \leq G(x)$ if and only if $P(X > x) \geq P(Y > x)$ so that X is stochastically larger than Y.

The simplest test of $H_0: F(x) = G(x)$ for all x is the median test, which was studied in Section 6.4. We recall that the median test is simply a test of the equality of medians $\tilde{\mu}_X$ and $\tilde{\mu}_Y$ of X and Y respectively. It is based on the test statistic U, the number of X's that are smaller than the sample median of the combined sample. Yet another test for $H_0: F(x) = G(x)$ for all x against the two-sided alternatives is the chi-square test of goodness of fit for multinomial distributions studied in Section 9.9. (See also Section 12.2.)

We now consider two more commonly used tests of the equality of two distributions. In view of the continuity assumption we note that $P(X_i = X_j) = 0$ for $i \neq j$ and $P(X_i = Y_j) = 0$ for all i, j.

11.6.1 The Mann–Whitney–Wilcoxon Rank Sum Test

Under H_0, the combined sample of $m + n$ observations $\{X_1, X_2, \ldots, X_m, Y_1, Y_2, \ldots, Y_n\}$ is a random sample of size $m + n$ from a single population. Let us order the combined sample from smallest (rank 1) to largest (rank $m + n$) and let

$$R(X_i) = \text{rank of } X_i \text{ in the combined sample}, \quad i = 1, 2, \ldots, m.$$

Then $1 \leq R(X_i) \leq m + n$ for $i = 1, 2, \ldots, m$. Under H_0 we expect the m ranks of the X's, namely, $R(X_1), \ldots, R(X_m)$ to be randomly spread out among the ranks $1, 2, \ldots, m + n$. The test is based on the rank sum statistic

$$T_X = \sum_{i=1}^{m} R(X_i).$$

Under H_0 we expect T_X to be proportional to the sample size m. A very large value of T_X indicates that the X's tends to be larger than the Y's so that a large value of T_X supports the alternative $H_1: F(x) \leq G(x)$. Similarly, a very small value of T_X supports the alternative $H_1: F(x) \geq G(x)$.

We note that

$$T_X + T_Y = \sum_{i=1}^{m} R(X_i) + \sum_{j=1}^{n} R(Y_j) = \frac{(m+n)(m+n+1)}{2}$$

so that T_X and T_Y offer equivalent test statistics. Although ties are theoretically impossible, they do occur in practice. In case of ties we assign average rank to the tied observations (as in Section 11.5). In the following test, t_x is the observed value of T_X.

Mann–Whitney–Wilcoxon Test for Testing $H_0: F = G$

Let $R(X_i)$ be the rank of X_i in the combined ordered sample, and $T_X = \sum_{i=1}^{m} R(X_i)$.

Alternative, H_1	Reject H_0 if	P-value
$F(x) \geq G(x)$	$T_X \leq c_1$ (T_X is too small)	$P_{H_0}(T_X \leq t_x)$
$F(x) \leq G(x)$	$T_X \geq c_2$ (T_X is too large)	$P_{H_0}(T_X \geq t_x)$
$F(x) \neq G(x)$	$T_X \geq c_3$ or $T_X \leq c_4$	2 (smaller tail
	(T_X is either too large	probability)
	or too small)	

Often the test is stated in terms of the statistic U_X where

$$U_X = \sum_{i=1}^{m} \left(\text{number of } Y_j\text{'s} < X_i \right).$$

Since, however,

$$R(X_i) = \left(\text{number of } Y_j\text{'s} < X_i \right) + \text{Rank of } X_i \text{ in the } X\text{'s}$$

we have

$$T_X = \sum_{i=1}^{m} R(X_i) = U_X + \sum_{i=1}^{m} i = U_X + \frac{m(m+1)}{2}.$$

It follows that

$$U_X = T_X - \frac{m(m+1)}{2}$$

and U_X and T_X are equivalent test statistics.

In order to compute the P-value (or to find the critical region if α is given), one needs the null distribution of T_X. Table A10 in the Appendix gives the cumulative probability in the tail of T_X. Since labeling of X and Y can be interchanged, we assume that the sample with fewer observations is the X sample so that $m \leq n$. Next note that the least value of T_X corresponds to the ranks $1, 2, \ldots, m$ and the maximum value corresponds to ranks $n + 1, n + 2, \ldots, n + m$. It follows that

$$\frac{m(m+1)}{2} \leq T_X \leq \sum_{k=1}^{m} (n+k) = \frac{m(m+2n+1)}{2}.$$

Finally, under H_0, T_X has a symmetric distribution about its mean $m(m + n + 1)/2$. It follows that a left tail cumulative probability equals the corresponding right tail cumulative probability. Table A10 gives the right tail probabilities for $T_X \geq m(m + n + 1)/2$ for values of $m \leq n \leq 10$. Left tail probabilities are obtained by symmetry. Indeed, if $t_x \leq m(m + n + 1)/2$, then

$$P_{H_0}(T_X \leq t_x) = P_{H_0}(T_X \geq m(m + n + 1) - t_x).$$

Suppose, for example, that $m = 5$, $n = 7$, and $t_x = 18$. Then $t_x = 18 < 5(5 + 7 + 1)/2$ so that

$$P_{H_0}(T_X \leq 18) = P_{H_0}(T_X \geq 5(5 + 7 + 1) - 18) = P_{H_0}(T_X \geq 47) = .009.$$

For larger values of m or n we use the fact that

$$\mathscr{E}_{H_0}(T_X) = \frac{m(m + n + 1)}{2}, \qquad \mathrm{var}_{H_0}(T_X) = \frac{mn(m + n + 1)}{12}$$

and

$$Z = \frac{\left(T_X - \mathscr{E}_{H_0}(T_X)\right)}{\sqrt{\mathrm{var}_{H_0}(T_X)}}$$

has approximately a standard normal distribution. Since T_X is integer valued, the application of continuity correction (Section 9.4) results in an improved approximation.

Example 1. Comparing Two Methods of Instruction. Seventeen students were randomly selected to participate in an educational research project. A group of eight students was asked to attend a traditional lecture course for four weeks. The remaining nine students were provided self-instructional material on video cassettes. At the end of four weeks all the students took the same test with the following results.

Lecture:	75,	82,	28,	82,	94,	78,	76,	64	
Self-Instruction:	78,	95,	63,	37,	48,	74,	65,	77,	63

At what level is the difference between the scores of the two groups significant?

Here we wish to test $H_0: F = G$ against the two-sided alternative $H_1: F \neq G$. Let the lecture scores represent X-values and self-instruction scores the Y-values, so that $m = 8$ and $n = 9$, $m < n$. Combining the scores, arranging them in increasing order, and assigning average rank to tied observations, we have:

Observation:	28	37	48	63	63	64	65	74	75
Rank:	1	2	3	4.5	4.5	6	7	8	9
X or Y:	X	Y	Y	Y	Y	X	Y	Y	X

Observation:	76	77	78	78	82	82	94	95
Rank:	10	11	12.5	12.5	14.5	14.5	16	17
X or Y:	X	Y	X	Y	X	X	X	Y

It follows that

$$t_x = 1 + 6 + 9 + 10 + 12.5 + 14.5 + 14.5 + 16 = 83.5.$$

Since the mean value is $8(8 + 9 + 1)/2 = 72$ we see that t_x is in the right tail so that the P-value associated with $T_X = 83.5$ is $2P_H(T_X \geq 83.5)$. Table A10 only gives values of cumulative tail probability for integral values of t_x. We note the

$$P_{H_0}(T_X \geq 83) = .161, \quad \text{and} \quad P_{H_0}(T_X \geq 84) = .138.$$

We estimate the right tail probability for 83.5 as the average $(.161 + .138)/2$, so that the P-value is .299. Since there is about a 30 percent chance of observing as large a value of T_X as 83.5 in random sampling under H_0, the data do not indicate a significant difference between the two sets of scores.

Suppose, on the other hand, that the level of significance is given to be $\alpha = .10$. In that case we wish to find the critical region $T_X \geq c_3$ or $T_X \leq c_4$. By symmetry we need to find c such that

$$P_{H_0}(T_X \geq c) \doteq \frac{\alpha}{2} = .05.$$

From Table A10, $c = 90 = c_3$ and $c_4 = 54$ since

$$P_{H_0}(T_X \geq 90) = P_{H_0}(T_X \leq 8(8 + 9 + 1) - 90) = P_{H_0}(T_X \leq 54) = .046. \qquad \square$$

Example 2.   Comparing Failure Times of Two Types of Light Bulbs.   The failure times of a certain type of light bulb manufactured by two different companies, X and Y, are given (in hundreds of hours) below:

X:	3.7,	2.8,	7.1,	8.4,	6.2,	2.7
Y:	6.4,	6.8,	9.1,	7.4,	6.9,	6.8.

Let us test the null hypothesis of equality of two medians against a one-sided alternative $H_1 : \tilde{\mu}_Y > \tilde{\mu}_X$. Combining the two samples and arranging them in increasing order of magnitude, we have

Observation:	2.7	2.8	3.7	6.2	6.4	6.8	6.8
Rank:	1	2	3	4	5	6.5	6.5
X or Y:	X	X	X	X	Y	Y	Y

Observation:	6.9	7.1	7.4	8.4	9.1
Rank:	8	9	10	11	12
X or Y:	Y	X	Y	X	Y

Thus $t_x = 1 + 2 + 3 + 4 + 9 + 11 = 30$ with P-value $P_{H_0}(T_X \leq 30) = P_{H_0}(T_X \geq 48)$ $= .09$. We conclude that the data do not substantiate the claim that $\tilde{\mu}_Y > \tilde{\mu}_X$ at the 5 percent level but they do substantiate it at the 10 percent level. $\qquad \square$

Example 3.   Breaking Strength of Two Types of Nylon Fiber.   In order to compare the breaking strength of nylon fiber produced by two different manufacturers, 10 measurements on one (say X) and 13 on the other (say Y) were taken with the following results.

Fiber X: 1.7,   1.9,   1.8,   1.1,   .7,   .9,   2.1,   1.6,   1.7,   1.3
Fiber Y: 2.1,   2.7,   1.6,   1.8,   1.7,   1.8,   1.6,   2.2,   2.4,   1.3,   1.9,   1.8,   2.0

Do the data indicate a significant difference between the breaking strengths?

Here $m = 10$, $n = 13$ and we have to use the normal approximation. Let us first find the observed value of T_X.

Observation:	.7	.9	1.1	1.3	1.3	1.6	1.6	1.6	1.7	1.7	1.7
Rank:	1	2	3	4.5	4.5	7	7	7	10	10	10
X or Y:	X	X	X	X	Y	X	Y	Y	X	X	Y

Observation:	1.8	1.8	1.8	1.8	1.9	1.9	2.0	2.1	2.1
Rank:	13.5	13.5	13.5	13.5	16.5	16.5	18	19.5	19.5
X or Y:	X	Y	Y	Y	X	Y	Y	X	Y

Observation:	2.2	2.4	2.7
Rank:	21	22	23
X or Y:	Y	Y	Y

Thus $t_x = 1 + 2 + 3 + 4.5 + 7 + 10 + 10 + 13.5 + 16.5 + 19.5 = 87$. We note that

$$\mathscr{E}_{H_0}(T_X) = \frac{m(m + n + 1)}{2} = \frac{10(10 + 13 + 1)}{2} = 120,$$

$$\text{var}_{H_0}(T_X) = \frac{mn(m + n + 1)}{12} = \frac{10(13)(24)}{12} = 260,$$

and since $t_x < 120$, the smaller tail probability is the left tail probability. Hence the P-value is given by

$$2P_{H_0}(T_X \leq 87) = 2P_{H_0}(T_X < 87.5) \doteq 2P\left(Z < \frac{87.5 - 120}{\sqrt{260}}\right)$$

$$= 2P(Z < -2.02) = 2(.0217) = .0434.$$

If we allow an error of no more than, say, .04 of rejecting a true null hypothesis, then we have to conclude that the brands of fiber have the same distribution of breaking strength. If, on the other hand, $\alpha = .05$, we will have to reject $H_0 : F = G$.   □

11.6.2   The Runs Test

Consider an ordered sequence of two types of symbols. A *run* is a succession of one or more identical symbols which are preceded and followed by a different symbol (or no symbol). The *length* of a run is the number of like symbols in a run.

Example 4.   Sex of Newborns at a Hospital.   The sexes of 15 children at a hospital are recorded in order of their birth with the following results.

$$\underline{GG}\ \underline{BBB}\ \underline{G}\ \underline{BB}\ \underline{GGGG}\ \underline{B}\ \underline{GG}$$

Here G stands for a girl and B for a boy. We underline groups of successive like symbols. Each underlined group is a run. Thus we have a run of length 2 (of G symbols), followed by a run of length 3 (of B symbols), and so on. The total number of runs in this sequence is seven.   □

The total number of runs can be used as a test statistic to test the equality of two distributions. Let $X_1, X_2, \ldots, X_m$ and $Y_1, Y_2, \ldots, Y_n$ be independent random samples with respective continuous distribution functions F and G. Let us arrange the combined sample in increasing order of magnitude (and assume for the moment that there are no ties between X and Y values). Replace the X-observations in this ordered arrangement by symbol x and the Y-observations by symbol y. Let R be the total number of runs in this ordered arrangement of $m + n$ symbols of type x and y. (We assume for the moment that there are no ties between x's and y's.)

Under H_0: $F(x) = G(x)$ for all x, we expect the symbols x and y to be well-mixed. If, for example, the x's tend to be larger than the y's, then most of the y's will precede the x's and we will have fewer runs. The same argument applies to the case when y's tend to be larger than the x's. It follows that a small value of R supports the alternative H_1: $F(x) \neq G(x)$. It is clear that a test based on R is appropriate only for two-sided alternatives.

Runs Test for Testing H_0: $F = G$

Let R be the total number of runs of x and y in the combined ordered sample.

Alternative, H_1	Reject H_0 if	P-value
$F(x) \neq G(x)$	$R < c$ (R is too small)	$P_{H_0}(R \leq r_0)$

Here r_0 is the observed value of R.

In case of ties between x's and y's, a conservative procedure is to break the ties in all possible ways. For each such resolution of ties, compute the value of R. Then use the largest possible value of R to give the largest P-value. For example, if one x and one y are tied there are two possible resolutions:

$$\ldots xy \ldots$$

and

$$\ldots yx \ldots$$

resulting in two values of R. Take the larger of the two R-values.

In order to perform the test (or to compute the P-value), we need the null distribution of R. Table A12 in the Appendix gives the left tail cumulative probabilities for all values of $m \leq n$ with $m + n \leq 20$. For larger values of m or n, we use the normal approximation. We have

$$\mathscr{E}_{H_0}(R) = 1 + \frac{2mn}{(m+n)}, \quad \text{var}_{H_0}(R) = \frac{2mn(2mn - m - n)}{(m+n-1)(m+n)^2},$$

and the statistic

$$Z = \frac{R - \mathscr{E}_{H_0}(R)}{\sqrt{\text{var}_{H_0}(R)}}$$

has approximately a standard normal distribution. Since R is integer valued, continuity correction should be applied to improve the normal approximation.

Example 5. **Comparing Two Methods of Instruction (Example 1).** We return to Example 1 where we used the Mann–Whitney–Wilcoxon test to compare two methods of instruction. We note that there is one tie between X and Y values at 78. Thus we have two ways of breaking the tie. Taking the X value corresponding to 78 first, we have

$$\underline{X} \ \underline{YYYY} \ \underline{X} \ \underline{YY} \ \underline{XX} \ \underline{Y} \ \underline{X} \ \underline{Y} \ \underline{XXX} \ \underline{Y}$$

so that $R = 10$. Taking the Y value corresponding to 78 first, we have

$$\underline{X} \ \underline{YYYY} \ \underline{X} \ \underline{YY} \ \underline{XX} \ \underline{YY} \ \underline{XXXX} \ \underline{Y}$$

so that $R = 8$. Thus we take the larger of the two values as the observed value of R. The P-value is therefore $P_{H_0}(R \leq 10)$. We note that $P_{H_0}(R \leq 10) > .5$, which is considerably larger than the P-value given by the Mann–Whitney–Wilcoxon test. In any case the data do substantiate the null hypothesis that there is no difference between the two methods of instruction. □

Example 6. **A Comparison of the Performance of Some Two-Samples Tests.** In order to compare the two-sample tests studied so far, we take two independent random samples from normal populations with means 0 and 1 respectively.

X values: 0.464, 0.060, 1.486, 1.022, 1.394, 0.906, 1.179, -1.501, -0.690
(mean 0)
Y-values: 0.672, 1.187, 1.785, 1.194, 0.742, 2.579, 2.090, 1.448, 0.543
(mean 1)

Let us test the null hypothesis H_0: $\tilde{\mu}_X = \tilde{\mu}_Y$, that the two samples have the same medians, against H_1: $\tilde{\mu}_X \neq \tilde{\mu}_Y$ in order to see which test is best able to detect the difference between the means (= medians).

t-test: We have $\bar{x} = 4.32/9 = 0.48$, $\bar{y} = 12.24/9 = 1.36$,

$$\sum_{i=1}^{9} x_i^2 = 10.3548, \quad \sum_{i=1}^{9} y_i^2 = 20.4339$$

and

$$s_p^2 = \frac{8.2812 + 3.7875}{9 + 9 - 2} = .7543.$$

Hence

$$t = \frac{0.48 - 1.36}{\sqrt{.7543}\sqrt{\frac{1}{9} + \frac{1}{9}}} = -\frac{.88}{.4094} = -2.1494,$$

so that the *P*-value is slightly less than .05. (For 16 d.f. $P(T \le -2.12) = .025$.)

Median test: We arrange the combined sample in increasing order, as shown.

Observation:	−1.501	−1.690	0.060	.464	.543	.672	.742	.906	1.022
X or *Y*:	*X*	*X*	*X*	*X*	*Y*	*Y*	*Y*	*X*	*X*
Rank:	1	2	3	4	5	6	7	8	9
Observation:	1.179	1.187	1.194	1.394	1.448	1.486	1.785	2.090	2.579
X or *Y*:	*X*	*Y*	*Y*	*X*	*Y*	*X*	*Y*	*Y*	*Y*
Rank:	10	11	12	13	14	15	16	17	18

Hence

$$U = \text{number of } X\text{'s that are smaller than the sample median} = 6.$$

(The sample median is any number between 1.022 and 1.179.) The smaller tail probability corresponds to $P_{H_0}(U \ge 6)$ so that the *P*-value is

$$2 P_{H_0}(U \ge 6) = 2 \sum_{k=6}^{9} \frac{\binom{9}{k}\binom{9}{9-k}}{\binom{18}{9}} = .3609.$$

It is clear that the median test performs rather poorly.

Mann–Whitney–Wilcoxon Test. We note that

$$t_x = 1 + 2 + 3 + 4 + 8 + 9 + 10 + 13 + 15 = 65$$

and since the mean value is $9(9 + 9 + 1)/2 = 85.5$, the *P*-value is two times the probability $P_{H_0}(T_X \le 65)$. Thus

$$P\text{-value} = 2(.039) = .078.$$

The Mann–Whitney–Wilcoxon test therefore rejects H_0 at the 10 percent level but not at the 5 percent level and compares rather well with the t-test even though it does not use the assumption of normality required for the t-test.

Runs test. We note that $R = 8$ so that the P-value is $P_{H_0} (R \leq 8) = .238$. Thus the runs test performs better than the median test but does much worse than the Mann–Whitney–Wilcoxon test.

A theoretical comparison of these two sample tests is beyond our scope. We state without proof that typically the runs test and the median test perform rather poorly in comparison with the Mann–Whitney–Wilcoxon test, which typically performs almost as well as the t-test if the underlying distribution is normal. If the underlying distribution is not normal the Mann–Whitney–Wilcoxon test is considered to be the best nonparametric test to detect shifts in location. □

Remark 1.   (The Runs Test for Randomness).   The statistic R is often used to test if an ordered sequence of two types of symbols is a random arrangement. If all possible distinguishable arrangements of the two types of symbols have the same chance, we say that the arrangement is random. Suppose for example that $x_1, x_2, \ldots, x_m$ and $y_1, y_2, \ldots, y_n$ are independent random samples from the same population. Then in the ordered arrangement of x and y symbols we expect the symbols to be randomly distributed. The alternative is usually taken to be two-sided, namely, that the sequence is nonrandom. If H_0 is rejected, then either the two samples do not come from the same population or they are not independent. Under the null hypothesis of randomness we expect the number of runs of the x symbols to be approximately the same as those of y symbols. If R is too small or too large then the two types of symbols are either clustered together or isolated, indicating nonrandomness. Hence we reject H_0 if R is either too large or too small.

Runs Test for Randomness

Let r_0 be the total number of runs in the combined ordered arrangement of x's and y's

H_0	H_1	Reject H_0 if	P-value
Sequence is Random	Sequence is Nonrandom	r_0 is either too large or too small	2 (smaller tail probability)

Example 7.   Tossing a Fair Coin.   Students were asked to toss a fair coin 20 times and report the results to the instructor. A student turned in the following sequence of heads (H) and tails (T):

$$H\ H\ T\ H\ T\ H\ T\ H\ T\ T\ H\ T\ H\ T\ H\ T\ T\ T\ H\ H\ T\ H$$

Does this sequence of 10 heads and 10 tails indicate departure from randomness? In other words, is there evidence to conclude that the student in question did not conduct the experiment? We note that the number of runs is given by $r = 15$. From Table A12 for $m = n = 10$,

$$P_{H_0}(R \geq 15) = 1 - P_{H_0}(R \leq 14) = P_{H_0}(R \leq 7) = .051$$

so that the P-value is $2(.051) = .102$ and we cannot reject H_0 at level $\alpha = .10$. That is, there does not appear to be an intentional mixing of symbols on the part of the student.

□

Problems for Section 11.6

1. The order in which test questions are asked affects a student's ability to answer them correctly and hence affects the student's total grade. In order to check this proposition, two tests were made. Test A had questions set in increasing order of difficulty; in Test B the order was reversed. A random sample of 20 students was selected in such a way that 10 pairs of students were matched in ability. From each pair, one student was assigned randomly to take Test A and the other Test B. The following scores were obtained.

Test A:	83,	82,	95,	92,	91,	60,	89,	69,	70,	72
Test B:	76,	62,	70,	74,	52,	63,	48,	80,	76,	74

   Is there evidence to indicate that the scores on Test B are lower than those on Test A? Use both Mann–Whitney–Wilcoxon and t-tests.

2. In a controlled laboratory environment, eight men and 10 women were tested to determine the room temperatures they found to be most comfortable. The following results were obtained.

Men:	72,	76,	70,	68,	66,	75,	73,	75		
Women:	75,	77,	77,	73,	79,	78,	80,	76,	74,	72

   Assuming that these samples are random samples from their respective populations, is the average comfortable temperature for men lower than that for women?

3. Nineteen pieces of flint were collected, nine from area A and ten from area B. The object of the study was to determine if the pieces of flint were of equal hardness. For the purposes of the study nineteen pieces of flint of equal hardness from a third area were brought in. Each of the sample pieces was then rubbed against a piece from the third area. The nineteen sample pieces were then ordered according to the amount of damage sustained from the softest (most damage) to the hardest (least damage): A A B B A A B B B B A A B A A A B B B. Is there evidence to suggest that the flints from areas A and B are of equal hardness? Use both Mann–Whitney–Wilcoxon and runs tests to analyze the data.

4. Fifteen three-year-old boys and 15 three-year-old girls were observed during two 30-minute sessions of recess in a nursery school. Each child's play was scored for incidence and degree of aggression as follows:

   Boys: 96, 65, 74, 78, 82, 121, 68, 79, 111, 48, 53, 92, 81, 31, 48
   Girls: 12, 47, 32, 59, 83, 14, 32, 15, 17, 82, 21, 34, 9, 15, 50

Is there evidence to suggest that there are sex differences in the incidence and amount of aggression? Use both Mann–Whitney–Wilcoxon and runs tests.

5. Two laboratory cultures are to be compared for difference in bacteria counts. Independent random samples of six from culture A and eight from culture B are taken and the number of bacteria per unit of volume are recorded as follows:

Culture A:　32,　29,　34,　47,　33,　27
Culture B:　38,　36,　33,　42,　34,　40,　39,　32.

Is there evidence to conclude that the bacteria counts for the two populations are not the same?

6. A true–false examination with 50 questions has the following sequence of answers.

$$F\ F\ T\ T\ F\ T\ T\ T\ T\ T\ T\ F\ T\ F\ T\ T\ F\ F\ F\ F\ T\ T\ F\ T\ T$$

$$T\ T\ T\ F\ F\ F\ T\ T\ F\ F\ T\ T\ T\ T\ F\ F\ T\ F\ T\ F\ F\ F\ F\ T\ F$$

Is there evidence to suggest that the instructor used intentional mixing of answers?

7. Twenty-five items emerging from a production line are tested and classified as defective (D) or nondefective (N). The following sequence is obtained:

$$N\ N\ N\ N\ N\ D\ D\ N\ D\ N\ N\ N\ N\ D\ D\ N\ N\ N\ N\ D\ N\ D\ N\ N\ N$$

How strong is the evidence that there is lack of randomness in the sequence?

8. An intelligence test was administered to 10 men and 12 women with the following test scores:

Men:　　150, 135, 132, 140, 147, 138, 129, 157, 134, 141
Women: 156, 138, 132, 144, 141, 147, 148, 124, 130, 132, 140, 129

Is it reasonable to conclude that there are no sex differences in average intelligence? Use the Mann–Whitney–Wilcoxon test as well as the t-test.

9. The following digits are chosen from Table A1, line 5.

69572, 68777, 39510, 35905, 14060, 40619, 29549, 69616, 33564, 60780

Use the runs test on odd and even digits to check if these numbers exhibit randomness.

10. A random sample of 18 married women is selected. Each woman is asked if she is contented or discontented with her marriage. The number of years she is married is also recorded with the following results.

Contented:　　3,　5,　4,　8,　2,　15,　17,　10
Discontented: 2,　20,　16,　18,　7,　1,　3,　5,　11,　4

Is there evidence to suggest that the two groups are different?

11. (i) Let U_X be the number of Y's that are smaller than the X's in independent random samples $X_1, X_2, \ldots, X_m$ and $Y_1, Y_2, \ldots, Y_n$. Find $\mathscr{E} U_X$ and $\mathrm{var}\,(U_X)$ under $H_0: F = G$ where X's have distribution F and Y's have distribution G.

(ii) Let $p_{m,n}(u) = P_{H_0}(U_X = u)$. Show that

$$p_{m,n}(u) = \frac{n}{m+n} p_{m,n-1}(u) + \frac{m}{m+n} p_{m-1,n}(u-n).$$

If $m = 0$, then for $n \geq 1$

$$p_{0,n}(u) = 1 \qquad \text{if } u = 0, \text{ and } 0 \text{ otherwise}$$

and if $n = 0$, then for $m \geq 1$

$$p_{m,0}(u) = 1 \qquad \text{if } u = 0, \text{ and } 0 \text{ otherwise.}$$

12*. Consider a sequence of two types of elements A and B with m of type A and n of type B. Let R_1 be the number of A runs in the sequence, R_2 the number of B runs, and $R = R_1 + R_2$. Show that under the hypothesis of randomness

$$P_{H_0}(R_1 = r_1, R_2 = r_2) = k \frac{\dbinom{m-1}{r_1-1}\dbinom{n-1}{r_2-1}}{\dbinom{m+n}{m}}$$

and

$$P_{H_0}(R = r) = \sum_{\{(r_1, r_2):\, r_1 + r_2 \leq r\}} P_{H_0}(R_1 = r_1, R_2 = r_2),$$

where $k = 2$ if $r_1 = r_2$ and $k = 1$ if $|r_1 - r_2| = 1, r_1 = 1, 2, \ldots, m$ and $r_2 = 1, 2, \ldots, n$. [Hint: Given $R_1 = r_1$, there are either r_1, $(r_1 - 1)$ or $(r_1 + 1)$ runs of type B elements. The maximum number of runs is $2m$ if $m = n$ and $2m + 1$ if $m < n$. Use Problem 2.5.18 to conclude that the number of ways of getting r_1 A runs is $\dbinom{m-1}{r_1-1}$.]

13. Show that the marginal distribution of R_1 is given by

$$P_{H_0}(R_1 = r_1) = \frac{\dbinom{m-1}{r_1-1}\dbinom{n+1}{r_1}}{\dbinom{m+n}{m}},$$

for $r_1 = 1, 2, \ldots, m$.

14. In Examples 1 and 3, apply the two-sample t-test and compare your results to those obtained in Examples 1 and 3.

11.7 CHI-SQUARE TEST OF GOODNESS OF FIT REVISITED

In Section 9.9 we introduced the chi-square test of goodness of fit when the cell probabilities are prespecified. As pointed out there, the case of most interest is not that the data can reasonably be fit by one specified distribution but rather that they can be fit by one of a family of distributions. Thus the question often is the following: Could the observations have come from some exponential distribution? Some normal distribution? Some Poisson distribution? This means, in particular, that H_0 is no longer a simple hypothesis. In other words, under H_0 the form of the probability or density function is specified except for one or more parameters.

Let $X_1, X_2, \ldots, X_n$ be a random sample from a discrete distribution given by

$$P(X = x_j) = p_j(\theta), \qquad j = 1, 2, \ldots, k \text{ and zero elsewhere,}$$

where $p_j(\theta) > 0$, $\sum_{j=1}^{k} p_j(\theta) = 1$ and θ is a scalar (or vector) parameter that is unknown. Under H_0 we know $p_j(\theta)$ except for θ. Let n_j be the number of X's in the sample that equal x_j. Then $\sum_{j=1}^{k} n_j = n$ and under H_0, $(n_1, n_2, \ldots, n_{k-1})$ has a multinomial distribution. Let

$$Q(\theta) = \sum_{j=1}^{k} \frac{\left[n_j - np_j(\theta)\right]^2}{np_j(\theta)}.$$

Our object is to test if for some value of θ, $p_j(\theta)$, $j = 1, 2, \ldots, k$, is a good fit to the data. In order to do so, we minimize $Q(\theta)$ for all possible values of θ in the parameter set. That is, we find that value of θ for which $p_j(\theta)$ best fits the data. Let $\hat{\theta}$ be the value of θ for which $Q(\theta)$ is minimum. Then $\hat{\theta}$ is called a *minimum chi-square* estimate of θ, and this method of estimation is called the *minimum chi-square method*. The following result holds and is stated without proof.

THEOREM 1. *Let $\hat{\theta}$ be the minimum chi-square estimate of θ. Then under certain conditions and for large n the minimum value $Q(\hat{\theta})$ of $Q(\theta)$ given by*

$$Q(\hat{\theta}) = \sum_{j=1}^{k} \frac{\left[n_j - np(\hat{\theta})\right]^2}{np_j(\hat{\theta})}$$

has approximately a chi-square distribution with $k - 1 - r$ d.f. where r is the dimensionality of θ. [If θ is scalar then $r = 1$, if $\theta = (\mu, \sigma^2)$ both μ and σ^2 unknown, then $r = 2$, and so on.]

The conditions left unspecified in Theorem 1 are satisfied for most commonly used models considered here. Unfortunately, however, it is not easy to estimate θ by minimizing $Q(\theta)$. Fortunately, it can be shown (Problem 7) that for sufficiently large n the θ that minimizes $Q(\theta)$ is approximately the maximum likelihood estimate of θ and, moreover, Theorem 1 holds when $\hat{\theta}$ is the maximum likelihood estimate of θ. It should be emphasized that Theorem 1 requires that the estimation be done by using the grouped data. Unfortunately, even this is often difficult to carry out. To appreciate the degree of difficulty involved, suppose X has *some* Poisson distribution under H_0 and we observe

j:	0	1	$\cdots$	k or more
Frequency:	n_0	n_1	$\cdots$	n_k

Then the likelihood function is

$$L(\lambda) = [p_0(\lambda)]^{n_0} [p_1(\lambda)]^{n_1} \ldots [p_k(\lambda)]^{n_k}$$

where $\mathscr{E}X = \lambda$ and

$$p_j(\lambda) = e^{-\lambda} \frac{\lambda^j}{j!} \qquad \text{for } j = 0, 1, \ldots, k-1,$$

$$p_k(\lambda) = 1 - \sum_{j=0}^{k-1} \frac{e^{-\lambda}\lambda^j}{j!}.$$

Taking logarithms, we have

$$\ln L(\lambda) = -\lambda \sum_{j=0}^{k-1} n_j + \ln \lambda \sum_{j=1}^{k-1} jn_j + n_k \ln\left(1 - \sum_{j=0}^{k-1} \frac{e^{-\lambda}\lambda^j}{j!}\right) + c,$$

where c is a constant. In order to maximize λ we need to solve the equation

$$\frac{d\ln(\lambda)}{d\lambda} = 0.$$

Given $n_0, n_1, \ldots, n_k$, this can only be done numerically. (If $n_k = 0$, then $\hat{\lambda} = \bar{x}$ is the maximum likelihood estimate.)

What we would like to do is use the estimate $\hat{\lambda} = \bar{x}$, the sample mean, since $\bar{x}$ is the maximum likelihood estimate of λ when the observations are not grouped. It is known that the approximate distribution function of Q lies between the chi-square distribution with $k - 2$ d.f. and the chi-square distribution with $k - 1$ d.f. In general, if θ is r dimensional and the r parameters are estimated from the ungrouped observations, then the approximate distribution of Q lies between the $\chi^2(k - 1)$ and the $\chi^2(k - r - 1)$ distributions. Since for a given α,

$$\chi_\alpha^2(k - 1) \geq \chi_\alpha^2(k - r - 1) \qquad \text{for } r \geq 1$$

where $\chi_\alpha^2(k - 1)$ is the $(1 - \alpha)$th percentile of the χ^2 distribution with $k - 1$ d.f., it is reasonable to adopt the following conservative procedure.

If θ is estimated from ungrouped data, then reject H_0 at level α if $Q(\hat{\theta}) \geq \chi_\alpha^2(k - 1)$, that is, if the P-value (for $k - 1$ d.f.) $\leq \alpha$.

In the continuous case the procedure is the same as described in Section 9.9. Thus if X is of the continuous type with density function $f(x, \theta)$ where θ is unknown, then we partition the real line into k intervals $I_1, I_2, \ldots, I_k$. Under H_0,

$$p_j(\theta) = \int_{I_j} f(x, \theta) \, dx,$$

and Theorem 1 applies. It should be emphasized again that the estimation of θ should be done from the grouped frequency distribution whenever possible. Moreover, we also lose one degree of freedom for each class pooled. If θ is estimated from the ungrouped data using maximum likelihood estimation procedure, then we use the conservative procedure described above.

Example 1.   Fatal Auto Accidents on a Holiday Weekend of Example 9.9.1.   In Example 9.9.1 we tested whether the following data on 340 fatal automobile accidents on a 72-hour weekend fits the Poisson distribution with mean $\lambda = 5$.

Accidents:	0 or 1	2	3	4	5	6	7	8 or more
Hours:	5	8	10	11	11	9	8	10

What we really want to test is whether the data could have come from *some* Poisson distribution. Let X be the number of fatal auto accidents per hour. We wish to test the composite null hypothesis

$$H_0 \colon P(X = x) = \frac{e^{-\lambda}\lambda^x}{x!}, \quad x = 0, 1, 2, \ldots \quad \text{for some } \lambda > 0$$

against all alternatives. Since λ is unknown, we compute the maximum likelihood estimate of λ. If we use the ungrouped data, then

$$\hat{\lambda} = \bar{x} = 340/72 = 4.72.$$

In order to apply the test we need

$$p_j(\hat{\lambda}) = \frac{e^{-\hat{\lambda}}\hat{\lambda}^j}{j!}, \quad \text{for } j = 0, 1, 2 \ldots.$$

Since

$$p_{j+1}(\hat{\lambda}) = e^{-\hat{\lambda}}\left(\frac{\hat{\lambda}^j}{j!}\right)\frac{\hat{\lambda}}{j+1} = \frac{\hat{\lambda}}{j+1}p_j(\hat{\lambda}),$$

and

$$p_0(\hat{\lambda}) = e^{-\hat{\lambda}} = .0089,$$

we can compute $p_j(\hat{\lambda})$ recursively. We have

$$p_1(\hat{\lambda}) = .0421, p_2(\hat{\lambda}) = .0993, p_3(\hat{\lambda}) = .1563, p_4(\hat{\lambda}) = .1844,$$

$$p_5(\hat{\lambda}) = .1741, p_6(\hat{\lambda}) = .1370, p_7(\hat{\lambda}) = .0923,$$

and

$$p_8(\hat{\lambda}) = 1 - \sum_{j=0}^{7} p_j(\hat{\lambda}) = .1056.$$

We can now compare the observed and expected frequencies as follows.

j:	0 or 1	2	3	4	5	6	7	8 or more
Observed:	5	8	10	11	11	9	8	10
Expected:	3.67	7.15	11.25	13.28	12.54	9.86	6.65	7.60

Pooling the first two cells, we get

$$q_0 = \sum_j \frac{\left[n_j - np_j(\hat{\lambda})\right]^2}{np_j(\hat{\lambda})} = 4.55.$$

The number of d.f. on the conservative side, is $7 - 1 = 6$ even though we estimated one parameter since we used the x's to estimate λ and not the grouped data. The P-value is given by

$$.70 > P_{H_0}(Q \geq 4.55) > .5$$

so that there is not much evidence to reject H_0. The data could have come from some Poisson distribution. □

Example 2. *Heights of Infants.* For the data on heights (in inches) of 60 infants of Problem 4.5.6, the following frequency distribution was constructed with eight classes of equal widths.

Class Interval	Frequency, f_i	Class Midpoint, x_i
(16.35, 17.25)	4	16.8
(17.25, 18.15)	7	17.7
(18.15, 19.05)	12	18.6
(19.05, 19.95)	8	19.5
(19.95, 20.85)	6	20.4
(20.85, 21.75)	11	21.3
(21.75, 22.65)	9	22.2
(22.65, 23.55)	3	23.1
Total	60	

Do the data come from a normal population? The maximum likelihood estimates for μ and σ^2 are given respectively by

$$\bar{x} = \sum_{i=1}^{8} \frac{f_i x_i}{60} = \frac{1196.1}{60} = 19.935,$$

and

$$s^2 = \frac{\sum_{i=1}^{8} f_i x_i^2 - 60\bar{x}^2}{60} = \frac{24039.41 - 23844.25}{60} = 3.2527.$$

Also $s = 1.80$. In order to compute $p_j(\hat{\theta})$ we standardize the upper class end points in order to read the standard normal distribution tables. We have

$$p_1(\hat{\theta}) = P\left(Z < \frac{17.25 - 19.94}{1.80}\right) = P(Z < -1.49) = .0681,$$

$$p_2(\hat{\theta}) = P\left(-1.49 < Z < \frac{18.15 - 19.94}{1.80}\right) = P(-1.49 < Z < -.99) = .0930,$$

$$p_3(\hat{\theta}) = P(-.99 < Z < -.49) = .1510, \qquad p_4(\hat{\theta}) = P(-.49 < Z < 0) = .1879,$$

$$p_5(\hat{\theta}) = P(0 < Z < .49) = .1879, \qquad p_6(\hat{\theta}) = P(.49 < Z < .99) = .1510,$$

$$p_7(\hat{\theta}) = P(.99 < Z < 1.49) = .0930, \qquad p_8(\hat{\theta}) = P(Z > 1.49) = .0681.$$

We can now compare the expected and observed frequencies as follows:

Cell Number:	1	2	3	4	5	6	7	8
Observed:	4	7	12	8	6	11	9	3
Expected:	4.08	5.58	9.06	11.27	11.27	9.08	5.58	4.08

If we do not pool the cells with expected frequencies < 5, $q_0 = 7.52$ with $8 - 2 - 1 = 5$ d.f. If we pool cells 1 and 2 and cells 7 and 8, then $q_0 = 5.14$ with $6 - 2 - 1 = 3$ d.f. For 5 d.f. we note that

$$.20 > P_{H_0}(Q \geq 7.52) > .10$$

and for 3 d.f.

$$.20 > P_{H_0}(Q \geq 5.14) > .10.$$

In either case the evidence against normality is not very strong. A look at the observed frequencies, however, shows that the data appear to come from a bimodal distribution. $\square$

Problems for Section 11.7

1. According to the Hardy–Weinburg stability principle (see Problem 10.4.5), the genotype probabilities after one generation should be in proportions $p_1 = p^2$, $p_2 = 2p(1 - p)$, and $p_3 = (1 - p)^2$, $0 < p < 1$ where p is unknown.
   (i) In a random sample of n offspring, suppose X_i are of type i, $X_1 + X_2 + X_3 = n$. Show that the maximum likelihood estimate of p is given by

   $$\hat{p} = \frac{2X_1 + X_2}{2n}.$$

   Is $\hat{p}$ unbiased for p? Find its variance. (See also Problem 10.5.16.)

(ii)   In a random sample of 75 offspring, the following data were observed:

Type:	1	2	3
Frequency:	15	45	15

Do the data support the Hardy–Weinburg principle?

2.   A random sample of size 50 showed the following results.

1.91,	.49,	.47,	1.60,	1.88,	.07,	2.68,	1.16,	−.89,	1.37,
2.18,	.06,	1.01,	1.77,	1.97,	1.71,	2.09,	.37,	.75,	.30,
−.50,	.51,	.84,	.86,	2.03,	1.20,	1.45,	1.75,	.58,	.57,
.31,	1.76,	−.62,	.65,	.49,	−1.05,	.54,	.78,	1.86,	.53,
2.37,	1.23,	1.38,	1.76,	1.18,	.26,	1.96,	−.53,	.74,	1.12.

(i)   Group the data in five intervals of equal length beginning with -1.025.
(ii)   Test the hypothesis that the sample comes from a standard normal population.
(iii)   Test the hypothesis that the sample comes from a normal population with mean $\mu = .5$ and variance 1.
(iv)   Test the hypothesis that the sample comes from a normal population with variance 1.
(v)   Test the hypothesis that the sample comes from some normal population.
(Estimate μ by $\bar{x}$ and σ by s where $\bar{x}$ and s are computed from the frequency distribution.)

3.   In Problem 9.9.10 the number of particles emitted in 150 consecutive 10-second intervals have the frequency distribution

Number of particles:	0	1	2	3	4	5
Number of intervals:	15	48	33	30	14	10

Could the data have come from some Poisson distribution?

4.   In Problem 9.9.11 the lifetimes of 80 rotary motors tested have the following frequency distribution.

Time to failure (years):	(0, 1)	[1, 2)	[2, 3)	[3, 4)	4 or more
Frequency of failure:	34	26	8	6	6

Could the time to failure be assumed to have an exponential distribution? (Assume that the last interval is [4, 5) in computing $\bar{x}$.)

5.   In Problem 9.9.7, the failure times of 100 transistor batteries (in hours) have the frequency distribution:

Hours:	[0, 25)	[25, 50)	[50, 75)	[75, 100)	[100,∞)
Frequency:	33	28	20	12	7

Could the data have come from some exponential distribution? (Assume that the last interval is $[100, 125)$ in order to estimate the mean.)

6. A random sample of 200 families, each with four children, has the following frequency distribution of the number of girls:

Number of girls:	0	1	2	3	4
Number of families:	5	32	65	75	23

Fit a binomial distribution and check the goodness of your fit.

7*. Show that the θ that minimizes $Q(\theta) = \sum_{j=1}^{k}[(n_j - np_j(\theta))^2/np_j(\theta)]$ is the maximum likelihood estimate of θ for sufficiently large n. [Hint: Write $Q(\theta) = \sum_{j=1}^{k}(n_j^2/np_j(\theta)) - n$ as in Problem 9.9.12 and differentiate. Use the law of large numbers.]

8. In Problem 2, test the hypothesis that the data come from some double exponential distribution with density

$$f(x) = (1/2)\exp(-|x - \mu|), \qquad x \in \mathbb{R}$$

where μ is unknown. [Hint: The maximum likelihood estimate of μ is the sample median. Take the midpoint of the class interval in which the median lies as the sample median.]

9. Word length, sentence length, and vocabulary used by an author can be distinctive measures of the author's style and may be used to discriminate among the writings of several authors. Word length counts have been carried out in the works of many well known authors and used to identify authors of disputed works. Suppose that in 250 passages of 2000 words in one author's works, the word "also" showed the following frequency distribution:

Frequency of "also":	0	1	2	3	4 or more	Total
Number of passages:	205	37	7	1	0	250

Is it reasonable to assume that the distribution of the number of times the author uses "also" in 2000-word passages is Poisson?

11.8* KOLMOGOROV–SMIRNOV GOODNESS OF FIT TEST

Let $X_1, X_2, \ldots, X_n$ be a random sample from a continuous distribution function F. In Sections 9.9 and 11.7 we considered the chi-square test of goodness of fit of the null hypothesis H_0: $F(x) = F_0(x)$ for all $x \in \mathbb{R}$ against H_1: $F(x) \neq F_0(x)$ for some x. Here F_0 is either completely specified or F_0 is specified except for a finite number of parameters. What we did was reduce the problem to the multinomial case by partitioning the real line (somewhat arbitrarily) into a finite number of intervals $I_1, I_2, \ldots, I_k$ and then compare the observed number of X_i's in each interval to that expected under H_0. The chi-square test is specifically designed for count data and is based on vertical deviations between the observed and hypothesized histograms.

Yet another test of goodness of fit, due to Kolmogorov and Smirnov, is based on vertical deviations between the empirical distribution function and the hypothesized distribution function. In order to apply this test we do not need any arbitrary grouping of the data. Rather the test is applied to the observations so that it permits a comparison at each of the observed values. Moreover, the test is exact in that the exact distribution of the test statistic under H_0 is known and tabulated. The Kolmogorov–Smirnov test is more flexible since it permits testing H_0 against both one-sided and two-sided alternatives.

We recall that the empirical distribution function of the sample $X_1, X_2, \ldots, X_n$ is defined by

$$\hat{F}_n(x) = \frac{\text{number of } X_i\text{'s} \le x}{n}.$$

Under H_0: $F(x) = F_0(x)$, $x \in \mathbb{R}$, we expect reasonable agreement between $F_0(x)$ and its estimate $\hat{F}_n(x)$. Suppose the alternative is H_1: $F(x) \ne F_0(x)$ for some x. Under H_0, the absolute deviations $|\hat{F}_n(x) - F_0(x)|$ should be small for all x. We define the two-sided Kolmogorov–Smirnov test statistic as follows:

(1)
$$D_n = \sup_{x \in \mathbb{R}} |\hat{F}_n(x) - F_0(x)|.$$

Thus D_n is the largest absolute deviation between $\hat{F}_n(x)$ and $F_0(x)$, and if D_n is small so are all the deviations. Therefore we reject H_0 in favor of H_1 if D_n is large.

The one-sided Kolmogorov–Smirnov statistics are defined analogously as follows:

(2)
$$D_n^+ = \sup_{x \in \mathbb{R}} \{ \hat{F}_n(x) - F_0(x) \},$$

(3)
$$D_n^- = \sup_{x \in \mathbb{R}} \{ F_0(x) - \hat{F}_n(x) \}.$$

The statistics D_n^+ and D_n^- are used to test H_0 against one-sided alternatives $F(x) \ge F_0(x)$ and $F(x) \le F_0(x)$ respectively. A large value of D_n^+ supports the alternative H_1: $F(x) \ge F_0(x)$ so that we reject H_0 in favor of H_1 if D_n^+ is large. Similar considerations apply to the alternative $F(x) \le F_0(x)$. If F_0 is continuous, the exact null distributions of D_n, D_n^+, D_n^- are known and tabulated.

Kolmogorov–Smirnov One-Sample Test of H_0: $F(x) = F_0(x)$ for all x

H_1	Reject H_0 if	P-value
$F(x) \ne F_0(x)$	$D_n > c_1$ (D_n is too large)	$P_{H_0}(D_n \ge d)$
$F(x) \ge F_0(x)$	$D_n^+ > c_2$ (D_n^+ is too large)	$P_{H_0}(D_n^+ \ge d^+)$
$F(x) \le F_0(x)$	$D_n^- > c_3$ (D_n^- is too large)	$P_{H_0}(D_n^- \ge d^-)$

Here d, d^+, d^- are the observed values of D_n, D_n^+ and D_n^- respectively.

How do we compute the observed value of D_n, D_n^+ or D_n^- from the sample? Let the sample order statistic be

$$x_{(1)} < x_{(2)} < \cdots < x_{(n)}.$$

First note that

(4) $$D_n = \max(D_n^+ , D_n^-).$$

We state without proof that

(5) $$D_n^+ = \max\left\{ \max_{1 \leq i \leq n}\left[\frac{i}{n} - F_0(x_{(i)}) \right], 0 \right\},$$

and

(6) $$D_n^- = \max\left\{ \max_{1 \leq i \leq n}\left[F_0(x_{(i)}) - \frac{i-1}{n} \right], 0 \right\}.$$

In view of (4), (5), and (6):

(7) $$D_n = \max_{1 \leq i \leq n} \left\{ \max\left[\frac{i}{n} - F_0(x_{(i)}), F_0(x_{(i)}) - \frac{i-1}{n} \right] \right\}$$

We note that $\hat{F}_n(x_{(i)}) = i/n$. Thus D_n^+ is the maximum (positive) vertical distance between $\hat{F}_n$ and F_0 taken over all the observations. Formulas (5), (6) and (7) simplify the computation of D_n, D_n^+, D_n^-. Table A9 in the Appendix gives the right tail under the distribution of Kolmogorov–Smirnov test statistic for $n \leq 40$.

   Example 1. Random Sampling from the Interval (0, 1). We return to Example 4.5.4, which was continued in Example 9.9.3. For convenience we take only the 16 observations in the first two rows:

.59	.72	.47	.43	.31	.56	.22	.90
.96	.78	.66	.18	.73	.43	.58	.11

Suppose we wish to test H_0 that the sample comes from the uniform distribution on $(0, 1)$ so that

$$F_0(x) = \begin{cases} 0 & x \leq 0 \\ x & 0 < x < 1 \\ 1 & 1 \leq x. \end{cases}$$

In the following table we compute the statistic D_n using formula (7).

Order Statistic, $x_{(i)}$	$F_0(x_{(i)})$	$\dfrac{i}{n} - F_0(x_{(i)})$	$F_0(x_{(i)}) - \dfrac{i-1}{n}$
.11	.11	−.0475	.1100
.18	.18	−.0550	.1175
.22	.22	−.0325	.0950
.31	.31	−.0600	.1225
.43 (2 observations)	.43	−.0550	.1175
.47	.47	−.0325	.0950
.56	.56	−.0600	.1225
.58	.58	−.0175	.0800
.59	.59	.0350	.0275
.66	.66	.0275	.0350
.72	.72	.0300	.0325
.73	.73	.0825	−.0200
.78	.78	.0950	−.0325
.90	.90	.0375	.0250
.96	.96	.0400	.0225

It follows that

$$D_{16} = .1225$$

and since from Table A9

$$P\text{-value} = P_{H_0}(D_{16} > .1225) > .20$$

we cannot reject the null hypothesis that the data come from the uniform distribution on $(0, 1)$.

Let us group the data in three intervals $(0, .33)$, $[.33, .67)$, and $[.67, 1.0)$ in order to apply the χ^2 goodness of fit test. The observed and expected frequencies are as follows.

	(0, .33)	[.33, .67)	[.67, 1.0)
Observed:	4	7	5
Expected:	5.28	5.44	5.28

Thus $q_0 = 1.53$ and for 2 d.f. the P-value is between .30 and .50. Again we cannot reject H_0. □

Remark 1. If F_0, the hypothesized distribution function, is discrete, then it is a step function. The same tests may be used with the same sampling distributions of D_n, D_n^+, D_n^- except that in this case the null distribution is no longer exact. The resulting test is conservative. The same tests may also be used if one or more parameters of F_0 are estimated, but once again the resulting tests are approximate (and conservative).

Remark 2. The statistic D_n can be used to compute the minimum sample size required in order to estimate F_0 by $\hat{F}_n$ with a prescribed degree of accuracy. Yet another application of D_n is to find confidence bands for $F(x)$, the unknown distribution function, based on $\hat{F}_n$ (see Example 3 below).

Remark 3. It is possible to extend the methods of Kolmogorov and Smirnov to compare two or more populations. For example, if we wish to test the hypothesis of equality of two distribution functions (the problem considered in Section 11.6) based on two independent samples, then we can use the two-sample analog of D_n defined by

$$D_{m,n} = \sup_{x \in \mathbf{R}} |\hat{F}_m(x) - \hat{G}_n(x)|$$

where $\hat{F}_m$ and $\hat{G}_n$ are the empirical distribution functions of the two samples. Under H_0 we expect $\hat{F}_m$ and $\hat{G}_n$ to be close so that large values of $D_{m,n}$ support the alternative hypothesis H_1: $F(x) \neq G(x)$ for some x. We will not consider the two-sample case any further.

Example 2. Heights of Infants. In Example 11.7.2 we analyzed the data on the heights of 60 infants given in Problem 4.5.6. For convenience we will consider only the 12 heights (in inches) given in the first row of Problem 4.5.6:

18.2,   21.4,   22.6,   17.4,   17.6,   16.7,   17.1,   21.4,   20.1,   17.9,   16.8,   23.1

We test the hypothesis that the data come from some normal population. Estimating μ and σ^2 by $\bar{x}$ and s^2 respectively where

$$\bar{x} = 230.3/12 = 19.19 \doteq 19.2,$$

$$s^2 = \frac{4482.01 - (230.3)^2/12}{12 - 1} = \frac{62.17}{11} = 5.65,$$

so that $s = 2.38$. Then

$$F_0(x_{(i)}) = P\left(Z \le \frac{x_{(i)} - 19.2}{2.38}\right), \qquad i = 1, 2, \ldots, 12.$$

Let us now compute D_{12}.

i	$x_{(i)}$	$F_0(x_{(i)})$	$\dfrac{i}{12} - F_0(x_{(i)})$	$F_0(x_{(i)}) - \dfrac{i-1}{12}$
1	16.7	.1469	− .0636	.1469
2	16.8	.1562	.0105	.0729
3	17.1	.1894	.0606	.0227
4	17.4	.2236	.1097	− .0264
5	17.6	.2514	.1653	− .0819
6	17.9	.2912	.2088	− .1255
7	18.2	.3372	.2461	− .1628
8	20.1	.6480	.0187	.0647
9	21.4	.8212	.0121	.0712
10	21.4			
11	22.6	.9236	− .0069	.0903
12	23.1	.9495	.0505	.0328

It follows that $D_{12} = .2461$. From Table A9 we note that the P-value $> .20$ so that normality cannot be rejected comfortably. □

Example 3. Confidence Bands for Unknown F. Suppose we wish to find a $1 - \alpha$ level confidence band for an unknown (continuous) distribution function F. That is, we should like to find two boundaries L and U for F such that

$$P(L(x) \le F(x) \le U(x) \text{ for all } x \in \mathbb{R}) = 1 - \alpha.$$

Here $L(x)$ is called the lower and $U(x)$ the upper boundary of F. Let $x_1, x_2, \ldots, x_n$ be a random sample from F. Given α and n, we find d_α from Table A9 such that

$$P(D_n > d_\alpha) = \alpha$$

using the two-sided values. We choose

$$L(x) = \max(0, \hat{F}_n(x) - d_\alpha)$$

and

$$U(x) = \min(1, \hat{F}_n(x) + d_\alpha).$$

Then

$$P\left(\sup_x |\hat{F}_n(x) - F(x)| \le d_\alpha \right) = 1 - \alpha$$

so that

$$P(\hat{F}_n(x) - d_\alpha \le F(x) \le \hat{F}_n(x) + d_\alpha \quad \text{for all } x) = 1 - \alpha,$$

and $L(x)$ and $U(x)$ have the required properties.

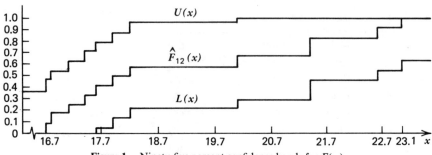

Figure 1. Ninety-five percent confidence bands for $F(x)$.

In particular, let us construct a 95 percent confidence band for the underlying distribution function of Example 2. We have $n = 12$, $\alpha = .05$, so that $d_\alpha = .375$. The confidence bands are now easily computed, as in the following table. Figure 1 gives a graph of $\hat{F}_n$, and $L(x)$ and $U(x)$.

x	16.7	16.8	17.1	17.4	17.6	17.9
$\hat{F}_n(x)$	.083	.167	.250	.333	.417	.500
$L(x)$	0	0	0	0	.042	.125
$U(x)$	.458	.542	.625	.708	.792	.875

x	18.2	20.1	21.4	22.6	23.1	
$\hat{F}_n(x)$	.583	.667	.833	.917	1.000	
$L(x)$	.208	.292	.458	.542	.625	
$U(x)$	.958	1.000	1.000	1.000	1.000	□

In conclusion, let us briefly compare the chi-square and the Kolmogorov–Smirnov tests of goodness of fit. We recall that the χ^2 test is specially designed for discrete distributions. For continuous distributions one needs to group the observations somewhat arbitrarily into cells, which results in loss of information. It can be applied when the parameters are estimated from the data. As we shall see, the χ^2 test can be used also for multidimensional distributions. The Kolmogorov–Smirnov test, on the other hand, is designed for continuous distributions and requires no grouping. It can be applied to discrete distributions but the resulting test is conservative. If the sample size is large, the test is computationally much more difficult to apply than the χ^2 test since it requires that the data be ordered.

Problems for Section 11.8

1. Test the null hypothesis that the following observations come from a uniform distribution on $(0,1)$:

.56 .22, .38, .32, .21, .94, .14, .24, .92, .10, .50, .86,
.46, .51, .38, .02, .19, .32, .65, .56, .86, .07, .48, .05, .39.

2. In Problem 1 test the hypothesis that the sample comes from the beta density

$$f(x) = 6x(1 - x), \qquad 0 < x < 1, \text{ and zero elsewhere.}$$

3. A random sample of 20 observations produced the following data:

    .46,   .14,   2.46,   −.32,   −.07,   .29,   −.29,   1.30, .24, −.96,
    .06,   −2.53,   −.53,   −.19,   .54, −1.56,   .19, −1.19, .02,   .53.

    Test the hypothesis that the sample comes from a standard normal distribution.

4. In Problem 1 find a 90 percent confidence band for the true underlying distribution function. Does the distribution function $F(x) = 0$ if $x < 0$, $= x$ if $0 \le x \le 1$, and $= 1$ if $x > 1$ lie in this band?

5. The following service times (in minutes) of a random sample of 15 customers were recorded at a bank:

    $$7, 3, 2, 7, 5, 4, 9, 15, 6, 1, 6, 8, 10, 1, 2.$$

    Test the hypothesis that the service times are exponentially distributed with mean 5 minutes.

6. Let X denote the error made when the digits to the right of the decimal point in a number are dropped. Then $X = \text{error} = N - N.n = -0.n$ where N is the integral part of the number $N.n$ when the digits n to the right of the decimal point are dropped. Clearly, $-1 < X \le 0$. Suppose a sample of 10 observations on X showed the following results:

    −.88,   −.15,   −.18,   −.52,   −.81,   −.33,   −.22,   −.90,   −.19,   −.59.

    Test the hypothesis that the distribution function of X is uniform on $(-1, 0]$.

7. In order to test the claim that the Scholastic Aptitude Test scores in mathematics for a certain group of students have a normal distribution with mean 500 and standard deviation 70, a random sample of 12 students from this group was taken and their SAT scores recorded as follows:

    580,   550,   485,   511,   570,   650,   425,   626,   460,   490,   558,   680.

    Is the claim justified?

8. Find a 95 percent confidence band for the distribution function of SAT mathematics scores in Problem 7.

9. The times to failure (in hours) of 10 randomly selected 9-volt transistor batteries of a certain brand were recorded as follows:

    28.9,   15.2,   28.7,   72.5,   48.6,   52.4,   37.6,   49.5,   62.1,   54.5.

    Test the hypothesis that the failure times are exponentially distributed with mean 45 minutes.

11.9   MEASURES OF ASSOCIATION FOR BIVARIATE DATA

In many practical problems involving pairs of measurements it is frequently of interest to study the relationship between the two sets of measurements. For example, are heights and weights of newborns dependent? Is there any relationship between the college entrance test scores and freshman grade point averages? Are dose level of an antihypersensitive drug and blood pressure dependent?

Let X and Y be two jointly distributed random variables. We have seen in Section 5.2 that the covariance between X and Y in a certain sense measures the interrelationship between X and Y. For example, the covariance is large and positive if large (small) values of X occur with large (small) values of Y with a high probability. We note that covariance is sensitive to the scales of measurement used. For this reason we defined the correlation coefficient between X and Y as

$$\rho(X, Y) = \text{cov}(X, Y)/\sqrt{\text{var}(X)\text{var}(Y)}$$

which is invariant under location and scale changes.

If X and Y are independent $\rho(X, Y) = 0$, but the converse is not true. If X and Y are independent there is no relationship or association between X and Y. On the other hand, if X and Y are associated in some way or dependent, ρ may still be zero [except in the case when (X, Y) has a bivariate normal distribution]. Thus, except in the bivariate normal case ρ may not be such a good measure of association between X and Y.

Often the null hypothesis of interest is that X and Y are not associated, to be tested against the alternative hypothesis that they are associated. Under H_0, X and Y are independent so $\rho = 0$ and it is tempting to test $H_0': \rho = 0$ instead. If then H_0' is accepted and we conclude that $\rho = 0$, it does not follow that H_0 is true. If, however, $H_1': \rho \neq 0$ is accepted then it does follow that H_1 is true, that is, that X and Y are associated.

In this section we consider two measures of association between X and Y which will also be used to test the existence of an association between X and Y. In the special case when (X, Y) has a bivariate normal distribution ρ does serve as a measure of association, and we have seen in Example 11.4.3 that the sample correlation coefficient given by

(1)
$$R = \frac{\sum_{i=1}^{n} (X_i - \bar{X})(Y_i - \bar{Y})}{\sqrt{\sum_{i=1}^{n} (X_i - \bar{X})^2 \sum_{i=1}^{n} (Y_i - \bar{Y})^2}}$$

may be used to test $H_0: \rho = 0$ against one-sided or two-sided alternatives. We record the results below for completeness.

Likelihood Ratio Test of $\rho = 0$ when (X, Y) Has a Bivariate Normal Distribution

H_1	Reject H_0 if	P-value
$\rho \neq 0$	$\lvert r_0 \rvert \geq c$ (r_0 has large numerical value)	$P_{H_0}(\lvert R \rvert \geq r_0)$
$\rho > 0$	$r_0 \geq c_1$ (r_0 is large positive)	$P_{H_0}(R \geq r_0)$
$\rho < 0$	$r_0 \leq c_2$ (r_0 is large negative)	$P_{H_0}(R \leq r_0)$

where r_0 is the observed value of R given by (1) and P-value is computed by using the t-statistic $T = R\sqrt{n-2}\,/\sqrt{1 - R^2}$ with $n - 2$ d.f.

Example 1. Testing for Independence. A random sample of 12 observations from a bivariate normal distribution yielded a sample correlation coefficient of $r_0 = .68$. Are X and Y independent? We wish to test $H_0 : \rho = 0$ against $H_1 : \rho \neq 0$. We have

$$t_0 = \frac{r_0\sqrt{n-2}}{\sqrt{1 - r_0^2}} = \frac{.68\sqrt{10}}{\sqrt{1 - .68^2}} = 2.93$$

so that P-value $= 2P_{H_0}(t(10) > 2.93)$, which is between .01 and .02. [Here $t(10)$ has t-distribution with 10 d.f.] Thus H_0 is rejected at levels $\alpha \geq .02$ and we conclude that ρ is significantly different from zero. That is, X and Y are not independent. $\square$

11.9.1 The Spearman's Rank Correlation Coefficient

The Spearman's rank correlation coefficient is essentially the sample correlation coefficient R defined in (1) with X's and Y's replaced by their ranks. We assume that there are no ties among the X's or among the Y's. Let $R(X_i)$ be the rank of X_i among the X's and $R(Y_i)$ be the rank of Y_i among the Y's, and let $D_i = R(X_i) - R(Y_i)$. Then

$$D_i^2 = R^2(X_i) + R^2(Y_i) - 2R(X_i)R(Y_i)$$

and

$$(2) \qquad \sum_{i=1}^{n} D_i^2 = \sum_{i=1}^{n} R^2(X_i) + \sum_{i=1}^{n} R^2(Y_i) - 2\sum_{i=1}^{n} R(X_i)R(Y_i).$$

Note, however, that

$$\sum_{i=1}^{n} R(X_i) = \sum_{i=1}^{n} R(Y_i) = \sum_{j=1}^{n} j = \frac{n(n+1)}{2},$$

and

$$\sum_{i=1}^{n} R^2(X_i) = \sum_{i=1}^{n} R^2(Y_i) = \sum_{j=1}^{n} j^2 = \frac{n(n+1)(2n+1)}{6}.$$

It follows from (2) that

(3)
$$\sum_{i=1}^{n} R(X_i)R(Y_i) = \frac{n(n+1)(2n+1)}{6} - \sum_{i=1}^{n} \frac{D_i^2}{2}.$$

Replacing X_i's and Y_i's by their ranks in (1) and writing R_S for the resulting correlation coefficient, we have

(4)
$$R_S = \frac{\sum_{i=1}^{n} R(X_i)R(Y_i) - n\bar{R}(X)\bar{R}(Y)}{\sqrt{\left[\sum_{i=1}^{n} R^2(X_i) - n\bar{R}^2(X)\right]\left[\sum_{i=1}^{n} R^2(Y_i) - n\bar{R}^2(Y)\right]}}$$

$$= \frac{n(n+1)(2n+1)/6 - (1/2)\sum_{i=1}^{n} D_i^2 - n(n+1)^2/4}{n(n+1)(2n+1)/6 - n(n+1)^2/4}$$

$$= \frac{(n^3-n)/12 - (1/2)\sum_{i=1}^{n} D_i^2}{(n^3-n)/12} = 1 - 6\sum_{i=1}^{n} \frac{D_i^2}{n^3-n}$$

where we have used the fact that

$$\bar{R}(X) = \sum_{i=1}^{n} \frac{R(X_i)}{n} = \frac{(n+1)}{2} = \bar{R}(Y).$$

Spearman's Rank Correlation Coefficient

Let $R(X_i) = $ Rank of X_i in the X's and $R(Y_i) = $ Rank of Y_i in the Y's. Then

$$R_S = 1 - 6\sum_{i=1}^{n} \frac{(R(X_i) - R(Y_i))^2}{n^3-n}.$$

Note that the rankings of the X's and Y's are in complete agreement if and only if $R(X_i) = R(Y_i)$ for each i, and in perfect disagreement if and only if

$R(X_i) = n + 1 - R(Y_i)$ for each i. The difference D_i between $R(X_i)$ and $R(Y_i)$ measures the degree of disagreement between $R(X_i)$ and $R(Y_i)$ in the sense that $D_i = 0$ when $R(X_i) = R(Y_i)$, and the larger the $|D_i|$ (or $\sum_{i=1}^{n} D_i^2$) the greater the disagreement. Thus $\sum_{i=1}^{n} D_i^2$ is a measure of the disagreement between the two sets of ranks. Since large values of $\sum_{i=1}^{n} D_i^2$ are associated with small values of R_S it follows that small values of R_S tend to support the null hypothesis of no association between X and Y. We note that $|R_S| = 1$ indicates a perfect association and $R_S = 0$ means no association. When $R_S = 1$ there is a direct perfect agreement and when $R_S = -1$ there is an inverse perfect disagreement. Hence large (positive) values of R_S tend to support the alternative that there is a direct association and large negative values of R_S support the alternative that there is an indirect association. If $|R_S|$ is too large then the two-sided alternative that some association exists is supported.

Spearman's Test for $H_0: X$ and Y Are Not Associated

H_1	Reject H_0	P-value
Direct Association	$R_S \geq c_1$ (R_S large positive)	$P_{H_0}(R_S \geq r_0)$
Indirect or Inverse Association	$R_S \leq -c_2$ (R_S large negative)	$P_{H_0}(R_S \leq r_0)$
Association	$|R_S| \geq c$ ($|R_S|$ is too large)	2 (smaller tail probability)

Here r_0 is the observed value of R_S.

In order to compute the P-value we note that R_S has a symmetric distribution under H_0. Table A13 in the Appendix gives the right tail probabilities $P_{H_0}(R_S \geq r)$ for $n \leq 10$ and all possible r values. The left tail probabilities are obtained by symmetry. For values of $n = 11, 12, \ldots, 30$, the table gives the smallest value of R (or the largest value of $-R$) for which the right tail (left tail) probability is at most .1, .05, .025, .01, .005.

For larger n (> 30) the statistic

$$Z = R_S\sqrt{n - 1}$$

has approximately a standard normal distribution. In case of a small number of ties one assigns the average rank to tied observations and uses the same null distribution to compute an approximate P-value.

Remark 1. The observations (X_i, Y_i) need not be numerical as long as it is possible to rank the X's and Y's. In many investigations two judges are asked to rank a group of subjects. In that case we can use R_S to assess the degree of agreement between the two rankings.

Example 2. *Dance Contest.* Two judges rated the participants in a dance contest as follows.

| | Contestant | | | | | | | |
Judge	A	B	C	D	E	F	G	H
1	5	2	6	3	4	1	8	7
2	3	1	7	4	5	2	6	8

A question of interest is, how closely do the two judges agree in their ratings? Computing the differences d_i between corresponding rankings we get

$$2, \quad 1, \quad -1, \quad -1, \quad -1, \quad -1, \quad 2, \quad -1$$

so that $\sum_{i=1}^{8} d_i^2 = 14$ and

$$r_S = 1 - \frac{6(14)}{512 - 8} = .833.$$

Since $r_S > 0$ we note that the relationship is direct. Suppose we wish to test H_0: no association, against H_1: direct association. Then the associated P-value is given by

$$P_{H_0}(R_S \geq .83) = .008$$

and we have to conclude that such a high value of R_S could not have been obtained by chance under H_0. That is, the data do support a direct association between the two rankings.   □

Example 3. *Relationship Between Age and Weight of Premature Infants.* Is there a direct relationship between the ages and weights of premature infants? In order to answer this question a random sample of 11 infants was taken and their ages and weights were recorded as follows:

| | Infant | | | | | | | | | | |
	1	2	3	4	5	6	7	8	9	10	11
Age (in weeks)	3	4	1	1	2	2	3	5	6	6	4
Weight (in pounds)	3.5	4	2.5	3	2.5	3	3	5	7	6.5	5

Since there are ties in ages as well as weights, we assign average ranks to tied observations.

| | Infant | | | | | | | | | | |
	1	2	3	4	5	6	7	8	9	10	11
$R(x_i)$	5.5	7.5	1.5	1.5	3.5	3.5	5.5	9	10.5	10.5	7.5
$R(y_i)$	6	7	1.5	4	1.5	4	4	8.5	11	10	8.5
d_i	-.5	.5	0	-2.5	2.0	-.5	1.5	.5	-.5	.5	-1

We have $\sum\limits_{i=1}^{11} d_i^2 = 15$ and

$$r_S = 1 - \frac{6(15)}{11(120)} = .932$$

which indicates a strong direct relationship between ages and weights of premature infants. Indeed,

$$P_{H_0}(R_S \geq .932) < .005.$$

It should be pointed out, however, that the number of ties in this example is not small. In general the value of R_S is underestimated by formula (4) in case of ties, and one needs to apply a correction for ties.[†] It is not necessary to do so in our example since it will only increase the value of r_S. We will not give a formula for this correction here.   □

11.9.2 Kendall's Tau Statistic

Kendall's tau statistic is also based on the ranks of X's and Y's. However, the test statistic is a function of the number of concordant and discordant pairs. We say that two pairs (x_i, y_i) and (x_j, y_j) $(i \neq j)$ are *concordant* if both members of one pair are larger than their respective members of the other pair. That is, (x_i, y_i) and (x_j, y_j) are concordant if and only if

$$(x_i - x_j)(y_i - y_j) > 0.$$

The pairs (x_i, y_i) and (x_j, y_j), $i \neq j$, are said to be *discordant* if and only if

$$(x_i - x_j)(y_i - y_j) < 0.$$

The pairs with ties between (one or both) respective members are neither concordant nor discordant. Thus the pairs $(1.2, 1.9)$, $(2.7, 2.0)$ and also the pairs $(2.4, 1.3)$ $(2.1, 1.1)$ are concordant, whereas the pairs $(1.2, 1.9)$ and $(2.7, 1.6)$ are discordant. The pairs $(1.2, 1.8)$ and $(1.3, 1.8)$ are neither concordant nor discordant. Let

$$N_c = \text{number of concordant pairs,}$$

$$N_d = \text{number of discordant pairs,}$$

and

$$N_0 = \text{number of pairs with ties.}$$

[†] For details, see J. D. Gibbons, *Nonparametric Methods for Quantitative Analysis*, Holt, Rinehart and Winston, New York, 1976, p. 279.

Then

$$N_c + N_d + N_0 = \text{total number of paired comparisons}$$

$$= \binom{n}{2}.$$

The measure of association proposed by Kendall is given by

(5)
$$\tau = \frac{N_c - N_d}{n(n-1)/2}.$$

If all pairs are concordant, then $N_c = \binom{n}{2}$, $N_d = 0$ and $\tau = 1$. If all pairs are discordant then $N_d = \binom{n}{2}$ and $\tau = -1$. If $\tau = 0$, then there is no association since the number of pairs that are concordant is the same as those that are discordant. Thus $|\tau| \leq 1$ and intermediate values of τ reflect the degree of association. The larger the numerical value of τ the greater the degree of association.

Under the assumption that (X, Y) has a continuous distribution, ties occur with probability zero so that $P(N_0 = 0) = 1$ and

$$N_c + N_d = \binom{n}{2}.$$

In that case

(6)
$$\tau = \frac{N_c - N_d}{\binom{n}{2}} = 1 - \frac{4N_d}{n(n-1)} = \frac{4N_c}{n(n-1)} - 1.$$

Whenever ties are present, we have to use (5) to compute τ. Otherwise one can use (6). The computation of τ in (6) is simplified by arranging (x_i, y_i) according to increasing values of x. Then each y is compared with those y's that are below it to determine the number N_d. The number N_c is similarly computed.

Kendall's Tau Statistic for Testing H_0: No Association Between X and Y

Compute $\tau = (N_c - N_d)/\binom{n}{2}$ where N_c is the number of concordant pairs and N_d is the number of discordant pairs.

H_1	Reject H_0 if	P-value
Direct association	$\tau \geq c_1$ (τ is too large)	$P_{H_0}(\tau \geq \tau_0)$
Inverse association	$\tau \leq c_2$ (τ is too small)	$P_{H_0}(\tau \leq \tau_0)$
Association	$|\tau| \geq c$ (τ is either too large or too small)	$2P_{H_0}(\tau \geq |\tau_0|)$

where τ_0 is the observed value of τ.

In order to compute the P-value one needs the null distribution of τ. Table A14 in the Appendix gives the right tail probabilities $P_{H_0}(\tau \geq \tau_0)$ for $n = 3, 4, \ldots, 20$ computed under the assumption that there are no ties. Since the distribution of τ is symmetric, the same table may be used for the left tail probabilities. In case of ties the same table may be used but the resulting test is approximate and conservative. For $n = 21, 22, \ldots, 40$ exact tables are available but not given here. We give only the exact quantile values of the null distribution of τ for some selected tail probabilities. For $n > 40$, the statistic

$$Z = \frac{3\tau\sqrt{n(n-1)}}{\sqrt{2(2n+5)}}$$

has approximately a standard normal distribution so that approximate P-values may be computed from a table of normal distribution.

Example 4. Dance Contest (Example 2). In Example 2, two judges in a dance contest rated the participants as follows:

Judge	\multicolumn{8}{c}{Contestant}
	A	B	C	D	E	F	G	H
1	5	2	6	3	4	1	8	7
2	3	1	7	4	5	2	6	8

In order to compute N_c let us arrange the ranks for Judge 1 in their natural order as follows.

Judge	\multicolumn{8}{c}{Contestant}
	F	B	D	E	A	C	H	G
1	1	2	3	4	5	6	7	8
2	2	1	4	5	3	7	8	6

We now compare each rank for Judge 2 with those that are above it. Thus there are six ranks larger than 2, six ranks larger than 1, and so on. It follows that

$$N_c = 6 + 6 + 4 + 3 + 3 + 1 + 0 + 0 = 23.$$

In view of (6)

$$\tau = \frac{4(23)}{8(7)} - 1 = 1.643 - 1 = .643.$$

Similarly, N_d is the total number of ranks for Judge 2 that are smaller than and to the right of each successive rank. Thus, there is one rank smaller than 2 (on its right), none smaller than 1, one smaller than 4, and so on, so that

$$N_d = 1 + 0 + 1 + 1 + 0 + 1 + 1 + 0 = 5.$$

Consequently,

$$\tau = 1 - \frac{4(5)}{8(7)} = .643.$$

as before. From Table A14

$$P_{H_0}(\tau \geq .643) = .016,$$

which is somewhat larger than the P-value that we obtained for the same data using the Spearman's rank correlation coefficient. The conclusion, however, is the same, that the data support a direct association between the two rankings.

In general, the numerical value of R_S is usually larger than that of τ for the same data. The two tests usually, though not always, produce the same inference. □

Example 5. Reading Ability and Arithmetic Performance of Third Graders. A random sample of nine third-graders received the following scores on tests in reading and arithmetic.

				Student					
Subject	1	2	3	4	5	6	7	8	9
Reading	65	94	82	66	72	84	92	72	80
Arithmetic	72	96	88	60	68	88	98	69	85

In order to compute Kendall's tau coefficient, let us rank the scores in reading and in arithmetic in increasing order (1 for the lowest score to 9 for the highest score). We assign average rank to tied scores.

				Student					
Subject	1	2	3	4	5	6	7	8	9
Reading	1	9	6	2	3.5	7	8	3.5	5
Arithmetic	4	8	6.5	1	2	6.5	9	3	5

In this case τ is computed from (5). We need to compute the difference between the number of concordant and discordant pairs. This is done by listing all the $\binom{n}{2} = \binom{9}{2} = 36$ X and Y pairs. We assign a score of $+1$ for a concordant pair, -1 for a discordant pair, and 0 for a pair where a tie occurs in either an X pair or a Y pair. Table 1 gives this comparison. Clearly $N_c - N_d$ = total number of "$+1$" scores − total number of "-1" scores = $30 - 4 = 26$ and

$$\tau = \frac{26}{\binom{9}{2}} = \frac{26}{36} = .722.$$

Actually the maximum value of $N_c - N_d$ is no longer $\binom{n}{2}$ and depends on the number of tied X ranks, tied Y ranks, and the tied ranks themselves. If the number of ties is large, a correction for ties is usually applied to the denominator, which results in a larger value of τ. We will not give this formula here.[†]

We note that $\tau > 0$, so there is direct association between reading and arithmetic. Since

$$P_{H_0}(\tau \geq .722) = .003$$

the evidence is strong against H_0: no association between reading and arithmetic. □

[†] For details, see J. D. Gibbons, *Nonparametric Methods for Quantitative Analysis*, Holt, Rinehart and Winston, New York, 1976, p. 289.

TABLE 1. Computation of Scores

X Pair	Y Pair	Score	X Pair	Y Pair	Score
1, 9	4, 8	+1	6, 8	6.5, 9	+1
1, 6	4, 6.5	+1	6, 3.5	6.5, 3	+1
1, 2	4, 1	−1	6, 5	6.5, 5	+1
1, 3.5	4, 2	−1	2, 3.5	1, 2	+1
1, 7	4, 6.5	+1	2, 7	1, 6.5	+1
1, 8	4, 9	+1	2, 8	1, 9	+1
1, 3.5	4, 3	−1	2, 3.5	1, 3	+1
1, 5	4, 5	+1	2, 5	1, 5	+1
9, 6	8, 6.5	+1	3.5, 7	2, 6.5	+1
9, 2	8, 1	+1	3.5, 8	2, 9	+1
9, 3.5	8, 2	+1	3.5, 3.5	2, 3	0
9, 7	8, 6.5	+1	3.5, 5	2, 5	+1
9, 8	8, 9	−1	7, 8	6.5, 9	+1
9, 3.5	8, 3	+1	7, 3.5	6.5, 3	+1
9, 5	8, 5	+1	7, 5	6.5, 5	+1
6, 2	6.5, 1	+1	8, 3.5	9, 3	+1
6, 3.5	6.5, 2	+1	8, 5	9, 5	+1
6, 7	6.5, 6.5	0	3.5, 5	3, 5	+1

Problems for Section 11.9

1. At what levels are the following values of the sample correlation coefficient significant? (Assume that the sample comes from a bivariate normal distribution.)
   (i) $r = .35$, $n = 15$, $H_0 : \rho = 0$, $H_1 : \rho > 0$.
   (ii) $r = .28$, $n = 17$, $H_0 : \rho = 0$, $H_1 : \rho \neq 0$.
   (iii) $r = -.64$, $n = 30$, $H_0 : \rho = 0$, $H_1 : \rho < 0$.

2. A random sample of eight students was selected from the freshman class at a state university. For each student the mathematics achievement test scores and the final score in the first calculus course were recorded.

Mathematics score:	49	50	79	63	47	37	55	64	52	59
Final Calculus score:	62	56	84	69	64	52	91	87	84	78

   Assuming that the mathematics achievement scores and the calculus scores have a bivariate normal distribution, is there evidence to conclude that the two scores are dependent?

3. Two judges ranked eight essays as follows:

		Essay							
Judge	1	2	3	4	5	6	7	8
1	3	1	2	6	5	4	8	7
2	4	3	2	5	6	4	7	8

Are the two rankings independent? Use both Kendall's τ and Spearman's rank correlation coefficient.

4. In order to test whether the heights of husband and wife are directly associated, a random sample of eight couples was taken and their heights recorded.

Heights of Married Couples (inches)

	Couple							
	1	2	3	4	5	6	7	8
Husband	62	73	68	74	69	67	78	70
Wife	63	61	66	68	65	64	72	67

(i) Compute the values of τ, R_S and R.
(ii) Test the claim that the heights are directly associated, using τ as well as R_S.
(iii) Assuming normality test $H_0 : \rho = 0$ against $H_1 : \rho > 0$.

5. Fifteen children were given two reading tests. Their relative rankings were as follows:

Child	Test 1	Test 2	Child	Test 1	Test 2
1	10	12	9	14	13
2	12	9	10	6	7
3	3	1	11	15	14
4	1	2	12	2	4
5	4	3	13	13	15
6	9	10	14	11	5
7	8	11	15	5	6
8	7	8			

(i) Compute R_S and τ.   (ii) Is there evidence of a direct association?

6. The heights and weights of eight athletes are as follows:

	Athlete							
	1	2	3	4	5	6	7	8
Height (inches)	68	72	66	69	72	74	71	73
Weight (pounds)	190	194	185	187	210	205	196	210

(i) Compute R, R_S, and τ.
(ii) Assuming that height and weight have a joint normal distribution, test $H_0 : \rho = 0$ against $H_1 : \rho > 0$.
(iii) Using R_S and τ test to check whether there is evidence of a direct association between the height and the weight of an athlete.

7. Seven applicants for the position of a probationary assistant professor were called in for interviews. Each applicant was assigned a rank by the chair in consultation with her advisory committee. Each applicant also saw the Dean, who also ranked the candidates. The following were the results of these two rankings.

	Applicant						
	1	2	3	4	5	6	7
Chair's ranking	3	4	1	2	7	6	5
Dean's ranking	3.5	1	6	5	3.5	2	7

(Average rank was assigned by the Dean to applicants 1 and 5 since in his view they tied.) Test the hypothesis that there is no relationship between the two rankings.

8. Two preseason polls (of coaches and sportswriters) ranked the eight football teams in the Mid-American Conference as follows.

	Team							
	A	B	C	D	E	F	G	H
Coaches	3	2	4	8	6	7	1	5
Sportswriters	2	3	8	7	5	4	1	6

Test the hypothesis that there is a direct association between the two polls.

9. The following table gives the ages at inauguration and at death of the last 20 deceased Presidents of the United States of America.

President	Age at Inauguration	Age at Death
Lincoln	52	56
Johnson, A.	56	66
Grant	46	63
Hayes	54	70
Garfield	49	49
Arthur	50	56
Cleveland[a]	47	71
Harrison	55	67
McKinley	54	58
Roosevelt, T.	42	60
Taft	51	72
Wilson	56	67
Harding	55	57
Coolidge	51	60
Hoover	54	90
Roosevelt, F.	51	63
Truman	60	88
Eisenhower	62	78
Kennedy	43	46
Johnson, L.	54	64

[a]Grover Cleveland had two nonconsecutive terms of office, 1885–1889 and 1893–1897. The age given is that at his first inauguration.

Test the hypothesis that there is a direct association between the age at inauguration and the age at death of U.S. Presidents.

11.10 REVIEW PROBLEMS

1. Find the Neyman–Pearson most powerful size α test of H_0: X is standard normal, against H_1: X has density $f(x) = (1/\pi)[1/(1 + x^2)]$, $x \in \mathbb{R}$, based on a sample of size 1.

2. A random sample of size n from probability function

$$P(X = x) = \binom{m}{x} p^x (1 - p)^{m-x}, \qquad x = 0, 1, \ldots, m$$

is taken in order to test the hypothesis H_0: $p = p_0$ against H_1: $p = p_1$, $p_1 < p_0$. Find the Neyman–Pearson most powerful test of its size.

3. In order to test the hypothesis H_0: $p = p_0$ against H_1: $p = p_1 (p_1 < p_0)$ where $p = P$ (heads) the coin is tossed until r heads are observed. Find the Neyman–Pearson most powerful test of its size. [Hint: The probability function of X, the number of trials needed to observe r heads, is given by

$$P(X = x) = \binom{x - 1}{r - 1} p^r (1 - p)^x, \qquad x = r, r + 1, \ldots .]$$

4. Let $X_1, X_2, \ldots, X_n$ be a random sample from an exponential density with mean β. Find the Neyman–Pearson most powerful size α test of H_0: $\beta = \beta_0$ against H_1: $\beta = \beta_1 (\beta_1 < \beta_0)$.

5*. In Problem 4 find a uniformly most powerful size α test of H_0: $\beta \geq \beta_0$ against H_1: $\beta < \beta_0$.

6*. In Problem 3 find a uniformly most powerful test of its size for testing H_0: $p \geq p_0$ against H_1: $p < p_0$.

7*. Let $X_1, X_2, \ldots, X_n$ be a random sample from the (folded normal) density function

$$f(x) = \frac{2}{\sqrt{2\pi}} \exp\left\{ -\frac{(x - \mu)^2}{2} \right\} \qquad \text{for } x > \mu \text{ and zero otherwise.}$$

Find a uniformly most powerful size α test of H_0: $\mu \leq \mu_0$ against H_1: $\mu > \mu_0$. Find the power function of your test.

8*. An urn contains N balls of which D are colored white. A random sample of size n is taken. Let X be the number of white balls in the sample. Find the form of the Neyman–Pearson most powerful test for testing H_0: $D = D_0$ against H_1: $D = D_1$ ($> D_0$). Find the form of the uniformly most powerful test of H_0': $D \leq D_0$ against H_1': $D > D_0$.

9*. Let $X_1, X_2, \ldots, X_n$ be a random sample from probability function

$$P(X = x) = p(1 - p)^{x-1}, \qquad x = 1, 2, \ldots .$$

Find the likelihood ratio test of H_0: $p = p_0$ against H_1: $p \neq p_0$.

10*. Let $X_1, X_2, \ldots, X_n$ be a random sample from the density function

$$f(x) = \frac{1}{\alpha} \exp\{-(x - \beta)/\alpha\}, \qquad x > \beta, \text{ and zero otherwise.}$$

Find the likelihood ratio test of $H_0: \alpha = \alpha_0$ against $H_1: \alpha < \alpha_0$.

11*. Let $\Omega = \cup_{j=1}^k A_j$ where the A_j's are disjoint and $P(A_j) = p_j > 0$, $j = 1, 2, \ldots, k$, $\sum_{j=1}^k p_j = 1$. Suppose the experiment is repeated n times. Let X_j = number of times A_j happens, $j = 1, 2, \ldots, k$. Then $n = \sum_{j=1}^k X_j$. Find the likelihood ratio test of $H_0: p_i = p_i^0$, $i = 1, 2, \ldots, k$ against all alternatives.

12*. In Problem 11 suppose n is large so that $-2 \ln \lambda(X_1, \ldots, X_k)$ has approximately a $\chi^2(k - 1)$ distribution. Show that the test based on $-2 \ln \lambda$ is equivalent to the χ^2 goodness of fit test based on the statistic

$$Q = \sum_{j=1}^k \frac{\left(X_j - np_j^0\right)^2}{np_j^0}.$$

[Hint: Expand the logarithm and show that only the leading terms are important. That is, show that

$$-2 \ln \lambda = \sum_{j=1}^k \frac{\left(X_j - np_j^0\right)^2}{np_j^0} + R_n$$

where $R_n \to 0$ (in probability) as $n \to \infty$.]

13*. In the Hardy–Weinburg stability principle (see Problems 11.7.1 and 10.4.5), suppose a random sample of n offspring shows that X_i offspring are of type i, $i = 1, 2, 3$, $\sum_{i=1}^3 X_i = n$. [According to the stability principle, the offspring should be in proportions $p_1 = p^2$, $p_2 = 2p(1 - p)$, and $p_3 = (1 - p)^2$.] Show that the likelihood ratio test of $H_0: p = p_0$ against $H_1: p \neq p_0$ is of the form: reject H_0 if

$$\frac{2X_1 + X_2}{2n} - p_0 < k_1 \quad \text{or} \quad \frac{2X_1 + X_2}{2n} - p_0 > k_2,$$

where k_1, k_2 are constants.

14.   The general manager of a discount department store chain of 12 stores is interested in finding out whether shoelaces sell best when placed in the shoe department or when located at the checkout counter. Each store manager is asked to try a display of shoelace packages for four weeks in the shoe department and then switch it to the checkout counter for the next four weeks. The following results were obtained.

						Store						
	1	2	3	4	5	6	7	8	9	10	11	12
Shoe department	100	104	92	110	84	73	84	95	86	72	90	104
Checkout counter	115	127	89	95	100	85	84	104	87	63	95	142

Is there evidence to conclude that the shoelaces sell better at the checkout counter? Use the sign test as well as the signed rank test to analyze the data.

15. Eighteen patients on a certain diet showed the following weight losses (in pounds):

$$7, \ 6, \ 9, \ 10, \ 1, \ -2, \ 4, \ 2, \ 1, \ 10, \ 8, \ 6, \ 5, \ 3, \ -1, \ 0, \ 8, \ 9.$$

Test the null hypothesis that the diet is ineffective (median weight loss is 0) against the alternative that the diet is effective (median weight loss is positive). You may assume that the distribution of weight loss is symmetric and continuous.

16. The median age of persons arrested for driving under the influence of alcohol (DUI) in a college town is believed to be 22 years. The sheriff believes that the median age of persons arrested for DUI is lower. She takes a random sample of 10 such arrest records and finds the following ages:

$$19, \quad 20, \quad 18, \quad 24, \quad 16, \quad 21, \quad 21, \quad 35, \quad 65, \quad 22.$$

Use the Wilcoxon signed rank test to test the claim that the median age of a DUI arrest is 22 years against the alternative hypothesis that the median age is lower. Also apply the sign test and compare your results.

17. A teacher is concerned about the poor performance of students in her introductory statistics course. She believes that this is due to the poor mathematics background of her students. She selects a random sample of 15 students in her class and divides them into two groups according to their mathematics background. The final scores of these 15 students are as follows:

Group 1 (Less than two years of school mathematics):	80, 54, 68, 53, 52, 67
Group 2 (More than 2 years of school mathematics):	92, 84, 48, 65, 86, 83, 58, 86, 72.

Test the null hypothesis that there is no difference in median performance of students with the two mathematics backgrounds against the appropriate alternative hypothesis.

18. Two auto mechanics charge different rates for the same job. Mechanic A, whose charge is lower, claims that this is because he can do the same job much faster than Mechanic B, whose rates are higher. Random samples of six cars each, which came for identical repairs to the two mechanics showed the following repair times, in minutes:

Mechanic A:	17,	28,	21,	16,	17,	22
Mechanic B:	19,	29,	30,	20,	25,	28.

Is Mechanic A justified in his claim?

19. Use a runs test in Problem 18 to test the null hypothesis that there is no difference in repair times of the two mechanics against the alternative that the repair times are different.

20. A student is hired to survey 20 randomly selected customers at a store. The ages of respondents are recorded in the order in which they were canvassed, as follows:

$$B\ B\ B\ A\ B\ B\ A\ A\ B\ A\ B\ B\ A\ A\ A\ A\ B\ B\ B\ A$$

where B represents aged 20 and below, and A represents aged 21 and above. Use the runs test to test the null hypothesis that the student selected the respondents at random (with respect to their age).

21. An arithmetic test is administered to a random sample of 10 sixth-graders from the Sylvania, Ohio, school district. The same test is administered to a random sample of 15 sixth-graders from the Toledo, Ohio, school district. The scores are as follows:

Sylvania: 70, 89, 74, 65, 76, 81, 82, 75, 89, 83
Toledo: 72, 69, 74, 82, 56, 66, 83, 75, 59, 62, 68, 73, 51, 61, 78.

Using the Mann–Whitney–Wilcoxon test, test the hypothesis that there is no difference in the performance of sixth graders in the two districts.

22. The runs test can be used to detect randomness of a sequence of measurements. To apply the test, first compute the sample median and then label each measurement in the ordered arrangement as being below (b) the median or above (a) the median. Count the number of runs. Reject H_0 that the sequence is random against the alternative that the sequence is nonrandom if the observed number of runs is either too large or too small.
    (i) In particular, consider a sample from the telephone book consisting of 15 single-digit numbers, using the next-to-the-last digit from each of the selected numbers:

$$5,\ 2,\ 3,\ 1,\ 1,\ 4,\ 7,\ 6,\ 6,\ 6,\ 9,\ 8,\ 0,\ 1,\ 2.$$

Is it a random sample?
    (ii) Suppose the distances of 25 pinholes in electrolytic tinplate, from one edge of a long strip of tinplate 20 inches wide, are as follows:

13.8,   4.6,   6.8,   9.2,   7.4,   2.0,   4.7,   13.2,   18.9,   15.4,   12.4,
11.8,   3.7,   2.8,   19.2,   18.4,   13.6,   11.2,   10.4,   4.6,   7.6,   6.4,
8.9,   12.6,   14.2.

Is this a random sample?

23. In 300 flips of four coins together, the following results were obtained.

Number of tails:	0	1	2	3	4
Frequency:	15	65	110	94	16

Test the hypothesis that all four coins are fair.

24. In Problem 22(ii), test the hypothesis that the sample comes from a uniform distribution:
    (i) Use the chi-square test.   (ii) Use the Kolmogorov–Smirnov test.

25. In the manufacture of explosives a certain number of ignitions may occur randomly. The number of ignitions per day is recorded for 150 days with the following results:

Number of Ignitions:	0	1	2	3	4	5	6
Number of Days:	45	53	33	12	4	2	1

Test the hypothesis that the number of ignitions per day is a Poisson random variable with mean λ for some $\lambda > 0$.

26. In a random sample of 60 cancer patients who were treated with a certain drug, 25 lived less than ten years, 12 lived ten or more but less than twenty years, 10 lived twenty or more but less than thirty years, 9 lived thirty or more but less than forty years, and 4 lived forty years or more.
    (i) Test the hypothesis that the lifetimes have an exponential distribution with mean 20 years.
    (ii) If in the 4 patients who lived forty years or more none lived for more than fifty years, test the hypothesis that the lifetimes have an exponential distribution.

27. A pediatrician tested the cholesterol level of 20 high-level young patients with the following results.

189	209	217	221	218	215	210	204	198	215
201	211	204	217	213	200	206	219	213	207

Test the hypothesis that the cholesterol level in high-level young patients is normally distributed.

28. The following data on the age of papaya trees and their heights were recorded.

X, Years:	3	1	2	5	4	5	2
Y, Feet:	8	5	6	14	11	12	7

Find the sample correlation coefficient. Is there evidence to conclude that the height and the age of a papaya tree are positively correlated?

29. The price and demand for gold in six different markets is given below.

Price (in hundreds of dollars):	1.9	2.8	4.9	8.5	3.8	6.5
Demand (in thousands of ounces):	50	42	24	20	40	22

Find the sample correlation coefficient between the price of gold and the demand for gold.

30. A customer who regularly visits a particular restaurant for lunch rates the quality of food and service over a period of 25 consecutive days as good (g) or bad (b) as follows:

$$g\ g\ b\ b\ g\ b\ b\ g\ g\ g\ b\ g\ g\ b\ b\ b\ g\ b\ g\ b\ g\ b\ g\ b\ b$$

Test the null hypothesis that the quality of food and service at this restaurant is random.

31. Fifty successive light bulbs produced by a certain machine are examined to see if they are defective (D) or nondefective (N) with the following results:

    D D D D N D N N D D D N D D N N N D N D D D N D N
    D D N D D D D D N D D D D D N N N D D D D D D D D

    Do the data indicate that the defective light bulbs occur randomly?

32. An educational scientist took a random sample of 15 fathers and sons and obtained the following data for their intelligence quotients.

	Family							
	1	2	3	4	5	6	7	8
Father	105	120	148	94	110	168	120	116
Son	115	132	154	100	108	149	135	98

	9	10	11	12	13	14	15
Father	132	143	101	128	124	112	135
Son	125	153	98	130	135	108	160

    Is there evidence of a direct association between a father's and son's I.Q.? Apply both Spearman's and Kendall's tests.

33. A random sample of nine middle-income families of four showed the following data on the monthly food cost and disposable income (in dollars).

	Family								
	1	2	3	4	5	6	7	8	9
Disposable income	1750	2800	1875	1900	2500	3000	2150	1900	2600
Food Cost	350	400	339	410	450	500	415	395	450

    Is there evidence of a direct relationship between the monthly disposable income and food costs?

CHAPTER 12

Analysis of Categorical Data

12.1 INTRODUCTION

Suppose each observation in a random sample is classified as belonging to one of a finite number of mutually exclusive classes or categories. For example, the family income may be classified as belonging to one of the five classes: poor ($10,000 or less), lower middle class (between 10 and 20 thousand dollars), middle class (20–30 thousand dollars), upper middle class (30–40 thousand dollars), and affluent (more than $40,000). Let $p_i = P$(an observation is in category i). The object may be to test the null hypothesis H_0: $p_i = p_i^0$, $i = 1, 2, \ldots, k$ when p_i^0 are completely specified. The reader will recognize that H_0 can be tested by using the χ^2 test of goodness of fit studied in Section 9.9 (when p_i^0's are prespecified) and Section 11.7 (when p_i^0's are specified except for some unknown parameter which has to be estimated from the data).

The chi-square test applies also to the case when we wish to test that

(i) the p_i's are the same for several different populations, or

(ii) family income is independent, say, of the geographical region in which a family resides.

In the former case we take independent random samples from the J-populations being compared, and the object is to test H_0: $p_{i1} = p_{i2} = \cdots = p_{iJ}$, $i = 1, 2, \ldots, k$, against all possible alternatives. The χ^2 test is specially designed for discrete data and it tests homogeneity of the J populations against general alternatives. This test is studied in Section 12.2. In the latter case we have one sample from a population categorized in two ways (for example, according to family incomes and according to geographic region). The resulting cross-categorized data can be put in a so-called *contingency table* with the object of testing H_0: The two categorical variables are independent. This test is studied in Section 12.3.

12.2 CHI-SQUARE TEST FOR HOMOGENEITY

Suppose we have two or more populations and we wish to test the null hypothesis that they all have the same distribution. In Section 11.6 we considered two well known tests for testing the equality of two distribution functions. These tests assume that the underlying distributions are continuous and do not easily extend to the case of more than two populations. The test that we consider here is specially designed for discrete populations although it applies also to continuous populations.

Let us begin with a simple case. Suppose we have two coins and we wish to test H_0: $p_1 = p_2 = p$ (known) against all possible alternatives where p_1, p_2 are the respective success probabilities. Suppose we toss the first coin n_1 times and observe X successes and the second coin n_2 times and observe Y successes. These results can now be put in a 2×2 contingency table as follows:

TABLE 1

	1st coin	2nd coin	Total
Successes	X	Y	$X + Y$
Failures	$n_1 - X$	$n_2 - Y$	$n - X - Y$
Sample size	n_1	n_2	$n = n_1 + n_2$

We assume that the $n_1 + n_2$ trials are independent. According to Section 9.9, under H_0, the statistics

$$\frac{(X - n_1 p)^2}{n_1 p} + \frac{(n_1 - X - n_1(1 - p))^2}{n_1(1 - p)}$$

and

$$\frac{(Y - n_2 p)^2}{n_2 p} + \frac{(n_2 - Y - n_2(1 - p))^2}{n_2(1 - p)}$$

are independent random variables each of which, for large n_1 and n_2, has approximately a chi-square distribution with $2 - 1 = 1$ d.f. Hence their sum is itself a χ^2 random variable with 2 d.f. Note, however, that

$$\frac{(X - n_1 p)^2}{n_1 p} + \frac{(n_1 - X - n_1(1 - p))^2}{n_1(1 - p)} = \frac{(X - n_1 p)^2}{n_1 p(1 - p)}$$

so that

(1)
$$Q = \frac{(X - n_1 p)^2}{n_1 p(1 - p)} + \frac{(Y - n_2 p)^2}{n_2 p(1 - p)}$$

has approximately a χ^2 distribution with 2 d.f. In complete analogy to the development in Section 9.9, therefore, we reject H_0: $p_1 = p_2 = p$ at level α if

$$Q > \chi^2_{2,\alpha}.$$

The case of most interest is when the common value of p_1 and p_2, under H_0, is unknown. In that case the maximum likelihood estimate of p is given by

(2)
$$\hat{p} = \frac{X + Y}{n_1 + n_2}.$$

According to Theorem 11.7.1 (and the discussion following it) if we substitute $\hat{p}$ for p in the expression for Q in (1), then the resulting test statistic $Q(\hat{p})$ has an approximate χ^2 distribution with $2 - 1 = 1$ d.f. since we estimated one parameter from the sample. We have

$$Q(\hat{p}) = \frac{n_2(X - n_1\hat{p})^2 + n_1(Y - n_2\hat{p})^2}{n_1 n_2 \hat{p}(1 - \hat{p})}$$

$$= \frac{n_2(n_2 X - n_1 Y)^2 + n_1(n_1 Y - n_2 X)^2}{n_1 n_2 (X + Y)(n_1 + n_2 - X - Y)}$$

(3)
$$= \frac{n(n_2 X - n_1 Y)^2}{n_1 n_2 (X + Y)(n - X - Y)}.$$

We therefore reject H_0: $p_1 = p_2$ at level α if $Q(\hat{p}) > \chi^2_{1,\alpha}$.

When n_1, n_2 are large, we saw in Section 9.7 that the statistic

(4)
$$Z = \frac{(X/n_1) - (Y/n_2)}{\sqrt{\hat{p}(1 - \hat{p})(1/n_1 + 1/n_2)}}$$

has, under H_0, approximately a standard normal distribution. Are the tests based on Z given in (4) and Q given in (3) equivalent? The answer is yes. Indeed, Z^2 has approximately a $\chi^2(1)$ distribution. Moreover,

$$Z^2 = n\frac{(n_2 X - n_1 Y)^2}{n_1 n_2 (X + Y)(n - X - Y)} = Q(\hat{p}).$$

We can rewrite Q in a more recognizable form:

(5)
$$Q(\hat{p}) = \frac{n\{X(n_2 - Y) - Y(n_1 - X)\}^2}{n_1 n_2(X + Y)(n - X - Y)}.$$

In terms of Table 1, it means that to compute Q we cross-multiply the table entries, take the difference and square it, multiply it by the total number of observations, and then divide by the product of row and column totals.

Example 1. Faculty Attitudes Toward Teaching Loads.   Independent random samples of 50 faculty members from the Colleges of Arts and Sciences and Business at a State University were asked whether they felt that the teaching loads were reasonable or unreasonable. The following results were obtained.

Response\College	Arts and Sciences	Business	Total
Reasonable	30	24	54
Unreasonable	20	26	46
Sample size	50	50	100

Let us test the hypothesis that the faculty in the two colleges have the same attitude toward teaching loads. We have

$$q_0 = \frac{100[30(26) - 20(24)]^2}{54(46)(50)(50)} = 1.45.$$

Under H_0, we note that

$$.20 < P_{H_0}\left(\chi^2(1) \geq 1.45\right) < .30$$

and there is sufficient evidence in support of H_0.
   Let us also apply the normal theory test of Section 8.7. We have

$$z = \frac{.6 - .48}{\sqrt{.54(.46)/25}} = 1.20$$

so that

$$P_{H_0}(Z \geq 1.20) = .1101$$

and the P-value is .2202. This result is consistent with the result based on the χ^2 test. Indeed, $q_0 = 1.45 \doteq z^2$, as it must. $\square$

The outline for the general case should now be evident. Let us suppose that we have J independent multinomial distributions. That is, suppose we have k

categories $A_1, A_2, \ldots, A_k$ and J independent samples of sizes $n_1, n_2, \ldots, n_J$ respectively. Let X_{ij} be the number of times A_i happens in the jth sample, $i = 1, 2, \ldots, k$, $j = 1, 2, \ldots, J$. The data can be put in a $k \times J$ contingency table as follows:

TABLE 2

Category	Sample						Row Total
	1	2	$\cdots$	j	$\cdots$	J	
A_1	X_{11}	X_{12}	$\cdots$	X_{1j}	$\cdots$	X_{1J}	$X_{1\cdot}$
A_2	X_{21}	X_{22}	$\cdots$	X_{2j}	$\cdots$	X_{2J}	$X_{2\cdot}$
$\vdots$	$\vdots$	$\vdots$	$\vdots$	$\vdots$	$\vdots$	$\vdots$	$\vdots$
A_i	X_{i1}	X_{i2}	$\cdots$	X_{ij}	$\cdots$	X_{iJ}	$X_{i\cdot}$
$\vdots$	$\vdots$	$\vdots$	$\vdots$	$\vdots$	$\vdots$	$\vdots$	$\vdots$
A_k	X_{k1}	X_{k2}	$\cdots$	X_{kj}	$\cdots$	X_{kJ}	$X_{k\cdot}$
Sample size	n_1	n_2	$\cdots$	n_j	$\cdots$	n_J	n

Let $p_{ij} = P_j(A_i)$, $i = 1, 2, \ldots, k$, $j = 1, 2, \ldots, J$ where P_j refers to the probability under the jth population. The hypothesis of homogeneity of the J populations that we wish to test is $H_0 : p_{i1} = p_{i2} = \cdots = p_{iJ}$, $i = 1, 2, \ldots, k$, against all possible alternatives. We consider the case of most interest where, under H_0, the common value p_i of the p_{ij}'s is unknown. The maximum likelihood estimates of the p_i's are easily shown to be

$$(6) \qquad \hat{p}_i = \frac{\sum_{j=1}^{J} X_{ij}}{\sum_{j=1}^{J} n_j} = \frac{X_{i\cdot}}{n}, \; i = 1, 2, \ldots, k-1$$

and $\hat{p}_k = 1 - \hat{p}_1 - \cdots - \hat{p}_{k-1}$, where $X_{i\cdot} = \sum_{j=1}^{J} X_{ij}$ and $n = \sum_{j=1}^{J} n_j$. We note that we have estimated $k-1$ parameters. In view of Theorem 11.7.1 we note that the statistic

$$(7) \qquad Q = \sum_{j=1}^{J} \sum_{i=1}^{k} \frac{\left(X_{ij} - n_j \hat{p}_i \right)^2}{n_j \hat{p}_i}$$

has, for large n, an approximate chi-square distribution with d.f. given by

$$\text{d.f.} = J(k-1) - (k-1) = (J-1)(k-1).$$

The argument used here is simple. If the p_i's are known, then

$$\sum_{j=1}^{J} \sum_{i=1}^{k} \frac{\left(X_{ij} - n_j p_i\right)^2}{n_j p_i}$$

is the sum of J independent (approximate) $\chi^2(k-1)$ random variables so that it has an approximate $\chi^2(J(k-1))$ distribution. Since we estimated $k-1$ parameters, we lose $k-1$ d.f.

Although the above procedure has been specially designed to test the equality of two or more multinomial distributions, it can be adapted to test the equality of two or more continuous distribution functions. Let us take $J = 2$ for convenience. We wish to test H_0: $F(x) = G(x)$ for all x against H_1: $F(x) \neq G(x)$ for some x. We have already considered (Section 11.6) some special tests designed to test this hypothesis. Let $X_1, X_2, \ldots, X_{n_1}$ and $Y_1, Y_2, \ldots, X_{n_2}$ be independent random samples from F and G respectively. Partition the real line into k disjoint intervals $A_1, A_2, \ldots, A_k$ and let

$$p_{i1} = P(X \in A_i), \quad p_{i2} = P(Y \in A_i), \quad i = 1, 2, \ldots, k.$$

Under H_0 we note that $p_{i1} = p_{i2}$ for $i = 1, 2, \ldots, k$. We test the less restrictive hypothesis H_0': $p_{i1} = p_{i2}$, $i = 1, 2, \ldots, k$ against all possible alternatives. Then the appropriate test statistic is

$$\sum_{j=1}^{2} \sum_{i=1}^{k} \frac{\left(X_{ij} - n_j \hat{p}_i\right)^2}{n_j \hat{p}_i}$$

where X_{i1} = number of X_j's that fall into A_i, and X_{i2} = number of Y_j's that fall into A_i, $i = 1, 2, \ldots, k$, and $\hat{p}_i = (X_{i1} + X_{i2})/(n_1 + n_2)$.

In applications, the question arises how the intervals A_i's are selected. This question has already been answered in Section 9.9.

Example 2.   Sex Differences in Incidence and Amount of Aggression.   In Problem 11.6.4 the reader was asked to test if there is evidence to suggest that there are sex differences in the incidence and amount of aggression in three-year-olds. The following scores were recorded:

Boys: 96,   65,   74,   78,   82,   121,   68,   79,   111,   48,   53,   92,   81,   31,   48
Girls: 12,   47,   32,   59,   83,   14,   32,   15,   17,   82,   21,   34,   9,   15,   50

Let us divide the scores in three classes of equal length, namely $A_1 = [8, 42]$, $A_2 = [43, 77]$, $A_3 = [78, 112]$. We have the following result:

Sex	Frequency			Sample Size
	A_1	A_2	A_3	
Boys	1	6	8	15
Girls	10	3	2	15
Total	11	9	10	30

Then the estimates of p_1, p_2, p_3 are given by

$$\hat{p}_1 = 11/30, \hat{p}_2 = 9/30, \hat{p}_3 = 10/30$$

and

$$q_0 = \frac{(1 - 11/2)^2}{5.5} + \frac{(6 - 9/2)^2}{4.5} + \frac{(8 - 5)^2}{5}$$

$$+ \frac{(10 - 11/2)^2}{5.5} + \frac{(3 - 9/2)^2}{4.5} + \frac{(2 - 5)^2}{5}$$

$$= 11.96.$$

For $(3 - 1)(2 - 1) = 2$ d.f. we note that

$$.001 < P_{H_0}(Q \geq 11.96) < .01$$

and the two distributions are significantly different at 1 percent level. This result is consistent with the result of Problem 11.6.4.　　　　　　　　　　□

Example 3.　Television Viewing Preference.　A market analyst believes that there is no difference in preferences of television viewers among the four Ohio cities of Toledo, Columbus, Cleveland, and Cincinnati. In order to test this belief, independent random samples of 150, 200, 250, and 200 persons were selected from the four cities and asked, "What type of program do you prefer most: Mystery, Soap, Comedy, or News Documentary?" The following responses were recorded.

Program Type	City			
	Toledo	Columbus	Cleveland	Cincinnati
Mystery	50	70	85	60
Soap	45	50	58	40
Comedy	35	50	72	67
News	20	30	35	33
Sample size	150	200	250	200

Under the null hypothesis that the proportions of viewers who prefer the four types of programs are the same in each city, the maximum likelihood estimates of p_i, $i = 1, 2, 3, 4$ are given by

$$\hat{p}_1 = \frac{50 + 70 + 85 + 60}{150 + 200 + 250 + 200} = \frac{265}{800} = .33, \hat{p}_3 = \frac{35 + 50 + 72 + 67}{800} = \frac{224}{800} = .28,$$

$$\hat{p}_2 = \frac{45 + 50 + 58 + 40}{800} = \frac{193}{800} = .24, \qquad \hat{p}_4 = \frac{20 + 30 + 35 + 33}{800} = \frac{118}{800} = .15.$$

Here p_1 = proportion of people who prefer mystery, and so on. The following table gives the expected frequencies under H_0.

Program Type	Expected Number of Responses Under H_0			
	Toledo	Columbus	Cleveland	Cincinnati
Mystery	$150 \times .33 = 49.5$	$200 \times .33 = 66$	$250 \times .33 = 82.5$	$200 \times .33 = 66$
Soap	$150 \times .24 = 36$	$200 \times .24 = 48$	$250 \times .24 = 60$	$200 \times .24 = 48$
Comedy	$150 \times .28 = 42$	$200 \times .28 = 56$	$250 \times .28 = 70$	$200 \times .28 = 56$
News	$150 \times .15 = 22.5$	$200 \times .15 = 30$	$250 \times .15 = 37.5$	$200 \times .15 = 30$
Sample size	150	200	250	200

It follows that

$$q_0 = \frac{(50 - 49.5)^2}{49.5} + \frac{(45 - 36)^2}{36} + \frac{(35 - 42)^2}{42} + \frac{(20 - 22.5)^2}{22.5}$$

$$+ \frac{(70 - 66)^2}{66} + \frac{(50 - 48)^2}{48} + \frac{(50 - 56)^2}{56} + \frac{(30 - 30)^2}{30}$$

$$+ \frac{(85 - 82.5)^2}{82.5} + \frac{(58 - 60)^2}{60} + \frac{(72 - 70)^2}{70} + \frac{(35 - 37.5)^2}{37.5}$$

$$+ \frac{(60 - 66)^2}{66} + \frac{(40 - 48)^2}{48} + \frac{(67 - 56)^2}{56} + \frac{(33 - 30)^2}{30}$$

$$= 9.37.$$

Since $J = 4$ and $k = 4$, the number of degrees of freedom is $(4 - 1)(4 - 1) = 9$ and we note that under H_0

$$.30 < P(Q \geq 9.37) < .50.$$

With such a large P-value we can hardly reject H_0. The data do not offer any evidence to conclude that the proportions in the four cities are different.   □

In conclusion, we emphasize that the χ^2 test studied in this section is a test designed for categorical data, and it tests the equality of J proportions against general alternatives. In a specific problem we may be more interested in testing the null hypothesis against restricted alternatives such as $p_{i1} > p_{i2} > \cdots > p_{iJ}$ for $i = 1, 2, \ldots, k - 1$. Any test specifically designed for this purpose will usually perform better than the χ^2 test of homogeneity. Similarly, tests designed for testing equality of several continuous distribution functions should perform better than the χ^2 test. (Such tests are studied in Chapter 13.)

Problems for Section 12.2

1. Three dice are rolled independently 360 times each with the following results.

Face Value	Die 1	Die 2	Die 3
1	50	62	38
2	48	55	60
3	69	61	64
4	45	54	58
5	71	78	73
6	77	50	67
Sample size	360	360	360

Are all the dice equally loaded? That is, test the hypothesis $H_0: p_{i1} = p_{i2} = p_{i3}$, $i = 1, 2, \ldots, 6$, where p_{i1} is the probability of getting an i with die 1, and so on.

2. Three different methods of instruction were used to teach a calculus course. In order to compare the effectiveness of the three methods, 45 comparable students were divided into three groups of 15 each. Each group was taught using one of these three methods. After the semester a standardized test was given and the following scores were recorded.

Method	Scores
I	92, 48, 37, 90, 87, 74, 83, 75, 54, 62, 82, 78, 81, 51, 68
II	77, 76, 81, 84, 65, 52, 87, 43, 61, 59, 71, 73, 82, 71, 50
III	73, 76, 75, 68, 52, 55, 79, 84, 63, 56, 49, 53, 72, 69, 89

Test the equality of the distributions of test scores. (Partition the combined sample of 45 scores into three equal parts, each part containing 15 scores by taking the intervals $[37, 62.5)$, $(62.5, 76.5)$, $(76.5, 92]$.)

3. Three independent random samples are taken with the following results:

Sample 1: 1.22, −.43, 1.29, .54, −1.66, .66, .34, .01, .11, 1.29, −.55, −1.18, −.51, .84, .58, −.43, −.13, −.73, 1.04, 2.04

Sample 2: −2.01, −.62, −.69, .48, −.58, −.57, −.12, .19, −.07, −3.00, .35, −.09, 1.5, .03, .40, .88, .45, −.79, 1.8, −.12

Sample 3: .00, −1.27, −1.79, −.98, −1.36, −.88, −.15, .83, −.81, −1.34, .50, −.31, 2.88, .19, −1.27, 1.26, −.28, 1.70, .58, −.43, 1.04, .73, 1.16, −.49, 1.00

(i) Test the hypothesis that the samples come from a standard normal distribution. You may use intervals $A_1 = (-\infty, -.75]$, $A_2 = (-.75, 0)$, $A_3 = [0, .75)$, $A_4 = [.75, \infty)$.

(ii) Using the same intervals as in part (i), test the hypothesis that the three samples come from the same population.

4. Independent random samples of 250 Democrats, 150 Republicans, and 100 Independent voters were selected one week before a nonpartisan election for mayor of a large city. Their preference for candidates Albert, Basu, and Chatfield were recorded as follows.

	Party Affiliation		
Preference	Democrat	Republican	Independent
Albert	160	70	90
Basu	32	45	25
Chatfield	30	23	15
Undecided	28	12	20
Sample size	250	150	150

Are the proportions of voters in favor of Albert, Basu, and Chatfield the same within each political affiliation?

5. An outpatient clinic at the County Health Center conducted an experiment to determine the degree of relief provided by three pain relievers. Each pain reliever was given to 75 patients with the following results:

	Pain Reliever		
Degree of Relief	A	B	C
Little or none	15	19	13
Moderate	42	32	40
Complete	18	24	22
Sample size	75	75	75

Is there evidence to conclude that the three pain relievers are equally effective?

6. Independent random samples of 50 students each were selected from the colleges of Arts and Sciences, Music, Business, Engineering, and Law at a large state university in order to determine whether there is a difference in proportions of students in the colleges who favor a proposal to have a student representative on the Board of Trustees. The following responses were recorded.

	College				
Response	Arts and Science	Music	Business	Engineering	Law
Favor	38	30	32	33	35
Oppose	12	20	18	17	15
Sample size	50	50	50	50	50

Is there evidence to conclude that the proportions of students in favor of representation is the same for these five colleges?

7. It is believed that the level of education that an 18-year-old man has received strongly influences his attitude toward the draft. Independent random samples were drawn of 100 men with only elementary school education, 100 with a high school diploma, and 75 with at least a year of college. The following data were collected.

Opinion on Draft	Education		
	Elementary	High School	College
Fair	62	53	20
Unfair	38	47	55
Sample size	100	100	75

Is there sufficient evidence to conclude that the distribution of the attitudes toward the draft differs among the three populations?

8. Independent random samples of 280 divorced males and 350 divorced females resulted in the following age distribution.

Age	Sex	
	Male	Female
30 or less	124	180
31–40	55	88
41–50	58	35
51 and more	43	47
Sample size	280	350

Is there sufficient evidence to conclude that the age distributions of divorced males and females are the same?

9. Independent random samples of 100 immigrants from Europe, 150 immigrants from the Far East, and 200 immigrants from Southeast Asia were taken after one year of their arrival into the United States. Their annual incomes were recorded as follows.

Annual Income (thousands of dollars)	Immigrants		
	Europe	Far East	Southeast Asia
Under 6	3	5	50
6–15	15	30	110
16–25	52	40	34
26 and over	30	75	6
Sample size	100	150	200

Is there sufficient evidence to conclude that the distribution of annual income is the same for each group of immigrants?

10*. Let $(X_{11}, X_{12}, \ldots, X_{k1}), \ldots, (X_{1J}, X_{2J}, \ldots, X_{kJ})$ be independent multinomial random variables with parameters $(n_1, p_{11}, p_{12}, \ldots, p_{k1}), \ldots, (n_J, p_{1J}, p_{2J}, \ldots, p_{kJ})$ respectively. Let $X_i = \sum_{j=1}^{J} X_{ij}$, $X_{\cdot j} = \sum_{i=1}^{k} X_{ij}$, and $\sum_{i=1}^{k}\sum_{j=1}^{J} X_{ij} = \sum_{i=1}^{k} X_i = \sum_{j=1}^{J} X_{\cdot j} = n$. Show that the likelihood ratio statistic for testing H_0: $p_{ij} = p_i$, $j = 1, 2, \ldots, J$, $i = 1, 2, \ldots, k$ where p_i are unknown against all alternatives is given by

$$\lambda(x) = \prod_{i=1}^{k} \left(\frac{x_i}{n}\right)^{x_i} \Big/ \prod_{i=1}^{k} \prod_{j=1}^{J} \left(\frac{x_{ij}}{n_j}\right)^{x_{ij}}$$

12.3 TESTING INDEPENDENCE IN CONTINGENCY TABLES

In Section 11.9 we defined some measures of association between two sets of measurements and used these to test the null hypothesis that the two sets of measurements are not associated and (hence independent). The χ^2 statistic can also be used to yield a test for independence. The test is specially designed for categorical data, although it can be adapted for use in the case of continuous data. The data for analysis therefore are frequency counts in each cross-category as against the rank data used in Section 11.9. The test answers questions of statistical interest such as: Is the color of a person's eyes independent of his or her hair color? Are the opinions of voters on a particular issue independent of their political views?

Consider a random experiment where each outcome can be categorized by two different attributes (such as height and weight, or level of education and annual income). Let $A_1, A_2, \ldots, A_r$ be the r mutually exclusive and exhaustive categories of the first attribute and let $B_1, B_2, \ldots, B_c$ be the c mutually exclusive and exhaustive categories of the second attribute. Write

$$p_{ij} = P(A_i \cap B_j), \qquad i = 1, 2, \ldots, r, j = 1, 2, \ldots, c.$$

Suppose that in n independent trials of the experiment X_{ij} is the frequency of the event $A_i \cap B_j$. Then $\sum_{i=1}^{r}\sum_{j=1}^{c} X_{ij} = n$ and $\sum_{i=1}^{r}\sum_{j=1}^{c} p_{ij} = 1$. In view of Section 9.9, the statistic

(1)
$$Q_1 = \sum_{i=1}^{r} \sum_{j=1}^{c} \frac{(X_{ij} - np_{ij})^2}{np_{ij}}$$

has, for sufficiently large n, an approximate chi-square distribution with $rc - 1$ d.f.

Suppose we wish to test the null hypothesis

$$H_0: p_{ij} = P(A_i)P(B_j), \qquad \text{for all } i, j$$

against all possible alternatives. Let

$$p_{i\cdot} = P(A_i) = \sum_{j=1}^{c} p_{ij}, \quad p_{\cdot j} = P(B_j) = \sum_{i=1}^{r} p_{ij}.$$

Then we can rewrite H_0 as follows:

$$H_0: p_{ij} = p_{i\cdot} p_{\cdot j}, \quad i = 1, 2, \ldots, r \quad \text{and} \quad j = 1, 2, \ldots, c.$$

In applications, $p_{i\cdot}$ and $p_{\cdot j}$ are usually unknown so that we cannot use the statistic Q_1 defined in (1) with p_{ij} replaced by $p_{i\cdot} p_{\cdot j}$. In that case we estimate $p_{i\cdot}$ and $p_{\cdot j}$ by their maximum likelihood estimates, namely,

$$\hat{p}_{i\cdot} = \sum_{j=1}^{c} \frac{X_{ij}}{n}, \quad i = 1, 2, \ldots, r-1, \quad \hat{p}_{r\cdot} = 1 - \sum_{i=1}^{r-1} \hat{p}_{i\cdot},$$

and

$$\hat{p}_{\cdot j} = \sum_{i=1}^{r} \frac{X_{ij}}{n}, \quad j = 1, 2, \ldots, c-1, \quad \hat{p}_{\cdot c} = 1 - \sum_{i=1}^{c-1} \hat{p}_{\cdot j}.$$

In view of Theorem 11.7.1 and the remarks following it, the statistic Q defined by

$$(2) \qquad Q = \sum_{i=1}^{r} \sum_{j=1}^{c} \frac{\left(X_{ij} - n\hat{p}_{i\cdot}\hat{p}_{\cdot j}\right)^2}{n\hat{p}_{i\cdot}\hat{p}_{\cdot j}}$$

has, under H_0 and for large n, an approximate χ^2 distribution with degrees of freedom given by

$$\text{d.f.} = rc - 1 - [(r-1) + (c-1)] = (r-1)(c-1).$$

This follows since we have estimated $r + c - 2$ parameters. Clearly, large values of Q indicate a large discrepancy between the observed frequencies X_{ij} and those expected under H_0 and, hence, lead to rejection of H_0. The approximate P-value corresponding to the computed value q_0 of Q is given by $P(\chi^2((r-1)(c-1)) \geq q_0)$. Bounds for this probability may be obtained from Table A7.

Let us examine the expression for Q in (2) a little more carefully. For convenience we write $X_{i\cdot} = \sum_{j=1}^{c} X_{ij}$ for the ith row total and $X_{\cdot j} = \sum_{i=1}^{r} X_{ij}$ for the jth column total. The $r \times c$ contingency table now looks as follows.

TABLE 1. $r \times c$ **Contingency Table**

A \ B	B_1	B_2	$\cdots$	B_j	$\cdots$	B_c	Row Total
A_1	X_{11}	X_{12}	$\cdots$	X_{1j}	$\cdots$	X_{1c}	$X_{1.}$
A_2	X_{21}	X_{22}	$\cdots$	X_{2j}	$\cdots$	X_{2c}	$X_{2.}$
$\vdots$	$\vdots$	$\vdots$	$\vdots$	$\vdots$	$\vdots$	$\vdots$	$\vdots$
A_i	X_{i1}	X_{i2}	$\cdots$	X_{ij}	$\cdots$	X_{ic}	$X_{i.}$
$\vdots$	$\vdots$	$\vdots$	$\vdots$	$\vdots$	$\vdots$	$\vdots$	$\vdots$
A_r	X_{r1}	X_{r2}	$\cdots$	X_{rj}	$\cdots$	X_{rc}	$X_{r.}$
Column Total	$X_{.1}$	$X_{.2}$	$\cdots$	$X_{.j}$		$X_{.c}$	n

We note that Table 1 looks much like Table 2 in Section 12.2. An important difference, however, is that in Table 2 of Section 12.2 the row totals are random variables while the column totals are fixed, whereas in Table 1 both the row and the column totals are random variables. Nevertheless, the computation of Q defined in (2) is exactly the same as that of Q defined in (12.2.7). Indeed,

$$\hat{p}_{i.} = \frac{X_{i.}}{n}, \hat{p}_{.j} = \frac{X_{.j}}{n}$$

so that Q defined in (2) can we rewritten as

$$(3) \qquad Q = \sum_{i=1}^{r} \sum_{j=1}^{c} \frac{\left[X_{ij} - \left(X_{i.} X_{.j} / n \right) \right]^2}{X_{i.} X_{.j} / n}.$$

This is precisely the expression for Q in (12.2.7) if we take $r = k$, $c = J$ and $X_{.j} = n_j$. From a computational point of view, therefore, the two expressions are the same. In fact, according to the law of large numbers for large n, the random column totals $X_{.j}$ in Table 1 will be approximately equal to $np_{.j} = n_j$, so that under the assumption of independence the cell frequencies in the jth column will have (approximately) a multinomial distribution with parameters n_j and $p_{1.}, p_{2.}, \ldots, p_{r.}$ for each j. This is precisely the assumption of homogeneity of the c populations.

The test can be easily adapted to test the independence of random variables X and Y when (X, Y) has a continuous bivariate distribution. We partition the ranges of X and Y into mutually disjoint intervals $A_1, A_2, \ldots, A_r$ and $B_1, B_2, \ldots, B_c$. Given the sample $(x_1, y_1), (x_2, y_2), \ldots, (x_n, y_n)$ it is then an easy matter to put the sample in the form of an $r \times c$ contingency table such as Table 1. The cell count x_{ij} represents the number of pairs (x_l, y_l) for which $x_l \in A_i$ and $y_l \in B_j$.

If, on the other hand, (X, Y) has a discrete distribution taking only a finite set of pairs of values, then the categories A_i and B_j are automatically determined respectively by the range of values of X and the range of values of Y.

   Example 1. Smoking and Cause of Death.   A random sample of 2100 death certificates of adults were examined in a large metropolitan area and showed the following results.

Death by Causes and Smoking Habits

Cause of Death	Smoking Habit			Total
	Heavy Smoker	Moderate Smoker	Nonsmoker	
Respiratory Disease	55	120	162	337
Heart Disease	49	388	315	752
Other	61	300	650	1011
Total	165	808	1127	2100

Is the cause of death independent of a person's smoking habit?

   We will use expression (3) to compute Q. Under H_0 that the cause of death is independent of a person's smoking habit, we expect $X_i. X_{.j}/n$ observations in cell (i, j). The following table shows the expected frequencies.

Expected Frequencies

Disease	Heavy Smoker	Moderate Smoker	Nonsmoker	Total
Respiratory	26.5	129.7	180.8	337.0
Heart	59.1	289.3	403.6	752.0
Others	79.4	389.0	542.6	1011.0
Total	165.0	808.0	1127.0	2100.0

Thus

$$q_0 = \frac{(55 - 26.5)^2}{26.5} + \frac{(120 - 129.7)^2}{129.7} + \frac{(162 - 180.8)^2}{180.8}$$

$$+ \frac{(49 - 59.1)^2}{59.1} + \frac{(388 - 289.3)^2}{289.3} + \frac{(315 - 403.6)^2}{403.6}$$

$$+ \frac{(61 - 79.4)^2}{79.4} + \frac{(300 - 389.0)^2}{389} + \frac{(650 - 542.6)^2}{542.6}$$

$$= 134.1$$

For $(3 - 1)(2 - 1) = 4$ d.f. we see that

$$P_{H_0}(Q \geq 134.1) < .001$$

so that we have to reject H_0 and conclude that the cause of death of a person depends on his or her smoking habit.   □

Example 2. Ages at Inauguration and at Death of U.S. Presidents.   The following pairs of observations represent the ages at inauguration and at death of the 35 U.S. Presidents.

(57, 67),	(61, 90),	(57, 83),	(57, 85),	(58, 73),	(57, 80),	(61, 78)
(54, 79),	(68, 68),	(51, 71),	(49, 53),	(64, 65),	(50, 74),	(48, 64)
(65, 77),	(52, 56),	(56, 66),	(46, 63),	(54, 70),	(49, 49),	(50, 56)
(47, 71),	(55, 67),	(54, 58),	(42, 60),	(51, 72),	(56, 67),	(55, 57)
(51, 60),	(54, 90),	(51, 63),	(60, 88),	(62, 78),	(43, 46),	(54, 64)

In Problem 11.9.9 we asked the reader to test the hypothesis that there is a direct association between the age at inauguration and the age at death (using a subset of 20 observations). The χ^2 test that we apply here will only test whether the two random variables are independent. We note that the ages at inauguration range from 42 to 68 and the ages at death range from 46 to 90. Let us choose the intervals

$$A_1 = (41.5, 50.5), \quad A_2 = (50.5, 59.5), \quad A_3 = (59.5, 68.5)$$

and

$$B_1 = (45.5, 60.5), \quad B_2 = (60.5, 75.5), \quad B_3 = (75.5, 90.5).$$

We get the following 3×3 contingency table where the numbers in parentheses give the expected cell counts under the hypothesis of independence.

	Age at Inauguration			
Age at Death	A_1	A_2	A_3	Total
B_1	5 (2.3)	4 (5.2)	0(1.5)	9
B_2	4 (4.1)	10 (9.1)	2 (2.8)	16
B_3	0 (2.6)	6 (5.7)	4 (1.7)	10
Total	9	20	6	35

Using the formula (3) we compute $q_0 = 10.99$, and for $(3 - 1)(3 - 1) = 4$ d.f. we see that

$$.025 < P\left(\chi^2(4) \geq 10.99\right) < .05.$$

Thus we reject H_0 at level .05 (but not at level .025) and conclude that the age at inauguration and age at death of U.S. Presidents are dependent. It should be noted that the number of observations is quite small and, moreover, that several cells have relatively small expected cell frequencies.   □

Problems for Section 12.3

1. A random sample of 1000 employed individuals was taken and the distribution of their salary by sex was found to be as follows:

Sex	Annual Income (dollars)			Total
	Less than 10,000	10,000–25,000	More than 25,000	
Male	103	195	240	538
Female	295	95	72	462
Total	398	290	312	1000

Is the annual income of an employee independent of his or her sex?

2. The following data represent the blood types and ethnic groups of a random sample of 550 minority citizens of the United States.

Ethnic Group	Blood Type				Total
	A	B	O	AB	
Hispanic	105	60	90	15	270
Oriental	50	40	42	26	158
Other Asians	55	22	38	7	122
Total	210	122	170	48	550

Test the hypothesis that the blood type of a minority citizen is independent of his or her ethnic group.

3. A random sample of 500 individuals was taken to investigate whether smoking and alcohol drinking are related. The following results were obtained.

	Heavy Smoker	Moderate Smoker	Nonsmoker	Total
Heavy Drinker	20	62	6	88
Moderate Drinker	40	8	159	207
Nondrinker	10	20	175	205
Total	70	90	340	500

Test the hypothesis that alcohol drinking and smoking are independent.

4. The score on the mathematics portion of the ACT (American College Test) examination and on the final examination in first semester Statistics course for a random sample of 150 students were classified as follows:

Score in Statistics	Mathematics Portion of ACT Scores			Total
	18 or less	19–27	28–36	
60 or less	10	17	3	30
61–85	5	88	4	97
86–100	6	5	12	23
Total	21	110	19	150

Is there sufficient evidence to conclude that the ACT mathematics score X of a student is independent of the statistics score Y?

5. In a study, a random sample of 15 racing days were selected at a race track and the results of all the races were classified according to the post position of the horse and the order in which it finished in the race.

Post Position	Order of Finish				Total
	1	2	3	4–10	
1–3	70	55	56	269	450
4–6	48	50	65	287	450
7–10	32	45	29	394	500
Total	150	150	150	950	1400

Is the order in which a horse finishes independent of its post position? (Note that marginal totals in this case are predetermined and not random variables.)

6. The final grades that students thought they would receive in five Elementary Statistics courses taught over five semesters were cross tabulated with the rating that the students gave the course. The following results were obtained.

Students' Rating of the Course	Predicted Grade				Total
	A	B	C	D or F	
Poor	2	7	17	9	35
Average	4	11	42	5	62
Better than average	3	20	27	3	53
Best	15	10	4	1	30
Total	24	48	90	18	180

Is the students' rating of the course independent of the predicted grade?

7. In a survey of 350 employees, conducted to determine whether the annual salary of an employee is dependent on the length of his or her service, the following results were tabulated.

Years of Service	Annual Salary			Total
	< $15,000	$15,000–$25,000	Over $25,000	
0–2	52	30	18	100
3–5	38	26	11	75
6–10	20	30	10	60
Over 10	25	59	31	115
Total	135	145	70	350

What is your conclusion?

8*. Consider an $r \times c$ contingency table with cell frequencies x_{ij}, $i = 1, 2, \ldots, r$ and $j = 1, 2, \ldots, c$. Let p_{ij} be the probability that a randomly selected observation falls in cell (i, j) and let $p_{i\cdot} = \sum_{j=1}^{c} p_{ij}$, $p_{\cdot j} = \sum_{i=1}^{r} p_{ij}$. The hypothesis of independence then can be written as:

$$H_0: p_{ij} = p_{i\cdot} p_{\cdot j}, \qquad i = 1, 2, \ldots, r, j = 1, 2, \ldots, c.$$

Show that the likelihood ratio test statistic for testing H_0 is given by

$$\lambda(x) = \frac{\left(\prod_{i=1}^{r} x_{i\cdot}^{x_{i\cdot}} \right)\left(\prod_{j=1}^{c} x_{\cdot j}^{x_{\cdot j}} \right)}{\left\{ n^n \prod_{i,j} x_{ij}^{x_{ij}} \right\}},$$

where $n = \sum_{i=1}^{r}\sum_{j=1}^{c} x_{ij}$, $x_{i\cdot} = \sum_{j=1}^{c} x_{ij}$ and $x_{\cdot j} = \sum_{i=1}^{r} x_{ij}$. For large n, the statistic $-2 \ln \lambda(X)$ has, under H_0, an approximate χ^2 distribution with $(r-1)(s-1)$ d.f.

12.4 REVIEW PROBLEMS

1. A small college is discussing the possibility of dropping its athletics program. A random sample of students, faculty, and administrative staff is taken in order to determine their reaction to the proposal. The results of the survey are as follows:

Opinion	Faculty	Students	Administration	Total
In favor	30	60	15	105
Opposed	17	108	10	135
No opinion	6	12	2	20
Total	53	180	27	260

Test the hypothesis that the attitude toward the proposal is independent of the status of the individual.

2. In Problem 1, suppose independent random samples of 50 faculty, 200 students, and 25 administrative staff members are taken in order to ascertain their opinion on whether or not the athletic program should be dropped. Test the hypothesis that the proportions of individuals who are in favor, opposed, or have no opinion among the three groups are the same. The results of the survey are as follows.

Opinion	Faculty	Students	Administration	Total
In favor	30	65	14	109
Opposed	15	115	9	139
No opinion	5	20	2	27
Total	50	200	25	275

3. A market opinion survey is conducted to determine if the weekly magazine that an individual trusts most is related to his or her annual income. The following table gives the results of a sample of 500 readers.

Magazine Trusted Most	Annual Income			Total
	< $20,000	$20,000–$35,000	Above $35,000	
News Time	42	60	43	145
Weekly News	38	49	38	125
U.S.A.	35	46	29	110
Others	20	50	50	120
Total	135	205	160	500

Test the hypothesis that the annual income of a reader is independent of the weekly magazine that he or she trusts most.

4. A statistical society sends a questionnaire to its members with the following results:

Type of Employment	Questionnaires		Total
	Returned	Not Returned	
Faculty	281	69	350
Government	102	70	172
Business	15	63	78
Total	398	202	600

Is the rate of response independent of the type of employment of the member?

5. Independent random samples of 500 freshmen in the state university system and 500 freshmen in the two-year colleges in the state are taken with the following distribution.

Institution Type	Age of Freshmen				Total
	17 and under	18–19	20–21	Over 21	
College	65	295	80	60	500
University	75	310	75	40	500
Total	140	605	155	100	1000

Test the null hypothesis that the age distribution of freshmen is the same in the two-year colleges and the state university system.

6. A survey is conducted to determine whether the incidence of certain types of crime in a large metropolitan area depends on the location where the crime is committed. The following table summarizes the 1205 crimes committed during a three-month period

by their location.

Location	Type of Crime				Total
	Assault	Burglary	Homicide	Larceny	
Northwest	55	130	15	35	235
Northeast	140	110	86	70	406
Southwest	39	150	10	26	225
Southeast	115	90	74	60	339
Total	349	480	185	191	1205

Does the type of crime depend on the location where it is committed?

CHAPTER 13

Analysis of Variance:
k-Sample Problems

13.1 INTRODUCTION

In the previous chapters, we have devoted several sections to the problem of testing the equality of two distribution functions. For example, the problem of testing the equality of two normal distribution functions was considered in Section 8.7. When the underlying populations are not normal, the same problem was considered in Section 11.6 and again in Section 12.2. The χ^2 test of homogeneity is the only test considered so far which can be used to test the equality of several distributions. This test, however, is designed specially for testing the equality of several proportions and is an approximate test. One needs a large sample size for the approximation to be good. Moreover, the test depends on how the observations are grouped.

In practical applications one often needs to compare more than two populations (for example, more than two brands of gasoline additives, or several methods of instruction, or several fertilizers, and so on). It is usual to call the different methods (or additives or fertilizers) *treatments*. The method of analysis depends very much on the design of the experiment and the type of assumption one is willing to make. Let us consider the following example.

Suppose we wish to compare the average number of units produced per day by four different machines (treatments). First suppose that the same worker is asked to operate the four machines so that there is no variation due to the manner in which the machines are operated (that is, due to the worker). Suppose we wish to have a total of n experimental units (that is, a total of n observations). If n_1 units are randomly assigned to treatment 1 (that is, n_1 observations are taken on machine 1), n_2 to treatment 2, n_3 to treatment 3, and n_4 to treatment 4, then we have what is known as a *completely randomized design*. The analysis of such a design depends on the type of assumptions we are willing to make about the observations. If we assume that each of the four samples comes from a normal population with the same (unknown) variance, then the data is analyzed by one-way analysis of variance, which is an extension of the two-sample t-test. One-way analysis of variance is considered in Section 13.2. In Section 13.3 we consider a procedure for constructing simultaneous confidence intervals for the difference $\mu_i - \mu_j$ in means μ_i and μ_j of the ith and the jth treatments for all $i \neq j$. This allows us to identify the means that are significantly different.

If, on the other hand, we assume only that the samples come from continuous distribution functions, then we can use either an extension of the two-sample median test or an extension of the Mann–Whitney–Wilcoxon test. Both these methods are considered in Section 13.5. We also consider a multiple comparison procedure, analogous to the one introduced in Section 13.3, that allows us to identify significantly different treatments.

Next suppose that different workers are asked to operate each of the four machines. In this case there is a variation due to the workers as well as one due to the machines. Indeed, there may also be variation due to interaction between the machines and the workers. A special case is when we have a total of n experimental units subdivided into b homogeneous groups called *blocks* (in our case each worker represents a block so that b = number of workers). Each block has four different units corresponding to four different treatments, so that $n = 4b$. This is the so-called *randomized complete block design*. The analysis of such a design in the special case when we assume normality (and also in the case when we have several observations in each block) is carried out by a two-way analysis of variance. Section 13.4 deals with this topic. If we drop the assumption of normality, then we need an extension of the paired comparison techniques of Section 11.9. This so-called Friedman test is studied in Section 13.6.

13.2 ONE-WAY ANALYSIS OF VARIANCE

Suppose three sections of the same Elementary Statistics course are taught by three instructors. The final grades of students in the three sections are recorded as follows:

Section 1: 95, 32, 47, 75, 83, 84, 73, 68,
Section 2: 85, 90, 79, 50, 32, 84, 78, 95, 65, 80
Section 3: 79, 92, 63, 68, 76, 20, 37, 74, 86

Is the distribution of grades the same in the three sections? Suppose we are willing to assume that the grades in each section are normally distributed and, moreover, the variances of the distributions of grades in the three sections are the same. Since a normal distribution is completely determined by its mean and the variance, the problem of testing the equality of the three distributions reduces to the problem of testing the equality of the three means. In mathematical language, therefore, the problem may be posed equivalently as follows:

Let $X_{11}, X_{12}, \ldots, X_{1n_1}, X_{21}, X_{22}, \ldots, X_{2n_2}$, and $X_{31}, X_{32}, \ldots, X_{3n_3}$ be independent random samples from three normal populations with respective parameters μ_1 and σ_1^2, μ_2 and σ_2^2, and μ_3 and σ_3^2. Suppose $\sigma_1 = \sigma_2 = \sigma_3$. How do we test the null hypothesis

$$H_0: \mu_1 = \mu_2 = \mu_3$$

against the alternatives that either μ_1 and μ_2, or μ_1 and μ_3, or μ_2 and μ_3, or μ_1 and μ_2 and μ_3 are unequal?

Let us first explain why the methods of Section 8.7 cannot be used here. According to Section 8.7 we can test the equality of μ_1 and μ_2 by using the t-statistic

$$
(1) \qquad T = \frac{\overline{X}_1 - \overline{X}_2}{S\sqrt{1/n_1 + 1/n_2}}
$$

where

$$
S^2 = \frac{\sum_{j=1}^{n_1} \left(X_{ij} - \overline{X}_1 \right)^2 + \sum_{j=1}^{n_2} \left(X_{2j} - \overline{X}_2 \right)^2}{n_1 + n_2 - 2}
$$

and

$$
\overline{X}_i = \frac{\sum_{j=1}^{n_i} X_{ij}}{n_i}, \qquad i = 1, 2.
$$

In order to apply this procedure we would have to run a sequence of t-tests on each of the null hypotheses $\mu_1 = \mu_2$, $\mu_2 = \mu_3$, and $\mu_3 = \mu_1$. If any of these hypotheses were rejected then we would conclude that the means are not all the same. This, however, is a very inefficient way to proceed. Suppose we ran the three t-tests at sizes α_1, α_2, and α_3 respectively. Let A_1, A_2, and A_3 respectively be the acceptance regions of these tests. Then

$$
P_{\mu_1 = \mu_2}(A_1) = 1 - \alpha_1, \quad P_{\mu_2 = \mu_3}(A_2) = 1 - \alpha_2, \quad P_{\mu_3 = \mu_1}(A_3) = 1 - \alpha_3.
$$

The probability of accepting H_0: $\mu_1 = \mu_2 = \mu_3$ is given by $P_{H_0}(A_1 \cap A_2 \cap A_3)$, so that the probability of rejecting H_0 when true is given by $1 - P_{H_0}(A_1 \cap A_2 \cap A_3)$. If A_1, A_2, and A_3 were independent events (which would be the case if the three test statistics were independent), then

$$
P\{\text{rejecting } H_0 | H_0 \text{ true}\} = 1 - P_{\mu_1 = \mu_2}(A_1) \, P_{\mu_2 = \mu_3}(A_2) \, P_{\mu_3 = \mu_1}(A_3)
$$

$$
= 1 - (1 - \alpha_1)(1 - \alpha_2)(1 - \alpha_3).
$$

If, for example, $\alpha_i = .05$, $i = 1, 2, 3$, then

$$
\alpha = P\{\text{rejecting } H_0 | H_0\} = 1 - (.95)^3 = .1424,
$$

so that the size of the combined test is considerably larger than the size of each of the three tests. In our case the A_i's are not independent, and we do not know what the overall α is. We do know from Boole's inequality (2.4.13) that

$$\alpha = 1 - P(A_1 \cap A_2 \cap A_3)$$

$$\leq 1 - \{1 - P(\bar{A_1}) - P(\bar{A_2}) - P(\bar{A_3})\}$$

$$= \alpha_1 + \alpha_2 + \alpha_3.$$

Moreover,

$$\alpha = 1 - P(A_1 \cap A_2 \cap A_3) = P(\overline{A_1 \cap A_2 \cap A_3})$$

$$\geq \max\{P(\bar{A_1}), P(\bar{A_2}), P(\bar{A_3})\}$$

$$= \max\{\alpha_1, \alpha_2, \alpha_3\}.$$

Note also that in $\binom{k}{2}$ comparisons of k samples taken two at a time, some will be significant by chance. For example, in 100 such comparisons by t-test we expect about five to be significant at level .05 even if there were no real differences among the means.

Thus the temptation to apply a sequence of t-tests to the data should be avoided. See also Remark 8.7.1.

The two-sample case, however, does provide a clue to how we should proceed in the (general) case under consideration. Let us therefore analyze the two-sample case a little more carefully.

Recall (Problem 8.8.27 or the definition of an F-statistic given in Example 8.4.6) that T^2 has an F distribution with $(1, n_1 + n_2 - 2)$ d.f. Hence the t-test is equivalent to the test based on the statistic

(2) $$F = T^2 = \frac{(n_1 n_2 / (n_1 + n_2))(\bar{X_1} - \bar{X_2})^2}{S^2}.$$

Let us write

$$\bar{X} = \frac{\displaystyle\sum_{i=1}^{2} \sum_{j=1}^{n_i} X_{ij}}{n_1 + n_2}$$

for the sample mean of the $n_1 + n_2$ observations in the combined sample. Then

(3) $$\bar{X} = \frac{n_1 \bar{X_1} + n_2 \bar{X_2}}{n_1 + n_2}.$$

Also,

$$\sum_{i=1}^{2} n_i (\bar{X}_i - \bar{X})^2 = n_1 \left(\bar{X}_1 - \frac{n_1 \bar{X}_1 + n_2 \bar{X}_2}{n_1 + n_2} \right)^2 + n_2 \left(\bar{X}_2 - \frac{n_1 \bar{X}_1 + n_2 \bar{X}_2}{n_1 + n_2} \right)^2$$

(4)
$$= \frac{n_1 n_2^2 + n_1^2 n_2}{(n_1 + n_2)^2} (\bar{X}_1 - \bar{X}_2)^2 = \frac{n_1 n_2}{n_1 + n_2} (\bar{X}_1 - \bar{X}_2)^2$$

so that the quantity $[n_1 n_2 / (n_1 + n_2)](\bar{X}_1 - \bar{X}_2)^2$ in the numerator of the expression for F in (2) is simply $\sum_{i=1}^{2} n_i (\bar{X}_i - \bar{X})^2$, the weighted sum of squares of deviations of sample means from the grand mean $\bar{X}$. Also, the quantity in the denominator of F can be written in the form

(5)
$$S^2 = \frac{\displaystyle\sum_{i=1}^{2} \sum_{j=1}^{n_i} (X_{ij} - \bar{X}_i)^2}{n_1 + n_2 - 2}.$$

Under H_0, $\mu_1 = \mu_2$, so that the two samples come from the same population and $\bar{X}$ is the best estimate of the common unknown mean. The total variation in the combined sample is therefore measured by the total sum of squares (TSS) given by

(6)
$$\text{TSS} = \sum_{i=1}^{2} \sum_{j=1}^{n_i} (X_{ij} - \bar{X})^2.$$

Let us now split up TSS as follows:

$$\text{TSS} = \sum_{i=1}^{2} \sum_{j=1}^{n_i} (X_{ij} - \bar{X}_i + \bar{X}_i - \bar{X})^2$$

$$= \sum_{i=1}^{2} \sum_{j=1}^{n_i} (X_{ij} - \bar{X}_i)^2 + \sum_{i=1}^{2} \sum_{j=1}^{n_i} (\bar{X}_i - \bar{X})^2$$

$$+ 2 \sum_{i=1}^{2} \sum_{j=1}^{n_i} (X_{ij} - \bar{X}_i)(\bar{X}_i - \bar{X})$$

(7)
$$= \sum_{i=1}^{2} \sum_{j=1}^{n_i} (X_{ij} - \bar{X}_i)^2 + \sum_{i=1}^{2} n_i (\bar{X}_i - \bar{X})^2$$

since $\sum_{j=1}^{n_i} (X_{ij} - \bar{X}_i) = 0$. In view of (4), (5), and (7) we have

(8)
$$\text{TSS} = (n_1 + n_2 - 2)S^2 + \frac{n_1 n_2}{n_1 + n_2} (\bar{X}_1 - \bar{X}_2)^2.$$

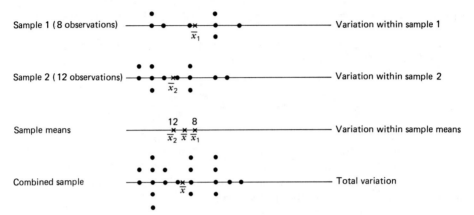

Figure 1. Partitioning the total variation. Dots represent individual observations.

In expression (8) we have partitioned the total variation in the combined sample into two parts (see Figure 1). The first part on the right side of (8) or of (7) is the so-called *error sum of squares* (SSE) which is used to give a pooled estimate of the common unknown variance. It measures the variation within samples. The second part,

$$\frac{n_1 n_2}{n_1 + n_2}\left(\overline{X}_1 - \overline{X}_2\right)^2 = \sum_{i=1}^{2} n_i\left(\overline{X}_i - \overline{X}\right)^2,$$

is the so-called *treatment sum of squares* (SST) which measures the variation among the samples. The larger the treatment sum of squares the greater the evidence against H_0. It follows that large values of the ratio

$$\frac{\text{SST}}{\text{SSE}} = \frac{\left(n_1 n_2/(n_1 + n_2)\right)\left(\overline{X}_1 - \overline{X}_2\right)^2}{(n_1 + n_2 - 2)S^2}$$

lead to the rejection of H_0. This is essentially the test based on T^2. The reason we have S^2 in the denominator of T^2 is to give an F distribution for the ratio that allows us to determine what values of the ratio will be significant at level α.

Yet another way to look at the same problem is to compare the two estimates of σ^2 (the common unknown variance). Under $H_0: \mu_1 = \mu_2$ we have

$$\mathscr{E}\frac{n_1 n_2}{n_1 + n_2}\left(\overline{X}_1 - \overline{X}_2\right)^2 = \mathscr{E}\sum_{i=1}^{2} n_i\left(\overline{X}_i - \overline{X}\right)^2 = \sigma^2,$$

so that SST provides an unbiased estimate of σ^2. Also [see (8.7.7) and the computation following it],

$$\mathscr{E}S^2 = \sigma^2$$

whether H_0 is true or not so that $SSE/(n_1 + n_2 - 2)$ is also an unbiased estimate of σ^2. When the null hypothesis is false we expect the estimate SST of σ^2 to exceed (Problem 9) the estimate $S^2 = SSE/(n_1 + n_2 - 2)$, and hence the null hypothesis will be rejected at level α if the ratio $F = T^2$ exceeds the $(1 - \alpha)$th quantile of the F distribution for $(1, n_1 + n_2 - 2)$ d.f.

This analysis of the two-sample case suggests how we should proceed in the general case: partition the total variation in the combined sample into two parts. One part explains the variation between the samples while the second part explains the variation within each sample. Compare the two. Equivalently, find two estimates of the common variance, one based on the deviations of $\overline{X}_i$'s from $\overline{X}$, and the second based on the pooled variance. Compare the two.

Let us first illustrate the procedure with the help of the example introduced at the beginning of this section.

Example 1.   Comparing Distribution of Grades. The following table gives the details of computation.

Section 1	Section 2	Section 3
95	85	79
32	90	92
47	79	63
75	50	68
83	32	76
84	84	20
73	78	37
68	95	74
	65	86
	80	

$$\overline{x}_1 = 557/8 = 69.6 \qquad \overline{x}_2 = 738/10 = 73.8 \qquad \overline{x}_3 = 595/9 = 66.1$$

$$\overline{x} = \frac{557 + 738 + 595}{8 + 10 + 9} = \frac{1890}{27} = 70, \quad \sum_{i=1}^{3} n_i(\overline{x}_i - \overline{x})^2 = 282.57$$

$$\sum_{j} x_{1j}^2 = 41781 \qquad \sum_{j} x_{2j}^2 = 57880 \qquad \sum_{j} x_{3j}^2 = 43716$$

$$s_1^2 = 428.55 \qquad s_2^2 = 379.51 \qquad s_3^2 = 547.36$$

$$s^2 = \frac{7s_1^2 + 9s_2^2 + 8s_3^2}{7 + 9 + 8} = \frac{10794.32}{24} = 449.76$$

Figure 2 shows the situation for each section in the table.

The estimate of σ^2 as obtained from variation between samples is given by

$$\sum_{i=1}^{3} \frac{n_i(x_i - \overline{x})^2}{3 - 1} = \frac{282.57}{2} = 141.29.$$

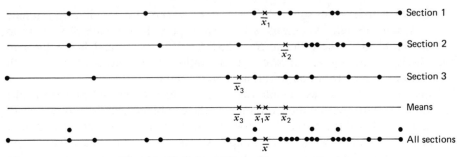

Figure 2. Variation within and between sections.

The estimate of σ^2 as obtained from the variation within samples is given by $s^2 = 449.76$. Hence

$$f = \frac{141.29}{449.76} = .31.$$

We will see that the d.f. for the numerator is 2 and for the denominator 24; there is not enough evidence to reject H_0. $\quad\square$

We are now ready to present the details of the general case. Suppose we have k independent random samples

$$x_{11}, x_{12}, \ldots, x_{1n_1}$$
$$x_{21}, x_{22}, \ldots, x_{2n_2}$$
$$\cdots \cdots \cdots \cdots$$
$$x_{k1}, x_{k2}, \ldots, x_{kn_k}$$

and we wish to test the null hypothesis

$$H_0: \mu_1 = \mu_2 = \cdots = \mu_k$$

against the alternative hypothesis

$$H_1: \text{At least two } \mu_i\text{'s are different.}$$

We will assume that the ith sample comes from a normal population with mean μ_i and variance σ^2, $i = 1, 2, \ldots, k$, where σ^2 is unknown. We can now express each x_{ij} as

(9) $$x_{ij} = \mu_i + \varepsilon_{ij}, \quad j = 1, 2, \ldots, n_i, i = 1, 2, \ldots, k$$

where ε_{ij} is random. Our assumption about normality of X_{ij} is equivalent to assuming that ε_{ij} is a normal random variable with mean 0 and variance σ^2. The model (9) is usually referred to as the *one-way analysis of variance* model because the k populations involve variation in only one *factor* (different sections, different methods of teaching, different fertilizers). The k populations or treatments are also called *levels* of the factor. It is called an analysis of variance model because, as we shall see, its analysis depends on the analysis of the total variation. The linear model (9) is usually written in the form

$$(10) \qquad x_{ij} = \mu + \alpha_i + \varepsilon_{ij}, \qquad j = 1, 2, \ldots, n_i, i = 1, 2, \ldots, k$$

where μ_i has been replaced by $\mu + \alpha_i$. Here μ is the mean of the μ_i's and, hence, $\sum_{i=1}^k \alpha_i = 0$ (see Problem 10). The null hypothesis H_0 can now be replaced by the null hypothesis that $\alpha_1 = \alpha_2 = \cdots = \alpha_k = 0$, and the alternative hypothesis states that $\alpha_i \neq 0$ for at least one i.

Let $\sum_{i=1}^k n_i = n$ and

$$\bar{x}_i = \sum_{j=1}^{n_i} \frac{x_{ij}}{n_i} = i\text{th sample mean}, i = 1, 2, \ldots, k,$$

$$\bar{x} = \sum_{i=1}^k \sum_{j=1}^{n_i} \frac{x_{ij}}{n} = \sum_{i=1}^k \frac{n_i \bar{x}_i}{n} = \text{grand sample mean},$$

$$s_i^2 = (n_i - 1)^{-1} \sum_{j=1}^{n_i} (x_{ij} - \bar{x}_i)^2 = i\text{th sample variance},$$

$$\text{SSE} = \sum_{i=1}^k (n_i - 1)s_i^2 = \text{error or within sum of squares},$$

and

$$\text{SST} = \sum_{i=1}^k n_i(\bar{x}_i - \bar{x})^2 = \text{treatment or between sum of squares}.$$

We will compare two estimates of the variance σ^2. One is based on the variation among the sample means and the other is based on the variation within the samples. First note that the total variation (TSS) is given by

$$\text{TSS} = \sum_{i=1}^k \sum_{j=1}^{n_i} (x_{ij} - \bar{x})^2 = \sum_{i=1}^k \sum_{j=1}^{n_i} (x_{ij} - \bar{x}_i + \bar{x}_i - \bar{x})^2$$

$$= \sum_{i=1}^k \sum_{j=1}^{n_i} (x_{ij} - \bar{x}_i)^2 + \sum_{i=1}^k n_i(\bar{x}_i - \bar{x})^2$$

so that

$$(11) \qquad \qquad \text{TSS} = \text{SSE} + \text{SST}.$$

Writing X_{ij} for the random variable that takes values x_{ij} we note also that

$$\mathscr{E} \sum_{i=1}^{k} \sum_{j=1}^{n_i} \left(X_{ij} - \overline{X}_i \right)^2 = \mathscr{E} \sum_{i=1}^{k} (n_i - 1)S_i^2 = (n - k)\sigma^2$$

so that

(12) $$\text{MSE} = \frac{\text{SSE}}{n - k}$$

is an unbiased estimate of σ^2, under H_0 as well as under H_1. Next we have

$$\mathscr{E}\,\text{SST} = \sum_{i=1}^{k} n_i \mathscr{E}\left(\overline{X}_i - \overline{X} \right)^2 = \sum_{i=1}^{k} n_i \mathscr{E} \overline{X}_i^2 - n\mathscr{E}\overline{X}^2$$

$$= \sum_{i=1}^{k} n_i \left(\frac{\sigma^2}{n_i} + \mu_i^2 \right) - n\left(\text{var}(\overline{X}) + (\mathscr{E}\overline{X})^2 \right)$$

$$= \sum_{i=1}^{k} \left(\sigma^2 + n_i \mu_i^2 \right) - n\left\{ \frac{\sigma^2}{n} + \left(\frac{1}{n} \sum_{i=1}^{k} n_i \mu_i \right)^2 \right\}$$

$$= (k - 1)\sigma^2 + \sum_{i=1}^{k} n_i \mu_i^2 - \frac{1}{n}\left(\sum_{i=1}^{k} n_i \mu_i \right)^2$$

$$= (k - 1)\sigma^2 + \sum_{i=1}^{n} n_i (\mu_i - \bar{\mu})^2$$

where $\bar{\mu} = \sum_{i=1}^{k} n_i \mu_i / n$. It follows that

$$\mathscr{E}\text{MST} = \mathscr{E}\frac{\text{SST}}{k - 1} = \sigma^2 + \frac{1}{k - 1} \sum_{i=1}^{k} n_i (\mu_i - \bar{\mu})^2 \geq \sigma^2$$

with equality if and only if $\mu_1 = \mu_2 = \cdots = \mu_k$. In particular,

(13) $$\text{MST} = \frac{\text{SST}}{k - 1}$$

is an unbiased estimate of σ^2 under H_0.

Consider the ratio

(14) $$F = \frac{\text{MST}}{\text{MSE}} = \frac{\text{SST}/(k - 1)}{\text{SSE}/(n - k)}$$

of the two estimates of σ^2. We note that the statistic F is an estimate of the ratio

(15) $$\frac{\mathscr{E}\text{MST}}{\mathscr{E}\text{MSE}} = 1 + \frac{1}{(k - 1)\sigma^2} \sum_{i=1}^{k} n_i (\mu_i - \bar{\mu})^2.$$

In view of the law of large numbers when each n_i is large, the ratio F in (14) is likely to be near the ratio (15), which takes the value 1 under H_0 and a value > 1 under H_1. Consequently, when the ratio in (15) is large, it means that the ratio $\sum_{i=1}^{k} n_i (\mu_i - \bar{\mu})^2 / \sigma^2$ is large and the test based on F will reject H_0 (with high probability).

How large should the value of F be in order to reject H_0 at level α? To answer this question we need the distribution of F in (14) under H_0. We first note that, in view of Theorem 8.7.1(i), SST and SSE are independent statistics. Moreover, $\sum_{i=1}^{k}(n_i - 1)S_i^2/\sigma^2$ is the sum of k independent χ^2 random variables with respective degrees of freedom $n_1 - 1, n_2 - 1, \ldots, n_k - 1$ so that SSE$/\sigma^2$ has a χ^2 distribution with $\sum_{i=1}^{k}n_i - k = n - k$ d.f. Also, $\bar{X}$ is the (weighted) sample mean of independent random variables $\bar{X}_1, \bar{X}_2, \ldots, \bar{X}_k$ and, under H_0, $\bar{X}_i$ has a normal distribution with mean $\mu(= \mu_1 = \cdots = \mu_k)$ and variance σ^2/n_i (see Section 8.7). It follows [by an argument similar to the one used in the proof of Theorem 8.7.1(ii)] that SST$/\sigma^2$ has a χ^2 distribution with $k - 1$ d.f. (see Problem 11). Consequently, it follows from the definition of an F-statistic (Example 8.4.6) that the ratio F defined in (14) has an F distribution, under H_0, with $(k - 1, n - k)$ d.f. We reject H_0 at level α if the value of F computed from (14) exceeds $F_{k-1, n-k, \alpha}$.

These results are usually summarized in the following so-called analysis of variance table.

One-Way Analysis of Variance

Source of Variation	Degrees of Freedom	Sum of Squares	Mean Square	F-Ratio
Treatment or between	$k - 1$	$\sum_{i=1}^{k} n_i(\bar{x}_i - \bar{x})^2$	MST $=$ SST$/(k - 1)$	$\dfrac{\text{MST}}{\text{MSE}}$
Error or Within	$n - k$	$\sum_{i=1}^{k}\sum_{j=1}^{n_i}(x_{ij} - \bar{x}_i)^2$	MSE $=$ SSE$/(n - k)$	
Total	$n - 1$	$\sum_{i=1}^{k}\sum_{j=1}^{n_i}(x_{ij} - \bar{x})^2$		

Example 2. *The Analysis of Variance for Example 1.* In Example 1, $k = 3$, $n_1 = 8$, $n_2 = 10$, $n_3 = 9$ and we can present the results in the following table.

Source	Degrees of Freedom	Sum of Squares	Mean Square	F-ratio
Sections	2	282.57	141.29	.31
Error	24	10,794.32	449.76	
Total	26	11,076.89		

Since $F_{2, 24, .05} = 3.4$, we cannot reject H_0 at $\alpha = .05$. $\square$

Example 3. *Comparing Yield with Different Fertilizers.* A farmer would like to test four fertilizers for his soybean field. He selects plots of the same size having similar soil, drainage, and exposure to the sun. He tries fertilizers A, B, C, and D respectively on 6, 8, 9 and 7 test plots with the following yields.

Fertilizer	Yield
A	47, 42, 43, 46, 44, 42
B	51, 58, 62, 49, 53, 51, 50, 59
C	37, 39, 41, 38, 39, 37, 42, 36, 40
D	42, 43, 42, 45, 47, 50, 48

Is there significant difference in the average yields?

Let $\mu_A, \mu_B, \mu_C, \mu_D$ be the mean yields. Then we wish to test $H_0: \mu_A = \mu_B = \mu_C = \mu_D$ against the alternative hypothesis that not all μ's are equal. We note that $k = 4$, $n_1 = 6$, $n_2 = 8$, $n_3 = 9$, $n_4 = 7$, and $n = 30$. We have

$$\bar{x}_1 = 264/6 = 44.0, \quad \bar{x}_2 = 433/8 = 54.1, \quad \bar{x}_3 = 349/9 = 38.8, \quad \bar{x}_4 = 317/7 = 45.3,$$

$$\bar{x} = 1363/30 = 45.4, s_1^2 = 22/5 = 4.4, s_2^2 = 164.88/7 = 23.6, s_3^2 = 31.56/8 = 3.9,$$

$$s_4^2 = 59.43/6 = 9.9.$$

Hence

$$\text{SST} = \sum_{i=1}^{4} n_i (\bar{x}_i - \bar{x})^2 = 1009.39,$$

$$\text{SSE} = \sum_{i=1}^{4} (n_i - 1) s_i^2 = 277.87.$$

Analysis of Variance

Source	D.F.	SS	Mean SS	F-Ratio
Fertilizers	3	1009.39	336.46	31.48
Error	26	277.87	10.69	
Total	29	1287.26		

Since $F_{3, 26, .05} = 2.89$, we reject H_0 at 5 percent level of significance and conclude that the fertilizers do not all have the same yield. □

Remark 1. The F-test obtained here can be shown to be the likelihood ratio test. We asked students to derive the likelihood ratio test for the $k = 3$ case in Problem 11.4.10(ii).

Remark 2. The assumption $\sigma_1 = \sigma_2 = \cdots = \sigma_k$ that we made in this section can, be tested, for example, by using the likelihood ratio procedure [see Problem 11.4.10(i)]. We advise once more against the use of a sequence of tests. The F-test that we developed in this section is *robust* in the sense that it is not too sensitive to departures from normality as well as from the assumption of equality of variances.

Remark 3. When the n_i's are large, one should avoid using the F-test. Indeed, when the n_i's are large the quantity $\sum_{i=1}^{k} n_i(\mu_i - \bar{\mu})^2/\sigma^2$ will be large even if the differences among the μ_i's are small. For sufficiently large values of n_i's the F-test will reject H_0 (with high probability) even if the differences among the μ_i's are practically negligible.

Remark 4. The case when we are not willing to assume that the X_{ij}'s have a normal distribution but only that they have a continuous distribution will be considered in Sections 13.5 and 13.6.

Remark 5. The one-way classification model of this section is also called a *completely randomized design*. Once the n subjects have been chosen they are allocated to the k different treatments in a completely random manner. Thus n_1 subjects are randomly chosen from n and allocated to treatment 1, next n_2 are randomly chosen from $n - n_1$ remaining subjects and allocated to treatment 2, and so on.

Remark 6. The one-way analysis of variance model (10) is called a *fixed effects* model because it is assumed that the levels chosen for the analysis are the only ones considered relevant. If we replace the fixed α_i's in (10) by random variables A_i, we get the so-called *random effects* model. In a random effects model we also need to make assumptions concerning the A_i's. We will not consider such models in this text.

Problems for Section 13.2

1. Five car models are chosen in a study to test the equality of average gasoline mileage per gallon that each model gives. Independent random samples of 5, 6, 4, 7, and 6 cars of each model were selected and each car was run on one gallon of gasoline until it ran out of gas. The miles traveled (rounded off to the nearest integer) by each car are given below.

Model	Distance Traveled (miles)
A	18, 17, 18, 21, 19
B	21, 24, 17, 23, 22, 23
C	15, 14, 16, 15
D	18, 20, 20, 24, 26, 23, 25
E	17, 16, 18, 17, 15, 17

Is there sufficient evidence to suggest that the five models give different mpg?

2. In order to compare the effectiveness of four methods of teaching young children a computer programming language, independent random samples of sizes 4, 7, 6, and 8 are taken from large groups of children taught by these four methods, and their

scores on a standardized achievement test are recorded as follows:

Method	Scores
A	75, 73, 68, 72
B	84, 92, 84, 82, 87, 85, 87
C	62, 65, 68, 67, 67, 66
D	74, 76, 73, 72, 76, 74, 75, 79

Is there evidence to conclude that there is significant difference among the four methods of instruction?

3. Five cigarette manufacturers claim that their product has low tar content. Independent random samples of cigarettes are taken from each manufacturer and the following tar levels (in milligrams) are recorded.

Brand	Tar Level (mg)
A	4.2, 4.8, 4.6, 4.0, 4.4
B	4.9, 4.8, 4.7, 5.0, 4.9, 5.2
C	5.4, 5.3, 5.4, 5.2, 5.5
D	5.8, 5.6, 5.5, 5.4, 5.6, 5.8
E	5.9, 6.2, 6.2, 6.8, 6.4, 6.3

Can the differences among the sample means be attributed to chance?

4. Four brands of bacon are tested for fat contents with the following sample results.

Brand	Fat Content(%)
1	41, 42, 40, 44, 43
2	38, 34, 36, 37, 38, 36
3	42, 45, 48, 46, 47, 48
4	54, 52, 51, 52, 53

Is the difference among the mean fat contents significant at level .05?

5. There are three banks in Bowling Green, Ohio. Customers are randomly selected from the main branch of each bank and their waiting times before obtaining service from a teller are recorded as follows:

Bank	Waiting Times (min.)
HB	2.5, 3.0, 3.5, 6.0, 3.5, 4.5
MA	4.5, 8.0, 6.5, 6.0, 7.5, 6.0, 7.5
TT	2.5, 3.5, 4.0, 3.5, 2.5

Do the data indicate a significant difference among the mean waiting times at these banks?

6. The quantity of oxygen dissolved in water is used as a measure of water pollution. Samples are taken at four locations in a lake and the quantity of dissolved oxygen is recorded as follows (lower reading corresponds to greater pollution):

Location	Quantity of Dissolved Oxygen (%)
A	7.8, 6.4, 8.2, 6.9
B	6.7, 6.8, 7.1, 6.9, 7.3
C	7.2, 7.4, 6.9, 6.4, 6.5
D	6.0, 7.4, 6.5, 6.9, 7.2, 6.8

Do the data indicate a significant difference in the average amount of dissolved oxygen for the four locations?

7. Four different brands of 9-volt batteries are to be compared to see if the mean life lengths are the same. Independent random samples were taken and the lifetimes of sample batteries were recorded (in hours) as follows:

Brand	Life Length (hr.)
A	42, 46, 43, 48, 47, 47
B	54, 53, 50, 58, 55
C	37, 35, 41, 32, 36, 37
D	49, 48, 48, 50, 49

What is your conclusion?

8. A psychologist is interested in comparing the mean reaction time of five types of laboratory animals to a particular stimulus. She obtains the following results:

	Animals				
	A	B	C	D	E
Sample Size	5	7	5	4	6
Mean (seconds)	5.2	4.8	6.1	4.6	5.9
Standard Deviation (seconds)	1.1	0.8	1.2	0.9	1.2

Do the data indicate that the mean reaction times are significantly different at a 5 percent level of significance?

9. Show that

$$\mathscr{E}\frac{n_1 n_2}{n_1 + n_2}\left(\bar{X}_1 - \bar{X}_2\right)^2 = \mathscr{E}\sum_{i=1}^{2} n_i\left(\bar{X}_i - \bar{X}\right)^2$$

$$= \sigma^2 + \frac{n_1 + n_2}{4}\left(\mu_1 - \mu_2\right)^2$$

and hence that

$$\frac{\mathscr{E}\left(n_1 n_2/(n_1 + n_2)\right)\left(\bar{X}_1 - \bar{X}_2\right)^2}{\mathscr{E}S^2} = 1 + \frac{n_1 n_2}{4\sigma^2}(\mu_1 - \mu_2)^2.$$

Conclude that when H_0: $\mu_1 = \mu_2$ is false (that is, when $\mu_1 \neq \mu_2$ is true) the estimate SST of σ^2 (with high probability) will exceed the estimate $S^2 = \text{SSE}/(n_1 + n_2 - 2)$ of σ^2.

10. Show that the model

$$x_{ij} = \mu_i + \varepsilon_{ij}, \quad j = 1, 2, \ldots, n_i, i = 1, 2, \ldots, k$$

can be written as

$$x_{ij} = \mu + \alpha_i + \varepsilon_{ij}, \quad j = 1, 2, \ldots, n_i, i = 1, 2, \ldots, k$$

where

$$\mu_i = \mu + \alpha_i, \quad \text{and} \quad \sum_{i=1}^{k} \alpha_i = 0.$$

11*. Show that the statistic $\sum_{i=1}^{k} n_i (\bar{X}_i - \bar{X})^2/\sigma^2$ has a $\chi^2 (k-1)$ distribution under H_0: $\mu_1 = \mu_2 = \cdots = \mu_k = \mu$ (unknown). {Hint: $\sum_{i=1}^{k} n_i (\bar{X}_i - \mu)^2/\sigma^2 = \sum_{i=1}^{k} [n_i (\bar{X}_i - \bar{X})^2/\sigma^2] + [n(\mu - \bar{X})^2/\sigma^2]$. Use an argument similar to the one used in the proof of Theorem 8.7.1(ii).}

12. Show that the F-test in one-way analysis of variance is equivalent to the test based on the statistic

$$\frac{\sum_{i=1}^{k} n_i (\bar{X}_i - \bar{X})^2}{\sum_{i=1}^{k} \sum_{j=1}^{n_i} (X_{ij} - \bar{X})^2}.$$

13.3 MULTIPLE COMPARISON OF MEANS

In Example 13.2.3 the null hypothesis, that the four fertilizers have the same mean yield, was rejected. From the farmer's point of view, the problems is not really solved. In order to determine which fertilizer to buy he needs to do a cost analysis and for that purpose he needs to be able to rank, if possible, the four fertilizers according to their average yield. The data of Example 13.2.3 suggest that $\mu_B > \mu_A = \mu_D > \mu_C$. What evidence does the data offer to make this assertion? In other words, with what degree of confidence can we assert that fertilizer B produces the largest average yield whereas the fertilizer C produces the least average yield?

More formally, suppose the F test in a one-way analysis of variance rejects the null hypothesis $H_0\colon \mu_1 = \mu_2 = \cdots = \mu_k$. How do we obtain simultaneous confidence intervals for the differences $\mu_i - \mu_j$ for each pair of population means (μ_i, μ_j) at level $1 - \alpha$? Several such multiple comparison procedures are available. We will only consider the following method, due to Scheffé:

In random sampling from k normal populations with common (unknown) variance, the probability is $1 - \alpha$ that

$$\sum_{i=1}^{k} c_i \overline{X}_i - l \le \sum_{i=1}^{k} c_i \mu_i \le \sum_{i=1}^{k} c_i \overline{X}_i + l$$

holds for all $c_1, c_2, \ldots, c_k$ such that $\sum_{i=1}^{k} c_i = 0$. Here

$$l^2 = (k-1) F_{k-1, n-k, \alpha} \left\{ \sum_{i=1}^{k} \frac{(n_i - 1) S_i^2}{n - k} \right\} \sum_{i=1}^{k} \frac{c_i^2}{n_i}.$$

We will not attempt to prove this result here. It should be noted that the result holds for *all comparisons* $\sum_{i=1}^{k} c_i \mu_i$ *simultaneously*. It should not be surprising, therefore, to find that the length of a confidence interval for a particular linear function such as $\mu_1 - \mu_2$ obtained by this method is much larger than the length of a $1 - \alpha$ level confidence interval for $\mu_1 - \mu_2$ based on the t-statistic (as obtained in Section 8.7).

We note that in view of (13.2.12),

$$\frac{\sum_{i=1}^{k} (n_i - 1) S_i^2}{n - k} = \frac{\text{SSE}}{n - k} = \text{MSE},$$

in the notation of Section 13.2, so that we have

$$P\left(\sum_{i=1}^{k} c_i \overline{X}_i - \sqrt{(k-1)\text{MSE}\left(\sum_{i=1}^{k} \frac{c_i^2}{n_i} \right) F_{k-1, n-k, \alpha}} \right.$$

$$\le \sum_{i=1}^{k} c_i \mu_i \le \sum_{i=1}^{k} c_i \overline{X}_i + \sqrt{(k-1)\text{MSE}\left(\sum_{i=1}^{k} \frac{c_i^2}{n_i} \right) F_{k-1, n-k, \alpha}} \, ,$$

$$\left. \text{for all } c_i \quad \text{such that} \quad \sum_{i=1}^{k} c_i = 0 \right) = 1 - \alpha.$$

In the particular case when $c_i = 1$, $c_j = -1$ and other c's are zero we have

$$P\left(\bar{X}_i - \bar{X}_j - \sqrt{(k-1)\text{MSE}\left(\frac{1}{n_i} + \frac{1}{n_j}\right) F_{k-1,\,n-k,\,\alpha}} \le \mu_i - \mu_j \right.$$

$$\left. \le \bar{X}_i - \bar{X}_j + \sqrt{(k-1)\text{MSE}\left(\frac{1}{n_j} + \frac{1}{n_j}\right) F_{k-1,\,n-k,\,\alpha}}\, , \text{ for all } i \ne j \right) = 1 - \alpha.$$

Example 1. Comparing Mean Yields with Different Fertilizers (Example 13.2.3 continued). In Example 13.2.3 we obtained the following sample results.

Fertilizer	Sample		
	Size	Mean	Variance
A	6	44.0	4.4
B	8	54.1	23.6
C	9	38.8	3.9
D	7	45.3	9.9

Also, MSE $= 277.87/26 = 10.69$ and $F_{3,\,26,\,.05} = 2.89$. Then

$$l = \sqrt{3(2.89)\left(\frac{1}{n_i} + \frac{1}{n_j}\right)(10.69)}\, .$$

The following table gives the six simultaneous confidence intervals for $\mu_i - \mu_j$, $i < j$, $i, j = 1, 2, 3, 4$. (For convenience, we have written $i = 1$ for A, $i = 2$, for B, and so on.)

Mean Difference	$\bar{x}_i - \bar{x}_j$	Lower Limit	Upper Limit
$\mu_A - \mu_B$	-10.1	-15.3	-4.9
$\mu_A - \mu_C$	5.2	-0.1	10.3
$\mu_A - \mu_D$	-1.3	-6.7	4.1
$\mu_B - \mu_C$	15.3	10.6	20.0
$\mu_B - \mu_D$	8.3	3.3	13.3
$\mu_C - \mu_D$	-6.5	-11.4	-1.6

The confidence intervals that do not contain zero are the ones for $\mu_A - \mu_B$, $\mu_B - \mu_C$, $\mu_B - \mu_D$ and $\mu_C - \mu_D$. Clearly we can assert with 95 percent confidence that

$$\mu_A - \mu_B < 0, \quad \mu_C - \mu_D < 0$$

and

$$\mu_B - \mu_C > 0, \quad \mu_B - \mu_D > 0.$$

Thus

$$\mu_B > \mu_D > \mu_C \quad \text{and} \quad \mu_B > \mu_A.$$

Consequently, with 95 percent confidence we can assert that fertilizer B results in the largest mean yield. It is not possible to make any assertion concerning the relative order of μ_A, μ_C and μ_D (except that $\mu_D > \mu_C$) since μ_A and μ_C, and μ_A and μ_D are not significantly different.

On the other hand, if our objective is to compare only μ_A and μ_C, then we can use a 95 percent confidence interval based on the t-statistic. This confidence interval is given by

$$\left[\bar{x}_1 - \bar{x}_3 - s\sqrt{\frac{1}{n_1} + \frac{1}{n_3}}\, t_{n_1 + n_3 - 2, .025}, \quad \bar{x}_1 - \bar{x}_3 + s\sqrt{\frac{1}{n_1} + \frac{1}{n_3}}\, t_{n_1 + n_3 - 2, .025} \right]$$

where

$$s^2 = \frac{(n_1 - 1)s_1^2 + (n_3 - 1)s_3^2}{n_1 + n_3 - 2} = 4.09.$$

For 13 d.f., $t_{13, .025} = 2.160$ and the confidence interval is given by

$$[5.2 - 2.3, 5.2 + 2.3] = [2.9, 7.5].$$

The length of this confidence interval is smaller than the length of the interval obtained by Scheffé's method. This confidence interval, however, is a statement concerning $\mu_A - \mu_C$ alone, whereas Scheffé's method makes a statement simultaneously concerning all the six possible differences. □

Example 2. Comparing the Average Number of Defects in Automobiles. An automobile assembly plant assembles several models. In order to investigate if more expensive models receive greater care than the less expensive models, several cars of the three models assembled at this plant are tested for defects with the following results. (Model A cars are the most expensive and Model C cars are the least expensive).

Model	Number of Defects
A	5, 4, 6, 6, 7
B	7, 8, 6, 7, 6, 5
C	9, 7, 8, 9, 10, 11, 10, 10

Writing $i = 1, 2, 3$ respectively for A, B, and C we see that

$$\bar{x}_1 = 28/5 = 5.6, \bar{x}_2 = 39/6 = 6.5, \bar{x}_3 = 74/8 = 9.25, \bar{x} = 7.42,$$

$$s_1^2 = 5.2/4 = 1.3, s_2^2 = 5.5/5 = 1.1, s_3^2 = 11.5/7 = 1.64.$$

Analysis of Variance

Source	D.F.	SS	Mean Square	F-Ratio
Models	2	48.43	24.22	17.45
Error	16	22.2	1.39	
Total	18	70.63		

Since $F_{2,16,.05} = 3.63$ we have to reject $H_0: \mu_A = \mu_B = \mu_C$ at level $\alpha = .05$ and conclude that the mean number of defects for the models are different.

Let us now apply Scheffé's method to investigate which differences are significantly different from zero. We note that

$$l = \sqrt{2(3.63)\left(\frac{1}{n_i} + \frac{1}{n_j}\right)(1.39)} = \sqrt{10.09\left(\frac{1}{n_i} + \frac{1}{n_j}\right)}.$$

The following table gives the confidence intervals for $\mu_i - \mu_j$, $i \neq j$

Mean Difference	Lower Limit	Upper Limit
$\mu_A - \mu_B$	-2.8	1.0
$\mu_A - \mu_C$	-5.5	-1.8
$\mu_B - \mu_C$	-4.5	-1.0

Since the confidence intervals for $\mu_A - \mu_C$ and $\mu_B - \mu_C$ are both to the left of zero, we can assert with 95 percent confidence that

$$\mu_A < \mu_C \text{ and } \mu_B < \mu_C$$

so that the average number of defects for the least expensive model is the greatest of the three models.   □

Problems for Section 13.3

1. For the data on distance traveled in miles per gallon of gasoline for the five models of automobiles in Problem 13.2.1, the sample results are as follows:

Model	Sample Size	Sample Mean	Sample Variance
A	5	18.60	2.30
B	6	21.67	6.27
C	4	15.00	0.67
D	7	22.29	8.90
E	6	16.67	1.07

Find simultaneous confidence intervals (for the differences between the means) at level .95. Which means are significantly different at level .05?

2. The following table gives the relevant sample statistics for the data of Problem 13.2.2.

Method	Sample Size	Sample Mean	Variance
A	4	72.0	8.67
B	7	85.9	10.48
C	6	65.8	4.57
D	8	74.9	4.70

Investigate significant differences in the true mean scores between different methods at level .05.

3. The following table gives the relevant sample statistics for the data on fat contents in four brands of bacon of Problem 13.2.4.

Brand	Size	Sample Mean	Variance
1	5	42.0	2.5
2	6	36.5	2.3
3	6	46.0	5.2
4	5	52.4	1.3

Identify differences in the true mean fat contents among the four brands of bacon at level .05.

4. In Problem 3, construct confidence intervals based on t-statistic for the six pairs of differences each at level .95, and compare your results with the simultaneous confidence intervals you obtained in Problem 3. Explain the difference.

5*. Let

$$X_{11}, X_{12}, \ldots, X_{1n_1},$$
$$X_{21}, X_{22}, \ldots, X_{2n_2},$$
$$\cdots$$
$$X_{k1}, X_{k2}, \ldots, X_{kn_k}$$

be independent random samples. For each i, suppose X_{ij} has a normal distribution with mean μ_i and variance σ^2, $i = 1, 2, \ldots, k$. Let $c_1, c_2, \ldots, c_k$ be known constants and consider the parametric function $\theta = \sum_{i=1}^{k} c_i \mu_i$. Construct a minimum $1 - \alpha$ level confidence interval for θ based on the unbiased estimate $\hat{\theta} = \sum_{i=1}^{k} c_i \bar{X}_i$ of θ where $\bar{X}_i = \sum_{j=1}^{n_i} X_{ij}/n_i$. How does this confidence interval differ from the simultaneous confidence interval obtained by Scheffé's method? [Hint: Each $\bar{X}_i$ has a normal $(\mu_i, \sigma^2/n_i)$ distribution, and $\bar{X}_1, \ldots, \bar{X}_k$ are independent of $\sum_{i=1}^{k} \sum_{j=1}^{n_i} (X_{ij} - \bar{X}_i)^2/\sigma^2$, which has a χ^2 distribution with $n - k$ d.f. Construct a t-statistic by taking the appropriate ratio.]

6. From the following sample data on lifetime (in hours) of three different brands of 9-volt batteries, construct simultaneous confidence intervals for the differences in

means at level .95. Identify significantly different means.

| | | Sample | |
Brand	Sample Size	Mean	Variance
X	5	40	100.00
Y	4	55	116.67
Z	6	60	170.00

13.4 TWO-WAY ANALYSIS OF VARIANCE

The methods of Section 13.2 are not appropriate when the data is classified according to two or more characteristics and the object is to study all the variables simultaneously. For each variable or factor of interest a number of levels may be chosen for study. This is the case, for example, when we have several drugs that are used at different dosage levels, or when several brands of automobiles tires are used under different road conditions, or when several brands of gasolines are to be compared with several different additives. All these are examples of the two-factor case, which is the only case that we will study in this text. We begin with an example.

Example 1. Comparing Workers on Two Machines. The data for this example have been selected to illustrate the basic concepts involved in a two-factor analysis of variance.

Suppose we wish to test whether there are any differences in two different machines that were used by two workers. Suppose, further, that four weeks are selected for this study, each week consisting of five working days. During the first and second weeks worker 1 is used and during third and fourth weeks worker 2 is used. Machine 1 is used during the first and third weeks and machine 2 is used during the second and fourth weeks. Suppose other conditions of work are identical. The data in Table 1 represent the number of units produced per day.

It would be inappropriate to apply a one-way analysis of variance to this data. There are two factors to be considered here. One is the effect of the workers and the other is the effect of the machines. There may also be an effect of the interaction between worker and machine. There are several questions of interest that we would like to answer. Do variations in the machines or in the worker affect the production? That is, which of these two factors makes a significant contribution to the number of units produced per day? Is a change in the number of units of production that can be attributed to a change in the machines the same for the two workers?

In order to show the interdependence between the two factors, we exhibit the data in Table 2.

TABLE 1.   Number of Units Produced per Day

Week	Number of Units Produced
1	69, 68, 72, 74, 75
2	95, 98, 100, 96, 97
3	81, 88, 84, 87, 88
4	105, 110, 107, 112, 118

TABLE 2

Workers	Machines		
	1	2	
1	69 68 72 74 75	95 98 100 96 97	Weeks 1 and 2
2	81 88 84 87 88	105 110 107 112 118	Weeks 3 and 4
	Weeks 1 and 3	Weeks 2 and 4	

Here we have two levels of factor A, workers 1 and 2, and two levels of factor B, machines 1 and 2. The four subsquares into which Table 2 is subdivided are usually referred to as *cells*. Let us call the cell for which factor A is at level i and factor B at level j, the (i, j)th cell, $i = 1, 2; j = 1, 2$. Each cell in Table 2 has five observations, all taken at the same set of factor levels. *We assume that the observations in the (i, j)th cell are a random sample from a normal population with mean μ_{ij} and variance σ^2, both unknown.*

The main features of a two-way analysis of variance are that:

(i) it allows us to compare the effects of two variables (factors),

(ii) it allows us to study the interaction effects, if any, of the two variables.

In order to study the simultaneous but separate effects of factors A and B one performs two F-tests, one for the worker "main" effect and another for the machine "main" effect. The most interesting aspect of the two-way analysis of variance (and also the most difficult to explain) is its ability to identify the *interaction* effects. It is easier to define "no interaction." We say that there are no interactions between two factors if the presence or absence of factor A has no effect on Factor B, and conversely. In that case the amount by which factor A raises the mean number of units produced is the same regardless of the level of Factor B, and conversely. Thus the cell mean when there are no interactions are as follows.

Factor A	Factor B	
	1	2
1	μ_{11}	μ_{12}
2	$\mu_{11} + \alpha$	$\mu_{12} + \alpha$

or

Factor A	Factor B	
	1	2
1	μ_{11}	$\mu_{11} + \beta$
2	μ_{21}	$\mu_{21} + \beta$

If we write $\mu_{12} = \mu_{11} + \beta$, we get the following table of means for the "no interaction" case.

Factor A	Factor B	
	1	2
1	μ_{11}	$\mu_{11} + \beta$
2	$\mu_{11} + \alpha$	$\mu_{11} + \alpha + \beta$

Thus "no interaction" means that the effects of the two factors are *additive* in the following sense. If a change in factor A by itself produces a change α in cell means, and a change in factor B by itself produces a change β in cell means, then changes in both factors produce a change of $\alpha + \beta$ in the cell means. This can be illustrated with the help of a diagram. Suppose the mean number of units produced per day are as follows.

	Machines	
Workers	1	2
1	72	97
2	85	110

Then $\mu_{11} = 72$, $\alpha = 13$, $\beta = 25$ and the two factors are additive. We can plot the means as in Figure 1. We see that there is no crossing over of the line segments depicting the two levels of each factor.

In order to illustrate an interaction effect, suppose the table of means was as follows.

	Machines	
Workers	1	2
1	85	97
2	72	110

Now we have two main effects (actually only one—see Problem 2) as well as an interaction effect. Both workers produce more units on Machine 2. Worker 1 produces less than worker 2 on Machine 2 but the reverse is true on Machine 1. This is seen more clearly in Figure 2.

In the case of additive factors, the graphs of cell means will look like those in Figure 1 (parallel line segments) and in the case of interaction the graphs will look like those in Figure 2 where the line segments cross over.

In practice one does not know the μ_{ij}'s. In that case one plots the corresponding sample means. The interpretation is the same. If there is little or no crossing over of line segments, it indicates that the two factors are additive. Otherwise there is also interaction between the two factors.   □

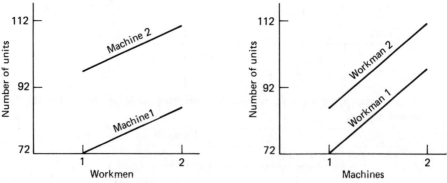

Figure 1.   Two main effects and no interaction.

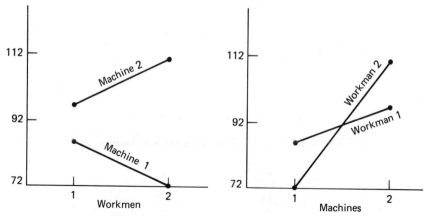

Figure 2. Two main effects and their interaction.

We are now ready to study the general case. Let $x_{ij1}, x_{ij2}, \ldots, x_{ijK}$, $i = 1, 2, \ldots, I$, $j = 1, 2, \ldots, J$ be independent random samples, each of size K, from normal populations with means μ_{ij} and a common (unknown) variance σ^2. The observations x_{ijk} can be presented in a two-way $I \times J$ table as follows.

Factor A	Factor B			
	1	2	$\cdots$	J
1	x_{111} x_{112} $\vdots$ x_{11K}	x_{121} x_{122} $\vdots$ x_{12K}	$\cdots$	x_{1J1} x_{1J2} $\vdots$ x_{1JK}
2	x_{211} x_{212} $\vdots$ x_{21K}	x_{221} x_{222} $\vdots$ x_{22K}	$\cdots$	x_{2J1} x_{2J2} $\vdots$ x_{2JK}
$\vdots$	$\vdots$	$\vdots$	$\vdots$	$\vdots$
I	x_{I11} x_{I12} $\vdots$ x_{I1K}	x_{I21} x_{I22} $\vdots$ x_{I2K}	$\cdots$	x_{IJ1} x_{IJ2} $\vdots$ x_{IJK}

Thus x_{ijk} is the kth observation, $k = 1, 2, \ldots, K$ in the (i, j)th cell; that is, it is the kth observation when factor A is held at level i and factor B is held at level j.

Let $I \geq 2$, $J \geq 2$, and $K \geq 2$. Set

$$\mu = \frac{1}{IJ} \sum_{i=1}^{I} \sum_{j=1}^{J} \mu_{ij}, \quad \mu_{i\cdot} = \frac{1}{J} \sum_{j=1}^{J} \mu_{ij}, \quad \mu_{\cdot j} = \frac{1}{I} \sum_{i=1}^{I} \mu_{ij}.$$

Let us call

$$\alpha_i = \mu_{i\cdot} - \mu = \text{effect of factor A at level } i$$

$$\beta_j = \mu_{\cdot j} - \mu = \text{effect of factor B at level } j$$

and

$$\gamma_{ij} = \mu_{ij} - (\mu + \alpha_i + \beta_j)$$

$$= \text{interaction effect of factor A at level } i \text{ and factor B at level } j.$$

Thus

(1) $$\mu_{ij} = \mu + \alpha_i + \beta_j + \gamma_{ij}.$$

We note that

$$\sum_{i=1}^{I} \alpha_i = \sum_{j=1}^{J} \beta_j = \sum_{i=1}^{I} \gamma_{ij} = \sum_{j=1}^{J} \gamma_{ij} = 0.$$

As usual let X_{ijk} denote the random variable, which takes values x_{ijk}, $k = 1, 2, \ldots, K$. Then (1) can be rewritten as

(2) $$\mathscr{E} X_{ijk} = \mu_{ij} = \mu + \alpha_i + \beta_j + \gamma_{ij},$$

for $k = 1, 2, \ldots, K$. We write

(3) $$X_{ijk} = \mu + \alpha_i + \beta_j + \gamma_{ij} + \varepsilon_{ijk}$$

where ε_{ijk}, for $k = 1, 2, \ldots, K$, are independent and identically distributed normal random variables with mean 0 and variance σ^2. Thus X_{ijk} for $k = 1, 2, \ldots, K$ are independent and identically distributed normal random variables with mean μ_{ij} and variance σ^2. Equation (3) is usually referred to as the (linear) *two-way analysis of variance* model.

The object of the analysis of variance is to test the following different hypotheses:

H_A: $\alpha_1 = \alpha_2 = \cdots = \alpha_I = 0$ (no effect due to factor A)

H_B: $\beta_1 = \beta_2 = \cdots = \beta_J = 0$ (no effect due to factor B)

H_{AB}: $\gamma_{ij} = 0$ for all i, j, $i = 1, 2, \ldots, I$, $j = 1, 2, \ldots, J$ (no interaction effects)

The null hypothesis H_{AB} simply states that the effects of factors A and B are additive. The procedure for simultaneously testing H_A, H_B, and H_{AB} is the same as in the one-way analysis of variance. We partition the total variation into four parts, one part attributable to factor A, the second part attributable to factor B, the third part attributable to interaction between factors A and B, and the fourth (remaining) part attributable to random error. Comparisons of each of the first three parts to the random error part allows us to test the three null hypotheses H_A, H_B, and H_{AB}. We define

$$\bar{X} = \frac{1}{IJK} \sum_{i=1}^{I} \sum_{j=1}^{J} \sum_{k=1}^{K} X_{ijk} = \text{grand mean,}$$

$$\bar{X}_{ij} = \frac{1}{K} \sum_{k=1}^{K} X_{ijk} = (i, j)\text{th cell mean,}$$

$$\bar{X}_{i\cdot} = \frac{1}{JK} \sum_{j=1}^{J} \sum_{k=1}^{K} X_{ijk} = \text{sample mean of observations at level } i \text{ of factor A,}$$

$$\bar{X}_{\cdot j} = \frac{1}{IK} \sum_{i=1}^{I} \sum_{k=1}^{K} X_{ijk} = \text{sample mean of observations at level } j \text{ of factor B.}$$

Clearly $\bar{X}, \bar{X}_{ij}, \bar{X}_{i\cdot}, \bar{X}_{\cdot j}$ respectively are the estimates of $\mu, \mu_{ij}, \mu_{i\cdot}, \mu_{\cdot j}$ so that the estimates of α_i, β_j and γ_{ij} are respectively given by $\bar{X}_{i\cdot} - \bar{X}$, $\bar{X}_{\cdot j} - \bar{X}$ and $\bar{X}_{ij} - \bar{X}_{i\cdot} - \bar{X}_{\cdot j} + \bar{X}$. The following table shows the respective sample means:

FACTOR A	FACTOR B				Row Means
	1	2	$\cdots$	J	
1	$\bar{x}_{11}$	$\bar{x}_{12}$	$\cdots$	$\bar{x}_{1J}$	$\bar{x}_{1\cdot}$
2	$\bar{x}_{21}$	$\bar{x}_{22}$	$\cdots$	$\bar{x}_{2J}$	$\bar{x}_{2\cdot}$
$\vdots$	$\vdots$	$\vdots$	$\vdots$	$\vdots$	$\vdots$
I	$\bar{x}_{I1}$	$\bar{x}_{I2}$	$\cdots$	$\bar{x}_{IJ}$	$\bar{x}_{I\cdot}$
Column Means	$\bar{x}_{\cdot 1}$	$\bar{x}_{\cdot 2}$	$\cdots$	$\bar{x}_{\cdot J}$	$\bar{x}$

Under H_A: $\alpha_1 = \alpha_2 = \cdots = \alpha_I = 0$ we expect all the row means to be close to the grand mean $\bar{x}$. Similarly, under H_B: $\beta_1 = \beta_2 = \cdots = \beta_J = 0$ and we expect all the column means to be close to $\bar{x}$. This explains the definitions of the variations due to factors A and B given below. Similar explanations may be given to the remaining sum of squares. Define

$$\text{TSS} = \sum_{i=1}^{I} \sum_{j=1}^{J} \sum_{k=1}^{K} \left(X_{ijk} - \bar{X} \right)^2 = \text{total sum of squares,}$$

$$\text{SSA} = \sum_{i=1}^{I} \sum_{j=1}^{J} \sum_{k=1}^{K} \left(\bar{X}_{i.} - \bar{X} \right)^2 = JK \sum_{i=1}^{I} \left(\bar{X}_{i.} - \bar{X} \right)^2$$

$$= \text{sum of squares due to factor A,}$$

$$\text{SSB} = IK \sum_{j=1}^{J} \left(\bar{X}_{.j} - \bar{X} \right)^2 = \text{sum of squares due to factor B,}$$

$$\text{SSI} = \sum_{i=1}^{I} \sum_{j=1}^{J} \sum_{k=1}^{K} \left(\bar{X}_{ij} - \bar{X}_{i.} - \bar{X}_{.j} + \bar{X} \right)^2$$

$$= K \sum_{i=1}^{I} \sum_{j=1}^{J} \left(\bar{X}_{ij} - \bar{X}_{i.} - \bar{X}_{.j} + \bar{X} \right)^2 = \text{sum of squares due to interaction,}$$

and

$$\text{SSE} = \sum_{i=1}^{I} \sum_{j=1}^{J} \sum_{k=1}^{K} \left(X_{ijk} - \bar{X}_{ij} \right)^2 = \text{sum of squares due to error.}$$

Clearly, SSA measures the total variation in sample means of factor A from $\bar{X}$, SSB measures the total variation in the sample means of factor B from $\bar{X}$, and so on. If factor A has little or no effect on μ, then SSA will be small. Similarly, SSB will be small if factor B has little or no effect, and SSI will be small if there is no interaction effect. The error sum of squares SSE is the variation that cannot be explained by the model (3). This is the variability in the data that is unaccounted for once the effects of factors A and B and their interaction have been accounted for. We note that

$$\text{TSS} = \sum_{i=1}^{I} \sum_{j=1}^{J} \sum_{k=1}^{K} \left(X_{ijk} - \bar{X} \right)^2$$

$$= \sum_{i=1}^{I} \sum_{j=1}^{J} \sum_{k=1}^{K} \left(X_{ijk} - \bar{X}_{ij} + \bar{X}_{ij} - \bar{X}_{i.} + \bar{X}_{i.} - \bar{X}_{.j} + \bar{X}_{.j} - \bar{X} \right)^2$$

$$= \sum_{i=1}^{I} \sum_{j=1}^{J} \sum_{k=1}^{K} \left\{ \left(\bar{X}_{i.} - \bar{X} \right) + \left(\bar{X}_{.j} - \bar{X} \right) + \left(\bar{X}_{ij} - \bar{X}_{i.} - \bar{X}_{.j} + \bar{X} \right) \right.$$

$$\left. + \left(X_{ijk} - \bar{X}_{ij} \right) \right\}^2,$$

so that on expanding we get

(4) $$\text{TSS} = \text{SSA} + \text{SSB} + \text{SSI} + \text{SSE}.$$

Relation (4) holds since all the six product terms vanish. For example,

$$\sum_{i=1}^{I} \sum_{j=1}^{J} \sum_{k=1}^{K} \left(\overline{X}_{i\cdot} - \overline{X} \right)\left(\overline{X}_{\cdot j} - \overline{X} \right) = K \sum_{i=1}^{I} \left(\overline{X}_{i\cdot} - \overline{X} \right) \sum_{j=1}^{J} \left(\overline{X}_{\cdot j} - \overline{X} \right) = 0,$$

$$\sum_{i=1}^{I} \sum_{j=1}^{J} \sum_{K=1}^{K} \left(\overline{X}_{\cdot j} - \overline{X} \right)\left(X_{ijk} - \overline{X}_{ij} \right) = \sum_{i=1}^{I} \sum_{j=1}^{J} \left(\overline{X}_{\cdot j} - \overline{X} \right) \sum_{k=1}^{K} \left(X_{ijk} - \overline{X}_{ij} \right)$$

$$= \sum_{i=1}^{I} \sum_{j=1}^{J} \left(\overline{X}_{\cdot j} - \overline{X} \right)\left(K\overline{X}_{ij} - K\overline{X}_{ij} \right) = 0,$$

and so on.

Exactly as in the one-way analysis of variance we can use each of the four sums of squares on the right side of (4) to estimate σ^2. Thus, we can show (Problem 1) that

$$\mathscr{E}\left(\frac{\text{SSA}}{I-1} \right) = \mathscr{E}(\text{MSA}) = \sigma^2 + \frac{JK}{I-1} \sum_{i=1}^{I} \alpha_i^2,$$

$$\mathscr{E}\left(\frac{\text{SSB}}{J-1} \right) = \mathscr{E}(\text{MSB}) = \sigma^2 + \frac{IK}{J-1} \sum_{j=1}^{J} \beta_j^2,$$

$$\mathscr{E}\left(\frac{\text{SSI}}{(I-1)(J-1)} \right) = \mathscr{E}(\text{MSI}) = \sigma^2 + \frac{K}{(I-1)(J-1)} \sum_{i=1}^{I} \sum_{j=1}^{J} \gamma_{ij}^2,$$

and

$$\mathscr{E}\left(\frac{\text{SSE}}{IJ(K-1)} \right) = \mathscr{E}(\text{MSE}) = \sigma^2,$$

where MSA is called the mean sum of squares due to factor A, MSB the mean sum of squares due to factor B, and so on. Under H_A, $\mathscr{E}(\text{MSA}) = \sigma^2$; under H_B, $\mathscr{E}(\text{MSB}) = \sigma^2$, and under H_{AB}, $\mathscr{E}(\text{MSI}) = \sigma^2$. Moreover, if H_A is false then

$$\mathscr{E}(\text{MSA}) > \sigma^2.$$

Now the ratio

$$F_A = \frac{\text{MSA}}{\text{MSE}}$$

is an estimate of the ratio

$$\frac{\mathscr{E}(\text{MSA})}{\mathscr{E}(\text{MSE})} = 1 + \frac{JK \sum_{I=1}^{I} \alpha_i^2}{(I-1)\sigma^2}.$$

If H_A is false then F_A will tend to be large. It is possible to show by using the normal theory of Section 8.7 (exactly as in the one-way analysis of variance case) that F_A has an $F(I-1, IJ(K-1))$ distribution. Hence we reject H_A at level α if

$$F_A > F_{I-1,\, IJ(K-1),\, \alpha}.$$

Similar arguments lead to the following level α tests of H_B and H_{AB}:

$$\text{Reject } H_B \text{ at level } \alpha \text{ if } F_B = \frac{\text{MSB}}{\text{MSE}} > F_{J-1,\, IJ(K-1),\, \alpha},$$

$$\text{Reject } H_{AB} \text{ at level } \alpha \text{ if } F_I = \frac{\text{MSI}}{\text{MSE}} > F_{(I-1)(J-1),\, IJ(K-1),\, \alpha}.$$

These results can now be summarized in the following two-way analysis of variance table:

Two-Way Analysis of Variance

Source of Variation	Degrees of Freedom	Sum of Squares	Mean Sum of Squares	F-Ratio
A	$I-1$	SSA	SSA/$(I-1)$	$F_A = \text{MSA}/\text{MSE}$
B	$J-1$	SSB	SSB/$(J-1)$	$F_B = \text{MSB}/\text{MSE}$
Interaction	$(I-1)(J-1)$	SSI	SSI/$(I-1)(J-1)$	$F_I = \text{MSI}/\text{MSE}$
Error	$IJ(K-1)$	SSE	SSE/$IJ(K-1)$	—
Total	$IJK-1$	TSS	—	—

The computation of SSA, SSB, SST, and SSE can be simplified be expanding each sum of squares. Thus

$$\text{SSA} = JK \sum_{i=1}^{I} \left(\overline{X}_{i.} - \overline{X} \right)^2 = JK \sum_{i=1}^{I} \overline{X}_{i.}^2 - IJK\overline{X}^2$$

$$= (JK)^{-1} \sum_{i=1}^{I} \left(\sum_{j=1}^{J} \sum_{k=1}^{K} X_{ijk} \right)^2 - (IJK)^{-1} \left(\sum_{i=1}^{I} \sum_{j=1}^{J} \sum_{k=1}^{K} X_{ijk} \right)^2.$$

For convenience, we write

$$T = \sum_{i=1}^{I} \sum_{j=1}^{J} \sum_{k=1}^{K} X_{ijk}, \; T_{i\cdot} = \sum_{j=1}^{J} \sum_{k=1}^{K} X_{ijk},$$

$$T_{\cdot j} = \sum_{i=1}^{I} \sum_{k=1}^{K} X_{ijk}, \; T_{ij} = \sum_{k=1}^{K} X_{ijk}.$$

Then

$$\text{SSA} = (JK)^{-1} \sum_{i=1}^{I} T_{i\cdot}^2 - (IJK)^{-1} T^2.$$

Similarly,

$$\text{SSB} = (IK)^{-1} \sum_{j=1}^{J} T_{\cdot j}^2 - (IJK)^{-1} T^2,$$

$$\text{SSE} = \sum_{i=1}^{I} \sum_{j=1}^{J} \sum_{k=1}^{K} X_{ijk}^2 - (K)^{-1} \sum_{i=1}^{I} \sum_{j=1}^{J} T_{ij}^2,$$

$$\text{TSS} = \sum_{i=1}^{I} \sum_{j=1}^{J} \sum_{k=1}^{K} X_{ijk}^2 - (IJK)^{-1} T^2,$$

and by subtraction

$$\text{SSI} = \text{TSS} - \text{SSA} - \text{SSB} - \text{SSE}.$$

Example 2.   Example 1 Revisited.   From Table 2 in Example 1 we get the following tables of sample sums and sample means

j	Sample Totals		
i	1	2	Total
1	358	486	$T_{1\cdot} = 844$
2	428	552	$T_{2\cdot} = 980$
Total	$T_{\cdot 1} = 786$	$T_{\cdot 2} = 1038$	$T = 1824$

j	Sample Means		
i	1	2	Workers
1	$\bar{x}_{11} = 71.6$	$\bar{x}_{12} = 97.2$	$\bar{x}_{1\cdot} = 84.4$
2	$\bar{x}_{21} = 85.6$	$\bar{x}_{22} = 110.4$	$\bar{x}_{2\cdot} = 98.0$
Machines	$\bar{x}_{\cdot 1} = 78.6$	$\bar{x}_{\cdot 2} = 103.8$	$\bar{x} = 91.2$

If the sample means are plotted as in Example 1, the plots will look like Figure 1. The line segments will not be parallel but almost parallel, the best that one can expect from the sample data. This suggests that there is no interaction between the factors and one could use the additive model

$$X_{ijk} = \mu + \alpha_i + \beta_j + \varepsilon_{ijk}$$

instead of the model (3). From the table for sample totals we have

$$\sum_{I=1}^{2} \sum_{j=1}^{2} T_{ij}^2 = 852{,}248, \quad \sum_{i=1}^{2} T_{i\cdot}^2 = 1{,}672{,}736,$$

$$\sum_{j=1}^{2} T_{\cdot j}^2 = 1{,}695{,}240, \quad T^2 = 3{,}326{,}976.$$

Also

$$\sum_{i=1}^{2} \sum_{j=1}^{2} \sum_{k=1}^{5} x_{ijk}^2 = 170{,}640.$$

Then

$$\text{SSA} = \frac{1672736}{10} - \frac{3326976}{20} = 924.8,$$

$$\text{SSB} = \frac{1695240}{10} - \frac{3326976}{20} = 3175.2,$$

$$\text{TSS} = 170640 - \frac{3326976}{20} = 4291.2,$$

$$\text{SSE} = 170640 - \frac{852248}{5} = 190.4,$$

so that

$$\text{SSI} = 4291.2 - 924.8 - 3175.2 - 190.4 = 0.8.$$

These results are summarized in the following analysis of variance table.

Analysis of Variance for Table 2

Source	D.F.	SS	Mean Square	F-ratio
Workers	1	924.8	924.8	77.7
Machines	1	3175.2	3175.2	266.8
Interaction	1	0.8	0.8	0.07
Error	16	190.4	11.9	
Total	19	4291.2		

We note that $F_{1,16,.01} = 8.53$, $F_{1,16,.05} = 4.49$ so that both H_A and H_B are rejected, whereas H_{AB} cannot be rejected. Thus there is an effect on production that can be attributed to differences in workers as well as to differences in machines, but there is no interaction between the workers and the machines. □

Example 3. ***Variety of Wheat vs. Location.*** Three varieties of wheat were grown at three different locations. A sample of each variety was taken from each location and milled into flour. Three loaves of bread were baked from each sample of flour. The following table gives the volumes of loaves measured in standard units.

	Variety of Wheat			
Location	1	2	3	Total
1	15.2	13.4	20.4	
	13.8	16.5	18.2	143.3
	14.3	15.2	16.3	
2	7.6	4.8	12.2	
	4.8	2.7	11.8	65.1
	3.9	3.9	13.4	
3	19.2	11.3	22.3	
	17.5	13.4	25.1	164.0
	18.4	12.6	24.2	
Total	114.7	93.8	163.9	372.4

We note that $I = J = K = 3$. The table of sample means is as follows.

i / j	Sample Means			
	1	2	3	$\bar{x}_{\cdot j}$
1	$T_{11} = 43.3$	$T_{21} = 45.1$	$T_{31} = 54.9$	15.92
	$\bar{x}_{11} = 14.43$	$\bar{x}_{21} = 15.03$	$\bar{x}_{31} = 18.3$	
2	$T_{12} = 16.3$	$T_{22} = 11.4$	$T_{32} = 37.4$	7.23
	$\bar{x}_{12} = 5.43$	$\bar{x}_{22} = 3.8$	$\bar{x}_{32} = 12.47$	
3	$T_{13} = 55.1$	$T_{23} = 37.3$	$T_{33} = 71.6$	18.22
	$\bar{x}_{13} = 18.37$	$\bar{x}_{23} = 12.43$	$\bar{x}_{33} = 23.87$	
$\bar{x}_{i\cdot}$	12.74	10.42	18.21	13.79

Figure 3 shows the graph of sample means indicating the presence of an interaction effect. We have

$$\sum_{i=1}^{3} \sum_{j=1}^{3} T_{ij}^2 = 18271.18, \qquad \sum_{i=1}^{3} T_{i\cdot}^2 = 48817.74, \qquad \sum_{j=1}^{3} T_{\cdot j}^2 = 51668.90,$$

and

$$\sum_{i=1}^{3} \sum_{j=1}^{3} \sum_{k=1}^{3} x_{ijk}^2 = 6123.5.$$

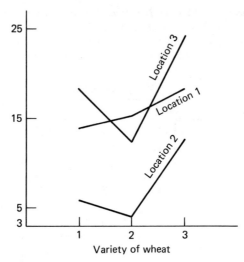

Figure 3. Interaction between variety and location.

Thus

$$\text{SSA} = \frac{48817.74}{9} - \frac{138681.76}{27} = 287.83, \qquad \text{SSB} = \frac{51668.9}{9} - \frac{138681.76}{27} = 604.66,$$

$$\text{TSS} = 6123.5 - \frac{138681.76}{27} = 987.17, \qquad \text{SSE} = 6123.5 - \frac{18271.18}{3} = 33.11,$$

and

$$\text{SST} = 987.17 - 287.83 - 604.66 - 33.11 = 61.57.$$

Source	D.F.	SS	Mean Square	F-ratio
Variety	2	287.83	143.92	79.08
Location	2	604.66	302.33	164.31
Interaction	4	61.57	15.39	8.36
Error	18	33.11	1.84	
Total	26	987.17		

Since $F_{2,18,.01} = 6.01$, and $F_{4,18,.01} = 4.58$ we have to reject H_A, H_B, and H_{AB} at level $\alpha = .01$ and conclude that both the location of the plot as well as the variety of wheat used affect the volume of loaves baked. Moreover, the volume is also affected by the interaction between the location and the variety of wheat used. $\quad\square$

13.4.1 Randomized Block Design

Suppose $I \geq 2$, $J \geq 2$, and $K = 1$ in a two-way analysis of variance so that we have exactly one observation per cell. In that case, $\text{SSE} = \Sigma\Sigma\Sigma(x_{ijk} - \bar{x}_{ij})^2 = 0$

since $x_{ijk} = \bar{x}_{ij}$ for all i, j and we cannot test for interaction between the two factors. Suppose the effects of the two factors are additive (the interactions γ_{ij} are all zero). In that case,

$$\mathscr{E}\left(\frac{\mathrm{SSI}}{(I-1)(J-1)}\right) = \mathscr{E}(\mathrm{MSI}) = \sigma^2$$

and we can use MSI as the denominator of the F-ratio for testing H_A or H_B. The analysis of variance table then reduces to the following table.

Two-way Analysis of Variance—One Observation per Cell

Source	D.F.	SS	Mean Square	F-Ratio
A	$I - 1$	SSA	$\mathrm{SSA}/(I-1)$	$F_A = \mathrm{MSA}/\mathrm{MSI}$
B	$J - 1$	SSB	$\mathrm{SSB}/(J-1)$	$F_B = \mathrm{MSB}/\mathrm{MSI}$
Error	$(I-1)(J-1)$	SSI	$\mathrm{SSI}/(I-1)(J-1)$	—
Total	$IJ - 1$	TSS	—	—

In many practical applications, an artifical factor B is created to reduce variability. Suppose we have $n = IJ$ experimental units available which can be divided into J homogeneous groups called *blocks*. Each block consists of I units. Within each homogeneous block, the I treatments are assigned randomly to the I units in the block. This arrangement of $n = IJ$ units in J blocks is called a *randomized* (complete) *block design*. We recall (Remark 13.2.5) that in a completely randomized design, on the other hand, the I treatments are *randomly allocated* to $n = IJ$ experimental units such that each treatment appears in J units (one can have repetitions of the I treatments).

The blocks in a randomized block design can be groups of plots of land, groups of animals of the same type (cows or rats or pigs), groups of locations, groups of students, and so on.

Even though the data in a randomized block design is classified according to two factors it does not necessarily follow that we are always interested in analyzing the differences among the block means. As pointed out earlier, blocks are introduced in order to eliminate heterogeneity caused by some extraneous factor; the object being to have experimental units that are as nearly identical as possible.

Example 45. Weight Gains by Swines. A swine-feeding experiment involving three different rations, A, B, and C, was laid out in a randomized block design using 15 animals in five groups of three each. The grouping was done on the basis of initial weight and litter. The litters are the blocks, since we expect the animals to be homogeneous within litter. The following table, arranged in five blocks, shows the weight gains in pounds. The ration fed is shown along with the observation.

C	B	B	A	A
8.2	15.2	20.2	10.8	14.2
A	A	C	B	C
6.7	16.2	13.8	9.9	14.4
B	C	A	C	B
13.3	16.3	13.4	15.8	19.6

We can regroup the data as follows.

Weight Gains of Swine (pounds)

Ration	Block					Total
	1	2	3	4	5	
A	6.7	16.2	13.4	10.8	14.2	61.3
B	13.3	15.2	20.2	9.9	19.6	78.2
C	8.2	16.3	13.8	15.8	14.4	68.5
Total	28.2	47.7	47.4	36.5	48.2	208.0

We note that $I = 3, J = 5$, and $K = 1$. Moreover,

$$SSA = \frac{14565.18}{5} - \frac{43264}{15} = 28.77,$$

$$SSB = \frac{8972.78}{3} - \frac{43264}{15} = 106.66,$$

$$TSS = 3083.68 - \frac{43264}{15} = 199.41,$$

so that

$$SSI = SSE = 199.41 - 28.77 - 106.66 = 63.98.$$

Analysis of Variance

Source	D.F.	SS	Mean Square	F-ratio
Blocks	4	106.66	26.665	3.33
Rations	2	28.77	14.385	1.80
Error	8	63.98	7.998	
Total	14	199.41		

Since $F_{4, 8, .05} = 3.84$, $F_{2, 8, .05} = 4.46$ we cannot reject the null hypotheses H_A: no differences in rations, and H_B: no differences in litters. $\square$

13.4.2 Multiple Comparisons in Two-Way Analysis of Variance

Whenever either H_A or H_B is rejected (and H_{AB} is not rejected), it is of interest to know the significantly different levels of the factor under study. The procedure is the same as in the one-way analysis of variance. In order to compare levels of factor A we use the simultaneous confidence intervals given by

$$P\left(\bar{X}_{i_1\cdot} - \bar{X}_{i_2\cdot} - l \le \mu_{i_1} - \mu_{i_2} \le \bar{X}_{i_1\cdot} - \bar{X}_{i_2\cdot} + l \quad \text{for all } i_1 \ne i_2 = 1, 2, \ldots, I\right)$$

$$= 1 - \alpha$$

where

$$l = \begin{cases} \sqrt{(I-1)F_{I-1,(I-1)(J-1),\,\alpha}\mathrm{MSE}(2/J)} & \text{if } K = 1 \\ \sqrt{(I-1)F_{I-1,(I-1)(J-1)+IJ(K-1),\,\alpha}\mathrm{MSE}(2/JK)} & \text{if } K > 1, \end{cases}$$

and

$$\mathrm{MSE} = \begin{cases} (\mathrm{TSS} - \mathrm{SSA} - \mathrm{SSB})/(I-1)(J-1) & \text{if } K = 1 \\ (\mathrm{TSS} - \mathrm{SSA} - \mathrm{SSB})/[(I-1)(J-1)+IJ(K-1)] & \text{if } K > 1. \end{cases}$$

Example 5. Comparing Four Brands of Tires.   In an experiment to compare tread wear of four different brands of tires after 20,000 miles of driving, a randomized block design was used with four cars serving as the four blocks. Each of the four sets of automobile tires is mounted in a random order on each car and then the car is run for 20,000 miles. The following table gives the tread wear (in units of 0.001 inch).

| | Tire Brand | | | | |
Automobile	A	B	C	D	Totals
1	10	11	15	11	47
2	10	9	12	10	41
3	8	10	11	10	39
4	8	8	11	8	35
Totals	36	38	49	39	162

Thus

$$\mathrm{SSA} = \text{treatment sum of squares} = \frac{6662}{4} - \frac{26244}{16} = 25.25,$$

$$\mathrm{SSB} = \text{block or car sum of squares} = \frac{6636}{4} - \frac{26244}{16} = 18.75,$$

and

$$\mathrm{TSS} = 1690 - \frac{26244}{16} = 49.75,$$

so that

$$SSE = 49.75 - 25.25 - 18.75 = 6.25.$$

The following table gives the two-way analysis of variance.

Source	D.F.	SS	Mean Square	F-ratio
Cars	3	18.75	6.25	9.00
Brands	3	25.25	8.42	12.12
Error	9	6.25	0.69	—
Total	15	49.75	—	—

Since $F_{3,9,.01} = 6.99$ we reject the null hypothesis H_A that the four brands of tires show the same mean tread wear after 20,000 miles of driving. Although the object was to compare differences in brands of tires, we can also test the null hypothesis H_B that average tread wear for all four cars is the same. We note that H_B is also rejected at 1 percent level.

Let us now compare the six pairs of brand means by using Scheffé's multiple comparison method. Since $I = J = 4$ and MSE = .694, we have with $\alpha = .05$

$$F_{I-1,(I-1)(J-1),.05} = F_{3,9,.05} = 3.86$$

so that

$$\sqrt{(I-1)MSE(2/J)F_{I-1,(I-1)(J-1),.05}} = \sqrt{3\left(\frac{6.25}{9}\right)\left(\frac{2}{4}\right)3.86} = 2.01.$$

We note that

$$\bar{x}_A = 9.0, \bar{x}_B = 9.5, \bar{x}_C = 12.25, \bar{x}_D = 9.75$$

so that at an overall level .05, the mean for Brand C is significantly different from those for Brands A, B, and D. Since the Brand C mean is the highest, this means that Brand C exhibits the greatest tread wear. A similar comparison may be made for the car means if so desired. □

Problems for Section 13.4

1. Recalling that

$$X_{ijk} = \mu + \alpha_i + \beta_j + \gamma_{ij} + \varepsilon_{ijk},$$

for $i = 1,2,\ldots,I, j = 1,2,\ldots,J, k = 1,2,\ldots,K$ where $\Sigma_{i=1}^I \alpha_i = \Sigma_{j=1}^J \beta_j = \Sigma_{i=1}^I \gamma_{ij} = \Sigma_{j=1}^J \gamma_{ij} = 0$ and the definitions of $\bar{X}, \bar{X}_{i.}, \bar{X}_{.j}$, and $\bar{X}_{ij}$ show that

$$\mathscr{E}\bar{X}_{i.}^2 = (JK)^{-1}\sigma^2 + (\mu + \alpha_i)^2, \qquad \mathscr{E}\bar{X}_{.j}^2 = (IK)^{-1}\sigma^2 + (\mu + \beta_j)^2,$$

$$\mathscr{E}\bar{X}_{ij}^2 = K^{-1}\sigma^2 + (\mu + \alpha_i + \beta_j + \gamma_{ij})^2, \qquad \mathscr{E}\bar{X}^2 = (IJK)^{-1}\sigma^2 + \mu^2.$$

Hence show that

$$\mathscr{E}(\mathrm{SSA}) = \sigma^2(I - 1) + JK \sum_{i=1}^{I} \alpha_i^2, \qquad \mathscr{E}(\mathrm{SSB}) = \sigma^2(J - 1) + IK \sum_{i=1}^{J} \beta_j^2,$$

$$\mathscr{E}(\mathrm{SSI}) = \sigma^2(I - 1)(J - 1) + K \sum_{i=1}^{I} \sum_{j=1}^{J} \gamma_{ij}^2, \qquad \mathscr{E}(\mathrm{SSE}) = IJ(K - 1)\sigma^2.$$

[Hint: Prove and use the relations $\bar{X}_{i.} = \mu + \alpha_i + (JK)^{-1}\varepsilon_{i.}$ where $\varepsilon_{i.} = \sum_{j=1}^{J}\sum_{k=1}^{K}\varepsilon_{ijk}$, and so on, and SSA $= JK\sum_{i=1}^{I}\bar{X}_{i.}^2 - IJK\bar{X}^2$, and so on.]

2. Let $K = 2$, $I = 2$, and $J = 2$ and consider the model in equation (2),

$$\mu_{ij} = \mu + \alpha_i + \beta_j + \gamma_{ij},$$

where

$$\alpha_1 + \alpha_2 = 0, \beta_1 + \beta_2 = 0, \gamma_{11} + \gamma_{12} = 0, \gamma_{12} + \gamma_{22} = 0, \text{ etc.}$$

Show that

$$\mu_{11} = \mu + \alpha_1 + \beta_1 + \gamma_{11}, \qquad \mu_{12} = \mu + \alpha_1 - \beta_1 - \gamma_{11},$$

$$\mu_{21} = \mu - \alpha_1 + \beta_1 - \gamma_{11}, \qquad \mu_{22} = \mu - \alpha_1 - \beta_1 + \gamma_{11}.$$

In particular, consider the table of means:

Workers	Machines	
	1	2
1	85	97
2	72	110

so that $\mu_{11} = 85$, $\mu_{12} = 97$, $\mu_{21} = 72$, and $\mu_{22} = 110$. Show that $\mu = 91$, $\gamma_{11} = 6.5$, $\beta_1 = -12.5$, and $\alpha_1 = 0$.

3. To test the effectiveness of three teaching methods, three instructors were randomly assigned 15 students each. The students were then randomly assigned to the different teaching methods (five to each method) and were taught exactly the same material. At the conclusion of the experiment, the same examination was administered to all the students with the following results.

Teaching Method	Instructor		
	A	B	C
1	75	90	85
	74	64	78
	86	80	77
	94	72	69
	78	68	82
2	92	90	82
	90	89	70
	84	91	73
	82	76	74
	75	84	83
3	85	69	89
	72	75	91
	80	98	73
	85	72	82
	76	73	84

(i) Assuming that the two main effects are additive, perform the two-way analysis of variance to test the hypotheses that the three methods are equally effective and that the three instructors are equally effective.

(ii) Construct the two-way analysis of variance table assuming interaction between the instructor and the teaching method. Test the three hypotheses H_A, H_B, and H_{AB}.

4. The following data represent the number of units produced by three workers, each working on the same machine for four different days.

Machine	Worker		
	A	B	C
M_1	15, 16, 15, 17	18, 19, 16, 18	19, 16, 20, 21
M_2	18, 18, 17, 18	16, 16, 15, 15	18, 23, 22, 23
M_3	16, 17, 18, 15	18, 16, 17, 17	19, 18, 18, 19

Test whether:
(i) The differences among the machines are significant.
(ii) The differences among the workmen are significant.
(iii) The interactions are significant.
Take $\alpha = .05$.

5. The following table shows the yield (pounds per plot) of four varieties of wheat, obtained with three different kinds of fertilizers.

Fertilizer	Variety of Wheat			
	A	B	C	D
α	8	3	6	7
β	10	4	5	8
γ	8	4	6	7

Test the hypotheses that the four varieties of wheat yield the same average yield and that the three fertilizers are equally effective.

6. In an experiment to determine whether four different makes of automobiles average the same gasoline mileage, a random sample of two cars of each make was taken from each of four cities. Each car was then test run on five gallons of gasoline of the same brand. The following table gives the number of miles traveled.

Cities	Automobile Make			
	A	B	C	D
Cleveland	92.3, 104.1	90.4, 103.8	110.2, 115.0	120.0, 125.4
Detroit	96.2, 98.6	91.8, 100.4	112.3, 111.7	124.1, 121.1
San Francisco	90.8, 96.2	90.3, 89.1	107.2, 103.8	118.4, 115.6
Denver	98.5, 97.3	96.8, 98.8	115.2, 110.2	126.2, 120.4

Construct the analysis of variance table. Test the hypotheses of no automobile effect, no city effect, and no interactions. Use $\alpha = .05$.

7. In a field trial four strains of wheat are planted in four blocks. The treatments (strains of wheat) are assigned to the four plots into which each block is divided at random. The following table shows the yield (in pounds per plot) by block.

Block 1		Block 2		Block 3		Block 4	
B	35.2	D	28.4	B	36.9	D	29.8
A	38.7	B	37.6	A	39.4	C	31.4
C	32.6	C	30.5	D	26.5	B	36.3
D	29.4	A	40.2	C	33.4	A	38.3

(i) Construct the analysis of variance table for this randomized block design experiment.
(ii) Test the hypothesis that there are no differences among the strains of wheat.
(iii) Construct .95 level simultaneous confidence intervals for differences among mean yields of the four strains of wheat. Identify significantly different strains of wheat.

8. An experiment is designed to test for differences in weight gains for steers produced by five different rations. Fifteen steers are allocated to three blocks on the basis of

their weight at the start of the experiment. The five rations (treatments) are assigned at random to the five steers within each block. The following data were obtained.

Weight Gains of Steers (pounds)

Ration	Block		
	1	2	3
A	6	2	4
B	5	5	4
C	10	8	9
D	3	4	6
E	8	7	8

Construct the two-way analysis of variance table and test the hypothesis that there are no differences among the five rations in producing weight gains for steers. Identify significantly different rations by using multiple comparisons. Use $\alpha = .05$.

9. In Problem 6 construct simultaneous confidence intervals at overall level .95 for differences among the means for the automobiles. Identify significantly different automobiles.

10. In Problem 3 construct simultaneous confidence intervals at overall level .95 for differences among the means of the three teaching methods.

11. The following data show the pulse rates of 10 subjects recorded before and again 90 minutes after the administration of a particular drug.

Subject	1	2	3	4	5	6	7	8	9	10
Before	74	80	86	95	92	98	74	77	89	87
After	65	74	71	73	74	68	75	65	68	69

(i) Apply Student's t-test to test the null hypothesis that the mean pulse rates before and after are the same.

(ii) Perform the two-way analysis of variance by using each subject as a block. Compare your result to that of part (i). (This is the data on the first 10 subjects of Problem 8.7.11.)

13.5   TESTING EQUALITY IN k-INDEPENDENT SAMPLES: NONNORMAL CASE

In Sections 13.2 and 13.3 we considered the problem of testing the null hypothesis

$$H_0: F_1(x) = F_2(x) = \cdots = F_k(x), \qquad \text{for all } x$$

based on k-independent random samples from normal distributions with common unknown variance σ^2. We now drop the assumption that the F_i's are normal distribution functions and assume only that each F_i is a continuous distribution function.

In Section 12.2 we showed how the χ^2 test for homogeneity of k proportions can be adapted to test equality of k distributions when the sample sizes are large. The median test of Section 6.4 can also be extended to test the equality of k distributions.

13.5.1   The Extended Median Test

Suppose we have k-independent random samples of sizes $n_1, n_2, \ldots, n_k$ from respective distribution functions $F_1, F_2, \ldots, F_k$ that are assumed to be continuous. The null hypothesis of interest is that all distributions have the same (unknown) median.

The method is simple enough. We compute the sample median of the combined sample. This is an estimate of the common population median under H_0. Under H_0, each observation is equally likely to be above or below the sample median of the combined sample. We therefore expect about half the observations in each sample to be above, and about half below, the sample median. If the samples show significant departure from what is expected under H_0, we reject H_0.

Let $X_1, X_2, \ldots, X_k$ be the number of observations in samples $1, 2, \ldots, k$ respectively that are above the sample median. Let $\sum_{i=1}^{k} n_i = n$ and $\sum_{i=1}^{k} X_i = t$. If the sample median is unique then

$$t = n/2 \text{ if } n \text{ is even, and } = (n-1)/2 \text{ if } n \text{ is odd,}$$

and under H_0

(1) $$P\left\{ X_1 = x_1, X_2 = x_2, \ldots, X_k = x_k \left| \sum_{i=1}^{k} X_i = t \right. \right\} = \frac{\prod_{j=1}^{k} \binom{n_j}{x_j}}{\binom{n}{t}}.$$

For an exact test we compute the P-value associated with an observation $x_1, x_2, \ldots, x_k$ by computing the probability of observing values at least as extreme as $x_1, x_2, \ldots, x_k$. Even for the $k = 2$ case we have seen that this is computationally a rather time-consuming exercise.

Example 1.   Comparing Weight Gains in Animals.   Three different feeds are to be compared to determine if they have the same distribution of weight gains on experimental animals (such as pigs). Suppose 12 animals are divided at random into three groups of four animals each and each group is given a different feed. The following results are obtained.

Feed	Weight Gains (pounds)
1	104, 108, 107, 106
2	112, 115, 118, 116
3	120, 124, 114, 112

Let us test the hypothesis that the three distributions of weight gains have the same median. Since $n = 12$, the sample median is the average of the sixth and seventh observations in the following set of ordered data:

$$104, 106, 107, 108, 112, 112, 114, 115, 116, 118, 120, 124.$$

The sample median is $(112 + 114)/2 = 113$. We note that $n_1 = n_2 = n_3 = 4$, $x_1 = 0$, $x_2 = 3$, $x_3 = 3$. Hence under H_0 the probability of observing triples (x_1, x_2, x_3) that are at least as extreme as the triple $(0, 3, 3)$ is given by

$$P\{ X_2 \geq 3, X_3 \geq 3 \mid t = 6\} = \sum_{\substack{k_1 + k_2 + k_3 = 6 \\ k_2 \geq 3, k_3 \geq 3}} P(X_1 = k_1, X_2 = k_2, X_3 = k_3)$$

$$= P(X_1 = 0, X_2 = 3, X_3 = 3) = \frac{\binom{4}{3}\binom{4}{3}}{\binom{12}{6}} = \frac{16}{924} = .017.$$

Thus we reject H_0 at levels $\alpha \geq .017$.

Let us now assume that the distributions of weight gains are normal and apply the one-way analysis of variance to the same data. We have

$$\bar{x}_1 = 425/4 = 106.25, \quad \bar{x}_2 = 461/4 = 115.25, \quad \bar{x}_3 = 470/4 = 117.5, \quad \bar{x} = 113,$$

and

$$s_1^2 = 8.75/3 = 2.92, \quad s_2^2 = 18.75/3 = 6.25, \quad s_3^2 = 91/3 = 30.33.$$

Analysis of Variance

Source	D.F.	S.S.	Mean Square	F-Ratio
Feeds	2	283.5	141.75	10.76
Error	9	118.5	13.17	
Total	11	402.0		

For 2 and 9 d.f. we have $F_{2,9,.01} = 8.02$ and we reject H_0 at level $\alpha = .01$. The result is the same although the normal theory test is able to detect the difference among means even at the 1 percent percent level. This is to be expected. Actually $F_{2,9,.005} = 10.1$ so that we reject H_0 even at level .005.   □

When n is moderately large $(n \geq 25)$ and each n_i is not too small $(n_i \geq 5)$ we can apply the χ^2 goodness of fit test, which is considerably simpler. Let us first settle the question of ties which, theoretically at least, occur with probability zero. In practice, several observations may be tied at the sample median. If the number of observations is large and there are only a few ties, one can drop the tied observations from consideration. However, we will adopt the following procedure. We will count the tied observations as being below the median. Accordingly, we replace each observation by the symbol *a* if it is above the sample median and by the symbol *b* otherwise. The resulting data can now be put into a $2 \times k$ contingency table as follows.

Number of	Sample				Total
	1	2	$\cdots$	*k*	
a's	X_1	X_2	$\cdots$	X_k	$\sum_{i=1}^{k} X_i$
b's	$n_1 - X_1$	$n_2 - X_2$	$\cdots$	$n_k - X_k$	$n - \sum_{i=1}^{k} X_i$
Total	n_1	n_2	$\cdots$	n_k	n

Let $\sum_{i=1}^{k} X_i = t$. Under H_0 we expect $n_i t/n$ observations in the ith sample to be above the median and $n_i(n - t)/n$ observations to be below the median. The statistic

$$(2) \qquad Q = \sum_{i=1}^{k} \frac{(X_i - n_i t/n)^2}{n_i t/n} + \sum_{i=1}^{k} \frac{(n_i - X_i - n_i(n - t)/n)^2}{n_i(n - t)/n}$$

$$= \sum_{i=1}^{k} \frac{(X_i - n_i t/n)^2}{n_i t(n - t)} n^2$$

has a χ^2 distribution with

$$\text{d.f.} = 2k - k - 1 = k - 1$$

since we have estimated k parameters. We reject H_0 at level (approximately) α if q_0, the computed value of Q, is $\geq \chi^2_{k-1, \alpha}$.

Example 2. Comparing Three Laboratories. In a study to ascertain whether or not three laboratories give essentially the same results, on the average, random samples of the same chemical mixture are sent to each laboratory for chemical analysis. The laboratories are asked to identify the percentage of chemical A in each sample.

Laboratory	Percentage of Chemical A
1	34.3, 32.4, 33.8, 35.6, 32.3, 34.5, 35.4, 33.9
2	36.4, 35.8, 37.2, 34.5, 36.1, 34.8, 35.6, 36.2, 36.4
3	35.3, 34.9, 35.1, 35.8, 36.1, 35.2, 35.8, 33.9

Since there are 25 observations in the pooled sample, the sample median is the 13th observation in the ordered arrangement. It is easy to see that the sample median if 35.3. The frequency counts of the number of observations below and above the median are as follows:

	Laboratory			
	1	2	3	Total
Above	2	7	3	12
Below	6	2	5	13
Total	8	9	8	25

Then

$$q_0 = \frac{n^2}{t(n-t)} \sum_{i=1}^{3} \frac{(x_i - n_i t/n)^2}{n_i}$$

$$= \frac{25^2}{12(13)} \left[\frac{(2 - 96/25)^2}{8} + \frac{(7 - 108/25)^2}{9} + \frac{(3 - 96/25)^2}{8} \right] = 5.25.$$

Since d.f. $= k - 1 = 3 - 1 = 2$, we see that the P-value satisfies

$$.05 < P_{H_0}(Q \geq 5.25) < .10.$$

We can therefore reject H_0 that the three medians are the same at level $\alpha = .1$ but not at level $\alpha = .05$. The evidence in favor of H_0 that the data offer is not very strong.

We leave the reader to show that the one-way analysis of variance yields the following results.

Source	D.F.	SS	Mean Square	F-ratio
Laboratories	2	14.98	7.49	8.57
Error	22	19.23	0.87	
Total	24	34.21		

For (2, 22) d.f. the F value 8.57 is significant even at .005 level.   □

13.5.2 The Kruskal–Wallis Test: Nonparametric One-way Analysis of Variance

The median test loses information since it compares each observation to the sample median. It does not use any information on the relative magnitude of each observation when compared with every other observation in the combined sample. The Kruskal–Wallis test effects this comparison by replacing each observation by its rank in the pooled sample. The smallest observation is replaced by its rank 1, the next smallest by rank 2, and so on, the largest by rank n. Since the test is an extension of the Mann–Whitney–Wilcoxon test (just as the one-way analysis of variance F-test is an extension of the two-sample t-test), it is instructive to analyze the Mann–Whitney–Wilcoxon test a little more carefully. Let

$$R_{ij} = \text{rank of } X_{ij}, \qquad j = 1, 2, \ldots, n_i, i = 1, 2,$$

and

$$R_i = \sum_{j=1}^{n_i} R_{ij}, i = 1, 2.$$

The Mann–Whitney–Wilcoxon test of $F_1 = F_2$ is based on the statistic R_1 (see Section 11.6.1). For large n_1, n_2 we know that

(3)
$$Z = \frac{R_1 - n_1(n + 1)/2}{\sqrt{n_1 n_2(n + 1)/12}}$$

has approximately a standard normal distribution, so that

(4) $$Z^2 = \frac{12}{n_1 n_2(n + 1)} \left[R_1 - \frac{n_1(n + 1)}{2} \right]^2 = \frac{12n_1}{n_2(n + 1)} \left(\frac{R_1}{n_1} - \frac{n + 1}{2} \right)^2$$

has approximately a $\chi^2(1)$ distribution. The first thing to note is that since $R_2 = [n(n + 1)/2] - R_1$, we can rewrite (after some simple algebra) Z^2 as

(5) $$Z^2 = \frac{12}{n(n + 1)} \left[n_1 \left(\frac{R_1}{n_1} - \frac{n + 1}{2} \right)^2 + n_2 \left(\frac{R_2}{n_2} - \frac{n + 1}{2} \right)^2 \right].$$

We note that R_i/n_i is the mean rank of the ith sample, whereas $(n + 1)/2$ is the mean rank of the combined sample. Thus Z^2 is essentially the weighted sum of squares of deviations of each sample mean rank from the combined sample mean rank. Under H_0, we have essentially one sample of n observations and we expect the sum of the ranks $n(n + 1)/2$ to be divided proportionally among the two samples according to their size. Thus statistic Z^2 makes good sense to consider even when $k \geq 3$. Accordingly, we define

(6) $$H = \frac{12}{n(n + 1)} \sum_{i=1}^{k} n_i \left(\frac{R_i}{n_i} - \frac{n + 1}{2} \right)^2.$$

This is the *Kruskal–Wallis statistic*. The larger the H the larger the disparity between the sample mean ranks and the grand mean rank so that the rejection region is of the form $H \geq c$.

The alert reader will notice that the quantity $\sum_{i=1}^{k} n_i[(R_i/n_i) - (n + 1)/2]^2$ is the sum of squares due to treatments (SST) in the one-way analysis of variance carried out with each X_{ij} replaced by its rank. Then why don't we have MSE as the divisor? The answer is that the F-test is equivalent to the test based on the ratio (Problem 13.2.12)

$$\frac{\sum_{i=1}^{k} n_i (\bar{X}_i - \bar{X})^2}{\sum_{i=1}^{k} \sum_{j=1}^{n_i} (X_{ij} - \bar{X})^2}.$$

When each X_{ij} is replaced by its rank, the denominator, which is the total sum of squares (TSS), is given by

$$\sum_{i=1}^{k} \sum_{j=1}^{n_i} \left(R_{ij} - \frac{n+1}{2} \right)^2 = \sum_{i=1}^{k} \sum_{j=1}^{n_i} R_{ij}^2 - n\left(\frac{n+1}{2} \right)^2$$

$$= \frac{n(n+1)(2n+1)}{6} - n\left(\frac{n+1}{2} \right)^2$$

$$= \frac{n^3 - n}{12}.$$

It follows that the test based on H is equivalent to the test based on the F-ratio (when each X_{ij} is replaced by its rank).

It is easy to show (Problem 8) that a computationally more convenient form of H given in (6) is the following

(7)
$$H = \frac{12}{n(n+1)} \sum_{i=1}^{k} \frac{R_i^2}{n_i} - 3(n+1).$$

The null distribution of H for $k = 3$ and $n_i \leq 5$ is extensively tabulated.[†] For values of n_i that are not too small, the null distribution of H may be approximated by the χ^2 $(k-1)$ distribution. We will use the χ^2 approximation whenever each $n_i > 5$.

When ties occur, we assign average rank to the tied observations. This does not affect the value of H when ties occur within the same sample. When ties occur across samples and the number of ties is small, we use the same test and the same distribution. When the number of ties is large it is possible to use a correction for ties, but we will not give this correction here.[‡]

> **Example 3. Comparing Weight Gains in Animals (Example 1).** For the data on weight gains of Example 1 let us assign ranks to each observation in the combined sample. The following table gives the ranks and the rank sums R_i.

	Feed	
1	2	3
1	5.5	11
4	8	12
3	10	7
2	9	5.5
10	32.5	35.5

[†]See, for example, W. J. Conover, *Practical Nonparametric Statistics*, Wiley, 1971, p. 393.
[‡]See J. D. Gibbons, *Nonparametric Methods for Quantitative Analysis*, Holt, Rinehart and Winston, 1976, pp. 178–179.

Then

$$h = \frac{12}{12(13)} \frac{10^2 + 32.5^2 + 35.5^2}{4} - 3(12 + 1) = 46.47 - 39 = 7.47.$$

For $k = 3$, $n_1 = n_2 = n_3 = 4$ we see from the table cited above that $P_{H_0}(H \geq 7.5385)$ = .011 and $P_{H_0}(H \geq 5.6923) = .049$ so that $.011 < P_{H_0}(H \geq 7.47) < .049$. In fact, the exact probability is closer to .011 than to .049. In any event we have to reject the null hypothesis of the equality of distributions at level .05.   □

*Example 4.   **Comparing Three Laboratories (Example 2).**   Let us order the 25 observations in increasing order of magnitude and assign ranks from 1 to 25. Tied observations are assigned average rank.

Observation	32.3	32.4	33.8	33.9	33.9	34.3	34.5
Rank	1	2	3	4.5	4.5	6	7.5
Laboratory	1	1	1	1	3	1	1

Observation	34.5	34.8	34.9	35.1	35.2	35.3	35.4
Rank	7.5	9	10	11	12	13	14
Laboratory	2	2	3	3	3	3	1

Observation	35.6	35.6	35.8	35.8	35.8	36.1	36.1
Rank	15.5	15.5	18	18	18	20.5	20.5
Laboratory	2	1	3	3	2	2	3

Observation	36.2	36.4	36.4	37.2
Rank	22	23.5	23.5	25
Laboratory	2	2	2	2

For convenience we rearrange the data in the following table.

Rank of X_{ij} in Laboratory		
1	2	3
6	23.5	13
2	18	10
3	25	11
15.5	7.5	18
1	22.5	20.5
7.5	9	12
14	15.5	18
4.5	22	4.5
	23.5	
Total 53.5	164.5	107.0

Thus

$$h = \frac{12}{25(26)} \left[\frac{53.5^2}{8} + \frac{164.5^2}{9} + \frac{107.0^2}{8} \right] - 3(26) = 10.53.$$

For $3 - 1 = 2$ d.f. we note that

$$.001 < P_{H_0}(H \geq 10.53) < .01$$

so that the Kruskal–Wallis test does better than the median test but somewhat worse than the F-test. This is to be expected, since F test uses more information whereas the median test loses information contained in the data. $\square$

13.5.3 Multiple Comparisons

In Example 4, the Kruskal–Wallis test rejects the equality of three medians so that we conclude that the three distributions are not the same. Often we wish to identify significantly different medians. As pointed out in Section 13.3, since the object is to obtain simultaneous confidence intervals, pairwise comparisons using the Mann–Whitney–Wilcoxon test are not appropriate.

We will follow the procedure proposed by Dunn. Let $\bar{R}_i = R_i / n_i$ be the sample mean of ranks assigned to the ith sample. Then

$$P\left\{ |\bar{R}_i - \bar{R}_j| \leq z_p \sqrt{\frac{n(n+1)}{12}\left(\frac{1}{n_i} + \frac{1}{n_j}\right)} \quad \text{for all } i, j, i \neq j \right\} \geq 1 - \alpha$$

where $p = \alpha / k(k-1)$ and z_p is the quantile of order $1 - p$ under the standard normal distribution.

> *Example 5. Significantly Different Laboratories in Example 4.* In Example 4, $n_1 = 8, n_2 = 9$, and $n_3 = 8$, so $n = 25, k = 3$, and $k(k-1) = 6$. Let us choose $\alpha = .15$ so that $p = .15/6 = .025$ and $z_p = 1.96$. In the following table we compute the critical values

$$c_{ij} = z_p \sqrt{\frac{n(n+1)}{12}\left(\frac{1}{n_i} + \frac{1}{n_j}\right)} = 1.96\sqrt{\frac{25(26)}{12}\left(\frac{1}{n_i} + \frac{1}{n_j}\right)}$$

for each pair (i, j). If $|\bar{R}_i - \bar{R}_j| > c_{ij}$ then the difference between laboratories i and j is considered significant at level α.

i \ j	1	2	3	$\bar{R}_i$
1	0			6.69
2	7.01	0		18.28
3	7.21	7.0	0	13.38

Thus

$$\bar{R}_1 - \bar{R}_2 = -11.59, \quad \bar{R}_1 - \bar{R}_3 = -6.69, \quad \bar{R}_2 - \bar{R}_3 = 4.9.$$

Comparing $|\bar{R}_i - \bar{R}_j|$ to c_{ij} we see that the difference between laboratories 1 and 2 is significant. Moreover, since $\bar{R}_1 - \bar{R}_2 < 0$ we conclude that Laboratory 1 reports a significantly smaller percentage of chemical A at level .15. □

Example 6. Comparing Four Diets. Suppose a dairy farmer wishes to compare the effect of four different diets on the amount of milk produced by cattle. He randomly selects 24 cows and assigns six cows randomly to each diet. After eight weeks, the amount of milk produced by each cow is recorded for one week. The following table gives the rank assigned to each cow's weekly production.

Diet	Rank	R_i	$\bar{R}_i$
1	$2, 4, 5, 8, 6.5, 9$	34.5	5.75
2	$6.5, 12, 10, 11, 13.5, 1$	54.0	9.00
3	$3, 23, 15, 19, 20, 21$	101.0	16.83
4	$13.5, 16, 18, 24, 17, 22$	110.5	18.42

Let us test the hypothesis that the diets are equally productive. Since $n_1 = n_2 = n_3 = n_4 = 6$,

$$h = \frac{12}{nn_1(n+1)} \sum_1^4 R_i^2 - 3(n+1)$$

$$= \frac{12}{24(6)(25)} (26517.5) - 3(25) = 13.39$$

and for 3 d.f. we see that

$$.001 < P_{H_0}(H \geq 13.39) < .01.$$

Thus we have to reject the equality of productivity at level $\alpha = .01$. Let us now identify significant differences. Let us choose $\alpha = .20$. Then $p = .20/12 = .0167$ and $z_p = 2.13$. Consequently

$$c_{ij} = 2.13 \sqrt{\frac{24(25)}{12} \left(\frac{1}{6} + \frac{1}{6} \right)} = 8.70.$$

It follows that the significant differences correspond to diets 1 and 3, 1 and 4, and 2 and 4. Moreover, in terms of yields

$$\text{Diet } 1 < \text{Diet } 3,$$

$$\text{Diet } 1 < \text{Diet } 4,$$

$$\text{Diet } 2 < \text{Diet } 4,$$

so with 80 percent confidence we can state that diet 4 is more productive than diets 1 and 2 and that diet 3 is more productive than diet 1. □

Problems for Section 13.5

1. The United Way would like to compare contributions to its annual appeal in three different towns. Independent random samples of contributors are selected from contribution records for each town and the annual contributions for each contributor in the sample are recorded.

Town	Annual Contributions (dollars)
1	35, 150, 200, 45, 75, 5
2	15, 110, 75, 25, 30, 115, 125
3	25, 35, 30, 75, 225, 160

   Is there sufficient evidence to conclude that the average (median) contribution in each town is the same? Analyze the data by using the large sample median test (even though $n = 19$) as well as the Kruskal–Wallis test.

2. In order to compare the amount of time it takes to recover from four different types of influenza, a random sample of 32 volunteers was selected. Eight subjects each were randomly assigned to each strain of virus. Each subject was then injected with the strain of the virus and all 32 subjects then received the same care. The number of days it took for each subject to recover completely was recorded as follows:

Influenza Type	Recovery Time (Days)
1	5, 7, 5, 8, 12, 6, 4, 5
2	4, 7, 3, 4, 6, 6, 5, 4
3	8, 12, 13, 10, 11, 10, 9, 11
4	12, 10, 13, 12, 14, 10, 13, 12

   (i) Is there sufficient evidence to indicate that the recovery times for the four types of influenza are the same? Apply the median test and the Kruskal–Wallis test and compare your results.

   (ii) If the recovery times are significantly different, identify the differences at level $\alpha = .15$.

3. In a study to compare the mean weight at four different heights, random samples of males from each height group are selected and their weights are recorded as follows:

Height Group (cm)	Weight (kg)
[150, 160)	50, 54, 52, 55, 52, 51
[160, 170)	55, 60, 62, 64, 61, 60
[170, 180)	65, 68, 67, 66, 67
[180, 190)	72, 75, 80, 79, 82, 81, 74

   (i) Is the distribution of weight the same for the four height groups? Compare the Kruskal–Wallis test and the one-way analysis of variance F-test.

   (ii) Identify significant differences if any at level $\alpha = .25$.

4. In order to check if there are significant differences in reading ability across the state, the State Board of Education classifies each school district as urban, suburban, or rural. From each district median reading scores of high school seniors are then recorded as follows:

District	Median Score
Urban	15.2, 14.7, 15.6, 14.8, 16.2, 13.4, 16.1
Suburban	17.8, 19.4, 16.8, 18.3, 20.3, 18.6, 19.8, 20.7
Rural	12.6, 14.3, 11.2, 13.4, 14.5, 15.1

Is there sufficient evidence to indicate that the median scores differ among the three classifications? If so, identify significant differences at level $\alpha = .25$.

5. Six different automobile models advertise the same fuel efficiency as measured by miles per gallon of gasoline. Randomly selected cars of each model were run on five gallons of the same brand of gasoline until the gas runs out. The road and weather conditions for each test run were identical. The following table gives the number of miles traveled by each car.

Model	Miles Traveled
A	66, 68, 67, 70, 69, 68, 71, 72
B	71, 75, 78, 79, 76, 78, 76
C	89, 90, 95, 90, 92, 96, 94
D	107, 105, 98, 106, 104, 95
E	115, 118, 122, 116, 118, 117
F	130, 132, 121, 129, 130, 126

Test the hypothesis that the distributions of the number of miles traveled are the same for each model. Identify significant differences if any at level $\alpha = .20$.

6. The following table gives the hourly wages of samples of floor workers in four different department stores in Toledo.

Department Store	Hourly Wage
A	3.35, 3.80, 3.55, 3.35, 3.80
B	3.80, 4.10, 4.10, 3.95, 4.25, 4.40
C	4.00, 4.50, 4.50, 4.75, 5.00
D	3.35, 3.80, 4.10, 3.95, 3.80

Test the hypothesis that the distributions of the hourly wages are the same for the four stores. Investigate significant differences if any at level $\alpha = .15$.

7. Show that the Kruskal–Wallis statistic and the Mann–Whitney–Wilcoxon test statistic offer equivalent criteria in the $k = 2$ case when n is large and each n_i is large. That is, show that (4) and (6) are equivalent when $k = 2$.

8. Show that expressions (6) and (7) for H are equivalent.

13.6　THE FRIEDMAN TEST FOR k-RELATED SAMPLES

In Section 13.5 we studied two methods of testing the equality of k distribution functions based on k-independent random samples. The Kruskal–Wallis test that we studied as an extension of the Mann–Whitney–Wilcoxon test is an analog, as we saw, of the one-way analysis of variance when we replace the normality assumption by the assumption that the distribution functions are all continuous. We now consider a statistic used in the analysis of data resulting from k-related samples. The problem we consider now may be regarded as an extension of the problem of dependent (or matched) pairs of observations considered in Section 11.5. The Friedman test statistic that is used to analyze k-related samples is, however, not an extension of the Wilcoxon signed rank test of Section 11.5. Indeed, it does not require the assumption of symmetry needed for the Wilcoxon test. The Friedman test is an extension of the two-way analysis of variance technique for a randomized block design when the assumption of normality is replaced by the assumption that the distributions are continuous.

Let $(X_{i1}, X_{i2}, \ldots, X_{ik})$, $i = 1, 2, \ldots, b$ be a random sample from a k-variate continuous type distribution function. The observations may be arranged in b rows (called blocks) and k columns (called treatments). The observations in different rows are independent and those in different columns are dependent. The observation x_{ij} then corresponds to the ith block and jth treatment, $i = 1, 2, \ldots, b$, and $j = 1, 2, \ldots, k$. The object is to test the null hypothesis H_0: treatment effects are all equal, against the alternative hypothesis H_1: treatment effects are not all equal.

Friedman suggested replacing each observation in a block by its rank. In the Kruskal–Wallis test we replace each observation by its rank in the combined sample in order to compare every observation with every other observation. In the present setup it does not make much sense to compare observations in two different blocks. Hence ranks are assigned separately for each block. We assume for now that there are no tied observations within each block.

Let $R_{ij} = R(X_{ij}) = $ rank of jth observation in ith block. Then $1 \leq R_{ij} \leq k$. Let

$$R_j = \sum_{i=1}^{b} R_{ij} = \text{sum of ranks for } j\text{th treatment}, j = 1, 2, \ldots, k.$$

Clearly

$$\sum_{j=1}^{k} R_{ij} = \frac{k(k + 1)}{2}, \qquad i = 1, 2, \ldots, b$$

whereas the R_j's are random. If the treatment effects are all the same then we expect each R_j to be equal to $b(k+1)/2$. That is, under H_0,

$$\mathscr{E}R_j = \frac{1}{k} \cdot \frac{bk(k+1)}{2} = \frac{b(k+1)}{2}.$$

The sum of squares of deviations of R_j's from $\mathscr{E}_{H_0}R_j$ therefore is a measure of the differences in treatment effects. Let

$$S = \sum_{j=1}^{k} \left\{ R_j - \frac{b(k+1)}{2} \right\}^2.$$

The Friedman test statistic is defined by

(1) $$F = \frac{12S}{bk(k+1)} = \frac{12}{bk(k+1)} \sum_{j=1}^{k} \left\{ R_j - \frac{b(k+1)}{2} \right\}^2.$$

We leave the reader to show (Problem 8) that (1) may be rewritten in the form

(2) $$F = \frac{12}{bk(k+1)} \sum_{j=1}^{k} R_j^2 - 3b(k+1)$$

which is more convenient for computational purposes.

The null distribution of F is extensively tabulated for small values of k and n. The null distribution of F can be approximated by a $\chi^2(k-1)$ distribution. The approximation gets better as b gets larger. In order to avoid giving long tables we will use this approximation. The null hypothesis is rejected at level α if the computed value of F exceeds $\chi^2_{k-1,\alpha}$, the quantile of order $1-\alpha$ for a $\chi^2(k-1)$ random variable.

Friedman Statistic for Testing Equality of *k* Treatment Effects

Let R_{ij} = rank of X_{ij}, $1 \le R_{ij} \le k$, and $R_j = \sum_{i=1}^{k} R_{ij}$ = sum of ranks for jth treatment. Compute

$$F = \frac{12}{bk(k+1)} \sum_{j=1}^{k} R_j^2 - 3b(k+1).$$

Reject H_0: no treatment differences at level α if the computed value f of F satisfies

$$f \ge \chi^2_{k-1,\alpha}.$$

Theoretically, ties within each block occur with probability zero. In practice, however, ties do occur. Tied observations are assigned average rank and the value of F is computed in the usual way.

Example 1. *Weight Gains by Swines (Example 13.4.4).* In Example 13.4.4 the weight gains (in pounds) by swines fed on three rations were as follows.

Blocks	Ration		
	A	B	C
1	6.7	13.3	8.2
2	16.2	15.2	16.3
3	13.4	20.2	13.8
4	10.8	9.9	15.8
5	14.2	19.6	14.4

We saw that we cannot reject H_0: no difference in rations at $\alpha = .05$ level. In order to apply the Friedman test we rank the rations in each block from 1 (least weight gain) to 3 (most gain) as follows.

Blocks	A	B	C
1	1	3	2
2	2	1	3
3	1	3	2
4	2	1	3
5	1	3	2
Total	7	11	12

Since $b = 5$, $k = 3$, we have

$$f = \frac{12}{15(4)}(49 + 121 + 144) - 3(5)(4) = 2.8$$

so that for $k - 1 = 2$ d.f. the approximate P-value satisfies $.20 < P\text{-value} = P(\chi^2(2) \geq 2.8) < .30$. Thus we cannot reject H_0. (The exact P-value in this case, is, in fact, .367.) $\square$

Example 2. *Comparing Tire Durability of Four Brands.* Four brands of automobile tires are to be tested for durability. Three drivers are selected for the experiment. Each driver is given a set of four tires. The tires are mounted in a random fashion on his car and rotated regularly. The car is run for 10,000 miles and the tread wear is measured. The experiment is then repeated with a second set of four tires of a different brand, and so on. Ranks are assigned by each driver to the four brands of tires according to the tread wear (rank 1 is assigned to the brand with the most tread wear, etc.). The following table shows the ranks and rank sums.

Driver	Brand			
	A	B	C	D
1	1.5	1.5	4	3
2	1	2	3.5	3.5
3	2	1	3	4
Total	4.5	4.5	10.5	10.5

Do the data indicate a significant difference in durability as measured by tread wear? Here $b = 3$ and $k = 4$, so that

$$f = \frac{12}{3(4)(5)} \left[2(4.5)^2 + 2(10.5)^2 \right] - 3(3)5 = 7.2.$$

For $k - 1 = 3$ d.f. we see that

$$.05 < P\text{-value} < .10$$

so that we reject H_0 at $\alpha = .10$ but not at $\alpha = .05$. Thus at significance level .10 we conclude that there are significant differences in tire durability. □

Remark 1. Suppose $k = 2$ so that two treatments are to be compared. In this case the randomized block design reduces to the paired comparisons design. The appropriate test for paired comparisons is the sign test of Section 6.3. How does Friedman test compare with the sign test in this special case? We note that each block has two ranks, 1 and 2. Suppose X = number of blocks in which treatment 1 receives rank 1 and treatment 2 rank 2. Then

$$R_1 = X + 2(b - X) = 2b - X, \quad \text{and} \quad R_2 = 3b - (2b - X) = b + X.$$

and

$$F = \frac{12}{6b} \left[(2b - X)^2 + (b + X)^2 \right] - 3b(3) = 4b \left(\frac{X}{b} - \frac{1}{2} \right)^2.$$

The null hypothesis is rejected when F is large, or equivalently, when $|X/b - 1/2|$ is large. Since X is the number of pairs for which the second treatment is "better" than the first we see that the F test is equivalent to the two-sided sign test based on the differences $X_{i1} - X_{i2}$, $i = 1, 2, \ldots, b$, provided b is large.

Remark 2. Suppose $b = 2$ so that there are only two blocks.

	Observations				Ranks			
Block 1	$X_{11}, X_{12}, \ldots, X_{1k}$				R_{11}	R_{12}	$\ldots$	R_{1k}
Block 2	$X_{21}, X_{22}, \ldots, X_{2k}$				R_{21}	R_{22}	$\ldots$	R_{2k}
	Total				R_1	R_2	$\ldots$	R_k

In Section 11.9.1 we used the Spearman rank correlation coefficient R_S given by

$$R_S = 1 - 6 \sum_{j=1}^{k} \frac{\left[R_{1j} - R_{2j} \right]^2}{k^3 - k}$$

as a measure of disagreement between the two sets of rankings. What, if any, is the relation between F and R_S? We leave the reader to show (Problem 9) that

$$F = (k - 1)(R_S + 1)$$

so that the Friedman test may also be used as a test of linear dependence in the two-sample case.

Remark 3. The Problem of b-Rankings. A statistic designed as a measure of agreement between b sets of rankings is closely related to the Friedman test statistic F. This is the so-called *Kendall's coefficient of concordance* defined by

$$W = \frac{F}{b(k - 1)} = \frac{12}{b^2 k(k^2 - 1)} \sum_{j=1}^{k} R_j^2 - \frac{3(k + 1)}{k - 1}.$$

One can show (Problem 10) that $0 \le W \le 1$. If all the b sets of rankings are in complete agreement, then $R_1, R_2, \ldots, R_k$ is a permutation of $b, 2b, \ldots, kb$ and it follows that $W = 1$. Thus values of W near 1 may be interpreted to mean that there is perfect agreement (or perfect association) among rankings. Similarly, values of W near zero may be interpreted to mean no agreement (or no association) among rankings. The statistic W may be used to test the null hypothesis of no association or independence among b rankings against the alternative that some association exists. (See Section 11.9 for the bivariate case.)

Example 3. Wine Testing. Six judges were asked to rank eight different brands of nonvintage Rosé wines all of which sell for approximately the same price. The following table shows their rankings. (Rank 1 means most preferred, rank 8 least preferred.)

	Wine Brand							
Judge	A	B	C	D	E	F	G	H
1	1	4	2	3	5	8	7	6
2	2	3	1	4	6	7	8	5
3	3	1	2	4	5	8	6	7
4	1	2	3	4	5	6	8	7
5	1	2	3	4	7	8	6	5
6	2	3	1	5	4	6	7	8
Total, R_j	10	15	12	24	32	43	42	38

Suppose first that we are only interested in measuring the association among the six sets of rankings. We compute

$$w = \frac{12}{36(8)(63)}(7126) - \frac{3(9)}{7} = .856$$

so that there is strong agreement among the rankings.

Let us test H_0: no association against H_1: association exists. We see that

$$f = b(k-1)w = 6(8-1)(.856) = 35.95$$

and for $k - 1 = 8 - 1 = 7$ d.f. we have

$$P(\chi^2(7) \geq 35.95) < .001.$$

Thus we reject H_0 and conclude that the judges agree reasonably well in their rankings of the wines.

If the (column) rank sums are used to estimate the ordering of judges' preferences then the preference ordering (from most preferred to least preferred) is as follows:

$$A\ C\ B\ D\ E\ H\ G\ F.$$

Thus the estimate of the rank of Brand A is 1, of Brand B is 3, of Brand C is 2, and so on.   ☐

Remark 4.   In the beginning of this section we said that Friedman's test is an extension of the two-way analysis of variance technique carried out by replacing each observation in a block by its rank in the block. We leave the reader to show (Problem 11) that

$$\text{TSS} = \text{Total sum of squares} = \sum_{i=1}^{b}\sum_{j=1}^{k}\left(R_{ij} - \frac{k+1}{2}\right)^2 = \frac{bk(k^2-1)}{12}$$

so that from (1)

$$F = \frac{(k-1)S}{\text{TSS}}.$$

Let $\bar{R}_j = R_j/b$ and $\bar{R} = \sum_{i=1}^{b}\sum_{j=1}^{k}R_{ij}/bk = (k+1)/2$. Then we can partition TSS as

$$\text{TSS} = \sum_{i=1}^{b}\sum_{j=1}^{k}\left(R_{ij} - \bar{R}\right)^2$$

$$= \sum_{i=1}^{b}\sum_{j=1}^{k}\left(R_{ij} - \bar{R}_j\right)^2 + \frac{1}{b}\sum_{j=1}^{k}\left\{R_j - \frac{b(k+1)}{2}\right\}^2$$

$$= \text{SSE} + \text{SST}.$$

The block sum of squares, SSB, is zero since the block sums are equal to $k(k + 1)/2$, so that there is no variation among the blocks. Thus the F-statistic in the two-way analysis of variance for testing H_0: no differences in treatments is the ratio

$$\frac{\text{MST}}{\text{MSE}} = \frac{\text{SST}/(k - 1)}{\text{SSE}/(b - 1)(k - 1)} = \frac{(b - 1)\text{SST}}{\text{SSE}}$$

$$= \frac{(b - 1)S/b}{\text{TSS} - S/b} = \frac{(b - 1)S}{b(\text{TSS}) - S}$$

where from (1)

$$\text{SST} = S/b.$$

Since TSS is a constant, the Friedman statistic is analogous to the F-ratio in the two-way analysis of variance.

Multiple Comparisons

When the Friedman test rejects the null hypothesis that the treatment effects are all the same, it is of interest to identify significantly different treatments. The procedure for multiple comparison is similar to the one outlined in Section 13.5.3. Indeed

$$P\left\{|R_i - R_j| \leq z_p\sqrt{\frac{bk(k + 1)}{6}} \quad \text{for all } i \neq j, 1 \leq i, j \leq k\right\} \geq 1 - \alpha$$

where z_p is the quantile of order $1 - p$ under the standard normal distribution and $p = \alpha/k(k - 1)$. Thus those pairs of treatment sums (R_i, R_j) are significantly different at level α for which

$$|R_i - R_j| > z_p\sqrt{bk(k + 1)/6}.$$

The sign of $R_i - R_j$ determines the direction of the difference.

Example 4. Multiple Comparisons in Example 2. In Example 2 we concluded that there were significant differences in tire durability at level .10. The data suggest that Brand C and D tires are more durable than Brand A and B tires. Let us apply the method of multiple comparisons to see if this is in fact the case. Let us choose $\alpha = .15$ so that $p = .15/4(3) = .0125$ and $z_{.0125} = 2.24$. Thus

$$z_p\sqrt{\frac{bk(k + 1)}{6}} = 2.24\sqrt{\frac{3(4)(5)}{6}} = 7.08.$$

Since none of the $|R_i - R_j|$, $i \neq j$, equals or exceeds 7.08 we cannot conclude at an overall level .15 that any pairs of treatments are significantly different.

If, on the other hand, we compare only Brands A and C, then with $k = 2$, $\alpha = .05$, $p = .05/2 = .025$, and $z_p = 1.96$ we have

$$z_p\sqrt{\frac{bk(k+1)}{6}} = 1.96\sqrt{\frac{3(2)(3)}{6}} = 3.39.$$

Since $|R_A - R_C| = |R_1 - R_3| = 6.5 > 3.39$ we conclude that Brand A and Brand C tires are significantly different at level .05. Moreover, since $R_1 < R_3$ we conclude that Brand C tires are more durable (that is, have less tread wear) than Brand A tires.   □

Example 5.   Comparing Five Gasoline Additives.   In an experiment to compare five gasoline additives three different car models are used. Each car is driven five times on the same course for 200 miles, once with each additive. All cars use the same gasoline. The following table gives the ranks assigned to gasoline mileage per gallon of gasoline for each car. (Rank 1 means lowest mpg.)

Car	Additive				
	A	B	C	D	E
1	2	4	3	1	5
2	2	5	3	1	4
3	1	3	4	2	5
Total	5	12	10	4	14

We note that $k = 5$, $b = 3$ and

$$f = \frac{12}{3(5)(6)}(25 + 144 + 100 + 16 + 196) - 3(3)(6) = 10.13.$$

For $k - 1 = 4$ d.f. we see that

$$.02 < P_{H_0}\left(\chi^2(4) \geq 10.13\right) < .05$$

and we reject H_0: no differences in additives at level $\alpha = .05$.

Let us now perform multiple comparisons to identify any significant differences. Choosing $\alpha = .20$ we see that $z_p = z_{.01} = 2.33$ and

$$z_p\sqrt{\frac{bk(k+1)}{6}} = 2.33\sqrt{\frac{3(5)(6)}{6}} = 9.02.$$

Writing $i = 1, 2, 3, 4, 5$ respectively for A, B, C, D, and E we see that

$$R_5 - R_4 = 10 > 9.02$$

so that of all the pairs the only pair that differs significantly at level .20 is the pair (D, E). Moreover, since $R_5 > R_4$ we conclude that the additive E is more effective in increasing gasoline mileage than additive D.   □

Problems for Section 13.6

1. In the randomized block experiment of Problem 13.4.5 the yield (pounds per plot) of four varieties of wheat obtained with three fertilizers were as follows.

	Variety of Wheat			
Fertilizer	A	B	C	D
α	8	3	6	7
β	10	4	5	8
γ	8	4	6	7

Using the Friedman test, test the hypothesis that the yield produced by the four varieties of wheat are not significantly different. Compare your result with the parametric case of Problem 13.4.5.

2. In a study of six groups under five different conditions, each group is assigned five matched subjects, one assigned to each of the five conditions. The following table gives the scores.

	Conditions				
Groups	1	2	3	4	5
A	10	2	4	6	5
B	8	2	3	8	6
C	9	1	5	7	5
D	7	3	4	8	4
E	9	1	3	7	4
F	10	1	5	5	3

Using Friedman's statistic, test the hypothesis that all the samples come from the same population.

3. In an experiment to study the effectiveness of five over-the-counter sleeping aids, eight subjects were selected. The experiment was conducted over a period of five weeks. Each subject received one of the five brands of sleeping aids for a week and was given a different brand for each subsequent week. The average number of hours of sleep was recorded as follows.

	Subject							
Sleeping Aid Brand	1	2	3	4	5	6	7	8
A	2.4	4.5	3.6	3.8	4.2	5.1	3.7	2.9
B	3.7	2.8	4.2	3.9	3.7	4.8	5.2	5.1
C	2.5	4.8	5.7	8.1	6.2	5.4	6.0	5.2
D	3.0	6.1	7.3	8.2	7.4	6.2	6.4	8.2
E	3.7	6.2	8.4	9.6	7.3	6.5	4.9	6.8

Test the null hypothesis that the five brands of sleeping aid medicines are equally effective.

4. In an experiment a psychologist asked 10 subjects to recall some material after five different time intervals. The degree of recall was then ranked for each subject from 1 to 5 (1 for least recall and 5 for most) with the following results.

	Time Interval				
Subject	A	B	C	D	E
1	2	1	4	3	5
2	1	2	5	3	4
3	3.5	1	2	3.5	5
4	2	1	3	5	4
5	2	1	5	3	4
6	2	3	4	1	5
7	3	1	4.5	2	4.5
8	1	2	4	3	5
9	2	1	3	4.5	4.5
10	2.5	1	2.5	5	4

Could the difference among rank sums be attributed to chance variation?

5. Suppose 10 randomly selected students are ranked in five subjects, English, Mathematics, History, Geography, and Economics from 1 (worst) to 10 (best) with the following results.

Subject	Student									
	A	B	C	D	E	F	G	H	I	J
English	1.5	10	1.5	3	6	4	7	5	9	8
Mathematics	4	9	1	3	5	2	6	7	8	10
History	2	9	3	1	5	4	6.5	6.5	10	8
Geography	1	6	4	3	7	2	5	8	9	10
Economics	3	8	2	1	4	5	6	10	7	9

Test the hypothesis of no agreement in the five sets of rankings. Estimate, if possible, the relative rank of each student in the five subjects.

6. In a dance contest, five judges are asked to rank each of the 10 finalists from 1 (best dancer) to 10 (worst dancer) with the following results.

	Contestant									
Judge	A	B	C	D	E	F	G	H	I	J
1	1	6	3	2	7	5	10	4	8	9
2	2	7	3	1	6	4	9	5	10	8
3	1	5	3	2	7	6	8	4	9	10
4	3	5	1	2	7	4	10	6	8	9
5	1	6	2	3	7	5	10	4	8	9

(i) Compute Kendall's coefficient of concordance.
(ii) Test the hypothesis that there is no association among the rankings.
(iii) If the null hypothesis of no association in part (ii) is rejected, estimate the rank of each contestant.

7. The following are the scores obtained by a group of fifth graders on tests in arithmetic, social studies, reading, and language.

				Student				
Subject	1	2	3	4	5	6	7	8
Arithmetic	92	41	86	56	55	65	96	72
Social Studies	89	32	82	49	58	59	98	78
Reading	98	38	84	47	52	49	100	71
Language	91	35	80	51	54	51	94	76

(i) Compute Kendall's coefficient of concordance.
(ii) Test the hypothesis that there is no agreement among the four sets of rankings of a student's performance.
(iii) If the null hypothesis in (ii) is rejected, estimate the relative ranks of the eight students.

8. Show that F can be written in the form

$$F = \frac{12}{bk(k+1)} \sum_{j=1}^{k} R_j^2 - 3b(k+1).$$

9. Show that in the $b = 2$ case the Friedman statistic and the Spearman's rank correlation coefficient are related according to the relation

$$F = (k-1)(R_S + 1).$$

10. Show that W, the Kendall coefficient of concordance, satisfies the inequality $0 \le W \le 1$. [Hint: W takes the least value when all the rank sums are equal to their mean value.]

11. Show that
(i) $\text{TSS} = \sum_{i=1}^{b}\sum_{j=1}^{k}\left(R_{ij} - \frac{k+1}{2}\right)^2 = \frac{bk(k^2-1)}{12}$,
(ii) $\text{TSS} = \sum_{i=1}^{b}\sum_{j=1}^{k}(R_{ij} - \bar{R}_j)^2 + \frac{S}{b}$. [Hint: $\text{TSS} = \sum_{i=1}^{b}\sum_{j=1}^{k}(R_{ij} - \bar{R}_j + \bar{R}_j - \bar{R})^2$.]

13.7 REVIEW PROBLEMS

1. In a study to determine whether there is a significant difference among the I.Q. scores of children of poor, lower middle class, middle class, upper middle class, and affluent

families the following results are obtained based on independent random samples:

Economic Status	I.Q. Scores
Poor	122, 115, 106, 112, 98, 105
Lower middle	120, 114, 97, 104, 112
Middle	115, 110, 100, 118, 98
Upper middle	124, 102, 99, 108, 118, 131
Affluent	104, 108, 110, 101, 115, 124, 116, 128

Is the difference among the sample mean I.Q. scores significant at the 1 percent level? Perform a one-way analysis of variance.

2. The following data give the service times (in minutes) that four tellers at a local bank spent serving their customers. The customers served by each teller were selected at random.

Teller	Service Times (minutes)
1	3.8, 1.7, 0.6, 2.3, 1.1, 1.4
2	4.5, 5.0, 3.8, 2.7, 4.2
3	3.2, 3.5, 4.0, 2.8, 4.6
4	1.6, 2.0, 2.8, 0.6, 1.2, 1.8, 1.2

(i) Is the difference among the sample mean service times of the four tellers significant at level .05? Perform a one-way analysis of variance.
(ii) If the answer to part (i) is yes, identify the significantly different true means at level .05.

3. In order to compare the effects of four stimuli on the reaction time of adult males, a completely randomized design was used. Twenty subjects were randomly assigned the four stimuli. The following reaction times (in seconds) were recorded.

Stimulus	Reaction Times
A	1.1, 1.4, 1.2, 1.3
B	.3, .4, .2, .4, .3
C	.8, .6, .7, .9, .8, .7
D	.6, .4, .5, .7, .3

(i) Test the null hypothesis that the mean reaction times to the four stimuli are equal. Construct simultaneous confidence intervals for all the six differences among the mean reaction times at overall level at least .95. Identify significantly different means.
(ii) Since the reaction time varies from individual to individual, the experimenter decides to redesign the experiment with subjects as blocks. Five subjects are used in this newly designed experiment. Each subject is randomly assigned to

each of the four stimuli. The following table shows the reaction times of the twenty subjects.

Stimulus	Subject				
	1	2	3	4	5
A	1.4	1.1	1.3	.9	1.2
B	.4	.3	.4	.5	.4
C	.9	.8	.8	.7	.6
D	.7	.6	.5	.5	.7

Construct the two-way analysis of variance table and test the null hypothesis that the mean reaction times to the four stimuli are all equal. Find simultaneous confidence intervals for the pairs of differences among the stimuli means at overall level .95.

(iii) In part (ii) apply Friedman's test to test the null hypothesis of no difference among the four stimulii.

4. Six brands of cigarettes are to be compared for tar content. Independent random samples of cigarettes of each brand are tested to yield the following tar levels (in milligrams).

Brand	Tar Content (mg)
A	4.2, 4.6, 4.4, 4.8, 4.4
B	6.4, 6.8, 6.2, 6.5, 7.0
C	8.1, 5.9, 7.3, 7.8, 6.6, 8.4
D	5.0, 5.4, 5.3, 5.2, 5.2
E	7.2, 6.9, 7.6, 7.4, 7.1
F	3.2, 3.1, 3.8, 3.6, 3.4

(i) Test the null hypothesis that the mean tar levels of each brand of cigarette are the same at level .05 using a one-way analysis of variance.

(ii) Do the same using a Kruskal–Wallis test.

5. Fifty female applicants for the positions of school bus drivers are given a psychological test in order to test their ability to stay calm and have self-control under trying conditions. The following table gives the scores of 50 applicants, classified according to their age group.

		Age of Applicant			
21–30	31–40	41–50	21–30	31–40	41–50
50	75	98	60	87	92
78	78	92	64	78	82
62	69	86	48	73	54
64	59	80	52	57	68
54	24	71	32	78	78
56	82	82	24	72	85
58	75	78	68	69	91
67	61	69	76		
48	95	89	41		

(i)   Are there significant differences among the age groups of applicants in the ability to stay calm and in control? Use one-way analysis of variance. If so, identify significantly different age groups at level $\alpha = .05$.

(ii) Apply the Kruskal–Wallis test to the same data to test the hypothesis that distribution of scores is the same for each age group. Identify significantly different age groups at level $\alpha = .15$.

(iii) State the difference in assumption in the two methods of analysis and compare your results.

6. The Internal Revenue Service wishes to compare the average time it takes to complete four different versions of the short form that it has developed. Twenty-five taxpayers who file the short form are selected and each is randomly assigned to one of the four forms. The amount of time it takes to complete the form (in minutes) is recorded as follows.

| | Forms | | |
A	B	C	D
19	15	31	42
24	21	34	39
26	23	28	48
18	17	36	50
21	12	22	41
23	16	31	
22		29	

Test the hypothesis that the four versions of the short form require the same average time to complete and identify significantly different means if any:
(i) Use the one-way analysis of variance.
(ii) Use the Kruskal–Wallis test.

7. The following data give the grade-point averages of 30 students by sex and major academic field of study.

	Sex	
Field	Male	Female
Mathematics	3.0, 2.6, 2.8	3.8, 4.0, 3.9
Business	3.9, 3.4, 3.6	2.6, 3.0, 2.9
Chemistry	2.7, 3.2, 3.8	4.0, 3.6, 3.9
History	3.5, 3.7, 3.8	3.4, 3.6, 3.4
Philosophy	4.0, 3.5, 3.8	3.5, 3.8, 3.7

Perform a two-way analysis of variance and test the null hypothesis of no sex effect, no field of study effect, and no interaction effect on grade-point averages.

8. Three doses of an antidiuretic substance are administered to rats. Three rats are taken from each of four litters, and one of the three doses is given to each rat randomly. Each experimental rat is given a standard water load at the beginning of

the experiment. The following table shows the amount of urine collected in the next 60 minutes as a percentage of the standard load.

Dose	Litter			
	1	2	3	4
A	18	15	16	20
B	9	11	7	10
C	4	5	3	3

  (i) Are the differences among the doses significant at the 5 percent level?
  (ii) Are the differences among the litters significant at the 5 percent level?
  (iii) Find simultaneous confidence intervals for the differences in doses at an overall level of at least .95.
  (iv) Use the Friedman test to test the null hypothesis of no dose effect.

9. In a consumer survey, seven consumers are asked to rank in order of preference four methods of packaging an over-the-counter pain reliever. The object of the study is to determine if there are differences among the four methods of packaging and if so, to estimate the relative order of preference of the four methods. The following table gives the ranks assigned (1 for the most preferred and 4 for the least preferred).

Consumer	Method			
	A	B	C	D
1	4	1	3	2
2	4	1	2	3
3	4	1	3	2
4	3	2	4	1
5	2	1	4	3
6	4	1	3	2
7	4	1	3	2

Analyze the data to carry out the objectives of the study.

10. Fifteen students were grouped according to their grades on a pretest in mathematics. Three sections of a new course were set up. The same teacher taught the three sections using the same method and the same material. The following table gives the average grade for each student in this new course.

Pretest Grade	Section		
	A	B	C
90–100	90.0	89.1	95.2
80–89	88.5	75.4	80.1
70–79	78.2	67.4	73.1
60–69	78.0	69.4	60.1
50–59	75.1	65.8	65.1

Rank the students in each of the five groups according to their grade (1 for lowest, 3 for highest). Compute Kendall's coefficient of concordance and test the null hypothesis of no association.

11. In a flower show the judges agreed that five exhibits were outstanding. These exhibits were numbered arbitrarily from 1 through 5 and presented to four judges for ranking in order of merit. The following rankings were obtained.

	Exhibit				
Judge	1	2	3	4	5
A	5	3	1	2	4
B	3	1	5	4	2
C	5	2	3	1	4
D	4	1	5	2	3

Compute Kendall's coefficient of concordance and test the null hypothesis of no association among the four rankings.

Appendix—Tables

TABLE A1. **Random Digits**[a]

Line No.	1–5	6–10	11–15	16–20	21–25	26–30	31–35	36–40	41–45	46–50
1	03991	10461	93716	16894	66083	24653	84609	58232	88618	19161
2	38555	95554	32886	59780	08355	60860	29735	47762	71299	23853
3	17546	73704	92052	46215	55121	29281	59076	07936	27954	58909
4	32643	52861	95819	06831	00911	98936	76355	93779	80863	00514
5	69572	68777	39510	35905	14060	40619	29549	69616	33564	60780
6	24122	66591	27699	06494	14845	46672	61958	77100	90899	75754
7	61196	30231	92962	61773	41839	55382	17267	70943	78038	70267
8	30532	21704	10274	12202	39685	23309	10061	68829	55986	66485
9	03788	97599	75867	20717	74416	53166	35208	33374	87539	08823
10	48228	63379	85783	47619	53152	67433	35663	52972	16818	60311
11	60365	94653	35075	33949	42614	29297	01918	28316	98953	73231
12	83799	42402	56623	34442	34994	41374	70071	14736	09958	18065
13	32960	07405	36409	83232	99385	41600	11133	07586	15917	06253
14	19322	53845	57620	52606	66497	68646	78138	66559	19640	99413
15	11220	94747	07399	37408	48509	23929	27482	45476	85244	35159
16	31751	57260	68980	05339	15470	48355	88651	22596	03152	19121
17	88492	99382	14454	04504	20094	98977	74843	93413	22109	78508
18	30934	47744	07481	83828	73788	06533	28597	20405	94205	20380
19	22888	48893	27499	98748	60530	45128	74022	84617	82037	10268
20	78212	16993	35902	91386	44372	15486	65741	14014	87481	37220
21	41849	84547	46850	52326	34677	58300	74910	64345	19325	81549
22	46352	33049	69248	93460	45305	07521	61318	31855	14413	70951
23	11087	96294	14013	31792	59747	67277	76503	34513	39663	77544
24	52701	08337	56303	87315	16520	69676	11654	99893	02181	68161
25	57275	36898	81304	48585	68652	27376	92852	55866	88448	03584

[a]From *A Million Random Digits*, pp. 3–4. © 1955 by The Rand Corporation. Used by permission.

Line No.	1–5	6–10	11–15	16–20	21–25	26–30	31–35	36–40	41–45	46–50
26	20857	73156	70284	24326	79375	95220	01159	63267	10622	48391
27	15633	84924	90415	93614	33521	26665	55823	47641	86225	31704
28	92694	48297	39904	02115	59589	49067	66821	41575	49767	04037
29	77613	19019	88152	00080	20554	91409	96277	48257	50816	97616
30	38688	32486	45134	63545	59404	72059	43947	51680	43852	59693
31	25163	01889	70014	15021	41290	67312	71857	15957	68971	11403
32	65251	07629	37239	33295	05870	01119	92784	26340	18477	65622
33	36815	43625	18637	37509	82444	99005	04921	73701	14707	93997
34	64397	11692	05327	82162	20247	81759	45197	25332	83745	22567
35	04515	25624	95096	67946	48460	85558	15191	18782	16930	33361
36	83761	60873	43253	84145	60833	25983	01291	41349	20368	07126
37	14387	06345	80854	09279	43529	06318	38384	74761	41196	37480
38	51321	92246	80088	77074	88722	56736	66164	49431	66919	31678
39	72472	00008	80890	18002	94813	31900	54155	83436	35352	54131
40	05466	55306	93128	18464	74457	90561	72848	11834	79982	68416
41	39528	72484	82474	25593	48545	35247	18619	13674	18611	19241
42	81616	18711	53342	44276	75122	11724	74627	73707	58319	15997
43	07586	16120	82641	22820	92904	13141	32392	19763	61199	67940
44	90767	04235	13574	17200	69902	63742	78464	22501	18627	90872
45	40188	28193	29593	88627	94972	11598	62095	36787	00441	58997
46	34414	82157	86887	55087	19152	00023	12302	80783	32624	68691
47	63439	75363	44989	16822	36024	00867	76378	41605	65961	73488
48	67049	09070	93399	45547	94458	74284	05041	49807	20288	34060
49	79495	04146	52162	90286	54158	34243	46978	35482	59362	95938
50	91704	30552	04737	21031	75051	93029	47665	64382	99782	93478
51	94015	46874	32444	48277	59820	96163	64654	25843	41145	42820
52	74108	88222	88570	74015	25704	91035	01755	14750	48968	38603
53	62880	87873	95160	59221	22304	90314	72877	17334	39283	04149
54	11748	12102	80580	41867	17710	59621	06554	07850	73950	79552
55	17944	05600	60478	03343	25852	58905	57216	39618	49856	99326
56	66067	42792	95043	52680	46780	56487	09971	59481	37006	22186
57	54244	91030	45547	70818	59849	96169	61459	21647	87417	17198
58	30945	57589	31732	57260	47670	07654	46376	25366	94746	49580
59	69170	37403	86995	90307	94304	71803	26825	05511	12459	91314
60	08345	88975	35841	85771	08105	59987	87112	21476	14713	71181
61	27767	43584	85301	88977	29490	69714	73035	41207	74699	09310
62	13025	14338	54066	15243	47724	66733	47431	43905	31048	56699
63	80217	36292	98525	24335	24432	24896	43277	58874	11466	16082

Line No.	1–5	6–10	11–15	16–20	21–25	26–30	31–35	36–40	41–45	46–50
64	10875	62004	90391	61105	57411	06368	53856	30743	08670	84741
65	54127	57326	26629	19087	24472	88779	30540	27886	61732	75454
66	60311	42824	37301	42678	45990	43242	17374	52003	70707	70214
67	49739	71484	92003	98086	76668	73209	59202	11973	02902	33250
68	78626	51594	16453	94614	39014	97066	83012	09832	25571	77628
69	66692	13986	99837	00582	81232	44987	09504	96412	90193	79568
70	44071	28091	07362	97703	76447	42537	98524	97831	65704	09514
71	41468	85149	49554	17994	14924	39650	95294	00556	70481	06905
72	94559	37559	49678	53119	70312	05682	66986	34099	74474	20740
73	41615	70360	64114	58660	90850	64618	80620	51790	11436	38072
74	50273	93113	41794	86861	24781	89683	55411	85667	77535	99892
75	41396	80504	90670	08289	40902	05069	95083	06783	28102	57816
76	25807	24260	71529	78920	72682	07385	90726	56166	98884	08583
77	06170	97965	88302	98041	21443	41808	68984	83620	89747	98882
78	60808	54444	74412	81105	01176	28838	36421	16489	18059	51061
79	80940	44893	10408	36222	80582	71944	92638	40333	67054	16067
80	19516	90120	46759	71643	13177	55292	21036	82808	77501	97427
81	49386	54480	23604	23554	21785	41101	91178	10174	29420	90438
82	06312	88940	15995	69321	47458	64809	98189	81851	29651	84215
83	60942	00307	11897	92674	40405	68032	96717	54244	10701	41393
84	92329	98932	78284	46347	71209	92061	39448	93136	25722	08564
85	77936	63574	31384	51924	85561	29671	58137	17820	22751	36518
86	38101	77756	11657	13897	95889	57067	47648	13885	70669	93406
87	39641	69457	91339	22502	92613	89719	11947	56203	19324	20504
88	84054	40455	99396	63680	67667	60631	69181	96845	38525	11600
89	47468	03577	57649	63266	24700	71594	14004	23153	69249	05747
90	43321	31370	28977	23896	76479	68562	62342	07589	08899	05985
91	64281	61826	18555	64937	13173	33365	78851	16499	87064	13075
92	66847	70495	32350	02985	86716	38746	26313	77463	55387	72681
93	72461	33230	21529	53424	92581	02262	78438	66276	18396	73538
94	21032	91050	13058	16218	12470	56500	15292	76139	59526	52113
95	95362	67011	06651	16136	01016	00857	55018	56374	35824	71708
96	49712	97380	10404	55452	34030	60726	75211	10271	36633	68424
97	58275	61764	97586	54716	50259	46345	87195	46092	26787	60939
98	89514	11788	68224	23417	73959	76145	30342	40277	11049	72049
99	15472	50669	48139	36732	46874	37088	73465	09819	58869	35220
100	12120	86124	51247	44302	60883	52109	21437	36786	49226	77837

TABLE A2. Squares and Square Roots

n	n^2	$\sqrt{n}$	$\sqrt{10n}$	n	n^2	$\sqrt{n}$	$\sqrt{10n}$
1	1	1.000 000	3.162 278	51	2 601	7.141 428	22.58318
2	4	1.414 214	4.472 136	52	2 704	7.211 103	22.80351
3	9	1.732 051	5.477 226	53	2 809	2.280 110	23.02173
4	16	2.000 000	6.324 555	54	2 916	7.348 469	23.23790
5	25	2.236 068	7.071 068	55	3 025	7.416 198	23.45208
6	36	2.449 490	7.745 967	56	3 136	7.483 315	23.66432
7	49	2.645 751	8.366 600	57	3 249	7.549 834	23.87467
8	64	2.828 427	8.944 272	58	3 364	7.615 773	24.08319
9	81	3.000 000	9.486 833	59	3 481	7.681 146	24.28992
10	100	3.162 278	10.00000	60	3 600	7.745 967	24.49490
11	121	3.316 625	10.48809	61	3 721	7.810 250	24.69818
12	144	3.464 102	10.95445	62	3 844	7.874 008	24.89980
13	169	3.605 551	11.40175	63	3 969	7.937 254	25.09980
14	196	3.741 657	11.83216	64	4 096	8.000 000	25.29822
15	225	3.872 983	12.24745	65	4 225	8.062 258	25.49510
16	256	4.000 000	12.64911	66	4 356	8.124 038	25.69047
17	289	4.123 106	13.03840	67	4 489	8.185 353	25.88436
18	324	4.242 641	13.41641	68	4 624	8.246 211	26.07681
19	361	4.358 899	13.78405	69	4 761	8.306 624	26.26785
20	400	4.472 136	14.14214	70	4 900	8.366 600	26.45751
21	441	4.582 576	14.49138	71	5 041	8.426 150	26.64583
22	484	4.690 416	14.83240	72	5 184	8.485 281	26.83282
23	529	4.795 832	15.16575	73	5 329	8.544 004	27.01851
24	576	4.898 979	15.49193	74	5 476	8.602 325	27.20294
25	625	5.000 000	15.81139	75	5 625	8.660 254	27.38613
26	676	5.099 020	16.12452	76	5 776	8.717 798	27.56810
27	729	5.196 152	16.43168	77	5 929	8.774 964	27.74887
28	784	5.291 503	16.73320	78	6 084	8.831 761	27.92848
29	841	5.385 165	17.02939	79	6 241	8.888 194	28.10694
30	900	5.477 226	17.32051	80	6 400	8.944 272	28.28427
31	961	5.567 764	17.60682	81	6 561	9.000 000	28.46050
32	1 024	5.656 854	17.88854	82	6 724	9.055 385	28.63564
33	1 089	5.744 563	18.16590	83	6 889	9.110 434	28.80972
34	1 156	5.830 952	18.43909	84	7 056	9.165 151	28.98275
35	1 225	5.916 080	18.70829	85	7 225	9.219 544	29.15476
36	1 296	6.000 000	18.97367	86	7 396	9.273 618	29.32576
37	1 369	6.082 763	19.23538	87	7 569	9.327 379	29.49576
38	1 444	6.164 414	19.49359	88	7 744	9.380 832	29.66479
39	1 521	6.244 998	19.74842	89	7 921	9.433 981	29.83287
40	1 600	6.324 555	20.00000	90	8 100	9.486 833	30.00000
41	1 681	6.403 124	20.24846	91	8 281	9.539 392	30.16621
42	1 764	6.480 741	20.49390	92	8 464	9.591 663	30.33150
43	1 849	6.557 439	20.73644	93	8 649	9.643 651	30.49590
44	1 936	6.633 250	20.97618	94	8 836	9.695 360	30.65942
45	2 025	6.708 204	21.21320	95	9 025	9.746 794	30.82207
46	2 116	6.782 330	21.44761	96	9 216	9.797 959	30.98387
47	2 209	6.855 655	21.67948	97	9 409	9.848 858	31.14482
48	2 304	6.928 203	21.90890	98	9 604	9.899 495	31.30495
49	2 401	7.000 000	22.13594	99	9 801	9.949 874	31.46427
50	2 500	7.071 068	22.36068	100	10 000	10.00000	31.62278

TABLE A3. Cumulative Binomial Probabilities[a]

$$P(X \leq k) = \sum_{j=0}^{k} \binom{n}{j} p^j (1-p)^{n-j}$$

n	k	.01	0.05	0.10	0.15	0.20	0.25	0.30	1/3	0.35	0.40	0.45	0.50
2	0	.9801	0.9025	0.8100	0.7225	0.6400	0.5625	0.4900	.4444	0.4225	0.3600	0.3025	0.2500
	1	.9999	0.9975	0.9900	0.9775	0.9600	0.9375	0.9100	.8889	0.8775	0.8400	0.7975	0.7500
	2	1.0000	1.0000	1.0000	1.0000	1.0000	1.0000	1.0000	1.0000	1.0000	1.0000	1.0000	1.0000
3	0	.9703	0.8574	0.7290	0.6141	0.5120	0.4219	0.3430	.2963	0.2746	0.2160	0.1664	0.1250
	1	.9997	0.9928	0.9720	0.9392	0.8960	0.8438	0.7840	.7407	0.7182	0.6480	0.5748	0.5000
	2	1.0000	0.9999	0.9990	0.9966	0.9920	0.9844	0.9730	.9630	0.9571	0.9360	0.9089	0.8750
	3		1.0000	1.0000	1.0000	1.0000	1.0000	1.0000	1.0000	1.0000	1.0000	1.0000	1.0000
4	0	.9606	0.8145	0.6561	0.5220	0.4096	0.3164	0.2401	.1975	0.1785	0.1296	0.0915	0.0625
	1	.9994	0.9860	0.9477	0.8905	0.8192	0.7383	0.6517	.5926	0.5630	0.4752	0.3910	0.3125
	2	1.0000	0.9995	0.9963	0.9880	0.9728	0.9492	0.9163	.8889	0.8735	0.8208	0.7585	0.6875
	3		1.0000	0.9999	0.9995	0.9984	0.9961	0.9919	.9876	0.9850	0.9744	0.9590	0.9375
	4			1.0000	1.0000	1.0000	1.0000	1.0000	1.0000	1.0000	1.0000	1.0000	1.0000
5	0	.9510	0.7738	0.5905	0.4437	0.3277	0.2373	0.1681	.1317	0.1160	0.0778	0.0503	0.0312
	1	.9990	0.9774	0.9185	0.8352	0.7373	0.6328	0.5282	.4609	0.4284	0.3370	0.2562	0.1875
	2	1.0000	0.9988	0.9914	0.9734	0.9421	0.8965	0.8369	.7901	0.7648	0.6826	0.5931	0.5000
	3		1.0000	0.9995	0.9978	0.9933	0.9844	0.9692	.9547	0.9460	0.9130	0.8688	0.8125
	4			0.9999	0.9997	0.9990	0.9976	.9959	0.9947	0.9898	0.9815	0.9688
	5			1.0000	1.0000	1.0000	1.0000	1.0000	1.0000	1.0000	1.0000	1.0000	1.0000
6	0	.9415	0.7351	0.5314	0.3771	0.2621	0.1780	0.1176	.0878	0.0754	0.0467	0.0277	0.0156
	1	.9986	0.9672	0.8857	0.7765	0.6553	0.5339	0.4202	.3512	0.3191	0.2333	0.1636	0.1094
	2	1.0000	0.9978	0.9842	0.9527	0.9011	0.8306	0.7443	.6804	0.6471	0.5443	0.4415	0.3438
	3		0.9999	0.9987	0.9941	0.9830	0.9624	0.9295	.8999	0.8826	0.8208	0.7447	0.6562
	4		1.0000	0.9999	0.9996	0.9984	0.9954	0.9891	.9822	0.9777	0.9590	0.9308	0.8906
	5			1.0000	1.0000	0.9999	0.9998	0.9993	.9986	0.9982	0.9959	0.9917	0.9844
	6					1.0000	1.0000	1.0000	1.0000	1.0000	1.0000	1.0000	1.0000
7.	0	.9321	0.6983	0.4783	0.3206	0.2097	0.1335	0.0824	.0585	0.0490	0.0280	0.0152	0.0078
	1	.9980	0.9556	0.8503	0.7166	0.5767	0.4449	0.3294	.2634	0.2338	0.1586	0.1024	0.0625
	2	1.0000	0.9962	0.9743	0.9262	0.8520	0.7564	0.6471	.5706	0.5323	0.4199	0.3164	0.2266
	3		0.9998	0.9973	0.9879	0.9667	0.9294	0.8740	.8267	0.8002	0.7102	0.6083	0.5000
	4		1.0000	0.9998	0.9988	0.9953	0.9871	0.9712	.9547	0.9444	0.9037	0.8471	0.7734
	5			1.0000	0.9999	0.9996	0.9987	0.9962	.9931	0.9910	0.9812	0.9643	0.9375
	6				1.0000	1.0000	0.9999	0.9998	.9995	0.9994	0.9984	0.9963	0.9922
	7						1.0000	1.0000	1.0000	1.0000	1.0000	1.0000	1.0000
8	0	.9227	0.6634	0.4305	0.2725	0.1678	0.1001	0.0576	.0390	0.0319	0.0168	0.0084	0.0039
	1	.9973	0.9428	0.8131	0.6572	0.5033	0.3671	0.2553	.1951	0.1691	0.1064	0.0632	0.0352

[a]For $n = 2$ to 20 and $n = 25$, from *Tables of Cumulative Binomial Probability Distribution*, Harvard University Press, Cambridge, Massachusetts, 1955, reprinted with permission. For $n = 100$, from *50–100 Binomial Tables* by H. G. Romig, copyright 1947, Bell Telephone Laboratories, Short Hills, N.J., reprinted with permission.

							p					
n	k	.01	0.05	0.10	0.15	0.20	0.25	0.30	1/3	0.35	0.40	0.45	0.50
	2	.9999	0.9942	0.9619	0.8948	0.7969	0.6785	0.5518	.4682	0.4278	0.3154	0.2201	0.1445
	3	1.0000	0.9996	0.9950	0.9786	0.9437	0.8862	0.8059	.7413	0.7064	0.5941	0.4770	0.3633
	4		1.0000	0.9996	0.9971	0.9896	0.9727	0.9420	.9121	0.8939	0.8263	0.7396	0.6367
	5			1.0000	0.9998	0.9988	0.9958	0.9887	.9803	0.9747	0.9502	0.9115	0.8555
	6				1.0000	0.9999	0.9996	0.9987	.9974	0.9964	0.9915	0.9819	0.9648
	7					1.0000	1.0000	0.9999	.9998	0.9998	0.9993	0.9983	0.9961
	8							1.0000	1.0000	1.0000	1.0000	1.0000	1.0000
9	0	.9135	0.6302	0.3874	0.2316	0.1342	0.0751	0.0404	.0260	0.0207	0.0101	0.0046	0.0020
	1	.9965	0.9288	0.7748	0.5995	0.4362	0.3003	0.1960	.1431	0.1211	0.0705	0.0385	0.0195
	2	.9999	0.9916	0.9470	0.8591	0.7382	0.6007	0.4628	.3772	0.3373	0.2318	0.1495	0.0898
	3	1.0000	0.9994	0.9917	0.9661	0.9144	0.8343	0.7297	.6503	0.6089	0.4826	0.3614	0.2539
	4		1.0000	0.9991	0.9944	0.9804	0.9511	0.9012	.1552	0.8283	0.7334	0.6214	0.5000
	5			0.9999	0.9994	0.9969	0.9900	0.9747	.9576	0.9464	0.9006	0.8342	0.7461
	6			1.0000	1.0000	0.9997	0.9987	0.9957	.9917	0.9888	0.9750	0.9502	0.9102
	7					1.0000	0.9999	0.9996	.9990	0.9986	0.9962	0.9909	0.9805
	8						1.0000	1.0000	.9999	0.9999	0.9997	0.9992	0.9980
	9								1.0000	1.0000	1.0000	1.0000	1.0000
10	0	.9044	0.5987	0.3487	0.1969	0.1074	0.0563	0.0282	.0173	0.0135	0.0060	0.0025	0.0010
	1	.9958	0.9139	0.7361	0.5443	0.3758	0.2440	0.1493	.1040	0.0860	0.0464	0.0233	0.0107
	2	1.0000	0.9885	0.9298	0.8202	0.6778	0.5256	0.3828	.2991	0.2616	0.1673	0.0996	0.0547
	3		0.9990	0.9872	0.9500	0.8791	0.7759	0.6496	.5593	0.5138	0.3823	0.2660	0.1719
	4		0.9999	0.9984	0.9901	0.9672	0.9219	0.8497	.7869	0.7515	0.6331	0.5044	0.3770
	5		1.0000	0.9999	0.9986	0.9936	0.9803	0.9527	.9234	0.9051	0.8338	0.7384	0.6230
	6			1.0000	0.9999	0.9991	0.9965	0.9894	.8803	0.9740	0.9452	0.8980	0.8281
	7				1.0000	0.9999	0.9996	0.9984	.9966	0.9952	0.9877	0.9726	0.9453
	8					1.0000	1.0000	0.9999	.9996	0.9995	0.9983	0.9955	0.9893
	9							1.0000	1.0000	1.0000	0.9999	0.9997	0.9990
	10										1.0000	1.0000	1.0000
11	0	.8954	0.5688	0.3138	0.1673	0.0859	0.0422	0.0198	.0116	0.0088	0.0036	0.0014	0.0005
	1	.9948	0.8981	0.6974	0.4922	0.3221	0.1971	0.1130	.0752	0.0606	0.0302	0.0139	0.0059
	2	.9998	0.9848	0.9104	0.7788	0.6174	0.4552	0.3127	.2341	0.2001	0.1189	0.0652	0.0327
	3	1.0000	0.9984	0.9815	0.9306	0.8389	0.7133	0.5696	.4726	0.4256	0.2963	0.1911	0.1133
	4		0.9999	0.9972	0.9841	0.9496	0.8854	0.7897	.7110	0.6683	0.5328	0.3971	0.2744
	5		1.0000	0.9997	0.9973	0.9883	0.9657	0.9218	.8779	0.8513	0.7535	0.6331	0.5000
	6			1.0000	0.9997	0.9980	0.9924	0.9784	.9614	0.9499	0.9006	0.8262	0.7256
	7				1.0000	0.9998	0.9988	0.9957	.9912	0.9878	0.9707	0.9390	0.8867
	8					1.0000	0.9999	0.9994	.9986	0.9980	0.9941	0.9852	0.9673
	9						1.0000	1.0000	.9999	0.9998	0.9993	0.9978	0.9941
	10								1.0000	1.0000	1.0000	0.9998	0.9995
	11											1.0000	1.0000
12	0	.8864	0.5404	0.2824	0.1422	0.0687	0.0317	0.0138	.0077	0.0057	0.0022	0.0008	0.0002
	1	.9938	0.8816	0.6590	0.4435	0.2749	0.1584	0.0850	.0540	0.0424	0.0196	0.0083	0.0032
	2	.9998	0.9804	0.8891	0.7358	0.5583	0.3907	0.2528	.1811	0.1513	0.0834	0.0421	0.0193
	3	1.0000	0.9978	0.9744	0.9078	0.7946	0.6488	0.4925	.3931	0.3467	0.2253	0.1345	0.0730
	4		0.9998	0.9957	0.9761	0.9274	0.8424	0.7237	.6315	0.5833	0.4382	0.3044	0.1938

875

TABLE A3. (*Continued*)

n	k	.01	0.05	0.10	0.15	0.20	0.25	0.30	1/3	0.35	0.40	0.45	0.50
	5		1.0000	0.9995	0.9954	0.9806	0.9456	0.8822	.8223	0.7873	0.6652	0.5269	0.3872
	6			0.9999	0.9993	0.9961	0.9857	0.9614	.9336	0.9154	0.8418	0.7393	0.6128
	7			1.0000	0.9999	0.9994	0.9972	0.9905	.9812	0.9745	0.9427	0.8883	0.8062
	8				1.0000	0.9999	0.9996	0.9983	.9961	0.9944	0.9847	0.9644	0.9270
	9					1.0000	1.0000	0.9998	.9995	0.9992	0.9972	0.9921	0.9807
	10							1.0000	.9999	0.9999	0.9997	0.9989	0.9968
	11								1.0000	1.0000	1.0000	0.9999	0.9998
	12											1.0000	1.0000
13	0	.8775	0.5133	0.2542	0.1209	0.0550	0.0238	0.0097	.0051	0.0037	0.0013	0.0004	0.0001
	1	.9928	0.8646	0.6213	0.3983	0.2336	0.1267	0.0637	.0358	0.0296	0.0126	0.0049	0.0017
	2	.9997	0.9755	0.8661	0.6920	0.5017	0.3326	0.2025	.1387	0.1132	0.0579	0.0269	0.0112
	3	1.0000	0.9969	0.9658	0.8820	0.7473	0.5843	0.4206	.3224	0.2783	0.1686	0.0929	0.0461
	4		0.9997	0.9935	0.9658	0.9009	0.7940	0.6543	.5520	0.5005	0.3530	0.2279	0.1334
	5		1.0000	0.9991	0.9924	0.9700	0.9198	0.8346	.7587	0.7159	0.5744	0.4268	0.2905
	6			0.9999	0.9987	0.9930	0.9757	0.9376	.8965	0.8705	0.7712	0.6437	0.5000
	7			1.0000	0.9998	0.9988	0.9944	0.9818	.9654	0.9538	0.9023	0.8212	0.7095
	8				1.0000	0.9998	0.9990	0.9960	.9912	0.9874	0.9679	0.9302	0.8666
	9					1.0000	0.9999	0.9993	.9984	0.9975	0.9922	0.9797	0.9539
	10						1.0000	0.9999	.9998	0.9997	0.9987	0.9959	0.9888
	11							1.0000	1.0000	1.0000	0.9999	0.9995	0.9983
	12										1.0000	1.0000	0.9999
	13												1.0000
14	0	.8687	0.4877	0.2288	0.1028	0.0440	0.0178	0.0068	.0034	0.0024	0.0008	0.0002	0.0001
	1	.9916	0.8470	0.5846	0.3567	0.1979	0.1010	0.0475	.0274	0.0205	0.0081	0.0029	0.0009
	2	.9997	0.9699	0.8416	0.6479	0.4481	0.2811	0.1608	.1053	0.0839	0.0398	0.0170	0.0065
	3	1.0000	0.9958	0.9559	0.8535	0.6982	0.5213	0.3552	.2612	0.2205	0.1243	0.0632	0.0287
	4		0.9996	0.9908	0.9533	0.8702	0.7415	0.5842	.4755	0.4227	0.2793	0.1672	0.0898
	5		1.0000	0.9985	0.9885	0.9561	0.8883	0.7805	.6899	0.6405	0.4859	0.3373	0.2120
	6			0.9998	0.9978	0.9884	0.9617	0.9067	.8505	0.8164	0.6925	0.5461	0.3953
	7			1.0000	0.9997	0.9976	0.9897	0.9685	.9424	0.9247	0.8499	0.7414	0.6047
	8				1.0000	0.9996	0.9978	0.9917	.9826	0.9757	0.9417	0.8811	0.7880
	9					1.0000	0.9997	0.9983	.9960	0.9940	0.9825	0.9574	0.9102
	10						1.0000	0.9998	.9993	0.9989	0.9961	0.9886	0.9713
	11							1.0000	.9999	0.9999	0.9994	0.9978	0.9935
	12								1.0000	1.0000	0.9999	0.9997	0.9991
	13										1.0000	1.0000	0.9999
	14												1.0000
15	0	.8601	0.4633	0.2059	0.0874	0.0352	0.0134	0.0047	.0023	0.0016	0.0005	0.0001	0.0000
	1	.9904	0.8290	0.5490	0.3186	0.1671	0.0802	0.0353	.0194	0.0142	0.0052	0.0017	0.0005
	2	.9996	0.9638	0.8159	0.6042	0.3980	0.2361	0.1268	.0794	0.0617	0.0271	0.0107	0.0037
	3	1.0000	0.9945	0.9444	0.8227	0.6482	0.4613	0.2969	.2092	0.1727	0.0905	0.0424	0.0176
	4		0.9994	0.9873	0.9383	0.8358	0.6865	0.5155	.4041	0.3519	0.2173	0.1204	0.0592
	5		0.9999	0.9978	0.9832	0.9389	0.8516	0.7216	.6184	0.5643	0.4032	0.2608	0.1509
	6		1.0000	0.9997	0.9964	0.9819	0.9434	0.8689	.8070	0.7548	0.6098	0.4522	0.3036
	7			1.0000	0.9994	0.9958	0.9827	0.9500	.9118	0.8868	0.7869	0.6535	0.5000
	8				0.9999	0.9992	0.9958	0.9848	.9692	0.9578	0.9050	0.8182	0.6964

								p					
n	k	.01	0.05	0.10	0.15	0.20	0.25	0.30	1/3	0.35	0.40	0.45	0.50
	9				1.0000	0.9999	0.9992	0.9963	.9915	0.9876	0.9662	0.9231	0.8491
	10					1.0000	0.9999	0.9993	.9982	0.9972	0.9907	0.9745	0.9408
	11						1.0000	0.9999	.9997	0.9995	0.9981	0.9937	0.9824
	12							1.0000	1.0000	0.9999	0.9997	0.9989	0.9963
	13									1.0000	1.0000	0.9999	0.9995
	14											1.0000	1.0000
16	0	.8515	.4401	.1853	.0743	.0281	.0100	.0033	.0015	.0010	.0003	.0001	.0000
	1	.9891	.8108	.5147	.2839	.1407	.0635	.0261	.0137	.0098	.0033	.0010	.0003
	2	.9995	.9571	.7892	.5614	.3518	.1971	.0994	.0594	.0451	.0183	.0066	.0021
	3	1.0000	.9930	.9316	.7899	.5981	.4050	.2459	.1660	.1339	.0651	.0281	.0106
	4		.9991	.9830	.9209	.7982	.6302	.4499	.3391	.2892	.1666	.0853	.0384
	5		.9999	.9967	.9765	.9183	.8103	.6598	.5469	.4900	.3288	.1976	.1051
	6		1.0000	.9995	.9944	.9733	.9204	.8247	.7374	.6881	.5272	.3660	.2272
	7			.9999	.9989	.9930	.9729	.9256	.8735	.8406	.7161	.5629	.4018
	8			1.0000	.9998	.9985	.9925	.9743	.9500	.9329	.8577	.7441	.5982
	9				1.0000	.9998	.9984	.9929	.9841	.9771	.9417	.8759	.7228
	10					1.0000	.9997	.9984	.9960	.9938	.9809	.9514	.8949
	11						1.0000	.9997	.9992	.9987	.9951	.9851	.9616
	12							1.0000	.9999	.9998	.9991	.9965	.9894
	13								1.0000	1.0000	.9999	.9994	.9979
	14										1.0000	.9999	.9997
	15											1.0000	1.0000
17	0	.8429	.4181	.1668	.0631	.0225	.0075	.0023	.0010	.0007	.0002	.0000	.0000
	1	.9877	.7922	.4818	.2525	.1182	.0501	.0193	.0096	.0067	.0021	.0006	.0001
	2	.9994	.9497	.7618	.5198	.3096	.1637	.0774	.0442	.0327	.0123	.0041	.0012
	3	1.0000	.9912	.9174	.7556	.5489	.3530	.2019	.1304	.1028	.0464	.0184	.0064
	4		.9988	.9779	.9013	.7582	.5739	.3887	.2814	.2348	.1260	.0596	.0245
	5		.9999	.9953	.9681	.8943	.7653	.5968	.4778	.4197	.2639	.1471	.0717
	6		1.0000	.9992	.9917	.9623	.8929	.7752	.6739	.6188	.4478	.2902	.1662
	7			.9999	.9983	.9891	.9598	.8954	.8281	.7872	.6405	.4743	.3145
	8			1.0000	.9997	.9974	.9876	.9597	.9245	.9006	.8011	.6626	.5000
	9				1.0000	.9995	.9969	.9873	.9727	.9617	.9081	.8166	.6855
	10					.9999	.9994	.9968	.9920	.9880	.9652	.9174	.8338
	11					1.0000	.9999	.9993	.9981	.9970	.9894	.9699	.9283
	12						1.0000	.9999	.9997	.9994	.9975	.9914	.9755
	13							1.0000	.9999	.9999	.9995	.9981	.9936
	14								1.0000	1.0000	.9999	.9997	.9988
	15										1.0000	1.0000	.9999
	16												1.0000
18	0	.8345	.3972	.1501	.0536	.0180	.0056	.0016	.0007	.0004	.0001	.0000	.0000
	1	.9862	.7735	.4503	.2241	.0991	.0395	.0142	.0068	.0046	.0013	.0003	.0001
	2	.9993	.9419	.7338	.4797	.2713	.1353	.0600	.0326	.0236	.0082	.0025	.0007
	3	1.0000	.9891	.9018	.7202	.5010	.3057	.1646	.1017	.0783	.0328	.0120	.0038

TABLE A3. (*Continued*)

n	k	.01	0.05	0.10	0.15	0.20	0.25	0.30	1/3	0.35	0.40	0.45	0.50
	4		.9985	.9718	.8794	.7164	.5187	.3327	.2311	.1886	.0942	.0411	.0154
	5		.9998	.9936	.9581	.8671	.7175	.5344	.4122	.3550	.2088	.1077	.0481
	6		1.0000	.9988	.9882	.9487	.8610	.7217	.6085	.5491	.3743	.2258	.1189
	7			.9998	.9973	.9837	.9431	.8593	.7767	.7283	.5634	.3915	.2403
	8			1.0000	.9995	.9957	.9807	.9404	.8924	.8609	.7368	.5778	.4073
	9				.9999	.9991	.9946	.9790	.9566	.9403	.8653	.7473	.5927
	10				1.0000	.9998	.9988	.9939	.9856	.9788	.9424	.8720	.7597
	11					1.0000	.9998	.9986	.9961	.9938	.9797	.9463	.8811
	12						1.0000	.9997	.9992	.9986	.9942	.9817	.9519
	13							1.0000	.9999	.9997	.9987	.9951	.9846
	14								1.0000	1.0000	.9998	.9990	.9962
	15										1.0000	.9999	.9993
	16											1.0000	.9999
	17												1.0000
19	0	.8262	.3774	.1351	.0456	.0144	.0042	.0011	.0004	.0003	.0001	.0000	
	1	.9847	.7547	.4203	.1985	.0829	.0310	.0104	.0047	.0031	.0008	.0002	.0000
	2	.9991	.9335	.7054	.4413	.2369	.1113	.0462	.0240	.0170	.0055	.0015	.0004
	3	1.0000	.9868	.8850	.6841	.4551	.2631	.1332	.0787	.0591	.0230	.0077	.0022
	4		.9980	.9648	.8556	.6733	.4654	.2822	.1879	.1500	.0696	.0280	.0096
	5		.9998	.9914	.9463	.8369	.6678	.4739	.3518	.2968	.1629	.0777	.0318
	6		1.0000	.9983	.9837	.9324	.8251	.6655	.5431	.4812	.3081	.1727	.0835
	7			.9997	.9959	.9767	.9225	.8180	.7207	.6656	.4878	.3169	.1796
	8			1.0000	.9992	.9933	.9713	.9161	.8538	.8145	.6675	.4940	.3238
	9				.9999	.9984	.9911	.9674	.9352	.9125	.8139	.6710	.5000
	10				1.0000	.9997	.9977	.9895	.9759	.9653	.9115	.8159	.6762
	11					1.0000	.9995	.9972	.9926	.9886	.9648	.9129	.8204
	12						.9999	.9994	.9981	.9969	.9884	.9658	.9165
	13						1.0000	.9999	.9996	.9993	.9969	.9891	.9682
	14							1.0000	.9999	.9999	.9994	.9972	.9904
	15								1.0000	1.0000	.9999	.9995	.9978
	16										1.0000	.9999	.9996
	17											1.0000	1.0000
20	0	.8179	.3585	.1216	.0388	.0115	.0032	.0008	.0003	.0002	.0000	.0000	
	1	.9831	.7358	.3917	.1756	.0692	.0243	.0076	.0033	.0021	.0005	.0001	.0000
	2	.9990	.9245	.6769	.4049	.2061	.0913	.0355	.0176	.0121	.0036	.0009	.0002
	3	1.0000	.9841	.8670	.6477	.4114	.2252	.1071	.0604	.0444	.0160	.0049	.0013
	4		.9974	.9568	.8298	.6296	.4148	.2375	.1515	.1182	.0510	.0189	.0059
	5		.9997	.9887	.9327	.8042	.6172	.4164	.2972	.2454	.1256	.0553	.0207
	6		1.0000	.9976	.9781	.9133	.7858	.6080	.4793	.4166	.2500	.1299	.0577
	7			.9996	.9941	.9679	.8982	.7723	.6615	.6010	.4159	.2520	.1316
	8			.9999	.9987	.9900	.9591	.8867	.8094	.7624	.5956	.4143	.2517
	9			1.0000	.9998	.9974	.9861	.9520	.9081	.8782	.7553	.5914	.4119
	10				1.0000	.9994	.9961	.9829	.9624	.9468	.8725	.7507	.5881
	11					.9999	.9991	.9949	.9870	.9804	.9435	.8692	.7483
	12					1.0000	.9998	.9987	.9963	.9940	.9790	.9420	.8684

							p					
n	k	.01	0.05	0.10	0.15	0.20	0.25	0.30	$1/3$	0.35	0.40	0.45	0.50
	13						1.0000	.9997	.9991	.9985	.9935	.9786	.9423
	14							1.0000	.9999	.9997	.9984	.9936	.9793
	15								1.0000	1.0000	.9997	.9985	.9941
	16										1.0000	.9997	.9987
	17											1.0000	.9998
	18												1.0000
25	0	.7778	.2774	.0718	.0172	.0038	.0008	.0001	.0000	.0000			
	1	.9742	.6424	.2712	.0931	.0274	.0070	.0016	.0005	.0003	.0000	.0000	
	2	.9980	.8729	.5371	.2537	.0982	.0321	.0090	.0035	.0021	.0004	.0005	.0000
	3	1.0000	.9659	.7636	.4711	.2340	.0962	.0332	.0149	.0097	.0024	.0023	.0001
	4		.9929	.9020	.6821	.4207	.2137	.0905	.0462	.0321	.0095	.0086	.0005
	5		.9988	.9667	.8385	.6167	.3783	.1935	.1120	.0826	.0294	.0258	.0020
	6		1.0000	.9905	.9305	.7800	.5611	.3406	.2215	.1734	.0736	.0638	.0073
	7			.9977	.9745	.8909	.7265	.5118	.3703	.3061	.1536	.1340	.0216
	8			.9995	.9920	.9532	.8506	.6769	.5376	.4662	.2735	.2424	.0539
	9			.9999	.9979	.9827	.9287	.8106	.6956	.6303	.4246	.3843	.1148
	10			1.0000	.9995	.9944	.9703	.9022	.8220	.7712	.5858	.5426	.2122
	11				.9999	.9985	.9803	.9558	.9082	.8746	.7323	.6937	.3450
	12				1.0000	.9996	.9966	.9825	.9585	.9396	.8462	.8173	.5000
	13					.9999	.9991	.9940	.9836	.9745	.9222	.9040	.6550
	14					1.0000	.9998	.9982	.9949	.9907	.9656	.9560	.7878
	15						1.0000	.9996	.9996	.9971	.9868	.9826	.8852
	16							.9999	.9999	.9992	.9957	.9942	.9461
	17							1.0000	1.0000	.9998	.9988	.9984	.9784
	18									1.0000	.9997	.9996	.9927
	19										.9999	.9999	.9980
	20										1.0000	1.0000	.9995
	21												.9999
	22												1.0000

				p					p	
n	k	.01	.05	.10	.20	.25	k	.30	.40	.50
100	0	.366032	.005921	.000027						
	1	.735762	.037081	.000322			10	.000002		
	2	.920627	.118263	.001945			11	.000006		
	3	.981626	.257839	.007836	.000001		12	.000019		
	4	.996568	.435981	.023711	.000004		13	.000057		
							14	.000157		
	5	.999465	.615999	.057577	.000019					
	6	.999929	.766014	.117156	.000078	.000001	15	.000405		
	7	.999992	.872039	.206051	.000277	.000003	16	.000969		

n	k	.01	.05	.10	.20	.25	k	.30	.40	.50
				p					*p*	
	8	.999999	.936910	.320874	.000855	.000012	17	.002163		
	9	1.000000	.971511	.451290	.002334	.000043	18	.004523	.000002	
							19	.008887	.000006	
	10		.988527	.583155	.005696	.000137				
	11		.995725	.703032	.012575	.000394	20	.016463	.000016	
	12		.998535	.801820	.025329	.001027	21	.028831	.000043	
	13		.999536	.876122	.046912	.002458	22	.047866	.000107	
	14		.999864	.927425	.080444	.005421	23	.075531	.000252	
							24	.113570	.000561	
	15		.999962	.960108	.128505	.011083				
	16		.999990	.979399	.192337	.021111	25	.163130	.001189	
	17		.999997	.989991	.271189	.037626	26	.224399	.002395	.000001
	18		.999999	.995417	.362087	.063011	27	.296365	.004600	.000002
	19		.999999	.998020	.460161	.099530	28	.376777	.008432	.000006
							29	.462338	.014775	.000016
	20			.999191	.559461	.148831				
	21			.999686	.654032	.211435	30	.549122	.024783	.000039
	22			.999884	.738931	.286370	31	.633106	.039848	.000091
	23			.999959	.810911	.371079	32	.710717	.061504	.000204
	24			.999985	.868645	.461670	33	.779256	.091253	.000437
							34	.837139	.130336	.000895
	25			.999994	.912523	.553470				
	26			.999997	.944166	.641738	35	.883919	.179469	.001758
	27			.999998	.965847	.722379	36	.920117	.238610	.003318
	28			.999998	.979978	.792459	37	.946952	.306810	.006016
	29				.988749	.850457	38	.966018	.382187	.010489
							39	.979009	.462075	.017600
	30				.993939	.896210				
	31				.996868	.930649	40	.987499	.543294	.028444
	32				.998448	.955401	41	.992824	.622533	.044313
	33				.999261	.972403	42	.996029	.696741	.066605
	34				.999662	.983571	43	.997883	.763470	.096674
							44	.998911	.821100	.135626
	35				.999851	.990590				
	36				.999936	.994815	45	.999460	.868911	.184101
	37				.999973	.997251	46	.999741	.907022	.242059
	38				.999988	.998597	47	.999880	.936213	.308649
	39				.999995	.999311	48	.999945	.957700	.382176
							49	.999975	.972903	.460205
	40					.999673				
	41					.999850	50	.999988	.983240	.539794
	42					.999933	51	.999994	.989997	.617823
	43					.999970	52		.994241	.691350
	44					.999986	53		.996804	.757940
							54		.998291	.815899

n	k	.01	.05	p .10	.20	.25	k	.30	p .40	.50
	45					.999993				
							55		.999120	.864373
							56		.999564	.903326
							57		.999793	.933394
							58		.999906	.955686
							59		.999960	.971556
							60		.999984	.982399
							61		.999995	.989510
							62			.993983
							63			.996681
							64			.998241
							65			.999105
							66			.999563
							67			.999795
							68			.999908
							69			.999960
							70			.999983
							71			.999993

TABLE A4. Cumulative Poisson Probabilities[a]

$$P(X \le k) = \sum_{x=0}^{k} e^{-\lambda}\frac{\lambda^x}{x!}$$

k	0.1	0.2	0.3	0.4	0.5	0.6	0.7	0.8	0.9	1.0	1.1	1.2	1.3	1.4
0	0.905	0.819	0.741	0.670	0.607	0.549	0.497	0.449	0.407	0.368	0.333	0.301	0.273	0.247
1	0.995	0.982	0.963	0.938	0.910	0.878	0.844	0.809	0.772	0.736	0.699	0.663	0.627	0.592
2	1.000	0.999	0.996	0.992	0.986	0.977	0.966	0.953	0.937	0.920	0.900	0.879	0.857	0.833
3		1.000	1.000	0.999	0.998	0.997	0.994	0.991	0.987	0.981	0.974	0.966	0.957	0.946
4				1.000	1.000	1.000	0.999	0.999	0.998	0.996	0.995	0.992	0.989	0.986
5							1.000	1.000	1.000	0.999	0.999	0.998	0.998	0.997
6										1.000	1.000	1.000	1.000	0.999
7														1.000

k	1.5	1.6	1.7	1.8	1.9	2.0	2.2	2.4	2.6	2.8	3.0	3.2	3.4	3.6
0	0.223	0.202	0.183	0.165	0.150	0.135	0.111	0.091	0.074	0.061	0.050	0.041	0.033	0.027
1	0.558	0.525	0.493	0.463	0.434	0.406	0.355	0.308	0.267	0.231	0.199	0.171	0.147	0.126
2	0.809	0.783	0.757	0.731	0.704	0.677	0.623	0.570	0.518	0.469	0.423	0.380	0.340	0.303
3	0.934	0.921	0.907	0.891	0.875	0.857	0.819	0.779	0.736	0.692	0.647	0.603	0.558	0.515
4	0.981	0.976	0.970	0.964	0.956	0.947	0.928	0.904	0.877	0.848	0.815	0.781	0.744	0.706
5	0.996	0.994	0.992	0.990	0.987	0.983	0.975	0.964	0.951	0.935	0.916	0.895	0.871	0.844
6	0.999	0.999	0.998	0.997	0.997	0.995	0.993	0.988	0.983	0.976	0.966	0.955	0.942	0.927
7	1.000	1.000	1.000	0.999	0.999	0.999	0.998	0.997	0.995	0.992	0.988	0.983	0.977	0.969
8				1.000	1.000	1.000	1.000	0.999	0.999	0.998	0.996	0.994	0.992	0.988
9								1.000	1.000	0.999	0.999	0.998	0.997	0.996
10										1.000	1.000	1.000	0.999	0.999
11													1.000	1.000

k	3.8	4.0	4.2	4.4	4.6	4.8	5.0	5.2	5.4	5.6	5.8	6.0
0	0.022	0.018	0.015	0.012	0.010	0.008	0.007	0.006	0.005	0.004	0.003	0.002
1	0.107	0.092	0.078	0.066	0.056	0.048	0.040	0.034	0.029	0.024	0.021	0.017
2	0.269	0.238	0.210	0.185	0.163	0.143	0.125	0.109	0.095	0.082	0.072	0.062
3	0.473	0.433	0.395	0.359	0.326	0.294	0.265	0.238	0.213	0.191	0.170	0.151
4	0.668	0.629	0.590	0.551	0.513	0.476	0.440	0.406	0.373	0.342	0.313	0.285
5	0.816	0.785	0.753	0.720	0.686	0.651	0.616	0.581	0.546	0.512	0.478	0.446
6	0.909	0.889	0.867	0.844	0.818	0.791	0.762	0.732	0.702	0.670	0.638	0.606
7	0.960	0.949	0.936	0.921	0.905	0.887	0.867	0.845	0.822	0.797	0.771	0.744
8	0.984	0.979	0.972	0.964	0.955	0.944	0.932	0.918	0.903	0.886	0.867	0.847
9	0.994	0.992	0.989	0.985	0.980	0.975	0.968	0.960	0.951	0.941	0.929	0.916
10	0.998	0.997	0.996	0.994	0.992	0.990	0.986	0.982	0.977	0.972	0.965	0.957
11	0.999	0.999	0.999	0.998	0.997	0.996	0.995	0.993	0.990	0.988	0.984	0.980

[a]Based on data in E. C. Molina, *Poisson's Exponential Binomial Limit*, 1942, with the permission of Wadsworth Publishing Company, Belmont, Calif. (original publisher, D. Van Nostrand).

TABLE A4. (*Continued*)

k	3.8	4.0	4.2	4.4	4.6	4.8	5.0	5.2	5.4	5.6	5.8	6.0
							λ					
12	1.000	1.000	1.000	0.999	0.999	0.999	0.998	0.997	0.996	0.995	0.993	0.991
13				1.000	1.000	1.000	0.999	0.999	0.999	0.998	0.997	0.996
14							1.000	1.000	0.999	0.999	0.999	0.999
15									1.000	1.000	1.000	0.999
16												1.000

k	6.5	7.0	7.5	8.0	8.5	9.0	9.5	10.0	10.5	11.0	11.5	12.0
0	0.002	0.001	0.001									
1	0.011	0.007	0.005	0.003	0.002	0.001	0.001					
2	0.043	0.030	0.020	0.014	0.009	0.006	0.004	0.003	0.002	0.001	0.001	0.001
3	0.112	0.082	0.059	0.042	0.030	0.021	0.015	0.010	0.007	0.005	0.003	0.002
4	0.224	0.173	0.132	0.100	0.074	0.055	0.040	0.029	0.021	0.015	0.011	0.008
5	0.369	0.301	0.241	0.191	0.150	0.116	0.089	0.067	0.050	0.038	0.028	0.020
6	0.527	0.450	0.378	0.313	0.256	0.207	0.165	0.130	0.102	0.079	0.060	0.046
7	0.673	0.599	0.525	0.453	0.386	0.324	0.269	0.220	0.179	0.143	0.114	0.090
8	0.792	0.729	0.662	0.593	0.523	0.456	0.392	0.333	0.279	0.232	0.191	0.155
9	0.877	0.830	0.776	0.717	0.653	0.587	0.522	0.458	0.397	0.341	0.289	0.242
10	0.933	0.901	0.862	0.816	0.763	0.706	0.645	0.583	0.521	0.460	0.402	0.347
11	0.966	0.947	0.921	0.888	0.849	0.803	0.752	0.697	0.639	0.579	0.520	0.462
12	0.984	0.973	0.957	0.936	0.909	0.876	0.836	0.792	0.742	0.689	0.633	0.576
13	0.993	0.987	0.978	0.966	0.949	0.926	0.898	0.864	0.825	0.781	0.733	0.682
14	0.997	0.994	0.990	0.983	0.973	0.959	0.940	0.917	0.888	0.854	0.815	0.772
15	0.999	0.998	0.995	0.992	0.986	0.978	0.967	0.951	0.932	0.907	0.878	0.844
16	1.000	0.999	0.998	0.996	0.993	0.989	0.982	0.973	0.960	0.944	0.924	0.899
17		1.000	0.999	0.998	0.997	0.995	0.991	0.986	0.978	0.968	0.954	0.937
18			1.000	0.999	0.999	0.998	0.996	0.993	0.988	0.982	0.974	0.963
19				1.000	0.999	0.999	0.998	0.997	0.994	0.991	0.986	0.979
20					1.000	1.000	0.999	0.998	0.997	0.995	0.992	0.989
21							1.000	0.999	0.999	0.998	0.996	0.994
22								1.000	0.999	0.999	0.998	0.997
23									1.000	1.000	0.999	0.999
24											1.000	0.999
25												1.000

TABLE A5. Standard Normal Distribution[a]
$P(0 \le Z \le z)$

z	.00	.01	.02	.03	.04	.05	.06	.07	.08	.09
0.0	.0000	.0040	.0080	.0120	.0160	.0199	.0239	.0279	.0319	.0359
0.1	.0398	.0438	.0478	.0517	.0557	.0596	.0636	.0675	.0714	.0753
0.2	.0793	.0832	.0871	.0910	.0948	.0987	.1026	.1064	.1103	.1141
0.3	.1179	.1217	.1255	.1293	.1331	.1368	.1406	.1443	.1480	.1517
0.4	.1554	.1591	.1628	.1664	.1700	.1736	.1772	.1808	.1844	.1879
0.5	.1915	.1950	.1985	.2019	.2054	.2088	.2123	.2157	.2190	.2224
0.6	.2257	.2291	.2324	.2357	.2389	.2422	.2454	.2486	.2517	.2549
0.7	.2580	.2611	.2642	.2673	.2703	.2734	.2764	.2794	.2823	.2852
0.8	.2881	.2910	.2939	.2967	.2995	.3023	.3051	.3078	.3106	.3133
0.9	.3159	.3186	.3212	.3238	.3264	.3289	.3315	.3340	.3365	.3389
1.0	.3413	.3438	.3461	.3485	.3508	.3531	.3554	.3577	.3599	.3621
1.1	.3643	.3665	.3686	.3708	.3729	.3749	.3770	.3790	.3810	.3830
1.2	.3849	.3869	.3888	.3907	.3925	.3944	.3962	.3980	.3997	.4015
1.3	.4032	.4049	.4066	.4082	.4099	.4115	.4131	.4147	.4162	.4177
1.4	.4192	.4207	.4222	.4236	.4251	.4265	.4279	.4292	.4306	.4319
1.5	.4332	.4345	.4357	.4370	.4382	.4394	.4406	.4418	.4429	.4441
1.6	.4452	.4463	.4474	.4484	.4495	.4505	.4515	.4525	.4535	.4545
1.7	.4554	.4564	.4573	.4582	.4591	.4599	.4608	.4616	.4625	.4633
1.8	.4641	.4649	.4656	.4664	.4671	.4678	.4686	.4693	.4699	.4706
1.9	.4713	.4719	.4726	.4732	.4738	.4744	.4750	.4756	.4761	.4767
2.0	.4772	.4778	.4783	.4788	.4793	.4798	.4803	.4808	.4812	.4817
2.1	.4821	.4826	.4830	.4834	.4838	.4842	.4846	.4850	.4854	.4857
2.2	.4861	.4864	.4868	.4871	.4875	.4878	.4881	.4884	.4887	.4890
2.3	.4893	.4896	.4898	.4901	.4904	.4906	.4909	.4911	.4913	.4916
2.4	.4918	.4920	.4922	.4925	.4927	.4929	.4931	.4932	.4934	.4936
2.5	.4938	.4940	.4941	.4943	.4945	.4946	.4948	.4949	.4951	.4952
2.6	.4953	.4955	.4956	.4957	.4959	.4960	.4961	.4962	.4963	.4964
2.7	.4965	.4966	.4967	.4968	.4969	.4970	.4971	.4972	.4973	.4974
2.8	.4974	.4975	.4976	.4977	.4977	.4978	.4979	.4979	.4980	.4981
2.9	.4981	.4982	.4982	.4983	.4984	.4984	.4985	.4985	.4986	.4986
3.0	.4987	.4987	.4987	.4988	.4988	.4989	.4989	.4989	.4990	.4990

[a]Reprinted from A. Hald, *Statistical Tables and Formulas*, Wiley, New York, 1953, with permission.

TABLE A6. Distribution of t^a

$$P(t(n) > t_{n,\alpha}) = \alpha$$

n \ α	.45	.40	.35	.3	.25	.2	.15	.1	.05	.025	.01	.005	.0005
1	.158	.325	.510	.727	1.000	1.376	1.963	3.078	6.314	12.706	31.821	63.657	636.619
2	.142	.289	.445	.617	.816	1.061	1.386	1.886	2.920	4.303	6.965	9.925	31.598
3	.137	.277	.424	.584	.765	.978	1.250	1.638	2.353	3.182	4.541	5.841	12.924
4	.134	.271	.414	.569	.741	.941	1.190	1.533	2.132	2.776	3.747	4.604	8.610
5	.132	.267	.408	.559	.727	.920	1.156	1.476	2.015	2.571	3.365	4.032	6.869
6	.131	.265	.404	.553	.718	.906	1.134	1.440	1.943	2.447	3.143	3.707	5.959
7	.130	.263	.402	.549	.711	.896	1.119	1.415	1.895	2.365	2.998	3.499	5.408
8	.130	.262	.399	.546	.706	.889	1.108	1.397	1.860	2.306	2.896	3.355	5.041
9	.129	.261	.398	.543	.703	.883	1.100	1.383	1.833	2.262	2.821	3.250	4.781
10	.129	.260	.397	.542	.700	.879	1.093	1.372	1.812	2.228	2.764	3.169	4.587
11	.129	.260	.396	.540	.697	.876	1.088	1.363	1.796	2.201	2.718	3.106	4.437
12	.128	.259	.395	.539	.695	.873	1.083	1.356	1.782	2.179	2.681	3.055	4.318
13	.128	.259	.394	.538	.694	.870	1.079	1.350	1.771	2.160	2.650	3.012	4.221
14	.128	.258	.393	.537	.692	.868	1.076	1.345	1.761	2.145	2.624	2.977	4.140
15	.128	.258	.393	.536	.691	.866	1.074	1.341	1.753	2.131	2.602	2.947	4.073
16	.128	.258	.392	.535	.690	.865	1.071	1.337	1.746	2.120	2.583	2.921	4.015
17	.128	.257	.392	.534	.689	.863	1.069	1.333	1.740	2.110	2.567	2.898	3.965
18	.127	.257	.392	.534	.688	.862	1.067	1.330	1.734	2.101	2.552	2.878	3.922
19	.127	.257	.391	.533	.688	.861	1.066	1.328	1.729	2.093	2.539	2.861	3.883
20	.127	.257	.391	.533	.687	.860	1.064	1.325	1.725	2.086	2.528	2.845	3.850
21	.127	.257	.391	.532	.686	.859	1.063	1.323	1.721	2.080	2.518	2.831	3.819
22	.127	.256	.390	.532	.686	.858	1.061	1.321	1.717	2.074	2.508	2.819	3.792
23	.127	.256	.390	.532	.685	.858	1.060	1.319	1.714	2.069	2.500	2.807	3.767
24	.127	.256	.390	.531	.685	.857	1.059	1.318	1.711	2.064	2.492	2.797	3.745
25	.127	.256	.390	.531	.684	.856	1.058	1.316	1.708	2.060	2.485	2.787	3.725
26	.127	.256	.390	.531	.684	.856	1.058	1.315	1.706	2.056	2.479	2.779	3.707
27	.127	.256	.389	.531	.684	.855	1.057	1.314	1.703	2.052	2.473	2.771	3.690
28	.127	.256	.389	.530	.683	.855	1.056	1.313	1.701	2.048	2.467	2.763	3.674
29	.127	.256	.389	.530	.683	.854	1.055	1.311	1.699	2.045	2.462	2.756	3.659
30	.127	.256	.389	.530	.683	.854	1.055	1.310	1.697	2.042	2.457	2.750	3.646
40	.126	.255	.388	.529	.681	.851	1.050	1.303	1.684	2.021	2.423	2.704	3.551
60	.126	.254	.387	.527	.679	.848	1.046	1.296	1.671	2.000	2.390	2.660	3.460
120	.126	.254	.386	.526	.677	.845	1.041	1.289	1.658	1.980	2.358	2.617	3.373
∞	.126	.253	.385	.524	.674	.842	1.036	1.282	1.645	1.960	2.326	2.576	3.291

^{a}Reprinted from Table III of Ronald A. Fisher and Frank Yates, *Statistical Tables for Biological, Agricultural, and Medical Research*, 6th ed., Longmans Group Ltd., London (previously published by Oliver and Boyd Ltd., Edinburgh), 1974, by permission of the authors and publishers.

TABLE A7. Distribution of $\chi^{2\,a}$
$$P(\chi^2(n) > \chi^2_{n,\alpha}) = \alpha$$

α n^c	.99	.975	.95	.90	.80	.70	.50	.30	.20	.10	.05	.025	.01	.001
1	.000157	0.00098	.00393	.0158	.0642	.148	.455	1.074	1.642	2.706	3.841	5.0238	6.635	10.827
2	.0201	0.0506	.103	.211	.446	.713	1.386	2.408	3.219	4.605	5.991	7.3780	9.210	13.815
3	.115	0.216	.352	.584	1.005	1.424	2.366	3.665	4.642	6.251	7.815	9.348	11.345	16.266
4	.297	0.484	.711	1.064	1.649	2.195	3.357	4.878	5.989	7.779	9.488	11.143	13.277	18.467
5	.554	0.831	1.145	1.610	2.343	3.000	4.351	6.064	7.289	9.236	11.070	12.832	15.086	20.515
6	.872	1.237	1.635	2.204	3.070	3.828	5.348	7.231	8.558	10.645	12.592	14.449	16.812	22.457
7	1.239	1.690	2.167	2.833	3.822	4.671	6.346	8.383	9.803	12.017	14.067	16.013	18.475	24.322
8	1.646	2.180	2.733	3.490	4.594	5.527	7.344	9.524	11.030	13.362	15.507	17.535	20.090	26.125
9	2.088	2.700	3.325	4.168	5.380	6.393	8.343	10.656	12.242	14.684	16.919	19.023	21.666	27.877
10	2.558	3.247	3.940	4.865	6.179	7.267	9.342	11.781	13.442	15.987	18.307	20.483	23.209	29.588
11	3.053	3.816	4.575	5.578	6.989	8.148	10.341	12.899	14.631	17.275	19.675	21.920	24.725	31.264
12	3.571	4.404	5.226	6.304	7.807	9.034	11.340	14.011	15.812	18.549	21.026	23.337	26.217	32.909
13	4.107	5.009	5.892	7.042	8.634	9.926	12.340	15.119	16.985	19.812	22.362	24.736	27.688	34.528
14	4.660	5.629	6.571	7.790	9.467	10.821	13.339	16.222	18.151	21.064	23.685	26.119	29.141	36.123
15	5.229	6.262	7.261	8.547	10.307	11.721	14.339	17.322	19.311	22.307	24.996	27.488	30.578	37.697

df														
16	5.812	6.908	7.962	9.312	11.152	12.624	15.338	18.418	20.465	23.542	26.296	28.845	32.000	39.252
17	6.408	7.564	8.672	10.085	12.002	13.531	16.338	19.511	21.615	24.769	27.587	30.191	33.409	40.790
18	7.015	8.231	9.390	10.865	12.857	14.440	17.338	20.601	22.760	25.989	28.869	31.526	34.805	42.312
19	7.633	8.907	10.117	11.651	13.716	15.352	18.338	21.689	23.900	27.204	30.144	32.852	36.191	43.820
20	8.260	9.591	10.851	12.443	14.578	16.266	19.337	22.775	25.038	28.412	31.410	34.170	37.566	45.315
21	8.897	10.283	11.591	13.240	15.445	17.182	20.337	23.858	26.171	29.615	32.671	35.479	38.932	46.797
22	9.542	10.982	12.338	14.041	16.314	18.101	21.337	24.939	27.301	30.813	33.924	36.781	40.289	48.268
23	10.196	11.689	13.091	14.848	17.187	19.021	22.337	26.018	28.429	32.007	35.172	38.076	41.638	49.728
24	10.856	12.401	13.848	15.659	18.062	19.943	23.337	27.096	29.553	33.196	36.415	39.364	42.980	51.179
25	11.524	13.120	14.611	16.473	18.940	20.867	24.337	28.172	30.675	34.382	37.652	40.646	44.314	52.620
26	12.198	13.844	15.379	17.292	19.820	21.792	25.336	29.246	31.795	35.563	38.885	41.923	45.642	54.052
27	12.879	14.573	16.151	18.114	20.703	22.719	26.336	30.319	32.912	36.741	40.113	43.194	46.963	55.476
28	13.565	15.308	16.928	18.939	21.588	23.647	27.336	31.391	34.027	37.916	41.337	44.461	48.278	56.893
29	14.256	16.047	17.708	19.768	22.475	24.577	28.336	32.461	35.139	39.087	42.557	45.722	49.588	58.302
30	14.953	16.791	18.493	20.599	23.364	25.508	29.336	33.530	36.250	40.256	43.773	46.979	50.892	59.703

[a] Reprinted (except for α = .975 and .025) from Table IV of Ronald A. Fisher and Frank Yates, *Statistical Tables for Biological, Agricultural, and Medical Research*, Longmans Group Ltd., London (previously published by Oliver and Boyd Ltd., Edinburgh), by permission of the authors and publishers.

[b] Values for α = .975 and .025 are taken from Table 3 of E. S. Pearson and H. O. Hartley, *Biometrika Tables for Statisticians*, Vol. II, 1972 with permission of the Biometrika Trustees.

[c] For larger values of n, the expression $\sqrt{2\chi^2} - \sqrt{2n-1}$ may be used as a normal deviate with unit variance, remembering that the probability for χ^2 corresponds with that of a single tail of the normal curve.

	Degrees of Freedom for the Numerator (*m*)											
n	1	2	3	4	5	6	7	8	9	10	11	12
1	161	200	216	225	230	234	237	239	241	242	243	244
	4,052	4,999	5,403	5,625	5,764	5,859	5,928	5,981	6,022	6,056	6,082	6,106
2	18.51	19.00	19.16	19.25	19.30	19.33	19.36	19.37	19.38	19.39	19.40	19.41
	98.49	99.00	99.17	99.25	99.30	99.33	99.36	99.37	99.39	99.40	99.41	99.42
3	10.13	9.55	9.28	9.12	9.01	8.94	8.88	8.84	8.81	8.78	8.76	8.74
	34.12	30.82	29.46	28.71	28.24	27.91	27.67	27.49	27.34	27.23	27.13	27.05
4	7.71	6.94	6.59	6.39	6.26	6.16	6.09	6.04	6.00	5.96	5.93	5.91
	21.20	18.00	16.69	15.98	15.52	15.21	14.98	14.80	14.66	14.54	14.45	14.37
5	6.61	5.79	5.41	5.19	5.05	4.95	4.88	4.82	4.78	4.74	4.70	4.68
	16.26	13.27	12.06	11.39	10.97	10.67	10.45	10.29	10.15	10.05	9.96	9.89
6	5.99	5.14	4.76	4.53	4.39	4.28	4.21	4.15	4.10	4.06	4.03	4.00
	13.74	10.92	9.78	9.15	8.75	8.47	8.26	8.10	7.98	7.87	7.79	7.72
7	5.59	4.74	4.34	4.12	3.97	3.87	3.79	3.73	3.68	3.63	3.60	3.57
	12.25	9.55	8.45	7.85	7.46	7.19	7.00	6.84	6.71	6.62	6.54	6.47
8	5.32	4.46	4.07	3.84	3.69	3.58	3.50	3.44	3.39	3.34	3.31	3.28 ·
	11.26	8.65	7.59	7.01	6.63	6.37	6.19	6.03	5.91	5.82	5.74	5.67
9	5.12	4.26	3.86	3.63	3.48	3.37	3.29	3.23	3.18	3.13	3.10	3.07
	10.56	8.02	6.99	6.42	6.06	5.80	5.62	5.47	5.35	5.26	5.18	5.11
10	4.96	4.10	3.71	3.48	3.33	3.22	3.14	3.07	3.02	2.97	2.94	2.91
	10.04	7.56	6.55	5.99	5.64	5.39	5.21	5.06	4.95	4.85	4.78	4.71
11	4.84	3.98	3.59	3.36	3.20	3.09	3.01	2.95	2.90	2.86	2.82	2.79
	9.65	7.20	6.22	5.67	5.32	5.07	4.88	4.74	4.63	4.54	4.46	4.40
12	4.75	3.88	3.49	3.26	3.11	3.00	2.92	2.85	2.80	2.76	2.72	2.69
	9.33	6.93	5.95	5.41	5.06	4.82	4.65	4.50	4.39	4.30	4.22	4.16
13	4.67	3.80	3.41	3.18	3.02	2.92	2.84	2.77	2.72	2.67	2.63	2.60
	9.07	6.70	5.74	5.20	4.86	4.62	4.44	4.30	4.19	4.10	4.02	3.96

1 Percent (Boldface Type) Points for the Distribution of F^a

			Degrees of Freedom for the Numerator (m)									
14	16	20	24	30	40	50	75	100	200	500	∞	n
245	246	248	249	250	251	252	253	253	254	254	254	1
6,142	**6,169**	**6,208**	**6,234**	**6,261**	**6,286**	**6,302**	**6,323**	**6,334**	**6,352**	**6,361**	**6,366**	
19.42	19.43	19.44	19.45	19.46	19.47	19.47	19.48	19.49	19.49	19.50	19.50	2
99.43	**99.44**	**99.45**	**99.46**	**99.47**	**99.48**	**99.48**	**99.49**	**99.49**	**99.49**	**99.50**	**99.50**	
8.71	8.69	8.66	8.64	8.62	8.60	8.58	8.57	8.56	8.54	8.54	8.53	3
26.92	**26.83**	**26.69**	**26.60**	**26.50**	**26.41**	**26.35**	**26.27**	**26.23**	**26.18**	**26.14**	**26.12**	
5.87	5.84	5.80	5.77	5.74	5.71	5.70	5.68	5.66	5.65	5.64	5.63	4
14.24	**14.15**	**14.02**	**13.93**	**13.83**	**13.74**	**13.69**	**13.61**	**13.57**	**13.52**	**13.48**	**13.46**	
4.64	4.60	4.56	4.53	4.50	4.46	4.44	4.42	4.40	4.38	4.37	4.36	5
9.77	**9.68**	**9.55**	**9.47**	**9.38**	**9.29**	**9.24**	**9.17**	**9.13**	**9.07**	**9.04**	**9.02**	
3.96	3.92	3.87	3.84	3.81	3.77	3.75	3.72	3.71	3.69	3.68	3.67	6
7.60	**7.52**	**7.39**	**7.31**	**7.23**	**7.14**	**7.09**	**7.02**	**6.99**	**6.94**	**6.90**	**6.88**	
3.52	3.49	3.44	3.41	3.38	3.34	3.32	3.29	3.28	3.25	3.24	3.23	7
6.35	**6.27**	**6.15**	**6.07**	**5.98**	**5.90**	**5.85**	**5.78**	**5.75**	**5.70**	**5.67**	**5.65**	
3.23	3.20	3.15	3.12	3.08	3.05	3.03	3.00	2.98	2.96	2.94	2.93	8
5.56	**5.48**	**5.36**	**5.28**	**5.20**	**5.11**	**5.06**	**5.00**	**4.96**	**4.91**	**4.88**	**4.86**	
3.02	2.98	2.93	2.90	2.86	2.82	2.80	2.77	2.76	2.73	2.72	2.71	9
5.00	**4.92**	**4.80**	**4.73**	**4.64**	**4.56**	**4.51**	**4.45**	**4.41**	**4.36**	**4.33**	**4.31**	
2.86	2.82	2.77	2.74	2.70	2.67	2.64	2.61	2.59	2.56	2.55	2.54	10
4.60	**4.52**	**4.41**	**4.33**	**4.25**	**4.17**	**4.12**	**4.05**	**4.01**	**3.96**	**3.93**	**3.91**	
2.74	2.70	2.65	2.61	2.57	2.53	2.50	2.47	2.45	2.42	2.41	2.40	11
4.29	**4.21**	**4.10**	**4.02**	**3.94**	**3.86**	**3.80**	**3.74**	**3.70**	**3.66**	**3.62**	**3.60**	
2.64	2.60	2.54	2.50	2.46	2.42	2.40	2.36	2.35	2.32	2.31	2.30	12
4.05	**3.98**	**3.86**	**3.78**	**3.70**	**3.61**	**3.56**	**3.49**	**3.46**	**3.41**	**3.38**	**3.36**	
2.55	2.51	2.46	2.42	2.38	2.34	2.32	2.28	2.26	2.24	2.22	2.21	13
3.85	**3.78**	**3.67**	**3.59**	**3.51**	**3.42**	**3.37**	**3.30**	**3.27**	**3.21**	**3.18**	**3.16**	

	Degrees of Freedom for the Numerator (m)											
n	1	2	3	4	5	6	7	8	9	10	11	12
14	4.60	3.74	3.34	3.11	2.96	2.85	2.77	2.70	2.65	2.60	2.56	2.53
	8.86	**6.51**	**5.56**	**5.03**	**4.69**	**4.46**	**4.28**	**4.14**	**4.03**	**3.94**	**3.86**	**3.80**
15	4.54	3.68	3.29	3.06	2.90	2.79	2.70	2.64	2.59	2.55	2.51	2.48
	8.68	**6.36**	**5.42**	**4.89**	**4.56**	**4.32**	**4.14**	**4.00**	**3.89**	**3.80**	**3.73**	**3.67**
16	4.49	3.63	3.24	3.01	2.85	2.74	2.66	2.59	2.54	2.49	2.45	2.42
	8.53	**6.23**	**5.29**	**4.77**	**4.44**	**4.20**	**4.03**	**3.89**	**3.78**	**3.69**	**3.61**	**3.55**
17	4.45	3.59	3.20	2.96	2.81	2.70	2.62	2.55	2.50	2.45	2.41	2.38
	8.40	**6.11**	**5.18**	**4.67**	**4.34**	**4.10**	**3.93**	**3.79**	**3.68**	**3.59**	**3.52**	**3.45**
18	4.41	3.55	3.16	2.93	2.77	3.66	2.58	2.51	2.46	2.41	2.37	2.34
	8.28	**6.01**	**5.09**	**4.58**	**4.25**	**4.01**	**3.85**	**3.71**	**3.60**	**3.51**	**3.44**	**3.37**
19	4.38	3.52	3.13	2.90	2.74	2.63	2.55	2.48	2.43	2.38	2.34	2.31
	8.18	**5.93**	**5.01**	**4.50**	**4.17**	**3.94**	**3.77**	**3.63**	**3.52**	**3.43**	**3.36**	**3.30**
20	4.35	3.49	3.10	2.87	2.71	2.60	2.52	2.45	2.40	2.35	2.31	2.28
	8.10	**5.85**	**4.94**	**4.43**	**4.10**	**3.87**	**3.71**	**3.56**	**3.45**	**3.37**	**3.30**	**3.23**
21	4.32	3.47	3.07	2.84	2.68	2.57	2.49	2.42	2.37	2.32	2.28	2.25
	8.02	**5.78**	**4.87**	**4.37**	**4.04**	**3.81**	**3.65**	**3.51**	**3.40**	**3.31**	**3.24**	**3.17**
22	4.30	3.44	3.05	2.82	2.66	2.55	2.47	2.40	2.35	2.30	2.26	2.23
	7.94	**5.72**	**4.82**	**4.31**	**3.99**	**3.76**	**3.59**	**3.45**	**3.35**	**3.26**	**3.18**	**3.12**
23	4.28	3.42	3.03	2.80	2.64	2.53	2.45	2.38	2.32	2.28	2.24	2.20
	7.88	**5.66**	**4.76**	**4.26**	**3.94**	**3.71**	**3.54**	**3.41**	**3.30**	**3.21**	**3.14**	**3.07**
24	4.26	3.40	3.01	2.78	2.62	2.51	2.43	2.36	2.30	2.26	2.22	2.18
	7.82	**5.61**	**4.72**	**4.22**	**3.90**	**3.67**	**3.50**	**3.36**	**3.25**	**3.17**	**3.09**	**3.03**
25	4.24	3.38	2.99	2.76	2.60	2.49	2.41	2.34	2.28	2.24	2.20	2.16
	7.77	**5.57**	**4.68**	**4.18**	**3.86**	**3.63**	**3.46**	**3.32**	**3.21**	**3.13**	**3.05**	**2.99**
26	4.22	3.37	2.98	2.74	2.59	2.47	2.39	2.32	2.27	2.22	2.18	2.15
	7.72	**5.53**	**4.64**	**4.14**	**3.82**	**3.59**	**3.42**	**3.29**	**3.17**	**3.09**	**3.02**	**2.96**

(*Continued*)

\			Degrees of Freedom for the Numerator (*m*)										
14	16	20	24	30	40	50	75	100	200	500	∞	*n*
2.48	2.44	2.39	2.35	2.31	2.27	2.24	2.21	2.19	2.16	2.14	2.13	14
3.70	**3.62**	**3.51**	**3.43**	**3.34**	**3.26**	**3.21**	**3.14**	**3.11**	**3.06**	**3.02**	**3.00**	
2.43	2.39	2.33	2.29	2.25	2.21	2.18	2.15	2.12	2.10	2.08	2.07	15
3.56	**3.48**	**3.36**	**3.29**	**3.20**	**3.12**	**3.07**	**3.00**	**2.97**	**2.92**	**2.89**	**2.87**	
2.37	2.33	2.28	2.24	2.20	2.16	2.13	2.09	2.07	2.04	2.02	2.01	16
3.45	**3.37**	**3.25**	**3.18**	**3.10**	**3.01**	**2.96**	**2.98**	**2.86**	**2.80**	**2.77**	**2.75**	
2.33	2.29	2.23	2.19	2.15	2.11	2.08	2.04	2.02	1.99	1.97	1.96	17
3.35	**3.27**	**3.16**	**3.08**	**3.00**	**2.92**	**2.86**	**2.79**	**2.76**	**2.70**	**2.67**	**2.65**	
2.29	2.25	2.19	2.15	2.11	2.07	2.04	2.00	1.98	1.95	1.93	1.92	18
3.27	**3.19**	**3.07**	**3.00**	**2.91**	**2.83**	**2.78**	**2.71**	**2.68**	**2.62**	**2.59**	**2.57**	
2.26	2.21	2.15	2.11	2.07	2.02	2.00	1.96	1.94	1.91	1.90	1.88	19
3.19	**3.12**	**3.00**	**2.92**	**2.84**	**2.76**	**2.70**	**2.63**	**2.60**	**2.54**	**2.51**	**2.49**	
2.23	2.18	2.12	2.08	2.04	1.99	1.96	1.92	1.90	1.87	1.85	1.84	20
3.13	**3.05**	**2.94**	**2.86**	**2.77**	**2.69**	**2.63**	**2.56**	**2.53**	**2.47**	**2.44**	**2.42**	
2.20	2.15	2.09	2.05	2.00	1.96	1.93	1.89	1.87	1.84	1.82	1.81	21
3.07	**2.99**	**2.88**	**2.80**	**2.72**	**2.63**	**2.58**	**2.51**	**2.47**	**2.42**	**2.38**	**2.36**	
2.18	2.13	2.07	2.03	1.98	1.93	1.91	1.87	1.84	1.81	1.80	1.78	22
3.02	**2.94**	**2.83**	**2.75**	**2.67**	**2.58**	**2.53**	**2.46**	**2.42**	**2.37**	**2.33**	**2.31**	
2.14	2.10	2.04	2.00	1.96	1.91	1.88	1.84	1.82	1.79	1.77	1.76	23
2.97	**2.89**	**2.78**	**2.70**	**2.62**	**2.53**	**2.48**	**2.41**	**2.37**	**2.32**	**2.28**	**2.26**	
2.13	2.09	2.02	1.98	1.94	1.89	1.86	1.82	1.80	1.76	1.74	1.73	24
2.93	**2.85**	**2.74**	**2.66**	**2.58**	**2.49**	**2.44**	**2.36**	**2.33**	**2.27**	**2.23**	**2.21**	
2.11	2.06	2.00	1.96	1.92	1.87	1.84	1.80	1.77	1.74	1.72	1.71	25
2.89	**2.81**	**2.70**	**2.62**	**2.54**	**2.45**	**2.40**	**2.32**	**2.29**	**2.23**	**2.19**	**2.17**	
2.10	2.05	1.99	1.95	1.90	1.85	1.82	1.78	1.76	1.72	1.70	1.69	26
2.86	**2.77**	**2.66**	**2.58**	**2.50**	**2.41**	**2.36**	**2.28**	**2.25**	**2.19**	**2.15**	**2.13**	

n	Degrees of Freedom for the Numerator (m)											
	1	2	3	4	5	6	7	8	9	10	11	12
27	4.21	3.35	2.96	2.73	2.57	2.46	2.37	2.30	2.25	2.20	2.16	2.13
	7.68	**5.49**	**4.60**	**4.11**	**3.79**	**3.56**	**3.39**	**3.26**	**3.14**	**3.06**	**2.98**	**2.93**
28	4.20	3.34	2.95	2.71	2.56	2.44	2.36	2.29	2.24	2.19	2.15	2.12
	7.64	**5.45**	**4.57**	**4.07**	**3.76**	**3.53**	**3.36**	**3.23**	**3.11**	**3.03**	**2.95**	**2.90**
29	4.18	3.33	2.93	2.70	2.54	2.43	2.35	2.28	2.22	2.18	2.14	2.10
	7.60	**5.42**	**4.54**	**4.04**	**3.73**	**3.50**	**3.33**	**3.20**	**3.08**	**3.00**	**2.92**	**2.87**
30	4.17	3.32	2.92	2.69	2.53	2.42	2.34	2.27	2.21	2.16	2.12	2.09
	7.56	**5.39**	**4.51**	**4.02**	**3.70**	**3.47**	**3.30**	**3.17**	**3.06**	**2.98**	**2.90**	**2.84**
32	4.15	3.30	2.90	2.67	2.51	2.40	2.32	2.25	2.19	2.14	2.10	2.07
	7.50	**5.34**	**4.46**	**3.97**	**3.66**	**3.42**	**3.25**	**3.12**	**3.01**	**2.94**	**2.86**	**2.80**
34	4.13	3.28	2.88	2.65	2.49	2.38	2.30	2.23	2.17	2.12	2.08	2.05
	7.44	**5.29**	**4.42**	**3.93**	**3.61**	**3.38**	**3.21**	**3.08**	**2.97**	**2.89**	**2.82**	**2.76**
36	4.11	3.26	2.86	2.63	2.48	2.36	2.28	2.21	2.15	2.10	2.06	2.03
	7.39	**5.25**	**4.38**	**3.89**	**3.58**	**3.35**	**3.18**	**3.04**	**2.94**	**2.86**	**2.78**	**2.72**
38	4.10	3.25	2.85	2.62	2.46	2.35	2.26	2.19	2.14	2.09	2.05	2.02
	7.35	**5.21**	**4.34**	**3.86**	**3.54**	**3.32**	**3.15**	**3.02**	**2.91**	**2.82**	**2.75**	**2.69**
40	4.07	3.23	2.84	2.61	2.45	2.34	2.25	2.18	2.12	2.07	2.04	2.00
	7.31	**5.18**	**4.31**	**3.83**	**3.51**	**3.29**	**3.12**	**2.99**	**2.88**	**2.80**	**2.73**	**2.66**
42	4.07	3.22	2.83	2.59	2.44	2.32	2.24	2.17	2.11	2.06	2.02	1.99
	7.27	**5.15**	**4.29**	**3.80**	**3.49**	**3.26**	**3.10**	**2.96**	**2.86**	**2.77**	**2.70**	**2.64**
44	4.06	3.21	2.82	2.58	2.43	2.31	2.23	2.16	2.10	2.05	2.01	1.98
	7.24	**5.12**	**4.26**	**3.78**	**3.46**	**3.24**	**3.07**	**2.94**	**2.84**	**2.75**	**2.68**	**2.62**
46	4.05	3.20	2.81	2.57	2.42	2.30	2.22	2.14	2.09	2.04	2.00	1.97
	7.21	**5.10**	**4.24**	**3.76**	**3.44**	**3.22**	**3.05**	**2.92**	**2.82**	**2.73**	**2.66**	**2.60**
48	4.04	3.19	2.80	2.56	2.41	2.30	2.21	2.14	2.08	2.03	1.99	1.96
	7.19	**5.08**	**4.22**	**3.74**	**3.42**	**3.20**	**3.04**	**2.90**	**2.80**	**2.71**	**2.64**	**2.58**

(Continued)

				Degrees of Freedom for the Numerator (m)								
14	16	20	24	30	40	50	75	100	200	500	∞	n
2.08	2.03	1.97	1.93	1.88	1.84	1.80	1.76	1.74	1.71	1.68	1.67	27
2.83	**2.74**	**2.63**	**2.55**	**2.47**	**2.38**	**2.33**	**2.25**	**2.21**	**2.16**	**2.12**	**2.10**	
2.06	2.02	1.96	1.91	1.87	1.81	1.78	1.75	1.72	1.69	1.67	1.65	28
2.80	**2.71**	**2.60**	**2.52**	**2.44**	**2.35**	**2.30**	**2.22**	**2.18**	**2.13**	**2.09**	**2.06**	
2.05	2.00	1.94	1.90	1.85	1.80	1.77	1.73	1.71	1.68	1.65	1.64	29
2.77	**2.68**	**2.57**	**2.49**	**2.41**	**2.32**	**2.27**	**2.19**	**2.15**	**2.10**	**2.06**	**2.03**	
2.04	1.99	1.93	1.89	1.84	1.79	1.76	1.72	1.69	1.66	1.64	1.62	30
2.74	**2.66**	**2.55**	**2.47**	**2.38**	**2.29**	**2.24**	**2.16**	**2.13**	**2.07**	**2.03**	**2.01**	
2.02	1.97	1.91	1.86	1.82	1.76	1.74	1.69	1.67	1.64	1.61	1.59	32
2.70	**2.62**	**2.51**	**2.42**	**2.34**	**2.25**	**2.20**	**2.12**	**2.08**	**2.02**	**1.98**	**1.96**	
2.00	1.95	1.89	1.84	1.80	1.74	1.71	1.67	1.64	1.61	1.59	1.57	34
2.66	**2.58**	**2.47**	**2.38**	**2.30**	**2.21**	**2.15**	**2.08**	**2.04**	**1.98**	**1.94**	**1.91**	
1.98	1.93	1.87	1.82	1.78	1.72	1.69	1.65	1.62	1.59	1.56	1.55	36
2.62	**2.54**	**2.43**	**2.35**	**2.26**	**2.17**	**2.12**	**2.04**	**2.00**	**1.94**	**1.90**	**1.87**	
1.96	1.92	1.85	1.80	1.76	1.71	1.67	1.63	1.60	1.57	1.54	1.53	38
2.59	**2.51**	**2.40**	**2.32**	**2.22**	**2.14**	**2.08**	**2.00**	**1.97**	**1.90**	**1.86**	**1.84**	
1.95	1.90	1.84	1.79	1.74	1.69	1.66	1.61	1.59	1.55	1.53	1.51	40
2.56	**2.49**	**2.37**	**2.29**	**2.20**	**2.11**	**2.05**	**1.97**	**1.94**	**1.88**	**1.84**	**1.81**	
1.94	1.89	1.82	1.78	1.73	1.68	1.64	1.60	1.57	1.54	1.51	1.49	42
2.54	**2.46**	**2.35**	**2.26**	**2.17**	**2.08**	**2.02**	**1.94**	**1.91**	**1.85**	**1.80**	**1.78**	
1.92	1.88	1.81	1.76	1.72	1.66	1.63	1.58	1.56	1.52	1.50	1.48	44
2.52	**2.44**	**2.32**	**2.24**	**2.15**	**2.06**	**2.00**	**1.92**	**1.88**	**1.82**	**1.78**	**1.75**	
1.91	1.87	1.80	1.75	1.71	1.65	1.62	1.57	1.54	1.51	1.48	1.46	46
2.50	**2.42**	**2.30**	**2.22**	**2.13**	**2.04**	**1.98**	**1.90**	**1.86**	**1.80**	**1.76**	**1.72**	
1.90	1.86	1.79	1.74	1.70	1.64	1.61	1.56	1.53	1.50	1.47	1.45	48
2.48	**2.40**	**2.28**	**2.20**	**2.11**	**2.02**	**1.96**	**1.88**	**1.84**	**1.78**	**1.73**	**1.70**	

	Degrees of Freedom for the Numerator (m)											
n	1	2	3	4	5	6	7	8	9	10	11	12
50	4.03	3.18	2.79	2.56	2.40	2.29	2.20	2.13	2.07	2.02	1.98	1.95
	7.17	5.06	4.20	3.72	3.41	3.18	3.02	2.88	2.78	2.70	2.62	2.56
55	4.02	3.17	2.78	2.54	2.38	2.27	2.18	2.11	2.05	2.00	1.97	1.93
	7.12	5.01	4.16	3.68	3.37	3.15	2.98	2.85	2.75	2.66	2.59	2.53
60	4.00	3.15	2.76	2.52	2.37	2.25	2.17	2.10	2.04	1.99	1.95	1.92
	7.08	4.98	4.13	3.65	3.34	3.12	2.95	2.82	2.72	2.63	2.56	2.50
65	3.99	3.14	2.75	2.51	2.36	2.24	2.15	2.08	2.02	1.98	1.94	1.90
	7.04	4.95	4.10	3.62	3.31	3.09	2.93	2.79	2.70	2.61	2.54	2.47
70	3.98	3.13	2.74	2.50	2.35	2.23	2.14	2.07	2.01	1.97	1.93	1.89
	7.01	4.92	4.08	3.60	3.29	3.07	2.91	2.77	2.67	2.59	2.51	2.45
80	3.96	3.11	2.72	2.48	2.33	2.21	2.12	2.05	1.99	1.95	1.91	1.88
	6.96	4.88	4.04	3.56	3.25	3.04	2.87	2.74	2.64	2.55	2.48	2.41
100	3.94	3.09	2.70	2.46	2.30	2.19	2.10	2.03	1.97	1.92	1.88	1.85
	6.90	4.82	3.98	3.51	3.20	2.99	2.82	2.69	2.59	2.51	2.43	2.36
125	3.92	3.07	2.68	2.44	2.29	2.17	2.08	2.01	1.95	1.90	1.86	1.83
	6.84	4.78	3.94	3.47	3.17	2.95	2.79	2.65	2.56	2.47	2.40	2.33
150	3.91	3.06	2.67	2.43	2.27	2.16	2.07	2.00	1.94	1.89	1.85	1.82
	6.81	4.75	3.91	3.44	3.14	2.92	2.76	2.62	2.53	2.44	2.37	2.30
200	3.89	3.04	2.65	2.41	2.26	2.14	2.05	1.98	1.92	1.87	1.83	1.80
	6.76	4.71	3.88	3.41	3.11	2.90	2.73	2.60	2.50	2.41	2.34	2.28
400	3.86	3.02	2.62	2.39	2.23	2.12	2.03	1.96	1.90	1.85	1.91	1.78
	6.70	4.66	3.83	3.36	3.06	2.85	2.69	2.55	2.46	2.37	2.29	2.23
1000	3.85	3.00	2.61	2.38	2.22	2.10	2.02	1.95	1.89	1.84	1.80	1.76
	6.66	4.62	3.80	3.34	3.04	2.82	2.66	2.53	2.43	2.34	2.26	2.20
∞	3.84	2.99	2.60	2.37	2.21	2.09	2.01	1.94	1.88	1.83	1.79	1.75
	6.64	4.60	3.78	3.32	3.02	2.80	2.64	2.51	2.41	2.32	2.24	2.18

(*Continued*)

	Degrees of Freedom for the Numerator (m)												
14	16	20	24	30	40	50	75	199	200	500	∞	n
1.90	1.85	1.78	1.74	1.69	1.63	1.60	1.55	1.52	1.48	1.46	1.44	50
2.46	**2.39**	**2.26**	**2.18**	**2.10**	**2.00**	**1.94**	**1.86**	**1.82**	**1.76**	**1.71**	**1.68**	
1.88	1.83	1.76	1.72	1.67	1.61	1.58	1.52	1.50	1.46	1.43	1.41	55
2.43	**2.35**	**2.23**	**2.15**	**2.06**	**1.96**	**1.90**	**1.82**	**1.78**	**1.71**	**1.66**	**1.64**	
1.86	1.81	1.75	1.70	1.65	1.59	1.56	1.50	1.48	1.44	1.41	1.39	60
2.40	**2.32**	**2.20**	**2.12**	**2.03**	**1.93**	**1.87**	**1.79**	**1.74**	**1.68**	**1.63**	**1.60**	
1.85	1.80	1.73	1.68	1.63	1.57	1.54	1.49	1.46	1.42	1.39	1.37	65
2.37	**2.30**	**2.18**	**2.09**	**2.00**	**1.90**	**1.84**	**1.76**	**1.71**	**1.64**	**1.60**	**1.56**	
1.84	1.79	1.72	1.67	1.62	1.56	1.53	1.47	1.45	1.40	1.37	1.35	70
2.35	**2.28**	**2.15**	**2.07**	**1.98**	**1.88**	**1.82**	**1.74**	**1.69**	**1.62**	**1.56**	**1.53**	
1.82	1.77	1.70	1.65	1.60	1.54	1.51	1.45	1.42	1.38	1.35	1.32	80
2.32	**2.24**	**2.11**	**2.03**	**1.94**	**1.84**	**1.78**	**1.70**	**1.65**	**1.57**	**1.52**	**1.49**	
1.79	1.75	1.68	1.63	1.57	1.51	1.48	1.42	1.39	1.34	1.30	1.28	100
2.26	**2.19**	**2.06**	**1.98**	**1.89**	**1.79**	**1.73**	**1.64**	**1.59**	**1.51**	**1.46**	**1.43**	
1.77	1.72	1.65	1.60	1.55	1.49	1.45	1.39	1.36	1.31	1.27	1.25	125
2.23	**2.15**	**2.03**	**1.94**	**1.85**	**1.75**	**1.68**	**1.59**	**1.54**	**1.46**	**1.40**	**1.37**	
1.76	1.71	1.64	1.59	1.54	1.47	1.44	1.37	1.34	1.29	1.25	1.22	150
2.20	**2.12**	**2.00**	**1.91**	**1.83**	**1.72**	**1.66**	**1.56**	**1.51**	**1.43**	**1.37**	**1.33**	
1.74	1.69	1.62	1.57	1.52	1.45	1.42	1.35	1.32	1.26	1.22	1.19	200
2.17	**2.09**	**1.97**	**1.88**	**1.79**	**1.69**	**1.62**	**1.53**	**1.48**	**1.39**	**1.33**	**1.28**	
1.72	1.67	1.60	1.54	1.49	1.42	1.38	1.32	1.28	1.22	1.16	1.13	400
2.12	**2.04**	**1.92**	**1.84**	**1.74**	**1.64**	**1.57**	**1.47**	**1.42**	**1.32**	**1.24**	**1.19**	
1.70	1.65	1.58	1.53	1.47	1.41	1.36	1.30	1.26	1.19	1.13	1.08	1000
2.09	**2.01**	**1.89**	**1.81**	**1.71**	**1.61**	**1.54**	**1.44**	**1.38**	**1.28**	**1.19**	**1.11**	
1.69	1.64	1.57	1.52	1.46	1.40	1.35	1.28	1.24	1.17	1.11	1.00	∞
2.07	**1.99**	**1.87**	**1.79**	**1.69**	**1.59**	**1.52**	**1.41**	**1.36**	**1.25**	**1.15**	**1.00**	

[a]Reprinted by permission from George W. Snedecor and William G. Cochran, *Statistical Methods*, 7th Ed., © 1980 by the Iowa State University Press, Ames, Iowa 50010.

TABLE A9. Right Tail Probabilities for One-Sample Kolmogorov–Smirnov Statistic[a]

		Right Tail Probability for Two-Sided Test									
n	.200	.100	.050	.020	.010	n	.200	.100	.050	.020	.010
1	.900	.950	.975	.990	.995	21	.226	.259	.287	.321	.344
2	.684	.776	.842	.900	.929	22	.221	.253	.281	.314	.337
3	.565	.636	.708	.785	.829	23	.216	.247	.275	.307	.330
4	.493	.565	.624	.689	.734	24	.212	.242	.269	.301	.323
5	.447	.509	.563	.627	.669	25	.208	.238	.264	.295	.317
6	.410	.468	.519	.577	.617	26	.204	.233	.259	.290	.311
7	.381	.436	.483	.538	.576	27	.200	.229	.254	.284	.305
8	.358	.410	.454	.507	.542	28	.197	.225	.250	.279	.300
9	.339	.387	.430	.480	.513	29	.193	.221	.246	.275	.295
10	.323	.369	.409	.457	.489	30	.190	.218	.242	.270	.290
11	.308	.352	.391	.437	.468	31	.187	.214	.238	.266	.285
12	.296	.338	.375	.419	.449	32	.184	.211	.234	.262	.281
13	.285	.325	.361	.404	.432	33	.182	.208	.231	.258	.277
14	.275	.314	.349	.390	.418	34	.179	.205	.227	.254	.273
15	.266	.304	.338	.377	.404	35	.177	.202	.224	.251	.269
16	.258	.295	.327	.366	.392	36	.174	.199	.221	.247	.265
17	.250	.286	.318	.355	.381	37	.172	.196	.218	.244	.262
18	.244	.279	.309	.346	.371	38	.170	.194	.215	.241	.258
19	.237	.271	.301	.337	.361	39	.168	.191	.213	.238	.255
20	.232	.265	.294	.329	.352	40	.165	.189	.210	.235	.252
	.100	.050	.025	.010	.005		.100	.050	.025	.010	.005

Right-Tail Probability for One-Sided Test

[a]Adapted from L. H. Miller, "Tables of Percentage Points of Kolmogorov Statistic," *Journal of the American Statistical Association*, **51** (1956), 111–121, with the permission of the American Statistical Association.

TABLE A10. Right Tail Probabilities $P(T_X \geq t_x)$ for Mann–Whitney–Wilcoxon Test Statistic[a]

Right tail probabilities are given for $t_x \geq m(m + n + 1)/2$ for $m \leq n$. Left tail probabilities are obtained from the relation $P(T_X \leq t_x) = P(T_X \geq m(m + n + 1) - t_x)$.

n	t_x	$P(T_X \geq t_x)$	n	t_x	$P(T_X \geq t_x)$	n	t_x	$P(T_X \geq t_x)$
	$m = 1$			**$m = 2$**			**$m = 2$**	
1	2	.500		7	.400		15	.291
2	2	.667		8	.200		16	.218
	3	.333		9	.100		17	.164
3	3	.500	4	7	.600		18	.109
	4	.250		8	.400		19	.073
4	3	.600		9	.267		20	.036
	4	.400		10	.133		21	.018
	5	.200		11	.067	10	13	.545
5	4	.500	5	8	.571		14	.455
	5	.333		9	.429		15	.379
	6	.167		10	.286		16	.303
6	4	.571		11	.190		17	.242
	5	.429		12	.095		18	.182
	6	.286		13	.048		19	.136
	7	.143	6	9	.571		20	.091
7	5	.500		10	.429		21	.061
	6	.375		11	.321		22	.030
	7	.250		12	.214		23	.015
	8	.125		13	.143			
8	5	.556		14	.071		**$m = 3$**	
	6	.444		15	.036	3	11	.500
	7	.333	7	10	.556		12	.350
	8	.222		11	.444		13	.200
	9	.111		12	.333		14	.100
9	6	.500		13	.250		15	.050
	7	.400		14	.167	4	12	.571
	8	.300		15	.111		13	.429
	9	.200		16	.056		14	.314
	10	.100		17	.028		15	.200
10	6	.545	8	11	.556		16	.114
	7	.455		12	.444		17	.057
	8	.364		13	.356		18	.029
	9	.273		14	.267	5	14	.500
	10	.182		15	.200		15	.393
	11	.091		16	.133		16	.286
				17	.089		17	.196
	$m = 2$			18	.044		18	.125
2	5	.667		19	.022		19	.071
	6	.333	9	12	.545		20	.036
	7	.167		13	.455		21	.018
3	6	.600		14	.364	6	15	.548

n	t_x	$P(T_X \geq t_x)$	n	t_x	$P(T_X \geq t_x)$	n	t_x	$P(T_X \geq t_x)$
	$m = 3$			$m = 3$			$m = 4$	
	16	.452		31	.018		25	.305
	17	.357		32	.009		26	.238
	18	.274		33	.005		27	.176
	19	.190	10	21	.531		28	.129
	20	.131		22	.469		29	.086
	21	.083		23	.406		30	.057
	22	.048		24	.346		31	.033
	23	.024		25	.287		32	.019
	24	.012		26	.234		33	.010
7	17	.500		27	.185		34	.005
	18	.417		28	.143	7	24	.536
	19	.333		29	.108		25	.464
	20	.258		30	.080		26	.394
	21	.192		31	.056		27	.324
	22	.133		32	.038		28	.264
	23	.092		33	.024		29	.206
	24	.058		34	.014		30	.158
	25	.033		35	.007		31	.115
	26	.017		36	.003		32	.082
	27	.008					33	.055
8	18	.539		$m = 4$			34	.036
	19	.461	4	18	.557		35	.021
	20	.388		19	.443		36	.012
	21	.315		20	.343		37	.006
	22	.248		21	.243		38	.003
	23	.188		22	.171	8	26	.533
	24	.139		23	.100		27	.467
	25	.097		24	.057		28	.404
	26	.067		25	.029		29	.341
	27	.042		26	.014		30	.285
	28	.024	5	20	.548		31	.230
	29	.012		21	.452		32	.184
	30	.006		22	.365		33	.141
9	20	.500		23	.278		34	.107
	21	.432		24	.206		35	.077
	22	.364		25	.143		36	.055
	23	.300		26	.095		37	.036
	24	.241		27	.056		38	.024
	25	.186		28	.032		39	.014
	26	.141		29	.016		40	.008
	27	.105		30	.008		41	.004
	28	.073	6	22	.543		42	.002
	29	.050		23	.457	9	28	.530
	30	.032		24	.381		29	.470

n	t_x	$P(T_X \geq t_x)$		n	t_x	$P(T_X \geq t_x)$		n	t_x	$P(T_X \geq t_x)$
		$m = 4$				$m = 5$				$m = 5$
	30	.413			32	.210			36	.472
	31	.355			33	.155			37	.416
	32	.302			34	.111			38	.362
	33	.252			35	.075			39	.311
	34	.207			36	.048			40	.262
	35	.165			37	.028			41	.218
	36	.130			38	.016			42	.177
	37	.099			39	.008			43	.142
	38	.074			40	.004			44	.111
	39	.053		6	30	.535			45	.085
	40	.038			31	.465			46	.064
	41	.025			32	.396			47	.047
	42	.017			33	.331			48	.033
	43	.010			34	.268			49	.023
	44	.006			35	.214			50	.015
	45	.003			36	.165			51	.009
	46	.001			37	.123			52	.005
10	30	.527			38	.089			53	.003
	31	.473			39	.063			54	.002
	32	.420			40	.041			55	.001
	33	.367			41	.026		9	38	.500
	34	.318			42	.015			39	.449
	35	.270			43	.009			40	.399
	36	.227			44	.004			41	.350
	37	.187			45	.002			42	.303
	38	.152		7	33	.500			43	.259
	39	.120			34	.438			44	.219
	40	.094			35	.378			45	.182
	41	.071			36	.319			46	.149
	42	.053			37	.265			47	.120
	43	.038			38	.216			48	.095
	44	.027			39	.172			49	.073
	45	.018			40	.134			50	.056
	46	.012			41	.101			51	.041
	47	.007			42	.074			52	.030
	48	.004			43	.053			53	.021
	49	.002			44	.037			54	.014
	50	.001			45	.024			55	.009
					46	.015			56	.006
		$m = 5$			47	.009			57	.003
5	28	.500			48	.005			58	.002
	29	.421			49	.003			59	.001
	30	.345			50	.001			60	.000
	31	.274		8	35	.528		10	40	.523

899

n	t_x	$P(T_X \geq t_x)$	n	t_x	$P(T_X \geq t_x)$	n	t_x	$P(T_X \geq t_x)$
		m = 5			**m = 6**			**m = 6**
	41	.477		57	.001		67	.001
	42	.430	7	42	.527		68	.001
	43	.384		43	.473		69	.000
	44	.339		44	.418	9	48	.523
	45	.297		45	.365		49	.477
	46	.257		46	.314		50	.432
	47	.220		47	.267		51	.388
	48	.185		48	.223		52	.344
	49	.155		49	.183		53	.303
	50	.127		50	.147		54	.264
	51	.103		51	.117		55	.228
	52	.082		52	.090		56	.194
	53	.065		53	.069		57	.164
	54	.050		54	.051		58	.136
	55	.038		55	.037		59	.112
	56	.028		56	.026		60	.091
	57	.020		57	.017		61	.072
	58	.014		58	.011		62	.057
	59	.010		59	.007		63	.044
	60	.006		60	.004		64	.033
	61	.004		61	.002		65	.025
	62	.002		62	.001		66	.018
	63	.001		63	.001		67	.013
	64	.001	8	45	.525		68	.009
	65	.000		46	.475		69	.006
				47	.426		70	.004
		m = 6		48	.377		71	.002
6	39	.531		49	.331		72	.001
	40	.469		50	.286		73	.001
	41	.409		51	.245		74	.000
	42	.350		52	.207		75	.000
	43	.294		53	.172	10	51	.521
	44	.242		54	.141		52	.479
	45	.197		55	.114		53	.437
	46	.155		56	.091		54	.396
	47	.120		57	.071		55	.356
	48	.090		58	.054		56	.318
	49	.066		59	.041		57	.281
	50	.047		60	.030		58	.246
	51	.032		61	.021		59	.214
	52	.021		62	.015		60	.184
	53	.013		63	.010		61	.157
	54	.008		64	.006		62	.132
	55	.004		65	.004		63	.110
	56	.002		66	.002		64	.090

n	t_x	$P(T_X \geq t_x)$	n	t_x	$P(T_X \geq t_x)$	n	t_x	$P(T_X \geq t_x)$
		$m = 6$			$m = 7$			$m = 7$
	65	.074		57	.478		77	.036
	66	.059		58	.433		78	.027
	67	.047		59	.389		79	.021
	68	.036		60	.347		80	.016
	69	.028		61	.306		81	.011
	70	.021		62	.268		82	.008
	71	.016		63	.232		83	.006
	72	.011		64	.198		84	.004
	73	.008		65	.168		85	.003
	74	.005		66	.140		86	.002
	75	.004		67	.116		87	.001
	76	.002		68	.095		88	.001
	77	.001		69	.076		89	.000
	78	.001		70	.060		90	.000
	79	.000		71	.047		91	.000
	80	.000		72	.036	10	63	.519
	81	.000		73	.027		64	.481
				74	.020		65	.443
		$m = 7$		75	.014		66	.406
7	53	.500		76	.010		67	.370
	54	.451		77	.007		68	.335
	55	.402		78	.005		69	.300
	56	.355		79	.003		70	.268
	57	.310		80	.002		71	.237
	58	.267		81	.001		72	.209
	59	.228		82	.001		73	.182
	60	.191		83	.000		74	.157
	61	.159		84	.000		75	.135
	62	.130	9	60	.500		76	.115
	63	.104		61	.459		77	.097
	64	.082		62	.419		78	.081
	65	.064		63	.379		79	.067
	66	.049		64	.340		80	.054
	67	.036		65	.303		81	.044
	68	.027		66	.268		82	.035
	69	.019		67	.235		83	.028
	70	.013		68	.204		84	.022
	71	.009		69	.176		85	.017
	72	.006		70	.150		86	.012
	73	.003		71	.126		87	.009
	74	.002		72	.105		88	.007
	75	.001		73	.087		89	.005
	76	.001		74	.071		90	.003
	77	.000		75	.057		91	.002
8	56	.522		76	.045		92	.002

n	t_x	$P(T_X \geq t_x)$	n	t_x	$P(T_X \geq t_x)$	n	t_x	$P(T_X \geq t_x)$
	m = 7			**m = 8**			**m = 8**	
	93	.001		76	.371		88	.158
	94	.001		77	.336		89	.137
	95	.000		78	.303		90	.118
	96	.000		79	.271		91	.102
	97	.000		80	.240		92	.086
	98	.000		81	.212		93	.073
				82	.185		94	.061
	m = 8			83	.161		95	.051
8	68	.520		84	.138		96	.042
	69	.480		85	.118		97	.034
	70	.439		86	.100		98	.027
	71	.399		87	.084		99	.022
	72	.360		88	.069		100	.017
	73	.323		89	.057		101	.013
	74	.287		90	.046		102	.010
	75	.253		91	.037		103	.008
	76	.221		92	.030		104	.006
	77	.191		93	.023		105	.004
	78	.164		94	.018		106	.003
	79	.139		95	.014		107	.002
	80	.117		96	.010		108	.002
	81	.097		97	.008		109	.001
	82	.080		98	.006		110	.001
	83	.065		99	.004		111	.000
	84	.052		100	.003		112	.000
	85	.041		101	.002		113	.000
	86	.032		102	.001		114	.000
	87	.025		103	.001		115	.000
	88	.019		104	.000		116	.000
	89	.014		105	.000				
	90	.010		106	.000		**m = 9**	
	91	.007		107	.000	9	86	.500
	92	.005		108	.000		87	.466
	93	.003	10	76	.517		88	.432
	94	.002		77	.483		89	.398
	95	.001		78	.448		90	.365
	96	.001		79	.414		91	.333
	97	.001		80	.381		92	.302
	98	.000		81	.348		93	.273
	99	.000		82	.317		94	.245
	100	.000		83	.286		95	.218
9	72	.519		84	.257		96	.193
	73	.481		85	.230		97	.170
	74	.444		86	.204		98	.149
	75	.407		87	.180		99	.129

n	t_x	$P(T_X \geq t_x)$	n	t_x	$P(T_X \geq t_x)$	n	t_x	$P(T_X \geq t_x)$
	m = 9			**m = 9**			**m = 10**	
	100	.111		105	.121		114	.264
	101	.095		106	.106		115	.241
	102	.081		107	.091		116	.218
	103	.068		108	.078		117	.197
	104	.057		109	.067		118	.176
	105	.047		110	.056		119	.157
	106	.039		111	.047		120	.140
	107	.031		112	.039		121	.124
	108	.025		113	.033		122	.109
	109	.020		114	.027		123	.095
	110	.016		115	.022		124	.083
	111	.012		116	.017		125	.072
	112	.009		117	.014		126	.062
	113	.007		118	.011		127	.053
	114	.005		119	.009		128	.045
	115	.004		120	.007		129	.038
	116	.003		121	.005		130	.032
	117	.002		122	.004		131	.026
	118	.001		123	.003		132	.022
	119	.001		124	.002		133	.018
	120	.001		125	.001		134	.014
	121	.000		126	.001		135	.012
	122	.000		127	.001		136	.009
	123	.000		128	.000		137	.007
	124	.000		129	.000		138	.006
	125	.000		130	.000		139	.004
	126	.000		131	.000		140	.003
10	90	.516		132	.000		141	.003
	91	.484		133	.000		142	.002
	92	.452		134	.000		143	.001
	93	.421		135	.000		144	.001
	94	.390						145	.001
	95	.360		**m = 10**			146	.001
	96	.330	10	105	.515		147	.000
	97	.302		106	.485		148	.000
	98	.274		107	.456		149	.000
	99	.248		108	.427		150	.000
	100	.223		109	.398		151	.000
	101	.200		110	.370		152	.000
	102	.178		111	.342		153	.000
	103	.158		112	.315		154	.000
	104	.139		113	.289		155	.000

TABLE A11. **Cumulative Right Tail Probabilities For Wilcoxon Signed Rank Statistic**[a]

Right tail probabilities $P(W \geq W_0)$ are given for $w_0 \geq n(n + 1)/4$. Here W is interpreted as either W_+ or W_-. Left tail probabilities are obtained from the relation $P(W_- \leq n(n + 1)/2 - w_0)$.

n	w_0	$P(W \geq w_0)$	n	w_0	$P(W \geq w_0)$	n	w_0	$P(W \geq w_0)$
2	2	.500		26	.023		44	.004
	3	.250		27	.016		45	.002
3	3	.625		28	.008	10	28	.500
	4	.375	8	18	.527		29	.461
	5	.250		19	.473		30	.423
	6	.125		20	.422		31	.385
4	5	.562		21	.371		32	.348
	6	.438		22	.320		33	.312
	7	.312		23	.273		34	.278
	8	.188		24	.230		35	.246
	9	.125		25	.191		36	.216
	10	.062		26	.156		37	.188
5	8	.500		27	.125		38	.161
	9	.406		28	.098		39	.138
	10	.312		29	.074		40	.116
	11	.219		30	.055		41	.097
	12	.156		31	.039		42	.080
	13	.094		32	.027		43	.065
	14	.062		33	.020		44	.053
	15	.031		34	.012		45	.042
6	11	.500		35	.008		46	.032
	12	.422		36	.004		47	.024
	13	.344	9	23	.500		48	.019
	14	.281		24	.455		49	.014
	15	.219		25	.410		50	.010
	16	.156		26	.367		51	.007
	17	.109		27	.326		52	.005
	18	.078		28	.285		53	.003
	19	.047		29	.248		54	.002
	20	.031		30	.213		55	.001
	21	.016		31	.180	11	33	.517
7	14	.531		32	.150		34	.483
	15	.469		33	.125		35	.449
	16	.406		34	.102		36	.416
	17	.344		35	.082		37	.382
	18	.289		36	.064		38	.350
	19	.234		37	.049		39	.319
	20	.188		38	.037		40	.289
	21	.148		39	.027		41	.260
	22	.109		40	.020		42	.232
	23	.078		41	.014		43	.207
	24	.055		42	.010		44	.183
	25	.039		43	.006		45	.160

n	w_0	$P(W \geq w_0)$	n	w_0	$P(W \geq w_0)$	n	w_0	$P(W \geq w_0)$
	46	.139		64	.026		77	.013
	47	.120		65	.021		78	.011
	48	.103		66	.017		79	.009
	49	.087		67	.013		80	.007
	50	.074		68	.010		81	.005
	51	.062		69	.008		82	.004
	52	.051		70	.006		83	.003
	53	.042		71	.005		84	.002
	54	.034		72	.003		85	.002
	55	.027		73	.002		86	.001
	56	.021		74	.002		87	.001
	57	.016		75	.001		88	.001
	58	.012		76	.001		89	.000
	59	.009		77	.000		90	.000
	60	.007		78	.000		91	.000
	61	.005	13	46	.500	14	53	.500
	62	.003		47	.473		54	.476
	63	.002		48	.446		55	.452
	64	.001		49	.420		56	.428
	65	.001		50	.393		57	.404
	66	.000		51	.368		58	.380
12	39	.515		52	.342		59	.357
	40	.485		53	.318		60	.335
	41	.455		54	.294		61	.313
	42	.425		55	.271		62	.292
	43	.396		56	.249		63	.271
	44	.367		57	.227		64	.251
	45	.339		58	.207		65	.232
	46	.311		59	.188		66	.213
	47	.285		60	.170		67	.196
	48	.259		61	.153		68	.179
	49	.235		62	.137		69	.163
	50	.212		63	.122		70	.148
	51	.190		64	.108		71	.134
	52	.170		65	.095		72	.121
	53	.151		66	.084		73	.108
	54	.133		67	.073		74	.097
	55	.117		68	.064		75	.086
	56	.102		69	.055		76	.077
	57	.088		70	.047		77	.068
	58	.076		71	.040		78	.059
	59	.065		72	.034		79	.052
	60	.055		73	.029		80	.045
	61	.046		74	.024		81	.039
	62	.039		75	.020		82	.034
	63	.032		76	.016		83	.029

n	w_0	$P(W \geq w_0)$	n	w_0	$P(W \geq w_0)$	n	w_0	$P(W \geq w_0)$
	84	.025	66	.381		94	.028
	85	.021	67	.360		95	.024
	86	.018	68	.339		96	.021
	87	.015	69	.319		97	.018
	88	.012	70	.300		98	.015
	89	.010	71	.281		99	.013
	90	.008	72	.262		100	.011
	91	.007	73	.244		101	.009
	92	.005	74	.227		102	.008
	93	.004	75	.211		103	.006
	94	.003	76	.195		104	.005
	95	.003	77	.180		105	.004
	96	.002	78	.165		106	.003
	97	.002	79	.151		107	.003
	98	.001	80	.138		108	.002
	99	.001	81	.126		109	.002
	100	.001	82	.115		110	.001
	101	.000	83	.104		111	.001
	102	.000	84	.094		112	.001
	103	.000	85	.084		113	.001
	104	.000	86	.076		114	.000
	105	.000	87	.068		115	.000
15	60	.511	88	.060		116	.000
	61	.489	89	.053		117	.000
	62	.467	90	.047		118	.000
	63	.445	91	.042		119	.000
	64	.423	92	.036		120	.000
	65	.402	93	.032			

[a] From Table II of H. L. Harter and D. B. Owen, Eds, *Selected Tables in Mathematical Statistics*, Vol. 1, Markham, Chicago, 1972, with the permission of the Institute of Mathematical Statistics.

TABLE A12. Left Tail Cumulative Probabilities, $P(R \le r)$, for Total Number of Runs R.[a]

(m, n)					r				
	2	3	4	5	6	7	8	9	10
(2, 3)	.200	.500	.900	1.000					
(2, 4)	.133	.400	.800	1.000					
(2, 5)	.095	.333	.714	1.000					
(2, 6)	.071	.286	.643	1.000					
(2, 7)	.056	.250	.583	1.000					
(2, 8)	.044	.222	.533	1.000					
(2, 9)	.036	.200	.491	1.000					
(2, 10)	.030	.182	.455	1.000					
(3, 3)	.100	.300	.700	.900	1.000				
(3, 4)	.057	.200	.543	.800	.971	1.000			
(3, 5)	.036	.143	.429	.714	.929	1.000			
(3, 6)	.024	.107	.345	.643	.881	1.000			
(3, 7)	.017	.083	.283	.583	.833	1.000			
(3, 8)	.012	.067	.236	.533	.788	1.000			
(3, 9)	.009	.055	.200	.491	.745	1.000			
(3, 10)	.007	.045	.171	.455	.706	1.000			
(4, 4)	.029	.114	.371	.629	.886	.971	1.000		
(4, 5)	.016	.071	.262	.500	.786	.929	.992	1.000	
(4, 6)	.010	.048	.190	.405	.690	.881	.976	1.000	
(4, 7)	.006	.033	.142	.333	.606	.833	.954	1.000	
(4, 8)	.004	.024	.109	.279	.533	.788	.929	1.000	
(4, 9)	.003	.018	.085	.236	.471	.745	.902	1.000	
(4, 10)	.002	.014	.068	.203	.419	.706	.874	1.000	
(5, 5)	.008	.040	.167	.357	.643	.833	.960	.992	1.000
(5, 6)	.004	.024	.110	.262	.522	.738	.911	.976	.998
(5, 7)	.003	.015	.076	.197	.424	.652	.854	.955	.992
(5, 8)	.002	.010	.054	.152	.347	.576	.793	.929	.984
(5, 9)	.001	.007	.039	.119	.287	.510	.734	.902	.972
(5, 10)	.001	.005	.029	.095	.239	.455	.678	.874	.958
(6, 6)	.002	.013	.067	.175	.392	.608	.825	.933	.987
(6, 7)	.001	.008	.043	.121	.296	.500	.733	.879	.966
(6, 8)	.001	.005	.028	.086	.226	.413	.646	.821	.937
(6, 9)	.000	.003	.019	.063	.175	.343	.566	.762	.902
(6, 10)	.000	.002	.013	.047	.137	.288	.497	.706	.864
(7, 7)	.001	.004	.025	.078	.209	.383	.617	.791	.922
(7, 8)	.000	.002	.015	.051	.149	.296	.514	.704	.867
(7, 9)	.000	.001	.010	.035	.108	.231	.427	.622	.806
(7, 10)	.000	.001	.006	.024	.080	.182	.355	.549	.743
(8, 8)	.000	.001	.009	.032	.100	.214	.405	.595	.786
(8, 9)	.000	.001	.005	.020	.069	.157	.319	.500	.702
(8, 10)	.000	.000	.003	.013	.048	.117	.251	.419	.621
(9, 9)	.000	.000	.003	.012	.044	.109	.238	.399	.601
(9, 10)	.000	.000	.002	.008	.029	.077	.179	.319	.510
(10, 10)	.000	.000	.001	.004	.019	.051	.128	.242	.41

(m, n)	11	12	13	14	15	16	17	18
(2, 3)								
(2, 4)								
(2, 5)								
(2, 6)								
(2, 7)								
(2, 8)								
(2, 9)								
(2, 10)								
(3, 3)								
(3, 4)								
(3, 5)								
(3, 6)								
(3, 7)								
(3, 8)								
(3, 9)								
(3, 10)								
(4, 4)								
(4, 5)								
(4, 6)								
(4, 7)								
(4, 8)								
(4, 9)								
(4, 10)								
(5, 5)								
(5, 6)	1.000							
(5, 7)	1.000							
(5, 8)	1.000							
(5, 9)	1.000							
(5, 10)	1.000							
(6, 6)	.998	1.000						
(6, 7)	.992	.999	1.000					
(6, 8)	.984	.998	1.000					
(6, 9)	.972	.994	1.000					
(6, 10)	.958	.990	1.000					
(7, 7)	.975	.996	.999	1.000				
(7, 8)	.949	.988	.998	1.000				
(7, 9)	.916	.975	.994	.999	1.000			
(7, 10)	.879	.957	.990	.998	1.000			
(8, 8)	.900	.968	.991	.999	1.000			
(8, 9)	.843	.939	.980	.996	.999	1.000		
(8, 10)	.782	.903	.964	.990	.998	1.000		
(9, 9)	.762	.891	.956	.988	.997	1.000		
(9, 10)	.681	.834	.923	.974	.992	.999	1.000	
(10, 10)	.586	.758	.872	.949	.981	.996	.999	1.000

[a]Reprinted from Frieda S. Swed and C. Eisenhart, "Tables for Testing Randomness of Grouping in a Sequence of Alternatives," *Annals of Mathematical Statistics*, **14** (1943), 66–87, with the permission of the Institute of Mathematical Statistics.

TABLE A13. Cumulative Probabilities $P(R \geq r)$ for Spearman's Rank Correlation Statistic $R_S = R$.[a]

Right tail probabilities are given for $r \geq 0$. Left tail probabilities
may be obtained from symmetry: For $r > 0$, $P(R \geq r) = P(-1 \leq R \leq -r)$.

r	$P(R \geq r)$	r	$P(R \geq r)$	r	$P(R \geq r)$	r	$P(R \geq r)$
n = 4		*n = 7*		*n = 8*		*n = 9*	
.000	.542	.071	.453	.310	.231	.167	.339
.200	.458	.107	.420	.333	.214	.183	.322
.400	.375	.143	.391	.357	.195	.200	.307
.600	.208	.179	.357	.381	.180	.217	.290
.800	.167	.214	.331	.405	.163	.233	.276
1.000	.042	.250	.297	.429	.150	.250	.260
		.286	.278	.452	.134	.267	.247
n = 5		.321	.249	.476	.122	.283	.231
.000	.525	.357	.222	.500	.108	.300	.218
.100	.475	.393	.198	.524	.098	.317	.205
.200	.392	.429	.177	.548	.085	.333	.193
.300	.342	.464	.151	.571	.076	.350	.179
.400	.258	.500	.133	.595	.066	.367	.168
.500	.225	.536	.118	.619	.057	.383	.156
.600	.175	.571	.100	.643	.048	.400	.146
.700	.117	.607	.083	.667	.042	.417	.135
.800	.067	.643	.069	.690	.035	.433	.125
.900	.042	.679	.055	.714	.029	.450	.115
1.000	.008	.714	.044	.738	.023	.467	.106
		.750	.033	.762	.018	.483	.097
n = 6		.786	.024	.786	.014	.500	.089
.029	.500	.821	.017	.810	.011	.517	.081
.086	.460	.857	.012	.833	.008	.533	.074
.143	.401	.893	.006	.857	.005	.550	.066
.200	.357	.929	.003	.881	.004	.567	.060
.257	.329	.964	.001	.905	.002	.583	.054
.314	.282	1.000	.000	.929	.001	.600	.048
.371	.249			.952	.001	.617	.043
.429	.210			.976	.000	.633	.038
.486	.178	*n = 8*		1.000	.000	.650	.033
.543	.149	.000	.512			.667	.029
.600	.121	.024	.488			.683	.025
.657	.088	.048	.467	*n = 9*		.700	.022
.714	.068	.071	.441	.000	.509	.717	.018
.771	.051	.095	.420	.017	.491	.733	.016
.829	.029	.119	.397	.033	.474	.750	.013
.886	.017	.143	.376	.050	.456	.767	.011
.943	.008	.167	.352	.067	.440	.783	.009
1.000	.001	.190	.332	.083	.422	.800	.007
		.214	.310	.100	.405	.817	.005
n = 7		.238	.291	.117	.388	.833	.004
.000	.518	.262	.268	.133	.372	.850	.003
.036	.482	.286	.250	.150	.354		

909

r	$P(R \geq r)$	r	$P(R \geq r)$	r	$P(R \geq r)$	r	$P(R \geq r)$
	$n = 9$		$n = 10$		$n = 10$		$n = 11$
.867	.002	.406	.124	.939	.000	.327	.163
.883	.002	.418	.116	.952	.000	.336	.157
.900	.001	.430	.109	.964	.000	.345	.150
.917	.001	.442	.102	.976	.000	.355	.143
.933	.000	.455	.096	.988	.000	.364	.137
.950	.000	.467	.089	1.000	.000	.373	.130
.967	.000	.479	.083			.382	.124
.983	.000	.491	.077		$n = 11$	.391	.118
1.000	.000	.503	.072	.000	.505	.400	.112
		.515	.067	.009	.495	.409	.107
	$n = 10$	.527	.062	.018	.484	.418	.102
.006	.500	.539	.057	.027	.473	.427	.096
.018	.486	.552	.052	.036	.462	.436	.091
.030	.473	.564	.048	.045	.452	.445	.087
.042	.459	.576	.044	.055	.441	.455	.082
.055	.446	.588	.040	.064	.430	.464	.077
.067	.433	.600	.037	.073	.419	.473	.073
.079	.419	.612	.033	.082	.409	.482	.069
.091	.406	.624	.030	.091	.398	.491	.065
.103	.393	.636	.027	.100	.388	.500	.061
.115	.379	.648	.024	.109	.377	.509	.057
.127	.367	.661	.022	.118	.367	.518	.054
.139	.354	.673	.019	.127	.357	.527	.050
.152	.341	.685	.017	.136	.347	.536	.047
.164	.328	.697	.015	.145	.337	.545	.044
.176	.316	.709	.013	.155	.327	.555	.041
.188	.304	.721	.012	.164	.317	.564	.038
.200	.292	.733	.010	.173	.307	.573	.035
.212	.280	.745	.009	.182	.298	.582	.033
.224	.268	.758	.008	.191	.288	.591	.030
.236	.257	.770	.006	.200	.279	.600	.028
.248	.246	.782	.005	.209	.270	.609	.026
.261	.235	.794	.004	.218	.260	.618	.024
.273	.224	.806	.004	.227	.252	.627	.022
.285	.214	.818	.003	.236	.243	.636	.020
.297	.203	.830	.002	.245	.234	.645	.018
.309	.193	.842	.002	.255	.226	.655	.017
.321	.184	.855	.001	.264	.217	.664	.015
.333	.174	.867	.001	.273	.209	.673	.014
.345	.165	.879	.001	.282	.201	.682	.013
.358	.156	.891	.001	.291	.193	.691	.011
.370	.148	.903	.000	.300	.186	.700	.010
.382	.139	.915	.000	.309	.178	.709	.009
.394	.132	.927	.000	.318	.171	.718	.008

r	$P(R \geq r)$	r	$P(R \geq r)$	r	$P(R \geq r)$	r	$P(R \geq r)$
	$n = 11$		$n = 11$		$n = 11$		$n = 11$
.727	.007	.800	.002	.873	.000	.945	.000
.736	.006	.809	.002	.882	.000	.955	.000
.745	.006	.818	.002	.891	.000	.964	.000
.755	.005	.827	.001	.900	.000	.973	.000
.764	.004	.836	.001	.909	.000	.982	.000
.773	.004	.845	.001	.918	.000	.991	.000
.782	.003	.855	.001	.927	.000	1.000	.000
.791	.003	.864	.001	.936	.000		

			Right-Tail Probability for One-Sided Test			
n	.100	.050	.025	.010	.005	.001
12	.406	.503	.587	.678	.734	.825
13	.385	.484	.560	.648	.703	.797
14	.367	.464	.538	.626	.679	.771
15	.354	.446	.521	.604	.657	.750
16	.341	.429	.503	.585	.635	.729
17	.329	.414	.488	.566	.618	.711
18	.317	.401	.474	.550	.600	.692
19	.309	.391	.460	.535	.584	.675
20	.299	.380	.447	.522	.570	.660
21	.292	.370	.436	.509	.556	.647
22	.284	.361	.425	.497	.544	.633
23	.278	.353	.416	.486	.532	.620
24	.275	.344	.407	.476	.521	.608
25	.265	.337	.398	.466	.511	.597
26	.260	.331	.390	.457	.501	.586
27	.255	.324	.383	.449	.492	.576
28	.250	.318	.376	.441	.483	.567
29	.245	.312	.369	.433	.475	.557
30	.241	.307	.363	.426	.467	.548
	.200	.100	.050	.020	.010	.002
			Tail Probability for Two-Sided Test			

[a]For $4 \leq n \leq 11$, adapted from Table 13.2 of Donald B. Owen, *Handbook of Statistical Tables*, © 1962, U.S. Department of Energy, published by Addison-Wesley Publishing Company, Inc., Reading, Massachusetts; reprinted with permission of the publisher.

For $12 \leq n \leq 30$, adapted from G. J. Glasser and R. F. Wintner, "Critical Values of the Rank Correlation Coefficient for Testing the Hypothesis of Independence," *Biometrika*, **48** (1961), 444–448, with permission of the Biometrika Trustees.

TABLE A14. Cumulative Probabilities $P(\tau \geq \tau_0)$ for Kendall's Tau Statistic[a]

Left tail probabilities may be obtained
from symmetry: For $\tau_0 > 0$, $P(\tau \leq -\tau_0) = P(\tau \geq \tau_0)$.

τ_0	$P(\tau \geq \tau_0)$	τ_0	$P(\tau \geq \tau_0)$	τ_0	$P(\tau \geq \tau_0)$
n = 2		**n = 8**		**n = 10**	
1.000	.500	.286	.199	.467	.036
n = 3		.357	.138	.511	.023
.333	.500	.429	.089	.556	.014
1.000	.167	.500	.054	.600	.008
n = 4		.571	.031	.644	.005
.000	.625	.643	.016	.689	.002
.333	.375	.714	.007	.733	.001
.667	.167	.786	.003	.778	.000
1.000	.042	.857	.001	.822	.000
n = 5		.929	.000	.867	.000
.000	.592	1.000	.000	.911	.000
.200	.408	**n = 9**		.956	.000
.400	.242	.000	.540	1.000	.000
.600	.117	.056	.460	**n = 11**	
.800	.042	.111	.381	.018	.500
1.000	.008	.167	.306	.054	.440
n = 6		.222	.238	.091	.381
.067	.500	.278	.179	.127	.324
.200	.360	.333	.130	.164	.271
.333	.235	.389	.090	.200	.223
.467	.136	.444	.060	.236	.179
.600	.068	.500	.038	.273	.141
.733	.028	.556	.022	.309	.109
.867	.008	.611	.012	.345	.082
1.000	.001	.667	.006	.382	.060
n = 7		.722	.003	.418	.043
.048	.500	.778	.001	.454	.030
.143	.386	.833	.000	.491	.020
.238	.281	.889	.000	.527	.013
.333	.191	.944	.000	.564	.008
.429	.119	1.000	.000	.600	.005
.524	.068	**n = 10**		.636	.003
.619	.035	.022	.500	.673	.002
.714	.015	.067	.431	.709	.001
.810	.005	.111	.364	.745	.000
.905	.001	.156	.300	.782	.000
1.000	.000	.200	.242	.818	.000
n = 8		.244	.190	.854	.000
.000	.548	.289	.146	.891	.000
.071	.452	.333	.108	.927	.000
.143	.360	.378	.078	.964	.000
.214	.274	.422	.054	1.000	.000

τ_0	$P(\tau \geq \tau_0)$	τ_0	$P(\tau \geq \tau_0)$	τ_0	$P(\tau \geq \tau_0)$
	n = 12		**n = 12**		**n = 12**
.000	.527	.364	.058	.727	.000
.030	.473	.394	.043	.758	.000
.061	.420	.424	.031	.788	.000
.091	.369	.454	.022	.818	.000
.121	.319	.485	.016	.848	.000
.152	.273	.515	.010	.879	.000
.182	.230	.545	.007	.909	.000
.212	.190	.578	.004	.939	.000
.242	.155	.606	.003	.970	.000
.273	.125	.636	.002	1.000	.000
.303	.098	.667	.001		
.333	.076	.696	.000		

α n	Critical Points for One-Sided Test				
	.10	.05	.025	.01	.005
13	.308	.359	.436	.513	.564
14	.275	.363	.407	.473	.516
15	.276	.333	.390	.467	.505
16	.250	.317	.383	.433	.483
17	.250	.309	.368	.426	.471
18	.242	.294	.346	.412	.451
19	.228	.287	.333	.392	.439
20	.221	.274	.326	.379	.421
21	.210	.267	.314	.371	.410
22	.195	.253	.295	.344	.378
23	.202	.257	.296	.352	.391
24	.196	.246	.290	.341	.377
25	.193	.240	.287	.333	.367
26	.188	.237	.280	.329	.360
27	.179	.231	.271	.322	.356
28	.180	.228	.265	.312	.344
29	.172	.222	.261	.310	.340
30	.172	.218	.255	.301	.333
n α	.20	.10	.05	.02	.01
			Two-Sided Test		

[a]For $2 \leq n \leq 12$, adapted from Table 13.1 of Donald B. Owen, *Handbook of Statistical Tables*, © 1962, U.S. Department of Energy, published by Addison-Wesley Publishing Company, Inc., Reading, Massachusetts; reprinted with permission of the publisher.

For $14 \leq n \leq 30$, adapted from L. Kaarsemaker and A. van Wijngaarden, "Tables for Use in Rank Correlation," *Statistica Neerlandica*, 7 (1953), 41–44, with permission.

Answers To Odd-Numbered Problems

SECTION 2.2

1. (i) $\{(i, j, k) : i, j, k \in \{H, T\}\}$; (iii) Each point of Ω is a subset of 13 cards from $\{AC, 2C, \ldots, KC, \ldots, AS, 2S, \ldots, KS\}$; (v) $\{0, 1, \ldots, n\}$; (vii) $\{2, 3, 4, 5, 6\}$; (ix) $\{(x, y, z) : x \geq 1 \text{ integral}, y > 0, z > 0\}$ (xi) $\{(x, y) : 0 \leq y \leq x \leq 1\}$; (xiii) $\{(i, j, k, l, m) : i, j, k, l, m \in \{L, R\}\}$.

3. (i) (a) $\bar{A} \cap \bar{C}$; (c) $(C - A) \cup (A - C)$.

5. $\{(X, Y, Z), (X, Z, Y), (Y, X, Z), (Y, Z, X), (Z, X, Y), (Z, Y, X)\}$; $\{(X, Y, Z), (X, Z, Y), (Z, X, Y)\}$; $\{(X, Y, Z), (X, Z, Y)\}$; $\{(Y, X, Z), (Z, X, Y)\}$.

7. (i) $\{(x, y) : 0 \leq x \leq y \leq 1\}$; (iii) $\{(x, y) : 0 \leq y \leq 3x, 0 \leq x \leq 1, 0 \leq y \leq 1\}$.

9. All subsets of three cities; (i) $\{\{D, P, L\}, \{D, P, H\}, \{P, L, H\}, \{D, L, H\}\}$, etc.

SECTION 2.3

1. In parts (ii), (v) and (vii).   3. 24/30; 28/30; 28/30.

5. 2/10; 7/10; 4/10.   7. 6/8.   9. 1/9.   11. 4/9; 5/9.

SECTION 2.4

1. 1/5; 4/5; 12/25; 1/10; 4/125.   3. .94; .18; .05; .95; .93, .82.

5. .0106.   7. 0.   9. $\geq .45$.   11. .75; .2; .05.

13. .05; .04; .15; .15; .16; .84; .40; .38; .60.

15. $.75 \leq p \leq .8$; $.1 \leq p \leq .55$; $.2 \leq p \leq .25$.

SECTION 2.5

1. 1680.   3. 10^9 (or $9(10)^8$ if number cannot begin with zero).

5. 4!; 4^4.   7. $11!/(2!)^3$.   9. 45321.   11. $\binom{8}{3}$; 7; 6.   13. $\binom{10}{2}$.

15. $\binom{13}{7}\binom{13}{3}\binom{13}{2}\binom{13}{1}$.   17. $20!/5!\,12!\,3!$.   19. $3/10^3(26)^3$.

21. Yes, $P(\text{all men}) = 6/143$.   23. 1/22.   25. 1/3; 2/5.

27. $(1 - 2/(N - 1))^{2^n - 2}$.   29. $\binom{13}{2} / \binom{15}{2}$; $1 - \binom{13}{2} / \binom{15}{2}$.

31. Yes, probability $\doteq .0151$.   33. $1 - 999990\,{}^P3 / 1000000\,{}^P3 = .00003$.

SECTION 2.6

1. 26/48; 32/52; 20/52; 22/48.   3. 100/126.   5. 8/13; .02.

7. .08.   9. 36/91.   11. 1/4; 1/3.   13. .08.

15. 173/480; 108/173, 15/173.   17. .0872.

SECTION 2.7

1. (vi).   3. Yes.   5. 0 or 5/7; .5.   9. .5 or .2; .2 or .5.

11. $1 - (1 - e^{-t_0/\lambda})^5$.   13. .001; .1; .05; .352; .216.

15. $1 - (.95)^{20} - (.95)^{19}$.

19. (i) 16807/279936; (ii) .463, .386, .1285, .0214, .0018, .0001 for $j = 1, 2, \ldots, 6$, etc.

21. $p_1 p_2 (2 - p_1)(2 - p_2)$.

SECTION 2.8

5. $10^P 5/10^5$.   7. 9; 9.

SECTION 2.9

1. Yes in each case.   5. $(n - r + 1)!/n!$.   9. $(1 - p)/(2 - p)$; $1/(2 - p)$.

11. 1/6; 1/3; 1/3.   13. $\pi/(\pi + p(1 - \pi))$.   15. 1/18; 1/17.

17. $.90 \le P(\cup A_i) \le .95$.   19. 13/16; 4/13; 12/13.   21. .00000064; .00043.

23. $1 - \pi/4$; $\pi/4$; $\pi/4$; 0.

SECTION 3.2

3. 1/2, 1/2; $1/2^9$.   7. 16/210, 8/210, 48/210, 48/210.

9. (ii) 16/27, 49/81, 5/9.

SECTION 3.3

1. No; yes; yes; no.

3. $P(X = 0) = .18$, $P(X = 1) = .54$, $P(X = 2) = .27$, $P(X = 3) = .02$; .82.

5. $c = 2/15$; 1/49; $2/N(N + 1)$; $P(X < 3) = 13/15$, 9/49, $6/N(N + 1)$; smallest $x = 1, 4, [\{1 + \sqrt{1 + 2N(N + 1)}\}/2]$; largest $x = 0, 3, [\{-1 + \sqrt{1 + 2N(N + 1)}\}/2]$.

7. .8647, .4059, .2706.

9. $1 - \exp\{-(x - \theta)/\lambda\}$ for $x \ge \theta$; $1/2 + (\arctan x)/\pi$.

11. Yes; 1/2 for $0 < x < 1$, 1/4 for $2 < x < 4$; $1/2\theta$, $|x| \le \theta$; xe^{-x}, $x > 0$; $(x - 1)/4$ for $1 \le x < 3$, and $P(X = 3) = 1/2$; $2xe^{-x^2}$, $x > 0$.

13. .3679; .141; .6988.   15. .2.   17. $20e^{-5}, 3/16$.

19. $[e^{-2/\lambda} + (1 - p)^2]/3; 0, e^{-2/\lambda}[e^{-2/\lambda} + (1 - p)^2]^{-1}$,
$(1 - p)^2[e^{-2/\lambda} + (1 - p)^2]^{-1}$.

SECTION 3.4

1. (ii) $11/36, 13/36$.   3. (ii) $5/16, 3/8$.   5. (ii) $23/60, 7/10, 23/60$.

9. $1/(2\pi\sqrt{x^2 + y^2})$, for $|x| \le 1, |y| \le 1$.   11. Yes, $1 - e^{-7}$.

13. (iii) $3/4, 9/16$.   15. $1/24; 15/16$.   17. $f(x, y) = 1, 0 \le x \le y \le \sqrt{2}$;
$1/4, (\sqrt{2} - 1)^2/2$.   19. $1/6, 2/3$.

SECTION 3.5

1. (ii) $X: 64/216, 48/108, 12/54, 1/27; \ Y: 27/216, 27/72, 9/24, 1/8$;
(iii) $P\{X = x | Y = y\} = \binom{3 - x}{y}(3/4)^y(1/4)^{3 - x - y}, \ P\{Y = y | X = x\}$
$= \binom{3 - y}{x}(2/3)^x(1/3)^{3 - y - x}$; (iv) $24/47$.

3. $X: 9/54, 8/48, 7/42, 6/36, 5/30, 4/24; \ Y: 1/54, 17/432, 191/3024, 257/3024$,
$1879/15120, 2509/15120, 2509/15120, 2509/15120, 2509/15120$.

5. (i) $\binom{3}{x}\binom{4}{y}\binom{2}{3 - x - y}/\binom{9}{3}$; (ii) $P(X = x) = \binom{3}{x}\binom{6}{3 - x}/\binom{9}{3}$,
$P(Y = y) = \binom{4}{y}\binom{5}{3 - y}/\binom{9}{3}$; (iv) $1/84, 4/84$; (v) $1/5, 1/10$.

7. (i) $3x/4 + 5/8; 3/8 + 5y/4$; (ii) $105/128, 81/128$;
(iii) $h(y|x) = 2(3x + 5y)/(6x + 5), g(x|y) = 2(3x + 5y)/(3 + 10y)$.

9. $f_2(y) = -\ln(1 - y), 0 < y < 1; \ g(x|y) = -1/[(1 - x)\ln(1 - y)], 0 \le x < y$;
$\ln(6/5)/\ln(8/5)$.

11. $P(N = n) = 1/2^{n+1}; h(t|n) = (2\lambda)^{n+1}t^n e^{-2\lambda t}/n!; P(N \ge 2) = 1/4$.

13. (iii) $e^{-x_n}x_3^2/2$, (iv) $2/x_3^2, 0 < x_1 < x_2 < x_3$, (v) $(x_3 - a)^2/x_3^2, x_3 > a$.

15. f_1 and f_2; $g_\alpha(x_1|x_2) = f_1(x_1)\{1 + \alpha[2F_1(x_1) - 1][2F_2(x_2) - 1]\}$.

SECTION 3.6

1. No.   3. No; Yes.   7. No closed form.   9. Yes in (ii) and (vi).

11. $1 - a/2b$ if $a < b, b/2a$ if $a > b$.   13. $2/(r_0 + 2)$.

15. $\lambda/(\lambda + \mu)$.   17. $1/8; .43$.

SECTION 3.7

1. $3, \$450$.   3. $1.5, 1/2$.   5. $a(p + .05)$.   7. $2/3; 55/3$.

9. 39.   11. $\alpha^2 + 4\alpha\beta/3 + \beta^2/2; (\beta^2/3)[2\alpha^2/3 + 4\alpha\beta/5 + \beta^2/4]$.

13. $p = 5$.   15. $3; 1/2; 1/5$.   17. At least 2.39 days in advance.

19. $.1975; .5556$.   27. $1/(1-p); [\ln(1/(1-p))]^{1/2}; p\theta; \ln(1-p)/\ln(1-\theta) \le \zeta_p \le 1 + \ln(1-p)/\ln(1-\theta); 1 - p = e^{-\zeta_p/\beta}\{1 + \zeta_p/\beta\}, b\{1 - (1-p)^{1/3}\}$.

SECTION 3.8

3. Yes, No.   5. $.02$.   7. $29/13, 4.74$.   13. $(98.6, 151.4); .28$.

SECTION 3.9

1. All.   3. $1 - e^{-2.5}; 3.5e^{-2.5}; 5/7$.   5. $12, 3/5, 1/25; 11/16$.

11. $5; 5/(1-p)^{1/(\alpha-1)}; 10$.   15. $P\{X \ge 1|Y \le 3\} = 23/33, P\{X \le 2|Y = 1\} = 5/8; 13/36, 23/36$.   17. $P(X - Y > 0) = 1/4$.   19. $e^{-\lambda^2}$.

SECTION 4.3

1. $\max\{X_1, \ldots, X_n\}$ or $\overline{X}/2$; $\overline{X}$; $\max X_i - \min X_i$; Sample quantile; $2\overline{X} - 1$ or $\max X_i$; $\Sigma X_i^2/n, \Sigma X_i^2/n - (\overline{X})^2$; $\exp\{-t/\overline{X}\}; \overline{X}^2, 2\overline{X} - 1, \overline{X} - \overline{X}^2$.

3. 1.   5. $3X + 2X^2$.   7. $15 + (9/2)X$.   9. (i) $[\alpha X/2, (1 - \alpha/2)X]$; (iii) $[\ln(1 - \alpha/2)/\ln X, \ln(\alpha/2)/\ln X]$; (v) $[X + \ln \alpha/2, X - \ln \alpha/2]$.

11. (i) $\mathscr{E}X$ does not exist; (v) $[X - (1/\sqrt{\alpha}), X + (1/\sqrt{\alpha})]$.

13. $[(X_1 + X_2)/(1 - \sqrt{\alpha}), (X_1 + X_2)/\sqrt{\alpha}]$

SECTION 4.4

3. (i) $H_0: \lambda = 1.25, H_1: \lambda > 1.25$; (ii) $\Sigma_{i=1}^{8} X_i > 14$; (iii) $.007$, yes.

5. $1 - 8/3e; 19/3e^2$.   7. $.173, \beta(\theta) = 1 - (\theta + 1.5)e^{-1.5/\theta}/\theta$.

9. $.33; .4, .6$.

11.

	C_0	C_1	C_2	C_3	C_4	C_5
Size	1	.5981	.1962	.0355	.0033	.0001
Power	1	.6723	.2627	.0579	.0067	.0003

13. $\{X < \theta_0 \ln[1/(1 - \alpha/2)]$ or $X > \theta_0 \ln[2/\alpha]\}; \beta(\theta) = 1 - \exp\{(\theta_0/\theta)\ln[2/(2 - \alpha)]\} + \exp\{-(\theta_0/\theta)\ln(2/\alpha)\}; [X/\ln(2/\alpha), X/\ln(2/(2 - \alpha))]$.

15. $\{|X - 1| > -\ln\alpha\}$ 17. (i) .4375, .2344; (ii) $\{X \geq 4\sqrt{1 - \alpha}\}$;
(iii) $1 - 16(1 - \alpha)/\theta^2$ for $\theta > 4\sqrt{1 - \alpha}$; (iv) $X/\sqrt{1 - \alpha}$.

SECTION 4.5

1. (i)

	Frequency	$\hat{f}(x)$	$\hat{F}(x)$
12.55–15.55	5	.033	.1
15.55–18.55	8	.053	.26
18.55–21.55	11	.073	.48
21.55–24.55	1	.007	.50
24.55–27.55	3	.020	.56
27.55–30.55	6	.040	.68
30.55–33.55	4	.027	.76
33.55–36.55	4	.027	.84
36.55–39.55	2	.013	.88
39.55–42.55	3	.020	.94
42.55–45.55	3	.020	1.00

(ii) $\bar{x} = 17.51$, Median 23.05.

3.

i	$\hat{P}(X = i)$
1	.142
2	.150
3	.200
4	.150
5	.175
6	.183

5.

$$\hat{F}_n(x) = \begin{cases} 0, & x < 0, \\ 2/9, & 1 \leq x < 2 \\ 4/9, & 2 \leq x < 3 \\ 11/18, & 3 \leq x < 4 \\ 5/6, & 4 \leq x < 5 \\ 1, & x \geq 5; \end{cases}$$

$\hat{P}(1 < X < 5) = .61$, $\hat{P}(X > 3) = .39$, $\hat{P}(X < 3) = .44$.

7. (i)

	Frequency	$\hat{f}(x)$	$\hat{F}(x)$
.005–.105	8	.8	.08
.105–.205	8	.8	.16
.205–.305	12	1.2	.28
.305–.405	14	1.4	.42
.405–.505	11	1.1	.53
.505–.605	12	1.2	.65
.605–.705	11	1.1	.76
.705–.805	6	0.6	.82
.805–.905	14	1.4	.96
.905–1.005	4	0.4	1.00

(iii) $\bar{x} = .489$, $\hat{\xi}_{.25} = .28$, $\hat{\xi}_{.5} = .43$, $\hat{\xi}_{.75} = .61$.

SECTION 4.6

1. (i)

	Frequency	$\hat{f}(x)$	$\hat{F}(x)$
13.95–18.95	8	.032	.16
18.95–23.95	11	.044	.38
23.95–28.95	11	.044	.60
28.95–33.95	14	.056	.88
33.95–38.9	6	.024	1.00

(iii) $\bar{x} = 26.35$ from frequency distribution and $\bar{x} = 26.398$ from sample observations.

3. (ii) $C_0 = \{(x_1, x_2) : x_1 \le \theta_0 \text{ or } x_2 \le \theta_0\}$.

5. $[\max(X_1, X_2)/\sqrt{1 - \alpha/2}, (\max(X_1, X_2) - 1)/\sqrt{\alpha/2}]$, $C = \{\max(X_1, X_2) > N_0\sqrt{1 - \alpha/2}$, or $\max(X_1, X_2) < 1 + N_0\sqrt{\alpha/2}\}$, $[28.5, 116.3]$.

7. (i) $C = \{|X| < \theta_0(1 - (1 - \alpha)^{1/3})\}$; (ii) $\beta(\theta) = 1 - [1 - (\theta_0/\theta)(1 - (1 - \alpha)^{1/3})]^3$; (iii) $\bar{X}/[1 - (1 - \alpha)^{1/3}]$.

9. (i) (a) $C = \{\bar{X} \le \theta_0\sqrt{\alpha}/2 \text{ or } \bar{X} > \theta_0 - \theta_0\sqrt{\alpha}/2\}$, (c) $[(X_1 + X_2)/(2 - \sqrt{\alpha})$, $(X_1 + X_2)/\sqrt{\alpha}]$; (ii) (a) $C = \{\max(X_1, X_2) \le \theta_0\sqrt{\alpha/2}$, or $\max(X_1, X_2) \ge \theta_0\sqrt{1 - \alpha/2}\}$, (c) $[\max(X_1, X_2)/\sqrt{1 - \alpha/2}]$, $\max(X_1, X_2)/\sqrt{\alpha/2}]$.

SECTION 5.2

1. $2\mathscr{E}X, (\mathscr{E}X)^2$.   3. $8/3$.   5. $2.4, .6, 1/8$.   9. 0.

SECTION 5.3

1. 2.   3. (i) $2\sigma^2$; (iii) $\sigma^2/2$; (v) σ^2/n.   5. 48, .025.

7. $46/25$.   9. $\alpha = \sigma_2^2/(\sigma_1^2 + \sigma_2^2)$.

11. (ii) $1/2$; (iii) $[\text{var}(X) - \text{var}(Y)]/[\text{var}(X) + \text{var}(Y)]$.   15. $(n - m)/n$.

21. S_p^2.   23. $p(1 - p)$ and $p(1 - p)(n - 1)/n$; $np(1 - 2p)^2(1 - p)/(n - 1)^2$ and $p(1 - p)(1 - 2p)^2/n$.   25. $(\mu_1/\mu_2)(1 + \sigma_2^2/\mu_2^2), (\mu_1^2/n\mu_2^2)(\sigma_1^2/\mu_1^2 + \sigma_2^2/\mu_2^2)$.

SECTION 5.4

1. Only in (ii).   9. 500; 2000; 8000.

SECTION 5.5

1. $\mathscr{E}\{X|y\} = 10/3$ for $y = 0$, $= 9/3$ for $y = 1$; $\text{var}\{X|y\} = 14/9$ for $y = 0$, $= 2$ for $y = 1$.   3. 75 minutes.   9. (i) $Y/2, Y^2/3$; (iii) $2(6 - Y)/3, (6 - Y)^2/2$.

11. $\propto \theta^2/3$.   19. Median of Y given X.

SECTION 5.6

1. $\mathscr{E}(Y - X) = 1, \mathscr{E}\{Y|x\} = 1 + x$.   3. $\hat{p} = .20, \hat{\sigma}_p^2 = .00365, [.0178, .4701]$.

5. $(3X + 2)/(6X + 3), 8/75$; $(6X^2 + 6X + 1)/18(2X + 1)^2$.   7. $\mathscr{E}\{X|Y\} = 0$, $\mathscr{E}XY = 0$.

13. $(ab/2)\{\sin\theta - (\theta\cos\theta)/2 - (\theta^2/6)\sin\theta\}$, $((ab)^2\theta^2/12)\sin^2\theta + [(ab)^2/4][(\theta/2)\cos\theta + (\theta^2/6)\sin\theta]^2$.

SECTION 6.2

5. $-1/38$.   7. (i) $P(N_n = k) = ((N - k + 1)/N)^n - ((N - k)/N)^n$, $\mathscr{E}N_n = N^{-n}\sum_{k=1}^{N}k^n$, $\mathscr{E}(M_n + N_n) = N + 1$; (ii) $P(N_n \geq k) = \binom{N - k + 1}{n}/\binom{N}{n}$.

SECTION 6.3

3. .0432.   5. (i) $p = 1 - \sigma^2/\mu$; $n = 25, p = .4$; $P(X \geq 10) = .2447$; (ii) No.

7. 8; 3; 6; 4.   9. .1133, no.   11. .0197, $(.0197)^4$.

13. $n = 13$ or 14.   15. .1091, 9.

25. $P(X = x) = \{[x(1 - q) + Mq(1 - p)]/Mq\}\binom{M}{x}(pq)^x(1 - pq)^{M-x-1}$; $p + pq(M - 1)$; $(M - 1)pq(1 - pq) + p(1 - p)$.   27. (i) .65; (iii) .6, .012; (v) 1.

29. (i) .4; (iii) $[0, .5]$; (v) .9568.   31. $[.18, .62]$.

33. P-value $= .5034$, power at .3, .4, .6, .7 respectively .416, .128, .128, .416.

35. P-value = .063.   37. (i) none; (ii) $\{X \le 2$ or $X \ge 10\}$; (iii) $\{X \ge 10\}$;
(iv) $\{X \le 18$ or $X \ge 32\}$.

39. $\ge .0002$ (for two-sided case).   41. P-value = .3436.   43. P-value = .7538.

45. P-value = .0004.   47. $X_1 X_2 / X_{12}$.

SECTION 6.4

1. $1/210$; $185/210$.   3. $7/9$; 3, 6.   5. $p = 1/42$, yes.

7. (i) $\left[\binom{58}{3} + \binom{42}{3}\right] / \binom{100}{3}$; (iii) $\left[58\binom{42}{2} + 42\binom{58}{2}\right] / \binom{100}{3}$.

11. No.   13. .29, 1740.   17. P-value = .1002.   19. P-value = 1.0.

21. P-value = .1849.   23. P-value = .5447.   25. P-value = .31.

SECTION 6.5

1. .067; .7209; 4.   3. .00001.   5. $1 - [1 - 3p(1 - p)]^{n-1}$; $n = 2$.

7. $1/4$; $p/(p + 3p_1)$.   9. $1/2^{x+1}$, $x = 0, 1, \ldots, n - 1$, $P(X = n) = 1/2^n$.

11. $(1 - p)^{k-1}$; .6561; 9 or 10.

13. Geometric with p^2; Geometric with $1 - (1 - p)^2$.

15. $2/3$.   17. .00032; .0044, .0149.   19. .4929; 3.38.

23. $P(X = 1) = 1 - p$, $P(X = k) = p^2(1 - p)^{k-2}$, $k \ge 2$; 2; $2(1 - p)/p$.

SECTION 6.6

1. .018; .195; .785.   3. $\lambda = 2.2$; .072; .244; .975.   5. .323; .135; .406.

7. .176; .440; .384.   9. (ii) .67 vs. .866; (iii) .916, .101.

11. $P(X \le 1|\lambda = 2) = .406$; $P(X \le 1|\lambda = 4) = .092$; $P(X \le 1|\lambda = 8) = .003$.

13. $\sum_1^{r-1} e^{-\lambda x}(\lambda x)^k / k!$.   15. $[\lambda]$.   17. (ii) $[.86, 29.14]$, $[2.1, 11.9]$.

19. (i) Poisson (λp); (ii) $e^{-\lambda(1-p)}[\lambda(1 - p)]^{x-k}/(x - k)!$, $x = k, k + 1, \ldots$.

SECTION 6.7

1. .1055.   3. .008; .115; .45.   5. .0073.   7. .042.   11. $\hat{p}_j = x_j/n$.

13. $\mathscr{E}\{X_1|x_2\} = (n - x_2)n_1/(N - n_2)$.   15. (i) $1/4, 1/2, 1/4$; (ii) $0, 1/2, 1/2$;
(iii) $\binom{6}{3}\left(\frac{1}{4}\right)^3 \left(\frac{3}{4}\right)^3$, $[6!/(2!)^3](1/4)^4(1/2)^2$.

SECTION 6.8

1. Binomial $(10, 1/5)$, 2, .006, 4.   3.   .442; .251; .934.

5. .2529; 9.   7. Poisson $(k + 1)$; .908.   11. Poisson (λp); λp.

15. P-value $= .377$.   17. P-value $= .0934$.

SECTION 7.2

1. .67, .27, .6; .3.   3. $(\ell + 1)/(k + \ell + 1)$.

5. $f(x) = 1$ on $[c - 3d/c, c + 3d/c]$; $[(3d/c) - 1/2]/2c$ for $6d > c$.

7. (i) $1/2, 1/5$; (iii) $1/3, 4/15$.   9. (i) $1, 1/9$; (ii) .0774; (iii) $\leq 4/9$.

11. $1/2$.   13. $(1/4) + (3/16)\ln 3 = .456$.

15. $2(1 - x), 0 \leq x \leq 1$, and $2y, 0 \leq y \leq 1$; $1/y, 0 \leq x \leq y$.   17. $7/9$.

19. $f(x, y, z) = 6, \ 0 \leq x \leq y \leq z \leq 1$; $f_{X,Y}(x, y) = 6(1 - y), \ 0 \leq x \leq y \leq 1$, $f_{Y,Z}(y, z) = 6y, 0 \leq y \leq z \leq 1$; $f_{X,Z}(x, z) = 6(z - x), 0 \leq x \leq z \leq 1$.

21. $4/9$.

SECTION 7.3

1. $\sqrt{\pi}/2, 105\sqrt{\pi}/16, 6, 9!$.   3. $9!/3^{10}$.

5. $1/60; 1/60; (1/n)\Gamma((m + 1)/n)(1/a)^{(m+1)/n}$.   9. $\pi a^6/32; 288\sqrt{3}/35$.

SECTION 7.4

1. $e^{-\lambda t}; e^{-\lambda t_1} - e^{-\lambda t_2}$.   3. $P(Y \geq n/2)$, Y binomial with $p = e^{-t/\lambda}$; $p = .6065$.

5. Process 1: $c + .8647d$; Process 2: $1.5c + .6363d$; Process 3: $2c + .6321d$.

7. $1 - e^{-1/\lambda}$.   9. $\$658.33$.   11. .8646, .5955, .6066.   13. .981; .228.

15. .003, .458.   17. (i) $\int_y^\infty x^{\alpha - 2} e^{-x/\beta} \, dx/[\Gamma(\alpha)\beta^\alpha], y > 0$.

19. $\alpha = \mu^2/\sigma^2, \ \beta = \sigma^2/\mu$.   25. $a\alpha/(\alpha - 1)$ for $\alpha > 1$, $a^2\alpha/[(\alpha - 2)(\alpha - 1)^2]$ for $\alpha > 2$.

SECTION 7.5

3. $1/6$; .3221; .2007.   5. .4449.   7. $\alpha = .08, \beta = .72$.

9. $f\{p|k\} = p^{k+\alpha-1}(1 - p)^{n-k+\beta-1}/B(k + \alpha, n - k + \beta), 0 < p < 1$.

11. $B(p + \alpha, q + \beta)/B(\alpha, \beta)$.

SECTION 7.6

1. (i) .7054; (iii) .8968; (v) .2362; (vii) .1949.

3. (i) .3707; (iii) .5899; (v) .6293.　　5. (i) 4.28; (iii) .38; (v) 2.67; (vii) .67.

7. .1151; 51.6.　　9. .6247.　　11. .0062, .1815, yes.

13. (i) .0618, (ii) 628, 664.　　15. 8 oz.　　17. $\sigma = .0667$ oz.

19. 2.28 percent; 81.2.　　21. $\mu = 95 + (\sigma^2/10)\ln\{(p_2 + p_3)/(p_2 - p_1)\}$.

23. $k^{-1/2}(2\pi)^{(1-k)/2}$; $c = k^{1/2}(2\pi)^{(k-1)/2}$; 0, $1/k$.

SECTION 7.7

1. (i) $\sqrt{3}/4\pi$, 0, 0, 4/3, 4/3, 1/2; (iii) $\sqrt{3}/8\pi$, -5, 0, 16/3, 4/3, $-1/2$.

3. 4, 15/4, 16, 9, $-3/4$; normal $(6 - 9x/16, 63/16)$; .3191.

5. (i) .4782; (ii) .1175; (iii) 1.5174.

7. Normal $(0, C/2(AC - B^2))$ and normal $(0, A/2(AC - B^2))$; $B = 0$, $c = \sqrt{AC}/2\pi$.

SECTION 7.8

1. 1/3, 1/6, 1/18, 7/180, 1/4, 1/2.　　3. $P(Y = y) = 1/(n + 1)$, $y = 0, 1, \ldots, n$.

11. 2.5, .0183.　　13. Normal $(.8x + 5, 2.54)$, 22.6, .5211.

15. $1/2^{x+1}$, $x = 0, 1, 2 \ldots$; .4859.　　17. 36.28; 8.04, 41.96; .8664 vs. .5556.

19. 2.5, 5, 7.5; $(15!/(6! 4! 5!))(1/6)^{10}(2/3)^5$.　　21. 53¢.

SECTION 8.2

1. (i) $\binom{n}{(y-b)/a} p^{(y-b)/a}(1 - p)^{n-(y-b)/a}$, $y = b, a + b, \ldots, na + b$.

3. $e^{-\lambda}\lambda^{\sqrt{y-4}}/(\sqrt{y-4})!$, $y = 4, 5, 8, \ldots$.　　5. $(2e^{-y^2/2})/\sqrt{2\pi}$, $y > 0$.

7. $9/(5\sigma\sqrt{2\pi})\exp\{-(\frac{9}{5}x - \mu + 32)^2/2\sigma^2\}$.　　9. $1/\pi(1 + x^2)$, $x \in \mathbb{R}$.

11. (i) 25/144, 60/144, 46/144, 12/144, 1/144 for $x = 0, 1, \ldots, 4$; (iii) 46/144, 72/144, 26/144 for $x = 0, 1, 2$.

13. Binomial $(n_1 + n_2, p)$.　　15. $1 - y/2$, $0 \le y < 2$.

17. $.4468/\sqrt{\pi x}$, $50\pi \le x \le 55.432\pi$.　　23. $1 - \sqrt{1 - 3X^2 + 2X^3}$.

SECTION 8.3

1. $(1/|a|)f((y - b)/a)$.

3. (i) $1/2\sqrt{y}$; (iii) $1/y^2$, $1 < y < \infty$; (v) $(1/\lambda)e^{-y/\lambda}$.

5. $(4y/\sqrt{2\pi})\exp(-y^4/2)$, $y > 0$.

9. e^y, $y < 0$; $[1/\lambda(1-y)^2]\exp(-y/\lambda(1-y))$, $0 < y < 1$.

19. $2z^3 e^{-z^2}$, $z > 0$. 21. $1/2$, $0 < u + v < 2$, $0 < u - v < 2$.

SECTION 8.4

1. (i) $1 - |y|$, $|y| < 1$; (iii) $1/2$, $0 < x < 1$, and $1/2x^2$ for $x \geq 1$.

5. $1/(1+y)^2$, $y \geq 0$.

7. $P(X_1 + X_2 = k) = (k-1)/N^2$, $k = 2,\ldots,N$ and $= 1/N$ for $k = N+1,\ldots,2N$.

17. $f_Z(z) = 2z^2 - 2z^3/3$, $0 < z < 1$, $= 4[(1-z)/2 + 1/3 + (z-1)^3/6]$, $1 < z < 2$; No.

SECTION 8.5

1. (i) $P(X_{(2)} = \ell) = (e^{-\lambda}\lambda^\ell/\ell!)^2 + 2(e^{-\lambda}\lambda^\ell/\ell!)\Sigma_0^{\ell-1}e^{-\lambda}\lambda^j/j!$; (ii) $P(X_{(2)} = \ell) = p(1-p)^{\ell-1}\{2 + p(1-p)^{\ell-1} - 2(1-p)^{\ell-1}\}$.

3. $P(X_{(1)} = x) = 7/12, 1/12, 4/12$, for $x = 1, 2, 3$.

5. .0989. 9. $[(n-1)/\lambda]e^{-r/\lambda}(1 - e^{-r/\lambda})^{n-2}$, $r > 0$.

11. (i) $\Sigma_{j=1}^n f_j(x)\prod_{i \neq j}\int_x^\infty f_i(y)\, dy$.

13. $1 - \prod_1^n(1/\lambda_i)(1 - e^{-t/\lambda_i})$; $\Sigma_1^n(1/\lambda_i)\int_0^\infty xe^{-x/\lambda_i}\prod_{j \neq i}(1 - e^{-x/\lambda_j})\, dx$.

15. $[(2m+1)/\lambda]\binom{2m}{m}e^{-(m+1)x/\lambda}(1 - e^{-x/\lambda})$, $x > 0$.

17. $[X_{(5)}, X_{(10)}]$. 19. 29.6; [27.9, 30.4] or [28.6, 30.8].

21. 9; 7; 11. 23. P-value $= .3214$. 25. P-value $= .754$.

SECTION 8.6

1. $sP(s)$; $P(s^m)$; $s^k P(s^m)$.

3. (i) $ps/(1 - s(1-p))$, $|s| \leq (1-p)^{-1}$; (iii) $[e^{-\lambda}/(1 - e^{-\lambda})](e^{s\lambda} - 1)$; (v) $f(s\theta)/f(\theta)$, $|s| \leq 1$.

9. (i) Binomial $(n, 1/2)$; (iii) $P(X = k) = p^k/[\ln(1-p)^{-k}]$, $k = 1, 2, \ldots$.

15. $\mathscr{E}X^k = (k+1)!$. 19. (i) Gamma (n, λ); (iii) Gamma $(\Sigma_1^n \alpha_i, \beta)$.

23. $M(s_1, s_2) = [(e^{s_2} - 1)/s_2 s_1^2](1 - e^{s_1} + s_1 e^{s_1}) + [(e^{s_1} - 1)/s_1 s_2^2](1 - e^{s_2} + s_2 e^{s_2})$.

25. $2[s_1 s_2(s_1 + s_2)]^{-1}\{s_1 e^{s_1}(e^{s_2} - 1) - s_2(e^{s_1} - 1)\}$.

SECTION 8.7

1. (i) .98; (iii) 2.086; (v) 1.74. 3. (i) .01; (iii) 14.067; (v) 3.94, 18.307.

5. (i) 8.45; (iii) 3.86; (v) .153. 7. [.082, .108]. 9. [206.1, 218.7].

11. $[9.53, 21.59]$.   13. 18.   15. $[6.21, 7.39]$.   17. $t_0 = 2.69$.

19. $[.49, 1.67]$.   21. $[-.04, 1.44]$.   23. $t_0 = 1.609$.   25. $t_0 = 6.39$.

27. $t_0 = 1.27$.   29. $t_0 = 2.36$.   31. $t_0 = 2.28$.   33. $v_0 = 15.36$.

35. $f_0 = .7398$, yes.   37. $f_0 = 1.25$, no.

SECTION 8.8

3. (i) $P(Y = 1) = 1 - F(0)$, $P(Y = -1) = F(0)$; (iii) $P(Y = c) = 1 - F(c)$, $P(Y = -c) = F(-c)$ and $f_Y(y) = f_X(y)$ for $|y| < c$.

5. (iii) $1/(1 - X)$; (v) $h(X) = \sqrt{2X}, 0 < X < 1$ and $= 2 + \sqrt{2(X - 1)}, 1 \le X < 2$.

9. $2/3$.   11. $z^3 e^{-z/\lambda}/6\lambda^4, z > 0$.

13. (i) $(18t - 12t^2)/5, 0 < t < 1$; (ii) $12t/(1 + t)^4, 0 < t < 1$.

15. $\sigma^2(\pi - 2)/2$.   17. (i) Binomial $(t, 1/n)$.

19. .375, .5625, .6797, .7617, .8218, .8664, .8999, .9249, .9437 for $n = 2, 3 \ldots, 10$ respectively.

21. (i) none; (ii) $[32,100, 47,600]$.   23. $[1, 33]$; P-value $= .5274$.

29. $t_0 = -2.04$.   31. $t_0 = .93$; $[-5.78, 11.98]$.

SECTION 9.2

3. $P(X = \beta) = 1$.

SECTION 9.3

1. (i) No; (iii) No; (v) Yes.   3. .5098.   5. .0021; .9958; .9236.

7. 271.   9. $z_0 = -3.6$.   11. $z_0 = 2.42$.   13. .5762; 385.

15. $\sigma = 60.79$.   17. (i) .33, .1922, .0409, 0 +, 0 +, limit 0.

23. Normal $(p_1 - p_2, [p_1(1 - p_1) + p_2(1 - p_2)]/n)$; normal $(p_1 - p_2, [p_1(1 - p_1) + 2p_1 p_2 + p_2(1 - p_2)]/n)$.

SECTION 9.4

1. $z_0 = -2.27$.   3. .8462; .9989.   5. 84.   7. 0^+; .0014.

9. 233.   11. .0357 (normal), .034 (Poisson).

13. .0104, .5693; .0043.   15. $n \doteq N\sigma^2 z_{\alpha/2}^2 [\varepsilon^2(N - 1) + \sigma^2 z_{\alpha/2}^2]^{-1}$.

17. Exact .2642, Poisson .264, normal .3085.   19. Binomial .0103, normal .0096.

SECTION 9.5

3. $2\hat{p}_1 - 1$; normal $(2p_1 - 1, [4p_1(1 - p_1)/n])$.   5. $\overline{X}$; $e^{-\overline{X}}$ or $\Sigma_1^n Y_i/n$, $Y_i = 1$ if $X_i = 0$, $Y_i = 0$ otherwise.

7. $1/\overline{X}$.   9. $\hat{\sigma}_2^2$ and $\hat{\sigma}_3^2$.   13. No.

SECTION 9.6

1. $[10.0035, 10.0165]$, No.   3. $[66.8, 69.6]$.   5. $[12.0, 12.06]$.

7. $[.32, .48]$.   9. $[.3967, .4783]$.   11. 96 or 97; 81.   13. 34.

15. $[29.8, 46.2]$; $[30.1, 45.9]$.   17. $[-.24, -.16]$.   19. $[-.138, .018]$.

21. $[7.25, 15.75]$.   23. $2\hat{p} - 1 \pm z_{\alpha/2}\sqrt{4\hat{p}(1 - \hat{p})/n}$; $[.258, .408]$.

25. $S_1 - S_2 \pm z_{\alpha/2}\left[\sqrt{S_1^2/2m + S_2^2/2n}\right]^{-1}$.

SECTION 9.7

1. $z_0 = -2.43$.   3. $z_0 = 3.39$; $z_0 = -4.93$.   5. $z_0 = 3.06$.

7. $z_0 = 3.03$.   9. $z_0 = .91$.   11. $z_0 = -1.75$.   13. $z_0 = -1.80$.

15. $z_0 = 5.06$.   17. 141.   19. (ii) $.63(\sigma_1^2 + \sigma_2^2)$; (iii) $2.71(\sigma_1^2 + \sigma_2^2)$.

SECTION 9.8

1. $[X_{(15)}, X_{(26)}]$.   3. $[X_{(18)}, X_{(33)}]$; $[X_{(19)}, X_{(32)}]$.   5. .0011.

7. (i) $\pi/2$, $X_{(n/2)} \pm \sigma z_{\alpha/2}\sqrt{\pi/2n}$; (ii) $n + 2$, $X_{(n/2)}/\{1 \pm (z_{\alpha/2}/\sqrt{n})\}$.

9. P-value = .0367.   11. P-value = .758.   13. $[X_{(10)}, X_{(21)}]$.   15. $X_{(38)}$.

SECTION 9.9

1. $q_0 = 10.06$.   3. $q_0 = 7.49$.   5. (i) $q_0 = 9.2$; (ii) $q_0 = 1.23$.

7. (i) $q_0 = 8.19$; (ii) $q_0 = 8.77$.   9. $q_0 = 7.3$.   11. $q_0 = 6.10$.

13. (i) $q_0 = 7.5$; (ii) $q_0 = 7.5$; (iii) $z_0 = -2.65$ or $q_0 = 7.5$.

SECTION 9.10

3. $P(X = 0) = 1$.   5. (i) No; (iii) No.   7. .0025, .0681.

9. .0082; .0228; 1/2.   11. $[24.1, 25.5]$; 139.

13. Proportion of 1's, or $\overline{X}e^{-\overline{X}}$.   15. $(X_i - X_j)/n$.

17. $[24.2, 31.8]$; $z_0 = -2.1$.   19. $z_0 = 3.65$; $[2.9, 7.5]$.

21. $[-.17, -.03]$; $z_0 = -2.18$.   23. $[20.6, 40.3]$; P-value = .6170; P-value = .401.

SECTION 10.2

3. No, yes. 5. $\{X_{(1)},\dots,X_{(n)}\}$; $\Pi_1^n X_j$; $(\Sigma_1^n X_j, \Sigma_1^n X_j^2)$.

7. (i) $\{X_{(1)},\dots,X_{(n)}\}$, $\Sigma_1^n X_j^\alpha$, $\{X_{(1)},\dots,X_{(n)}\}$; (iii) ΠX_j, $\Pi(1 - X_j)$, $(\Pi X_j, \Pi(1 - X_j))$; (v) $\{X_{(1)}, X_{(n)}\}$, (vii) $\Sigma_1^n |X_j - \mu|$, $\{X_{(1)},\dots,X_{(n)}\}$, $\{X_{(1)},\dots,X_n)\}$; (ix) $\Sigma \ln X_j$, $\Sigma(\ln X_j)^2$, $(\Sigma \ln X_j, \Sigma(\ln X_j)^2)$.

SECTION 10.3

1. $(n - X)/n$; $(2X - n)/n$. 3. (i) Not possible; (ii) $(T - b)/a$.

5. $\mathscr{E}X^2$; $(X^2 - X)N(N - 1)/n(n - 1) + NX/n$.

7. pX, $r(1 - p)$; $p^2 X^2 + p^2 X - pX$. 9. 6.71, .96.

11. 2.04, $1/50$. 13. $3\overline{X} - 2b$. 15. $3\overline{X}$.

19. $\overline{X} + z_{1-p}\sqrt{(n - 1)/2}\,[\Gamma((n - 1)/2)/\Gamma(n/2)]S$. 21. $X_i X_j$; $X_i X_j X_k$; $\Pi_{j=1}^k X_j$.

23. $((n - 1)/n)^{\Sigma X_j}$. 25. ΣX_i^2 sufficient; neither is best.

SECTION 10.4

1. (ii) $\overline{X}(1 - \overline{X})$; (iii) $(n - 1)S^2/n$. 3. X/p. 5. $\sqrt{X_1/n}$ or $1 - \sqrt{X_3/n}$.

7. $\hat{\alpha} = \overline{X}^2/\hat{\sigma}^2$, $\hat{\beta} = \hat{\sigma}^2/\overline{X}$, $\hat{\sigma}^2 = n^{-1}\Sigma(X_i - \overline{X})^2$.

9. $\hat{p} = \overline{X}/(\overline{X} + \hat{\sigma}^2)$, $r = \overline{X}^2/(\overline{X} + \hat{\sigma}^2)$, $\hat{\sigma}^2 = n^{-1}\Sigma(X_i - \overline{X})^2$.

11. $\hat{\alpha} = \overline{X}/\sqrt{2\pi}$, $\hat{S}(t) = e^{-t^2/2\hat{\alpha}^2}$, or proportion of X_i's $> t$.

SECTION 10.5

1. $X_{(n)}$.

3. $l(x_i) =$ Number of x_i in sample, $\hat{\theta} = l(x_3)/(l(x_1) + l(x_3))$ if $l(x_1), l(x_3) > 0$, if $l(x_1) = 0 = l(x_3)$ any value in $(0, 1)$; if $l(x_1) = 0$, $\hat{\theta} = 1$; if $l(x_3) = 0$, $\hat{\theta} = 0$.

5. $\hat{\theta}_1 = X_{(1)}$, $\hat{\theta}_2 = X_{(n)}$. 7. $\overline{X} + z_{1-p}\sqrt{\Sigma(X_i - \overline{X})^2/n}$.

9. (i) $e^{-t/\overline{X}}$; (ii) $e^{-t^{\hat{\alpha}}/\hat{\beta}}$ where $\hat{\alpha}$, $\hat{\beta}$ are *mles*; (iii) $\exp\{-nt^2/\Sigma X_j^2\}$; $\lambda(t)$: (i) $1/\overline{X}$; (ii) $\hat{\alpha}t^{\hat{\alpha}-1}/\hat{\beta}$; (ii) $2nt/\Sigma X_j^2$.

11. $\overline{x} = (n - m)\hat{\mu} + m\phi(-\hat{\mu})/\Phi(-\hat{\mu})$, ϕ, Φ density and distribution function of standard normal.

13. $\overline{X}$, $\overline{Y}$, $\Sigma(X_i - \overline{X})^2/n$, $\Sigma(Y_i - \overline{Y})^2/n$, $\Sigma(X_i - \overline{X})(Y_i - \overline{Y})/\sqrt{\Sigma(X_i - \overline{X})^2 \Sigma(Y_i - \overline{Y})^2}$.

15. $\hat{p}_j = X_j/n$.

SECTION 10.6

1. (ii) $(k + 1)/(n + 2)$; (c) $1/6(n + 2)$. 3. $xe^{-\theta x}$; $1/X$.

7. $[(n + 1)/(n + 2)]^{\Sigma X_i + 1}$. 9. (ii) $(\Sigma X_i + 1)/(n + 2)$; (iii) $1/6(n + 2)$.

11. $(\Sigma X_i + 1)(\Sigma X_i + 2)/(n + 1)^2$.

13. (ii) $(\alpha n\overline{X} + \sigma^2\mu_0)/(n\alpha + \sigma^2)$; (iv) $\overline{x} = 1302.71$, $\theta_n = 1303.48$, risk ratio 1.008.

15. (ii) $\sqrt{X}$; (iii) $(n\overline{X} + \mu_0/\alpha)/(n + 1/\alpha)$. 17. .104. 19. 4601.8; .8291.

SECTION 10.7

1. $(\theta - X)/\theta$; $X - \theta$; X/θ; X/θ; X/θ; $(X - \mu_1, Y - \mu_2)$.

3. $[2\Sigma X_i/\lambda_2, 2\Sigma X_i/\lambda_1]$ where λ_1 and λ_2 are solutions of $1 - \alpha = \int_{\lambda_1}^{\lambda_2}\chi_{2n}^2(t)\,dt$ and $\lambda_1^2\chi_{2n}^2(\lambda_1) = \lambda_2^2\chi_{2n}^2(\lambda_2)$.

5. $[X/(1 - \lambda_1), X/(1 - \lambda_2)]$ where $1 - \alpha = \lambda_2^2 - \lambda_1^2$ and $(\lambda_1 + \lambda_2 - 1)^2 = \lambda_1\lambda_2$.

7. $\alpha^{-1/n}X_{(n)}$. 9. $\overline{X} - \overline{Y} \pm z_{\alpha/2}\sigma\sqrt{(1/m) + (1/n)}$.

11. $\overline{X} - \overline{Y} \pm z_{\alpha/2}\sqrt{(\sigma_1^2/m) + (\sigma_2^2/n)}$.

17. $h(\theta|x) = [(\theta - x)/4]\exp(-(\theta - x)/2)$, $0 < x < \theta$.

SECTION 10.8

1. (a) (i) ΣX_j^2 and $\sqrt{\Sigma X_j^2/n}$; (iii) $X_{(n)}$ and $X_{(n)}$; (b) $[(3n + 1)/3n]X_{(n)}$ and $\overline{X}$.

3. Yes, yes. 5. No. 7. $R(\hat{\theta}_1, \hat{\theta}_2) = n$. 11. $2X$; No.

13. $\hat{p} = 1 - \hat{\sigma}^2/\overline{X}$, $\hat{n} = \overline{X}^2/(\overline{X} - \hat{\sigma}^2)$, $\sigma^2 = \Sigma(X_i - \overline{X})^2/m$.

15. Choose c from $\alpha = c^{n-1}[n - (n - 1)c]$, $\mathscr{E}((R/c) - R) = [(n - 1)/(n + 1)]\theta$
$[(1 - c)/c]$, $\mathscr{E}(\alpha^{-1/n}X_{(n)} - X_{(n)}) = (\alpha^{-1/n} - 1)n\theta/(n + 1)$.

17. $n \doteq (1/\varepsilon^2\alpha)(1/16(p - 1/2)^2)$; $\hat{\pi} \pm z_{\alpha/2}\hat{\sigma}_{\hat{\pi}}$; .5625 and $[.52, .62]$.

19. $[(n + \alpha)/(n + \alpha - 1)]\max(a, x_1, \ldots, x_n)$; $2^{1/(n+\alpha)}\max(a, x_1, \ldots, x_n)$; $[\max(a, x_1, \ldots, x_n), \alpha^{-1/(n+\alpha)}\max(a, x_1, \ldots, x_n)]$.

21. r/n. 23. $n\overline{X}^2/(n + 1)$.

SECTION 11.2

1. $\{\Sigma X_i \geq k\}$. 3. $\{X \geq k\}$. 5. $\{X \geq k\}$.

7. $\{|X| \leq k\}$. 9. $\{\Sigma X_j^2 \geq k\}$.

11. (i) $\{X_{(1)} \geq \theta_0 + (1/n)\ln(1/\alpha)\}$, $\{X \geq 1 - \alpha\}$, $\{X_{(1)} \geq \theta_0\alpha^{-1/n}\}$,
$\{\overline{X} \leq \theta_0 - z_\alpha/\sqrt{n}\}$, $\{\Sigma X_j^2 \geq \theta_0^2\chi_{n,\alpha}^2\}$; (ii) $\exp\{n(\theta_1 - \theta_0 - (1/n)\ln(1/\alpha))\}$,
$1 - (1 - \alpha)^{\theta_1}$, $\alpha(\theta_1/\theta_0)^n$, $\Phi((\theta_0 - \theta_1)\sqrt{n} - z_\alpha)$, $P\{\chi^2(n) \geq (\theta_0/\theta_1)^2\chi_{n,\alpha}^2\}$.

SECTION 11.3

1. $\{\Sigma X_i \geq k\}$, P-value $= .0539$.   3. $\{\Sigma X_j \geq k\}$.   5. $\{\overline{X} \leq k\}$.

7. (i) $\{\Pi X_j \geq k\}$; (ii) $\{\Pi X_j \geq k\}$; (iii) $\{X_{(1)} \geq k\}$.

SECTION 11.4

1. (i) $\{X_{(n)} \geq N_0\}$; (ii) $\{X_{(n)} > N_0 \text{ or } X_{(n)} < c\}$.

3. $\{\overline{X} \leq c_1 \text{ or } \overline{X} \geq c_2\}$.   5. $\{X \leq c_1 \text{ or } X \geq c_2\}$.

7. (i) $\{[\Sigma(X_i - \overline{X})^2/\Sigma(Y_i - \overline{Y})^2] \geq c_1 \text{ or } [\Sigma(X_i - \overline{X})^2/\Sigma(Y_i - \overline{Y})^2] \leq c_2\}$;
(ii) $\{|(\overline{X} - \overline{Y})/S_p\sqrt{(1/m + 1/n)}| \geq c\}$, $S_p^2 = [\Sigma(X_i - \overline{X})^2 + \Sigma(Y_i - \overline{Y})^2]/(m + n - 2)$.

9. $\{(n/k)^n/(\Pi x_j^{x_j}) < c\}$ or $\{\Sigma_1^k x_j \ln x_j > c_1\}$.

SECTION 11.5

1. $w_+ = 29$, $t_0 = 1.424$.   3. $w_- = 49$, $t_0 = -3.76$.

5. $w_- = 37.5$.   7. $W_+ = 111$.

9. (i) P-values of sign, Wilcoxon and t-tests respectively .8238, .2802, between .3 and .4.

SECTION 11.6

1. $t_x = 130.5$.   3. $t_x = 82$, $R = 8$.   5. $t_x = 33.5$.

7. $R = 11$.   9. $R = 23$.   11. $\mathscr{E}_{H_0} U_x = mn/2$, $\text{var}(U_x) = mn(m + n + 1)/12$.

SECTION 11.7

1. $\mathscr{E}\hat{p} = p$, $\text{var}(\hat{p}) = p(1 - p)/2n$; $\hat{p} = .5$, $q_0 = 3$.   3. $q_0 = 8.41$.

5. $q_0 = 8.41$.   9. $q_0 = 2.71$.

SECTION 11.8

1. $D_n = .24$.   3. $D_n = .1946$.   5. $D_n = .2321$.   7. $D_n = .3444$.

9. $D_n = .3715$.

SECTION 11.9

1. (i) $t_0 = 1.182$, (ii) $t_0 = 1.13$; (ii) $t_0 = -4.407$.

3. $r_S = .8571$, $\tau = .7857$.   5. $r_S = .8607$, $\tau = .7143$.

7. $r_S = -.3482$, $\tau = -.2857$.   9. $r = .5647$ and $t_0 = 2.903$, $r_S = .5135$.

SECTION 11.10

1. $|X| > c_1$ or $|X| < c_2$.   3. $\{X \le k\}$.   5. $\{\Sigma X_i < k\}$.   7. $\{X_{(1)} > k\}$.

9. $\{\Sigma X_i < c_1 \text{ or } \Sigma X_i > c_2\}$.   11. $\{\Sigma X_i \ln(np_i^0 / X_i) < k\}$.   15. $w_+ = 146.5$.

17. $t_X = 34$, $r_0 = 7$.   19. $r_0 = 6$.

21. $t_x = 173$.   23. $q_0 = 7.356$.   25. $q_0 = .669$.

27. $D_{20} = .1049$ 29. $r = -.9228$.   31. $r_0 = 21$.   33. $r_S = .80$, $\tau = .67$.

SECTION 12.2

1. $q_0 = 15.41$.   3. (i) $q_0 = 6.7$; (ii) $q_0 = 6.3$.   5. $q_0 = 3.54$.

7. $q_0 = 21.81$.   9. $q_0 = 219.1$.

SECTION 12.3

1. $q_0 = 217.6$.   3. $q_0 = 256.4$.   5. $q_0 = 58.7$.   7. $q_0 = 28.47$.

SECTION 12.4

1. $q_0 = 15.74$.   3. $q_0 = 11.46$.   5. $q_0 = 5.25$.

SECTION 13.2

1. $f = 12.17$.   3. $f = 56.45$.   5. $f = 16.46$.   7. $f = 49.13$.

SECTION 13.3

1. $\mu_B > \mu_C$, $\mu_B > \mu_E$, $\mu_D > \mu_C$, $\mu_D > \mu_E$.   3. $\mu_4 > \mu_3 > \mu_1 > \mu_2$.

5. $\hat{\theta} \mp t_{n-k,\,\alpha/2} \sqrt{\left[\Sigma_{i=1}^{k}\Sigma_{j=1}^{n_i}\left(X_{ij} - \overline{X}_i\right)^2 / (n-k)\right]\Sigma_{i=1}^{k} c_i^2 / n_i}$.

SECTION 13.4

3. (i) $f_{\text{Methods}} = 1.0$, $f_{\text{Instructors}} = .45$; (ii) $f_{\text{Methods}} = 1.08$, $f_{\text{Instructors}} = .48$, $f_I = 1.73$.

5. $f_{\text{Variety}} = 24.00$.   7. $f_{\text{Strains}} = 48.61$; $\mu_A > \mu_C > \mu_D$, $\mu_B > \mu_C > \mu_D$.

9. $\mu_D > \mu_C > \mu_A$, $\mu_D > \mu_C > \mu_B$.   11. $t_0 = 5.34$; $f_{\text{Treatments}} = 28.66$.

SECTION 13.5

1. $q_0 = .1719$, $h = .2598$.   3. $h = 21.42$, $f = 93.52$; $\mu_1 < \mu_3$, $\mu_1 < \mu_4$, $\mu_2 < \mu_4$.

5. $h = 37.52$; $\mu_A < \mu_D$, $\mu_A < \mu_E$, $\mu_A < \mu_F$, $\mu_B < \mu_E$, $\mu_B < \mu_F$, $\mu_C < \mu_F$.

SECTION 13.6

1. $f = 9$.   3. $f = 21.775$.   5. $w = .8443$, $f = 37.996$.

7. $w = .9665$, $f = 27.0625$.

SECTION 13.7

1. $f = .3747$.

3. (i) $f = 46.34$, $\mu_A > \mu_C > \mu_D$, $\mu_C > \mu_B$; (ii) $f_{\text{Stimuli}} = 38.27$, $f_{\text{Subjects}} = 1.53$, $\mu_A > \mu_C > \mu_D > \mu_B$; (iii) $f = 13.38$.

5. $f = 28.82$, $\mu_1 < \mu_2$, $\mu_1 < \mu_3$; $h = 22.06$, $\mu_1 < \mu_2$, $\mu_1 < \mu_3$.

7. $f_{\text{Fields}} = 3.45$, $f_{\text{Sex}} = 1.72$, $f_I = 13.09$.   9. $w = .706$, $f = 14.83$.

11. $w = .4$, $f = 6.4$.

INDEX